To Lee,

Wishing you every success in the future.

Best wishes,

Chris.

MAY 2007

Methods in Cell Biology

VOLUME 80

Mitochondria, 2nd Edition

Series Editors

Leslie Wilson

Department of Molecular, Cellular and Developmental Biology
University of California
Santa Barbara, California

Paul Matsudaira

Whitehead Institute for Biomedical Research
Department of Biology
Division of Biological Engineering
Massachusetts Institute of Technology
Cambridge, Massachusetts

Methods in Cell Biology

VOLUME 80

Mitochondria, 2nd Edition

Edited by

Liza A. Pon

Department of Anatomy and Cell Biology
College of Physicians and Surgeons
Columbia University
New York, New York

Eric A. Schon

Department of Neurology and Department of
Genetics and Development
Columbia University
New York, New York

ELSEVIER

AMSTERDAM • BOSTON • HEIDELBERG • LONDON
NEW YORK • OXFORD • PARIS • SAN DIEGO
SAN FRANCISCO • SINGAPORE • SYDNEY • TOKYO
Academic Press is an imprint of Elsevier

Cover Photo Credit: Laser scanning confocal image of human fibroblasts stained for mitochondria using Mito Tracker (red) and microtubules (green) using a monoclonal anti-tubulin antibody and indirect immunofluorescence. Image provided by Istvan Boldogh (Columbia University) and Michael Yaffe (University of California, San Diego).

Academic Press is an imprint of Elsevier
525 B Street, Suite 1900, San Diego, California 92101-4495, USA
84 Theobald's Road, London WC1X 8RR, UK

For information on all Academic Press publications visit our Web site at www.books.elsevier.com

ISBN-13: 978-0-12-544173-5
ISBN-10: 0-12-544173-8

PRINTED IN THE UNITED STATES OF AMERICA
07   08   09   10   9   8   7   6   5   4   3   2   1

Working together to grow
libraries in developing countries

www.elsevier.com | www.bookaid.org | www.sabre.org

ELSEVIER   BOOK AID International   Sabre Foundation

CONTENTS

PART II   Biochemical Assays of Mitochondrial Activity

11. Assays of Cardiolipin Levels

Michael Schlame

12. Measurement of VDAC Permeability in Intact Mitochondria
and in Reconstituted Systems

Marco Colombini

13. Methods for Studying Iron Metabolism in Yeast Mitochondria

Sabine Molik, Roland Lill, and Ulrich Mühlenhoff

PART IV   Oxidative Stress Measurements

PART V   Mitochondrial Genes and Gene Expression

PART VI   Assays for Mitochondrial Morphology and Motility

30. Visualization and Quantification of Mitochondrial Dynamics in Living Animal Cells

Kurt J. De Vos and Michael P. Sheetz

31. Cell-Free Assays for Mitochondria–Cytoskeleton Interactions

Istvan R. Boldogh, Liza A. Pon, Michael P. Sheetz, and Kurt J. De Vos

32. *In Vitro* Assays for Mitochondrial Fusion and Division

Elena Ingerman, Shelly Meeusen, Rachel DeVay, and Jodi Nunnari

PART VII   Methods to Determine Protein Localization to Mitochondria

PART VIII Appendices

CONTRIBUTORS

Numbers in parentheses indicate the pages on which the authors' contributions begin.

R. Acín-Pérez (571), Departamento de Bioquímica y Biología Molecular y Celular, Universidad de Zaragoza, Zaragoza 50013, Spain

J. S. Armstrong (355), Department of Biochemistry, Yong Loo Lin School of Medicine, National University of Singapore, Republic of Singapore 117597, Singapore

Giuseppe Attardi (121), Division of Biology, California Institute of Technology, Pasadena, California 91125

Sandra R. Bacman (503), Department of Neurology, University of Miami Miller School of Medicine, Miami, Florida 33136

Christopher J. Bell (341), Department of Biochemistry, Henry Wellcome Signalling Laboratories, School of Medical Sciences, University Walk, University of Bristol, Bristol BS8 1TD, United Kingdom

Michael J. Bennett (179), Department of Pathology and Laboratory Medicine, University of Pennsylvania, Metabolic Disease Laboratory, The Children's Hospital of Philadelphia, Philadelphia 19104

Istvan R. Boldogh (45, 683), Department of Anatomy and Cell Biology, College of Physicians and Surgeons, Columbia University, New York, New York 10032

Eduardo Bonilla (135), Department of Pathology and Department of Neurology, Columbia University, New York, New York 10032

Nathalie Bonnefoy (525), Centre de Génétique Moléculaire, CNRS UPR2167, Avenue de la Terrasse, 91198 Gif-sur-Yvette cedex, France

Paul S. Brookes (395), Department of Anesthesiology, University of Rochester Medical Center, Rochester, New York 14642

Marco Colombini (241), Department of Biology, University of Maryland, College Park, Maryland 20742

Victor M. Darley-Usmar (395), Cellular and Molecular Pathology, University of Alabama at Birmingham, Birmingham, Alabama 35294

Kurt J. De Vos (627, 683), Department of Neuroscience, MRC Centre for Neurodegeneration Research, The Institute of Psychiatry, King's College London, De Crespigny Park, Denmark Hill, London SE5 8AF, United Kingdom

Rachel DeVay (707), Department of Molecular and Cellular Biology, University of California Davis, Davis, California 95616

Orly Elpeleg (199), The Metabolic Disease Unit, Hadassah-Hebrew University Medical Centre, Jerusalem 91120, Israel

J. A. Enriquez (571), Departamento de Bioquímica y Biología Molecular y Celular, Universidad de Zaragoza, Zaragoza 50013, Spain

P. Fernández-Silva (571), Departamento de Bioquímica y Biología Molecular y Celular, Universidad de Zaragoza, Zaragoza 50013, Spain

E. Fernández-Vizarra (571), Departamento de Bioquímica y Biología Molecular y Celular, Universidad de Zaragoza, Zaragoza 50013, Spain

Thomas D. Fox (525), Department of Molecular Biology and Genetics, Cornell University, Ithaca, New York 14853

Elena Garcia-Arumi (379), Centre d'Investigacions en Bioquímica i Biología Molecular (CIBBIM), Institut de Recerca Hospital Universitari Vall d'Hebrón, Barcelona, Spain

Anna C. Gay (591), Department of Anatomy and Cell Biology, College of Physicians and Surgeons, Columbia University, New York, New York 10032

Elinor J. Griffiths (341), Department of Biochemistry, Henry Wellcome Signalling Laboratories, School of Medical Sciences, University Walk, University of Bristol, Bristol BS8 1TD, United Kingdom

Shukry J. Habib (761), Institut für Physiologische Chemie, Universität München, D-81377 Munich, Germany

Johannes M. Herrmann (743), Institut für Zellbiologie, Universität Kaiserslautern, 67663 Kaiserslautern, Germany

Elena Ingerman (707), Department of Molecular and Cellular Biology, University of California Davis, Davis, California 95616

Michelle S. Johnson (417), Department of Pathology, University of Alabama at Birmingham, Birmingham, Alabama 35294; Center for Free Radical Biology, University of Alabama at Birmingham, Birmingham, Alabama 35294

Denise M. Kirby (93), Mitochondrial Research Group, School of Neurology, Neurobiology and Psychiatry, The Medical School, Framlington Place, Newcastle University, Newcastle upon Tyne, NE2 4HH, United Kingdom; Mitochondrial Research Laboratory, Murdoch Childrens Research Institute, Royal Children's Hospital, Parkville, Victoria 3052, Australia

Guido Kroemer (327), INSERM U848, Université Paris Sud, Institut Gustave Roussy, F-94805 Villejuif, France

Aimee Landar (417), Department of Pathology, University of Alabama at Birmingham, Birmingham, Alabama 35294; Center for Free Radical Biology, University of Alabama at Birmingham, Birmingham, Alabama 35294

C. J. Leaver (65), Department of Plant Sciences, University of Oxford, Oxford OX1 3RB, United Kingdom

John J. Lemasters (283), Center for Cell Death, Injury and Regeneration, and Departments of Pharmaceutical Sciences and Biochemistry & Molecular Biology, Medical University of South Carolina, Charleston, South Carolina 29425

Giorgio Lenaz (3), Dipartimento di Biochimica "G. Moruzzi", Università degli Studi di Bologna, Bologna, Italy

A. Liddell (65), Department of Plant Sciences, University of Oxford, Oxford OX1 3RB, United Kingdom

Roland Lill (261), Institut für Zytobiologie und Zytopathologie, Philipps-Universität Marburg, 35033 Marburg, Germany

Nevi Mackay (173), Department of Pediatric Laboratory Medicine and the Research Institute, Hospital for Sick Children, University of Toronto, Toronto, Ontario, Canada; Department of Biochemistry, University of Toronto, Toronto, Ontario, Canada

Carine Maisse (327), INSERM U848, Université Paris Sud, Institut Gustave Roussy, F-94805 Villejuif, France

Giovanni Manfredi (155, 341), Department of Neurology and Neuroscience, Weill Medical College of Cornell University, New York, New York 10021

Shelly Meeusen (707), Department of Molecular and Cellular Biology, University of California Davis, Davis, California 95616

Didier Métivier (327), INSERM U848, Université Paris Sud, Institut Gustave Roussy, F-94805 Villejuif, France

A. H. Millar (65), ARC Centre of Excellence in Plant Energy Biology, The University of Western Australia, Crawley 6009, Western Australia, Australia

Sabine Molik (261), Institut für Zytobiologie und Zytopathologie, Philipps-Universität Marburg, 35033 Marburg, Germany

Carlos T. Moraes (481, 503), Department of Neurology, University of Miami Miller School of Medicine, Miami, Florida 33136

Ulrich Mühlenhoff (261), Institut für Zytobiologie und Zytopathologie, Philipps-Universität Marburg, 35033 Marburg, Germany

Ali Naini (437), H. Houston Merritt Clinical Research Center for Muscular Dystrophy and Related disorders, Department of Neurology, College of Physicians and Surgeons, Columbia University, New York, New York 10032

Walter Neupert (761), Institut für Physiologische Chemie, Universität München, D-81377 Munich, Germany

Jodi Nunnari (707), Department of Molecular and Cellular Biology, University of California Davis, Davis, California 95616

Joo Yeun Oh (417), Department of Pathology, University of Alabama at Birmingham, Birmingham, Alabama 35294; Center for Free Radical Biology, University of Alabama at Birmingham, Birmingham, Alabama 35294

A. Pérez-Martos (571), Departamento de Bioquímica y Biología Molecular y Celular, Universidad de Zaragoza, Zaragoza 50013, Spain

Francesco Pallotti (3), Dipartimento di Scienze Biomediche Sperimentali e Cliniche, Università degli Studi dell'Insubria, 21100 Varese, Italy

Nikolaus Pfanner (783), Institut für Biochemie und Molekularbiologie, Zentrum für Biochemie und Malekulare Zellforschung, Universität Freiburg, Hermann-Herder-Straße 7, D-79104 Freiburg, Germany

Carl A. Pinkert (549), Department of Pathobiology, College of Veterinary Medicine, Auburn University, Auburn, Alabama 36849

Paolo Pinton (297), Department of Experimental and Diagnostic Medicine, Section of General Pathology, Interdisciplinary Center for the Study of Inflammation (ICSI) and Emilia Romagna Laboratory for Genomics and Biotechnology (ER-Gentech), University of Ferrara, I-44100 Ferrara, Italy

Liza A. Pon (45, 591, 683), Department of Anatomy and Cell Biology, College of Physicians and Surgeons, Columbia University, New York, New York 10032

Andrea Prandini (297), Department of Experimental and Diagnostic Medicine, Section of General Pathology, Interdisciplinary Center for the Study of Inflammation (ICSI) and Emilia Romagna Laboratory for Genomics and Biotechnology (ER-Gentech), University of Ferrara, I-44100 Ferrara, Italy

Venkat K. Ramshesh (283), Center for Cell Death, Injury and Regeneration, and Departments of Pharmaceutical Sciences and Biochemistry & Molecular Biology, Medical University of South Carolina, Charleston, South Carolina 29425

Doron Rapaport (761), Institut für Physiologische Chemie, Universität München, D-81377 Munich, Germany

Ann Saada Reisch (199), The Metabolic Disease Unit, Hadassah-Hebrew University Medical Centre, Jerusalem 91120, Israel; Ben Gurion University of the Negev, Beer Sheva 84105, Israel

Claire Remacle (525), Laboratoire de Génétique des Microorganismes, Département des Sciences de la vie, Université de Liège, B-4000 Liège, Belgium

Alessandro Rimessi (297), Department of Experimental and Diagnostic Medicine, Section of General Pathology, Interdisciplinary Center for the Study of Inflammation (ICSI) and Emilia Romagna Laboratory for Genomics and Biotechnology (ER-Gentech), University of Ferrara, I-44100 Ferrara, Italy

Rosario Rizzuto (297), Department of Experimental and Diagnostic Medicine, Section of General Pathology, Interdisciplinary Center for the Study of Inflammation (ICSI) and Emilia Romagna Laboratory for Genomics and Biotechnology (ER-Gentech), University of Ferrara, I-44100 Ferrara, Italy

Brian H. Robinson (173), Department of Pediatric Laboratory Medicine and the Research Institute, Hospital for Sick Children, University of Toronto, Toronto, Ontario, Canada; Department of Biochemistry, University of Toronto, Toronto, Ontario, Canada

Anna Romagnoli (297), Department of Experimental and Diagnostic Medicine, Section of General Pathology, Interdisciplinary Center for the Study of Inflammation (ICSI) and Emilia Romagna Laboratory for Genomics and Biotechnology (ER-Gentech), University of Ferrara, I-44100 Ferrara, Italy

Guy A. Rutter (341), Department of Biochemistry, Henry Wellcome Signalling Laboratories, School of Medical Sciences, University Walk, University of Bristol, Bristol BS8 1TD, United Kingdom; Department of Cell Biology, Faculty of Medicine, Imperial College London, London SW7 2AZ, United Kingdom

Hermann Schägger (723), Molekulare Bioenergetik, ZBC, Universitätsklinikum Frankfurt, 60590 Frankfurt am Main, Germany

Michael Schlame (223), Department of Anesthesiology and Department of Cell Biology, New York University School of Medicine, New York, New York 10016

Bonnie L. Seidel-Rogol (465), Department of Biochemistry, Emory University School of Medicine, Atlanta, Georgia 30322

Gerald S. Shadel (465), Department of Biochemistry, Emory University School of Medicine, Atlanta, Georgia 30322

Sara Shanske (437), H. Houston Merritt Clinical Research Center for Muscular Dystrophy and Related disorders, Department of Neurology, College of Physicians and Surgeons, Columbia University, New York, New York 10032

Michael P. Sheetz (627, 683), Department of Biological Sciences, Columbia University, New York, New York 10025

Sruti Shiva (395), Vascular Medicine Branch, NHLBI, NIH, Bethesda, Maryland 20892

Anatoly Starkov (379), Department of Neurology and Neuroscience, Weill Medical College, Cornell University, New York, New York 10021

Diana Stojanovski (783), Institut für Biochemie und Molekularbiologie, Zentrum für Biochemie und Molekulare Zellforschung, Universität Freiburg, Hermann-Herder-Straße 7, D-79104 Freiburg, Germany

Theresa C. Swayne (591), Department of Anatomy and Cell Biology, College of Physicians and Surgeons, Columbia University, New York, New York 10032

Kurenai Tanji (135), Department of Pathology, Columbia University, New York, New York 10032

Robert W. Taylor (93), Mitochondrial Research Group, School of Neurology, Neurobiology and Psychiatry, The Medical School, Framlington Place, Newcastle University, Newcastle upon Tyne, NE2 4HH, United Kingdom

David R. Thorburn (93), Mitochondrial Research Laboratory, Murdoch Childrens Research Institute, Royal Children's Hospital, Parkville, Victoria 3052, Australia

Ian A. Trounce (549), Centre for Neuroscience, University of Melbourne, Victoria, 3010 Australia

Douglass M. Turnbull (93), Mitochondrial Research Group, School of Neurology, Neurobiology and Psychiatry, The Medical School, Framlington Place, Newcastle University, Newcastle upon Tyne, NE2 4HH, United Kingdom

Gaetano Villani* (121), Division of Biology, California Institute of Technology, Pasadena, California 91125

Cristofol Vives-Bauza (155, 379), Department of Neurology and Neuroscience, Weill Medical College of Cornell University, New York, New York 10021; Centre d'Investigacions en Bioquímica i Biología Molecular (CIBBIM), Institut de Recerca Hospital Universitari Vall d'Hebrón, Barcelona, Spain

Benedikt Westermann (743), Institut für Zellbiologie, Universität Bayreuth, 95440 Bayreuth, Germany

M. Whiteman (355), Department of Biochemistry, Yong Loo Lin School of Medicine, National University of Singapore, Republic of Singapore 117597, Singapore

Nils Wiedemann (783), Institut für Biochemie und Molekularbiologie, Zentrum für Biochemie und Molekulare Zellforschung, Universität Freiburg, Hermann-Herder-Straße 7, D-79104 Freiburg, Germany

Sion L. Williams (481), Department of Neurology, University of Miami Miller School of Medicine, Miami, Florida 33136

Ilka Wittig (723), Molekulare Bioenergetik, ZBC, Universitätsklinikum Frankfurt, 60590 Frankfurt am Main, Germany

Lichuan Yang (155), Department of Neurology and Neuroscience, Weill Medical College of Cornell University, New York, New York 10021

Naoufal Zamzami (327), INSERM U848, Université Paris Sud, Institut Gustave Roussy, F-94805 Villejuif, France

* Current address: Department of Medical Biochemistry and Biology, University of Bari, Bari, Italy.

PREFACE

The field of "mitochondriology" has undergone a renaissance in the last 15 years. Compared to the state of the field as recently as 1985, entirely new areas of investigation have emerged, such as the role of mitochondrial function in apoptosis, aging, and mitochondrial dysfunction in human disease. Furthermore, there have been landmark discoveries and tremendous advances in our understanding of basic phenomena, including mitochondrial import, movement, fission, fusion, inheritance, and interactions with the nucleus and other organelles.

Concomitant with these conceptual advances has been the remarkable increase in the types of tools available to study mitochondrial structure and function. In addition to the traditional methods for isolating mitochondria and assaying their biochemical properties, we now have new and powerful ways to visualize, monitor, and perturb mitochondrial function and to assess the genetic consequences of those perturbations.

In 2001, we assembled the first volume in the Methods in Cell Biology series to bring together those methods—both "classic" and modern—that would enable anyone to study this organelle, be it from a biochemical, morphological, or genetic point of view. The popularity of that volume has inspired us to assemble a second edition that would reflect technical advances of the last few years, as well as new or emerging areas of investigation. Not only has the number of chapters grown from 25 to 36, but much of the material presented in the first volume has been revamped. A few chapters have been eliminated. Fourteen chapters are completely new and cover methods for assaying lactate, pyruvate, β-oxidation, TCA cycle intermediates, cardiolipin, iron–sulfur clusters, and the permeability transition. We have also added a new section on measurements of oxidative stress (assays of reactive oxygen species, free radical scavengers, nitric oxide, and protein thiol modification) and new chapters on examining mitochondrial fission/fusion and on measuring mitochondrial movement in living cells—two related emerging "hot" topics.

A noteworthy, and we believe useful, adjunct to the methods are the appendices, which bring together fundamental information regarding mitochondria that are not found in any single source. These include lists of every known or suspected mitochondrial protein (with approximately 50% more entries in this edition compared to the last), maps of mitochondrial genomes from several commonly-used organisms, and information on agents that perturb respiratory function. Reflecting the boom in "omics," a new appendix has been added describing microarray and proteomic analyses of alterations in mitochondrial function under various conditions.

We dedicate this volume to Giuseppe Attardi, a pioneer in the field of mitochondria in general and mammalian mitochondrial genetics in particular. Among his many notable achievements are the elucidation of the transcription of the human mitochondrial genome and the development of mammalian ρ^0 cells and cybrids. Giuseppe is an inspiration to us all.

Liza A. Pon
Eric A. Schon

PART I

Isolation and Subfractionation
of Mitochondria

CHAPTER 1

Isolation and Subfractionation of Mitochondria from Animal Cells and Tissue Culture Lines

Francesco Pallotti★ and Giorgio Lenaz†

★Dipartimento di Scienze Biomediche Sperimentali e Cliniche
Università degli Studi dell'Insubria, 21100 Varese, Italy

†Dipartimento di Biochimica "G. Moruzzi"
Università degli Studi di Bologna, Bologna, Italy

I. Introduction

Past and current mitochondrial research has been performed on mitochondria prepared from rat liver, rat heart, or beef heart, since these tissues can be obtained readily and in quantity. Toward the end of the 1980s, a new branch of human pathology began with the discovery of human disorders linked to mitochondrial dysfunction (mitochondrial encephalomyopathies). Therefore, it became necessary to investigate and to study the status of mitochondria in human tissue. As it is not always possible to obtain large amounts of human tissues for extracting mitochondria, interest arose in developing smaller scale mitochondrial isolation methods to determine mitochondrial enzyme profiles. Moreover, the discovery of a deep involvement of mitochondria in many aspects of human pathology (Lenaz *et al.*, 2006; Wallace, 2005) has prompted the creation of novel animal models requiring short times for reproduction, such as insects and worms, and of methods to isolate mitochondria from these sources.

Since our previous edition of the chapter on isolation and subfractionation of mitochondria from animal tissues, the field has developed a need for fast and reliable methods for isolation of mitochondria from various tissues that produce pure and reasonable amount of functional mitochondria for further investigation.

In this chapter, we update the methods described in the previous edition in the attempt to address the broad interests of research groups on mitochondria derived from animal cells and tissues. The reader is also directed to Chapter 4 by Kirby *et al.*, this volume for methods to isolate mitochondria from human biopsy samples and from cultured human cells and cell lines.

II. General Features of Mitochondrial Preparations

All methods for mitochondrial isolation are laboratory-specific and were developed to obtain preparations of mitochondria suitable for different applications. However, all isolation procedures for preparation of mitochondria must be carried out on ice or in a cold room, and follow a standardized flowchart, as shown in Fig. 1.

A. Cell Rupturing by Mechanical and/or Chemical Means and Resuspension in Isolation Medium

1. Starting Tissues

The choice of starting tissue or cell line is critical for the yield of the final preparation. Amounts of starting material determine the scale of preparation. Usually, for a large scale preparation of beef heart mitochondria (BHM), 500 g of starting material is optimal, giving typical yields of 1 g of mitochondria. Small-scale preparations of tissues, such as rat or mouse tissues, require a starting amount of about 0.5 g to obtain a yield of 5 mg of mitochondria.

The yield of mitochondria from cell lines is influenced by the culture conditions. Cells that are grown in conditions of low bioenergetics requirements usually contain few mitochondria. It is difficult to produce high quantities of mitochondria from cultured cells that adhere to a solid surface, in part because adherent cells are not ruptured efficiently using conventional mechanical or chemical methods. A small-scale preparation from cultured fibroblasts requires a starting amount of 30-mg fibroblasts, equivalent to about fifteen 150-mm plates of confluent cells, in order to obtain 1 mg of mitochondria.

2. Homogenization

The method of choice for rupturing intercellular connections, cell walls, and plasma membranes is dictated by cell type and their intercellular connections. For fibrous tissues, such as muscle, the use of a tissue blender (Waring, Ultraturrax) is necessary. Soft tissue, like liver and cells from suspension cultures can be disrupted using Dounce or Potter-Elvehjem homogenizers. Suspension cultures are disrupted more efficiently using Dounce homogenizers. Dounce homogenizers consist of a glass pestle that is manually driven into a glass vessel. Potter-Elvehjem homogenizers consist of Teflon pestles that are driven into a glass vessel either manually or mechanically. For both of these homogenizers, the degree of homogenization is controlled by the clearance between the pestle and walls of the glass vessel, the rotational speed of the pestle, and the number of compression strokes made. Two pestles are necessary to perform a two-step-based homogenization: a large clearance pestle (0.12 mm) is used for the initial sample reduction, and a small clearance pestle (0.05 mm) is used to form the final homogenates.

Localized heating within a sample can occur during homogenization, leading to protein denaturation and aggregation. To avoid this problem, it is essential to prechill

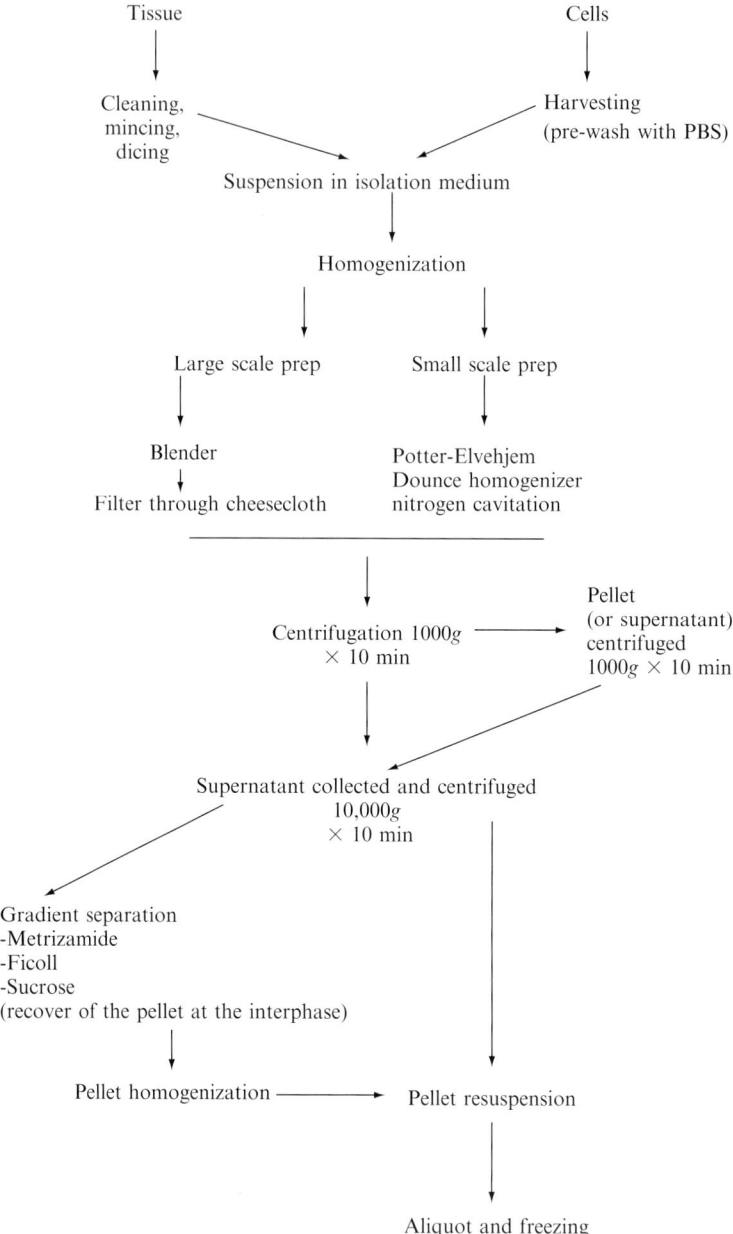

Fig. 1 General flowchart for the isolation of mitochondria.

equipment and keep samples on ice at all times. Moreover, protease inhibitors should be added to all samples prior to lysis. Proteases generally belong to one of four evolutionarily distinct enzyme families, based on the functional groups involved in cleavage of the peptide bond. Therefore, several different types of inhibitors are generally required to protect proteins from proteolysis during cells homogenization. We recommend the following cocktail of protease inhibitors: phenylmethylsulfonyl fluoride (PMSF), an inhibitor of serine proteases; leupeptin and Pepstatin A, inhibitors of thiol and acid proteases, respectively; and EDTA, a metalloprotease inhibitor. We use a 1000× stock solution of a protease inhibitor cocktail containing 1-mM PMSF, 1-mg/ml leupeptin, and 1-mg/ml pepstatin proteases may be used to soften the tissues, especially trypsin (0.3 mg/ml) or Nagarse, on muscle tissue prior to homogenization.

3. Isolation Medium

Isolation media can be distinguished as either nonionic or ionic. Sucrose, mannitol, or sorbitol are nonionic osmotic supports for mammalian tissues, usually at 0.25 M. Ionic osmotic supports should be used for those tissues, such as muscle, which develop a gelatinous consistency on homogenization in isotonic sucrose. We recommend 100- to 150-mM KCl for ionic osmotic support. Hypotonic KCl media is used to swell mitochondria during preparation of mitoplasts as an alternative to incubation in digitonin, as is 1-mM EDTA, to chelate Ca^{2+} ions; 0.1–1% bovine serum albumin (BSA), which quenches proteolytic activity of trypsin, removes free fatty acids, and maintains mitochondrial respiration. Buffer, such as 5- to 20-mM Tris–HCl or Tris–acetate is usually added.

B. Differential Centrifugation

Differential centrifugation consists of a two-step centrifugation carried out at low and at high speed, consecutively. The low-speed centrifugation is usually carried out at $1000 \times g$ for 10 min and the subsequent high-speed centrifugation at $10,000 \times g$ for 10 min. The low-speed centrifugation is necessary to remove intact cells, cellular debris, and nuclei. Since mitochondria can be trapped in the low-speed pellet, resuspension of the pellet and recentrifugation at low speed increases the yield of mitochondria. The two supernatants obtained from the low-speed centrifugation undergo a subsequent high-speed centrifugation to sediment mitochondria. In many tissues, a light-colored pellet or "fluffy" layer may sediment over the darker brown mitochondrial pellet after high-speed centrifugation. This layer consists of broken mitochondrial membranes and of mitochondria with structural alterations. The fluffy layer can be removed by gentle shaking in the presence of few drops of medium and discarded. In some tissues, two distinct mitochondrial fractions can be isolated. Usually, the heavier pellet contains the most intact, metabolically active mitochondria. This pellet is usually considered "crude," and in some cases a purification based on size and density is necessary. This can be achieved by using discontinuous gradients (sucrose, Ficoll, or metrizamide).

C. Storage

Mitochondrial pellets should be suspended in a minimal volume of isolation buffer. It is preferable to use nonionic media to prevent the loss of peripheral membrane proteins. Prior to storage or to use, the protein concentration of the mitochondrial preparation should be determined, using either the biuret (Gornall *et al.*, 1949) or Lowry method (Lowry *et al.*, 1951). The biuret method is recommended if high yields of mitochondria are obtained, while the Lowry method is more sensitive for determining small protein concentrations. Mitochondria should be stored at concentrations no lower than 40-mg protein/ml of resuspension buffer, and should be kept frozen at $-70\,^\circ$C.

D. Criteria of Purity and Intactness

Mitochondria free from contaminant membranes should have negligible activities of the marker enzymes for other subcellular fractions, such as glucose-6-phosphatase for endoplasmic reticulum, acid hydrolases for lysosomes, and catalase and d-amino acid oxidase for peroxisomes. On the other hand, mitochondria should be highly enriched for cytochrome *c* oxidase (COX) and succinate dehydrogenase.

COX activity is determined spectrophotometrically following ferrocytochrome *c* oxidation at 550 nm at $25\,^\circ$C (Wharton and Tzagaloff, 1967). Determination of COX activity is performed in the presence or absence of a detergent such as lauryl maltoside (Stieglerovà *et al.*, 2000) or *n*-dodecyl-β-d-maltoside (Musatov *et al.*, 2000). The ratio of the two activities provides a measure of the integrity of the outer membrane (OM). Citrate synthase is a stable mitochondrial enzyme whose activity is not subject to fluctuations and pathological changes. For this reason, when homogenates or impure mitochondrial fractions are used for enzymatic determinations, activities are best compared by normalization to citrate synthase, in order to prevent artifacts due to differences in the mitochondrial content (Chapter 4 by Kirby *et al.*, this volume for details).

Alternatively, it is possible to evaluate the content of cytochromes following the method described by Vanneste (1966) and Nicholls (1976). Briefly, the cytochrome content is evaluated by differential spectra (dithionite-reduced minus ferricyanide-oxidized) in a sample of 5- to 6-mg mitochondrial protein diluted in 1 ml of 25-mM KH_2PO_4, 1-mM EDTA, pH 7.4 in the presence of 1% deoxycholate, in a double-beam spectrophotometer.

III. Mitochondria from Beef Heart

Mitochondria prepared from heart muscle present some advantages over those from other mammalian tissues. They are stable with respect to oxidation and phosphorylation for up to a week when stored at $4\,^\circ$C, and for up to a year at $-20\,^\circ$C. Generally, mitochondria prepared from slaughterhouse material are functionally intact. While beef heart is the tissue of choice for this preparation, porcine heart may also be used.

A. Small–Scale Preparation

The method used is essentially that described by Smith (1967), with modifications to allow large-scale isolation (i.e., from one or two beef hearts). Hearts are obtained from a slaughterhouse within 1–2 h after the animal is slaughtered, and placed on ice. All subsequent procedures are carried out at 4°C. Fat and connective tissues are removed and the heart tissue is cut into small cubes. Two-hundred grams of tissues are passed through a meat grinder and placed in 400 ml of sucrose buffer (0.25-M sucrose, 0.01-M Tris–HCl, pH 7.8). The suspension is homogenized in a Waring blender for 5 sec at low speed, followed by 25 sec at high speed. At this stage, the pH of the suspension must be adjusted to 7.5 with 1-M Tris. The homogenate is centrifuged for 20 min at $1200 \times g$ to remove unruptured muscle tissue and nuclei. The supernatant is filtered through two layers of cheesecloth to remove lipid granules, and then centrifuged for 15 min at $26,000 \times g$. The mitochondrial pellet obtained is resuspended in 4 volumes of sucrose buffer, homogenized in a tight-fitting Teflon-glass Potter-Elvejhem homogenizer (clearance of 0.02 mm), and then centrifuged at $12,000 \times g$ for 30 min. The pellet is resuspended in the sucrose buffer and stored at –80°C, at a protein concentration of 40 mg/ml [determined using the biuret method (Gornall *et al.*, 1949)].

The heavy fraction of BHM obtained by this procedure was reported to have a high content of mitochondrial components, with an excess of cytochrome oxidase over the other complexes (Capaldi, 1982) (Table I) and with high respiratory activities accompanied by high RCR and by P/O ratios approaching the theoretical values (Table II). In our experience, mitochondria obtained using the above isolation method do not exhibit good respiratory control. However, they have rates of substrate oxidation comparable to those described by Smith (1967). This is likely to be due to the quality of distilled water used for the preparation, as well as to the extent of mitochondrial rupture resulting from the harsh treatment. Therefore, we only use BHM for manipulations that do not require coupled mitochondria.

When the activity of individual enzymes, and not overall respiration, is determined, BHM preparations should be subjected to 2- to 3-freeze-thaw cycles. This disrupts the outer and the inner mitochondrial membranes, allowing impermeable substrates, such as NADH, to enter the matrix, and allows cytochrome c to become available at the outer side of the inner membrane (IM). Alternatively, small amounts of detergents, such as deoxycholate, may be added to breach the permeability barrier. Many individual and combined activities can be assayed, such as NADH CoQ reductase (Complex I), succinate CoQ reductase (Complex II), ubiquinol cytochrome c reductase (Complex III), COX (Complex IV), NADH-cytochrome c reductase (Complexes I + III), and succinate cytochrome c reductase (Complexes II + III) (These assays are desribed in Chapter 4 by Kirby *et al.*, this volume).

1. A Rapid Method to Prepare BHM

Up to 500 g of beef heart are dissected into 25-mm sections and added to 1.5 volumes of 0.25-M sucrose, 0.01-M Tris–HCl, pH 7.8, 0.2-mM EDTA, and homogenized in an Ultraturrax blender with a rheostat set at 5–6 for 90 sec in

Table I

The Main Components of the Oxidative Phosphorylation System in the Bovine Heart Mitochondrial Membrane[a]

	Concentration range		Molecular weight (kDa)	Number of polypeptides	Prosthetic groups[c]
	nmol/mg protein	μM in lipids[b]			
Complex I	0.06–0.13	0.12–0.26	700	46	FMN, 7Fe–S
Complex II	0.19	0.38	200	4–5	FAD, 3Fe–S
Complex III	0.25–0.53	0.50–1.06	250	12	2b, c_1, Fe–S
Complex IV	0.60–1.0	1.20–2.0	160	12	a, a_3, 2Cu
Cytochrome c	0.80–1.02	1.60–2.04	12	1	c
Ubiquinone-10	3.0–8.0	6.0–16.0	0.75	–	–
NADH-NADP			120	1	–
Transhydrogenase	0.05	0.1			
ATP synthase	0.52–0.54	1.04–1.08	500	23[d]	–
ADP/ATP translocator	3.40–3.60	6.8–9.2	30	1	–
PL	440–587	–	0.7–1.0	–	–

[a]Modified from Capaldi, 1982.
[b]Assumes PL to be 0.5-mg/mg protein.
[c]Fe–S = iron–sulfur clusters; b, c_1, c, a, a_3 are the corresponding cytochromes.
[d]14 types of subunits.

Table II

Oxidation of Various Substrates by BHM[a]

Substrate	Rate of oxidation[b]	P:O
Pyruvate + malate	0.234	2.9
Glutamate	0.181	3.1
2-Oxoglutarate	0.190	3.8
3-Hydroxybutyrate	0.123	3.0
Succinate	0.050	2.0

[a]Modified from Hatefi et al., 1961.
[b]Rate of oxidation is expressed as μg atoms oxygen taken up per minute per millgram mitochondrial protein.

30-sec bursts plus 30-sec pause until smooth. The pH of the homogenate is adjusted to 7.8 with 1-M Tris–HCl. After this first homogenization step, a second homogenization is carried out using a Teflon-glass Potter-Elvehjem homogenizer and the pH is adjusted to 7.8 with 1-M Tris–HCl. The homogenate is centrifuged at $1000 \times g$ for 10 min at $4°C$; the supernatant is collected, passed through a four layers of cheesecloth, and is centrifuged again at $1000 \times g$ for 15 min. The supernatant is collected and centrifuged at $12,000 \times g$ for 15 min. The pellet is recovered, resuspended in 4 volumes of 0.25-M sucrose, 0.01-M Tris–HCl, pH 7.8,

0.2-mM EDTA, and is recentrifuged at 12,000 × *g* for 15 min. The pellet is resuspended in a small volume of 0.25-M sucrose, 0.01-M Tris–HCl, pH 7.8, 0.2-mM EDTA, and is stored at –80°C.

B. Preparation of Coupled Submitochondrial Particles

BHM preparations obtained with the method described above are not maximally efficient for the coupling of oxidation to phosphorylation. However, further manipulations of BHM can yield a preparation of submitochondrial particles (SMP) fully capable of undergoing coupled oxidative phosphorylation. These are known as phosphorylating electron transfer particles derived from BHM (ETPH).

ETPH are inside-out vesicles formed by pinching and resealing of the cristae during sonication (Fig. 2) and are obtained using the methods of Beyer (1967) and Hansen and Smith (1964), with minor modifications. These particles are often better sealed than their parent mitochondria, with a more intact permeability barrier, and therefore have higher coupling capacity.

BHM are prepared as described in Section III.A.1. However, the mitochondrial pellet, either freshly prepared or thawed from –80°C storage, is resuspended in STAMS buffer (0.25-M sucrose, 0.01-M Tris–HCl, pH 7.8, 1-mM ATP, 5-mM $MgCl_2$, 10-mM $MnCl_2$, and 1-mM potassium succinate). The final protein concentration is adjusted to 40 mg/ml with STAMS. Aliquots of 20–25 ml of this mitochondrial suspension are sonicated for 30 sec using a probe sonicator. We use either a Branson sonicator at 20 kc or a Braun Sonifier set at 150. The suspension is then centrifuged at 20,000 × *g* for 7 min in order to remove big particles. The supernatant is decanted and centrifuged at 152,000 × *g* for 25 min, and the pellet is rinsed with STAMS buffer and resuspended to a final protein concentration of 20 mg/ml in a preserving mix containing 0.25-M sucrose, 10-mM Tris–HCl, pH 7.5, 5-mM $MgCl_2$, 2-mM glutathione, 2-mM ATP, and 1-mM succinate (Linnane and Titchener, 1960).

ETPH are loosely coupled: they exhibit high rates of NADH and succinate oxidation, but no respiratory control. However, respiratory control can be detected after the addition of oligomycin. Since this ATPase inhibitor slows down respiration rates, the lack of respiratory control is probably due to backflow of protons through the ATPase membrane sector.

Since EPTH are closed inverted vesicles derived from the inner mitochondrial membrane, a pH gradient (ΔpH) can form in coupled EPTH preparations. The actual ΔpH in ETPH preparations can be determined by analysis of the distribution of 9-aminoacridine, an amine that is membrane permeable only when uncharged (Casadio *et al.*, 1974). The ΔpH of ETPH obtained using succinate as a substrate is typically 3.15 (Lenaz *et al.*, 1982). Quenching of 9-aminoacridine and derivatives, like atebrine and ACMA (9-amino-6-chloro-2-methoxy acridine) in the presence of substrates and valinomycin plus potassium (added in order to

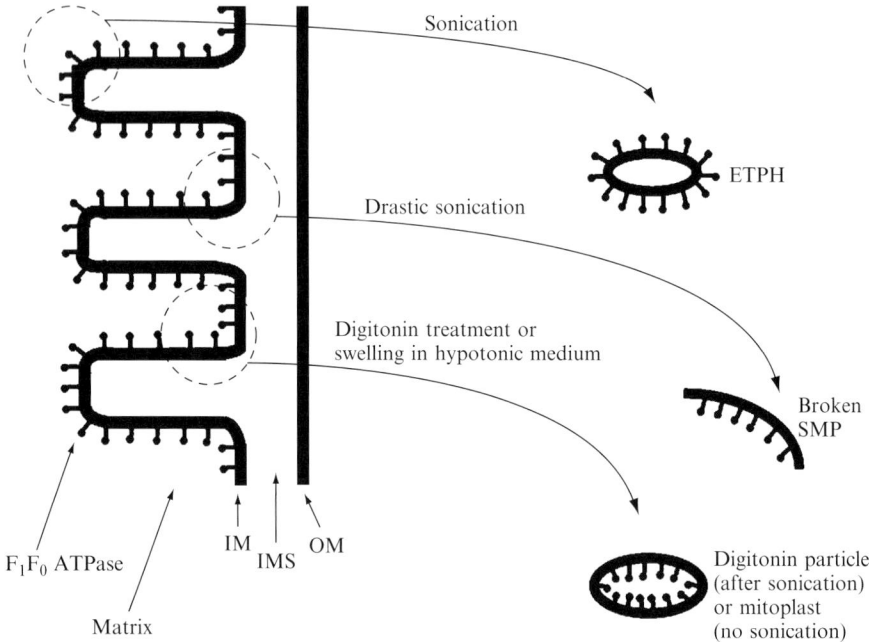

Fig. 2 Derivation of ETPH, digitonin particles, mitoplasts, and SMP from mitochondrial cristae. IM, inner membrane; IMS, intermembrane space; and OM, outer membrane.

collapse the membrane potential), can also be used to detect ΔpH in ETPH preparations (Fig. 3).

C. Broken Submitochondrial Particles

SMP are broken membrane fragments that can react with exogenously added cytochrome c. They are not coupled, but they have good rates of individual enzymatic activities, such as NADH-coenzyme Q (CoQ) oxidoreductase and ubiquinol-cytochrome c oxidoreductase. SMP have been used in our laboratory for the kinetic characterization of Complex I (Estornell $et\ al.$, 1993) and Complex III (Fato $et\ al.$, 1993).

SMP are prepared by sonication of 10-mg/ml BHM suspension as obtained in Section III.A, in a Braun (Labsonic U) sonicator equipped with a 9-mm tip at 150 W for a total of 5 min, in 30-sec bursts followed by 30-sec off, thus allowing the preparation to cool down. Preparations are kept on ice and under nitrogen in order to avoid lipid peroxidation. The mitochondrial suspension is then centrifuged at 20,000 $\times$ g for 10 min; the supernatant is collected and ultracentrifuged at 152,000 $\times$ g for 40 min. SMP are resuspended in sucrose buffer and kept frozen at $-80\,^{\circ}$C at a protein concentration of 40–50 mg/ml, until needed.

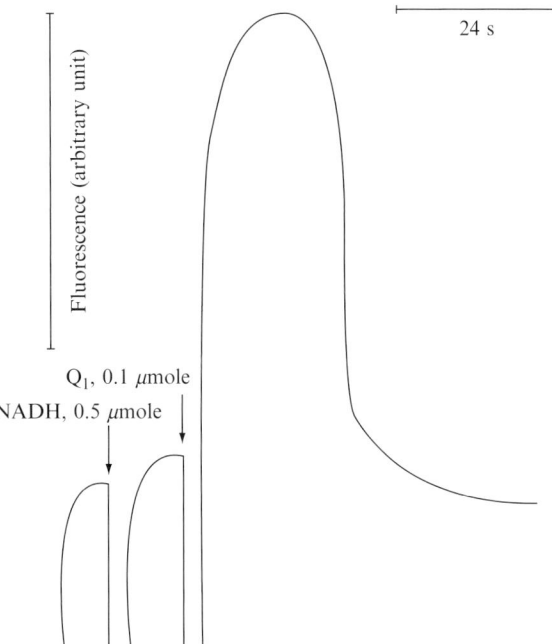

Fig. 3 Quenching of the fluorescence of atebrine in an ETPH preparation, induced by the oxidation of NADH by coenzyme Q_1 (modified from Melandri *et al.*, 1974). The assay is performed in a final volume of 2.5 ml containing: 100-μmol glycylglycine, pH 8.0; 1.25-mmol sucrose; 12.5-μmol $MgCl_2$; 250-μmol KCl; 2.5-μmol EDTA; 4-μg valinomycin; 10-nmol atebrine; and ETPH corresponding to 480 μg of protein.

In these particles, succinate evokes a slight quenching of 9-aminoacridine fluorescence that was roughly calculated to correspond to <5–8% sealed inverted vesicles (Casadio *et al.*, 1974). SMP can be used to assay NADH-, succinate-, and ubiquinol-cytochrome *c* reductase (or COX) without adding detergents (Table III). The use of detergents (i.e., deoxycholate) is important in evaluating the status of the particles; if detergents stimulate cytochrome *c* reduction by upstream substrates, the preparation contains high levels of closed particles (Degli Esposti and Lenaz, 1982). On the other hand, ETPH must have very low cytochrome *c* reductase activities that are strongly stimulated by adding detergents.

D. Keilin–Hartree Heart Muscle Preparation

The Keilin-Hartree heart muscle preparation consists of disrupted mitochondrial IMs. The original procedure (Keilin and Hartree, 1940) has been modified and adapted several times, and these preparations are still used with success for kinetic studies of mitochondrial activities. Here we describe the method as modified by

Table III
Some Individual Respiratory Chain Activities in Broken SMP

Activity	Substrates	V_{max} (μmol/min/mg)	k_{cat} (s^{-1})
NADH-CoQ[a]	NADH	0.98	380
	NADH	0.58	225
Ubiquinol-cytochrome c[b]	Ubiquinol-1 cytochrome c	2.77	220
	Ubiquinol-2 cytochrome c	4.65	370
Succinate-CoQ[c]	Succinate, CoQ_1	1.08	63

[a]From Fato *et al.*, 1996.
[b]From Fato *et al.*, 1993.
[c]Unpublished data from R. Fato and G. Lenaz.

Vinogradov and King (1979) that is particularly suitable for small-scale preparations, not only from beef heart but also from the heart of various organisms; it can be scaled down to 3 g of heart tissue.

Briefly, the ground heart muscle mince is washed in tap water and squeezed through a double layer of cheesecloth to remove excess liquid. The tissue is then washed overnight in 0.1-M phosphate buffer, pH 7.4, with efficient stirring, then squeezed vigorously through cheesecloth and washed in distilled water. The mince is then placed in a porcelain mortar with 350 ml of 0.02-M phosphate buffer, pH 7.4, and ground twice with 500 g of well-washed dry sand for 45–60 min at room temperature. The mixture is then centrifuged for 20 min at 2000 × g. The supernatant layer is collected. The fluffy layer that is recovered above the pellet is discarded and the pellet is subjected to another round of grinding and centrifugation, as above. The combined supernatants are centrifuged at 40,000 × g for 1.5 h and the supernatant is discarded. The pellets is suspended in 0.15-M borate-phosphate buffer, pH 7.8, and stored at 0–2 °C.

This preparation is largely a mixture of broken IM fragments which are mainly in the inside-out orientation. It has good respiratory activities that can be elevated by addition of exogenous cytochrome c. It has oligomycin-sensitive ATPase, but no oxidative phosphorylation activity.

E. Cytochrome c-Depleted and Cytochrome c-Reconstituted Mitochondria and ETPH

Cytochrome c is a mobile component of the respiratory chain, must be released from the organelle in order to estimate its amount. The quantitative extraction of cytochrome c can be accomplished with salts, yielding a good relative amount of non-denatured mitochondria. The method for cytochrome c extraction from mitochondrial preparations has been described previously by MacLennan *et al.* (1966). This procedure usually removes about 85% of the cytochrome c content from mitochondria; the rates of substrate oxidation of these mitochondria are 15%

of the rates of the mitochondrial preparation before extraction procedure. It is summarized briefly in this section.

Suspensions of freshly isolated BHM are suspended in a solution of 0.015-M KCl (Jacobs and Sanadi, 1960), a hypotonic medium for intact mitochondria, to a protein concentration of 20 mg/ml. Mitochondria are allowed to swell for 10 min on ice before centrifugation at $105,000 \times g$ for 15 min. The colorless supernatant is discarded and the pellet is resuspended in 0.15-M KCl, an isotonic medium for intact mitochondria. The mixture is left on ice for 10 min and then subjected to centrifugation step at $105,000 \times g$ for 15 min. The resulting supernatant is red in color. The mitochondrial pellet is then extracted twice in isotonic medium, as described above. The supernatants contain the extracted cytochrome c (hence the red color). The pellet of cytochrome c-depleted mitochondria is resuspended in 0.25-M sucrose and 0.01-M Tris–acetate, pH 7.5.

Preparation of cytochrome c-depleted ETPH is achieved by sonication of cytochrome c-depleted mitochondria (Lenaz and MacLennan, 1967), using the conditions for sonication described in Section III.A. Usually, the ETPH preparation undergoes two sonication cycles: after the first low-speed centrifugation ($26,000 \times g$ for 15 min) the supernatant is collected and the pellet is resonicated and recentrifuged at low speed. The resulting supernatants are pooled and centrifuged at high speed ($105,000 \times g$ for 45 min). Cytochrome c-depleted ETPH preparations show reductions in substrate oxidation and in P/O ratios in comparison with normal ETPH.

Reincorporation of cytochrome c into cytochrome c-depleted ETPH is complicated by the fact that ETPH are inside-out vesicles that cannot react with exogenous cytochrome c. For this reason, KCl-extracted BHM are resuspended in the preserving mix for ETPH, except that cytochrome c is added at a concentration of 10 μg/mg of protein before the BHM are disrupted by sonication. The same procedure as for preparation of the cytochrome c-depleted ETPH is then followed.

Table IV summarizes the properties of cytochrome c-depleted and reconstituted preparations.

F. CoQ-Depleted and CoQ-Reconstituted Mitochondria

Coenzyme Q can be extracted from BHM preparations using organic solvents. CoQ-depleted mitochondria are usually reconstituted with CoQ homologues and analogues. Polar solvents, such as acetone, were first used for CoQ extraction, but they irreversibly damage Complex I; thus nonpolar solvents are preferred, but they can extract neutral lipids, such as CoQ, only from dry material. The extraction is performed on lyophilized BHM preparations using a modification of the method of Szarkowska (1966).

The BHM preparation (Section III.A) is thawed on ice and diluted to a final protein concentration of 20 mg/ml in sucrose buffer (0.25-M sucrose, 0.01-M Tris–HCl, pH 7.5). The suspension is centrifuged at $35,000 \times g$ for 10 min (17,000 rpm in a

Table IV
Oxidative and Phosphorylating Properties of Cytochrome c-Depleted and -Reconstituted Mitochondria and ETPH

	A. BHM[a]				
Substrate	Cytochrome c removed oxygen uptake[b]	P/O ratio	Cytochrome c added oxygen uptake[b]	P/O ratio
Pyruvate + malate	0.022	1.14	0.189	2.40
Succinate	0.035	0.71	0.136	1.06
Ascorbate + TMPD	0.037	0.24	0.169	0.63

	B. ETPH[c]	
Treatment	Oxygen uptake with succinate[a]	P/O ratio
None	0.163	1.02
ETPH(– cytochrome c)[d]	0.064	0.77
ETPH(+ cytochrome c)[e]	0.155	1.15
ETPH(– cytochrome c) + cytochrome c added after sonication	0.093	0.55

[a]Data obtained from MacLennan et al., 1966.
[b]Expressed as μg atoms/min/mg protein.
[c]Data obtained from Lenaz and MacLennan, 1966.
[d]By sonication of cytochrome c-depleted mitochondria.
[e]Cytochrome c added before sonication.

Sorvall SS34 rotor). The pellet is resuspended in 0.15-M KCl, frozen at –80 °C, and lyophilized.

To extract CoQ, 10 ml of pentane is added to about 0.5 mg of lyophilized BHM. The mixture is homogenized in a Potter-Elvejhem homogenizer with Teflon pestle and then centrifuged at $1100 \times g$ for 10 min (3000 rpm in a Sorvall SS34 rotor). The supernatant, containing pentane, is removed and the extraction–homogenization is repeated four times. Finally, pentane is removed from the extracted mitochondria, first in a rotary evaporator under reduced pressure at 30 °C, and then under high vacuum at room temperature, for 2 h. Extracted mitochondria are homogenized in sucrose buffer, centrifuged at $35,000 \times g$ for 10 min at 4 °C, and resuspended in the same buffer. This suspension can either be used immediately or stored at –80 °C for later use.

Reconstitution of CoQ-depleted mitochondria is attained by treating the dry-extracted mitochondrial powder with either the pentane extract or CoQ homologues and analogues dissolved in pentane, using a modification of the method of Norling (Norling et al., 1974). Depleted particles are homogenized in a small volume of pentane (usually 2 ml) in the presence of a known amount of CoQ. The amount of CoQ is usually adjusted to a final concentration between 0.4 and 25-nmol CoQ/mg

Table V

Kinetic Constants of NADH- and Succinate–Cytochrome c Reductase in Lyophilized and Pentane–Extracted BHM Reconstituted with Some Representative CoQ Homologues Having Different Lengths of the Isoprenoid Side Chain

Quinone	NADH-cytochrome c reductase		Succinate-cytochrome c reductase	
	V_{max}[a]	K_m[b]	V_{max}[a]	K_m[b]
CoQ_{10}	0.64 ± 0.46	1.53 ± 1.16	0.26 ± 0.17	0.42 ± 0.57[c]
CoQ_3	0.12 ± 0.04	1.42 ± 0.92	0.20 ± 0.01	0.53 ± 0.33[c]
CoQ_5	0.15 ± 0.02	1.46 ± 0.38	0.13 ± 0.05	0.37 ± 0.24[d]
6-Decylubiquinone (DB)	0.55	25	0.52	3.8

[a]Expressed as μmol/min/mg protein.

[b]Expressed as nmol quinone/mg protein, except for DB, whose K_m was calculated as mM quinone in the PL.

[c]Statistically significant with respect to the corresponding K_m of NADH-cytochrome c reductase, $p < 0.05$.

[d]As above, $p < 0.001$.

mitochondrial protein. The protein content of CoQ-depleted mitochondria is estimated from the dry weight; 3-mg dry weight usually corresponds to 1-mg mitochondrial protein, as measured by the biuret method.

The particle suspension is transferred in a rotary evaporator at 4°C under a slightly reduced pressure. Gradual removal of pentane is essential in order to get a good rate of incorporation (pentane should evaporate over a period of 30 min). After complete removal of the pentane, particles are dried at 4°C for additional 30 min under vacuum and resuspended in sucrose buffer.

CoQ-depleted mitochondria have negligible contents of residual CoQ (<20-pmol/ mg protein) and exhibit very low rates of both NADH and succinate oxidation in comparison with controls (either intact or lyophilized mitochondria). Reconstitution with long-chain ubiquinones (CoQ5 to CoQ10) reconstitutes maximal rates of NADH and succinate oxidation. Reconstitution with short-chain homologues results in reconstitution of succinate oxidation. Thus, short-chain homologues behave like Complex I inhibitors.

Some properties of these mitochondria are shown in Table V.

IV. Mitochondria from Rat Liver

Liver mitochondria prepared from rats are usually suitable for biochemical assays in studies on the pharmacological effects of different drugs or on effects of specific diets on mitochondrial membrane composition. We used rat liver mitochondria in studies after perfusion and in studies where quinones were incorporated in liver fractions. Rat liver contains a considerable amount of mitochondria and is easier to manipulate than is skeletal muscle. Generally, a high-mitochondrial yield is obtained from rat liver.

A. Standard Preparation

Our method is a modification of that described by Kun *et al.* (1979). This method is based on the typical differential centrifugation procedure used for other mitochondrial preparations.

Fresh tissue is chilled on ice and washed in 0.22-M mannitol, 0.07-M sucrose, 0.02-M HEPES, 2-mM Tris–HCl, pH 7.2, 1-mM EDTA (Solution A). Subsequently, it is minced with scissors and washed thrice in solution A containing 0.4% BSA to remove blood and connective tissue, and weighed in a prechilled glass Petri dish. About 10 g of mince are suspended in 200–250 ml of solution A containing 0.4% BSA and then homogenized in a prechilled Potter-Elvehjem glass homogenizer using a Teflon pestle and filtered. The homogenate is centrifuged at 3000 × *g* for 1.5 min (5000 rpm in a Sorvall SS34 rotor). The supernatant is decanted. The pellet is resuspended in Solution A, and subjected to a second centrifugation step.

The two supernatants are combined and centrifuged at 17,500 × *g* for 2.5 min (12,000 rpm in a Sorvall SS-34 rotor). The resulting pellet is washed in Solution A and then centrifuged at 17,500 × *g* for 4.5 min. The pellet is resuspended in 0.22-M mannitol, 0.07-M sucrose, 0.01-M Tris–HCl, pH 7.2, 1-mM EDTA (Solution B) and is centrifuged at 17,500 × *g* for 4.5 min. The pellet is finally resuspended in Solution B at a ratio of 10-ml Solution B per 10 g of starting material, in order to standardize the protein content of the mitochondrial fraction.

Mitochondria obtained by this method also contain a lysosomal fraction, but if used immediately they show a good respiratory control with glutamate-malate and with succinate (Fig. 4), and good enzymatic activities of the four mitochondrial complexes (after membrane permeabilization by freezing-thawing cycles or by detergents).

B. Gradient–Purified Rat Liver Mitochondria

The preparation of rat liver mitochondria free from lysosomal contamination involves purification of mitochondria in metrizamide gradient. The method described here is a modification of Kalen *et al.* (1990). This procedure for purification of mitochondria has been used by us for monitoring the incorporation of exogenous CoQ by rat liver fractions (Genova *et al.*, 1994).

The liver is removed, chopped, weighed, and washed at least 10 times in ice-cold Solution A (0.33-M sucrose, 0.01-M Tris–HCl, pH 7.4, 1-mM EDTA, 0.4% BSA). The tissue is then homogenized in a prechilled Potter-Elvehjem glass homogenizer with a Teflon pestle to a ratio 1:4 (w/v) tissue to Solution A.

The homogenate is centrifuged at 310 × *g* for 10 min. The supernatant is decanted and the pellet resuspended in Solution B (0.33-M sucrose, 0.01-M

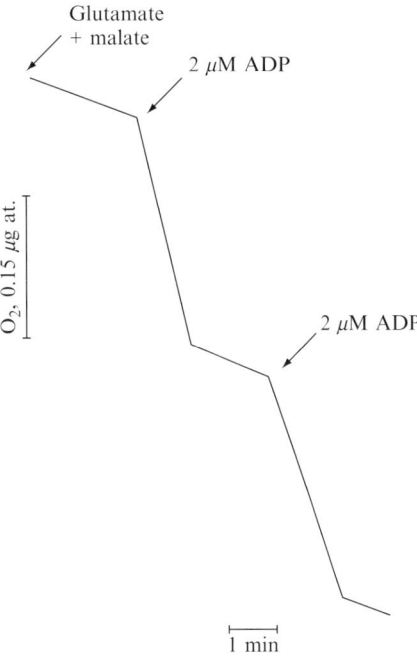

Fig. 4 Respiratory control in the presence of glutamate + malate in a mitochondrial preparation from rat liver. This preparation had an RCR equal to 11.

Tris–HCl, pH 7.4, 1-M EDTA) and recentrifuged in order to obtain a clean nuclear fraction. The supernatant, previously collected, is centrifuged at 2800 × g for 20 min and the pellet (washed twice) is recovered for resuspension in 3 volumes as specified below; volumes for washes are not generally stated or specified. This pellet contains a fraction of mitochondria defined as "heavy mitochondria" (HM). The supernatant is then centrifuged at 11,400 × g for 20 min, thus giving a pellet of mitochondria defined as "light mitochondria" (LM) containing lysosomes. The resulting supernatant, containing microsomes and cytosol, is discarded. The LM and HM fractions are resuspended in Solution B at a volume three times the weight of the starting rat tissue.

Aliquots of HM and LM are then added to 2 volumes of metrizamide solution as described (Wattiaux *et al.*, 1978), usually 0.4 ml of mitochondria in 0.8-ml metrizamide 85.6% (w/v). One milliliter of HM/LM-metrizamide solution is then layered at the bottom of a 5-ml ultracentrifuge tube, and a discontinuous metrizamide gradient is stratified on top of it by adding 0.6-ml 32.82% metrizamide, 0.6-ml 26.34% metrizamide, 0.6-ml 24.53% metrizamide, and 0.6-ml 19.78% metrizamide. Centrifugation is then performed in a swinging bucket rotor at 95,000 × g for 2 h at 4°C.

Mitochondrial subfractions can be collected using a Pasteur pipette. Usually, three subfractions from HM preparations and four subfractions from LM preparations

(Genova *et al.*, 1994) are observed. Mitochondria prepared with this procedure exhibit good activities of the respiratory chain complexes, but little respiratory control.

C. Liver SMP

It is also possible to obtain SMP from rat liver. Usually SMP from rat liver are prepared after intensive sonication of mitochondria; thus they are not coupled. However, they can be used for studying NADH oxidation, since intact mitochondria are not permeable to NADH.

The method used in our experiments is essentially the one described by Gregg (1967). Mitochondria are prepared from 20 to 30 g of rat liver and the buffer used is the same as that used for the preparation of broken SMP particles from BHM (Section III.C), with the addition of 0.4% BSA from the beginning of the isolation procedure.

D. Mitochondria from Rat Hepatocytes

Preparation of isolated rat liver cells is usually performed using the two-step collagenase liver perfusion technique of Seglen (1976). The purpose of this technique is to obtain intact hepatocytes separated from nonparenchymal cells (up to 40% of total liver tissue). We have used this method to treat isolated cells with adriamycin in order to induce oxidative stress by reduction of this potent anticancer agent to semiquinone, which releases superoxide anion and hydrogen peroxide. In order to evaluate cellular integrity (or viability), the trypan blue exclusion test is performed.

Mitochondria are prepared from hepatocytes (30×10^6 cells); 1×10^6 cells/ml are resuspended in Krebs-Henseleit medium (pH 7.4) supplemented with 5-mM glucose, incubated under an atmosphere of 95% O_2/5% CO_2 in a shaking bath at $37°C$ for 2 h (Barogi *et al.*, 2000; Wells *et al.*, 1987). Cells are then pelleted (300 g for 3 min) and resuspended in 10 ml of Solution A (0.25-M sucrose, 0.01-M tricine, 1-mM EDTA, 10-mM NaH_2PO_4, 2-mM $MgCl_2$, pH 8) supplemented with 0.4% BSA and frozen at $-80°C$ for 10 min to break the plasma membrane, and centrifuged at $760 \times g$ for 5 min. The supernatant is kept while the pellet is homogenized, using Ultra-turrax homogenizer for 10 min, followed by centrifugation at $760 \times g$ for 5 min. The supernatants from the previous two steps are combined and centrifuged for 20 min at $8000 \times g$. The mitochondrial pellet is washed once with Solution A and finally resuspended in the same buffer. Mitochondria obtained with this method show good respiratory activities, but the NAD-dependent activities are rather low (Barogi *et al.*, 2000).

E. Isolation of Mitochondria from Frozen Tissues

Generally, mitochondrial isolations are performed on fresh tissues. Under these conditions, isolated mitochondria are pure and suitable for most of the biochemical assays and for further modifications of mitochondria to be used in particular assays. However, in cases when fresh tissues are not available, it is often necessary to prepare mitochondria from frozen tissues.

The most important step in the procedure is the freeze-thawing of the tissue. Mitochondrial membranes are very sensitive to temperature variations. Rupture of the membranes (cellular and mitochondrial) prior to homogenization makes it impossible to separate the subcellular fractions, and also modifies biochemical parameters.

In our experience, the most suitable method to preserve organs prior to extraction of mitochondria is that described by Fleischer and Kervina (1974) for liver tissue. We have also applied the procedure for extractions from other organs, such as heart, muscle, and kidney (Barogi et al., 1995; Castelluccio et al., 1994). This method allows for tissue storage for prolonged periods prior to the extraction of mitochondria.

After sacrificing the animal, organs are weighed and 1 volume of storage medium [0.21-M mannitol, 0.07-M sucrose, 20% dimethyl sulfoxide (DMSO), pH 7.5] is added. DMSO, used as an antifreeze agent, has the advantage over glycerol in that it is less viscous and penetrates the tissues rapidly. It is important to prevent ice crystal formation; in order to minimize this phenomenon, the organs should be frozen rapidly, and their thawing should be performed as quickly as possible. Once in storage medium, the organ can either be diced into small pieces and homogenized, or it can be maintained intact before quick freezing in liquid nitrogen. In soft tissues, such as liver, the storage medium rapidly diffuses through the organ. However, it is recommended to cut more fibrous tissues, such as kidney, heart, and muscle, into two to three pieces (1 cm^3) to facilitate medium penetration into the tissue. After rapid freezing in liquid nitrogen, the organs are stored in liquid nitrogen until mitochondrial isolation is required. In our experience, intact organs can be stored in liquid nitrogen with no apparent alterations to mitochondrial function.

The thawing procedure should be performed quickly. It is necessary to preheat the thawing medium (0.25-M sucrose, 0.01-M Tris–HCl, pH 7.5) at 45 °C before adding it to the frozen tissue at a 4:1 ratio of medium to tissue. The thawing medium for liver should also contain 0.4% BSA. The extraction of mitochondria is then performed using the method described in Section III.A, which is suitable for heart, kidney, and muscles as well.

F. Preparation of Rat Liver Mitoplasts

Mitoplasts, that is, mitochondria with ruptured OMs, may be prepared from rat liver mitochondria by treatment with osmotic shock or nonionic detergent. Mitoplasts are often used to determine the submitochondrial localization of proteins or protein activities.

One approach to disruption of OMs is by osmotic shock produced by suspending the mitochondria in a hypotonic medium (Jacobs and Sanadi, 1960; MacLennan et al., 1966). Mitochondria are prepared from rat liver as described in Section IV.A. The mitochondrial pellet, however, is resuspended in 0.22-M mannitol, 0.07-M sucrose, 2-mM HEPES, pH 7.4 (Solution A). After centrifugation at $20,500 \times g$ for 20 min, the pellet is resuspended in 15-mM KCl to a final concentration of 20 mg/ml and kept on ice for 10 min, thus allowing the external membrane to swell. The suspension is centrifuged at $31,000 \times g$ for 45 min and resuspended in Solution A to a final concentration of 40–50 mg/ml. Alternatively, mitochondria from rat liver are washed in Solution A in the presence of 0.5-mg/ml BSA, centrifuged at $4500 \times g$, and resuspended in Solution A.

An alternative method for removing the OM and for purifying the IM-matrix fraction is given by the "controlled digitonin incubation" method (Schnaitman and Greenawalt, 1968). Digitonin is diluted in Solution A at a concentration of 2% (w/v) and 0.5-mg/ml BSA is added after the digitonin is dissolved. The rat liver mitochondrial suspension is then treated with 1.1 mg of digitonin solution per 10 mg of mitochondrial protein. The suspension is stirred gently for 15 min and then diluted in 3 volumes of Solution A. This suspension is then centrifuged at $31,000 \times g$ for 20 min and then resuspended in Solution A.

G. Phospholipid-Enriched Mitoplasts and SMP with or Without Excess CoQ

Addition of phospholipid (PL) to mitoplasts or SMP increases the inner mitochondrial membrane surface area and dilutes intramembrane proteins. In the case of SMP, PL enrichment facilitates study of diffusion control of electron transfer by ubiquinone.

Unilamellar vesicles are prepared from soybean PL (asolectin) by suspending the PL in Solution A (0.22-M mannitol, 0.07-M sucrose, 2-mM HEPES, pH 7.4) at a concentration of 200-mg PL/ml Solution A. The suspension is then sonicated using a Braun (Labsonic U) sonicator for a total time of 30 min, with cycles of 30 sec on followed by 30 sec off, at 150 W.

To prepare unilamellar PL vesicles enriched with coenzyme Q_{10}, CoQ_{10} suspended in ethanol and PL suspended in chloroform/methanol (ratio 2:1) are mixed and dried under nitrogen gas. Dried PL and CoQ_{10} are then resuspended in Solution A and sonicated with a Braun sonicator for as described above.

Reincorporating mitoplasts or SMP with PL- or PL/CoQ_{10}-enriched vesicles is carried out according to Schneider $et\ al.$ (1980). Three milligrams of mitochondrial protein are mixed with 1.5-ml PL or PL + CoQ_{10} vesicles (150 mg/ml). The mixture is frozen in liquid nitrogen and then is allowed to thaw at room temperature. This step is repeated thrice. The suspension is then layered on a discontinuous sucrose density gradient (0.6, 0.75, 1, 1.25-M sucrose) and centrifuged in swinging bucket rotor (SW 28) at $90,000 \times g$ for 16 h. The fractions obtained are diluted in 0.25-M sucrose, 0.01-M Tris–HCl, pH 7.8 (SMP solution) for PL-enriched SMP, and in Solution A for PL-enriched mitoplasts. Fractions are then centrifuged twice at $160,000 \times g$ for 45 min and resuspended in the appropriate buffers (SMP solution and Solution A). Protein content is assayed by the biuret method (Gornall $et\ al.$, 1949) and lipid content is measured by the method of Marinetti (1962).

V. Mitochondria from Skeletal Muscle

Biochemical analysis of mitochondria isolated from muscle tissue provides valuable insights into the pathological and physiological characteristics of human diseases related to impaired mitochondrial metabolism (mitochondrial encephalomyopathies). Muscle tissue is a postmitotic tissue, and becomes damaged over time, making it of particular interest for studies on aging. Finally, this tissue is one of the best characterized tissues bioenergetically.

Skeletal muscle mitochondria are usually isolated from several grams of tissues. This may not be a problem if the animal is sufficiently large or if several muscle samples can be pooled. However, human samples are usually obtained from surgical biopsies; needle biopsies yield <1 g of muscle tissues.

When dealing with experimental animals, the method described by Kun *et al.* (1979) is generally used for the isolation of skeletal muscle mitochondria. We find this method to be suitable for the isolation of mitochondria from human muscle biopsies (Zucchini *et al.*, 1995) and from rat gastrocnemius muscle (Barogi *et al.*, 1995). We have exploited this method to investigate some individual respiratory chain activities of muscle mitochondria from young and old individuals. The activities could be transformed into actual turnover numbers (TN) related to Complex III content, established on the basis of the antimycin A inhibition titer (Zucchini *et al.*, 1995) (Table VI).

The method is based on the use of mannitol and sucrose medium, BSA, and EDTA as complexing agent. Essentially, 25–100 mg of muscle tissue is freed from collagen and nerves, weighed and homogenized in Teflon-glass Potter-Elvehjem homogenizer in Solution A (0.22-M mannitol, 0.07-M sucrose, 2-mM Tris, 1-mM EDTA, and 20-mM HEPES, pH 7.2) containing 0.4% BSA, and centrifuged at $600 \times g$ for 80 sec. The supernatant is collected and the crude nuclear fraction in the pellet is subjected to a second round of homogenization and centrifugation. The two supernatants are combined and centrifuged at $17,000 \times g$ for 2.5 min. The mitochondrial pellet is then washed twice in Solution A and resuspended in the same solution.

Lee *et al.* (1993) have described a valid alternative method for the isolation of mitochondria from 3 to 5 g of skeletal muscle, using KCl medium and proteinase K to soften the tissue and facilitate cell disruption during homogenization. The muscle is trimmed, minced, and placed in Medium A (0.1-M KCl, 50-mM Tris–HCl, pH 7.5, 5-mM $MgCl_2$, 1-mM EDTA, 1-mM ATP), in a proportion of 10-ml Medium A/g of muscle. Between 3- and 5-mg proteinase K/g of muscle tissue is then added to the muscle suspension. After 5 min of incubation (2 min is sufficient for human muscle) at room temperature, the mixture is diluted 1:2 with medium A and homogenized for 15 sec by Ultraturrax. The homogenate (pH 7.3) is centrifuged at $600 \times g$ for 10 min. The pellet is discarded and the supernatant is filtered through a two-layer cheese cloth and centrifuged at $14,000 \times g$ for 10 min. The resulting pellet is resuspended in Medium B (0.1-M KCl, 50-mM Tris–HCl, pH 7.5, 1-mM $MgCl_2$, 0.2-mM

Table VI
Enzymatic Activities of Skeletal Muscle Mitochondria from Young and Old Individuals, Expressed as Specific Activities and Turnover Numbers[a]

Age range (years)	NADH-DB Reductase		Succinate-cytochrome c reductase		Ubiquinol-cytochrome c reductase	
	nmol/min.mg protein	TN (s^{-1})	nmol/min.mg protein	TN (s^{-1})	nmol/min.mg protein	TN (s^{-1})
18–29 (n = 5)[b]	52.44 ± 25.35	8.78 ± 1.86	102.82 ± 57.97	16.44 ± 4.92	876.88 ± 437.93	142.16 ± 18.37
69–90 (n = 9)[b]	49.31 ± 31.92	8.45 ± 7.04	90.79 ± 48.86	12.70 ± 8.31	939.72 ± 449.47	124.65 ± 37.29

[a]From Zucchini et al., 1995.
[b]n = number of samples in the age range.
Mann-Whitney U nonparametric test: $p > 0.05$ for all parameters evaluated.

EDTA, 0.2-mM ATP) containing 1% BSA and centrifuged at 7000 $\times$ g for 10 min. The pellet is resuspended in Medium B and centrifuged at 3500 $\times$ g for 10 min and resuspended in 0.25-M sucrose, to give an approximate content of 40- to 50-mg mitochondrial protein/ml.

Another method, which uses special equipment to isolate mitochondria from 25 to 100 mg of skeletal muscle tissue, has been described by Rasmussen *et al.* (1997). Briefly, muscle tissue is weighed and incubated for 2 min in 500 μl ATP Medium (100-mM KCl, 50-mM Tris, 5-mM $MgSO_4$, 1-mM ATP, 1-mM EDTA, pH 7.4, 0.5% BSA) containing 2-mg proteinase K/ml. The proteinase is then diluted with 3-ml ATP medium and the liquid is discarded. ATP medium is added to the digested muscle and the tissue is homogenized (Rasmussen *et al.*, 1997). The homogenate is centrifuged at 300 $\times$ g for 5 min (1600 rpm in a Sorvall SS-34 rotor) and the supernatant obtained is centrifuged at 4500 $\times$ g (6000 rpm in a Sorvall SS-34 rotor) for 10 min. The pellet is washed with 100-mM KCl, 50-mM Tris, 5-mM $MgSO_4$, 1-mM EDTA, pH 7.4 and centrifuged at 7000 $\times$ g for 10 min (6800 rpm in a Sorvall HB-4 swing-out rotor). The pellet is finally resuspended in 0.225-M mannitol, 75-mM sucrose. The relative yield is 40–50% and the mitochondria obtained with this method are well coupled and exhibit high rates of phosphorylating respiration.

Skeletal muscle mitochondria consist of two separate and distinct subfractions, the subsarcolemmal (SS) and the intermyofibrillar (IMF) mitochondria, located beneath the sarcolemma and between myofibrils, respectively. IMF mitochondria contain a higher amount of cardiolipin and a more elevated state-3 respiration rate (Cogswell *et al.*, 1993).

SS and IMF mitochondria can be prepared following the procedure adopted by Jimenez *et al.* (2002). Briefly, muscle tissue (50–250 mg) is minced in 5 ml homogenization buffer (100-mM sucrose, 180-mM KCl, 10-mM EDTA, 5-mM $MgCl_2$, 1-mM ATP, 50-mM Tris–HCl, pH 7.4) with 0.06% protease inhibitor cocktail, and homogenized in a Teflon-glass Potter-Elvehjem homogenizer, using a loose-fitting pestle followed by tight-fittting pestle step. Homogenates are centrifuged at 1600 $\times$ g for 10 min and the pellet is saved for subsequent extraction of IMF mitochondria. The supernatant is filtered through two layers of surgical gauze and centrifuged at 9200 $\times$ g for 10 min; the resulting pellet consists of SS mitochondria.

The pellet saved from the low-speed centrifugation of the homogenate is resuspended in a homogenization buffer containing 100-mM KCl, 1-mM EDTA, 5-mM $MgSO_4$, 1-mM ATP, 50-mM Tris–HCl, pH 7.4 supplemented with 0.06% protease inhibitor cocktail (Sigma, St. Louis, MO), and homogenized using a tight-fitting Teflon pestle in a glass Potter-Elvehjem, in order to avoid the use of protease to soften the muscle tissue. The homogenate is centifuged and processed as described above and the final pellet, containing IMF mitochondria, is obtained by the mean of a centrifugation step at 15,000 $\times$ g for 15 min.

VI. Synaptic and Nonsynaptic Mitochondria from Different Rat Brain Regions

Mitochondrial preparations from brain are often heterogeneous and a number of different methods have been developed to separate the populations of brain mitochondria. These methods, however, involve lengthy centrifugation in hypertonic sucrose gradients (Clark and Nicklas, 1970), and are not compatible with some metabolic studies.

The method for the isolation and the characterization of functional mitochondria from rat brain was first described by Lai and colleagues (Lai *et al.*, 1977). Three distinct populations of mitochondria from rat brain can be isolated from a single homogenate preparation, two from the synaptosomal fraction (HM and LM, for heavy and light mitochondria, respectively) and one from nonsynaptic origin, the so-called "free mitochondria" (FM).

It is possible to separate the different types of mitochondria from the cerebral cortex, hippocampus, and striatum. The protocol that we use was described by Battino *et al.* (1995) and modified by our group (Genova *et al.*, 1997; Pallotti *et al.*, 1998). The animal is sacrificed by decapitation and the skull is opened rapidly. The brain is placed on an ice-chilled glass plate and dissected according to the procedure described by Glowinski and Iversen (1966). First, the posterior region containing the pons and cerebellum is removed. The remainder is divided into two hemispheres. The two brain cortexes are freed from the hippocampus and the striatum, and each one is cut into two pieces. Brain areas should be dissected rapidly (<20 sec) and placed in Buffer A (0.32-M sucrose, 1-mM EDTA, 10-mM Tris–HCl, pH 7.4). The tissue is then homogenized using a Teflon-glass homogenizer at 800 rpm by five up-and-down passes of the pestle (Villa *et al.*, 1989). The homogenate is centrifuged by gradually increasing the centrifugation rate to $1000 \times g$ over a period of 4 min, with centrifugation at $1000 \times g$ for an additional 11 sec. The pellet is resuspended in Buffer A and centrifuged; this step is repeated once again. The three supernatants recovered from the three rounds of centrifugation are pooled and centrifuged at $15,000 \times g$ for 20 min. Mitochondria and synaptosomes are enriched in this "crude" mitochondrial pellet. Mitochondria can be separated from synaptosomes by layering the pellet, resuspended in Buffer A, on a discontinuous Ficoll-sucrose two-step gradient (12% and 7.5% w/w Ficoll in 0.32-M sucrose, 50 μM EDTA, 10-mM Tris–HCl, pH 7.4). The gradient is then centrifuged in a swinging bucket rotor (Sorvall SW 50.1) at $73,000 \times g$ for 24 min (24,000 rpm). Myelin and synaptosomes are recovered in two bands in the gradient. Crude free mitochondria (CFM) are recovered in the pellet. The myelin is removed by aspiration. The synaptosomal band, at the 7.5–12% (w/w) Ficoll interphase, is collected by aspiration, diluted in Buffer A, and centrifuged at $15,000 \times g$ for 20 min. In the original method (Battino *et al.*, 1991) Buffer A used for harvesting the synaptosomes contained protease inhibitors.

We omit protease inhibitors in our buffers because they interfere with Complex I activity.

The pellet obtained from the isopycnic gradient is resuspended in lysis buffer (6-mM Tris–HCl, pH 8.1), homogenized, and centrifuged at $14,000 \times g$ for 30 min. The pellet is resuspended in 3% Ficoll in 0.12-M mannitol, 30-mM sucrose, 25 μM EDTA, 5-mM Tris–HCl, pH 7.4, and layered on a discontinuous Ficoll gradient consisting of two layers of 6% and 4.5% Ficoll in 0.24-M mannitol, 60-mM sucrose, 50-μM EDTA, 10-mM Tris–HCl, pH 7.4. After centrifugation at $10,000 \times g$ for 30 min in a swinging bucket rotor, the HM fraction is recovered in the pellet. The intermediate fraction is diluted in Buffer A and centrifuged at $15,000 \times g$ for 30 min to pellet the LM fraction. FM, LM, and HM pellets are resuspended in minimal volumes of 0.22-M mannitol, 0.07-M sucrose, 50-mM Tris–HCl, 1-mM EDTA, pH 7.2, and stored at $-80\,^{\circ}$C. Several respiratory chain enzymatic activities from three different regions of the brain are reported in Tables VII and VIII.

Total brain mitochondria can be isolated following the procedure described by Kudin *et al.* (2004). Total brain from a single Wistar rat is transferred to ice-cold MSE solution (225-mM mannitol, 75-mM sucrose, 1-mM EGTA, 5-mM HEPES, 1-mg/mL BSA, pH 7.4), minced, and transferred to 10 ml of MSE solution containing 0.05% nagarse for homogenization using a Potter-Elvehjem homogenizer. The resulting homogenate is diluted with an additional 20 ml of MSE-nagarse solution and centrifuged at $2000 \times g$ for 4 min. The supernatant is collected, filtered through cheesecloth, and then centrifuged at $12,000 \times g$ for 9 min. Synaptosomes are enriched in the pellet from this centrifugation. To permeabilize the synaptosomes, the pellet is resuspended in MSE containing 0.02% digitonin, hand homogenized

Table VII

Respiratory Chain Activities of Nonsynaptic and Synaptic Light and Heavy Mitochondria from Rat Brain Regions[a]

	Cortex	Hippocampus	Striatum
Succinate-cytochrome c			
FM	0.196 ± 0.076	0.159 ± 0.048	0.134 ± 0.039
LM	0.171 ± 0.050	0.154 ± 0.051	0.159 ± 0.045
HM	0.075 ± 0.019	0.062 ± 0.011	0.063 ± 0.019
CoQ_2H_2-cytochrome c			
FM	2.397 ± 0.445	1.513 ± 0.175	1.776 ± 0.666
LM	2.647 ± 1.097	2.153 ± 0.995	2.028 ± 0.823
HM	1.137 ± 0.471	0.505 ± 0.129	0.766 ± 0.415
COX			
FM	2.217 ± 0.342	1.677 ± 0.259	2.021 ± 0.380
LM	2.328 ± 0.498	1.870 ± 0.427	1.819 ± 0.688
HM	1.125 ± 0.307	0.698 ± 0.203	0.762 ± 0.219

[a]Enzymatic activities are expressed in μM/min/mg mitochondrial protein. Results are means $\pm$ S.D. for number of rats >10.

Table VIII

Biochemical Parameters in Nonsynaptic Mitochondria (FM), and in Synaptic Light and Heavy Mitochondria from Rat Brain Cortex

	FM		LM		HM	
	4 months	24 months	4 months	24 months	4 months	24 months
Mitochondrial yield[c]	4.46 ± 1.20	3.40 ± 1.82	3.06 ± 0.45	3.26 ± 0.82	5.83 ± 2.04	5.42 ± 1.58
NADH oxidase activity[d]	203 ± 72^b	130 ± 51^b	109 ± 41	100 ± 55	104 ± 18^b	83 ± 18^b
NADH-ferricyanide reductase[d]	4.29 ± 1.26	3.49 ± 0.73	3.91 ± 0.81	4.36 ± 2.36	3.89 ± 0.47	3.78 ± 0.58
I_{50} of rotenone[e]	29 ± 15	41 ± 26	40 ± 14	33 ± 11	25 ± 8	26 ± 9
I_{50} of rotenone (corrected)[a]	6.8 ± 1.7^b	10.8 ± 4.8^b	9.4 ± 2.8	7.9 ± 4.6	6.3 ± 2.4	6.6 ± 2.4

[a](pmol rotenone/mg protein)/NADH-ferricyanide reductase.
[b]$p < 0.05$.
[c]mg protein/g tissue.
[d]$nmol.min^{-1}mg^{-1}$.
[e]pmol rotenone/mg protein.

with 8–10 strokes in a glass homogenizer, and centrifuged at $12,000 \times g$ for 11 min. The pellet is resuspended in a minimal amount of MSE solution to obtain a protein concentration of 20 mg/ml. The method is suitable also for isolation of human mitochondria from biopsies (150–200 mg).

An efficient method for the isolation of a rat mitochondrial fraction freed from synaptosomes and related structures by Percoll density gradient centrifugation has been described by Anderson and Sims (2000). Rats are fasted overnight. They are euthanized by decapitation. Brains are removed as above. Forebrains are separated, minced, and washed thrice in isolation buffer (0.32-M sucrose, 1-mM EDTA, 10-mM Tris–HCl, pH 7.4). The tissue is then homogenized in isolation buffer (10% w/v) using a Dounce homogenizer with 4 strokes using a loose-fitting pestle followed by 8 strokes with a tight-fitting pestle. The homogenate is diluted 1:1 with Percoll (24%) in isolation buffer and centrifuged at $30,700 \times g$ for 5 min. The upper half of the mixture, which is depleted of myelin, is removed and diluted with isolation buffer until the final concentration of Percoll is 12%. Thereafter it is homogenized in a loose-fitting pestle (4 strokes) and recentrifuged at $30,700 \times g$ for 5 min. After centrifugation, the upper part is discarded and the lower part is retained, mixed with the lower half obtained from the first centrifugation, and treated with 125 μl/g initial wet weight of digitonin stock solution (50 mg/ml in 100% methanol). The resulting mixture is layered on a Percoll step gradient (19% and 40%) and centrifuged at $30,700 \times g$ for 5–10 min. The fraction recovered at the interface between the two lower Percoll layers is collected, diluted 1:4 with isolation buffer, and centrifuged at $16,700 \times g$ for 10 min. The loose pellet from this centrifugation is resuspended in 0.5 ml BSA (10 mg/ml) and up to 5 ml of isolation buffer, and centrifuged at $6900 \times g$ for 10 min. The resulting pellet is resuspended in 200 μl of isolation buffer, and homogenized with 4 strokes using a

Dounce homogenizer with a loose-fitting pestle. The described method is suitable for preparation of whole forebrain (100 mg or more) from rats, but it can be used, with minor modifications, for small samples. The mitochondrial fraction obtained using this method is metabolically highly active and well coupled.

VII. Mitochondria from Kidney and Testis

A recent application of the subcellular fractionation method has been introduced by Hoffmann *et al.* (2005) to isolate purified mitochondria from human kidney and testis. Briefly, tissue is chopped and minced with a scalpel and resuspended in 10 volumes of homogenization medium (0.25-M sucrose, 1-mM EDTA, 10-mM Tris–HCl, pH 7.5). The suspension is then homogenized in an Ultraturrax for 5 sec and subsequently with 6 strokes in a Teflon-glass Potter-Elvehjem homogenizer. The homogenate is separated from unbroken tissue lumps by filtration through a coarse metal sieve, and is centrifuged at $1000 \times g$ for 10 min (for kidney) or 20 min (for testis). The supernatant is collected, and centrifuged at $39,000 \times g$ for 45 min. The pellet obtained from this centrifugation is resuspended in 1.38-M sucrose and homogenized with 3 strokes using a Potter-Elvehjem homogenizer. The homogenate is transferred to an ultracentrifuge tube. Sucrose (0.25 M) is layered on top of the homogenate and the discontinuous gradient is centrifuged at $57,000 \times g$ for 90 min. The pellet contains mitochondria contaminated mainly by peroxisomes. Mitochondria prepared using this method should be stored at $-80\,^{\circ}$C at high protein concentration in a medium containing protease inhibitor cocktail.

VIII. Mitochondria from Hamster Brown Adipose Tissue

Glycerol-3-phosphate dehydrogenase is tightly bound to the outer surface of the inner mitochondrial membrane. The amount of this enzyme varies in mitochondrial preparations from different tissues. The highest activity of this enzyme was found in insect flight muscle (Estabrook and Sacktor, 1958), but it is also present in brown adipose tissue of either newborn or cold-adapted adult mammals (Chaffee *et al.*, 1964).

We have studied the activities of glycerol-3-phosphate dehydrogenase and glycerol-3-phosphate cytochrome *c* reductase in brown adipose tissue mitochondria from cold-adapted hamsters (*Mesocricetus auratus*) (Rauchova *et al.*, 1992, 1997). Mitochondria are prepared following the method of Hittelman *et al.* (1969). Brown fat is rapidly excised, placed in ice-cold 0.25-M sucrose, 10-mM Tris–HCl, and 1-mM EDTA, pH 7.4 (STE Buffer). Extraneous tissue is removed and the brown adipose tissue is homogenized using a Teflon-glass homogenizer in 10 volumes of STE buffer. The homogenate is subjected to high-speed centrifugation ($14,000 \times g$ for 10 min) (Smith *et al.*, 1966). The supernatant is carefully aspirated from beneath the overlying lipid layer, and the latter removed. The pellet is resuspended in the original volume of STE buffer and centrifuged at 8500 $\times$ g for 10 min (Schneider, 1948). The final resuspension of mitochondrial pellet is then carried out in STE buffer.

IX. Mitochondria from Insect Flight Muscle

Mitochondria isolated from cockroach flight muscle have been studied for their high content in glycerol-3-phosphate dehydrogenase activity (Rauchova *et al.*, 1997). Cockroach (*Periplaneta americana*) flight muscle mitochondria are prepared according to Novak *et al.* (1979). Red metathoracic muscles are minced and added to 0.32-M sucrose, 0.01-M EDTA, pH 7.4, using 10 ml of medium per 3 g of tissue, and homogenized by twelve up-and-down strokes in a 1-min period in a Teflon-glass homogenizer. Fractions are then separated by low-speed centrifugation (2000 $\times$ g for 20 min) followed by high-speed centrifugation (10,000 $\times$ g for 10 min). The fractions obtained from the high-speed centrifugations are composed mainly of mitochondria. Centrifugation of the homogenate at 18,500 $\times$ g for 20 min yields a fraction containing mitochondria and membranes of other origin.

X. Mitochondria from Porcine Adrenal Cortex

Mitochondria from adrenal cortex contain three monooxygenase systems involved in corticosteroidogenesis: 11β- and 18-hydroxylation of deoxycorticosterone (DOC) to corticosterone, and side chain cleavage of cholesterol to pregnenolone (Simpson and Boyd, 1971; Simpson and Estabrook, 1969). These monooxygenase systems use a second electron-transport chain whose primary electron donor is NADPH. Electrons are then transferred from NADPH to a flavoprotein, to adrenodoxin, and ultimately to cytochrome P-450, where hydroxylation of the steroid precursor occurs. The two electron transport chains are not independent, as steroid hydroxylation can inhibit ATP synthesis.

Mitochondria are extracted from porcine adrenal cortex according to Popinigis *et al.* (1990). The adrenal glands are separated from all connective and fatty tissue. The central medulla is scraped away and the cortex is cut up and resuspended in 0.33-M sucrose, 20-mM Tris–HCl, pH 7.4, 2-mM EDTA, and 0.2% BSA. The suspension is homogenized using first a loose-fitting glass-glass homogenizer and then a tight-fitting Teflon-glass homogenizer. The homogenate is centrifuged at 600 $\times$ g for 8 min, and the resulting supernatant is further centrifuged at 2000 $\times$ g for 8 min. The mitochondrial pellet is resuspended in 0.33-M sucrose, 20-mM Tris–HCl, pH 7.4, and 0.5-mM EDTA and washed thrice. The final pellet is resuspended in the same washing medium at a final concentration of ~50–60 mg/ml. Mitochondria prepared using the method described above have respiratory control with both succinate and glutamate, but ADP/O ratios are not very high (Popinigis *et al.*, 1990).

XI. Mitochondria from Human Platelets

A. Crude Mitochondrial Membranes

Preparation of crude mitochondrial membranes from platelets has an advantage in that a large amount of blood is not required. The platelet-enriched fraction is prepared according to Blass *et al.* (1977). Blood is obtained by venipuncture; 40–60 ml of blood yields quantities of mitochondria that are sufficient for a few respiratory activities (Merlo Pich *et al.*, 1996). For more complex experiments, involving kinetic determinations or inhibitor titrations (e.g., rotenone titration), 100 ml of blood is required (Degli Esposti *et al.*, 1994).

Erythrocytes (from 100 ml venous blood) are aggregated by the addition of 25 ml of 5% Dextran 250,000, 0.12-M NaCl, 10-mM EDTA, pH 7.4 at 4°C for 45 min (Blass *et al.*, 1977, Degli Esposti *et al.*, 1994). The upper phase is centrifuged at $5000 \times g$ for 10 min. Platelets that are recovered in the pellet obtained from the centrifugation are lysed by suspension in 3-ml H_2O and incubation for 60 sec. One milliliter of 0.6-M NaCl is then added to the mixture to restore osmolarity. After centrifugation at $5000 \times g$ for 10 min, the pellet is subjected to a second round of lysis and centrifugation, as described above. The pellet is resuspended in 3 ml of phosphate buffer (10-mM NaH_2PO_4, 10-mM Na_2HPO_4, 0.12-M NaCl, pH 7.4) and washed twice at $15,000 \times g$ for 20 min. The platelet pellet is resuspended in 0.1-M Tris–HCl, 1-mM EDTA, pH 7.4, and sonicated in a Braun sonicator (Labsonic U) five times for 10-sec each at 150 Hz, at 50-sec intervals. After sonication, the homogenate is centrifuged at $8000 \times g$ for 10 min (10,000 rpm in a tabletop Eppendorf centrifuge) and the supernatant is diluted 1:1 with 0.25-M sucrose, 50-mM Tricine-Cl, 2.5-mM $MgCl_2$, 40-mM KCl, 0.5% BSA, pH 8, and centrifuged at $100,000 \times g$ for 40 min to pellet mitochondrial particles. The pellet is resuspended in 0.125-M sucrose, 50-mM Tricine-Cl, 2.5-mM $MgCl_2$, 40-mM KCl, pH 8.

Alternatively, Merlo Pich *et al.* (1996) have described a modification of this method. Forty to sixty milliliters of blood is mixed with 5% Dextran 70,000, 0.9% NaCl at room temperature for 30 min, and the upper platelet-containing phase is centrifuged at $3000 \times g$ for 3 min. The upper phase is collected, centrifuged at $4000 \times g$ for 25 min, and then washed in the phosphate buffer as described above. After sonication, as described above, platelet membrane fragments are diluted 1:1 in 0.25-M sucrose, 30-mM Tris, 1-mM EDTA, pH 7.7 (STE buffer), and separated from the heavier cell debris by centrifugation at $33,000 \times g$ for 10 min. The supernatant is then ultracentrifuged at $100,000 \times g$ for 40 min and the pellet recovered from the ultracentrifugation is resuspended in STE buffer. This crude mitochondrial fraction has been used to study Complex I activity in patients affected by Leber's Hereditary Optic Neuropathy

(Carelli *et al.*, 1999; Degli Esposti *et al.*, 1994) and in aged individuals (Merlo Pich *et al.*, 1996).

B. Coupled Mitochondrial Particles from Platelet Mitochondria

This method involves the preparation of mitochondria followed by sonication in order to obtain coupled inside-out SMP of the ETPH type (Section III.B). Baracca *et al.* (1997) have described a method for preparation of coupled SMP from horse platelets. This method is also suitable for human platelets with minor modification.

Horse platelets are isolated and purified from 200–500 ml of venous blood according to Blass *et al.* (1977). To isolate mitochondria, platelets are suspended in a hypotonic medium (10-mM Tris–HCl, pH 7.6) for 7 min. The osmotic shock is stopped by addition of 0.25-M sucrose, and the suspension is centrifuged at $600 \times g$ for 10 min in an Eppendorf microcentrifuge. The pellet undergoes a second round of osmotic shock and centrifugation. The obtained supernatant is pooled with the one obtained from the first osmotic step, and centrifuged at $12,000 \times g$ (12,000 rpm in a tabletop Eppendorf centrifuge) for 20 min to sediment mitochondria. The mitochondrial pellet is then suspended in 0.25-M sucrose, 2-mM EDTA, pH 8, and subjected to sonic oscillation under partial N_2 atmosphere for 1 min at low output (40 W) (Braun Labsonic U sonicator). The suspension is centrifuged at $12,000 \times g$ for 10 min. The supernatant obtained is decanted and centrifuged at $105,000 \times g$ for 40 min to sediment the particles. The pellet is finally suspended in 0.25-M sucrose to give a protein concentration of 10 mg/ml as assayed by the Lowry method (Lowry *et al.*, 1951) in the presence of 1% deoxycholate. These inverted vesicles can be used to oxidize NADH and to measure ΔpH formation by fluorescence quenching of suitable acridine probes (Baracca *et al.*, 1997).

The method has been adapted for preparation of coupled SMP from human platelets. Platelets are isolated from 100 ml of venous blood. The osmotic shock is carried out for 4 min instead of 7 min, and the sonic oscillation step lasts 20 sec rather than 1 min. In SMPs from human platelets, we were not able to detect a ΔpH formed by NADH. However, we have measured the protonophoric activity of ATP synthase in these particles. We have also assayed ATP-dependent activities in SMPs from platelets from patients with mutations in ATP synthase (Baracca *et al.*, 2000).

XII. Mitochondria from Fish Liver

We have used the method described for the preparation and the characterization of liver mitochondria and SMP from eel (*Anguilla anguilla*) (Baracca *et al.*, 1992). It has also been applied to isolate mitochondria from sea bass (*Morone labrax*) liver (Ventrella *et al.*, 1982), and from other types of fish and amphibians (Degli Esposti *et al.*, 1992). The liver is removed, blotted dry, and immediately washed in 0.25-M sucrose, 0.01-M Tris–HCl, 1-mM EDTA, pH 7.4, and homogenized in 0.25-M sucrose, 24-mM Tris–HCl, 1-mM EDTA, pH 7.4 containing 0.5-mg/ml BSA (homogenizing solution), using an Ultraturrax homogenizer. The homogenate is then centrifuged at $750 \times g$ for 10 min. The supernatant is filtered through cheesecloth and centrifuged at $13,000 \times g$ for 10 min. The obtained pellet is resuspended in the homogenizing solution and is centrifuged at $11,000 \times g$ for 10 min. The mitochondrial pellet is resuspended in the homogenizing solution at a concentration of 40 mg mitochondrial protein/ml of solution.

XIII. Mitochondria from Sea Urchin Eggs

Mitochondria can be extracted from eggs of the sea urchin *Paracentrotus lividus* (Cantatore *et al.*, 1974). Eggs are fertilized and stored in sea water until the mesenchymatic blastula stage (about 15 h at 18 °C); the embryos are collected and suspended in 0.25-M sucrose, 0.1-M Tris–HCl, pH 7.6, 1-mM EDTA, 0.24-M KCl (TEK buffer) in a ratio 1:10 and homogenized in Teflon-glass Potter-Elvehjem homogenizer. The homogenate is centrifuged at $600 \times g$ for 10 min, and the supernatant obtained is centrifuged at $2600 \times g$ for 10 min. Finally, mitochondria are sedimented by centrifugation at $7500 \times g$ for 10 min. The pellet obtained is suspended in TEK buffer, and is centrifuged for 10 min at $15,000 \times g$. The pellet obtained is suspended in TEK buffer and layered on top of a sucrose discontinuous gradient consisting of 1-M and 1.5-M steps of sucrose in 0.1-M Tris–HCl, pH 7.6, 1-mM EDTA, 0.24-M KCl. The gradient is centrifuged at $50,000 \times g$ for 3 h. The band obtained in the 1.5-M sucrose region is collected, diluted with TEK buffer, and centrifuged at $15,000 \times g$ for 10 min. Finally, the pellet obtained is suspended in TEK buffer. This method has been used by Degli Esposti *et al.* (1990) to study the natural resistance of the sea urchin mitochondrial Complex III to the cytochrome *b* inhibitors.

XIV. Mitochondria and Kinetoplasts from Protozoa

Mitochondria have been isolated from ciliate and trypanosome protozoans and have been tested for their sensitivity to inhibitors of ubiquinol: cytochrome *c* reductase (Ghelli *et al.*, 1992). Mitochondria of ciliates, such as *Tetrahymena piriformis*, are resistant to antimycin A and rotenone, whereas mitochondria of trypanosomes are quite resistant to stigmatellin. Mitochondria from both ciliates and trypanosomes are highly resistant to myxothiazol.

Mitochondria from *Tetrahymena piriformis* are prepared according to Kilpatrick and Erecinska (1977). Cultured cells are harvested by low-speed centrifugation and washed once in cold 0.25-M sucrose. The pellet is then resuspended in 6 volumes of 0.25-M sucrose, 10-mM KCl, 5-mM MOPS, and 0.2-mM EDTA, pH 7.2 (Solution A), and homogenized in a Teflon-glass Potter-Elvehjem homogenizer. The cell homogenate is centrifuged at $600 \times g$ for 6 min, and the supernatant obtained is collected. The pellet is resuspended in half the previous volume, homogenized, and centrifuged as described above. The supernatant obtained is collected and pooled from the previous steps. The pooled supernatants are centrifuged at 5000 $\times$ g for 10 min to sediment the mitochondria. The mitochondrial pellet is then resuspended in Solution A containing 0.2% BSA and sedimented by centrifugation at 8000 $\times$ g for 10 min. The pellet obtained consists of a colorless layer over the mitochondrial fraction, and black sediment that adheres firmly to the bottom of the centrifuge tube. Discard the colorless layer and the black sediment, and resuspend the mitochondrial fraction in Solution A containing 0.2% BSA. Centrifuge the suspension as above and resuspend the mitochondrial fraction in the pellet obtained in Solution A. According to Kilpatrick and Erecinska (1977), the pellet obtained after the second extraction contains mitochondria with higher P/O ratios.

Mitochondria from *Paramecium tetraurelia* are prepared essentially as described by Doussiere *et al.* (1979). Cells are harvested by centrifugation at 20 °C for 3 min at 100 $\times$ g, and washed once in 0.5-M mannitol and 5-mM MOPS. Packed cells are then resuspended in 0.5-M mannitol, 5-mM MOPS, and 1-mM EDTA, pH 7.3 containing 0.5% BSA (homogenization medium), and homogenized in a Potter-Elvehjem Teflon-glass homogenizer.

The homogenate is centrifuged at 600 $\times$ g for 5 min. The supernatant obtained is collected and centrifuged at 600 $\times$ g for 10 min. Mitochondria in the pellet are resuspended in the homogenization medium and centrifuged at 600 $\times$ g for 4 min to eliminate trichocystis. The supernatant obtained is centrifuged at 5000 $\times$ g for 10 min to sediment a crude mitochondrial fraction. The latter is resuspended in homogenization medium and purified by centrifugation through a discontinuous gradient of sorbitol (at 40 and 60%, w/w) in 5-mM MOPS and 1-mM EDTA, pH 7.3 at 150,000 $\times$ g for 1 h. Mitochondrial particles, which are recovered at the interphase between the two sorbitol steps, are removed and diluted in 10 volumes of homogenization medium. The suspension is then centrifuged at 10,000 $\times$ g for 15 min. Finally, the pellet, which contains mitochondria, is resuspended in the homogenization medium.

Mitochondria from *Crithidia luciliae* and *Leishmania infantum* are prepared following the method of Renger and Wolstenholme (1972). Protozoans of the order Kinetoplastida, which includes *Trypanosoma*, *Leishmania*, and *Crithidia* genera, all contain kinetoplasts. This organelle is a modified mitochondrion. Therefore, mitochondria from trypanosomes are purified from kinetoplast-enriched fractions.

Cells are harvested by centrifugation at 1000 $\times$ g for 10 min and washed thrice in SSC buffer, pH 7.5. Cells are then resuspended in 0.3-M sucrose, 10-mM Tris–HCl, pH 7.4, and 1-mM EDTA (STE buffer), homogenized in a Waring blender for 15–20 sec at high speed, and centrifuged at 700 $\times$ g for 10 min. This centrifugation step should be repeated until all the cells are removed from the supernatant. The supernatant is then centrifuged at 8000 $\times$ g for 10 min. The pellet obtained is resuspended in STE buffer, and incubated with 200 μg/ml of DNase I at 37 °C for 30 min in the presence of 7-mM MgCl$_2$. The DNase I is then removed by washing thrice with 40-mM EDTA. The kinetoplast-enriched fraction is then resuspended in 0.15-M NaCl, 0.1-M EDTA, and 0.05-M Na$_2$HPO$_4$.

XV. Mitochondria from Fish Erythrocytes

In nonmammalian vertebrates, mature erythrocytes retain the nucleus and cytoplasmic organelles. Fish erythrocytes have a relatively long life and are a convenient source of mitochondria. The isolation of mitochondria from red blood cells of *Torpedo marmorata* has been described by Pica *et al.* (2001). The red blood cell pellet from 200 ml of blood is diluted threefold in sucrose buffer (800-mM sucrose, 2-mM K-EDTA, 10-mM Tris, pH 7.4) and incubated with 200 μg/ml digitonin for 15 min at 4°C. After centrifugation to remove hemoglobin, the pellet is incubated with 400 μg/ml lysozyme and 200 μg/ml nagarse for 20 min at 4°C and homogenized. The homogenate is centrifuged at 1500 $\times$ g and the supernatant is filtered and recentrifuged twice at 7000 $\times$ g. The pellet is resuspended and centrifuged at 22,000 $\times$ g for 10 min, and the mitochondrial pellet is collected at 5–6-mg protein/ml. The isolated mitochondria are intact and exhibit low content of respiratory enzymes and low respiratory activities in comparison with mammalian mitochondria.

XVI. Mitochondria from *Caenorhabditis elegans*

The soil nematode *C. elegans* offers numerous advantages to study mitochondrial dysfunction (Lewis and Fleming 1995). The organism grows rapidly, and has a life cycle of 3–4 days at 20°C. Moreover, the sequences of its nuclear and mitochondrial genome have been determined (*C. elegans* sequencing consortium, 1998). Finally, the mitochondrial bioenergetics of *C. eleagans* is similar to that of mammals.

Here, we describe the method published by Grad and Lemire (2004) for the isolation of mitochondria from whole worms. Worms are harvested from liquid culture, washed to remove bacteria, and suspended in mannitol–sucrose buffer (0.2-M mannitol, 70-mM sucrose, 0.1-M EDTA, pH 7.4 and 1-mM PMSF). Worms are disrupted with 0.1-mm acid-washed glass beads in a Bead-Beater homogenizer (Biospec Products, Bartlesville, OK) by three 30-sec pulses, and then homogenized by 10 strokes of a glass-glass homogenizer. The homogenate is centrifuged at 1000 $\times$ g for 10 min to remove debris. The supernatant obtained is centrifuged at 19,200 $\times$ g for 10 min to sediment mitochondria. The mitochondrial pellet is resuspended in the mannitol–sucrose buffer and centrifuged as described above. The final pellet is resuspended in the same medium. Respiratory activities can be determined in the mitochondrial fraction.

XVII. Mitochondria from *Drosophila*

Drosophila is a well-established model organism in cellular and developmental biology. Mitochondrial function during aging in *Drosophila* has been investigated in several studies (Sohal *et al*, 2002). Here, we report the method for the isolation of mitochondria from *Drosophila* described by Miwa and Brand (2005). Briefly, about 200 flies are immobilized by chilling on ice. They are then gently pressed with a pestle in a chilled mortar containing isolation medium (250-mM sucrose, 5-mM Tris–HCl, 2-mM EGTA, 1% BSA, pH 7.4) at 4°C. The homogenate is filtered through muslin, and centrifuged at $150 \times g$ for 3 min. The supernatant obtained is filtered through muslin and centrifuged at $9000 \times g$ for 10 min. The pellet obtained is resuspended in isolation medium at 30-mg protein/ml. The yield of mitochondria is about 8- to 9-mg/g wet weight flies. The mitochondria are coupled and stable for some hours on storage on ice.

XVIII. Mitochondria and Mitoplasts from Cultured Cells

Human cultured cells represent a valid experimental model for investigating mitochondria function, both in physiological and pathological states. Cells are usually easy to cultivate and to obtain in large amounts, and are therefore a good source for the isolation of intact mitochondria for biochemical and genetic analyses.

Mitochondria from cell lines have been isolated as described by Yang *et al.* (1997) with minor modifications. Cells are harvested by centrifugation at $600 \times g$ for 10 min, washed with PBS, and resuspended with 5 volumes of Solution A (0.25-M sucrose, 20-mM HEPES-KOH, pH 7.5, 10-mM KCl, 1.5-mM $MgCl_2$, 1-mM EDTA, 1-mM EGTA, 1-mM dithiothreitol, 0.1-mM PMSF). The cellular suspension is homogenized with a Teflon-glass homogenizer with 20 up and down passes of the pestle. In our experience, the mitochondrial yield is increased by the use of a glass-glass pestle. The homogenate is then centrifuged at $750 \times g$ for 10 min. The resulting supernatant is collected, and the pellet is resuspended in Solution A and centrifuged as described above. The supernatants obtained from the two low-speed spins are pooled and then centrifuged at $10,000 \times g$ for 15 min. Crude mitochondria, which are recovered in the pellet, are resuspended in Solution A.

Mitochondria of higher purity can be prepared using a sucrose gradient. A valid method, which is a modification of the "two-step" procedure described by Tapper *et al.* (1983), has been described by Magalhães *et al.* (1998). Cells are harvested by low-speed centrifugation and resuspended in 10-mM NaCl, 1.5-mM $MgCl_2$, and 10-mM Tris–HCl, pH 7.5. Cells are allowed to swell for 4–5 min on ice and briefly homogenized in a Teflon-glass Potter-Elvehjem homogenizer. The sucrose concentration is then adjusted to 250 mM by adding 2-M sucrose in 10-mM Tris–HCl, and 1-mM EDTA, pH 7.6 ($T_{10}E_{20}$ buffer). The suspension is then centrifuged at $1300 \times g$ for 3 min. The supernatant obtained is subjected to a second round of low-speed centrifugation. Mitochondria are sedimented from the supernatant of the low-speed centrifugation by centrifugation at $15,000 \times g$ for 15 min. The pellet obtained is washed thrice with 250-mM sucrose in $T_{10}E_{20}$ buffer, and resuspended in the same solution. The mitochondrial suspension is layered on a discontinuous sucrose gradient (1.0 and 1.7 M) in $T_{10}E_{20}$ buffer and centrifuged at $70,000 \times g$ for 40 min. The mitochondrial fraction is recovered from the interface between the two sucrose steps, diluted with an equal volume of 250-mM sucrose in $T_{10}E_{20}$ buffer, and washed twice in the same solution. Finally, the mitochondrial pellet is resuspended in 250-mM sucrose in $T_{10}E_{20}$ buffer and protein concentration is determined by the Lowry method.

An alternative method to this procedure consists in the "one-step" procedure described by Bogenhagen and Clayton (1974). Cells are harvested and resuspended as described above. However, in this method there is no osmotic swelling. Therefore, the sucrose concentration is adjusted to 250-mM using 2-M sucrose in $T_{10}E_{20}$ buffer immediately after resuspension. Cell debris and nuclei are sedimented and removed as described in the "two-step" procedure, and the supernatant obtained is layered on top of 3 volumes of 1.7-M sucrose in $T_{10}E_{20}$ buffer in an ultracentrifuge tube. The mixture is centrifuged at 70,000 × g for 40 min. The mitochondrial fraction that is recovered on top of the sucrose cushion is harvested, washed, and resuspended as above.

A "no gradient procedure" method is based on slight modifications to the method described by Schneider (1948). Instead of isotonic sucrose used in the original method, the buffer used consists of 0.21-M mannitol, 0.07-M sucrose, 5-mM Tris–HCl, 5-mM EDTA, pH 7.5 (mannitol–sucrose buffer) (Bogenhagen and Clayton, 1974). Cells are homogenized and the nuclei are pelleted as described in "one-step" procedure. Cell debris and nuclei are sedimented by centrifugation at 400 × g for 4 min, and a crude mitochondrial pellet is obtained from centrifugation of the supernatant at 27,000 × g for 10 min (Bestwick et al., 1982).

An alternative method to mechanical homogenization of cells and tissues is disruption represented by nitrogen cavitation (Hunter and Commerford, 1961). This method limits the shear stress due to mechanical forces. Kristian et al. (2006) have used this technique to isolate mitochondria from primary cultures of neurons and astrocytes. The mitochondria obtained from this preparation are tightly coupled, as demonstrated by respiratory control ratios (RCR) higher than 9 in the presence of glutamate/malate.

Briefly, 2×10^8 cells are washed with isolation medium (225-mM mannitol, 75-mM sucrose, 5-mM HEPES–Tris, 1-mM EGTA, pH 7.4), scraped in 3 ml of isolation medium, and collected into a precooled cavitation chamber. The cell suspension is then subjected to 1500 psi (atmospheric pressure) for 15 min with subsequent decay of the pressure at 800 psi. The cell suspension is released from the chamber through outflow tubing attached to the valve localized at the bottom of the cavitation chamber, with immediate pressure falling from 800 psi to 14.7 psi. The suspension is centrifuged at 1500 × g for 3 min. The supernatant obtained is collected and centrifuged at 20,000 × g for 10 min. The pellet obtained contains crude mitochondria. It is resuspended in 0.8 ml of 15% Percoll and layered on top of a Percoll step gradient. The bottom of the step gradient is 50% Percoll. The top step of the gradient is 21% Percoll (for neuronal mitochondria) or 23% (for astrocyte mitochondria). The suspension is centrifuged at 30,700 × g for 6 min and the mitochondrial fraction (between the top and bottom Percoll layers) is collected, diluted with isolation medium (1:8), and centrifuged at 17,000 × g for 10 min. The pellet obtained contains mitochondria.

It is suspended in isolation medium (1.5 ml) and centrifuged at $7000 \times g$ for 10 min. The final pellet is resuspended in 30 μl isolation medium without EGTA.

Mitoplasts from gradient purified mitochondria can be obtained by modifications of the original method, as described by Magalhães *et al.* (1998). In the modification of the "swell-contract" method (Murthy and Pande, 1987), gradient-purified mitochondria are resuspended in 20-mM potassium phosphate (pH 7.2) containing 0.02% BSA, and are allowed to swell for 20 min on ice. After addition of 1-mM ATP and 1-mM $MgCl_2$, incubation is carried out for an additional 5 min, and mitoplasts are collected by centrifugation at $15,000 \times g$ for 10 min.

The modification of the "digitonin method" is an alternative method for the production of mitoplasts (Greenawalt, 1974). The resuspended gradient-purified mitochondria are incubated with 0.1-mg digitonin/mg of mitochondrial protein, with gentle stirring on ice for 15 min. Thereafter, 3 volumes of 250-mM sucrose in $T_{10}E_{20}$ buffer are added, and mitoplasts are sedimented after centrifugation at $15,000 \times g$ for 15 min. Mitoplasts, which are recovered in the pellet, are washed once in 250-mM sucrose in $T_{10}E_{20}$ buffer and finally resuspended in 250-mM sucrose, 10-mM Tris–HCl, pH 7.6.

Generally, methods presented in this section are suitable for various types of cell lines, and usually the isolation of a crude mitochondrial pellet from cultured cells is sufficient for protein analysis, such as Western blotting, ELISA, and proteomic research. For *in vitro* protein synthesis purposes, a method of isolation is recommended that emphasizes the preparation of intact, coupled mitochondria rather than the yield (i.e., the "two-step procedure" using a discontinuous gradient). Moreover, since contamination of mitochondrial fraction by other cellular fractions capable of *in vitro* protein synthesis is possible, the use of cycloheximide (usually 5 μg/ml) in incubation solutions is recommended, in order to inhibit the activity of cytoplasmic 80S ribosomes.

Care must also be taken in the isolation of mitochondria from human fibroblasts. As stated previously, fibroblast cell membranes are difficult to break open; for this reason, it is recommended to freeze-thaw cells once harvested. Usually, cells are recovered and centrifuged (after a brief wash in PBS) at maximum speed in a tabletop centrifuge for 3 min; the pellet is then stored at $-80\,^{\circ}$C until the next isolation step (homogenization).

The homogenization step is another critical point for the final yield of mitochondrial preparations. There are no stated rules, but experience obtained through direct observation will facilitate the standardization for single cell lines.

The use of a glass-glass homogenizer increases the yield of mitochondrial preparations: usually 10 strokes are sufficient, but also 5 strokes in a loose-fitting homogenizer followed by 5 strokes in tight-fitting homogenizer can be used. The starting amount of cells can dictate the number of strokes necessary to break open cell membranes: from a starting amount of 5×10^7–3×10^8 cells, the number of recommended strokes ranges from 30 to 50. The freeze-thawing step of the cell pellet can reduce the number down to 15–20 strokes.

The homogenization step should lyse about 80% of the cells. To check the efficiency of homogenization, 2–3 μl of homogenization suspension are layered on a coverslip and observed under a microscope. A shiny ring around the cells indicates that cells are still intact. If more than 20% of cells are intact, 10 additional passes should be performed and the homogenization solution should be checked thereafter until 80% of cells are broken.

Different cell lines require different numbers of strokes in the homogenization step to have about 80% of lysis efficiency: C6 cells require 40 strokes, HeLa cells 60 strokes, and NIH 3T3 cells about 80 strokes (as suggested by the "mitochondrial isolation kit" from Pierce, Rockford, IL), but standardized conditions must be determined by direct observation under a microscope. It should be also borne in mind that excessive homogenization can cause damage to the mitochondrial membrane, thus triggering the release of mitochondrial components.

Acknowledgments

The authors thank Romana Fato for critical revision of the manuscript and for helpful comments and suggestions. This work is supported by grants from the "Fondo di Ateneo per la Ricerca" (FAR) (to F.P.) and from MURST, Rome, Italy (to G.L.).

References

Anderson, M. F., and Sims, N. R. (2000). Improved recovery of highly enriched mitochondrial fractions from small brain tissue samples. *Brain Res. Brain Res. Protoc.* **5**, 95–101.

Baracca, A., Degli Esposti, M., Parenti Castelli, G., and Solaini, G. (1992). Purification and characterization of adenosine-triphosphatase from eel liver mitochondria. *Comp. Biochem Physiol. B.* **101**, 421–426.

Baracca, A., Bucchi, L., Ghelli, A., and Lenaz, G. (1997). Protonophoric activity of NADH coenzyme Q reductase and ATP synthase in coupled submitochondrial particles from horse platelets. *Biochem. Biophys. Res. Commun.* **235**, 469–473.

Baracca, A., Barogi, S., Carelli, V., Lenaz, G., and Solaini, G. (2000). Catalytic activities of the mitochondrial ATP synthase in patients with the mtDNA T8993G mutation in the ATPase 6 gene encoding for the a subunit. *J. Biol. Chem.* **275**, 4177–4182.

Barogi, S., Baracca, A., Parenti Castelli, G., Bovina, C., Formiggini, G., Marchetti, M., Solaini, G., and Lenaz, G. (1995). Lack of major changes in mitochondria from liver, heart, and skeletal muscle of rats upon ageing. *Mech. Ageing Dev.* **84**, 139–150.

Barogi, S., Baracca, A., Cavazzoni, M., Parenti Castelli, G., and Lenaz, G. (2000). Effect of oxidative stress induced by adriamycin on rat hepatocyte bioenergetics during ageing. *Mech. Ageing Dev.* **113**, 1–21.

Battino, M., Bertoli, E., Formiggini, G., Sassi, S., Gorini, A., Villa, R. F., and Lenaz, G. (1991). Structural and functional aspects of the respiratory chain of synaptic and nonsynaptic mitochondria derived from selected brain regions. *J. Bioenerg. Biomembr.* **23**, 345–363.

Battino, M., Gorini, A., Villa, R. F., Genova, M. L., Bovina, C., Sassi, S., Littarru, G. P., and Lenaz, G. (1995). Coenzyme Q content in synaptic and non-synaptic mitochondria from different brain regions in the ageing rat. *Mech. Ageing Dev.* **78**, 173–187.

Bestwick, R. K., Moffett, G. L., and Mathews, C. K. (1982). Selective expansion of mitochondrial nucleoside triphosphate pools in antimetabolite-treated HeLa cells. *J. Biol. Chem.* **257**, 9300–9304.

Beyer, R. E. (1967). Preparation, properties, and conditions for assay of phosphorylating electron transport particles (ETPH) and its variations. *In* "Methods in Enzymology" (R. W. Estabrook, and M. E. Pullman, eds.), Vol. 10, pp. 186–194. Academic Press, San Diego.

Blass, J. P., Cederbaum, S. D., and Kark, R. A. P. (1977). Rapid diagnosis of pyruvate and ketoglutarate dehydrogenase deficiencies in platelet-enriched preparations from blood. *Clin. Chim. Acta* **75,** 21–30.

Bogenhagen, D., and Clayton, D. A. (1974). The number of mitochondrial deoxyribonucleic acid genomes in mouse L and human HeLa cells. Quantitative isolation of mitochondrial deoxyribonucleic acid. *J. Biol. Chem.* **249,** 7991–7995.

Caenorhabditis elegans Sequencing Consortium (1998). Genome sequence of the nematode *C. elegans*: A platform for investigating bilogy. *Science* **282,** 2012–2018.

Cantatore, P., Nicotra, A., Loria, P., and Saccone, C. (1974). RNA synthesis in isolated mitochondria from sea urchin embryos. *Cell Differ.* **3,** 45–53.

Capaldi, R. A. (1982). Arrangement of proteins in the mitochondrial inner membrane. *Biochim. Biophys. Acta* **694,** 291–306.

Carelli, V., Ghelli, A., Bucchi, L., Montagna, P., De Negri, A., Leuzzi, V., Carducci, C., Lenaz, G., Lugaresi, E., and Degli Esposti, M. (1999). Biochemical features of mtDNA 14484 (ND6/M64V) point mutation associated with Leber's hereditary optic neuropathy (LHON). *Ann. Neurol.* **45,** 320–328.

Casadio, R., Baccarini-Melandri, A., and Melandri, B. A. (1974). On the determination of the transmembrane pH difference in bacterial chromatophores using 9-aminoacridine. *Eur. J. Biochem.* **47,** 121–128.

Castelluccio, C., Baracca, A., Fato, R., Pallotti, F., Maranesi, M., Barzanti, V., Gorini, A., Villa, R. F., Parenti Castelli, G., Marchetti, M., and Lenaz, G. (1994). Mitochondrial activities of rat heart during ageing. *Mech. Ageing Dev.* **76,** 73–88.

Chaffee, R. R. J., Allen, J. R., Cassuto, Y., and Smith, R. E. (1964). Biochemistry of brown fat and liver of cold-acclimated hamsters. *Am. J. Physiol.* **207,** 1211–1214.

Clark, J. B., and Nicklas, W. J. (1970). The metabolism of rat brain mitochondria. Preparation and characterization. *J. Biol. Chem.* **245,** 4724–4731.

Cogswell, A. M., Stevens, R. J., and Hood, D. A. (1993). Properties of skeletal muscle mitochondria isolated from subsarcolemmal and intermyofibrillar regions. *Am. J. Physiol.* **264,** C383–C389.

Degli Esposti, M., and Lenaz, G. (1982). Kinetics of ubiquinol-1-cytochrome *c* reductase in bovine heart mitochondria and submitochondrial particles. *Biochim. Biophys. Acta* **682,** 189–200.

Degli Esposti, M., Ghelli, A., Butler, G., Roberti, M., Mustich, A., and Cantatore, P. (1990). The cytochrome *c* of the sea urchin *Paracentrotus lividus* is naturally resistant to myxothiazol and mucidin. *FEBS Lett.* **263,** 245–247.

Degli Esposti, M., Ghelli, A., Crimi, M., Baracca, A., Solaini, G., Tron, T., and Meyer, A. (1992). Cytochrome *b* of fish mitochondria is strongly resistant to funiculosin, a powerful inhibitor of respiration. *Arch. Biochem. Biophys.* **295,** 198–204.

Degli Esposti, M., Carelli, V., Ghelli, A., Ratta, M., Crimi, M., Sangiorgi, S., Montagna, P., Lenaz, G., Lugaresi, E., and Cortelli, P. (1994). Functional alterations of the mitochondrially encoded ND4 subunit associated with Leber's hereditary optic neuropathy. *FEBS Lett.* **352,** 375–379.

Doussiere, J., Sainsard-Chanet, A., and Vignais, P. V. (1979). The respiratory chain of *Paramecium tetraurelia* in wild type and the mutant. C_1. I. Spectral properties and redox potentials. *Biochim. Biophys. Acta* **548,** 224–235.

Estabrook, R. W., and Sacktor, B. (1958). α-Glycerophosphate oxidase of flight muscle mitochondria. *J. Biol. Chem.* **233,** 1014–1019.

Estornell, E., Fato, R., Pallotti, F., and Lenaz, G. (1993). Assay conditions for the mitochondrial NADH: Coenzyme Q oxidoreductase. *FEBS Lett.* **332,** 127–131.

Fato, R., Cavazzoni, M., Castelluccio, C., Parenti Castelli, G., Palmer, G., Degli Esposti, M., and Lenaz, G. (1993). Steady-state kinetics of ubiquinol—cytochrome *c* reductase in bovine heart submitochondrial particles: Diffusional effects. *Biochem. J.* **290,** 225–236.

Fato, R., Estornell, E., Di Bernardo, S., Pallotti, F., Parenti Castelli, G., and Lenaz, G. (1996). Steady-state kinetics of the reduction of Coenzyme Q analogs by Complex I (NADH: Ubiquinone oxidoreductase) in bovine heart mitochondria and submitochondrial particles. *Biochemistry (US)* **35**, 2705–2716.

Fleischer, S., and Kervina, M. (1974). Long-term preservation of liver for subcellular fractionation. *In* "Methods in Enzymology" (S. Fleischer, and L. Packer, eds.), Vol. 31, pp. 3–6. Academic Press, San Diego.

Genova, M. L., Bovina, C., Formiggini, G., Ottani, V., Sassi, S., and Marchetti, M. (1994). Uptake and distribution of exogenous CoQ in the mitochondrial fraction of perfused rat liver. *Mol. Aspects Med.* **15**(Suppl.), s47–s55.

Genova, M. L., Bovina, C., Marchetti, M., Pallotti, F., Tietz, C., Biagini, G., Pugnaloni, A., Viticchi, C., Gorini, A., Villa, R. F., and Lenaz, G. (1997). Decrease of rotenone inhibition is a sensitive parameter of complex I damage in brain non-synaptic mitochondria of aged rats. *FEBS Lett.* **410**, 467–469.

Ghelli, A., Crimi, M., Orsini, S., Gradoni, L., Zannotti, M., Lenaz, G., and Degli Esposti, M. (1992). Cytochrome *b* of protozoan mitochondria: Relationships between function and structure. *Comp. Biochem. Physiol. B* **103**, 329–338.

Glowinski, J., and Iversen, L. L. (1966). Regional studies of catecholamines in the rat brain. I. The disposition of [³H]norepinephrine, [³H]dopamine and [³H]dopa in various regions of the brain. *J. Neurochem.* **13**, 655–669.

Gornall, A. G., Bardawill, C. J., and David, M. M. (1949). Determination of serum proteins by means of the biuret reaction. *J. Biol. Chem.* **177**, 751–766.

Grad, L. I., and Lemire, B. D. (2004). Mitochondrial complex I mutations in *Caenorhabditis elegans* produce cytochrome c oxidase deficiency, oxidative stress and vitamin-responsive lactic acidosis. *Hum. Mol. Genet.* **13**, 303–314.

Greenawalt, J. W. (1974). The isolation of outer and inner mitochondrial membranes. *In* "Methods in Enzymology" (S. Fleisher, and L. Packer, eds.), Vol. 31, pp. 310–323. Academic Press, San Diego.

Gregg, C. T. (1967). Preparation and assay of phosphorylating submitochondrial particles: Particles from rat liver prepared by drastic sonication. *In* "Methods in Enzymology" (R. W. Estabrook, and M. E. Pullman, eds.), Vol. 10, pp. 181–185. Academic Press, San Diego.

Hansen, M., and Smith, A. L. (1964). Studies on the mechanism of oxidative phosphorylation. VII. Preparation of a submitochondrial particle (ETP$_H$) which is capable of fully coupled oxidative phosphorylation. *Biochim. Biophys. Acta* **81**, 214–222.

Hatefi, Y., Jurtshuk, P., and Haavik, A. G. (1961). Studies on the electron transport system. XXXII. Respiratory control in beef heart mitochondria. *Arch. Biochem. Biophys.* **94**, 148–155.

Hittelman, K. J., Lindberg, O., and Cannon, B. (1969). Oxidative phosphorylation and compartmentation of fatty acid metabolism in brown fat mitochondria. *Eur. J. Biochem.* **11**, 183–192.

Hoffmann, K., Blaudszun, J., Brunken, C., Hopker, W. W., Tauber, R., and Steinhart, H. (2005). New application of a subcellular fractionation method to kidney and testis for the determination of conjugated linoleic acid in selected cell organelles of healthy and cancerous human tissues. *Anal. Bioanal. Chem.* **381**, 1138–1144.

Hunter, M. J., and Commerford, S. L. (1961). Pressure homogenization of mammalian tissues. *Biochim. Biophys. Acta* **47**, 580–586.

Jacobs, E. E., and Sanadi, D. R. (1960). The reversible removal of cytochrome *c* from mitochondria. *J. Biol. Chem.* **235**, 531–534.

Jimenez, M., Yvon, C., Lehr, L., Leger, B., Keller, P., Russell, A., Kuhne, F., Flandin, P., Giacobino, J-P., and Muzzin, P. (2002). Expression of uncoupling protein-3 in subsarcolemmal and intermyofibrillar mitochondria of various mouse muscle types and its modulation by fasting. *Eur. J. Biochem.* **269**, 2878–2884.

Kalen, A., Soderberg, N., Elmberger, P. G., and Dallner, G. (1990). Uptake and metabolism of dolichol and cholesterol in perfused rat liver. *Lipids* **25**, 93–99.

Keilin, D., and Hartree, E. F. (1940). Succinic dehydrogenase-cytochrome system of cells. Intracellular respiratory system catalysing aerobic oxidation of succinic acid. *Proc. R. Soc. Lond. B* **129**, 277–306.

Kilpatrick, L., and Erecinska, M. (1977). Mitochondrial respiratory chain of *Tetrahymena piriformis*. The termodynamic and spectral properties. *Biochim. Biophys. Acta* **460,** 346–363.

Kristian, T., Hopkins, I. B., McKenna, M. C., and Fiskum, G. (2006). Isolation of mitochondria with high respiratory control from primary cultures of neurons and astrocytes using nitrogen cavitation. *J. Neurosci. Methods* **152,** 136–143.

Kudin, A. P., Bimpong-Buta, N. Y., Vielhaber, S., Elger, C. E., and Kunz, W. S. (2004). Characterization of superoxide-producing sites in isolated brain mitochondria. *J. Biol. Chem.* **279,** 4127–4135.

Kun, E., Kirsten, E., and Piper, W. N. (1979). Stabilization of mitochondrial functions with digitonin. *In* "Methods in Enzymology" (S. Fleischer, and L. Packer, eds.), Vol. 55, pp. 115–118. Academic Press, San Diego.

Lai, J. C. K., Walsh, J. M., Dennis, S. C., and Clark, J. B. (1977). Synaptic and non-synaptic mitochondria from rat brain: Isolation and characterization. *J. Neurochem.* **28,** 625–631.

Lee, C. P., Martens, M. E., and Tsang, S. H. (1993). Small scale preparation of skeletal muscle mitochondria and its application in the study of human disease. *In* "Methods in Toxicology" (D. Jones, and L. Lash, eds.), Vol. 2, pp. 70–83. Academic Press, New York.

Lenaz, G., and MacLennan, D. H. (1966). Studies on the mechanism of oxidative phosphorylation. X. The effect of cytochrome *c* on energy-linked processes in submitochondrial particles. *J. Biol. Chem.* **241,** 5260–5265.

Lenaz, G., and MacLennan, D. H. (1967). Extraction and estimation of cytochrome *c* from mitochondria and submitochondrial particles. *In* "Methods in Enzymology" (R. W. Estabrook, and M. E. Pullman, eds.), Vol. 10, pp. 499–504. Academic Press, San Diego.

Lenaz, G., Degli Esposti, M., and Parenti Castelli, G. (1982). DCCD inhibits proton translocation and electron flow at the second site of the mitochondrial respiratory chain. *Biochem. Biophys. Res. Commun.* **105,** 589–595.

Lenaz, G., Baracca, A., Fato, R., Genova, M. L., and Solaini, G. (2006). New insights into structure and function of mitochondria and their role in aging and disease. *Antioxid. Redox Signal.* **8,** 417–437.

Linnane, A. W., and Titchener, E. B. (1960). Studies on the mechanism of oxidative phosphorylation. VI. A factor for coupled oxidation in the electron transport particle. *Biochim. Biophys. Acta* **39,** 469–478.

Lewis, J. A., and Fleming, J. T. (1995). Basic culture methods. *Methods Cell Biol.* **48,** 3–29.

Lowry, O. H., Rosenbrough, N. J., Farr, A. L., and Randall, R. J. (1951). Protein measurement with the Folin phenol reagent. *J. Biol. Chem.* **193,** 265–275.

MacLennan, D. H., Lenaz, G., and Szarkowska, L. (1966). Studies on the mechanism of oxidative phosphorylation. IX. Effect of cytochrome *c* on energy-linked processes. *J. Biol. Chem.* **241,** 5251–5259.

Magalhães, P. J., Andreu, A. L., and Schon, E. A. (1998). Evidence for the presence of 5S rRNA in mammalian mitochondria. *Mol. Biol. Cell* **9,** 2375–2382.

Marinetti, G. V. (1962). Chromatographic separation, identification and analysis of phosphatides. *J. Lipid Res.* **3,** 1–20.

Melandri, B. A., Baccarini Melandri, A., Lenaz, G., Bartoli, E., and Masotti, L. (1974). The site of inhibition of dibromothymoquinone in mitochondrial respiration. *J. Bioenerg. Biomembr.* **6,** 125–133.

Merlo Pich, M., Bovina, C., Formiggini, G., Cometti, G. G., Ghelli, A., Parenti Castelli, G., Genova, M. L., Marchetti, M., Semeraro, S., and Lenaz, G. (1996). Inhibitor sensitivity of respiratory complex I in human platelets: A possible biomarker of ageing. *FEBS Lett.* **380,** 176–178.

Miwa, S., and Brand, M. D. (2005). The topology of superoxide production by Complex III and glycerol 3 phosphate dehydrogenase in Drosophila mitochondria. *Biochim. Biophys. Acta* **1709,** 214–219.

Murthy, M. S. R., and Pande, S. V. (1987). Malonyl-CoA binding site and the overt carnitine palmitoyltransferase activity reside on the opposite sides of the outer mitochondrial membrane. *Proc. Natl. Acad. Sci. USA* **84,** 378–382.

Musatov, A., Ortega-Lopez, J., and Robinson, N. C. (2000). Detergent-solubilized bovine Cytochrome *c* Oxidase: Dimerization depends on the amphiphilic environment. *Biochemistry* **39,** 12996–13004.

Nicholls, P. (1976). Catalytic activity of cytochromes c and c_1 in mitochondria and submitochondrial particles. *Biochim. Biophys. Acta* **430**, 30–45.

Norling, B., Glazek, E., Nelson, B. D., and Ernster, L. (1974). Studies with ubiquinone-depleted submitochondrial particles. Quantitative incorporation of small amounts of ubiquinone and its effects on the NADH and succinate oxidase activities. *Eur. J. Biochem.* **47**, 475–482.

Novak, F., Novakova, O., and Kubista, V. (1979). Heterogeneity of insect flight muscle mitochondria as demonstrated by different phospholipid turnover *in vivo*. *Insect. Biochem.* **9**, 389–396.

Pallotti, F., Genova, M. L., Merlo Pich, M., Zucchini, C., Carraro, S., Tesei, M., Bovina, C., and Lenaz, G. (1998). Mitochondrial dysfunction and brain disorders. *Arch. Gerontol. Geriatr.* **6**(Suppl.), 385–392.

Pica, A., Scacco, S., Papa, F., De Nitto, E., and Papa, S. (2001). Morphological and biochemical characterization of mitochondria in Torpedo red blood cells. *Comp. Biochem. Physiol. B Biochem. Mol. Biol.* **128**, 213–219.

Popinigis, J., Antosiewicz, J., Mazzanti, L., Bertoli, E., Lenaz, G., and Cambria, A. (1990). Direct oxidation of glutamate by mitochondria from porcine adrenal cortex. *Biochem. Int.* **21**, 441–451.

Rasmussen, H. N., Andersen, A. J., and Rasmussen, U. F. (1997). Optimization of preparation of mitochondria from 25–100 mg skeletal muscle. *Anal. Biochem.* **252**, 153–159.

Rauchova, H., Battino, M., Fato, R., Lenaz, G., and Drahota, Z. (1992). Coenzyme Q-pool function in glycerol-3-phosphate oxidation in hamster brown adipose tissue mitochondria. *J. Bioenerg. Biomembr.* **24**, 235–241.

Rauchova, H., Fato, R., Drahota, Z., and Lenaz, G. (1997). Steady-state kinetics of reduction of coenzyme Q analogs by glycerol-3-phospate dehydrogenase in brown adipose tissue mitochondria. *Arch. Biochem. Biophys.* **344**, 235–241.

Renger, H. C., and Wolstenholme, D. R. (1972). The form and structure of kinetoplast DNA of *Crithidia*. *J. Cell Biol.* **54**, 346–364.

Schnaitman, C., and Greenawalt, J. W. (1968). Enzymatic properties of the inner and outer membranes of rat liver mitochondria. *J. Cell Biol.* **38**, 158–175.

Schneider, H., Lemasters, J. J., Hochli, J. J., and Hackenbrock, C. R. (1980). Fusion of liposomes with mitochondrial inner membranes. *Proc. Natl. Acad. Sci. USA* **77**, 442–446.

Schneider, W. C. (1948). Intracellular distribution of enzymes. III. The oxidation of octanoic acid by rat liver fractions. *J. Biol. Chem.* **176**, 259–267.

Seglen, P. O. (1976). Preparation of isolated rat liver cells. *In* "Methods in Cell Biology" (D. M. Prescott, ed.), Vol. 13, pp. 29–83. Academic Press, New York.

Simpson, E. R., and Boyd, G. S. (1971). The metabolism of pyruvate by bovine-adrenal-cortex mitochondria. *Eur. J. Biochem.* **22**, 489–499.

Simpson, E. R., and Estabrook, R. W. (1969). Mitochondrial malic enzyme: The source of reduced nicotinamide adenine dinucleotide phosphate for steroid hydroxylation in bovine adrenal cortex mitochondria. *Arch. Biochem. Biophys.* **129**, 384–395.

Smith, A. L. (1967). Preparation, properties, and conditions for assay of mitochondria: Slaughter-house material, small-scale. *In* "Methods in Enzymology" (R. W. Estabrook, and M. E. Pullman, eds.), Vol. 10, pp. 81–86. Academic Press, San Diego.

Smith, R. E., Roberts, J. C., and Hittelman, K. J. (1966). Nonphosphorylating respiration of mitochondria from brown adipose tissue. *Science* **154**, 653–654.

Sohal, R. S., Mockett, R. J., and Orr, W. C. (2002). Mechanisms of aging: An appraisal of the oxidative stress hypothesis. *Free Radic. Biol. Med.* **33**, 575–586.

Stieglerovà, A., Drahota, Z., Ostadal, B., and Houstek, J. (2000). Optimal conditions for determination of cytochrome c oxidase activity in the rat heart. *Physiol. Res.* **49**, 245–250.

Szarkowska, L. (1966). The restoration of DPNH oxidase activity by Coenzyme Q (ubiquinone). *Arch. Biochem. Biophys.* **113**, 519–525.

Tapper, D. P., Van Etten, R. A., and Clayton, D. A. (1983). Isolation of mammalian mitochondrial DNA and RNA and cloning of the mitochondrial genome. *In* "Methods in Enzymology" (S. Fleischer, and B. Fleisher, eds.), Vol. 97, pp. 426–434. Academic Press, San Diego.

Vanneste, V. H. (1966). Molecular properties of the fixed cytochrome components of the respiratory chain of Keilin-Hartree particles beef heart mitocondria. *Biochim. Biophys. Acta* **113,** 175–178.

Ventrella, V., Pagliarani, A., Trigari, R., and Borgatti, A. R. (1982). Respirazione e fosforilazione ossidativa in mitocondri epatici di branzino (*Morone labrax*) e loro dipendenza dalla temperatura. *Boll. Soc. Ital. Biol. Sper.* **58,** 1509–1515.

Villa, R. F., Gorini, A., Geroldi, D., Lo Faro, A., and Dell'Orbo, C. (1989). Enzyme activities in perikaryal and synaptic mitochondrial fractions from rat hippocampus during development. *Mech. Ageing Dev.* **49,** 211–225.

Vinogradov, A. D., and King, T. E. (1979). The Keilin-Hartree heart muscle preparation. *Methods Enzymol.* **55,** 118–127.

Wallace, D. C. (2005). A mitochondrial paradigm of metabolic and degenerative diseases, aging and cancer: A dawn for evolutionary medicine. *Annu. Rev. Genet.* **39,** 359–407.

Wharton, D. C., and Tzagaloff, A. (1967). Cytochrome oxidase from beef heart mitochondria. *Methods Enzymol.* **10,** 245–252.

Wattiaux, R., Wattiaux-De Coninck, , Ronveaux-Dupal, M. F., and Dubois, F. (1978). Isolation of rat liver lysosomes by isopycnic centrifugation in a metrizamide gradient. *J. Cell Biol.* **78,** 349–368.

Wells, W. W., Seyfred, M. A., Smith, C. D., and Sakai, M. (1987). Measurement of subcellular sites of polyphosphoinositide metabolism in isolated rat hepatocytes. *In* "Methods in Enzymology" (P. M. Conn, and A. R. Means, eds.), Vol. 141, pp. 92–99. Academic Press, San Diego.

Yang, J., Liu, X., Bhalla, K., Kim, C. N., Ibrado, A. M., Cai, J., Peng, T. I., Jones, D. P., and Wang, X. (1997). Prevention of apoptosis by Bcl-2: Release of cytochrome *c* from mitochondria blocked. *Science* **275,** 1129–1132.

Zucchini, C., Pugnaloni, A., Pallotti, F., Solmi, R., Crimi, M., Castaldini, C., Biagini, G., and Lenaz, G. (1995). Human skeletal muscle mitochondria in aging: Lack of detectable morphological and enzymic defects. *Biochem. Mol. Biol. Int.* **37,** 607–616.

CHAPTER 2

Purification and Subfractionation of Mitochondria from the Yeast *Saccharomyces cerevisiae*

Istvan R. Boldogh and Liza A. Pon

Department of Anatomy and Cell Biology, College of Physicians and Surgeons
Columbia University, New York, New York 10032

I. Introduction

The yeast *Saccharomyces cerevisiae* has emerged as an important model for the study of cell physiology and pathophysiology. The simple life cycle of yeast, its essentially trouble-free culture conditions, and its rapid growth allow for the preparation of quantities of material that are sufficient for biochemical and biophysical studies. Moreover, since the budding yeast is genetically manipulatable and is the first eukaryote to have its genome sequenced, many fundamental genomic and proteomic studies have been conducted on it.

Budding yeast is also widely used as a source for the preparation of mitochondria for characterization of various functions, including mitochondrial protein import, cytoskeletal interactions, DNA maintenance, and membrane fusion (Chapter 10 by Saada Reisch and Elpeleg, this volume; Boldogh *et al.*, 1998; Kaufman *et al.*, 2000; Meeusen *et al.*, 2004). Since mitochondrial function and protein composition are conserved (Prokisch *et al.*, 2006), studies of yeast mitochondria have been instrumental for understanding mitochondrial functions in higher organisms. Finally, the budding yeast has an additional advantage over other eukaryotic organisms for studying mitochondrial protein functions: it is a facultative aerobe that can tolerate mutations in mitochondrial respiration that are lethal in other organisms. Thus, the budding yeast has become the organism of choice for the study of mitochondrial genetics, inheritance, and function.

Early methods for the isolation of yeast mitochondria were based on removing the cell wall (i.e., converting intact yeast cells to spheroplasts), followed by cell lysis and subcellular fractionation by differential centrifugation (Daum *et al.*, 1982). Mitochondria obtained from these methods contain substantial amounts of other cellular organelles. These crude mitochondria are of sufficient purity for some applications, including the analysis of the activity of known mitochondrial proteins. However, other studies, including analysis of proteins that localize to more than one cellular location, or determining the localization of a protein, require the use of purer mitochondrial fractions. More recent methods for mitochondrial preparation have employed sucrose or Nycodenz density gradient centrifugations to decrease the amount of contaminating membranes and to increase the purity of isolated mitochondria (Diekert *et al.*, 2001; Glick and Pon, 1995; Meisinger *et al.*, 2000). Both sucrose and Nycodenz are nonionic gradient-forming materials. The advantage of Nycodenz over sucrose is that Nycodenz does not affect osmolarity. We find that, compared to sucrose, the yield of intact mitochondria is significantly higher in Nycodenz gradients containing sorbitol as an osmotic stabilizer.

In this chapter, we describe well-established protocols for isolating yeast mitochondria, purifying the organelle using Nycodenz gradients, determining the integrity of isolated mitochondria, and fractionating mitochondria into their subcompartments. We also describe how these methods may be used to determine whether a protein localizes to mitochondria and to subcompartments (outer membrane, inner membrane, intermembrane space, and matrix) within the organelle. Finally, we describe methods to determine the disposition of proteins on mitochondrial membranes.

II. Isolation of Mitochondria

This protocol refers to a 10-l culture, grown in lactate medium in a stirred-tank fermentor (BioFlo 2000, New Brunswick Scientific, Edison, NJ). Under these conditions, the yeast strain D273–10B will yield 40–60 g of yeast, which in turn will yield 200–250 mg of crude mitochondria and 40–80 mg of Nycodenz-purified mitochondria. The yield of cells and/or mitochondria can be substantially lower when using other strains, especially mutant strains, or cells grown in fermentable and/or selective media (e.g., glucose-based rich media, YPD, or glucose-based synthetic media, SC). In these cases, the amount of materials used for organelle preparation should be scaled down.

Stock solutions

We use deionized and Millipore-purified water for making buffers. For preparation of media, deionized water is adequate.

1-M Tris–SO_4, pH 9.4. Autoclave. Store at 4°C.

1-M DTT. Dissolve in water. Store at −20°C.

1-M $MgCl_2$. Autoclave. Store at 4°C.

0.5-M EGTA. Dissolve in water and adjust to pH 8.0 with KOH. Autoclave. Store at 4°C.

2.4-M Sorbitol. Autoclave. Store at 4°C.

200-mM Phenylmethylsulfonylfluoride (PMSF) (Roche, Indianapolis, IN). Dissolve in 100% ethanol. This stock is stable for several months of storage at −20°C.

1-M HEPES–KOH, pH 7.4. Adjust pH with KOH. Autoclave. Store at 4°C.

1.0-M KPi, pH 7.4. Mix 80.2 ml of 1-M KH_2PO_4 and 19.8 ml of 1-M K_2HPO_4. Autoclave. Store at 4°C.

Protease Inhibitor (PI) cocktails (all reagents are from Sigma, St. Louis, MO).

PI-1: 1000× stock (Sigma catalog numbers in parentheses)

0.5-mg/ml Pepstatin A (P4265)

0.5-mg/ml Chymostatin (C7268)

0.5-mg/ml Antipain (A6191)

0.5-mg/ml Leupeptin (L2884)

0.5-mg/ml Aprotinin (A1153)

Prepare 1000× stock from five starting solutions. Dissolve pepstatin A in dimethyl sulfoxide (DMSO), chymostatin in DMSO, antipain in water, leupeptin in water, and aprotinin in water, each at a final concentration of 10 mg/ml. Mix equal volumes of all five solutions and add water so that the final concentration of each PI in the mixture is 0.5 mg/ml.

PI-2: 1000× stock

10-mM Benzamidine–HCl (B6506)

1-mg/ml 1,10-Phenanthroline (P9375) in 100% ethanol

Both cocktails are divided into small aliquots and can be stored for several months at −20°C.

2× Lactate media without glucose

Make 2× concentrated lactate medium but omit glucose (see below for protocol preparation of lactate medium). Autoclave. Store at 4°C.

Nycodenz 50% (w/v)

Dissolve by slow addition of Nycodenz powder (Sigma, St. Louis, MO) to water with constant stirring. Nycodenz powder has a tendency to form clumps and therefore care must be taken to dissolve it completely before use. Store in aliquots at −20°C.

Working solutions

Lactate medium for yeast cell growth:

> 3-g/l Yeast extract (Difco)
> 0.5-g/l Glucose
> 0.5-g/l $CaCl_2 \cdot 2H_2O$
> 0.5-g/l NaCl
> 0.6-g/l $MgCl_2 \cdot 6H_2O$
> 1-g/l KH_2PO_4
> 1-g/l NH_4Cl
> 22 ml/l of 90% Lactic acid
> 7.5-g/l NaOH

Adjust pH to 5.5 with NaOH pellets.

Tris–DTT buffer (200 ml):

> 0.1-M Tris–SO_4, pH 9.4
> 10-mM DTT

SP buffer (300 ml):

> 1.2-M Sorbitol
> 20-mM KPi, pH 7.4

Regeneration solution (200 ml):

> 1.2-M Sorbitol
> 1× Lactate media without glucose

SEH buffer (500 ml):

> 0.6-M Sorbitol
> 20-mM HEPES–KOH, pH 7.4
> 2-mM $MgCl_2$
> 1-mM EGTA, pH 8.0
> 1× PI-1
> 1× PI-2
> 1-mM PMSF

Make 500 ml ice-cold SEH buffer without PMSF. Immediately before each Dounce homogenization step (see below), add PMSF to the SEH buffer with constant stirring, and filter using Whatman No. 4 filter paper.

SEM buffer (use this buffer instead of SEH buffer to carry out mitochondrial purification at pH 6.0):

0.6-M Sorbitol

20-mM MES–KOH, pH 6.0

2-mM $MgCl_2$

1-mM EGTA, pH 8.0

1× PI-1

1× PI-2

1-mM PMSF

Make 500 ml ice-cold SEM buffer without PMSF. Immediately before each Dounce homogenization step, add PMSF to the SEM buffer with constant stirring, and filter using Whatman No. 4 filter paper.

A. Growth of Yeast Cells

Generate a stationary phase preculture of D273–10B cells, by inoculating 50 ml of lactate medium in a shake flask with a colony from a fresh plate. Incubate cells with vigorous shaking for 2 days at 30 °C. Inoculate 10 liter of lactate medium in a fermentor with the 50-ml preculture (1:200). Grow yeast at 30 °C for 16–17 h with good aeration. At the time of harvest, yeast should be in mid-log phase and at a density (OD_{600}) of three.

B. Isolation of Crude Mitochondria

1. Concentrate cells by centrifugation at 2000 × g for 5 min. (Sorvall GS-3 rotor 3500 rpm.)

2. Resuspend all of the cell pellets obtained from the centrifugation in a total volume of ∼200 ml of water. For quick resuspension, use a rubber bulb attached to the end of a pipette or a rubber policeman. Wash the tubes thoroughly to minimize losses. Pool all resuspended cell pellets into one tared centrifuge tube. Centrifuge the suspension for 5 min at 2000 × g. Discard the supernatant obtained from the centrifugation and determine the weight of the "wet" cell pellet. For a mid-log phase culture of D273–10B cells grown in 10 liter of lactate medium, a typical wet weight for the cell pellet is 40–60 g.

3. Resuspend the cell pellet in 180 ml of Tris–DTT buffer. Incubate for 15 min at 30 °C, with shaking. Transfer the cell suspension into six Sorvall SS34 rotor tubes (35 ml/tube), and centrifuge for 5 min at 1500 × g (3500 rpm).

4. Resuspend cells from the pellet obtained in 180 ml of SP buffer, and centrifuge the suspension at $1500 \times g$ for 5 min. Resuspend the cell pellet obtained in 70–80 ml of SP buffer, and transfer the suspension into a 500-ml Erlenmeyer flask.

5. To convert yeasts to spheroplasts (i.e., to remove yeast cell walls), dissolve 7.5 mg of zymolyase 20T (Seikagaku Corporation, Tokyo, Japan) per gram of wet cells in 15 ml of SP buffer. Add the zymolyase solution to the cell suspension, and shake gently (200 rpm) at $30\,^{\circ}$C for 40 min. After incubation, centrifuge the mixture at $4500 \times g$ (6000 rpm in a Sorvall SS34 rotor) for 5 min at room temperature (RT).

Note: Conditions for spheroplast formation vary according to the batch of zymolyase and/or the strains used. The progress of cell wall removal can be monitored by visual inspection using a light microscope. Cells with intact cell walls have highly refractile borders, while spheroplasts are less refractile. Also, the pellet of spheroplasts is fluffier than that of intact yeast cells.

6. Resuspend the spheroplast pellet obtained in step 5 in 180 ml of Regeneration Solution. For efficient resuspension, add a small volume of liquid to the pellet and resuspend gently using a rubber policeman. Repeat until all of the Regeneration Solution is added and the entire spheroplast pellet is resuspended. Transfer the suspension to a 500-ml Erlenmeyer flask and shake gently (200 rpm) for 40 min at $30\,^{\circ}$C. Transfer the suspension to six Sorvall SS34 tubes and centrifuge at $4500 \times g$ for 5 min.

Note: Early studies (Baker and Schekman, 1989) revealed that yeast that are harvested and converted to spheroplasts using the methods described exhibit reduced levels of protein synthesis activity. The regeneration step described above results in restoration of translation activity in spheroplasts without regeneration of the cell wall. In our experience, mitochondria recovered from "regenerated" spheroplasts are more robust than those obtained from untreated spheroplasts.

7. Add PMSF to ice-cold SEH buffer, as described above. Resuspend each pellet of regenerated spheroplasts in 25 ml of ice-cold SEH buffer. From this step forward, all tubes and solutions are kept on ice. Centrifuge this suspension at $4500 \times g$ for 5 min at $4\,^{\circ}$C.

8. Resuspend the pellet obtained in each tube in 10 ml of ice-cold SEH buffer. Pour suspension from two tubes into a prechilled 40-ml glass/glass Dounce homogenizer (Wheaton Science Products, Millville, NJ). Wash centrifuge tubes with SEH buffer, and transfer the wash solution and residual cells to the Dounce homogenizer. The total volume for each round of Dounce homogenization should be $\sim$35 ml. Using the tight-fitting pestle, homogenize the suspension with 15 forceful strokes. Repeat homogenization with the remaining four tubes. Transfer homogenates to SS34 tubes.

9. Centrifuge the homogenate at low speed, $1500 \times g$ for 5 min, at $4\,^{\circ}$C. Pour the "low-speed" supernatant obtained into fresh SS34 tubes, using care to avoid

disrupting the loose pellet. Save the "low-speed" pellets, which contain nuclei, cell debris, and unbroken cells. Centrifuge the "low-speed" supernatant at high speed, 12,000 × g (10,000 rpm) for 10 min, at 4°C. Save the "high-speed" pellet, which contains mitochondria, and discard the "high-speed" supernatant.

10. Subject the low-speed pellet obtained in step 9 to a second round of homogenization followed by a low-speed and then a high-speed centrifugation. Do the second high-speed centrifugation on top of the first high-speed pellet.

11. Resuspend the high-speed pellets obtained in steps 9 and 10 in a total of 70 ml of ice-cold SEH buffer, using a 40-ml Dounce homogenizer with a loose-fitting pestle. Transfer the suspension to SS34 tubes and centrifuge at low speed (1500 × g for 5 min) at 4°C. Transfer the supernatants obtained to fresh SS34 tubes, and centrifuge at high speed (12,000 × g for 10 min) at 4°C. Resuspend the high-speed pellet in 35-ml SEH buffer; repeat low- and high-speed centrifugations, as above.

12. Resuspend high-speed pellets obtained from step 11 to a final volume of 3–4 ml in ice-cold SEH buffer, using a 7-ml glass/glass Dounce homogenizer with a loose-fitting pestle. This product is the crude mitochondrial fraction. The darker brown the pellet, the higher the mitochondrial content.

13. To estimate the concentration of protein in crude mitochondria, mix 10 μl of the crude mitochondria with 990 μl 0.6% SDS. Measure OD at 280 nm. A value of 0.21 corresponds to a protein concentration of 10 mg/ml in the undiluted crude mitochondria.

C. Purification of Crude Mitochondria Using Continuous Nycodenz Gradient Centrifugation

The volumes described are for the preparation of two 10.5-ml gradients. Two gradients are usually sufficient for purification of mitochondria from 200 to 250 mg of crude mitochondria.

Working solutions
2× SEH buffer:

1.2-M Sorbitol
40-mM HEPES–KOH, pH 7.4
4-mM $MgCl_2$
1-mM EGTA, pH 8.0
2× PI-1
2× PI-2
2-mM PMSF

Protocol
Prepare the two Nycodenz gradients in Beckman 14 × 89 mm Ultra-Clear Centrifuge tubes (344059) 5–6 h before use.

1. Prepare 5 ml of each Nycodenz gradient step (5, 10, 15, 20, and 25% Nycodenz), using 50% Nycodenz, 2× SEH buffer and water, according to the following scheme:

Final Nycodenz concentration (%)	Volume 50% Nycodenz (ml)	Volume 2× SEH buffer (ml)	Volume water (ml)
5	0.5	2.5	2.0
10	1.0	2.5	1.5
15	1.5	2.5	1.0
20	2.0	2.5	0.5
25	2.5	2.5	0

2. For each gradient, transfer 2.1 ml of the 25% Nycodenz solution into the ultracentrifuge tube. Overlay this step with 2.1 ml of the 20% Nycodenz solution, using a 3-ml syringe and 21-gauge needle. Use care to avoid mixing the 20 and 25% Nycodenz solutions, and strive to produce a defined interface between each step. Overlay the 20% Nycodenz step with 2.1 ml of the 15% Nycodenz solution, as described above. Repeat the overlay with the 10% Nycodenz solution and then with the 5% Nycodenz solution.

3. Allow gradients to diffuse for 3–4 h at RT. Chill them at 4 °C for 1 h before use.

4. Apply 1.9-ml crude mitochondrial solution (50 mg/ml) onto the top of each gradient using care not to disrupt the gradient's structure.

5. Ultracentrifuge gradients in a swinging bucket rotor (Beckman SW41 rotor) at $100,000 \times g$ (30,000 rpm) for 60 min at 4 °C. This centrifugation results in separation of crude mitochondria into four to five bands, which contain lipids, mitoplasts (mitochondria with ruptured outer membranes), intact mitochondria, and cell debris (Fig. 1).

6. Aspirate and discard the solution above the mitochondrial band (Fig. 1), using care to avoid disturbing the mitochondrial band itself. Transfer mitochondria from the Nycodenz gradient to a fresh, prechilled tube, using a Pasteur pipette. Separate the Nycodenz-purified mitochondria into aliquots, as needed. For long-term storage, aliquots can be frozen and kept in liquid nitrogen.

7. To determine the recovery of protein in the Nycodenz-purified mitochondria, dilute 10 μl of purified mitochondria in 0.5 ml SEH buffer, centrifuge the suspension for 5 min at $12,000 \times g$, resuspend pellet in 1 ml 0.6% SDS, and measure OD at 280 nm. A value of 0.12 corresponds to 10 mg/ml mitochondrial protein in the undiluted mitochondrial suspension.

Note: For subsequent use of mitochondria, remove Nycodenz by diluting the mitochondrial suspension five-fold with SEH buffer. Centrifuge diluted mito-chondria for 5 min at $12,000 \times g$ at 4 °C. Remove the supernatant, and resuspend the mitochondrial pellet in the buffer of choice for further experiments. If the mitochondrial protein concentration falls below 1 mg/ml, the *in vitro* protein import activity of the organelle decreases. This decrease can be prevented by the addition of 1-mg/ml fatty acid-free bovine serum albumin to the mitochondrial suspension.

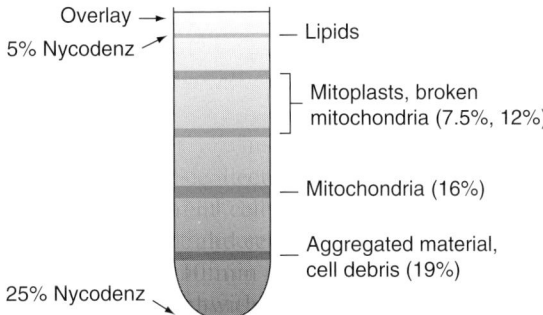

Fig. 1   Separation pattern of crude mitochondria on a 5–25% Nycodenz gradient. Purified mitochondria appear as a wide, brownish band at around 16% Nycodenz concentration. Mitochondrial fragments and mitoplasts can be seen at around 7.5 and 12% Nycodenz concentration as lighter brownish bands. Aggregated mitochondria and cell debris are recovered at around 19% Nycodenz concentration. If resuspension of mitochondria is not sufficient, aggregated material may appear at the bottom of the tube as well.

III. Analysis of Isolated Mitochondria

A. Assessing the Purity of Mitochondrial Preparations

Isolated mitochondria are used routinely to study activities of resident mitochondrial proteins and to determine whether a protein localizes to mitochondria. For both applications, it is critical to document the purity and integrity of the mitochondrial preparation. The simplest way to do so is to perform SDS-PAGE and Western blot analysis, using antibodies raised against proteins of different membrane organelles (Fig. 2).

Quantitation of the fractionation behavior of marker proteins during preparation of mitochondria (Table I) reveals that differential centrifugation results in the enrichment of a mitochondrial marker protein in the crude mitochondrial fraction compared to whole cell lysates, as expected. Surprisingly, differential centrifugation does not separate crude mitochondria from the plasma membrane. Moreover, endoplasmic reticulum (ER) membranes are enriched in the crude mitochondrial fraction. Thus, ER, plasma membrane, and other membranes are contaminants in mitochondria isolated by differential centrifugation. This quantitation also reveals that Nycodenz gradient purification reduces, but does not completely eliminate, these contaminating membranes (Table I). Specifically, Nycodenz purification leads to a significant enrichment of the mitochondrial marker, and depletion of both ER and plasma membrane contaminants, in the mitochondrial fraction.

ER is present in the mitochondrial fraction, in part because ER and mitochondria interact physically and functionally during cellular processes, including phospholipid

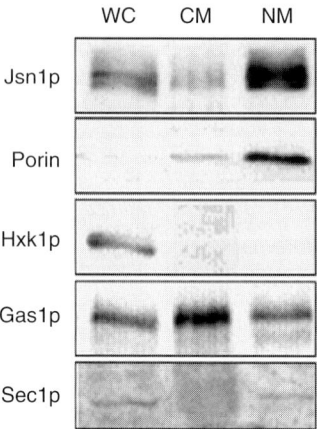

Fig. 2 Purification of mitochondria on a Nycodenz gradient. Aliquots of the different fractions (WC, whole cells; CM, crude mitochondria; NM, Nycodenz-purified mitochondria) containing equal amounts of protein were analyzed by SDS-PAGE and Western blot, using antibodies specific for proteins that are markers for mitochondria (porin), the cytosol (hexokinase, Hxk1p), the plasma membrane (Gas1p), and the ER (Sec1p). Porin and the PUF family protein Jsn1p are enriched in crude mitochondria and in Nycodenz-purified mitochondria.

Table I
Analysis of Marker Proteins in Different Fractions During Mitochondrial Preparation[a]

	Relative levels of marker protein		
		Mitochondria	
Marker protein (protein localization)	Spheroplasts	Crude	Gradient purified
Cytochrome b_2 (mitochondria)	1	3.77	4.36
Gas1p (plasma membrane)	1	1.01	0.31
NPC (nucleus)	1	0.61	0.10
Sec61 (ER)	1	2.36	0.48
Hexokinase (cytosol)	1	0.09	0.02

[a]Relative levels of marker proteins in spheroplast, crude mitochondria, and Nycodenz gradient-purified mitochondria. Aliquots of the different fractions were centrifuged at $12,000 \times g$ for 5 min. The pellets were resuspended in buffer SEH. The protein concentrations of these samples were determined using BCA assay (Pierce, Rockford, Illinois, IL) with bovine serum albumin as the standard. Different amounts (5–50 μg/lane) of each fraction were analyzed by SDS-PAGE and Western blot using specific antibodies for each marker protein. The antibodies were detected using chemiluminescence and quantified by the public domain software, ImageJ. Gray values were used to normalize the relative amount of membrane markers in these fractions: the gray values per microgram of total protein were determined for each marker protein and expressed relative to the level found in the spheroplast.

Table II
Differences in Membrane Contaminations in Mitochondria Purified
at pH 7.4 or 6.0[a]

Protein (marker)	pH 7.4	pH 6.0
Cytochrome b_2 (mitochondria)	1	0.98
Gas1p (plasma membrane)	1	0.61
NPC (nucleus)	1	0.21
Sec61 (ER)	1	0.59
Hexokinase (cytosol)	1	0.17

[a]Values in the table represent relative levels of marker proteins in gradient-purified mitochondria. Nycodenz-purifed mitochondria were collected and diluted in SEH (pH 7.4) or SEM (pH 6.0) buffer. The diluted samples were centrifuged at $12,000 \times g$ for 5 min. The pellets obtained were resuspended in buffer SEH. For each marker protein, the gray values per microgram of total protein were determined and expressed relative to the level found in mitochondria isolated at pH 7.4. SDS-PAGE, Western blot analysis, and quantitation of proteins were carried out using the same procedure as for Table I.

biosynthesis and calcium regulation (Zinser and Daum, 1995). Therefore, methods have been developed to remove ER from mitochondrial fractions. Previous studies have shown that incubation of crude mitochondria at pH 6.0 results in the release of microsomal (ER) membranes from mitochondria (Gaigg et al., 1995). Alternatively, mitochondria can be isolated using a buffer with pH value of 6.0 (SEM) instead of 7.4 (SEH) (Zinser and Daum, 1995). Table II shows the level of membrane contamination in Nycodenz-purified yeast mitochondria, where cell homogenization and all subsequent steps were carried out at pH 7.4 in SEH buffer or at pH 6.0 using SEM buffer.

B. Determining the Integrity of Mitochondrial Preparations

Below we describe methods to assess the integrity of the outer and inner membranes in isolated mitochondria. These methods are also used to determine the submitochondrial localization of a protein and the disposition of proteins on mitochondrial membranes. In this section, we provide an overview of the methods used and the outcomes expected in assessing the quality of mitochondrial preparations.

The integrity of the mitochondrial outer membrane can be determined by evaluating the recovery of an intermembrane-space marker protein (e.g., cytochrome b_2) relative to an integral outer or inner membrane protein (e.g., porin) by SDS-PAGE and Western blot analysis, as described above. Alternatively, mitochondria can be treated with protease under conditions that allow for degradation of accessible proteins, without disrupting membrane integrity. Figure 3 shows results obtained following protease treatment, in an assay for the integrity of the mitochondrial outer membrane. Here, Nycodenz-purified mitochondria were treated with protease and the levels of protease-sensitive outer membrane (Tom70p) and intermembrane-space

Protease

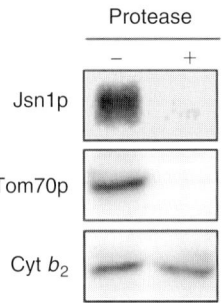

Fig. 3 Protease sensitivity of selected proteins in Nycodenz-purified mitochondria. Mitochondria were incubated in the presence or absence of Trypsin/Chymotrypsin mixture and were concentrated by centrifugation, as described in the text. Mitochondria recovered in the pellet of the centrifugation were analyzed by SDS-PAGE and Western blots. Since cytochrome b_2 (cyt b_2), an intermembrane-space protein, is protease-resistant, it is clear that the mitochondria are intact. Tom70p, a mitochondrial outer membrane protein, is protease-sensitive. Thus, the protease conditions used degraded mitochondrial surface proteins without affecting the integrity of the organelle. Since Jsn1p is protease-sensitive, it localizes to the surface of the organelle.

(cytochrome b_2) proteins were monitored by Western blot analysis. The observed protease sensitivity of Tom70p indicates that the conditions used allowed for the degradation of mitochondrial surface proteins. Moreover, the observed protease resistance of cytochrome b_2 demonstrates that the mitochondrial outer membrane was intact.

The integrity of the mitochondrial inner membrane can be determined by monitoring mitochondrial respiratory activity, as described in Chapters 4 and 5, this volume. An alternative method is to monitor the ability of the organelle to respond to changes in osmolarity of the medium. If mitochondrial inner membranes are intact, then incubation of the organelle in hypotonic media will result in swelling of the organelle, which in turn results in rupture of the outer membrane. Swelling of mitochondria and rupture of the mitochondrial outer membrane, which produces mitoplasts, can be monitored by protease treatment. If mitochondria swell and mitoplast formation occurs, then proteins in the intermembrane space will be sensitive to externally added protease, while proteins in the matrix will not.

1. Methods to Produce Mitoplasts and Assess Membrane Integrity in Intact Mitochondria and Mitoplasts

We describe here a protocol to assess the integrity of both outer and inner mitochondrial membranes. In the protocol described, mitochondria and mitoplasts, which are produced by incubation of mitochondria in hypotonic buffer, are treated with protease, and the levels of mitochondrial marker proteins are determined by Western blot analysis. A control, in which mitochondria are solubilized with a nonionic detergent, is included, to demonstrate that the marker proteins analyzed are protease-sensitive when accessible to externally added protease.

Working solutions

20-mM HEPES–KOH, pH 7.4 (hypotonic buffer)

SH–PI buffer

   0.6-M Sorbitol

   20-mM HEPES–KOH, pH 7.4

Protease (Trypsin/Chymotrypsin mixture or Proteinase K). Dissolve protease(s) to a final concentration of 1 mg/ml each in water immediately before use.

Trypsin inhibitor: 30 mg/ml. Dissolve in water.

1000× PI-1 (Section II)

1000× PI-2 (Section II)

200-mM PMSF (Section II)

Triton X-100 (20%)

100% (w/v) trichloroacetic acid (TCA) solution. Dissolve in water. Store in an airtight, dark bottle at RT. This solution is very corrosive and should be used in a fume hood.

Samples and controls

1. Mitochondria
2. Protease-treated mitochondria
3. Mitoplasts (mitochondria treated with hypotonic media)
4. Protease-treated mitochondria
5. Detergent-treated mitochondria
6. Detergent- and protease-treated mitochondria, a control to document protease sensitivity of marker proteins

Reagent	1	2	3	4	5	6
Mitochondria (5 mg/ml)	10 μl	10 μl	10 μl	10 μl	10 μl	10 μl
SH–PI buffer	90 μl	90 μl	–	–	90 μl	90 μl
20-mM HEPES–KOH, pH 7.4	–	–	90 μl	90 μl	–	–
Triton X-100	–	–	–	–	1 μl	1 μl
Protease	–	1 μl	–	1 μl	–	1 μl

Protocol

1. To remove PI that is present in mitochondrial preparations, dilute mitochondria five-fold with ice-cold SH–PI buffer, and centrifuge the mixture at $12,000 \times g$ for 10 min at 4°C. Discard the supernatant obtained, and resuspend the mitochondrial pellet again in ice-cold SH–PI buffer. Centrifuge the mixture, as

described above. Discard supernatant and resuspend the mitochondrial pellet in ice-cold SH–PI buffer at a concentration of 5 mg/ml.

2. Transfer washed mitochondria (10 μl) to prechilled, labeled Eppendorf tubes.

3. Add SH–PI buffer (90 μl) to samples 1, 2 and 5, 6, and add 20-mM HEPES–KOH, pH 7.4 (90 μl) to samples 3–4. Mix gently.

4. To initiate protease and/or detergent treatment, add protease (1 μl) to samples 2, 4, and 6, and add Triton X-100 (1 μl) to samples 5 and 6. Mix gently.

5. Incubate for 20 min on ice. Thereafter, add 0.5 μl of each 1000$\times$ PI-1, 1000$\times$ PI-2, 200-mM PMSF, and 30 mg/ml trypsin inhibitor to each tube.

6. Centrifuge at 12,500 $\times$ g for 5 min at 4°C. Analyze pellets by SDS-PAGE and Western blots.

C. Assessment of Protein Localization to Mitochondria and to Subcompartments in the Organelle

Since proteins that do not localize to mitochondria are recovered in mitochondrial fractions even after purification using Nycodenz gradients, detection of a protein in crude or Nycodenz-purified mitochondria is not a sufficient criterion to establish the localization of a protein. Instead, it is critical to use methods, including Western blot analysis, that compare the fractionation behavior of the protein of interest with that of other marker proteins during mitochondrial preparation. In the fractionation study shown in Fig. 2, Jsn1p, a PUF (Pumilio) family protein, is enriched in crude and Nycodenz-purified mitochondria to the same extent as is the mitochondrial marker protein cytochrome b_2 (Fehrenbacher et al., 2005). This finding, combined with localization of the protein by indirect immunofluorescence, indicates that Jsn1p is a mitochondrial protein.

The protease-sensitivity treatments described above can also be used to study the submitochondrial localization of a mitochondrial protein. Proteins that localize to the surface of the organelle are typically degraded on treatment of intact mitochondria with protease, if the conditions used result in digestion of a mitochondrial outer membrane protein (e.g., Tom70p) without affecting the integrity of the outer membrane. In contrast, proteins that localize to the inner leaflet of the outer membrane, the intermembrane space, or the outer leaflet of the inner membrane typically remain intact on treatment of intact mitochondria with protease, but are degraded on treatment of mitoplasts with protease.

Finally, proteins that localize to the inner leaflet of the inner membrane or to the mitochondrial matrix typically remain intact on treatment of mitochondria or mitoplasts with protease, but are degraded after detergent permeabilization of mitochondrial outer and inner membranes followed by protease treatment. These experiments are only informative if the proteins of interest are known to be sensitive to the protease used for digestion. Thus, controls showing that the protein of interest is digested with protein on disruption of mitochondrial membranes with a nonionic detergent are critical for interpretation of protease-sensitivity studies.

D. Assessment of the Disposition of Proteins on Mitochondrial Membranes

Treatments with high concentration of salt and alkali (sodium carbonate) are routinely used to determine the disposition of proteins on mitochondrial membranes. Peripheral membrane proteins are released from membranes by alkali treatment. In contrast, integral membrane proteins remain associated with membranes after washes with high salt and alkali treatment. Treatment of intact mitochondria with high salt can release peripheral membrane proteins that are on the surface of the outer membrane, while treatment of mitoplasts with high salt can release peripheral membrane proteins from the inner and outer leaflet of the outer membrane, and from the outer leaflet of the inner membrane. In contrast, since alkali treatment disrupts all mitochondrial membranes, carbonate extracts and releases peripheral membrane proteins from all of the membranes of isolated yeast mitochondria.

Figure 4 illustrates the results of salt and alkali treatment of Nycodenz-purified mitochondria. Cytochrome b_2, a peripheral inner membrane protein, is extracted from mitochondrial membranes with alkali treatment. However, since salt treatment does not disrupt the integrity of the mitochondrial outer membrane, cytochrome b_2 is not released from the organelle by high salt treatment. Porin is an integral mitochondrial

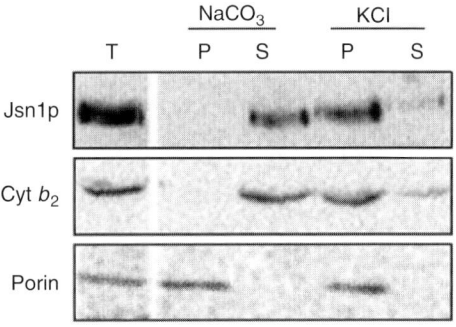

Fig. 4 Carbonate and salt extraction of Nycodenz-purified yeast mitochondria. Mitochondria were incubated in the presence of sodium carbonate or 1-M KCl, as described in the text. After separation of extracted and unextracted material by centrifugation, pellets and supernatants were analyzed by SDS-PAGE and Western blots. Cyt b_2, cytochrome b_2, is a peripheral inner mitochondrial membrane protein. Porin is an integral mitochondrial outer membrane protein. The PUF family protein Jsn1p behaves like a peripheral membrane protein. T, total; P, pellet; S, supernatant.

outer membrane protein that is not released from mitochondrial membranes by either alkali or salt treatment. Finally, the PUF family protein Jsn1p is released by treatment of mitochondria with alkali. Thus, Jsn1p behaves like a peripheral membrane protein, and based on protease sensitivity studies (see Fig. 3) it resides on the outer leaflet of the outer mitochondrial membrane.

1. Carbonate and Salt Extraction of Isolated Yeast Mitochondria

Working solutions
SEH buffer (Section II)

0.2-M NaCl. Dissolve in water.

0.2-M Na_2CO_3. Dissolve in water. Prepare immediately before use.

SEH + salt: SEH buffer containing 1-M KCl

100% (w/v) TCA solution. Dissolve in water. Store in an airtight, dark bottle at RT. This solution is very corrosive and should be used in a fume hood.

Protocol

1. Mix solutions according to the following table, in the order shown:

Reagent	Carbonate treatment (μl)		Salt treatment (μl)	
	1—control	2—treated	3—control	4—treated
SEH buffer	90	90	190	—
0.2-M NaCl	100	—	—	—
0.2-M Na_2CO_3	—	100	—	—
SEH + salt	—	—	—	190
Mitochondria (5 mg/ml)	10	10	10	10

2. Incubate for 20 min on ice.

3. Centrifuge samples 1 and 2 at $100,000 \times g$ for 20 min at 4°C. Separate supernatant and pellet obtained from the centrifugation. Save the supernatant and resuspend the pellet in 200-μl buffer SEH.

4. Centrifuge samples 3 and 4 at $12,500 \times g$ for 20 min at 4°C. Separate supernatant and pellet obtained from the centrifugation. Save the supernatant and resuspend the pellet in 200-μl buffer SEH.

5. Add 1/10 volume (20 μl) of the TCA solution to the supernatant and to the resuspended pellet. Incubate for 15 min on ice. Centrifuge the TCA-treated samples at $12,500 \times g$ for 15 min at 4°C. Aspirate and discard supernatants, taking care not to disrupt the loose pellets.

6. Resuspend the pellets obtained from TCA precipitation in 20- to 40-μl SDS-PAGE loading buffer. If the solution is yellow, neutralize residual TCA by addition of an amount of 1-M Tris–base (1–5 μl) that allows the solution to turn

blue. Mix by vortexing and complete solubilization by heating the samples to 95°C for 3–4 min. Analyze samples by SDS-PAGE and Western blot.

IV. Isolation of Mitochondrial Outer Membranes, Contact Sites, and Inner Membranes

Isolated mitochondria can be fractionated further into inner and outer membranes, as well as contact sites, which are sites of close contact between the outer and inner membranes that are implicated as sites for translocation of protein and phospholipids across, and between, the mitochondrial outer and inner membranes (Simbeni *et al.*, 1991). For submitochondrial fractionation, mitochondria are disrupted by hypotonic treatment followed by sonication. Small unilamellar vesicles produced from these procedures are separated from intact mitochondria and mitoplasts by centrifugation. Disruption is carried out in the presence of EDTA, which dissociates cytoplasmic ribosomes from contact sites and shifts the density of contact site-enriched vesicles to one that is distinct from those of submitochondrial membrane vesicles derived from the outer or inner membrane. The vesicles obtained are then separated on the basis of density, using a continuous sucrose gradient. Outer and inner membranes separate as low- and high-density bands, respectively, while contact sites sediment at a density that is intermediate between those of the outer and inner membranes (Pon *et al.*, 1989) (Fig. 5). The yield of submitochondrial membrane vesicles obtained using this procedure is low. However, the vesicles obtained are of suitable purity to allow for the analysis of the activity or submitochondrial membrane localization of a protein of interest.

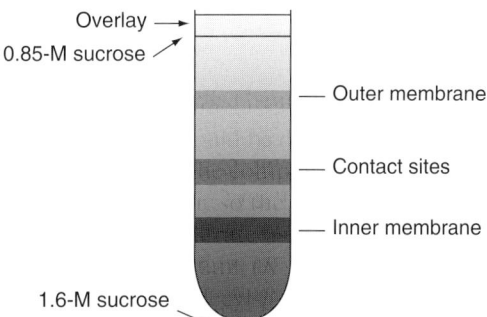

Fig. 5 Separation pattern of mitochondrial membrane fractions on a 0.85- to 1.6-M sucrose gradient. Submitochondrial membrane vesicles separate into three bands. Outer membrane fractions, contact sites, and inner membrane fractions appear at around 1-, 1.2-, and 1.36-M sucrose concentrations, respectively.

Working solutions

HE buffer
   20-mM HEPES–KOH, pH 7.4
   0.5-mM EDTA
   1-mM PMSF
100× HK buffer
   0.5-M HEPES–KOH, pH 7.4
   1-M KCl
Buffer A
   1.6-M Sucrose, 5-mM HEPES–KOH, pH 7.4
   10-mM KCl
Buffer B
   0.85-M Sucrose 5-mM HEPES–KOH, pH 7.4
   10-mM KCl

Protocol

A. Preparation of sucrose gradients:

The volumes described are for the preparation of four 10 ml sucrose gradients. Four gradients are usually sufficient to isolated submitochondrial membrane vesicles from 200 to 250 mg of crude mitochondria.

1. Prepare four gradients in Beckman 14 × 89 mm Ultra-Clear Centrifuge tubes (344059). The sucrose solutions should be prepared as described below:

Step [sucrose] (M)	Buffer A (ml)	Buffer B (ml)
0.85	0	12
1.1	4	8
1.35	8	4
1.6	12	0

2. For each gradient, transfer 2.5 ml of the 1.6-M sucrose solution into the ultracentrifuge tube. Overlay this step with 2.5 ml of the 1.35-M sucrose solution, using a 3-ml syringe and 21-gauge needle. Use care to avoid mixing the 1.6- and 1.35-M sucrose solutions, and strive to produce a defined interface between each step. Overlay the 1.35-M sucrose step with 2.5 ml of the 1.1-M sucrose solution, as described above. Repeat the overlay with the 0.85-M sucrose solution.

3. Allow gradients to diffuse for 6 h at RT. Chill them at 4 °C for 1 h before use.

B. Formation of Submitochondrial Membrane Vesicles for Isolated Yeast mitochondria:

1. To subject mitochondria to hypotonic shock, which converts mitochondria to mitoplasts, resuspend 200–250 mg of crude mitochondria to a final concentration of 10-mg/ml HE buffer. Incubate this suspension on ice for 30 min with occasional mixing.

2. Add sucrose to the mitoplasts to a final concentration of 0.45 M and incubate the suspension for 10 min on ice. Transfer the mixture to a prechilled sonication flask. Add fresh PMSF to a final concentration of 1 mM. Sonicate the mixture on ice, using a cell disruptor (Heat Systems-Ultrasonic, Farmingdale, NY) for $3\times$ 30-sec pulses at 80% duty cycle and maximum power, with 30-sec intervals between the pulses.

3. To separate submitochondrial membrane vesicles from mitochondria and mitoplasts, transfer the sonicated material to fresh centrifuge tubes and centrifuge for 15 min at 34,000 $\times$ g (17,000 rpm in a SS34 rotor).

5. To concentrate the submitochondrial membrane vesicles, transfer the supernatant obtained to fresh ultracentrifuge tubes, and centrifuge in a fixed-angle rotor at 200,000 $\times$ g for 90 min at 4°C (Beckman Ti60 rotor at 45,000 rpm).

6. Discard the resultant supernatant, and resuspend the pellet in 2 ml of prechilled, filtered buffer B containing DTT and 1-mM PMSF, using a 7.0-ml Dounce homogenizer and a loose-fitting pestle.

7. Apply the resuspended material onto the sucrose gradients. Centrifuge in a swinging bucket rotor at 100,000 $\times$ g for 14 h at 4°C (Beckman SW41 rotor at 29,000 rpm). This centrifugation results in separation of submitochondrial membrane vesicles into three bands, which contain outer membrane, contact sites, and inner membrane (Fig. 5).

8. Aspirate and discard the solution above the outer membrane band, using care to avoid disturbing it. Transfer the outer membrane band to a fresh, prechilled tube using a Pasteur pipette. Remove the contact site- and inner membrane-containing vesicles as above. Separate the outer membrane, contact site, and inner membrane vesicles into aliquots, as needed. For long-term storage, aliquots can be frozen and kept in liquid nitrogen.

Acknowledgments

We thank the members of the Pon laboratory for the critical evaluation of the manuscript. The work was supported by research grants to L.A.P. from the National Institutes of Health (GM45735).

References

Baker, D., and Schekman, R. (1989). Reconstitution of protein transport using broken yeast spheroplasts. *Methods Cell Biol.* **31,** 127–141.

Boldogh, I., Vojtov, N., Karmon, S., and Pon, L. A. (1998). Interaction between mitochondria and the actin cytoskeleton in budding yeast requires two integral mitochondrial outer membrane proteins, Mmm1p and Mdm10p. *J. Cell Biol.* **141,** 1371–1381.

Daum, G., Bohni, P. C., and Schatz, G. (1982). Import of proteins into mitochondria. Cytochrome b2 and cytochrome c peroxidase are located in the intermembrane space of yeast mitochondria. *J. Biol. Chem.* **257,** 13028–13033.

Diekert, K., de Kroon, A. I., Kispal, G., and Lill, R. (2001). Isolation and subfractionation of mitochondria from the yeast *Saccharomyces cerevisiae. Methods Cell Biol.* **65,** 37–51.

Fehrenbacher, K. L., Boldogh, I. R., and Pon, L. A. (2005). A role for Jsn1p in recruiting the Arp2/3 complex to mitochondria in budding yeast. *Mol. Biol. Cell* **16,** 5094–5102.

Gaigg, B., Simbeni, R., Hrastnik, C., Paltauf, F., and Daum, G. (1995). Characterization of a micro-somal subfraction associated with mitochondria of the yeast, *Saccharomyces cerevisiae*. Involvement in synthesis and import of phospholipids into mitochondria. *Biochim. Biophys. Acta* **1234**, 214–220.

Glick, B. S., and Pon, L. A. (1995). Isolation of highly purified mitochondria from *Saccharomyces cerevisiae*. *Methods Enzymol.* **260**, 213–223.

Kaufman, B. A., Newman, S. M., Hallberg, R. L., Slaughter, C. A., Perlman, P. S., and Butow, R. A. (2000). In organello formaldehyde crosslinking of proteins to mtDNA: Identification of bifunctional proteins. *Proc. Natl. Acad. Sci. USA* **97**, 7772–7777.

Meeusen, S., McCaffery, J. M., and Nunnari, J. (2004). Mitochondrial fusion intermediates revealed *in vitro*. *Science* **305**, 1747–1752.

Meisinger, C., Sommer, T., and Pfanner, N. (2000). Purification of *Saccharomcyes cerevisiae* mitochon-dria devoid of microsomal and cytosolic contaminations. *Anal. Biochem.* **287**, 339–342.

Pon, L., Moll, T., Vestweber, D., Marshallsay, B., and Schatz, G. (1989). Protein import into mito-chondria: ATP-dependent protein translocation activity in a submitochondrial fraction enriched in membrane contact sites and specific proteins. *J. Cell Biol.* **109**, 2603–2616.

Prokisch, H., Andreoli, C., Ahting, U., Heiss, K., Ruepp, A., Scharfe, C., and Meitinger, T. (2006). MitoP2: The mitochondrial proteome database—now including mouse data. *Nucleic Acids Res.* **34**, D705–D711.

Simbeni, R., Pon, L., Zinser, E., Paltauf, F., and Daum, G. (1991). Mitochondrial membrane contact sites of yeast. Characterization of lipid components and possible involvement in intramitochondrial translocation of phospholipids. *J. Biol. Chem.* **266**, 10047–10049.

Zinser, E., and Daum, G. (1995). Isolation and biochemical characterization of organelles from the yeast, *Saccharomyces cerevisiae*. *Yeast* **11**, 493–536.

CHAPTER 3

Isolation and Subfractionation of Mitochondria from Plants

A. H. Millar,* A. Liddell,† and C. J. Leaver†

*ARC Centre of Excellence in Plant Energy Biology
The University of Western Australia, Crawley 6009, Western Australia, Australia

†Department of Plant Sciences, University of Oxford, Oxford OX1 3RB, United Kingdom

METHODS IN CELL BIOLOGY, VOL. 80
Copyright 2007, Elsevier Inc. All rights reserved.

0091-679X/07 $35.00
DOI: 10.1016/S0091-679X(06)80003-8

I. Introduction

Mitochondria from plants share significant similarities with those from other eukaryotic organisms in terms of both their structure and function. Plant mitochondria are the main site of the synthesis of ATP, provide a high flux of metabolic precursors in the form of tricarboxylic acids to the rest of the cell for nitrogen assimilation and biosynthesis of amino acids, and have roles in plant development, productivity, fertility, and resistance to disease. To engage in novel pathways, plant mitochondria also contain a range of enzymatic functions that are not found in the mammalian organelles and which influence assays of mitochondrial function. The presence of a rigid cell wall and different intra- and extracellular structures also influence methods for mitochondrial purification from plants.

Techniques for the isolation of plant mitochondria, while maintaining their morphological structure observed *in vivo* and their functional characteristics, have required methods that avoid osmotic rupture of membranes and that protect organelles from harmful products released from other cellular compartments. Several extensive methodology reviews (Douce, 1985; Neuburger, 1985) and more specific methodology papers (Day *et al.*, 1985; Hausmann *et al.*, 2003; Keech *et al.*, 2005; Leaver *et al.*, 1983; Millar *et al.*, 2001; Neuburger *et al.*, 1982) are now available on plant mitochondrial purification. Here, we aim to provide appraise of these classic methods for the reader and some more detailed examples of recent protocols for specialized purposes.

Study of the composition, metabolism, transport processes, and biogenesis of plant mitochondria require an array of procedures for the subfractionation of mitochondria to provide information on localization and association of proteins or enzymatic activities. This chapter also presents subfractionation procedures to separate mitochondria into four compartments. Further techniques for separation and partial purification of membrane protein complexes are outlined. Finally, profiling of the mitochondrial proteome and identification of proteins by the use of mass spectrometry (MS), antibodies, and *in organello* [^{35}S] methionine translation assays are discussed with examples.

II. Growth and Preparation of Plant Material

Mitochondria can be isolated from virtually any plant tissue; however, depending on the requirements of amount, purity, and yield, particular tissues have their advantages. (1) Nonphotosynthetic fleshy root and tuber tissues allow for large scale, but low percentage yield, plant mitochondrial isolation ($\sim$300-mg mitochondrial protein from 5- to 10-kg FW). Typically, potato, sweet potato, turnip, and sugar beet have been used for this purpose. (2) Etiolated seedling tissues such as hypocotyls, cotyledons, roots, or coleoptiles are used due to the lack of dense thylakoid membranes. Etiolated tissues also have lower phenolic content than green tissues and high yields of functional mitochondria ($\sim$20- to 50-mg mitochondrial protein from 100- to 200-g FW). (3) Interest in the function of mitochondria during photosynthesis and interest in mitochondria from plant species without abundant storage organs has led to methods for the isolation of chlorophyll-free mitochondria preparations from light-grown leaves and cotyledons (10- to 25-mg mitochondrial protein from 50- to 100-g FW).

Once the plant material has been chosen, the tissue must be free from obvious fungal or bacterial contamination and should be washed in cold H_2O, or in cases of greater contamination, surface sterilized for 5 min in 1:20 dilution of a 14% (w/v) sodium hypochlorite stock solution. These steps ensure that minimal contamination is carried into the further steps of the protocol. All tissues should then be cooled to 2–4°C before proceeding.

III. Isolation of Mitochondria

Tissues should be processed as soon as possible following harvesting and cooling in order to maintain cell turgor and thus ensure maximal mitochondrial yields following homogenization. A basic homogenization medium consists of 0.3–0.4 M of an osmoticum (sucrose or mannitol), 1–5 mM of a divalent cation chelator (EDTA or EGTA), 25–50 mM of a pH buffer (MOPS, TES or Na-pyrophosphate), and 5–10 mM of a reductant (cysteine or ascorbate) which is added just prior to homogenization. The osmoticum maintains the mitochondrial structure and prevents physical swelling and rupture of membranes, the buffer prevents acidification from the contents of ruptured vacuoles, and the EDTA inhibits the function of phospholipases and various proteases requiring Ca^{2+} or Mg^{2+}. The reductant prevents damage from oxidants present in the tissue or produced on homogenization. This basic medium is sufficient for tuber tissues, but a variety of additions have been suggested to improve yields and protect mitochondria from damage during isolation in some tissues. We have found that addition of tricine (5–10 mM) to all solutions aids integrity and yield of mitochondria from seedling tissues. Etiolated seedling tissues and green tissues also often require addition of 0.1–1% (w/v) BSA to remove free fatty acids, and 1–5% (w/v) PVP to remove phenolics that damage organelles in the initial homogenate. As a result we have routinely used a homogenization medium consisting of 0.4-M mannitol, 1-mM EGTA, 25-mM MOPS–KOH (pH 7.8), 10-mM tricine, 8-mM cysteine, 0.1% BSA, and 1% PVP-40 with wide success in a variety of plant tissues.

Following preparation of the medium and the plant material, the three most critical steps in yield of mitochondria are often found to be (1) the method of homogenization, (2) the pH of the medium following homogenization, and (3) the ratio of homogenization medium to plant tissue.

Depending on the tissue, homogenization can be accomplished by one of several methods:

a. Grinding in a precooled mortar and pestle
b. Homogenization in a square-form beaker with either a Moulinex mixer for 20–60 sec or a Polytron, Ultra-Turrax blender at 50% full-speed for 5×2 sec
c. Homogenization in a Waring blender for 3×5 sec at low speed
d. Juicing tuber material directly into $5 \times$ homogenization medium using a commercial vegetable/fruit juice extractor
e. Grating of tissue by hand using a vegetable grater submerged under homogenization medium (useful for mitochondrial extraction from soft-fruits)

The pH value of the homogenization medium should be adjusted to ~ 7.8 before homogenization to compensate for the further acidification of the suspension following cell breakage. For some plant tissues, however, the amount of buffer may need to be increased to ensure that the pH of the final homogenate remains at or above pH 7.0. Alternatively, the pH may need to be adjusted following homogenization with a stock of 5-M KOH or NaOH.

Homogenizing medium should be added in a ratio of at least 2-ml/g FW of nongreen tissue or at least 4-ml/g FW of green tissue. Decreases in these ratios can result in dramatic losses in yield. However, smaller ratios of medium to tissue (as low as 0.25-ml/g FW) do work in the case of tuber tissues extracted by method (d).

Once homogenization has been completed, the final suspension must be filtered to remove the majority of the starch and cell debris, and kept at $4\,^{\circ}$C. Filtration through four to eight layers of muslin, Miracloth (Calbiochem, La Jolla, CA) or disposable clinical sheeting (available from most medical suppliers) is effective. Filtrate is directed via a funnel into a cooled beaker or conical flask and the remainder of the fluid in the filter is then collected by wringing the filtration cloth from the top into the funnel. The filtered homogenate is then poured into 50-, 250-, or 500-ml centrifuge tubes, depending on the volume of the preparation, and centrifuged for 5 min at $1000 \times g$ in a fixed-angle rotor fitted to a preparative centrifuge at $4\,^{\circ}$C. The supernatant following centrifugation is gently decanted into another set of centrifuge tubes taking care not to transfer the pellet material which contains starch, nuclei, and cell debris. The supernatant is then centrifuged for 15–20 min at $12,000 \times g$ at $4\,^{\circ}$C and the resulting high-speed supernatant is discarded. The tan- or green-colored pellet in each tube is resuspended in 5–10 ml of a standard wash medium with the aid of a clean, soft bristle paint brush. A standard wash medium consists of 0.4-M mannitol, 1-mM EGTA, 10-mM MOPS–KOH, pH 7.2, and 0.1% (w/v) BSA. If sucrose is the osmoticum used in the preparation, then sucrose can be used to replace mannitol in the standard wash medium. The resuspended organelles are transferred to 50-ml centrifuge tubes and the volume in each tube adjusted to 40 ml with wash medium, and the samples are centrifuged at $1000 \times g$ for 5 min. The supernatants are transferred into another set of tubes and the organelles are sedimented by centrifugation at $12,000 \times g$ for 15–20 min. The high-speed supernatant is once again discarded and the washed organelles are resuspended uniformly in ~2 ml of wash medium, as above. Preparation of the mitochondrial fraction to this point should be undertaken as quickly as is possible given the steps involved, without samples warming above 4°C, and without storage for extended periods between centrifugation runs.

IV. Density Gradient Purification of Mitochondria

The mitochondrial preparation described above is often adequate for a variety of respiratory measurements; however, depending on the plant tissue, it is frequently contaminated by thylakoid membranes, plastids, peroxisomes and/or glyoxysomes, and occasionally by bacteria. Further purification can be carried out using density gradients of either sucrose or Percoll (Amersham Biosciences, Uppsala, Sweden). Many of the early isolations of plant mitochondria were performed on sucrose gradients. Sucrose offers the advantage that it is inexpensive and gradients of sucrose are more effective in separation of bacteria from mitochondria than are the Percoll equivalents. However, Percoll gradients offer more rapid purification, allow separation of mitochondria and thylakoid membranes from green tissues, and also ensure iso-osmotic conditions in the gradient. The latter eliminates the potential of extreme osmotic shock encountered in sucrose gradients that can rupture mitochondria from some plant sources.

A. Sucrose

Washed mitochondria in 1–2 ml of wash medium (from 75-g tissue) are layered on a 13-ml or 30-ml continuous sucrose gradient (0.6–1.8 M) and centrifuged in a swing-out rotor (e.g., Beckman SW28 or SW28.1) at $72,000$–$77,000 \times g$ for 60 min (Fig. 1A). The gradients can be prepared either using a gradient maker with equal volumes of solution A [0.6-M sucrose, 10-mM tricine, 1-mM EGTA, pH 7.2, 0.1% (w/v) BSA] and B [1.8-M sucrose, 10-mM tricine, 1-mM EGTA, pH 7.2, 0.1% (w/v) BSA]. Alternatively, five or more steps of different sucrose concentration (1.8, 1.45, 1.2, 0.9, and 0.6 M) can be layered in centrifuge tubes and left overnight at 4°C to equilibrate. After centrifugation, the mitochondria form a band at ~1.25- to 1.35-M sucrose and can be recovered by aspirating with a pasteur pipette into a graduated tube on ice (Fig. 1A). The molarity of the suspension is then estimated using a refractometer, and the suspension is slowly diluted over 15–20 min by addition of 0.2-M mannitol, 10-mM tricine, 1-mM EGTA (pH 7.2) to give a final osmotic concentration of 0.6 M. The diluted suspension is centrifuged at $12,000 \times g$ for 15 min, the supernatant discarded, and the purified mitochondrial pellet resuspended in resuspension buffer (0.4-M mannitol, 10-mM tricine, 1-mM EGTA, pH 7.2) at 5- to 20-mg mitochondrial protein/ml.

B. Percoll

A versatile colloidal silica sol such as Percoll allows the formation of iso-osmotic gradients and, through isopycnic centrifugation, facilitates a range of

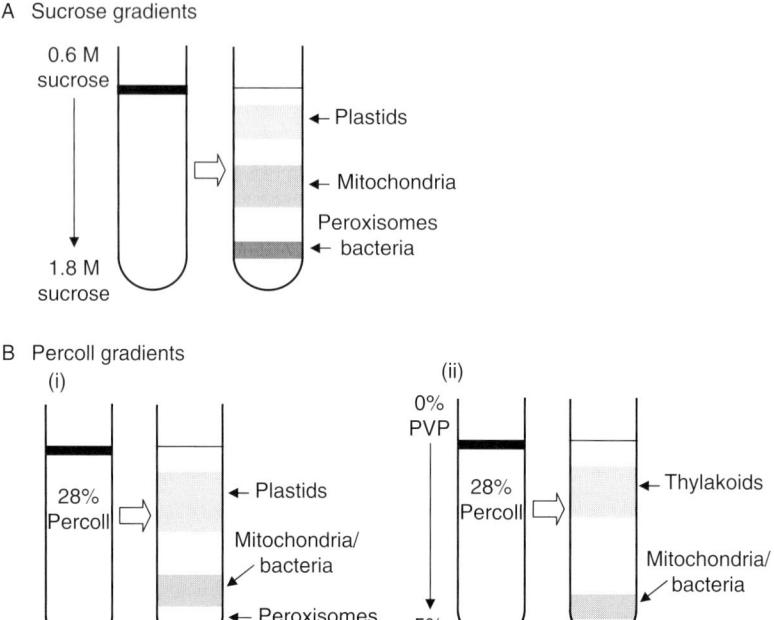

Fig. 1 Sucrose and Percoll gradients for purification of plant mitochondria. (A) Crude mitochondria sample loaded onto a 0.6- to 1.8-M sucrose gradient is separated by centrifugation at 75,000 × g for 60 min, yielding plastid, mitochondrial, and peroxisome/bacterial fractions. (B) Crude mitochondria samples are loaded on either (i) 28% (v/v) Percoll medium or (ii) 28% (v/v) Percoll medium with a 0–5% (w/v) PVP-40 gradient, separated by centrifugation at 40,000 × g for 45 min, and yielding mitochondrial fractions separated from plastid and peroxisome contamination.

methods for the density purification of mitochondria. The most common method is the sigmoidal, self-generating gradient obtained by centrifugation of a Percoll solution in a fixed-angle rotor. The density gradient is formed during centrifugation at >10,000 × g due to the sedimentation of the polydispersed colloid (average particle size is 29-nm diameter, average density $\rho = 2.2$ g/ml). The concentration of Percoll in the starting solution and the time of centrifugation can be varied to optimize a particular separation. A solution of 28% (v/v) Percoll in 0.4-M mannitol or 28% (v/v) Percoll in 0.3-M sucrose supplemented with 10-mM MOPS, pH 7.2, 0.1% (w/v) BSA is prepared on the day of use. Washed mitochondria from up to 80 g of etiolated plant tissue are layered over 35 ml of the gradient solution in a 50-ml centrifuge tube. Gradients are centrifuged at 40,000 × g for 45 min in a fixed-angle rotor of a Sorvall or Beckman preparative centrifuge. After centrifugation, mitochondria form a buff-colored band below the yellow-orange plastid membranes (Fig. 1Bi). The mitochondria are aspirated with a pasteur pipette avoiding collection of the yellow or green plastid fractions. The suspension is diluted with at least 4 volumes of standard wash medium and centrifuged at 15,000 × g for 15–20 min in 50-ml centrifuge tubes.

The resultant loose pellets are resuspended in wash medium and centrifuged again at 15,000 × g for 15–20 min and the mitochondrial pellet is resuspended in resuspension medium at a concentration of 5- to 20-mg mitochondrial protein/ml.

Separation of mitochondria from green plant tissues is greatly aided by inclusion of a 0–5% PVP-40 preformed gradient in the Percoll solution (Day et al., 1985). Such gradients can be formed using aliquots of the Percoll wash buffer solution with or without addition of 5% PVP-40 and a standard linear gradient former. This technique allows much greater separation of mitochondria and thylakoid membranes (Fig. 1Bii).

Percoll step gradients can also be used and these aid the concentration of mitochondria fractions on a gradient at an interface between Percoll concentrations. Step gradients can be formed easily by setting up a series of inverted 20-ml syringes (fitted with 19-gauge needles) strapped to a flat block of wood, clamped to a retort stand at an angle of 45° over a rack containing the centrifuge tubes. The needles are lowered to touch the bevels against the inside, lower edge of the tubes. The step gradient solutions are then added (from bottom to top) to the empty inverted syringe bodies and each allowed to drain through in turn before the addition of the next step solution. Step gradients should, however, only be used after the density of mitochondria from a particular tissue and the density of contaminating components have been determined using continuous gradients. A range of density marker beads is available from Amersham Biosciences for standardizing running conditions and establishing new protocols.

C. Mini–Mitochondrial Preparations for Screening Transgenic Lines

Rapid, high percentage yield preparations of mitochondria from a large number of small samples can sometimes be required. This is especially the case in the screening of transgenic plants with mitochondrial-targeted changes in protein expression or in experimentation on small amounts of material. In potato lines expressing antisense constructs to tricarboxylic acid cycle enzyme subunits, we have used a Percoll step-gradient protocol that yields 25–35% of the mitochondria present in 3- to 5-g FW tuber samples. This procedure allows for the purification of eight samples at once using a standard fixed-angle rotor housing 8 × 50-ml tubes. Tuber samples are cooled to 4 °C, diced finely (2- to 3-mm squares), and ground in a cooled mortar and pestle in 8–10 ml of standard grinding medium. After filtering through muslin, a 1000 × g centrifugation for 5 min removes starch and cell debris. The 10- to 12-ml supernatant is then layered directly onto a Percoll step gradient made in a standard mannitol wash buffer (2-ml 40%, 15-ml 28%, 5-ml 20% Percoll in a 50-ml tube). After centrifugation at 40,000 × g for 45 min, mitochondria are removed from the 28%/40% interface and washed twice in 50-ml tubes diluted with wash medium (Fig. 2A). Approximately 0.2–0.5 mg of mitochondria protein is obtained per sample. These preparations have enzyme specific activities similar to those from samples of conventional potato tuber mitochondrial preparations.

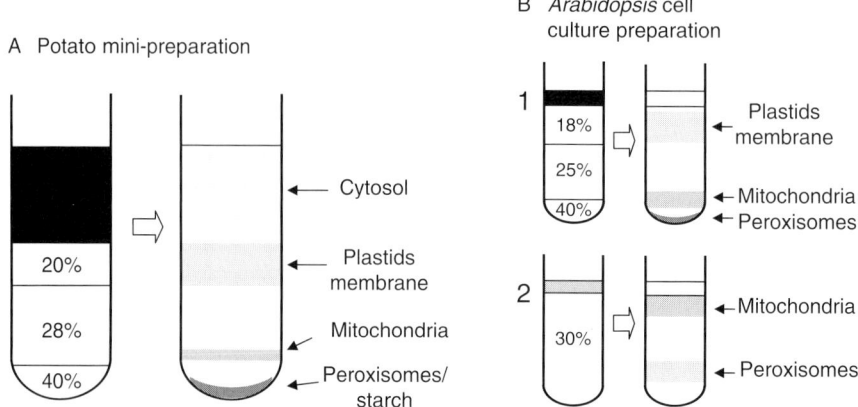

Fig. 2 Percoll density gradients for the purification of potato tuber mitochondria and *Arabidopsis* cell culture mitochondria. (A) Minipreparations of mitochondria from 5- to 10-g FW of potato tuber using 20%/28%/40% step gradients. (B) Two-stage Percoll gradient separation of *Arabidopsis* cell culture mitochondria using a 18%/25%/40% step gradient following by layering the mitochondrial fraction onto a 30% self-forming gradient to maximize removal of peroxisomal material.

Similar protocols can be developed for screening transgenic tissue culture or callus samples.

D. Cell Culture/Callus Mitochondria Preparation

Plant cell culture and callus material can be notoriously difficult for the purification of mitochondria. This is due to two main factors: (1) the difficulties in breaking open the cell walls of cells in small clusters, and (2) the densities of mitochondria from nondifferentiated cells tend to be lower than those of mitochondria from whole plant tissues, and thus sediment to different regions of density gradients. Cells from cultures are not easily broken by homogenizers or blenders due to the strong cell wall structure. We have found a mortar and pestle with added sand or glass beads (0.4-mm diameter) is reasonably effective in breaking cells (Millar *et al.*, 2001), while other groups have utilized enzymatic digestion of the cell wall to release protoplasts from which mitochondria are then isolated (Vanemmerik *et al.*, 1992; Zhang *et al.*, 1999). The advantage of the mortar and pestle method is its speed of isolation; however, yields are not high (5- to 8-mg mitochondrial protein/100-g FW cells). The disadvantage of the enzymatic approach is the need to incubate cells for more than 1 h in digestion enzymes which may alter mitochondrial functions of interest, but this approach does tend to give higher yields of purified mitochondria (15- to 25-mg mitochondrial protein/100-g FW cells).

We have used the following method to isolate plant mitochondria from an *Arabidopsis thaliana* cell culture. Approximately 100-g FW cells are harvested by collecting material from ten flasks of 100-ml culture by filtration through a $20 \times 20 \ cm^2$ of muslin. In batches of 20-g FW, cells are ground in standard homogenization buffer in the presence of 1–2 g of glass beads (0.4-mm diameter). Grinding commences on addition of 10 ml of homogenization buffer and continues with subsequent additions to a total of 50 ml. Once all the batches are ground, the homogenate is filtered and centrifuged in 50-ml tubes and centrifuged at $2500 \times g$ for 5 min to remove cell debris. The resultant supernatant is centrifuged at $15,000 \times g$ for 15 min. Following resuspension of the pellets in standard mannitol wash buffer, the two centrifugation steps are repeated. The final pellet is resuspended in several milliliters of mannitol wash buffer and layered over a Percoll step gradient in a 50-ml tube, comprising, from bottom to top, 5-ml 40%, 15-ml 25%, 15-ml 20% Percoll in mannitol wash buffer. After centrifugation at $40,000 \times g$ for 45 min, mitochondria and peroxisomes are found at the 25%/40% interface, while plastid membrane remains in the 20% Percoll step (Fig. 2B). After washing in five times the volume of wash buffer, the mitochondria and peroxisome pellet is resuspended in mannitol wash buffer and layered over 30 ml of 30% Percoll in 0.3-M sucrose wash buffer. After centrifugation at $40,000 \times g$ for 45 min, mitochondria are found just below the start of the 25% Percoll step, while peroxisomes are found toward the bottom of this step (Fig. 2B). If sufficient material is present, the peroxisomal layer will have a translucent yellow color while the mitochondria will be a white-buff, opaque band. It is important to note that preparation of mitochondria from other cell cultures may require even lower Percoll concentrations on the first gradient. In the case of some plant cell cultures, gradients of 5-ml 25%, 15-ml 20%, 15-ml 15% Percoll have been used with some success, with mitochondria found at the 20–25% Percoll interface.

V. Mitochondrial Yield, Purity, Integrity, Storage, and Function

A. Yield Calculations

The yield of mitochondria from plant tissues are successively decreased through the purification process due to incomplete grinding of plant tissue, removal in discarded supernatant and pellet fractions, and removal of damaged mitochondria on a density basis in the final gradient steps. Typically, 45–75% of total mitochondria membrane marker enzyme activity in a tissue is found in the first high-speed pellet, 15–30% is loaded onto the density gradient after washing, and only 5–15% is found in the washed mitochondrial sample at the end of the purification procedure. In nongreen tissues, cytochrome c oxidase can be used as a marker enzyme, but in green tissues a variety of oxidases can prevent the use of this assay. Cytochrome c oxidase activity can be measured using an O_2 electrode as 0.1-mM KCN-sensitive O_2 consumption in the presence of 10-mM ascorbate, 50-μM cytochrome c, 0.05% Triton X-100 in 10-mM TES–KOH (pH 7.2), or it can be measured spectrophotometrically as 0.1-mM KCN-sensitive reduced cytochrome c oxidation at 550 nm ($\varepsilon = 28,000 \ M^{-1}$). In cases where this assay does not work, fumarase is the best candidate for a mitochondrially specific enzyme, but its low activity can hamper the use of this marker in some tissues. Fumarase can be assayed spectrophotometrically as fumarate formation at 240 nm ($\varepsilon = 4880$ M^{-1}) by addition of 50-mM malate to samples in 0.1-M K$_2$PO$_4$–KOH (pH 7.7). In potato tuber tissue, we have successfully used assays for mitochondrial citrate synthase, NAD-dependent malic enzyme, fumarase, and cytochrome *c* oxidase to provide an estimate of the mitochondrial content, which averages 0.26 ± 0.04-mg mitochondrial protein/g FW potato tuber.

B. Purity Determinations

Marker enzymes for contaminants commonly found in plant mitochondrial samples can be used to access the purity of preparations. Peroxisomes can be identified by catalase, hydroxy-pyruvate reductase, or glycolate oxidase activities. Chloroplasts can be identified by chlorophyll content, etioplasts by carotenoid content and/or alkaline pyrophosphatase activity, and glyoxysomes by isocitrate lyase activity. Endoplasmic reticulum can be identified by antimycin-A-insensitive cytochrome *c* reductase activity, and plasma membranes by K+-ATPase activity. Cytosolic contamination is rare if density centrifugation is performed properly and care is taken in removal of mitochondrial fractions, but such contamination can be measured easily as alcohol dehydrogenase or lactate dehydrogenase activities. References to and methods for the assay of these enzymes are summarized by Quail (1979), with several additions by Neuburger (1985).

C. Integrity Determinations

A variety of assays can be used to determine the integrity of the outer and inner membranes of plant mitochondrial samples, thus providing information on the structural damage caused by the isolation procedure. Two latency tests are commonly used to estimate the proportion of broken mitochondria in a sample; both rely on the impermeable nature of an intact outer mitochondrial membrane (OM) to added cytochrome *c*. The ratio of cytochrome *c* oxidase activity (measured as in Section V.A) before and after addition of a nonionic detergent [0.05% (w/v) Triton X-100] or before and after osmotic shock (induced by dilution of mitochondria in H$_2$O followed by addition of 2× reaction buffer) gives an estimate of the proportion of ruptured mitochondria in a sample. Alternatively, succinate: cytochrome *c* oxidoreductase can be assayed as cytochrome *c* reduction at 550 nm ($\varepsilon = 21,000$ M^{-1}) in wash medium supplemented with 0.5-mM ATP, 50-μM cytochrome *c*, 1-mM KCN, and 10-mM succinate. The ratio of the activity of a control sample over the activity of a sample diluted in H$_2$O followed by addition of 2× reaction buffer gives an estimate of the proportion of ruptured mitochondria in the sample.

Inner membrane integrity can be assayed by the latency of matrix enzyme activities with or without detergent or osmotic shock. Transport of organic acids and cofactors across the inner mitochondrial membrane does hamper accurate measurements of matrix enzyme activity latency; however, malate dehydrogenase activity is so limited by transport in intact mitochondria that assaying the latency of its activity gives a good indication of inner membrane integrity (Douce, 1985). Finally, the ability of mitochondria samples to maintain a proton-motive force across the inner membrane for ATP synthesis can be assessed by the ratio of respiratory rates in the presence and absence of added ADP and by measurement of the ratio of ADP consumed:O_2 consumed. Both these latter assays will be affected by the presence and operation of the nonproton pumping plant alternative oxidase (AOX), which should be inhibited by 0.1-mM n-propyl gallate or 1-mM salicylhydroxamic acid during experiments to assess coupling of O_2 consumption and ADP synthesis rate.

D. Storage

Once isolated by density gradient purification, mitochondria from most plant tissues can be kept on ice for 5–6 h without significant losses in membrane integrity and respiratory function. Pure preparations of some storage tuber mitochondria can be kept at 4 °C for several days without loss of function, although cofactors such as NAD, TPP, and CoA often need to be added back for full tricarboxylic acid cycle function. Longer-term storage of functional mitochondria can be achieved by addition of DMSO to 5% (v/v) or ethylene glycol to 7.5% (v/v), followed by rapid-freezing of small volumes of mitochondrial samples (<0.5 ml) in liquid N_2 (Schieber et al., 1994; Zhang et al., 1999). Frozen samples can then be kept at −80 °C. In our hands, potato mitochondrial samples prepared in this way can be stored for over a year, and following thawing of these samples on ice, no significant losses could be detected in respiratory rate, coupling to ATP production, or outer membrane integrity.

E. Assays of Mitochondria Function

The Clark-type polarographic O_2 electrode utilizing platinum–silver electrodes has been the routine method of assaying function of isolated plant mitochondria for many years, and has formed the basis of much of the characterization of the novel mitochondrial functions found in plants. Commercial designs of such dissolved O_2 electrodes are available from Hansatech (Kings Lynn, Norfolk, UK) and Rank Brothers Ltd (Bottisham, Cambridgeshire, UK) and allow rapid measurements of O_2 consumption rate in 1- to 2-ml volumes containing 0.1- to 0.3-mg mitochondrial protein. Using a basic reaction medium [0.3-M sucrose, 10-mM NaCl, 5-mM KH_2PO_4, 3-mM $MgCl_2$, 0.1 V% (W/V) BSA, 10-mM TES–NaOH, pH 7.2] and adding substrates, cofactors and effectors of O_2 consumption rate, a large amount of functional information can be obtained.

In addition to the assays used for studying mitochondrial function from other sources, the following assays can be undertaken in plant mitochondria due to their unique features. NADH (1 mM) or NADPH (1 mM) can be supplied exogenously as respiratory substrates due to the function of cytosolic-facing NADH and

NADPH rotenone-insensitive nonproton pumping dehydrogenases that bypass the function of Complex I of the respiratory chain and directly reduce ubiquionone in the plant inner mitochondrial membrane. Pyruvate (10 mM) rarely provides a rapid respiratory rate, due to the lack of an endogenous malate/OAA pool in isolated plant mitochondria, and thus a "sparker" addition of malate (50 μM) is required. Malate (10 mM) itself is a good respiratory substrate, due to the presence of an NAD-dependent malic enzyme in the plant mitochondrial matrix that catalyses the decarboxylation of malate to pyruvate, providing, along with malate dehydrogenase, the substrates for further TCA cycle operation via citrate synthase. Large amounts of glycine decarboxylase are found in mitochondria from photosynthetic sources, allowing glycine (10 mM) to be used as a substrate to support high rates of O_2 consumption in mitochondria from these tissues.

Due to the presence of novel electron transport bypasses, respiratory inhibitors often have unexpected effects on respiratory rates in plant mitochondria. Rotenone does not fully inhibit O_2 consumption mediated via the generation of NADH in the matrix of mitochondria, due to the presence of matrix-facing rotenone-insensitive NADH dehydrogenases in plant mitochondria. Cytochrome pathway inhibitors (antimycin A, myxothiazol, KCN, and CO) do not fully inhibit O_2 consumption, due to the presence of an AOX that oxidizes the ubiquinone pool directly and reduces O_2 to H_2O. The AOX can be inhibited by salicylhydroxamic acid ($\sim$250 μM) or by

Table I

Methods Optimized for the Assessment of Enzyme Activities and for Partial or Complete Purification of the Enzymes from Plant Mitochondria

Enzyme	References
Pyruvate dehydrogenase complex	Millar *et al.*, 1998; Randall and Miernyk, 1990
Citrate synthase	Stevens *et al.*, 1997; Stitt, 1984
NAD(P)-isocitrate dehydrogenase	McIntosh and Oliver, 1992; Rasmusson and Moller, 1990
Aconitase	Verniquet *et al.*, 1991
2-Oxoglutarate dehydrogenase complex	Millar *et al.*, 1999; Poulsen and Wedding, 1970
Succinyl CoA synthase	Palmer and Wedding, 1966
Fumarase	Behal and Oliver, 1997; Behal *et al.*, 1996
Malate dehydrogenase	Hayes *et al.*, 1991; Walk *et al.*, 1977
NAD-malic enzyme	Grover and Wedding, 1984; Grover *et al.*, 1981; Hatch *et al.*, 1982
Glycine decarboxylase	Neuburger *et al.*, 1986; Oliver, 1994; Walker and Oliver, 1986
Complex I	Herz *et al.*, 1994; Leterme and Boutry, 1993; Rasmusson *et al.*, 1998
Non-Phos NADH dehydrogenases	Moller *et al.*, 1993; Soole and Menz, 1995
Complex II	Hattori and Asahi, 1982; Igamberdiev and Falaleeva, 1994
Complex III	Braun and Schmitz, 1992, 1995; Emmermann *et al.*, 1993
Complex IV	Devirville *et al.*, 1994; Maeshima and Asahi, 1978
Alternative oxidases	Hoefnagel *et al.*, 1995; Vanlerberghe and McIntosh, 1997; Zhang *et al.*, 1996
Complex V	Glaser and Norling, 1983; Hamasur and Glaser, 1992; Hamasur *et al.*, 1992

gallates such as *n*-propyl gallate (~50 μM) and octyl gallate (~5 μM). The AOX is also activated by short 2-oxo acids (such as pyruvate and glyoxylate) and by reduction of a disulphide bond. The latter can be catalyzed via reduction of the NADPH pool in the mitochondrial matrix or direct application of chemical reductants, via such as DTT (Vanlerberghe and McIntosh, 1997). Thus, assessment of the activity of AOX can vary depending on the substrate used, and the relevant literature should be studied carefully if investigation of this activity is to be undertaken. Specific spectrophotometric assays optimized for enzymes from plant mitochondria and protocols for their partial or complete purification have been published in a variety of reviews and research papers, and a list in Table I provides the reader with a ready entry to this material.

VI. Subfractionation of Mitochondrial Compartments

A. Separation of Inner Mitochondrial Membrane, Outer Mitochondrial Membrane, Matrix, and Intermembrane Space

Using a combination of osmotic shock and differential centrifugation, plant mitochondrial samples can be fractionated into four components. These comprise the two aqueous compartments—the matrix (MA) and the intermembrane space (IMS)—and the two membrane compartments—the inner mitochondrial membrane (IM) and the OM. While separation of the IM and MA fractions from each other is relatively trivial, removal of the IMS and OM fractions is a greater challenge and much is owed in the development of these techniques to the work of Mannella and Bonner (1975) and Mannella (1985) on the plant mitochondrial outer membrane.

To perform a subfractionation, a purified mitochondrial sample that has not been frozen is separated into two aliquots and each is added to 49 times the volume of two separate solutions. One solution contains 10-mM sucrose and the other 86-mM sucrose, both in 10-mM MOPS, pH 7.2 (depending on the source of the mitochondria to subfractionated, these osmotic strengths can be altered to significantly improve yields and purity of final fractions). The two solutions are slowly stirred on ice for 15 min. The two osmotically shocked mitochondrial samples (10 and 86 mM) are then kept separate for the following procedures, except for the OM pellets, which are combined from the two samples at the end of the protocol. Using a stock solution of 2-M sucrose in 100-mM Tris–HCl, pH 7.4, the sucrose concentration in each solution is brought to a final concentration of 0.3 M and stirring is continued for a further 15 min. Mitoplasts (IM + MA) are removed from both samples by centrifugation at 15,000 × *g* for 15 min at 4°C.

The supernatant (OM + IMS) is then transferred to clean tubes and centrifuged at 100,000 × *g* for 90 min. The supernatants of this high-speed centrifugation contain IMS-10 and IMS-86 and can be frozen separately for later analysis. The pellets are OM, but due to the small amount of this membrane obtained, a further purification is required to remove contaminating IM and mitoplasts. This is achieved by combining the OM-10 and OM-86 pellets, resuspended in a minimum volume of 100-mM Tris–HCl, pH 7.4, and layering the combined sample over a 4-ml of 0.6/0.9-M step sucrose gradient. This gradient is centrifuged in a swing-out rotor at $50,000 \times g$ for 60 min. OM is collected from the 0.6/0.9 interface and is only visible when large amounts of mitochondria (>100-mg protein) are being fractionated. Visible bands at other locations are probably contaminants and should not be collected.

The mitoplast pellet of the first centrifugation, which may be frozen before this step, is then diluted to 15 mg/ml and sonicated at full power for 3×5-sec bursts with 20-sec rest periods on ice. Centrifugation at $80,000 \times g$ for 60-min pellets the IM-10 and IM-86 fractions. Retain the supernatants separately as MA-10 and MA-86 and freeze all remaining fractions.

Depending on the tissue and the preparation of mitochondria, the 10- or 86-mM samples will give higher purity separations of some fractions. The 86-mM sample will yield a higher purity IMS fraction due to lower contamination of MA, while the 10-mM sample will yield a higher purity IM and MA fraction by ensuring a more complete rupture of the whole mitochondria sample, yielding mitoplasts. Sample purity can be assayed by measuring: cytochrome *c* oxidase for IM, fumarase for MA, and antimycin A-insensitive NADH: cytochrome *c* oxidoreductase for OM. In mammalian mitochondria, adenylate kinase activity is used as a marker for IMS, but in plants this enzyme activity is found most abundantly associated with the outer face of the IM, and therefore is not a reliable marker of IMS. In Fig. 3, sodium dodecyl sulfate polyacrylamide gel electrophoresis (SDS-PAGE) separations of the proteins of these four mitochondrial subcompartments are presented from subfractionation experiments performed on mitochondria isolated from a dicotyledonous plant (potato) and from a monocotyledonous plant (maize). Similarities between the apparent molecular masses of proteins in each compartment between these divergent species is evident. Typically, the enzymatic analysis of cross-contamination shows that less than 10% of the protein in any compartment is derived from contaminating proteins from the three other compartments.

B. Separation of Mitochondrial Inner Membrane Components by MonoQ

Samples of IM can be separated further by two simple chromatography columns to provide relatively high purity fractions containing functional components of the electron transport chain (ETC) and the mitochondrial superfamily of transporters (Fig. 4). Samples of mitochondrial membrane (up to 25 mg of protein) are dissolved in 1% (w/v) *n*-dodecyl-β-D-maltoside and 25-mM MOPS (pH 7.5). Approximately 1.5 ml of solution is required for each 10 mg of mitochondrial membrane. The dissolved membrane is then diluted to 0.1% (w/v) *n*-dodecyl-β-D-maltoside and 25-mM MOPS (pH 7.5) and loaded on a MonoQ anion-exchange column (Amersham Biosciences) pre-equilibrated with the same solution and eluted using a 0- to 0.6-M NaCl gradient over 30 ml at 1 ml/min with 1-ml fractions collected. The inner mitochondrial transporter superfamily (IMTS)

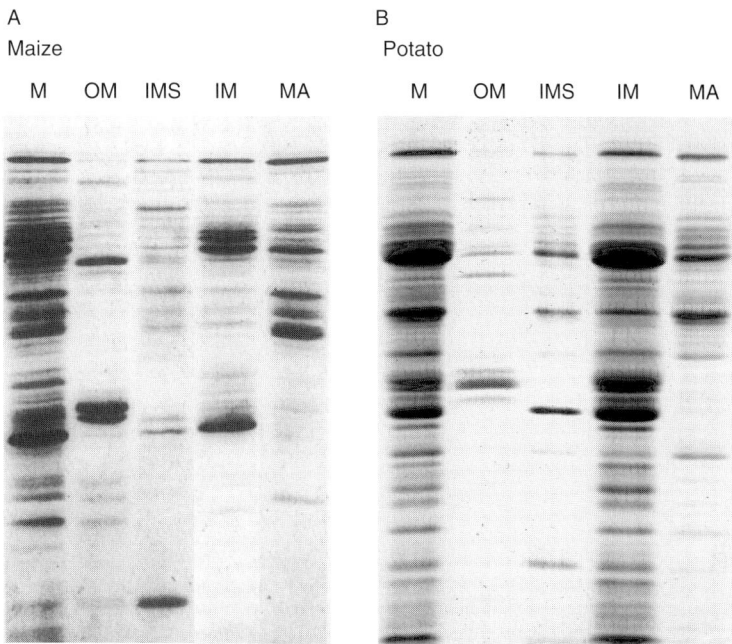

Fig. 3 SDS-PAGE separations of the four compartments of plant mitochondria isolated from (A) maize coleoptiles and (B) potato tubers. M, intact mitochondria; OM, outer mitochondrial membrane; IMS, intermembrane space; IM, inner mitochondrial membrane; MA, matrix space.

(which all have p*I* values of greater than 9.0) are eluted in the void and appear as proteins of 30–35 kDa. During the gradient elution, ETC components appear in the following order: Complex V, ATP synthetase (peak fractions 6–8); Complex I, NADH-Q oxidoreductase (peak fractions 9 and 10); Complex IV, cytochrome *c* oxidase (peak fractions 11 and 12, green colored from bound Cu), Complex III, Q-cytochrome *c* oxidoreductase (peak fractions 13 and 14, red colored from bound Fe). The oxo-glutarate dehydrogenase complex (OGDC) which is bound to the IM is eluted in fractions 10–12 and can be identified by its prominent 100-kDa subunit (Millar *et al.*, 1999). The ETC components can be separated further by size using a gel filtration column. In this case, fractions 10–12 are concentrated to 250 μl and loaded on a Superose 6 gel filtration column (Amersham Biosciences) pre-equilibrated with 0.1% (w/v) *n*-dodecyl-β-D-maltoside, 100-mM NaCl, and 25-mM MOPS–NaOH (pH 7.5). Complexes are eluted in the same buffer by size; OGDC ~2500 kDa (peak fractions 2–6), Complex I ~1000 kDa (peak fractions 7–10), Complex III ~650 kDa (peak fractions 11–13), Complex IV ~300 kDa (peak fractions 14–16), and some Complex V fragments ~200–300 kDa (fractions 17–18).

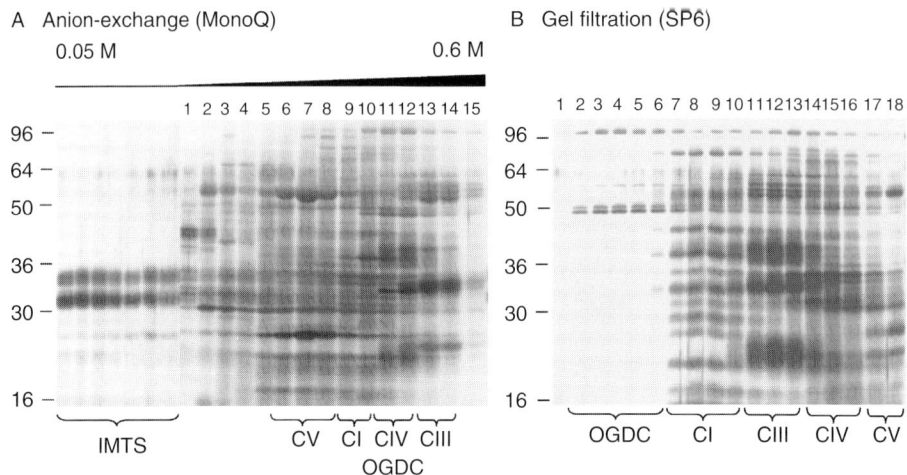

Fig. 4 Separation of potato inner mitochondrial membrane protein complexes using anion-exchange (MonoQ) and gel filtration (Superose 6, SP6) chromatography. (A) 25 mg of IM protein dissolved in *n*-dodecyl-β-D-maltoside was absorbed on a MonoQ column and protein complexes were eluted using a linear NaCl gradient (0.05–0.6 M). (B) Fractions 10–12 from the MonoQ run were separated by size on a Superose 6 column revealing partial purifications of OGDC (2-oxoglutarate dehydrogenase complex) and respiratory complexes I, III, IV, and V.

C. Isoelectric Focusing and SDS-PAGE

Advances in two-dimensional (2D) gel electrophoresis, utilizing fixed pH gradients for the first isoelectric focusing (IEF) dimension, have greatly enhanced the ease and reproducibility of such gels for proteome analysis (Link, 1999). Using this approach, a plant mitochondrial sample can be separated to reveal 400- to 550-protein features with p*I* values ranging between values of 3 and 10 and apparent molecular masses between 100 and 10 kDa (Fig. 5). Samples ranging from 150 μg to 1 mg of mitochondrial protein are acetone-extracted for 2D analysis. In a 1.5-ml tube, 400 μl of absolute acetone cooled to $-80\,^{\circ}$C is added to a sample of mitochondria made to 100 μl with dH$_2$O. The sample is stored at $-20\,^{\circ}$C for 1 h and centrifuged at 20,000 $\times$ g for 15 min. The supernatant is removed and the sample is air-dried by placing the open tube in a heating block set at 30 $^{\circ}$C for 10 min. A total of 300 μl of an IEF sample buffer [6-M urea, 2-M thiourea, 2% (w/v) CHAPS, 0.3% (w/v) DTT, 0.5% (v/v) Amersham Biosciences IPG-buffer, and 0.005% (w/v) Bromophenol blue] is added and the pellet is resuspended by pipette, vortexing, and heating to 30 $^{\circ}$C. After a further centrifugation to remove insoluble material, the 200-μl sample in IEF buffer is added to a reswell chamber, an ampholyte (Immobiline) strip is placed over the solution, and mineral oil is layered over the top to allow for re-swelling overnight. Re-swollen IEF gels are run using the Amersham Multiphor II flat-bed electrophoresis tank or the Amersham IPGphor II dedicated IPG separator, using a program

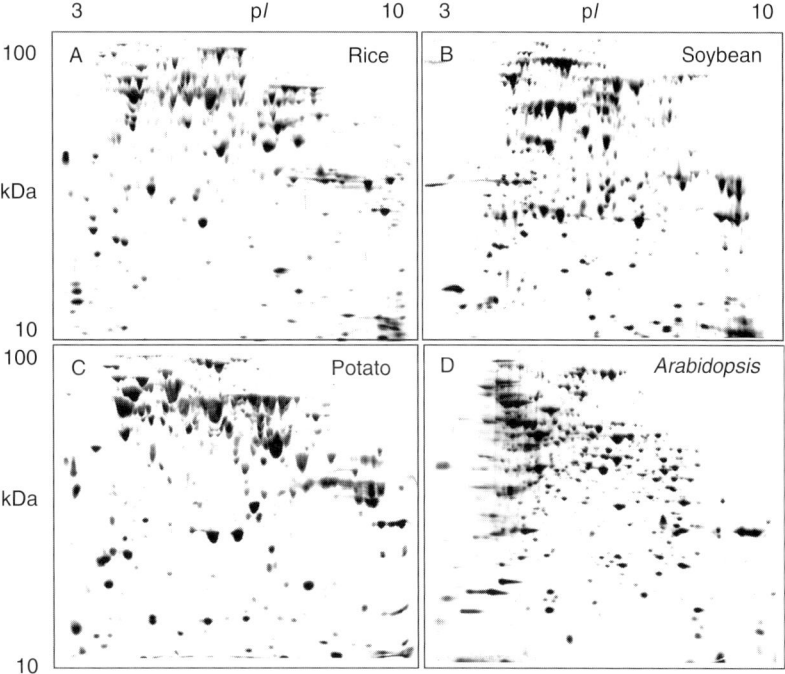

Fig. 5   Two-dimensional gel electrophoresis of mitochondrial proteins from (A) rice, (B) soybean, (C) potato, and (D) *Arabidopsis*. Isoelectric focusing (left to right) on a 3- to 10-pH nonlinear immobilized pH gradient strip followed by SDS-PAGE (top to bottom) on a 12% acrylamide gel. Approximate p*I* and MW of proteins are shown on each axis.

of ramped voltages according to the manufacturer's instructions. IEF strips are then slotted into a central single well of a 4% acrylamide stacking gel, formed using a single-well comb, above a 0.1 cm × 18.5 cm × 20 cm, 12% (w/v) acrylamide, 0.1% (w/v) SDS-PAGE. Gel electrophoresis is performed at 30 mA and completed in 6 h. Figure 5 shows typical separations of plant mitochondria from a variety of sources using a commercial 3- to 10-pH nonlinear 180-mm IPG strip and a 12% SDS-PAGE.

D. BN–PAGE of Intact ETC Components

Blue native PAGE (BN-PAGE) has been used very successfully for separation of mammalian and plant mitochondrial electron transport protein complexes. This electrophoresis technique utilizes Coomassie Serva Blue G-250 to bind and provide a net negative charge to protein complexes under nondenaturing conditions and can resolve protein complexes in the 200- to 1000-kDa range. A combination of first dimension BN-PAGE and second dimension SDS-PAGE and Western blotting also allows for the immunodetection of proteins in mitochondria and provides evidence of their native state incorporation or association with mitochondrial complexes (see review by Eubel *et al.*, 2005).

We have used a protocol largely based on the modifications of the method of Schägger and von Jagow (1991) which were introduced by Jansch *et al.* (1996) to adapt BN-PAGE to separation of plant mitochondrial proteins. Gels consist of a separating gel [12% (w/v) acrylamide] and a stacking gel [4% (w/v) acrylamide] formed in a solution of 0.25-M ε-amino-*n*-caproic acid (ACA) and 25-mM Bis-Tris–HCl (pH 7.0) using APS and TEMED for polymerization. The anode buffer consists of 50-mM tricine, 15-mM Bis-Tris–HCl, 0.02% (w/v) Serva Blue G-250 (pH 7.0) and the cathode buffer of 50-mM Bis-Tris–HCl (pH 7.0). Sample preparation involves resuspension of 1 mg of membrane protein in 75 μl of an ACA buffer solution (containing 0.75-M ACA, 0.5-mM Na$_2$EDTA, 50-mM Bis-Tris–HCl, pH 7.0) followed by addition of 15 μl of a freshly prepared solution of 10%

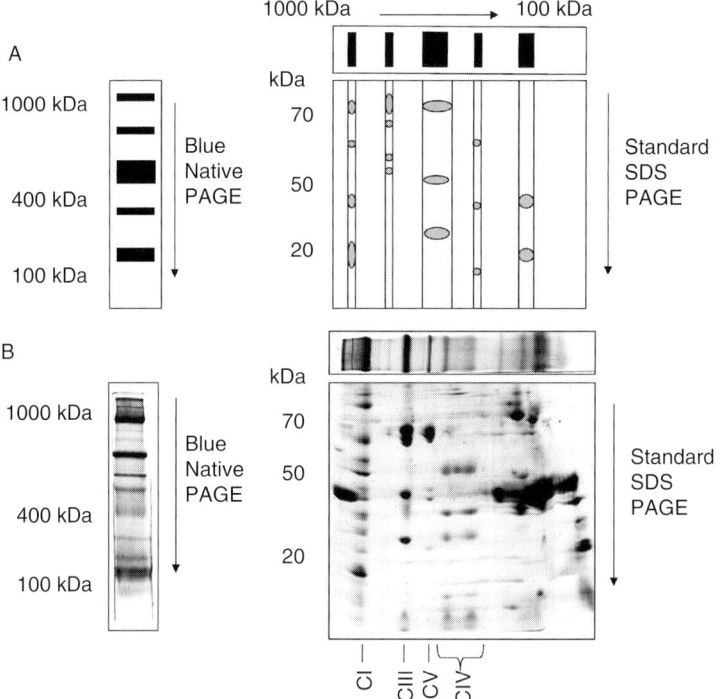

Fig. 6 Blue native gel electrophoresis of protein complexes. (A) Cartoon of BN-PAGE separation of complexes that is then affixed at 90 °C to an SDS-PAGE separation of the components in each complex in a series of tracks. (B) BN-PAGE separation of potato mitochondrial protein complexes and SDS-PAGE separation of the components of respiratory chain complexes (CI, CIII, and CIV) and of the ATP Synthase (CV). Protein spots are visualized with Coomassie blue staining.

(w/v) *n*-dodecyl maltoside. Following centrifugation for 10 min at 20,000 × *g*, supernatants are transferred to microtubes containing 15 μl of solution comprising 5% (w/v) Serva Blue G-250 dissolved in ACA buffer solution. Unused protein tracks in gels should be filled with a solution containing ACA buffer, 10% (w/v) *n*-dodecyl maltoside, and 5% Serva Blue solutions in a ratio of 0.7:0.15:0.15, respectively.

Gels are run at 4°C in a precooled apparatus with all samples and buffers precooled to 4°C. Approximately 30 min before the run, 0.03% (w/v) *n*-dodecylmaltoside is added to the cathode buffer only. Electrophoresis is commenced at 100 V constant voltage. After 45 min, the current is limited to 15 mA and the voltage limit removed. Voltage will rise from 100 to ~500 V during the 5 h of the electrophoresis run time. A cartoon of the process and a typical gel and the subsequent separation of the components of each complex by SDS-PAGE are shown in Fig. 6.

VII. *In Organello* Translation Analysis

The mitochondrial genome encodes a small percentage of the proteins found in mitochondria and the translation of these proteins in isolated mitochondria can be analyzed directly by *in organello* translation assays coupled to 2D profiling of the translated products. Mitochondria need to be essentially free of contaminating bacteria for successful *in organello* translation assays. To aid this requirement, sucrose gradients are often used for purification of mitochondria. However, Percoll can be used in purification from sterile plant material (e.g., tissue or cell culture) or easily sterilized plant material. All the buffers used in the preparation of mitochondria can be sterilized by autoclave or passage through 0.45-μm filters. Percoll must be sterilized on its own and not in the presence of salts and sugars. Following the completion of isolation, mitochondria are suspended at 20 mg/ml in a solution containing 0.3-M mannitol and 20-mM MOPS–NaOH (pH 7.5).

A set of four incubations are performed for each mitochondrial sample in order to assess bacterial contamination and to test the effect of an extramitochondrial and intramitochondrial energy generation systems for ^{35}S-Met incorporation. Sodium acetate is used in the first incubation as an energy source which is utilized by bacteria but not mitochondria; ATP is used in the second to provide energy directly to mitochondria and bypass oxidative phosphorylation; succinate/ADP is used in the third to generate ATP by oxidative phosphorylation; a water control is used in the fourth showing incorporation of ^{35}S-Met in the absence of an energy source.

A reaction mix containing GTP (2.0 mM), ^{35}S-Met (0.75 mCi/ml), and 0.3-M mannitol, KCl (60 mM), $MgCl_2$ (10 mM), KH_2PO_4 (15 mM) and MOPS–NaOH (20 mM) or HEPES (50 mM, pH 7.2) and amino acids (50 μM each of aspartic acid, glutamine, lysine, serine, valine, asparagine hydrate, glutamic acid, leucine, proline, tyrosine, arginine HCl, glycine, isoleucine, phenylalanine, tryptophan, alanine, cysteine, histidine HCl, and threonine) is prepared. The salts and amino acids can be prepared as 1.66× and 20× stocks, respectively, and stored at −20°C in aliquots.

Substrate mixes are also prepared in sterilized H_2O containing the following: (1) 200-mM sodium acetate, (2) 120-mM ATP, (3) 100-mM succinate and 10-mM ADP, and (4) H_2O only. A single incubation will utilize a 40-μl aliquot of the reaction mix, 5 μl of an energy mix, and 5 μl of a 20-mg/ml mitochondrial suspension.

Aliquots of reaction mix (40 μl) and substrate mixes (5 μl) are combined for each incubation, shaken at 25 °C on an orbital shaker, and the reaction started by addition of mitochondrial suspension (5 μl). Following incubation for 10–90 min, reactions are stopped by addition of 4 volumes (200 μl) of an ice-cold solution containing 0.3-M mannitol, 20-mM MOPS–NaOH, pH 7.2, and 10-mM l-methionine and placed on ice. To estimate TCA precipitable counts, 5-μl samples of each reaction are spotted onto 2-cm Whatman 3MM chromatography paper discs and the filters are dropped into ice-cold 10% (w/v) trichloroacetic acid (TCA) and left on ice overnight at 4 °C. The solution is discarded and replaced with 5% (w/v) TCA heated to 90–99 °C for 5 min. The filters are then washed four times in 5% (w/v) TCA at room temperature, washed in absolute ethanol for 15 min, then acetone for 5 min, and dried for scintillation counting. For analysis of protein translation products, mitochondria are pelleted from the reaction solutions by centrifugation at 20,000 × g for 5 min. These pelleted samples can be frozen for later analysis or solubilized in SDS-PAGE sample loading-buffer or IEF sample buffer. Following separation by one-dimensional (1D) or 2D electrophoresis, gels are dried for subsequent autoradiography to visualize ^{35}S-Met incorporation into polypeptides. In Fig. 7, a sample from an *in organello* translation of maize mitochondria has been separated by 1D and 2D electrophoresis and total proteins and ^{35}S-Met-labeled proteins presented. Translated products from the ATP synthase F_1 α-subunit gene can be seen prominently at 55 kDa.

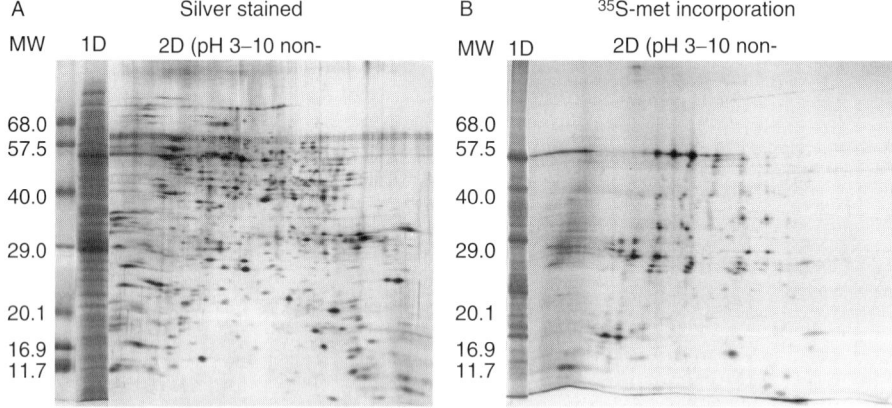

Fig. 7 One- and two-dimensional electrophoresis of a protein sample from an *in organello* translation assay performed on maize coleoptile mitochondria. (A) Silver-stained gel of total maize mitochondrial proteome. (B) Labeled proteins translated in a 1-h incubation with ^{35}S-Met.

VIII. Proteome Analysis

The plant mitochondrial proteome is likely to contain over 1500 different gene products, most encoded in the nuclear genome. In the past 5 years, a series of studies have used the gel separation techniques outlined above coupled to MS identification from in-gel digestions, along with 1D and 2D liquid chromatography linked online with tandem MS, to identify well over 400 mitochondrial proteins in *Arabidopsis* and over 200 proteins in rice mitochondria (Heazlewood *et al.*, 2003b, 2004). The results have provided new information on the composition of the respiratory chain complexes. They have also highlighted a range of new mitochondrial proteins, new mitochondrial functions, and possible new mechanisms for regulating mitochondrial metabolism (Millar *et al.*, 2005). Many identified proteins in both rice and *Arabidopsis* mitochondrial samples lack similarity to any protein of known function, suggesting that a range of novel activities in plant mitochondria await discovery. These discoveries have occurred alongside wider scale use of targeting prediction programs to suggest location based on primary sequence data, and *in situ* imaging using fluorescent tags linked to protein of interest that are then expressed and imaged in living cells.

A. Experimental Analysis of Mitochondrial Proteomes

Systematic analysis of the proteins separated on IEF, SDS-PAGE 2D gels (Section VI.C) has been undertaken by in-gel digestion of protein spot plugs with trypsin. The resulting peptides have then been analyzed by MALDI-TOF peptide mass fingerprinting, N-terminal sequencing, and tandem MS of selected peptides to provide insight into many of the more soluble proteins in *Arabidopsis* (Kruft *et al.*, 2001; Millar *et al.*, 2001) and rice (Heazlewood *et al.*, 2003b). This has provided a list of over 120 mitochondrial proteins and some assessment of their relative abundance in mitochondrial extracts.

The use of BN gels (Section VI.D) has aided in the systematic analysis of the components of the classical complexes I, II, IV, and V of the respiratory chain, and the complexes of the import apparatus. Proteomic analysis of BN-separated Complex I components in rice and *Arabidopsis* identified a set of 20 proteins with sequence similarity to known Complex I subunits, and also identified a further series of plant-specific subunits (Heazlewood *et al.*, 2003a). The newly identified components include a series of carbonic anhydrase-like proteins (Perales *et al.*, 2004), and the final enzyme in ascorbate synthesis pathway in plants, galactonolactone dehydrogenase (Millar *et al.*, 2003). Complex II (succinate dehydrogenase), while classically composed of four subunits (SDHI-4), contains up to four additional proteins of unknown function in plants (Eubel *et al.*, 2003; Millar *et al.*, 2004). Complex IV (cytochrome *c* oxidase), in mammals contains 13 nonidentical subunits. Early purifications from plants found only seven or eight subunits in the plant complex, but recently, a COX complex containing up to 12-proteins bands has been isolated and the proteins identified (Eubel *et al.*, 2003; Millar *et al.*, 2004). Complex V, the F_1F_0-type H + ATP synthase, has a general structure highly conserved in prokaryotes and eukaryotes. Analysis of the complex from *Arabidopsis*, sunflower, and protists has identified two mitochondrial-encoded ORFs as components of the F_0 subcomplex, likely representing the equivalents of the AL6 (or subunit 8) and the β-subunit (or subunit 4), and encoded by orfB and orf25 in plant mitochondrial genomes (Burger *et al.*, 2003; Heazlewood *et al.*, 2003c; Sabar *et al.*, 2003). The other major protein complexes within and associated with the mitochondrial membranes belong to the protein import apparatus. The translocase of the outer mitochondrial membrane (TOM) complex has been purified and analyzed from *Arabidopsis* (Werhahn *et al.*, 2001, 2003). This 390-kDa complex contains six conserved subunits with similarity to yeast counterparts, but key components are missing or modified, providing the potential for separate reception of different classes of proteins for import and/or differentiation between chloroplast and mitochondrial-targeted proteins in plants (Lister *et al.*, 2004; Werhahn *et al.*, 2001, 2003).

Use of HPLC separation of peptides from trypsination of intact mitochondrial samples has yielded an even deeper insight into the mitochondrial proteome (Heazlewood *et al.*, 2004). The identified proteins reflect the many well-known roles these organelles play in primary metabolism. For example, some 30 proteins from the tricarboxylic acid (TCA) cycle, 78 proteins from the ETC, and over 20 proteins from amino acid metabolism pathways, including the mitochondrial steps in photorespiration, have been identified in *Arabidopsis*. Additionally, a range of inner membrane carriers and components of the mitochondrial protein import and assembly pathways are well represented in this list. A significant number of low abundance proteins involved in DNA synthesis, transcriptional regulation, protein complex assembly, and cellular signaling were also discovered (Heazlewood *et al.*, 2004), which make this list a rich source for further experimental analysis of mitochondrial function.

Analysis of these different datasets shows the distribution of molecular mass, isoelectric point and hydrophobicity of the identified proteins were dependent on the technology used for proteome analyses. The gel-free strategies were more successful in identifying very large, small, and basic proteins as well as proteins of low abundance.

B. Prediction of Mitochondrial Proteomes

The experimental avenue for enlarging the proteome is aided by a range of algorithms that use both defined characteristics and machine-learning techniques to predict subcellular localization specifically on the basis of the N-terminal region of proteins that often contain presequence targeting information. This has led to a number of publicly available programs that can be used to generally predict protein subcellular localization, and most notably, mitochondrial localization. MitoProt II, PSORT, and iPSORT utilize predefined parameters of signal peptides to predict subcellular localization. The programs TargetP, Predotar, and SubLoc utilize machine-learning techniques for subcellular predictions. Such programs typically predict 5–10% of *Arabidopsis* proteins as mitochondrial targeted (1500–3000). Comparisons of the prediction sets obtained from each predictor program indicate the consensus sets between programs are much smaller (~3% of all *Arabidopsis* proteins), indicating there are relatively large nonoverlapping sets of positives within these predicted sets (Heazlewood *et al.*, 2004). However, despite these difficulties, proteins that are widely predicted by these programs to be mitochondrial represent a useful set for further analysis.

Several databases are available online to give access to these prediction sets and experimentally determined sets of proteins based on MS and GFP fluorescence studies:

Arabidopsis Mitochondrial Protein Database   http://www.ampdb.bcs.uwa.edu.au/
SUBA—Subcellular Location in *Arabidopsis*   http://www.suba.bcs.uwa.edu.au/
The *Arabidopsis* Mitochondrial Proteome   http://www.gartenbau.uni-hannover.de/genetik/AMPP
   Project

IX. Conclusion

These techniques outline a basis for studying plant mitochondria at the level of overall metabolic function, at the level of individual enzyme and protein complex composition and operation, and at the level of each polypeptide chain in the organelle proteome. Integration of these approaches will be essential tools in future identification and characterization of the mitochondrial-targeted products of nuclear-encoded genes sequenced in plant genome projects.

References

Behal, R. H., and Oliver, D. J. (1997). Biochemical and molecular characterization of fumarase from plants: Purification and characterization of the enzyme—cloning, sequencing, and expression of the gene. *Arch. Biochem. Biophys.* **348,** 65–74.

Behal, R. H., Neal, W., and Oliver, D. J. (1996). Biochemical and molecular characterization of NAD-dependent isocitrate dehydrogenase and fumarase from oil seed plants. *Plant Physiol.* **111,** 601.

Braun, H. P., and Schmitz, U. K. (1992). Affinity purification of cytochrome-c deductase from potato mitochondria. *Eur. J. Biochem.* **208,** 761–767.

Braun, H. P., and Schmitz, U. K. (1995). Cytochrome-c reductase/processing peptidase complex from potato mitochondria. *Methods Enzymol.* **260,** 70–82.

Burger, G., Lang, B. F., Braun, H. P., and Marx, S. (2003). The enigmatic mitochondrial ORF ymf39 codes for ATP synthase chain b. *Nucleic Acids Res.* **31,** 2353–2360.

Day, D. A., Neuburger, M., and Douce, R. (1985). Biochemical characterization of chlorophyll-free mitochondria from pea leaves. *Aust. J. Plant Physiol.* **12,** 219–228.

Devirville, J. D., Moreau, F., and Denis, M. (1994). Proton to electron stoichiometry of the cytochrome-c-oxidase proton pump in plant mitoplasts. *Plant Physiol. Biochem.* **32,** 85–92.

Douce, R. (1985). "Mitochondria in Higher Plants: Structure, Function and Biogenesis." Academic Press, London.

Emmermann, M., Braun, H. P., Arretz, M., and Schmitz, U. K. (1993). Characterization of the bifunctional cytochrome-c reductase processing-peptidase complex from potato mitochondria. *J. Biol. Chem.* **268**, 18936–18942.

Eubel, H., Jänsch, L., and Braun, H. P. (2003). New insights into the respiratory chain of plant mitochondria. Supercomplexes and a unique composition of complex II. *Plant Physiol.* **133**, 274–286.

Eubel, H., Braun, H. P., and Millar, A. H. (2005). Blue-Native PAGE in plants: A tool in analysis of protein-protein interactions. *Plant Methods* **1**, 11.

Glaser, E., and Norling, B. (1983). Kinetics of interaction between the H+-translocating component of the mitochondrial ATPase complex and oligomycin or dicyclohexylcarbodiimide. *Biochem. Biophys. Res. Commun.* **111**, 333–339.

Grover, S. D., and Wedding, R. T. (1984). Modulation of the activity of NAD-malic enzyme from Solanum tuberosum by changes in oligomeric state. *Arch. Biochem. Biophys.* **234**, 418–425.

Grover, S. D., Canellas, P. F., and Wedding, R. T. (1981). Purification of NAD-malic enzyme from potato and investigation of some physical and kinetic properties. *Arch. Biochem. Biophys.* **209**, 396–407.

Hamasur, B., and Glaser, E. (1992). Plant mitochondrial F0F1 ATP synthase—identification of the individual subunits and properties of the purified spinach leaf mitochondrial ATP synthase. *Eur. J. Biochem.* **205**, 409–416.

Hamasur, B., Guerrieri, F., Zanotti, F., and Glaser, E. (1992). F0F1-ATP synthase of potato tuber mitochondria—structural and functional characterization by resolution and reconstitution studies. *Biochim. Biophys. Acta* **1101**, 339–344.

Hatch, M. D., Tsuzuki, M., and Edwards, G. E. (1982). Determination of NAD-malic enzyme in leaves of C4 plants—effects of malate dehydrogenase and other factors. *Plant Physiol.* **69**, 483–491.

Hattori, T., and Asahi, T. (1982). The presence of two forms of succinate dehydrogenase in sweet potato root mitochondria. *Plant Cell Physiol.* **23**, 515–523.

Hausmann, N., Werhahn, W., Huchzermeyer, B., Braun, H. P., and Papenbrock, J. (2003). How to document the purity of mitochondria prepared from green tissue of pea, tobacco and Arabidopsis thaliana. *Phyton (Annales Rei Botanicae)* **43**, 215–229.

Hayes, M. K., Luethy, M. H., and Elthon, T. E. (1991). Mitochondrial malate dehydrogenase from corn—purification of multiple forms. *Plant Physiol.* **97**, 1381–1387.

Heazlewood, J. A., Howell, K. A., and Millar, A. H. (2003a). Mitochondrial complex I from Arabidopsis and rice: Orthologs of mammalian and yeast components coupled to plant-specific subunits. *Biochim. Biophys. Acta Bioenerg.* **1604**, 159–169.

Heazlewood, J. L., Howell, K. A., Whelan, J., and Millar, A. H. (2003b). Towards an analysis of the rice mitochondrial proteome. *Plant Physiol.* **132**, 230–242.

Heazlewood, J. L., Whelan, J., and Millar, A. H. (2003c). The products of the mitochondrial ORF25 and ORFB genes are FO components of the plant F1FO ATP synthase. *FEBS Lett.* **540**, 201–205.

Heazlewood, J. L., Tonti-Filippini, J. S., Gout, A., Day, D. A., Whelan, J., and Millar, A. H. (2004). Experimental analysis of the Arabidopsis mitochondrial proteome highlights signaling and regulatory components, provides assessment of targeting prediction programs, and indicates plant-specific mitochondrial proteins. *Plant Cell* **16**, 241–256.

Herz, U., Schroder, W., Liddell, A., Leaver, C. J., Brennicke, A., and Grohmann, L. (1994). Purification of the NADH ubiquinone oxidoreductase (complex-I) of the respiratory chain from the inner mitochondrial membrane of Solanum tuberosum. *J. Biol. Chem.* **269**, 2263–2269.

Hoefnagel, M. H. N., Wiskich, J. T., Madgwick, S. A., Patterson, Z., Oettmeier, W., and Rich, P. R. (1995). New inhibitors of the ubiquinol oxidase of higher plant mitochondria. *Eur. J. Biochem.* **233**, 531–537.

Igamberdiev, A. U., and Falaleeva, M. I. (1994). Isolation and characterization of the succinate-dehydrogenase complex from plant mitochondria. *Biochemistry (Moscow)* **59**, 895–900.

Jansch, L., Kruft, V., Schmitz, U. K., and Braun, H. P. (1996). New insights into the composition, molecular-mass and stoichiometry of the protein complexes of plant mitochondria. *Plant J.* **9**, 357–368.

Keech, O, Dizengremel, P., and Gardestrom, P. (2005). Preparation of leaf mitochondria from *Arabidopsis thaliana*. *Physiol. Plant.* **124**, 403–409.

Kruft, V., Eubel, H., Jansch, L., Werhahn, W., and Braun, H. P. (2001). Proteomic approach to identify novel mitochondrial proteins in Arabidopsis. *Plant Physiol.* **127**, 1694–1710.

Leaver, C. J., Hack, E., and Forde, B. G. (1983). Protein synthesis by isolated plant mitochondria. *Methods Enzymol.* **97**, 476–484.

Leterme, S., and Boutry, M. (1993). Purification and preliminary characterization of mitochondrial complex I (NADH-ubiquinone reductase) from broad bean (*Vicia faba* L). *Plant Physiol.* **102**, 435–443.

Link, A. J. (1999). "2D Proteome Analysis Protocols." Humana Press, Totowa, New Jersey.

Lister, R., Chew, O., Heazlewood, J. L., Lee, M.-N., Millar, A. H., and Whelan, J. (2004). A transcriptomic and proteomic characterization of the Arabidopsis mitochondrial protein import apparatus and its response to mitochondrial dysfunction. *Plant Physiol.* **134**, 777–789.

Maeshima, M., and Asahi, T. (1978). Purification and characterisation of sweet potato cytochrome c oxidase. *Arch. Biochem. Biophys.* **187**, 423–430.

Mannella, C. A. (1985). The outer membrane of plant mitochondria. *In* "Higher Plant Cell Respiration" (R. Douce, and D. A. Day, eds.), Vol. 18, pp. 106–133. Springer-Verlag, Berlin.

Mannella, C. A., and Bonner, W. D. (1975). Biochemical characteristics of the outer membranes of plant mitochondria. *Biochim. Biophys. Acta* **413**, 213–225.

McIntosh, C. A., and Oliver, D. J. (1992). NAD-linked isocitrate dehydrogenase—isolation, purification, and characterization of the protein from pea mitochondria. *Plant Physiol.* **100**, 69–75.

Millar, A. H., Knorpp, C., Leaver, C. J., and Hill, S. A. (1998). Plant mitochondrial pyruvate dehydrogenase complex: Purification and identification of catalytic components in potato. *Biochem. J.* **334**, 571–576.

Millar, A. H., Hill, S. A., and Leaver, C. J. (1999). The plant mitochondrial 2-oxoglutarate dehydrogenase complex: Purification and characterisation in potato. *Biochem. J.* **343**, 327–334.

Millar, A. H., Sweetlove, L. J., Giegé, P., and Leaver, C. J. (2001). Analysis of the Arabidopsis mitochondrial proteome. *Plant Physiol.* **127**, 1711–1727.

Millar, A. H., Mittova, V., Kiddle, G., Heazlewood, J. L., Bartoli, C. G., Guiamet, J. J., Theodoulou, F. L., and Foyer, C. H. (2003). Control of ascorbate synthesis by respiration and its implications for stress responses. *Plant Physiol.* **133**, 443–447.

Millar, A. H., Eubel, H., Jänsch, L., Kruft, V., Heazlewood, J. L., and Braun, H. P. (2004). Mitochondrial cytochrome c oxidase and succinate dehydrogenase complexes contain plant-specific subunits. *Plant Mol. Biol.* **56**, 77–89.

Millar, A. H., Heazlewood, J. L., Kristensen, B., Bruan, H. P., and Moller, I. M. (2005). The plant mitochondrial proteome. *Trends Plant Sci.* **10**, 36–43.

Moller, I. M., Rasmusson, A. G., and Fredlund, K. M. (1993). NAD(P)H-ubiquinone oxidoreductases in plant mitochondria. *J. Bioenerg. Biomembr.* **25**, 377–384.

Neuburger, M. (1985). Preparation of plant mitochondria, criterea for assessment of mitochondrial integerity, and purity, survival *in vitro*. *In* "Higher Plant Cell Respiration" (R. Douce, and D. A. Day, eds.), Vol. 18, pp. 7–24. Springer-Verlag, Berlin.

Neuburger, M., Journet, E., Bligny, R., Carde, J., and Douce, R. (1982). Purification of plant mitochondria by isopycnic centrifugation in density gradients of Percoll. *Arch. Biochem. Biophys.* **217**, 312–323.

Neuburger, M., Bourguignon, J., and Douce, R. (1986). Isolation of a large complex from the matrix of pea leaf mitochondria involved in the rapid transformation of glycine into serine. *FEBS Lett.* **207**, 18–22.

Oliver, D. J. (1994). The glycine decarboxylase complex from plant mitochondria. *Annu. Rev. Plant Physiol. Plant Mol. Biol.* **45**, 323–337.

Palmer, J. M., and Wedding, R. T. (1966). Purification and properties of succinyl-CoA synthetase from Jerusalem artichoke mitochondria. *Biochim. Biophys. Acta* **113**, 167–174.

Perales, M., Parisi, G., Fornasari, M. S., Colaneri, A., Villarreal, F., Gonzalez-Schain, N., Echave, J., Gomez-Casati, D., Braun, H. P., Araya, A., and Zabaleta, E. (2004). Gamma carbonic anhydrase like complex interact with plant mitochondrial complex I. *Plant Mol. Biol.* **56**, 947–957.

Poulsen, L. L., and Wedding, R. T. (1970). Purification and properties of the a-ketoglutarate dehydrogenase complex of cauliflower mitochondria. *J. Biol. Chem.* **245**, 5709–5717.

Quail, P. H. (1979). Plant cell fractionation. *Annu. Rev. Plant Physiol.* **30**, 425–484.

Randall, D. D., and Miernyk, J. A. (1990). The mitochondrial pyruvate dehydrogenase comples. *In* "Methods in Plant Biochemistry" (P. J. Lea, ed.), Vol. 3, pp. 175–192. Academic Press, New York.

Rasmusson, A. G., and Moller, I. M. (1990). NADP-utilizing enzymes in the matrix of plant-mitochondria. *Plant Physiol.* **94**, 1012–1018.

Rasmusson, A. G., Heiser, V., Zabaleta, E., Brennicke, A., and Grohmann, L. (1998). Physiological, biochemical and molecular aspects of mitochondrial complex I in plants. *Biochim. Biophys Acta-Bioenerg.* **1364**, 101–111.

Sabar, M., Gagliardi, D., Balk, J., and Leaver, C. J. (2003). ORFB is a subunit of F1F(O)-ATP synthase: Insight into the basis of cytoplasmic male sterility in sunflower. *EMBO Rep.* **4**, 381–386.

Schägger, H., and von Jagow, G. (1991). Blue native electrophoresis for isolation of membrane complexes in enzymatically active form. *Anal. Biochem.* **199**, 223–231.

Schieber, O., Dietrich, A., and Marechal-Drouard, L. (1994). Cryopreservation of plant mitochondria as a tool for protein import or in organello protein synthesis studies. *Plant Physiol.* **106**, 159–164.

Soole, K. L., and Menz, R. I. (1995). Functional molecular aspects of the NADH dehydrogenases of plant mitochondria. *J. Bioenerg. Biomembr.* **27**, 397–406.

Stevens, F. J., Li, A. D., Lateef, S. S., and Anderson, L. E. (1997). Identification of potential inter-domain disulfides in three higher plant mitochondrial citrate synthases: Paradoxical differences in redox-sensitivity as compared with the animal enzyme. *Photosynth. Res.* **54**, 185–197.

Stitt, M. (1984). Citrate synthase. *In* "Methods of Enzymatic Analysis" (H. U. Bergmeyer, J. Bergmeyer, and M. Grabl, eds.), Vol. 4, pp. 353–358. Verlag Chemie, Weinheim.

Vanemmerik, W. A. M., Wagner, A. M., and Vanderplas, L. H. W. (1992). A quantitative comparison of respiration in cells and isolated mitochondria from Petunia hybrida suspension cultures—a high yield isolation procedure. *J. Plant Physiol.* **139**, 390–396.

Vanlerberghe, G. C., and McIntosh, L. (1997). Alternative oxidase: From gene to function. *Annu. Rev. Plant Physiol. Plant Mol. Biol.* **48**, 703–734.

Verniquet, F., Gallard, J., Neuburger, M., and Douce, R. (1991). Rapid inactivation of plant aconitase by hydrogen peroxide. *Biochem. J.* **276**, 643–648.

Walk, R. A., Michaeli, S., and Hock, B. (1977). Glyoxysomal and mitochondrial malate dehydrogenase of watermelon (Citrullus vulgaris) cotyledons. I. Molecular properties of the purified isoenzymes. *Planta* **136**, 211–220.

Walker, J. L., and Oliver, D. J. (1986). Glycine decarboxylase multienzyme complex—purification and partial characterization from pea leaf mitochondria. *J. Biol. Chem.* **261**, 2214–2221.

Werhahn, W., Niemeyer, A., Jänsch, L., Kruft, V., Schmitz, U. K., and Braun, H. P. (2001). Purification and characterization of the preprotein translocase of the outer mitochondrial membrane from Arabidopsis. Identification of multiple forms of TOM20. *Plant Physiol.* **125**, 943–954.

Werhahn, W., Jänsch, L., and Braun, H. P. (2003). Identification of novel subunits of the TOM complex from Arabidopsis thaliana. *Plant Physiol. Biochem.* **41**, 407–416.

Zhang, Q. S., Hoefnagel, M. H. N., and Wiskich, J. T. (1996). Alternative oxidase from Arum and soybean: Its stabilization during purification. *Physiol. Plant.* **96**, 551–558.

Zhang, Q., Wiskich, J. T., and Soole, K. L. (1999). Respiratory activities in chloramphenicol-treated tobacco cells. *Physiol. Plant.* **105**, 224–232.

PART II

Biochemical Assays of
Mitochondrial Activity

CHAPTER 4

Biochemical Assays of Respiratory Chain Complex Activity

Denise M. Kirby,[*,†] David R. Thorburn,[†] Douglass M. Turnbull,[*] and Robert W. Taylor[*]

[*]Mitochondrial Research Group, School of Neurology
Neurobiology and Psychiatry, The Medical School
Framlington Place, Newcastle University
Newcastle upon Tyne, NE2 4HH, United Kingdom

[†]Mitochondrial Research Laboratory
Murdoch Childrens Research Institute
Royal Children's Hospital
Parkville, Victoria 3052, Australia

I. Introduction

Mitochondrial disorders are estimated to affect at least 1 in 5000 of the population (Skladal *et al.*, 2003). These are usually multisystem diseases; indeed, mitochondrial disease "can give rise to any symptom, in any tissue, at any age, and with any mode of inheritance" (Munnich and Rustin, 2001). Skeletal and cardiac muscle and the central nervous system are predominantly affected. Many mitochondrial disorders are caused by deficiencies of the activities of one or more of the enzyme complexes of the respiratory chain (RC), thus RC biochemistry is an important part of the diagnostic work up of a patient suspected of mitochondrial disease. This is especially true for children, where defects caused by mutations in mitochondrial

METHODS IN CELL BIOLOGY, VOL. 80
Copyright 2007, Elsevier Inc. All rights reserved.

0091-679X/07 $35.00
DOI: 10.1016/S0091-679X(06)80004-X

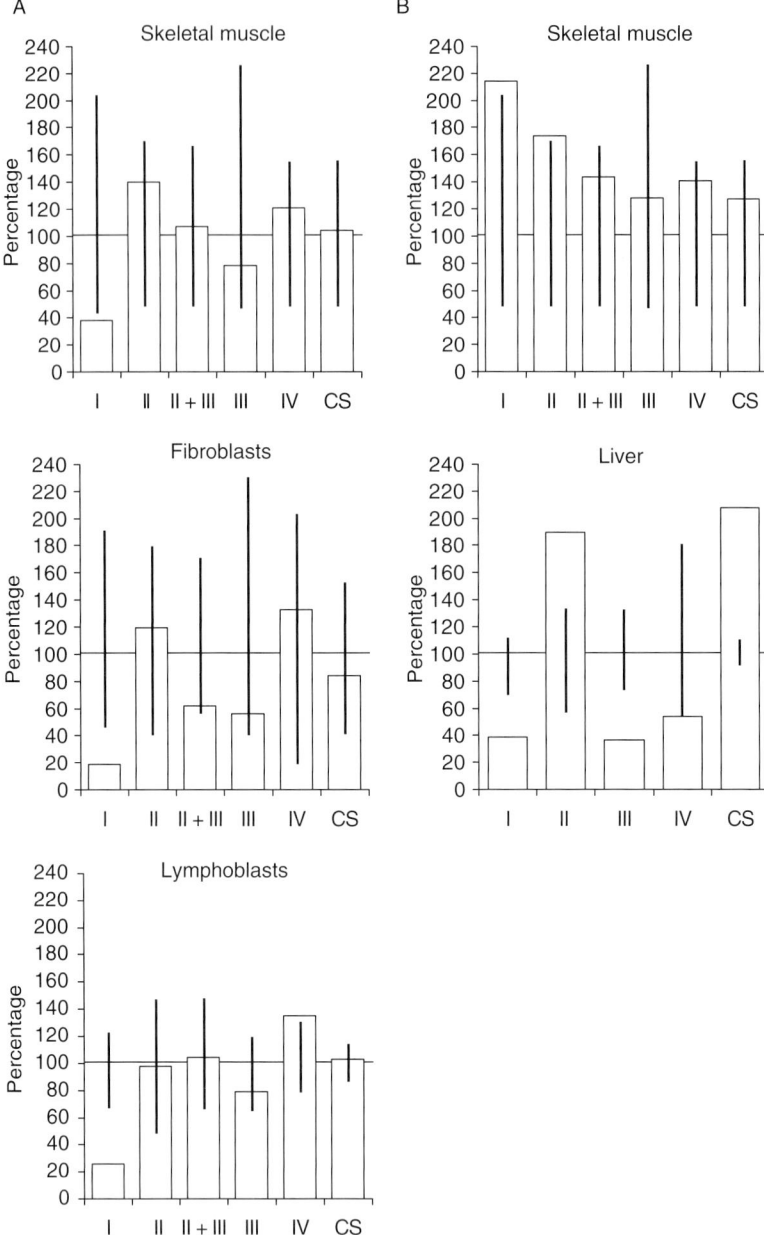

Fig. 1 Respiratory chain enzyme deficiencies may be isolated, affecting only one enzyme complex, or they may affect multiple enzyme activities. They may be systemic, affecting all tissues, or tissues specific, affecting only one tissue. (A) Respiratory chain complexes I, II, II+III, III, IV, and citrate synthase activities in skeletal muscle, fibroblasts, and transformed lymphoblasts from a patient with systemic isolated complex I deficiency and Leigh-like syndrome. (B) Respiratory chain complexes I, II, II+III, III, IV, and citrate

DNA (mtDNA) are less common than in adults. RC enzyme deficiencies may be isolated, that is, affect only one of the enzyme complexes (Fig. 1A), or they may affect several, particularly those with mtDNA-encoded subunits (Fig. 1B). Mutations in mitochondrial transfer RNA (tRNA) genes, and in nuclear genes encoding enzymes involved in regulation and metabolism of mitochondrial nucleoside pools, such as the mtDNA polymerase $\gamma 1$ (*POLG1*) (Davidzon *et al.*, 2005; Naviaux and Nguyen 2004) and mitochondrial deoxyguanosine kinase (*DUOGK*) (Mandel *et al.*, 2001; Slama *et al.*, 2005) genes, can cause multiple RC enzyme defects. Accurate identification of such biochemical defects is desirable in order to direct efforts in identifying the molecular basis of the disease in a particular patient. It is therefore important to be able to measure individual RC complexes to enable correlation with both nuclear and mtDNA mutations, to direct screening of candidate genes (e.g., a biochemical complex I deficiency would direct one toward sequencing the mitochondrial *MTND* genes, or some of the many nuclear DNA-encoded complex I subunits, while an isolated complex IV deficiency would more likely lead to investigation of genes involved in complex IV assembly such as *SURF1*), and to narrow down defects detected by other methods [e.g., cytochrome *c* oxidase (COX) histochemistry, substrate oxidation, adenosine triphosphate (ATP) synthesis].

There are many methods described to measure RC enzyme activities, and there is no one "correct" way. This chapter is written collaboratively by representatives from two laboratories that provide diagnostic services for RC enzymology using spectrophotometric assays. It is notable that although the precise methods differ slightly, the general approach in the two centers is remarkably similar. We find that linked assays are less useful in the diagnosis of RC deficiencies than the isolated activities. For example, the combined activity of complex II + III (succinate: cytochrome *c* reductase) is not very sensitive for the diagnosis of complex III defects, since the flux through these enzymes relies mostly on the complex II activity (Taylor *et al.*, 1994). However, it can be useful in detecting complex II deficiency, and since it relies on endogenous coenzyme Q, complex II + III activity is useful in delineating coenzyme Q deficiency. Thus, if the isolated complexes II and III activities are normal, but the combined II + III activity is reduced, the RC abnormality may well be a defect in coenzyme Q synthesis or recycling.

Measurement of RC enzymes in cultured cells is useful to confirm a defect found in another tissue, although it is our experience, and that of others, that only about 50% of patients with an RC defect detected in skeletal muscle, liver or heart will express that defect in cultured fibroblasts (Faivre *et al.*, 2000; Niers *et al.*, 2003). In adults, the proportion of patients expressing a defect in fibroblasts is probably

synthase activities in skeletal muscle, and complexes I, II, III, IV, and citrate synthase activities in liver from a patient with Alpers syndrome caused by mutations in the *POLG1* gene. This patient shows liver-specific deficiencies of multiple enzyme complexes. Enzyme activities are expressed as percentages of the mean of controls ($n = 17$ for skeletal muscle, $n = 6$ for liver, $n = 35$ for fibroblasts, and $n=7$ for transformed lymphoblasts). The observed range for the controls is represented by the vertical lines. CS, citrate synthase.

even less than this. Since fibroblasts, chorionic villus sampling (CVS) cells, and amniocytes have a common embryonic origin (Niers *et al.*, 2003), expression of an enzyme defect in fibroblasts suggests that the deficiency will also be expressed in CVS and amniocytes, implying that enzyme-based prenatal diagnosis may be possible in cases where a molecular defect has not been demonstrated. The ability to measure the RC complexes in cultured cells is also useful in cell biology studies to demonstrate the pathogenicity of a putative mtDNA mutation. Such studies may include correlation of mutant load with residual enzyme activity in cybrids (transmitochondrial cytoplasmic hybrids) or lack of phenotypic rescue (restoration of enzyme activity) in hybrids between RC enzyme-deficient patient fibroblasts and mtDNA-less cells (ρ^0 cells).

Here, we describe the preparation of mitochondrial fractions from tissues and cultured cells for RC enzymology, and measurement of activity of the individual complexes I, II, III, IV, V, the mitochondrial matrix marker enzyme citrate synthase, and the combined activity of complexes II + III. Citrate synthase is not part of the RC, but RC enzyme activities are frequently expressed relative to its activity. Such ratios are more robust than absolute activities due to (1) the variability inherent in cell culture conditions such as passage number and the degree of confluence and (2) the proliferation of mitochondria seen in tissues of many patients with mitochondrial disease.

II. Materials and Methods

A. Sample Preparation

1. Reagents for Sample Preparation

Medium A: 120-mM KCl; 20-mM 4-(2-hydroxyethyl)piperazine-1-(2-ethanesulfonic acid) (HEPES); 2-mM $MgCl_2$; 1-mM ethyleneglycolbis(β-aminoethyl ether)-*N,N,N',N'*-tetraacetic acid (EGTA); 5-mg/ml fatty acid free bovine serum albumin, Fraction V (BSA), pH 7.4

Medium B: 300-mM sucrose; 2-mM HEPES; 0.1-mM EGTA, pH 7.4

Medium C: 250-mM sucrose; 2-mM HEPES; 0.1-mM EGTA, pH 7.4

Medium D: 25-mM potassium phosphate; 5-mM $MgCl_2$, pH 7.2

Medium E: 210-mM mannitol; 70-mM sucrose; 5-mM HEPES; 1-mM EGTA, pH 7.2 (0.1-mg/ml digitonin added as specified; see page 101)

2. Tissues

It is generally considered that the ideal way to investigate a patient with a suspected RC defect is to obtain a fresh biopsy from an affected tissue, for example skeletal muscle, liver, cardiac muscle, kidney, and so on (Thorburn and Smeitink, 2001). This allows for functional studies (e.g., substrate oxidation, oxygen consumption) as well as for enzyme assays to be performed on isolated mitochondria or homogenates. However, a number of practical issues impinge on the approach taken by different centers. Often biopsies are very small and isolation of sufficient mitochondria to perform RC enzyme analysis and functional assays is not possible. This applies particularly to samples from children and to tissues other than skeletal muscle. Obtaining fresh tissues may not be feasible where biopsy or autopsy samples are collected at a time or location that prevents immediate processing by a specialist center. Both of our centers receive samples from widespread regions, and these geographic constraints generally preclude collection of fresh tissue and expedient preparation of mitochondria. In these cases, RC enzymes can be assayed in homogenates (post-600 $\times$ g_{av} supernatants) prepared from frozen tissues, using as little as 20 mg of skeletal muscle and 10 mg of liver or cardiac muscle.

Mitochondrial fractions can be prepared by differential centrifugation of homogenates from a variety of tissues. Skeletal muscle is the tissue traditionally used for RC enzymology, but we have found that RC enzyme analysis in cardiac muscle and liver is very useful in the diagnosis of mitochondrial disease, particularly in pediatric patients. Often, the residual RC enzyme activity is not definitively low in skeletal muscle, but more markedly reduced in liver or heart. In fact, patients may have tissue-specific enzyme deficiencies, and express the defect in liver or heart, but not in skeletal muscle. For example, Fig. 1B shows normal RC enzymes in skeletal muscle but liver-specific deficiencies of multiple RC enzymes in a patient with Alpers syndrome who was a compound heterozygote for two mutations (G848S and W748S) in the *POLG1* gene (Davidzon *et al.*, 2005).

There is still debate as to the suitability of frozen tissues for RC enzymology. There are few data on the stability of the RC enzymes in tissues frozen and stored at $-70\,^{\circ}$C, particularly on long-term storage of several years. In Melbourne, the practice has been that each batch of sample assays includes a homogenate prepared freshly from a single donated normal skeletal muscle or liver sample. Although there is some day-to-day variation in RC enzyme activities, we observed no significant decrease in activity over 4–5 years for skeletal muscle or ~9 years for liver (Thorburn *et al.*, 2004). The data for complex I are shown in Fig. 2. We have also stored tissue homogenates at $-70\,^{\circ}$C for periods up to 3 months and seen no significant decline in enzyme activities. Thus, the activities of RC complexes I, II, III, and IV (assayed as individual complexes) and citrate synthase are stable in tissue samples and homogenates stored at $-70\,^{\circ}$C for extended periods.

The use of tissue homogenates prepared from frozen tissues has limitations. Janssen *et al.* (2003) estimate that only ~50% of patients showing defects in substrate oxidation in fresh tissues show an RC enzyme defect in frozen tissues. Therefore, many RC defects may be missed by assaying frozen tissues. However, not all patients with substrate oxidation defects prove to have RC enzyme complex defects. Complex V and adenine nucleotide translocator (ANT) defects cannot be detected in frozen tissues, and there is potential for artifacts (Scholte and Trijbels 1995; Zheng *et al.*, 1990). The complex V assay described in this chapter has a very high background in homogenates prepared from frozen tissues and is unreliable in such preparations.

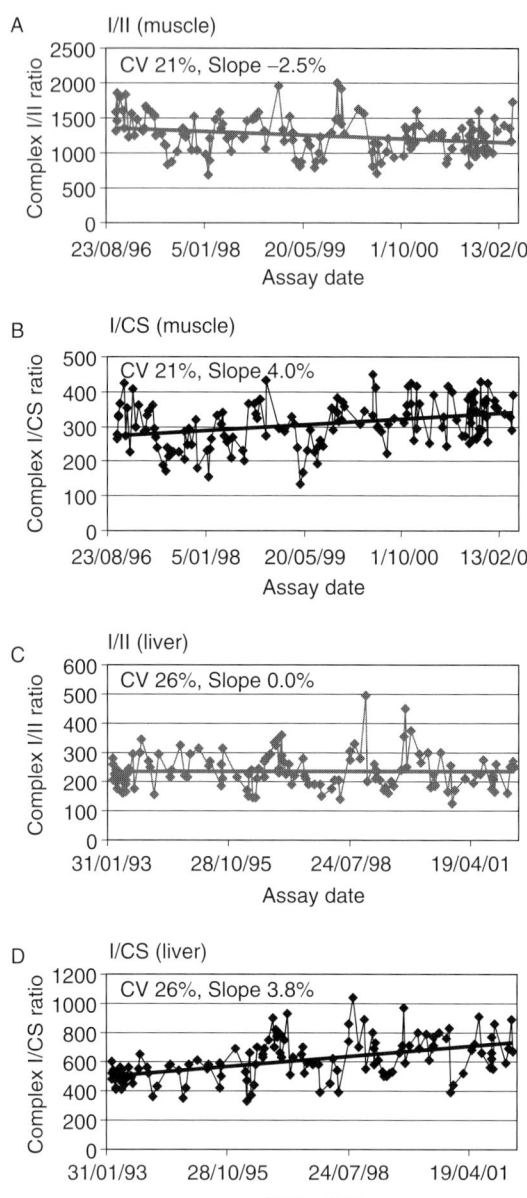

Fig. 2 Complex I ratios in single large samples of skeletal muscle (A, B) and liver (C, D) that had been stored at −70°C. The tissues were used as "in batch" controls for respiratory chain enzymology. Muscle was homogenized and analyzed on 138 occasions between 1996 and 2002, and the liver on 127 occasions between 1993 and 2002. Results are expressed as complex I/II ratio (A, C) and complex I/CS ratio (B, D); the coefficient of variation (CV) and change in activity per year (slope) is also shown. CS, citrate synthase. Reproduced from Thorburn *et al.*, 2004, with permission.

We have studied the effects of postmortem delay on RC enzymes (Thorburn *et al.*, 2004). Liver samples frozen at varying times after death were analyzed to assess the stability of RC enzyme activities postmortem. There can be considerable loss of RC enzyme activity postmortem, particularly in liver, but our observations suggest that muscle collected and frozen at $-70\,^{\circ}$C within 6 h of death and liver within 2 h remain suitable for RC enzyme analysis. Storage at $-20\,^{\circ}$C results in a dramatic decrease in the activities of all RC complexes, with relative preservation of citrate synthase activity. Citrate synthase activity appears to be more robust than the RC enzymes. Complex II loses activity most rapidly after death and on inappropriate storage. Complexes II + III and III are the next most labile and complexes I and IV have intermediate lability. The complex II activity can therefore be considered a marker of the integrity of the tissue and the reliability of RC enzymology, particularly if the tissues have been taken at longer or unknown times postmortem. Generally, we consider that complex II activity of greater than or equal to 50% of normal means that RC enzymology is reliable, but we are wary of the results if complex II is less than 50%. The reduced complex II activity may be due to tissue pathology, poor sample processing, or postmortem delay in freezing of tissues at $-70\,^{\circ}$C.

a. Preparation of Tissue Homogenates

All procedures are carried out on ice or at $4\,^{\circ}$C. The tissue (50–150 mg of skeletal muscle or 10–50 mg of cardiac muscle) is dissected out, weighed, and placed in ice-cold Medium A. It is trimmed of fat and connective tissues, chopped finely with a pair of scissors, and rinsed in Medium A to remove any blood. The tissue is chopped further using a scalpel blade and forceps to minimize loss of tissue. The tissue is then homogenized on ice using a T25 Ultra-Turrax homogenizer (9500 rpm for 5 sec) in a total volume of 1 ml of Medium A. The homogenate is centrifuged at $600 \times g_{av}$ for 10 min at $4\,^{\circ}$C to remove nuclear debris. The pellet obtained at this stage can be retained; it contains nuclei and some mitochondria and is suitable for molecular genetic studies (Southern blotting, mtDNA sequencing, restriction fragment length polymorphism (RFLP) analysis of specific mutations) other than analysis for mtDNA depletion. The supernatant is centrifuged again at $600 \times g_{av}$ for 10 min at $4\,^{\circ}$C. Enzyme measurements are performed on this second supernatant.

Liver and kidney biopsies are usually small (<50 mg), especially those from children. For these and other small specimens, including muscle, the tissue is chopped finely, then resuspended in 9 volumes (with respect to the original wet weight of the tissue) of Medium E, and homogenized in a hand-held glass/glass homogenizer to minimize sample loss. The homogenizer is rinsed with a small volume (50–100 μl) of Medium E and this is added to the homogenate. The homogenate is centrifuged at $600 \times g_{av}$ for 10 min at $4\,^{\circ}$C as before, and the RC enzyme assays are performed on the supernatant. Again, the pellet obtained at this stage can be retained for molecular genetic studies.

b. Preparation of Muscle Mitochondrial Fraction

For muscle samples >150 mg, a mitochondrial fraction may be prepared. All procedures are carried out on ice or at 4°C. The muscle is chopped finely as described above. The disrupted muscle is then made up to 20 volumes with respect to the original wet weight of tissue with Medium A and homogenized with the T25 Ultra-Turrax homogenizer (9500 rpm for 5 sec). The homogenate is centrifuged at $600 \times g_{av}$ for 10 min and the supernatant is filtered through four layers of cheesecloth to remove fat and fibrous tissue. The pellet is resuspended in 8 volumes of Medium A with a hand-held Teflon-glass homogenizer (Potter-Elvehjem) (5–6 strokes) and centrifuged at $600 \times g_{av}$ for 10 min. The second supernatant is filtered through four layers of cheesecloth and combined with the first. The pellet obtained at this stage is retained; it contains nuclei and some mitochondria and is suitable for molecular genetic studies. The combined supernatants are centrifuged ($17,000 \times g_{av}$ for 10 min), and the pellet, which contains the mitochondria, is resuspended in 10 volumes of Medium A and centrifuged at $7000 \times g_{av}$ for 10 min. The pellet is resuspended in 10 volumes of Medium B and centrifuged at $3500 \times g_{av}$ for 10 min, and the mitochondrial fraction is finally suspended in a small volume of Medium B ($\sim$25–50 mg protein/ml).

If large amounts of animal mitochondria are being prepared, resuspension is performed using a hand-held Teflon-glass homogenizer. However, if human muscle is used, resuspension is performed by gently detaching the pellet and mixing mitochondria using a plastic Pasteur pipette, which minimizes any loss of mitochondria. In addition, we wash the centrifuge tube further with 1.25 ml of Medium B, transfer this suspension to a microcentrifuge tube, and centrifuge at $11,000 \times g_{av}$ for 10 min. Any mitochondria pelleted during this procedure are kept, but are only used for enzyme assays if the original mitochondrial pellet is fully used. The mitochondrial suspension is divided into aliquots and stored at $-70°C$ for further biochemical and molecular studies.

3. Preparation of Mitochondrial Fraction from Cultured Cells

An enriched mitochondrial fraction can be isolated from cultured cells (fibroblasts, CVS, amniocytes, EBV-transformed lymphoblasts, myoblasts, cybrids, hybrids) by physical homogenization and treatment with hypotonic buffer. For fibroblasts and myoblasts, we typically use one roller bottle (850 cm^2 surface area) or six to eight 80-cm^2 flasks; for CVS, amniocytes, cybrids and hybrids, two 80-cm^2 flasks; and for EBV-transformed lymphoblasts, 100 ml of suspension culture.

Attached cells are harvested by trypsinization, followed by inactivation of trypsin with culture medium containing bovine serum. The cells are then washed three times with phosphate buffered saline (PBS). Nonadherent cells such as EBV-transformed lymphoblasts are disaggregated in culture medium by trituration, centrifuged at $800 \times g_{av}$ for 10 min to pellet them, and washed three times with PBS.

All subsequent procedures are carried out on ice or at 4°C. For adherent cells, the washed cell pellet is suspended in 1–2 ml of ice-cold Medium C, and transferred to a 1–2 ml capacity smooth-surfaced glass homogenizer. For nonadherent cells, the washed cell pellet is suspended in 1 ml of Medium E containing 0.1-mg/ml digitonin. For both cell types, the cells are disrupted by 20 passes in the homogenizer with a tight-fitting power-driven Teflon plunger. The homogenate is then centrifuged for 10 min at $600 \times g_{av}$ in a refrigerated centrifuge at $4°C$, and the mitochondria-rich supernatant is collected. The cell debris pellet is resuspended in 800 μl of Medium C for adherent cells or in Medium E without digitonin for nonadherent cells, homogenized, and centrifuged as before. The two supernatants are pooled and centrifuged for 10 min at $11,000 \times g_{av}$ at $4°C$. The pellet (mitochondrial fraction) is then washed with 1 ml of hypotonic buffer (Medium D) and centrifuged for 10 min at $11,000 \times g_{av}$ at $4°C$. The resulting mitochondrial pellet is suspended in 400 μl of Medium D and subjected to three rounds of freeze/thaw prior to use in the enzyme assays. The mitochondrial fraction suspended in Medium D may be frozen at $-70°C$ until required for the enzyme assays. Since the combined complex II + III activity relies on the integrity of the mitochondrial membrane, the hypotonically treated mitochondria are not suitable for this assay.

B. Enzyme Assays

The methods for measuring enzyme activities are very similar between both laboratories, with minor variations in the assay conditions. We both use a Varian Cary 300 Bio spectrophotometer with a Cary temperature controller, sample transporter, and multicell holder, driven by Varian WinUV software. The assays are performed in duplicate or triplicate at $30°C$ in either 1- or 0.5-ml volumes; multiple cuvettes may be assayed at the same time. Assaying in 0.5-ml volumes can effectively double the protein concentration and is particularly useful when the sample is limited. The assay mixtures are allowed to equilibrate at $30°C$ for 5–10 min before starting the reaction. In Newcastle, usually only one cuvette is assayed at a time, and a reaction mix is made up in each individual cuvette. In Melbourne, typically four to six cuvettes are assayed at once: one cuvette without mitochondrial extract as a blank, the others with samples. A reaction mix sufficient for all samples to be assayed is prepared and aliquoted into the individual cuvettes. For the assay of complex IV, only two cuvettes are assayed together (see the following sections).

1. Stock Solutions for Enzyme Assays

250-mM potassium phosphate buffer, 50-mM $MgCl_2$, pH 7.2

200-mM potassium phosphate, pH 7.0

500-mM potassium phosphate, pH 7.4

50-mM potassium phosphate, 0.1-mM ethylenediaminetetraacetic acid (EDTA), pH 7.4

0.4-M Tris-HCO_3, 10-mM EGTA, pH 8.0

1-M Tris-HCl, pH 8.0

1-M potassium cyanide (KCN)

1-mg/ml antimycin A in ethanol

1-mg/ml rotenone in ethanol

1-mM oligomycin in ethanol

25-mg/ml BSA (fatty acid free, Fraction V)

13-mM reduced nicotinamide adenine dinucleotide disodium salt (NADH)

65-mM ubiquinone$_1$ in ethanol

1-M sodium succinate

5-mM 2,6-dichlorophenolindophenol (DCPIP)

2.5-mM cytochrome *c* from horse heart, type VI

30-mM *n*-dodecyl-β-D-maltoside

0.1-M phosphoenolpyruvate, monopotassium salt (PEP)

0.1-M ATP, adjusted to pH 7.2 with 2.0-M KOH

1-M magnesium chloride (MgCl$_2$)

10-mg/ml lactate dehydrogenase (LDH) from hog muscle

10-mg/ml pyruvate kinase (PK) from rabbit muscle

L-ascorbic acid, solid

Potassium hexacyanoferrate (K$_3$Fe(CN)$_6$), solid

5-mM acetyl coenzyme A

10-mM 5,5′-dithiobis-(2-nitrobenzoic acid) (DTNB)

50-mM oxaloacetate, pH 7.2 (freshly prepared)

10% Triton X-100 (w/v)

35-mM ubiquinol$_2$ (UQ$_2$·H$_2$)

  Ubiquinol$_2$ is prepared by dissolving ubiquinone$_2$ (10 μmol) in 1 ml of ethanol, acidified to pH 2 with 6-M HCl. The quinone is reduced with excess solid sodium borohydride. Ubiquinol is then extracted into diethylether:cyclohexane (2:1, v/v) and evaporated to dryness under nitrogen gas, dissolved in 1 ml of ethanol acidified to pH 2 with HCl. This solution is stable at $-70\,^{\circ}$C or lower for 1 year.

10-mM decylbenzylquinol (DB·H$_2$)

  Reduced decylbenzylquinone can be prepared just prior to assay. Add 250 μl of 10-mM decylbenzylquinone (3.22 mg/ml in ethanol) to a few grains of potassium borohydride (KBH$_4$). Acidify the mixture by adding 10 μl of 0.1-M HCl in ethanol. Mix on a vortex mixer until the yellow color disappears (3–4 min). Add 10 μl of 3-M HCl in ethanol to stabilize the DB·H$_2$, centrifuge briefly at top speed in a benchtop centrifuge to pellet the excess KBH$_4$, and then transfer the DB·H$_2$ to a fresh tube. Use on the day of preparation.

Reduced cytochrome *c*

  This can be prepared by reducing oxidized cytochrome *c*. Add a few grains of ascorbic acid to 0.5 ml of 2.5-mM cytochrome *c*. The ascorbic acid is then removed from the mixture by gel filtration on a Sephadex G-25 column equilibrated with 50-mM potassium phosphate/0.1-mM EDTA, pH 7.4. The colored fraction is collected and stored frozen at $-20\,^{\circ}C$ under nitrogen. This preparation is stable for at least 1 month. Just prior to performing the assays, the concentration of reduced cytochrome c can be estimated by measuring the absorbance at 550 nm (extinction coefficient for cytochrome c at 550 nm is $18.7\ mM^{-1}cm^{-1}$). To determine if the preparation is still usable, that is not too oxidized, the absorbance at 550 nm and at 565 nm is read and the ratio of A_{550}/A_{565} calculated. If this is <6, the preparation is too oxidized to use, and a fresh preparation must be made.

2. Complex I

Complex I deficiency is the most common RC defect and frequently presents as Leigh syndrome or lethal infantile mitochondrial disease (Kirby *et al.*, 1999). Causative mutations in all 7 mtDNA genes (Kirby *et al.*, 2004a; Lebon *et al.*, 2003; McFarland *et al.*, 2004), in 9 of the 36 nuclear DNA-encoded subunit genes [discussed in Kirby *et al.* (2004b)], and in an assembly factor gene, *B17.2L* (Ogilvie *et al.*, 2005) have been reported. Mutations in the *MTTL1* gene (tRNA$^{\text{Leu(UUR)}}$) are also associated with complex I deficiency. Mitochondrial NADH:ubiquinone oxidoreductase (complex I) has been considered difficult to assay, largely because of the inaccessibility of the complex in the inner mitochondrial membrane and the insolubility of the natural substrate, ubiquinone (coenzyme Q). Coenzyme Q is a member of a group of quinine-containing compounds that have long hydrophobic side chains consisting of varying numbers of isoprenoid groups. The most common form in mammals has 10 isoprenoid units (coenzyme Q_{10}). The hydrocarbon tail makes it largely insoluble in aqueous *in vitro* assay systems. The availability of soluble short chain coenzyme Q analogues as substrates, along with improved methods of sample preparation, has enhanced the reliability of complex I assays. The gentle disruption techniques needed for assay of other RC enzymes do not achieve maximal activity of complex I, and harsher methods of lysis, such as sonication or hypotonic shock, are necessary. Illustrative data comparing methods to disrupt mitochondrial membranes for the assay of complex I activity are shown in Table I (Birch-Machin and Turnbull, 2001).

Complex I is inhibited by rotenone, but there is considerable rotenone-insensitive "background" activity in many tissues (Birch-Machin *et al.*, 1994). It has been considered in the past that complex I cannot be measured reliably in cultured cells due to high rotenone-insensitive activity (Chretien *et al.*, 1994; Faivre *et al.*, 2000). However, the use of hypotonic lysis in the preparation of isolated mitochondria from cultured cells minimizes the rotenone-insensitive activity (Lowerson *et al.*, 1992). The hypotonic treatment involves a wash with 25-mM potassium phosphate/5-mM $MgCl_2$, pH 7.2, followed by centrifugation at $11,000 \times g_{av}$. Crude mitochondrial extracts are resuspended in the hypotonic wash buffer and frozen and thawed three times prior to assay. For tissue homogenates, hypotonic treatment or sonication

Table I

Assessment of Methods to Achieve Mitochondrial Membrane Disruption: Effect on Complex I Activity in Human Muscle Mitochondria

Mitochondrial membrane treatment	Complex I activity[a]
Freeze-thaw in hypotonic media	100 (90)
Freeze-thaw in hypotonic media + detergent treatment	94.8 ± 14.8 (91)
Sonication in isotonic media + detergent treatment	92.0 ± 9.0 (87)
Sonication in isotonic media	71.5 ± 17.1 (80)
Hypotonic shock	60.0 ± 25.3 (83)
Freeze-thaw in isotonic media	44.0 ± 13.4 (64)

[a]Complex I activities are expressed as percentage values, with those obtained by freeze-thawing in hypotonic media representing 100%. The results given represent mean ± SD for five samples, with the figures in parentheses representing the average percentage of total NADH:ubiquinone oxidoreductase activity that is rotenone sensitive (Birch-Machin and Turnbull, 2001).

can be used to disrupt the mitochondrial membranes for complex I measurement. In Newcastle, tissue homogenates are diluted in hypotonic buffer (Medium D) and assayed in this same buffer. In Melbourne, the homogenates are typically sonicated with a microtip, 5 × 6 pulses at setting 3, 30% duty cycle on a Branson sonifier. Sonication decreases the activities of the other RC enzymes, so a separate aliquot is reserved for sonication and complex I measurement. Data relating to rotenone sensitivity of total NADH:ubiquinone oxidoreductase activity have been published previously by both laboratories and are comparable, with reported rotenone sensitivities of 78–84% for skeletal muscle mitochondrial fractions, 48–55% for liver, and 52–54% for fibroblasts (Birch-Machin and Turnbull 2001; Kirby *et al.*, 1999).

It is worth emphasizing the importance of a reliable spectrophotometric assay of complex I activity since there is no available histochemical assay to assess enzyme activity in tissue sections.

Complex I is measured as NADH:ubiquinone oxidoreductase.

Reaction

$$NADH + H^+ + Ubiquinone_1 \longrightarrow NAD^+ + Ubiquinone_1 \cdot H_2$$

Reaction conditions

25-mM potassium phosphate buffer, 5-mM $MgCl_2$, pH 7.2
2-mM KCN
2.5-mg/ml BSA
0.13-mM NADH
2-μg/ml antimycin A
65-μM ubiquinone$_1$
±2 μg/ml rotenone in ethanol

Sample

Mitochondrial extract diluted in hypotonic buffer (Medium D) or sonicated as described above and frozen in liquid nitrogen or dry ice/ethanol slurry and subjected to three rounds of freeze/thaw.

Procedure

Make up an assay reagent sufficient for ten 1-ml cuvettes consisting of:

1 ml of 250-mM potassium phosphate, 50-mM $MgCl_2$, pH 7.2

1 ml of 25-mg/ml BSA

1 ml of 13-mM NADH

20 μl of 1-M KCN

20 μl of 1-mg/ml antimycin A

10 μl of 65-mM ubiquinone$_1$

Complex I-specific activity is measured by following the decrease in absorbance due to the oxidation of NADH at 340 nm with 425 nm as the reference wavelength (extinction coefficient for NADH $= 6.81$ mM^{-1}cm^{-1}, to account for the contribution of ubiquinone$_1$ to the absorbance at 340 nm). Add 305 (for 1-ml reaction) or 152.5 μl (for 0.5-ml reaction) reagent mix, make up to 1 or 0.5 ml with water in each cuvette, allowing for addition of sample volume. Equilibrate cuvettes at 30 °C for 5–10 min and then record the absorbance change for about 1 min to ensure that the baseline is stable. Add mitochondria (20–50 μg of protein) and measure the NADH:ubiquinone oxidoreductase activity for 3–5 min before the addition of rotenone (2 μl of 2-μg/ml rotenone/ml). Then, measure the activity for an additional 3 min. If there is sufficient sample, this assay can also be performed using two separate cuvettes in parallel, one with and one without rotenone. Complex I activity is the rotenone-sensitive NADH:ubiquinone oxidoreductase activity.

3. Complex II

Deficiency of complex II is rare, but several cases have been reported (Rustin *et al.*, 2002). All four subunits of complex II are encoded by nuclear DNA. Mutations in subunit A are associated predominantly with Leigh syndrome (Bourgeron *et al.*, 1995). The activities of the other RC complexes are often referenced to complex II activity. As noted earlier, complex II loses activity most rapidly after death and on inappropriate storage, and we consider its activity as a marker of the integrity of the tissue and the reliability of RC enzymology (Thorburn *et al.*, 2004). As for complex I, hypotonic lysis maximizes activity.

Complex II can be measured as succinate:ubiquinone$_1$ oxidoreductase, linked to the artificial electron acceptor DCPIP.

Reactions

$$\text{Succinate} + \text{Ubiquinone}_1 \longrightarrow \text{Fumarate} + \text{Ubiquinone}_1 \cdot H_2$$

$$\text{Ubiquinone}_1 \cdot H_2 + \text{DCPIP} \longrightarrow \text{Ubiquinone}_1 + \text{DCPIP} \cdot H_2$$

Reaction conditions

25-mM potassium phosphate buffer, 5-mM MgCl$_2$, pH 7.2

20-mM sodium succinate

50-μm DCPIP

2-mM KCN

2-μg/ml antimycin A

2-μg/ml rotenone in ethanol

65-μM ubiquinone$_1$

Sample

Mitochondrial extract diluted in hypotonic buffer (Medium D) and frozen in liquid nitrogen or dry ice/ethanol slurry and subjected to three rounds of freeze/thaw.

Procedure

Make up the assay reagent in a 1-ml cuvette. If using multiple cuvettes, make up an assay reagent sufficient for ten 1-ml cuvettes consisting of:

1 ml of 250-mM potassium phosphate, 50-mM MgCl$_2$, pH 7.2

200 μl of 1-M sodium succinate

100 μl of 5-mM DCPIP

20 μl of 1-M KCN

20 μl of 1-mg/ml antimycin A

20 μl of 1-mg/ml rotenone

Complex II-specific activity is measured by following the decrease in absorbance due to the oxidation of DCPIP at 600 nm (extinction coefficient for DCPIP = 19.1 mM^{-1}cm^{-1}). Add 136 (for 1-ml reaction) or 68 μl (for 0.5-ml reaction) assay reagent, and bring the volume up to 1 ml or 0.5 ml in each cuvette with water, allowing for addition of sample volume. Equilibrate cuvettes at 30°C for 5–10 min, and then preincubate the mitochondria (10–50 μg protein) for 10 min at 30°C. Record the baseline rate for 3 min. The reaction is started by adding 1 μl/ml of 65-mM ubiquinone$_1$, and the enzyme-catalyzed reduction of DCPIP is measured for 3–5 min.

4. Complex III

Isolated complex III deficiency is also rare. Mutations in the mtDNA-encoded cytochrome *b* (*MTCYB*) gene are associated with complex III deficiency (Schon, 2000), with patients typically presenting in adulthood with myopathy, rhabdomyolysis, and exercise intolerance (Andreu *et al.*, 1999). A mutation in the nuclear-encoded *UQCRB* subunit gene causes hypoglycemia and lactic acidosis (Haut *et al.*, 2003). Mutations in BCS1L, a complex III assembly factor, are associated with complex III deficiency and GRACILE syndrome (growth retardation, Fanconi-type aminoaciduria, cholestasis, iron overload, lactic acidosis, early death) (de Lonlay *et al.*, 2001; Visapaa *et al.*, 2002). Studies using both human and mouse cell lines harboring deleterious *MTCYB* mutations have demonstrated a structural dependence among complexes I and III, highlighting the need for a fully assembled respiratory complex III for the stability and activity of complex I (Acin-Perez *et al.*, 2004). Some patients with MTCYB mutations have very low activities of both complexes I and III, again highlighting the need to measure complex activities in isolation rather than using linked assays (Blakely *et al.*, 2005).

Each of the contributing laboratories measures complex III slightly differently. The main difference is the use different ubiquinol analogues (ubiquinol$_2$ in Newcastle and decylbenzylquinol in Melbourne) as a substrate. We will describe both methods. For the complex III assay in both laboratories, the K_m for the substrate and the K_i for the product are similar, so the reaction rate decays pseudoexponentially during the assay as the substrate is oxidized. Therefore, the activity is described by a first-order rate constant ($K \cdot min^{-1} \cdot mg^{-1}$) rather than an initial rate. The use of the detergent *n*-dodecyl-β-D-maltoside in the reaction mixtures maximizes enzyme activity.

a. Ubiquinol$_2$:Cytochrome c Reductase
Reaction

$$UQ_2 \cdot H_2 + \text{Cytochrome } c \text{ (oxidized)} \longrightarrow UQ_2 + \text{Cytochrome } c \text{ (reduced)}$$

Reaction conditions

25-mM potassium phosphate buffer, 5-mM MgCl$_2$, pH 7.2
2.5-mg/ml BSA
2-mM KCN
2-μg/ml rotenone
0.6-mM *n*-dodecyl-β-D-maltoside
15-μM cytochrome *c* (oxidized)
35-μM ubiquinol$_2$

Sample

Mitochondrial extract frozen in liquid nitrogen or dry ice/ethanol slurry and subjected to three rounds of freeze/thaw.

Procedure

Make up the assay reagent in a 1-ml cuvette. If using multiple cuvettes, make up an assay reagent sufficient for ten 1-ml cuvettes consisting of:

1 ml of 250-mM potassium phosphate, 50-mM MgCl$_2$, pH 7.2
1 ml of 25-mg/ml BSA
200 μl of 30-mM *n*-dodecyl-β-D-maltoside
60 μl of 2.5-mM cytochrome *c*
20 μl of 1-M KCN

20 μl of 1-mg/ml rotenone

10 μl of 35-mM ubiquinol$_2$

Complex III-specific activity is measured by following the increase in absorbance due to the reduction of cytochrome c at 550 nm, with 580 nm as the reference wavelength (extinction coefficient for cytochrome c = 18.7 mM^{-1}cm^{-1}). Add 241 (for 1-ml reaction) or 120.5-μl (for 0.5-ml reaction) assay reagent, and bring the volume up to 1 or 0.5 ml in each cuvette with water, allowing for addition of sample volume. Equilibrate cuvettes at 30°C for 5–10 min, and then record the baseline rate for 1 min. The reaction is started by adding mitochondria (5–20 μg of protein) and the enzyme-catalyzed reduction of cytchrome c is measured for 3 min. The increase in absorbance rapidly becomes nonlinear and activity is expressed as an apparent first-order rate constant after reduction of the remaining cytochrome c with a few grains of ascorbate (see below).

b. Decylbenzylquinol:Cytochrome c Reductase
Reaction

$$DB \cdot H_2 + \text{Cytochrome } c \text{ (oxidized)} \longrightarrow DB + \text{Cytochrome } c \text{ (reduced)}$$

Reaction conditions

50-mM potassium phosphate, pH 7.2

1-mM *n*-dodecylmaltoside

1-mM KCN

1-μg/ml rotenone

100-μM DB·H$_2$

15-μM cytochrome c (oxidized)

0.1% BSA

Sample

Mitochondrial extract frozen in liquid nitrogen or dry ice/ethanol slurry and subjected to three rounds of freeze/thaw.

Procedure

Make up the assay reagent without DB·H$_2$ and cytochrome c in a 1-ml cuvette. If using multiple cuvettes, make up an assay reagent sufficient for ten 1-ml cuvettes consisting of:

1 ml of 500-mM potassium phosphate, pH 7.4

400 μl of 25-mg/ml BSA

330 μl of 30-mM *n*-dodecyl-β-D-maltoside

100 μl of 1-mg/ml rotenone

10 μl of 1-M KCN

Complex III-specific activity is measured by following the increase in absorbance due to the reduction of cytochrome c at 550 nm (extinction coefficient for cytochrome $c = 18.7 \, \text{mM}^{-1}\text{cm}^{-1}$). Add 184 (for 1-ml reaction) or 92-μl (for 0.5-ml reaction) assay reagent, and bring the volume up to 1 or 0.5 ml in each cuvette with water, allowing for addition of sample volume. After equilibration, start the reaction by addition of mitochondrial extract (5–20 μg of protein), then 10-mM DB·H_2 (10 μl/ml) and 2.5-mM cytochrome c (20 μl/ml). The enzyme-catalyzed reduction of cytchrome c is measured for 3 min. The increase in absorbance rapidly becomes nonlinear and activity is expressed as an apparent first-order rate constant after reduction of the remaining cytochrome c with a few grains of ascorbate. At the completion of the time course, add a few grains of L-ascorbic acid to the cuvette, mix, and read the absorbance at 550 nm after ~1 min. The final absorbance is used in the rate constant calculation. There is a significant blank rate constant, which is determined by running the reaction without mitochondrial extract. The rate constant for the blank then must be subtracted from the rate constants determined for the samples.

5. Complex IV

Isolated Complex IV deficiency is the second most prevalent RC defect described (Shoubridge, 2001) and is a frequent cause of Leigh syndrome (Rahman *et al.*, 1996). No causative mutations have been reported in the nuclear DNA-encoded subunits, and mutations in mtDNA-encoded subunits are rare causes (Shoubridge, 2001). Most cases are due to defects in the assembly of the multisubunit complex IV molecule, with mutations in genes encoding assembly factors such as *SURF1* (Tiranti *et al.*, 1998; Zhu *et al.*, 1998), *SCO1* (Valnot *et al.*, 2000a), *SCO2* (Papadopoulou *et al.*, 1999), *COX10* (Valnot *et al.*, 2000b), and *COX15* (Antonicka *et al.*, 2003) or in *LRPPRC*, a gene putatively involved in mtDNA expression (Mootha *et al.*, 2003). Complex IV deficiency can be detected histochemically in muscle as COX-deficient fibers. Heteroplasmic tRNA mutations and single, large-scale mtDNA deletions often express a mosaic pattern of COX-deficient fibers in muscle. Since we are measuring an average activity, histochemically apparent complex IV deficiency is not always detectable by direct enzyme assay in tissue homogenates. In contrast, global histochemical COX defects should be detected by the biochemical assay.

As with the complex III assay, in the complex IV assay, the K_m for the substrate and the K_i for the product are similar, so the reaction rate decays pseudoexponentially during the assay as the substrate is oxidized. Therefore the activity is described by a first-order rate constant ($K \cdot \text{min}^{-1} \cdot \text{mg}^{-1}$) rather than an initial rate. The use of the detergent n-dodecyl-β-D-maltoside in the reaction mixtures maximizes enzyme activity.

Reaction

$$\text{Cytochrome } c \text{ (reduced)} + O_2 \longrightarrow \text{Cytochrome } c \text{ (oxidized)} + H_2O$$

Reaction conditions

   20-mM potassium phosphate, pH 7.0

   0.45-mM *n*-dodecyl-β-D-maltoside

   15-μM cytochrome *c* (reduced)

Sample

   Mitochondrial extract frozen in liquid nitrogen or dry ice/ethanol slurry and subjected to three rounds of freeze/thaw.

Procedure

   For 1-ml samples, add 100-μl buffer and 15-μl *n*-dodecyl-β-D-maltoside to a 1.0-ml cuvette and bring volume to 1 ml with water, allowing for the sample volume. For 0.5-ml samples, add 50-μl buffer and 7.5-μl *n*-dodecyl-β-D-maltoside to a 0.5-ml cuvette, and bring the volume to 0.5 ml with water, allowing for sample volume. Equilibrate cuvettes at 30°C for 5–10 min. Add cytochrome *c* (reduced) and record the nonenzymatic (blank) rate. Start the reaction by addition of mitochondrial extract and follow the decrease in absorbance at 550 nm for 2 min. The absorbance declines exponentially as the reaction rapidly proceeds, and 2 min is usually sufficient time to gather enough data to calculate the first-order rate constant. Because the reaction proceeds rapidly, it is best not to use all six cuvettes, but to restrict the analysis to two cuvettes at one time. At the completion of the time course, add a few grains of $K_3Fe(CN)_6$ to the cuvette, mix, and read the absorbance at 550 nm after $\sim$1 min. The final absorbance is used in the rate constant calculation.

   Assay blanks are determined by running the reaction with cytochrome *c* but without mitochondrial extract, and with mitochondrial extract but without cytochrome *c*. In contrast with the complex III assay, the blank rate constant is usually insignificant and it is not necessary to subtract it from the rate constants determined for the samples.

6. Complex V

   Complex V (F_1-ATP synthase) is less routinely assayed, although a number of deficiencies of its activity have been reported. The most frequent of these are associated with mutations in mtDNA, most often at nucleotide 8993 of the *MTATP6* gene. Leigh syndrome is a frequent presentation (de Vries *et al.*, 1993; Santorelli *et al.*, 1993; Tatuch *et al.*, 1992). A mutation in a nuclear-encoded assembly factor, ATP12, has been reported to cause complex V deficiency (De Meirleir *et al.*, 2004).

   Complex V is assayed in the reverse direction, as F_1-ATPase. The assay relies on linking the ATPase activity to NADH oxidation via conversion of phosphoenolpyruvate to pyruvate by PK and then pyruvate to lactate by LDH. Other ATPase activities can contribute a significant blank rate in some tissues, which can be minimized by using mitochondrial preparations rather than homogenates, by inhibiting Ca^{2+}-ATPase with EGTA and by determining the oligomycin-sensitive activity (see later).

Reactions

$$Mg \cdot ATP \xrightarrow{\text{ATPase}} Mg \cdot ADP + P_i$$

$$Mg \cdot ADP + Phosphonenolpyruvate \xrightarrow{\text{PK}} Mg \cdot ATP + Pyruvate$$

$$Pyruvate + NADH + H^+ \xrightarrow{\text{LDH}} Lactate + NAD^+$$

Reaction conditions

40-mM Tris–HCO$_3$, 10-mM EGTA, pH 8.0

0.2-mM NADH

2.5-mM PEP

25-μg/ml antimycin A

50-mM MgCl$_2$

0.5-mg/ml LDH

0.5-mg/ml PK

2.5-mM ATP

Sample

Mitochondrial extract frozen in liquid nitrogen or dry ice/ethanol slurry and subjected to three rounds of freeze/thaw.

Procedure

Make up the assay reagent without ATP in a 1-ml cuvette. For multiple cuvettes, prepare an assay reagent mix (sufficient for ten 1-ml cuvettes) consisting of:

1 ml of 0.4-M Tris–HCO$_3$, 10-mM EGTA pH 8.0

250 μl of 0.1-M PEP

154 μl of 13-mM NADH

25 μl of 1-mg/ml antimycin A

50 μl of 1-M MgCl$_2$

50 μl of 10-mg/ml LDH

50 μl of 10-mg/ml PK

For multiple cuvettes, add 153 (for 1-ml reaction) or 76.5 μl (for 0.5-ml reaction) reagent mix, make up to 1 or 0.5 ml in each cuvette with water, allowing for addition of ATP and sample volume. Equilibrate cuvettes at 30°C for 5–10 min.

Add ATP to a final concentration of 2.5 mM (25 μl/ml of 0.1-M ATP) and leave for 2 min to allow phosphorylation of any ADP present in the ATP solution. Start the reaction by addition of mitochondrial extract (5–20 μg). Follow the change in absorbance over 3–5 min at 340 nm (extinction coefficient for NADH = 6.22 mM^{-1}cm^{-1}). Appropriate controls for this assay are a sample blank (cuvette with ATP replaced by water), and an ATP blank (cuvette without sample).

F_1-ATPase is inhibited by oligomycin. Muscle mitochondria do not have a significant oligomycin-insensitive activity, but other samples may. The oligomycin-sensitive rate can be determined by assaying the sample with and without 2-μM oligomycin (2 μl/ml of 1-mM oligomycin). This assay is not reliable for post-600 $\times$ g$_{av}$ supernatants prepared from frozen tissues, since oligomycin sensitivity can be highly variable in such samples.

7. Complex II + III

As noted in the Introduction, the combined complex II + III (succinate: cytochrome c reductase) assay can be useful in the diagnosis of coenzyme Q deficiency. The complex II + III assay is unreliable in liver homogenates, with high and variable antimycin A-insensitive background activities.

Reaction

Succinate + Cytochrome c (oxidized) $\longrightarrow$ Fumarate + Cytochrome c (reduced)

Reaction conditions

25-mM potassium phosphate buffer, 5-mM MgCl$_2$, pH 7.2
20-mM sodium succinate
2-mM KCN
2-μg/ml rotenone
37.5-μM cytochrome c (oxidized)

Sample
Mitochondrial extract frozen in liquid nitrogen or dry ice/ethanol slurry and subjected to three rounds of freeze/thaw.
Procedure
Make up the assay reagent without cytochrome c in a 1-ml cuvette, or if using multiple cuvettes, make up an assay reagent sufficient for ten 1-ml cuvettes consisting of:

1 ml of 250-mM potassium phosphate, 50-mM MgCl$_2$, pH 7.2
200 μl of 1-M sodium succinate
20 μl of 1-M KCN
10 μl of 1-mg/ml rotenone

Complex II + III-specific activity is measured by following the increase in absorbance due to the reduction of cytochrome c at 550 nm, with 580 nm as the reference wavelength (extinction coefficient for cytochrome $c = 18.7$ mM^{-1}cm^{-1}). Add 123 (for 1-ml reaction) or 61.5-μl (for 0.5-ml reaction) assay reagent, make up to 1 or 0.5 ml with water in each cuvette, allowing for addition of sample volume and cytochrome c. Equilibrate cuvettes at 30 °C for 5–10 min then preincubate the mitochondria (10–50 μg protein) for 10 min at 30 °C. This ensures that complex II is fully activated. Record the baseline rate for 3 min. The reaction is started by adding 7.5 μl/ml of 5-mM cytochrome c (oxidized) and the enzyme-catalyzed reduction of cytochrome c is measured for 3–5 min.

8. Citrate Synthase

Citrate synthase is assayed by measuring the rate of production coenzyme A (CoA.SH) from oxaloacetate by measuring free sulfhydryl groups using the thiol reagent DTNB. DTNB reacts spontaneously with sulfhydryl groups to produce free 5-thio-2-nitrobenzoate anions, which have a yellow color and can be monitored at 412 nm.

Reactions

$$\text{Oxaolacetate} + \text{Acetyl coenzyme A} \xrightarrow{\text{citrate synthase}} \text{Citrate} + \text{CoA} \cdot \text{SH}$$

$$\text{CoA} \cdot \text{SH} + \text{DTNB} \longrightarrow \text{CoA} \cdot \text{S-}S\text{-nitrobenzoate} + \text{5-thio-2-nitrobenzoate anion}$$

Reaction conditions

100-mM Tris–HCl, pH 8.0
100-μM DTNB
50-μM acetyl coenzyme A
0.1% (w/v) Triton X-100
250-μM oxaloacetate

Sample

Mitochondrial extract frozen in liquid nitrogen or dry ice/ethanol slurry and subjected to three rounds of freeze/thaw.

Procedure

Make up the assay reagent without oxaloacetate in a 1-ml cuvette. If using multiple cuvettes, make up an assay reagent sufficient for ten 1-ml cuvettes consisting of:

1 ml of 1-M Tris–HCl, pH 8.0
100 μl of 5-mM acetyl coenzyme A
100 μl of 10-mM DTNB
100 μl of 10% Triton X-100 (w/v)

Citrate synthase-specific activity is measured by following the increase in absorbance due to the formation of 5-thio-2-nitrobenzoate anion at 412 nm (extinction coefficient = 13.6 mM^{-1} cm^{-1}). For assaying multiple cuvettes, add 130 (for 1-ml reaction) or 65-μl (for 0.5-ml reaction) assay reagent, make up to 1 ml or 0.5 ml with water in each cuvette, allowing for addition of sample volume and oxaloacetate. Equilibrate cuvettes at 30 °C for 5–10 min, and then add mitochondria (10–20 μg protein). The reaction is started by the addition of excess (250 μM) oxaloacetate (5 μl/ml of 50 mM) and followed over 3 min.

In liver, but not in skeletal muscle, cardiac muscle, or cultured cells, there is a significant background activity for citrate synthase, and its activity must be measured with and without oxaloacetate to correct for nonspecific thiolase activity. This can be significant in liver homogenates, representing up to 35% of total activity, but is negligible in other tissues (Kirby *et al.*, 1999).

III. Interpreting Results of RC Enzymology

In both our centers, the results of RC enzymology are calculated as initial rates (nmol min^{-1} mg^{-1} for complexes I, II, II + III, V, and citrate synthase) or first-order rate constants (K, min^{-1} mg^{-1} for complexes III and IV). However, in interpreting results of RC enzymology, reliance on activity alone can be misleading (Thorburn *et al.*, 2004). The results of all activities must be viewed together, for often tissue pathology may cause all activities to be reduced. Increased activities due to mitochondrial proliferation may also be misleading. Thus, for interpretation, all activities are related to citrate synthase activity and/or complex II activity rather than activity relative to protein or wet weight alone. Figure 3 shows the results of RC enzymology in a boy with isolated complex I deficiency and a mutation in the *MTTL1* gene (tRNA$^{Leu(UUR)}$) (3271T > C). Complex I is 90% of control mean relative to protein, and all other RC enzymes and citrate synthase are elevated, reflecting mitochondrial proliferation. When expressed relative to complex II, complex I is only 20% of control mean. Interpretation of results depends on the interplay of many factors, including clinical, histological, histochemical, functional, and molecular findings. There are several published diagnostic schemes (Bernier *et al.*, 2002; Walker *et al.*, 1996; Wolf and Smeitink 2002) in which RC enzyme data are incorporated into the scheme for interpretation and allocation of the patient to a definite, probable, or possible diagnosis of an RC defect.

We have studied the effects of tissue pathology by comparing RC enzymes in tissues from patients without RC complexes I–IV defects with normal controls (Hui *et al.*, 2006; Thorburn *et al.*, 2004). These "disease" controls included tissue samples from children later diagnosed with other inborn errors of metabolism (IEM) and patients with end-stage liver disease of nonmetabolic causes. The rationale for studying such samples was that they may have pathological processes

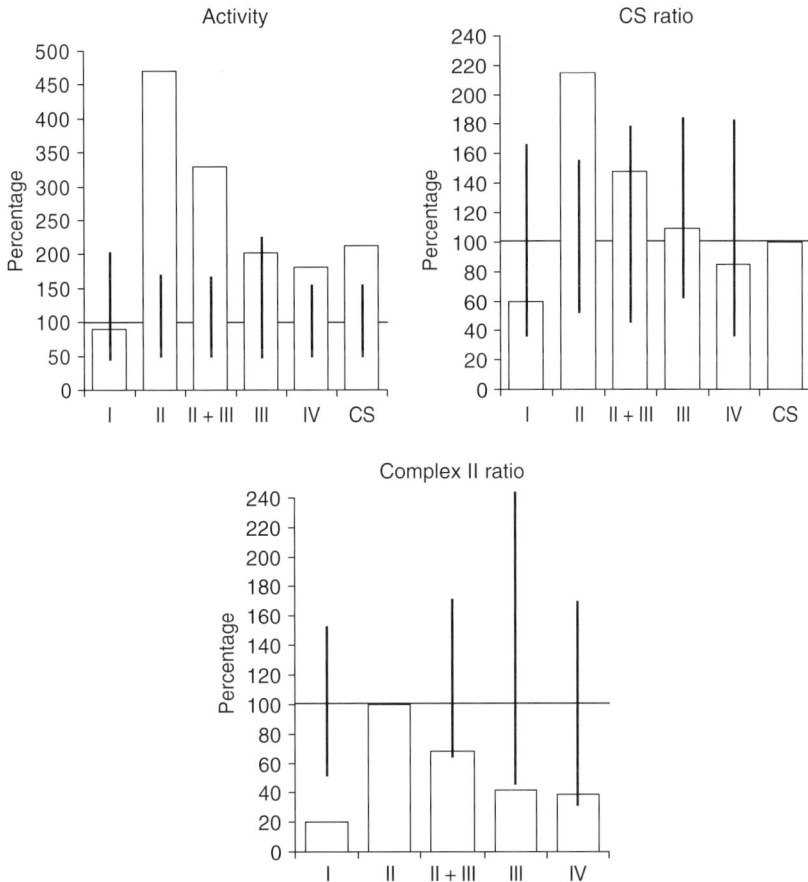

Fig. 3 Respiratory chain enzymes in skeletal muscle from a patient with the 3271T > C mutation in the *MTTL1* gene (tRNA^Leu(UUR)). Relative to protein and citrate synthase, complex I activity is within the normal range, but is 20% of control mean when expressed relative to complex II. Enzyme activities, CS ratios, and complex II ratios are expressed as percentages of the mean of controls ($n = 17$ for skeletal muscle). The observed range for the controls is represented by the vertical lines. CS, citrate synthase.

that overlap those in patients with *bona fide* RC complex I–IV defects, and secondary abnormalities of RC enzymes may be present. The RC enzymes in the IEM control tissues were compared with those in normal control tissues. The normal control muscles were from children undergoing orthopedic procedures, and the normal control livers were from pediatric organ donors. From the data obtained from the RC enzyme profiles of tissues from patients with IEMs other than RC complex I–IV defects or with nonmetabolic liver failure, it is apparent that tissue pathology can extend the normal ranges for RC enzymes, even when expressed as enzyme ratios. The possibility of secondary decreases in enzyme activity and the broadening of reference ranges in the presence of tissue pathology should be considered in interpreting RC enzyme profiles.

Acknowledgments

D.M.K. holds a National Health and Medical Research Council of Australia (NHMRC) CJ Martin Postdoctoral Training Fellowship. D.R.T. is supported by an NHMRC Senior Research Fellowship and grants from NHMRC and the Muscular Dystrophy Association (MDA) of United States. D.M.T. and R.W.T. are grateful for the financial support of the Wellcome Trust, Muscular Dystrophy Campaign, and Newcastle upon Tyne Hospitals NHS Foundation Trust.

References

Acin-Perez, R., Bayona-Bafaluy, M. P., Fernandez-Silva, P., Moreno-Loshuertos, R., Perez-Martos, A., Bruno, C., Moraes, C. T., and Enriquez, J. A. (2004). Respiratory complex III is required to maintain complex I in mammalian mitochondria. *Mol. Cell* **13,** 805–815.

Andreu, A. L., Hanna, M. G., Reichmann, H., Bruno, C., Penn, A. S., Tanji, K., Pallotti, F., Iwata, S., Bonilla, E., Lach, B., Morgan-Hughes, J., and DiMauro, S. (1999). Exercise intolerance due to mutations in the cytochrome b gene of mitochondrial DNA. *N. Engl. J. Med.* **341,** 1037–1044.

Antonicka, H., Mattman, A., Carlson, C. G., Glerum, D. M., Hoffbuhr, K. C., Leary, S. C., Kennaway, N. G., and Shoubridge, E. A. (2003). Mutations in COX15 produce a defect in the mitochondrial heme biosynthetic pathway, causing early-onset fatal hypertrophic cardiomyopathy. *Am. J. Hum. Genet.* **72,** 101–114.

Bernier, F. P., Boneh, A., Dennett, X., Chow, C. W., Cleary, M. A., and Thorburn, D. R. (2002). Diagnostic criteria for respiratory chain disorders in adults and children. *Neurology* **59,** 1406–1411.

Birch-Machin, M. A., and Turnbull, D. M. (2001). Assaying mitochondrial respiratory complex activity in mitochondria isolated from human cells and tissues. *Methods Cell Biol.* **65,** 97–117.

Birch-Machin, M. A., Briggs, H. L., Saborido, A. A., Bindoff, L. A., and Turnbull, D. M. (1994). An evaluation of the measurement of the activities of complexes I-IV in the respiratory chain of human skeletal muscle mitochondria. *Biochem. Med. Metab. Biol.* **51,** 35–42.

Blakely, E. L., Mitchell, A. L., Fisher, N., Meunier, B., Nijtmans, L. G., Schaefer, A. M., Jackson, M. J., Turnbull, D. M., and Taylor, R. W. (2005). A mitochondrial cytochrome *b* mutation causing severe respiratory chain enzyme deficiency in humans and yeast. *FEBS J.* **272,** 3583–3592.

Bourgeron, T., Rustin, P., Chretien, D., Birch-Machin, M., Bourgeois, M., Viegas-Pequignot, E., Munnich, A., and Rotig, A. (1995). Mutation of a nuclear succinate dehydrogenase gene results in mitochondrial respiratory chain deficiency. *Nat. Genet.* **11,** 144–149.

Chretien, D., Rustin, P., Bourgeron, T., Rotig, A., Saudubray, J. M., and Munnich, A. (1994). Reference charts for respiratory chain activities in human tissues. *Clin. Chim. Acta* **228,** 53–70.

Davidzon, G., Mancuso, M., Ferraris, S., Quinzii, C., Hirano, M., Peters, H. L., Kirby, D., Thorburn, D. R., and DiMauro, S. (2005). POLG mutations and Alpers syndrome. *Ann. Neurol.* **57,** 921–923.

de Lonlay, P., Valnot, I., Barrientos, A., Gorbatyuk, M., Tzagoloff, A., Taanman, J. W., Benayoun, E., Chretien, D., Kadhom, N., Lombes, A., de Baulny, H. O., Niaudet, P., *et al.* (2001). A mutant mitochondrial respiratory chain assembly protein causes complex III deficiency in patients with tubulopathy, encephalopathy and liver failure. *Nat. Genet.* **29,** 57–60.

De Meirleir, L., Seneca, S., Lissens, W., De Clercq, I., Eyskens, F., Gerlo, E., Smet, J., and Van Coster, R. (2004). Respiratory chain complex V deficiency due to a mutation in the assembly gene ATP12. *J. Med. Genet.* **41,** 120–124.

de Vries, D. D., van Engelen, B. G., Gabreels, F. J., Ruitenbeek, W., and van Oost, B. A. (1993). A second missense mutation in the mitochondrial ATPase 6 gene in Leigh's syndrome. *Ann. Neurol.* **34,** 410–412.

Faivre, L., Cormier-Daire, V., Chretien, D., Christoph von Kleist-Retzow, J., Amiel, J., Dommergues, M., Saudubray, J. M., Dumez, Y., Rotig, A., Rustin, P., and Munnich, A. (2000). Determination of enzyme activities for prenatal diagnosis of respiratory chain deficiency. *Prenat. Diagn.* **20,** 732–737.

Haut, S., Brivet, M., Touati, G., Rustin, P., Lebon, S., Garcia-Cazorla, A., Saudubray, J. M., Boutron, A., Legrand, A., and Slama, A. (2003). A deletion in the human QP-C gene causes a complex III deficiency resulting in hypoglycaemia and lactic acidosis. *Hum. Genet.* **113,** 118–122.

Hui, J., Kirby, D. M., Thorburn, D. R., and Boneh, A. (2006). Decreased activities of mitochondrial respiratory chain complexes in non-mitochondrial respiratory chain diseases. *Dev. Med. Child Neurol.* **48,** 132–136.

Janssen, A. J., Smeitink, J. A., and van den Heuvel, L. P. (2003). Some practical aspects of providing a diagnostic service for respiratory chain defects. *Ann. Clin. Biochem.* **40,** 3–8.

Kirby, D. M., Crawford, M., Cleary, M. A., Dahl, H. H., Dennett, X., and Thorburn, D. R. (1999). Respiratory chain complex I deficiency: An underdiagnosed energy generation disorder. *Neurology* **52,** 1255–1264.

Kirby, D. M., McFarland, R., Ohtake, A., Dunning, C., Ryan, M. T., Wilson, C., Ketteridge, D., Turnbull, D. M., Thorburn, D. R., and Taylor, R. W. (2004a). Mutations of the mitochondrial ND1 gene as a cause of MELAS. *J. Med. Genet.* **41,** 784–789.

Kirby, D. M., Salemi, R., Sugiana, C., Ohtake, A., Parry, L., Bell, K. M., Kirk, E. P., Boneh, A., Taylor, R. W., Dahl, H. H., Ryan, M. T., and Thorburn, D. R. (2004b). NDUFS6 mutations are a novel cause of lethal neonatal mitochondrial complex I deficiency. *J. Clin. Invest.* **114,** 837–845.

Lebon, S., Chol, M., Benit, P., Mugnier, C., Chretien, D., Giurgea, I., Kern, I., Girardin, E., Hertz-Pannier, L., de Lonlay, P., Rotig, A., Rustin, P., *et al.* (2003). Recurrent de novo mitochondrial DNA mutations in respiratory chain deficiency. *J. Med. Genet.* **40,** 896–899.

Lowerson, S. A., Taylor, L., Briggs, H. L., and Turnbull, D. M. (1992). Measurement of the activity of individual respiratory chain complexes in isolated fibroblast mitochondria. *Anal. Biochem.* **205,** 372–374.

Mandel, H., Szargel, R., Labay, V., Elpeleg, O., Saada, A., Shalata, A., Anbinder, Y., Berkowitz, D., Hartman, C., Barak, M., Eriksson, S., and Cohen, N. (2001). The deoxyguanosine kinase gene is mutated in individuals with depleted hepatocerebral mitochondrial DNA. *Nat. Genet.* **29,** 337–341.

McFarland, R., Kirby, D. M., Fowler, K. J., Ohtake, A., Ryan, M. T., Amor, D. J., Fletcher, J. M., Dixon, J. W., Collins, F. A., Turnbull, D. M., Taylor, R. W., and Thorburn, D. R. (2004). *De novo* mutations in the mitochondrial ND3 gene as a cause of infantile mitochondrial encephalopathy and complex I deficiency. *Ann. Neurol.* **55,** 58–64.

Mootha, V. K., Lepage, P., Miller, K., Bunkenborg, J., Reich, M., Hjerrild, M., Delmonte, T., Villeneuve, A., Sladek, R., Xu, F., Mitchell, G. A., Morin, C., *et al.* (2003). Identification of a gene causing human cytochrome c oxidase deficiency by integrative genomics. *Proc. Natl. Acad. Sci. USA* **100,** 605–610.

Munnich, A., and Rustin, P. (2001). Clinical spectrum and diagnosis of mitochondrial disorders. *Am. J. Med. Genet.* **106,** 4–17.

Naviaux, R. K., and Nguyen, K. V. (2004). POLG mutations associated with Alpers' syndrome and mitochondrial DNA depletion. *Ann. Neurol.* **55,** 706–712.

Niers, L., van den Heuvel, L., Trijbels, F., Sengers, R., and Smeitink, J. (2003). Prerequisites and strategies for prenatal diagnosis of respiratory chain deficiency in chorionic villi. *J. Inherit. Metab. Dis.* **26,** 647–658.

Ogilvie, I., Kennaway, N. G., and Shoubridge, E. A. (2005). A molecular chaperone for mitochondrial complex I assembly is mutated in a progressive encephalopathy. *J. Clin. Invest.* **115,** 2784–2792.

Papadopoulou, L. C., Sue, C. M., Davidson, M. M., Tanji, K., Nishino, I., Sadlock, J. E., Krishna, S., Walker, W., Selby, J., Glerum, D. M., Coster, R. V., Lyon, G., *et al.* (1999). Fatal infantile cardioencephalomyopathy with COX deficiency and mutations in SCO2, a COX assembly gene. *Nat. Genet.* **23,** 333–337.

Rahman, S., Blok, R. B., Dahl, H. H., Danks, D. M., Kirby, D. M., Chow, C. W., Christodoulou, J., and Thorburn, D. R. (1996). Leigh syndrome: Clinical features and biochemical and DNA abnormalities. *Ann. Neurol.* **39,** 343–351.

Rustin, P., Munnich, A., and Rotig, A. (2002). Succinate dehydrogenase and human diseases: New insights into a well-known enzyme. *Eur. J. Hum. Genet.* **10,** 289–291.

Santorelli, F. M., Shanske, S., Macaya, A., DeVivo, D. C., and DiMauro, S. (1993). The mutation at nt 8993 of mitochondrial DNA is a common cause of Leigh's syndrome. *Ann. Neurol.* **34,** 827–834.

Scholte, H. R., and Trijbels, J. M. (1995). Isolated mitochondria from frozen muscle have limited value in diagnostics. *Eur. J. Pediatr.* **154,** 80.

Schon, E. A. (2000). Mitochondrial genetics and disease. *Trends Biochem. Sci.* **25,** 555–560.

Shoubridge, E. A. (2001). Cytochrome c oxidase deficiency. *Am. J. Med. Genet.* **106,** 46–52.

Skladal, D., Halliday, J., and Thorburn, D. R. (2003). Minimum birth prevalence of mitochondrial respiratory chain disorders in children. *Brain* **126,** 1905–1912.

Slama, A., Giurgea, I., Debrey, D., Bridoux, D., de Lonlay, P., Levy, P., Chretien, D., Brivet, M., Legrand, A., Rustin, P., Munnich, A., and Rotig, A. (2005). Deoxyguanosine kinase mutations and combined deficiencies of the mitochondrial respiratory chain in patients with hepatic involvement. *Mol. Genet. Metab.* **86,** 462–465.

Tatuch, Y., Christodoulou, J., Feigenbaum, A., Clarke, J. T., Wherret, J., Smith, C., Rudd, N., Petrova-Benedict, R., and Robinson, B. H. (1992). Heteroplasmic mtDNA mutation (T—G) at 8993 can cause Leigh disease when the percentage of abnormal mtDNA is high. *Am. J. Hum. Genet.* **50,** 852–858.

Taylor, R. W., Birch-Machin, M. A., Bartlett, K., Lowerson, S. A., and Turnbull, D. M. (1994). The control of mitochondrial oxidations by complex III in rat muscle and liver mitochondria. Implications for our understanding of mitochondrial cytopathies in man. *J. Biol. Chem.* **269,** 3523–3528.

Thorburn, D. R., and Smeitink, J. (2001). Diagnosis of mitochondrial disorders: Clinical and biochemical approach. *J. Inherit. Metab. Dis.* **24,** 312–316.

Thorburn, D. R., Chow, C. W., and Kirby, D. M. (2004). Respiratory chain enzyme analysis in muscle and liver. *Mitochondrion* **4,** 363–375.

Tiranti, V., Hoertnagel, K., Carrozzo, R., Galimberti, C., Munaro, M., Granatiero, M., Zelante, L., Gasparini, P., Marzella, R., Rocchi, M., Bayona-Bafaluy, M. P., Enriquez, J. A., *et al.* (1998). Mutations of SURF-1 in Leigh disease associated with cytochrome c oxidase deficiency. *Am. J. Hum. Genet.* **63,** 1609–1621.

Valnot, I., Osmond, S., Gigarel, N., Mehaye, B., Amiel, J., Cormier-Daire, V., Munnich, A., Bonnefont, J. P., Rustin, P., and Rotig, A. (2000a). Mutations of the SCO1 gene in mitochondrial cytochrome c oxidase deficiency with neonatal-onset hepatic failure and encephalopathy. *Am. J. Hum. Genet.* **67,** 1104–1109.

Valnot, I., von Kleist-Retzow, J. C., Barrientos, A., Gorbatyuk, M., Taanman, J. W., Mehaye, B., Rustin, P., Tzagoloff, A., Munnich, A., and Rotig, A. (2000b). A mutation in the human heme A: Farnesyltransferase gene (COX10) causes cytochrome c oxidase deficiency. *Hum. Mol. Genet.* **9,** 1245–1249.

Visapaa, I., Fellman, V., Vesa, J., Dasvarma, A., Hutton, J. L., Kumar, V., Payne, G. S., Makarow, M., Van Coster, R., Taylor, R. W., Turnbull, D. M., Suomalainen, A., *et al.* (2002). GRACILE syndrome, a lethal metabolic disorder with iron overload, is caused by a point mutation in BCS1L. *Am. J. Hum. Genet.* **71,** 863–876.

Walker, U. A., Collins, S., and Byrne, E. (1996). Respiratory chain encephalomyopathies: A diagnostic classification. *Eur. Neurol.* **36,** 260–267.

Wolf, N. I., and Smeitink, J. A. (2002). Mitochondrial disorders: A proposal for consensus diagnostic criteria in infants and children. *Neurology* **59,** 1402–1405.

Zheng, X. X., Shoffner, J. M., Voljavec, A. S., and Wallace, D. C. (1990). Evaluation of procedures for assaying oxidative phosphorylation enzyme activities in mitochondrial myopathy muscle biopsies. *Biochim. Biophys. Acta* **1019,** 1–10.

Zhu, Z., Yao, J., Johns, T., Fu, K., De Bie, I., Macmillan, C., Cuthbert, A. P., Newbold, R. F., Wang, J., Chevrette, M., Brown, G. K., Brown, R. M., *et al.* (1998). SURF1, encoding a factor involved in the biogenesis of cytochrome c oxidase, is mutated in Leigh syndrome. *Nat. Genet.* **20,** 337–343.

CHAPTER 5

Polarographic Assays of Respiratory Chain Complex Activity*

Gaetano Villani[†] and Giuseppe Attardi

Division of Biology
California Institute of Technology
Pasadena, California 91125

I. Introduction

Traditionally, studies on oxidative phosphorylation (OXPHOS) have been carried out on isolated mitochondria. In recent years, however, the increasing evidence

*Dedicated to the memory of Ferruccio Guerrieri.
[†]Current address: Department of Medical Biochemistry and Biology, University of Bari, Bari, Italy.

METHODS IN CELL BIOLOGY, VOL. 80
Copyright 2001, Elsevier Inc. All rights reserved.

0091-679X/07 $35.00
DOI: 10.1016/S0091-679X(06)80005-1

that interactions of these organelles with the cytoskeleton, other organelles, and the cytosol play a significant role in the various OXPHOS reactions has created the need for new methods that can probe these reactions under conditions approximating more closely the *in vivo* situation. This need has been accentuated by the discovery since the late 1980s of a large variety of mitochondrial DNA mutations that cause diseases in humans and by the increasing evidence that a reduction in the rate of assembly and in the activities of the OXPHOS apparatus is associated with aging or neurodegenerative diseases (Schon *et al.*, 1997; Wallace, 1992). In particular, the recognition of threshold effects in the capacity of a mitochondrial DNA (mtDNA) mutation to produce a dysfunction in the presence of varying amounts of wild-type mtDNA has stimulated a strong interest in determining the degree of control that a particular OXPHOS step exerts on the rate of mitochondrial respiration and/or ATP production *in vivo*. This chapter describes an approach that has been developed to determine the control that cytochrome *c* oxidase (COX) exerts on the rate of endogenous respiration in intact cells (Villani and Attardi, 1997; Villani *et al.*, 1998), as well as a method for analysis of OXPHOS in permeabilized cells.

The experimental procedures described are based on polarographic measurements of oxygen consumption by means of Clark-type oxygen electrodes. The equipment and its utilization have been described in detail (Hofhaus *et al.*, 1996; Trounce *et al.*, 1996), and new generations of oxygen electrodes and oxygraphic chambers, as well as of software for the automatic analysis of respiratory kinetics, are now available. With any chosen reaction chamber, it is important to measure the exact mixing volume. This corresponds to the minimal volume required to fill the chamber up to the bottom of the channel port of the stopper. This value is important for subsequent calculations. The volume introduced into the chamber has to exceed slightly the mixing volume in order to avoid formation of air bubbles from the vortex created by stirring.

II. Measurements of Endogenous Respiration in Intact Cells

A. Media and Reagents

Dulbecco's modified Eagle's medium (DMEM) lacking glucose, supplemented with 5% fetal bovine serum (FBS)

Tris-based, Mg^{2+}-, Ca^{2+}-deficient (TD) buffer: 0.137 M NaCl, 5 mM KCl, 0.7 mM Na_2HPO_4, 25 mM Tris–HCl, pH 7.4, at 25°C

B. Procedure

The measurements are performed at 37°C on exponentially growing cell cultures change the media in cultured cells the day before measurements are made. Allow the temperature of the measurement medium to equilibrate under stirring in the reaction chamber: this will result in a constant horizontal tracing on the recorder paper. Wash with TD buffer and collect cells by centrifugation (for cells growing in suspension) or by trypsinization and centrifugation (for cells growing on plates). The final cell pellet is resuspended in a portion of the equilibrated medium from the chamber by gently pipetting. Cells are then transferred into the oxygraphic chamber. Close the chamber and record the endogenous oxygen consumption rate. After obtaining a reliable rate, open the chamber and remove one sample for cell counting and four 20- to 50-μl samples for total protein determination.

C. Comments

It has been reported (Villani and Attardi, 1997) that the osteosarcoma-derived 143B. TK$^-$ cell line respires in TD buffer at the same rate as in DMEM lacking glucose (Fig. 1A). When the effects of inhibitors or uncouplers of OXPHOS have to be analyzed, the use of TD buffer is recommended.

The oxygen consumption rate, expressed relative to the cell number, tends to decrease with increasing cell density in the culture. This is due, at least in part, to a decrease in the size of the cells, as shown by the fact that the respiratory rate,

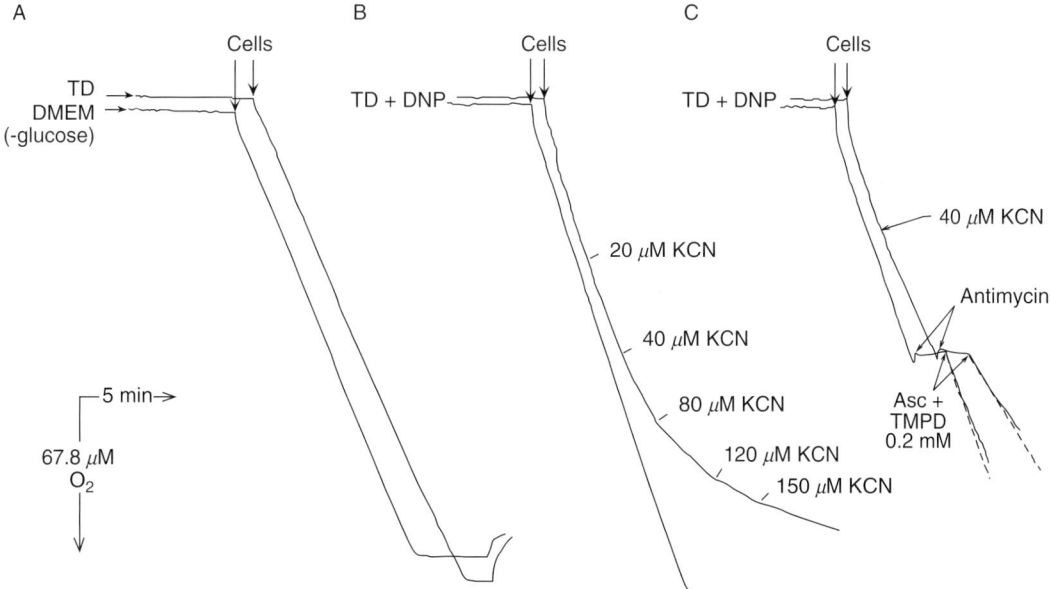

Fig. 1 Polarographic tracings of oxygen consumption in 143B. TK$^-$ cells. (A) Endogenous respiration in TD buffer or DMEM-glucose medium. (B) Cyanide titration of uncoupled endogenous respiration; KCN was added sequentially at different concentrations in one of the two chambers, as indicated. (C) Inhibitory effect of 40 μM KCN on the uncoupled endogenous respiration rate and on the initial rate (dashed lines) of 10 mM ascorbate (Asc) + 0.2 mM TMPD-supported oxygen consumption.

normalized per milligram of protein, remains fairly constant at different cell densities (Villani *et al.*, 1998).

The mitochondrial respiration rate should be corrected for KCN (2–6 mM)-resistant oxygen consumption rate (see next section for preparation of KCN solution), which reflects oxygen metabolism from other pathways. This includes superoxide radical formation that normally does not exceed 1–2% of total oxygen consumption (Ernster, 1986). An increase beyond such a value could be taken as indicative of cellular oxidative stress.

Additional information can be obtained by the analysis of the endogenous respiration rate of intact cells in the presence of membrane-permeant inhibitors or uncouplers. As an example, the degree of coupling of endogenous respiration to ATP production can be estimated by inhibiting the mitochondrial ATPase with oligomycin. This will result in a decrease in the respiratory rate due to the establishment of a maximal steady-state electrochemical membrane potential. The respiration rate can then be restored by the addition of mitochondrial uncouplers. *Note*: Oligomycin should be used at the minimal concentration that results in the maximal ratio between uncoupled and oligomycin-inhibited respiration rate.

III. KCN Titration of COX Activity in Intact Cells

A. Principle

The application of the metabolic control theory (Heinrich and Rapoport, 1974; Kacser and Burns, 1973) to the study of the mitochondrial metabolism can be a valuable approach for determining the degree of control exerted by different OXPHOS steps on the rate of mitochondrial respiration (Groen *et al.*, 1982) and for identifying and quantifying enzymatic defects in the OXPHOS machinery caused by mitochondrial or nuclear DNA mutations. The method used most often to analyze the control that a given enzymatic step exerts on the overall metabolic flux is the inhibitor titration technique (Groen *et al.*, 1982). In this approach, a specific inhibitor is utilized to titrate an enzymatic activity both as an isolated step and as a step integrated in the metabolic pathway. In this way, one can measure, for each concentration of inhibitor used, the percentage of inhibition of the isolated step and to what degree this specific inhibition affects the overall pathway.

This method can be applied to intact cells to study the control exerted by cytochrome *c* oxidase on endogenous respiration. In particular, the isolated COX activity is titrated with KCN after inhibiting the upstream segment of the respiratory chain with antimycin A and reducing the endogenous cytochrome *c* pool with tetramethyl-*p*-phenylene diamine (TMPD), a membrane-permeant, one-equivalent electron donor, in the presence of an excess of ascorbate as the primary reducing agent. The titration curve thus determined is then compared to the KCN titration curve of the endogenous respiration, where the COX activity participates as a respiratory chain-integrated step (Villani and Attardi, 1997).

To avoid any influence of proton cycling and OXPHOS on the respiratory flux, the mitochondrial uncoupler dinitrophenol (DNP) is routinely utilized in these experiments at the concentration that produces the highest stimulation of respiration rate. For DNP titration, after recording the coupled endogenous respiration rate, as described earlier, DNP is added at increasing concentrations, and, at each concentration, the uncoupled respiratory rate is measured and expressed as a percentage of the starting coupled endogenous respiration rate (Villani et al., 1998).

B. Media and Reagents

TD buffer

50 mM DNP (Sigma) in absolute ethanol; the stock solution can be stored at $-20°C$

4 μM antimycin A (Sigma) in absolute ethanol (stored at $-20°C$)

1 M ascorbate: adjust to pH 7.4 by gradual addition of NaOH tablets. Distribute in microfuge tubes and store at $-80°C$.

300 mM TMPD: distribute in tubes and store at $-80°C$. Note that the solution can turn blue due to the oxidized form of the TMPD, but this will not affect the results, as $TMPD^+$ will be reduced by ascorbate rapidly.

1 M KCN: this solution is adjusted to pH 8 by adding concentrated HCl. The stock solution is then distributed in tubes and stored for long periods at $-80°C$. Note: Preparation of the KCN stock must be carried out in a fume hood, as highly toxic volatile HCN may be produced, especially when an excess of added HCl rapidly drops the pH to acidic values. The pH adjustment can lower the effective titer of the solution without affecting the following experimental results, as relative inhibition rates have to be measured. However, if the solution is exposed to excess acid during preparation, it should be discarded.

C. Procedure

For this type of experiment, the parallel use of two oxygraphic chambers equipped with electrodes is recommended. These should be connected to a double-channel oxygraph and recorder. Calibrate the instrument to produce the same 0–100% oxygen range in the two channels. This allows one to correct for small decreases in respiratory rates occurring during the oxygraphic run.

1. KCN Titration of Integrated COX Activity

1. Allow the temperature of the measurement medium to equilibrate under stirring in the reaction chambers.

2. At the same time, wash with TD and collect exponentially growing cell cultures in which the medium was changed the day before.

3. Resuspend the final cell pellet in a volume of measured medium (one-fifth of the chamber volume), which is made by combining two equal samples taken from the two chambers. Gently pipette up and down, and split the resulting suspension between the two chambers to which DNP has been added. As a rule, DNP should be used at the minimal concentration needed to produce maximal stimulation of the endogenous respiration. In the case of 143B.TK$^-$ cells, DNP is added to final concentration of 25–30 μM. The concentration of cells in the chamber is adjusted to ensure that the uninhibited endogenous respiration rate remains fairly constant during the titration experiment, and the TMPD autooxidation rate represents a small percentage of the overall isolated COX activity (see later).

4. Close the chambers and measure the initial oxygen consumption rate. You should obtain two parallel lines on the chart recorder (see Fig. 1B).

5. Sequentially introduce small amounts of a 10 mM KCN solution at appropriate intervals near the bottom of one of the chambers using a Hamilton syringe fitted with a 2.5-in. (6.3 cm)-long needle (Hamilton). The intervals of addition are chosen to allow adequate time for reaching constant slopes in the tracing and for constructing detailed inhibition curves (see Fig. 2A).

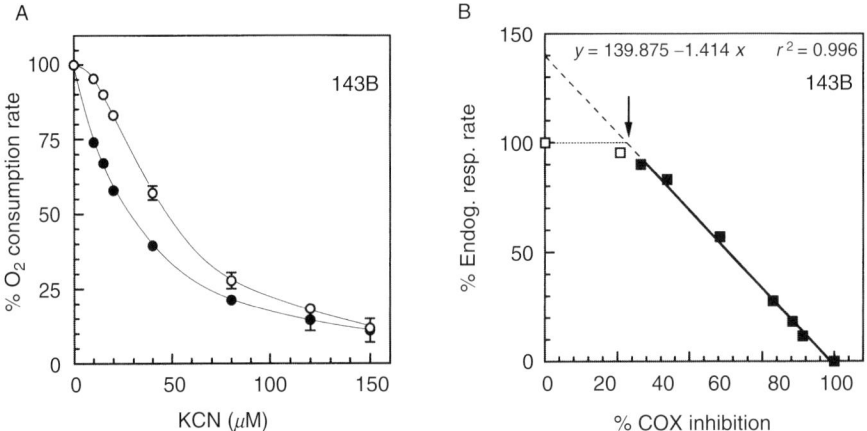

Fig. 2 Inhibition by KCN of endogenous respiration rate in TD buffer in the presence of DNP (○) or ascorbate + TMPD-dependent respiration in the presence of DNP and antimycin A (●) in intact 143B. TK$^-$ cells (A), and percentage of endogenous respiration rate as a function of percentage of isolated COX inhibition (threshold plot) in 143B.TK$^-$ cells, with determination of relative maximum COX capacity (COX$_{Rmax}$) (B). Adapted from Villani and Attardi (1997). (A) Data shown represent the means $\pm$SE (the error bars that fall within the individual data symbols are not shown) obtained in six or seven determinations on 143B.TK$^-$ cells. The TMPD concentration used for KCN titration of the O$_2$ consumption rate of antimycin-treated cells was 0.2 mM. (B) Percentage rates of endogenous respiration at different KCN concentrations in the experiments illustrated in (A) are plotted against percentages of COX inhibition by the same KCN concentrations, and the least-square regression line through the filled symbols beyond the inflection point (threshold) in the curve (arrow) is extended to zero COX inhibition. The equation describing the extrapolated line is shown. For further details, see text and Villani and Attardi (1997).

2. KCN Titration of Isolated COX Activity

1. Same as in Section III.C.1, step 1.

2. Same as in Section III.C.1, step 2.

3. Split the cell suspension, prepared as in Section III.C.1, step 3, between the two chambers, to which DNP and antimycin A have been added. Antimycin should be used at the minimal concentration needed to produce maximal inhibition of respiration (20 nM for 143B.TK$^-$ cells).

4. After closing the chambers, add ascorbate and TMPD to both chambers and measure the initial oxygen consumption rate. You should obtain two parallel lines on the chart recorder. Ascorbate is used at 10 mM. For 143B.TK$^-$ cells, addition of ascorbate alone does not result in any significant oxygen consumption. TMPD, which starts oxygen consumption, should be used at a concentration that produces oxygen consumption rates comparable to the endogenous respiration rate. Under these conditions, cytochrome c oxidase can be titrated both as an integrated step and as an isolated step under similar electron fluxes.

5. Same as in Section III.C.1, step 5.

D. Analysis of Data

In the titration of both integrated and isolated COX activity, the KCN-inhibited activity measured after each addition of KCN is expressed as a percentage of the uninhibited activity measured [in parallel (i.e., at the same O_2 level)] in the other chamber. If the starting oxygen consumption rates are not the same, the percentage difference should be used to correct the subsequent calculation of the inhibited activities.

For the measurement of isolated COX activity, the oxygen consumption rates must be corrected for the KCN-insensitive oxygen consumption due to TMPD autooxidation. For this purpose, in a separate experiment, DNP, antimycin, ascorbate, and TMPD are added to TD buffer at the same concentrations used in the assays performed with cells, and the oxygen consumption rate is measured in the oxygen concentration range in which the cyanide titrations are performed. This is important because the rate of oxygen consumption depends on the O_2 concentration in the chamber. In a similar experiment performed in the presence of mtDNA-less cells, we have observed no difference in TMPD oxidation rate. Practically, when measuring the slopes corresponding to the reference and KCN-inhibited isolated COX activities, data obtained should be corrected by substracting the slope corresponding to the TMPD autooxidation rate in the same O_2 concentration range.

An alternative way to perform the titration at each KCN concentration is presented in Fig. 1C. The experiment is set up as before, with cells resuspended in TD plus DNP. After recording the uncoupled respiration rate in both chambers, KCN is added to one of the chambers at a given concentration and the oxygen consumption is recorded until a constant slope is obtained. Then, antimycin is added to both chambers to stop the respiration completely. Finally, ascorbate and TMPD are added to both chambers, and the two isolated COX activities (control and KCN-inhibited) are measured as initial rates and then corrected for the TMPD autooxidation rate as determined in a separate experiment. This method is preferable in the case of unstable oxygen consumption rates.

Two main parameters can be determined using the KCN titration technique: (i) the inhibition "threshold" value, defined as the percentage inhibition of the isolated COX activity where COX activity becomes rate limiting for the overall flux, and (ii) the "maximum relative capacity," which indicates the percentage excess of COX activity over that needed to support the maximal rate of endogenous respiration.

As shown in Fig. 2A, the KCN titration of the O_2 consumption rate, in the presence of ascorbate and TMPD, in antimycin A-inhibited 143B.TK$^-$ cells (isolated step), produces a curve that is quasilinear at nonsaturating concentrations of the inhibitor. However, the variation of the endogenous respiration rate (overall flux) over the same range of KCN concentrations produces a curve that is sigmoidal. This difference is consistent with the occurrence in 143B.TK$^-$ cells of an excess of COX activity over that required to support a normal respiratory rate.

In Fig. 2B, the difference in relative KCN sensitivity between the endogenous respiration rate and the isolated COX activity in 143.TK$^-$ cells is illustrated in the form of a threshold plot; that is, a plot of the relative endogenous respiration rates against percentage inhibition of isolated COX activity by the same KCN concentrations. In this plot, the endogenous respiration rate remains fairly constant with increasing inhibition of the isolated enzyme, up to the percentage inhibition threshold (COX_T), that is, ~28%, at which a further decrease in enzyme activity has a marked effect on the rate of endogenous respiration. When the least-square regression line through the filled symbols (beyond the inflexion point in the curve) is extrapolated to zero COX inhibition, the y intercept gives an estimate of the maximum COX capacity as an integrated step, expressed relative to the uninhibited endogenous respiration rate (Taylor et $al.$, 1994; Villani and Attardi, 1997).

This method has been applied to a variety of human cell types and has revealed generally low values of reserve COX capacity with interesting differences, which can be helpful in understanding the tissue specificity of mitochondrial diseases (Villani et $al.$, 1998). Furthermore, by means of the same strategy, the impact of the apoptotic release of cytochrome c on cell respiration has been investigated (Hájek et $al.$, 2001).

IV. *In Situ* Analysis of Mitochondrial OXPHOS

One of the most important bioenergetic parameters to be determined when dealing with mitochondrial metabolism is the efficiency of oxidative phosphorylation. This is usually measured as coupling of mitochondrial respiration to ATP production, namely as the P/O ratio [defined as nanomoles of ATP produced per natoms of oxygen consumed during ADP-stimulated (state III) respiration (Chance and Williams, 1955)]. This classic methodology has been used on isolated mitochondria [for an application to mitochondria isolated from cultured cells, see Trounce *et al.* (1996)].

A. Media and Reagents

Buffer A: 75 mM sucrose, 5 mM potassium phosphate (monobasic), 40 mM KCl, 0.5 mM EDTA, 3 mM $MgCl_2$, 30 mM Tris–HCl (pH 7.4 at 25°C)

Permeabilization buffer: buffer A + 1 mM phenylmethylsulfonyl fluoride (PMSF)

Measurement buffer: buffer A + 0.35% bovine serum albumin (BSA; Sigma)

Substrate stock solutions: 1 M glutamate, 1 M malate, 1 M succinate. Adjust to pH 7.0–7.4 with KOH or NaOH, distribute in tubes, and store at −20°C.

10% digitonin (Calbiochem) in dimethyl sulfoxide

2 mM rotenone in absolute ethanol (stored at −20°C)

50 mM ADP (Sigma): adjust pH to 6.5–6.8 with diluted KOH and add $MgCl_2$ to 20 mM. Distribute in tubes and store at −80°C.50 mM p^1, p^5-di(adenosine-5') pentaphosphate (Ap_5A) (Sigma) in H_2O (stored at −80°C)

B. Digitonin Titration

The amount of digitonin needed for optimal cell permeabilization must be titrated for each cell type analyzed. A simple functional test of optimal permeabilization can be performed using succinate as the respiratory substrate. Succinate is normally not utilized for endogenous respiration and can enter intact cells only poorly, as tested in a variety of human cell types (Kunz *et al.*, 1995; Villani *et al.*, 1998). Alternatively, a previously described method based on the trypan blue exclusion test (Hofhaus *et al.*, 1996) can be utilized to determine the extent of permeabilization. The exclusion test is particularly useful when working with cells showing a high contribution of rotenone-insensitive endogenous respiration.

Collect cells, resuspend the final pellet, and transfer it to the oxygraphic chamber, as described in the previous sections, by using buffer A equilibrated at 37°C and supplemented with 0.5 mM ADP just before adding the cell suspension. Take a sample for cell counting. After recording endogenous respiration, add rotenone to a final concentration of 100 nM and wait until maximal inhibition is reached. Then, add digitonin and, after 2 min, add 5 mM succinate and measure respiration. The optimal amount of digitonin gives the highest succinate-dependent respiration. This rate should not increase on the addition of exogenous cytochrome *c* if the integrity of the mitochondrial outer membrane is preserved.

C. P/O Ratio Assay

A calibration of the oxygraph is needed at the beginning and at the end of the experiments. If many samples are analyzed, additional dithionite calibrations (Hofhaus *et al.*, 1996) are recommended during the course of the experiment.

The experiments are run at $37\,^{\circ}C$. At this temperature the concentration of O_2 in distilled water under saturating conditions is 217 nmol/ml (Cooper, 1977). All the buffers must be kept at the measurement temperature during the experiment.

Collect cells, wash once in buffer A, and remove a sample for cell counting. Resuspend the final pellet in permeabilization buffer at a cell concentration similar to the one utilized for digitonin titration. Add the optimal amount of digitonin and wait for 1–2 min, occasionally shaking the tube. Then, add 10 volumes of measurement buffer and collect permeabilized cells by centrifugation. Resuspend the final pellet in a sample of measurement buffer previously equilibrated at $37\,^{\circ}C$, by gentle pipetting and transfer the suspension into the chamber under stirring. After supplying the mixture with 0.3 mM Ap_5A, close the chamber and record the baseline. At this point, the oxygen consumption rate is undetectable or very low and produced by endogenous substrates within mitochondria. If the starting oxygen consumption rate is too high, it indicates that the permeabilization procedure has probably not worked properly, and a higher digitonin concentration is usually required. Add glutamate and malate at a concentration of 5 mM each and record state IV respiration. When a stable rate is obtained, add ADP (100–200 nmol for a 1.5-ml chamber); this will result in a transient increase in respiration rate (state III), which will return to state IV respiration when all the ADP has been phosphorylated. After obtaining a linear state IV respiration, a second addition of ADP can be made to obtain a second measurement of the P/O ratio. Thereafter, block respiration by adding rotenone to 100 nM, add 5 mM succinate, and, after obtaining a stable state IV respiration, add ADP (50–100 nmol); and record the transient state III respiratory rate and the subsequent return to state IV.

D. Analysis of Data

Figure 3 shows the oxygraphic tracing obtained for 143B.TK⁻-permeabilized cells. Values of respiratory control ratio [defined as ratio between respiration rate in the presence of added ADP (state III) and state IV respiration rate obtained following its consumption (Chance, 1959; see also Estabrook, 1967)] of 3–4 have been measured for these cells, with glutamate and malate as respiratory substrates. These results are very similar to previous results obtained on mitochondria isolated from the same cells (Trounce et al., 1996). These values are also very close to the ratio between the endogenous respiration rate [>95% rotenone sensitive (Villani and Attardi, 1997)] and the oligomycin-inhibited respiration rate, as measured in intact cells (data not shown). Furthermore, the glutamate/malate-dependent respiration rate in the presence of ADP is very similar to the coupled endogenous respiration rate of intact cells (data not shown), thus allowing one to measure the ATP production efficiency under physiological respiratory fluxes. This is especially important when the dependence of the H^+/e^- stoichiometry of cytochrome c oxidase (Capitanio et al., 1991, 1996) or, in general, of the efficiency of ATP production on the electron transfer rate (Fitton et al., 1994) is considered. P/O ratios can be calculated as ratios between the amount (nanomoles) of added ADP

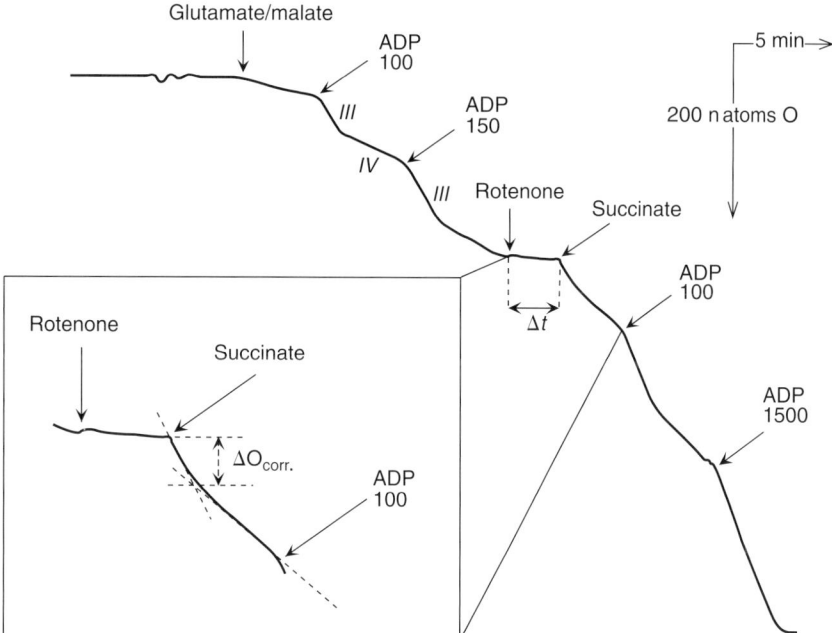

Fig. 3 Polarographic tracings of an experiment for P/O ratio measurements in digitonin-permeabilized 143B.TK⁻ cells. The additions of ADP are expressed in nanomoles. *III*, ADP-stimulated state III respiration; *IV*, state IV respiration. (Inset) An enlargement of the tracing corresponding to the transient state III respiration obtained on succinate addition. See text for further details.

and the oxygen (natoms) consumed during the ADP-induced state III respiration (Chance and Williams, 1955; Estabrook, 1967). In this way, P/O ratios up to about 2.3 and 1.3 have been measured in 143B.TK⁻ digitonin-permeabilized cells respiring, respectively, in the presence of glutamate/malate and succinate.

The first addition of succinate after inhibition with rotenone induces a transient stimulation of respiration. This may be due to the ADP that has accumulated as a result of ATP-consuming systems operating in permeabilized cells during the time of rotenone inhibition (Δt in Fig. 3). In fact, the ADP produced under these conditions cannot be rephosphorylated to ATP because of the block of respiration and consequent collapse of the electrochemical transmembrane potential that drives the ATP synthesis by mitochondrial ATP synthase. The rate of ADP production by ATP-consuming processes can be estimated by measuring the amount of oxygen consumed during the transient state III cited earlier (ΔO_{corr} in the inset of Fig. 3), multiplying this amount by the P/O ratio, as measured for succinate respiration, and then dividing the value thus obtained by the time of rotenone inhibition (Δt). The final value thus calculated (ADP/s) can then be used to correct the P/O ratio measurements for state III respiration for the additional amount of ADP produced and then converted to ATP. In the experiment shown in Fig. 3, for instance, ATP consumption has been estimated to account for the cycling of

0.23 nmol ADP/sec. The final P/O ratio values, corrected by this factor, would then be 2.78 and 2.52 (instead of 2.33 and 2.16) for the first and the second addition of ADP to glutamate/malate-oxidizing cells, respectively, and 1.37 (instead of 1.1) for succinate-driven respiration.

In the case of 143B.TK$^-$ cells, no P/O ratio could be measured if glucose was present in the measurement buffer due to ADP production by membrane-associated hexokinase activity. Therefore, the presence of traces of glucose could result in a lower P/O ratio. Furthermore, it was observed that the presence of PMSF in the measurement buffer resulted in the formation of aggregates in the oxygraphic chamber.

The procedure described in this section is very useful when dealing with a limited amount of sample, as much information can be collected within a single experiment. In fact, substrate-dependent respiration rates, respiratory control ratios, and P/O ratios can be measured at different energy conservation sites with as few as $2-4 \times 10^6$ cells/ml, as has been done with 143B.TK$^-$ cells and its derivative cell lines.

Acknowledgments

This work was supported by National Institutes of Health Grant GM-11726 (to G.A.). We are very grateful to M. Greco for help in some of the experiments, to the late F. Guerrieri for helpful discussions and A. Chomyn for her critical reading of the manuscript.

References

Capitanio, N., Capitanio, G., Demarinis, D. A., De Nitto, E., Massari, S., and Papa, S. (1996). Factors affecting the H$^+$/e$^-$ stoichiometry in mitochondrial cytochrome c oxidase: Influence of the rate of electron flow and transmembrane delta pH. *Biochemistry* **35**, 10800–10806.

Capitanio, N., Capitanio, G., De Nitto, E., Villani, G., and Papa, S. (1991). H$^+$/e$^-$ stoichiometry of mitochondrial cytochrome complexes reconstituted in liposomes. Rate-dependent changes of the stoichiometry in the cytochrome c oxidase vesicles. *FEBS Lett.* **288**, 179–182.

Chance, B. (1959). "Ciba Found. Symp. Regulation Cell Metabolism," p. 91. Little Brown, Boston, MA.

Chance, B., and Williams, G. R. (1955). Respiratory enzymes in oxidative phosphorylation. I. Kinetics of oxygen utilization. *J. Biol. Chem.* **217**, 383–393.

Cooper, T. G. (1977). "The Tools of Biochemistry." Wiley, New York.

Ernster, L. (1986). Oxygen as an environmental poison. *Chem. Scripta* **26**, 525–534.

Estabrook, R. W. (1967). Mitochondrial respiratory control and the polarographic measurement of ADP:O ratios. *Methods Enzymol.* **10**, 41–47.

Fitton, V., Rigoulet, M., Ouhabi, R., and Guerin, B. (1994). Mechanistic stoichiometry of yeast mitochondrial oxidative phosphorylation. *Biochemistry* **33**, 9692–9698.

Groen, A. K., Wanders, R. J., Westerhoff, H. V., van der Meer, R., and Tager, J. M. (1982). Quantification of the contribution of various steps to the control of mitochondrial respiration. *J. Biol. Chem.* **257**, 2754–2757.

Hájek, P., Villani, G., and Attardi, G. (2001). Rate-limiting step preceding cytochrome c release in cells primed for Fas-mediated apoptosis is revealed by cell sorting and respiration analysis. *J. Biol. Chem.* **276**, 606–615.

Heinrich, R., and Rapoport, T. A. (1974). A linear steady-state treatment of enzymatic chains: General properties, control and effector strength. *Eur. J. Biochem.* **42**, 89–95.

Hofhaus, G., Shakeley, R. M., and Attardi, G. (1996). Use of polarography to detect respiration defects in cell cultures. *Methods Enzymol.* **264,** 476–483.

Kacser, H., and Burns, J. A. (1973). *In* "Rate Control of Biological Processes" (D. D. Davies, ed.), pp. 65–104. Cambridge University Press, London.

Kunz, D., Luley, C., Fritz, S., Bohnensack, R., Winkler, K., Kunz, W. S., and Wallesch, C. W. (1995). Oxygraphic evaluation of mitochondrial function in digitonin-permeabilized mononuclear cells and cultured skin fibroblasts of patients with chronic progressive external ophthalmoplegia. *Biochem. Mol. Med.* **54,** 105–111.

Schon, E. A., Bonilla, E., and Di Mauro, S. (1997). Mitochondrial DNA mutations and pathogenesis. *J. Bioenerg. Biomembr.* **29,** 131–149.

Taylor, R. W., Birch-Machin, M. A., Bartlett, K., Lowerson, S. A., and Turnbull, D. M. (1994). The control of mitochondrial oxidations by complex III in rat muscle and liver mitochondria: Implications for our understanding of mitochondrial cytopathies in man. *J. Biol. Chem.* **269,** 3523–3528.

Trounce, I. A., Kim, Y. L., Jun, A. S., and Wallace, D. C. (1996). Assessment of mitochondrial oxidative phosphorylation in patient muscle biopsies, lymphoblasts, and transmitochondrial cell lines. *Methods Enzymol.* **264,** 484–509.

Villani, G., and Attardi, G. (1997). *In vivo* control of respiration by cytochrome *c* oxidase in wild-type and mitochondrial DNA mutation-carrying human cells. *Proc. Natl. Acad. Sci. USA* **94,** 1166–1171.

Villani, G., Greco, M., Papa, S., and Attardi, G. (1998). Low reserve of cytochrome *c* oxidase capacity *in vivo* in the respiratory chain of a variety of human cell types. *J. Biol. Chem.* **273,** 31829–31836.

Wallace, D. C. (1992). Diseases of the mitochondrial DNA. *Annu. Rev. Biochem.* **61,** 1175–1212.

CHAPTER 6

Optical Imaging Techniques (Histochemical, Immunohistochemical, and *In Situ* Hybridization Staining Methods) to Visualize Mitochondria

Kurenai Tanji* and Eduardo Bonilla*,†

*Department of Pathology, Columbia University
New York, New York 10032

†Department of Neurology, Columbia University
New York, New York 10032

I. Introduction

The mitochondria are the primary ATP-generating organelles in all mammalian cells. ATP is produced via oxidative phosphorylation through five respiratory complexes located in the inner mitochondrial membrane. Mitochondria contain their own DNA (mtDNA), which is maternally inherited (Anderson *et al.*, 1981; Giles *et al.*, 1980). The human mitochondrial genome is a 16,569-bp double-stranded circle. It is highly compact and contains only 37 genes: 2 genes encode ribosomal RNAs (rRNAs), 22 encode transfer RNAs (tRNAs), and 13 encode polypeptides. All 13 polypeptides are components of the respiratory chain, including 7 subunits of complex I or NADH dehydrogenase-ubiquinone oxidoreductase, 1 subunit of complex III, or ubiquinone-cytochrome *c* oxidoreductase, 3 subunits of complex IV, or cytochrome *c* oxidase (COX), and 2 subunits of complex V, or ATP synthase (Attardi and Schatz, 1988). The respiratory complexes also contain nuclear DNA (nDNA)-encoded sububits, which are imported into the organelle from the cytosol and are assembled, together, with the mtDNA-encoded subunits, into the respective holoenzymes in the mitochondrial inner membrane (Attardi and Schatz, 1988; Neupert, 1997). Complex II, or succinate dehydrogenase-ubiquinone oxidoreductase, contains only nDNA-encoded subunits.

The mitochondrial disorders encompass a heterogeneous group of diseases in which mitochondrial dysfunction produces clinical manifestations. Because of the dual genetic make up of mitochondria (Attardi and Schatz, 1988), these diseases are typically caused by genetic errors either in mtDNA or nDNA.

Pathogenic mtDNA mutations have been identified in three of the "prototypical" mitochondrial disorders. First, large-scale mtDNA rearrangements [i.e., deletions (Δ-mtDNA) and/or duplications (dup-mtDNA)] have been associated with sporadic Kearns-Sayre syndrome (KSS) and are often seen in patients with isolated ocular myopathy (OM) (Moraes *et al.*, 1989; Zeviani *et al.*, 1988). Second, myoclonus epilepsy with ragged-red fibers (MERRF) has been frequently associated with two different point mutations, both in the $tRNA^{Lys}$ gene (Shoffner *et al.*, 1990; Silvestri *et al.*, 1992). Third, mitochondrial encephalopathy with lactic acidosis and stroke-like episodes (MELAS) has been mainly associated with two different point mutations, both in the $tRNA^{Leu(UUR)}$ gene (Goto *et al.*, 1990, 1991). Other point mutations in tRNAs and protein-coding genes of the mitochondrial genome, as well as disorders leading to generation of multiple Δ-mtDNAs or to depletion of mtDNA that are due to primary defects in nDNA, have now been described (DiMauro and Bonilla, 2004). Because mitochondria are inherited only from the mother (Giles *et al.*, 1980), pedigrees harboring defects in mtDNA genes should exhibit maternal inheritance. Moreover, because there are hundreds or even thousands of mitochondria in each cell, with an average of five mtDNAs per organelle (Bogenhagen and Clayton, 1984; Satoh and Kuroiwa, 1991), mutations in mtDNA result in two populations of mtDNAs—mutated and wild type—a condition known as heteroplasmy. The phenotypic expression of an mtDNA mutation is regulated by the threshold effect, that is, the mutant phenotype is expressed in the heteroplasmic cells only when the relative proportion of mutant mtDNAs exceeds a minimum value (Wallace, 1992).

Research on mitochondrial diseases has also uncovered an increasing number of disorders that are caused by mutations in nuclear genes encoding the subunits of the respiratory chain, or other proteins that are essential for the biosynthesis of specific cofactors or assembly of the complexes. Interestingly, defects in complexes I and II are associated with mutations in genes encoding the subunits of the complexes, while defects in complexes III, IV, and V are caused by mutations in specific assembly proteins or ancillary biosynthetic factors (Shoubridge, 2001).

Brain and muscle, whose functions are highly dependent on oxidative metabolism, are the tissues most severely affected in mitochondrial disorders (Wallace, 1992). Consequently, genetic as well as morphologic studies of muscle and brain from patients with mitochondrial diseases have been proven to be fundamental to understanding the pathogenesis of mitochondrial dysfunction at the level of individual muscle fibers and in different neuronal nuclei (Moraes *et al.*, 1992; Sparaco *et al.*, 1995; Tanji *et al.*, 1999).

The purpose of this chapter is to present the histochemical, immunohistochemical, and *in situ* hybridization (ISH) methods that, in our experience, appear to be the most reliable for the correct identification of mitochondria on frozen or paraffin-embedded tissue sections. We illustrate here their potential using specific pathological samples, and provide an updated version of the methods. While the described protocols refer to skeletal muscle or brain mitochondria, the methods described can be applied to any cell type (Szabolcs *et al.*, 1994; Tanji *et al.*, 2003). It is not our intention to cover every study or method related to morphological aspects of mitochondria, but rather to provide enough information to allow investigators to apply these selected light microscopy tools to a particular scientific or diagnostic question.

II. Histochemistry

The visualization of normal and pathological mitochondria on frozen tissue sections can be carried out using a number of cytochemical techniques. These include the modified Gomori trichrome and hematoxylin-eosin stains, and histochemical methods for the demonstration of oxidative enzyme activity.

The most informative histochemical alteration of mitochondria in skeletal muscle are the ragged-red fibers (RRF), observed on frozen sections stained with the modified trichrome method of Engel and Cunningham (1963). The name derives from the reddish appearance of the trichrome-stained muscle fiber as a result of subsarcolemmal and/or intermyofibrillar proliferation of the mitochondria. The fibers harboring abnormal deposits of mitochondria are most often type I (i.e., oxidative) myofibers, and they may also contain increased numbers of lipid droplets. Since accumulations of materials other than mitochondria may simulate RRF formation, the identification of deposits suspected of being mitochondrial proliferation should be confirmed histochemically by the application of oxidative enzyme stains.

In our experience, enzyme histochemistry for the activity of succinate dehydrogenase (SDH) and COX have proven to be the most reliable methods for the correct visualization of normal mitochondria, and for the interpretation and ultimately the diagnosis of some of the mitochondrial disorders affecting skeletal muscle (DiMauro and Bonilla, 2004).

A. Succinate Dehydrogenase

SDH is the enzyme that catalyzes the conversion of succinate to fumarate in the tricarboxylic acid cycle. It consists of four subunits. The two large subunits—a 70-kDa flavoprotein (SDHA) and a 30-kDa iron-sulfur-containing protein (SDHB)—form complex II of the mitochondrial respiratory chain, while the two smaller subunits (SDHC and SDHD) are responsible for attaching SDH to the inner mitochondrial membrane (Ackrell, 2002). Because complex II is the only component of the respiratory chain whose subunits are all encoded by nDNA, SDH histochemistry is extremely useful for detecting any variation in the fiber distribution of mitochondria, independently of any alteration affecting the mtDNA.

The histochemical method for the microscopic demonstration of SDH activity on frozen tissue sections is based on the use of a tetrazolium salt (nitro blue tetrazolium, NBT) as an electron acceptor, with phenazine methosulfate (PMS) serving as the intermediate electron donor to NBT (Pette, 1981; Seligman and Rutenburg, 1951). The specificity of the method may be tested by performing control experiments in which an SDH inhibitor, sodium malonate (0.01 M), is added to the incubation medium.

Using this method for detecting SDH activity in normal muscle sections, two populations of fibers are seen resulting in a checkerboard pattern. Type II fibers, which rely on glycolytic metabolism, show a light blue networklike stain. Type I fibers, whose metabolism is highly oxidative and therefore contain more mitochondria, show a more elaborate and darker mitochondrial network (Fig. 1A).

In samples with pathological proliferation of mitochondria (RRF), the RRF show an intense blue SDH reaction corresponding to the distribution of the mitochondria within the fiber (marked with an asterisk in Fig. 1C). This proliferation of mitochondria is associated with most mtDNA defects (deletions and tRNA point mutations), but RRF can also be observed in other disorders that are thought to be due to defects of nDNA, such as the depletion of muscle mtDNA, and the fatal and "benign" COX-deficient myopathies of infancy (DiMauro et al., 1994).

SDH histochemistry is also useful for the diagnosis of complex II deficiency. Several patients with myopathy and complex II deficiency have been reported. In agreement with the biochemical observations, SDH histochemistry showed complete lack of reaction in muscle (Haller et al., 1991; Taylor et al., 1996).

1. Method

1. Collect 8-μm-thick cryostat sections on poly (L-lysine)-coated (0.1%) coverslips. Dissolve the following in 10 ml of 5-mM phosphate buffer, pH 7.4:

   5-mM Ethylenediaminetetraacetic (EDTA)

   1-mM Potassium cyanide (KCN)

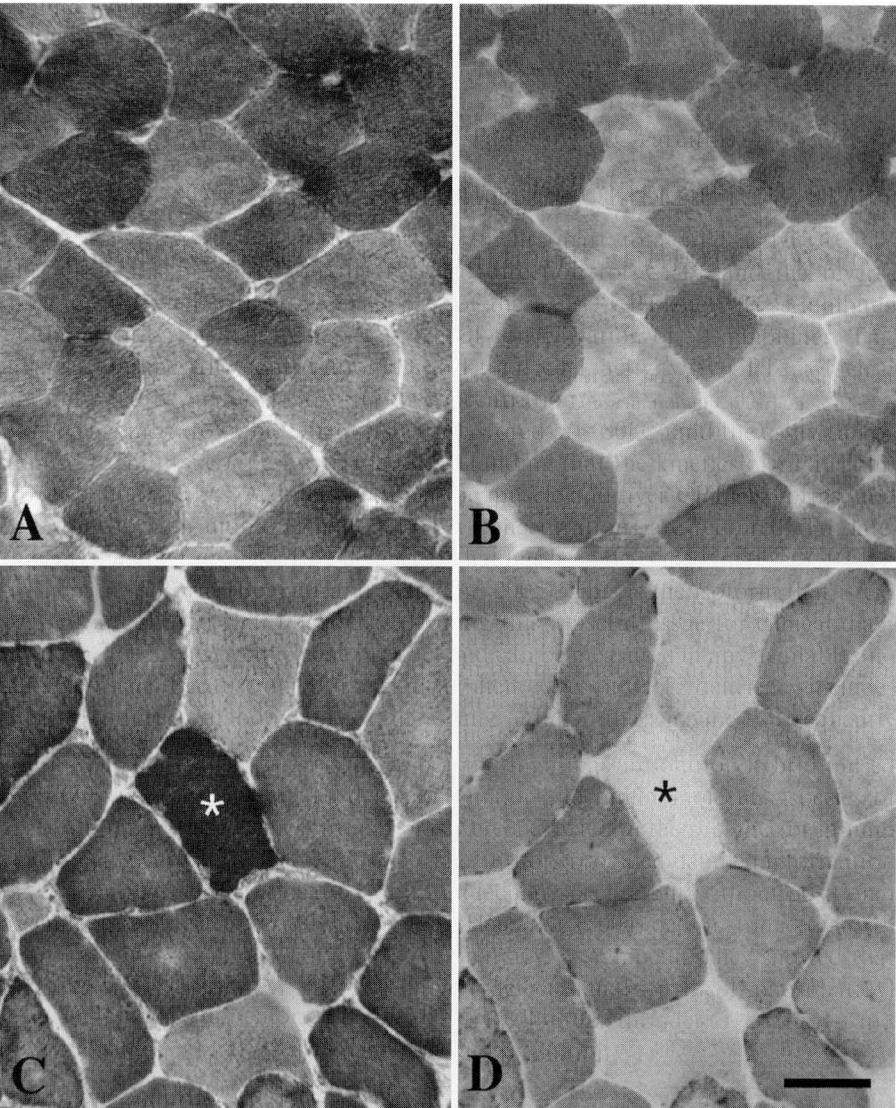

Fig. 1 Histochemical stains for SDH and COX activities on serial muscle sections from a normal subject and a patient with KSS. The normal subject shows a checkerboard pattern with both enzymes (A and B). The KSS samples show, in one section, one RRF (asterisk) by SDH stain (C), and the same fiber on the serial section (asterisk) shows lack of COX activity (D). Bar = 50 μ. (See Color Insert.)

0.2-mM PMS

50-mM Succinic acid

1.5-mM NBT

Adjust pH to 7.6.

2. Filter solution through filter paper (Whatman No. 1).

3. Incubate sections for 20 min at 37°C. For control sections, sodium malonate (0.01 M) is added to the incubation medium.

4. Rinse with distilled water (dH$_2$O) three times for 5 min each, at RT.

5. Mount on glass slides with warm glycerin gel.

B. Cytochrome *c* Oxidase

COX, the last component of the respiratory chain, catalyzes the transfer of reducing equivalents from cytochrome *c* to molecular oxygen (Hatefi, 1985). In mammals, the apoprotein is composed of 13 different subunits embedded in the mitochondrial inner membrane (Cooper *et al.*, 1991; Taanman, 1997). The three largest polypeptides (I, II, and III), which are encoded by mtDNA and are synthesized within the mitochondria, confer the catalytic and proton pumping activities to the enzyme. The 10 smaller subunits are synthesized in the cytoplasm under the control of nuclear genes and are presumed to confer tissue specificity, thus adjusting the enzymatic activity to the metabolic demands of different tissues (Capaldi, 1990; Kadenbach *et al.*, 1987). The holoenzyme contains two heme *a* moieties (*a* and *a₃*) and three copper atoms (two in the Cu$_A$ site in subunit II and one in the Cu$_A$ site in subunit I). Additional nDNA-encoded factors are required for the assembly of COX, including those involved in the synthesis of hemes *a* and *a₃*, transport and insertion of copper atoms, and proper coassembly of the mtDNA- and nDNA-encoded subunits. Several COX assembly genes have been identified in yeast (Kloeckener-Gruissem *et al.*, 1987; Pel *et al.*, 1992), and pathogenic mutations in the human homologues of two of these genes, SURF1 and SCO2, have been discovered in patients with COX-deficient Leigh syndrome (Tiranti *et al.*, 1998; Zhu *et al.*, 1998) and in patients with a cardioencephalomyopathy characterized by COX deficiency (Papadopoulou *et al.*, 1999).

The dual genetic make up of COX and the availability of a reliable histochemical method to visualize its activity have made COX one of the ideal tools for basic investigations of mitochondrial biogenesis, nDNA–mtDNA interactions, and for the study of human mitochondrial disorders at both the light and electron microscopic levels (Bonilla *et al.*, 1975; Johnson *et al.*, 1983; Wong-Riley, 1989).

The histochemical method to visualize COX activity is based on the use of 3,3′-diaminobenzidine tetrahydrochloride (DAB) as an electron donor for cytochrome *c* (Seligman *et al.*, 1968). On oxidation, DAB forms a brown pigmentation whose distribution corresponds to the distribution of mitochondria in the tissue. The specificity of the method may be tested by performing control experiments in which the COX inhibitor, KCN (0.01 M), is added to the incubation medium.

As in the case of SDH, staining of normal muscle for COX activity also shows a checkerboard pattern. Type I fibers stain darker due to their mainly oxidative metabolism and more abundant mitochondria content, and type II fibers show a finer and less intensely stained mitochondria network (Fig. 1B).

The application of COX histochemistry to the investigation of KSS, MERRF, and MELAS has revealed one of the most important clues for the study of pathogenesis in these disorders. Muscle from KSS and MERRF patients shows a mosaic expression of COX, consisting of a variable number of COX-deficient and COX-positive fibers (Johnson *et al.*, 1983) (asterisk in Fig. 1D). Before the advent of molecular genetics, it was difficult to understand the reason for the appearance of this mosaic, but when it was discovered that these patients harbored mutations of mtDNA in their muscles, it became evident that the mosaic was an indicator of the heteroplasmic nature of the genetic defect. The mosaic pattern of COX expression in mitochondrial disorders is now considered the "histochemical signature" of a heteroplasmic mtDNA mutation affecting the expression of mtDNA-encoded genes in skeletal muscle (Bonilla *et al.*, 1992; Shoubridge, 1993).

Muscle biopsies from patients with MELAS also show COX-deficient fibers. Almost unique to this disorder, RRF are COX-*positive*, but the activity is decreased in the center of the fibers although it is largely preserved in the subsarcolemmal regions.

It should also be noted that a more generalized pattern of COX deficiency as well as COX-negative RRF are also observed in infants with depletion of muscle mtDNA or in patients with either the fatal or "benign" COX-deficient myopathies of infancy (DiMauro and Bonilla, 2004). In addition, a generalized pattern of COX deficiency, including extrafusal muscle fibers, intrafusal fibers, and arterial walls of blood vessels, is seen in children affected with COX-deficient Leigh syndrome resulting from mutations in either SURF1 or SCO2 (Papadopoulou *et al.*, 1999; Tiranti *et al.*, 1998; Zhu *et al.*, 1998).

These observations indicate that histochemical studies in mitochondrial disorders provide significant information about both the nature and the pathogenesis of mitochondrial disorders. Moreover, they provide useful clues as to which molecular testing is needed to provide a specific diagnosis.

1. Method

1. Collect 8-μm-thick cryostat sections on poly (L-lysine)-coated (0.1%) coverslips. Dissolve the following in 10 ml of 5-mM phosphate buffer, pH 7.4:

   0.1% (10 mg) DAB

   0.1% (10 mg) Cytochrome *c* (from horse heart)

   0.02% (2 mg) Catalase

   Adjust pH to 7.4

2. Do not expose solution to light.

3. Filter solution through filter paper (Whatman No. 1).

4. Incubate sections for 1 h at 37°C. For control sections, KCN (0.01 M) is added to the incubation medium.

5. Rinse with dH$_2$O three times for 5 min each, at RT.

6. Mount on glass slides with warm glycerin gel.

III. Immunohistochemistry

The unique ability of immunohistochemistry to allow for the detection of specific proteins in single cells makes it the method of choice to study the expression of both mtDNA and nDNA genes in mitochondria of small and heterogeneous tissue samples. Recent technical advances have increased greatly the scope of immuno-histochemistry and have made it accessible to a variety of investigators with minimal expertise in immunology.

Several immunological probes are presently available to perform immunohisto-chemical studies of mitochondria on frozen tissue sections. These include anti-bodies directed against mtDNA- and nDNA-encoded subunits of the respiratory chain complexes, antibodies against other mitochondrial proteins, and antibodies against DNA that allow for the detection of mtDNA (Andreetta *et al.*, 1991; Bonilla *et al.*, 1992; Papadopoulou *et al.*, 1999). Because the entire mitochondrial genome has been sequenced, any mtDNA-encoded respiratory chain subunit is potentially available for immunocytochemical studies, and it is anticipated that the same will soon be true for all the nDNA-encoded subunits of the respiratory chain (Taanman *et al.*, 1996).

There are several immunocytochemical methods to study of mitochondria on tissue sections. These include enzyme-linked methods (peroxidase, alkaline phos-phatase, and glucose oxidase) and methods based on the application of fluoro-chromes. For studies on frozen tissue sections, we favor the use of fluorochromes because they allow for the direct visualization of the antigen-antibody binding sites, and because they are more flexible for double-labeling experiments. For studies of mitochondria on formalin-fixed and paraffin-embedded samples, we routinely employ the avidin-biotin-peroxidase complex (ABC) method (Bedetti, 1985; Hsu *et al.*, 1981) for detection.

A. Immunolocalization of nDNA- and mtDNA-Encoded Subunits of the Respiratory Chain in Frozen Samples

As mentioned earlier, we prefer immunolocalization via immunofluorescence on serial frozen sections, using different fluorochromes for double-labeling studies. The main advantage of this approach is that it allows for the visualization of two different probes in the same mitochondrion and in the same plane of section. These methods also eliminate the inferences that must be made with studies on serial sections, and they are particularly useful in immunocytochemical investigations of mitochondria in nonsynctial tissues such as heart, kidney, and brain.

In our laboratory, we routinely use a monoclonal antibody against COX IV as probe for an nDNA-encoded mitochondrial protein and a polyclonal antibody against COX II as probe for an mtDNA-encoded protein. For these studies, the sections are first incubated with both the polyclonal and the monoclonal antibodies at optimal dilution, that is, the lowest concentration of the antibody giving a clear particulate immunostain corresponding to the localization of the mitochondria in normal muscle fibers. Subsequently, the sections are incubated with goat anti-rabbit IgG-fluorescein (to visualize the mtDNA probe in "green") and goat anti-mouse IgG-Texas red (to visualize the nDNA probe in "red"). We carry out these studies with unfixed frozen sections, but with some antibodies it may be required to permeabilize mitochondrial membranes to uncover the antigenic epitopes or to facilitate the penetration of the probes into the inner mitochondrial compartment. In agreement with Johnson *et al.* (1988), we have also found that fixation of fresh frozen sections with 4% formaldehyde in 0.1-M $CaCl_2$, pH 7, followed by dehydration in serial alcohols (outlined in Section III. A.1) provides the most reproducible and successful results.

Using unfixed serial muscle sections from normal samples, a checkerboard pattern resembling the one described for histochemistry is usually observed, with type I fibers appearing brighter due to their higher mitochondria content (Fig. 2A and B).

In muscle serial sections from patients with KSS harboring a documented Δ-mtDNA, COX-deficient RRF typically show an absence of or a marked reduction of COX II (asterisk in Fig. 2C), whereas using antibodies against COX IV the immunostain is typically normal not only in the COX-positive fibers, but in the COX-deficient fibers as well (asterisk in Fig. 2D). Presumably, the loss of COX II immunostain is due to the fact that even the smallest deletion eliminates essential tRNAs that are required for translation of *all* mtDNA-encoded mRNAs, irrespective of whether the corresponding genes on the mtDNA are, or are not, located within the deleted region of the Δ-mtDNA (Moraes *et al.*, 1992; Nakase *et al.*, 1990). Cell culture studies have confirmed this hypothesis: transfer of Δ-mtDNA to mtDNA-less human cell lines produces a severe defect in the synthesis of the mtDNA-encoded subunits of the respiratory chain (Hayashi *et al.*, 1991).

1. Method

Simultaneous visualization of mtDNA- and nDNA-encoded subunits of the respiratory chain using different fluorochromes:

1. Collect 4-μm-thick cryostat sections on poly(L-lysine)-coated (0.1%) coverslips.
2. Incubate the sections for 2 h at RT (in a wet chamber) with anti-COX II polyclonal antibody and with anti-COX IV monoclonal antibody at optimal dilutions (1:100–1:500; Molecular Probes, Eugene, OR) in phosphate buffer saline containing 1% bovine serum albumin (PBS/BSA). Control sections are incubated without the primary antibodies.
3. Rinse the samples with PBS three times for 5 min each, at RT.
4. Incubate the sections for 1 h at RT (in a wet chamber) with anti-rabbit IgG-fluorescein and with anti-mouse IgG-Texas red diluted 1:100 in 1% BSA/PBS.

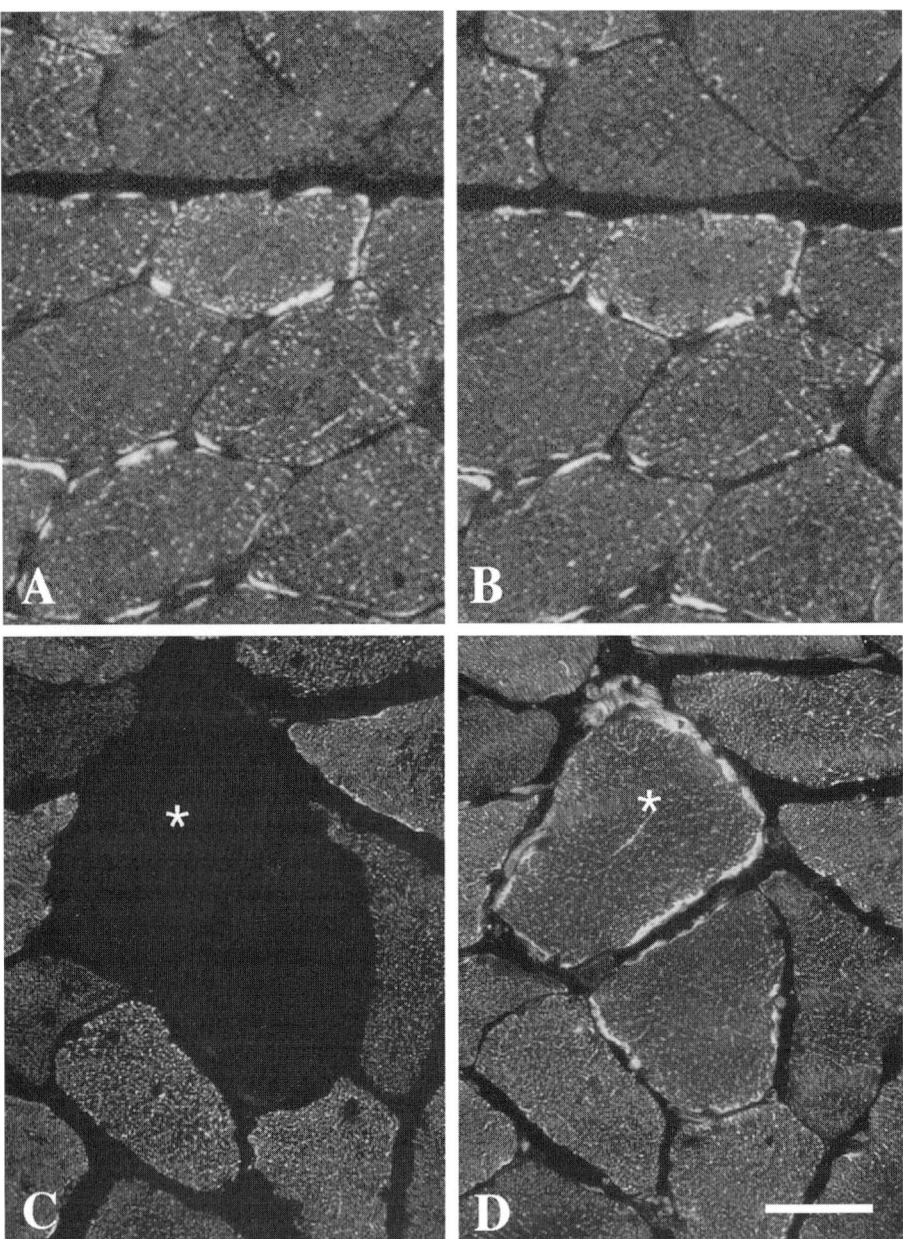

Fig. 2 Immunolocalization of COX II and COX IV on serial muscle sections from a normal subject and a patient with KSS. The normal muscle shows an almost identical mitochondrial network for COX II (A) and for COX IV (B). The KSS samples show one RRF (asterisk) that lacks COX II immunostain in one section (C), and the same fiber on the serial section (asterisk) shows enhanced stain for COX IV (D). Bar = 50 μ. (See Color Insert.)

5. Rinse the samples with PBS three times for 5 min each, at RT.

6. Mount on slides with 50% glycerol in PBS.

B. Immunolocalization of mtDNA on Frozen Samples

Immunohistochemistry using anti-DNA antibodies has been applied as an alternative method to ISH (see below) for the studies of the localization and distribution of mtDNA in normal and pathological conditions (Andreetta *et al.*, 1991; Tritschler *et al.*, 1992). The advantages of this method are that both mitochondrial and nDNA are detected simultaneously, at the single cell level, and that the nuclear signal can be used as an internal control.

For detection of mtDNA using immunological probes, we use monoclonal antibodies against DNA, on frozen sections. In muscle sections from normal controls, these antibodies show an intense staining of the nuclei and a particulate stain of the mitochondrial network (Andreetta *et al.*, 1991). When muscle sections from patients with KSS are studied, the mitochondrial network is intensely stained in RRF (Andreetta *et al.*, 1991). Conversely, when muscle biopsies from patients with mtDNA depletion are analyzed, the particulate immunostaining of the mitochondrial network is not detectable or is only present in a small number fibers, but the intensity of immunostaining in nuclei shows no alteration (Bonilla *et al.*, 1992).

Immunohistochemistry utilizing antibodies against DNA is a useful method for the rapid evaluation of the distribution of mtDNA in normal cells and for the detection of depletion of muscle mtDNA. The method is particularly precise for the diagnosis of mtDNA depletion when it is confined to only a subpopulation of fibers (Vu *et al.*, 1998).

1. Method

1. Collect 4-μm-thick cryostat sections on poly(L-lysine)-coated (0.1%) coverslips.

2. Fix the sections in 4% formaldehyde in 0.1-M $CaCl_2$, pH 7 for 1 h at RT.

3. Dehydrate the sections in 70, 80, 90% ethyl alcohol, 5 min each, and in 100% ethyl alcohol for 15 min.

4. Rinse the samples with PBS three times for 5 min each, at RT.

5. Incubate the sections for 2 h at RT (in a wet chamber) with anti-DNA monoclonal antibody (1:100–1:500; Chemicon, Temecula, CA) in 1% BSA/PBS. Control sections are incubated without the primary antibodies.

6. Rinse the samples with PBS three times for 5 min each, at RT.

7. Incubate the sections for 30 min at RT (in a wet chamber) with biotinylated anti-mouse IgG (1:100) in 1% BSA/PBS.

8. Rinse the samples with PBS three times for 5 min each, at RT.

9. Incubate the sections for 30 min at RT (wet chamber) with streptavidin-fluorescein or streptavidin-Texas Red (1:250) in 1% BSA/PBS.

10. Rinse the samples with PBS three times for 5 min each, at RT.

11. Mount on slides with 50% glycerol in PBS.

C. Immunolocalization of Mitochondrial Proteins on Paraffin-Embedded Samples

As mentioned earlier, we prefer immunoperoxidase for the localization of mitochondria on paraffin-embedded brain samples, in particular the ABC method (Bedetti, 1985; Hsu *et al.*, 1981). This method is based on the high affinity of avidin, an egg white protein for the vitamin biotin. In this technique, avidin can be viewed as an antibody against the biotin-labeled peroxidase. A reliable ABC kit can be obtained from a commercial source (e.g., Vector Laboratories Inc., Burlingame, CA); alternatively, the ABC reagents can be prepared according to a previously published method (Hsu *et al.*, 1981).

Studies of the mitochondrial respiratory chain on paraffin-embedded samples of brain, using the ABC method, have provided significant information regarding the pathogenesis of neuronal dysfunction in mitochondrial disorders and of the role of mitochondria in neurodegenerative disorders of the central nervous system (CNS), including Parkinson and Alzheimer disease (Bonilla *et al.*, 1999; Hattori *et al.*, 1990; Sparaco *et al.*, 1995; Tanji *et al.*, 1999). For example, in patients with MERRF, a syndrome clinically characterized by myoclonic epilepsy, cerebellar ataxia and myopathy, studies of the mitochondrial respiratory chain have shown a severe defect in the expression of the mtDNA-encoded subunit II of COX in neurons of the cerebellar system, including cerebellar cortex, and the dentate, and olivary nuclei (Fig. 3). Based on these observations, we have proposed that in MERRF, mitochondrial dysfunction in neurons of the cerebellar system play a significant role in the genesis of cerebellar ataxia (Sparaco *et al.*, 1995).

1. Method (Using ABC)

1. Collect 4-μm-thick paraffin-embedded sections on poly(L-lysine)-coated (0.1%) slides.
2. Deparaffinize the sections by incubation in xylene followed by a descending ethanol series (100, 95, 80, and 75%). Each incubation step is carried out at 5 min at RT.
3. Incubate the sections in methanol containing 5% H_2O_2 for 30 min at RT.
4. Place the slides in PBS, and then incubate the samples with 5% normal serum (from the same species as the host of the second antibody) for 1 h at RT.
5. Incubate the slides with the primary antibody (1:1000–1:2000) overnight at 4°C.
6. Rinse the slides with PBS three times for 5 min each, at RT.
7. Incubate the slides with the biotinylated second antibody at the optimal conditions (1:100–1:300) for 1 h at RT.
8. Rinse the slides with PBS three times for 5 min each, at RT.
9. Incubate the slides with ABC complex (prepared 30 min—1 h prior to use).
10. Rinse the slides with PBS three times for 5 min each, at RT.
11. Incubate the slides with DAB–H_2O_2 solution [40 mg of 3,3'-diaminobenzidine tetrahydrochloride dissolved in 100 ml of either PBS or 0.05-M Tris–HCl buffer (pH 7.6) containing 0.005% H_2O_2] for 1–3 min at RT.

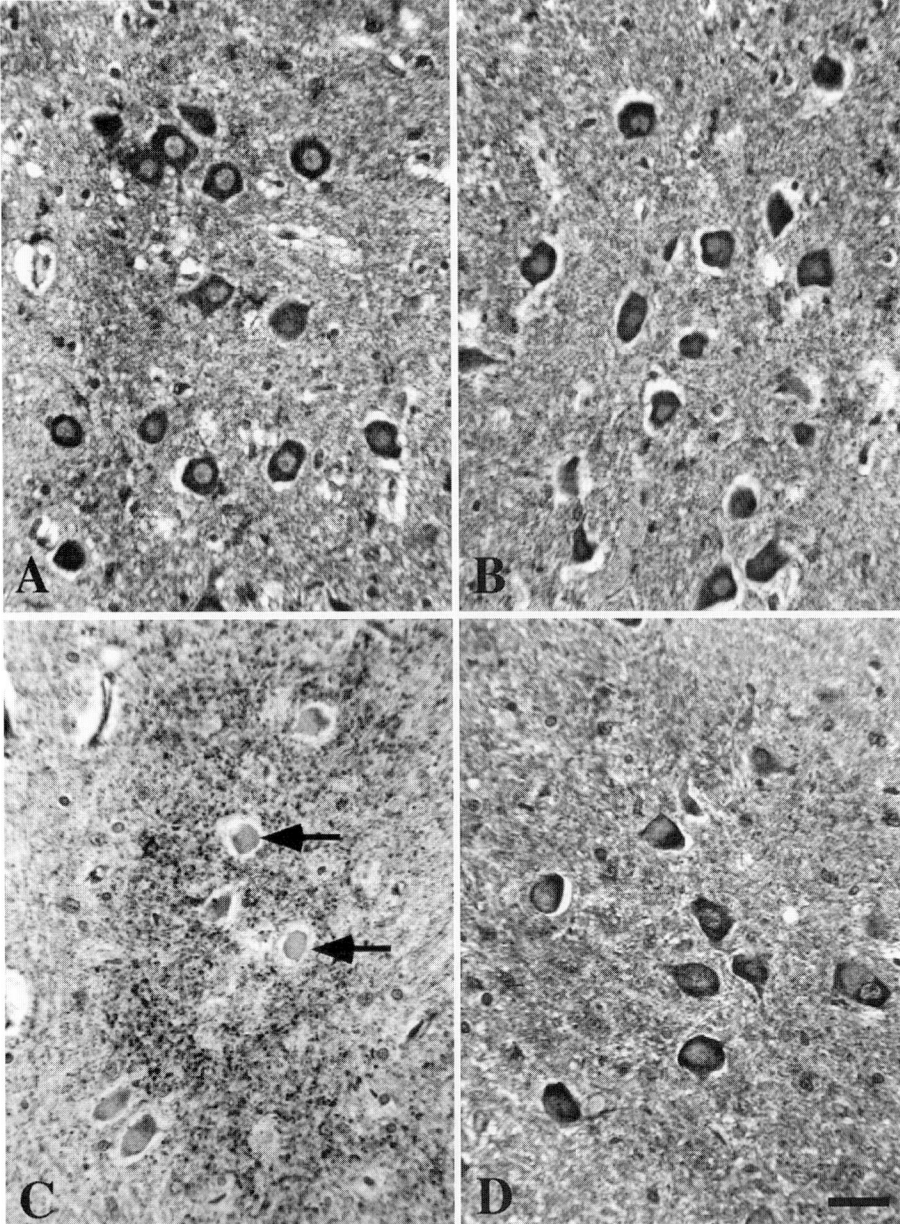

Fig. 3   Immunostaining of sections of the olivary nucleus from a control (A and B) and from a patient with MERRF (C and D) for the localization of the mtDNA-encoded COX II subunit of complex IV (A and C) and the nDNA-encoded FeS subunit of complex III (B and D). The patient shows a marked decrease of immunostain in neurons for COX II (arrows), but normal stain for FeS. Bar = 50 μ. (See Color Insert.)

12. Rinse the slides with dH_2O several times.

13. Counterstain the slides briefly with hematoxylin.

14. Rinse the slides with dH_2O, dehydrate through ascending ethanol series, and clear in xylene.

15. Mount the slides with synthetic resin (Permount).

IV. *In Situ* Hybridization to mtDNA

ISH is a technique that permits the precise cellular localization and identification of cells that express a particular nucleic acid sequence. The essence of this method is the hybridization of a nucleic acid probe with a specific nucleic acid sequence found in a tissue section. ISH has been used extensively in human pathologic conditions to correlate mitochondrial abnormalities with the presence of mutated mtDNAs, an analysis that provides strong support for a pathogenic role of a specific mitochondrial genotype. As the method relies on sequence homology, it has been applied mainly to differentiate mtDNAs with large-scale deletions from wild-type sequences in samples from patients with KSS or isolated OM. In these studies, two probes have been used to determine the distribution of wild-type and Δ-mtDNAs: one probe, in the undeleted region, hybridizes to both wild-type and Δ-mtDNAs (the "common" probe); the other, correspond to mtDNA sequences located within the deletion, hybridizes only to wild-type mtDNAs (the "wild-type" probe). Typical results on serial muscle sections showed abundant hybridization signal (focal accumulations of mtDNAs) with the common probe, but not with the wild-type probe, in COX-deficient RRF of the patients (Bonilla *et al.*, 1992; Moraes *et al.*, 1992). These observations indicated that the predominant species of mtDNAs in COX-deficient RRF of KSS and OM patients are Δ-mtDNAs. Furthermore, the concentration of Δ-mtDNAs in these RRF reached the required threshold level to impair the translation of the mitochondrial genome because they were characterized immunohistochemically by a lack or a marked reduction of the mtDNA-encoded COX II polypeptide (Bonilla *et al.*, 1992; Moraes *et al.*, 1992).

An important extension of ISH to mtDNA is regional ISH, which has been applied to determine the spatial distribution of multiple Δ-mtDNAs in samples from patients with Mendelian-inherited OM. These are disorders characterized by progressive external ophthalmoplegia (PEO) and mitochondrial myopathy. In Mendelian OM, hundreds of different deletions coexist within the same muscle in affected family members. Thus, as opposed to the single deletions found in sporadic KSS and OM, Mendelian OM is associated with multiple Δ-mtDNAs that are apparently generated over the life span of the individual. The genetic defects in these disorders cause mutations in nuclear gene products that regulate the mitochondrial nucleotide pool (DiMauro and Bonilla, 2004). ISH of serial muscle sections from these patients, in which a different mtDNA regional probe was used on each section, showed specific, and different, RRFs losing hybridization signal with each specific probe, while the remaining RRFs hybridized intensely. These observations have provided the strongest evidence to date that each RRF in Mendelian-inherited OM contains a clonal expansion of a single species of Δ-mtDNA (Moslemi *et al.*, 1996; Vu *et al.*, 2000).

Although ISH utilizing RNA probes is more widely used in typical cell and molecular biology applications, most of our recent experience derives from the use of digoxigenin (DIG)-labeled DNA probes to visualize mtDNA (Manfredi *et al.*, 1997; Vu *et al.*, 2000). We describe here a method that we employ for the identification of RRF (Fig. 4) and for the detection of depletion of mtDNA on muscle-frozen sections. Although tissue samples can be frozen or fixed and paraffin-embedded, we described a procedure optimized for frozen sections that has been tested thoroughly on muscle sections from patients with mitochondrial myopathies. Because of the focal pattern of distribution of mutant and wild-type mtDNAs in muscle fibers, one must make sure that serial sections above and below the ones used for ISH are characterized histochemically by staining for COX and SDH activities (Fig. 4).

The size of the labeled probe is very important and should be optimized to promote specificity and tissue penetration. Sizes between 300 and 400 nucleotides are optimum, even though we have obtained excellent results with 500-nucleotide probes. DNA probes can be prepared in different ways, but we routinely use PCR-generated DIG-labeled mtDNA probes. Kits and polymerases for DNA labeling are available from most molecular biology companies. It is important that prior to performing ISH experiments, the concentration of the probe be determined by dot blot and that the specificity of the probe be tested by Southern blot.

1. Method

1. Collect 8-μm-thick frozen sections on poly(L-lysine)-coated (0.1%) slides.
2. Fix the sections with 4% paraformaldehyde for 30 min at RT.
3. Rinse the slides with PBS containing 5-mM $MgCl_2$ (PBS–$MgCl_2$, pH 7.4) three times for 5 min at RT.
4. Incubate the sections with 5-mg/ml proteinase K for 1 h at 37°C.
5. Place the slides in PBS–$MgCl_2$ at 4°C for 5 min to remove the proteinase K.
6. Acetylate the sections by incubating in 0.1-M triethanolamine containing 0.25% acetic anhydride for 10 min at RT.
7. Treat the slides with 5-mg/ml RNase (DNase-free) in RNase buffer (50-mM NaCl and 10-mM Tris–HCl, pH 8) for 30 min at 37°C.
8. Rinse the slides briefly with PBS–$MgCl_2$.
9. Dehydrate the sections through ascending series of ethanol (optional).
10. Incubate with hybridization buffer without probe (prehybridization) for 1–2 h at 37°C. (In our experience, this step can be eliminated.) Hybridization solution: 50% deionized formamide, 20-mM Tris–HCl, 0.5-mM NaCl, 10-mM EDTA, 0.02% Ficoll, 0.02% polyvinylpyrrollidone, 0.12% BSA,

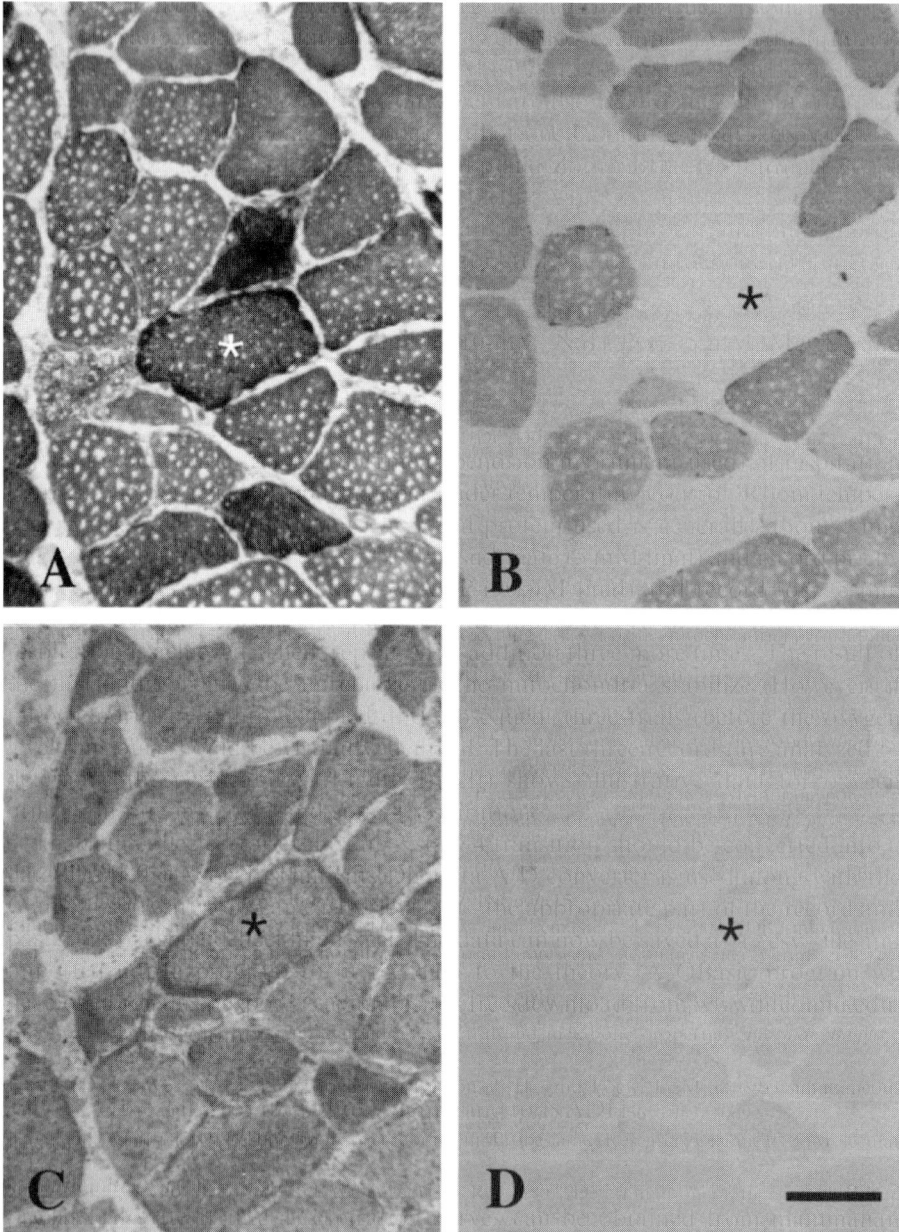

Fig. 4 Cellular localization of mtDNA by ISH. Serial muscle sections from a patient with MERRF were stained for SDH (A) and COX (B) activity, and for mtDNA localized by ISH using digoxigenin-labeled probes (C). The ISH signal is seen as the red material. A control section subjected to ISH without the denaturing step (D) shows no hybridization signal. Note the strong mtDNA signal in an RRF (asterisk) characterized by increased SDH activity and lack of COX activity. Bar = 50 μ. (See Color Insert.)

0.05% salmon sperm DNA, 0.05% total yeast RNA, 0.01% yeast tRNA, and 10% dextran sulfate.

11. Denature the probe/hybridization solution (20 ng/ml) at 92 °C for 10 min and immediately place the solution on ice until it is applied to the sections.

12. Blot off excess hybridization solution and apply probe/hybridization solution to the sections.

13. Denature the sections covered with the hybridization solution at 92 °C for 10–15 min.

14. Hybridize overnight at 42 °C.

15. Rinse the slides briefly with 2× SSC (3-M NaCl, 0.3-M sodium citrate, pH 7.0) at RT followed by rinsing twice with 1× SSC for 15 min at 45 °C, and once with 0.2× SSC for 30 min at 45 °C.

16. Rinse the slides with PBS for 5 min at RT.

17. Incubate the slides with 1% BSA for 1 h at RT.

18. Incubate the slides with alkaline phosphatase-conjugated anti-DIG antibody (1:1000–1:5000) for 1 h at RT.

19. Rinse the slides with PBS three times for 5 min each, at RT.

20. Incubate the slides with color-substrate solution (SIGMA FAST™ Fast Red TR/Naphthol AS-MX Alkaline phosphatase substrate tablets) until the desired strength of the signal is obtained.

21. Mount the sections with glycerol-PBS, 1:1.

Acknowledgments

This work was supported by grants from the National Institutes of Health (NS11766 and PO1HD32062).

References

Ackrell, B. A. C. (2002). Cytopathies involving mitochondrial complex II. *Mol. Aspects Med.* **23,** 369–384.

Anderson, S., Bankier, A. T., Barrel, B. G., de Bruin, N. H. L., Coulson, A. R., Drouin, J., Operon, I. C., Nierlich, D. P., Roe, B. A., Sanger, F., Schreier, P. H., Smith, A. J., *et al.* (1981). Sequence and organization of the human mitochondrial genome. *Nature* **290,** 457–465.

Andreetta, F., Tritschler, H. J., Schon, E. A., DiMauro, S., and Bonilla, E. (1991). Localization of mitochondrial DNA using immunological probes: A new approach for the study of mitochondrial myopathies. *J. Neurol. Sci.* **105,** 88–92.

Attardi, G., and Schatz, G. (1988). Biogenesis of mitochondria. *Annu. Rev. Cell Biol.* **4,** 289–333.

Bedetti, C. D. (1985). Immunocytological demonstration of cytochrome *c* oxidase with an immuno-peroxidase method. *J. Histochem. Cytochem.* **33,** 446–452.

Bogenhagen, D., and Clayton, D. A. (1984). The number of mitochondrial DNA genomes in mouse and human HeLa cells. *J. Biol. Chem.* **249,** 7991–7995.

Bonilla, E., Schotland, D. L., DiMauro, S., and Aldover, B. (1975). Electron cytochemistry of crystalline inclusions in human skeletal muscle mitochondria. *J. Ultrastruct. Res.* **51,** 404–408.

Bonilla, E., Sciacco, M., Tanji, K., Sparaco, M., Petruzzella, V., and Moraes, C. T. (1992). New morphological approaches for the study of mitochondrial encephalopathies. *Brain Pathol.* **2,** 113–119.

Bonilla, E., Tanji, K., Hirano, M., Vu, T. H., DiMauro, S., and Schon, E. A. (1999). Mitochondrial involvement in Alzheimer's disease. *Biochem. Biophys. Acta* **1410,** 171–182.

Capaldi, R. A. (1990). Structure and assembly of cytochrome *c* oxidase. *Arch. Biochem. Biophys.* **280,** 252–262.

Cooper, C. E., Nicholls, P., and Freedman, J. A. (1991). Cytochrome *c* oxidase: Structure, function, and membrane topology of the polypeptide subunits. *Biochem. Cell Biol.* **69,** 596–607.

DiMauro, S., and Bonilla, E. (2004). Mitochondrial encephalomyopathies. *In* "Myology Third Edition" (A. G. Engel, and C. Franzini-Armstrong, eds.), pp. 1623–1662. McGraw-Hill, New York.

DiMauro, S., Hirano, M., Bonilla, E., Moraes, C. T., and Schon, E. A. (1994). Cytochrome oxidase deficiency: Progress and problems. *In* "Mitochondrial Disorders in Neurology" (A. H. V. Shapira, and S. DiMauro, eds.), pp. 91–115. Butterworth-Heiemann, Oxford.

Engel, W. K., and Cunningham, G. G. (1963). Rapid examination of muscle tissue: An improved trichrome stain method for fresh frozen biopsy sections. *Neurology* **13,** 919–926.

Giles, R. E., Blanc, H., Cann, H. M., and Wallace, D. C. (1980). Maternal inheritance of human mitochondrial DNA. *Proc. Natl. Acad. Sci. USA* **77,** 6715–6719.

Goto, Y. I., Nonaka, I., and Horai, S. (1990). A mutation in the tRNA$^{Leu(UUR)}$ gene associated with the MELAS subgroup of mitochondrial encephalopathies. *Nature* **348,** 651–653.

Goto, Y. I., Nonaka, I., and Horai, S. (1991). A new mtDNA mutation associated with mitochondrial myopathy, encephalopathy, lactic acidosis, and stroke-like episodes. *Biochem. Biophys. Acta* **1097,** 238–240.

Haller, R. G., Henriksson, K. G., Jorfeldt, L., Hultman, E., Wibom, R., Sahlin, K., Areskog, N. H., Gunder, M., Ayyad, K., Blomqvist, C. G., Hall, R. E., Thuiller, P., *et al.* (1991). Deficiency of skeletal muscle SDH and aconitase. *J. Clin. Invest.* **88,** 1197–1206.

Hatefi, Y. (1985). The mitochondrial electron transport and oxidative phosphorylation system. *Annu. Rev. Biochem.* **54,** 1015–1076.

Hattori, N., Tanaka, M., Ozawa, T., and Mizuno, Y. (1990). Immunohistochemical studies of complexes I, II, III and IV of mitochondria in Parkinson disease. *Ann. Neurol.* **30,** 563–571.

Hayashi, J.-I., Ohta, S., Kikuchi, A., Takemitsu, M., Goto, Y. I., and Nonaka, I. (1991). Introduction of disease-related mitochondrial DNA deletions into HeLa cells lacking mitochondrial DNA results in mitochondrial dysfunction. *Proc. Natl. Acad. Sci. USA* **88,** 10614–10618.

Hsu, S. M., Raine, L., and Fanger, H. (1981). Use of avidin-peroxidase complex (ABC) in immuno-peroxidase techniques: A comparison between ABC and unlabeled antibody (PAP) procedures. *J. Histochem. Cytochem.* **29,** 577–580.

Johnson, M. A., Turnbull, D. M., Dick, D. J., and Sherratt, H. S. (1983). A partial deficiency of cytochrome *c* oxidase in chronic progressive external ophtalmoplegia. *J. Neurol. Sci.* **60,** 31–53.

Johnson, M. A., Kadenbach, B., Droste, M., Old, S. L., and Turnbull, D. M. (1988). Immunocytochemical studies of cytochrome c oxidase subunits in skeletal muscle of patients with partial cytochrome oxidase deficiency. *J. Neurol. Sci.* **87,** 75–90.

Kadenbach, B., Kuhn-Nentwig, L., and Buge, U. (1987). Evolution of a regulatory enzyme: Cytochrome *c* oxidase (Complex IV). *Curr. Topics Bioenerg.* **15,** 113–161.

Kloeckener-Gruissem, B., McEwen, J. E., and Poyton, R. O. (1987). Nuclear functions required for cytochrome *c* oxidase biogenesis in *Saccharomyces cerevisiae*: Multiple trans-acting nuclear genes exert specific effects on expression of each of the cytochrome *c* oxidase subunits encoded on mitochondrial DNA. *Curr. Genet.* **12,** 311–322.

Manfredi, G., Vu, T. H., Bonilla, E., Schon, E. A., DiMauro, S., Arnaudo, E., Zhang, L., Rowland, L. P., and Hirano, M. (1997). Association of myopathy with large-scale mitochondrial DNA duplications and deletions: Which is pathogenic? *Ann. Neurol.* **42,** 180–188.

Moraes, C. T., DiMauro, S., Zeviani, M., Lombes, A., Shanske, S., Miranda, A. F., Nakase, H., Bonilla, E., Werneck, L. C., Servidei, S., *et al.* (1989). Mitochondrial DNA deletions in progressive external ophthalmoplegia and Kearns-Sayre syndrome. *N. Engl. J. Med.* **320,** 1293–1299.

Moraes, C. T., Ricci, E., Petruzzella, V., Shanske, S., DiMauro, S., Schon, E. A., and Bonilla, E. (1992). Molecular analysis of the muscle pathology associated with mitochondrial DNA deletions. *Nat. Genet.* **1,** 359–367.

Moslemi, A. R., Melberg, A., Holme, E., and Oldfors, A. (1996). Clonal expansion of mitochondrial DNA with multiple deletions in autosomal dominant progressive external ophthalmoplegia. *Ann. Neurol.* **48,** 707–713.

Nakase, H., Moraes, C. T., Rizzuto, R., Lombes, A., DiMauoro, S., and Schon, E. A. (1990). Transcription and translation of deleted mitochondrial genomes in Kearns-Sayre syndrome: Implications for pathogenesis. *Am. J. Hum. Genet.* **46,** 418–427.

Neupert, W. (1997). Protein import into mitochondria. *Ann. Rev. Biochem.* **66,** 863–917.

Papadopoulou, L. C., Sue, C. M., Davidson, M. M., Tanji, K., Nishino, I., Sadlock, J. E., Krishna, S., Walker, W., Selby, J., Moira Glerum, D., Van Coster, R., Lyon, G., *et al.* (1999). Fatal infantile cardioencephalopathy with cytochrome *c* oxidase (COX) deficiency and mutations in *SCO2*, a human COX assembly gene. *Nat. Genet.* **23,** 333–337.

Pel, H. J., Tzagoloff, A., and Grivell, L. A. (1992). The identification of 18 nuclear genes required for the expression of the yeast mitochondrial gene encoding cytochrome *c* oxidase subunit I. *Curr. Genet.* **21,** 139–146.

Pette, D. (1981). Microphotometric measurement of initial maximum reaction rates in quantitative enzyme histochemistry *in situ*. *Histochem. J.* **13,** 319–327.

Satoh, M., and Kuroiwa, T. (1991). Organization of multiple nucleotides of DNA molecules in mitochondria of a human cell. *Exp. Cell Res.* **196,** 137–140.

Seligman, A. M., and Rutenburg, R. M. (1951). The histochemical demonstration of succinic dehydrogenase. *Science* **113,** 317–320.

Seligman, A. M., Karnovsky, M. J., Wasserkrug, H. L., and Hanker, J. S. (1968). Non-droplet ultrastructural demonstration of cytochrome oxidase activity with a polymerizing osmiophilic reagent, diaminobenzidine (DAB). *J. Cell Biol.* **38,** 1–15.

Shoffner, J. M., Lott, M. T., Lezza, A. M. S., Seibel, P., Ballinger, S. W., and Wallace, D. C. (1990). Myoclonic epilepsy and ragged-red fiber disease (MERRF) is associated with a mitochondrial DNA tRNA[Lys] mutation. *Cell* **61,** 931–937.

Shoubridge, E. A. (1993). Molecular histology of mitochondrial diseases. *In* "Mitochondrial DNA in Human Pathology" (S. DiMauro, and D. C. Wallace, eds.), pp. 109–123. Raven Press, New York.

Shoubridge, E. A. (2001). Nuclear genetic defects of oxidative phosphorylation. *Hum. Mol. Genet.* **10,** 2277–2284.

Silvestri, G., Moraes, C. T., Shanske, S., Oh, S. J., and DiMauro, S. (1992). A new mtDNA mutation in the tRNA[Lys] gene associated with myoclonic epilepsy and ragged-red fibers (MERRF). *Am. J. Hum. Genet.* **51,** 1213–1217.

Sparaco, M., Schon, E. A., DiMauro, S., and Bonilla, E. (1995). Myoclonus epilepsy with ragged-red fibers (MERRF): An immunohistochemical study of the brain. *Brain Pathol.* **5,** 125–133.

Szabolcs, M. J., Seigle, R., Shanske, S., Bonilla, E., DiMauro, S., and D'Agati, V. (1994). Mitochondrial DNA deletion: A cause of chronic tubulointerstitial nephropathy. *Kidney Int.* **45,** 1388–1396.

Taanman, J. W. (1997). Human cytochrome c oxidase: Structure, function and deficiency. *J. Bioenerg. Biomembr.* **29,** 151–163.

Taanman, J. W., Burton, M. D., Marusich, M. F., Kennaway, N. G., and Capaldi, R. A. (1996). Subunit specific monoclonal antibodies show different steady-state levels of various cytochrome c oxidase subunits in chronic external ophthalmoplegia. *Biochem. Biophys. Acta* **1315,** 199–207.

Tanji, K., Vu, T. H., Schon, E. A., DiMauro, S., and Bonilla, E. (1999). Kearns-Sayre syndrome: Unusual pattern of expression of subunits of the respiratory chain in the cerebellar system. *Ann. Neurol.* **45,** 377–383.

Tanji, K., Gamez, J., Cervera, C., Mearin, F., Ortega, A., de la Torre, J., Montoya, J., Andreu, A. L., DiMauro, S., and Bonilla, E. (2003). The A8344G mutation in mitochondrial DNA associated with stroke-like episodes and gastrointestinal dysfunction. *Acta Neuropathol.* **105,** 69–75.

Taylor, R. W., Birch-Machin, M. A., Schaefer, J., Taylor, L., Shakir, R., Ackrell, B. A., Cochran, B., Bindoff, L. A., Jackson, M. J., Griffiths, P., and Turnbull, D. M. (1996). Deficiency of complex II of the mitochondrial respiratory chain in late-onset optic atrophy and ataxia. *Ann. Neurol.* **39**, 224–232.

Tiranti, V., Hoertnagel, K., Carrozo, R., Galimberti, C., Munaro, M., Granatiero, M., Zelante, L., Gasparini, P., Marzella, R., Rocchi, M., Bayona-Bafaluy, M. P., Enriquez, J. A., *et al.* (1998). Mutations of *SURF-1* in Leigh disease associated with cytochrome c oxidase deficiency. *Am. J. Hum. Genet.* **63**, 1609–1621.

Tritschler, H. J., Andreetta, F., Moraes, C. T., Bonilla, E., Arnaudo, E., Danon, M. J., Glass, S., Zalaya, B. M., Vamos, E., Shanske, S., Kadenbach, B., DiMauro, S., *et al.* (1992). Mitochondrial myopathy of childhood with depletion of mitochondrial DNA. *Neurology* **42**, 209–217.

Vu, T. H., Tanji, K., Valsamis, H., DiMauro, S., and Bonilla, E. (1998). Mitochondrial DNA depletion in a patient with long survival. *Neurology* **51**, 1190–1193.

Vu, T. H., Tanji, K., Palotti, F., Golzi, V., Hirano, M., DiMauro, S., and Bonilla, E. (2000). Analysis of muscle from patients with multiple mtDNA deletions by *in situ* hybridization. *Muscle Nerve* **23**, 80–85.

Wallace, D. C. (1992). Diseases of the mitochondrial DNA. *Annu. Rev. Biochem.* **61**, 1175–1212.

Wong-Riley, M. T. T. (1989). Cytochrome oxidase: An endogenous metabolic marker for neuronal activity. *Trends Neurosci.* **12**, 94–101.

Zeviani, M., Moraes, C. T., DiMauro, S., Nakase, H., Bonilla, E., Schon, E. A., and Rowland, L. P. (1988). Deletions of mitochondrial DNA in Kearns-Sayre syndrome. *Neurology* **38**, 1339–1346.

Zhu, Z., Yao, J., Johns, T., Fu, K., De Bie, I., Macmillan, C., Cuthber, A. P., Newbold, R. F., Wang, J., Chevrette, M., Brown, G. K., Brown, R. M., *et al.* (1998). *SURF-1*, encoding a factor involved in the biogenesis of cytochrome c oxidase, is mutated in Leigh syndrome. *Nat. Genet.* **20**, 337–343.

CHAPTER 7

Assay of Mitochondrial ATP Synthesis in Animal Cells and Tissues

Cristofol Vives–Bauza, Lichuan Yang, and Giovanni Manfredi

Department of Neurology and Neuroscience
Weill Medical College of Cornell University
New York, New York 10021

I. Introduction

Mitochondria are the major cellular source of adenosine triphosphate (ATP) synthesis. Its oxidative phosphorylation system generates 36 molecules of ATP per molecule of glucose, as opposed to only 2 molecules of ATP generated by

0091-679X/07 $35.00
DOI: 10.1016/S0091-679X(06)80007-5

glycolysis in the cytoplasm. Therefore, the measurement of mitochondrial ATP synthesis can be considered a pivotal tool in understanding many important characteristics of cellular energy metabolism, both in the normal physiological state and in pathological conditions such as mitochondrial disorders. In the first section of this chapter, we will discuss approaches to measure ATP synthesis from mammalian cells and tissues. In the second section, we will discuss procedures to estimate the steady-state content of ATP and other high-energy phosphates by high-performance liquid chromatography (HPLC).

II. ATP Synthesis Assays

Two methodological issues need to be addressed when measuring mitochondrial ATP synthesis. The first one is how to deliver reaction substrates to the mitochondria. Because of the low permeability of the plasma membrane to some hydrophilic substrates, such as adenosine diphosphate (ADP), some investigators prefer to analyze isolated mitochondria (Tatuch and Robinson, 1993; Tuena de Gomez-Puyou *et al.*, 1984; Vazquez-Memije *et al.*, 1996). However, for ADP phosphorylation to take place, mitochondria need to be maintained intact and in a coupled state. Therefore, ATP synthesis can only be measured on freshly isolated mitochondria. Moreover, the isolation of highly coupled mitochondria from cultured cells involves delicate and time-consuming procedures, and the results are sometimes inconsistent, leading to a potential lack of reproducibility. Measurement of ATP synthesis on whole cells requires a permeabilization step to allow for the penetration of hydrophilic substrates through biological membranes. Cell membranes can be permeabilized with detergents such as saponin (Kunz *et al.*, 1993) or digitonin (Houstek *et al.*, 1995; Wanders *et al.*, 1993, 1994, 1996). In addition, the use of permeabilized cells rather than isolated mitochondria reduces substantially the number of cells required for each assay.

The second issue is how to detect and quantify ATP. Fluorimetry is a commonly used method to detect ATP produced by isolated mitochondria (Houstek *et al.*, 1995; Tatuch and Robinson, 1993; Wanders *et al.*, 1993, 1994, 1996). Another method employed on isolated mitochondria is the incorporation of [32]Pi into ADP and its subsequent transfer to glucose-6-phosphate by hexokinase, followed by extraction of unincorporated [32]Pi and measurement of radioactivity in a scintillation counter (Tuena de Gomez-Puyou *et al.*, 1984).

Typically, for steady-state ATP measurements based on fluorescence or radio-assays, isolated mitochondria or permeabilized cells are incubated with appropriate substrates followed by a lysis extraction in strong acid conditions to inactivate cellular ATPases. With this approach, in order to obtain kinetic measurements of ATP synthesis, replicate tests have to be run at various time intervals. Alternatively, it would be convenient to perform measurements of ATP synthesis, which do not require repeated sampling and which allow for kinetic measurements to be performed on single samples. In this chapter, we describe a rapid kinetic approach to monitor continuous ATP synthesis in mammalian cells that takes advantage of the luciferase–luciferin system.

Firefly luciferase is widely used as a reporter for gene expression to study promoter regulation in mammalian cells, but its bioluminescence properties have also been used to measure ATP content in isolated mitochondria (Lemasters and Hackenbrock, 1973; Strehler and Totter, 1952; Wibom *et al.*, 1990, 1991) and in permeabilized cells (James *et al.*, 1999; Maechler *et al.*, 1998; Ouhabi *et al.*, 1998). The reaction catalyzed by luciferase is:

$$\text{Luciferase} + \text{Luciferin} + \text{ATP} \rightarrow \text{Luciferase - Luciferyl - AMP} + \text{PPi}$$

$$\text{Luciferase - Luciferyl - AMP} + O_2 \rightarrow \text{Luciferase} + \text{Oxyluciferin} + \text{AMP} + CO_2 + hv$$

The reaction produces a flash of yellow-green light, with a peak emission at 560 nm, whose intensity is proportional to the amount of substrates in the reaction mixture (Deluca, 1976).

III. Methodological Considerations

A. Cell Permeabilization (Detergent Titration in Cultured Cells)

Although more convenient than isolating coupled mitochondria, permeabilization procedures also require standardization. Insufficient permeabilization could result in an underestimation of ATP synthesis due to lack of available substrates. Conversely, excessive permeabilization leads to mitochondrial membrane damage and uncoupling. For these reasons, it is important to establish the optimal amount of detergent needed per unit of cell protein. We use digitonin as the detergent of choice because it results in plasma membrane permeabilization at concentrations that do not affect mitochondrial membranes significantly. Alternatively, saponin may be used as a detergent (Kunz *et al.*, 1993) particularly in muscle cells. Due to variation in membrane cholesterol content, the parameters for digitonin treatment need to be optimized for each cell type. In our experience, the digitonin concentration at which HeLa and COS-7 cells (1-mg/ml cell proteins) become permeable to substrates and achieve the maximal ATP synthesis rate is 50 μg/ml, while N2A mouse neuroblastoma cells require 25 μg/ml and HEK 293T (HEK, human embryonic kidney cells) cells require 75 μg/ml (Fig. 1).

B. ATP Detection by Luciferase–Luciferin

A number of variables need to be taken into consideration in setting up a luminescence assay. First, depending on the ATP concentrations, firefly luciferase shows two different rates of light production, possibly due to the binding of substrates at two different catalytic sites. At high concentrations of ATP, a short flash

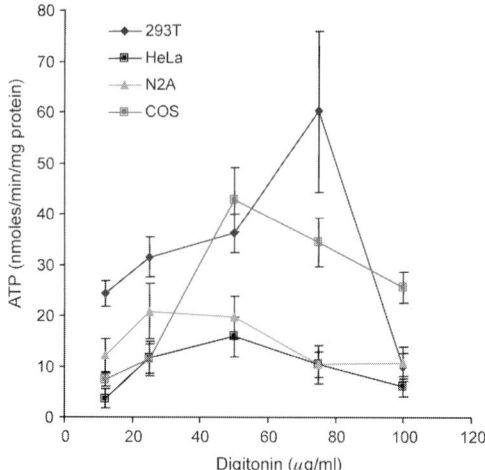

Fig. 1 Titration of digitonin. HeLa, COS-7, HEK 293T, and N2A cells were incubated with increasing concentrations of digitonin. All measurements were performed after 1 min of incubation with digitonin, followed by a wash in buffer A. ATP synthesis was performed using malate/pyruvate as substrates. The maximal ATP synthesis in HeLa and COS-7 cells is obtained with 50-μg/ml digitonin. In contrast, N2A cells show maximum ATP production after being permeabilized with 25-μg/ml digitonin, while HEK 293T require 75 μg/ml.

of light is produced followed by enzyme inactivation. At low concentrations of ATP, the flash is less intense but more prolonged (DeLuca and McElroy, 1974). We observed that in the range of ATP concentrations normally present in cultured cells, the initial flash and the ensuing inactivation of luciferase could be prevented by preincubating luciferase with luciferin for 10 min on ice, prior to the assay (see buffer B in Table I). Second, the kinetics of luciferase is unstable at low concentrations of ATP and high concentrations of luciferin (Lembert and Idahl, 1995), and its activity decreases over time. Therefore, the concentration of luciferin in the reaction mixture has to be appropriate for the levels of ATP present in cells. Furthermore, buffer B (containing the luciferase–luciferin mix) has to be protected from ambient light and kept on ice until the beginning of the measurement. In order to test for the stability of the luciferin–luciferase complex in buffer B, measure the luminescence derived from a fixed ATP standard in between measurements on test samples.

C. Specificity of the Assay

ATP is formed not only through mitochondrial oxidative phosphorylation, but also through other metabolic pathways such as glycolysis and adenylate kinase. Therefore, to obtain a specific measurement of mitochondrial ATP synthesis, these other sources of ATP must be either accounted for or, preferably, eliminated.

Table I
Media and Reagents[a]

	Volume (final concentration)	Stock solutions	Notes
Buffer A (cell suspension)	160 μl	150-mM KCl, 25-mM Tris–HCl, 2-mM EDTA, 0.1% BSA, 10-mM potassium phosphate, 0.1-mM MgCl$_2$, pH 7.4	Store at $-20\,^{\circ}$C; at room temperature for the assay
Digitonin	5 μl (see Fig. 1 for the appropriate concentration)	10 mg/ml in DMSO	Store at $-20\,^{\circ}$C; at room temperature for the assay
Di(adenosine pentaphosphate)	5 μl (0.15 mM)	6 mM in water	Store at $-20\,^{\circ}$C; thaw and keep on ice
Malate[b]	2.5 μl (1 mM)	80 mM in buffer A	1-M stock in water; store at $-20\,^{\circ}$C; thaw and dilute
Pyruvate[b]	2.5 μl (1 mM)	80 mM in buffer A	1-M stock in water; store at $-20\,^{\circ}$C; thaw and dilute
Succinate[b]	5 μl (5 mM)	200 mM in buffer A	1-M stock in water; store at $-20\,^{\circ}$C; thaw and dilute
Rotenone	2 μg/ml	400 μg/ml in ethanol	Store at $-20\,^{\circ}$C; thaw and keep on ice
Atractyloside[b]	10 μl (0.78 mM)	12.5 mM in water	Prepare the same day; keep on ice
Oligomycin[b]	2 μl (2 μg/ml)	0.2 mg/ml in ethanol	Store at $-20\,^{\circ}$C; thaw and keep on ice
ADP	5 μl (0.1 mM)	4 mM in buffer A	Prepare the same day; keep on ice
Luciferin	See buffer B	100 mM in water	Store at $-20\,^{\circ}$C; thaw and keep on ice
Luciferase	See buffer B	1 mg/ml in 0.5-M Tris–acetate, pH 7.75	Store at $-20\,^{\circ}$C; thaw and keep on ice
Buffer B	10 μl	0.5-M Tris–acetate, pH 7.75, 0.8-mM luciferin, 20-μg/ml luciferase	Prepare the same day; keep on ice
Total volume	To 200 μl with buffer A		

[a]All reagents other than *Photinus pyralis* luciferase (Roche Molecular Biochemicals, Indianapolis, IN) and luciferin (Promega Life Sciences, Madison, WI) are from Sigma Biochemicals, St. Louis, MO.

[b]Malate plus pyruvate or succinate is used alternatively as substrate; atractyloside or olygomycin is used alternatively as inhibitor.

Adenylate kinase catalyzes the following reversible reaction:

$$2ADP \rightleftarrows ATP + AMP$$

When ADP is added to the cells, the luminescence response shows an initial rapid phase due to adenylate kinase activity and a second slow phase derived from oxidative phosphorylation. For these reasons, P^1,P^5-di(adenosine) pentaphosphate, an inhibitor of adenylate kinase (Kurebayashi *et al.*, 1980), is added to the reaction mixture to eliminate the initial rapid phase of light emission. Moreover, prior to the measurement, the cells are rinsed and maintained in glucose-free buffer to deplete most of the residual intracellular glucose. By assaying luminescence with and without inhibitors of mitochondrial ATP synthesis, such as atractyloside (an inhibitor of the adenine nucleotide translocator) or oligomycin (an inhibitor of the ATP synthase), it is possible to determine how much luminescence derives from the ATP that is not produced by oxidative phosphorylation. The inhibitor-insensitive ATP synthesis is subtracted from the rate of synthesis of total ATP measured.

IV. Experimental Procedures

This section describes the assay of ATP synthesis in five cell types: HeLa, COS-7, N2A, HEK 293T, and 143B-derived cytoplasmic hybrids (cybrids; King and Attardi, 1989) harboring either wild-type mitochondrial DNA (mtDNA) or the T8993G mtDNA mutation in the ATPase 6 gene, which is responsible for a mitochondrial disorder characterized by neuropathy, ataxia, and retinitis pigmentosa (NARP; Holt *et al.*, 1990).

A. Measurement of ATP Synthesis in Cultured Cells

1. Cell Culture

Cells are grown in 100-mm culture dishes in Dulbecco's Modified Eagle's Medium (DMEM) containing high glucose (4.5 mg/ml), 2-mM L-glutamine, 110-mg/liter sodium pyruvate, and supplemented with 5% fetal bovine serum (FBS) at 37°C and 5% CO_2. Cells are harvested by trypsinization when they reach ~80% confluence. Cells are then sedimented by centrifugation at $800 \times g$ at RT, and the cell pellet is washed with glucose-free, serum-free DMEM. Cells are counted in a hemocytometer or an automated cell counter, and resuspended in glucose-free, serum-free DMEM at a concentration of 1.5×10^6 cells/ml.

2. Measurement of ATP Synthesis

Media and reagents used for these assays are shown in Table I. For HeLa, COS-7, and 143B-derived cybrids, 1.5×10^6 cells are used in each assay. Cells are concentrated by centrifugation at $800 \times g$ at RT, and the cell pellet is resuspended in 160 μl of buffer A (Table I) at RT. In the case of HEK 293T and N2A cells, 2×10^6 cells are used in each assay. Buffer A-containing cells are incubated with the appropriate amount of digitonin (Fig. 1) for 1 min at RT, with gentle agitation. Digitonin is removed by washing cells with 1 ml of buffer A. Cells are concentrated by centrifugation at $800 \times g$ at RT, and the cell pellet is resuspended in 160 μl of buffer A and P^1,P^5-di(adenosine) pentaphosphate (to 0.15 mM). Add 10 μl of buffer B (containing luciferin and luciferase), ADP (to 0.1 mM), and either malate plus pyruvate (both to 1 mM) or succinate (to 5 mM) plus 2-μg/ml rotenone to the cell suspension to obtain the baseline luminescence corresponding to nonmitochondrial ATP production, one replicate tube for each sample is prepared containing the components described above plus either 1-μg/ml oligomycin or 1-mM atractyloside. These inhibitors have no effect on luciferase activity.

Cells are transferred to a luminometer cuvette and, after a gentle mixing with a vortexer for 2 sec, they are placed in a counting luminometer and light emission is recorded. We use an Optocomp I luminometer (MGM Instruments, Inc., Hamden, CT), which allows for multiple recordings in the kinetic mode. The integration time for each reading is set at 1 sec and the interval between readings at 15 sec, for a total recording time of 3.5 min (15 readings). Other luminometers can be used, but those that allow for kinetic protocols are preferable. It is also useful to interface the instrument with a computer to facilitate data collection and analysis. Total cellular protein content is measured on the digitonized cells using the Bio-Rad DC protein assay kit (Bio-Rad Laboratories, Hercules, CA). Serial dilutions of bovine serum albumin (BSA) are used as standards.

Figure 2 shows a representative assay on HeLa cells. The luminescence curves describe the kinetics of ATP synthesis in cells energized with malate plus pyruvate in the presence and absence of oligomycin. Luminescence increases linearly for ~3.5 min. It reaches a peak at 4–5 min, followed by a progressive decrease (not shown). The decrease in luminescence is probably due to the chemical properties of luciferin, which is converted progressively into the inactive derivate deoxyluciferin (Lembert and Idahl, 1995). The linear portion of the curve is used to extrapolate the variation in luminescence per unit of time (change in relative light units, ΔRLU). The ΔRLU measured in the presence of the inhibitor is subtracted from the total ΔRLU in order to obtain the proportion of ΔRLU derived from mitochondrial ATP synthesis. The change in luminescence is then converted to ATP concentration

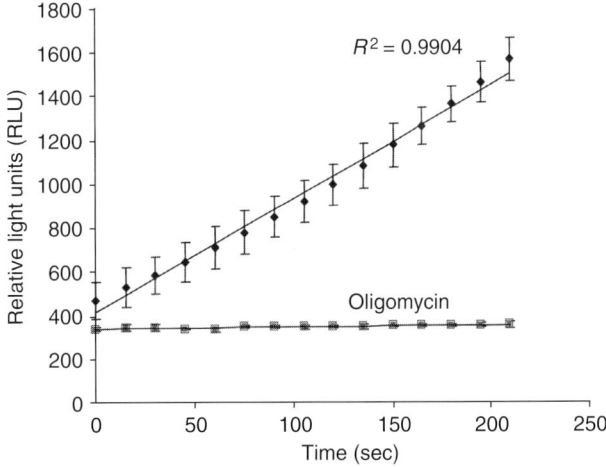

Fig. 2 Linear luminescence curve obtained from 1×10^6 HeLa cells permeabilized with 50-μg/ml digitonin, using malate/pyruvate as substrates, with and without the addition of 0.5 μg of oligomycin. Luminescence (expressed as RLU) was recorded at 15-sec intervals with a 1-sec integration time.

based on an ATP standard curve, since the luminescence is directly proportional to the ATP concentration in the physiological concentration range (Strehler, 1968). A standard ATP–luminescence curve is constructed by measuring flash luminescence derived from ATP solutions containing 0-, 0.05-, 0.1-, 0.5-, 1-, 5-, and 10-mM ATP in buffer A plus 10 μl of buffer B, using a single point reading (Fig. 3). ATP synthesis rates with malate plus pyruvate in HeLa, COS-7, N2A, HEK 293T are shown in Fig. 4A, and ATP synthesis in 143B-derived cybrid cells in Fig. 4B.

B. ATP Synthesis in Mitochondria Isolated from Animal Tissues

We use the luciferin–luciferase method to measure ATP synthesis in mitochondria freshly isolated from mouse liver and brain.

1. Isolation of Mitochondria

Reagents
Buffer H

0.22-M D-mannitol
0.007-M sucrose
20-mM HEPES
1-mM EGTA
1% BSA
pH 7.2

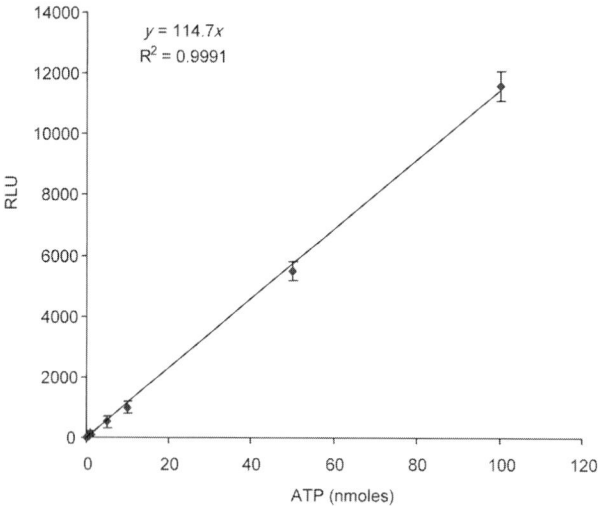

Fig. 3   ATP standard assay. Linear regression analysis for the correlation between ATP concentration and luciferin/luciferase-dependent luminescence (RLU). Each point indicates the value corresponding to the mean luminescence $\pm$ SD ($n = 5$); $R^2 = 0.9991$.

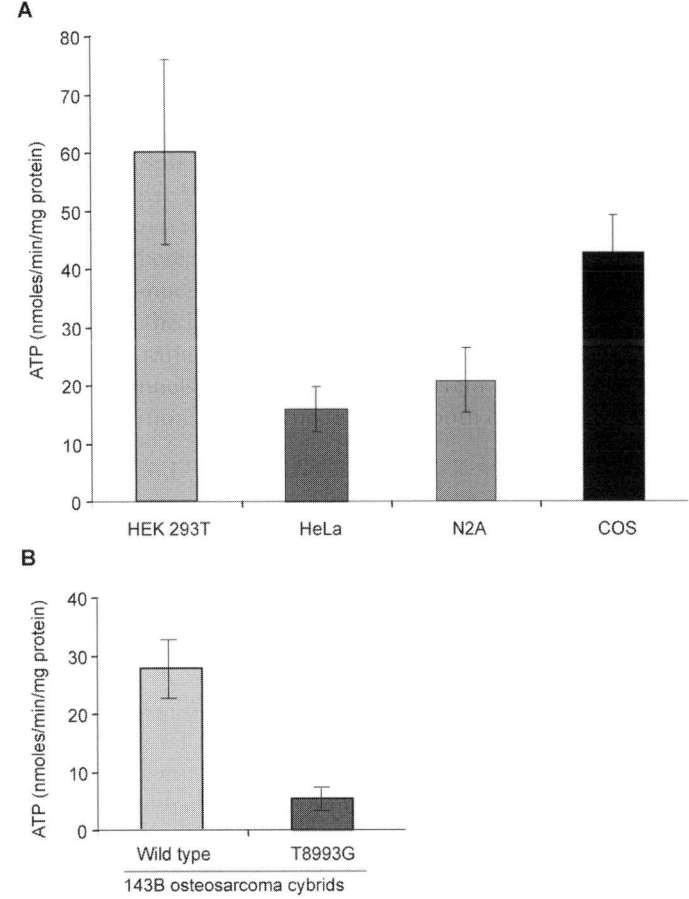

Fig. 4 (A) ATP synthesis rates in HeLa, COS-7, HEK 293T, and N2A cells using malate/pyruvate as substrates. Bars indicate the values corresponding to the mean activity $\pm$ SD ($n = 3$). (B) ATP synthesis in wild-type and T8993G mutant cytoplasmic hybrids (cybrids) resulting from fusion of mtDNA-less 143B osteosarcoma cells (ρ^0 cells) with platelets from individuals affected by heteroplasmic T8993G mtDNA mutations associated with the mitochondrial disease NARP (neuropathy, ataxia, and retinitis pigmentosa). ATP synthesis was measured using either malate/pyruvate as substrates. Bars indicate the values corresponding to the mean activity $\pm$ SD ($n = 3$).

Protocol

A whole mouse liver is homogenized with 8 ml of ice-cold buffer H, containing 0.22-M D-mannitol, 0.07-M sucrose, 20-mM HEPES, 1-mM EGTA, and 1% BSA, pH 7.2 in a glass-teflon pestle, gently by hand with approximately five strokes. The homogenate is centrifuged at 1500 $\times$ g for 5 min at 4°C. The supernatant is centrifuged at 10,000 $\times$ g for 10 min at 4°C. The mitochondria-rich pellet is resuspended in a small volume ($\sim$250 μl) of buffer H and kept on ice. The final protein concentration is 40–60 mg/ml. Higher dilutions are not recommended because they tend to cause uncoupling of mitochondrial respiration.

Mitochondria from whole mouse brain are isolated by homogenization in 2 ml of ice-cold buffer H. The homogenate is centrifuged at $1500 \times g$ for 5 min at $4\,^\circ$C. The supernatant is kept on ice, while the resulting pellet is resuspended in 1 ml of buffer H and subjected to a second centrifugation of $1500 \times g$ at $4\,^\circ$C. The two supernatants are combined and centrifuged at $13{,}500 \times g$ for 10 min at $4\,^\circ$C. The mitochondria-rich pellet is resuspended in 50–100 μl of buffer H and kept on ice. The final protein concentration is 20–40 mg/ml.

2. Measurement of ATP Synthesis

Reagents
Buffer R

  0.25-M sucrose
  50-mM HEPES
  2-mM $MgCl_2$
  1-mM EGTA
  10-mM KH_2PO_4
  pH 7.4
  30-mM glutamate
  30-mM malate
  300-μM ADP
  1-μg/ml oligomycin

Protocol
Before starting to measure the ATP synthesis, we test the coupling state of mitochondria by polarography. Approximately 600 μg of liver and 400 μg of brain mitochondrial proteins are resuspended in 0.3 ml of buffer R. Oxygen consumption is measured in a Clark-type electrode oxygraph (Hansatech, UK) using either 20-mM succinate or 30-mM glutamate plus 30-mM malate in the absence of exogenous ADP (state 2 respiration) and after addition of 300-μM ADP (state 3 respiration). The addition of 1-μg/ml oligomycin inhibits the mitochondrial ATP synthase and reduces oxygen consumption to a rate similar to that preceding the addition of ADP. Respiratory control ratios (RCRs, the ratio between state 2 and state 3 respirations) should be above 3 (with succinate) and 7 (with glutamate/malate) in liver, and 2.5 (with succinate) and 4 (with glutamate/malate) in brain. Lower RCRs suggest mitochondrial uncoupling.

The protocol used to measure ATP synthesis is similar to that described above for permeabilized cells, with the omission of the digitonization step, and using 30 μg of liver or 100 μg of brain mitochondrial proteins resuspended in buffer A (Table I).

V. Measurement of High–Energy Phosphates in Animal Tissue and Cultured Cells by HPLC

Among the different methods used to assay high-energy phosphates in biological samples, HPLC has the advantage of high sensitivity and efficiency because it allows for the simultaneous analyses of all species of phosphorylated nucleotides in one analysis.

Different HPLC techniques are commonly used to separate nucleotides (Zakaria and Brown, 1981). Nucleotide phosphates are charged molecules that can be separated well by ion-exchange HPLC. However, the separation time is relatively long, and the columns are less stable than are reversed-phase ones. Reversed-phase HPLC separates compounds based on hydrophobic interactions and is less suitable for charged nucleotide phosphates. However, ion-pair reagents, such as tetrabutyl-ammonium hydrogen sulfate and tetrabutylammonium dihydrogen phosphate, when added to the phosphate-buffered mobile phase can complex the charged molecules and enable reversed-phase HPLC to be used in nucleotide separation, thereby allowing for shorter separation times and a better reproducibility than ion-exchange HPLC (Meyer et al., 1999). After adding an ion-pair reagent, the elution order of nucleotides is reversed compared with reversed-phase HPLC without ion-pair reagent (Uesugi et al., 1997).

Because of the high rate of conversion of ATP to ADP and adenosine mono-phosphate (AMP) in fresh samples by cellular ATPases, protocols for the quick inactivation of phosphatases are essential for ATP measurement. Rapid-heating or flash-freezing procedures have been used to preserve high-energy phosphates in preparing biological samples. The method of focused microwave irradiation, which can rapidly kill small rodents and irreversibly inactivate enzymes, has been used in sample preparations for investigating rapidly modulated neuro-chemicals such as neurotransmitters and high-energy phosphorylated nucleotides (McCandless et al., 1984; Schneider et al., 1982). The microwave method can also be used for the extraction of cellular ATP (Tsai, 1986). This method is also useful for preserving the in vivo protein phosphorylation state of many phosphoproteins (O'Callaghan and Sriram, 2004). Due to the resistance of phosphorylated com-pounds to heat, denaturing tissue culture samples in boiling distilled water can produce high yield and good reproducibility for assays of phosphorylated nucleo-tides (Ozer et al., 2000). Alternatively, flash-freezing of tissues or small animals in liquid nitrogen is, in our experience, a simple but effective way to instantly prevent the inactivation of ATPases. However, because the inactivation of enzymes is reversible, maintaining subfreezing conditions during dissection and homogeniza-tion of the samples is critical. A cold chamber with a cryoplate can be used to dissect target tissues from the frozen animal body. For enzyme inactivation in cell culture samples, following quick removal of the culture medium, concentrated perchloric acid or alkaline KOH can be added directly to cells. A proteinase K-based extraction technique has been reported to generate consistently higher adenylate yields from a broad range of cellular samples compared to perchloric acid or the boiling method (Napolitano and Shain, 2005).

We have tested different sample preparation procedures on rat and mouse tissues as well as on cultured cells, and have developed an ion-pair reversed-phase HPLC system, using a phosphate-buffered acetonitrile gradient mobile phase, to simultaneously determine the levels of creatine, its high-energy phosphate, and all of the three adenylates (AMP, ADP, and ATP). With this system, high-energy phosphates separated into well-resolved, tight, reproducible peaks, allowing for a reliable quantification.

A. Apparatus and Reagents

A Perkin–Elmer (Norwalk, CT) M-250 binary LC pump, a Waters (Milford, MA) 717 plus autosampler, a Waters 490 programmable multiwavelength UV detector, and an ESA (Chelmsford, MA) 501 chromatography data process system are used in the HPLC assay. A mechanically refrigerated thermal platform (Sigma, San Diego, CA) is used for frozen tissue dissection. All chemicals are from Sigma (St. Louis, MO).

B. Preparation of Biological Samples

Rats are anesthetized with pentobarbital (64.8 mg/ml, 50 mg/kg) and decapitated (according to institutional protocols), and the heads are immediately snap-frozen in liquid nitrogen. Mice are anesthetized with isoflurane delivered by a vaporizer (VetEquip, Inc., CA) and the animal is immediately frozen in liquid nitrogen. Striatum, cortex, and cerebellum are quickly dissected on a cold plate at $-20\,^{\circ}$C. Frozen tissues are transferred to a 1.5-ml microfuge tube and 10 μl of ice-cold 0.4-M perchloric acid are added per milligram wet weight. The tissue is immediately homogenized with a pellet pestle. The acidic homogenate is kept on ice for 30 min and then centrifuged at $18,000 \times g$ in a microfuge at $4\,^{\circ}$C for 10 min. An aliquot of the pellets is set aside for protein measurements. The supernatant (100 μl) is neutralized with 10 μl of 4-M K_2CO_3, kept on ice for 10 min, and then at $-80\,^{\circ}$C for 1–2 h, to promote precipitation of the perchlorate. Finally, the mixture is centrifuged, as described above. Supernatants are stored at $-80\,^{\circ}$C until HPLC assay.

Cultured cells are grown in six-well dishes, as described above. Culture medium is removed by aspiration, followed by immediate addition of ice-cold 0.4-M perchloric acid (500 μl per 1×10^6 cells). The culture dish is sealed tightly with Parafilm and cooled to $-80\,^{\circ}$C. Cell lysates are thawed on ice, scraped off thoroughly from the wells, and transferred to 1.5-ml microfuge tubes. Samples are centrifuged at $18,000 \times g$ at $4\,^{\circ}$C for 10 min and supernatants are neutralized with K_2CO_3, as described above.

External standards are prepared in 0.4-M perchloric acid, neutralized, and treated in exactly the same way as the tissue and cells samples. Protein measurements are performed using a Bio-Rad protein analysis protocol (Bio-Rad Laboratories).

C. Chromatography

Reagents
Buffer A

25-mM NaH_2PO_4

100-mg/liter tetrabutylammonium, pH 5

Buffer B

10% (v/v) acetonitrile in 200-mM NaH_2PO_4

100-mg/liter tetrabutylammonium, pH 4.0

Buffers are filtered through a Rainin 0.2-μm, Nylon-66 filter (Woburn, MA) and degassed in a flask linked to a vacuum pipe.

Procedure

The gradient elution is performed on a 4.6-mm i.d. $\times$ 250-mm, 3-μm-particle-size YMC C18 HPLC column (Waters, Milford, MA) with buffer A and B, at a rate of 1 ml/min. The gradient is:

100% buffer A from 0 to 5 min

100% buffer A to 100% buffer B from 5 to 30 min

100% buffer B to 100% buffer A from 30 to 31 min

100% buffer A from 31 to 45 min for column reequilibration, which is sufficient to achieve stable baseline conditions.

Fifty microliters of prepared sample or standard mixture are autoinjected and monitored by UV at 210 nm from 0 to 15 min (for detection of creatine and phosphocreatine) and at 260 nm from 15 to 45 min (for phosphorylated nucleotides). Peaks are identified by their retention times and by using co-chromatography with standards.

D. Standard Curves

Each standard of interest is first subjected individually to chromatography to determine its retention time and later identify each compound in a standard mixture. A standard curve for each compound is constructed by plotting peak heights (μV) versus concentration (10–1000 μM for creatine and phosphorylated creatine; 5–500 μM for phosphorylated nucleotides). Linear curves are obtained (R^2 values are 0.98) from which the detection limits of the HPLC method are estimated to be fivefold above baseline noise. The quantification of creatine, phosphocreatine, and phosphorylated nucleotides in the sample is carried out using the external standard calibration (i.e., by co-chromatography of the mixed standard solution and samples), integrating sample peak heights against corresponding standard curves.

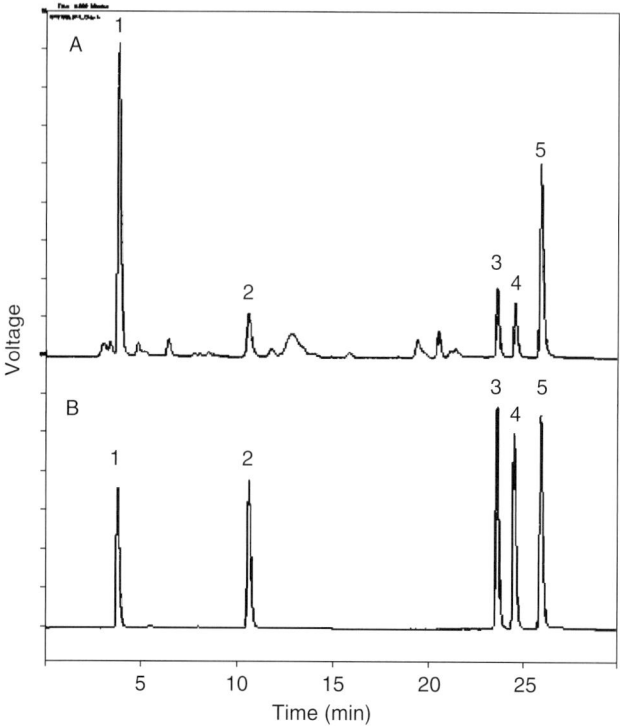

Fig. 5 Representative chromatograms showing the separation of creatine, phosphocreatine, AMP, ADP, and ATP from a rat cortex tissue sample (A) and a standard mixture (B). Running time is 45 min. Creatine, 500 μM (1); phosphocreatine, 500 μM (2); AMP, 100 μM (3); ADP, 100 μM (4); and ATP, 100 μM (5).

E. Measurement of Creatine, Phosphocreatine, and Phosphorylated Nucleotides

Figure 5 shows the separation of creatine, phosphorylated creatine, AMP, ADP, and ATP from a rat brain tissue sample (Fig. 5A) and a standard mixture (Fig. 5B). One single run takes 45 min; an average of 30 min is sufficient for the separation of adenosine nucleotides. Retention times are: creatine 3.5 min, phosphocreatine 11 min, AMP 22.5 min, ADP 24.5 min, and ATP 26 min. All the compounds of interest in the tissue samples are resolved clearly and have very sharp peaks. In cultured cell samples, there is no detectable creatine and phosphocreatine peak, since the cells do not synthesize creatine (not shown). Figure 6 shows an example of a standard curve for ATP. Detection limits for the three phosphorylated nucleotides are $\sim$200 pmol and are $\sim$500 pmol for creatine and phosphocreatine. Calculations are performed by peak height calibrations and values are expressed as μmol/g of wet tissue or nmol/mg of protein. The following values are obtained from rat brain (μmol/g wet wt.): creatine 9.07 $\pm$ 0.52, phosphocreatine 2.2 $\pm$ 0.23, AMP 0.91 $\pm$ 0.08, ADP 0.71 $\pm$ 0.03, and ATP 1.23 $\pm$ 0.18;

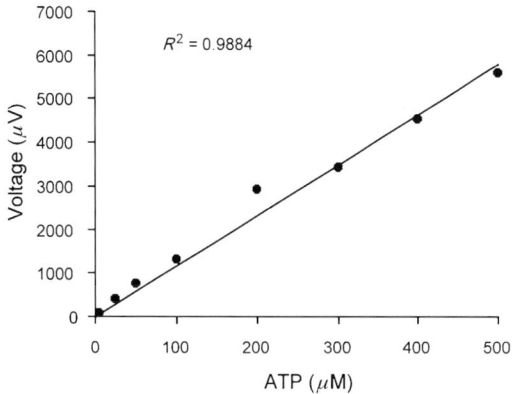

Fig. 6 ATP standard curve obtained by HPLC. ATP concentrations range from 5 to 500 μM.

from mouse brain (nmol/mg protein): creatine 68.0 ± 4.9, phosphocreatine 18.2 ± 1.5, AMP 14.4 ± 2.1, ADP 7.3 ± 1.1, and ATP 16.0 ± 2.8; from cell culture samples (nmol/mg protein): AMP 14.8 ± 1.2, ADP 9.1 ± 0.5, and ATP 21.5 ± 1.3. All these values are in good agreement with those reported in the literature (Carter and Muller, 1990; Horn *et al.*, 1998; Mitani *et al.*, 1994; Pissarek *et al.*, 1999).

Acknowledgments

This work was supported by NIH grant K02-NS47306 and by the Muscular Dystrophy Association.

References

Carter, A. J., and Muller, R. E. (1990). Application and validation of an ion-exchange high-performance liquid chromatographic method for measuring adenine nucleotides, creatine and creatine phosphate in mouse brain. *J. Chromatogr.* **527,** 31–39.

Deluca, M. (1976). Firefly luciferase. *Adv. Enzymol. Relat. Areas. Mol. Biol.* **44,** 37–68.

DeLuca, M., and McElroy, W. D. (1974). Kinetics of the firefly luciferase catalyzed reactions. *Biochemistry* **13,** 921–925.

Holt, I. J., Harding, A. E., Petty, R. K., and Morgan-Hughes, J. A. (1990). A new mitochondrial disease associated with mitochondrial DNA heteroplasmy. *Am. J. Hum. Genet.* **46,** 428–433.

Horn, M., Frantz, S., Remkes, H., Laser, A., Urban, B., Mettenleiter, A., Schnackerz, K., and Neubauer, S. (1998). Effects of chronic dietary creatine feeding on cardiac energy metabolism and on creatine content in heart, skeletal muscle, brain, liver and kidney. *J. Mol. Cell. Cardiol.* **30,** 277–284.

Houstek, J., Klement, P., Hermanska, J., Houstkova, H., Hansikova, H., Van den Bogert, C., and Zeman, J. (1995). Altered properties of mitochondrial ATP-synthase in patients with a T→G mutation in the ATPase 6 (subunit a) gene at position 8993 of mtDNA. *Biochim. Biophys. Acta* **1271,** 349–357.

James, A. M., Sheard, P. W., Wei, Y. H., and Murphy, M. P. (1999). Decreased ATP synthesis is phenotypically expressed during increased energy demand in fibroblasts containing mitochondrial tRNA mutations. *Eur. J. Biochem.* **259,** 462–469.

King, M. P., and Attardi, G. (1989). Human cells lacking mtDNA: Repopulation with exogenous mitochondria by complementation. *Science* **246,** 500–503.

Kunz, W. S., Kuznetsov, A. V., Schulze, W., Eichhorn, K., Schild, L., Striggow, F., Bohnensack, R., Neuhof, S., Grasshoff, H., Neumann, H. W., and Gelleric, F. N. (1993). Functional characterization of mitochondrial oxidative phosphorylation in saponin-skinned human muscle fibers. *Biochim. Biophys. Acta* **1144,** 46–53.

Kurebayashi, N., Kodama, T., and Ogawa, Y. (1980). P1, P5-Di (adenosine-5′) pentaphosphate (Ap5A) as an inhibitor of adenylate kinase in studies of fragmented sarcoplasmic reticulum from bullfrog skeletal muscle. *J. Biochem. (Tokyo)* **88,** 871–876.

Lemasters, J. J., and Hackenbrock, C. E. (1973). Adenosine triphosphate: Continuous measurement in mitochondrial suspension by firefly luciferase luminescence. *Biochem. Biophys. Res. Commun.* **55,** 1262–1270.

Lembert, N., and Idahl, L. A. (1995). Regulatory effects of ATP and luciferin on firefly luciferase activity. *Biochem. J.* **305**(Pt. 3), 929–933.

Maechler, P., Wang, H., and Wollheim, C. B. (1998). Continuous monitoring of ATP levels in living insulin secreting cells expressing cytosolic firefly luciferase. *FEBS Lett.* **422,** 328–332.

McCandless, D. W., Stavinoha, W. B., and Abel, M. S. (1984). Maintenance of regional chemical integrity for energy metabolites in microwave heat inactivated mouse brain. *Brain Res. Bull.* **13,** 253–255.

Meyer, S., Noisommit-Rizzi, N., Reuss, M., and Neubauer, P. (1999). Optimized analysis of intra-cellular adenosine and guanosine phosphates in *Escherichia coli. Anal. Biochem.* **271,** 43–52.

Mitani, A., Takeyasu, S., Yanase, H., Nakamura, Y., and Kataoka, K. (1994). Changes in intracellular Ca^{2+} and energy levels during *in vitro* ischemia in the gerbil hippocampal slice. *J. Neurochem.* **62,** 626–634.

Napolitano, M. J., and Shain, D. H. (2005). Quantitating adenylate nucleotides in diverse organisms. *J. Biochem. Biophys. Methods* **63,** 69–77.

O'Callaghan, J. P., and Sriram, K. (2004). Focused microwave irradiation of the brain preserves *in vivo* protein phosphorylation: Comparison with other methods of sacrifice and analysis of multiple phosphoproteins. *J. Neurosci. Methods* **135,** 159–168.

Ouhabi, R., Boue-Grabot, M., and Mazat, J. P. (1998). Mitochondrial ATP synthesis in permeabilized cells: Assessment of the ATP/O values *in situ. Anal. Biochem.* **263,** 169–175.

Ozer, N., Aksoy, Y., and Ogus, I. H. (2000). New sample preparation method for the capillary electrophoretic determination of adenylate energy charge in human erythrocytes. *J. Biochem. Biophys. Methods* **45,** 141–146.

Pissarek, M., Reinhardt, R., Reichelt, C., Manaenko, A., Krauss, G., and Illes, P. (1999). Rapid assay for one-run determination of purine and pyrimidine nucleotide contents in neocortical slices and cell cultures. *Brain Res. Brain Res. Protoc.* **4,** 314–321.

Schneider, D. R., Felt, B. T., Rappaport, M. S., and Goldman, H. (1982). Development and use of a nonrestraining waveguide chamber for rapid microwave radiation killing of the mouse and neonate rat. *J. Pharmacol. Methods* **8,** 265–274.

Strehler, B. L. (1968). Bioluminescence assay: Principles and practice. *Methods Biochem. Anal.* **16,** 99–181.

Strehler, B. L., and Totter, J. R. (1952). Firefly luminescence in the study of energy transfer mechanism. I. Substrates and enzyme determinations. *Arch. Biochem. Byophys.* **40,** 28–41.

Tatuch, Y., and Robinson, B. H. (1993). The mitochondrial DNA mutation at 8993 associated with NARP slows the rate of ATP synthesis in isolated lymphoblast mitochondria. *Biochem. Biophys. Res. Commun.* **192,** 124–128.

Tsai, T. L. (1986). A microwave method for the extraction of cellular ATP. *J. Biochem. Biophys. Methods* **13,** 343–345.

Tuena de Gomez-Puyou, M., Ayala, G., Darszon, A., and Gomez-Puyou, A. (1984). Oxidative phosphorylation and the Pi-ATP exchange reaction of submitochondrial particles under the influence of organic solvents. *J. Biol. Chem.* **259**, 9472–9478.

Uesugi, T., Sano, K., Uesawa, Y., Ikegami, Y., and Mohri, K. (1997). Ion-pair reversed-phase high-performance liquid chromatography of adenine nucleotides and nucleoside using triethylamine as a counterion. *J. Chromatogr. B Biomed. Sci. Appl.* **703**, 63–74.

Vazquez-Memije, M. E., Shanske, S., Santorelli, F. M., Kranz-Eble, P., Davidson, E., DeVivo, D. C., and DiMauro, S. (1996). Comparative biochemical studies in fibroblasts from patients with different forms of Leigh syndrome. *J. Inherit. Metab. Dis.* **19**, 43–50.

Wanders, R. J., Ruiter, J. P., and Wijburg, F. A. (1993). Studies on mitochondrial oxidative phosphorylation in permeabilized human skin fibroblasts: Application to mitochondrial encephalomyopathies. *Biochim. Biophys. Acta* **1181**, 219–222.

Wanders, R. J., Ruiter, J. P., and Wijburg, F. A. (1994). Mitochondrial oxidative phosphorylation in digitonin-permeabilized chorionic villus fibroblasts: A new method with potential for prenatal diagnosis. *J. Inherit. Metab. Dis.* **17**, 304–306.

Wanders, R. J., Ruiter, J. P., Wijburg, F. A., Zeman, J., Klement, P., and Houstek, J. (1996). Prenatal diagnosis of systemic disorders of the respiratory chain in cultured chorionic villus fibroblasts by study of ATP-synthesis in digitonin-permeabilized cells. *J. Inherit. Metab. Dis.* **19**, 133–136.

Wibom, R., Lundin, A., and Hultman, E. (1990). A sensitive method for measuring ATP-formation in rat muscle mitochondria. *Scand. J. Clin. Lab. Invest.* **50**, 143–152.

Wibom, R., Soderlund, K., Lundin, A., and Hultman, E. (1991). A luminometric method for the determination of ATP and phosphocreatine in single human skeletal muscle fibres. *J. Biolumin. Chemilumin.* **6**, 123–129.

Zakaria, M., and Brown, P. R. (1981). High-performance liquid chromatography of nucleotides, nucleosides and bases. *J. Chromatogr.* **226**, 267–290.

CHAPTER 8

Measurement of the Ratio of Lactate to Pyruvate in Skin Fibroblast Cultures

Nevi Mackay and Brian H. Robinson

Department of Pediatric Laboratory Medicine and the Research Institute
Hospital for Sick Children
University of Toronto
Toronto, Ontario, Canada

Department of Biochemistry
University of Toronto
Toronto, Ontario, Canada

I. Introduction

A group of inborn errors of metabolism exists that result in the condition of chronic lactic academia of childhood. Nearly all of the defects that can be identified occur in mitochondrial proteins, and many can be demonstrated in skin fibroblast cultures established from the patients concerned. One approach to diagnose these defects is to measure production of lactate and pyruvate from fibroblast cultures after incubation in glucose-containing medium. The total amount of lactate and pyruvate and the ratio between them is different in cells from patients with defects in the pyruvate dehydrogenase complex (PDH) or the respiratory chain compared to cells from unaffected individuals.

METHODS IN CELL BIOLOGY, VOL. 80
Copyright 2007, Elsevier Inc. All rights reserved.
0091-679X/07 $35.00
DOI: 10.1016/S0091-679X(06)80008-7

II. Principle

When glucose is metabolized by skin fibroblasts, the end product of glycolysis is pyruvic acid. This molecule has two major fates in most oxidative tissues. It can be oxidized to acetyl coenzyme A (CoA) through the PDH, or it can be reduced to form lactic acid by NADH. The extent to which this latter reaction takes place is governed by two elements: the rate of flux of pyruvate through PDH into the citric acid cycle, and the extent of reduction of the NADH/NAD couple in the cytosol. The cytosolic redox couple is again in equilibrium with the mitochondrial NADH/NAD couple through the glutamate/aspartate shuttle system (Fig. 1).

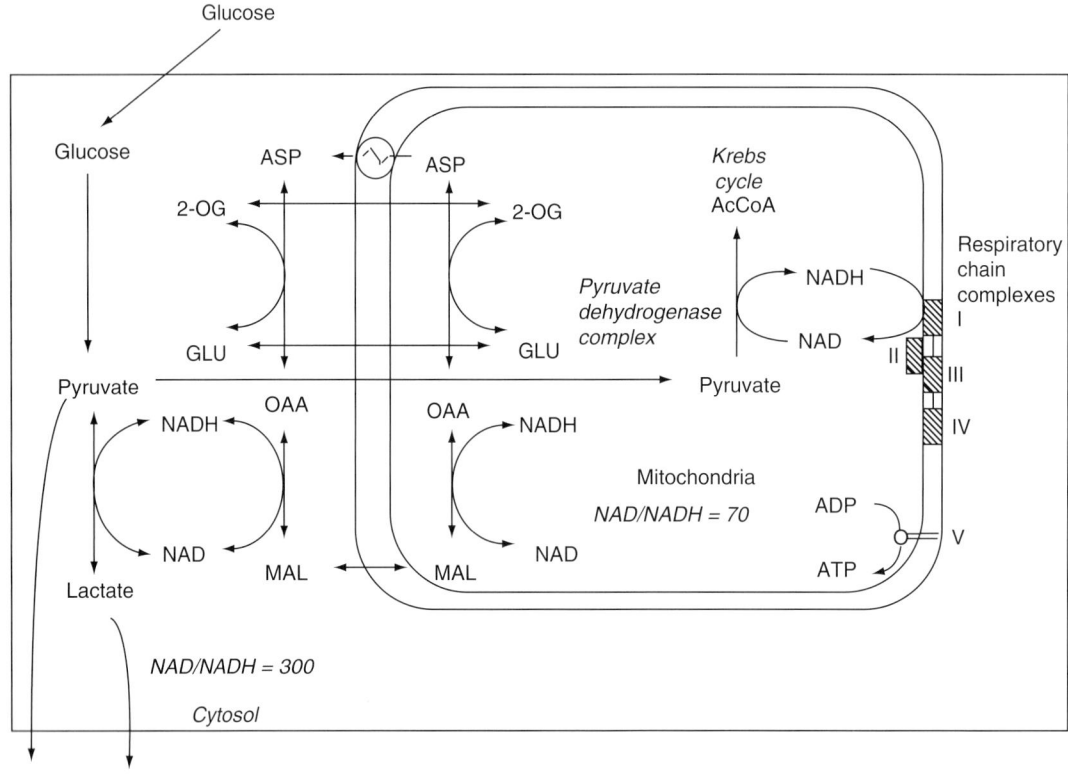

Fig. 1 The relationship between the NADH/NAD redox state of mitochondria and the cytoplasm. The systems are kept in equilibrium via the metabolites involved in the malate dehydrogenase and aspartate aminotransferase equilibria in mitochondrial and cytosolic compartments. The cytosolic NADH/NAD couple is maintained in a more oxidized state by the electrogenic expulsion of aspartate from mitochondria. Glucose taken up by cells is metabolized to pyruvate. Pyruvate is then reduced to lactate by LDH or converted to acetyl CoA by the PDH complex to produce NADH in the mitochondria. As a result, respiratory chain defects lead to an increased L/P ratio, because electron transport from NADH is hampered. In contrast, defects in the PDH complex result in decreased production of NADH in mitochondria and lower L/P ratios.

The lactate dehydrogenase (LDH) equilibrium in the cytosolic compartment is described by:

$$H^+ + Pyruvate + NADH \longleftrightarrow Lactate + NAD^+$$

This emphasizes the fact that the equilibrium of the LDH lies to the right, even more so in acidic conditions. The redox ratio of the $[NADH]_c/[NAD]_c$ couple in the cytosol can be calculated from:

$$\frac{[Lactate]}{[Pyruvate]} \times K_{LDH} = \frac{[NADH]_c}{[NAD]_c}$$

where K_{LDH} is the equilibrium constant for the LDH reaction, having a value of 1.11×10^{-4} (Krebs, 1973). The lactate/pyruvate (L/P) ratio both in the intracellular compartment and in the surrounding body fluids or tissue culture medium is usually between 10:1 and 25:1 in favor of lactate. By calculation, the NADH/NAD ratio in the cytosol is about 1:300 in most cell types. In contrast, the NADH/NAD ratio in mitochondria of most cell types is 1:10 (Williamson et al., 1973).

The measurement of lactate and pyruvate accumulated after incubating cells with glucose can give an indication of the absolute rate of production of lactate. It is also a measure whether the flux into the respiratory chain is compromised by a defect in pyruvate dehydrogenase (normal L/P ratio) or by a defect in the respiratory chain (high L/P ratio) (Robinson, 1996; Robinson et al., 1985, 1986). This latter application has been documented for defects in complex I, complex IV, and multiple defects of the respiratory chain.

III. Procedure

Fibroblast cells are grown on 9-cm Petri dish tissue culture plates or 25-ml tissue culture flasks (T25). Use one plate or flask per sample. The cells should be subcultured 1:4 from a confluent culture and their growth should be monitored so that the cells are used once confluence is reached. The test should be done no later than 2 weeks after the cell have been subcultured. Cells in the growth phase are not suitable for this assay, as they have somewhat lower L/P ratios, depending on the growth rate.

A. Sample Preparation

1. Remove as much medium from the sample as possible by vacuum suction. Thereafter, tilt plates again and remove residual medium by aspiration.
2. Wash cells with 0.5-ml phosphate-buffered saline (PBS). Remove PBS by aspiration, as described above.

3. Add 1-ml PBS (sterile) to each plate, and incubate for 1 h at 37°C. Thereafter, remove medium by aspiration, as above.

4. Add 1-ml PBS and 1-mM glucose to each plate and incubate for 1 h.

5. Add 50 ml of 1.6-M perchloric acid (PCA), to stop metabolism.

6. Pipette extract from the plate into one Eppendorf tube.

7. Add 1 ml of biuret reagent to each plate and determine the protein content of the sample, using 1 and 2 mg/ml protein standards.

B. Determination of Lactate

1. For 20 samples, combine:

   20-ml hydrazine buffer (0.1-M Tris, 0.4-M hydrazine, 0.4-mM EDTA, 10-mM $MgSO_4$, pH 8.5)

   200-μl NAD, 80 mg/ml

   200-μl LDH, 50 mg/10 ml in distilled water

2. Pipette 1 ml of the above mixture into a set of tubes (Eppendorf 1 ml).

3. Add 100 μl of cell extract (Section III.A, step 6), vortex, and incubate at RT for 1 h. For controls, which should be carried out in duplicate, use 100-μl PBS instead of cell extract. For standards, add L-Lactate to final concentrations of 25 and 50 nmol, instead of cell extract.

4. Measure the amount of lactate produced in each sample using a spectrophotometer at a wavelength of 340 nm. The blanks can be subtracted automatically if a split-beam spectrophotometer is used. If only a single-beam spectrophotometer is available, measure the blank samples against water and subtract from experimental samples.

C. Determination of Pyruvate

1. Adjust a fluorimeter to the following settings:

   excitation wavelength: 340 nm

   emission wavelength: >400 nm

   sensitivity (for 2- to 4-nmol NADH): full scale, using either computerized recordings or a chart recorder.

2. Transfer 1-ml 0.1-M potassium phosphate buffer pH 7.0 to a fluorimeter cuvette.

3. Add 100 μl of acidified cell extract (Section III.A, step 6) to the cuvette.

4. Add 4-nmol NADH, offset the fluorescence background, and read baseline fluorescence.

5. Add 20-μl LDH, close the lid tightly, and read fluorescence for about 3 min. Measure the change in fluorescence produced in response to added NADH.

Calculate the amount of pyruvate, which is almost stoichiometric under these conditions.

IV. Results

In the set of measurements shown in Fig. 2, the amount of lactate produced by control cells is 400–500 nmol/min/mg protein. This rate is increased in PDH deficiency, COX deficiency, and complex I deficiency. Since the pyruvate production rate in control cells is 20–25 nmol/min/mg protein, the L/P ratio for control cells is typically 20:1. In PDH deficiency, pyruvate production is increased and L/P ratios are normal or low. In COX deficiency and complex I deficiency, pyruvate production is markedly decreased, and L/P ratios are elevated.

The change in redox state, as measured by L/P ratios, is usually proportional to the severity of the defect. This can be seen in Fig. 3, where the measured L/P ratios for a series of cell lines with COX deficiency are plotted against residual cytochrome oxidase activity. It would seem that little change in redox state occurs until the COX activity has fallen to ~40% of control values. This implies that cytochrome oxidase activity is present in a roughly 2.5-fold excess over the rate limiting step, so that overall oxidation is little affected by residual values higher than this.

Since redox state and demands for ATP production are quite different in quiescent and actively growing cells (Erecinska and Wilson, 1982), care should be taken to ensure that cells are confluent when these measurements are made. Cells that have not reached confluency tend to have an abnormally low L/P ratio. In contrast, cells that are over confluence have an abnormally high L/P ratio.

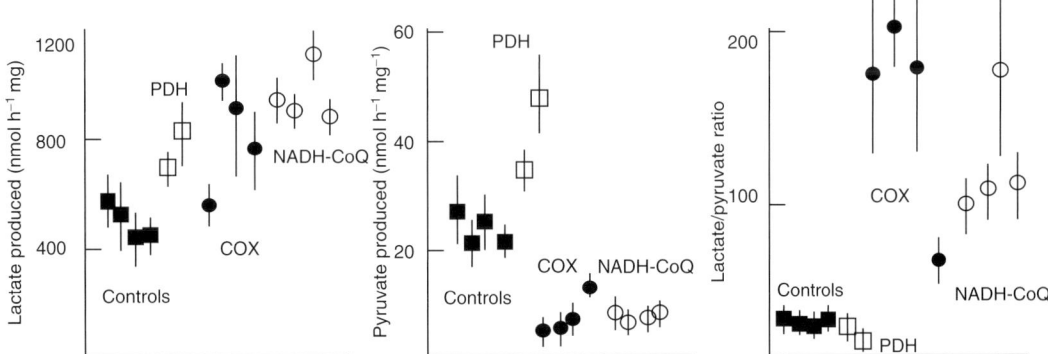

Fig. 2 The Lactate and pyruvate produced in confluent fibroblast cultures after 1 h of incubation in glucose-containing medium. The lactate and pyruvate production rates are documented for control (v), PDH-deficient (θ), cytochrome oxidase-deficient (λ), and NADH-CoQ reductase-deficient (μ) cell lines. Values are given as the mean $\pm$ SEM for at least four determinations.

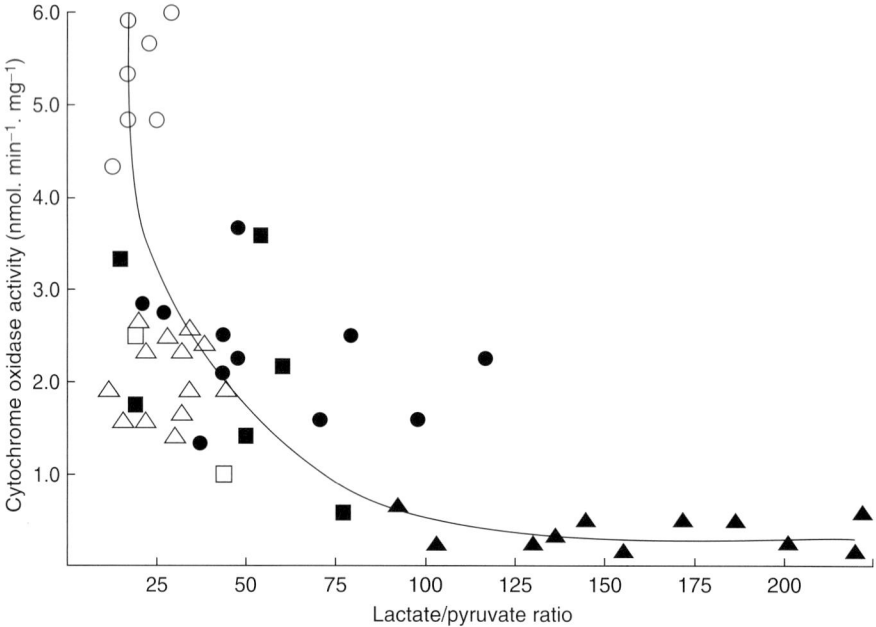

Fig. 3   Relationship between cytochrome oxidase activity and L/P ratios in cultured skin fibroblasts. The residual cytochrome oxidase activity is plotted against the L/P ratios measured in skin fibroblast cultures. The groups are: control (μ), SURF1-deficient COX deficiency (σ), Leigh syndrome COX deficiency, non-SURF1 COX deficiency (λ), COX deficiency with cardiomyopathy (ν), and LRPPRC-deficient (i.e., Leigh Syndrome of the French Canadian type) COX deficiency (Δ).

References

Erecinska, M., and Wilson, D. F. (1982). Regulation of cellular energy metabolism. *J. Membr. Biol.* **70**(1), 1–14.

Krebs, H. A. (1973). Pyridine nucleotides and rate control. *Symp. Soc. Exp. Biol.* **27**, 299–318.

Robinson, B. (1996). Use of fibroblast and lymphoblast cultures for detection of respiratory chain defects. *Methods Enzymol.* **264**, 454–464.

Robinson, B. H., McKay, N., Goodyer, P., and Lancaster, G. (1985). Defective intramitochondrial NADH oxidation in skin fibroblasts from an infant with fatal neonatal lacticacidemia. *Am. J. Hum. Genet.* **37**(5), 938–946.

Robinson, B. H., Ward, J., Goodyer, P., and Baudet, A. (1986). Respiratory chain defects in the mitochondria of cultured skin fibroblasts from three patients with lacticacidemia. *J. Clin. Invest.* **77**(5), 1422–1427.

Williamson, J. R., Safer, B., LaNoue, K. F., Smith, C. M., and Walajtys, E. (1973). Mitochondrial-cytosolic interactions in cardiac tissue: Role of the malate-aspartate cycle in the removal of glycolytic NADH from the cytosol. *Symp. Soc. Exp. Biol.* **27**, 241–281.

CHAPTER 9

Assays of Fatty Acid β-Oxidation Activity

Michael J. Bennett

Department of Pathology and Laboratory Medicine
University of Pennsylvania
Metabolic Disease Laboratory
The Children's Hospital of Philadelphia
Philadelphia 19104

I. Introduction

Mitochondrial fatty acid β-oxidation represents an essential pathway of energy metabolism for tissues such as liver, muscle, and kidney during periods of increased demand due to fasting, febrile illness, reduced caloric intake due to gastrointestinal illness, and when cold. In heart muscle, fatty acid oxidation may be as important as carbohydrate metabolism for normal cardiac function. The end products of fatty acid oxidation in the liver are ketone bodies, which are transported in the circulation to tissues such as brain, which has minimal β-oxidation capacity. In nonketogenic tissues, such as muscle and kidney, the end product of β-oxidation is acetyl-coenzyme A (CoA), which is used directly as an energy source within the TCA cycle (Rinaldo *et al.*, 2002; Stanley *et al.*, 2006). In recent years, it has also been recognized that gastrointestinal, fetal, and placental metabolism also uses the fatty acid β-oxidation pathway as a source of fuel (Oey *et al.*, 2005; Shekhawat *et al.*, 2003). However, the extent and necessity of use of this pathway in these tissues remains to be established.

The fatty acid oxidation pathway is divided into a number of cellular components (Fig. 1). Typically, long-chain fatty acids of chain lengths C_{16}–C_{18} are released at the plasma membrane following the action of lipoprotein lipase activity on circulating triglycerides, or are derived from albumin-bound nonesterified fatty acids. The fatty acids are then transported across the plasma membrane and activated to their acyl CoA derivatives at the outer mitochondrial membrane. Evidence suggests that CD36, the fatty acid transporter (also known as FAT) may

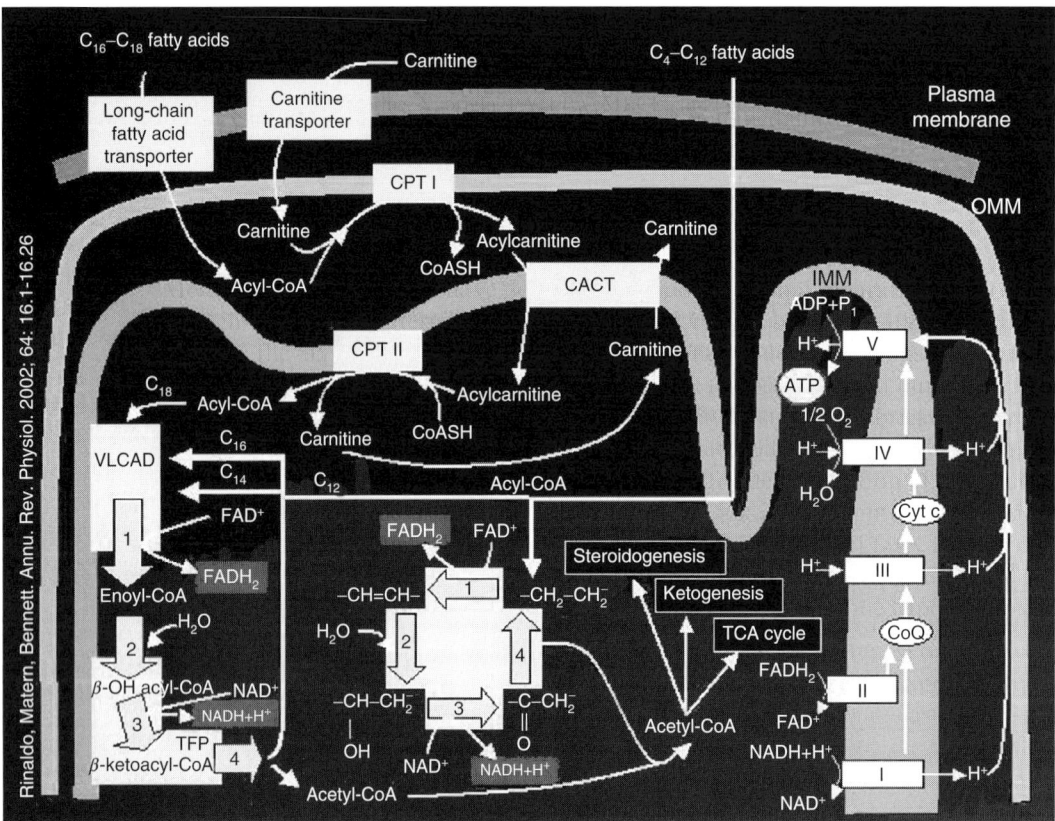

Fig. 1 Interactions of mitochondrial fatty acid oxidation and the respiratory chain. OMM, outer mitochondrial membrane; IMM inner mitochondrial membrane. [Reprinted with permission from Rinaldo *et al.* (2002) © 2002 by Annual Reviews www.annualreviews.org]

play an important role in this process, and provide a direct connection with the outer mitochondrial membrane, where the fatty acids are activated to become acyl CoAs (Campbell *et al.*, 2004). The plasma membrane carnitine transporter, which provides carnitine, an essential cofactor, is also necessary for normal fatty acid oxidation.

Long-chain acyl CoA species cannot be transported directly across the mitochondrial membrane. Substrate transfer is dependent on the carnitine cycle (McGarry and Brown, 1997). In this cycle, long-chain acyl CoAs are first converted to their respective acylcarnitine species by the action of carnitine palmitoyltransferase 1 (CPT 1) at the outer mitochondrial membrane. The acylcarnitine species is transported through the inner mitochondrial membrane by the bidirectional carnitine:acylcarnitine translocase (CACT), a carrier that also transports free carnitine and acylcarnitine out of the mitochondria. Finally, the acylcarnitine is reconverted to the acyl CoA species by the action of carnitine palmitoyltransferase 2 (CPT2) at the inner mitochondrial membrane. Free carnitine is recycled through CACT to provide substrate for CPT1. Medium-chain length fatty acids (C_6–C_{10}) appear to enter into the mitochondrial matrix by a carnitine-independent process that is not yet characterized. Medium-chain fatty acids are frequently provided as a nutritional supplement in the form of medium-chain triglycerides, and are presumably metabolized entirely within the mitochondrial matrix.

The process of acyl CoA β-oxidation involves four enzymatic steps. First, a double bond is inserted into the 2,3 position by an acyl CoA dehydrogenase (ACD), to generate a 2,3-enoyl-CoA species. This is an energy-generating step and electrons are transferred by electron transfer flavoprotein (ETF), an FAD-dependent pathway, to ETF: coenzyme Q (CoQ) oxidoreductase in the respiratory chain. The second step involves hydration across this double bond by a hydratase, which generates a stereospecific L-3-hydroxyacyl CoA species. The third enzymatic reaction involves further reduction in the 3-position by an NAD-requiring L-3-hydroxyacyl CoA dehydrogenase. Electrons from this step are transferred to the respiratory chain through Complex I, which recycles NAD for further metabolic use. The product of this third reaction, a 3-ketoacyl CoA species, is then thiolytically cleaved by a 3-ketoacyl CoA thiolase into acetyl-CoA and a two-carbon chain-shortened acyl CoA, which continues to reenter the spiral until all of the original fatty acid has been converted to acetyl-CoA.

The process of complete oxidation of a long-chain fatty acid requires the participation of a series of genetically similar enzymes, each with different chain-length specificities. Long-chain acyl CoAs are metabolized near the inner mitochondrial membrane by a family of long-chain-specific membrane-bound enzymes, while medium- and short-chain intermediates are metabolized in the mitochondrial matrix by a group of soluble enzymes (Table I). Three of the enzymes involved in membrane-associated metabolism of long-chain intermediates are part of a multifunctional enzyme complex known as the mitochondrial trifunctional protein (TFP). This complex is a hetero-octamer consisting of four α-subunits and four β-subunits. The α-subunit has both long-chain 2,3 dienoyl-CoA hydratase

Table I
Characterized Components of Mitochondrial Fatty Acid Oxidation[a]

	Gene name and locus	Protein structure
Carnitine uptake and cycle		
Carnitine transporter	*OCTN2* 5q 31.1–32	Monomer 63 kDa
CPT1A	*CPTI A* 22q 13.2–13.3	Monomer 88 kDa
CACT	*SLC25A20/CACT* 3p21–31	Monomer 33 kDa
CPT2	*CPT II* 1p32	Homotetramer each 71 kDa
Inner mitochondrial membrane enzymes (long-chain fatty acids)		
VLCAD	*ACADVL/VLCAD* 17p 11.2	Homodimer each 66 kDa
LHYD[α]	*HADHA* 2p 23	Hetero-octamer of four α79 kDa
LCHAD[α]	*HADHA* 2p 23	and four β 47-kDa
LKAT[β]	*HADHB* 2p.23	subunits
Mitochondrial matrix enzymes (medium- and short-chain fatty acids)		
MCAD	*ACADM/MCAD* 1p31	Homotetramer each of 44 kDa
M/SCHAD	*HAD 1/SCHAD* 4q 22-26	Homodimer each of 33 kDa
SCAD	*ACADS/SCAD* 12q22-ter	Homotetramer each of 44 kDa
SKAT	*ACAT 1* 11q22.3-23.1	Homotetramer each of 42 kDa

[a]CPT1A, carnitine palmitoyltransferase 1A (hepatic); CACT, carnitine: acylcarnitine translocase; CPT2, carnitine palmitoyltransferase 2; VLCAD, very-long-chain acyl CoA dehydrogenase; LHYD, long-chain enoyl-CoA hydratase; LCHAD, long-chain L-3-hydroxyacyl CoA dehydrogenase; LKAT, long-chain 3-ketoacyl CoA thiolase; MCAD, medium-chain acyl CoA dehydrogenase; M/SCHAD, medium and short-chain l-3-hydroxyacyl CoA dehydrogenase; SCAD, short-chain acyl CoA dehydrogenase; SKAT, short-chain 3-keto acyl CoA thiolase (also known as β-ketothiolase). LHYD and LCHAD are both components of the α-subunit of the mitochondrial trifunctional protein; LKAT is a component of the β-subunit.

(LHYD) and long-chain-L-3-hydroxyacyl CoA dehydrogenase (LCHAD) activities, while the β-subunit has long-chain-3-ketoacyl CoA thiolase (LKAT) activity. Little is known about the organization of the enzymes within the mitochondrial matrix, but there is increasing evidence that the enzymes are organized as a uniquely structured and closely associated metabolome, which is in a required order for the necessary efficient substrate flow.

Genetic defects that impair flux through the pathway have been identified for most of the well-characterized steps in fatty acid oxidation. The clinical phenotypes of the genetic disorders have been described in a number of recent reviews (Rinaldo *et al.*, 2002; Roe and Ding, 2001; Stanley *et al.*, 2006). In this chapter, we describe methods to measure unique metabolites, specific enzymes, and genes in fatty acid oxidation, and to monitor metabolic flux through the entire pathway. We also describe methods for the diagnosis of diseases of fatty acid oxidation. Please refer to reviews that have been cited for full clinical descriptions of the diseases.

================ **II. Metabolite Measurements**

A. Acylcarnitine Profiles and Total and Free Carnitine Levels

Transport of long-chain fatty acids into the mitochondrion for fatty acid oxidation depends on the essential cofactor carnitine. The carnitine transporter that is present in muscle and kidney is a member of the organic cation transporter family, OCTN2, which is a high-affinity sodium cotransporter that can concentrate carnitine within the cell to levels that are 50 times higher than those observed in the circulation. OCTN2 is competitively inhibited by certain acylcarnitine species, which results in renal tubular loss of carnitine, and eventual depletion of tissue carnitine levels (Stanley *et al.*, 1993). In genetic OCTN2 deficiency, there is massive renal loss and marked deficiency in tissues, leading to symptoms of cardiac and skeletal muscle dysfunction similar to those seen in patients with primary defects which, if untreated, are fatal. Measurement of total and free carnitine levels is important in the evaluation of patients who may have a fatty acid oxidation defect or an OCTN2 deficiency. Moreover, evaluation of acylcarnitine profiles and identification of specific acylcarnitine species may provide clues to the site of a fatty acid oxidation defect. For example, elevations of the level of medium-chain C_8, C_{10}, and $C_{10:1}$ acylcarnitine species is diagnostic for medium-chain acyl CoA dehydrogenase (MCAD) deficiency. The majority of newborns in the United States and other western countries are screened for this disorder and other disorders of fatty acid oxidation at birth. In this screen, acylcarnitine levels are measured in blood samples using tandem mass spectrometry. The sensitivity of this technique is sufficient for acylcarnitine analysis in small blood samples or in blood that has been collected as spots onto filter paper. The method is used in whole population screening (Chace *et al.*, 2003; Frazier *et al.*, 2006).

For the measurement of total and free carnitine, a small blood sample is required. The sample is divided analyzed in two aliquots. The level of free carnitine is quantified in one aliquot by the same technology as for the acylcarnitine analysis. The second aliquot is subjected to alkaline hydrolysis to liberate any acylated carnitine, and the level of total carnitine is measured. The difference between the total and free carnitine represents the acylcarnitine fraction. Individuals with a carnitine transporter defect have markedly reduced plasma total carnitine levels (to less than 10% of normal), and carriers for the defect have levels that are intermediate. Most patients with disorders of fatty acid oxidation have reduced total carnitine levels.

Protocol for measurement of total and free carnitine and acylcarnitine in plasma

1. Separate plasma into 20- to 50-μl aliquots.
2. Add a series of stable-isotope-labeled acylcarnitine internal standards (including free carnitine) to each sample. (Deuterated standards are available from Cambridge Isotope Laboratories Inc., Product No. NSKB.)
3. Dilute each sample with 450 μl of ethanol and vortex mixed for 30 sec.

4. Centrifugation at 13,000 × g for 10 min at RT to separate the deproteinized plasma.

5. Transfer 300 μl of the supernatant from each sample into wells in 96-well plates, and evaporate to dryness under a steady stream of nitrogen.

6. Prepare butyl-derivatives by the addition of 100 μl of butanolic hydrochloric acid, heating at 65 °C for 15 min. (Not all laboratories choose to make butylated acylcarnitine and carnitine derivatives, claiming that nonderivatization is a simpler process. We find that butylation is preferable, particularly for measurement of dicarboxylic acid carnitine conjugates, such as glutarylcarnitine.)

7. Thereafter, evaporate to dryness under a steady stream of nitrogen at RT.

8. Resuspend the material in each well with 100-μl acetonitrile:water (50:50).

9. Ten microliters of this material is injected directly into the tandem mass spectrometer via a column-free HPLC system.

Analysis of the mass spectroscopy data

Analyze the parent compounds for a fragment of *m/z* 85. These masses are usually characteristic of carnitine and carnitine-conjugated compounds that can be associated with disease. Occasionally, two acylcarnitine compounds have the same mass but are associated with different conditions. These isobaric compounds cannot be differentiated using tandem mass spectrometry. In this case, alternate methods for analysis of the metabolite profile, including urine organic acid analysis, are required to differentiate the species and identify the disease.

Determine the level of these carnitine fragments using known concentrations of the stable isotope-labeled internal standards that are added at the outset. Isotope-labeled internal standards are not commercially available for all acylcarnitine species. When an immediate internal standard is not available for a particular compound, it is acceptable, but not ideal, to quantitate using the nearest mass equivalent internal standard.

B. Measurement of Urine Organic Acids and Acylglycine

Methods for the analysis of urine organic acids and acylglycines have been well established as a clinical tool, and usually involve acidification of urine, addition of suitable internal standards, extraction into an organic solvent, and derivatization and analysis by gas chromatography/electron impact mass spectrometry, with data collection in the mass range 50–600 amu. Please refer to an online document for procedures for this assay: http://www.aacc.org/AACC/members/nacb/LMPG/OnlineGuide/PublishedGuidelines/Maternal_Fetal_Risk/

1. Organic Acids

The analysis of urine organic acids has become a standard tool for investigation of suspected metabolic disease patients showing a wide variety of clinical signs (Kumps *et al.*, 2002). In mitochondrial fatty acid oxidation defects, a number of specific and nonspecific metabolites can be detected in urine, using gas chromatography/mass spectrometry. In the event that flux through the fatty acid oxidation pathway is impaired, long-chain fatty acids can be partially oxidized within peroxisomes to chain lengths C_6–C_{10}. Further metabolism of these medium-chain species takes place in the microsomes, where ω and ω-1 oxidation products are produced. Thus, the ω-oxidation products adipic, suberic, and sebacic acids, which are also known as medium-chain dicarboxylic acids, are prominent but nonspecific markers of impaired mitochondrial β-oxidation, as are the ω-1 products of 5-hydroxyhexanoic, 7-hydroxyoctanoic, and 9-hydroxydecanoic acids. In addition, a number of 3-hydroxydicarboxylic acids up to chain length C_{14} may be detected by this process. Interpretation of 3-hydroxydicarboxylic aciduria may include defects at the level of 3-hydroxyacyl CoA dehydrogenase and 3-ketoacyl CoA thiolase. Longer chain-length dicarboxylic acids are not found in urine. They are probably circulated as protein-bound metabolites, which are not filtered by the renal glomeruli.

Protocol

1. Add 50 μl of an internal standard of hendecanedioic acid (1 mg/ml) and 50 μl of methoxylamine hydrochloride (25 g/liter) to a volume of urine that contains 2-mg creatinine. (The methoxylamine protects ketoacids from being converted to their respective hydroxy acids.)

2. Bring volume up to 3 ml with distilled deionized water, alkalinized with — three to four drops of 2-M sodium hydroxide, and heat at 70°C for 30 min to complete the oximation.

3. Cool the samples and then acidify them with 5-M HCl to pH 1–2.

4. Extract sample three times with ethylacetate. Pool the organic phases and dry them at under a steady stream of nitrogen at RT.

5. Trimethylsilyl-derivatives are made by adding 150-μl bis (trimethylsilyl)-trifluoroacetamide:trimethylchlorosilane (BSTMS:TMCS, 99:1 obtained from Pearce Chemical Company) and heating for 30 min at 70°C.

6. Inject the derivatized extract into the gas chromatograph-mass spectrometer and generate a total ion chromatogram (mass range 50–600 amu) for the organic acid profile.

2. Acylglycines

The extraction procedure that isolates and measures organic acids in urine also isolates acylglycines. These compounds are the products of mitochondrial glycine conjugation of acyl CoA intermediates, including hexanoylglycine in MCAD deficiency and butyrylglycine in SCAD deficiency. These markers are probably pathognomic for the particular disorders, but are present at lower levels, necessitating the approach of stable isotope dilution and selected ion monitoring mass spectrometry for accurate quantitation. The acylglycine internal standards for this procedure are $[^{13}C,^{15}N]$-labeled acylglycine species, which generate a mass that is 2 amu greater than that of the naturally occurring species (Rinaldo *et al.*, 1988).

<hr>

III. Enzyme and Transporter Assays

A. Carnitine Uptake

Carnitine uptake can only be measured in living intact cells. The muscle and kidney carnitine transporter, which presently represents the only genetic carnitine transporter defect, is present in cultured skin fibroblasts, and this represents the major tissue in which carnitine transport is evaluated. There is a second carnitine transporter that is expressed primarily in liver, and in the case of a suspected transporter defect for this isoform, the assay would need to be performed in cultured hepatocytes. As with all assays that require the use of cultured skin fibroblasts, testing should be performed at the lowest possible cell passage number, since fibroblasts can transform over time and certain properties regarding metabolism may change. The assay monitors intracellular uptake of radiolabeled $[^{14}C]$-carnitine and is capable of differentiating homozygous deficient, normal, and heterozygous carriers (Treem *et al.*, 1988).

Protocol

1. Transfer fibroblasts to 24-well plates and grow to confluence under normal cell culture conditions.
2. On the day of the assay, wash fibroblasts twice with PBS, and incubate fibroblasts for 90 min at $37\,^{\circ}C$ in medium consisting of 26-mM Tris (pH 7.4), 116-mM NaCl, 5.4-mM KCl, 1.8-mM $CaCl_2$, 1-mM $NaH_2\,PO_4$, 0.8-mM $MgSO_4$, 5.5-mM D-glucose, and 1% bovine serum albumin (BSA) to deplete the intracellular stores of carnitine.
3. Replace the medium with the same medium also containing 0.5-μM $[^3H]$-carnitine (0.5 μCi/ml).
4. At 0, 1, 2, and 4 h time points, terminate the reaction by repeated washing 4 times with ice-cold 0.1-M $MgCl_2$.
5. Extract intracellular carnitine from each well with 0.5-ml ethanol, and determine the amount of radiolabeled carnitine in the extract using a scintillation counter. The rate of carnitine uptake is determined by kinetic analysis of the labeled intracellular carnitine at the various time points.

B. Carnitine Cycle

The individual elements of the carnitine cycle include carnitine palmitoyltransferase 1 (CPT1), carnitine: acylcarnitine transferase (CACT), and carnitine palmitoyltransferase 2 (CPT2). CPT1 and CPT2 activities can be measured simultaneously using the method of McGarry and Brown (1997). In this assay, the conversion of palmitoyl-CoA to palmitoyl carnitine is monitored using radiolabeled carnitine.

$$\text{Palmitoyl-CoA} + [^{14}\text{C-carnitine}] \longrightarrow \text{Palmitoyl-}[^{14}\text{C-carnitine}] + \text{CoA}$$

Protocol for measurement of CPT1 and CPT2 activity

The assay is performed in two parts. In the first part, total CPT activity (CPT1 plus CPT2) is measured, in the forward direction for CPT1 and the reverse direction for CPT2. In the second part of the assay, CPT1 activity is inhibited by the addition of malonyl-CoA at a final concentration of 50 mmol/liter. This assay determines the contribution toward total CPT activity by CPT2, which is not inhibited by malonyl-CoA at this concentration.

The enzymes are stable when tissues are stored at $-80\,^{\circ}\text{C}$, and the assay can be applied to all tissue types, with the understanding that the kinetics of CPT1 inhibition by malonyl-CoA differs between the muscle and liver CPT1 isoforms, and that appropriate inhibiting malonyl-CoA concentrations are required. To date, the only genetic cause of CPT1 deficiency that has been described involves CPT1A, the hepatic isoform (Bennett and Narayan, 2005). This is the isoform that is expressed in cultured skin fibroblasts, making this a useful tissue for clinical diagnostics. CPT2 is a systemic gene product and can be measured in any available and suitable tissue. Because one of the clinical phenotypes of CPT2 deficiency includes exercise-induced rhabdomyolysis in young adults, the enzyme is frequently measured in biopsied muscle tissue.

Assay cocktail

> BSA (200 mg)
> Glutathione (1.5 mg)
> KCN (2.6 mg)
> ATP (480 μl of a 100-mg/ml stock solution)
> 14-ml Trizma pH 7.4, 0.15 M
> 34.2-μl magnesium chloride (2.45-M stock solution)
> 20-μl rotenone (400 mg/10 ml acetone stock solution)

> This cocktail is made fresh daily.

Substrate mixture

> 320-μl L-carnitine (100 mM)
> 20-μCi [^{14}C]-carnitine
> 400-μl palmitoyl-CoA (20 mM)
> Bring up to 15 ml with distilled water.

> This is stable for 6 months and can be aliquoted and frozen.

Protocol

1. Bring ~0.2-mg protein of the cell preparation to a final volume of 100 μl.
2. Mix 100 μl of the substrate mixture, 100-μl of 0.15-M KCl, and 700 μl of the assay cocktail.
3. Start the reaction by the addition of the cell preparation to the substrate/ assay cocktail/KCl mixture.
4. Terminate the reaction by the addition of 1 ml of 1-M HCl.
5. Extract palmitoyl carnitine using butanol and determine the amount of radioactivity in the extract using a liquid scintillation counter.
6. Determine the CPT2 activity by performing the same assay on samples after pretreatment with malonyl-CoA at a final concentration of 50 mmol/liter.

Protocol for measurement of CACT activity

CACT activity is most conveniently measured in tissues using the method described by Murthy *et al.* (1986) and modified by Stanley *et al.* (1992). In the clinical assay, the conversion of [2-^{14}C]-pyruvate to [^{14}C]-acetyl-carnitine by pyruvate dehydrogenase (PDH) and carnitine acyltransferase (CAT) is monitored in intact skin fibroblasts. The assay is performed in the presence of saturating concentrations of carnitine (using digitonin to permeabilize the outer mitochondrial membrane and to remove any potential errors due to abnormal pyruvate transporter activity) and dichloroacetate to activate PDH. Sulfobetaine, a specific inhibitor of CACT, is added in a separate reaction to determine a blank activity. The radiolabeled acetylcarnitine product is isolated by ion exchange chromatography.

1. 2-^{14}C-Pyruvate $\xrightarrow{\text{PDH}}$ ^{14}Acetyl-CoA

2. ^{14}C-Acetyl-CoA + Carnitine $\xrightarrow{\text{CAT}}$ ^{14}C-Acetylcarnitine (intramitochondrial)

3. ^{14}C-Acetylcarnitine (intramitochondrial) $\xrightarrow{\text{CACT}}$ ^{14}C-Acetylcarnitine (in media)

Reagents
Homogenization medium

210-mM mannitol
70-mM sucrose
0.1-mM EDTA
10-mM Tris–HCl (pH 7.4)

Assay buffer

250-mM mannitol
25-mM HEPES (pH 7.4)
5-mM $K_2 PO_4$
0.05-mM EDTA

Preincubation mixture (final volume 200 μl)

3-mM ADP
2-mM dichloroacetate
0.5-mM L-carnitine
0.004-mg digitonin
25-mM sulfobetaine

Protocol

1. Wash fibroblast and suspend in ice-cold homogenizing media to an approximate protein concentration of 2 mg/ml. Homogenization takes place with the sample on ice.
2. Add 200 μl of preincubation mixture and preincubate for 15 min at 30 °C.
3. Start the reaction by addition of 2-mM malonate and 0.2-mM pyruvate (0.05 μCi/ml of 2-[^{14}C]-pyruvate) and incubate at 30 °C for 30 min.
4. Terminate the reaction by addition of 800-μl methanol and transfer sample to ice.
5. The mixture is centrifuged at 2500 rpm for 5 min to separate the protein.
6. The supernatant from the reaction is passed through an anion exchange column to isolate the product [^{14}C]-acetyl carnitine which is eluted with 60% methanol.

C. Intramitochondrial Enzymes

1. Acyl CoA Dehydrogenases

Presently, three members of the nine-member acyl CoA dehydrogenase (ACAD) genetic superfamily have been defined to play a significant chain-length-specific role in the complete oxidation of long-chain fatty acids within the mitochondrion. Very long-chain-acyl CoA dehydrogenase (VLCAD) is bound to the inner mitochondrial membrane. It is responsible for the first cycles of long-chain fatty acid oxidation, providing chain shortening of C_{16}–C_{18} precursors to around C_{10}–C_{12}. The oxidative process is then targeted, by an unknown mechanism, to MCAD in the mitochondrial matrix, which shortens the chains of intermediates to C_4. The final cycle of oxidation requires SCAD, also a matrix enzyme. The roles of additional acyl CoA dehydrogenases, including LCAD and ACD9, have not yet been fully defined, but may involve differential tissue or substrate preferences (Ensenauer *et al.*, 2005).

VLCAD, MCAD, and SCAD activities are all measured spectrometrically or fluorimetrically by monitoring the transfer of electrons from appropriate chain length straight-chain acyl CoAs to an appropriate electron acceptor. Typically, palmitoyl-CoA is used as a substrate for VLCAD measurement, and octanoyl-CoA and butyryl-CoA are substrates for MCAD and SCAD, respectively (Hale *et al.*, 1990). A number of artificial electron acceptors have been used in attempts to measure VLCAD, MCAD, and SCAD activities, including phenazine methosulfate as an initial acceptor followed by transfer to indole dyes, such as dichloropheno-lindophenol (DCIP) and the ferricenium ion. However, the improved kinetics of the reaction when using the flavoprotein ETF, which is the natural electron acceptor, has made this the favored electron acceptor for enzyme measurement in most diagnostic laboratories. The human *ETF* gene has been cloned, and recombinant ETF is available for the assay.

Procedure

1. Prepare assay cocktail:

    20-mM Tris–HCl, pH 8.0

    variable amounts of electron transfer flavoprotein (depending on the yield from purification)

    18.5-mM D-glucose

    20 units D-glucose

    0.5 units catalase (to remove oxygen from the reaction vessel)

    50-μM substrate (palmitoyl-CoA for VLCAD, octanoyl-CoA for MCAD, and butyryl-CoA for SCAD).

2. Degas and flush an anaerobic fluorimeter cuvette with oxygen-free nitrogen. (Under aerobic conditions, ETF rapidly auto-oxidizes, creating excessive noise in the analytical system. Therefore, this fluorimetric assay is performed using an anaerobic cuvette that has been exposed to oxygen-free nitrogen prior to sealing.)

3. Transfer 400 μl of the assay cocktail to the cuvette.

4. Start reaction by addition of 20- to 200-μg protein.

5. Monitor fluorescence at an excitation wavelength of 405 nm and emission wavelength of 490 nm.

There are a number of pitfalls and additional steps required in the performance and interpretation of these *in vitro* enzyme assays. *In vitro*, ~50% of the activity that is measured using butyryl-CoA as the substrate for the determination of SCAD activity, is derived from the MCAD that is present in any tissue preparation. MCAD has to be removed prior to activity measurement by immunoprecipitation. Although, there is a distinctive enzymatic overlap between SCAD and MCAD activities in disrupted cells, this probably does not occur in intact cells or *in vivo*, where these two activities are distinct and organized to facilitate substrate channeling.

2. 2,3–Enoyl–CoA Hydratases

Two hydratase enzymes are thought to be responsible for the complete oxidation of a long-chain fatty acid. Long-chain enoyl-CoA hydratase (LHYD) is part of the inner-mitochondrial trifunctional protein (TFP), and can be assayed using ultraviolet spectroscopy by monitoring the loss of the double bond from long-chain (C_{16}) 2,3-enoyl-CoAs at 280 nm. A short-chain hydratase (SHYD), also known as crotonase, can be assayed using crotonyl-CoA (C_4) as substrate. Neither of these hydratases are measured routinely in clinical samples. Mitochondrial TFP deficiency is usually detected by analysis of its α- and β-subunit components, LCHAD and LKAT, respectively. A specific defect of LHYD can occur, due to missense mutations in the LHYD coding region of the α-subunit, which do not impact LCHAD or the octameric TFP formation, and membrane attachment. However, these defects have not yet been identified. There are multiple additional short-chain hydratase activities within the mitochondrion, including one that is situated on the catabolic pathway of the amino acid lysine. These other short-chain hydratases can interfere with the assay for the SHYD (crotonase), particularly when disrupted cells are used as a source or when crotonyl-CoA is used as a substrate.

3. L-3-Hydroxyacyl CoA Dehydrogenases

Two L-3-hydroxyacyl CoA dehydrogenases are known, with differing chain-length specificities. LCHAD is part of the mitochondrial TFP, and as described above, measurement of LCHAD activity is frequently used to determine the integrity of the α-subunit. A second enzyme has been identified which has broad-chain-length specificity, but because the substrate used to monitor its activity was a short-chain species (C_4), this enzyme was initially called short-chain-L-3-hydroxyacyl CoA dehydrogenase (SCHAD). We prefer to use the term medium- and short-chain (M/SCHAD) to define more accurately the chain-length specificity and to distinguish this enzyme from a second hydroxyacyl CoA dehydrogenase that utilizes sterols as substrates and is metabolically unrelated to the fatty acid oxidation pathway. There is confusion in the literature regarding these two different enzymes. For clarification, *HAD1* is the gene assignment for M/SCHAD while *HAD2* is the gene assignment for the second enzyme (Bennett *et al.*, 2006; Yang *et al.*, 2005).

Both 3-hydroxyacyl CoA dehydrogenase enzymes are routinely measured in the reverse direction, by monitoring the conversion of the 3-ketoacyl CoA intermediate to the 3-hydroxyacyl CoA, with NADH being converted to NAD, and monitoring the change in optical density at 340 nm. The substrate that is routinely used to determine LCHAD activity is 3-ketopalmitoyl-CoA, and that for M/SCHAD is 3-ketobutyryl-CoA (acetoacetyl-CoA) (Wanders *et al.*, 1992). 3-Ketopalmitoyl-CoA is not commercially available and requires synthesis in house. In our laboratory, this process was achieved enzymatically starting with palmitoyl-CoA, which is converted to the 2,3-enoyl-CoA by acyl CoA oxidase, then to the 3-hydroxyacyl CoA by crotonase, and finally to the 3-ketopalmitoyl-CoA by the addition of recombinant SCHAD. The product is purified by HPLC prior to use. Acyl CoA oxidase is commercially available, and lines expressing crotonase and SCHAD are available. Other substrates of different chain lengths, including 3-keto-octanoyl-CoA, can also be synthesized using this procedure. The advantage of using the 3-ketoacyl CoA substrate is that it can also be used in the forward direction to measure long-chain 3-ketoacyl CoA thiolase (LKAT) activity. Acetoacety-CoA for the measurement of M/SCHAD activity can be obtained commercially.

Protocol

1. Prepare assay mixture (final volume 350 μl):
   100-mM potassium phosphate buffer
   pH 6.3, 0.1-mM dithiothreitol
   0.1% (w/v) Triton X-100
   100-μM NADH
   0.1-mg/ml tissue protein
2. Preincubate assay mixture at 37 °C in a cuvette.
3. Start the reaction by addition of substrate at a concentration of 40–50 μM.
4. Monitor the reaction spectrophotometrically at 340 nm.

4. 3–Ketoacyl CoA Thiolases

Two or three 3-ketoacyl CoA thiolases are responsible for the complete oxidation of a long-chain fatty acid. LKAT, the product of the β-subunit of the mitochondrial TFP, is responsible for the thiolytic cleavage of long-chain species at the inner mitochondrial membrane. A short-chain 3-ketothiolase, which is also responsible for the thiolytic cleavage of methylacetoacetyl-CoA from the isoleucine degradative pathway, is probably also responsible for short-chain thiolytic cleavage of fatty acid oxidative intermediates. This enzyme is also known as β-ketothiolase. The single report of a medium-chain thiolase (MKAT), which is active in the fatty acid pathway, has not yet been confirmed by additional reports or by the identification of a gene encoding this activity.

MKAT activity is monitored in the forward direction and measured as the loss of the double bond in the 3-keto position of the same substrates that are used above for 3-hydroxyacyl CoA dehydrogenase, as they are thiolytically cleaved to form a two-carbon chain-shortened saturated acyl CoA species and acetyl-CoA. A similar assay is used to measure SKAT activity. The only differences are (1) the BSA and Triton X-100 are omitted and (2) acetoacetyl-CoA is used as a substrate at a final concentration of 50 μM.

These assays can be performed on a variety of cell types, including liver, skeletal muscle, cardiac muscle, skin fibroblasts, amniocytes, and chorionic villus cells. There is preliminary but unconfirmed evidence that M/SCHAD may have different tissue distributions, with hepatic and myopathic clinical phenotypes described. However, as with MKAT, which is discussed above, the genetic basis for different tissue expression of M/SCHAD activities is not well understood.

Protocol to measure MKAT activity

1. Prepare the assay cocktail:
   100-mM Tris–HCl, pH 8.3

50-mM KCl

25-mM $MgCl_2$

0.1% (w/v) Triton X-100

0.2-mg/ml BSA (fatty acid free)

10-μM 3-ketopalmitoyl-CoA

0.1-mg/ml protein extract

2. Preincubate the cocktail at 37°C in a cuvette.

3. Start the reaction by addition of CoA to a final concentration of 70 μM.

4. Monitor the reaction spectrophotometrically at 280 nm.

IV. Metabolic Flux Studies

Frequently, patients are identified in whom there are clinical hallmarks of impaired fatty acid oxidation but in whom there are no clues from metabolite or enzyme analysis. In these situations, there are a number of ways in which flux through the entire fatty acid oxidative pathway can be studied. The end products of complete oxidation of, and electron transfer from, fatty acids, are CO_2 and H_2O. Using appropriately radiolabeled fatty acid substrates, it is possible to monitor the production of radiolabeled CO_2 and H_2O, and test the entire pathway in intact living cells. Alternate methods to study the whole pathway have been developed which monitor the production of fatty acid intermediates, including acylcarnitines and free fatty acids. The relative merits of each of these approaches are discussed below.

A. Whole-Flux Measurement

The initial approach to whole-flux measurement of fatty acid oxidation uses fatty acids labeled with ^{14}C, usually in the 1-^{14}C or terminal position (16-^{14}C for palmitate), with the assay monitoring the production of [14]CO_2 (Saudubray *et al.,* 1982). Because of the pH of the reaction mixture, the CO_2 is present in solution as bicarbonate. At the end of the measurement, the CO_2 is released by acidification of the reaction mixture, and the CO_2 gas is captured on a filter paper by an acidic trapping agent containing concentrated HCl. Alternate approaches have monitored the production of trichloroacetic acid-precipitable radiolabeled intermediates in the culture media (Olpin *et al.*, 1997). The limitation of using 1-^{14}C substrates is that they can only be used to study flux relative to specific chain length within the pathway. For instance, 1-^{14}C-palmitate can be used for the study of long-chain defects. In order to study medium-chain or short-chain fatty acid oxidative flux, additional experiments using 1-^{14}C-labeled octanoate and butyrate are required.

Olpin *et al.* (1997) developed an assay that monitors the production of tritiated water from fatty acids labeled with ^{3}H in the 9,10 position. In the assay, confluent cells in monolayer are incubated for with the 9,10-radiolabeled fatty acid (usually palmitate or myristate) in 24-well plates, and the tritiated water is isolated from the unused labeled substrate by use of an ion exchange resin. Isolation of tritiated water is simpler than CO_2 trapping, and appears to be more precise. This assay, which is used routinely in our laboratory, can be used to detect all long- and medium-chain fatty acid oxidation defects, and tritium-labeled fatty acid is an excellent substrate with which to monitor metabolic flux.

The assay can also be applied to any cells in culture and has been adapted for the analysis of fatty acid oxidative flux in isolated white blood cells and in whole blood. (Seargeant et al., 1999; Wanders et al., 1991). These latter two applications have an advantage in not requiring tissue culture, but do require that analysis be performed immediately on sample collection, as intact, fully metabolic cells are required to perform the assay. The only drawback of the methods is that it does not detect SCAD deficiency.

All of the above metabolic flux assays monitor the rate of fatty acid oxidation in both the mitochondrial and peroxisomal compartments. Peroxisomal fatty acid oxidation can contribute up to 30% of the total metabolic flux, particularly under conditions when the mitochondrial pathway is inhibited or deficient. We modified the assay by addition of a cocktail that inhibits mitochondrial fatty acid oxidation. Comparison of activity in the presence and absence of the inhibitor cocktail provides an index of peroxisomal contribution to the overall rate of fatty acid oxidation and a more accurate evaluation of the mitochondrial fatty acid oxidation activity (Narayan et al., 2005).

Reagents

PBS (pH 6.9)

Krebs-Ringer bicarbonate buffer containing 0.5-g/liter BSA (fatty acid free)

Fatty acid oxidation inhibitor cocktail:

  0.45-mM malonyl-CoA

  0.02-mM glyburide

  0.79-mM 3-mercaptopropionic acid

^{3}H fatty acid (9,10-radiolabeled palmitate or myristate)

Protocol

1. Low passage fibroblasts (<P8) are harvested from a confluent T25 flask and placed into each of 2 wells in the 24-well plate.

2. Incubate the 24-well plate at 37°C in a CO_2 incubator for 48 h until the cells are attached and confluent.

3. Thereafter, wash the cell monolayer with PBS.

4. Preincubate one well with Krebs-Ringer bicarbonate medium containing BSA. Preincubate the other well with Krebs-Ringer bicarbonate medium containing BSA and the fatty acid oxidation inhibitor cocktail.

5. Add radiolabeled fatty acid (110 μM, 16.7 μCi/ml) to each well and incubate for 2–3 h at 37°C in a CO_2 incubator.

6. Apply the culture medium to Dowex-1 minicolums. Unused labeled fatty acids are bound to the ion-exchange resin, and water, which passes through the column, is collected. Radioactivity of tritiated water in the column eluate is then measured by liquid scintillation counting.

7. Hydrolyze cells with sodium hydroxide for total protein measurement.

B. Metabolite Accumulation Assays

Tandem mass spectrometry has been applied to measure metabolic intermediates, at low concentrations or in small sample volumes, in media from cells in tissue culture. This has provided an excellent complementary tool to the whole-flux assays that are described above.

In this technique, cells in culture are incubated with fatty acids (0.2 mM in the tissue culture medium), which are either unlabeled or labeled with a stable isotope (e.g., deuterium), and are incubated either with or without carnitine (usually 0.4 mM). At the end of the incubation period, acylcarnitine profiles are analyzed by tandem mass spectrometry, as described for plasma acylcarnitine profiling. The profiles are then compared to those obtained from patients with well-characterized fatty acid oxidation defects (Nada *et al.*, 1995; Shen *et al.*, 2000). These assays analyze as little as 100 μl of culture medium and have the advantage of preserving the cultured cells for additional experimentation.

V. Discussion

Fatty acid oxidation can be performed using a variety of different and usually complementary techniques, including metabolite measurement, classical enzymology, and studies of metabolic flux, using both product formation and intermediate accumulation. These assays have an important role to play in the diagnosis of genetic diseases of the pathway. Metabolite measurement, most specifically acylcarnitine measurement, is rapidly developing into the method of choice for whole population newborn screening for these disorders in the Western world, such that diagnosis can be established prior to the development of potentially fatal clinical signs. The tools that are available for metabolic flux studies can be applied in any experimental system in which alterations in metabolic flux are required to be evaluated.

References

Bennett, M. J., and Narayan, S. B. (2005). Carnitine palmitoyltransferase 1A. *In* "Gene Reviews." Available at www.genetests.org.

Bennett, M. J., Russell, L. K., Tokunada, C., Narayan, S. B., Tan, L., Seegmiller, A., Boriack, R. L., and Strauss, A. W. (2006). Reye-like syndrome resulting from novel missense mutations in mitochondrial medium- and short-chain L-3-hydroxyacyl-CoA dehydrogenase. *Mol. Genet. Metab.* **88,** 74–79.

Campbell, S. E., Tandon, N. T., Woldegiorgis, G., Luiken, J. J. F. P., Glatz, J. F. C., and Bonen, A. (2004). A novel function for fatty acid translocase (FAT)/CD36. *J. Biol. Chem.* **279,** 36235–36241.

Chace, D. H., Kalas, T. A., and Naylor, E. W. (2003). Use of tandem mass spectrometry for multianalyte screening of dried blood specimens from newborns. *Clin. Chem.* **49,** 1797–1817.

Ensenauer, R., He, M., Willard, J. M., Goetzman, E. S., Corydon, T. J., Vandahl, B. B., Mohsen, A. W., and Vockley, J. (2005). Human acyl-CoA dehydrogenase-9 plays a novel role in the mitochondrial beta-oxidation of unsaturated fatty acids. *J. Biol. Chem.* **280,** 32309–32316.

Frazier, D. M., Millington, D. S., McCandless, S. E., Koeberl, D. D., Weavil, S. D., Chaing, S. H., and Muenzer, J. (2006). The tandem mass spectrometry newborn screening experience in North Carolina: 1997–2005. *J. Inherit. Metab. Dis.* **29,** 76–85.

Hale, D. E., Stanley, C. A., and Coates, P. M. (1990). Genetic defects of acyl-CoA dehydrogenases: Studies using an electron transfer flavoprotein reduction assay. *Prog. Clin. Biol. Res.* **321,** 333–348.

Kumps, A., Duez, P., and Mardens, Y. (2002). Metabolic, nutritional, iatrogenic, and artifactual sources of urinary organic acids: A comprehensive table. *Clin. Chem.* **48,** 708–717.

McGarry, J. D., and Brown, N. F. (1997). The mitochondrial carnitine palmitoyl-transferase system. From concept to molecular analysis. *Eur. J. Biochem.* **244,** 1–14.

Murthy, M. S., Kamanna, V. S., and Pande, S. V. (1986). A carnitine/acylcarnitine translocase assay applicable to biopsied specimens without requiring mitochondrial isolation. *Biochem. J.* **236,** 143–148.

Nada, M. A., Chace, D. H., Sprecher, H., and Roe, C. R. (1995). Investigation of β-oxidation in normal and MCAD-deficient human fibroblasts using tandem mass spectrometry. *Biochem. Mol. Med.* **54,** 59–66.

Narayan, S. B., Boriack, R. L., Messmer, B., and Bennett, M. J. (2005). Establishing a normal range for measurement of flux through the mitochondrial fatty acid oxidation pathway. *Clin. Chem.* **51,** 644–646.

Oey, N. A., Den Boer, M. E. J., Wijberg, F. A., Vekemans, M., Auge, J., Steiner, C., Wanders, R. J. A., Waterham, H. R., Ruiter, J. P. N., and Attie-Bitach, T. (2005). Long-chain fatty acid oxidation during early human development. *Pediatr. Res.* **56,** 755–759.

Olpin, S. E., Manning, N. J., Pollitt, R. J., and Clarke, S. (1997). Improved detection of long-chain fatty acid oxidation defects in intact cells using [9,10-^{3}H] Oleic acid. *J. Inherit. Metab. Dis.* **20,** 415–419.

Rinaldo, P., Matern, D., and Bennett, M. J. (2002). Fatty acid oxidation disorders. *Annu. Rev. Physiol.* **64,** 477–502.

Rinaldo, P., O'Shea, J. J., Coates, P. M., Hale, D. E., Stanley, C. A., and Tanaka, K. (1988). Medium-chain acyl-CoA dehydrogenase deficiency: Diagnosis by stable isotope dilution measurement of urinary n-hexanoylglycine and 3-phenyl propionylglycine. *N. Engl. J. Med.* **319,** 1308–1313.

Roe, C. R., and Ding, J. (2001). Mitochondrial fatty acid oxidation disorders. *In* "The Metabolic & Molecular Bases of Inherited Disease" (C. R. Scriver, A. L. Beaudet, W. S. Sly, and D. Valle, eds.), pp. 2297–2326, 8th edn. McGraw-Hill, New York.

Saudubray, J.-M., Coude, F.-X., Demaugre, F., Johnson, C., Gibson, K. M., and Nyhan, W. L. (1982). Oxidation of fatty acids in cultured fibroblasts: A model system for the detection and study of defects in oxidation. *Pediatr. Res.* **16,** 877–881.

Seargeant, L. E., Balachandra, K., Malory, C., Dilling, L. A., and Greenberg, C. R. (1999). A simple screening test for fatty acid oxidation defects using whole blood palmitate oxidation. *J. Inherit. Metab. Dis.* **22,** 740–746.

Shekhawat, P., Bennett, M. J., Sadovsky, Y., Nelson, D. M., Rakheja, D., and Strauss, A. W. (2003). Human placenta metabolizes fatty acids: Implications for fetal fatty acid oxidation disorders and maternal liver disease. *Am. J. Physiol. Endocrinol. Metab.* **284,** E1098–E1105.

Shen, J. J., Matern, D., Millington, D. S., Hillman, S., Feezor, M. D., Bennett, M. J., Qumsiyeh, M., Kahler, S. G., Chen, Y.-T., and Van Hove, J. L. K. (2000). Acylcarnitines in fibroblasts of patients with long-chain 3-hydroxyacyl-CoA dehydrogenase deficiency and other fatty acid oxidation defects. *J. Inherit. Metab. Dis.* **23,** 27–44.

Stanley, C. A., Bennett, M. J., and Mayatepek, E. (2006). Disorders of mitochondrial fatty acid oxidation and related metabolic pathways. *In* "Inborn Metabolic Diseases: Diagnosis and Treatment" (J. Fernandes, J.-M. Saudubray, G. Van den Berghe, and J. H. Walter, eds.), pp. 175–190. Springer, Heidelberg.

Stanley, C. A., Berry, G., Hale, D. E., Bennett, M. J., Willi, S., and Treem, W. (1993). Renal handling of carnitine in secondary carnitine deficiency disorders. *Pediatr. Res.* **34,** 89–97.

Stanley, C. A., Hale, D. E., Berry, G. T., Deleeuw, S., Boxer, J., and Bonnefont, J.-P. (1992). A deficiency of carnitine-acylcarnitine translocase in the inner mitochondrial membrane. *N. Engl. J. Med.* **327,** 19–23.

Treem, W. R., Stanley, C. A., Finegold, D. N., Hale, D. E., and Coates, P. M. (1988). Primary carnitine deficiency due to a failure of carnitine transport in kidney, muscle, and fibroblasts. *N. Engl. J. Med.* **319,** 1331–1336.

Wanders, R. J. A., IJlst, L., van Elk, E., de Klerk, J. B. C., and Przyrembel, H. (1991). Octanoate and palmitate β-oxidation in human leukocytes: Implications for rapid diagnosis of fatty acid β-oxidation. *J. Inherit. Metab. Dis.* **14,** 317–320.

Wanders, R. J. A., IJlst, L., Poggi, F., Bonnefont, J.-P., Munnich, A., Rabier, D., and Saudubray, J.-M. (1992). Human trifunctional protein deficiency: A new disorder of mitochondrial fatty aid beta-oxidation. *Biochem. Biophys. Res. Commun.* **188,** 1139–1145.

Yang, S.-Y., He, X.-J., and Schulz, H. (2005). 3-Hydroxyacyl-CoA dehydrogenase and short-chain 3-hydroxyacyl-CoA dehydrogenase in human health and disease. *FEBS J.* **272,** 4874–4883.

CHAPTER 10

Biochemical Assays for Mitochondrial Activity: Assays of TCA Cycle Enzymes and PDHc

Ann Saada Reisch[*,†] and Orly Elpeleg[*]

[*]The Metabolic Disease Unit
Hadassah-Hebrew University Medical Centre
Jerusalem 91120, Israel

[†]Ben Gurion University of the Negev
Beer Sheva 84105, Israel

I. Introduction

The tricarboxylic acid cycle (TCA), also called the citric acid cycle or Krebs cycle, is a central pathway of metabolism. Its main function is the oxidation of acetyl coenzyme A (CoA) derived from carbohydrates, amino acids, and fatty acids. The pyruvate dehydrogenase complex (PDHc) provides a link between glycolysis and the TCA.

Both TCA and PDHc generate reducing equivalents in the form of NADH and FADH$_2$, which comprises the first step of oxygen-dependent energy production (Fig. 1). The TCA cycle also functions in a biosynthetic capacity, primarily in the synthesis of glucose, heme, and amino acids. The committed step in the synthesis of heme is the condensation of succinyl CoA and glycine. Glutamate and aspartate are products of the transamination reaction of α-ketoglutarate and oxaloacetate, respectively. TCA cycle flux appears to be limited by the availability

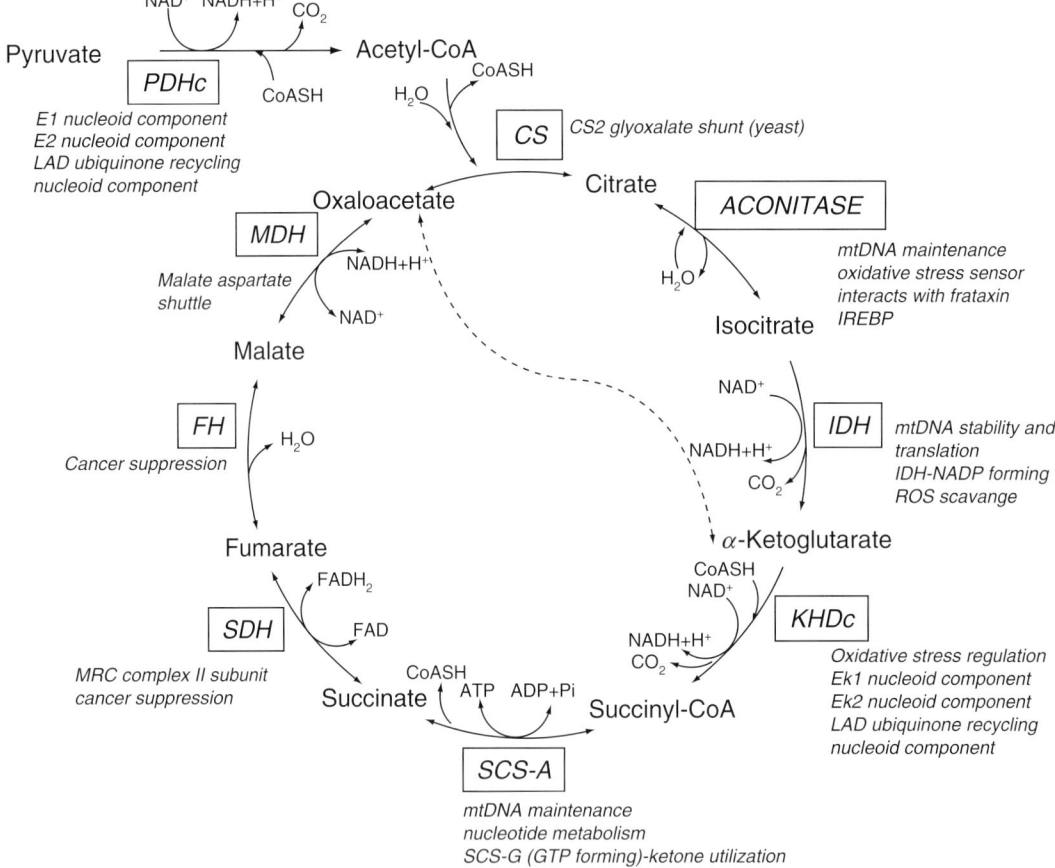

Fig. 1 A schematic representation of the reactions of pyruvate dehydrogenase and the TCA. "Moonlighting" functions of the enzymes and subunits are outlined in *italics*. PDHc, pyruvate dehydrogenase; LAD, lipoamide dehydrogenase; CS, citrate synthase; IDH, isocitrate dehydrogenase; KHDc, α-ketoglutarate dehydrogenase; SCS-A, succinyl CoA synthase, ATP forming; SCS-G, succinyl CoA synthase, GTP forming; SDH, succinate dehydrogenase; FH, fumarase; MDH, malate dehydrogenase.

of oxaloacetate and α-ketoglutarate. This has led to the proposal of a model dividing the TCA cycle into two minicycles that are interconnected by these substrates and their transamination products (Rustin *et al.*, 1997; Yudkoff *et al.*, 1994). The TCA enzymes are located in the mitochondrial matrix, organized into a supramolecular complex—the "Krebs metabolon"—that interacts with mitochondrial membranes and the mitochondrial respiratory chain (MRC) (Velot and Srere, 2000).

Many of the TCA enzymes—aconitase, isocitrate dehydrogenase (IDH), fumarase (FH), malate dehydrogenase (MDH), and yeast citrate synthase (CS)—have cytosolic counterparts that participate in extramitochondrial reactions. In addition to their roles in the TCA, most enzymes perform additional "moonlighting" functions (Sriram *et al.*, 2005). Aconitase, α-ketoglutarate subunits, succinyl CoA synthase (SCS), and IDH are associated with the mitochondrial nucleoid, and may play a role in stabilizing the mitochondrial DNA (mtDNA). While IDH, lipoamide dehydrogenase (LAD), aconitase, succinate dehydrogenase (SDH), and FH are associated with mitochondrial mRNA translation, oxidative stress, iron metabolism, or tumor suppression (Fig. 1).

Congenital defects of TCA enzymes were previously thought to be rare, overshadowed by the more common defects in PDHc and the MRC. However, since the first report of fumaric aciduria about 20 years ago, an increasing number of defects has been identified, affecting not only the TCA cycle but also their "moonlighting" functions (Pithukpakorn, 2005; Rustin *et al.*, 1997; Whelan *et al.*, 1983). It should be noted that all subunits of PDHc and TCA enzymes are encoded by the nuclear genome and are targeted to mitochondria by N-terminal presequences, which are removed by mitochondrial matrix peptidases (MPPs). As opposed to the significant proportion of MRC disorders, which are maternally inherited, all PDHc and TCA defects are inherited as Mendelian traits.

The characterization of these defects requires the accurate measurement of the respective enzymatic activities. The assays may be performed on tissue homogenates or in isolated mitochondria. Disruption of the sample before assay is a crucial step. The most suitable disruption method depends on the enzyme and the material to be assayed. Several methods are employed, including treatment with various detergents such as Triton X-100, cholic acid, or dodecyl maltoside. The protein:detergent ratio should be kept fairly constant, as too much detergent could interfere with the enzymatic assay, whereas too little may lead to insufficient disruption. "Fine tuning" is also warranted for mechanical means of disruption such as freeze thawing, sonication, and osmotic shock (Robinson *et al.*, 1987). A number of assays have been the subject of modifications by many distinguished laboratories, and several spectrophotometric and radiochemical methods, differing in their sensitivity, have been described for each enzyme. Thus, each laboratory should adapt the preferred method, optimizing sample preparation, disruption, and assay conditions to aim for a linear reaction with increased signal:background ratio. These individual modifications dictate the establishment of normal reference ranges by each laboratory for each enzyme.

II. Pyruvate Dehydrogenase Complex

The PDHc (EC 1.2.4.1) catalyzes the irreversible conversion of pyruvate to acetyl CoA. It regulates the metabolic flux of glycolytic metabolites into the TCA cycle. PDHc is a 9-MDa multisubunit enzyme complex consisting of multiple copies of its three component enzymes: E1 (EC 1.4.1.1), composed of pyruvate decarboxylase, which is a tetramer of 2 α- and 2 β-subunits; E2 (EC 2.3.1.12), composed of dihydrolipoyl transacetylase; and E3 (EC 1.8.1.4), composed of LAD, a dihydro-lipoyl dehydrogenase (Reed and Willms, 1966). Mammalian PDHc contains an additional component, the E3 binding protein (E3BP; PDX1; X). PDHc components E1α, E1β, and LAD have been identified in yeast, and E2 has been found in the *Xenopus* nucleoid (Bogenhagen *et al.*, 2003; Chen *et al.*, 2005).

The mammalian PDHc requires five different coenzymes that act as carriers or oxidants of the intermediates of the reaction. These include three prosthetic groups—thiamine pyrophosphate (TPP), lipoic acid, and FAD—which are tightly bound, and two substrates, NADH and CoA.

PDHc activity is modulated by numerous factors. It is inhibited by its products, acetyl CoA and NADH, and is dependent on cofactor availability. The E1 subunit, which catalyzes the TPP-dependent decarboxylation of pyruvate, is highly regulated by its state of phosphorylation. It is inactivated by a PDHc-specific protein kinase (EC 2.7.1.99) and activated by the PDHc protein phosphatase (EC 3.1.3.43). Isoforms of this phosphatase are expressed differentially in various tissues. The kinases and phosphatases are regulated by ADP/ATP ratio, Ca^{2+}, Mg^{2+}, and insulin (Holness and Sugden 2003). Defects in each of the PHDc subunits and in the phosphatases cause disease (Maj *et al.*, 2005).

PDHc deficiency is a relatively common cause of primary lactic acidosis (Robinson *et al.*, 1980; Stromme *et al.*, 1976). The clinical spectrum of PDHc deficiency is broad, ranging from fatal, infantile lactic acidosis to a mild, intermittent ataxia, to a fairly asymptomatic course. The vast majority of reported patients are those suffering from mutations in the E1 α-subunit, transmitted as an X-linked dominant trait. Some of these patients are thiamine-responsive. Mutations in E1β and E3BP have been reported only rarely; the clinical course was invariably severe and most patients expired early in life (Aral *et al.*, 1997; De Meirleir, 2002; Robinson *et al.*, 1990). E2 defects are equally rare, but the clinical outcome is more favorable, albeit the patients' neurological condition is severely compromised (Head *et al.*, 2005).

A. Pyruvate Dehydrogenase Assays

The activity of the whole PDHc is measured in the presence of CoA, NAD^+, and TPP. Spectophotometric as well as radiometric assays have been described. Generally, the radiochemical methods, which are based on the release of radioactive CO_2 from labeled pyruvate, are more sensitive. Radiolabeled pyruvate is also used for the determination of E1 activity, which is significantly lower than the activity of the whole complex. The obvious drawbacks are high costs, complications associated with CO_2 collection, and high background values. Spectrophotometric methods, which measure the formation of NADH or acetyl CoA, are more convenient but are less sensitive, especially in homogenates of tissues with low PDH activity such as fibroblasts. For both approaches, care must be taken for the proper storage of pyruvate, which is labile. The actual concentration of the pyruvate substrate solution should therefore be checked on a regular basis, using methods including the lactate dehydrogenase (LDH) assay. High concentrations of pyruvate should be avoided, as branched-chain 2-oxo acid dehydrogenases may interfere with the measurements (Goodwin et al., 1986).

PDHc phosphatases are activated by calcium and are dependent on magnesium. Therefore, both must be present in the assay. Frequently, a preincubation step with calcium and magnesium and/or dichloroacetate, an inhibitor of the PDH kinase, is performed. Oxamic acid, an LDH inhibitor, is present in the more recently described spectrophotometric assays (Chretien et al., 1995; Schwab et al., 2005). It should be noted that PDHc activity assayed in patients with mitochondrial complex I (NADH CoQ reductase) deficiency is frequently low because of the inhibitory effect of the accumulated NADH. Decreased PDHc activity may also result from defective PDHc phosphatase. Thus, in order to advance molecular diagnosis, separate assays for the most common PDHc disorders, E1 and LAD deficiency, are justified.

A radiometric PDHc assay suitable for freeze-thawed fibroblast homogenates is described in Table I (Robinson et al., 1980; Seals and Jarett, 1980). Activity of PDHc in muscle homogenates or mitochondria may be measured by a modified assay, independent of mitochondrial complex I activity, also as described in Table I (Sperl et al., 1993). E1 activity is measured radiochemically in the presence of TPP and 2,6-dichlorophenol indophenol (DCIP), an artificial electron acceptor, without the addition of CoA and NAD, as detailed in Table II (van Laack et al., 1988). The specific radioactivity activity of the [^{14}C] pyruvate used as the E1 substrate is higher than that of the PDHc substrate because of the lower activity of E1 compared to that of the whole complex. The measurement of E2 activity is rarely performed, as the assay requires labeled acetyl CoA and the preparation of dihydrolipoamide (Butterworth et al., 1975).

Deficiency of the various PDHc components may be confirmed by Western blot with anti-PDH antibodies, which are now available commercially. Western blot is currently the only method to identify a deficiency in the E3BP component of PDHc. It should be noted that some polyclonal antisera are devoid of anti-LAD antibodies (Elpeleg et al., 1997).

B. LAD Assay

LAD (E3; dihydrolipoyl dehydrogenase; EC 1.6.4.3) is the third catalytic component of PDHc. It binds to the core of E2 and catalyzes the transfer of hydrogen from reduced lipoyl groups to NAD via FAD. LAD is also a component of three additional enzyme complexes: the α-ketoglutarate dehydrogenase complex

Table I
A Selection of Radiometric Assays for Pyruvate Dehydrogenase Complex

	Reagents and assay conditions					Samples and ranges		
Assay	Reagent (stock solution)	Assay mix	Concentration	Conditions	Sample	Normal values (nmol/min/mg)	Deficient values (nmol/min/mg)	References
Pyruvate dehydrogenase complex (PDHc) assay I	Tris (100 mM), EDTA (0.5 mM), pH 7.4 buffer	1.4 ml	100 mM, 0.5 mM	Mix buffer, NAD, TPP, CaCl$_2$, MgSO$_4$ Adjust to pH 7.4 with 0.4-M KOH	Frozen homogenate from human muscle 2–7 mg/ml Freeze thawed 3×	5.7 ± 1.5	0.9	Aptowitzer et al., 1997; Seals and Jarett, 1980; Sperl et al., 1993
	NAD (5 mM) in buffer	1 ml	0.5 mM	Add CoA, cytochrome c, and cytochrome c reductase				
	TPP (10 mM) in buffer	1 ml	1 mM	Divide into 0.4-ml reaction/aliquots.				
	Carnitine (50 mM) in buffer	1 ml	5 mM	Add 50-μl sample/reaction Preincubate 5 min 37°C				
	CaCl$_2$ (100 mM)	0.2 ml	2 mM	Start reaction with 50-μl pyruvate/reaction				
	MgSO$_4$ (100 mM)	0.4 ml	4 mM	Insert hyamine hydroxide				
	CoA (10 mM) in buffer	1 ml	1 mM	(1 M in methanol) 0.2 ml in vial/filter paper (avoid contact with assay mixture)				
	Cytochrome c (20 mg/ml) buffer	1 ml	2 mg/ml	Seal tube immediately. Stop assay after 20 min by injecting 0.2 ml of 3% perchloric acid.				
	Cytochrome c reductase (10 U/ml)	1 ml	0.1 U/ml	Leave on ice 1 h. Carefully remove hyamine hydroxide and count by liquid scintillation. Subtract background reaction lacking coenzymes Work in ventilated hood				
	Pyruvate (5 mM), 3 μCi/ml ^{14}C	250 nmol + 0.15 μCi/0.5 ml reaction						
	Sample	50 μl/reaction						

Pyruvate dhehydrogenase complex (PDHc) assay II	Potassium phosphate (10 mM), EDTA (1 mM), 1% fatty acid free bovine serum albumin, pH 7.4	7.24 ml	10 mM	Mix buffer. H$_2$O, albumin. NAD, TPP, CaCl$_2$, MgCl$_2$, DTT	Human skin fibroblasts 1.35 mg/ml In 10-mM phosphate buffer, 1-mM EDTA, pH 7.4, 1% bovine albumin sonicated 10″ × 2	0.450–0.703	0.085–0.087	Robinson et al., 1980
				Adjust to pH 7.4 with 0.4-M KOH				
				Add CoA				
	MgCl$_2$ (100 mM)	200 μl	2 mM	Divide into 0.4-ml aliquots.				
	CaCl$_2$ (100 mM)	200 μl	2 mM	Add 50-μl sample/aliquot.				
	Dithiothreitol DTT (1 M)	10 μl	1 mM	Preincubate 5 min 37°C				
	CoA (10 mM) in buffer	150 μl	0.15 mM	Start reaction with pyruvate 50-μl/aliquot. Insert hyamine hydroxide and stop reaction as described above				
	TPP(10 mM) in buffer	200 μl	0.1 mM	Subtract background reaction lacking coenzymes				
	NAD	10.6 mg	1.6 mM	Work in ventilated hood				
	Pyruvate (2.5 mM), 2.5 μCi/ml ^{14}C	125 nmol + 0.125 μCi/ 0.5 ml reaction						
	Sample	50 μl/reaction						

Table II
Radiometric Assays for Pyruvate Decarboxylase and α-Ketoglutarate Dehydrogenase Complex

Assay	Reagents and assay conditions					Samples and ranges		
	Reagent (stock solution) [storage]	Assay mix	Concentration	Conditions	Sample	Normal values (nmol/min/mg)	Deficient values (nmol/min/mg)	References
Pyruvate decarboxylase E1	Tris (100 mM), EDTA (0.5 mM), pH 6.2 buffer	5.6 ml	100 mM, 0.5 mm	Mix buffer. H_2O, NAD, TPP, $CaCl_2$, $MgCl_2$, DTT	Frozen homogenate from human muscle 2–7 mg/ml	0.083–0.24	0.009–0.0019	van Laack et al., 1988
	EDTA (100 mM)	20 μl	0.2 mM	Adjust to pH 6.2 with 0.4-M KOH	Freeze thawed 3×			
	$MgCl_2$ (100 mM)	200 μl	2 mM	Add DCIP				
	$CaCl_2$ (100 mM)	200 μl	2 mM	Divide into 0.4-ml aliquots. Add 50 μl sample/aliquot				
	TPP (10 mM) in buffer	200 μl	0.1 mM	Preincubate 5 min 37°C				
	Dichloroindophenol, DCIP (5 mM) [fresh]	1 ml	0.5 mM	Start reaction with pyruvate 50 μl/aliquot. Insert hyamine hydroxide and stop reaction as described for PDHC				
	Pyruvate (2.5 mM), 5 μCi/ml ^{14}C	250 nmol + 0.25 μCi/0.5 ml reaction		Subtract background reaction lacking TPP Work in ventilated hood				
	Sample	50 μl/reaction						
α-Ketoglutarate complex	Potassium phosphate (0.1 M), pH 7.4 buffer	0.75 ml	10 mM	Mix buffer. H_2O, NAD, TPP, $CaCl_2$, $MgCl_2$, DTT	Frozen homogenate from human muscle 2–7 mg/ml	3.7 ± 1.6	0.7 (LAD deficient)	Aptowitzer et al., 1997; Robinson et al., 1977, 1980
	H_2O	5.57 ml		Adjust to pH 7.4 with 0.4-M KOH Add CoA	Freeze thawed 3×			
	EDTA (100 mM)	20 μl	0.2 mM	Divide into 0.4-ml aliquots. Add 50 μl sample/aliquot				
	$MgCl_2$ (100 mM)	200 μl	2 mM	Preincubate 5 min 37°C				
	$CaCl_2$ (100 mM)	200 μl	2 mM	Start reaction with α-ketoglutarate 50 μl/aliquot. Insert hyamine hydroxide and stop reaction as described for PDHC				
	Dithiothreitol, DTT (1 M)	10 μl	1 mM	Subtract background reaction lacking coenzymes Work in ventilated hood				
	CoA (10 mM) in buffer	150 μl	0.15 mM					
	TPP (10 mM) in buffer	200 μl	0.1 mM					
	NAD	10.6 mg	1.6 mM					
	α-Ketoglutarate (2.5 mM), 2.5 μCi/ml ^{14}C	125 nmol + 0.25 μCi/0.5 ml reaction						
	Sample	50 μl/reaction						

(KGDHc), the branched-chain keto acid dehydrogenase, and the glycine cleavage system (Reed and Willms, 1966; Yoshino *et al.*, 1986). Finally, LAD is instrumental in cellular defenses against lipid oxidation, since it is able to reduce ubiquinone (CoQ) to ubiquinol, with either NADH or NADPH as cofactors. The reaction is stimulated by zinc at acidic pH. Ubiquinol functions as not only an electron carrier in the MRC, but has also an important role in the protection of polyunsaturated fatty acids from damage by reactive oxygen species (ROS). Ubiquinol is the only endogenous lipid-soluble antioxidant presently known, and its regeneration both within the mitochondria and in the cytosol is mediated by LAD (Xia *et al.*, 2001).

LAD deficiency results in numerous metabolic disturbances, TCA cycle dysfunction, and impaired branched-chain amino acid degradation. The age at onset and the clinical course are related directly to the genotype, ranging from stormy lactic acidemia immediately after birth, with severe neurological disease in infancy, to relapsing liver failure or intermittent myoglobinuria (Robinson *et al.*, 1977; Shaag *et al.*, 1999). The residual enzymatic activity does not reflect the disease severity, but the ATP production rate in the patient's fibroblasts does (Saada *et al.*, 2000).

LAD is most conveniently measured in the reaction's reverse direction (Berger *et al.*, 1996), in lymphocytes, fibroblasts, or muscle homogenates, as detailed in Table III.

III. Citrate Synthase

CS (EC 2.3.3.1) a homodimer that catalyzes the condensation of acetyl CoA and oxaloacetate to form citrate. The rate of the reaction is largely determined by the availability of its substrates, while succinyl CoA competes with the acetyl CoA. CS activity is also inhibited by ATP, NADH, and derivatives of fatty acids (Kispal *et al.*, 1989; Srere, 1974).

The enzyme is localized in the mitochondrial matrix and is frequently used as a mitochondrial marker in cells that do not contain the glyoxalate shunt. CS activity is commonly used as a measure of mitochondrial disruption during treatment with detergents, sonication, or freeze thawing (Robinson *et al.*, 1987).

To date, no deficiency of CS in humans has been reported. This CS is used frequently to estimate mitochondrial content and to normalize enzymatic activities of the MRC in diagnostics. A mutation in a conserved amino acid in CS detected in "bang-sensitive" fruit flies, suggested that CS mutations may be compatible with life, and the possibility for human disease deficiency does exist (Fergestad *et al.*, 2006).

CS activity is measured by monitoring the liberation of CoASH by coupling to 5,5'-dithiobis(2-nitrobenzoic) acid (DTNB), with release of a mercaptide ion. The assay is described in Table III.

Table III
A Selection of Spectrophotometric Assays for TCA Enzymes

	Reagents and assay conditions					Samples and ranges		
Assay	Reagent (stock solution) [storage]	Volume in assay	Amount in assay	Conditions ($\varepsilon = \mu M^{-1}\,\mu^{-1}$; t = temperature)	Sample (pretreatment)	Normal values (nmol/min/mg)	Deficient values (nmol/min/mg)	References
Lipoamide dehydrogenase	Potassium phosphate buffer (60.6 mM), EDTA (1.2 mM), pH 6.5	825 μl	50 mM, 1 mM	A_{340} $\varepsilon = 6.22$ $t = 37\,^{\circ}$C Preincubate until stable baseline Start assay with lipoamide	Frozen homogenate from fresh human skeletal muscle 1- to 5-mg protein/ml (0.5% Triton X-100, 0°C, 5 min)	121 ± 9 84–239	10	Aptowitzer et al., 1997; Berger et al., 1996; Reed and Willms, 1966
	NADH (2 mM) [fresh]	75 μl	0.15 mM		Homogenate from frozen lymphocyte pellet. 1- to 2-mg protein/ml (sonication 5 sec, 0.5% Triton X-100, 0°C, 5 min)	81 ± 24 35–139	13 ± 24 7–21	
	Sample	50 μl	50–25 μg					
	Lipoamide-thioctic acid amide (40 mM in 80% ethanol) [fresh]	50 μl	2 mM					
Citrate synthase	DTNB (10 mM in 1-M Tris–HCl pH 8.1) [fresh]	100 μl	1 mM	A_{412} $\varepsilon = 13.6$ $t = 37\,^{\circ}$C Preincubate until stable baseline Start assay with oxaloacetate	Frozen mitochondria from fresh human skeletal muscle/ 0.1-mg protein/ml (1% sodium cholate, 0°C, 10 min)	1880 ± 340 820–2900		Berger et al., 1996; Saada et al., 2001; Shaag et al., 1997; Srere et al., 1963
	H$_2$O	Up to 1 ml						
	Acetyl CoA (10 mM) [frozen aliquots]	30 μl	0.3 mM		Homogenate from frozen lymphocyte pellet. 1- to 2-mg protein/ml (sonication 5″ + 0.5% Triton X-100, 5 min, 0°C)	75.5 ± 16 56–88		
	Sample	10–50 μl	1–50 μg					
	Oxaloacetate (10 mM in 0.1-M Tris–HCl, pH 8.1) [fresh]	50 μl	0.5 mM					
Aconitase	Tris–HCl 0.1 M, pH 7.4	200 μl	20 mM	A_{240} (quartz) $\varepsilon = 4.28$ $t = 37\,^{\circ}$C Preincubate until stable baseline Start assay with sample	Mitochondria from human skeletal muscle 1–3 mg/ml (freeze thawed)	348 ± 24 156–467	113	Fansler and Lowenstein, 1969; Hall et al., 1993
	NaCl 0.5 M	200 μl	100 mM					
	Cis-aconitic acid (1 mM) Neutralized with NaOH	100 μl	0.1 mM					
	H$_2$O	475 μl						
	Sample	25 μl	50–100 μg					

IV. Aconitase

Aconitase (aconitate hydratase; citrate hydro-lyase 2; ACO2), EC 4.2.1.3, catalyzes the reversible interconversion of citrate and isocitrate via the enzyme-bound intermediate *cis*-aconitate. Aconitase belongs to the family of iron–sulfur-containing dehydratases. This enzyme undergoes reversible, citrate-dependent inactivation induced by pro-oxidants, which cause disassembly of the iron–sulfur cluster (Rose and O'Connell, 1967). Aconitase activity is commonly used as a biomarker for oxidative stress and has been suggested to serve as an intramitochondrial sensor of redox status. Numerous degenerative mitochondrial disorders are associated with elevated levels of pro-oxidants with subsequent decline of aconitase activity. This decline is frequently reported, together with decreased SDH and complex III of the MRC, suggesting a general defect in the assembly of mitochondrial iron–sulfur clusters (Haile *et al.*, 1992).

Aconitase activity is modulated by the mitochondrial matrix protein frataxin, which acts as an iron chaperone. Frataxin plays a prominent role in mitochondrial iron storage and promotes the maturation of cellular iron–sulfur-containing and heme-containing proteins (Bulteau *et al.,* 2004). Frataxin is deficient in Friedrich's ataxia, a neurodegenerative and cardiac disorder characterized by intramitochondrial iron accumulation and diminished activities of mitochondrial iron–sulfur cluster-containing enzymes (Patel and Isaya, 2001; Rotig *et al.*, 1997).

An additional aconitase activity is present in the cytosol, thus its activity is not a suitable mitochondrial marker. The cytosolic aconitase activity in higher eukaryotes is attributed to the iron regulatory protein IRP1 (IREBP), which possesses dual functions. It can either bind to stem-loop mRNA structures in iron-responsive elements, modulating their expression, or it can assemble iron–sulfur clusters and function as aconitase (ACO1). In yeast, a single nuclear gene encodes the mitochondrial and cytosolic aconitase, participating in the Krebs cycle, mtDNA maintenance, and glyoxalate shunt, respectively (Chen *et al.,* 2005; Regev-Rudzki *et al.,* 2005; Shadel, 2005).

Aconitase enzymatic activity is assayed by measuring the conversion of citrate to isocitrate, or of *cis*-aconitate to isocitrate (Fansler and Lowenstein, 1969). The former assay is coupled with IDH and is measured by the formation of NADPH. The latter assay monitors the disappearance of *cis*-aconitate as measured spectrophotometrically in a quartz cuvette, as detailed in Table III. The enzyme is activated by ferrous ions and reducing agents, and a preincubation step may be of advantage.

V. Isocitrate Dehydrogenase

IDH (IDH3; NAD-specific mitochondrial isocitrate dehydrogenase; IDHm), EC 1.1.1.41, catalyzes the irreversible decarboxylation of isocitrate to α-ketoglutarate, with the formation of NADH and CO_2. In mammalian tissues two additional IDH isoenzymes are present, the cytosolic NADP(+)-specific IDH1

(EC 1.1.4.2.) and the mitochondrial NADP(+)-specific IDH2. The mitochondrial NAD(+)-specific IDH3 is a heterotetramer composed of 2 α-subunits and single β- and γ-subunits (Nichols *et al.*, 1995). Four IDHs are present in yeast. The three NADP(+)-specific enzymes are localized to the cytosol, mitochondria, and peroxisomes. The fourth enzyme is the NAD(+)-specific mitochondrial enzyme.

Among the isoenzymes, the NAD(+)-specific IDH is presumed to play a major role in regulating the TCA cycle flux, since its activity is modulated by numerous allosteric effectors. The eukaryotic enzyme displays cooperative binding to iso-citrate and is activated allosterically by adenine dinucleotides. IDH3 is activated by ADP, whereas the yeast enzyme is activated by AMP. The activity is inhibited by ATP and NADH. The mammalian enzyme is also activated by Ca^{2+}. IDH activity is one of the rate-limiting steps of the TCA cycle and contributes to its regulation according to the cell's energy status. In yeast, IDH mutations were associated with the loss of mtDNA and decreased stability of mitochondrial translation products (de Jong *et al.*, 2000; McCammon and McAlister-Henn, 2003).

It is unclear whether the mitochondrial NADP(+)-specific enzyme IDH2 is a part of the TCA cycle. This enzyme produces NADPH, which is required for the regeneration of glutathione. Decreased expression of IDH2 results in increased the production of ROS, with concurrent mitochondrial damage. Conversely, over-production of IDH2 led to protection against oxidative damage. Thus, IDH2 plays a key role in controlling the redox balance and cellular defenses against ROS (Jo *et al.*, 2001).

NAD(+)-specific IDH activity is measured by following the formation of NADH, as described in Table IV. When the yeast enzyme is measured, AMP is substituted ADP.

VI. α-Ketoglutarate Dehydrogenase Complex

The α-ketoglutarate dehydrogenase complex (KGDHc), 2-oxoglutarate dehy-drogenase, catalyzes the irreversible decarboxylation of α-ketoglutarate, yielding succinyl CoA, NADH, and CO_2. This reaction is also a critical step in the removal of glutamate, a potentially toxic neurotransmitter. Similar to PDHc, KGDHc is a multienzyme complex consisting of multiple copies of three protein subunits: the TPP containing oxoglutarate decarboxylase (E1k; EC 1.2.4.2); the core enzyme dihydrolipoyl succinyltransferase (E2k; EC 2.3.1.61), and LAD (E3). E1k and E2k are specific for KGDHc, while LAD is also a component of PDHc, branched-chain keto acid dehydrogenase, and the glycine cleavage system (Sheu and Blass, 1999). All three components of the KGDHc have been identified in the mitochondrial nucleoid (Bogenhagen *et al.*, 2003; Chen *et al.*, 2005).

Severe deficiency of KGDHc is rare and presents with infantile lactic acidosis and severe psychomotor retardation (Bonnefont *et al.*, 1992). Mutated LAD is a major cause of KGDHc deficiency; only rarely does the mutation affects KGDHc activity without of PDHc dysfunction (Odièvre *et al.*, 2005). Since all KGDHc components

Table IV
A Selection of Spectrophotometric Assays for TCA Enzymes

	Reagents and assay conditions				Samples and ranges			
Assay	Reagent (stock solution) [storage]	Volume in assay	Amount in assay	Conditions $\varepsilon = \mu M^{-1} \mu^{-1}$ t = temperature	Sample (pretreatment)	Normal values (nmol/min/mg)	Deficient values (nmol/min/mg)	References
Isocitrate dehydrogenase	Tris–HCl 1 M, pH 7.2	70 μl	70 mM	A_{340} $\varepsilon = 6.22$ $t = 30°C$	Mitochondria from human placenta 25-mg protein/ml (0.2% Triton X-100, sonication)	98–102		Alp *et al.*, 1976; Scislowski *et al.*, 1983
	H$_2$O	Up to 1 ml						
	NAD (20 mM) [fresh]	100 μl	2 mM	Preincubate until stable baseline				
	ADP (20 mM)	100 μl	2 mM	Start assay with citrate and isocitrate				
	MgCl$_2$ (100 mM)	80 μl	8 mM					
	MnCl$_2$ (20 mM)	50 μl	1 mM					
	Antimycin A [5% in ethanol]	5 μl	5 μl/ml					
	Sample	20–25 μl						
	Citrate (100 mM)	100 μl	10 mM					
	Threo-D$_s$-isocitrate (20 mM)	75 μl	1.5 mM					
Succinyl CoA Synthase	Tris–HCl (1 M) pH 7.4	100 μl	100 mM	A_{235} (quartz) $\varepsilon = 13.6$ $t = 37°C$	Frozen mitochondria from fresh human skeletal muscle/1- to 2-mg protein/ml (1% sodium cholate)	0.21 0.18–0.25 U/mg SCS-A 0.39 0.31–0.46 U/mg SCS-G	0.038 U/mg SCS-A	Cha and Parks, 1964; Elpeleg *et al.*, 2005
	H$_2$O	Up to 1 ml						
	MgCl$_2$ (100 mM)	100 μl	10 mM	Preincubate until stable baseline				
	CoA [fresh] 10 mM	10 μl	0.1 mM	Start assay with sample				
	ATP (or GTP) (10 mM) [neutralized, frozen aliquots]	10 μl	0.1 mM					
	Succinate (0.5 M)	100 μl	50 mM					
	Sample	20–40 μl	20–80 μg					
Fumarase	HEPES KOH, pH 7.6 (250 mM)	100 μl	25 mM	A_{340} $\varepsilon = 6.22$ $t = 37°C$	Mitochondrial extracts from human muscle (1 mg/ml in 1% sodium cholate)	1997 ± 24	23	Elpeleg *et al.*, 1992; Hatch, 1978; Wei *et al.*, 1996; Zinn *et al.* 1986
	H$_2$O	Up to 1 ml			Skin fibroblasts	103 ± 19.5	11	
	KHPO$_4$ (100 mM)	50 μl	5 mM	Preincubate until stable baseline				
	MgCl$_2$ (100 mM)	40 μl	4 mM	Start assay with fumarate	Lymphoblastoid cells 1 × 10^6 cells/ml in (sonication 5″)	513	118,182	
	NADP (10 mM)	400 μl	0.4 mM					
	Malic enzyme-NADP		0.2 U					
	Sample	10–50 μl	10–50 μg					
	Fumarate (200 mM)	50 μl	10 mM					

are present in the nucleoid, KGDHc deficiency may also cause mtDNA depletion. This is not the case for E3 deficiency, whose defects were previously shown not to be associated with quantitative defects of the mtDNA (Shaag *et al.,* 1999). Other neurogenerative diseases, such as Alzheimer disease, have also been associated with decreased KGDHc activity. Human brain contains low, presumably rate-limiting, amounts of KGDHc activity relative to other enzymes involved in energy metabolism. Therefore, this tissue is especially vulnerable to KGDHc inactivation.

KGDHc is sensitive to inactivation by ROS and neurotoxins. However, studies in knockout mice indicate that KGDHc is not only a target of oxidative stress but is also a primary site for the generation of ROS. Thus, KGDHc functions as a mediator between mitochondria and oxidative stress in neurodegeneration (McCammon *et al.,* 2003; Starkov *et al.,* 2004).

KGDH activity can be measured spectrophotometrically by monitoring the production of NADH (Gibson *et al.,* 2005; Robinson *et al.,* 1987). When activity and sample sizes are small, radiometric methods are preferred, measuring the formation of CO_2 from radioactive α-ketoglutarate (Robinson *et al.,* 1977, 1980). The assay is detailed in Table II.

VII. Succinyl CoA Synthetase

SCS (succinyl CoA ligase; succinic thiokinase) is a heterodimer composed of α- and β-subunits. SCS catalyzes the cleavage of the thioester bond of succinyl CoA forming succinate and CoASH. The reaction is coupled to the formation of a ribonucleoside triphosphate. Two distinct SCSs are present in mitochondria, one ATP-forming (SCS-A; EC 6.2.1.5) and one GTP-forming (SCS-G; EC 6.2.1.4). The GTP produced by substrate-level phosphorylation by SCS-G may be converted to ATP by the nucleoside diphosphate kinase (NDPK). The β-subunit determines the nucleotide specificity of the enzyme; the α-subunit is identical in the two isoenzymes. It is not clear which of the two is the main TCA cycle enzyme. It has been proposed that SCS-A participates in the TCA cycle and that SCS-G catalyzes the reverse reaction and contributes to the synthesis of succinyl CoA for ketone utilization and heme synthesis (Furuyama and Sassa, 2000). Nevertheless, their tissue expression pattern suggests that these two enzymes have overlapping but nonredundant activities: SCS-G is predominantly expressed in anabolic tissues, such as liver and kidney, while SCS-A predominates in testis and brain (Lambeth *et al.,* 2004). In support of this, we detected severe SCS-A deficiency in a patient with isolated encephalopathy. The case also establishes the importance of SCS for mtDNA maintenance. The deficiency of SCS-A, due to a mutated β-subunit, was associated with mtDNA depletion, most likely because of the tight association between the mitochondrial NDPK and SCS-A (Elpeleg *et al.,* 2005). NDPK is a key component in the mitochondrial deoxyribonu-cleoside salvage pathway, which provides precursors for mtDNA synthesis (Kowluru *et al.,* 2002; Saada, 2004).

SCS activity is reversible and assays have been developed to measure activity in both directions. The assay of the reaction in the forward direction measures the release of CoASH from succinyl CoA by DTNB. In the reverse direction, the formation of GDP may be monitored by coupling with pyruvate kinase and LDH. Alternatively, the formation of succinyl CoA may be followed directly by measuring the formation of the thioester bond at 235 nm (Cha and Parks, 1964; Lambeth et al., 2004). The latter reaction is described in Table IV. Measurements are performed in a quartz cuvette. Solubilization in Triton X-100 should be avoided, as this detergent interferes with the spectrophotometric measurements. One unit is defined as the amount of enzyme that gives an absorbance of 1.000 per minute.

VIII. Succinate Dehydrogenase

SDH (EC 1.3.5.1) catalyzes the conversion of succinate to fumarate. SDH is a component of the TCA cycle as well as of the MRC, where it functions as complex II. SDH is composed of four subunits. The largest subunit A (SDHA) is a flavoprotein containing the active site. Subunit B (SDHB) is an iron sulfur protein. Together with subunits C (SDHC) and D (SDHD), the four subunits comprise complex II, the succinate-ubiquinone oxidoreductase activity of the MRC. Complex II is embedded/located in the inner mitochondrial membrane and carries electrons from $FADH_2$ to the ubiquinone pool. Subunits A and B are hydrophilic and face the mitochondrial matrix, while subunits C and D anchor the complex to the mitochondrial inner membrane. Contrary to other MRC complexes, all complex II subunits are nuclear encoded (Brièrea et al., 2005; Rustin and Rotig, 2002). Hence, SDH activity can be used to normalize other MRC activities when a defect in mtDNA is suspected (Saada et al., 2001).

Mutations in all SDH subunits have been associated with a wide spectrum of clinical presentations. Mutations in SDHA are a rare cause of mitochondrial encephalomyopathy (Bourgeron et al., 1995). Heterozygosity for mutations in the other subunits (SDHB, SDHC, and SDHD) is one of the causes for familial paraganglioma (PLG) and pheochromocytoma. Loss of heterozygosity with preservation of the mutant allele has been observed in the tumors. The accumulated data suggest that SDHB, SDHC, and SDHD genes act as tumor suppressors (Astuti et al., 2001). SDH deficiency has also been reported in combination with deficiencies of other iron–sulfur proteins such as aconitase in Friedrich ataxia (Hall et al., 1993; Rotig et al., 1997).

Dysfunction of the MRC is commonly known to cause ATP depletion followed by tissue necrosis. This pathogenic mechanism negates the proliferative nature of the SDH-associated tumors. Although increased oxidative stress and the resulting ROS production may promote tumorigenesis, a specific link between the type of metabolic blockade and tumor formation was suggested by the observation that mutations in the FH gene are also associated with tumor formation (mainly uterine leiomyomas). Recent data support the hypothesis that increased succinate and/or fumarate causes stabilization of hypoxia-inducible factor-1α (HIF-1α), which initiates the transcription of genes coding for proangiogenic proteins, such as VEGF (Pollard *et al.*, 2005).

SDH is measured spectrophotometrically, based on the CoQ- or phenazine methosulfate (PMS)-mediated reduction of DCIP, as described in Table V (Fischer *et al.*, 1986; Rustin *et al.*, 1994). SDH activity is activated by preincubation of the sample with the succinate. For research purposes, an assay of SDH based on the reduction of methyl-thiazolyl-tetrazolium (MTT) is widely used as a cell viability proliferative marker (Loveland *et al.*, 1992).

IX. Fumarase

FH (fumarate hydratase; l-malate hydrolyase), EC 4.2.1.2, is a homotetramer that catalyzes the conversion of fumarate to malate. FH exists in both cytosolic and mitochondrial forms; both isoforms are encoded by a single gene. In rat, the two isoenzymes arise from a single mRNA transcript, whereas in humans, two different transcripts are formed (Doonan *et al.*, 1984). On the basis of the study in yeast, it seems that the targeting of the cytosolic and mitochondrial FH follow a unique mechanism whereby the mitochondrial FH proteins are fully imported into the matrix, but the cytosolic form is translocated only partially. The amino terminus is then processed by cleavage of the presequence by MPP and the rapidly folded cytosolic isoform is released back into the cytosol (Sass *et al.*, 2003; Stein *et al.*, 1994).

FH deficiency was the first TCA disorder to be identified, about 20 years ago, perhaps because this is the only TCA deficiency associated with a diagnostic organic aciduria (Whelan *et al.*, 1983; Zinn *et al.*, 1986). Clinically, the defect is typically characterized by static encephalopathy with psychomotor retardation and hypotonia (Elpeleg *et al.*, 1992). FH deficiency was also the first TCA disorder to be characterized at the molecular level (Bourgeron *et al.*, 1994).

FH also acts as a tumor suppressor, and mutations are associated with leiomyomata and renal cell cancer; its measured activity is very low or absent in tumors. Possibly due to loss of heterozygosity, leiomyomata were found in some of the parents of patients with congenital FH deficiency (Pithukpakorn *et al.*, 2006; Tomlinson *et al.*, 2002). The mechanism for carcinogenesis is similar to that proposed for SDH mutations, that is, the accumulation of Krebs cycle intermediates. In this case, loss of FH leads to overexpression of HIF-1α (Isaacs *et al.*, 2005; Pollard *et al.*, 2005).

FH activity may be assayed by the direct measurement of the production or disappearance of fumarate, following the absorption at 240 nm or 290 mm. These assays are limited by the fumarate's relatively low absorption index and by high background values in tissue extracts. A more convenient method couples the reaction to NADP malic enzyme, as described in Table IV (Hatch, 1978).

Table V
A Selection of Spectrophotometric Assays for TCA Enzymes

	Reagents and assay conditions				Samples and ranges		
Assay	Reagent (stock solution) [storage]	Volume in assay	Amount in assay	Conditions $\varepsilon = \mu M^{-1} \mu^{-1}$ t = temperature	Sample (pretreatment)	Normal values (nmol/min/mg)	Deficient values (nmol/ min/mg)	References
Succinate dehydrogenase assay I	Tris–HCl (1 M), pH 8.0	50 μl	50 mM	A_{600} $\varepsilon = 21$ $t = 37°C$	Mitochondria from human skeletal muscle	50–140	5–12	Rustin *et al.*, 1994; Bourgeron *et al.*, 1995
	H$_2$O	Up to 1 ml		Preincubate with succinate ATP, KCN, and rotenone. Add DCIP and antimycin.				
	ATP 10 mM [neutralized] [aliquots]	20 μl	0.2 mM	Start assay with DUQ				
	Succinate 0.2 M	50 μl	10 mM	Read background after the	Human lymphocytes	12–33	5	
	KCN 0.2 M [neutralized]	1.5 μl	0.3 mM	addition of malonate				
	Rotenone 0.6 mM [fresh in ethanol]	5 μl	3 μM					
	Sample	10 μl	0.5 μg		Human fibroblasts	13–31	4.5–5	
	Antimycin A [5% in ethanol]	10 μl	1 μM					
	Dichloroindophenol, DCIP (5 mM) [fresh]	16 μl	80 μM					
	Decyl ubiquinone, DUQ 2.5 mM [ethanol]	20 μl	50 μM					
	Malonate 0.5 M		10 mM					

(continues)

Table V (*continued*)

Assay	Reagent (stock solution) [storage]	Volume in assay	Amount in assay	Conditions $\varepsilon = \mu M^{-1}\,\mu l^{-1}$ t = temperature	Sample (pretreatment)	Normal values (nmol/min/mg)	Deficient values (nmol/ min/mg)	References
Succinate Dehydrogenase assay II	Potassium phosphate buffer KPB, pH 7.4 0.1 M	500 μl	100 mM	A_{600} $\varepsilon = 21$ $t = 25\,^\circ C$ Preincubate in buffer with succinate, 20 min, 25 °C, add KCN, DCIP, and PMS	Mitochondrial extracts from human muscle (0.2–0.5 mg/ml in 1% sodium cholate)	258 ± 52 165–442		Fischer et al., 1986; Saada et al., 2001
	H₂O	Up to 1 ml						
	Succinate 0.2 M	100 μl	20 mM					
	Sample	10 μl	5–10 μg					
	KCN 0.2 M	10 μl	2 mM					
	DCIP 5 mM [fresh]	10 μl	50 μM					
	Phenzine methosulfate PMS (65 mM) [fresh]	25 μl	1.6 mM					
Malate dehydrogenase m	Potassium phosphate buffer pH 7.4 100 mM	875 μl	100 mM	A_{340} $\varepsilon = 6.22$ $t = 25\,^\circ C$ Preincubate until stable baseline Start assay with oxaloacetate	Human fibroblast mitochondria	816 ± 628		Kitto and Lewis, 1969; Robinson et al., 1987; Stumpf et al., 1982
	NADH (2 mM) [fresh in buffer]	75 μl	0.15 mM					
	Triton X-100 10% [fresh in H₂O]-optional	50 μl	0.5%	Human leukocyte mitochondria		393 ± 146		
	Sample	10 μl	10–50 μg					
	Oxaloacetate 10 mM [in KPB]	50 μl	0.50 mM					

X. Malate Dehydrogenase

Mitochondrial malate dehydrogenase (MDH2; M-MDH), EC 1.1.1.37, a homodimer, catalyzes the NAD^+/NADH-dependent interconversion of the substrates malate and oxaloacetate. The mitochondrial enzyme participates in the TCA cycle and, together with its cytosolic counterpart, contributes to the malate/aspartate shuttle. This shunt transfers NADH into the mitochondria, allowing cytosolic reducing equivalents to access the MRC. The mitochondrial and cytosolic forms are encoded by two different genes, which likely originated from the same ancestor (Birktoft *et al.*, 1982; Minarik *et al.*, 2002; Schantz and Henriksson, 1987). To date, no MDH deficiency has been reported in humans, suggesting that these enzymes are either indispensable for human life or are function in an unexpectedly critical way.

MDH activity is measured by the disappearance of NADH following the conversion of oxaloacetate to malate (Kitto and Lewis, 1969), as described in Table V.

XI. Conclusion

The TCA cycle was discovered more than half a century ago. The study of its various components has been extensive, and a wide range of methods for the assay of TCA enzymes have been described over the years. It is therefore surprising to realize that relatively few human disorders are associated with a deficiency of TCA enzymes. This should not be confused with the scarce reports of increased excretion of TCA metabolites, which, with the exception of fumaric aciduria in FH deficiency, is a nonspecific phenomenon that is usually secondary to a number of other metabolic conditions. We speculate that because of their indispensability, most TCA components have counterparts in the form of isoenzymes or other yet-to-be identified factors. Defects in one of counterpart is likely compatible with life, but present clinically in a manner which seems far from a TCA cycle defect, as in mtDNA replication defects or tumor formation. With the advance of linkage analysis techniques, it is expected that molecular defects of most TCA components will be identified. Enzymatic methodologies will therefore be indispensable for confirming pathogenicity and for the study of disease mechanisms.

Acknowledgments

The authors are grateful to Corinne Belaiche for expert technical assistance, and to Drs. Elisabeth Holme, Avraham Shaag, Ophry Pines, and Wim Ruitenbeek for helpful discussions over the years.

References

Alp, P. R., Newsholme, E. R., and Zammit, V. A. (1976). Activities of citrate synthase and NAD^+-linked and $NADP^+$-linked isocitrate dehydrogenase in muscle from vertebrates and invertebrates. *Biochem. J.* **154,** 689–700.

Aptowitzer, I., Saada, A., Faber, J., Kleid, D., and Elpeleg, O. N. (1997). Liver disease in the Ashkenazi-Jewish lipoamide dehydrogenase deficiency. *J. Pediatr. Gastroenterol. Nutr.* **24,** 599–601.

Aral, B., Benelli, C., Ait-Ghezala, G., Amessou, M., Fouque, F., Maunoury, C., Creau, N., Kamoun, P., and Marsac, C. (1997). Mutations in PDX1, the human lipoyl-containing component X of the pyruvate dehydrogenase-complex gene on chromosome 11p1, in congenital lactic acidosis. *Am. J. Hum. Genet.* **61,** 1318–1326.

Astuti, D., Latif, F., Dallol, A., Dahia, P. L. M., Douglas, F., George, E., Idberg, F. S., Husebye, E. S., Eng, C., and Maher, E. R. (2001). Gene mutations in the succinate dehydrogenase subunit SDHB cause susceptibility to familial pheochromocytoma. *Am. J. Hum. Genet.* **69,** 49–54.

Berger, I., Elpeleg, O. N., and Saada, A. (1996). Lipoamide dehydrogenase activity in lymphocytes. *Clin. Chim. Acta* **256,** 197–201.

Birktoft, J. J., Fernley, R. T., Bradshaw, R. A., and Banaszak, L. J. (1982). Amino acid sequence homology among the 2-hydroxy acid dehydrogenases: Mitochondrial and cytoplasmic malate dehydrogenases form a homologous system with lactate dehydrogenase. *Proc. Natl. Acad. Sci. USA* **79,** 6166–6170.

Bogenhagen, D. F., Wang, Y., Shen, E.L, and Kobayashi, R. (2003). Protein components of mitochondrial DNA nucleoids in higher eukaryotes. *Mol. Cell Proteomics* **2,** 1205–1216.

Bonnefont, J.-P., Chretien, D., Rustin, P., Robinson, B., Vassault, A., Aupetit, J., Charpentier, C., Rabier, D., Saudubray, J.-M., and Munnich, A. (1992). Alpha-ketoglutarate dehydrogenase deficiency presenting as congenital lactic acidosis. *J. Pediatr.* **121,** 255–258.

Bourgeron, T., Chretien, D., Poggi-Bach, J., Doonan, S., Rabier, D., Letouze, P., Munnich, A., Rotig, A., Landrieu, P., and Rustin, P. (1994). Mutation of the fumarase gene in two siblings with progressive encephalopathy and fumarase deficiency. *J. Clin. Invest.* **93,** 2514–2518.

Bourgeron, T., Rustin, P., Chretien, D., Birch-Machin, M., Bourgeois, M., Viegas-Pequignot, E., Munnich, A., and Rotig, A. (1995). Mutation of a nuclear succinate dehydrogenase gene results in mitochondrial respiratory chain deficiency. *Nat. Genet.* **11,** 144–149.

Brièrea, J. J., Favierb, J., El Ghouzzia, V., Djouadic, F., Bénita, P., Gimenezc, A. P., and Rustin, P. (2005). Succinate dehydrogenase deficiency in human. *Cell. Mol. Life Sci.* **62,** 2317–2324.

Bulteau, A. L., O'Neill, H. A., Kennedy, M. C., Ikeda-Saito, M., Isaya, G., and Szweda, L. I. (2004). Frataxin acts as an iron chaperone protein to modulate mitochondrial aconitase activity. *Science* **305,** 242–245.

Butterworth, P. J., Tsai, S. T., Eley, M. H., Roche, T. E., and Reed, L. (1975). A kinetic study of dihydrolipoyl transacetylase from bovine kidney. *J. Biol. Chem.* **250,** 1921–1925.

Cha, S., and Parks, R. E. (1964). Succinic thiokinase. I. Purification of the enzyme from pig heart. *J. Biol. Chem.* **239,** 1961–1967.

Chen, X. J., Wang, X., Kaufman, B. A., and Butow, R. A. (2005). Aconitase couples metabolic regulation to mitochondrial DNA maintenance. *Science* **307,** 714–717.

Chretien, D., Pourrier, M., Bourgeron, T., Sene, M., Rotig, A., Munnich, A., and Rustin, P. (1995). An improved spectrophotometric assay of pyruvate dehydrogenase in lactate dehydrogenase contaminated mitochondrial preparations from human skeletal muscle. *Clin. Chim. Acta* **240,** 129–136.

de Jong, L., Elzinga, S. D., McCammon, M.T, Grivell, L. A., and van der Spek, H. (2000). Increased synthesis and decreased stability of mitochondrial translation products in yeast as a result of loss of mitochondrial (NAD(+))-dependent isocitrate dehydrogenase. *FEBS Lett.* **483,** 62–66.

De Meirleir, L. (2002). Defects of pyruvate metabolism and the Krebs cycle. *J. Child Neurol.* **17**(Suppl. 33), S26–S33.

Doonan, S., Barra, D., and Bossa, F. (1984). Structural and genetic relationships between cytosolic and mitochondrial isoenzymes. *Int. J. Biochem.* **16,** 1193–1199.

Elpeleg, O., Miller, C., Hershkovitz, E., Bitner-Glindzicz, M., Bondi-Rubinstein, G., Rahman, S., Pagnamenta, A., Eshhar, S., and Saada, A. (2005). Deficiency of the ADP-forming succinyl-CoA synthase activity is associated with encephalomyopathy and mitochondrial DNA depletion. *Am. J. Hum. Genet.* **76,** 1081–1086.

Elpeleg, O. N., Amir, N., and Christensen, E. (1992). Variability of clinical presentation in fumarate hydratase deficiency. *J. Pediatr.* **121**, 752–754.

Elpeleg, O. N., Saada, A. B., Shaag, A., Glustein, J. Z., Ruitenbeek, W., Tein, I., and Halevy, J. (1997). Lipoamide dehydrogenase deficiency: A new cause for recurrent myoglobinuria. *Muscle Nerve* **20**, 238–240.

Fansler, B., and Lowenstein, J. M. (1969). Aconitase from pig heart. *Methods Enzymol.* **13**, 26–30.

Fergestad, T., Bostwick, B., and Ganetzky, B. (2006). Metabolic disruption in *Drosophila* bang-sensitive seizure mutants. *Genetics* **173**, 1357–1364.

Fischer, J. C., Ruitenbeek, W., Gabreels, F. J., Janssen, A. J., Renier, W. O., Sengers, R. C., Stadhouders, A. M., ter Laak, H. J., Trijbels, J. M., and Veerkamp, J. H. (1986). A mitochondrial encephalomyopathy: The first case with an established defect at the level of coenzyme Q. *Eur. J. Pediatr.* **144**, 441–444.

Furuyama, K., and Sassa, S. (2000). Interaction between succinyl CoA synthetase and the heme-biosynthetic enzyme ALAS-E is disrupted in sideroblastic anemia. *J. Clin. Invest.* **105**, 757–764.

Gibson, G. E., Blass, J. P., Beal, M. F., and Bunik, V. (2005). The alpha-ketoglutarate-dehydrogenase complex: A mediator between mitochondria and oxidative stress in neurodegeneration. *Mol. Neurobiol.* **31**, 43–63.

Goodwin, G. W., Paxton, R., Gillim, S. E., and Harris, R. A. (1986). Branched-chain 2-oxo acid dehydrogenase interferes with the measurement of the activity and activity state of hepatic pyruvate dehydrogenase. *Biochem. J.* **236**, 111–114.

Haile, D. J., Rouault, T. A., Tang, C. K., Chin, J., Harford, J. B., and Klausner, R. D. (1992). Reciprocal control of RNA-binding and aconitase activity in the regulation of the iron-responsive element binding protein: Role of the iron-sulfur cluster. *Proc. Natl. Acad. Sci. USA* **89**, 7536–7540.

Hall, R. E., Henriksson, K. G., Lewis, S. F., Haller, R. G., and Kennaway, N. G. (1993). Mitochondrial myopathy with succinate dehydrogenase and aconitase deficiency. Abnormalities of several iron-sulfur proteins. *J. Clin. Invest.* **92**, 2660–2666.

Hatch, M. D. (1978). A simple spectrophotometric assay for fumarate hydratase in crude tissue extracts. *Anal. Biochem.* **85**, 271–275.

Head, R. A., Brown, R. M., Zolkipli, Z., Shahdadpuri, R., King, M. D., Clayton, P. T., and Brown, G. K. (2005). Clinical and genetic spectrum of pyruvate dehydrogenase deficiency: Dihydrolipoamide acetyltransferase (E2) deficiency. *Ann. Neurol.* **58**, 234–241.

Holness, M. J., and Sugden, M. C. (2003). Regulation of pyruvate dehydrogenase complex activity by reversible phosphorylation. *Biochem. Soc. Trans.* **31**, 1143–1151.

Isaacs, J. S., Jung, Y. J., Mole, D. R., Lee, S., Torres-Cabala, C., Chung, Y. L., Merino, M., Trepel, J., Zbar, B., Toro, J., Ratcliffe, P. J., Linehan, W. M., *et al.* (2005). HIF overexpression correlates with biallelic loss of fumarate hydratase in renal cancer: Novel role of fumarate in regulation of HIF stability. *Cancer Cells* **8**, 143–153.

Jo, S. H., Son, M. K., Koh, H. J., Lee, S. M., Song, I. H., Kim, Y. O., Lee, Y. S., Jeong, K. S., Kim, W. B., Park, J. W., Song, B. J., and Huh, T. L. (2001). Control of mitochondrial redox balance and cellular defense against oxidative damage by mitochondrial NADP+-dependent isocitrate dehydrogenase. *J. Biol. Chem.* **276**, 16168–16176.

Kispal, G., Evans, C. T., Malloy, C., and Srere, P. A. (1989). Metabolic studies on citrate synthase mutants of yeast. A change in phenotype following transformation with an inactive enzyme. *J. Biol. Chem.* **264**, 11204–11210.

Kitto, G. B., and Lewis, R. G. (1969). Purification and properties of tuna supernatant and mitochondrial malate dehydrogenases. *Biochim. Biophys. Acta* **139**, 1–15.

Kowluru, A., Tannous, M., and Chen, H. Q. (2002). Localization and characterization of the mitochondrial isoform of the nucleoside diphosphate kinase in the pancreatic beta cell: Evidence for its complexation with mitochondrial succinyl-CoA synthetase. *Arch. Biochem. Biophys.* **398**, 160–169.

Lambeth, D. O., Tews, K. N., Adkins, S., Frohlich, D., and Milavetz, B. I. (2004). Expression of two succinyl-CoA synthetases with different nucleotide specificities in mammalian tissues. *J. Biol. Chem.* **279**, 36621–36624.

Loveland, B. E., Johns, T. G., Mackay, I. R., Vaillant, F., Wang, Z. X., and Hertzog, P. J. (1992). Validation of the MTT dye assay for enumeration of cells in proliferative and antiproliferative, assays. *Biochem. Int.* **27,** 501–510.

Maj, M. C., MacKay, N., Levandovskiy, V., Addis, J., Baumgartner, E. R., Baumgartner, M. R., Robinson, B. H., and Cameron, J. M. (2005). Pyruvate dehydrogenase phosphatase deficiency: Identification of the first mutation in two brothers and restoration of activity by protein complementation. *J. Clin. Endocrinol. Metab.* **90,** 4101–4107.

McCammon, M. T., and McAlister-Henn, L. (2003). Multiple cellular consequences of isocitrate dehydrogenase isozyme dysfunction. *Arch. Biochem. Biophys.* **419,** 222–233.

McCammon, M. T., Epstein, C. B., Przybyla-Zawislak, B., McAlister-Henn, L., and Butow, R. A. (2003). Global transcription analysis of Krebs tricarboxylic acid cycle mutants reveals an alternating pattern of gene expression and effects on hypoxic and oxidative genes. *Mol. Biol. Cell* **14,** 958–972.

Minarik, P., Tomaskova, N., Kollarova, M., and Antalik, M. (2002). Malate dehydrogenases—structure and function. *Gen. Physiol. Biophys.* **21,** 257–265.

Nichols, B. J., Perry, A. C., Hall, L., and Denton, R. M. (1995). Molecular cloning and deduced amino acid sequences of the alpha- and beta-subunits of mammalian NAD(+)-isocitrate dehydrogenase. *Biochem. J.* **310,** 917–922.

Odièvre, M. H., Chretien, D., Munnich, A., Robinson, B. H., Dumoulin, R., Masmoudi, S., Kadhom, N., Rotig, A., Rustin, P., and Bonnefont, J. P. (2005). A novel mutation in the dihydrolipoamide dehydrogenase E3 subunit gene (DLD) resulting in an atypical form of alpha-ketoglutarate dehydrogenase deficiency. *Hum. Mutat.* **25,** 323–324.

Patel, P. I., and Isaya, G. (2001). Friedreich ataxia: From GAA triplet-repeat expansion to frataxin deficiency. *Am. J. Hum. Genet.* **69,** 15–24.

Pithukpakorn, M. (2005). Disorders of pyruvate metabolism and the tricarboxylic acid cycle. *Mol. Genet. Metab.* **85,** 243–246.

Pithukpakorn, M., Wei, M. H., Toure, O., Steinbach, P. J., Glenn, G. M., Zbar, B., Linehan, W. M., and Toro, J. R. (2006). Fumarate Hydratase enzyme activity in lymphoblastoid cells and fibroblasts of individuals in families with hereditary leiomyomatosis and renal cell cancer. *J. Med. Genet.* **43,** 755–762.

Pollard, P. J., Brière, J. J., Alam, N. A., Barwell, J., Barclay, E., Wortham, N. J., Hunt, T., Mitchell, M., Olpin, S., Moat, S. J., Hargreaves, I. P., Heales, S. J., *et al.* (2005). Accumulation of Krebs cycle intermediates and over-expression of HIF1a in tumours which result from germline FH and SDH mutations. *Hum. Mol. Genet.* **14,** 2231–2239.

Reed, L. J., and Willms, C. R. (1966). Purification and resolution of the pyruvate dehydrogenase complex. *Methods Enzymol.* **9,** 247–265.

Regev-Rudzki, N., Karniely, S., Ben-Haim, N. N., and Pines, O. (2005). Yeast aconitase in two locations and two metabolic pathways: Seeing small amounts is believing. *Mol. Biol. Cell* **16,** 4163–4171.

Robinson, B. H., Taylor, J., and Sheerwood, W. G. (1977). Deficiency of dihydrolipoyl dehydrogenase (a component of the pyruvate and α-ketoglutarate dehydrogenase complexes): A cause of congenital chronic lactic acidosis in infancy. *Pediatr. Res.* **11,** 1198–1202.

Robinson, B. H., Taylor, J., and Sherwood, W. G. (1980). The genetic heterogeneity of lactic acidosis: Occurrence of recognizable inborn errors of metabolism in a pediatric population with lactic acidosis. *Pediatr. Res.* **14,** 956–962.

Robinson, J. B., Brent, L. G., Sumegi, B., and Srere, P. A. (1987). An enzymatic approach to the study of the Krebs tricarboxylic acid cycle. *In* "Mitochondrial a Practical Approach" (V. M. Darley-Usmar, and M. T. Wilson, eds.). IRL Press Oxford, Washington, DC.

Robinson, B. H., MacKay, N., Petrova-Benedict, R., Ozalp, I., Coskun, T., and Stacpoole, P. W. (1990). Defects in the E(2) lipoyl transacetylase and the X-lipoyl containing component of the pyruvate dehydrogenase complex in patients with lactic acidemia. *J. Clin. Invest.* **85,** 1821–1824.

Rose, I. A., and O'Connell, E. L. (1967). Mechanism of aconitase action. I. The hydrogen transfer reaction. *J. Biol. Chem.* **242,** 1870–1879.

Rotig, A., de Lonlay, P., Chretien, D., Foury, F., Koenig, M., Sidi, D., Munnich, A., and Rustin, P. (1997). Aconitase and mitochondrial iron-sulphur protein deficiency in Friedreich ataxia. *Nat. Genet.* **17,** 215–217.

Rustin, P., and Rotig, A. (2002). Inborn errors of complex II Unusual human mitochondrial diseases. *Biochim. Biophys. Acta* **1553,** 117–122.

Rustin, P., Chretien, D., Burgeron, T., Rotig, A., Saudubray, J. M., and Munnich, A. (1994). Biochemical and molecular investigations in respiratory chain deficiencies. *Clin. Chim. Acta.* **1,** 35–51.

Rustin, P., Bourgeron, T., Parfait, B., Chretien, D., Munnich, A., and Rotig, A. (1997). Inborn errors of the Krebs cycle: A group of unusual mitochondrial diseases in human. *Biochim. Biophys. Acta* **1361,** 185–197.

Saada, A. (2004). Deoxyribonucleotides and disorders of mitochondrial DNA integrity. *DNA Cell Biol.* **23,** 797–806.

Saada, A., Aptowitzer, I., Link, G., and Elpeleg, O. N. (2000). ATP synthesis in lipoamide dehydrogenase deficiency. *Biochem. Biophys. Res. Commun.* **269,** 382–386.

Saada, A., Shaag, A., Mandel, H., Nevo, Y., Eriksson, S., and Elpeleg, O. (2001). Mutant mitochondrial thymidine kinase in mitochondrial DNA depletion myopathy. *Nat. Genet.* **29,** 342–344.

Sass, E., Karniely, S., and Pines, O. (2003). Folding of fumarase during mitochondrial import determines its dual targeting in yeast. *J. Biol. Chem.* **278,** 45109–45116.

Schantz, P. G., and Henriksson, J. (1987). Enzyme levels of the NADH shuttle systems: Measurements in isolated muscle fibres from humans of differing physical activity. *Acta Physiol. Scand.* **129,** 505–515.

Scislowski, P. W., Zolnierowicz, S., and Zelewski, L. (1983). Subcellular distribution of isocitrate dehydrogenase in early and term human placenta. *Biochem. J.* **214,** 339–343.

Schwab, M. A., Kolker, S., van den Heuvel, L. P., Sauer, S., Wolf, N. I., Rating, D., Hoffmann, G. F., Smeitink, J. A., and Okun, J. G. (2005). Optimized spectrophotometric assay for the completely activated pyruvate dehydrogenase complex in fibroblasts. *Clin. Chem.* **51,** 151–160.

Seals, J. R., and Jarett, L. (1980). Activation of pyruvate dehydrogenase by direct addition of insulin to an isolated plasma membrane/mitochondria mixture: Evidence for generation of insulin's second messenger in a subcellular system. *Proc. Natl. Acad. Sci. USA* **77,** 77–81.

Shaag, A., Saada, A., Steinberg, A., Navon, P., and Elpeleg, O. N. (1997). Mitochondrial encephalomyopathy associated with a novel mutation in the mitochondrial tRNA(leu)(UUR) gene (A3243T). *Biochem. Biophys. Res. Commun.* **233,** 637–639.

Shaag, A., Saada, A., Berger, I., Mandel, H., Joseph, A., Feigenbaum, A., and Elpeleg, O. N. (1999). Molecular basis of lipoamide dehydrogenase deficiency in Ashkenazi Jews. *Am. J. Med. Genet.* **82,** 177–182.

Shadel, G. S. (2005). Mitochondrial DNA, aconitase 'wraps' it up. *Trends Biochem. Sci.* **30,** 294–296.

Sheu, K. F., and Blass, J. P. (1999). The alpha-ketoglutarate dehydrogenase complex. *Ann. NY Acad. Sci.* **893,** 61–78.

Sperl, W., Trijbels, J. M., Ruitenbeek, W., van Laack, H. L., Janssen, A. J., Kerkhof, C. M., and Sengers, R. C. (1993). Measurement of totally activated pyruvate dehydrogenase complex activity in human muscle: Evaluation of a useful assay. *Enzyme Protein* **47,** 37–46.

Srere, P. A. (1974). Controls of citrate synthase. *Life Sci.* **15,** 1695–1710.

Srere, P. A., Brazil, H., and Gonen, L. (1963). Citrate condensing enzyme of pigeon breast muscle and moth flight muscle. *Acta Chem. Scand.* **17,** S129–S134.

Sriram, G., Martinez, J. A., McCabe, E. R., Liao, J. C., and Dipple, K. M. (2005). Single-gene disorders: What role could moonlighting enzymes play? *Am. J. Hum. Genet.* **76,** 911–924.

Starkov, A. A., Fiskum, G., Chinopoulos, C., Lorenzo, B. J., Browne, S. E., Patel, M. S., and Beal, M. F. (2004). Mitochondrial alpha-ketoglutarate dehydrogenase complex generates reactive oxygen species. *J. Neurosci.* **24,** 7779–7788.

Stein, I., Peleg, Y., Even-Ram, S., and Pines, O. (1994). The single translation product of the FUM1 gene (fumarase) is processed in mitochondria before being distributed between the cytosol and mitochondria in Saccharomyces cerevisiae. *Mol. Cell. Biol.* **14,** 4770–4778.

Stromme, J. H., Borud, O., and Moe, P. J. (1976). Fatal lactic acidosis in a newborn attributable to a congenital defect of pyruvate dehydrogenase. *Pediatr. Res.* **10,** 62–66.

Stumpf, D. A., Parks, J. K., Eguren, L. A., and Haas, R. (1982). Friedreich ataxia: III. Mitochondrial malic enzyme deficiency. *Neurology* **32,** 221–227.

Tomlinson, I. P., Alam, N. A., Rowan, A. J., Barclay, E., Jaeger, E. E., Kelsell, D., Leigh, I., Gorman, P., Lamlum, H., Rahman, S., Roylance, R. R., Olpin, S., *et al.* (2002). Germline mutations in FH predispose to dominantly inherited uterine fibroids, skin leiomyomata and papillary renal cell cancer. *Nat. Genet.* **30,** 406–410.

van Laack, H. L., Ruitenbeek, W., Trijbels, J. M., Sengers, R. C., Gabreels, F. J., Janssen, A. J., and Kerkhof, C. M. (1988). Estimation of pyruvate dehydrogenase (E1) activity in human skeletal muscle; three cases with E1 deficiency. *Clin. Chim. Acta* **171,** 109–118.

Velot, C., and Srere, P. A. (2000). Reversible transdominant inhibition of a metabolic pathway. *In vivo* evidence of interaction between two sequential tricarboxylic acid cycle enzymes in yeast. *J. Biol. Chem.* **275,** 12926–12933.

Wei, W. H., Toure, O., Glenn, G. M., Pithukpakorn, M., Neckers, L., Stolle, C., Choyke, P., Grubb, R., Middelton, L., Turner, M. L., Walther, M. M., *et al.* (1996). Novel mutations in FH and expansion of the spectrum of phenotypes expressed in families with hereditary leiomyomatosis and renal cell cancer. *J. Med. Genet.* **43,** 18–27.

Whelan, D. T., Hill, R. E., and McClorry, S. (1983). Fumaric aciduria: A new organic aciduria, associated with mental retardation and speech impairment. *Clin. Chim. Acta* **132,** 301–308.

Xia, L., Bjornstedt, M., Nordman, T., Eriksson, L. C., and Olsson, J. M. (2001). Reduction of ubiquinone by lipoamide dehydrogenase. An antioxidant regenerating pathway. *Eur. J. Biochem.* **268,** 1486–1490.

Yoshino, M., Koga, Y., and Yamashita, F. (1986). A decrease in glycine cleavage activity in the liver of a patient with dihydrolipoyl dehydrogenase deficiency. *J. Inherit. Metab. Dis.* **9,** 399–400.

Yudkoff, M., Nelson, D., Daikhin, Y., and Erecinska, M. (1994). Tricarboxylic acid cycle in rat brain synaptosomes. Fluxes and interactions with aspartate aminotransferase and malate/aspartate shuttle. *J. Biol. Chem.* **269,** 27414–27420.

Zinn, A. B., Kerr, D. S., and Hoppel, C. L. (1986). Fumarase deficiency: A new cause of mitochondrial encephalomyopathy. *N. Engl. J. Med.* **315,** s469–s475.

CHAPTER 11

Assays of Cardiolipin Levels

Michael Schlame

Department of Anesthesiology
New York University School of Medicine, New York, New York 10016

Department of Cell Biology
New York University School of Medicine, New York, New York 10016

I. Introduction

Cardiolipin is a dimeric phospholipid that is specifically localized in mitochondria. As an organelle marker, cardiolipin can be used to estimate the tissue content of mitochondria or to assess the purity of subcellular preparations. Cardiolipin is associated with the inner mitochondrial membrane, where its biosynthesis takes place at the matrix interface (Schlame and Haldar, 1993). It is likely that cardiolipin is also present in the outer membrane and in outer–inner membrane contact zones, but this issue has not been resolved unequivocally (for a review see Schlame *et al.*, 2000). The biological function of cardiolipin is beginning to emerge. In particular, it has become clear that cardiolipin plays a role in the supramolecular organization of the multimeric complexes of oxidative phosphorylation (Pfeiffer *et al.*, 2003; Zhang *et al.*, 2002), a task that is presumably related to its ubiquitous protein affinity (Schlame *et al.*, 2000). Deficiency of cardiolipin in yeast and Chinese Hamster Ovary mutants and in human Barth syndrome, causes a number of pleiotropic changes, affecting respiratory coupling, protein import, and crista morphology (Jiang *et al.*, 2000; Ohtsuka *et al.*, 1993; Xu *et al.*, 2005).

Several techniques are available to quantify cardiolipin (Table I). The method of choice depends on the required sensitivity, the availability of equipment, and whether or not information on the molecular composition of cardiolipin is desired. The least sensitive technique is ^{31}P-nuclear magnetic resonance (^{31}P-NMR) spectroscopy. Although rarely employed for the purpose of cardiolipin assay, NMR may be an attractive option if sufficient material is present because the method requires little sample work-up (Beyer and Klingenberg, 1985). At the other end of the scale, mass spectrometry (MS) is the most sensitive technique to measure cardiolipin. Encouraged by the recent boom of affordable instruments, this method has become more and more popular.

When the need arises for cardiolipin quantification in a laboratory without lipid expertise, I highly recommend the classical phosphorus assay in combination with thin-layer chromatography (TLC). This method is robust, does not require specialized equipment, and produces reliable results as long as at least a few nanomoles of cardiolipin are present. What I do not recommend is cardiolipin quantification with the fluorescence dye 10-*N*-nonyl acridine orange (NAO). Although the content of cardiolipin correlates with the NAO fluorescence signal in isolated mitochondria (Petit *et al.*, 1992), this correlation is lost in more complex systems, such as cells or tissues, presumably because NAO binding is not sufficiently specific for cardiolipin (Gohil *et al.*, 2005). Unfortunately, the easy applicability of the NAO assay to flow cytometry has prompted its misuse in a number of studies.

Table I
Overview of Quantitative Cardiolipin Assays

Technique of quantification	Analytical range (mol/sample)	Qualitative information[a]
^{31}P-NMR spectrometry	10^{-7}–10^{-5}	No
TLC/phosphorus assay	10^{-9}–10^{-6}	No
TLC/fatty acid assay	10^{-9}–10^{-6}	Yes
UV-HPLC	10^{-9}–10^{-7}	Yes
Fluorescence-HPLC	10^{-10}–10^{-7}	Yes
Mass spectrometry	10^{-12}–10^{-9}	Yes

[a]Composition of fatty acids or molecular species.

In the following, I will describe the most useful methods for cardiolipin analysis and discuss their respective merits and shortcomings. Regardless of the technical details, key to each method is the separation of cardiolipin from the other lipids that are present in biological tissues. This can be achieved by TLC, high-performance liquid chromatography (HPLC), or MS. However, before describing these methods in detail, I will review the structure of cardiolipin, in which multiple fatty acid combinations and their positional isomers provide a unique challenge for comprehensive compositional analysis.

II. Molecular Species of Cardiolipin

What is commonly referred to as cardiolipin is almost always a mixture of different molecular species, that is, cardiolipins with specific combinations of fatty acids. A huge number of such molecular species may exist because the structure of cardiolipin is 1′,3′-bis(1,2-diacylglycero-3-phospho)glycerol; thus it includes four acyl residues, each attached to a stereochemically unique hydroxyl group (Powell and Jacobus, 1974). The uniqueness of each ester site is caused by the prochirality of the central glycerol, which makes one glycerophosphate *pro-S* (1′-linked) and the other one *pro-R* (3′-linked). As a result, there are four distinguishable ester sites, namely the 1′-(1-glycero-), 1′-(2-glycero-), 3′-(1-glycero-), and 3′-(2-glycero-) position (Fig. 1). This implies that N^4 positional permutations are possible in a cardiolipin with N types of fatty acids. For example, 625 cardiolipin species could occur in yeast that contains 5 fatty acids, and 38,416 cardiolipin species could occur in humans with 14 fatty acids.

However, the actual number of molecular species in mitochondria seems to be much smaller. This is because only one or two kinds of fatty acids are usually selected for the assembly of cardiolipins. For instance, in many mammalian tissues, cardiolipin has a strong preference for linoleic acid, which makes tetralinoleoyl-cardiolipin the dominant molecular species that may account for up to 80% of total cardiolipin. Selective incorporation of fatty acids into cardiolipin is found throughout the eukaryotic world, albeit with varied fatty acids. Yeast cardiolipin has a preference for oleic acid, in bivalves it is docosahexaenoic acid, and in *Drosophila* it is both linoleic and palmitoleic acid (Schlame *et al.*, 2005). Selection of two fatty acids, as in *Drosophila*, creates 16 molecular species (Fig. 2). This example demonstrates the complexity of cardiolipin species even in the presence of only two acyl types.

The functional implication of fatty acid selection is not known. One consequence is limitation of the variety of molecular species, that is, generation of structural uniformity. Another consequence is molecular symmetry, that is, the formation of molecular species with identical 1,2-diacylglycerol moieties. When the process of fatty acid selection is impaired, as in patients with Barth syndrome, multiple, mostly asymmetric cardiolipin species are formed. In asymmetric species, the prochiral carbon atom of the central glycerol group becomes a true chiral center

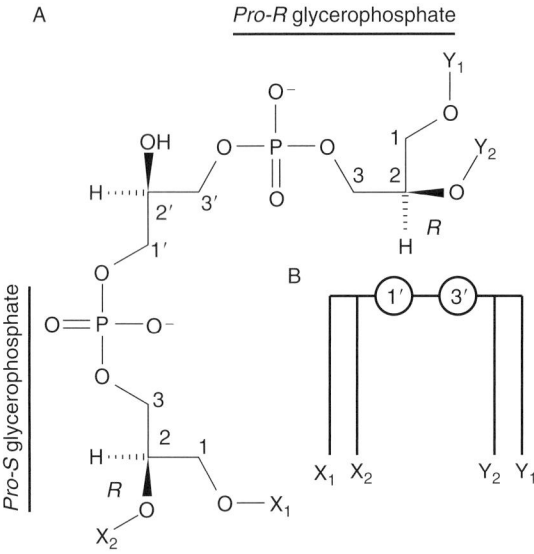

Fig. 1 Structure of cardiolipin. (A) Stereochemical relations. The structure consists of two *sn*-glycero-3-phosphate moieties linked by a glycerol group. Four acyl groups (X_1, X_2, Y_1, and Y_2) are attached to the glycerophosphates. Both glycerophosphates carry obligate chiral centers in *R* conformation. The central glycerol carries a prochiral center (if $X_1 = Y_1$ and $X_2 = Y_2$) or a true chiral center (if $X_1 \neq Y_1$ or $X_2 \neq Y_2$). As a result, the two glycerophosphates are stereochemically nonequivalent. (B) Shorthand of cardiolipin structure. Horizontal lines represent glycerol moieties, vertical lines represent fatty acids, and circles represent phosphate groups.

(Schlame *et al.*, 2005), but whether or not the chirality of cardiolipin has any functional relevance remains to be shown.

To date, no analytical method is able to resolve all molecular species of cardiolipin. Partial resolution can be achieved by either reversed-phase HPLC or MS, and significant improvement may result from combining the two techniques (Fig. 2). In general, HPLC is superior in resolving homologous cardiolipins with variable double bond numbers, while MS is superior in resolving cardiolipins with different carbon numbers. The most promising approach to comprehensive cardiolipin analysis is multidimensional MS, in which molecular species are separated, trapped, and fragmented in a stereo-specific manner.

III. Cardiolipin Analysis by TLC

A. Overview

TLC is a powerful method to separate phospholipids (Fig. 3). A large number of techniques, using various solvents and stationary phases have been described. Since many kinds of TLC plates have become available from commercial sources,

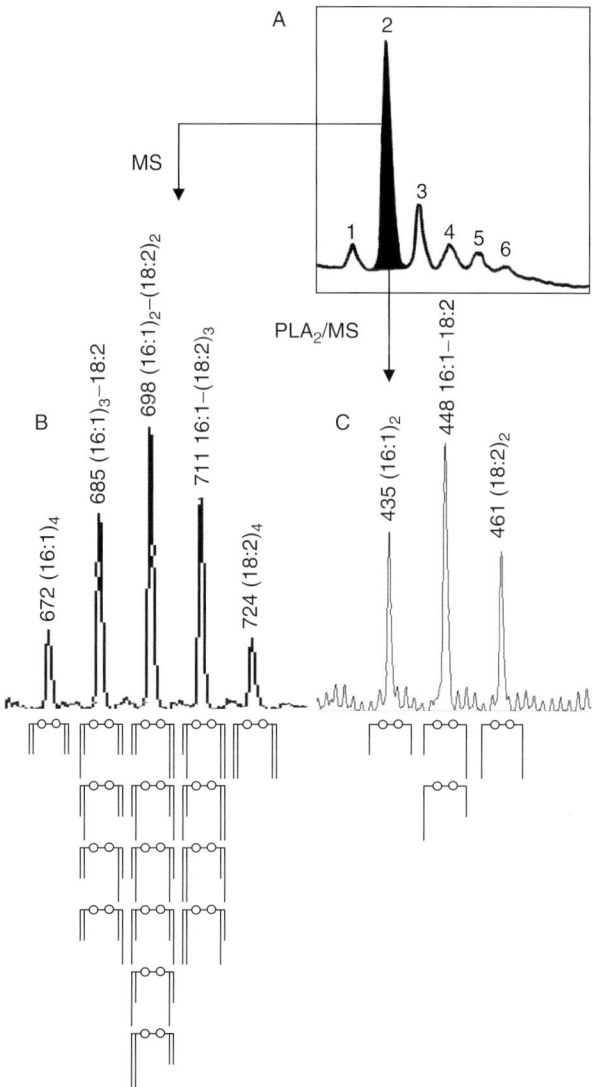

Fig. 2 Molecular species of cardiolipin from *Drosophila melanogaster*. Experimental details are described elsewhere (Schlame *et al.*, 2005). (A) Reversed-phase HPLC of cardiolipin. Peak 2 was collected for MS analysis. An aliquot was treated with phospholipase A_2 (PLA_2). (B) MS of cardiolipin from peak 2. The spectrum shows five mass signals, each corresponding to a specific combination of palmitoleoyl (16:1) and linoleoyl (18:2) residues. The mass intensity correlates roughly to the number of possible positional isomers, suggesting that all of them occur with equal likelihood. Isomers are depicted below the peaks as schematic drawings according to the shorthand in Fig. 1B (short vertical lines, 16:1; long vertical lines, 18:2). (C) MS of di-(2-lyso)-cardiolipin formed by PLA_2 treatment. The spectrum shows the three mass signals that are expected for a mixture of di-(2-lyso)-cardiolipins containing 16:1 and 18:2 residues.

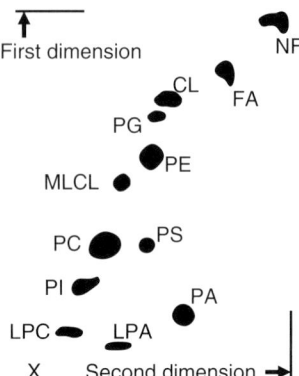

Fig. 3 2D TLC of a mitochondrial lipid extract. Silica gel 60 TLC plates were developed with chloroform-methanol-20% ammonia [65/30/5 (v/v/v)] in the first dimension and with chloroform-acetone-methanol-acetic acid-water [50/20/10/10/5 (v/v/v/v/v)] in the second dimension. Lipids were visualized by iodine vapor. CL, cardiolipin; FA, fatty acids; LPC, lysophosphatidylcholine; LPA, lysophosphatidic acid; MLCL, monolyso-cardiolipin; NF, neutral fats; PA, phosphatidic acid; PC, phosphatidylcholine; PE, phosphatidylethanolamine; PG, phosphatidylglycerol; PI, phosphatidylinositol; PS, phosphatidylserine; X, origin. PA, LPA, MLCL, and PG were added as standard lipids; they are usually not visible in mitochondrial extracts.

the practice of preparing self-made silica phases on glass plates has largely fallen out of favor. I routinely perform two-dimensional (2D) TLC on silica gel 60 plates from Merck. Although authors have described monodimensional separation techniques on high-resolution material, I believe that 2D TLC gives better results, especially in view of the day-to-day variations in performance. Such variations are caused mainly by changes in weather conditions, notably temperature and humidity.

In order to visualize the separated lipids, they may be stained by brief exposure to iodine vapor in order to mark them on the TLC plate. Silica spots are then scraped into test tubes for determination of phosphorus or for determination of fatty acids. For the phosphorus assay, a standard technique (Cook and Daughton, 1981) and a microtechnique (Zhou and Arthur, 1992) have been described. If fatty acids are to be measured, TLC plates can be stained by 8-anilino-1-naphthenesulfonic acid spray or by 2',7'-dichlorofluorescin spray as an alternative to iodine vapor. Sprayed lipids may be visualized under UV light.

B. Protocol for TLC with Phosphorus Assay

Step 1
Prepare samples: Typical samples include membrane preparations, cell pellets, or tissue homogenates, containing 0.5- to 5-mg protein in a volume up to 0.5 ml.

The total amount of lipid should not exceed 1 μmol to guarantee good chromatographic resolution.

Step 2

Extract lipids: Add 2-ml methanol and 1-ml chloroform, vortex, and incubate 20 min at 37°C to facilitate denaturation of proteins. Then, add 0.4-ml normal saline and vortex; add 1-ml chloroform and vortex; add 1-ml normal saline and vortex. Spin at 500 × g for 5 min to achieve phase separation. Collect the lower phase and add 2 ml of chloroform to repeat the extraction cycle once. Dry the lipid extract under a gentle stream of nitrogen. The tube should be placed in a heating block set at 40°C. Redissolve lipids in 0.3-ml chloroform/methanol (2:1).

Step 3

Perform TLC: Spot the lipid solution in the left lower corner of a silica gel 60 plate from Merck (10 × 20 cm^2). Spots should not exceed 3/8 in. Use a stream of air or nitrogen to facilitate evaporation of the solvent. For separation in the first dimension, place plates in a solvent tank with chloroform-methanol-20% ammonia [65/30/5 (v/v/v)]. Let the solvent run up to about 5 in. Remove plate and dry for about 20 min. For separation in the second dimension, place plate, perpendicular to the first dimension, in a solvent tank containing chloroform-acetone-methanol-acetic acid-water [50/20/10/10/5 (v/v/v/v/v)]. Let the solvent run up to about 3 in. Place the TLC plate briefly in a chamber that contains iodine crystals. Exposure to iodine vapor should be terminated immediately after yellow lipid spots become visible (Fig. 3). Mark lipids by pencil and allow unbound iodine to dissipate off the plate in a fume hood. Scrape silica spots that contain phospholipids into appropriate test tubes for either phosphorus assay or fatty acid assay.

Step 4

Perform phosphorus assay: This procedure involves digestion of the phospholipid followed by colorimetric determination of the liberated phosphate. The test has to be calibrated by a series of 10- to 200-μl aliquots of a 1-mM potassium phosphate solution. Transfer standards and silica gel spots into borosilicate tubes that are at least $4^3/_4$ in. long. Add 0.5- to 1.0-ml 70% perchloric acid, cap the tubes with aluminum foil, and incubate overnight in a heating block set at 160–180°C. This creates conditions for reflux distillation. Let samples cool. Add 2.5-ml 0.568% (w/v) ammonium molybdate-4-hydrate, followed immediately by 0.05-ml 11.4% (w/v) ascorbic acid, and vortex. Incubate for 1 h in a water bath at 60–70°C. Spin samples to sediment silica gel. Read absorbance of supernatants at 705 nm. This signal is proportional to the amount of phosphorus.

A more sensitive variation of this assay was developed by Zhou and Arthur (1992). In this protocol, aliquots of the perchloric acid digests (0.05–0.2 ml) are incubated for 10 min with 2 ml of a solution containing 0.3% (w/v) malachite green base, 1% (w/v) ammonium molybdate-4-hydrate, and 0.05% (w/v) Tween 20 in 1.25-M hydrochloric acid. Absorbance at 660 nm is proportional to the phosphorus concentration. However, the malachite green solution is prone to precipitations that can cause errors in absorbance readings.

C. Protocol for TLC with Fatty Acid Assay

Steps 1–3
Perform these steps as described above (Section III.B).

Step 4
Measure the content of fatty acids: Transfer silica gel spots into 13 × 100-mm glass tubes that can be sealed with screw caps. Add internal standard (e.g., 50-nmol heptadecanoic acid) and 1 ml of 0.5-M hydrochloric acid in dry methanol (methanolic-HCl from Supelco, Bellefonte, PA). Incubate overnight in a heating block set at 90 °C. Cool samples and add 1-ml saturated sodium bicarbonate solution. Add 2-ml hexane, vortex, and spin to achieve phase separation. Collect the upper phase into a fresh tube that contains a few grains of anhydrous sodium sulfate, and vortex briefly. Transfer hexane solution into a new tube and evaporate solvent under a stream of nitrogen. Redissolve fatty acid methyl esters in 30- to 50-μl hexane. Inject 0.5 μl of this solution into a gas chromatograph equipped with an SP-2330 capillary column (length, 30 m; internal diameter, 0.32 mm; film, 0.2 μm; Supelco, Bellefonte, PA), set at a temperature of 185 °C. Phospholipids are quantified by measuring the surface areas of all endogenous fatty acid peaks relative to the standard peak.

IV. Cardiolipin Analysis by HPLC

A. Overview

The presence of four fatty acids makes cardiolipin ideal for reversed-phase HPLC. This method resolves molecular species of cardiolipin because the chromatographic retention time increases proportional to the length of fatty acids and it decreases proportional to the number of double bonds (Schlame *et al.*, 2005). Figure 4 shows the chromatographic properties of a series of cardiolipin species with four C_{18} chains, demonstrating a quasilinear, inverse relationship between the retention time and the number of double bonds.

HPLC may be performed either with derivatized or with nonderivatized cardiolipin. Teng and Smith (1985) developed an HPLC technique for native cardiolipin that can be used in connection with UV absorbance recording. However, since the UV signal is not specific, cardiolipin needs to be purified before HPLC. Also, quantification of cardiolipin species may be difficult because the absorbance coefficient varies among individual peaks, depending on their double bond content. An alternative way of quantification is the determination of cardiolipin immunoreactivity in collected HPLC fractions (Schlame *et al.*, 2001). This "immunochromatogram" is relatively specific for cardiolipin and, within limits, may yield quantitative information. Measurement of immunoreactivity is applicable to highly saturated cardiolipins that have very low UV absorbance (Fig. 4D). When samples contain radioactive cardiolipin, the HPLC technique by Teng and Smith (1985) may be combined with scintillation counting (Xu *et al.*, 2003).

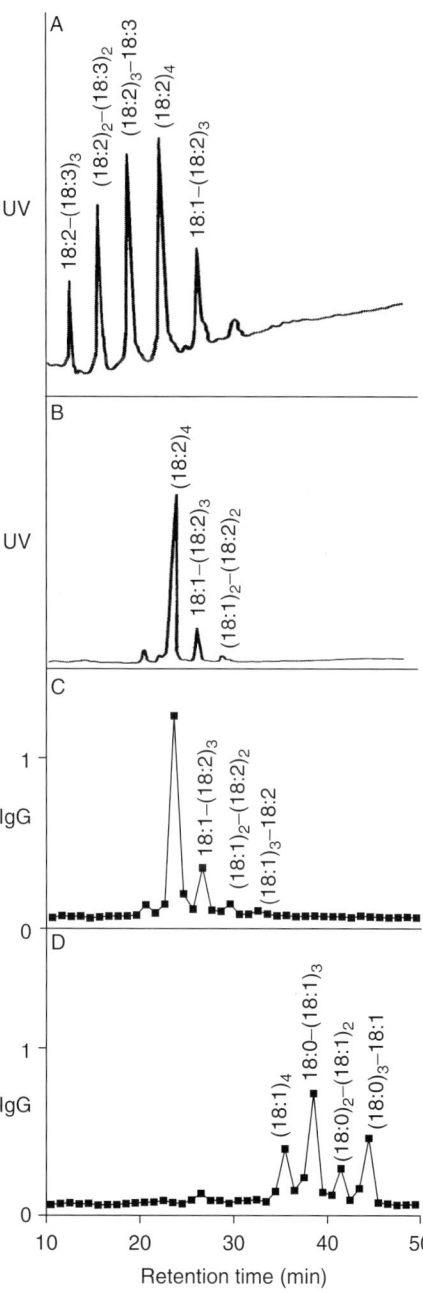

Fig. 4 HPLC of cardiolipin species with C_{18} fatty acids. Chromatograms were recorded by absorbance at 205 nm (UV) or by immunoassay using serum from patients with antiphospholipid syndrome (IgG). Experimental details are described elsewhere (Schlame *et al.*, 2001, 2005). (A) Cardiolipin from mung bean hypocotyls. (B, C) Cardiolipin from bovine heart. (D) Cardiolipin from bovine heart after hydrogenation with PtO_2/H_2. 18:0, stearic acid; 18:1, oleic acid; 18:2, linoleic acid; 18:3, linolenic acid.

For precise HPLC quantification, cardiolipin has to be chemically modified. First, cardiolipin is methylated in order to render the phosphate groups inert. Then, the only residue amenable to derivatization is the central hydroxyl group that can be linked to chromophores like benzoic acid (Schlame and Otten, 1991) or naphthyl-1-acetic acid (Schlame *et al.*, 1999). This approach allows for specific detection, either by 228-nm absorbance or by naphthyl-specific fluorescence. The fluorescence signal is proportional to the molar concentration of cardiolipin and it is sensitive enough to detect less than 1 nmol of cardiolipin.

B. Protocol for Nonderivatized Cardiolipin

Step 1

Isolate cardiolipin by TLC as described above. Extract cardiolipin from silica gel into chloroform/methanol [1/1 (v/v)]. Remove silica gel by centrifugation. Add normal saline, vortex, spin to achieve phase separation, and collect the lower phase. This step removes any remaining traces of silica gel. Evaporate the solvent and redissolve sample in 0.2-ml methanol.

Step 2

Prepare HPLC solvents. Solvent A contains 300-ml methanol, 600-ml acetonitrile, and 100-ml aqueous potassium phosphate (10 mM, pH 7.5). It is important to adjust the pH with KOH rather than NaOH because sodium salts will precipitate in the HPLC pump. Solvent B contains 400-ml methanol and 600-ml acetonitrile.

Step 3

Inject sample into analytical C_{18} reversed-phase column (e.g., ODS Hypersil, 5 μm, column dimensions: 4.6×250 mm^2). The amount of cardiolipin should not exceed 300 nmol. Run a gradient from 50% solvent B to 100% solvent B in 50 min at a flow rate of 1 ml/min. Record absorbance at 205 nm.

Step 4

To measure cardiolipin by solid-phase immunoassay, collect 1-ml fractions of the HPLC effluent. Transfer 10- to 50-μl aliquots of the HPLC fractions onto 96-well microtiter plates and let the solvent evaporate. Perform standard anticardiolipin ELISA as described (Schlame *et al.*, 2001).

C. Protocol for Fluorescence-Labeled Cardiolipin

Step 1

Prepare the internal standard, oleoyl-tristearoyl-cardiolipin, by catalytic hydrogenation of commercial bovine heart cardiolipin. Add 7-mg platinum (IV) oxide to 1-ml methanol and bubble with hydrogen. Add 10-mg bovine heart cardiolipin in 0.5-ml methanol and continue to bubble with hydrogen gas for another 45 min. Remove the catalyst by centrifugation, add normal saline, and purify the hydrogenated cardiolipin by chloroform extraction. Most of the hydrogenation product is oleoyl-tristearoyl-cardiolipin. The absolute concentration of the standard can be determined by phosphorus assay (Section III.B).

Step 2

Extract lipids from the biological sample with chloroform/methanol, as described above (Section III.B). At the beginning of the extraction, add 1–2 nmol of the internal standard. Dry the lipid extract under a stream of nitrogen (40°C).

Step 3

Place samples on ice and add 2-ml ice-cold methanol, 1-ml ice-cold chloroform, and 1-ml ice-cold 0.1-M hydrochloric acid. Vortex and incubate on ice for 5 min. Add 1 ml of chloroform and 1 ml of 0.1-M hydrochloric acid, vortex, and wait for phase separation. Collect the lower phase and evaporate the solvent under a stream of nitrogen (40°C).

Step 4

Produce diazomethane from 1-g *N*-methyl-*N*-nitroso-p-toluenesulfonamide by adding 2-ml ethanol and 0.3-ml 16-M potassium hydroxide. Trap the diazomethane gas in 16-ml ice-cold chloroform. Add 1 ml of the diazomethane/chloroform solution to each sample and incubate on ice for 15 min. Evaporate the solvent and dissolve the residue in 0.2-ml chloroform.

Step 5

Equilibrate silica solid-phase extraction tubes (Supelclean LC-Si, 1-ml tubes from Supelco, Bellefonte, PA) in diethylether-ethanol [9/1 (v/v)]. Load samples onto extraction tubes and elute with 2-ml diethylether-ethanol [9/1 (v/v)]. Collect the eluate and evaporate the solvent under nitrogen.

Step 6

Add 0.25 ml of 0.1-M 1-naphthylacetic anhydride in anhydrous pyridine. Incubate at 40°C for 2 h. This step is highly sensitive to moisture. The reagent needs to be made fresh each time.

Step 7

Add 6 ml of hexane and spin tubes ($500 \times g$, 5 min) to sediment precipitated material. Collect supernatant and extract with 4-ml water. Recover the hexane phase and remove traces of water by adding a few grains of anhydrous sodium sulfate. Transfer to a clean tube, evaporate hexane, and redissolve the sample in 1-ml hexane-diethylether [1/1 (v/v)].

Step 8

Equilibrate silica solid-phase extraction tubes (Supelclean LC-Si, 1 ml tubes from Supelco, Bellefonte, PA) in hexane-diethylether [1/1 (v/v)]. Load samples onto extraction tubes, elute with 2-ml hexane-diethylether [1/1 (v/v)], and discard this eluate. Next, elute with 2-ml diethylether-ethanol [9/1 (v/v)] and collect the eluate. Evaporate solvent and redissolve residue in 0.2-ml hexane–ethanol [1/1 (v/v)].

Step 9

Inject 20 μl of the hexane–ethanol solution into an ODS Hypersil column (particle size 5 μm; column dimensions, 4.6×150 mm^2). Elute with 20% 2-propanol in acetonitrile for 10 min at a flow rate of 2 ml/min. From 10 to 60 min, increase the concentration of 2-propanol from 20% to 70%, while maintaining a flow rate of 2 ml/min. Monitor eluate with a fluorescence detector set at an excitation wavelength of 280 nm and an emission wavelength of 360 nm. Molecular species of cardiolipin are quantified by determination of their peak area relative to the peak area of the internal standard. The internal standard elutes at a retention time of about 50 min.

V. Cardiolipin Analysis by MS

A. Overview

Advances in MS technology have made this method very attractive for cardiolipin analysis. MS is not only the most sensitive of the cardiolipin methods, it is also most amenable to automation and it can potentially provide a wealth of structural information. However, the costs associated with the purchase of a mass spectrometer are substantial and its successful use requires special training. Most MS core facilities have focused on protein analysis and are often reluctant to expose their instruments to lipids.

Analysis of cardiolipin has been accomplished by fast atom bombardment MS (Peter-Katalinic and Fischer, 1998) and by electrospray ionization (ESI) MS (Lesnefsky et al., 2001; Schlame et al., 2001). The most widely employed technique is ESI-MS in negative ion mode, in which a single cardiolipin species typically produces two signals, representing the double-charged ion and the single-charged ion, respectively. In addition, the sodium adduct of the double-charged ion may give rise to a third mass (Fig. 5). The double-charged ion is most suitable for quantification because it has the highest intensity (Valianpour et al., 2002). However, resolution of cardiolipin species with different double bond content is better among single-charged ions because their m/z values are further apart (Sparagna et al., 2005). Excellent resolution can be obtained, even in the double-charged ion cluster, when cardiolipin species differ by carbon chain length (Fig. 2C). Quantification of cardiolipin by MS requires an internal standard. Tetramyristoyl-cardiolipin has been used for this purpose because it is a commercially available, nonnatural species with a mass far removed from that of native cardiolipins (Sparagna et al., 2005; Valianpour et al., 2002).

To improve performance, MS may be combined with HPLC. This increases the signal to noise ratio, minimizes "ghost peaks" from impurities, and enhances resolution. Both reversed-phase and normal-phase HPLC have been employed. Combination of MS with reversed-phase HPLC is most suitable for experiments in which a comprehensive analysis of the cardiolipin composition is the primary goal (Fig. 2). However, since the reversed-phase solvent system by Teng and Smith (1985) contains a high concentration of potassium phosphate, cardiolipin needs to be extracted into a solvent that is compatible with MS (Schlame et al., 2001). Alternatively, the phosphate salt can be replaced by ammonium acetate to allow online coupling of the HPLC system to the mass spectrometer (Lesnefsky et al., 2001). MS can also be combined with normal-phase HPLC, especially when a high throughput of samples is desired (Sparagna et al., 2005; Valianpour et al., 2002). This technique eliminates the requirement for extensive preanalytical preparation because normal-phase HPLC

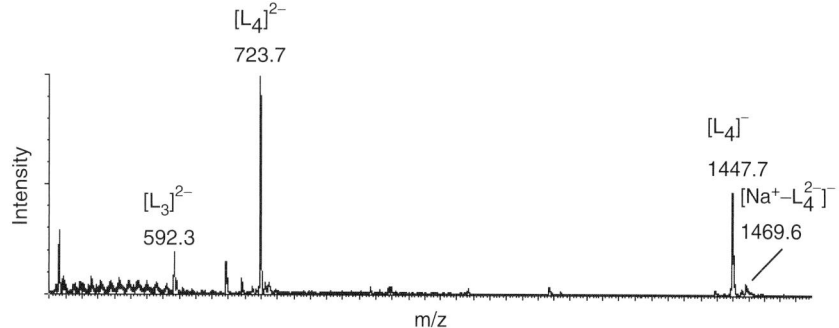

Fig. 5 Mass spectrum of bovine heart cardiolipin. A commercial preparation of bovine heart cardiolipin was dissolved in chloroform-methanol-8.4 mM aqueous ammonium bicarbonate [11/8/2 (v/v/v)] to a final concentration of 10 μM. ESI-MS analysis was performed on a Quattro II triple quadrupole mass spectrometer (Micromass, Beverly, MA) in negative ion mode with the capillary voltage set to 3.2 kV, the cone voltage set to 50 V, and the temperature set to 65°C. L_4, tetralinoleoyl-cardiolipin; L_3, trilinoleoyl-monolyso-cardiolipin.

results in sufficient cardiolipin purification. Thus, samples may be injected into the HPLC-MS instrument immediately following lipid extraction.

Novel MS technologies have been driven by the lipidomics approach, in which comprehensive compositional and structural analysis is achieved solely on the basis of multidimensional mass-spectrometric arrays. Han *et al.* (2006) used continuous ion-transmission (QqQ type) and high-mass resolution hybrid pulsed (QqTOF type) instruments to determine major and minor molecular species of cardiolipin in mouse tissues. Total lipids were extracted in the presence of lithium chloride and directly subjected to MS. The authors provided detailed information on comprehensive quantification of all molecular species, but without positional assignment of the acyl residues. However, Hsu *et al.* (2004) demonstrated that MS is, at least in principle, capable of positional acyl assignment. They showed that multiple stage tandem MS may selectively detect signals of the 1'-phosphatidyl and 3'-phosphatidyl moieties respectively, and reveal the identity of *sn*-1- and *sn*-2-linked fatty acids.

B. Method

1. MS Coupled to Reversed-Phase HPLC

Extract lipids into chloroform/methanol and isolate cardiolipin by TLC as described in Section III.B. Resolve cardiolipin by reversed-phase HPLC, essentially as described in Section IV.B. In solvent A, replace potassium phosphate with 15-mM ammonium acetate, pH 7.4. HPLC may be coupled to ESI-MS as described by Lesnefsky *et al.* (2001). Alternatively, HPLC fractions may be collected and cardiolipin may be extracted into chloroform. Evaporate solvent and redissolve in chloroform-methanol-8.4-mM aqueous ammonium bicarbonate [11/8/2 (v/v)]. Perform ESI-MS as described by Schlame *et al.* (2001, 2005).

2. MS Coupled to Normal–Phase HPLC

Extract lipids into chloroform/methanol and analyze lipid extracts by normal-phase HPLC on a silica column coupled to a mass spectrometer operated in negative ESI mode. Both chloroform-methanol (Valianpour *et al.*, 2002) and hexane/2-propanol (Sparagna *et al.*, 2005) have been employed as HPLC solvent systems.

3. MS Array Analysis

Extract lipids into chloroform/methanol in the presence of LiCl as described by Han *et al.* (2006). Infuse lipid extracts directly into a triple-quadrupole (QqQ), Qq time-of-flight, or quadrupole ion-trap instrument. A detailed description of instrument settings and data interpretation can be found elsewhere (Han *et al.*, 2006; Hsu *et al.*, 2004).

VI. Applications of Cardiolipin Measurements

Earlier studies of cardiolipin focused on occurrence, tissue concentration, and fatty acid composition. Typically, these studies used TLC followed either by phosphorus assay or by gas chromatographic analysis of fatty acids. Cardiolipin metabolism has also been studied by TLC, mostly after incorporation of radioactive precursors. Such precursors may be detected by scintillation counting in lipid fractions separated by TLC.

Involvement of cardiolipin in disease processes has been investigated in animal models of thyroid dysfunction (Paradies and Ruggiero, 1990), aging (Paradies and Ruggiero, 1991), heart failure (Sparagna *et al.*, 2005), ischemia (Lesnefsky *et al.*, 2001), and diabetes (Han *et al.*, 2005). Some of these studies have used HPLC and MS, providing information not only on cardiolipin content, but also on the composition of cardiolipin molecular species. Cardiolipin has also been measured in human specimens in the context of clinical studies. Cardiac and skeletal muscle biopsies were obtained from patients with heart failure (Heerdt *et al.*, 2002) and from patients with mitochondrial disease (Schlame *et al.*, 1999). Fluorescence-HPLC was used in these studies because of the small amount of material that is available in clinical samples.

Barth syndrome is an X-linked recessive human disorder of cardiolipin metabolism (Barth *et al.*, 2004). Affected patients have a reduced content of cardiolipin and an altered composition of cardiolipin molecular species (Fig. 6). They present with

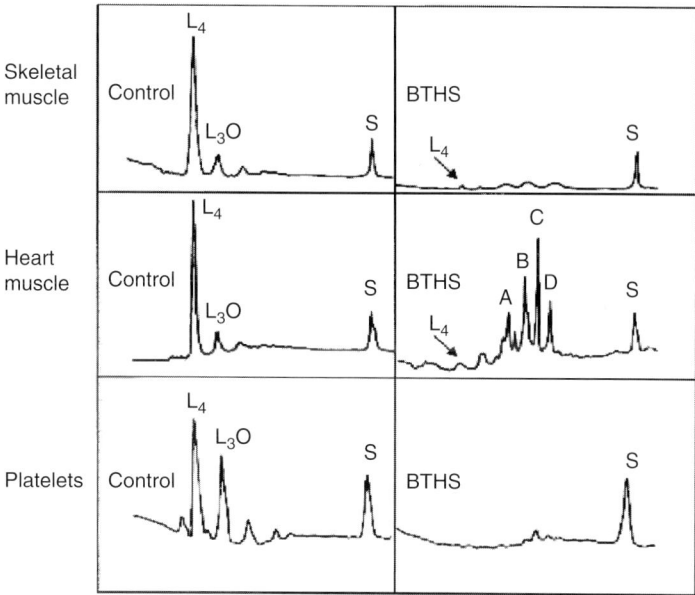

Fig. 6   HPLC of cardiolipin from control subjects and patients with Barth syndrome (BTHS). Human specimens were extracted with chloroform, derivatized, and analyzed by fluorescence HPLC. Each box shows a chromatogram in which the fluorescence yield is plotted against retention time. In BTHS, the content of cardiolipin is reduced (especially in platelets and skeletal muscle) and the composition of cardiolipin is altered. L_4, tetralinoleoyl-cardiolipin; L_3O, trilinoleoyl-oleoyl-cardiolipin; S, internal standard (tristearoyl-oleoyl-cardiolipin); A–D, other molecular species of cardiolipin. [Reprinted with permission from Schlame *et al.* (2002).]

multisystemic symptoms, including cardiomyopathy, skeletal muscle weakness, neutropenia, and growth retardation. The absence of tetralinoleoyl-cardiolipin in human platelets is diagnostic for Barth syndrome, as it differentiates this disease from related cardiomyopathies and skeletal muscle disorders (Schlame *et al.*, 2003). The platelet test for tetralinoleoyl-cardiolipin is now routinely employed in the work-up of patients who meet the clinical criteria of Barth syndrome. Platelet cardiolipin can be analyzed either by fluorescence-HPLC (Schlame *et al.*, 2002) or by MS (Valianpour *et al.*, 2002). The MS technique is faster and more convenient, and may be automated to accommodate a large number of samples.

Apoptosis is associated with a specific decline of the cardiolipin content of cells. This has been documented *in vitro*, using TLC (Ostrander *et al.*, 2001), HPLC (Petrosillo *et al.*, 2004), and MS (Sorice *et al.*, 2004). Since free radicals are involved in apoptosis, oxidative damage of cardiolipin has been proposed as the main mechanism that leads to the degradation of cardiolipin. In principle, oxidized cardiolipin can be measured by HPLC with UV detection at 232 nm (Parinandi *et al.*, 1988;

Schlame *et al.*, 2001) or by HPLC with chemiluminescence detection (Nomura *et al.*, 2000). However, it has been difficult to define the oxidation products of cardiolipin that presumably form during apoptosis. Besides HPLC, ESI-MS has the capability to identify oxidized species of cardiolipin; for example, we detected a series of oxidatively modified cardiolipins, containing two to eight extra oxygen atoms, by ESI-MS (Schlame *et al.*, 2001).

Acknowledgments

This work was supported in part by grants from the Barth Syndrome Foundation, the American Heart Association, and the National Institutes of Health.

References

Barth, P. G., Valianpour, F., Bowen, V. M., Lam, J., Duran, M., Vaz, F. M., and Wanders, R. J. A. (2004). X-linked cardioskeletal myopathy and neutropenia (Barth syndrome): An update. *Am. J. Med. Genet.* **126A,** 349–354.

Beyer, K., and Klingenberg, M. (1985). ADP/ATP carrier protein from beef heart mitochondria has high amounts of tightly bound cardiolipin, as revealed by 31P nuclear magnetic resonance. *Biochemistry* **24,** 3821–3826.

Cook, A. M., and Daughton, C. G. (1981). Total phosphorus determination by spectrophotometry. *Methods Enzymol.* **72,** 292–295.

Gohil, V. M., Gvozdenovic-Jeremic, J., Schlame, M., and Greenberg, M. L. (2005). Binding of 10-N-nonyl acridine orange to cardiolipin-deficient yeast cells: Implications for assay of cardiolipin. *Anal. Biochem.* **343,** 350–352.

Han, X., Yang, J., Cheng, H., Yang, K., Abendschein, D. R., and Gross, R. W. (2005). Shotgun lipidomics identifies cardiolipin depletion in diabetic myocardium linking altered substrate utilization with mitochondrial dysfunction. *Biochemistry* **44,** 16684–16694.

Han, X., Yang, K., Yang, J., Cheng, H., and Gross, R. W. (2006). Shotgun lipidomics of cardiolipin molecular species in lipid extracts of biological samples. *J. Lipid Res.* **47,** 864–879.

Heerdt, P. M., Schlame, M., Jehle, R., Barbone, A., Burkhoff, D., and Blanck, T. J. J. (2002). Disease-specific remodeling of cardiac mitochondria after left ventricular assist device. *Ann. Thorac. Surg.* **73,** 1216–1221.

Hsu, F.-F., Turk, J., Rhoades, E. R., Russell, D. G., Shi, Y., and Groisman, E. A. (2004). Structural characterization of cardiolipin by tandem quadrupole and multiple-stage quadrupole ion-trap mass spectrometry with electrospray ionization. *J. Am. Soc. Mass Spectrom.* **16,** 491–504.

Jiang, F., Ryan, M. T., Schlame, M., Zhao, M., Gu, Z., Klingenberg, M., Pfanner, N., and Greenberg, M. L. (2000). Absence of cardiolipin in the crd1 null mutant results in decreased mitochondrial membrane potential and reduced mitochondrial function. *J. Biol. Chem.* **275,** 22387–22394.

Lesnefsky, E. J., Slabe, T. J., Stoll, M. S. K., Minkler, P. E., and Hoppel, C. L. (2001). Myocardial ischemia selectively depletes cardiolipin in rabbit heart subsarcolemmal mitochondria. *Am. J. Physiol.* **280,** H2770–H2778.

Nomura, K., Imai, H., Koumura, T., Kobayashi, T., and Nakagawa, Y. (2000). Mitochondrial phospholipids hydroperoxide glutathione peroxidase inhibits the release of cytochrome c from mitochondria by suppressing the peroxidation of cardiolipin in hypoglycemia-induced apoptosis. *Biochem. J.* **351,** 183–193.

Ohtsuka, T., Nishijima, M., Suzuki, K., and Akamatsu, Y. (1993). Mitochondrial dysfunction of a cultured Chinese hamster ovary cell mutant deficient in cardiolipin. *J. Biol. Chem.* **268,** 22914–22919.

Ostrander, D. B., Sparagna, G. C., Amoscato, A. A., McMillin, J. B., and Dowhan, W. (2001). Decreased cardiolipin synthesis corresponds with cytochrome c release in palmitate-induced cardiomyocyte apoptosis. *J. Biol. Chem.* **276,** 38061–38067.

Paradies, G., and Ruggiero, F. M. (1990). Stimulation of phosphate transport in rat liver mitochondria by thyroid hormones. *Biochim. Biophys. Acta* **1019,** 133–136.

Paradies, G., and Ruggiero, F. M. (1991). Effect of aging on the activity of the phosphate carrier and on the lipid composition in rat liver mitochondria. *Arch. Biochem. Biophys.* **284,** 332–337.

Parinandi, N. L., Weis, B. K., and Schmid, H. H. (1988). Assay of cardiolipin peroxidation by high-performance liquid chromatography. *Chem. Phys. Lipids* **49,** 215–220.

Peter-Katalinic, J., and Fischer, W. (1998). α-D-Glucopyranosyl-, D-alanyl- and L-lysylcardiolipin from gram-positive bacteria: analysis by fast atom bombardment mass spectrometry. *J. Lipid Res.* **39,** 2286–2292.

Petit, J. M., Maftah, A., Ratinaud, M. H., and Julien, R. (1992). 10N-nonyl-acridine orange interacts with cardiolipin and allows the quantification of this phospholipids in isolated mitochondria. *Eur. J. Biochem.* **209,** 267–273.

Petrosillo, G., Ruggiero, F. M., Pistolese, M., and Paradies, G. (2004). Ca^{2+}-induced reactive oxygen species production promotes cytochrome c release from rat liver mitochondria via mitochondrial permeability transition (MPT)-dependent and MPT-independent mechanisms. *J. Biol. Chem.* **279,** 53103–53108.

Pfeiffer, K., Gohil, V., Stuart, R. A., Hunte, C., Brandt, U., Greenberg, M. L., and Schagger, H. (2003). Cardiolipin stabilizes respiratory chain supercomplexes. *J. Biol. Chem.* **278,** 52873–52880.

Powell, G. L., and Jacobus, J. (1974). The non-equivalence of the phosphorus atoms in cardiolipin. *Biochemistry* **13,** 4024–4026.

Schlame, M., and Haldar, D. (1993). Cardiolipin is synthesized on the matrix side of the inner membrane in rat liver mitochondria. *J. Biol. Chem.* **268,** 74–79.

Schlame, M., Haller, I., Sammaritano, L. R., and Blanck, T. J. J. (2001). Effect of cardiolipin oxidation on solid-phase immunoassay for antiphospholipid antibodies. *Thromb. Haemost.* **86,** 1475–1482.

Schlame, M., Kelley, R. I., Feigenbaum, A., Towbin, J. A., Heerdt, P. M., Schieble, T., Wanders, R. J. A., DiMauro, S., and Blanck, T. J. J. (2003). Phospholipid abnormalities in children with Barth syndrome. *J. Am. Coll. Cardiol.* **42,** 1994–1999.

Schlame, M., and Otten, D. (1991). Analysis of cardiolipin molecular species by high-performance liquid chromatography of its derivative 1,3-bisphosphatidyl-2-benzoyl-sn-glycerol dimethyl ester. *Anal. Biochem.* **195,** 290–295.

Schlame, M., Ren, M., Xu, Y., Greenberg, M. L., and Haller, I. (2005). Molecular symmetry in mitochondrial cardiolipins. *Chem. Phys. Lipids* **138,** 38–49.

Schlame, M., Rua, D., and Greenberg, M. L. (2000). The biosynthesis and functional role of cardiolipin. *Prog. Lipid Res.* **39,** 257–288.

Schlame, M., Shanske, S., Doty, S., Konig, T., Sculco, T., DiMauro, S., and Blanck, T. J. J. (1999). Microanalysis of cardiolipin in small biopsies including skeletal muscle from patients with mitochondrial disease. *J. Lipid Res.* **40,** 1585–1592.

Schlame, M., Towbin, J. A., Heerdt, P. M., Jehle, R., DiMauro, S., and Blanck, T. J. J. (2002). Deficiency of tetralinoleoyl-cardiolipin in Barth syndrome. *Ann. Neurol.* **51,** 634–637.

Sorice, M., Circella, A., Cristea, I. M., Garofalo, T., Di Renzo, L., Alessandri, C., Valesini, G., and Degli Esposti, M. (2004). Cardiolipin and its metabolites move from mitochondria to other cellular membranes during death receptor-mediated apoptosis. *Cell Death Differ.* **11,** 1133–1145.

Sparagna, G. C., Johnson, C. A., McCune, S. A., Moore, R. L., and Murphy, R. C. (2005). Quantification of cardiolipin molecular species in spontaneously hypertensive heart failure rats using electrospray ionization mass spectrometry. *J. Lipid Res.* **46,** 1196–1204.

Teng, J. L., and Smith, L. L. (1985). High-performance liquid chromatography of cardiolipin. *J. Chromatogr.* **339,** 35–44.

Valianpour, F., Wanders, R. J. A., Barth, P. G., Overmars, H., and van Gennip, A. H. (2002). Quantitative and compositional study of cardiolipin in platelets by electrospray ionization mass spectrometry: Application for the identification of Barth syndrome patients. *Clin. Chem.* **48,** 1390–1397.

Xu, Y., Kelley, R. I., Blanck, T. J. J., and Schlame, M. (2003). Remodeling of cardiolipin by phospholipids transacylation. *J. Biol. Chem.* **278,** 51380–51385.

Xu, Y., Sutachan, J. J., Plesken, H., Kelley, R. I., and Schlame, M. (2005). Characterization of lymphoblast mitochondria from patients with Barth syndrome. *Lab. Invest.* **85,** 823–830.

Zhang, M., Mileykovskaya, E., and Dowhan, W. (2002). Gluing the respiratory chain together. Cardiolipin is required for supercomplex formation in the inner mitochondrial membrane. *J. Biol. Chem.* **277,** 43553–43556.

Zhou, X., and Arthur, G. (1992). Improved procedures for the determination of lipid phosphorus by malachite green. *J. Lipid Res.* **33,** 1233–1236.

CHAPTER 12

Measurement of VDAC Permeability in Intact Mitochondria and in Reconstituted Systems

Marco Colombini

Department of Biology, University of Maryland
College Park, Maryland 20742

I. Introduction

Since the early work of Werkheiser and Bartley (1957), mitochondria have been known to consist of two membranes: the outer and the inner membrane. In the early days, mitochondrial research focused primarily on energy transduction, and the characterization of isolated mitochondria focused on criteria related to coupled energy transduction, such as the respiratory control ratio. In those days, the function of the outer membrane and the molecular basis for its permeability were

Copyright 2007, Elsevier Inc. All rights reserved.

0091-679X/07 $35.00
DOI: 10.1016/S0091-679X(06)80012-9

not understood or appreciated. However, times have changed and the importance of the outer membrane is no longer in question. It is literally a question of life and death for the cell. Among the roles of the outer membrane is the regulation of the permeability of the outer membrane to metabolites. This role is the function of VDAC (Colombini, 1979), the channel by which metabolites cross the outer membrane.

VDAC is a 30–32-kDa protein that forms monomeric channels in the mitochondrial outer membrane of all eukaryotes tested (Colombini, 1994; Mannella *et al.*, 1992). Where multiple isoforms exist, at least one isoform, called VDAC1, has properties that are virtually identical, irrespective of the source (Colombini, 1989). This includes two voltage-gating processes, preference for anions, and single-channel conductance. The variation in the properties of other isoforms is believed to be related to specialized functions in that organism or cell type.

Isolation of mitochondria typically involves the homogenization of the cell. This traumatic experience results in some damage to organelles, and mitochondria are no exception. If mitochondria in that cell exist as a mitochondrial reticulum (one large reticulated mitochondrion), the structure will be fragmented. When many mitochondria are present, less damage is likely. In any case, the outer membrane is most vulnerable, not only because it is the mitochondrial surface but also because its surface area is much smaller than that of the inner membrane. Swelling of the inner membrane can damage the outer membrane without breaching the integrity of the inner membrane. Thus, assessing the intactness of the outer membrane is essential.

Beyond assessment of mitochondrial integrity, characterization of mitochondria should include an assessment of the status of the VDAC channels in the intact mitochondrion. VDAC channels are the pathway by which metabolites cross the outer membrane and thus the permeability of the outer membrane to metabolites reflects the state of VDAC. A variety of conditions can influence the probability of VDAC closure (Colombini, 2004). Thus, characterizing the permeability of the outer membrane to metabolites provides information on the state and history of the mitochondrion.

Permeability measurements must go beyond the determination of absolute permeability to the determination of rates of permeation. Mitochondria are dynamic structures capable of high rates of metabolism. Thus, the rate at which metabolites can cross the outer membrane will limit mitochondria metabolic rates. This, in turn, will influence metabolic rates in the rest of the cell because of the highly intertwined nature of the two metabolic systems. The extensive impact of altering the permeability of the outer membrane has been explored in a recent publication (Lemasters and Holmuhamedov, 2006).

Three methods are described to measure the permeability of the outer membrane to metabolites. These all take advantage of mitochondrial enzymatic activity and the rate limitation of the outer membrane. In order to characterize the detailed properties of VDAC, the protein needs to be purified and reconstituted into phospholipid membranes for electrophysiological analysis. A discussion of this method is presented, pointing out the pitfalls to avoid in making these measurements.

II. Determining the Permeability of the Mitochondrial Outer Membrane to Metabolites

A. Rationale

Since VDAC is responsible for metabolite flux across the outer membrane, determining the permeability of the outer membrane to metabolites also determines the state and function of VDAC in the intact organelle. However, the small volume of the intermembrane space and the high permeability of the outer membrane to metabolites make metabolite flux measurements across the outer membrane very difficult. A solution is to take advantage of an enzymatic activity to consume the metabolite once it has crossed the outer membrane. If the rate of the enzymatic reaction were limited by the rate of metabolite translocation across the outer membrane, then the permeability of that membrane to the metabolite could be calculated. Changes in this permeability following any treatment would provide quantitative information of the influence of the treatment on the open-channel probability of VDAC.

B. Assay of Outer Membrane Intactness

Outer membrane intactness must be assessed first to estimate the fraction of mitochondria with damaged outer membranes. This is measured by the method of Douce *et al.* (1987). Intact outer membranes are impermeable to cytochrome *c*. Following outer membrane damage, exogenously added cytochrome *c* can have access to enzymes in the inner membrane. The measurement of KCN-sensitive cytochrome *c* oxidation by complex IV (cytochrome oxidase) is convenient. This can be done either by following the drop in absorbance of reduced cytochrome *c* at 550 nm or by measuring the rate of oxygen consumption using an oxygen electrode (Lee and Colombini, 1997).

C. Methods

1. Determination of the Permeability of the Outer Membrane to ADP from Respiration Measurements

a. Theory

The permeability of the outer membrane to ADP can be extracted from measurements of ADP-dependent respiration (Lee and Colombini, 1997; Lee *et al.*, 1994). ADP-dependent respiration by isolated mitochondria requires four processes in series:

1. ADP diffusion through VDAC channels in the outer membrane
2. ADP/ATP exchange through the inner membrane
3. Proton pumping through the respiratory chain coupled to oxygen consumption.
4. Phosphorylation by the ATP synthase

At steady-state, steps (1) and (2) must proceed at the same rate. The experiments must be run under conditions where steps (3) and (4) are not rate limiting. If this condition does not hold true, then the measured values of the permeability are only apparent but are still useful for comparison purposes with appropriate controls.

Thus, at steady state, the rate of depletion of medium [ADP]:

$$\frac{d[ADP]}{dt} = -\frac{P([ADP]_o - [ADP]_i)}{Vol} = -\frac{V_{max}[ADP]_i}{K_M + [ADP]_i} \qquad (1)$$

where P is the total permeability of the outer membrane, Vol is the volume of the medium, and K_M and V_{max} are the kinetic parameters of the adenine nucleotide translocator, and "o" and "i" refer to the outside compartment (the medium) and the intermembrane space, respectively. Note that the units of V_{max} are not the standard units but concentration per unit time. Solving for $[ADP]_i$:

$$[ADP]_i = \frac{-\left(\frac{V_{max}Vol}{P} + K_M - [ADP]_o\right)}{2} + \sqrt{\left(\frac{\frac{V_{max}Vol}{P} + K_M - [ADP]_o}{2}\right) + K_M[ADP]_o}$$

$$(2)$$

This equation needs to be fit to the experimental results to extract a value for the permeability P. The experimental results begin by measuring the medium [oxygen] as a function of time following the addition of a fixed amount of ADP (Fig. 1, left panel). By subtracting state 4 from the respiration curve, one achieves a relationship for the ADP-dependent oxygen consumption. We must assume that the ADP-independent oxygen consumption is not altered by the addition of ADP. This may introduce a small error because of the change in membrane potential on ADP addition. However the error should be small, especially since state 4 respiration is much lower than state 3. Using the P/O ratio (2 for succinate), one can convert the curve from [oxygen] as a function of time to [ADP] as a function of time (Fig. 1, right panel). Now the theoretical relation can be used to fit the theory to the data.

A theoretical curve can be generated for arbitrary values of P, V_{max}, and K_M by calculating new values of $[ADP]_o$ for each time increment dt (matching the time increments in the digitized data). Thus:

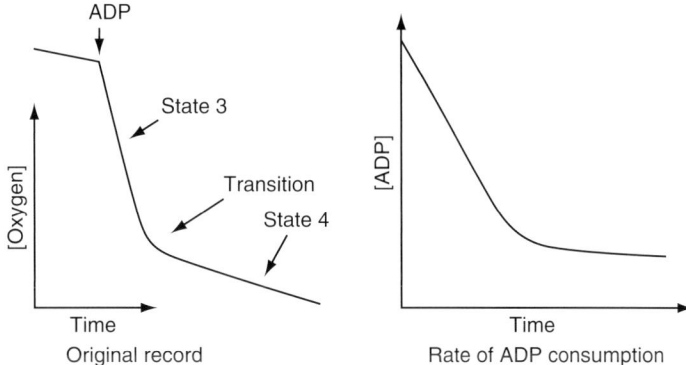

Fig. 1 Illustration of respiration data used to obtain a quantitative measure of the permeability of the mitochondrial outer membrane. The left panel shows a drawing of a typical mitochondrial respiration curve obtained after the addition of ADP. The data needs to be collected directly via an A/D converter with at least 10-bit accuracy. The state 4 respiration rate is subtracted from the recording by fitting a straight line to the state 4 recording (use the asymptote to the end part of the data) and subtracting this from all the data. Only the record after ADP addition is used for further calculations. Using the standard P/O ratio, the [oxygen] is converted to the [ADP] (right panel). Each state 3/state 4 transition takes ~5 min. See text for further explanations.

$$[ADP]_o^{t+dt} = [ADP]_o^t + \frac{d[ADP]_o}{dt}\,dt = [ADP]_o^t - \frac{V_{max}[ADP]_i^t}{K_M + [ADP]_i^t}\,dt \qquad (3)$$

For details on two methods of fitting the data see Lee and Colombini (1997). A unique fit can sometimes be achieved without the need for additional information. However, the K_M value can be obtained on the same set of mitochondria in a separate experiment on hypotonically shocked mitochondria. The same fitting is performed, except that the P value is given a very large number, say 1000. In our experience, this worked well for mitochondria from potatoes but not from mitochondria from rat liver. Damage of the outer membrane of rat liver mitochondria changes the respiration properties. Even rather mild osmotic shock (100 mOsmolar for 10 min on ice) caused changes that cannot be attributed merely to outer membrane damage. An alternative way to reduce the number of variables is to set the K_M to the published value of 4 μM (Pfaff et al., 1969) for the adenine nucleotide translocator. This will yield relative values of permeability, before and after some experimental treatment.

b. Equipment

Clark oxygen electrode setup: If this unit has an analogue output, then the data should be digitized into a good-quality analog-to-digital (A/D) converter (at least 10 bits). If it has a digital output, then this signal can be fed directly into a computer and processed using the manufacturer's software. For example: Yellow Springs Instruments; Mitocell S200 from Strathkelvin Instruments.

A/D converter (optional, see above): A good quality unit such as one used for electrophysiological recordings will work well. For example: Axon Instruments Digidata 1322A.

Chart recorder (optional): It is convenient to also record the output from the oxygen electrode into a chart recorder so that one can more easily keep track of the experimental results. For example: Kipp and Zonen BD12E.

c. Reagents

10-mM ADP

100-mM disodium succinate

Respiration buffer: 0.3-M mannitol, 10-mM NaH_2PO_4, 5-mM $MgCl_2$, and 10-mM KCl (pH 7.2)

d. Protocol

Add mitochondria to 3 ml (volume depends on instrument used) of respiration buffer in the oxygen electrode chamber under temperature control. Room temperature works well. The total mitochondrial protein used is generally about 1 mg. Begin recording oxygen level. Add 150-μl succinate (5 mM final) and record briefly to see a measurable rate. Add 27-μl ADP (90 μM final) and record for $\sim$5 min to detect the state 3/state 4 respiration transition. Record until constant state 4 respiration is achieved. Then, repeat ADP addition three more times. The result of the first addition may be unreliable as the mitochondria stabilize. However, if mitochondria are tightly coupled, the subsequent three trials (before the oxygen level runs out) should be virtually identical. The last three records are analyzed to determine the permeability. Figure 1 (left) shows illustrative data for oxygen utilization rate in state 3 and state 4 respiration.

The data collected for each state 3/state 4 transition should be converted into a data file. If the Axon Instruments Digidata A/D converter is used along with the Clampfit software, one can easily "cut-out" the appropriate part of the record and paste it into a spreadsheet like Excel. The data can now be saved as a "csv" file and imported into a program to fit the data to the theory. A QBasic program we developed for this purpose is available for free download from www.life.umd.edu/biology/faculty/colombini/.

2. Measurement of the Permeability of the Outer Membrane to NADH from Measurements of NADH Oxidation

a. Rationale

Classical state 3/state 4 respiration curves can be obtained from mammalian mitochondria and from those of higher plants but these cannot be obtained by using mitochondria from the yeast, *Saccharomyces cerevisiae*. Addition of ADP to mitochondria isolated from this yeast results in an increase in respiration that does not return to baseline for the length of the experiment. Addition of ATP yields similar results. Thus, the method described in Section II.C.1 cannot be used to determine the permeability of the outer membrane to metabolites. Instead, one can take advantage of another peculiarity of yeast mitochondria to measure the permeability of the outer membrane to the metabolite NADH. Unlike mammalian mitochondria, yeast mitochondria have an NADH dehydrogenase on the outer surface of their inner membrane and thus can use cytosolic NADH to deliver the electrons to cytochrome oxidase and ultimately to oxygen.

b. Procedure

The [NADH] is measured as function of time by recording the absorbance at 340 nm (Lee *et al.*, 1998). This is done for both intact mitochondria and hypotonically shocked mitochondria (Fig. 2, left panel). Simultaneous recordings at 400 nm can be used to detect any volume changes in the mitochondria that could affect the measured values at 340 nm. The rate of NADH oxidation is obtained from these oxidation curves by calculating the slope of the tangent at each point. This can be done by determining the least squares fit on all groups of adjacent 13 points by processing the data in a spreadsheet. The slope m is given by:

$$m = \frac{\sum xy - n\overline{xy}}{\sum x^2 - n\overline{x}} \tag{4}$$

where x and y are the [NADH] and time for each data point, $\overline{x}$ and $\overline{y}$ are the mean values, and n is the number of points used to fit the line. This equation can be

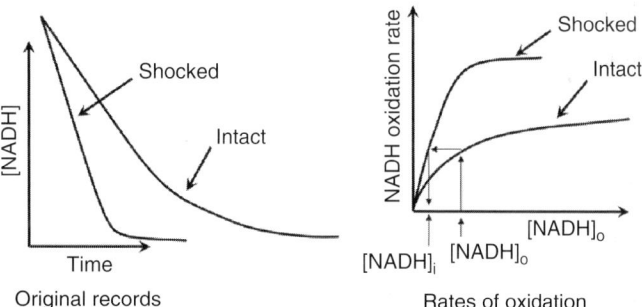

Fig. 2 Illustration of NADH oxidation data used to obtain a quantitative measure of the permeability of the mitochondrial outer membrane. The panel on the left shows an illustration of typical recordings of NADH oxidation by intact and hypotonically shocked mitochondria obtained from the yeast, *S. cerevisiae*. The absorbance at 340 nm was converted to [NADH] by using the extinction coefficient $(6.22 \times 10^3$ molar absorbancy). Rates of NADH oxidation were obtained by determining the tangents at each point from the data in the left panel (see text). These rates were plotted as a function of the [NADH]. For each point on the curve for intact mitochondria one can obtain the medium [NADH] (subscript "o") by dropping a perpendicular and the intermembrane space [NADH] (subscript "i") by finding the point on the shocked mitochondria curve and dropping a perpendicular from that point. This determination was not obtained graphically (as explained in the text) but it is hoped that the illustration with clarify how the information can be obtained. Each experiment takes 10–30 min depending on the permeability of the outer membrane.

written in individual cells in the spreadsheet to obtain the slope for each point (except the first and last seven points) and thus the corresponding rates of NADH oxidation. Note that these rates are in concentration per unit time. The number of points to be used to determine the slopes can be determined by trial and error, although 13 worked well for us. The greater the number of points the less the noise, but if too many points are chosen it is equivalent to filtering the data and the values of the rates are reduced. We chose a number of points that did not reduce the calculated rates.

The rate of NADH oxidation depends on the medium [NADH] and thus the calculated rates are plotted against the [NADH] in the medium (Fig. 2, right panel). From these plots, one can obtain the permeability of the outer membrane. The relationship is as follows:

$$\frac{d[NADH]_o}{dt} = \frac{P([NADH]_o - [NADH]_i)}{Vol} \tag{5}$$

where "o" and "i" refer to the outside compartment (the medium) and the intermembrane space, respectively. Thus, the only value needed in order to calculate the permeability P is the intermembrane space concentration of NADH. This can be obtained using the results from the shocked mitochondria (damaged outer membrane) if one assumes that the rate of NADH oxidation depends on the [NADH] in the vicinity of the enzyme.

A ninth-order polynomial fit of the data relating the measured NADH oxidation rate of shocked mitochondria and the measured medium [NADH] yields an equation that converts oxidation rates to [NADH] close to the enzyme. This equation can then be used to convert the rates of oxidation measured with intact mitochondria to the intermembrane concentration of NADH. This is illustrated in Fig. 2, right panel. Each point on the curve for intact mitochondria has its associated [NADH]_o but the [NADH]_o that resulted in the same respiration rate in shocked mitochondria is the [NADH]_i for the intact mitochondria. Ideally, one could just determine this graphically as shown in the figure. However, because of noise in the data, it must be obtained from the ninth-order polynomial equation. This is also much faster that determining each [NADH]_i graphically. With these values, one has all the parameters needed to calculate a value of P for each data point using Eq. (5).

c. Procedures for Isolating Mitochondria and Obtaining Shocked Mitochondria from Yeast

Mitochondria were isolated from yeast by methods designed to maximize the intactness of the outer membrane (Lee et al., 1998). The method described results in disruption of the mitochondrial outer membrane under conditions that minimize sample degradation (Lee et al., 1998). An aliquot of a suspension of intact mitochondria is diluted with 2 volumes of ice-cold distilled water and allowed to incubate on ice for 10 min. A volume of cold medium that is equal to 5 volumes of the original mitochondrial aliquot is then added. Finally, a volume of ice-cold twofold concentrated medium that is equal to two times the volume of the original mitochondrial aliquot is added. The final volume of these shocked mitochondia is 10 times the original volume.

Reagents

3-mM NADH

Oxidation buffer: 0.65-M sucrose, 10-mM HEPES, 10-mM phosphate, 5-mM KCl, 5-mM MgCl$_2$, pH 7.2 with KOH

Protocol

Add mitochondria to 1 ml (volume depends on cuvette used) of oxidation buffer in a cuvette (25-μg mitochondrial protein). Allow to equilibrate at room temperature for about 20 min. (The time for equilibration may vary. To determine optimal equilibration times, monitor absorbance of the sample at 600 nm until there is no change in absorbance.) Add 40 μl of NADH (30 μM final) and record absorbance at 340 nm. Continue recording until there is no further decline in absorbance. Repeat the experiment with the shocked mitochondria.

3. Measurement of the Permeability of the Outer Membrane to ATP in Yeast Mitochondria Using Adenylate Kinase Activity

a. Principle of the Method

Adenylate kinase catalyzes the reaction (Komarov *et al.*, 2005):

$$2ADP \rightarrow ATP + AMP$$

This enzyme is located in the mitochondrial intermembrane space and thus the rate of reaction depends on the permeability of adjacent membranes to substrates and products. If the adenine nucleotide translocator in the inner membrane is inhibited using atractyloside (20 μg/ml), then VDAC channels in the outer membrane are the only pathway for metabolite access to this enzyme.

Adenylate kinase activity is measured by coupled assays using the enzymes hexokinase and glucose-6-phosphate dehydrogenase, which ultimately produce NADPH in response to added ADP. To generate one molecule of NADPH, two molecules of ADP must move through VDAC from the medium and be converted to AMP and ATP in the intermembrane space. The ATP must return to the medium to be used by hexokinase to phosphorylate glucose. The resulting glucose-6-phosphate is then used by glucose-6-phosphate dehydrogenase to convert NADP to NADPH. The latter is detected by an increase in absorbance at 340 nm. To inhibit oxidation of NADPH, the medium is supplemented with 0.2-mM KCN.

The system must be run in such a way that the coupled enzyme reactions do not limit the rate of adenylate kinase activity. Under these conditions, the reaction can be limited by the rate of entry ADP into and exit of ATP from the intermembrane space.

b. Determination of the Permeability to ATP

At steady state, the rate of NADPH production must be equal to the net efflux of ATP from the mitochondrion and also equal to one-half the net influx of ADP (note the stoichiometry of adenylate kinase). The $[ADP]_o$ at any time during the reaction is known because it must be equal to the initial concentration minus the [NADPH] that is measured continuously (absorbance at 340 nm). Therefore:

$$\frac{d[NADPH]}{dt} = -\frac{d[ADP]}{dt} = \frac{0.5P_D([ADP]_i - [ADP]_o)}{Vol} = \frac{P_T([ATP]_o - [ATP]_i)}{Vol}$$

(6)

Vol is the volume of the medium. To determine the permeability of the outer membrane to ADP, P_D, we need to know $[ADP]_i$, the [ADP] in the intermembrane space. The medium [ATP] is kept essentially at zero through of the action of hexokinase. Thus, to obtain the permeability of the outer membrane to ATP, P_T, we need to know the [ATP] in the intermembrane space $[ATP]_i$. The approach used in Section II does not work here because the rate of the reaction depends on both substrate and product concentrations. Damaging the outer membrane results in obtaining a rate in the presence of medium [ADP] in which [ATP] is also virtually zero (due to the action of hexokinase). This rate cannot be used to measure the $[ADP]_i$ because the [ATP] is changing.

However, the unknown concentrations can be estimated from the kinetic properties of adenylate kinase. The net forward rate of this enzymatic activity will be determined by the $[ADP]_i$ and reduced by $[ATP]_i$. These effects must be consistent, and indeed are constrained by the known kinetic properties of the enzyme. In order to reduce the unknowns, one assumes that the permeability of the outer membrane is the same for ADP, ATP, and AMP. Since the goal is to assess the permeability of the outer membrane to metabolites, this assumption is reasonable. With this assumption and Eq. (6) it follows that:

$$[ADP]_i - [ADP]_o = [ATP]_i = [AMP]_i$$

(7)

This is correct for the $[AMP]_i$ at initial times. Therefore, initial rates must be used.

Kinetic studies of yeast adenylate kinase (Su and Russell, 1968) have shown that the reaction is limited by the rate of conversion of the enzyme from E-ADP$_2$ to E-AMP-ATP. Thus, the binding steps in the adenylate kinase reaction can be assumed to be at equilibrium. Thus:

$$v_0 = k_f[E(ADP)_2] - k_r[E(ATP)(AMP)]$$

(8)

for the reaction in intact mitochondria, where v_0 is the initial rate and k_f and k_r are the forward and reverse rate constants, respectively. For the shocked mitochondria, the second term is zero.

Substituting for the unknowns:

$$v_0 = \frac{V_{max}^f K_D^2 [ADP]^2 - V_{max}^r K_M K_T [ATP]^2}{1 + K_D[ADP] + K_D^2[ADP]^2 + K_T[ATP] + K_M[ATP] + K_M K_T[ATP]^2} \quad (9)$$

where K_D, K_T, and K_M are the equilibrium binding constants for ADP, ATP, and AMP, respectively. For yeast mitochondria these are 3700, 19,000, and 17,000, respectively. V_{max}^f and V_{max}^r are the maximal velocities in the forward and reverse directions, respectively. The equilibrium constants are obtained from the reciprocal of the Michaelis constants (Su and Russell, 1968); the ratio of the V_{max} values is known to be 0.6 at neutral pH (Chiu *et al.*, 1967). Additional expressions needed to solve for the needed concentrations numerically are:

$$\text{Ratio} \quad = \frac{V_{max}^f}{V_{max}^r} \quad (10)$$

$$V_{max}^f = \frac{v_0^{shock}(1 + K_D[ADP]_o + K_D^2[ADP]_o^2)}{K_D^2[ADP]_o^2} \quad (11)$$

$$v_0 = \frac{d[NADPH]}{dt}Vol \quad (12)$$

where v_0^{shock} is the initial velocity measured for shocked mitochondria. Note that here the [ADP] is different from that in Eq. (8) because the mitochondria are shocked and thus the enzyme is in the medium.

The equations are easily solved with the aid of a spreadsheet program or math program. The unknown $[ADP]_i$ can be replaced, using Eq. (6), by the known medium concentration and the $[ATP]_i$, thus eliminating one of the unknowns. The initial rate for shocked mitochondria is used to calculate V_{max}^f using Eq. (11). Then an $[ATP]_i$ is selected for Eq. (8) that yields an initial velocity for adenylate kinase in intact mitochondria that matches the measured value. The other concentrations can be calculated from this value, along with the permeability to nucleotides.

c. Protocol to Measure Adenylate Kinase Activity in Isolated Yeast Mitochondria
Reagents

12.5-mM ADP hexokinase and glucose-6-phosphate dehydrogenase, each dissolved in water at 1 unit per microliter (can be stored frozen; survives freeze thawing).

Reaction medium: 0.6-M sucrose, 50-mM TrisCl, 5-mM $MgSO_4$, 10-mM glucose, 0.2-mM NADP, 20-mg/ml atractylosides, 0.2-mM KCN, pH 7.5.

Twofold reaction medium: 1.2-M sucrose, 100-mM TrisCl, 10-mM $MgSO_4$, 20-mM glucose, 0.4-mM NADP, 40-mg/ml atractylosides, 0.4-mM KCN, pH 7.5.

Protocol

Mitochondria are isolated from yeast by methods designed to maximize the intactness of the outer membrane (Lee *et al.*, 1998). These mitochondria (10- to 14-mg protein/ml) are diluted 40-fold with reaction medium (room temperature) just before use. Shocked mitochondria are generated by mixing the initial mitochondrial suspension with 9 volumes of ice-cold distilled water and incubating on ice for 5 min. Thereafter, 10 volumes of twofold reaction medium (at room temperature) are added followed by 20 volumes of reaction medium (at room temperature). This results in the same 40-fold dilution and restores the original solute concentrations.

To 1 ml of mitochondrial suspension, add 20 μl of ADP (250 μM final concentration) and 10 μl of the hexokinase/glucose-6-phosphate dehydrogenase mixture (10 units each). The increase in [NADPH] is followed by recording the absorbance at 340 nm. There may be an initial rapid increase in absorbance due to the presence of a small amount of ATP in the ADP solution (Fig. 3). A 1% contamination produces an obvious effect. Thus, the initial rate is assessed after the initial rapid increase. By overlapping the curves obtained with intact and shocked mitochondria one can immediately get a sense of the level of permeability of the outer membrane to ATP. Figure 3 shows an example of data collected with intact and shocked mitochondria from the yeast, *S. cerevisiae*, lacking VDAC (nul) and yeast expressing VDAC1 (POR1). The theory extracts the numerical values.

D. Controls

The methods described above are designed to assess the permeability of the mitochondrial outer membrane to metabolites and, in doing so, to probe the

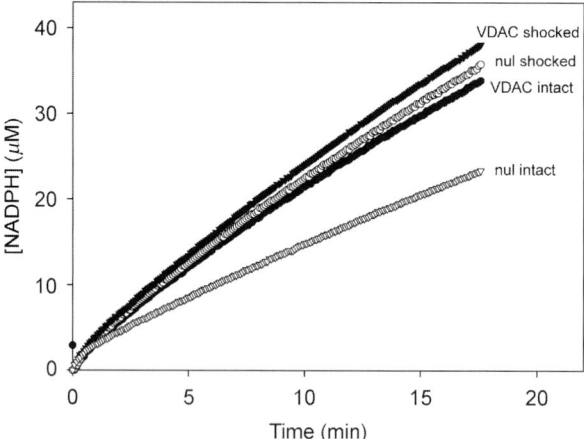

Fig. 3 Sample results of experiments to measure the permeability of the mitochondrial outer membrane by measuring adenylate kinase activity. Mitochondria were isolated from *S. cerevisiae* cells, both wild type (VDAC containing) and lacking VDAC1 (POR1). The experiment was run as described in the text.

functional state of VDAC. The changes that occur on outer membrane damage provide some confidence that the measurements provide the desired information. However, further evidence exists. The knockout of VDAC1(POR1) in yeast drops the permeability measured by a factor of 20 (Lee *et al.*, 1998). The remaining permeability could be attributed to damaged mitochondria or to VDAC2. A recently discovered blocker of VDAC drops the permeability of the outer membrane (measured by method 1) of rat liver mitochondria by a factor of 6–7 (Tan *et al.*, 2007). These findings provide further evidence that VDAC is the pathway responsible for the permeability of the outer membrane. They also help to validate the methods.

III. VDAC Activity After Reconstitution into Phospholipid Membranes

A. Rationale

The detailed properties of VDAC can only be determined after purification and reconstitution into phospholipid membranes. Electrophysiological recordings of single channels allow one to determine single-channel conductance and selectivity. Recordings on small populations of VDAC channels yield the best estimates of voltage-dependent parameters. Effective pore size is best estimated from experiments with liposomes. Studies on reconstituted VDAC channels also allow one to study the influence of test molecules or macromolecules on the properties of VDAC.

B. Purification of VDAC

VDACs have been successfully purified from a wide variety of mitochondria by the method of Freitag *et al.* (1983). This rapid, one-step procedure yields material of sufficient purity for functional, reconstitution experiments. Higher levels of purity can be obtained by additional purification steps (Gincel *et al.*, 2000).

At this writing, I am unaware of a satisfactory way of obtaining VDAC with normal function by refolding denatured protein obtained from inclusion bodies. One can obtain channel-forming activity but the properties are different from those observed for VDAC purified from mitochondria. The properties change markedly depending on the method used to refold the protein.

1. Reagents

Shock medium: 1-mM KCl, 1-mM HEPES, pH 7.5

Solubilization medium: 2.5% Triton X-100, 50-mM KCl, 10-mM Tris, 1-mM EDTA, 15% DMSO, pH 7.0

Siliconizing solution such as Sigmacote

1:1 dry mixture of hydroxyapatite and celite

2. Protocol

Mitochondria are hypotonically shocked to remove soluble proteins from the intermembrane and matrix space. The mitochondrial pellet is suspended in about 50-fold excess of ice-cold shock medium and the membranes are concentrated by centrifugation at $24,000 \times g$ for 20 min. These membranes can be stored for extended periods at $-80\,^{\circ}$C if suspended in the shock medium supplemented with 15% DMSO. The pellet (about 1 mg of protein) is resuspended in 0.7 ml solubilization medium and incubated on ice for 30 min with occasional vortexing. The suspension is centrifuged (generally $14,000 \times g$ for 30 min in a microfuge) at $4\,^{\circ}$C to remove nonsolubilized material. The supernatant is applied to 1 ml of dry hydroxyapatite/celite in a Pasteur pipette (end plugged with a small amount of glass wool and siliconized) at $4\,^{\circ}$C. The effluent is collected and another 0.7 ml of solubilization medium is applied to the column. The second effluent is mixed with the first and the solution is aliquotted and stored at $-80\,^{\circ}$C. Some VDAC proteins have cysteine residues and so 1-mM DTT could be added to the above solutions.

C. Reconstitution of VDAC into Planar Membranes

1. General Considerations

Electrophysiological recordings of VDAC channels have traditionally been made on channels reconstituted into planar phospholipid membranes (Colombini, 1979; Schein *et al.*, 1976). Recordings with patch-clamp techniques of VDAC in liposomes have been made (Wunder and Colombini, 1991) but VDAC channels have a tendency to rapidly disappear from the patch. Patch-clamp recordings of mitochondrial outer membranes reveal channels with properties that differ from those of purified VDAC (Jonas *et al.*, 2005; Tedeschi *et al.*, 1989). The reason for this is unclear but may have to do with the presence of associated proteins. In one case, VDAC was isolated along with what appeared to be a tightly bound protein. Protease treatment converted the properties of this reconstituted channel to more classical VDAC behavior (Elkeles *et al.*, 1997).

Generally, VDAC needs to be solubilized in a mild nonionic detergent to be reconstituted into planar membranes. However, reconstitution without detergent results in properties that are indistinguishable from those seen when detergents are used (Schein *et al.*, 1976).

2. Equipment

Teflon chamber: Although some chambers are available commercially (e.g., Warner Instruments), the ones used to make membranes from monolayers are made by special order. Contact the author for details.

Electrodes: Mini calomel electrodes from Fisher Scientific work well. They need to be matched so that their electrode asymmetry is 1 mV or less. They can also be made from silver wire by coating them with AgCl and then generating a KCl bridge with 3-M KCl.

Dissecting microscope mounted on stand with side arm to observe the formation of the membrane.

Stimulator, recording amplifier, filters, and so on: Various systems are commercially available from Axon Instruments, Warner Instruments, and World Precision Instruments.

Faraday cage: Wire mesh cage either home made or purchased commercially.

Air table: Optional for VDAC recordings because of their large single-channel conductance.

3. Reagents

Lipids must be of high quality. All lipids should be purchased from Avanti Polar Lipids. For VDAC studies, a lipid mixture that resembles the lipid composition of the mitochondrial outer membrane is the polar extract of soybean phospholipids (aka asolectin). Cholesterol is added separately. Often diphytanoyl phosphatidylcholine (DPhPC) is also added to improve membrane stability. A good mixture for easy membrane formation is 1% phospholipid (DPhPC:asolectin; 1:1) and 0.1% cholesterol in hexane.

Coating solution: 20% petrolatum in petroleum ether.

4. Protocol

Planar membranes can be made by a variety of published methods. Variations of the monolayer method of Montal and Mueller (1972) work very reliably (Colombini, 1987). A schematic of the setup is shown in Fig. 4. Two identical Teflon hemi-chambers are clamped together and squeeze between them a thin partition made of Teflon or Saran. There is a hole in the Teflon or Saran partition that is ca. 0.1 mm in diameter. The partition is coated with a thin layer of petrolatum (high-molecular-weight hydrocarbon) from a solution of petrolatum in petroleum ether. The petroleum ether is allowed to evaporate for 5 min. Syringes are used to inject the aqueous solution into each hemi-chamber to just completely cover the floor of the chamber. To form the monolayers, ca. 25 μl of the lipid solution is layered on top of the aqueous solution in each hemi-chamber and the hexane is allowed to evaporate, resulting in a small visible nodule of excess lipid floating on the surface of the water. The level of the *trans* compartment is raised by injecting solution into the subphase, until the monolayer is just above the hole. Then, the level of the *cis* compartment is raised to the same level by injecting solution into the subphase. This generates a bilayer, the planar membrane, in the hole in the Telfon or Saran partition.

VDAC purified as described in Section III.B.2 is added directly to the aqueous phase bathing the planar phospholipid membrane. Typically 1–4 μl of the VDAC solution is added to 5 ml of aqueous solution while the solution is being stirred with a Teflon-coated magnetic bar. VDAC channels insert spontaneously into the planar

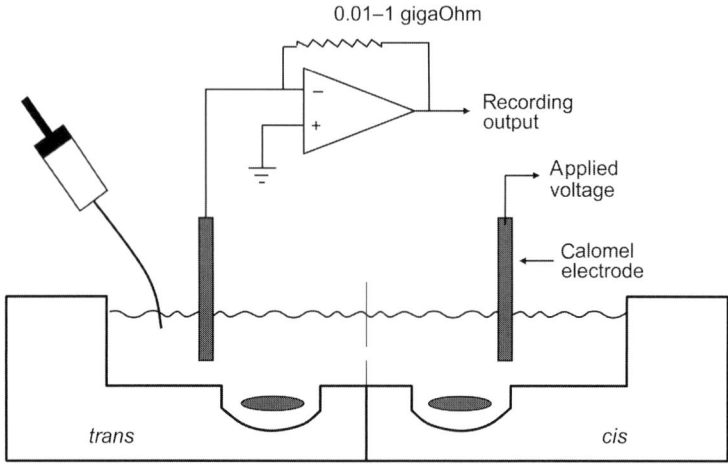

Fig. 4 Experimental setup for recording the conductance of VDAC channels reconstituted into planar membranes. The Teflon chamber consists of two hemi-chambers separated by a thin partition containing the small hole (0.1mm in diameter) across which a phospholipid membrane is formed. The aqueous solutions on the two sides are shown after both levels have been raised and allowing for the formation of the planar membrane. The bottom of each hemi-chamber has a small well containing the Teflon-coated stir bar for mixing the solutions. A DC powered geared motor is used to rotate a magnetic bar below the chamber and this bar rotates both stir bars simultaneously. The rate of mixing is varied by varying the voltage applied to the motor. The electrodes are made from calomel and contain salt bridges. Syringes are used to control the level of the solution. Only one syringe is illustrated for clarity. No needles are used; rather Tygon tubing is used to connect the syringe to the solution. The electronics illustrated is that of a high-quality operational amplifier in the inverted mode. One can use a commercial amplifier instead. The voltage is clamped and the current measured. The compartment held at virtual ground by the amplifier is labeled "*trans*" and the compartment whose voltage is controlled by the operator is labeled "*cis*."

membrane. Warming the VDAC solution to room temperature before the addition helps the insertion process. Typically, the voltage is clamped at 10 mV during the insertion period. Depending on the source of VDAC and its concentration, insertion of channels can occur almost immediately or may require an hour or more.

The insertion efficacy is influenced by many variables. Some sources of VDAC insert more readily than others. VDAC tends to insert more easily in the presence of high salt concentrations such as 1-M salt. The amount of detergent added also influences the probability of insertion. The detergent Triton X-100 is particularly effective at facilitating VDAC insertion. VDAC in the planar membrane facilitates further insertion of VDAC, and affects the orientation of VDAC insertion (Xu and Colombini, 1996; Zizi *et al.*, 1995).

D. Electrophysiological Recordings

1. General Considerations

Basic electrophysiological characterization of VDAC involves three measurements: single-channel conductance, reversal potential as a measure of selectivity, and voltage dependence parameters. All these measurements can be made under an almost limitless variety of conditions. If one wants to compare with literature values (Colombini, 1989), then the most common conditions are single-channel conductance and voltage-dependent measurements in 1.0-M KCl; reversal potential measurement in the presence of 1.0-M KCl versus 0.10-M KCl. Generally, the solutions also contain some buffer and divalent salts. VDAC selectivity values (permeability ratios for anions over cations) are much higher if measured in lower salt concentrations and in shallower salt gradients. Selectivity values are also influenced by the choice of membrane lipid, especially the presence of charge. The voltage-dependent parameters, the steepness of the voltage dependence n, and the voltage at which half the channels are closed, V_0, are strongly influenced by the presence of a salt gradient (Zizi *et al.*, 1998).

The voltage gating of VDAC involves two distinct gating processes, one at positive potentials and the other at negative potentials. The channels are open at low potentials and closed at high potentials, but closed states at positive potentials are different from those at negative potentials. This is not explained by a simple electrostriction phenomenon. One cannot go from a positive to a negative closed state without reopening the channel (Colombini, 1986). More detailed information on the molecular basis for gating in VDAC is available in the publications of Song *et al.* (1998a,b).

There are two ways to measure the voltage-dependent parameters for VDAC: the use of distinct steps of voltage and the use of triangular voltage waves. When steps of voltage are applied to a multichannel membrane, one typically begins with a voltage at which the channel is open (typically between 10 and -10 mV) and then steps the voltage up to the desired value. The conductance is allowed to relax to a steady value and that value is assumed to be an equilibrium value reflecting the conductance through some closed channels and some open channels in dynamic equilibrium. This process is repeated until a voltage range is spanned from voltages where essentially all the channels are open to voltages where they are essentially all closed. The problem with this approach stems from the complexity of the gating of VDAC. There is only one high-conductance, anion-selective, open state and a large set of lower-conducting and usually cation-preferring states called closed states. Some of these are readily accessible but others are only accessible after prolonged application of an elevated potential. The voltage step approach results in VDAC occupying a different population of states at low and high voltages. This becomes evident when single channels are analyzed.

The use of triangular voltages is not the ideal method but the best available method. It takes advantage of the large difference in rates between the voltage-dependent closing process (milliseconds to minutes) and the largely voltage-independent opening process (submillisecond). It also probes the system in the time scale of 1 min (for channel reopening) by applying triangular waves or 2–5 mHz. Thus, one is probing the voltage dependence of rapidly accessible closed states. (Physiologically, the other states are going to be important as well.) One analyzes the part of the recording where the channels are reopening so that the record is close to an equilibrium between rapidly accessible states. This approach yields results that are essentially independent of the speed of the triangular wave from 1 to 10 mHz.

The results are fit to a Boltzmann distribution following the fundamental work of Hodgkin and Huxley (1952).

2. Simple Protocol for Observing Basic Properties of VDAC

a. Reagents
Chamber solution: 1.0-M KCl, 1-mM $MgCl_2$, 5-mM HEPES, pH 7.5

b. Protocol
Generate a planar membrane using the desired lipid mixture in the presence of the chamber solution. Begin recording the current through the membrane. Insert VDAC as described in Section III.C.2 with an applied voltage of 10 mV. If one channel inserts (4-nS conductance), step the voltage to 40 mV and you should observe a fourfold increase in current (ohmic behavior). With time the current should tend to drop to a lower value, about 40% of the original. The current should move stepwise between discrete levels. There should be one high-current level corresponding to a conductance of 4 nS (the open state) and a variety of low-conducting levels (closed states). When the voltage is returned to 10 mV, the channel should reopen quickly and return to its original value. Usually this opening transition is only detected at high time resolution (a few milliseconds). The closing rate at 40 mV is in the seconds time range. The closing rate increases as the applied voltage is increased. Voltages greater that 70 mV are quite likely to result in membrane breakdown and this likelihood increases with the magnitude of the applied voltage. If the channel is held closed for several minutes, the reopening rate should be much slower and the application of an opposite potential may be

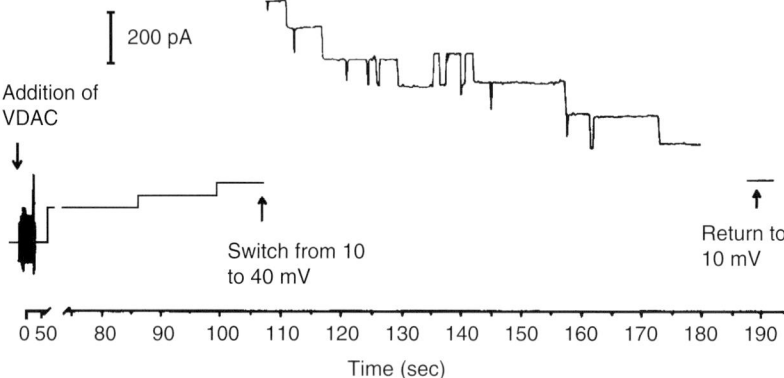

Fig. 5 An example of a recording of the insertion of 5 VDACs and their voltage-dependent closure. These channels were isolated from *Neurospora crassa* mitochondria. *N. crassa* VDAC channels often insert as groups of three channels (see first insertion). Addition of the VDAC-containing sample to the *cis* compartment while stirring produces a noisy record. When the stirring is stopped, the noise is no longer there. During the addition and channel-insertion period, the applied voltage was 10 mV. At the indicated time point, the voltage was raised to 40 mV. After all the channels closed, the voltage was reduced to 10 mV.

needed. VDAC develops a memory of the applied voltage that becomes erased when the channel reopens. This voltage gating should be detected at both positive and negative applied potentials. During the experiment, more channels may insert, making the recordings more complex.

If many channels insert, the same voltage step protocol can be used, but one sees the current changes resulting from the action of many channels. Figure 5 shows an experiment in which five channels inserted. A group of three channels inserted in unison (first current step then two single channels inserted). When the voltage was increased to 40 mV, there were clearly five levels of current drop corresponding to the closure of five channels. When the voltage was returned to 10 mV, the current returned instantly to the level it was prior to the transition to 40 mV because all five channels opened rapidly.

If hundreds of channels have inserted, the current changes will look like exponentials as the population is large enough to smooth out the stochastic variation.

IV. Summary

Quantitative measurements of the permeability of the mitochondrial outer membrane to metabolites allow one to assess the ability of mitochondria to exchange metabolites with the cytosol of the cell. This can be used to determine the influence of treatments or conditions that alter this permeability to regulate cellular metabolic rate or control the initiation of the apoptotic process by closing VDAC. By comparing effects on pure VDAC with those in the outer membrane, one can discover the existence of endogenous factors that influence VDAC's response. Naturally the isolated mitochondrion lacks any cytosolic influences but cytosolic factors could be added back to determine their mode of action.

Acknowledgments

This work was supported by NIH grants: AI49388, NS42319, and NS42025.

References

Chiu, C., Su, S., and Russel, P. J. (1967). Adenylate kinase from baker's yeast: I. Purification and intracellular location. *Biochim. Biophys. Acta* **132**, 361–369.

Colombini, M. (1979). A candidate for the permeability pathway of the outer mitochondrial membrane. *Nature* **279**, 643–645.

Colombini, M. (1986). Voltage gating in VDAC: Toward a molecular mechanism. *In* "Ion Channel Reconstitution" (C. Miller, ed.), Section 4 , Chapter 10. Plenum Press, New York.

Colombini, M. (1987). Characterization of channels isolated from plant mitochondria. *Methods Enzymol.* **148**, 465–475.

Colombini, M. (1989). Voltage gating in the mitochondrial channel, VDAC. *J. Membr. Biol.* **111**, 103–111.

Colombini, M. (1994). Anion channels in the mitochondrial outer membrane. *In* "Current Topics in Membranes" (W. G. Guggino, ed.), Vol. 42, Chapter 4, pp. 73–101. Academic Press, Inc., San Diego, CA.

Colombini, M. (2004). VDAC: The channel at the interface between mitochondria and the cytosol. *Mol. Cell. Biochem.* **256**, 107–115.

Douce, R., Bourguignon, J., Brouquisse, R., and Neuburger, M. (1987). Isolation of plant mitochondria: General principles and criteria of integrity. *Methods Enzymol.* **148**, 403–415.

Elkeles, A., Breiman, A., and Zizi, M. (1997). Functional differences among wheat voltage-dependent anion channel (VDAC) isoforms expressed in yeast. Indication for the presence of a novel VDAC-modulating protein? *J. Biol. Chem.* **272**, 6252–6260.

Freitag, H., Benz, R., and Neupert, W. (1983). Isolation and properties of the porin of the outer mitochondrial membrane from neurospora crassa. *Methods Enzymol.* **97**, 286–294.

Gincel, D., Silberberg, S. D., and Shoshan-Barmatz, V. (2000). Modulation of the voltage-dependent anion channel (VDAC) by glutamate. *J. Bioenerg. Biomembr.* **32**, 571–583.

Hodgkin, A. L., and Huxley, A. F. (1952). A quantitative description of membrane current and its application to conduction and excitation in nerve. *J. Physiol. (Lond.)* **117**, 500–544.

Jonas, E. A., Hickman, J. A., Hardwick, J. M., and Kaczmarek, L. K. (2005). Exposure to hypoxia rapidly induces mitochondrial channel activity within a living synapse. *J. Biol. Chem.* **280**, 4491–4497.

Komarov, A. G., Deng, D., Craigen, W. J., and Colombini, M. (2005). New insights into the mechanism of permeation through large channels. *Biophys. J.* **89**, 3950–3959.

Lee, A., and Colombini, M. (1997). A unique method for determining the permeability of the mitochondrial outer membrane. *Methods Cell Sci.* **19**, 71–81.

Lee, A., Xu, X., Blachly-Dyson, E., Forte, M., and Colombini, M. (1998). The role of yeast VDAC genes on the permeability of the mitochondrial outer membrane. *J. Membr. Biol.* **161**, 173–181.

Lee, A., Zizi, M., and Colombini, M. (1994). β-NADH decreases the permeability of the mitochondrial outer membrane to ADP by a factor of 6. *J. Biol. Chem.* **269**, 30974–30980.

Lemasters, J. J., and Holmuhamedov, E. (2006). Voltage-dependent anion channel (VDAC) as mitochondrial governator-thinking outside the box. *Biochim. Biophys. Acta* **1762**, 181–190.

Mannella, C. A., Forte, M., and Colombini, M. (1992). Toward the molecular structure of the mitochondrial channel, VDAC. *J. Bioenerg. Biomembr.* **24**, 7–19.

Montal, M., and Mueller, P. (1972). Formation of bimolecular membranes from lipid monolayers and a study of their electrical properties. *Proc. Natl. Acad. Sci. USA* **69**, 3561–3566.

Pfaff, E., Heldt, H. W., and Klingenberg, M. (1969). Adenine nucleotide translocation of mitochondria. Kinetics of the adenine nucleotide exchange. *Eur. J. Biochem.* **10**, 484–493.

Schein, S. J., Colombini, M., and Finkelstein, A. (1976). Reconstitution in planar lipid bilayers of a voltage-dependent anion-selective channel obtained from *Paramecium* mitochondria. *J. Membr. Biol.* **30**, 99–120.

Song, J., Midson, C., Blachly-Dyson, E., Forte, M., and Colombini, M. (1998). The sensor regions of VDAC are translocated from within the membrane to the surface during the gating processes. *Biophys. J.* **74**, 2926–2944.

Song, J., Midson, C., Blachly-Dyson, E., Forte, M., and Colombini, M. (1998). The topology of VDAC as probed by biotin modification. *J. Biol. Chem.* **273**, 24406–24413.

Su, S., and Russell, P. J. (1968). Adenylate kinase from baker's yeast. 3. Equilibria: Equilibrium exchange and mechanism. *J. Biol. Chem.* **243**, 3826–3833.

Tedeschi, H., Kinnally, K. W., and Mannella, C. A. (1989). Properties of channels in the mitochondrial outer membrane. *J. Bioenerg. Biomembr.* **21**, 451–459.

Tan, W., Lai, J. C., Miller, P., Stein, C. A., and Colombini, M. (2007). Phosphorothioate oligonucleotides reduce mitochondrial outer membrane permeability to ADP. *Am. J. Physiol.* (in press).

Werkheiser, W. C., and Bartley, W. (1957). The study of the steady-state concentration of internal solutes of mitochondria by rapid centrifugal transfer to a fixation medium. *Biochem. J.* **66**, 79–91.

Wunder, U. R., and Colombini, M. (1991). Patch-clamping VDAC in liposomes containing whole mitochondrial membranes. *J. Membr. Biol.* **123**, 83–91.

Xu, X., and Colombini, M. (1996). Self-catalyzed insertion of proteins into phospholipid membranes. *J. Biol. Chem.* **271**, 23675–23682.

Zizi, M., Byrd, C., Boxus, R., and Colombini, M. (1998). The voltage-gating process of the voltage-dependent anion channel is sensitive to ion flow. *Biophys. J.* **75**, 704–713.

Zizi, M., Thomas, L., Blachly-Dyson, E., Forte, M., and Colombini, M. (1995). Oriented channel insertion reveals the motion of a transmembrane beta strand during voltage gating of VDAC. *J. Membr. Biol.* **144**, 121–129.

CHAPTER 13

Methods for Studying Iron Metabolism in Yeast Mitochondria

Sabine Molik, Roland Lill, and Ulrich Mühlenhoff

Institut für Zytobiologie und Zytopathologie
Philipps-Universität Marburg, 35033 Marburg, Germany

I. Introduction

Iron–sulfur (Fe/S) clusters are versatile, ancient cofactors of proteins that are involved in electron transport, enzyme catalysis, and regulation of gene expression (Beinert *et al.*, 1997). Recent years have shown that the synthesis of these cofactors and their insertion into apoproteins involve the function of complex cellular machineries in all kingdoms of life (reviewed by Balk and Lill, 2004; Johnson and Dean, 2004; Lill and Muhlenhoff, 2005, 2006). The budding yeast *Saccharomyces cerevisiae* serves as a key model organism for the study of this novel biochemical pathway in eukaryotes. Mitochondria play a central role in this pathway, as they are essential for the maturation of all cellular Fe/S proteins. The so-called mitochondrial iron–sulfur cluster (ISC) assembly machinery is responsible for the *de novo* synthesis of Fe/S clusters and for the insertion of these cofactors into mitochondrial Fe/S apoproteins (Fig. 1). This system is also involved in the maturation of Fe/S proteins that are located outside the mitochondria, in the cytosol or in the nucleus. A mitochondrial export system and a recently discovered cytosolic Fe/S protein assembly (CIA) system specifically participate in the maturation of cytosolic and nuclear Fe/S proteins (Fig. 1). Of the ∼20 assembly components known to date, many are encoded by essential genes, including several components of mitochondria. This indicates that the process is indispensable for life. In fact, the maturation of cellular Fe/S proteins is the only mitochondrial function known so far that is essential for viability of eukaryotes (Lill and Muhlenhoff, 2005). In *S. cerevisiae*, defects in Fe/S protein maturation result in respiratory deficiency due to a collapse of the respiratory chain and/or the citric acid cycle, and auxotrophies for certain amino acids and vitamins, which require Fe/S proteins for their biosynthesis (Jensen and Culotta, 2000; Kispal *et al.*, 1999). Severe defects result in cell death. The first essential Fe/S protein that was identified in budding yeast (and is not a member of the ISC or CIA machineries) is Rli1p, which plays a key role in ribosome biogenesis (Kispal *et al.*, 2005; Yarunin *et al.*, 2005). The finding that this protein is essential for cell viability indicates that Fe/S biogenesis is also an essential process.

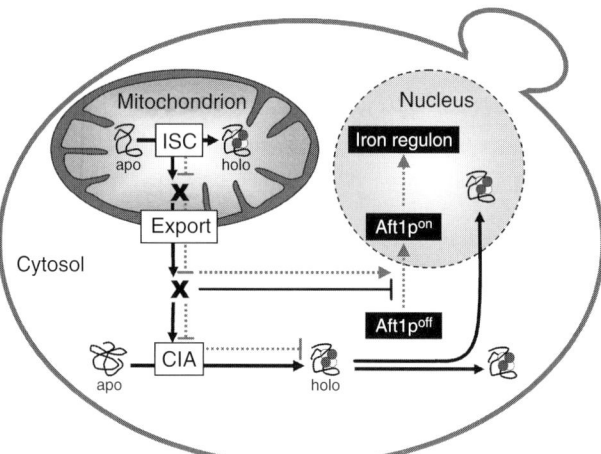

Fig. 1   Fe/S protein biogenesis and Aft1/2p-dependent gene regulation in *S. cerevisiae*. In yeast and other eukaryotes, the mitochondrial ISC assembly system is required for the maturation of mitochondrial Fe/S protein. An unknown component (X), which is produced and exported by the ISC assembly and ISC export systems (export), is required for the maturation of cytosolic and nuclear Fe/S proteins by the cytosolic Fe/S protein assembly system (CIA). In addition, this component is required for attenuation of the iron regulon by the transcriptional activators Aft1p/Aft2p. Defects in the ISC assembly and export systems induce the import of Aft1p into the nucleus and a constitutive transcriptional activation of many genes involved in iron uptake and storage. The connection between Fe/S protein biogenesis, the mechanism for iron sensing, and the transcriptional regulation of iron-responsive genes are not well understood.

In *S. cerevisiae*, defects in the mitochondrial ISC assembly and export apparatus are associated with low levels of cellular heme (the second most abundant iron-containing cofactor of the cell), mitochondrial iron overload, and a massive constitutive transcriptional deregulation of iron-responsive genes that is similar to the transcriptional response of yeast to iron deprivation (Shakoury-Elizeh *et al.*, 2004). These phenotypes strongly suggest that the mitochondrial Fe/S protein assembly system plays a crucial role in the sensing of iron, in the transcriptional regulation of cellular iron homeostasis, and in the control of cellular heme levels (Lange *et al.*, 2004). However, the underlying regulatory mechanisms for these processes are unknown. In yeast, defects in the mitochondrial ISC assembly and export apparatus elicit the induction of iron-uptake genes, the so-called "iron regulon." The iron regulon comprises a set of ~30 genes with products that function in ionic iron uptake at the cell surface (*FET3*, *FTR1*, *FRE1*, and *FRE2*), siderophore uptake (*ARN1–4*, *FIT1–3*), iron transport across the vacuolar and mitochondrial membranes (*FET5*, *FTH1*, and *MRS4*), and the modulation of the mRNA stability of iron-dependent genes (Rutherford and Bird, 2004; Rutherford *et al.*, 2003; Shakoury-Elizeh *et al.*, 2004). The induction of these genes is a physiological response by yeast to iron deprivation that contributes significantly to the coordinated adaptation of cellular iron usage and uptake to low iron conditions. The expression of the iron regulon is controlled by the iron-sensing transcriptional activators Aft1p and Aft2p (Rutherford and Bird, 2004). Aft1p resides in the cytoplasm under iron-replete conditions, but shuttles to the nucleus on iron deprivation (Yamaguchi-Iwai *et al.*, 2002). The precise mechanism of the iron-responsive activation of Aft1p is unclear, since it is unknown what this transcription factor is actually sensing. Analyses have shown that Aft1 (and possibly Aft2p) are regulated by an unidentified key molecule that is produced and sequestered by the mitochondrial ISC systems, and which functions as a regulatory signal for the sensing of iron and for the expression of iron-regulated genes in the nucleus (Fig. 1) (Rutherford *et al.*, 2005). The constitutive activation of Aft1/2p in ISC assembly- or ISC export-deficient cells results in an increased iron uptake linked with mitochondrial iron accumulation (Kispal *et al.*, 1997; Lill and Muhlenhoff, 2005; Rutherford *et al.*, 2005).

The present compendium includes assays for the analysis of *de novo* synthesis of Fe/S clusters and heme formation in *S. cerevisiae* under *in vivo* and *in vitro* conditions in Section II. Methods to determine the mitochondrial iron content and reporter assays to analyze iron-dependent gene expression are described in Sections III and IV. These approaches are crucial to elucidate the mechanisms underlying the maturation of Fe/S proteins and may aid in the identification of new members of this evolutionary ancient process.

II. Determination of Cellular Fe/S Protein Formation and Heme Synthesis

The phenotypes outlined above show that cellular Fe/S protein maturation is tightly connected to mitochondrial respiration, heme synthesis, and cellular iron homeostasis in *S. cerevisiae*. A full understanding of Fe/S cluster maturation thus requires an integrative approach that takes all of these processes into account. This chapter provides a comprehensive compilation of the most important routine methods used for the analysis of these processes in *S. cerevisiae* in our laboratory.

We first describe standard methods for analyzing the *de novo* maturation of Fe/S proteins by radiolabeling iron-starved yeast cells with ^{55}Fe *in vivo* (Fig. 2). Following the radiolabeling, cells are lysed, and the ^{55}Fe/S proteins are immunoprecipitated with specific antibodies. The amount of radioactive iron that is copurified with the protein is quantified by liquid scintillation counting. The uptake of ^{55}Fe by the cells during the labeling reaction may serve as an indication of cell viability. This semi-quantitative method can be used for the analysis of endogenously synthesized proteins (e.g., aconitase and Leu1p) and of overproduced proteins (e.g., Bio2p, Yah1p, Isu1, Nbp35p, Nar1p). If suitable antibodies are not available, appropriate tags can be used. In this context, we have good experience with the use of hemagglutinin (HA)- or tandem affinity purification (TAP)-tagged proteins f analysis of both mitochondrial and cytosolic Fe/S proteins. Finally, the formation of heme can be analyzed by this *in vivo* labeling assay. Due to its high solubility in organic solvents, radiolabeled heme can be extracted quantitatively from a cell extract into the organic phase. The degree of incorporation of ^{55}Fe into heme is quantified by liquid scintillation counting of the organic phase.

In addition, Section II includes two assays for the analysis of Fe/S cluster synthesis activities in mitochondrial extracts *in vitro*. The first assay is based on radiolabeling with radioactive ^{55}Fe or ^{35}S-cysteine of yeast mitochondria overproducing an endogenous Fe/S protein (Muhlenhoff *et al.*, 2002a). Mitochondria are lysed with detergent and incubated with cysteine, iron, NADH, and DTT under anaerobic conditions. Radiolabeled Fe/S proteins are immunoprecipitated with specific antibodies and are quantified by scintillation counting. To obtain mitochondria with a large portion of Fe/S proteins in the apoform, organelles prepared from yeast cells cultivated in iron-free medium are optimal. In the second assay, a soluble apoferredoxin is added to the mitochondrial extract and the reconstituted holoferredoxin is subsequently purified either by anion exchange or by native gel electrophoresis (Lutz *et al.*, 2001). This method can be used with or without radiolabel and is suitable for mitochondria or cell extracts from a variety of organisms with only minor adjustments. However, in the case of mammalian tissue or cell cultures or of pathogenic protists, experiments may be complicated by low sample amounts.

A. Determination of Cellular Fe/S Cluster Formation by Radiolabeling of Yeast Cells *In Vivo*

Materials

"Iron-free" minimal medium for growth of *S. cerevisiae*. This medium corresponds to regular synthetic complete (SC) medium (Sherman, 2002), lacking iron chloride. A ready-made powder is commercially available (Formedium, United Kingdom).

^{55}FeCl$_3$ (NEN/Perkin-Elmer)

Citrate buffer: 50-mM sodium citrate; 1-mM EDTA, pH 7.0

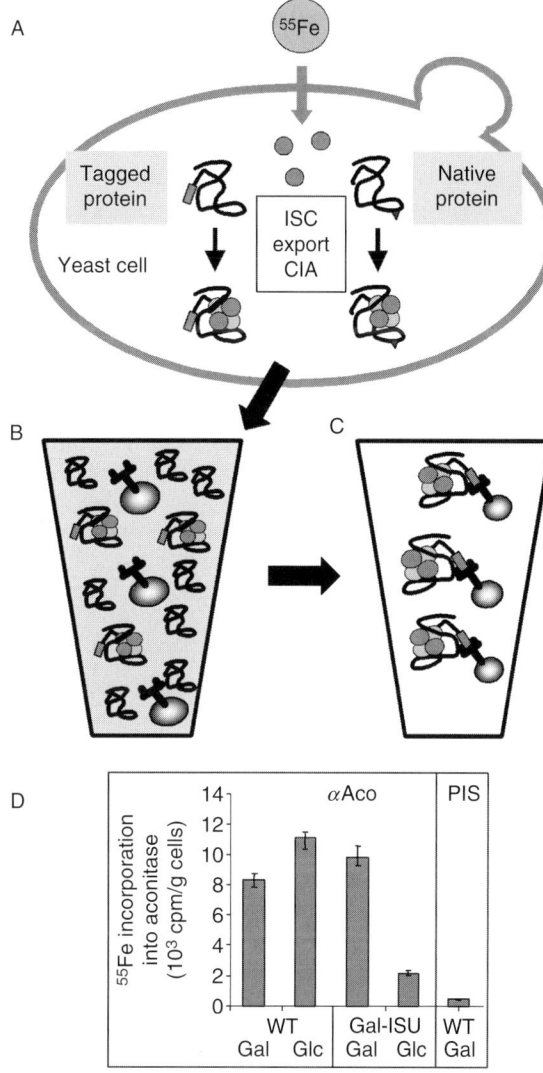

Fig. 2 Flow sheet for the analysis of *de novo* Fe/S protein formation by radiolabeling of yeast cells *in vivo*. (A) Yeast cells that were cultivated in iron-free medium are incubated with radioactive ^{55}Fe for at least 1 h. During this time, the radioactive iron is taken up and incorporated into Fe/S apoproteins by the mitochondrial ISC assembly, ISC export, and CIA systems. (B) Subsequently, a cell extract is prepared and the labeled ^{55}Fe/S reporter proteins are immunoprecipitated with specific antibodies coupled to protein-A Sepharose. (C) The radioactive ^{55}Fe that is copurified with the immunobeads is quantified by scintillation counting. Both tagged and untagged (native) proteins can be analyzed in this assay. (D) The conditional ISC mutant Gal-ISU1/Δisu2 (Gal-ISU) was radiolabeled under permissive conditions in the presence of galactose (Gal) and under repressive conditions in the presence of glucose (Glc), and the amount of iron incorporation into the mitochondrial aconitase was determined. Wild-type cells (WT) were analyzed in parallel. Antibodies against aconitase (αAco) and preimmune serum (PIS) were used for immunoprecipitation.

TNETG buffer: 20-mM Tris–HCl, pH 7.4; 2.5-mM EDTA; 150-mM NaCl; 10% (w/v) glycerol; 0.5% (w/v) Triton X-100

20-mM HEPES-KOH, pH 7.4; saturated PMSF in ethanol; 0.5-mm glass beads; 25% trichloroacetic acid (TCA) (optional)

Appropriate antibodies coupled to protein-A Sepharose. Antibodies are coupled to protein-A Sepharose for immunoprecipitation as follows:

Resuspend 50 mg of dried protein-A Sepharose (GE Healthcare, Chalfont St. Giles, UK) in cold 500-μl TNETG. The beads are swollen by incubation for at least 30 min at 4 °C with occasional mixing. Do not vortex immunobeads.

Collect the beads by centrifugation at 850 $\times$ g for 5 min. Add 500 μl of antiserum and incubate at 4 °C with gentle mixing using a rotating shaker for at least 1 h.

Collect the beads by centrifugation. Wash five times in 500-μl TNETG, spin down in between washes. The beads are resuspended in 500-μl TNETG and stored at 4 °C.

Procedure

1. Harvest yeast cells from a 50-ml overnight culture by centrifugation (1500 $\times$ g for 5 min), wash once in 10-ml sterile water, and dilute cells in 100-ml "iron-free" SC medium supplemented with the appropriate carbon source. Continue growth until cell density reaches an OD_{600} of 0.2. Incubate overnight in a shaking incubator at 30 °C.

2. Collect cells by centrifugation (1500 $\times$ g for 5 min), wash once in 10-ml sterile water and determine the wet weight. Resuspend 0.5-g cells in 10-ml iron-free medium in a 50-ml culture flask and incubate for 10 min at 30 °C in a shaking incubator.

3. Dilute 10-μCi ^{55}FeCl$_3$ in 100 μl of 0.1-M sodium ascorbate, and add mixture to the preincubated cells. The final concentration of ascorbate in the medium is 1 mM. Incubate for 1–2 h at 30 °C.

4. Transfer the radiolabeled cells to a 15-ml Falcon tube and harvest by centrifugation (1500 $\times$ g for 5 min). Wash the cells once in 10-ml citrate buffer, and once in 1 ml of 20-mM Hepes-KOH, pH 7.4. Concentrate the cells by centrifugation (1500 $\times$ g for 5 min).

5. Resuspend the cell pellet in 0.5-ml TNETG buffer. Add 10-μl saturated PMSF and 1/2 volume of glass beads. Lyse the cells by three bursts of 1 min on a vortex at maximum speed with intermediate cooling for 1 min on ice. Tubes are best vortexed upside down. All subsequent steps are carried out at 4 °C.

6. Remove the coarse cell debris by centrifugation for 5 min at 1500 $\times$ g. Transfer the supernatant to a 1.5-ml reaction tube and centrifuge at 17,000 $\times$ g for 10 min. Transfer the supernatant to a fresh tube. At this point, carefully avoid carrying over any membrane debris. Remove 5 μl of the extract for scintillation counting. This sample serves as a crude measure of the cellular iron uptake. Another 25 μl of the extract are precipitated with TCA for immunoblotting (optional). About 250 μl

are used for immunoprecipitation of the proteins of interest. Usually, two immuno-precipitations can be carried out with one labeling reaction.

7. For immunoprecipitation, add 20–50 μl of IgG-coupled immunobeads or 10 μl of commercially available coupled anti-hemagglutinin A (HA) or anti-Myc beads (Santa Cruz) to 250 μl of cell extract. Cut off the end of the pipette tip when handling immunobeads. Do not vortex the beads. Incubate the reaction tubes in a rotating shaker at 4 °C for 1 h.

8. Collect the beads by centrifugation at 1500 × g for 5 min. Remove the supernatant completely with a syringe. Wash the beads three times in 500-μl TNETG buffer and collect the beads by centrifugation. Virtually no labeled supernatant should remain in the reaction tubes after each washing step.

9. Add 50 μl of water and 1 ml of scintillation cocktail to the beads, vortex briefly, and determine the radioactivity associated with the beads in a scintillation counter using the counter settings for ^{3}H (Muhlenhoff *et al.*, 2002b).

Comments

1. In all labeling experiments involving ^{55}Fe, it is essential that all solutions and glassware are iron free. Standard dishwasher detergent and laboratory glassware frequently contain iron. Glassware should be acid-washed in 1-M HCl. Double-distilled water of the highest quality should be used throughout. The contaminated glass flasks used for *in vivo* radiolabeling of yeast are incubated with citrate buffer and washed in distilled water to remove remnant radioactivity. The flasks are rinsed with 70% ethanol for sterilization.

2. Reduction of ^{55}FeCl$_3$ is essential, as oxidized Fe^{3+} is virtually insoluble at neutral pH. Therefore, labeling reactions with ^{55}FeCl$_3$ *in vivo* or *in vitro* are always carried out in the presence of 1-mM fresh ascorbate to avoid precipitation of ferric iron. The radiation safety conditions for ^{55}Fe (an electron capture radiation) are similar to those for radioactive ^{3}H. For the quantification of ^{55}Fe, the counter settings for ^{3}H are appropriate.

3. For yeast, best results are obtained with cells overproducing the Fe/S protein of interest from a high-copy plasmid under the control of a strong promoter. If antibodies are not available, HA-, Myc-, or TAP-tagged versions of the Fe/S protein can frequently be used. In *S. cerevisiae*, the endogenous levels of aconitase, Yah1p (ferredoxin), and Leu1p are sufficient for analysis without overexpression. For other organisms, a suitable reporter protein has to be determined empirically.

B. Determination of Cellular Heme Levels by Radiolabeling of Yeast Cells *In Vivo*

Materials

Stop solution: 100-mM FeCl$_2$ in 5-M HCl
Butyl acetate

Procedure

1. Perform steps 1–4 of an *in vivo* [55]Fe radiolabeling assay as described in Section II.A.

2. Resuspend the washed, radiolabeled cells in 0.5-ml water and split the suspension in two halves.

3. To each half add 25-μl Stop solution, 800-μl butyl acetate, and 1/2 volume of glass beads at $0\,^{\circ}C$. Lyse the cells by three bursts of 1 min each on a vortex at high speed with intermediate cooling for 1 min on ice. The tubes should be vortexed upside down.

4. Separate the organic phase by centrifugation at $10,500 \times g$ for 10 min at $20\,^{\circ}C$. Transfer two 250-μl aliquots (for duplicate values) of the upper organic phase into two reaction tubes. Carefully avoid touching the membrane debris or parts of the lower aqueous phase.

5. Add 1 ml of scintillation cocktail to the beads, mix, and determine the [55]Fe radioactivity in the organic phase by scintillation counting, using the standard settings for [3]H.

Comments

1. The two assays described in Sections II.A and II.B can be performed in parallel on the same sample.

2. The heme biosynthesis assay takes advantage of the high solubility of protonated heme in organic solvents at low pH. Butyl acetate is the preferred solvent as it does not interfere with scintillation counting.

3. Due to the short incubation times (usually 60 min), this *in vivo* heme formation assay gives a measure mainly for the activity of ferrochelatase. The steady-state levels of heme in yeast cells *in vivo* can be determined using cells cultivated in medium supplemented with [55]$FeCl_2$ for longer periods. To this end, cells from a preculture are diluted in 50-ml "iron-free" SC medium supplemented with the appropriate carbon source, 1-mM ascorbate, and 10-μCi [55]Fe, at a density of $OD_{600} = 0.1$. The cultures are incubated overnight at $30\,^{\circ}C$ in a shaking incubator and the amount of radioactive heme is determined as described above. For details see Lange *et al.* (2004). This assay is highly sensitive and works well with mammalian cell cultures.

C. Determination of Fe/S Cluster Formation in Isolated Mitochondria *In Vitro*

Materials

In vitro experiments for the *de novo* formation of Fe/S proteins essentially require anoxic conditions and oxygen-free buffers. We work in an anaerobic chamber (Coy Laboratory Products Inc., Grass Lake, MI) filled with 95% (v/v) nitrogen and 5% (v/v) hydrogen gas. Stock solutions are introduced to anaerobic conditions at least 1 week in advance. They can be stored under anaerobic conditions indefinitely.

Anaerobic solutions of the following chemicals:

2 × Mitobuffer

40-mM HEPES, pH 7.4

100-mM KCl

2-mM MgSO$_4$

1.2-M Sorbitol

Other reagents:

10% (w/v) Triton X-100

2-Mercaptoethanol

25% (v/v) HCl

1-M Tris–HCl, pH 8.3

0.5-M EDTA

The following stock solutions are prepared freshly with oxygen-free water:

0.1-M Sodium ascorbate

0.1-M DTT

10-mM Cysteine

1-mM Pyridoxal phosphate

0.1-M NADH

30-mM Ferric ammonium citrate

30-mM FeCl$_2$

^{55}FeCl$_3$ (NEN/Perkin-Elmer) or ^{35}S-cysteine (GE Healthcare, Chalfont St. Giles, UK)

Procedure

Steps 1–4 are carried out under anaerobic conditions.

1. Store isolated mitochondria and all other solutions in an anaerobic chamber with open caps for 1 h on ice to remove oxygen.

2. In a standard reaction tube, add 125 μl of 2× mitobuffer, 2.5 μl of 100-mM sodium ascorbate, 2.5 μl of 0.1-M DTT, 2.5 μl of 0.1-M NADH (1-mM final), 2.5 μl of 1-mM pyridoxal phosphate (10-μM final), 5 μl of 10-mM cysteine (0.2-mM final), 2.5 μl of 10% (w/v) Triton X-100, and bring volume to 250 μl (minus the volume of mitochondria) with anaerobic water.

3. Add mitochondria (100-μg protein) isolated from iron-starved cells, 5-μCi ^{55}FeCl$_3$ (reduced in 10-mM ascorbate), and incubate with gentle shaking for 2.5 h at 25 °C under anaerobic conditions. [For labeling with ^{35}S, add 10-μCi ^{35}S cysteine and FeCl$_2$ (0.3-mM final concentration) instead of non-radioactive cysteine and FeCl$_3$.]

4. Terminate the labeling reaction by addition of 2.5 μl of 0.5-M EDTA on ice. All further steps are carried out aerobically at 4 °C.

5. Remove membranes by centrifugation at 17,000 × *g* for 10 min at 4 °C. Transfer the supernatant to a fresh tube. Using a Pipetman with the ends of the

pipette tips cut off, add 20–40 μl of immunobeads. Incubate the tubes in a rotating shaker for 1 h.

6. Collect the beads by centrifugation at 3000 $\times$ *g* for 5 min at 4°C. Carefully remove all of the supernatant with a syringe. Wash the beads three times in 500 μl of ice-cold TNETG buffer. In between the washing steps, collect the beads by centrifugation. It is essential that virtually no supernatant remains in the reaction tubes after each wash.

7. Add 50 μl of water and 1-ml scintillation cocktail, vortex, and count the radioactivity associated with the beads in a scintillation counter.

Comments

Mitochondria are isolated from iron-starved yeast cells grown in iron-free medium (Diekert *et al.*, 2001). We obtained the best results with mitochondria isolated from cells that overproduce biotin synthase (Bio2p) for this type of experiment (Muhlenhoff *et al.*, 2002a).

D. Analysis of Fe/S Cluster Formation *In Vitro* Using Recombinant Ferredoxins

Materials

Materials as in Section II.C.

An acidic, low molecular weight [2Fe-2S] ferredoxin. These can be obtained in recombinant form from *E. coli* or are commercially available.

Additional reagents requiring no deoxygenation.

20-mM HEPES-KOH, pH 7.4; saturated PMSF in ethanol; 25% TCA (optional); appropriate antibodies coupled to protein-A Sepharose; Bromophenol blue (0.2% w/v).

Procedure

Preparation of apoferredoxin

1. Incubate a recombinant ferredoxin (4–10 mg/ml concentration) under anaerobic conditions for at least 1 h on ice to remove oxygen.

2. In a standard reaction tube, add 250-μl ferredoxin, 10-μl 2-mercaptoethanol, and bring volume up to 1 ml with anaerobic water. Cool the sample on ice, add 25% HCl to a final concentration of 0.5 M, mix gently, and incubate on ice for 5 min. A white precipitate forms immediately.

3. Collect the precipitated protein by centrifugation at 12,000 $\times$ *g* for 10 min at 4°C. Remove the supernatant completely and rinse the pellet briefly with 500 μl of ice-cold water containing 0.1% 2-mercaptoethanol. Remove water completely.

4. Add 250 μl of 50-mM Tris–HCl, pH 8.3. Resuspend the pellet carefully with a pipette and store the sample on ice. Do not vortex. The protein should dissolve within 10–15 min. If necessary, add 1 μl drops of unbuffered 1-M Tris until the solution is completely clear.

5. Repeat steps 1–4. For details see comment 2 and Meyer *et al.* (1986).

Reconstitution assay

1. Under anaerobic conditions, combine 125 μl of 2× mitobuffer, 2.5 μl of 100-mM sodium ascorbate, 2.5 μl of 0.1-M DTT, 2.5 μl of 0.1-M NADH (1-mM final), 2.5 μl of 1-mM pyridoxal phosphate (10-μM final), 2.5 μl of 10% Triton X-100, and bring volume up to 250 μl with anaerobic water. In cases where holoferredoxin formation is detected by native PAGE, smaller reaction volumes (50–100 μl) should be used.

2. Add isolated mitochondria or cell extracts corresponding to at least 100 μg of protein and 20 μg of apoferredoxin. For radioassays, add either 5 μl of 10-mM cysteine (0.2-mM final) and 5-μCi ^{55}FeCl$_3$ (reduced in 10-mM ascorbate) or 2.5 μl of 30-mM FeCl$_3$ (0.3-mM final) and 10-μCi ^{35}S-cysteine. Control reactions without mitochondria and added apoferredoxin should be analyzed in parallel. For nonradioactive assays, add ferric ammonium citrate to a final concentration of 0.3 mM and cysteine to 4 mM. Higher amounts of ferredoxin (50 μg) may be used. Incubate with gentle shaking for 2 h at 25 °C under anaerobic conditions.

3. Terminate the reconstitution reaction by addition of 2.5 μl of 0.5-M EDTA on ice.

4. Remove membrane debris and aggregates by centrifugation at 17,000 × g for 10 min at 20 °C. Transfer the supernatant to a fresh tube. For analysis of holoferredoxin formation continue with one of the two following protocols. All further steps are carried out aerobically at 4 °C.

Isolation of radiolabeled holoferredoxin by binding to ion exchange resins

1. Add 25 μl of a 1:1 slurry of anion exchanger beads (Q-Sepharose, GE Healthcare, Chalfont St. Giles, UK) in TNETG-buffer containing 200-mM NaCl (TNETG-200) and incubate in a rotating shaker for 10 min at 4 °C.

2. Collect the beads by centrifugation at 3000 × g for 5 min. Carefully remove all of the supernatant with a syringe. Wash the beads three times in 500-μl TNETG-200 buffer containing 1-mM ascorbate. Between the washing steps, collect the beads by centrifugation. Remove the supernatant completely.

3. Add 50 μl of water and 1-ml scintillation cocktail, vortex, and count the radioactivity associated with the beads in a scintillation counter.

Separation by native PAGE

1. Prepare a 17.5% polyacrylamide gel with a 6% stacking gel and cold standard electrophoresis buffer according to the standard Laemmli procedure, with the exception that SDS is omitted.

2. Add 1 μl of 0.2% bromophenol blue to each 100 μl of reaction mix, load 50 μl of the samples onto the chilled native polyacrylamide gel. Electrophoresis is carried out at 4 °C at 30 mA and 200 V until the dye front reaches the bottom of the gel.

3. For detection of radiolabeled ferredoxin, the gel is fixed by gentle shaking in cold 20% ethanol at 4 °C for 30 min, dried under vacuum, and analyzed by

autoradiography. Since Fe/S cofactors are acid-labile, the gels should not be stained or fixed with acetic acid. In case of nonradioactive assays, the red color of the reconstituted holoferredoxin may be visible by eye. The gel is stained with Coomassie Brilliant Blue. Holoferredoxin forms a sharp band slightly above the dye front that is well separated from the majority of proteins.

Comments

1. The preparation of apoferredoxin and Fe/S cluster reconstitution necessitate anaerobic conditions and oxygen-free solutions.

2. These protocols take advantage of the fact that most low molecular mass [2Fe-2S] ferredoxins are soluble in their apoform, which can be generated by acid precipitation. Care should be taken that the Fe/S cluster is removed completely and no oxidation occurs. In an optimal case, the UV/Vis spectrum of the apoferredoxin lacks any absorption above 300 nm. If this is not the case, the procedure should be repeated.

3. Biochemical reconstitution of mitochondrial Fe/S protein assembly requires ATP. For mitochondria from *S. cerevisiae*, the endogenous ATP levels are sufficient and the addition of ATP is not recommended since ATP is an effective chelator of iron. For mitochondria from other sources, however, the addition of low amounts of ATP (0.2–0.5 mM) may improve the reconstitution. In any case, control reactions without added ATP, apoferredoxins or mitochondria should be analyzed in parallel.

4. The quantitative estimation of holoferredoxin formation takes advantage of the acidic character of this type of ferredoxin, which allows its binding to either an anion exchange resin at relatively high ionic strength or a rapid mobility in an electric field. It is therefore essential that a model protein with a low pI is used. We use either yeast ferredoxin Yah1p or plant-type ferredoxins.

5. Original references: Lutz *et al.* (2001), Muhlenhoff *et al.* (2002a), Suzuki *et al.* (1991), and Takahashi *et al.* (1991).

III. Determination of Mitochondrial Iron Contents

Defects in mitochondrial Fe/S cluster formation or in the ISC export machinery result in an increased uptake of iron that is eventually accumulated within the mitochondria. The latter is an unphysiological event, as the vacuole usually serves as the physiological storage organelle for iron. Section III includes two methods for the determination of the iron content of isolated mitochondria or of cell lysates, which are based on the formation of colored iron complexes with the chelators bathophenantroline and nitro-PAPS, respectively. The bathophenantroline assay is a rapid assay useful for the determination of artificially increased mitochondrial iron levels. The nitro-PAPS assay is more sensitive but more time consuming, as labile iron is quantitatively released from proteins by acid prior to quantification. This assay may be used to determine physiological iron levels of cell lysates.

A. Bathophenantroline Assay

Materials

100-mM Bathophenantroline (prepare freshly)
1-M Sodium dithionite (prepare freshly)
10% (w/v) SDS
1-M Tris–HCl, pH 7.4

Procedure

1. Mix 100 μl of 1-M Tris–HCl, pH 7.4 with 60 μl of 10% SDS (0.6% final), 20 μl of 1-M dithionite, 100-μl bathophenantroline (10-mM final) and isolated mitochondria (at least 200-μg protein), and bring to a final volume of 1 ml with distilled water. Incubate at 20°C for 5 min.

2. Remove membrane debris by centrifugation for 5 min at 10,000 $\times$ g. Record the absorption spectrum of the sample between 500 and 700 nm using a reference sample lacking mitochondria. Determine the difference between OD_{540} minus OD_{700} (to account for light scattering). The absorption coefficient $\varepsilon_{540 \ nm}$ is $\sim$23,500 M^{-1} cm^{-1} (Li *et al.*, 1999).

B. Nitro-PAPS Assay

Materials

2.5% (w/v) SDS
1% HCl
7.5% (w/v) Ammonium acetate solution
4% (w/v) Ascorbic acid (prepare fresh)
1.3-mM 2-(5-nitro-2-pyridylazo)-5-(*N*-propyl-*N*-sulfopropylamino)phenol (Nitro-
 PAPS)
Recommended iron standard: 0.2-mM $(NH_4)_2Fe(SO_4)_2\cdot6H_2O$; FW = 392.14
 [Mohr's Salt, 0.008% (w/v), $\sim$39 mg/500 ml, prepare fresh]

Procedure

1. Dilute samples and a blank to 100 μl with water in standard reaction tubes.

2. Add 100 μl of 1% HCl, mix by gentle shaking, and incubate at 80°C for 10 min. Allow the tubes to cool down (keep closed) and centrifuge for 1 min at 9000 $\times$ g.

3. In the following order: add 500 μl of 7.5% ammonium acetate, 100 μl of 4% ascorbic acid, and 100 μl of 2.5% SDS. Vortex after each addition.

4. Centrifuge for 5 min at 9000 $\times$ g, transfer supernatant (855 μl) into a fresh tube, and add 95 μl of the iron chelator nitro-PAPS (final concentration 130 μM).

Determine the absorbance at 585 nm against the blank sample in quartz cuvettes. The absorption coefficient $\varepsilon_{585\ nm}$ is ~94,000 M^{-1} cm^{-1} (Makino *et al.*, 1988).

Comments

The same protocol may also be used for the iron chelator Ferene (3-(2-pyridyl)-5,6-bis(2-[5-furylsulfonic acid])-1,2,4-triazine). The absorption coefficient $\varepsilon_{593\ nm}$ is 35,000 M^{-1} cm^{-1}.

IV. Reporter Assays for Analysis of Iron–Dependent Gene Expression in *S. cerevisiae*

Defects in mitochondrial Fe/S cluster formation or in ISC export systems induce the constitutive transcriptional deregulation of iron-dependent genes, a response that resembles the physiological reaction of yeast to iron deprivation (Shakoury-Elizeh *et al.*, 2004). The transcriptional (de)regulation of iron-responsive genes can be studied conveniently by determining promoter activities in reporter assays. In *S. cerevisiae*, reporter assays based on the β-galactosidase (LacZ) are most frequently used. In Section IV, we describe alternative reporter assays that take advantage of green fluorescent protein (GFP) or luciferase for monitoring gene expression in *S. cerevisiae*. Both GFP and luciferase from *Renilla reniformis* can be measured directly in whole yeast cells and do not require cell lysis. The use of these whole cell assays, however, is limited to the analysis of strong promoters. Weak promoters can be routinely analyzed in cell extracts using firefly or *Renilla* luciferase as reporters. Both luciferase systems are highly sensitive and can be analyzed in parallel in a single cell (Harger and Dinman, 2003; McNabb *et al.*, 2005; Szittner *et al.*, 2003). Although we use these reporter assays in the context of iron-responsive transcriptional regulation, these assays can be used generally for monitoring gene expression in *S. cerevisiae*.

Our GFP reporter constructs are based on the CEN/ARS plasmid pPS1372 that carries two GFP open reading frames in tandem, under the control of the *ADH2* promoter (Taura *et al.*, 1998). The *ADH2* promoter can be excised by double digests with *Sac*I and *Sma*I or *Sac*I and *Eco*RI, and replaced by the promoter of interest. Our luciferase reporter constructs are based on the CEN/ARS plasmids p414-MET25 and p416-MET25 (Funk *et al.*, 2002) (Fig. 3). Coding regions for luciferase from firefly or *Renilla reniformis* were taken from the original vectors pGL3 and pGL4.70 (Promega Corporation, Madison, WI), respectively, and inserted into the *Sma*I and *Sal*I sites downstream of the *MET25* promoter. The latter can be replaced by the promoter of interest by directional cloning. For details see Fig. 3. For an example of iron-responsive genes, we choose the promoter of *CYC1*, encoding cytochrome *c*, and *FET3*, a member of the iron regulon that is frequently used for the analysis of iron-responsive transcription in *S. cerevisiae*. These genes are constitutively deregulated in cells with defects in mitochondrial Fe/S protein biogenesis (Fig. 3).

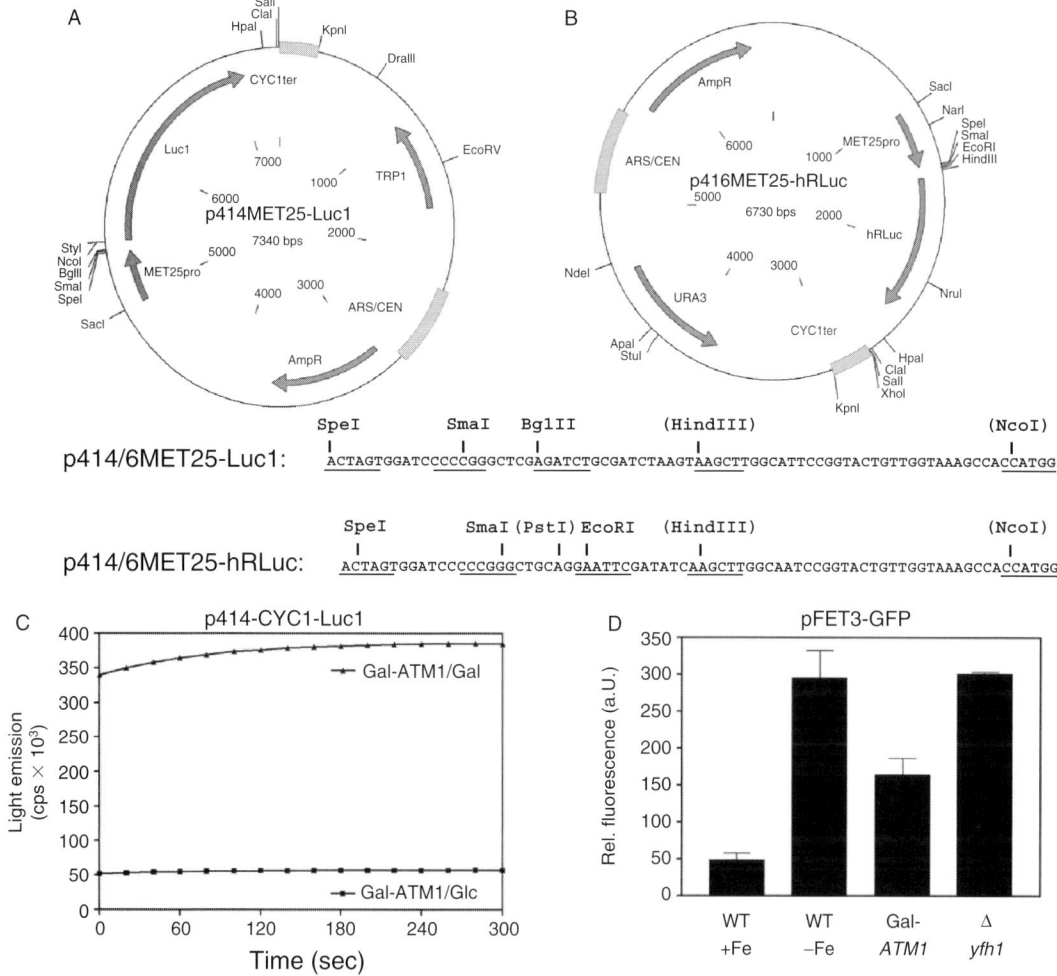

Fig. 3 Reporter assays based on luciferase and GFP in *S. cerevisiae*. Physical map of the luciferase reporter plasmids p414-MET25-Luc1 (A) and p416-MET25-hRLuc (B) carrying the firefly (luc1) or *Renilla* luciferase (hRluc) gene. The sequence and restriction sites of the poly-linker regions between the *MET25* promoter and the start codons of the luciferase open reading frames are shown below the vector maps. (C) The conditional ISC export mutant Gal-ATM1 harboring the reporter plasmid p414-CYC1-Luc1 was grown under permissive conditions in the presence of galactose (Gal) and under repressive conditions in the presence of glucose (Glc), and the transcriptional activity of the *CYC1* promoter was determined by recording the luciferase-specific luminescence of cell extracts. Five microliters of cell extracts at a protein concentration of 0.1 $\mu g/\mu l$ were analyzed over a 5-min interval. The maximal activity was observed ∼3 min after the addition of substrate. (D) Regular and iron-starved (-Fe) wild-type cells (WT) and the ISC mutants Δ*yfh1* and Gal-ATM1 harboring plasmid pFET3-GFP were grown in the presence of glucose, and the transcriptional activity of the *FET3* promoter was determined by recording the GFP-specific fluorescence emission of whole cells.

A. Promoter Assay Based on Green Fluorescent Protein

Procedure

1. Grow the yeast cells carrying the *GFP* reporter plasmid in a small volume of minimal medium supplemented with the required amino acids and carbon source overnight at 30°C. Collect the cells by centrifugation (1500 × g for 5 min), dilute in 25 ml of the desired growth medium to a final cell density of $OD_{600} = 0.2$, and incubate at 30°C in a shaking incubator. For a single end point measurement, cells are grown up to density of OD_{600} of ~0.5. Otherwise, aliquots are withdrawn at the desired time points.

2. Harvest the cells by centrifugation (1500 × g for 5 min), resuspend the cell pellet in 3 ml of distilled water, and determine the optical density at 600 nm.

3. For the determination of the GFP-specific fluorescence, cells are diluted in a 3-ml fluorescence cuvette to an OD_{600} of 1 with distilled water. The fluorescence emission of the cells is set to zero at 600 nm and the emission spectrum is recorded between 500 and 550 nm. The maximum of the emission is at ~512 nm. For an accurate readout, the spectrum of a sample lacking the reporter plasmid is subtracted. We use the following fluorescence spectrometer settings. Excitation wavelength: 480 nm with 10 nm bandwidth; emission: 5 nm or 10 nm bandwidth; detector at medium amplification. We use a FP6300 fluorescence spectrometer (Jasco Ltd., Tokyo, Japan).

Comments

1. Since yeast cells settle quickly, the use of a built-in magnetic stirrer during fluorescence measurements is recommended.

2. Since the GFP fluorescence is partially quenched by the growth medium, medium must be removed completely prior to the measurement. The fluorescence is also reduced in cells at stationary phase.

3. The assay is much simpler, but also less sensitive, than reporter assays based on β-galactosidase or luciferase. This assay is useful for the analysis of strong promoters only.

4. The assay can be carried out in microplate formats and quantified in microplate readers equipped with fluorescence detection.

5. *FET3* and promoters of other genes of the yeast iron regulon are induced on iron starvation, which can be mimicked by growth of yeast in the presence of up to 50-μM bathophenantroline, an iron chelator. The FET3 promoter is constitutively induced in cells with defects in the mitochondrial ISC assembly or export systems and may be used for identification of these types of mutations in yeast (Fig. 3).

B. Promoter Assays Based on Luciferase Reporters

1. Analysis in Cell Extracts

Procedure
Sample preparation

1. Grow yeast cells carrying the luciferase reporter plasmids in a small volume of minimal media with the required amino acids and carbon sources overnight at

30°C in a shaking incubator. Collect the cells by centrifugation ($1500 \times g$), dilute in 5 ml of the desired growth medium until the cell density reaches an OD_{600} of 0.2, and incubate at 30°C in a shaking incubator until the cell density reaches ~0.5.

2. Collect the cells by centrifugation ($1500 \times g$ for 5 min), resuspend in 1-ml water, and transfer the cells to a 2.2-ml reaction tube. Collect the cells by centrifugation ($1500 \times g$ for 5 min). At this point the cells may be shock frozen.

3. Resuspend the cells in 300-μl cell culture lysis reagent (provided by the manufacturer with the luciferase substrates). Add 50–100 μl of 0.5-mm glass beads and 2.5-μl saturated PMSF and cool the cells on ice. Vortex two times for 3 min with intermediate cooling on ice.

4. Sediment the glass beads and membrane debris by centrifugation at $17,000 \times g$ for 10 min at 4°C. Transfer the supernatant to a fresh tube and determine the protein concentration (typical yields are 0.2–0.5 μg/μl). At this point the extracts may be shock frozen.

Luminescence measurements
Firefly luciferase:

1. Add 10-ml Assay Reagent buffer to the lyophilized luciferase substrate (Luciferase Assay System, Promega). Store the working solution in aliquots at -20°C.

2. Mix 5-μl yeast extract with 20-μl substrate working solution in a microplate, mix well, and incubate for 1 min. Determine the luminescence in a microplate reader equipped with luminescence detection over 2–5 min with an integration time of 1000 μsec. Dilution of the cell extracts may be required.

Renilla *luciferase:*

1. Dilute *Renilla* substrate stock solution 1:100 in Assay Reagent buffer (*Renilla* Luciferase Assay System, Promega). Prepared fresh.

2. In a microplate, mix 10-μl yeast extract with 50 μl of substrate working solution in *Renilla* Assay buffer. Mix well and determine the luminescence in a microplate reader equipped with luminescence detection over 1 min with an integration time of 1000 μsec. Serial dilutions of the extracts may be required.

Comments

1. To minimize plasmid loss, cells must be grown with selection for the auxotrophic marker of the plasmid throughout the analysis.

2. Both luciferase systems give comparable results and can be measured in parallel in the same sample (McNabb *et al.*, 2005). In general, the firefly system is more sensitive. Extracts are stable on ice for extended times. Specialized luminometers are not necessary. The luminescence can be quantified in microplate readers equipped with luminescence detection or in scintillation counters, provided the coincidence mode is switched off. We routinely use a microplate reader (Infinite M200, Tecan AG, Männedorf, Switzerland) or a scintillation counter. The protocols described above are for microplate readers, but they can be easily adjusted for scintillation counters. Since the latter are far more sensitive, dilutions of the extracts up to 1:100,000 may be required.

3. The luminescence of luciferase is time dependent. In yeast extracts, the luminescence of firefly luciferase peaks after $\sim$2–3 min, while that of the *Rhenilla* enzyme is maximal immediately after addition of the substrate. Hence, it is necessary to monitor the time course of the luminescence. As a result, only a restricted number of samples can be analyzed in parallel in a microplate reader. Luciferase assays are thus difficult to adjust for high through-put analysis.

2. Promoter Assay with Whole Cells Using *Renilla* Luciferase

Procedure

1. Grow the yeast cells carrying the luciferase reporter plasmid in a small volume of the desired growth medium overnight at 30 °C in a shaking incubator. Collect the cells by centrifugation (1500 $\times$ *g* for 5 min), dilute in 5 ml of the desired growth medium until cells reach a density of $OD_{600} = 0.2$, and incubate at 30 °C in a shaking incubator until the cell density reaches $\sim$0.5. Dilute the cells to an OD_{600} of 0.1 in 1-ml water.

2. Dilute the *Renilla* luciferase substrate (*Viviren*, Promega) 1:20 in DMSO.

3. Transfer 2 μl of diluted substrate to a fresh tube and bring final volume to 100 μl, accounting for the volume of cells that will be assayed. Mix gently.

4. Record the luminescence of the sample for 1 min. This is necessary for preequilibration and to get a background reading.

5. Add the cells, mix gently, and monitor the luminescence of the sample for 2–3 min.

6. Alternatively, cells and substrate are mixed in 100-μl water and the luminescence is recorded directly for 2–4 min. A control lacking cells expressing luciferase is analyzed separately.

Comments

1. The luminescence of *Renilla* luciferase in whole yeast cells is much weaker than in cell extracts. The analysis in luminometers or microplate readers frequently gives unsatisfactory results. We routinely quantify the luminescence in a standard scintillation counter with the coincidence detection mode switched off. Scintillation counters are much more sensitive than luminescence readers and are generally recommended for highly sensitive luminescence assays or simply when a luminescence device is not available.

2. Store the *Viviren* substrate in small aliquots at -80 °C. Avoid repetitive freeze thawing. The working solution is best prepared a day or two in advance. For reasons that are unclear, preparation of the substrate in advance reduces the autofluorescence of the substrate and increases the luminescence signal. The working solution can be stored in the refrigerator for at least 1 week.

3. All samples and controls have to be prepared freshly and analyzed with the same substrate-working solution. Cells should not be frozen.

Acknowledgments

This work was supported by grants of the Deutsche Forschungsgemeinschaft (SFB593 and Gottfried-Wilhelm-Leibniz program), Fonds der Chemischen Industrie, Deutsches Humangenomprojekt, and the Fritz-Thyssen-Stiftung.

References

Balk, J., and Lill, R. (2004). The cell's cookbook for iron-sulfur clusters: Recipes for fool's gold? *Chembiochem* **5,** 1044–1049.

Beinert, H., Holm, R. H., and Munck, E. (1997). Iron-sulfur clusters: Nature's modular, multipurpose structures. *Science* **277,** 653–659.

Diekert, K., de Kroon, A. I., Kispal, G., and Lill, R. (2001). Isolation and subfractionation of mitochondria from the yeast *Saccharomyces cerevisiae*. *Methods Cell Biol.* **65,** 37–51.

Funk, M., Niedenthal, R., Mumberg, D., Brinkmann, K., Ronicke, V., and Henkel, T. (2002). Vector systems for heterologous expression of proteins in *Saccharomyces cerevisiae*. *Methods Enzymol.* **350,** 248–257.

Harger, J. W., and Dinman, J. D. (2003). An *in vivo* dual-luciferase assay system for studying translational recoding in the yeast *Saccharomyces cerevisiae*. *RNA* **9,** 1019–1024.

Jensen, L. T., and Culotta, V. C. (2000). Role of *Saccharomyces cerevisiae* ISA1 and ISA2 in iron homeostasis. *Mol. Cell. Biol.* **20,** 3918–3927.

Johnson, D., and Dean, D. R. (2004). Structure, function, and formation of biological iron-sulfur clusters. *Annu. Rev. Biochem.* **19,** 19.

Kispal, G., Csere, P., Guiard, B., and Lill, R. (1997). The ABC transporter Atm1p is required for mitochondrial iron homeostasis. *FEBS Lett.* **418,** 346–350.

Kispal, G., Csere, P., Prohl, C., and Lill, R. (1999). The mitochondrial proteins Atm1p and Nfs1p are essential for biogenesis of cytosolic Fe/S proteins. *EMBO J.* **18,** 3981–3989.

Kispal, G., Sipos, K., Lange, H., Fekete, Z., Bedekovics, T., Janaky, T., Bassler, J., Aguilar Netz, D. J., Balk, J., Rotte, C., and Lill, R. (2005). Biogenesis of cytosolic ribosomes requires the essential iron-sulphur protein Rli1p and mitochondria. *EMBO J.* **24,** 589–598.

Lange, H., Muhlenhoff, U., Denzel, M., Kispal, G., and Lill, R. (2004). The heme synthesis defect of mutants impaired in mitochondrial iron-sulfur protein biogenesis is caused by reversible inhibition of ferrochelatase. *J. Biol. Chem.* **279,** 29101–29108.

Li, J., Kogan, M., Knight, S. A., Pain, D., and Dancis, A. (1999). Yeast mitochondrial protein, Nfs1p, coordinately regulates iron–sulfur cluster proteins, cellular iron uptake, and iron distribution. *J. Biol. Chem.* **274,** 33025–33034.

Lill, R., and Mühlenhoff, U. (2005). Iron-sulfur protein biogenesis in eukaryotes. *Trends Biochem. Sci.* **30,** 133–141.

Lill, R., and Mühlenhoff, U. (2006). Iron-sulfur protein biogenesis in eukaryotes: Components and mechanisms. *Annu. Rev. Cell Dev. Biol.* **22,** 457–486.

Lutz, T., Westermann, B., Neupert, W., and Herrmann, J. M. (2001). The mitochondrial proteins Ssq1 and Jac1 are required for the assembly of iron-sulfur clusters in mitochondria. *J. Mol. Biol.* **307,** 815–825.

Makino, T., Kiyonaga, M., and Kina, K. (1988). A sensitive, direct colorimetric assay of serum iron using the chromogen, nitro-PAPS. *Clin. Chim. Acta* **171,** 19–27.

McNabb, D. S., Reed, R., and Marciniak, R. A. (2005). Dual luciferase assay system for rapid assessment of gene expression in *Saccharomyces cerevisiae*. *Eukaryot. Cell* **4,** 1539–1549.

Meyer, J., Moulis, J. M., and Lutz, M. (1986). High-yield chemical assembly of [2Fe-2X] (X = S, Se) clusters into spinach apoferredoxin, product characterisation by Raman spectroscopy. *Biochim. Biophys. Acta* **871,** 243–249.

Muhlenhoff, U., Richhardt, N., Gerber, J., and Lill, R. (2002a). Characterization of iron-sulfur protein assembly in isolated mitochondria. A requirement for ATP, NADH, and reduced iron. *J. Biol. Chem.* **277,** 29810–29816. Epub 2002 Jun 13.

Muhlenhoff, U., Richhardt, N., Ristow, M., Kispal, G., and Lill, R. (2002b). The yeast frataxin homolog Yfh1p plays a specific role in the maturation of cellular Fe/S proteins. *Hum. Mol. Genet.* **11,** 2025–2036.

Rutherford, J. C., and Bird, A. J. (2004). Metal-responsive transcription factors that regulate iron, zinc, and copper homeostasis in eukaryotic cells. *Eukaryot. Cell* **3,** 1–13.

Rutherford, J. C., Jaron, S., and Winge, D. R. (2003). Aft1p and Aft2p mediate iron-responsive gene expression in yeast through related promoter elements. *J. Biol. Chem.* **278,** 27636–27643.

Rutherford, J. C., Ojeda, L., Balk, J., Muhlenhoff, U., Lill, R., and Winge, D. R. (2005). Activation of the iron regulon by the yeast Aft1/Aft2 transcription factors depends on mitochondrial but not cytosolic iron-sulfur protein biogenesis. *J. Biol. Chem.* **280,** 10135–10140.

Shakoury-Elizeh, M., Tiedeman, J., Rashford, J., Ferea, T., Demeter, J., Garcia, E., Rolfes, R., Brown, P. O., Botstein, D., and Philpott, C. C. (2004). Transcriptional remodeling in response to iron deprivation in *Saccharomyces cerevisiae. Mol. Biol. Cell* **15,** 1233–1243.

Sherman, F. (2002). Getting started with yeast. *Methods Enzymol.* **350,** 3–41.

Suzuki, S., Izumihara, K., and Hase, T. (1991). Plastid import and iron-sulphur cluster assembly of photosynthetic and nonphotosynthetic ferredoxin isoproteins in maize. *Plant Physiol.* **97,** 375–380.

Szittner, R., Jansen, G., Thomas, D. Y., and Meighen, E. (2003). Bright stable luminescent yeast using bacterial luciferase as a sensor. *Biochem. Biophys. Res. Commun.* **309,** 66–70.

Takahashi, Y., Mitsui, A., and Matsubara, H. (1991). Formation of the Fe-S cluster of ferredoxin in lysed spinach chloroplasts. *Plant Physiol.* **95,** 97–103.

Taura, T., Krebber, H., and Silver, P. A. (1998). A member of the Ran-binding protein family, Yrb2p, is involved in nuclear protein export. *Proc. Natl. Acad. Sci. USA* **95,** 7427–7432.

Yamaguchi-Iwai, Y., Ueta, R., Fukunaka, A., and Sasaki, R. (2002). Subcellular localization of Aft1 transcription factor responds to iron status in *Saccharomyces cerevisiae. J. Biol. Chem.* **277,** 18914–18918.

Yarunin, A., Panse, V. G., Petfalski, E., Dez, C., Tollervey, D., and Hurt, E. C. (2005). Functional link between ribosome formation and biogenesis of iron-sulfur proteins. *EMBO J.* **24,** 580–588.

PART III

Assays for Mitochondrial Respiratory Activity and Permeability in Living Cells

CHAPTER 14

Imaging of Mitochondrial Polarization and Depolarization with Cationic Fluorophores

John J. Lemasters and Venkat K. Ramshesh

Center for Cell Death, Injury and Regeneration, and
Departments of Pharmaceutical Sciences and Biochemistry & Molecular Biology
Medical University of South Carolina, Charleston, South Carolina 29425

I. Introduction

Mitochondrial respiration generates an electrochemical gradient of protons made up mostly of a negative electrical potential difference ($\Delta\Psi$) across the mitochondrial inner membrane. In living cells, the plasma membrane also generates a negative inside $\Delta\Psi$. Electrophysiological techniques have long been used to measure plasmalemmal $\Delta\Psi$, but mitochondria are in general too small to impale with a microelectrode. Instead, mitochondrial $\Delta\Psi$ must be estimated from the equilibrium distribution of

0091-679X/07 $35.00
DOI: 10.1016/S0091-679X(06)80014-2

membrane-permeant cations, which accumulate electrophoretically into polarized cells and mitochondria (Hoek *et al.*, 1980). By using fluorescent lipophilic cations, $\Delta\Psi$ in individual cells and mitochondria can also be visualized and quantified (Chacon *et al.*, 1994; Ehrenberg *et al.*, 1988; Farkas *et al.*, 1989; Zahrebelski *et al.*, 1995). By coupling fluorescent cation uptake with fluorescence resonance energy transfer (FRET), mitochondria undergoing depolarization can also be revealed selectively (Elmore *et al.*, 2001, 2004).

II. Quantitative Imaging of $\Delta\Psi$ with Fluorescent Cations

A. Nernstian Distribution of Cationic Fluorophores

Membrane-permeant monovalent cationic fluorophores, such as rhodamine 123, tetramethylrhodamine methyl ester (TMRM), and many others, accumulate electrophoretically within cells in response to the electrical potential Ψ of subcellular compartments (Ehrenberg *et al.*, 1988; Emaus *et al.*, 1986; Johnson *et al.*, 1981). At equilibrium, Ψ is related to fluorescent cation uptake by the Nernst equation:

$$\Psi = -59 \log\left(\frac{F_{in}}{F_{out}}\right) \qquad (1)$$

where Ψ is electrical potential in millivolts, F_{out} is the monovalent cationic fluorophore concentration in the extracellular space (electrical ground), and F_{in} is the fluorophore concentration at any point within the cell. By using confocal/multiphoton microscopy to quantify intracellular fluorophore distribution, maps of intracellular Ψ can be generated using Eq. (1). From differences of Ψ between compartments, membrane potentials ($\Delta\Psi$) can be determined, specifically plasmalemmal $\Delta\Psi$ (cytosolic minus extracellular Ψ) and mitochondrial $\Delta\Psi$ (mitochondrial minus cytosolic Ψ).

Under normal conditions in most cells, plasmalemmal $\Delta\Psi$ ranges from -30 to -100 mV and mitochondrial $\Delta\Psi$ ranges from -120 to -160 mV. Since these $\Delta\Psi$'s are additive, mitochondria are 150–260 mV more negative than the extracellular space. From Eq. (1), such large $\Delta\Psi$'s correspond to cation concentration ratios between mitochondria and the cell exterior that can exceed 10,000:1. Such large gradients cannot be captured in digital images using a conventional linear scale of 256 gray levels per pixel (8-bit pixel memory). Instead, 12- to 16-bit memory, sequential imaging at different laser powers, or a nonlinear logarithmic (gamma) scale must be used (Chacon *et al.*, 1994). Gamma scales, commonly used in scanning electron microscopy, compress input signals logarithmically into the available 256 gray levels of pixel memory. Early confocal microscopes, such as the Bio-Rad MRC-600, included gamma imaging circuits, but gamma circuitry is generally not available in current commercial laser scanning confocal/multiphoton microscope systems. Accordingly, this chapter addresses measurement of $\Delta\Psi$'s using high bit memory and sequential imaging.

B. Cellular Loading of Potential-Indicating Fluorophores

Cells plated on glass coverslips are loaded by simple incubation with 50–500 nM of TMRM, rhodamine 123, or other indicator, for 15–30 min at 37 °C. No special loading buffer is required, and cells can be loaded with equal success in complete culture medium or in a simple Ringer's or saline solution. However, when the loading buffer is replaced, the new medium should contain a third to a fourth of the initial loading concentration of the potential-indicating probe in order to maintain its equilibrium distribution. Otherwise, with each buffer change, some fluorophore will leach from the cells irrespective of changes of intracellular Ψ.

Loading concentrations above 1 μM should be avoided. Even with loading below 1 μM, fluorophore accumulation can reach millimolar concentrations inside the mitochondrial matrix space. Such high concentrations can cause metabolic inhibition. For example, rhodamine 123 at high matrix concentrations inhibits the mitochondrial ATP synthase of oxidative phosphorylation (Emaus et al., 1986), and DiOC (6), a cationic fluorophore still used in flow cytometry, strongly inhibits mitochondrial respiration even when loaded at nanomolar concentrations (Rottenberg and Wu, 1997). Of commonly used fluorophores, TMRM exhibits the least mitochondrial toxicity (Scaduto and Grotyohann, 1999). In general, fluorophore concentrations exceeding 500 nM should be avoided unless an adequate fluorescent signal cannot otherwise be produced.

C. Image Acquisition and Processing

After loading, confocal and multiphoton images may then be collected. At least two images must be collected. The first is an optical section through the specimen of interest. The second is an image obtained after refocusing inside the glass coverslip. The latter image serves as a background in an area devoid of fluorophore. Both images should be collected using identical instrumental settings of laser power, gain, and brightness. Image oversaturation (pixels at highest gray level) and undersaturation (pixels with a zero gray level) should be negligible. A low laser power setting ($\leq$1%) should be used to minimize photodamage, especially if serial imaging over time of the same portion of the specimen will be performed. If the microscope can collect images with pixel depth of 12 or 16 bits (4096 and 65,536 gray levels, respectively), then additional images may not be needed. However, if pixel depth is only 8 bits (256 gray levels) or if the ratio of fluorescence intensity between mitochondria and the extracellular space exceeds the gray level range, then additional images at $\sim$10 times greater laser power ($\sim$3.3 times laser power for two-photon microscopy since two-photon excitation varies with the square of laser intensity) should be collected both in the cell and within the glass coverslip.

With multitracking systems, lower and higher laser power images can be collected simultaneously one line at a time. Each line (row of pixels) in the images is collected in sequence. A first line scan is performed at lower laser power and is followed immediately by a second line scan at higher laser power, a process that is repeated with each succeeding line in the image. Multitracking permits the two images to be acquired simultaneously and eliminates problems associated with specimen movement and focus drift. When images at higher laser power are collected, oversaturation of mitochondria will be marked, but oversaturation in higher power images does not pose a problem, since these images will only be used to quantify areas of weak fluorescence intensity in the extracellular space.

By processing images collected through the sample and within the glass coverslip, maps of the intracellular distribution of Ψ may be generated by a three-step procedure of: (1) background subtraction, (2) quantitation of extracellular fluorescence, and (3) calculation of Ψ for each pixel gray level to create a pseudocolor map of Ψ.

1. Background Subtraction

Unless photon counting or an equivalent procedure is performed, light detectors such as photomultiplier tubes generate signals even in the absence of light. This background signal must be subtracted from the signal collected in the presence of light to obtain an output truly proportional to fluorescence intensity. In confocal/ multiphoton microscopy, images collected in the plane of the coverslip represent this background, since added fluorophore cannot penetrate into the glass. A mean pixel value is then calculated for all the pixels of each background image using utilities in ImageJ, Adobe Photoshop, MetaMorph, or other image analysis software. This average value is then subtracted from every pixel of the corresponding specimen image. Such background-corrected images represent the true relative distribution of fluorescence intensity within the images.

2. Fluorescence of the Extracellular Space

The next step is to estimate extracellular fluorescence. Since fluorescence should be the same everywhere in the extracellular space, all areas in the cell-free extracellular space in background-subtracted specimen images are selected, and average pixel intensity is determined. If extracellular fluorescence is too weak, then extracellular fluorescence can be measured in the same way in a higher laser power image. Division (after background subtraction) of mean extracellular fluorescence from the higher laser power image by the power ratio (or its square for two-photon excitation) between the higher and lower power images then yields an estimate of extracellular fluorescence for the lower power image. The latter estimate is valid even if it is less than one.

3. Pixel–by–Pixel Calculation of Ψ

Using Eq. (1), a value for Ψ can be calculated for every pixel of background-corrected images on the assumption that fluorescence intensity is proportional to monovalent cationic fluorophore concentration. To display the intracellular distribution of Ψ, colors can be assigned to specific millivolt ranges of Ψ. To determine the pixel value corresponding to a specific millivolt value of Ψ, Eq. (1) is rearranged:

$$P_i = \text{antilog}\,(\log P_{\text{out}} - \Psi/59) \qquad (2)$$

where P_{out} is average background-subtracted pixel intensity in the extracellular space, and P_i is the pixel value representing a particular millivolt value of Ψ. Using Eq. (2), ranges of gray level values corresponding to a specific range of Ψ can be calculated, to which individual colors are assigned.

Figures 1 and 2 illustrate the pseudocolored images that result for a TMRM-loaded mouse hepatocyte and adult feline cardiac myocyte imaged by confocal microscopy. The difference of Ψ between the extracellular space (where Ψ is zero) and the cytosol and nucleus represents plasmalemmal $\Delta\Psi$, whereas the difference between the cytosol/nucleus and mitochondria represents mitochondrial $\Delta\Psi$. In the hepatocyte, plasmalemmal $\Delta\Psi$ was about -30 mV, and mitochondrial $\Delta\Psi$ was as

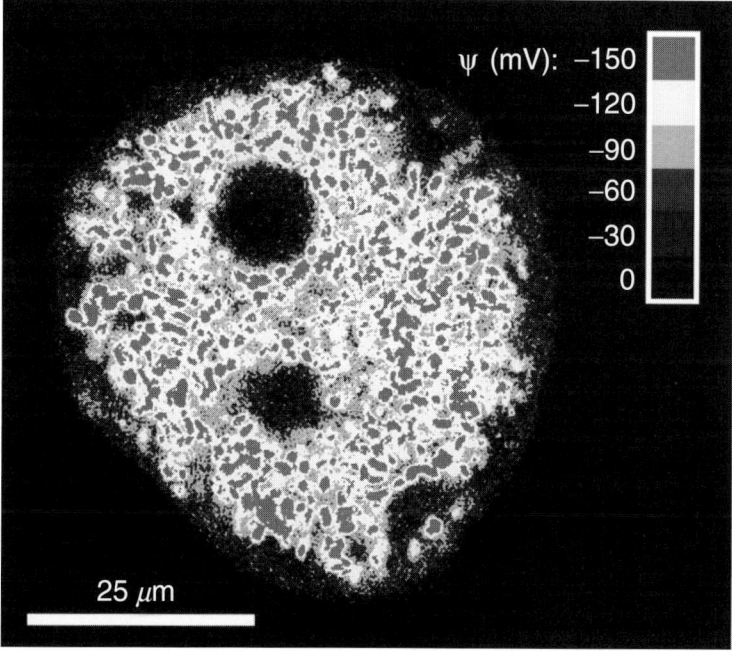

Fig. 1 Distribution of electrical potential in a mouse hepatocyte. A mouse hepatocyte plated on a Type 1 collagen-coated coverslip for 5 h was loaded with 200 nM TMRM for 20 min in KRH at 37 °C and imaged in KRH containing 50 nM TMRM. The distribution of Ψ was determined by laser scanning confocal microscopy using 543-nm excitation from a helium–neon laser and a 565- to 615-nm emission barrier filter. (See Color Insert.)

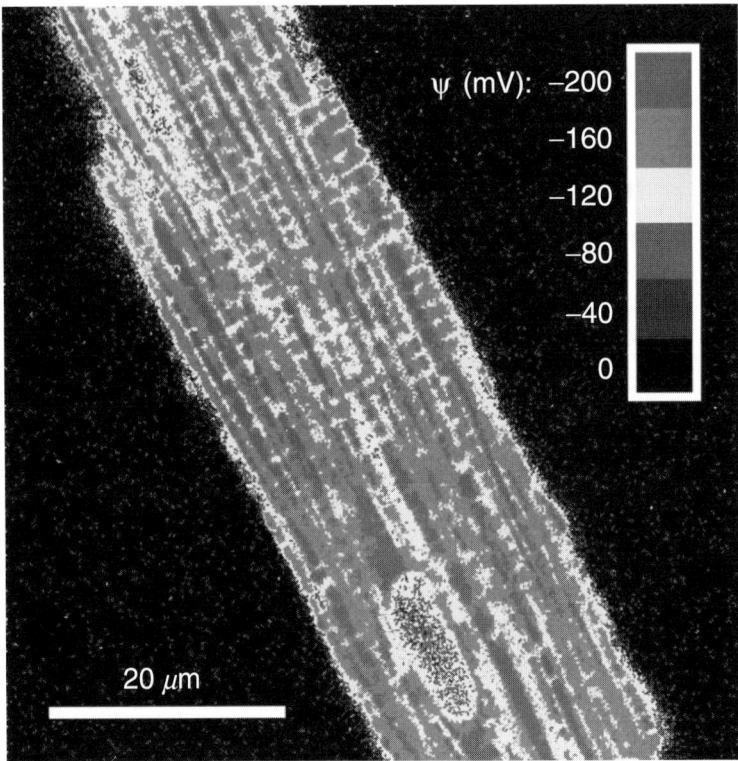

Fig. 2 Electrical potential of an adult feline cardiac myocyte. An adult feline cardiac myocyte plated on a Matrigel-coated coverslip for 5 h was loaded with 200 nM TMRM for 20 min in Medium 199 at 37°C and imaged in Medium 199 containing 50 nM TMRM, as described in Fig. 1. (See Color Insert.)

great as -120 mV. In the cardiac myocyte, the cytoplasm was so densely packed with mitochondria that cytosolic Ψ needs to be estimated from the nucleus, which is equipotential with the cytosol. From nuclear Ψ, plasmalemmal (sarcolemmal) $\Delta\Psi$ was estimated to be about -100 mV. The greater plasmalemmal $\Delta\Psi$ of the cardiac myocyte is consistent with the greater plasmalemmal polarization of excitable cells such as myocytes. However despite the greater plasmalemmal $\Delta\Psi$, mitochondrial $\Delta\Psi$ of the myocyte was similar to that of the hepatocytes.

In hepatocytes and cardiac myocytes, mitochondrial diameter is about 1 μm. Since the thickness of confocal and multiphoton optical sections imaged with a high numerical aperture lens is slightly less than 1 μm, some mitochondria occupy the entire thickness of the optical slices, but other mitochondria will occupy only part of the thickness of the section. The presence of mitochondria that are only partially included in confocal/multiphoton optical sections causes an apparent heterogeneity and underestimation of mitochondrial $\Delta\Psi$. Consequently, the highest values of mitochondrial Ψ most likely reflect true mitochondrial $\Delta\Psi$.

D. Nonideal Characteristics of Fluorophores

The Nernst equation [Eq. (1)] describes ideal electrophoretic distribution of membrane-permeant monovalent cations, but probes may not behave ideally. Rhodamine 123, for example, can accumulate in mitochondria to levels that exceed those predicted by Eq. (1) (Emaus et al., 1986). Excess accumulation may in part be due to concentration-dependent fluorophore stacking and formation of so-called J-aggregates. For fluorophores like rhodamine 123, TMRM, and others, J-aggregate formation causes fluorescence quenching and a redshift of absorbance. In cuvette and multiwell assays, absorbance changes and fluorescence quenching can be used to monitor mitochondrial fluorophore accumulation and hence mitochondrial $\Delta\Psi$ (Blattner et al., 2001; Emaus et al., 1986; Scaduto and Grotyohann, 1999). J-aggregate formation and quenching are dependent on fluorophore concentration and may be minimized by using smaller loading concentrations.

With some monovalent cationic fluorescent probes, such as JC-1, J-aggregate formation leads to a shift from green to red fluorescence emission rather than to fluorescence quenching (Smiley et al., 1991). As more JC-1 accumulates in mitochondria, more and larger J-aggregates form. Thus, increased red fluorescence of JC-1 signifies increased mitochondrial polarization. Confocal images of JC-1-loaded hepatocytes visualize these aggregates of JC-1 within the matrix of individual green-fluorescing mitochondria (Fig. 3). These J-aggregates are literally microprecipitates and characteristically do not fill the mitochondrial matrix. In Fig. 3, J-aggregates are seen to localize to the lateral margins of mitochondria, presumably in association with mitochondrial membranes. Heterogeneity of red and green fluorescence in single JC-1-loaded mitochondria represents the physical distribution of J-aggregate microprecipitates and does not signify variation of $\Delta\Psi$ within the single organelles. Formation of physical aggregates of JC-1 inside mitochondria limits the probe's usefulness for high-resolution imaging of cellular and mitochondrial $\Delta\Psi$.

III. Visualization of Depolarized Mitochondria

A. Covalent Adduct Formation by MitoTracker Probes

When mitochondria depolarize, they release their potential-indicating fluorophores. As a result, the organelles disappear from fluorescence images. To circumvent this problem, membrane-permeant cationic fluorophores have been developed that incorporate reactive chloromethyl groups that form covalent bonds with protein sulfhydryls (Haugland, 1999). These MitoTracker probes, such as MitoTracker Green and MitoTracker Red, accumulate electrophoretically into mitochondria in the same fashion as do other potential-indicating cationic fluorophores (Fig. 4, left panel). After intramitochondrial accumulation, adducts form between MitoTracker probes and mitochondrial matrix proteins, although adduct formation may require many minutes to go to completion. After adduct formation, MitoTracker fluorescence

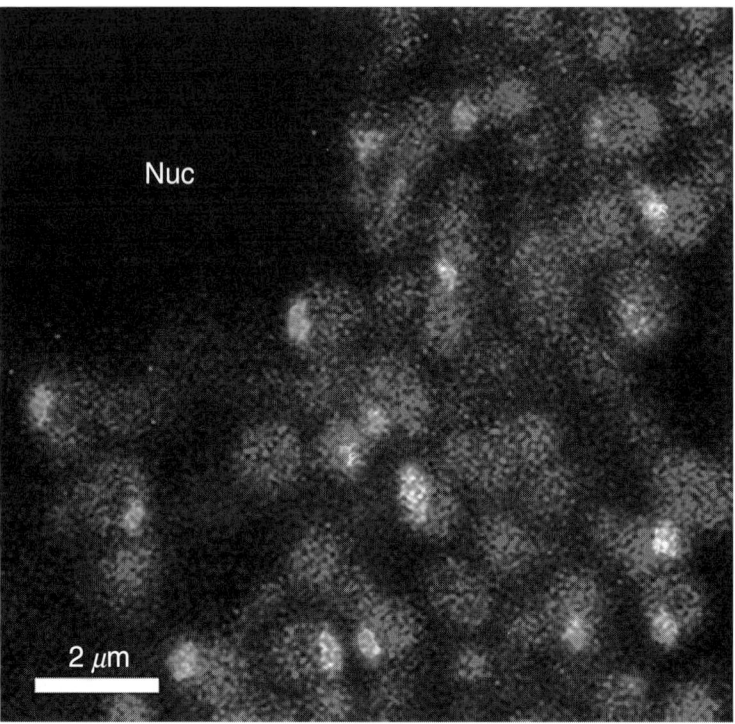

Fig. 3 Red and green mitochondrial fluorescence after hepatocyte loading with JC-1. A mouse hepatocyte plated as in Fig. 1 was loaded with 100 nM JC-1 for 30 min in KRH at 37°C and imaged by multitrack confocal microscopy. Shown is an overlay of green fluorescence excited with 488-nm light and red fluorescence excited with 543-nm light. Note that many green-fluorescing mitochondria contain peripheral red J-aggregate inclusions. Nuc, nucleus. (See Color Insert.)

becomes independent of mitochondrial polarization and is retained even if mitochondria subsequently depolarize (Fig. 4, right panel). MitoTracker fluorescence may even survive chemical fixation with aldehydes and other chemical preservatives. Thus, MitoTracker labeling of mitochondria has been valuable for use in conjunction with immunocytochemistry and related procedures that require some type of fixation.

B. FRET Between Cationic Fluorophores

When fluorophores with different spectral characteristics come into proximity, the phenomenon of FRET can occur (Stryer, 1978). In FRET, photons excite a donor fluorophore to an excited state. If an acceptor fluorophore is nearby whose excitation spectrum overlaps the donor emission spectrum, then nonradiative energy transfer can occur from the donor to the acceptor, leading to fluorescence

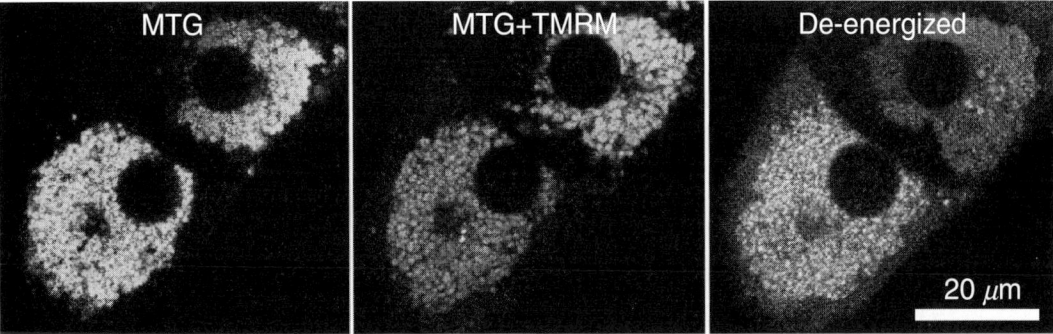

Fig. 4   FRET-dependent quenching of MTG by TMRM in cultured rat hepatocytes. Overnight cultured rat hepatocytes were loaded with 0.5 μM MTG for 60 min in Waymouth's growth medium, and green and red fluorescence was imaged by confocal microscopy using blue (488-nm) excitation from an argon–krypton laser (left panel). Subsequently, 1 μM TMRM was added (middle panel), followed by de-energization with 2.5 mM KCN and 10 μM CCCP in the presence of 20 mM fructose and 1 μg/ml oligomycin (right panel). After MTG loading, mitochondria fluoresced green but not red (left panel). After subsequent TMRM loading, green fluorescence was quenched, and mitochondria fluoresced red after blue excitation (middle panel). De-energization restored green MTG fluorescence and simultaneously caused red TMRM fluorescence to disappear. (See Color Insert.)

emission at wavelengths characteristic of the acceptor rather than the donor. When FRET occurs, ordinary donor fluorescence becomes quenched. Instead, longer wavelength photons are emitted with wavelength characteristic of the acceptor. FRET varies with the sixth power of the distance between donor and acceptor. For FRET to occur to any appreciable extent, donor and acceptor molecules must be within 10 nm of one another.

C. FRET Between MitoTracker Green FM and TMRM

MitoTracker Green (MTG) FM is an example of a cationic MitoTracker probe that accumulates into mitochondria and binds covalently to mitochondrial proteins. Green-fluorescing MTG can serve as a donor for FRET with TMRM as the acceptor, a phenomenon likely promoted by stacking interactions between the two fluorophores. When cultured rat hepatocytes are loaded with 0.5 μM MTG for 1 h in culture medium, green MTG fluorescence excited with blue (488 nm) light shows a characteristic mitochondrial pattern in confocal images (Fig. 4, left panel). Red fluorescence is absent. After incubation with 1 μM TMRM, green fluorescence excited with blue light becomes quenched and red fluorescence occurs instead (Fig. 4, middle panel). Since blue light does not excite red TMRM fluorescence, the observed shift from green to red emission represents FRET. This shift of emission is reversible, since de-energization of mitochondria with KCN (a respiratory inhibitor), CCCP (an uncoupler), and oligomycin (a mitochondrial ATPase inhibitor) in the presence of fructose (a glycolytic substrate to maintain ATP) restores green fluorescence as red fluorescence is lost (Fig. 4, right panel). TMRM alone in the absence of MTG produces little red fluorescence after blue excitation until MTG is loaded (data not shown) (Elmore *et al.*, 2004). Radiative colorimetric quenching in which green fluorescence emitted by MTG is reabsorbed by TMRM, leading to red fluorescence, cannot account for these observations, since photon absorbance over the optical path length involved (i.e., the diameter of a mitochondrion or 1 μm) is negligible (Elmore *et al.*, 2001, 2004).

D. Use of FRET to Distinguish Depolarized from Polarized Mitochondria

Mitochondrial depolarization frequently contributes to necrotic and apoptotic killing of cells from liver, heart, and other organs. The cellular fate of depolarized mitochondria is difficult to study because after depolarization, labeling of mitochondria with potential-indicating probes like rhodamine 123 and TMRM simply disappears. Because MitoTracker probes are retained after depolarization, MitoTracker dyes do not distinguish mitochondria that depolarize from those that remain polarized. A strategy to visualize depolarized mitochondria selectively is to coload cells with MTG and TMRM. After coloading, TMRM quenches MTG fluorescence by FRET when the two fluorophores reside together inside polarized mitochondria. When an individual mitochondrion depolarizes, it releases TMRM, and MTG becomes unquenched. As a result, the fluorescence of the depolarized mitochondrion changes from red to green. Such green fluorescence allows discrimination of a single depolarized mitochondrion against a background of hundreds of polarized mitochondria.

For coloading, cells cultured on coverslips or glass-bottomed Petri dishes are incubated at 37 °C with 200 to 500 nM MTG FM for 1 h in either complete culture medium or a saline solution such as Krebs-Ringer-HEPES solution (KRH containing: 25 mM HEPES, 115 mM NaCl, 5 mM KCl, 1 mM KH_2PO_4, 1.2 mM $MgSO_4$, and 2 mM $CaCl_2$, pH 7.4). After mounting the cells on the microscope, TMRM (0.5–1 μM) is added for 30 min followed by baseline imaging, as shown for cultured rat hepatocytes in Fig. 5. For effective MTG quenching, TMRM must be used at a higher concentration than when TMRM is used alone, e.g. 1 μM. After coloading with TMRM and MTG, green MTG fluorescence excited with blue light was quenched in nearly all mitochondria, but due to FRET, nearly all mitochondria fluoresced red (Fig. 5). A few mitochondria, however, did fluoresce green (circles), but none were yellow in the overlay of the red and green fluorescence images, signifying that no individual structures within the cells were simultaneously emitting red and green fluorescence (Fig. 5, right). Since cationic TMRM and MTG are both initially taken up only by polarized mitochondria, recovery of the green fluorescence of covalently bound MTG and loss of red fluorescence indicates spontaneous depolarization of these mitochondria. When the hepatocytes were then incubated in nutrient-free KRH plus the hormone glucagon, individual

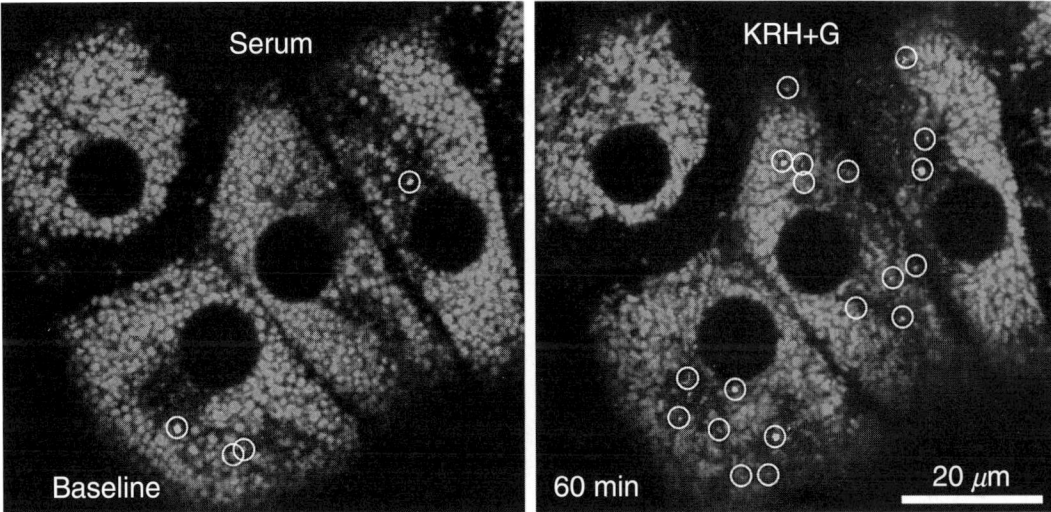

Fig. 5 Spontaneous mitochondrial depolarization in rat hepatocytes during nutrient deprivation plus glucagon. Cultured hepatocytes were loaded sequentially with MTG (0.5 μM) and TMRM (1 μM) in serum containing complete growth medium and mounted on the stage in growth medium. After collection of a baseline image (left panel), the cells were incubated in KRH plus 1 μM glucagon (KRH + G) for 60 min (right panel). Shown are overlay confocal images of green and red fluorescence after blue (488 nm) excitation. Green-fluorescing structures increased when the medium was changed to KRH plus glucagon (circles). (See Color Insert.)

mitochondria reverted from red to green fluorescence (Fig. 5, right panel, circles). Most of these mitochondria were undergoing lysosomal digestion by the process of mitochondrial autophagy or mitophagy (Elmore *et al.*, 2004; Rodriguez-Enriquez *et al.*, 2006).

IV. Conclusion

Quantitative confocal and multiphoton microscopy of the intracellular distribution of membrane-permeant cationic fluorophores provides a minimally perturbing means to measure dynamically both mitochondrial and plasmalemmal $\Delta\Psi$ in cultured cells. A limiting factor is time resolution, since several seconds are required for cationic fluorophores to reestablish a new steady-state equilibrium after a change of $\Delta\Psi$. A further shortcoming is that mitochondria lose their potential-indicating fluorophores and become invisible in fluorescence images after depolarization. The latter problem can be overcome by coloading with TMRM and MTG. With coloading, $\Delta\Psi$ can still be monitored by red TMRM fluorescence excited with green light, but additionally TMRM quenches green MTG fluorescence excited with blue light via a FRET mechanism. Consequently, when a mitochondrion depolarizes, TMRM is released, and the green fluorescence of covalently bound MTR recovers, which reveals depolarized mitochondria as individual green structures against a background of red-fluorescing polarized mitochondria. Unlike virtually any other technique, confocal and multiphoton microscopy permits nondestructive serial observation of the $\Delta\Psi$'s of populations of cells and mitochondria. Live cell, three-dimensionally resolved confocal and multiphoton microscopy is now an indispensable tool for studying the mitochondrial function and pathophysiology.

Acknowledgments

We thank Insil Kim for mouse hepatocytes and Dr. Donald R. Menick for adult feline cardiac myocytes. This work was supported, in part, by Grants DK37034 and DK070195 from the National Institutes of Health.

References

Blattner, J. R., He, L., and Lemasters, J. J. (2001). Screening assays for the mitochondrial permeability transition using a fluorescence multiwell plate reader. *Anal. Biochem.* **295,** 220–226.

Chacon, E., Reece, J. M., Nieminen, A. L., Zahrebelski, G., Herman, B., and Lemasters, J. J. (1994). Distribution of electrical potential, pH, free Ca^{2+}, and volume inside cultured adult rabbit cardiac myocytes during chemical hypoxia: A multiparameter digitized confocal microscopic study. *Biophys. J.* **66,** 942–952.

Ehrenberg, B., Montana, V., Wei, M.-D., Wuskell, J. P., and Loew, L. M. (1988). Membrane potential can be determined in individual cells from the Nernstian distribution of cationic dyes. *Biophys. J.* **53,** 785–794.

Elmore, S. P., Qian, T., Grissom, S. F., and Lemasters, J. J. (2001). The mitochondrial permeability transition initiates autophagy in rat hepatocytes. *FASEB J.* **15,** 2286–2287.

Elmore, S. P., Nishimura, Y., Qian, T., Herman, B., and Lemasters, J. J. (2004). Discrimination of depolarized from polarized mitochondria by confocal fluorescence resonance energy transfer. *Arch. Biochem. Biophys.* **422,** 145–152.

Emaus, R. K., Grunwald, R., and Lemasters, J. J. (1986). Rhodamine 123 as a probe of transmembrane potential in isolated rat-liver mitochondria: Spectral and metabolic properties. *Biochim. Biophys. Acta* **850,** 436–448.

Farkas, D. L., Wei, M.-D., Febbroriello, P., Carson, J. H., and Loew, L. M. (1989). Simultaneous imaging of cell and mitochondrial membrane potentials. *Biophys. J.* **56,** 1053–1069.

Haugland, R. P. (1999). "Handbook of Fluorescent Probes and Research Chemicals." Molecular Probes, Inc., Eugene, Oregon.

Hoek, J. B., Nicholls, D. G., and Williamson, J. R. (1980). Determination of the mitochondrial protonmotive force in isolated hepatocytes. *J. Biol. Chem.* **255,** 1458–1464.

Johnson, L. V., Walsh, M. L., Bockus, B. J., and Chen, L. B. (1981). Monitoring of relative mitochondrial membrane potential in living cells by fluorescence microscopy. *J. Cell Biol.* **88,** 526–535.

Rodriguez-Enriquez, S., Kim, I., Currin, R. T., and Lemasters, J. J. (2006). Tracker dyes to probe mitochondrial autophagy (mitophagy) in rat hepatocytes. *Autophagy* **2,** 39–46.

Rottenberg, H., and Wu, S. (1997). Mitochondrial dysfunction in lymphocytes from old mice: Enhanced activation of the permeability transition. *Biochem. Biophys. Res. Commun.* **240,** 68–74.

Scaduto, R. C., Jr., and Grotyohann, L. W. (1999). Measurement of mitochondrial membrane potential using fluorescent rhodamine derivatives. *Biophys. J.* **76,** 469–477.

Smiley, S. T., Reers, M., Mottola-Hartshorn, C., Lin, M., Chen, A., Smith, T. W., Steele, G. D., Jr., and Chen, L. B. (1991). Intracellular heterogeneity in mitochondrial membrane potentials revealed by a J-aggregate-forming lipophilic cation JC-1. *Proc. Natl. Acad. Sci. USA* **88,** 3671–3675.

Stryer, L. (1978). Fluorescence energy transfer as a spectroscopic ruler. *Annu. Rev. Biochem.* **47,** 819–846.

Zahrebelski, G., Nieminen, A. L., al Ghoul, K., Qian, T., Herman, B., and Lemasters, J. J. (1995). Progression of subcellular changes during chemical hypoxia to cultured rat hepatocytes: A laser scanning confocal microscopic study. *Hepatology* **21,** 1361–1372.

CHAPTER 15

Biosensors for the Detection of Calcium and pH

Paolo Pinton, Alessandro Rimessi, Anna Romagnoli, Andrea Prandini, and Rosario Rizzuto

Department of Experimental and Diagnostic Medicine, Section of General Pathology
Interdisciplinary Center for the Study of Inflammation (ICSI) and
Emilia Romagna Laboratory for Genomics and Biotechnology (ER-Gentech)
University of Ferrara, I-44100 Ferrara, Italy

I. Introduction

The development of diverse molecular biology techniques, which can modify and express exogenous cDNAs in virtually all cell types, has been responsible for the recent expansion in the use of protein probes for imaging studies in cell biology. Two groups of visualizable reporter proteins are currently employed: the chemiluminescent proteins (e.g., aequorin from *Aequorea victoria* and luciferase from *Photinus pyralis*), and the fluorescent proteins (e.g., green fluorescent protein [GFP] from *A. victoria* and red fluorescent protein [DsRFP] from *Discosoma* sp.).

The study of isolated mitochondria, dating back to the 1960s, has provided a wealth of information on the biochemical routes that allow these organelles, derived from the adaptation of primordial symbionts, to couple oxidation of substrates to the production of ATP. Moreover, recent work has highlighted the role of signals reaching the mitochondria in the activation of apoptosis. In this context, it is an exciting task to study mitochondrial function in living cells. For this purpose, new tools are needed, that combine specific targeting to mitochondria and the sensitivity to detect the parameter of interest. Recombinant reporter proteins are emerging as the tools of choice, since their light output can be robust, and they can be targeting to compartments of interest using signal sequences.

In particular, we describe the development and use of protein chimeras (derived from proteins naturally present in the medusa, *A. victoria*) that are specifically targeted to the mitochondria, either to the matrix or to the intermembrane space and can be used to monitor calcium and pH. Aequorin, the pioneer of the targeted recombinant probes, is a Ca^{2+}-sensitive photoprotein that emits light on binding of Ca^{2+} to its three high affinity Ca^{2+}-binding sites. We also describe the use of mutants of GFP as Ca^{2+} and pH probes.

II. Targeting Strategies and Transfection

The strategy used to achieve the correct delivery of recombinant probes to mitochondria takes advantage of the naturally occurring signal peptides that are present in nuclear-encoded mitochondrial proteins (Fig. 1). The vast majority of mitochondrial proteins are encoded in nuclear DNA, synthesized in the cytosol, and imported into the organelle. Correct targeting of these nuclear-encoded mitochondrial proteins is mediated by signal peptides, which are usually present at the N-terminus. Mitochondrial signal sequences have no consensus sequence but bear some common amino acid compositions, charge distribution and in some cases, secondary structure (Gavel and von Heijne, 1990) (for details on import of proteins into mitochondria, please refer to Chapter 35 by Habib *et al.* and Chapter 36 by Stojanovski *et al.*, this volume).

A. Targeting to the Mitochondrial Matrix

Subunit VIII of human cytochrome *c* oxidase possesses a 25-amino acid signal peptide at its N-terminus, which is cleaved by the matrix protease on import. It is presumed that the action of the protease is dependent on a motif present within the first few amino acids of the mature protein (Rizzuto *et al.*, 1989). To target functional reporter probes to the mitochondrial matrix, a chimera is constructed which consists of the first 31 amino acids of COX Subunit VIII, which contains the

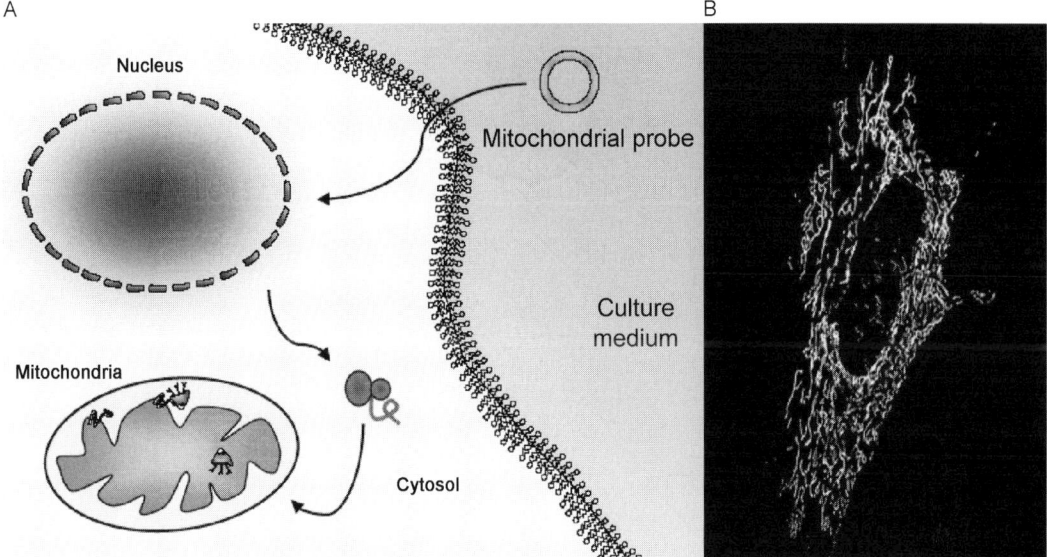

Fig. 1 Transfection of reporter constructs. (A) Recombinant expression of a mitochondrial probe. The cDNA of the reporter protein is introduced into cells by transfection. The precursor protein is synthesized by the cellular machinery and imported into mitochondria. For the mitochondrial matrix, HA1-tagged probe is fused to the cleavable-targeting sequence (mitochondrial presequence) of a subunit (e.g., VIII) of cytochrome c oxidase, the terminal complex of the respiratory chain. For the MIMS, the gene-encoded probe is fused to the C-terminus of GDP, an integral protein of the mitochondrial inner membrane that has a large C-terminal domain protruding into the intermembrane space. (B) Immunolocalization of mtAEQ 36 h after transfection was carried out by standard procedures.

matrix-targeting signal peptide and the first 6 amino acids of the mature protein, fused to the N-terminus of the reporter protein of interest (e.g., aequorin or GFP mutant) (De Giorgi *et al.*, 1996; Rizzuto *et al.*, 1992).

B. Targeting to the Mitochondrial Intermembrane Space

For the delivery of probes to the mitochondrial intermembrane space (MIMS), we exploited the characteristics of glycerol phosphate dehydrogenase (GPD), an enzyme that is integrated into the mitochondrial inner membrane with its C-terminal domain protruding into the MIMS. To target aequorin and GFP to this space, we fused the photoprotein to the C-terminal portion of GPD. This allowed for efficient targeting of the reporter protein to the intermembrane space, without altering its C-terminus, which is essential for its luminescent properties (Pinton *et al.*, 1998; Porcelli *et al.*, 2005; Rizzuto *et al.*, 1998b). In all cases, an epitope tag (9-aa-long HA1 hemagglutinin) is added to the coding region of the reporter protein, to allow immunolocalization of the transfected protein.

Our experience has been that mitochondrial reporter constructs are sorted to the expected location efficiently and quantitatively. However, when a series of experiments is initiated in a new cell type, it is prudent, especially with photoproteins that cannot be directly visualized under the microscope, to confirm the correct sorting. Figure 1B shows the immunolocalization of one of these constructs (mitochondria-targeted aequorin, mtAEQ). The typical rod-like appearance of mitochondria can be easily appreciated.

The first step in mitochondrial measurements with recombinant reporter probes is the introduction of the encoding cDNA into the cell type of interest (Fig. 1A). Introduction of foreign DNA into cells is generally feasible using the calcium phosphate method. Certain cell types, however, are particularly resistant to this procedure. In such cases, the solution has been to seek alternative methods for the introduction of cDNA into that particular cell type. Since detailed protocols are supplied by the manufacturers of both kits and specific instrumentation, these are not be discussed in this chapter. In our experiment, we have determined empirically that all cell types in which we have tried to express chimeric probes can be transfected successfully using one method or another.

III. Calcium Reporters

The contribution of mitochondria to intracellular Ca^{2+} signaling and the role of mitochondrial Ca^{2+} uptake, both in shaping the cytoplasmic response and in controlling mitochondrial function, are currently areas of intense investigation (Bernardi, 1999; Budd and Nicholls, 1996a,b; Duchen, 2000a,b; Nicholls, 2005; Pozzan and Rizzuto, 2000; Rizzuto and Pozzan, 2006; Rizzuto *et al.*, 2000). These studies rely to a large extent on the appropriate use of emerging techniques combined with judicious data interpretation. The outer and inner mitochondrial membranes (OMMs and IMMs, respectively) are markedly different, architecturally as well as functionally. It is noteworthy that the IMM is very impermeable to ions in general, including Ca^{2+}.

Mitochondrial Ca^{2+} traffic takes place essentially through two pathways: (1) a ruthenium red (RR)-sensitive electrogenic Ca^{2+} uniporter that exhibits a very low affinity ($K_m \sim 5$–10 μM) for the cation, allows for the transport of Ca^{2+} down its electrochemical gradient (~ 200 mV on the side of the matrix) and (2) an electroneutral antiporter that, by exchanging Ca^{2+} with either Na^+ or H^+, prevents the attainment of an electrochemical equilibrium (Fig. 2) (Carafoli, 1987; Gunter and Gunter, 1994, 2001; Gunter *et al.*, 2000; Pozzan *et al.*, 1994).

Although the biochemical properties of these transport pathways have been known for over three decades, the molecular events that regulate the dynamics of this transport system were not well understood. Nevertheless, a role for mitochondria in Ca^{2+} signaling is widely accepted and work was initiated to clarify the precise function of mitochondria in these signal transduction events (Berridge *et al.*, 2000, 2003; Duchen, 1999, 2000a; Hajnoczky *et al.*, 1995; Rizzuto and Pozzan, 2006;

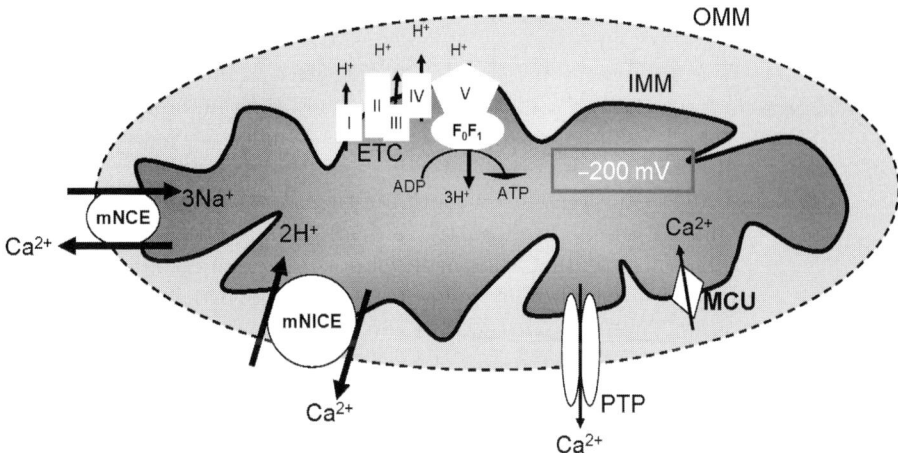

OMM = outer mitochondrial membrane, IMM = inner mitochondrial membrane, mNICE = mitochondrial sodium independent calcium exchanger, mNCE = mitochondrial sodium calcium exchanger, MCU = mitochondrial calcium uniporter, PTP = permeability transition pore, ETC = electron transport chain, F_0F_1 = ATP synthase

Fig. 2 Calcium trafficking in mitochondria. Ca^{2+} entry takes place via a low affinity uniporter (which is inhibited by RR), due to the high electronegative potential (which is collapsed by FCCP or CCCP) in the mitochondrial matrix. Extrusion of Ca^{2+} takes place through an electro-neutral antiporter (in exchange with either Na^+ or H^+). In the matrix, Ca^{2+} stimulates the activity of three Ca^{2+}-sensitive dehydrogenases of the Krebs cycle (NAD^+-isocitrate-, 2-oxoglutarate-, and pyruvate-dehydrogenase), thus promoting electron flow through the electron transport chain (ETC).

Rizzuto *et al.*, 2004). An obvious function for mitochondrial Ca^{2+} homeostasis was soon found, namely, to rapidly adapt aerobic metabolism to the changing needs of a cell. However, subsequent work revealed a much broader picture. It showed that mitochondrial Ca^{2+} uptake influences the kinetics and spatial properties of the cytoplasmic concentration of Ca^{2+} or $[Ca^{2+}]_c$ (Budd and Nicholls, 1996a; Drummond and Fay, 1996; Herrington *et al.*, 1996), and that within the mitochondrion a Ca^{2+}-mediated signal can induce a radically different effect, that is, triggers the onset of apoptosis (Duchen, 2000b; Green and Kroemer, 2004; Kroemer and Reed, 2000).

We describe here two methodological approaches to measure the concentration of Ca^{2+} within the mitochondrial context, that is, $[Ca^{2+}]_m$. In the first part, we focus on the use of aequorin (a luminescent protein). The second part is devoted to measurements using chimeric versions of GFP (a fluorescent protein).

A. Mitochondrial Calcium Measurements Using Aequorin

Aequorin is a 21-kda protein from various *Aequorea* species (Shimomura, 1986; Shimomura *et al.*, 1962) which, in the active form, consists of an apoprotein and a covalently bound prosthetic group (coelenterazine). When Ca^{2+} ions bind to three high-affinity sites (EF-hand type), aequorin undergoes an irreversible reaction in which a photon is emitted (Fig. 3). The use of aequorin as a Ca^{2+} probe is based on the fact that, at $[Ca^{2+}]$ between 10^{-7} and 10^{-5} M, the concentrations normally occurring in the cytoplasm of living cells, there is a relationship between $[Ca^{2+}]$ and the fractional rate of aequorin consumption (i.e., L/L_{max}, where L_{max} is the maximal rate of light production at saturating Ca^{2+} concentrations) (Allen and Blinks, 1978; Cobbold, 1980; Ridgway and Ashley, 1967). Figure 3B shows the Ca^{2+} response curve of expressed recombinant aequorin under physiological conditions of pH, temperature, and ionic strength. It is apparent that, due to the cooperativity

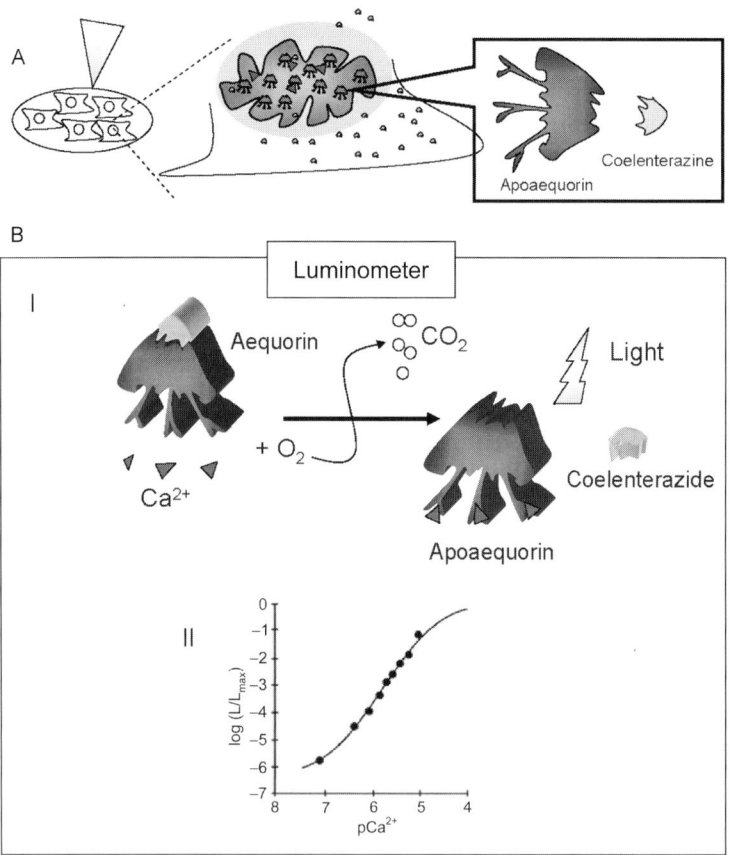

Fig. 3 Aequorin reporters. (A) Schematic representation of the Ca^{2+}-dependent photon emission in cells transfected with aequorin cDNA. The functional probe is composed of the apoaequorin moiety (synthesized by the cell) and coelenterazine (imported from the surrounding medium). (B) (I) Binding of calcium ions disrupts the binding of the prosthetic group, with concomitant light emission. (II) Relationship between calcium concentration and the rate of light emission by active aequorin (L and L_{max} are, respectively, the instant and maximal rates of light emission).

between the three binding sites, light emission is proportional to the second to third power of $[Ca^{2+}]$. This accounts for the excellent signal-to-noise ratio of aequorin. On the other, however, if the $[Ca^{2+}]$ is not homogeneous, aequorin it tends to bias the average luminosity toward the higher values (see below).

Aequorin is well suited for measuring $[Ca^{2+}]$ between 0.5 and 10 μM. However, it is possible to extend the range of measurable $[Ca^{2+}]$ by reducing the affinity of aequorin for Ca^{2+}. There are at least three different ways to reduce this affinity: (1) introducing point-mutations in the Ca^{2+}-binding sites of the apoprotein (Kendall *et al.*, 1992); (2) using surrogate cations that elicit a lower rate of photo-protein consumption than Ca^{2+}, for example, Sr^{2+} (Montero *et al.*, 1995); and (3) using modified prosthetic groups that alter the Ca^{2+}-dependent reaction. These three approaches can be combined to obtain a clear shift in the Ca^{2+} affinity of the photoprotein. Because of the cooperativity between the three Ca^{2+}-binding sites of aequorin, the aequorin mutant that we generated (Ala119 $\rightarrow$ Asp), which affects the second EF-hand domain, is suited for measuring $[Ca^{2+}]$ in the range of 10–100 μM. The range of aequorin sensitivity can be expanded further by employing divalent cations other than Ca^{2+}. We have used Sr^{2+}, a common Ca^{2+} surrogate: Sr^{2+} permeates across Ca^{2+} channels and is actively transported, although with a low affinity, by both the plasma membrane and the sarco-endoplasmic Ca^{2+} ATPases (SERCAs). By combining the two approaches, an aequorin-based ER probe can measure $[cation^{2+}]$ ranging from the μM to the mM (Montero *et al.*, 1995). To avoid possible discrepancies between the behavior of the two cations and to provide a more accurate estimate of the $[Ca^{2+}]$ in compartments with high $[Ca^{2+}]$, it is now possible to measure Ca^{2+} directly using a low-affinity coelenter-azine analog (coelenterazine n) (Montero *et al.*, 1997).

While the experimental protocols will be discussed in more detail below, we will review here briefly the main advantages and disadvantages of the recombinant aequorin approach.

1. The Advantages of Aequorin

1. *Selective intracellular distribution.* The ability to engineer the protein (e.g., add highly specific-targeting sequences or introduce substitutions that alter Ca^{2+}-binding affinity) is a key advantage of genetically encoded Ca^{2+} probes, such as aequorin and fluorescent indicator based on GFP.

2. *High signal-to-noise ratio.* Due to the low luminescence background of cells and the steepness of the Ca^{2+} response curve of aequorin, minor variations in the amplitude of the agonist-induced $[Ca^{2+}]$ changes can be easily detected with aequorin.

3. *Low Ca^{2+} buffering effect.* Although the binding of Ca^{2+} by aequorin may, in principle, affect intracellular Ca^{2+} homeostasis, this undesired effect is less delete-rious for aequorin compared to other fluorescent indicators. In fact, thanks to the excellent signal-to-noise ratio, aequorin is loaded at a concentration that is 2–3

orders of magnitude lower than other dyes, that is, usually from <0.1 μM (for the recombinantly expressed photoprotein) to ~1 μM (in the case of microinjected photoprotein for single cell studies) (Brini *et al.*, 1995).

4. *Wide dynamic range.* As is evident in Fig. 3B, aequorin can measure $[Ca^{2+}]$ accurately in the range of 0.5–10 μM, concentrations at which most fluorescent indicators are saturated. Indeed, thanks to these properties and low buffering effect, it is possible to estimate the large $[Ca^{2+}]$ rises that occur, for example, in neurons. Moreover, as described above, the range of $[Ca^{2+}]$ measurement can be increased to more than 100 μM.

5. *Possibility of coexpression with proteins of interest.* The ability to modify the molecular repertoire of a cell is one of the most powerful tools for dissecting complex, often interconnected signaling pathways. This is also true for calcium signaling, and one of the experimental tasks we often face is to measure Ca^{2+} concentrations in cells expressing a normal or mutated signaling component. It is not possible to load indicator dyes in the subset of transfected cells. Therefore, calcium measurements using indicator dyes are usually carried out using clones stably expressing the transgene (which may have altered behaviors as a result of natural variability of cell clones) or using laborious single-cell analysis of trans- fected cell populations. Aequorin provides an easy solution to the problem, because it can be coexpressed with the protein of interest. In transient expression studies, the Ca^{2+} probe is localized exclusively to the fraction of transfected cells (depending on the cell type), which are thus very representative of the behavior of the parental population (Bastianutto *et al.*, 1995; Lievremont *et al.*, 1997; Pinton *et al.*, 2000; Rapizzi *et al.*, 2002).

2. The Disadvantages of Aequorin

1. *Low-light emission.* The major disadvantage in the use of aequorin is the low amount of light emitted by the photoprotein. Each aequorin molecule emits only one photon, and only a small fraction of the pool of photoproteins ($<10^{-3}$) emits light throughout the experiment. This is not a major problem when light output form aequorin from an entire coverslip of transfected cells is measured, as described above. However, it is often desirable to carry out single cell-imaging experiments (e.g., to detect Ca^{2+} waves or local changes in Ca^{2+} concentration). With aequorin, such analyses are quite difficult. Special imaging systems are needed, with enhanced sensitivity at the cost of lower spatial resolution. Thus, although single cell-imaging experiments with recombinant aequorin have been carried out and have provided interesting biological information (Rutter *et al.*, 1996), this approach is technically difficult and yield low image quality.

2. *Overestimation of the average rise in cells (or compartments) with nonhomo- geneous behavior.* Due to the steep Ca^{2+} response curve of aequorin, if the probe is distributed between areas of high and low $[Ca^{2+}]$, the former undergoes a much

greater emission of light. The total signal will be calibrated as an "average" $[Ca^{2+}]$ increase that will be severely biased by the region at high Ca^{2+}. In other words, given that an increase in 1 pCa unit causes a 100 to 1000-fold increase in light emission, a tiny volume undergoing a very large increase in $[Ca^{2+}]$ will increase total light output drastically and thus will be "interpreted" by the calibration algorithms as a moderate Ca^{2+} rise throughout the cells. Conditions can be envisaged in which this effect can become very significant. For example, a Ca^{2+} rise of 3 μM detected in neurons could be, in fact, lower, but severely biased by hotspots occurring throughout the dendritic tree.

3. *The loading procedure*. The obvious requirement of the recombinant reporter approach is that a cell must be amenable to transfection. Significant improvements in transfection techniques have made this problem less severe. In our experiment, all cell types can be transfected, with either using simple calcium phosphate procedure (which proves effective in most cell types) or using other techniques (liposomes, gene gun, electroporation). Conversely, the need to maintain the cells in culture for enough time to produce the recombinant protein can be an important limitation, rendering the extension of the approach to interesting cell types (e.g., pancreatic acinar cells, spermatozoa, and so on) quite difficult.

B. Procedure

We have employed a wide variety of cell types (e.g., HeLa, CHO, COS, neurons, myocytes) with good results in terms of aequorin expression. In all cases, the cells are seeded in coverslips (13-mm diameter) and allowed to grow until about 50% confluency. At this point, the cells are transfected with 4 μg of mtAEQ or 0.5 μg of mimsAEQ, using the calcium phosphate procedure. With mimsAEQ, we noticed that in a few cases high levels of expression had deleterious effects on the transfected cell population. This problem was solved by reducing the quantity of DNA used for transfection.

After 36 h, aequorin is reconstituted (Fig. 3A) by adding the prosthetic group to the incubation medium (5-μM coelenterazine for 2 h in DMEM supplemented with 1% FCS at 37 °C in 5% CO_2 atmosphere). The cells are then washed and transferred to the perfusion chamber of the measuring system.

After reconstitution ($\sim$2 h), the coverslip is placed in the chamber of the measuring system, where it is perfused with modified Krebs-Ringer buffer saline solution (modified KRB: 125-mM NaCl, 5-mM KCl, 1-mM Na_3PO_4, 1-mM $MgSO_4$, 5.5-mM glucose, 20-mM HEPES, pH 7.4, 37 °C).

The schematic representation of the measuring system is depicted in Fig. 4. In this system, the perfusion chamber, on top of a hollow cylinder (with temperature control maintained by a thermostatted by water jacket), is perfused continuously with buffer using a peristaltic pump. Agonists and drugs are added to the same KRB medium.

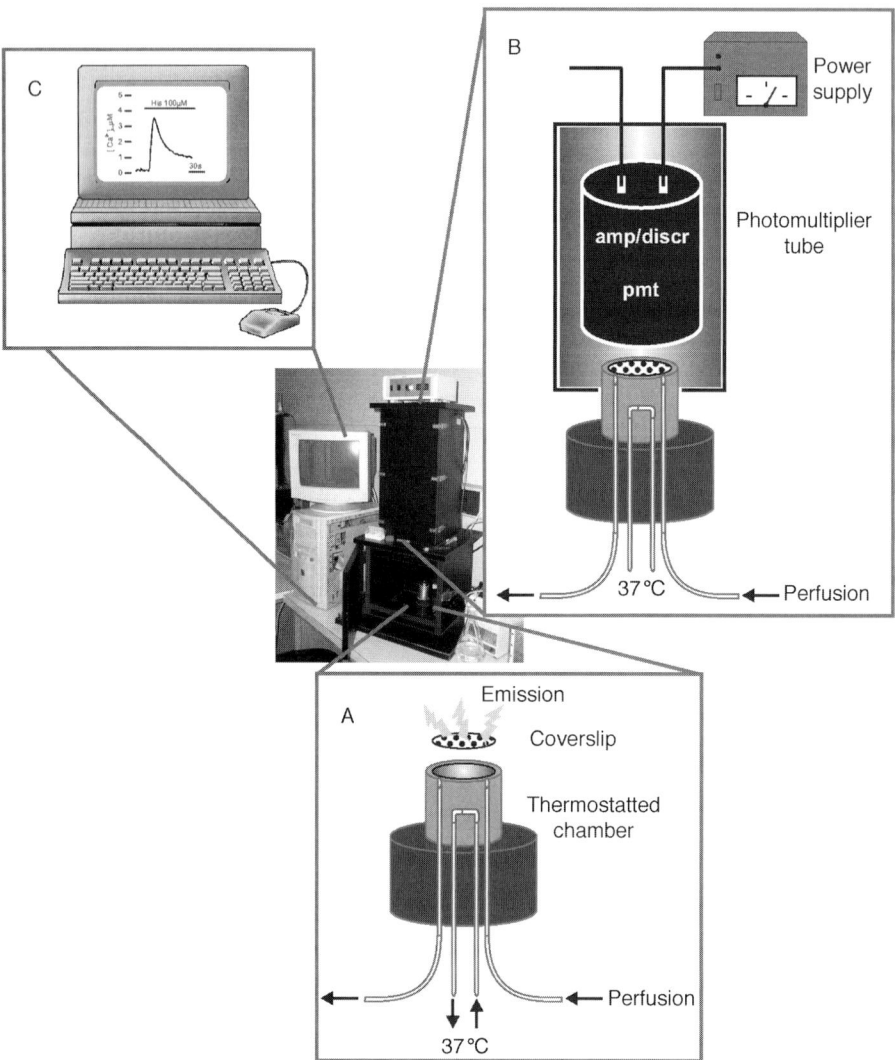

Fig. 4 Experimental setup to measure light production. (A) A thermostatted (37 °C) chamber, engineered with a perfusion system, is the support for cells expressing the functional aequorin probe. Cells are plated on a coverslip which is fit into a notch positioned at the top of the chamber and then sealed with a larger coverslip. (B) The chamber is in proximity to a photomultiplier tube (pmt), with a built-in amplifier/discriminator (amp/discr) that is able to capture the light emitted in the reaction. The complete assemblage is in the dark to minimize extraneous signals. (C) An analysis system (IBM-compatible computer) records and captures the photon emission data and converts them into Ca^{2+} concentration values, using the aequorin calibration curve.

The cell coverslip is placed at few millimeter away from the surface of a low-noise phototube. The photomultiplier is kept in a dark, refrigerated box. An amplifier discriminator is built into the photomultiplier housing. The pulses generated by the discriminator are captured by a Thorn EMI photon counting board, connected to an IBM-compatible computer. The board allows the storing of the data in the computer memory for further analyses.

To calibrate the crude luminescent signal in terms of $[Ca^{2+}]$, we have developed an algorithm that takes into account the instantaneous rate of photon emission and the total number of photons that can be emitted by the aequorin in the sample (Brini *et al.*, 1995). To determine the latter, emitted light is measured in the sample, after cells used in an experiment are lyzed by perfusion with hypo-osmotic medium containing 10-mM $CaCl_2$ and detergent (100-μM digitonin). This discharges all the aequorin that was not consumed during the experiment.

1. Agonist Stimulation Causes Large and Rapid Mitochondrial Ca^{2+} Uptake in Most Cell Types

To stimulate an increase in $[Ca^{2+}]_c$, we use an agonist—histamine or ATP—that, acting on a G-protein-coupled receptor, leads to the generation of inositol 1,4,5-trisphosphate (IP_3) (Fig. 5A). Briefly, histamine or ATP (100 μM, final concentration) is added to the perfusion medium (modified KRB buffer) for 1 min, followed by rinsing in the same buffer. The biphasic kinetics of $[Ca^{2+}]_c$ in HeLa cells transfected with wild-type aequorin (without any targeting sequences and thus localized to the cytosol) is shown in Fig. 5C. The release of Ca^{2+} from intracellular stores causes an initial rapid, but transient, increase of $[Ca^{2+}]_c$ (to values of $\sim$2.5 μM), followed by a sustained increase of $[Ca^{2+}]_c$ above normal basal levels throughout the stimulation period.

Figure 5B shows the measurements obtained in HeLa cells transiently expressing mtAEQ or mtAEQmut. On stimulation with histamine, there is a rapid increase in $[Ca^{2+}]_m$ up to $\sim$10 μM (mtAEQ) or 100 μM (mtAEQmut), followed by the return, within 2 min, to almost basal values. In turn, the rise in $[Ca^{2+}]_m$ stimulates the activity of Ca^{2+}-dependent enzymes of the Krebs cycle, and hence increases mitochondrial ATP production (Jouaville *et al.*, 1999).

Similar studies can be conducted in virtually any cell type capable of being transfected. Indeed, large $[Ca^{2+}]_m$ changes were observed in a variety of cell lines and in different primary cultures (Pinton *et al.*, 1998).

2. The mimsAEQ Chimera Reveals Ca^{2+} Rises Higher Than Those of the Bulk Cytosol

The fact that alterations in $[Ca^{2+}]_m$ are much larger than those observed for $[Ca^{2+}]_c$ were unexpected, given the low affinity of the mitochondrial uniporter for Ca^{2+}. However, this apparent contradiction could be explained if the endoplasmic reticulum (ER) and the mitochondria have a close physical relationship, allowing mitochondria be exposed to microdomains of high $[Ca^{2+}]$ generate by release of Ca^{2+} from the ER (Rizzuto *et al.*, 1993, 1998b).

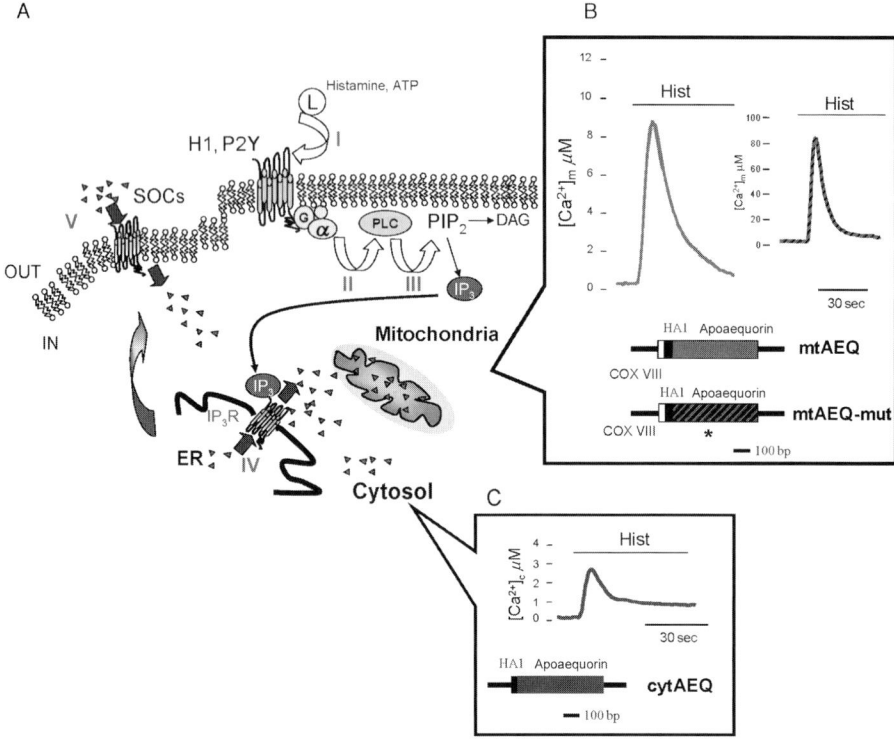

Fig. 5 Calcium measurements in mitochondria. (A) When an extracellular agonist, for example, histamine or ATP, binds to its receptor (I), a G-protein–coupled receptor is activated (II), with consequent stimulation of phospholipase C (III). Thus, phosphatidylinositol-4,5-bisphosphate is hydrolyzed to diacylglycerol and IP₃, which diffuses through the cytosol and binds to its IP₃ receptors on internal stores of Ca^{2+}. Ca^{2+} is released (IV) and an increase of $[Ca^{2+}]_c$ and $[Ca^{2+}]_m$ occurs. Emptying of the store is detected and is transmitted to the store-operated channels (SOCs) to induce Ca^{2+}-influx across the plasma membrane (V). (B) Schematic maps of the mt-AEQ-wt and -mut cDNAs. Coding and noncoding regions are represented as boxes and lines, respectively. In the coding region, the portions encoding mitochondrial targeting sequence (subunit of cytochrome *c* oxidase), the HA1 epitope (HA1) and apoaequorin are white, black and gray–gray/black, respectively. For mtAEQ-mut, the asterisk indicates a mutation in the Ca^{2+}-binding sites (Asp-119 → Ala), which causes ~20-fold decrease in the Ca^{2+} affinity of the photoprotein. Traces show the monitoring of $[Ca^{2+}]_m$ in HeLa cells transfected transiently with the probes. Where indicated, the cells were treated with 100-μM histamine (Hist), added to KRB. (C) Schematic map of the cyt-AEQ cDNA. In the coding region, the HA1 epitope (HA1) and apoaequorin are black and dark gray, respectively. The figure also includes a representative measurement showing the $[Ca^{2+}]_c$ rise evoked by histamine stimulation.

To analyze the possible existence of such high Ca^{2+} microdomains, we used a different aequorin chimera which is targeted to the mitochondrial intermembrane space (mimsAEQ). Since the OMM is freely permeable to ions and small molecules, aequorin molecules present between the two mitochondrial membranes are located in a region that is in rapid equilibrium with the cytosolic portion in contact with the organelle. Such a chimera is thus sensitive to changes in Ca^{2+} in the cytosolic region immediately adjacent to mitochondria (Rizzuto *et al.*, 1998b).

HeLa cells transfected with mimsAEQ and stimulated with histamine show a biphasic response, as shown in Fig. 6A. An initial rise in $[Ca^{2+}]_{mims}$ to ~3.5 μM is followed by a rapid decrease, that gradually levels out to values above the initial $[Ca^{2+}]$. This type of response is due to the two mechanisms of action of the agonist, namely the release of Ca^{2+} from intracellular stores and the entry of Ca^{2+} from the extracellular medium. The comparison of the responses of $[Ca^{2+}]_c$ and $[Ca^{2+}]_{mims}$ to histamine shows a clear difference only in the initial phase (the phase that is due to the release of Ca^{2+} from intracellular stores). These data support the hypothesis that the opening of IP_3-sensitive channels in the ER that are in proximity to mitochondria generate microdomains of high $[Ca^{2+}]$. Indeed, the $[Ca^{2+}]$ in such microdomains is much higher than the average $[Ca^{2+}]_c$, and thus the low affinity Ca^{2+} uptake systems present in mitochondria are capable of accumulating Ca^{2+} efficiently in the matrix of the organelle. Conversely, depletion of Ca^{2+} in the ER by a different mechanism, such as passive diffusion after blocking of the ER Ca^{2+} accumulation via ER Ca^{2+}-ATPase, caused the same rise (Fig. 6B).

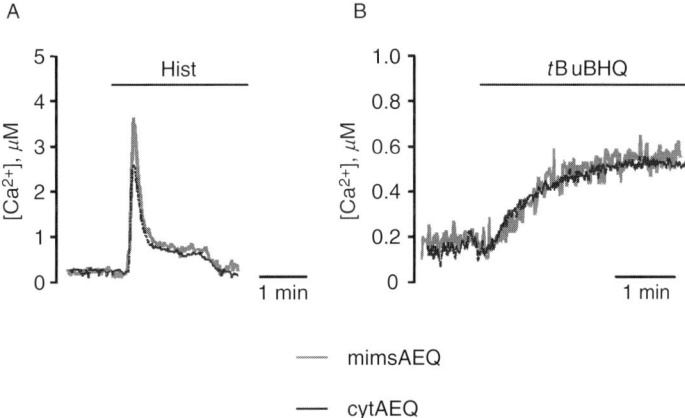

Fig. 6 Calcium measurement in the MIMS. (A) The effect of histamine on the $[Ca^{2+}]$ of the cytosol ($[Ca^{2+}]_c$) and MIMS ($[Ca^{2+}]_{mims}$). The traces show $[Ca^{2+}]$ of the two compartments in parallel batches of HeLa cells, transfected transiently with the appropriate aequorin chimera (cytAEQ or mimsAEQ). All conditions are as in Fig. 5. Where indicated, the cells were treated with 100-μM histamine (Hist), added to KRB. (B) The effect of inhibition of the sarco-endoplasmic Ca^{2+}-ATPase (SERCA) on $[Ca^{2+}]_c$ and $[Ca^{2+}]_{mims}$. All conditions are as in Fig. 5. Where indicated, the cells were challenged with 10-μM 2,5-di(*tert*-butyl)-1,4-benzohydroquinone (*t*BuBHQ), an inhibitor of the SERCAs (Kass *et al.*, 1989).

3. A More Detailed Analysis of the Mitochondrial Localization of Aequorin Probes

Given the targeting signal, chimeric protein mtAEQ is expected to be localized to the mitochondrial matrix. However, the immunofluorescence data (see above) indicated only a general mitochondrial localization of mtAEQ, and it was not possible to discriminate its exact localization within the organelle. To ensure that our $[Ca^{2+}]$ measurements were indeed due to changes in the mitochondrial matrix, we analyzed the effect of uncouplers on the $[Ca^{2+}]$ values generated by our chimeric probe. Since Ca^{2+} entry into the mitochondrial matrix is driven by the electrochemical gradient across the IMM, collapse of this gradient with an uncoupler should abolish the changes in $[Ca^{2+}]$ measured by our probe. Using FCCP, a compound that dissipates the proton gradient across the IMM, the response to the histamine stimulus was abolished, as shown in Fig. 7A. This confirms that mtAEQ is localized to the mitochondrial matrix and that calcium accumulation occurs via the known uptake mechanism (the Ca^{2+} uniporter). Similar results were obtained using RR which inhibits calcium channels. Since cells are not permeable to RR,

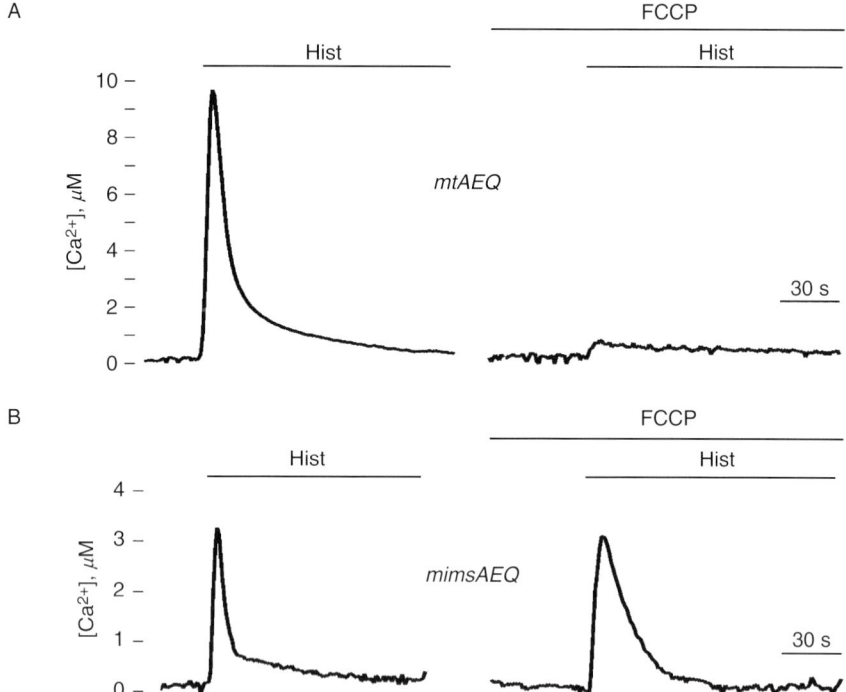

Fig. 7 Effect of FCCP on mitochondrial calcium measurements. (A) $[Ca^{2+}]_m$ changes on stimulation with histamine, in the absence and presence of FCCP (left and right panels, respectively). (B) $[Ca^{2+}]_{mims}$ changes on stimulation with histamine, in the absence and presence of FCCP (left and right panels, respectively).

cells were first permeabilized to all RR to gain access to the Ca^{2+} uniporter in mitochondria. Calcium uptake was then initiated by supplementing the medium with known concentrations of Ca^{2+}. On challenging with 2-μM free Ca^{2+}, $[Ca^{2+}]_m$ increased to over 3 μM, but in the presence of RR, uptake was almost completely abolished (data not shown).

To determine the exact location of mimsAEQ, we performed a similar series of experiments. As noted above, we had previously determined that the collapse of the proton gradient drastically reduces the accumulation of Ca^{2+} in the matrix, as visualized using mtAEQ. If mimsAEQ were mistargeted to the matrix, the rise in $[Ca^{2+}]$ should be abolished in the presence of uncouplers. Conversely, if mims-AEQ were localized in the intermembrane space, then the presence of uncouplers should not affect the $[Ca^{2+}]$ values obtained. As shown in Fig. 7B, HeLa cells transfected with mimsAEQ and treated with FCCP did not show a significant difference in $[Ca^{2+}]$ dynamics, when compared to similar, coupled cells, strongly suggesting that the aequorin moiety of mimsAEQ lies in the MIMS.

Taken as a whole, these experiments demonstrate that aequorin is useful as a calcium probe in a wide range of circumstances. Aequorin can be targeted to different subcellular locations (including different organelle compartments), where it can sense specific calcium changes (Chiesa *et al.*, 2001). Moreover, the use of different mitochondrial drugs can be used to analyze calcium changes in a variety of conditions, as exemplified by, but not limited to, the cases documented above.

C. Mitochondrial Calcium Measurements Using GFP

In nature, GFP binds to aequorin (Chalfie, 1995; Chalfie *et al.*, 1994; Rizzuto *et al.*, 1996). Indeed, it is produced by the same jellyfish (*A. victoria*) and is situated in close association with the photoprotein, acting as a natural fluorophore that absorbs the blue light emitted by the aequorin and reemits photons of a longer wavelength (thus accounting for its own name and for the greenish hue of the jellyfish luminescence) (Ormo *et al.*, 1996). In research applications, GFP retains its fluorescence properties, and thus can be added to the long list of probes in the cell biologist's toolbox. Some of its unique properties account for its explosive success, and has made it a powerful and versatile tool for investigating virtually all fields of cell biology, ranging from the study of gene expression to protein sorting, organelle structure, and measurements of physiological parameters in living cells (Rizzuto *et al.*, 1996; Zhang *et al.*, 2002).

The main reason for the success of GFP is its own nature: the fluorescent moiety is a gene product (with no need for cofactors) that is open to molecular engineering and transient or stable expression in virtually every cell type. Moreover, its mutagenesis has allowed for the adaptation of its fluorescence properties to different experimental needs (for a detailed description see the review by Zhang *et al.*, 2002). Equally important, with the notable exception of the blue mutant, GFP is strongly resistant to photobleaching. As a result, it can be used for applications including prolonged laser illumination during confocal microscopy. For these reasons, GFP has replaced available probes for many applications.

GFPs can also be used as sensors for physiological parameters. To overcome the problems of single cell-imaging experiments using aequorin probes, researchers have tried to exploit GFP to measure intracellular $[Ca^{2+}]$. Different Ca^{2+} indicators have been developed, by fusing of one or two GFP derivatives with a Ca^{2+}-binding protein, such as calmodulin (CaM) and/or a CaM-binding peptide. These chimeric probes are called cameleons, camgaroos, and pericams (Filippin *et al.*, 2005; Rudolf *et al.*, 2003; Zhang *et al.*, 2002). In general, these indicators can be classified depending on the process that causes the calcium signal. Cameleons are based on a fluorescent resonance energy transfer (FRET) process between two variants of GFP. Camgaroos and pericams can be classified as environment-sensitive GFPs. There are also fusion proteins (fusion aequorin-GFP and aequorin-RFP [red fluorescent protein]) that exhibit chemiluminescence resonance energy transfer (CRET or BRET) on binding calcium to aequorin.

1. Cameleons

FRET occurs only if the fluorescence emission spectrum of the "donor" GFP overlaps with the excitation spectrum of the "acceptor" GFP, and if the two fluorophores are located within few nanometers of each other and in a favorable orientation. Any alterations of these parameters can drastically alter the efficiency of FRET. Indeed, the rate of energy transfer is proportional to the sixth power of the distance $E = [1 + (R \backslash R_0)^6]^{-1}$, where R/R_0 is the fractional change in distance, and thus becomes negligible when the two fluorophores are >5–6 nm apart. The earliest use of FRET between GFP mutants was reported by Heim and Tsien (1996), who linked two fluorophores via a sequence of 25 amino acids, including a trypsin cleavage site. After addition of trypsin, the short peptide was cleaved and FRET was reduced drastically.

Cameleons are genetically encoded fluorescent indicators for Ca^{2+} based on chimeric proteins composed of four domains: a blue (BFP) or cyan (CFP) variant of GFP (the FRET donor), CaM, a glycylglycine linker, the CaM-binding domain of myosin light chain kinase (M13), and a green (GFP) or yellow (YFP) version of GFP (the FRET acceptor) (Miyawaki *et al.*, 1997, 1999). Modifications to improve and/or modulate their response have been performed in the GFP units and also in the calcium switch; as a result, different cameleons have been developed.

Cameleons change their structure on Ca^{2+} binding. The fluorescent proteins joined at both ends stand away from each other in the absence of calcium ion, and under these conditions thus excitation of CFP yields light emission mostly from CFP itself. This conformation has very low FRET probability. However, in the presence of calcium ion, the activated CaM wraps around the M13 protein. The structure of Cameleon then changes and reduces the distance between CFP and YFP. This conformational chkange makes CFP and YFP come together, and both units adopt an orientation that induces an increase in FRET (FRET depends on distance between donor and acceptor and also on their relative orientation).

Yellow cameleons (YCs) are improved versions of cameleons composed of enhanced CFPs and YFPs. These constructs are more efficient than the original one, composed of BFP and GFP. BFP bleaches quickly, has a very low quantum yield, and requires UV excitation (excitation at ~430 nm), which might be cyto-toxic and should be avoided. Conversely, YCs are brighter, more robust, and are excited by illumination that is less damaging.

YCs are classified in different groups, depending on their binding domains. Those in the YC2 group contain an intact CaM with high affinity for Ca^{2+}, while YC3 and YC4 have a mutation in CaM that lowers its affinity for calcium. For this reason, YC2 cameleons are high-affinity indicators ($K_d \sim 1.2\ \mu M$) and YC3 and YC4 are low-affinity indicators ($K_d \sim 4\ \mu M$ and $K_d \sim 105\ \mu M$, respectively). Different modifications in the YFP have been introduced to improve the properties of YCs. The original YCs were named as YC2, YC3, and YC4. Their first variants, YC2.1 and YC3.1, contain a mutation on YFP that decreases its sensitivity to acidification (rendering the signal unaffected at pH ≥ 6.5). Other mutations introduced into YFP allow for quicker maturation (Venus is a brighter version of YFP) or improve the cameleon's dynamic range.

The main drawback of most YCs is their poor dynamic range: the maximum variation between R_{max} and R_{min} obtained with YC3.12 is about 20%. However, the latest reported variant, YC3.60 (Nagai *et al.*, 2004), has a promising dynamic range of nearly 600% which dramatically improves signal-to-noise ratio.

Cameleons are used as ratiometric indicators to increase the signal-to-noise ratio (in the case of the CFP–YFP pair, emission is measured at 480 and 535 nm). An example of mitochondrial $[Ca^{2+}]$ measurements using targeted cameleons is shown in Fig. 8A.

2. Camgaroos and Pericams

Camgaroos and pericams are based on the fact that GFP and its mutants do not lose their ability to fold and form the chromophore if they are modified at tyrosine-145. This allows insertion of peptides between positions 145 and 146, and is the origin of camgaroos and pericams. In the case of camgaroos (Fig. 8B), a CaM unit is bound to positions 145 and 146 of YFP. When calcium interacts with CaM, the protein changes its conformation and induces an increase in fluores-cence of YFP.

Pericams (Fig. 8C) are obtained by a more complex approach; the linear sequence of YFP is cleaved between positions 145 and 146 (generating new C and N termini at the cleavage site), while the original C and N termini are fused. This modification is called circular permutation. The linkage of CaM and CaM-binding domain of myosin light chain kinase (M13) to the new C and N termini makes pericams sensitive to calcium. Three types of pericams have been developed by mutating several amino acids adjacent to the chromophore.

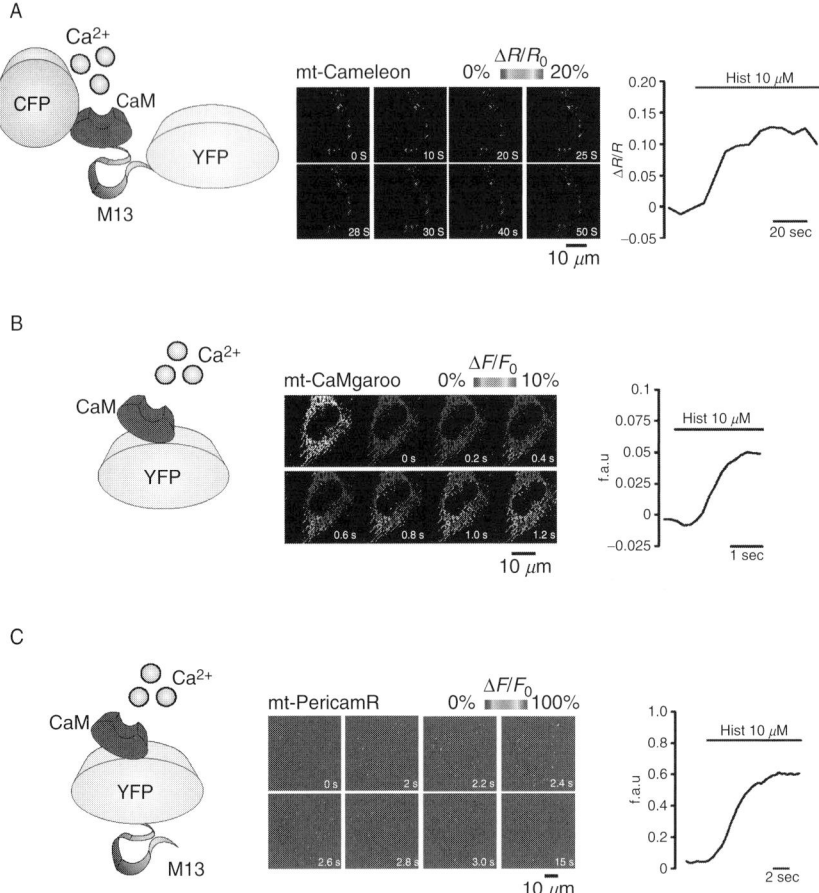

Fig. 8 Camgaroos and pericams. (A) $[Ca^{2+}]$ measurement using mtCaMgaroo-2. On the *left*, the schematic structure of CaMgaroo is shown. On the *right* the pseudocolor-rendered images of ratio values (F/F_0) show the changes in fluorescence after a 5-sec puff of histamine (10 μM). At each time point (200 msec apart), the $[Ca^{2+}]$ is measured in mitochondria. The trace representing the changes in fluorescence (f.a.u., fluorescence arbitrary units) of $[Ca^{2+}]_m$ during histamine challenge. (B) $[Ca^{2+}]$ measurement using mtPericamR. On the *left*, the schematic structure of mitochondrial ratiometric-pericam is shown; on the *right* is a representative experiment. The cells were illuminated at 410 nm at 100-msec intervals. The pseudocolor images show the changes in fluorescence after stimulation with 10-μM histamine. Ca^{2+} concentrations are color coded, with a basal Ca^{2+} concentration in blue and a high Ca^{2+} concentration in red. The *black trace* represents the kinetic changes of the $\Delta F/F_0$ ratio over large mitochondrial region of a single representative cell. (C) Mitochondrial calcium response measurement using mtYC2.1. On the *left*, the schematic structure of mitochondrial targeting Cameleon (mtYC 2.1) is shown; on the *right* is a representative experiment. Emission ratio images were obtained at 440 nm excitation and 510/20 nm (YFP) and 460/20 nm (CFP) emission, and are rendered in pseudocolor. The normalized trace represents the $[Ca^{2+}]_m$ change in HeLa cells stimulated with 100-μM histamine. (See Color Insert.)

Of these, "flash-pericam" became brighter with Ca^{2+}, whereas "inverse-pericam" became dimmer. On the other hand, "ratiometric-pericam" had an excitation wavelength that changed in a Ca^{2+}-dependent manner (Nagai *et al.*, 2001).

Compared with cameleons, pericams show greater Ca^{2+} responses and improved signal-to-noise ratio (Nagai *et al.*, 2001). By contrast to low Ca^{2+}-affinity camgaroos ($K_d = 7$ μmol/l) (Baird *et al.*, 1999), pericams have higher Ca^{2+} affinity ($K_d = 0.7$ μmol/l), which is favorable for sensing physiological Ca^{2+} changes.

Both kind of indicators are usually employed in nonratiometric measurements and are affected by changes in pH. An example of mitochondrial $[Ca^{2+}]$ measurements using targeted camgaroos and pericams is reported in Fig. 8B and C, respectively.

Figure 9 shows an example of the use of those probes for high-speed single cell imaging of mitochondrial Ca^{2+} changes simultaneously with cytosolic measurements

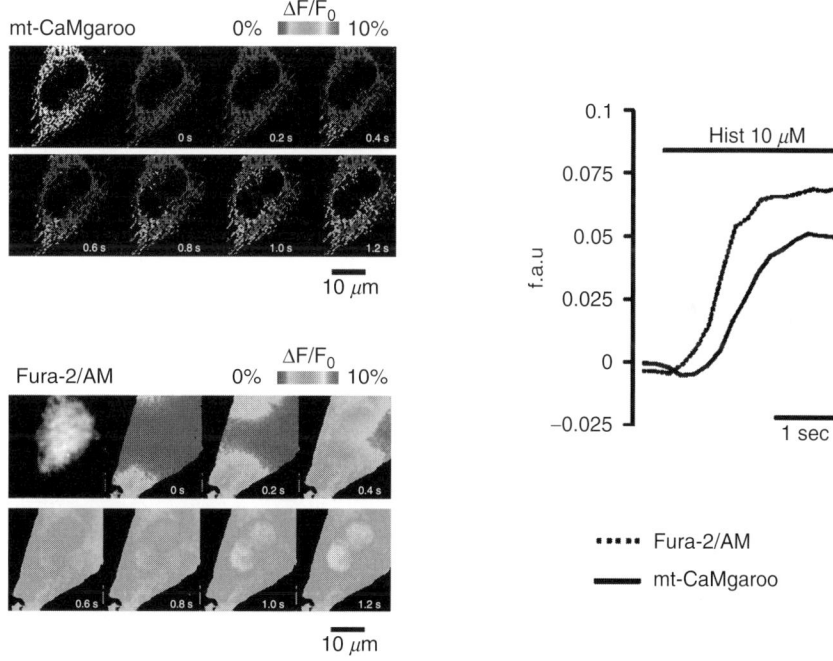

Fig. 9 Simultaneous single cell imaging of mitochondrial and cytosolic $[Ca^{2+}]$ changes in single cell. The imaging of Fura-2/AM (bottom) and mt-CaMgaroo (top) is shown. The black and white images show Fura-2/AM and three-dimensional CaMgaroo distribution within the cells. The pseudocolored images (350 nm/380 nm ratio for Fura-2/AM and 514 excitation for CaMgaroo) show the changes in fluorescence after a 5-sec puff of histamine (Hist 10 μM). At each time point (200 msec apart), the $[Ca^{2+}]$ is measured simultaneously in the cytosol and mitochondria. On the right, the traces representing the changes in fluorescence (f.a.u., fluorescence arbitrary units) of $[Ca^{2+}]_c$ (dotted lines) and $[Ca^{2+}]_m$ (continuous lines) are shown. (See Color Insert.)

(using an organelle-targeted camgaroo for the mitochondria and the Fura-2 for the cytosol). The results demonstrated that the $[Ca^{2+}]_c$ rise evoked by cell stimulation is followed by a delayed upstroke of $[Ca^{2+}]_m$ (Rapizzi *et al.*, 2002).

D. Procedure

Cells grown on 24-mm diameter coverslips are transfected with 8 μg of DNA bearing the GFP-based sensor of interest. After 36 h, the coverslip is placed on a thermostatted chamber in KRB saline solution and $[Ca^{2+}]$ changes are determined using a high-speed, wide-field digital-imaging microscope. A microperfusion system (ALA DVD-12, Ala Scientific, NY), allowing rapid solution exchange, can be used to apply stimuli. Perfusion and image acquisition are controlled by MetaFluor 5.0 software (Universal Imaging Corporation, PA).

The emission ratio of cameleon (excitation wavelength of 430 nm and emission at 480 and 535 nm) is acquired at 2 frames per second using a CoolSNAPHQ interline CCD camera (Roper Scientific, NJ) using a 440/20-nm excitation filter, 455DCLP dichroic, and 485/40-nm or 535/30-nm emission filters (Chroma Technology, VT) placed in an emission filter wheel controlled by Lambda 10–2 (Sutter Instrument, Novato, CA). For fast kinetic imaging of $[Ca^{2+}]_m$, the probe is excited at 410 nm, using a random access monochromator (Photon Technology International, NJ). Images are acquired at 5–10 frames per second using a BFT512 back-illuminated camera (Princeton Instruments, AZ). Dual-excitation imaging with ratiometric pericam or camgaroo (two excitation wavelengths at 415 and 490 nm, and emission at 525 nm) use two excitation filters (480DF10 and 410DF10), which are alternated by a filter changer (Lambda 10–2, Sutter Instruments), a 505 DRLP-XR dichroic mirror, and a 535DF25 emission filter (Rizzuto *et al.*, 1998a).

IV. pH Reporters

All cells contain an intracellular fluid with a defined and highly controlled pH value—this is known as the intracellular pH (pHi). The pHi plays a critical role in cellular function, and tight regulation of pH is required for cell survival. There are numerous mechanisms that can cause pHi to change, including metabolic acid production, leakage of acid across plasma and organelle membranes, and membrane transport processes. pHi regulates many cellular processes, including cell metabolism (Roos and Boron, 1981), gene expression (Isfort *et al.*, 1993, 1996), cell–cell coupling (Orchard and Kentish, 1990), and cell death (Gottlieb *et al.*, 1996).

In mitochondria, it is universally accepted that the H^+ electrochemical potential ($\Delta\mu_{H^+}$), generated by electron transport across the inner membrane coupled to H^+ ejection on the redox H^+ pumps, is used to drive ATP synthesis by the F_oF_1-ATP synthase. In addition to ATP synthesis, $\Delta\mu_{H^+}$ supports a variety of mitochondrial processes, some of which are a prerequisite for respiration and ATP synthesis, such as the uptake of respiratory substrates and phosphate, the uptake of ADP in exchange for ATP and of Ca^{2+} ions, the transhydrogenation reaction, and the import of respiratory chain and ATP synthase subunits encoded by nuclear genes. $\Delta\mu_{H^+}$ comprises both the electrical ($\Delta\psi$) and the chemical component of the H^+ gradient (ΔpH). It is well established that under physiological conditions, $\Delta\psi$ represents the dominant component of $\Delta\mu_{H^+}$, whereas the ΔpH gradient is small. The relative contribution of $\Delta\psi$ and ΔpH across the mitochondrial membrane can be changed by the redistribution of permeant ions, such as phosphate, Ca^{2+}, or K^+, in the presence of the ionophore valinomycin. Several techniques have been developed for the determination of $\Delta\psi$ and ΔpH in isolated mitochondria. Attempts have been also made to measure $\Delta\psi_m$ in intact cells (for a critical evaluation of these methods, see Bernardi et al., 1999). However, measurements of matrix pH have been elusive, due to the difficulty of separating the mitochondrial signal from that in the surrounding cytoplasm (Chacon et al., 1994). One of the most promising tools developed to overcome this difficulty the use of recombinant pH-sensitive fluorescent proteins targeted specifically to the mitochondrial compartments.

A. Mitochondrial pH–Sensitive Fluorescent Proteins

The pH sensitivity of GFP, and in particular of some of its mutants, is usually a disadvantage, confounding the interpretation of other data (e.g., FRET-based measurements of $[Ca^{2+}]$). In the case of the measurement of pH, however, this disadvantage can be turned to an advantage. Both purified and recombinant wild-type GFP, and most of the GFP mutants, are influenced by pH, suggesting that pHi sensitivity is a general property of this entire class of proteins (Patterson et al., 1997; Ward and Bokman, 1982). Early after the first reports on the use of GFP as a general reporter, GFP mutants appeared promising as pH sensors in living cells. This possibility was first investigated by Kneen et al. (1998), who reported that mutations S65T and F64L/S65T in wild-type GFP exhibited fluorescent spectra and pH titration data indistinguishable from the unmutated polypeptide, with a pK_a value of 5.98. In addition, the finding that the titration curve for GFP-S65T was similar to that of fluorescein suggested the involvement of only a single amino acid residue in its pH-sensitive mechanism (Kneen et al., 1998). Crystallographic analysis of this GFP mutant at both basic (pH 8.0) and acidic (pH 4.6) pH identified a model, likely valid for other mutants as well, where the phenolic hydroxyl of Tyr-66 is the site of protonation (Elsliger et al., 1999).

The same GFP mutant (F64L/S65T), termed GFPmut1, was expressed heterologously in the cytosol and nuclear compartments of BS-C-1 rabbit proximal tubule cells. In this study, comparison of GFPmut1 with the pH-sensitive dye BCECF fluorescence showed a uniform agreement between pHi estimates using the two methods (Robey et al., 1998).

Another pH-sensitive GFP mutant, named enhanced yellow fluorescent protein (EYFP), was made by introducing the amino acid substitutions S65G/S72A/T203Y. The absorbance spectrum of EYFP in vitro (Fig. 10A) shows that the

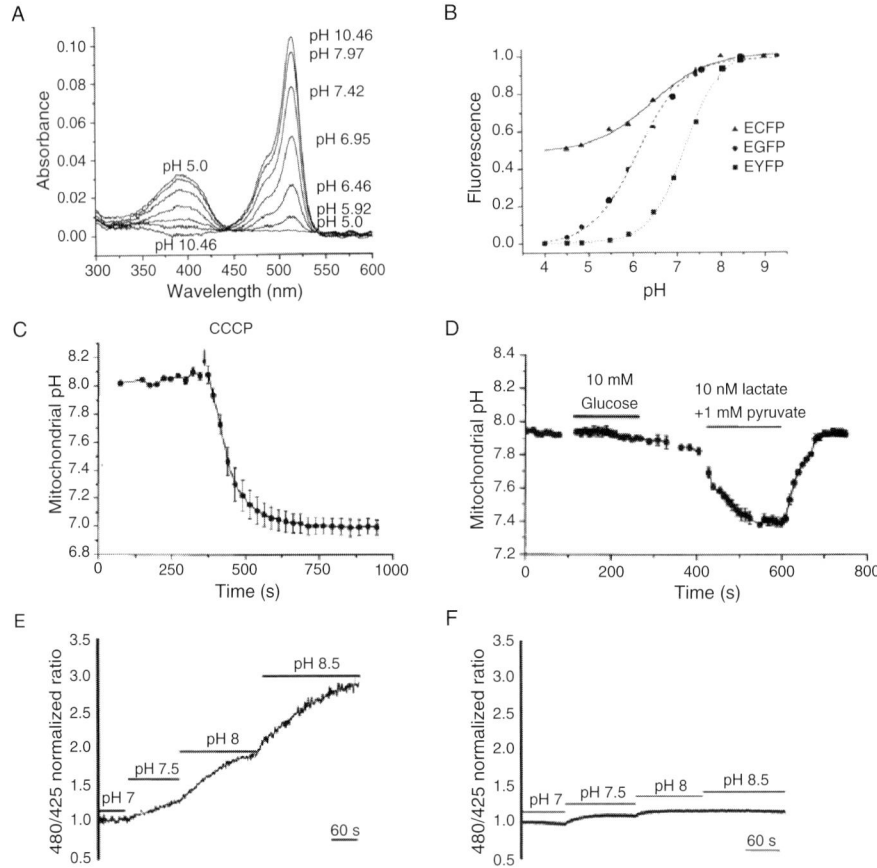

Fig. 10 pH reporters. (A) pH-dependent absorbance of EYFP (modified from Llopis *et al.*, 1998). (B) pH dependency of fluorescence of various GFP mutants. The fluorescence intensity of purified recombinant GFP mutant protein as a function of pH was measured in a microplate fluorometer (modified from Llopis *et al.*, 1998). (C) The uncoupler CCCP collapses mitochondrial pH (modified from Llopis *et al.*, 1998). (D) Effect of glucose and lactate/pyruvate perfusion on mitochondrial pH (modified from Llopis *et al.*, 1998). (E and F) HeLa cells coexpressing mtAlpHi and mtECFP were perfused with mKRB, supplemented with the ionophores nigericin and monensin (5 μM). The pH was increased in 0.5-unit steps, from 7.0 to 8.5, and the fluorescence of both probes was monitored. For comparison, HeLa cells coexpressing mtEYFP and mtECFP were subjected to the same protocol. Whereas mtAlpHi yields a signal that increases by threefold between pH 7 and 8.5, mtEYFP shows an increase of only <15% (modified from Abad *et al.*, 2004).

absorbance of the peak at 514 nm increased from pH 5–8 (Llopis *et al.*, 1998). The apparent pK_a (pK_a') of EYFP is 7.1, compared to a pK_a' of EGFP of 6.15. The change in fluorescence of enhanced cyan fluorescent protein (ECFP) with pH (pK_a', 6.4) is smaller than that of EGFP or EYFP (Fig. 10B).

The suitability of these GFP mutants as pH indicators in the mitochondrial matrix of living cells was evaluated in CHO (Kneen *et al.*, 1998), HeLa cells, and in rat neonatal cardiomyocytes (Llopis *et al.*, 1998). In CHO cells expressing recombinant GFPmut1 targeted to the mitochondrial matrix, a qualitative reversible acidification of mitochondria was observed after addition of the protonophore CCCP. However, the pH value of the mitochondrial matrix could not be determined accurately due to the low pK_a of this GFP mutant. Thus, GFP-mut1 is not a suitable pH indicator in this compartment, but it would be advantageous in more acidic organelles, such as the cytoplasm and trans-Golgi cisternae (Kneen *et al.*, 1998). Conversely, expression of EYFP allowed for the quantitative determination of pH changes in the mitochondrial matrix. The resting value, which was 7.98 in HeLa cells and 7.91 in rat neonatal cardiomyocytes, collapsed rapidly to about pH 7 following addition of a protonophore (Fig. 10C), indicating that this GFP mutant is a good tool for measurement of mitochondrial matrix pH (Llopis *et al.*, 1998).

In resting cells, the matrix pH did not change following addition of medium containing 10-mM glucose (an oxidizable substrate), but growth in medium containing 10-mM lactate plus 1-mM pyruvate caused an acidification (Fig. 10D), which was reversible on washout. This can be accounted for by diffusion of the protonated acid or by cotransport of pyruvate$^-$/H$^+$ through the IMM (Llopis *et al.*, 1998).

A new GFP-based pH indicator with a pK_a' in the alkaline region (i.e., with an ideal sensitivity to monitor pH variations within the mitochondrial matrix), named mtAlpHi, for mitochondrial alkaline pH indicator, has been developed (Abad *et al.*, 2004). This probe appears even more suitable than mtEYFP for studying the pH dynamics of mitochondrial matrix; it responds rapidly and reversibly to changes in pH, it has a pK_a' around 8.5, and it lacks toxicity or evident interference with normal cellular functions. Moreover the dynamic range of mtAlpHi fluorescence is markedly different from that of mtEYFP, whereas the former increases approximately threefold between pH 7 and 8.5, the latter shows a much smaller change ($<15\%$) (Fig. 10E and F). With mtAlpHi, it has been calculated that the mitochondrial matrix pH in intact cells in steady state conditions is 8.05 ± 0.11, and on addition of FCCP this value drops to about 7.50 ± 0.22.

An equally important compartment is the intermembrane space of mitochondria, that is, the other side of the membrane across which the respiratory chain pumps H$^+$. Given the much smaller volume (compensated by the diffusion to the cytosol through the OMM) pH values could significantly differ from those of the neighboring regions. To get insight into this important bioenergetic parameter, we developed a specific YFP-based pH sensor. For this purpose, the same cloning strategy employed for mimsAEQ (Pinton *et al.*, 1998; Rizzuto *et al.*, 1998b) was followed, and mimsYFP was generated (Porcelli *et al.*, 2005). When transfected in HeLa cells, mimsYFP showed a classical mitochondrial distribution (i.e., typical rod-like structures distributed throughout the cytoplasm). The choice of a GFP mutant with a pK_a value around 7 as a suitable pH sensor in this compartment derives from the assumption that the pH of the MIMS is in equilibrium with that of cytosol. Indeed, the outer membrane has not been considered a barrier to transport of ions and molecules up to 5 kDa (Mannella *et al.*, 1998). However, the MIMS might be influenced by pH changes occurring in the matrix as a result of variations of the $\Delta\psi$, and microdomains exhibiting different $[H^+]$ from cytosol might exist under physiological and pathological conditions.

In order to generate a titration curve of mimsYFP, we quantitated the fluorescence of mimsEYFP as a function of pH (Fig. 11A). Coverslips with mimsEYFP-transfected cells were mounted in the thermostatted chamber (37 °C) of a digital-imaging system, and were incubated in a KCl-based saline solution containing 125-mM KCl, 20-mM NaCl, 5.5-mM D-glucose, 1-mM CaCl$_2$, 1-mM MgSO$_4$, 1-mM K$_2$HPO$_4$, 20-mM Na-HEPES (pH 6.5), and the ionophores nigericin plus monensin (5 μM each) to equilibrate intracellular and extracellular pH. pH was increased by the addition of small amounts of NaOH while measuring the pH with a semimicro-combination pH electrode. Fluorescence of mimsEYFP in selected areas of cells was linear with respect to pH over the range between pH 6.5 and 8. When the average values of fluorescence in the selected areas were plotted against pH, we obtained the calibration curve shown in Fig. 11B (Porcelli *et al.*, 2005).

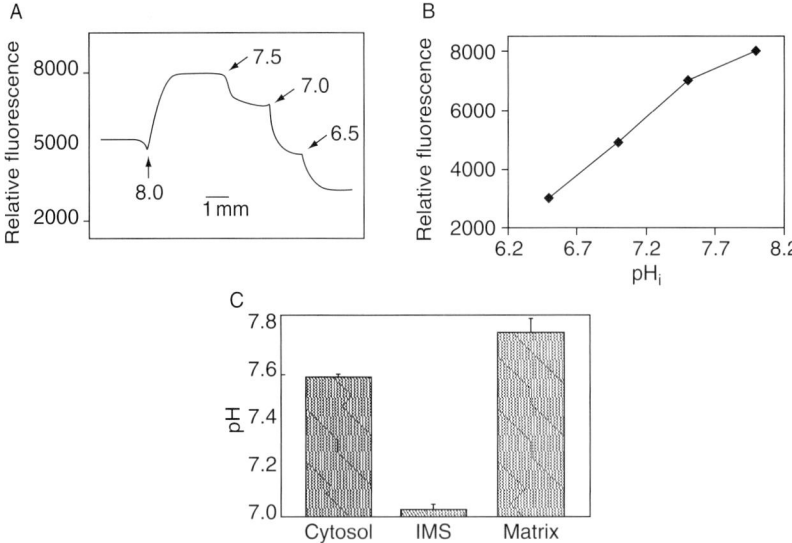

Fig. 11 pH measurements in mitochondria. (A) MIMS-EYFP-transfected ECV304 cells were first incubated with KBS, then perfused with high KCl saline solution containing nigericin and monensin (5 μM) at the indicated pH, as detailed in the figure (modified from Porcelli *et al.*, 2005). (B) Normalized values of fluorescence intensity obtained in (A) are plotted against pH. Data are means $\pm$ SD (modified from Porcelli *et al.*, 2005). (C) Values of pH in the cytosol, MIMS, and mitochondrial matrix. ECV304 cells were transfected with cytosolic EYFP, mimsEYFP, or mtEYFP, incubated with KBS, and then perfused as described in (A) (modified from Porcelli *et al.*, 2005).

Using the mimsYFP, mtYFP, and cytYFP probes with this calibration curve, the following pH values were obtained: MIMS 6.88 ± 0.09; mitochondrial matrix 7.78 ± 0.17, and cytosol 7.59 ± 0.01 (Fig. 11C) (Porcelli et $al.$, 2005).

B. Procedure

The procedures employed for expressing mtYFP and mimsYFP are the same as those described previously for the corresponding aequorin chimeras, that is, in most cases, transection can be accomplished by a simple transfection with the calcium phosphate protocol. The only difference is that the cells are seeded onto 24-mm coverslips and are transfected with 8-μg DNA/coverslip. In the case of mimsYFP, the amount of DNA can be reduced to 2–4 μg per coverslip to prevent mitochondrial damage. It is our experience that in many cell types high levels of expression of a recombinant protein retained in the intermembrane space (i.e., not only YFP, but also aequorin, see above) cause a change in the morphology of mitochondria. After 36 h of transfection, the coverslip is transferred to the thermostatted stage of a digital-imaging system, as described previously for the GFP-based calcium probes. For YFP chimeras, dedicated YFP filter sets (500/542 nm) are usually used, but comparable results can also be obtained with a traditional FITC set (482/536 nm). For mtAlpHi, the maximal peak is at 498 nm for the excitation and 522 nm for the emission light.

V. Conclusions

Genetically encoded biosensors (chimeric proteins) are intracellular probes of physiological parameters endowed with unique advantages. These advantages include the ability to measure intracellular parameters (such as Ca^{2+} and pH) at specific intracellular sites, with little background signal, and no probe leakage. Additionally, the use of biosensors is not associated with the toxicity observed with many chemical probes nor with the need of invasive loading procedures. Furthermore, an ideal intracellular probe should be stable, have high parameter sensitivity and specificity, rapid signal response to parameter changes, and good luminescent/optical properties. The probes described here fulfill many of these requirements.

Acknowledgments

The authors are deeply indebted to past and present collaborators. We thank the Italian University Ministry (MURST), Telethon-Italy (Grants no. GGP05284 and GTF02013), the Italian Association for Cancer Research (AIRC), the Italian Space Agency (ASI), EU (fondi strutturali Obiettivo 2), and the PRRIITT program of the Emilia Romagna Region for financial support.

References

Abad, M. F., Di Benedetto, G., Magalhaes, P. J., Filippin, L., and Pozzan, T. (2004). Mitochondrial pH monitored by a new engineered green fluorescent protein mutant. *J. Biol. Chem.* **279**, 11521–11529.

Allen, D. G., and Blinks, J. R. (1978). Calcium transients in aequorin-injected frog cardiac muscle. *Nature* **273**, 509–513.

Baird, G. S., Zacharias, D. A., and Tsien, R. Y. (1999). Circular permutation and receptor insertion within green fluorescent proteins. *Proc. Natl. Acad. Sci. USA* **96**, 11241–11246.

Bastianutto, C., Clementi, E., Codazzi, F., Podini, P., De Giorgi, F., Rizzuto, R., Meldolesi, J., and Pozzan, T. (1995). Overexpression of calreticulin increases the Ca2+ capacity of rapidly exchanging Ca2+ stores and reveals aspects of their lumenal microenvironment and function. *J. Cell Biol.* **130**, 847–855.

Bernardi, P. (1999). Mitochondrial transport of cations: Channels, exchangers, and permeability transition. *Physiol. Rev.* **79**, 1127–1155.

Bernardi, P., Scorrano, L., Colonna, R., Petronilli, V., and Di Lisa, F. (1999). Mitochondria and cell death. Mechanistic aspects and methodological issues. *Eur. J. Biochem.* **264**, 687–701.

Berridge, M. J., Lipp, P., and Bootman, M. D. (2000). The versatility and universality of calcium signalling. *Nat. Rev. Mol. Cell Biol.* **1**, 11–21.

Berridge, M. J., Bootman, M. D., and Roderick, H. L. (2003). Calcium signalling: Dynamics, homeostasis and remodelling. *Nat. Rev. Mol. Cell Biol.* **4**, 517–529.

Brini, M., Marsault, R., Bastianutto, C., Alvarez, J., Pozzan, T., and Rizzuto, R. (1995). Transfected aequorin in the measurement of cytosolic Ca2+ concentration ([Ca2+]c). A critical evaluation. *J. Biol. Chem.* **270**, 9896–9903.

Budd, S. L., and Nicholls, D. G. (1996a). A reevaluation of the role of mitochondria in neuronal Ca2+ homeostasis. *J. Neurochem.* **66**, 403–411.

Budd, S. L., and Nicholls, D. G. (1996b). Mitochondria, calcium regulation, and acute glutamate excitotoxicity in cultured cerebellar granule cells. *J. Neurochem.* **67**, 2282–2291.

Carafoli, E. (1987). Intracellular calcium homeostasis. *Annu. Rev. Biochem.* **56**, 395–433.

Chacon, E., Reece, J. M., Nieminen, A. L., Zahrebelski, G., Herman, B., and Lemasters, J. J. (1994). Distribution of electrical potential, pH, free Ca2+, and volume inside cultured adult rabbit cardiac myocytes during chemical hypoxia: A multiparameter digitized confocal microscopic study. *Biophys. J.* **66**, 942–952.

Chalfie, M. (1995). Green fluorescent protein. *Photochem. Photobiol.* **62**, 651–656.

Chalfie, M., Tu, Y., Euskirchen, G., Ward, W. W., and Prasher, D. C. (1994). Green fluorescent protein as a marker for gene expression. *Science* **263**, 802–805.

Chiesa, A., Rapizzi, E., Tosello, V., Pinton, P., de Virgilio, M., Fogarty, K. E., and Rizzuto, R. (2001). Recombinant aequorin and green fluorescent protein as valuable tools in the study of cell signalling. *Biochem. J.* **355**, 1–12.

Cobbold, P. H. (1980). Cytoplasmic free calcium and amoeboid movement. *Nature* **285**, 441–446.

De Giorgi, F., Brini, M., Bastianutto, C., Marsault, R., Montero, M., Pizzo, P., Rossi, R., and Rizzuto, R. (1996). Targeting aequorin and green fluorescent protein to intracellular organelles. *Gene* **173**, 113–117.

Drummond, R. M., and Fay, F. S. (1996). Mitochondria contribute to Ca2+ removal in smooth muscle cells. *Pflugers Arch.* **431**, 473–482.

Duchen, M. R. (1999). Contributions of mitochondria to animal physiology: From homeostatic sensor to calcium signalling and cell death. *J. Physiol.* **516**(Pt. 1), 1–17.

Duchen, M. R. (2000a). Mitochondria and Ca(2+) in cell physiology and pathophysiology. *Cell Calcium* **28**, 339–348.

Duchen, M. R. (2000b). Mitochondria and calcium: From cell signalling to cell death. *J. Physiol.* **529** (Pt. 1), 57–68.

Elsliger, M. A., Wachter, R. M., Hanson, G. T., Kallio, K., and Remington, S. J. (1999). Structural and spectral response of green fluorescent protein variants to changes in pH. *Biochemistry* **38**, 5296–5301.

Filippin, L., Abad, M. C., Gastaldello, S., Magalhaes, P. J., Sandona, D., and Pozzan, T. (2005). Improved strategies for the delivery of GFP-based Ca2+ sensors into the mitochondrial matrix. *Cell Calcium* **37**, 129–136.

Gavel, Y., and von Heijne, G. (1990). Cleavage-site motifs in mitochondrial targeting peptides. *Protein Eng.* **4**, 33–37.

Gottlieb, R. A., Gruol, D. L., Zhu, J. Y., and Engler, R. L. (1996). Preconditioning rabbit cardiomyocytes: Role of pH, vacuolar proton ATPase, and apoptosis. *J. Clin. Invest.* **97**, 2391–2398.

Green, D. R., and Kroemer, G. (2004). The pathophysiology of mitochondrial cell death. *Science* **305**, 626–629.

Gunter, K. K., and Gunter, T. E. (1994). Transport of calcium by mitochondria. *J. Bioenerg. Biomembr.* **26**, 471–485.

Gunter, T. E., Buntinas, L., Sparagna, G., Eliseev, R., and Gunter, K. (2000). Mitochondrial calcium transport: Mechanisms and functions. *Cell Calcium* **28**, 285–296.

Gunter, T. E., and Gunter, K. K. (2001). Uptake of calcium by mitochondria: Transport and possible function. *IUBMB Life* **52**, 197–204.

Hajnoczky, G., Robb-Gaspers, L. D., Seitz, M. B., and Thomas, A. P. (1995). Decoding of cytosolic calcium oscillations in the mitochondria. *Cell* **82**, 415–424.

Heim, R., and Tsien, R. Y. (1996). Engineering green fluorescent protein for improved brightness, longer wavelengths and fluorescence resonance energy transfer. *Curr. Biol.* **6**, 178–182.

Herrington, J., Park, Y. B., Babcock, D. F., and Hille, B. (1996). Dominant role of mitochondria in clearance of large Ca2+ loads from rat adrenal chromaffin cells. *Neuron* **16**, 219–228.

Isfort, R. J., Cody, D. B., Asquith, T. N., Ridder, G. M., Stuard, S. B., and LeBoeuf, R. A. (1993). Induction of protein phosphorylation, protein synthesis, immediate-early-gene expression and cellular proliferation by intracellular pH modulation. Implications for the role of hydrogen ions in signal transduction. *Eur. J. Biochem.* **213**, 349–357.

Isfort, R. J., Cody, D. B., Stuard, S. B., Ridder, G. M., and LeBoeuf, R. A. (1996). Calcium functions as a transcriptional and mitogenic repressor in Syrian hamster embryo cells: Roles of intracellular pH and calcium in controlling embryonic cell differentiation and proliferation. *Exp. Cell Res.* **226**, 363–371.

Jouaville, L. S., Pinton, P., Bastianutto, C., Rutter, G. A., and Rizzuto, R. (1999). Regulation of mitochondrial ATP synthesis by calcium: Evidence for a long-term metabolic priming. *Proc. Natl. Acad. Sci. USA* **96**, 13807–13812.

Kass, G. E., Duddy, S. K., Moore, G. A., and Orrenius, S. (1989). 1,2-Di-(tert-butyl)-1,4-benzohydroquinone rapidly elevates cytosolic Ca^{2+} concentration by mobilizing the inositol 1,4,5-trisphosphate-sensitive pool. *J. Biol. Chem.* **264**, 15192–15198.

Kendall, J. M., Sala-Newby, G., Ghalaut, V., Dormer, R. L., and Campbell, A. K. (1992). Engineering the CA(2+)-activated photoprotein aequorin with reduced affinity for calcium. *Biochem. Biophys. Res. Commun.* **187**, 1091–1097.

Kneen, M., Farinas, J., Li, Y., and Verkman, A. S. (1998). Green fluorescent protein as a noninvasive intracellular pH indicator. *Biophys. J.* **74**, 1591–1599.

Kroemer, G., and Reed, J. C. (2000). Mitochondrial control of cell death. *Nat. Med.* **6**, 513–519.

Lievremont, J. P., Rizzuto, R., Hendershot, L., and Meldolesi, J. (1997). BiP, a major chaperone protein of the endoplasmic reticulum lumen, plays a direct and important role in the storage of the rapidly exchanging pool of Ca2+. *J. Biol. Chem.* **272**, 30873–30879.

Llopis, J., McCaffery, J. M., Miyawaki, A., Farquhar, M. G., and Tsien, R. Y. (1998). Measurement of cytosolic, mitochondrial, and Golgi pH in single living cells with green fluorescent proteins. *Proc. Natl. Acad. Sci. USA* **95**, 6803–6808.

Mannella, C. A., Buttle, K., Rath, B. K., and Marko, M. (1998). Electron microscopic tomography of rat-liver mitochondria and their interaction with the endoplasmic reticulum. *Biofactors* **8**, 225–228.

Miyawaki, A., Llopis, J., Heim, R., McCaffery, J. M., Adams, J. A., Ikura, M., and Tsien, R. Y. (1997). Fluorescent indicators for Ca2+ based on green fluorescent proteins and calmodulin. *Nature* **388**, 882–887.

Miyawaki, A., Griesbeck, O., Heim, R., and Tsien, R. Y. (1999). Dynamic and quantitative Ca2+ measurements using improved cameleons. *Proc. Natl. Acad. Sci. USA* **96,** 2135–2140.

Montero, M., Brini, M., Marsault, R., Alvarez, J., Sitia, R., Pozzan, T., and Rizzuto, R. (1995). Monitoring dynamic changes in free Ca2+ concentration in the endoplasmic reticulum of intact cells. *EMBO J.* **14,** 5467–5475.

Montero, M., Barrero, M. J., and Alvarez, J. (1997). [Ca2+] microdomains control agonist-induced Ca2+ release in intact HeLa cells. *FASEB J.* **11,** 881–885.

Nagai, T., Sawano, A., Park, E. S., and Miyawaki, A. (2001). Circularly permuted green fluorescent proteins engineered to sense Ca2+. *Proc. Natl. Acad. Sci. USA* **98,** 3197–3202.

Nagai, T., Yamada, S., Tominaga, T., Ichikawa, M., and Miyawaki, A. (2004). Expanded dynamic range of fluorescent indicators for Ca(2+) by circularly permuted yellow fluorescent proteins. *Proc. Natl. Acad. Sci. USA* **101,** 10554–10559.

Nicholls, D. G. (2005). Mitochondria and calcium signaling. *Cell Calcium* **38,** 311–317.

Orchard, C. H., and Kentish, J. C. (1990). Effects of changes of pH on the contractile function of cardiac muscle. *Am. J. Physiol.* **258,** C967–C981.

Ormo, M., Cubitt, A. B., Kallio, K., Gross, L. A., Tsien, R. Y., and Remington, S. J. (1996). Crystal structure of the Aequorea victoria green fluorescent protein. *Science* **273,** 1392–1395.

Patterson, G. H., Knobel, S. M., Sharif, W. D., Kain, S. R., and Piston, D. W. (1997). Use of the green fluorescent protein and its mutants in quantitative fluorescence microscopy. *Biophys. J.* **73,** 2782–2790.

Pinton, P., Brini, M., Bastianutto, C., Tuft, R. A., Pozzan, T., and Rizzuto, R. (1998). New light on mitochondrial calcium. *Biofactors* **8,** 243–253.

Pinton, P., Ferrari, D., Magalhaes, P., Schulze-Osthoff, K., Di Virgilio, F., Pozzan, T., and Rizzuto, R. (2000). Reduced loading of intracellular Ca(2+) stores and downregulation of capacitative Ca(2+) influx in Bcl-2-overexpressing cells. *J. Cell Biol.* **148,** 857–862.

Porcelli, A. M., Ghelli, A., Zanna, C., Pinton, P., Rizzuto, R., and Rugolo, M. (2005). pH difference across the outer mitochondrial membrane measured with a green fluorescent protein mutant. *Biochem. Biophys. Res. Commun.* **326,** 799–804.

Pozzan, T., and Rizzuto, R. (2000). The renaissance of mitochondrial calcium transport. *Eur. J. Biochem.* **267,** 5269–5273.

Pozzan, T., Rizzuto, R., Volpe, P., and Meldolesi, J. (1994). Molecular and cellular physiology of intracellular calcium stores. *Physiol. Rev.* **74,** 595–636.

Rapizzi, E., Pinton, P., Szabadkai, G., Wieckowski, M. R., Vandecasteele, G., Baird, G., Tuft, R. A., Fogarty, K. E., and Rizzuto, R. (2002). Recombinant expression of the voltage-dependent anion channel enhances the transfer of Ca2+ microdomains to mitochondria. *J. Cell Biol.* **159,** 613–624.

Ridgway, E. B., and Ashley, C. C. (1967). Calcium transients in single muscle fibers. *Biochem. Biophys. Res. Commun.* **29,** 229–234.

Rizzuto, R., and Pozzan, T. (2006). Microdomains of intracellular Ca2+: Molecular determinants and functional consequences. *Physiol. Rev.* **86,** 369–408.

Rizzuto, R., Nakase, H., Darras, B., Francke, U., Fabrizi, G. M., Mengel, T., Walsh, F., Kadenbach, B., DiMauro, S., and Schon, E. A. (1989). A gene specifying subunit VIII of human cytochrome *c* oxidase is localized to chromosome 11 and is expressed in both muscle and non-muscle tissues. *J. Biol. Chem.* **264,** 10595–10600.

Rizzuto, R., Simpson, A. W., Brini, M., and Pozzan, T. (1992). Rapid changes of mitochondrial Ca2+ revealed by specifically targeted recombinant aequorin. *Nature* **358,** 325–327.

Rizzuto, R., Brini, M., Murgia, M., and Pozzan, T. (1993). Microdomains with high Ca2+ close to IP3-sensitive channels that are sensed by neighboring mitochondria. *Science* **262,** 744–747.

Rizzuto, R., Brini, M., De Giorgi, F., Rossi, R., Heim, R., Tsien, R. Y., and Pozzan, T. (1996). Double labelling of subcellular structures with organelle-targeted GFP mutants *in vivo*. *Curr. Biol.* **6,** 183–188.

Rizzuto, R., Carrington, W., and Tuft, R. A. (1998a). Digital imaging microscopy of living cells. *Trends Cell Biol.* **8,** 288–292.

Rizzuto, R., Pinton, P., Carrington, W., Fay, F. S., Fogarty, K. E., Lifshitz, L. M., Tuft, R. A., and Pozzan, T. (1998b). Close contacts with the endoplasmic reticulum as determinants of mitochondrial Ca2+ responses. *Science* **280,** 1763–1766.

Rizzuto, R., Bernardi, P., and Pozzan, T. (2000). Mitochondria as all-round players of the calcium game. *J. Physiol.* **529**(Pt. 1), 37–47.

Rizzuto, R., Duchen, M. R., and Pozzan, T. (2004). Flirting in little space: The ER/mitochondria Ca2+ liaison. *Sci. STKE* **2004**(215), re1.

Robey, R. B., Ruiz, O., Santos, A. V., Ma, J., Kear, F., Wang, L. J., Li, C. J., Bernardo, A. A., and Arruda, J. A. (1998). pH-dependent fluorescence of a heterologously expressed Aequorea green fluorescent protein mutant: *In situ* spectral characteristics and applicability to intracellular pH estimation. *Biochemistry* **37,** 9894–9901.

Roos, A., and Boron, W. F. (1981). Intracellular pH. *Physiol. Rev.* **61,** 296–434.

Rudolf, R., Mongillo, M., Rizzuto, R., and Pozzan, T. (2003). Looking forward to seeing calcium. *Nat. Rev. Mol. Cell Biol.* **4,** 579–586.

Rutter, G. A., Burnett, P., Rizzuto, R., Brini, M., Murgia, M., Pozzan, T., Tavare, J. M., and Denton, R. M. (1996). Subcellular imaging of intramitochondrial Ca2+ with recombinant targeted aequorin: Significance for the regulation of pyruvate dehydrogenase activity. *Proc. Natl. Acad. Sci. USA* **93,** 5489–5494.

Shimomura, O. (1986). Isolation and properties of various molecular forms of aequorin. *Biochem. J.* **234,** 271–277.

Shimomura, O., Johnson, F. H., and Saiga, Y. (1962). Extraction, purification and properties of aequorin, a bioluminescent protein from the luminous hydromedusan, Aequorea. *J. Cell. Comp. Physiol.* **59,** 223–239.

Ward, W. W., and Bokman, S. H. (1982). Reversible denaturation of Aequorea green-fluorescent protein: Physical separation and characterization of the renatured protein. *Biochemistry* **21,** 4535–4540.

Zhang, J., Campbell, R. E., Ting, A. Y., and Tsien, R. Y. (2002). Creating new fluorescent probes for cell biology. *Nat. Rev. Mol. Cell Biol.* **3,** 906–918.

CHAPTER 16

Measurement of Membrane Permeability and the Permeability Transition of Mitochondria

Naoufal Zamzami, Carine Maisse, Didier Métivier, and Guido Kroemer

INSERM U848, Université Paris Sud
Institut Gustave Roussy
F-94805 Villejuif, France

I. Introduction

Mitochondria are essential for the evolution of complex animals in aerobic conditions. The mitochondrion carries out most cellular oxidations and produces the bulk of the animal cell's ATP. Recently, this organelle has also been implicated in the regulation of cell death. Indeed, a major site of the activity of (pro/anti)-apoptotic proteins is the mitochondrion. Following a variety of death signals, mitochondria show early alterations in their function, which at least in some instances may be explained by the opening of the permeability transition (PT) pore. The PT pore, also called mitochondrial megachannel or multiple conductance channel (Bernardi, 1996; Kinnally et al., 1996; Zoratti and Szabò, 1995), is a dynamic multiprotein complex located at the contact site between the inner and the outer mitochondrial membranes, one of the critical sites of metabolic coordination between the cytosol, the mitochondrial intermembrane space, and the matrix. The PT pore participates in the regulation of matrix Ca^{2+}, pH, $\Delta\Psi_m$, and volume.

0091-679X/07 $35.00
DOI: 10.1016/S0091-679X(06)80016-6

It also functions as a Ca^{2+}-, voltage-, pH-, and redox-gated channel, with several levels of conductance and little if any ion selectivity (Bernardi, 1996; Beutner *et al.*, 1996; Ichas *et al.*, 1997; Kinnally *et al.*, 1996; Marzo *et al.*, 1998; Zoratti and Szabò, 1995). In isolated mitochondria, opening of the PT pore causes matrix swelling with consequent distension and local disruption of the mitochondrial outer membrane (whose surface is smaller than that of the inner membrane), release of soluble products from the intermembrane space, dissipation of the mitochondrial inner transmembrane potential, and release of small molecules (up to 1500 Da) from the matrix through the inner membrane (Bernardi, 1996; Kantrow and Piantadosi, 1997; Petit *et al.*, 1998; Zoratti and Szabò, 1995). Similar changes are found in apoptotic cells, perhaps with the exception of matrix swelling, which is only observed in a transient fashion before cells shrink (Kroemer *et al.*, 1995, 1997, 1998; Liu *et al.*, 1996; Vander Heiden *et al.*, 1997; Obeid *et al.*, 2007). In several models of apoptosis, pharmacological inhibition of the PT pore is cytoprotective, suggesting that opening of the PT pore can be rate-limiting for the death process (Kroemer *et al.*, 1998; Marchetti *et al.*, 1996; Zamzami *et al.*, 1996). Moreover, the cytoprotective oncoprotein Bcl-2 has been shown to function as an endogenous inhibitor of the PT pore (Kroemer, 1997; Susin *et al.*, 1996; Zamzami *et al.*, 1996). Some authors have suggested that other mechanisms not involving the PT pore may account for mitochondrial membrane permeabilization during apoptosis (Bossy-Wetzel *et al.*, 1998; Kluck *et al.*, 1997). Irrespective of this possibility, mitochondrial membrane permeabilization is a general feature of apoptosis. Here we will detail experimental procedures designed to measure apoptosis-associated mitochondrial membrane permeabilization, either in intact cells or in isolated mitochondria.

II. Procedures

A. Mitochondrial Alterations in Intact Cells

Mitochondrial inner membrane depolarization is one of the major alterations observed in apoptotic cells. Accurate quantitation of the mitochondrial inner transmembrane potential ($\Delta\Psi_m$), using appropriate potentiometric fluorochromes, reveals a $\Delta\Psi_m$ decrease in most models of apoptosis. This $\Delta\Psi_m$ decrease is an early event and thus mostly precedes the activation of downstream caspases and nucleases. Our current data are compatible with the notion that a $\Delta\Psi_m$ decrease mediated by the opening of the PT pore constitutes an irreversible event of the apoptotic process (Susin *et al.*, 1998; Zamzami *et al.*, 1998). Therefore, the determination of the $\Delta\Psi_m$ allows for the detection of early apoptosis both *in vitro* and *ex vivo*. Here, we propose several protocols for quantitation of $\Delta\Psi_m$ and other mitochondrial alterations in purified mitochondria as well as in intact cells.

B. Cytometric Analysis: Lipophilic Cationic Dyes

Lipophilic cations accumulate in the mitochondrial matrix, driven by the electrochemical gradient following the Nernst equation; every 61.5-mV increase in membrane potential (usually 120–170 mV) corresponds to a 10-fold increase in cation concentration in mitochondria. Therefore, the concentration of such cations is two to three logs higher in the mitochondrial matrix than the cytosol. Several different cationic fluorochromes can be employed to measure mitochondrial transmembrane potentials. These markers include 3,3'-dihexyloxacarbocyanine iodide ($DiOC_6$) (green fluorescence) (Petit *et al.*, 1990), chloromethyl-X-rosamine (CMXRos; also called MitoTracker Red^{TM}) (red fluorescence) (Castedo *et al.*, 1996; Macho *et al.*, 1996), and 5,5',6,6'-tetrachloro-1,1',3,3'-tetraethylbenzimidazolcarbocyanine iodide (JC-1) (red and green fluorescence) (Smiley, 1991). As compared to rhodamine 123, which we do not recommend for cytofluorometric analyses (Metivier *et al.*, 1998), $DiOC_6(3)$ offers the important advantage that it does not show major quenching effects. JC-1 incorporates into mitochondria, where it either forms monomers (green fluorescence, 527 nm) or, at high transmembrane potentials, aggregates (red fluorescence, 590 nm) (Smiley, 1991). Thus, the quotient between green and red JC-1 fluorescence provides an estimate of $\Delta\Psi_m$ that is (relatively) independent of mitochondrial mass.

1. Single Staining of Mitochondrial Parameters for Flow Cytometric Ananlysis

Materials

- Stock solutions of fluorochromes: $DiOC_6$ should be diluted to 40 μM in DMSO, CMXRos to 1 mM in DMSO, and JC-1 to 0.76 mM in DMSO. All three fluorochromes can be purchased from Molecular Probes (France) and should be stored, once diluted, at $-20\,°C$ in the dark.
- Working solutions: Dilute $DiOC_6(3)$ to 400 nM (10-μl stock solution + 1-ml PBS), CMXRos to 2 μM (2-μl stock solution + 1-ml PBS or mitochondrial resuspension buffer), and JC-1 to 7.6 μM (10-μl stock solution + 1-ml PBS). These solutions should be prepared freshly for each series of stainings.
- Carbonyl cyanide *m*-chlorophenylhydrazone (CCCP) diluted in ethanol (stock at 20 mM), a protonophore required for control purposes ($\Delta\Psi_m$ disruption).
- Cytofluorometer with appropriate filters.

Staining protocol

- Cells (5–10 $\times$ 10^6 in 0.5-ml culture medium or PBS) should be kept on ice until the staining. If necessary, cells can be labeled with specific antibodies conjugated to compatible fluorochromes [e.g., phycoerythrin for $DiOC_6$, fluorescein isothiocyanate (FITC) for CMXRos] before determination of mitochondrial potential.
- For staining, add the following amounts of working solutions to 0.5 ml of cell suspension: 25-μl $DiOC_6(3)$ (final concentration: 20 nM), 10-μl CMXRos (final concentration: 40 nM), or 6.5-μl JC-1 (final concentration 0.1 μM), and transfer tubes to a water bath kept at 37$°C$. Incubate cells for 20 min at 37$°C$. Do not wash cells. As a negative control, in each experiment, aliquots of cells should be labeled in the presence of the protonophore CCCP (100 μM).

• Perform cytofluorometric analysis within 10 min, while gating the forward and sideward scatters on viable, normal-sized cells. When large series of tubes are to be analyzed (>10 tubes), the interval between labeling and cytofluorometric analysis should be kept constant. When using an Epics Profile cytofluorometer (Coulter, France), $DiOC_6$ should be monitored in FL1, CMXRos in FL2 (excitation, 488 nM; emission, 599) (Fig. 1), and JC-1 in FL1 versus FL3 (excitation, 488 nM; emission at 527 and 590 nM). The following compensations are recommended for JC-1: 10% of FL2 in FL1 and 21% of FL1 in FL2 (indicative values). Note that the incorporation of these fluorochromes may be influenced by parameters not determined by mitochondria (cell size, plasma membrane permeability, efficacy of the multiple drug resistance pump, and so on). Results can only be interpreted when the staining profiles obtained in different experimental conditions (e.g., controls versus apoptosis) are identical in the presence of CCCP (Fig. 1).

2. Combined Detection of Different Mitochondrial Parameters in Intact Cells

One of the interesting aspects of flow cytometry is the possibility to perform a multiparametric study. It is well known that mitochondrial depolarization is followed by other mitochondrial and nonmitochondrial perturbations. Among

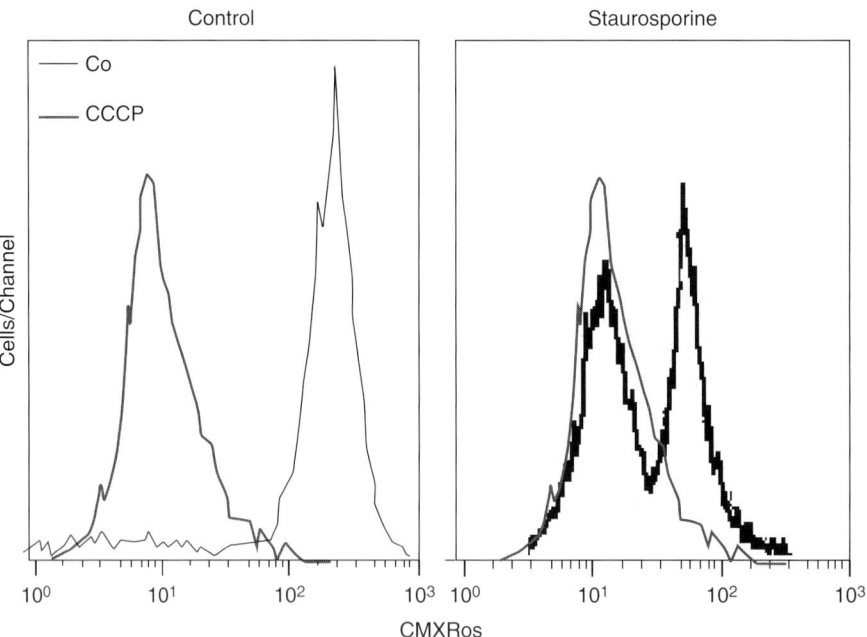

Fig. 1 Representative examples for $\Delta\Psi_m$ measurements. $\Delta\Psi_m$ measurement in Rat1 cells which were either left untreated (control) or treated for 4 h with the apoptosis stimulator staurosporine 0.5 μM, followed by staining with CMXRos (40 nM) in the presence (full line) or absence (dotted line) of CCCP. Similar results are obtained using similar $\Delta\Psi_m$ sensitive probes.

the consequences of depolarization, mitochondria hyperproduce reactive oxygen species (ROS), which in turn oxidize mitochondrial cardiolipins. It is possible to determine the production of ROS by dihydroethidine (HE), a substance that is oxidized by superoxide anion to become ethidium bromide (EthBr) and emits a red fluorescence (Rothe and Valet, 1990). HE is more sensitive to superoxide anion than is 2′,7′-dichlorofluorescein diacetate, which is used to measure H_2O_2 formation (Carter *et al.*, 1994; Rothe and Valet, 1990). Thus, enhanced HE → EthBr conversion can be observed in cells that cannot be labeled with 2′,7′-dichlorofluorescein diacetate. Alternatively, the damage produced by ROS in mitochondria can be determined indirectly by assessing the oxidation state of cardiolipin, a molecule restricted to the inner mitochondrial membrane. Nonyl acridine orange (NAO) interacts stoichiometrically with intact, nonoxidized cardiolipin (Petit *et al.*, 1992). Thus a reduction in NAO fluorescence indicates a decrease in cardiolipin content (see caveats on using NAO in Chapter 11 by Schlame, this volume).

Materials

- Stock solutions of fluorochromes: HE (4.73 mg/ml) should be stored at −20°C; 3.15-mg/ml (10 mM) NAO in DMSO should be stored at 4°C. Both fluorochromes, available from Molecular Probes, are light sensitive.

- Working solutions: Dilute 5 μl of either of the stock solutions in 5-ml PBS (final concentration: 10 μM). Prepare freshly for each series of stainings.

Staining protocol

- Cells (5–10 × 10^6 in 1-ml culture medium or PBS) should be kept on ice until the staining. Before HE staining, cells may be labeled with specific antibodies conjugated to either FITC or phycoerythrine (PE).

- For staining, add the following amounts of working solutions to 1 ml of cell suspension: 25-μl HE (final concentration: 2.5 μM), 10-μl NAO (final concentration: 100 nM), or 25-μl JC-1 (final concentration: 1 μM), and transfer tubes to a water bath kept at 37°C. Incubate cells for 20 min at 37°C. Do not wash cells.

- Perform cytofluorometric analysis within 10 min, while gating forward and sideward scatters on viable, normal-sized cells. When using an Epics Profile cytofluorometer (Coulter), HE should be monitored in FL3 and NAO in FL1 or FL3. For double staining, compensations have to be adjusted in accordance with the apparatus. If HE is combined with NAO, we recommend to compensate FL3 − FL1 = 27% and FL1 − FL3 = 1%. In this case, NAO is measured in FL3. For double staining with HE and $DiOC_6$ the compensation should be approximately FL1 − FL3 = 5% and FL3 − FL1 = 2%.

3. Detection of Soluble Intermembrane Mitochondrial Proteins by Fluorescence Microscopy

Many apoptotic regulatory proteins are associated with mitochondrial membranes or translocate from cytosol to mitochondria during the apoptotic process. In addition, mitochondria do release apoptogenic factors, called here soluble intermembrane mitochondrial proteins (SIMPs), for soluble intermembrane proteins (e.g., AIF, cytochrome *c*, caspases 9 and 2, DDF; Patterson *et al.*, 2000) when they are damaged during apoptosis. Therefore, the detection of the release of such proteins into the cytosol is a complementary indication of the mitochondrial implication in different cell death pathways. Two strategies may be employed to detect the mitochondrial release of such proteins. First, cells may be disrupted and subcellular fractions can be prepared, followed by Western blot analysis of the subcellular redistribution of SIMPs. Second, the localization of SIMPs can be detected by immunofluorescence *in situ* on fixed intact cells (Fig. 2). We recommend this latter method because it is more direct and less prone to artifacts.

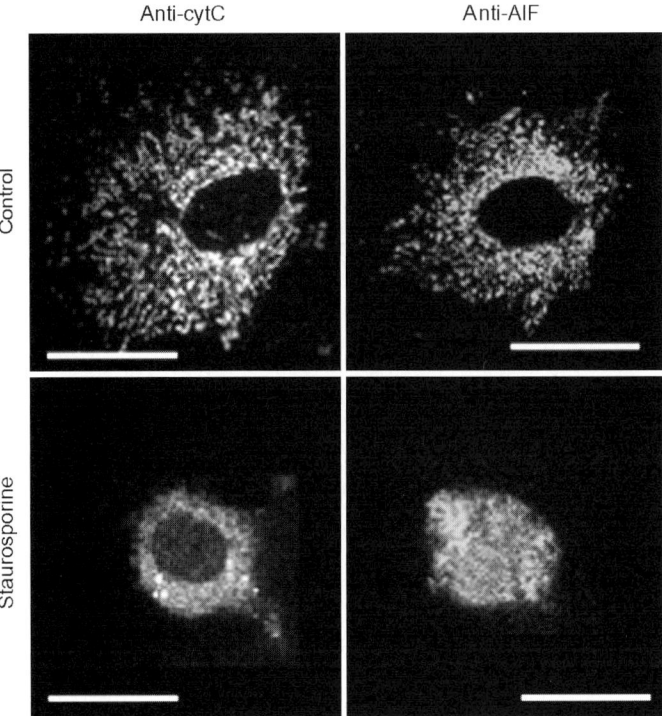

Fig. 2 Analysis of AIF and cytochrome *c* release from mitochondria. MEF cells, treated with or without staurosporin (8 h, 2 μM), were stained for AIF (red fluorescence) or cytochrome *c* (green fluorescence) followed by confocal analysis. In these experimental conditions, staurosporin-treated cells have released both AIF and cytochrome *c* from the punctate mitochondrial localization. Note that only AIF translocates to the nucleus, where it induces fragmentation of the DNA in high molecular weight fragments (not shown). Immunostaining protocol: A rabbit antiserum was generated against a mixture of 3 peptides derived from the mouse AIF aa sequence (aa 151–170, 166–185, 181–200, coupled to keyhole limpet haemocyanine, generated by Syntem, Nîmes, France). This antiserum (ELISA titer ~10.000) was used (diluted 1/1000) on paraformaldehyde (4% w/v) and picric acid-fixed (0.19% v/v) cells (cultured on 100-μm cover slips; 18 mm Ø; Superior, Germany), and revealed with a goat anti-rabbit IgG conjugated to PE (Southern Biotechnology, Birmingham, AL). Cells were counterstained for the detection of cytochrome *c* (mAb 6H2.B4 from Pharmingen, revealed by a goat anti-mouse IgG FITC conjugate; Southern Biotechnology).

Immunofluorescence analysis
Materials

- Adherent cells cultured and treated on cover glasses, in 12-well dishes.
- Slides.
- PBS.
- Paraformaldehyde 4% in PBS + 0.19% (v/v) of a saturated solution of picric acid.
- SDS 0.1% in PBS (v/v) from a 10% SDS stock.
- FCS 10% in PBS.
- PBS-BSA 1 mg/ml.
- First antibody (anti-SIMPs).
- Corresponding second antibody (anti-isotype to the first antibody), labeled with fluorochrome (PE or FITC).

Protocol

All steps are performed at room temperature.

- Withdraw the culture medium.
- Wash once with PBS.
- Fix with the fixative solution (paraformaldehyde 4% + picric acid 0.19% in PBS), 30–60 min.
- Wash three times with PBS. Gentle pipetting is recommended to avoid cell detachment.
- Permeabilize with SDS 0.1% in PBS, 10 min.
- Wash three times with PBS.
- Block with FCS 10% in PBS, 20 min.
- Wash once with PBS.
- Incubate with the first antibody (anti-SIMPs) in PBS-BSA 1 mg/ml, 60 min.
- Wash three times with PBS.

From this step onward samples should be protected from light.

- Incubate with the fluorescent second antibody (anti-isotype to the first antibody) diluted in PBS-BSA 1 mg/ml, 30 min.
- Wash three times with PBS.
- Place the coverslip on a glass slide and examine using a fluorescence microscope.

4. Monitoring Changes in Isolated Mitochondria

In cell-free systems of apoptosis, purified mitochondria can be exposed to apoptogenic molecules, such as Ca^{2+}, ganglioside GD3, or recombinant Bax protein, which function as endogenous-permeabilizing agents. Similarly, purified mitochondria can be exposed to cytotoxic xenobiotics or viral proteins, which in some cases have a direct membrane-permeabilizing effects (Jacotot *et al.*, 2000; Larochette *et al.*, 1999; Marchetti *et al.*, 1999; Ravagnan *et al.*, 1999). The quantitation of $\Delta\Psi_m$ in isolated mitochondria is useful, because it detects changes in the inner membrane permeability which usually are accompanied by an increase of outer membrane permeability, leading to the release of the so-called SIMPs normally stored in the intermembrane space. Here, we provide a protocol for mitochondrial preparation and three different methods for the measurement of mitochondrial membrane permeability.

a. Purification of Mitochondria
Principle of protocol

Tissues or cells are mechanically disrupted in a buffer compatible with mito-chondrial integrity, followed by several rounds of differential centrifugation to separate mitochondria from other organelles and debris. For a better separation of mitochondria, particularly from endoplasmic reticulum and ribosomes, enriched mitochondria can be purified on a discontinuous density gradient. This latter step can be omitted if a low degree of purity is acceptable (see also Chapter 1 by Pallotti and Lenoz, this volume).

Required equipments

- Thomas potter with Teflon inlet.
- Hamilton 500-μl syringe.
- Corex 15-ml tubes.
- Cylinder made of cork loosely fitting into the Corex tube (height: ~2 mm), attached in the center to a nylon sewing cotton. This cork cylinder can be easily produced from the cork of a wine bottle.

Reagents

- Homogenization (H) buffer: 300-mM saccharose, 5-mM N-tris(hydroxymethyl) methyl-2-aminoethanesulfonic acid (TES), 200-μM EGTA, pH 7.2, to be stored at 4°C.
- Percoll (P) buffer: 300-mM saccharose, 10-mM TES, 200-μM EGTA, pH 6.9, to be stored at 4°C.
- Solution A: 90-ml P buffer + 2.05-g saccharose.
- Solution B: 90-ml P buffer + 3.96-g saccharose.
- Solution C: 90-ml P buffer + 13.86-g saccharose.
- Percoll stock solution (100%, Pharmacia).
- 18% Percoll: 6.55-ml solution A + 1.45-ml Percoll solution (prepare fresh).
- 30% Percoll: 5.6-ml solution B + 2.4-ml Percoll solution (prepare fresh).
- 60% Percoll: 3.2-ml solution C + 4.8-ml Percoll solution (prepare fresh).

Protocol for purification of mitochondria from mouse liver

- After sacrifice, mouse liver is rapidly removed, rinsed with cold H buffer and cut into small pieces using a pair of scissors. Note that all steps are performed at 4°C or on wet ice.
- Homogenize tissue in the Potter-Thomas homogenizer (~20 strokes, ~500 revolutions/min).
- Distribute homogenate in two Corex tubes and centrifuge for 10 min at $760 \times g$.
- Recover supernatant. Resuspend pellet in H buffer, centrifuge as above, and recover supernatant. Join the two supernatants.
- Centrifuge for 10 min at $8740 \times g$. In the meantime, prepare Percoll gradient. Transfer 4-ml 60% Percoll solution into a Corex glass tube, then introduce the cork cylinder, which floats on the solution. Keep tube vertical and add the 30% Percoll solution, then the 18% solution (4 ml each) directly onto the cork using a 5-ml plastic pipette. Remove cork by pulling the attached string.
- Recover pellet and resuspend in 1-ml H buffer, carefully add the mitochondria-containing solution on top of the 60%/30%/18% Percoll gradient while inclining the tube 45°.
- Centrifuge for 10 min at $8740 \times g$.
- Remove the lower interface (between 60% and 30% Percoll) with the Hamilton syringe. Dilute 10× in H buffer.
- Centrifuge for 10 min at $6800 \times g$ to remove Percoll, which is toxic to mitochondria.
- Discard supernatant and resuspend in the appropriate buffer, for example, SB (0.2-M sucrose; 10-mM Tris–MOPS, pH 7.4; 5-mM succinate; 1-mM Pi; 2-μM rotenone; 10-μM EGTA–Tris) for the measurement of $\Delta\Psi_m$ or the mitochondrial swelling subsequent to PTP opening.

Note that this protocol is optimized for mouse tissues. Check purity of preparations (which should be higher than 90%) by electron microscopy. For cell lines, an alternative protocol of disruption—nitrogen cavitation—can be used to liberate mitochondria from the cell: 3×10^7–1×10^8 cells are suspended in H buffer, washed ($600 \times g$, 10 min, RT) twice in this buffer, and exposed to a nitrogen decompression of 150 psi during 30 min using a "cell disruption bomb" (Parr Instrument Company, Moline, IL). Thereafter the cell lysate is subjected to differential centrifugation as described above. Note that the resuspension buffer of mitochondria depends on the intended use of the mitochondria (cell-free system, determination of large amplitude swelling, and so on).

b. Volume Changes of Isolated Mitochondria: "Large Amplitude Swelling"
Reagents for swelling assay

- Instrument for absorbance analysis (spectrophotometer or spectrofluorimeter).
- Freshly prepared mitochondria.

- Swelling buffer (SB): 0.2-M sucrose; 10-mM Tris–MOPS, pH 7.4; 5-mM succinate; 1-mM Pi; 2-μM rotenone; 10-μM EGTA–Tris.
- Note that the SB solution should be prepared fresh before each use.

Principle

Rat or mouse liver mitochondria are prepared by standard differential centrifugation as described above. Mitochondrial swelling following to PTP opening is followed at 545 nm by the variation of the absorbance when a spectrophotometer is used or by the variation of 90° light scattering with a spectrofluorimeter, in which both excitation and emission wavelengths are fixed at 545 nm. When mitochondria swell, there is a decrease in the absorbance at 545 nm, which is measured by recording the kinetics of absorbance.

Measurement

- Dilute freshly prepared mitochondria in SB at 0.5 mg/ml and incubate them in a thermostated, magnetically stirred cuvette in a final volume of 1 ml.
- Two minutes after starting the record the reagents to be tested are added.

A useful positive control consists in the addition of 100-μM Ca^{2+}, which opens the PT pore and causes large amplitude swelling. Cyclosporine A (CsA: 1 μM) added before Ca^{2+} prevents this swelling, and CsA-mediated inhibition of mitochondrial volume change is generally interpreted to mean that the CsA-sensitive PT pore mediates this reaction (Fig. 3).

c. Mitochondrial Transmembrane Potential Measurement in Isolated Mitochondria

The $\Delta\Psi_m$ of isolated mitochondria can be quantified by multiple different methods. Here we propose two of them. One is based on the cytofluorometric

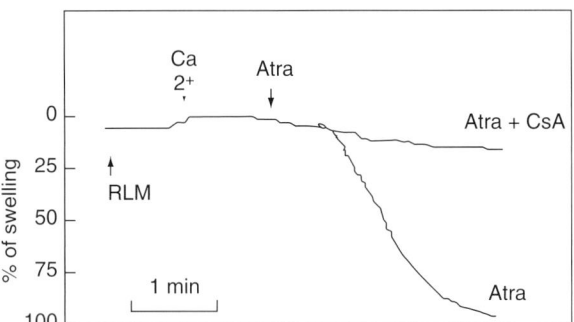

Fig. 3 Induction of PT pore opening in purified mitochondria by atractyloside. The incubation medium contained 0.2-M sucrose; 10-mM Tris–MOPS, pH 7.4; 5-mM succinate–Tris; 1-mM Pi; 2-mM rotenone; and 10-mM EGTA–Tris. Final volume 1 ml, 25 °C. All experiments were started by the addition under stirring of 1 mg of mouse or rat liver mitochondria (not shown). Where indicated, 25-μM Ca^{2+}, 100-μM actratyloside, and mitochondria were preincubated with 1-μM cyclosporin A.

analysis of purified mitochondria on a per-mitochondrion-basis. In this case, the incorporation of the dye CMXRos is measured and low levels of CMXRos incorporation indicate a low $\Delta\Psi_m$. The second protocol is performed as a bulk measurement, based on the quantitation of rhodamine 123 quenching. At a high $\Delta\Psi_m$ level, most of the rhodamine 123 is concentrated in the mitochondrial matrix and quenches. At lower $\Delta\Psi_m$ levels, rhodamine 123 is released, causing dequenching and an increase in rhodamine 123 fluorescence. Thus, a low $\Delta\Psi_m$ level corresponds to a higher value of rhodamine 123 fluorescence.

Required materials

- M buffer: 220-mM sucrose; 68-mM mannitol; 10-mM KCl; 5-mM KH_2PO_4; 2-mM $MgCl_2$; 500-μM EGTA; 5-mM succinate; 2-μM rotenone; 10-mM HEPES, pH 7.2.
- Mitochondria are purified as described above, kept on ice for a maximum of 4 h, and resuspended in M buffer (or similar buffers).
- CMXRos stock solution is prepared as described above. Rhodamine 123 stock solution (10 mM in ethanol) should be kept at $-20\,^\circ$C, protected against light.
- Advanced cytofluorometer capable of detecting isolated mitochondria.
- Spectrofluorometer.

Protocol

Mitochondria are incubated in the presence of the indicated reagent for 30 min at 20 $^\circ$C. Use 100 μM of the protonophore CCCP as a control. Determine the $\Delta\Psi_m$ using the potential-sensitive fluorochrome chloromethyl-X-rosamine (100 nM, 15 min, RT) and analyze in a FACS Vantage cytofluorometer (Becton Dickinson, San Jose, CA) or similar, while gating on single-mitochondrion events in the forward and side scatters (Fig. 4).

Alternatively, mitochondria (1-mg protein per ml) are incubated in a buffer supplemented with 5-μM rhodamine 123 for 5 min, and the $\Delta\Psi_m$-dependent quenching of rhodamine fluorescence (excitation 490 nm, emission 535 nm) is measured continuously in a fluorometer (Shimizu *et al.*, 1998) as shown in Fig. 4.

Anticipated results and pitfalls

The cytofluorometric determination of the $\Delta\Psi_m$ in purified mitochondria is a delicate procedure and requires an advanced cytofluorometer capable of detecting isolated mitochondria. Note that the quality of mitochondrial preparation is very important and that for each experiment controls must be performed to assess background incorporation of fluorochromes in the presence of CCCP, a protonophore causing a complete disruption of the $\Delta\Psi_m$. Working with isolated mitochondria requires that the mitochondrial preparations are optimal and fresh (<4 h). Rhodamine 123 fluorescence measurements in a spectrofluorometer generate less problems than cytofluorometric measurements of isolated mitochondria.

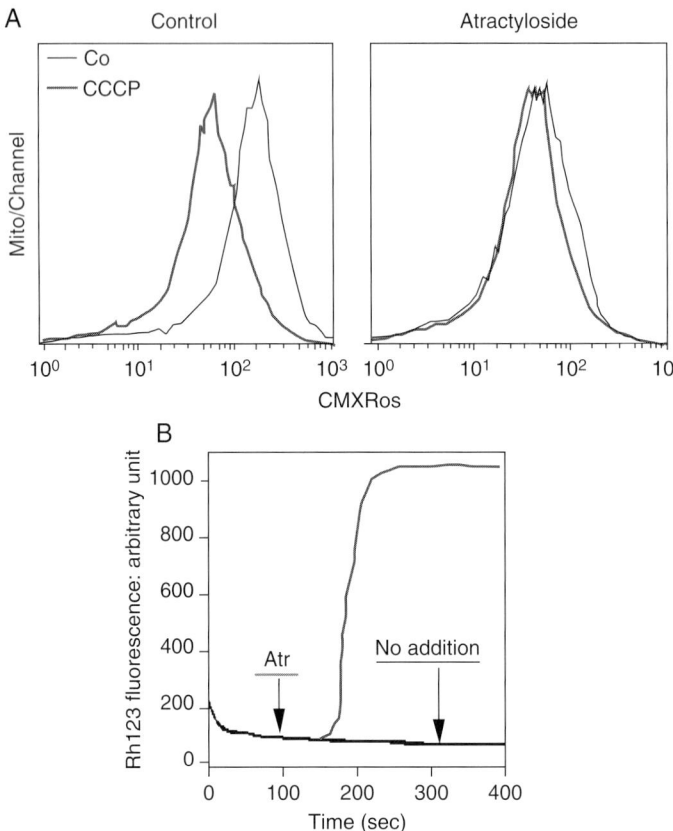

Fig. 4 (A) $\Delta\Psi_m$ measurements in isolated mouse liver mitochondria treated in the presence or absence of 100-μM atractyloside, followed by staining with CMXRos in the presence or absence of CCCP 100 μM and cytofluorometric evaluation of the CMXRos-dependent fluorescence. Note that higher fluorescence values imply a higher $\Delta\Psi_m$. (B) $\Delta\Psi_m$ measurement using Rh123. The same samples as in A were stained with Rh123 and the kinetics of rhodamine dequenching was evaluated in a fluorometer.

References

Bernardi, P. (1996). The permeability transition pore. Control points of a cyclosporin A-sensitive mitochondrial channel involved in cell death. *Biochim. Biophys. Acta* **1275,** 5–9.

Beutner, G., Ruck, A., Riede, B., Welte, W., and Brdiczka, D. (1996). Complexes between kinases, mitochondrial porin, and adenylate translocator in rat brain resemble the permeability transition pore. *FEBS Lett.* **396,** 189–195.

Bossy-Wetzel, E., Newmeyer, D. D., and Green, D. R. (1998). Mitochondrial cytochrome c release in apoptosis occurs upstream of DEVD-specific caspase activation and independently of mitochondrial transmembrane depolarization. *EMBO J.* **17,** 37–49.

Carter, W. O., Narayanan, P. K., and Robinson, J. P. (1994). Intracellular hydrogen peroxide and superoxide anion detection in endothelial cells. *J. Leukoc. Biol.* **55,** 253–258.

Castedo, M., Hirsch, T., Susin, S. A., Zamzami, N., Marchetti, P., Macho, A., and Kroemer, G. (1996). Sequential acquisition of mitochondrial and plasma membrane alterations during early lymphocyte apoptosis. *J. Immunol.* **157,** 512–521.

Ichas, F., Jouavill, L. S., and Mazat, J.-P. (1997). Mitochondria are excitable organelles capable of generating and conveying electric and calcium currents. *Cell* **89,** 1145–1153.

Jacotot, E., Ravagnan, L., Loeffler, M., Ferri, K., Vieira, H. L. A., Zamzami, N., Costantini, P., Druillennec, S., Hoebeke, J., Brian, J. P., Irinopoulos, T., Daugas, E., *et al.* (2000). The HIV-1 viral protein R induces apoptosis via a direct effect on the mitochondrial permeability transition pore. *J. Exp. Med.* **191,** 33–46.

Kantrow, S. P., and Piantadosi, C. A. (1997). Release of cytochrome c from liver mitochondria during permeability transition. *Biochem. Biophys. Res. Comm.* **232,** 669–671.

Kinnally, K. W., Lohret, T. A., Campo, M. L., and Mannella, C. A. (1996). Perspectives on the mitochondrial multiple conductance channel. *J. Bioenerg. Biomembr.* **28,** 115–123.

Kluck, R. M., Bossy-Wetzel, E., Green, D. R., and Newmeyer, D. D. (1997). The release of cytochrome c from mitochondria: A primary site for Bcl-2 regulation of apoptosis. *Science* **275,** 1132–1136.

Kroemer, G. (1997). The proto-oncogene Bcl-2 and its role in regulating apoptosis. *Nat. Med.* **3,** 614–620.

Kroemer, G., Dallaporta, B., and Resche-Rigon, M. (1998). The mitochondrial death/life regulator in apoptosis and necrosis. *Annu. Rev. Physiol.* **60,** 619–642.

Kroemer, G., Petit, P. X., Zamzami, N., Vayssière, J.-L., and Mignotte, B. (1995). The biochemistry of apoptosis. *FASEB J.* **9,** 1277–1287.

Kroemer, G., Zamzami, N., and Susin, S. A. (1997). Mitochondrial control of apoptosis. *Immunol. Today* **18,** 44–51.

Larochette, N., Decaudin, D., Jacotot, E., Brenner, C., Marzo, I., Susin, S. A., Zamzami, N., Xie, Z., Reed, J. C., and Kroemer, G. (1999). Arsenite induces apoptosis via a direct effect on the mitochondrial permeability transition pore. *Exp. Cell Res.* **249,** 413–421.

Liu, X. S., Kim, C. N., Yang, J., Jemmerson, R., and Wang, X. (1996). Induction of apoptotic program in cell-free extracts: Requirement for dATP and cytochrome C. *Cell* **86,** 147–157.

Macho, A., Decaudin, D., Castedo, M., Hirsch, T., Susin, S. A., Zamzami, N., and Kroemer, G. (1996). Chloromethyl-X-rosamine is an aldehyde-fixable potential-sensitive fluorochrome for the detection of early apoptosis. *Cytometry* **25,** 333–340.

Marchetti, P., Castedo, M., Susin, S. A., Zamzami, N., Hirsch, T., Haeffner, A., Hirsch, F., Geuskens, M., and Kroemer, G. (1996). Mitochondrial permeability transition is a central coordinating event of apoptosis. *J. Exp. Med.* **184,** 1155–1160.

Marchetti, P., Zamzami, N., Josph, B., Schraen-Maschke, S., Mereau-Richard, C., Costantini, P., Metivier, D., Susin, S. A., Kroemer, G., and Formstecher, P. (1999). The novel retinoid AHPN/CD437 triggers apoptosis through a mitochondrial pathway independent of the nucleus. *Cancer Res.* **59**(24), 6257–6266.

Marzo, I., Brenner, C., Zamzami, N., Susin, S. A., Beutner, G., Brdiczka, D., Rémy, R., Xie, Z.-H., Reed, J. C., and Kroemer, G. (1998). The permeability transition pore complex: A target for apoptosis regulation by caspases and Bcl-2 related proteins. *J. Exp. Med.* **187,** 1261–1271.

Metivier, D., Dallaporta, B., Zamzami, N., Larochette, N., Susin, S. A., Marzo, I., and Kroemer, G. (1998). Cytofluorometric detection of mitochondrial alterations in early CD95/Fas/APO-1-triggered apoptosis of Jurkat T lymphoma cells. Comparison of seven mitochondrion-specific fluorochromes. *Immunol. Lett.* **61,** 157–163.

Obeid, M., Tesniere, A., Ghiringhelli, F., Fimia, G. M., Apetoh, L., Perfettini, J. L., Castedo, M., Mignot, G., Panaretakis, T., Casares, N., Metivier, D., Larochette, N., *et al.* (2007). Calreticulin exposure dictates the immunogenicity of cancer cell death. *Nat. Med.* **13**(1), 54–61.

Patterson, S., Spahr, C. S., Daugas, E., Susin, S. A., Irinopoulos, T., Koehler, C., and Kroemer, G. (2000). Mass spectrometric identification of proteins released from mitochondria undergoing permeability transition. *Cell Death Differ.* **7,** 137–144.

Petit, J. M., Maftah, A., Ratinaud, M. H., and Julien, R. (1992). 10 N-nonyclacridine orange interacts with cardiolipin and allows for the quantification of phospholipids in isolated mitochondria. *Eur. J. Biochem.* **209,** 267–273.

Petit, P. X., Goubern, M., Diolez, P., Susin, S. A., Zamzami, N., and Kroemer, G. (1998). Disruption of the outer mitochondrial membrane as a result of mitochondrial swelling. The impact of irreversible permeability transition. *FEBS Lett.* **426,** 111–116.

Petit, P. X., O'Connor, J. E., Grunwald, D., and Brown, S. C. (1990). Analysis of the membrane potential of rat- and mouse-liver mitochondria by flow cytometry and possible applications. *Eur. J. Biochem.* **220,** 389–397.

Ravagnan, L., Marzo, I., Costantini, P., Susin, S. A., Zamzami, N., Petit, P. X., Hirsch, F., Poupon, M. F., Miccoli, L., Xie, Z., Reed, J. C., and Kroemer, G. (1999). Lonidamine triggers apoptosis via a direct, Bcl-2-inhibited effect on the mitochondrial permeability transition pore. *Oncogene* **18,** 2537–2546.

Rothe, G., and Valet, G. (1990). Flow cytometric analysis of respiratory burst activity in phagocytes with hydroethidine and $2',7'$-dichlorofluorescein. *J. Leukoc. Biol.* **47,** 440–448.

Shimizu, S., Eguchi, Y., Kamiike, W., Funahashi, Y., Mignon, A., Lacronique, V., Matsuda, H., and Tsujimoto, Y. (1998). Bcl-2 prevents apoptotic mitochondrial dysfunction by regulating proton flux. *Proc. Natl. Acad. Sci. USA* **95,** 1455–1459.

Smiley, S. T. (1991). Intracellular heterogeneity in mitochondrial membrane potential revealed by a J-aggregate-forming lipophilic cation JC-1. *Proc. Natl. Acad. Sci. USA* **88,** 3671–3675.

Susin, S. A., Zamzami, N., Castedo, M., Hirsch, T., Marchetti, P., Macho, A., Daugas, E., Geuskens, M., and Kroemer, G. (1996). Bcl-2 inhibits the mitochondrial release of an apoptogenic protease. *J. Exp. Med.* **184,** 1331–1342.

Susin, S. A., Zamzami, N., and Kroemer, G. (1998). Mitochondrial regulation of apoptosis. Doubt no more. *Biochim. Biophys. Acta* **1366,** 151–165.

Vander Heiden, M. G., Chandal, N. S., Williamson, E. K., Schumacker, P. T., and Thompson, C. B. (1997). Bcl-XL regulates the membrane potential and volume homeostasis of mitochondria. *Cell* **91,** 627–637.

Zamzami, N., Brenner, C., Marzo, I., Susin, S. A., and Kroemer, G. (1998). Subcellular and submito-chondrial mechanisms of apoptosis inhibition by Bcl-2-related proteins. *Oncogene* **16,** 2265–2282.

Zamzami, N., Susin, S. A., Marchetti, P., Hirsch, T., Gómez-Monterrey, I., Castedo, M., and Kroemer, G. (1996). Mitochondrial control of nuclear apoptosis. *J. Exp. Med.* **183,** 1533–1544.

Zoratti, M., and Szabò, I. (1995). The mitochondrial permeability transition. *Biochem. Biophys. Acta— Rev. Biomembranes* **1241,** 139–176.

CHAPTER 17

Luciferase Expression for ATP Imaging: Application to Cardiac Myocytes

Christopher J. Bell,★ Giovanni Manfredi,† Elinor J. Griffiths,★ and Guy A. Rutter★,‡

★Department of Biochemistry
Henry Wellcome Signalling Laboratories
School of Medical Sciences, University Walk
University of Bristol, Bristol BS8 1TD, United Kingdom

†Department of Neurology and Neuroscience, Weill Medical College
Cornell University, New York, New York 10021

‡Department of Cell Biology, Faculty of Medicine
Imperial College London
London SW7 2AZ, United Kingdom

I. Introduction

Targeted luciferase is a powerful tool for measuring ATP levels within various compartments of living cells. Here we describe some potential applications of this system, with particular reference to the measurement of mitochondrial and cytosolic ATP changes in working cardiac myocytes.

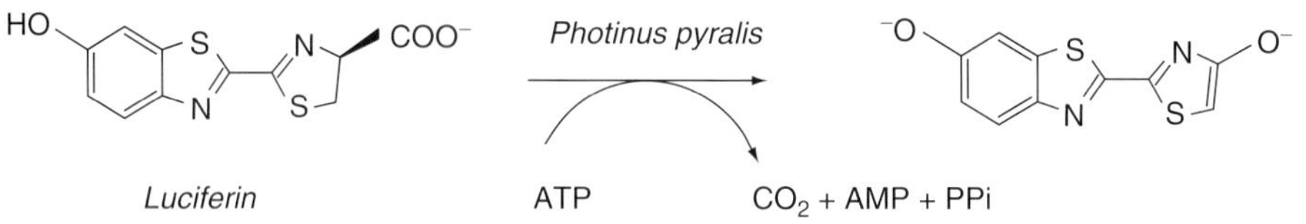

Fig. 1 Luciferase chemistry.

At present, the best available biosensor for ATP is firefly luciferase because it is capable of producing a detectable signal in response to ATP within individual compartments of living cells. The luciferase gene was cloned in 1987 (De Wet *et al.*, 1987) and has been used as a reporter gene in gene expression studies (Rutter *et al.*, 1998). Versions of the gene were then generated that were optimized for thermostability and expression in mammalian cells. Luciferase can be used to detect ATP by virtue of the fact that a photon of light is emitted during the ATP-dependent oxidation of a substrate (luciferin), generating AMP and CO_2 (Fig. 1). Importantly, fusing the luciferase gene to a cDNA encoding a localization sequence, for example, the mitochondrial import sequence of cytochrome *c* oxidase subunit VIII (Jouaville *et al.*, 1999; Kennedy *et al.*, 1999), can also target the enzyme to individual organelles.

A cDNA encoding luciferase can readily be introduced into most mammalian cell types by conventional transfection techniques, more sophisticated techniques including microinjection (Rutter *et al.*, 1995), or the use of adenoviral vectors (Ainscow and Rutter, 2001, 2002; Ravier and Rutter, 2005). Use of adenoviral vectors is extremely effective, since >95% of cells of any mammalian species/embryological origin can be infected at low (nontoxic) viral titers, compared with 2–10% of cells using conventional expression systems. Adenoviral transfection is particularly valuable when studying primary cells or tissue samples (e.g., myocytes or islet β cells), where conventional transfection is essentially impossible.

Luciferase light output can be quantitated from single living cells, after the addition of the (reasonably cell-permeant) cofactor, luciferin. Under most conditions, O_2 and cofactors other than ATP are not limiting. Furthermore, it can be calculated that the contribution of ATP consumption by luciferase represents only a tiny fraction of total cellular ATP turnover (<0.1%, even at relatively high levels of luciferase expression, e.g., 1×10^6 molecules/cell) and is thus nonperturbing for normal cellular ATP homeostasis. Light output from single cells is clearly detectable even after the expression from weak promoters, using highly sensitive photon–counting devices and long integration times (minutes) (Alekseev *et al.*, 1997; Rutter *et al.*, 1995, 1998; White *et al.*, 1994).

Wild-type luciferase has previously been used to study changes in ATP in a variety of different cell types, for example HeLa cells (Jouaville *et al.*, 1999), islet β and α cells (Kennedy *et al.*, 1999; Ravier and Rutter, 2005), and cardiac myocytes (see below). More recently, mutated forms of luciferase have also been developed with markedly decreased affinity for ATP (Guyot, J., Rutter, G. A., and Hanrahan, J., unpublished observations), which may provide more precise measurements of ATP in the millimolar range that may exist in subcellular organelles, including secretory vesicles.

II. ATP Homeostasis in the Heart

Aerobic energy production in the heart is essential for the maintenance of normal contractility, but the mechanisms through which ATP homeostasis is achieved are incompletely understood. Early studies on isolated mitochondria showed that the main regulator of ATP production was likely to be ADP (Chance and Williams, 1956), so it seemed probable that this parameter had to increase, and ATP levels fall, before any stimulation of respiration occurred (reviewed by Brown, 1992). However, subsequent studies using ^{31}P-NMR showed that ATP levels were maintained during increased workload (Katz et $al.$, 1989), implicating alternative regulatory mechanisms in the control of mitochondrial respiration. It is now generally well accepted that this control is provided by Ca^{2+}: in many mammalian cell types, changes in the free Ca^{2+} concentration in the mitochondrial matrix ($[Ca^{2+}]_m$) (Denton and McCormack, 1980), by agonists that increase cytosolic free $[Ca^{2+}]$ ($[Ca^{2+}]_c$), stimulate ATP production and also modulate both the amplitude and duration of cytosolic Ca^{2+} waves (Rutter and Rizzuto, 2000). However, in heart cells, normal excitation–contraction coupling involves very rapid changes in $[Ca^{2+}]_c$, and it remains controversial as to whether these rapid changes also occur in mitochondria, and if so, what the consequences are for regulation of ATP production.

ATP levels had never previously been observed directly in living myocytes, and our aim was to use luciferase to monitor ATP changes in isolated adult rat myocytes.

III. Measurement of Mitochondrial and Cytosolic ATP in Cardiac Myocytes

We measured levels of ATP directly in beating isolated cardiac myocytes from the adult rat, using either untargeted (cytosolic) or mitochondrially targeted luciferase, in order to investigate how changes in workload might affect ATP levels in each compartment. Since primary myocytes are not amenable to transfection using conventional methods, we used an adenoviral transfection method, where myocytes were isolated and then cultured in the presence of adenovirus at a multiplicity of infection (MOI) of 50–100 for ~24 h prior to experimentation. Adenoviruses were generated using the pAdEasy system as described previously (Ainscow and Rutter, 2001). This method allowed close to 100% of myocytes to be infected. Correct localization was confirmed by immunocytochemistry in both neonatal and adult myocytes (which have distinct morphologies), using an anti-luciferase antibody (Fig. 2). Mitochondrially targeted luciferase (mLuc) showed a characteristic mitochondrial distribution and was absent from the nuclei, while untargeted luciferase (cLuc) was more widespread throughout the cell, including the nuclei.

The changes in ATP that occurred within cardiac myocytes stimulated to contract at varying rates were then measured. Adult rat myocytes were isolated and cultured in the presence of adenoviruses encoding mLuc or cLuc for ~24 h and then transferred to modified Krebs buffer (in mmol/liter: 4.2 HEPES, pH 7.6; 130 NaCl; 5.4 KCl; 1.4 $MgCl_2$; 0.4 NaH_2PO_4). The luciferase molecule was then reconstituted by the addition of 1-mmol/liter luciferin shortly (minutes) before measurement. Culture dishes were placed onto a heated (37 °C) microscope

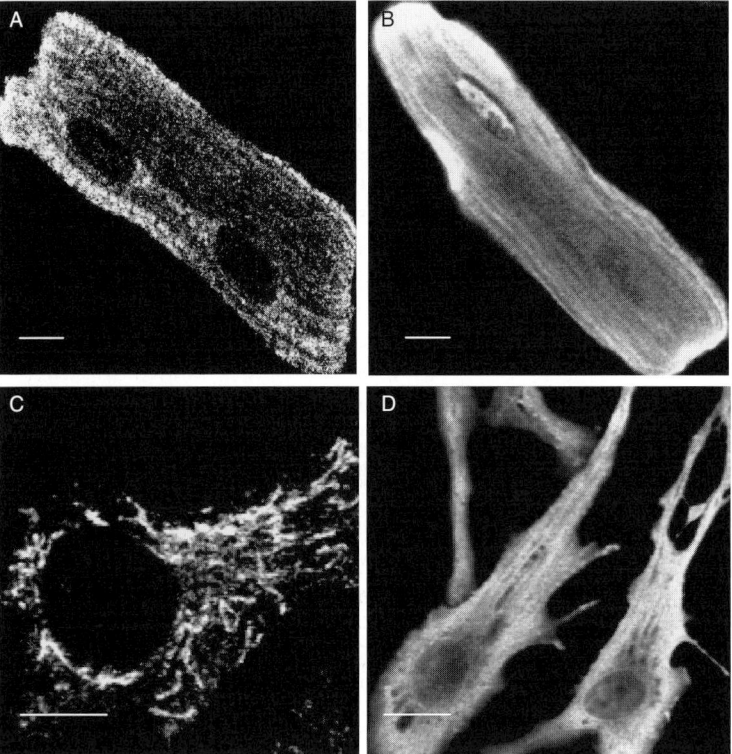

Fig. 2 Immunocytochemistry showing localization of adenovirally expressed luciferase in adult and neonatal rat cardiac myocytes. Primary adult (A, B) and neonatal (C, D) rat myocytes were isolated by Langendorf perfusion (Mitcheson *et al.*, 1998) and enzymatic digestion, respectively, and then cultured for ~24 h in the presence of adenoviruses encoding either mitochondrially targeted (A, C) or untargeted (B, D) luciferase (MOI = 50–100). Luciferase was then detected by immunofluorescence (Saghir *et al.*, 2001), using a 1:50 dilution of goat polyclonal anti-luciferase (Promega) as primary antibody and 1:500 dilution of an Alexa 568 conjugated anti-goat secondary antibody (Molecular Probes, www.invitrogen. com). Scale bars represent 10 microns.

stage, and light emission from the luciferase could be observed easily using the
$10\times$ objective of an Olympus IX-70 inverted microscope coupled to a triply
intensified charge coupled camera (ICCD218; Photek, Lewes, East Sussex) in
time-resolved imaging mode (60 frames/sec). The light emission from a typical
field of view is shown in Fig. 3, in the absence (Fig. 3A) or presence (Fig. 3B) of the
mitochondrial poison sodium azide, which reduces ATP levels, and thus luciferase
activity. Changes in ATP levels were measured as the percentage change in light
output compared to the original baseline light output due to the difficulty in
determining absolute [ATP] (Kennedy *et al.*, 1999).

To measure ATP changes in beating myocytes, cells were stimulated to contract
at 2 Hz in the presence of 2-mM extracellular Ca^{2+} (see Figs. 3–5 for further
details). No changes in ATP were observed in either compartment on the addition
of the β-adrenergic agonist isoproterenol (which increases workload, Fig. 4) or on
addition of increased extracellular $[Ca^{2+}]$ or increases in stimulation rate. These
results suggest that ATP production is extremely well coupled to changes in
demand in both compartments when cells are already beating. In addition, no
changes in ATP were observed during the contractile cycle, indicating that even if
Ca^{2+} enters the mitochondria on a beat-to-beat basis (Bell *et al.*, 2005; Robert
et al., 2001), this is not translated into ATP changes on this timescale.

In order to see whether changes in ATP levels could be observed during very
large changes in workload, cells were stimulated to beat rapidly from rest: ATP
levels were measured before and after stimulation of cells at 2 Hz in presence of
isoproterenol (Fig. 5). Under these conditions, mitochondrial ATP dropped
initially ($\approx$10%) and then rose to a level that was $\sim$20% higher than in the resting
state. This delay in the rise in ATP is postulated to be due to a lag time in the

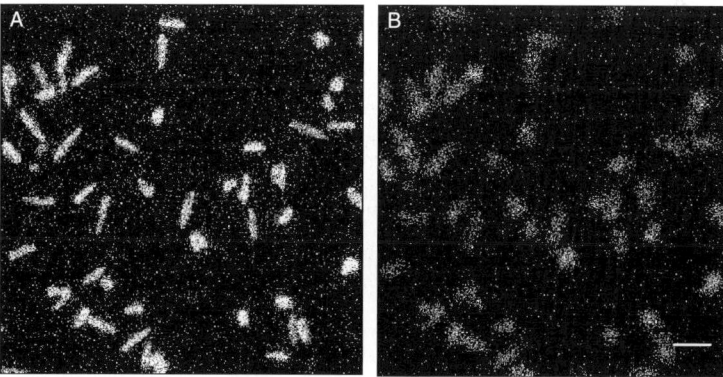

Fig. 3 Light emission from a typical field of view of adult myocytes expressing mLuc. Adult rat
myocytes were isolated and infected with mLuc as described in Fig. 2 before being reconstituted with
1-mM luciferin $\sim$1 min prior to imaging. The image shows light captured from a field of view of beating
myocytes in the absence (A) or presence (B) of 20-mM sodium azide. Integration time was 15 sec and
scale bar represents 100 microns.

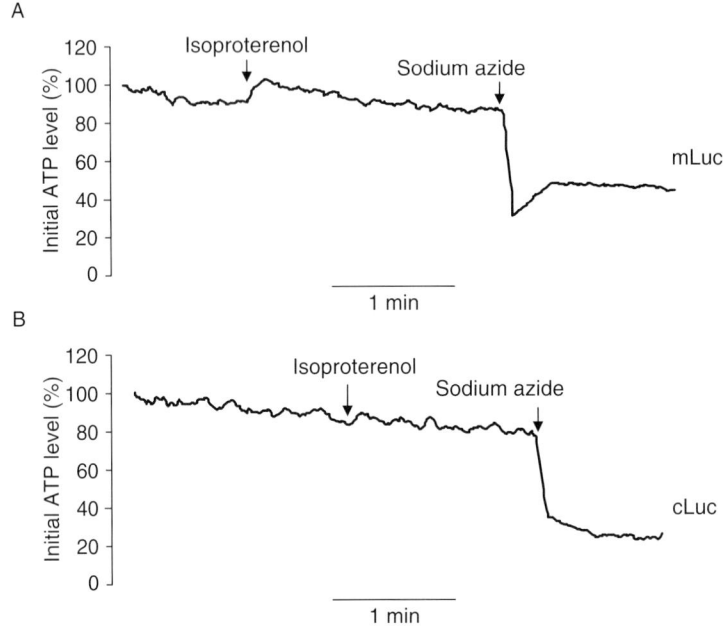

Fig. 4 No effect of isoproterenol on ATP levels in mitochondria or cytosol of already-beating cells. Beating adult myocytes expressing either mLuc (A) or cLuc (B) were stimulated at 2 Hz in the presence of 2-mM extracellular Ca^{2+}. Five micromolar isoproterenol was added at the time indicated on each trace, followed by 20-mM sodium azide, as indicated, to inhibit ATP production.

activation of mitochondrial oxidative metabolism. On return to rest, the ATP level dropped to its original level. In the cytosol, the rise in ATP level was significantly less than in mitochondria ($\sim$7%), suggesting that cytosolic ATP supply and demand are extremely well matched, even on sudden, large increases in demand.

The above experiments demonstrate that adenovirally expressed luciferase is a useful tool for measuring changes in ATP in different compartments of living primary cells such as single rat cardiac myocytes. Although clear changes in ATP levels could be observed on large changes in workload in these cells, it can be concluded that in normal circumstances ATP generation is extremely well matched to demand, in particular that ATP levels in the cytosol are extremely well buffered.

IV. Measurement of ATP in Subcellular Compartments by Luminometery

In addition to the procedures described above, the relative ATP concentration within individual cell compartments can be estimated using a "bulk approach," where luminescence from mammalian cells expressing the targeted luciferase

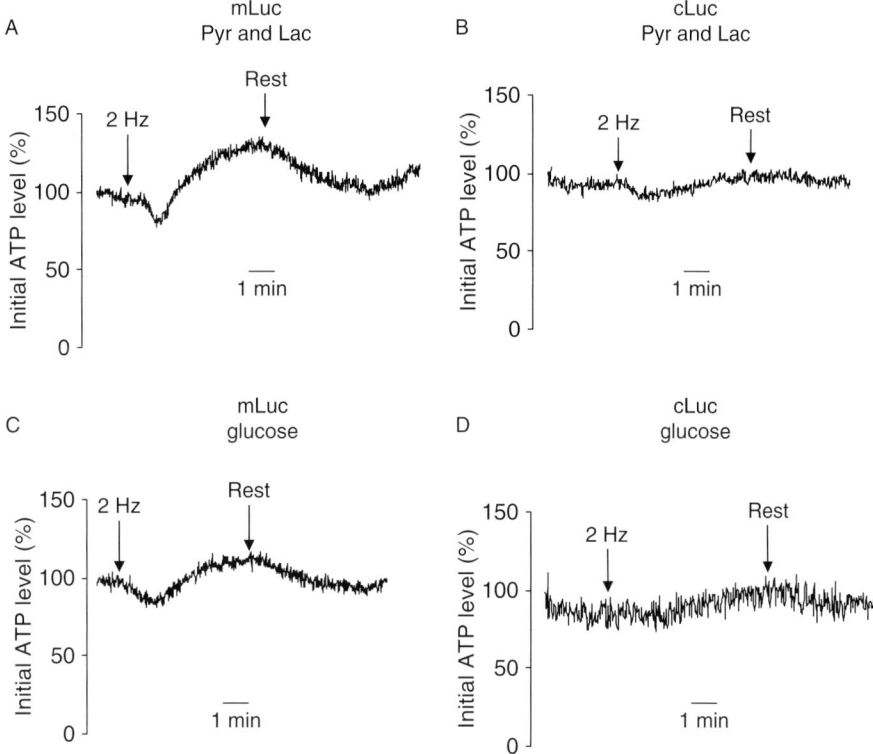

Fig. 5 Effect of stimulation from rest in the presence of isoproterenol on ATP levels in mitochondria and cytosol of adult cardiac myocytes. Adult myocytes expressing either mLuc (A, C) or cLuc (B, D) were allowed to rest in the presence of 2-mM extracellular Ca^{2+} and 5-μM isoproterenol, with either pyruvate and lactate (A, B) or glucose (C, D) as fuel. Cells were stimulated to beat at the point indicated on each trace and then left to beat for several minutes before being returned to rest at the point indicated.

reporters is measured in a luminometer. We have tested this approach by expressing various targeted recombinant luciferases by standard transient transfection with plasmid vectors.

A. Luciferase Vectors

For a detailed description of all the luciferase vectors see Gajewski *et al.* (2003).

1. *Cytosolic luciferase*: Luciferase cDNA with a C-terminal amino acid substitution L550V was cloned in the expression vector PCDNA3.0 (Invitrogen, Inc., Carlsbad, CA). The L/V substitution eliminates the natural peroxisomal targeting signal (Gould *et al.*, 1989), and therefore luciferase is untargeted and resides diffusely in the cytoplasm. (NB this mutation is also present in the humanized

luciferase vectors provided by Promega and was used in the live cells imaging studies discussed above.)

2. *Mitochondrial luciferase*: The mitochondrial-targeting signal for COX8 is appended at the N-terminus of L550V luciferase. The COX8-luciferase protein is directed to the mitochondrial matrix (Gajewski *et al.*, 2003; Jouaville *et al.*, 1999; Kennedy *et al.*, 1999; Porcelli *et al.*, 2001).

In addition to the cytosolic and mitochondrial luciferases, which are also described in the first part of this chapter, we have constructed reporters directed to the plasma membrane or the nucleus.

3. *Nuclear luciferase*: The SV-40 large T antigen nuclear localization sequence (NLS) is appended at the N-terminus of L550V luciferase and directs luciferase to the nucleus (Gajewski *et al.*, 2003).

4. *Plasma membrane luciferase*: The second external arm of the HT1a receptor, which belongs to the family of G-linked protein (Singer, 1990) is appended to the N-terminus of L550V luciferase and directs it to the internal side of the plasma membrane (Gajewski *et al.*, 2003).

B. Cell Transfection and ATP Assays

Attached cells are transfected transiently with plasmid vectors expressing the recombinant targeted luciferase gene, using FuGENE6 (Roche Diagnostics Ltd., West Sussex, UK) following the manufacturer's recommendations. Local intra-cellular ATP content is measured in intact cells 48 h after transfection to allow for sufficient protein expression. We have also tested other transfection reagents, such as lipofectamine (Invitrogen), and obtained similar results. Successful expression of enzymatically functional targeted luciferase has been achieved in several cell lines, including COS-1, 143B osteosarcoma, and 293T HEK cells.

Transfected cells are collected by trypsinization or by scraping, resuspended by trituration, diluted at the concentration of 0.2×10^6 cell/ml in culture media, and 1 ml/well of cell suspension is placed in 24-well plastic plates (preferably noncoated, to prevent attachment). Cells are incubated for variable lengths of time, generally 30 min to 1 h, in medium containing different combinations of substrates and inhibitors, according to the metabolic conditions being tested in the assay, with gentle rocking to prevent cells from attaching to the plate, at $37\,^\circ$C with 5% CO_2. For example, glucose and oligomycin, a specific inhibitor of the mitochondrial ATP synthetase, can be combined to test glycolytic ATP production, whereas pyruvate and 2-deoxyglucose, an inhibitor of glycolysis, can be used to test mitochondrial ATP production. After an appropriate incubation time, cells are collected with a pipette and spun down at $800 \times$ g for 2 min, rinsed in phosphate saline buffer (PBS), and resuspended in 90 μl of a buffer containing 25-mM tricine, 150-mM NaCl, pH 7.4. Ten microliters of 20-mM beetle luciferin (Promega, Inc., Madison, WI) (final concentration 2 mm) are added to the cell suspension and light emission is measured immediately in a counting luminometer, using a kinetic protocol with repeated measurements at 5-second intervals, until the maximum value of relative luminescence is achieved (typically, after 30 sec). We employ an Optocomp I luminometer (MGM Instruments, Hamden, CT), in the kinetic measurement setting, but other type of cuvette or multiwell plate luminometers can be used.

Because only cells expressing the luciferase reporter are measured, untransfected cells are irrelevant to the assay. However, due to the inherent variability of targeted luciferase expression levels in transiently transfected cells, it is necessary to utilize a correction factor to express the ATP levels based on the intensity of the emitted light. To this end, the maximal relative luminescence can be expressed as a ratio of the "total potential luminescence" measured on an aliquot of cells from the same transfection. This second aliquot of cells is lysed and light emission is measured in the presence of excess ATP (0.5 mm, final concentration), using a standard luciferase assay kit (Promega, Inc.) according to the manufacturer's protocol. Using this correction factor, ATP levels can be compared across multiple transfections and different incubation conditions.

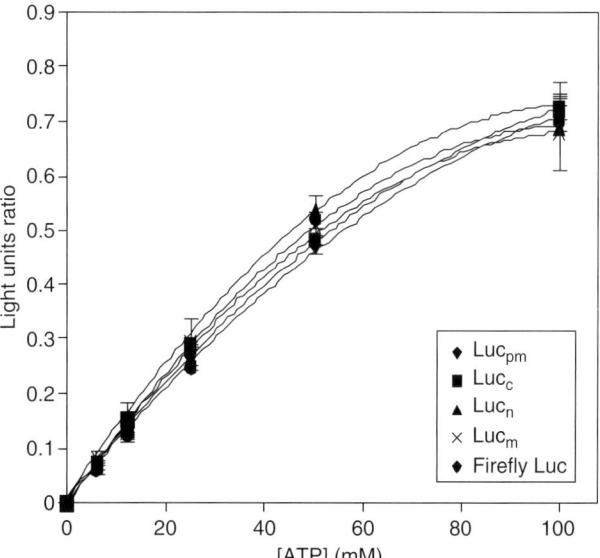

Fig. 6 Calibration of the luciferase ATP reporter system. Kinetics of the affinity of recombinant chimeric luciferases for ATP in tricine buffer. Luminescence with the indicated recombinant luciferases was measured in aliquots of 2×10^5 lysed transfected cells resuspended in tricine buffer containing 2-mM luciferin and 100 pg of luciferase. The "light units ratio" is the ratio of luminescence obtained from the cells in tricine buffer and the "total potential luminescence" obtained from an identical aliquot of cells measured with a luciferase detection kit. Scale bars represent standard deviations of three replicate samples.

C. ATP Standards

When the goal of the experiment is to determine the absolute concentrations of "free" ATP available for interaction with luciferase in the various cell compartments, ATP-luminescence standards curves can be constructed, either using lysates from cells expressing the targeted luciferase or with purified firefly luciferase. In the first case, 2×10^5 cells transfected with recombinant luciferase are lysed in 20 μl of a buffer containing 20-mM HEPES and 0.1% Triton X-100, pH 7.2. In the second approach, 150 pg of firefly luciferase (Roche Diagnostics Ltd., West Sussex, UK) are dissolved in 20 μl of the lysis buffer described above. In both approaches, luminescence is measured from the 20 μl of lysis buffer containing cell extracts or purified luciferase, after the addition of 80 μl of the "physiological" luciferin buffer (20-mM HEPES, 140-mM KCl, 10.2-mM EGTA, 6.7-mM $CaCl_2$, 2-mM luciferin, 20-μg/ml digitonin, 0.5-mM $MgCl_2$, and 0.01-mM CoA), that we have previously described (Jouaville *et al.*, 1999; Kennedy *et al.*, 1999), plus ATP standards ranging from 0 to 100 μM (Gajewski *et al.*, 2003) (Fig. 6).

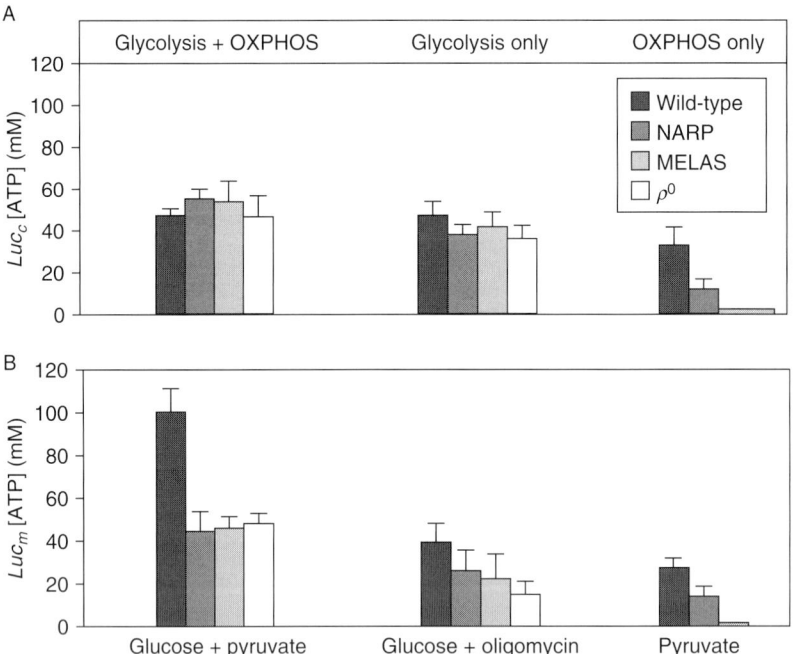

Fig. 7 ATP concentration in intracellular compartments. Wild-type, NARP, MELAS, and r^0 cells expressing luciferases targeted to the cytosol (A) and the mitochondrial matrix (B) were incubated for 2 h in medium containing the indicated combinations of substrate and inhibitors. Luminescence was measured on aliquots of 2×10^5 intact cells in the presence of 2-mM luciferin. ATP concentrations were extrapolated based on the light units ratio curves obtained in the ATP titration experiment shown in Fig. 6. Scale bars represent standard deviations of 20 replicate samples.

Figure 7 shows an example of the use of ATP reporters to measure ATP in the cytosol and the mitochondrial matrix (Gajewski *et al.*, 2003).

Acknowledgments

We thank the British Heart Foundation and Wellcome Trust for financial support. GAR was a Wellcome Trust Research Leave Fellow.

References

Ainscow, E. K., and Rutter, G. A. (2001). Mitochondrial priming modifies Ca^{2+} oscillations and insulin secretion in pancreatic islets. *Biochem. J.* **353,** 175–180.

Ainscow, E. K., and Rutter, G. A. (2002). Glucose stimulated oscillations in free cytosolic ATP concentration imaged in single islet beta-cells: Evidence for a Ca^{2+} dependent mechanism. *Diabetes* **51**(Suppl. 1), S162–S170.

Alekseev, A. E., Kennedy, M. E., Navarro, B., and Terzic, A. (1997). Burst kinetics of co-expressed Kir6.2/SUR1 clones: Comparison of recombinant with native ATP-sensitive K^+ channel behaviour. *J. Membr. Biol.* **159,** 161–168.

Bell, C. J., Rutter, G. A., and Griffiths, E. J. (2005). Mitochondrial and cytosolic calcium transients in adult cardiomyocytes detected using targeted aequorin. *J. Physiol.* **567P,** C96 (Abstract).

Brown, G. C. (1992). Control of respiration and ATP synthesis in mammalian mitochondria and cells. *Biochem. J.* **284,** 1–13.

Chance, B., and Williams, G. R. (1956). The respiratory chain and oxidative phosphorylation. *Adv. Enzymol. Relat. Subj. Biochem.* **17,** 65–134.

De Wet, J. R., Wood, K. V., DeLuca, M., Helinski, D. R., and Subramani, S. (1987). Firefly luciferase gene: Structure and expression in mammalian cells. *Mol. Cell. Biol.* **7,** 725–737.

Denton, R. M., and McCormack, J. G. (1980). On the role of the calcium transport cycle in heart and other mammalian mitochondria. *FEBS Lett.* **119,** 1–8.

Gajewski, C. D., Yang, L., Schon, E. A., and Manfredi, G. (2003). New insights into the bioenergetics of mitochondrial disorders using intracellular ATP reporters. *Mol. Biol. Cell* **14,** 3628–3635.

Gould, S. J., Keller, G. A., Hosken, N., Wilkinson, J., and Subramani, S. (1989). A conserved tripeptide sorts proteins to peroxisomes. *J. Cell Biol.* **108,** 1657–1664.

Jouaville, L. S., Bastianutto, C., Rutter, G. A., and Rizzuto, R. (1999). Regulation of mitochondrial ATP synthesis by calcium: Evidence for a long term, Ca^{2+} triggered, mitochondrial activation. *Proc. Natl. Acad. Sci. USA* **96,** 13807–13812.

Katz, L. A., Swain, J. A., Portman, M. A., and Balaban, R. S. (1989). Relation between phosphate metabolites and oxygen consumption of heart *in vivo. Am. J. Physiol.* **256,** H265–H274.

Kennedy, H. J., Pouli, A. E., Jouaville, L. S., Rizzuto, R., and Rutter, G. A. (1999). Glucose-induced ATP microdomains in single islet beta-cells. *J. Biol. Chem.* **274,** 13281–13291.

Mitcheson, J. S., Hancox, J. C., and Levi, A. J. (1998). Cultured adult cardiac myocytes: Future applications, culture methods, morphological and electrophysiological properties. *Cardiovasc. Res.* **39,** 280–300.

Porcelli, A. M., Pinton, P., Ainscow, E. K., Chiesa, A., Rugolo, M., Rutter, G. A., and Rizzuto, R. (2001). Targeting of reporter molecules to mitochondria to measure calcium, ATP, and pH. *Methods Cell Biol.* **65,** 353–380.

Ravier, M. A., and Rutter, G. A. (2005). Glucose or insulin, but not zinc ions, inhibit glucagon secretion from mouse pancreatic alpha-cells. *Diabetes* **54,** 1789–1797.

Robert, V., Gurlini, P., Tosello, V., Nagai, T., Miyawaki, A., Di Lisa, F., and Pozzan, T. (2001). Beat-to-beat oscillations of mitochondrial $[Ca^{2+}]$ in cardiac cells. *EMBO J.* **20,** 4998–5007.

Rutter, G. A., and Rizzuto, R. (2000). Regulation of mitochondrial metabolism by ER Ca^{2+} release: An intimate connection. *Trends Biochem. Sci.* **25,** 215–221.

Rutter, G. A., Kennedy, H. J., Wood, C. D., White, M. R. H., and Tavare, J. M. (1998). Quantitative real-time imaging of gene expression in single cells using multiple luciferase reporters. *Chem. Biol.* **5**, R285–R290 (Abstract).

Rutter, G. A., White, M. R. H, and Tavare, J. M. (1995). Non-invasive imaging of luciferase gene expression in single living cells reveals the involvement of MAP kinase in insulin signalling. *Curr. Biol.* **5**, 890–899.

Saghir, A. N., Tuxworth, W. J., Jr., Hagedorn, C. H., and McDermott, P. J. (2001). Modifications of eukaryotic initiation factor 4F (eIF4F) in adult cardiocytes by adenoviral gene transfer: Differential effects on eIF4F activity and total protein synthesis rates. *Biochem. J.* **356**, 557–566.

Singer, S. J. (1990). The structure and insertion of integral proteins in membranes. *Annu. Rev. Cell Biol.* **6**, 247–296.

White, M. R. H., Masuko, M., Amet, L., Elliott, G., Braddock, M., Kingsman, A. J., and Kingsman, S. M. (1994). Real-time analysis of the transcriptional regulation of HIV and hCMV promoters in single mammalian cells. *J. Cell Sci.* **108**, 441–455.

PART IV

Oxidative Stress Measurements

CHAPTER 18

Measurement of Reactive Oxygen Species in Cells and Mitochondria

J. S. Armstrong and M. Whiteman

Department of Biochemistry
Yong Loo Lin School of Medicine
National University of Singapore
Republic of Singapore 117597, Singapore

I. Introduction

Reactive oxygen species (ROS), including superoxide anion ($O_2^{\bullet-}$), hydroxyl radicals (OH$^{\bullet}$), and hydrogen peroxide (H_2O_2), are generated during aerobic metabolism. The mitochondrion is the most common source of ROS production, although there are other important cellular sources, including enzymes, such as cytochrome P450 in the endoplasmic reticulum (ER), lipoxygenases, cyclooxy-genases, xanthine oxidase (XO), and the NADPH oxidase of white blood cells (WBCs).

Under normal physiological conditions, it is estimated that ~1–3% of electrons carried by the mitochondrial electron transport chain (ETC) leak out of the pathway and pass directly to oxygen, generating $O_2^{\bullet-}$ (Halliwell and Gutteridge, 1999). Mitochondrial respiratory complex I, also known as the NADH-ubiqui-none oxidoreductase, is an important source of ROS production, especially in the reverse transport mode (Lambert and Brand, 2004). Respiratory complex III, also known as the ubiquinol-cytochrome c oxidoreductase or cytochrome bc_1, as well as the coenzyme Q radical generated during the Q cycle, is probably quanti-tatively the most important site of mitochondrial $O_2^{\bullet-}$ production (Halliwell and Gutteridge, 1999; Raha and Robinson, 2000; Turrens, 1997).

High concentrations of the enzyme manganese superoxide dismutase (MnSOD) in the mitochondrial matrix ensure that the basal level of $O_2^{\bullet-}$ formed during normal electron transport is kept at a level whereby it does not damage proteins of the matrix involved in the regulation of metabolism, including the iron sulfur proteins aconitase and succinate dehydrogenase (Halliwell and Gutteridge, 1999; Li *et al.*, 1995). MnSOD catalyzes the breakdown or dismutation of $O_2^{\bullet-}$ to yield H_2O_2, which is an important signaling molecule generated throughout the plant and animal kingdoms. Normally, H_2O_2 generated by the dismutation of mito-chondrial $O_2^{\bullet-}$ is broken down by a glutathione (GSH)-dependent peroxidase enzyme (GSH/Px) to yield water, whereas in the cytosol H_2O_2, which is generated by peroxisomal β-oxidation of fatty acids or by the enzyme XO, it is broken down by catalase (CAT) and GSH/Px, respectively (see Fig. 1 for a schematic represen-tation of cellular and mitochondrial ROS production pathways). However, when

Sources of reactive oxygen species

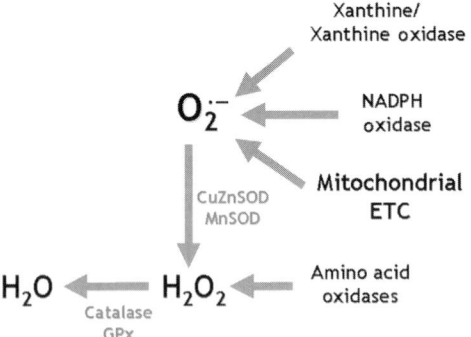

Fig. 1 Cellular and mitochondrial sources of ROS production and breakdown. The mitochondrial ETC, NADPH oxidases, and X/XO systems are shown as important sources of $O_2^{\bullet-}$. SODs convert $O_2^{\bullet-}$ to H_2O_2 enzymatically, which is then broken down to H_2O by CATs and the GSH/GSHPx systems.

H_2O_2 is produced in excess, it is also a toxic molecule which can react with ferrous iron (Fe^{2+}) to form $OH^{\bullet}$ (via the Fenton reaction or the metal-catalyzed Haber–Weiss reaction (Halliwell and Gutteridge, 1999). $OH^{\bullet}$ is a short-lived species that can react with molecules such as DNA.

In the mitochondrion, the cooperative action of MnSOD and the GSH/GSHPx system ensures that ROS ($O_2^{\bullet-}$ and H_2O_2) generated during normal aerobic metabolism are kept at a nontoxic level. Although, the cell possesses an extensive antioxidant defense system to combat ROS toxicity, including CAT, SODs, and GSH and GSH/GSHPx systems, when ROS overcome the antioxidant defense systems of the cell and redox homeostasis is lost, the result is oxidative stress which can lead to cell damage and death. The importance of being able to make qualitative and quantitative measurements of ROS production by cells is under-scored by the fact that oxidative stress is involved in the pathogenesis or progres-sion of numerous diseases, including neurodegenerative diseases (Huntington, Parkinson, and Alzheimer diseases), retinal degenerative disorders, AIDS, cancer, and the aging process in general.

This chapter discusses (1) the cellular sources of ROS and their enzymatic detoxification; (2) common methods used to determine cellular and mitochondrial ROS, including chemiluminescence (CL), spin trapping, and fluorescence and enzymatic techniques; (3) common problems associated with these assays and the interpretation of data; and (4) simple protocols for the determination of ROS in cells and mitochondria. When measuring ROS in cells and mitochondria we emphasize the need for thorough understanding of results obtained and their interpretation.

II. Measurement of ROS in Cells

A. CL-Based Characterization of ROS Generated by HX/XO and Phorbol Myristate Acetate (PMA)–Stimulated WBCs

In 1960, Totter and colleagues investigated the use of luminol as a detector of radicals formed during the reaction of XO with hypoxanthine (HX). They re-ported that the light produced by luminol and XO/HX was directly related to the velocity of the XO/HX reaction, and suggested that this system could be used as a general test for oxidizing radicals (Totter *et al.*, 1960). Lucigenin (bis-*N*-methyla-cridinium) has also frequently been used for the luminescent detection of the radical $O_2^{\bullet-}$ by the XO/HX system or by activated phagocytes (Storch and Ferber, 1988). Luminol and lucigenin are often used to detect the production of ROS by activated phagocytes, although they have also been used to determine ROS in other cell types (Faulkner and Fridovich, 1993). More recently, several other compounds have been used for chemiluminescent detection of $O_2^{\bullet-}$, including coelenterazine [2-(4-hydroxybenzyl)-6-(4-hydroxyphenyl)-8-benzyl-3,7-dihydroi-midazo [1,2-α]pyrazin-3-one] and its analogues CLA (2-methyl-6-phenyl-3,7-dihy-droimidazo [1,2-α]pyrazin-3-one) and MCLA [2-methyl-6-(4-methoxyphenyl)-3,

7-dihydroimidazo[1,2-α]pyrazin-3-one]. We provide an example of the use of luminol to detect ROS generated by (1) HX/XO and (2) PMA-stimulated WBCs.

Reagents

1-mM HX (Sigma-Aldrich, St. Louis, MO)
Copper and zinc-containing superoxide dismutase (CuZnSOD) (Sigma-Aldrich)
Dimethyl sulfoxide (DMSO) (Sigma-Aldrich)
Luminol (Sigma-Aldrich)
XO (Sigma-Aldrich)
Ethylene diamine tetraacetic acid (EDTA) (Sigma-Aldrich)
CAT (Sigma-Aldrich)
Phosphate-buffered saline (PBS), pH 7.4

Protocol for the luminol-dependent characterization of ROS generated by the XO/ HX system

1. 100 μl of 1-mM HX in PBS are added to 50 μl of either CuZnSOD (100 U/ml), CAT (4×10^3 U/ml), or DMSO (100 mM), and 100 μl of CL probe (1-mM luminol) are added to the test wells of a white 96-well microtiter plate (in triplicate).

2. The reaction is started by the addition of 50-μl XO (0.25 U/ml) using the automated luminometer dispenser.

3. The luminometer is operated in the integration mode for a time of 10 sec using "Ascent" software from LabSystems, MN.

4. Results are expressed as total integrated CL signal [relative light units (RLU)] (Table I).

5. This protocol may require substantial optimization to achieve maximum sensitivity and reliability of results.

Protocol for the luminol-dependent characterization of ROS generated by white blood cells (WBCs)

1. EDTA-anticoagulated whole blood is allowed to sediment under gravity for 1 h at 37°C and the plasma is removed and centrifuged at $\sim400 \times g$ for 10 min to pellet out the WBCs.

2. WBCs are washed in PBS at 4°C and resuspended in PBS to a concentration of $2–5 \times 10^6$ cells/ml.

3. WBCs (100 μl) are added to a white 96-well microtiter plate in triplicate together with 50 μl of PMA (2 ng/ml) and 50 μl of test compounds, including either CuZnSOD (100 U/ml), CAT (4×10^3 U/ml), or DMSO (100 mM).

4. 100 μl of 1-mM luminol is added, giving a final total volume of 300 μl.

5. The luminometer is operated in the kinetic mode at 10 sec intervals for a total of 90 min using "Ascent" software (Fig. 2) and in the integration mode (Table I).

Table I

Effect of Antioxidants on Superoxide Radicals Generated by HX + XO and PMA-Stimulated Leukocytes, Measured by CL[a]

Source of ROS	Control	SOD	Catalase	DMSO
[HX + XO]				
Mean RLU	1590 ± 75.5	37.9 ± 4.7	941 ± 35.3	1605 ± 46.9
Inhibition (%)		$98\% \pm 0.6\%$	$41\%^{*} \pm 4\%$	$-1\% \pm 6\%$
[PMA + WBCs]				
Mean RLU	192 ± 6.4	19.5 ± 0.27	107 ± 9.9	195 ± 11
Inhibition (%)		$90\%^{*} \pm 0.3\%$	$44\%^{*} \pm 10\%$	$-1\% \pm 11\%$

[a]Luminol-dependent CL generated by HX/XO and PMA-stimulated WBCs in the presence and absence of antioxidants SOD, CAT, and DMSO. CL was monitored over 30 min (mean RLU of three experiments) and inhibition (%) of CL signal by SOD, CAT, and DMSO is shown (reproduced from Armstrong *et al.*, 1999, with permission).

*$P < .05$ compared to control.

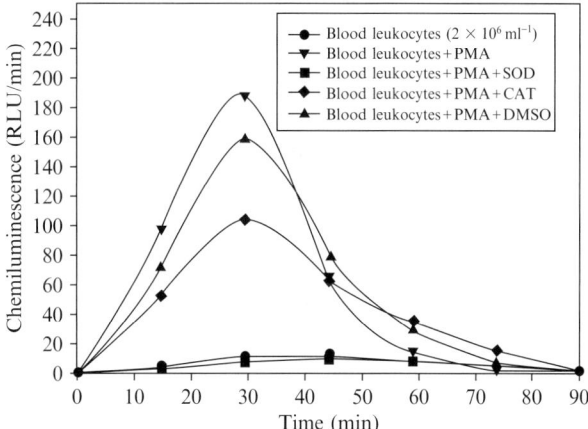

Fig. 2 Luminol-dependent CL generated by PMA-stimulated WBCs in the presence and absence of antioxidants SOD, CAT, and DMSO. CL was monitored kinetically at 10-sec intervals for a total time of 90 min (see text for details) (reproduced with permission Armstrong *et al.*, 1999).

6. The antioxidants SOD, CAT, and DMSO are used to measure the relative proportions of $O_2^{\bullet-}$, H_2O_2 and $OH^{\bullet}$ generated by PMA-stimulated WBCs.

7. The results are expressed either as the total integrated CL signal (RLU) (Table I; reproduced with permission of Armstrong *et al.*, 1999) or as RLU/min (Fig. 2; reproduced with permission of Armstrong *et al.*, 1999).

Comments

Although luminol and lucigenin are both used for the determination of $O_2^{\bullet-}$, the two reactions are different, in that luminol requires univalent oxidation and lucigenin requires univalent reduction, before either can react with $O_2^{\bullet-}$ and produce luminescence that is, luminol or lucigenin, per se, do not react with $O_2^{\bullet-}$ (Faulkner and Fridovich, 1993; Spasojevic *et al.*, 2000).

A major problem with lucigenin is that the radical cation, formed by its reduction, can auto-oxidize and generate $O_2^{\bullet-}$ artifactually. Because of this, it has been suggested that lucigenin can be used for assaying SOD activity, but it should not be used for measuring $O_2^{\bullet-}$ (Liochev and Fridovich, 1997). The extent to which this artifact can interfere with the accurate measurement of $O_2^{\bullet-}$ by lucigenin is controversial (Munzel *et al.*, 2002), but it appears to be significant in some cell systems (Sohn *et al.*, 2000; Spasojevic *et al.*, 2000; Tarpey *et al.*, 1999; Wardman *et al.*, 2002). It has been suggested that modulating the concentrations of lucigenin may circumvent the problems associated with the overestimation of $O_2^{\bullet-}$. However, since it is not possible to predict the extent of redox cycling, this may be difficult to determine accurately (Tarpey and Fridovich, 2001). There are other problems associated with use of lucigenin. For example, the conversion of lucigenin to the radical by $O_2^{\bullet-}$ is not rapid and requires other cellular-reducing systems (e.g., XO, the mitochondrial ETC, or the phagocyte NADPH oxidase). This introduces an obvious complexity in interpreting results (Sohn *et al.*, 2000; Spasojevic *et al.*, 2000; Tarpey and Fridovich, 2001; Tarpey *et al.*, 1999).

There are similar problems with the use of luminol for the detection of radicals such as $O_2^{\bullet-}$, since the radical intermediate formed on oxidation also auto-oxidizes (Tarpey and Fridovich, 2001). Furthermore, luminol luminescence is also not a reliable indicator of $O_2^{\bullet-}$, even when $O_2^{\bullet-}$ is involved in the reaction leading to light emission, because it can mediate $O_2^{\bullet-}$ formation by a variety of oxidants, including ferricyanide, hypochlorite, and XO/HX (Hodgson and Fridovich, 1973).

The problems of luminol and lucigenin associated with the artifactual generation of $O_2^{\bullet-}$ can be circumvented by the use of more specific alternative nonredox-cycling compounds for the determination of $O_2^{\bullet-}$, including coelenterazine, a luminophore isolated from the coelenterate *Aequorea* (Munzel *et al.*, 2002; Tarpey and Fridovich, 2001; Tarpey *et al.*, 1999; Teranishi and Shimomura, 1997), and the luciferin analogues CLA and MCLA from *Cypridina*, of which MCLA is the most sensitive probe for $O_2^{\bullet-}$. Although the intensity of light emitted from the interaction of co-elenterazine with $O_2^{\bullet-}$ is greater than that from either lucigenin or luminol, coelenterazine-dependent CL is not entirely specific for $O_2^{\bullet-}$, because $ONOO^-$ will also cause coelenterazine to luminescence (Tarpey *et al.*, 1999). A similar problem is found with MCLA, which reacts with peroxyl radicals (Tampo *et al.*, 1998). Our investigations with MCLA have shown that this compound, although more sensitive to $O_2^{\bullet-}$ than luminol, has a high level of background luminescence, is both light- and temperature-sensitive, and auto-oxidizes, adding to the practical problems associated with nonspecificity.

B. Electron Spin Resonance Spectroscopy

The only technique that detects free radicals specifically is electron spin resonance (ESR) spectroscopy, because it detects the presence of unpaired electrons. However, unpaired electrons of species such as $O_2^{\bullet-}$, $OH^{\bullet}$, and $NO^{\bullet}$ are highly reactive radicals and generally do not accumulate to high-enough levels to be measured. One solution to the problem has been to use "spin traps" or "probes" that intercept reactive radicals and form more stable, longer-lasting radical adducts with characteristic ESR signatures (Halliwell and Gutteridge, 1999). These adducts accumulate to levels that can be detected using ESR spectroscopy, and provide information enabling the identification of the originating free radical species. A wide range of spin traps is available for use both in whole animal studies and in cell culture systems including *N*-tert butyl-*p*-phenylnitrone (PBN) and 5,5-dimethyl-1-pyrroline *N*-oxide (DMPO) (Khan *et al.*, 2003). Newer compounds, including 1,1,3-trimethyl-isoindole *N*-oxide (TMINO) (Bottle *et al.*, 2003), *N*-2-(2-ethoxycarbonyl-propyl)-α-phenylnitrone (EPPN) (Stolze *et al.*, 2003), and 5-diethoxyphosphoryl-5-methyl-1-pyrroline *N*-oxide (DEPMPO) (Fig. 3), have been reported to be more $O_2^{\bullet-}$ specific and stable (Frejaville *et al.*, 1995; Khan *et al.*, 2003; Liu *et al.*, 1999).

Reagents

1-g/100-ml Chelex-100 (Biorad, Hercules, CA)
PBS, pH 7.4
50-mM DEPMPO (Oxis International, Inc., Portland, OR)
2-ng/ml PMA (Sigma-Aldrich)

Equipment

Aqueous quartz flat cell (Wilmad, Buena, NJ)
Bruker 200D EPR spectrometer

Protocol for ESR spectroscopy characterization of ROS generated by activated WBCs using the spin trap DEPMPO

1. PBS buffer is pretreated with the ion-exchange resin Chelex-100 (1 g/100 ml) to remove adventitious metals in the buffer (Armstrong *et al.*, 1999).

2. WBCs (2×10^5 cells/ml) in Chelex-treated PBS are treated with PMA (2 ng/ml) in the presence of the spin trap DEPMPO (50 mM).

3. The mixtures are added to an aqueous quartz flat cell (Wilmad) which is then centered in a TE011 cavity.

4. EPR spectra are recorded with a Bruker 200D spectrometer operated at 9.7 GHz with a 1 GHz modulation frequency.

5. The data are transferred to a computer for simulation analysis (Armstrong *et al.*, 1999).

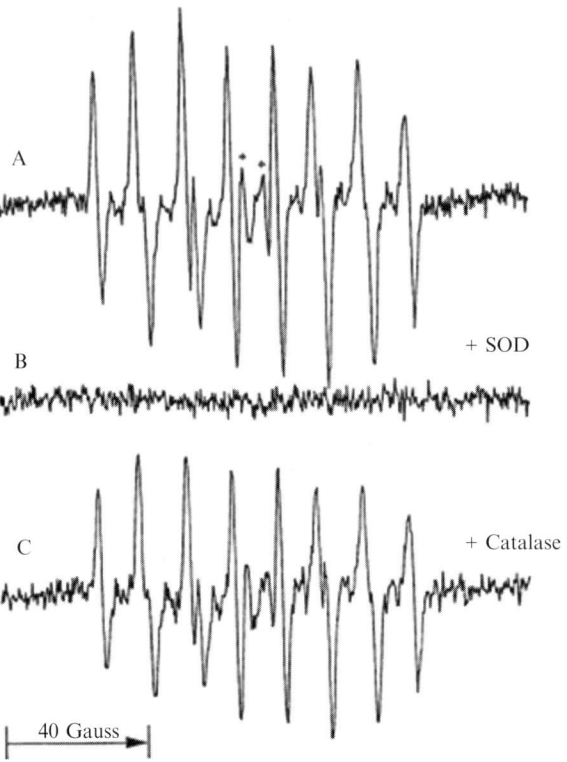

Fig. 3 ESR determination of $O_2^{\bullet-}$ and $OH^\bullet$ generated by PMA-stimulated WBCs $\pm$ SOD and CAT, using the spin trap DEPMPO. See text for details (reproduced with permission Armstrong *et al.*, 1999).

Comments

The determination of radical species such as $O_2^{\bullet-}$, $OH^\bullet$, and $NO^\bullet$ by ESR spectroscopy necessitates the use of employing spin traps to generate species with longer half-lives and to facilitate their detection by ESR spectroscopy. However, the technique can detect more stable free radical-derived species, including ascorbyl and tocopheroxyl radicals, and heme–nitrosyl complexes produced in vascular tissues during oxidative injury and inflammation (Laurindo *et al.*, 1994).

There are a number of potential problems with the use of spin traps (Halliwell and Gutteridge, 1999; Khan *et al.*, 2003; Rosen *et al.*, 1999). For example, one problem is that the reaction products giving the ESR signal can be removed rapidly *in vivo* and in cultured cells by enzymatic metabolism and also by direct reduction by agents such as ascorbate. For example, when DMPO is used to trap the $OH^\bullet$, ascorbate can directly reduce the DMPO-hydroxyl radical adduct to an ESR-silent species (Halliwell and Gutteridge, 1999). DMPO can also be oxidized by ferric ions, which generates the 4-line spectrum normally produced by $OH^\bullet$ (Makino *et al.*, 1990). Papers reporting on the use of spin traps to detect radical species often fail to show use of the proper controls (Halliwell, 1995); for example, does the added compound interfere with the radical-generating system (e.g., decomposing H_2O_2 or chelating iron in the Fe^{2+}/H_2O_2 system), or does it interact directly with the trap spin adduct, reducing it to an ESR-silent species?

Rizzi et al. (2003) have introduced triarylmethyl free radical, TAM OX063, as a probe for the detection of $O_2^{\bullet-}$ in aqueous solution. In this case, the $O_2^{\bullet-}$ reacts with the probe to cause the loss of the ESR signal. One advantage of TAM OX063 is that it is not subjected to reduction by such agents as ascorbate or reduced GSH.

Valgimigli et al. (2001) described an ESR method for the measurement of the oxidative stress status in biological systems. The method was based on the X-band ESR detection of a nitroxide generated under physiological conditions by oxidation of bis(1-hydroxy-2,2,6,6-tetramethyl-4-piperidinyl)decandioate, which is administrated as hydrochloride salt. Since the probe is reported to react rapidly with the majority of radical species involved in oxidative stress and to cross cell membranes easily, it was suggested to be applicable in the clinical setting (Valgimigli et al., 2001).

C. Fluorescence Dye Techniques for the Determination of Cellular ROS

Dichlorofluorescein diacetate (DCFH-DA) is taken up by cells and is hydrolyzed to 2′,7′-dichlorofluorescein (DCFH) which is trapped inside cells. Intracellular DCFH, a nonfluorescent fluorescein analogue, is oxidized by ROS to highly fluorescent 2′,7′-dichlorofluorescein (DCF). DCFH-DA has been used quantitatively to detect ROS produced by PMA-stimulated phagocytes, as well as in cultured cells. The fact that DCFH is also oxidized by H_2O_2 or H_2O_2-generating systems, such as glucose oxidase and glucose, and is inhibited by CAT but not by SOD, implied that DCFH oxidation is relatively specific for H_2O_2 formation (Bass et al., 1983). However, it is now known that this is not the case, and that DCF yields fluorescence in response to a variety of ROS, including not only H_2O_2, but also other peroxides, ONOO⁻ (Whiteman et al., 2004), and hypochlorous acid (HOCl) (Whiteman et al., 2005), and after intracellular oxidation that is mediated by GSH depletion (Armstrong and Jones, 2002; Armstrong et al., 2004a) (see example in Fig. 4 from Armstrong et al., 2004b). DCFH oxidation does not require a functional ETC (see example in Fig. 6 and Whiteman et al., 2006), indicating that the ROS generated under these conditions are not mitochondrial in origin (Fig. 5). Because of this, *specific scavengers* of ROS must be used to understand the source of the DCF fluorescence signal.

Reagents and equipment

DCFH-DA (Molecular Probes, Eugene, Oregon)

10-mM glucose in PBS, pH 7.4

Fluorescence activated cell sorting (FACS) machine or fluorescence plate reader (excitation wavelength 498 nm and emission wavelength 522 nm)

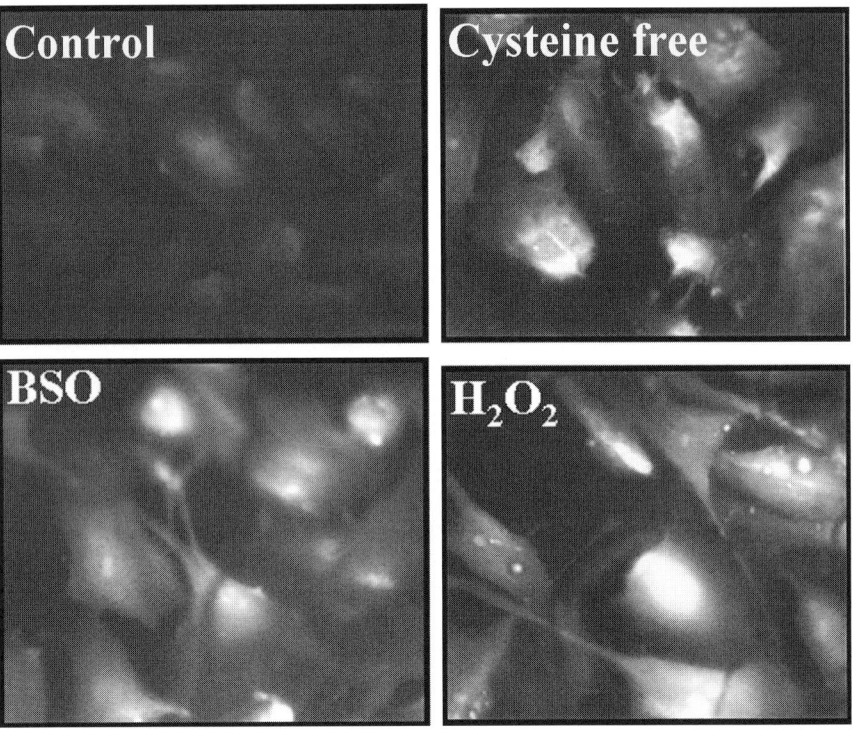

Fig. 4 Confocal microscopy fluorescence determination of ROS produced by retinal pigment epithe-lial cells treated with agents to deplete GSH, including L-BSO and cysteine-free media. Authentic H_2O_2 (10 mM) was used as a positive control for ROS (reproduced from Armstrong *et al.*, 2004b, with permission).

Glucose 4 days

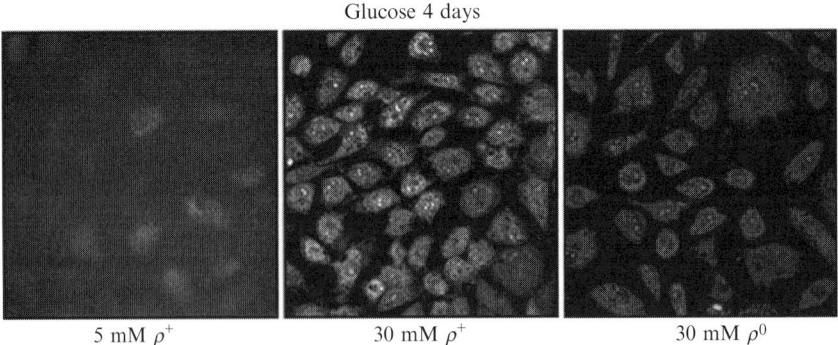

Fig. 5 ROS measurements in BAEC. BAEC parental (ρ^+) cells and BAEC cells lacking mtDNA (ρ^0) were incubated in media containing either 5-mM or 30-mM glucose for 4 days. Cells were stained with 10-μM HE and E fluorescence was viewed by confocal microscopy.

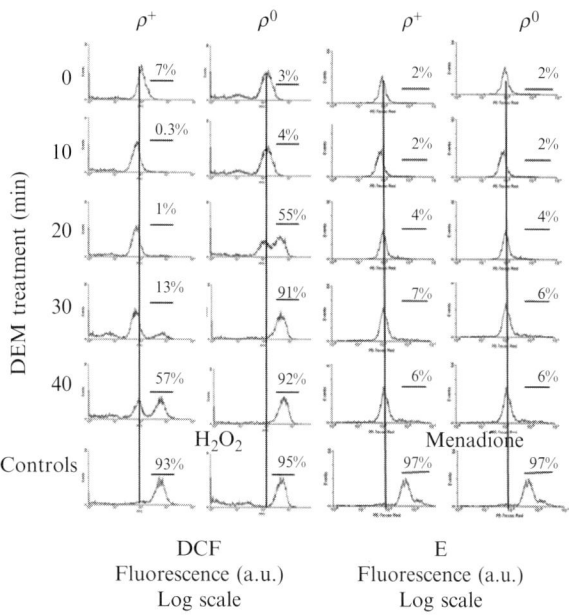

Fig. 6 ROS measurements in lymphoblastic leukemic CEM cells. Lymphoblastic leukemic CEM parental (ρ^+) cells and CEM cells lacking mtDNA (ρ^0) were treated with the GSH-depleting agent diethylmaleate (DEM), and DCF and E fluorescence were determined at the times indicated by FACS analysis. Figure (left) shows that DCF fluorescence increased in a time-dependent manner over 40 min in both cell lines after treatment with DEM (5 mM). However, DCF fluorescence increased earlier in the ρ^0 cell line compared to the ρ^+ cell line. This is probably explained by the finding that control CEM ρ^0 cells had a more oxidized GSH redox potential compared to control CEM ρ^+ cells (Chua *et al.*, 2005) (authentic H_2O_2 was used as a positive control for ROS). Figure (right) shows that E fluorescence does not increase after DEM treatment (menadione was used as a positive control for $O_2^{\bullet-}$) (see text for interpretation).

Protocol to determine ROS using DCFDA and FACS analysis

1. Leukemic CEM suspension cells or HL60 suspension cells are cultured in RPMI with 10% FCS and supplements. Cells are maintained in log phase and are kept at a concentration between 2.5 and 5×10^5 cells/ml.

2. After cell treatments (e.g., with H_2O_2), cells are centrifuged gently to remove culture media and are washed twice by centrifugation in PBS at $100 \times g$.

3. Cells are suspended in PBS by light vortexing and are loaded with 5–10-μM DCFDA by incubation for 15–20 min at 37°C.

4. Cells are washed 2× in PBS (centrifuge for 1 min at $100 \times g$) and suspended in PBS containing 10-mM glucose.

5. Since this is a viable cell assay, immediate FACS analysis is required, using appropriate software for analysis.

6. The FACS setting is FL-2 (FITC) using log mode, after cell debris have been electronically gated out.

Comments

The use of DCFH for the measurement of H_2O_2 is associated with a number of problems. First, it has been reported that the H_2O_2-dependent oxidation of DCFH to DCF occurs slowly in the absence of ferrous iron and can be completely inhibited by desferroxamine (LeBel *et al.*, 1992). Second, cellular fluorescence of DCFH itself may also auto-oxidize to form hydrogen peroxide. Third, numerous other substances are capable of directly inducing DCF formation in the absence of H_2O_2, including peroxidases (Rota *et al.*, 1999), $ONOO^-$, and HOCl (see above for references). The reaction of DCFH with peroxidase forms DCF radicals, with the subsequent generation of $O_2^{\bullet-}$, suggesting that the use of DCFH to measure ROS may be problematic (Halliwell and Whiteman, 2004). Because of the multiple pathways that can lead to DCF fluorescence, and the inherent uncertainty relating to endogenous versus artifactual oxidant generation, this assay may best be applied as a qualitative marker of *cellular oxidant stress* rather than as a precise indicator of rates of H_2O_2 formation.

Hydroethidine (HE) is cell-permeable and reacts with $O_2^{\bullet-}$ to form ethidium (E), which in turn intercalates with DNA, providing nuclear fluorescence at an excitation wavelength of 520 nm and an emission wavelength of 610 nm (Rothe and Valet, 1990). Bindokas *et al.* (1996) reported that oxidation of HE by $O_2^{\bullet-}$ was specific since the reaction did not occur in the presence of other ROS including $OH^{\bullet}$, H_2O_2, or $ONOO^-$. Because of this apparent selectivity, HE has frequently been used to detect intracellular $O_2^{\bullet-}$ (Rothe and Valet, 1990).

D. The Determination of ROS Using HE and FACS Analyses

Reagents and equipment

HE (Molecular Probes)

10-mM glucose in PBS, pH 7.4

FACS machine or fluorescence plate reader (excitation wavelength 520 nm and emission wavelength 610 nm)

Protocol

1. Leukemic CEM suspension cells or HL60 suspension cells are cultured in RPMI with 10% FCS and supplements. Cells are maintained in log phase and kept at a concentration between 2.5 and 5×10^5 ml^{-1}.

2. After treatments (e.g., with the redox-cycling compound menadione, which generates $O_2^{\bullet-}$), cells are centrifuged gently to remove culture media and are washed twice by centrifugation in PBS (centrifuge for 1 min at $100 \times g$).

3. Cells are suspended in PBS by light vortexing and are loaded with 5–10-μM HE by incubation for 15–20 min at 37 °C.

4. Cells are washed $2\times$ in PBS (centrifuge for 1 min at $100 \times g$) and are suspended in PBS containing 10-mM glucose.

5. Since this is a viable cell assay, immediate FACS analysis is required, using appropriate software for analysis.

6. The FACS setting is FL-3 (PE) in log mode, after cell debris have been electronically gated out.

Comments

The intracellular oxidation of HE to E by $O_2^{\bullet-}$ has previously been analyzed using flow cytometry, and also by visualization of adherent neuronal cells and brain tissue using digital-imaging microfluorometry (Bindokas *et al.*, 1996). HE has also been used to study the respiratory burst in immune cells (Rothe and Valet, 1990) and the redox state in tumor cells (Olive, 1989). To show the potential use of this probe, we have included an example of oxidation of HE to E by mitochondrial $O_2^{\bullet-}$ generated by treating bovine aortic endothelial cells (BAEC) with high glucose for 4 days (Fig. 5). However, there are a number of problems associated with the use of HE as a quantitative marker of $O_2^{\bullet-}$ production. First, the amount of E produced by the oxidation of HE decreases with increasing $O_2^{\bullet-}$ flux, suggesting that HE catalyzes the dismutation of $O_2^{\bullet-}$, leading to an under-estimation of $O_2^{\bullet-}$ (Benov *et al.*, 1998). Second, HE can also be oxidized by a variety of heme proteins, including mitochondrial cytochromes, hemoglobin, and myoglobin (Papapostolou *et al.*, 2004). Third, and probably most important, recent work has indicated that $O_2^{\bullet-}$ does not oxidize HE to E, since it was found that $O_2^{\bullet-}$ generated by a variety of enzymatic and chemical systems [e.g., xanthine/ xanthine oxidase (X/XO), endothelial nitric oxide synthase, or potassium super-oxide] oxidized HE to a fluorescent product (excitation, 480 nm; emission, 567 nm) that was different from E (Zhao *et al.*, 2003). The authors concluded that the reaction between $O_2^{\bullet-}$ and HE formed a fluorescent marker product which was not E. The method is still widely used and reported to detect $O_2^{\bullet-}$ specifically. Although not entirely specific for $O_2^{\bullet-}$, the method continues to be used widely for the determination of this radical species. We suggest that when HE is used for the determination of $O_2^{\bullet-}$ production, the combined use of pharmacological inhibitors of the mitochondrial ETC together with the use of appropriate cell lines lacking mtDNA (i.e., ρ^0 cells, without a functional ETC) be used to clarify the subcellular source of the $O_2^{\bullet-}$ signal.

E. Spectrophotometric Methods for Determination of ROS

The spectrophotometric methods outlined below are often widely used and well-accepted techniques for measurement of extracellular $O_2^{\bullet-}$ production by isolated enzymes, cell homogenates, and neutrophils. In general, $O_2^{\bullet-}$ is measured as the SOD-sensitive reduction of substrate. However, as described below, many of the methods in common use are not highly specific for $O_2^{\bullet-}$.

1. Nitro Blue Tetrazolium

Nitro blue tetrazolium (NBT) is a nitro-substituted aromatic compound that can be reduced by $O_2^{\bullet-}$ to the monoformazan, whose formation can be monitored spectrophotometrically at 550–560 nm. The reaction is a two-step process that proceeds via the formation of an intermediate NBT radical, which can then undergo either further reduction or dismutation to the monoformazan (Munzel et al., 2002; Tarpey and Fridovich, 2001). Similar to DHE, NBT detects intracellular $O_2^{\bullet-}$; however, it is less sensitive and specific for $O_2^{\bullet-}$ than is DHE. The NBT radical intermediate can also react with molecular oxygen under aerobic conditions and, in doing so, generate $O_2^{\bullet-}$ artifactually, which can further reduce the NBT (Auclair and Voisin, 1985). Importantly, although this formazan product is SOD-inhibitable, this technique does not confirm unequivocally that SOD-inhibitable NBT reduction is due to $O_2^{\bullet-}$. NBT is also susceptible to reduction by several tissue reductases and, in fact, is used as the substrate in the blue formazan reaction used to detect NO synthase (Grozdanovic et al., 1995). For these reasons, detection of $O_2^{\bullet-}$ in biological samples should not rely exclusively on NBT reduction but should be supplemented with an additional, independent, measure of $O_2^{\bullet-}$.

2. Cytochrome c Reduction

Reduction of cytochrome c is a widely used, accepted, technique for measurement of $O_2^{\bullet-}$ production by isolated enzymes, cell homogenates, and neutrophils. In general, $O_2^{\bullet-}$ is measured spectrophotometrically by the increase in absorbance at 550 nm that results from the SOD-inhibitable reduction of cytochrome c. Because of the inability of cytochrome c to penetrate cells, it can be used only to measure extracellular $O_2^{\bullet-}$.

There are several precautions when using this reaction to detect $O_2^{\bullet-}$. First, cytochrome c reduction is nonspecific for $O_2^{\bullet-}$; compounds such as ascorbate and GSH, in addition to cellular reductases, can catalyze cytochrome c reduction. In addition, cytochrome c can be reoxidized by cytochrome oxidase (COX), peroxidases, and a variety of oxidants (including H_2O_2 and $ONOO^-$), thereby underestimating the rate of $O_2^{\bullet-}$ production (Thomson et al., 1995). Cyanide (CN^-) can be added to the reaction mixture to inhibit COX activity, and ROS scavengers, such as CAT, will block oxidation by H_2O_2, while urate will prevent oxidation by $ONOO^-$. Alternatively, improved specificity of the cytochrome c assay for $O_2^{\bullet-}$ can be achieved by either measuring SOD-inhibitable cytochrome c reduction or by using acetylated or succinoylated forms of cytochrome c, which minimize artifactual reduction and oxidation without affecting the ability of $O_2^{\bullet-}$ to reduce cytochrome c (Azzi et al., 1975; Kuthan et al., 1982).

3. Aconitase

Aconitase is a citric acid cycle enzyme belonging to the family of dehydro-genases containing iron–sulfur (4Fe–4S) centers that catalyze the conversion of citrate to isocitrate. The mitochondrial and the cytosolic forms of aconitase are inactivated by $O_2^{\bullet-}$, so that its activity has been proposed to reflect intracellular levels of $O_2^{\bullet-}$ production, with low levels of enzyme activity reflecting high levels of $O_2^{\bullet-}$ (Gardner and Fridovich, 1991). Inactivation of the enzyme occurs because of the oxidation of the enzyme and the subsequent loss of iron from its [4Fe–4S] cluster. Unfortunately, numerous ROS, and even NO, have been shown to inactivate aconitase; therefore, the assay is not specific for $O_2^{\bullet-}$ (Hausladen and Fridovich, 1994). Also, because inactivation of the enzyme occurs over several hours or days, it is impossible to intervene with specific scavengers to restore its activity.

III. Measurement of ROS in Mitochondria

A. Isolation of Mitochondria from Cells and Tissue

1. Animal Cells

Concentrate $\sim 2 \times 10^7$ cells by centrifugation in a microcentrifuge tube at $1000 \times g$ for 3 min. Carefully remove and discard the supernatant. Cells are washed in MSHE buffer (70-mM sucrose, 220-mM mannitol, 2-mM HEPES, 0.5-mM EGTA, 0.1% BSA, pH 7.4) and centrifuged at $1000 \times g$ for 3 min. Supernatant is removed and the cell pellet is suspended in $\sim 150 \, \mu l$ of fresh, ice-cold MSHE buffer. Cells are homogenized by passage through a syringe needle (G $27^{1/2}$) until 80–90% cells are blue by trypan blue exclusion. The homogenate is then centrifuged at $1000 \times g$ for 10 min at $4\,^\circ$C. The supernatant obtained is trans-ferred to a fresh microfuge tube and centrifuged at $10,000 \times g$ for 20 min at $4\,^\circ$C. The supernatant obtained from the $10,000 \times g$ centrifugation contains the cyto-solic fraction and is transferred to a new microcentrifuge tube. The pellet obtained from the $10,000 \times g$ centrifugation contains the isolated mitochondrial fraction. The mitochondrial pellet is resuspended in MSHE buffer on ice prior to processing. Mitochondrial fractions should support coupled respiration.

2. Animal Tissue: Isolation of Rat Liver Mitochondria

Male Sprague Dawley rats weighing 180–220 g are used for isolation of liver mitochondria. Briefly, fresh liver tissues are chopped finely and are homogenized in ice-cold isolation medium consisting of 220-mM mannitol, 70-mM sucrose, 2-mM HEPES, 0.5-mM EGTA, 0.1% BSA (fatty acid free), pH 7.4, using a prechilled Dounce homogenizer. The homogenate is centrifuged at $1000 \times g$ for 10 min at $4\,^\circ$C in a Beckman JA20 rotor. The supernatant obtained is transferred into a fresh tube and centrifuged as described above. The supernatant obtained is centrifuged at $10,000 \times g$ for 10 min at $4\,^\circ$C. The pellet obtained, which is enriched in mitochondria, is retained. The dark brown layer in the middle of the pellet is removed from the pellet and is transferred to a fresh tube. This material is suspended in a small volume of the isolation buffer. This sample may now be assay for protein concentration. The condition of intact mitochondria can be tested by measuring oxygen consumption in the presence of succinate and ADP and determining respiratory control ratio (RCR).

B. Measurement of Mitochondrial ROS (H_2O_2)

In the presence of horseradish peroxidase (HRP), Amplex red reagent (10-acetyl-3,7-dihydroxyphenoxazine) reacts with H_2O_2 with a 1:1 stoichiometry to produce a highly fluorescent resorufin product. Amplex red/HRP has been observed to detect levels of H_2O_2 produced by the NADPH oxidase and other oxidases (Zhou *et al.*, 1997).

Reagents

Amplex red (Molecular Probes)
HRP (Molecular Probes)
Krebs-Ringer phosphate (KRP)
145-mM NaCl
5.7-mM Sodium phosphate
4.86-mM KCl
0.54-mM $CaCl_2$
1.22-mM $MgSO_4$
5.5-mM Glucose
pH 7.35
RPMI 1640 media
Fetal calf serum (FCS)

Fluorescence plate reader (e.g., SpectraMax Plus spectrophotometer, Molecular Devices Corporation, Sunnyvale, CA) set at excitation wavelength 560 nm and emission wavelength 590 nm.

Protocol

1. Leukemic CEM suspension cells are cultured in RPMI 1640 with 10% FCS and supplements and maintained in log phase and kept at a concentration between 2.5 and $5 \times 10^5 \mathrm{ml}^{-1}$.

2. The mitochondrial pellet obtained from $\sim$40 ml (1.5×10^6) cells is suspended in 200 μl of KRP buffer and is kept on ice prior to use.

3. 100 μl of a solution containing HRP and Amplex red (0.1 unit/ml and 50 μM, respectively, in KRP buffer) is added to each well in a microtiter plate and the

reaction is initiated by adding 20 μl of the mitochondrial mixture. For negative control, add 20 μl of KRP alone. A standard curve is constructed by adding a known amount of H_2O_2 to the assay medium in the absence of mitochondria. Typical standard curves are linear to $\sim$2-μM H_2O_2.

4. Fluorescence is measured in a fluorescence microplate reader, prewarmed to 37°C, at 15sec intervals, for a total time of 30 min. The initial linear phase of increase in absorbance is used to calculate the rate of H_2O_2 production per milligram of mitochondrial protein. Background fluorescence is measured in the absence of mitochondria and results are calculated as relative fluorescence units (RFU) in the mitochondrial sample minus background RFU, and are expressed as pmol H_2O_2/mg of protein/30 min).

5. Under the above assay conditions SOD does not increase the rate of production of H_2O_2, indicating that the nonenzymatic dismutation of $O_2^{\bullet-}$ to H_2O_2 is rapid and complete.

C. Measurement of Mitochondrial ROS ($O_2^{\bullet-}$)

DHE can be used to measure mitochondrial $O_2^{\bullet-}$ produced by isolated mitochondria.

Reagents

MSE buffer

220-mM Mannitol

70-mM Sucrose

2-mM HEPES

0.5-mM EGTA

0.1% BSA (fatty acid free)

pH 7.4

Rotenone (2 μM) (Sigma-Aldrich)

Succinate (10 mM) (Sigma-Aldrich)

Antimycin A (10 μM) (Sigma-Aldrich)

ADP (250 μM) (Sigma-Aldrich)

KH_2PO_4 (5 mM)

Fluorescence is measured in black 96-well microplates on a Gemini XS dual monochromatic reader at an excitation/emission wavelength of 520/610 nm.
Protocol for determining mitochondrial ROS using DHE and a Gemini XS dual monochromatic fluoresecence plate reader

1. For the determination of mitochondrial ROS under state 4 (resting) conditions, reagents are added in the following sequence (total volume 200 μl): MSE buffer, mitochondria (40 μg), succinate (10 mM)/rotenone (2 μM) mix, and DHE (1 μM).

2. For experiments with mitochondria under state 3 (ADP-stimulated) conditions, ADP (250 μM) and KH_2PO_4 (5 mM) are added after the succinate/rotenone mix.

3. The assay is carried out in triplicate and the fluorescence is followed kinetically over 15 min (at 37°C) in a Gemini XS dual monochromatic reader at an excitation/emission wavelength of 520/610 nm, respectively. For mitochondria, antimycin A (10 μM) is used as a positive control.

IV. General Problems Associated with the Measurement of ROS

In addition to the specific methodological problems and caveats associated with measuring ROS (briefly described in the relevant sections above), there are a number of general problems that should be taken into account. For example, many methods allegedly measuring ROS have become widely used, but precise information on what they really measure is conspicuously lacking. An important first consideration is that cell culture itself induces oxidative stress, both by facilitating generation of ROS and by hindering the adaptive upregulation of cellular antioxidants (reviewed by Halliwell, 2003). For example, the "Hayflick limit" in fibroblasts (replicative senescence after a certain number of cell divisions) may be an artifact of oxidative stress imposed during cell culture (Wright and Shay, 2002). In addition, results can be confounded by free radical reactions taking place in the culture media (Grzelak et al., 2000; Halliwell and Whiteman, 2004; Roques et al., 2002; Wee et al., 2003). For example, a number of reports of effects of ascorbate and polyphenolic compounds (e.g., flavonoids) on cells in culture appear to be largely artifactual, caused by the oxidation of these compounds in the culture media (Clement et al., 2001; Long and Halliwell, 2001). Therefore, it is important to consider what reaction takes place in the cell culture medium alone when compounds are added to it (Clement et al., 2001, 2002). Also, the presence of cells can often suppress free radical reactions occurring in the medium (Halliwell, 2003).

Some simple principles can be used as guidelines in understanding oxidative stress/oxidative damage in cell culture. Hydrogen peroxide generally crosses cell membranes readily, probably via aquaporins (Henzler and Steudle, 2000). Thus, CAT added outside cells can exert both intracellular and extracellular effects on H_2O_2 level, the former by "draining" H_2O_2 out of the cell by removing extracellular H_2O_2 and thus establishing a concentration gradient (Halliwell, 2003). In contrast, $O_2^{\bullet-}$ does not readily cross cell membranes (Lynch and Fridovich, 1978; Marla et al., 1997). Thus, if externally added SOD is protective against an event in cell culture, be wary of what this means: it could be indicative of extracellular $O_2^{\bullet-}$ generating reactions. To further complicate this issue, if externally added SOD is damaging to cells in culture, this could also indicate that an extracellular reaction generating $O_2^{\bullet-}$ is occurring, since the SOD can increase levels of H_2O_2, which will rapidly enter the cell and inhibit cellular ATP production (Armstrong *et al.*, 1999).

Similarly, neither the iron-chelating agent desferroxamine (which suppresses most, but not all, iron-dependent free radical reactions) nor the thiol antioxidant GSH enters cells easily. So again, there is a need for caution if they have protective effects in short-term experiments that are suggestive of extracellular effects (Halliwell, 1989; Marla *et al.*, 1997; Meister and Anderson, 1983). As an example, Clement *et al.* (2002) showed that GSH protected against the cytotoxicity of dopamine simply because it reacts with dopamine oxidation products generated in the cell culture medium. Given long enough, however, everything can enter cells, including desferroxamine, superoxide dismutase (SOD), and GSH (Doulias *et al.*, 2003; Rius *et al.*, 2003).

Studies of ROS production by mitochondria is an area of particular interest, but also of great difficulty. DHE is often used to determine mitochondrial production of $O_2^{\bullet-}$ *in situ*, but it is clear that DHE also responds to extra-mitochondrially produced $O_2^{\bullet-}$ by redox-cycling agents such as menadione or paraquat. Therefore, the use of DHE to indicate mitochondrial $O_2^{\bullet-}$ production specifically may often be erroneous. More specificity will be attained by the use of pharmacological inhibitors of mitochondrial electron transport, as well as the use of genetically modified cells lacking mitochondrial DNA (mtDNA), to determine whether the $O_2^{\bullet-}$ is mitochondrial or extramitochondrial in origin (Fig. 5).

Studies of ROS production in cells can, of course, be achieved directly by using aromatic compounds or spin traps as described above. Spin traps have been used successfully in many cell studies, since a wider range of traps at higher concentrations can be used than could ever be employed *in vivo*. Again, one must be aware of the possibility of rapid reduction of free radical–spin trap adducts to ESR-silent species by nonenzymatic antioxidants (such as ascorbate) and cellular enzymatic-reducing systems. An interesting combination is 5-[(2-carboxy)phenyl]-5-hydroxy-1-(2,2,5,5-tetramethyl-1-oxypyrrolidin-3-yl)methyl-3-phenyl-2-pyrrolin-4-one sodium salt, a nitroxide that is nonfluorescent. When it combines with a ROS, the nitroxide is removed, the ESR signal is lost, and the fluorescence is restored (Pou *et al.*, 1993).

V. Conclusions

Whatever method is used to measure ROS, it is necessary to consider the methodological limitations and problems of specificity. Therefore, investigators are urged to think carefully about the method's mechanism, potential problems with specificity, and whether the method is quantitative. With care and attention, and the judicious use of pharmacological inhibitors and genetically modified cells, erroneous interpretations can be minimized and the vibrant field of free radical research can continue to move forward.

Acknowledgments

This chapter was supported by research grants from National of University Singapore (NUS) Academic Research Fund Grant R183–000–127–112 and Biomedical Research Council (BMRC) of Singapore R183–000–132–305 to JSA. We would like to acknowledge Dr. Urs Boelsterli for the protocol on measurement of mitochondrial ROS using HE.

References

Armstrong, J. S., and Jones, D. P. (2002). Glutathione depletion enforces mitochondrial permeability transition and apoptosis in HL60 cells overexpressing Bcl-2. *FASEB J.* **16,** 1263–1265.

Armstrong, J. S., Rajasekaran, M., Chamulitrat, W., Gatti, P. J., Hellstrom, W. J., and Sikka, S. C. (1999). The effects of reactive oxygen intermediates on human spermatozoa movement and energy metabolism. *Free Radic. Biol. Med.* **26,** 869–880.

Armstrong, J. S., Whiteman, M., Yang, H., Jones, D. P., and Sternberg, P. (2004a). Cysteine-starvation activates the redox-dependent mitochondrial permeability transition in retinal pigment epithelial cells. *Invest. Ophthalmol. Vis. Sci.* **45,** 4183–4189.

Armstrong, J. S., Yang, H., Duan, W., Chua, Y., and Whiteman, M. (2004b). Cytochrome bc_1 regulates the mitochondrial permeability transition by two distinct pathways. *J. Biol. Chem.* **279,** 50420–50428.

Auclair, C., and Voisin, E. (1985). Nitroblue tetrazolium reduction. *In* "CRC Handbook of Methods for Oxygen Radical Research" (R. A. Greenwald, ed.), pp. 123–132. CRC Press, Boca Raton, Fla.

Azzi, A., Montecucco, C., and Richter, C. (1975). The use of acetylated ferricytochrome c for the detection of superoxide radicals produced in biological membranes. *Biochem. Biophys. Res. Commun.* **65,** 597–603.

Bass, D. A., Parce, J. W., Dechatelet, L. R., Szejda, P., Seeds, M. C., and Thomas, M. (1983). Flow cytometric studies of oxidative product formation by neutrophils: A graded response to membrane stimulation. *J. Immunol.* **130,** 1910–1917.

Benov, L., Sztejnberg, L., and Fridovich, I. (1998). Critical evaluation of the use of hydroethidine as a measure of superoxide anion radical. *Free Radic. Biol. Med.* **25,** 826–831.

Bindokas, V. P., Jordan, J., Lee, C. C., and Miller, R. J. (1996). Superoxide production in rat hippocampal neurons: Selective imaging with hydroethidine. *J. Neurosci.* **16,** 1324–1326.

Bottle, S. E., Hanson, G. R., and Micallef, A. S. (2003). Application of the new EPR spin trap 1,1,3-trimethylisoindole N-oxide (TMINO) in trapping HO and related biologically important radicals. *Org. Biomol. Chem.* **1,** 2585–2589.

Chua, Y. L., Zhang, D., Boelsterli, U., Moore, P. K., Whiteman, M., and Armstrong, J. S. (2005). Oltipraz-induced phase 2 enzyme response conserved in cells lacking mitochondrial DNA. *Biochem. Biophys. Res. Commun.* **337,** 375–381.

Clement, M. V., Long, L. H., Ramalingam, J., and Halliwell, B. (2002). The cytotoxicity of dopamine may be an artefact of cell culture. *J. Neurochem.* **81,** 414–421.

Clement, M. V., Ramalingam, J., Long, L. H., and Halliwell, B. (2001). The *in vitro* cytotoxicity of ascorbate depends on the culture medium used to perform the assay and involves hydrogen peroxide. *Antioxid. Redox Signal.* **3,** 157–163.

Doulias, P. T., Christoforidis, S., Brunk, U. T., and Galaris, D. (2003). Endosomal and lysosomal effects of desferrioxamine: Protection of HeLa cells from hydrogen peroxide-induced DNA damage and induction of cell-cycle arrest. *Free Radic. Biol. Med.* **35,** 719–728.

Faulkner, K., and Fridovich, I. (1993). Luminol and lucigenin as detectors for $O_2^{\bullet-}$. *Free Radic. Biol. Med.* **15,** 447–451.

Frejaville, C., Karoui, H., Tuccio, B., Le Moigne, F., Culcasi, M., Pietri, S., Lauricella, R., and Tordo, P. (1995). 5-(Diethoxyphosphoryl)-5-methyl-1-pyrroline N-oxide: A new efficient phosphorylated nitrone for the *in vitro* and *in vivo* spin trapping of oxygen-centered radicals. *J. Med. Chem.* **38,** 258–265.

Gardner, P. R., and Fridovich, I. (1991). Superoxide sensitivity of the Escherichia coli aconitase. *J. Biol. Chem.* **266,** 19328–19333.

Grozdanovic, Z., Nakos, G., Christova, T., Nikolova, Z., Mayer, B., and Gossrau, R. (1995). Demonstration of nitric oxide synthase (NOS) in marmosets by NADPH diaphorase (NADPH-d) histochemistry and NOS immunoreactivity. *Acta Histochem.* **97,** 321–331.

Grzelak, A., Rychlik, B., and Bartosz, G. (2000). Reactive oxygen species are formed in cell culture media. *Acta Biochim. Pol.* **47,** 1197–1198.

Halliwell, B. (1989). Protection against tissue damage *in vivo* by desferrioxamine: What is its mechanism of action? *Free Radic. Biol. Med.* **7,** 645–651.

Halliwell, B. (1995). Antioxidant characterization. Methodology and mechanism. *Biochem. Pharmacol.* **49,** 1341–1348.

Halliwell, B. (2003). Oxidative stress in cell culture: An under-appreciated problem? *FEBS Lett.* **540,** 3–6.

Halliwell, B., and Gutteridge, J. M. (1999). "Free Radicals in Biology and Medicine." Oxford University Press, UK.

Halliwell, B., and Whiteman, M. (2004). Measuring reactive species and oxidative damage *in vivo* and in cell culture: How should you do it and what do the results mean? *Br. J. Pharmacol.* **142,** 231–255.

Hausladen, A., and Fridovich, I. (1994). Superoxide and peroxynitrite inactivate aconitases, but nitric oxide does not. *J. Biol. Chem.* **269,** 29405–29408.

Henzler, T., and Steudle, E. (2000). Transport and metabolic degradation of hydrogen peroxide in Chara corallina: Model calculations and measurements with the pressure probe suggest transport of $H(2)O(2)$ across water channels. *J. Exp. Bot.* **51,** 2053–2066.

Hodgson, E. K., and Fridovich, I. (1973). The role of O_2^- in the chemiluminescence of luminol. *Photochem. Photobiol.* **18,** 451–455.

Khan, N., Wilmot, C. M., Rosen, G. M., Demidenko, E., Sun, J., Joseph, J., O'Hara, J., Kalyanaraman, B., and Swartz, H. M. (2003). Spin traps: *In vitro* toxicity and stability of radical adducts. *Free Radic. Biol. Med.* **34,** 1473–1481.

Kuthan, H., Ullrich, V., and Estabrook, R. W. (1982). A quantitative test for superoxide radicals produced in biological systems. *Biochem. J.* **203,** 551–558.

Lambert, A. J., and Brand, M. D. (2004). Inhibitors of the quinone-binding site allow rapid superoxide production from mitochondrial NADH: Ubiquinone oxidoreductase (complex I). *J. Biol. Chem.* **279,** 39414–39420.

Laurindo, F. R., Pedro Mde, A., Barbeiro, H. V., Pileggi, F., Carvalho, M. H., Augusto, O., and da Luz, P. L. (1994). Vascular free radical release: *Ex vivo* and *in vivo* evidence for a flow-dependent endothelial mechanism. *Circ. Res.* **74,** 700–709.

LeBel, C. P., Ischiropoulos, H., and Bondy, S. C. (1992). Evaluation of the probe 2′,7′-dichlorofluorescin as an indicator of reactive oxygen species formation and oxidative stress. *Chem. Res. Toxicol.* **5,** 227–231.

Li, Y., Huang, T. T., Carlson, E. J., Melov, S., Ursell, P. C., Olson, J. L., Noble, L. J., Yoshimura, M. P., Berger, C., Chan, P. H., Wallace, D. C., and Epstein, C. J. (1995). Dilated cardiomyopathy and neonatal lethality in mutant mice lacking manganese superoxide dismutase. *Nat. Genet.* **11,** 376–381.

Liochev, S. I., and Fridovich, I. (1997). Lucigenin (bis-N-methylacridinium) as a mediator of superoxide anion production. *Arch. Biochem. Biophys.* **337**(1), 115–120. Related Articles, Links.

Liu, K. J., Miyake, M., Panz, T., and Swartz, H. (1999). Evaluation of DEPMPO as a spin trapping agent in biological systems. *Free Radic. Biol. Med.* **26,** 714–721.

Long, L. H., and Halliwell, B. (2001). Antioxidant and prooxidant abilities of foods and beverages. *Methods Enzymol.* **335,** 181–190. No abstract available.

Lynch, R. E., and Fridovich, I. (1978). Permeation of the erythrocyte stroma by superoxide radical. *J. Biol. Chem.* **253,** 4697–4699.

Makino, K., Hagiwara, T., Hagi, A., Nishi, M., and Murakami, A. (1990). Cautionary note for DMPO spin trapping in the presence of iron ion. *Biochem. Biophys. Res. Commun.* **172,** 1073–1080.

Marla, S. S., Lee, J., and Groves, J. T. (1997). Peroxynitrite rapidly permeates phospholipid membranes. *Proc. Natl. Acad. Sci. USA* **94**, 14243–14248.

Meister, A., and Anderson, M. E. (1983). Glutathione. *Annu. Rev. Biochem.* **52**, 711–760.

Munzel, T., Afanas'ev, I. B., Kleschyov, A. L., and Harrison, D. G. (2002). Detection of superoxide in vascular tissue. *Arterioscler. Thromb. Vasc. Biol.* **22**, 1761–1768.

Olive, P. L. (1989). Hydroethidine: A fluorescent redox probe for locating hypoxic cells in spheroids and murine tumours. *Br. J. Cancer* **160**, 328–332.

Papapostolou, I., Patsoukis, N., and Georgiou, C. D. (2004). The fluorescence detection of superoxide radical using hydroethidine could be complicated by the presence of heme proteins. *Anal. Biochem.* **332**, 290–298.

Pou, S., Huang, Y. I., Bhan, A., Bhadti, V. S., Hosmane, R. S., Wu, S. Y., Cao, G. L., and Rosen, G. M. (1993). A fluorophore-containing nitroxide as a probe to detect superoxide and hydroxyl radical generated by stimulated neutrophils. *Anal. Biochem.* **212**, 85–90.

Raha, S., and Robinson, B. H. (2000). Mitochondria, oxygen free radicals, disease and ageing. *Trends Biochem. Sci.* **25**, 502–508.

Rius, M., Nies, A. T., Hummel-Eisenbeiss, J., Jedlitschky, G., and Keppler, D. (2003). Cotransport of reduced glutathione with bile salts by MRP4 (ABCC4) localized to the basolateral hepatocyte membrane. *Hepatology* **38**, 374–384.

Rizzi, C., Samouilov, A., Kutala, V. K., Parinandi, N. L., Zweier, J. L., and Kuppusamy, P. (2003). Application of a trityl-based radical probe for measuring superoxide. *Free Radic. Biol. Med.* **35**, 1608–1618.

Roques, S. C., Landrault, N., Teissedre, P. L., Laurent, C., Besancon, P., Rouane, J. M., and Caporiccio, B. (2002). Hydrogen peroxide generation in caco-2 cell culture medium by addition of phenolic compounds: Effect of ascorbic acid. *Free Radic. Res.* **36**, 593–599.

Rosen, G. M., Britigan, B., Halpern, H., and Pou, S. (1999). "Free Radicals. Biology and Detection by Spin Trapping." Oxford University Press, Oxford, UK.

Rota, C., Chignell, C. F., and Mason, R. P. (1999). Evidence for free radical formation during the oxidation of 2'-7'-dichlorofluorescin to the fluorescent dye 2'-7'-dichlorofluorescein by horseradish peroxidase: Possible implications for oxidative stress measurements. *Free Radic. Biol. Med.* **27**, 873–881.

Rothe, G., and Valet, G. (1990). Flow cytometric analysis of respiratory burst activity in phagocytes with hydroethidine and 2,7-dichlorofluorescin. *J. Leukoc. Biol.* **47**, 440–448.

Sohn, H. Y., Keller, M., Gloe, T., Crause, P., and Pohl, U. (2000). Pitfalls of using lucigenin in endothelial cells: Implications for NAD(P)H dependent superoxide formation. *Free Radic. Res.* **32**, 265–272.

Spasojevic, I., Liochev, S. I., and Fridovich, I. (2000). Lucigenin: Redox potential in aqueous media and redox cycling with O-(2) production. *Arch. Biochem. Biophys.* **373**, 447–450.

Stolze, K., Udilova, N., Rosenau, T., Hofinger, A., and Nohl, H. (2003). Spin trapping of superoxide, alkyl- and lipid-derived radicals with derivatives of the spin trap EPPN. *Biochem. Pharmacol.* **66**, 1717–1726.

Storch, J., and Ferber, E. (1988). Detergent-amplified chemiluminescence of lucigenin for determination of superoxide anion production by NADPH oxidase and xanthine oxidase. *Anal. Biochem.* **169**, 262–267.

Tampo, Y., Tsukamoto, M., and Yonaha, M. (1998). The antioxidant action of 2-methyl-6-(p-methoxyphenyl)-3,7-dihydroimidazo[1,2-alpha]pyra z in-3-one (MCLA), a chemiluminescence probe to detect superoxide anions. *FEBS Lett.* **430**, 348–352.

Tarpey, M. M., and Fridovich, I. (2001). Methods of detection of vascular reactive species: Nitric oxide, superoxide, hydrogen peroxide, and peroxynitrite. *Circ. Res.* **89**, 224–236.

Tarpey, M. M., White, C. R., Suarez, E., Richardson, G., Radi, R., and Freeman, B. A. (1999). Chemiluminescent detection of oxidants in vascular tissue. Lucigenin but not coelenterazine enhances superoxide formation. *Circ. Res.* **84**, 1203–1211.

Teranishi, K., and Shimomura, O. (1997). Coelenterazine analogs as chemiluminescent probe for superoxide anion. *Anal. Biochem.* **249**, 37–43.

Thomson, L., Trujillo, M., Telleri, R., and Radi, R. (1995). Kinetics of cytochrome c oxidation by peroxynitrite: Implications for superoxide measurements in nitric oxide-producing biological systems. *Arch. Biochem. Biophys.* **319**, 491–497.

Totter, J. R., de Dugros, E. C., and Riveiro, C. (1960). The use of chemiluminescent compounds as possible indicators of radical production during xanthine oxidase action. *J. Biol. Chem.* **235**, 1839–1842.

Turrens, J. F. (1997). Superoxide production by the mitochondrial respiratory chain. *Biosci. Rep.* **17**, 3–8.

Valgimigli, L., Pedulli, G. F., and Paolini, M. (2001). Measurement of oxidative stress by EPR radical-probe technique. *Free Radic. Biol. Med.* **31**(6), 708–716.

Wardman, P., Burkitt, M. J., Patel, K. B., Lawrence, A., Jones, C. M., Everett, S. A., and Vojnovic, B. (2002). Pitfalls in the use of common luminescent probes for oxidative and nitrosative stress. *J. Fluoresc.* **12**, 65–68.

Wee, L. M., Long, L. H., Whiteman, M., and Halliwell, B. (2003). Factors affecting the ascorbate- and phenolic-dependent generation of hydrogen peroxide in Dulbecco's Modified Eagles Medium. *Free Radic. Res.* **37**, 1123–1130.

Whiteman, M., Armstrong, J. S., Jones, D. P., and Halliwell, B. (2004). Peroxynitrite mediates calcium-dependent mitochondrial dysfunction and cell death via activation of calpains. *FASEB J.* **18**, 1395–1397.

Whiteman, M., Chua, Y. L., Zhang, D., Duan, W., Liou, Y. C., and Armstrong, J. S. (2006). Nitric oxide blocks glutathione-dependent cell death independently of mitochondrial reactive oxygen species: Potential role of s-nitrosylation? *Biochem. Biophys. Res. Commun.* **339**, 255–262.

Whiteman, M., Rose, P., Siau, J. L., Cheung, N. S., Tan, G. S., Halliwell, B., and Armstrong, J. S. (2005). Hypochlorous acid-mediated mitochondrial dysfunction and apoptosis in human hepatoma Hepg2 and human fetal liver cells: Role of mitochondrial permeability transition. *Free Radic. Biol. Med.* **38**, 1571–1584.

Wright, W. E., and Shay, J. W. (2002). Historical claims and current interpretations of replicative aging. *Nat. Biotechnol.* **20**, 682–688. Review.

Zhao, H., Kalivendi, S., Zhang, H., Joseph, J., Nithipatikom, K., Vasquez-Vivar, J., and Kalyanaraman, B. (2003). Superoxide reacts with hydroethidine but forms a fluorescent product that is distinctly different from ethidium: Potential implications in intracellular fluorescence detection of superoxide. *Free Radic. Biol. Med.* **34**, 1359–1368.

Zhou, M., Diwu, Z., Panchuk-Voloshina, N., and Haugland, R. P. (1997). A stable nonfluorescent derivative of resorufin for the fluorometric determination of trace hydrogen peroxide: Applications in detecting the activity of phagocyte NADPH oxidase and other oxidases. *Anal. Biochem.* **253**, 162–168.

CHAPTER 19

Measurements of the Antioxidant Enzyme Activities of Superoxide Dismutase, Catalase, and Glutathione Peroxidase

Cristofol Vives-Bauza,[*,†] Anatoly Starkov,[*] and Elena Garcia-Arumi[†]

[*]Department of Neurology and Neuroscience
Weill Medical College, Cornell University, New York, New York 10021

[†]Centre d'Investigacions en Bioquímica i Biología Molecular (CIBBIM)
Institut de Recerca Hospital Universitari Vall d'Hebrón, Barcelona, Spain

I. Introduction

The mitochondrial electron transport chain and a variety of cellular oxidases are the main sources of reactive oxygen species (ROS), such as superoxide anion ($O_2^{\bullet-}$), hydrogen peroxide (H_2O_2), and hydroxyl radical ($^{\bullet}OH$), which are continuously generated as by-products of various intracellular redox reactions (Andreyev *et al.*, 2005; Halliwell and Gutteridge, 1990). To cope with the damaging actions of ROS, organisms have evolved a sophisticated ROS defense system (RDS), consisting of low-molecular-weight antioxidants, such as glutathione, ascorbic acid, tocopherol, and uric acid, and specialized ROS-detoxifying enzymes, such

0091-679X/07 $35.00
DOI: 10.1016/S0091-679X(06)80019-1

as superoxide dismutases (SODs), catalase (CAT), glutathione peroxidases (GPxs), and various thio-, peroxi-, and glutaredoxins (Andreyev *et al.*, 2005; Rhee *et al.*, 2005). These enzymes represent the primary line of ROS defense. Defense enzymes can remove ROS directly or can repair the damage to other macromolecules caused by ROS. Other enzymes are involved in the renewal of the reducing power of defense enzymes.

Failure of RDS to cope with the intracellular ROS production results in oxidative stress, which contributes to the damage and death of cells. Therefore, measuring the activity of RDS enzymes is a valuable diagnostic tool to determine the role of the oxidative stress in the pathology of a particular disease. This chapter describes the activity assays of the major RDS enzymes—SOD, GPx, and CAT—in cultured cells, tissue homogenates, and mammalian mitochondria.

II. Experimental Procedures

A. Sample Preparation

1. Cultured Cells

Cells are grown in 150-mm culture dishes in Dulbecco's Modified Eagle's Medium (DMEM) containing high levels of glucose (4.5 mg/ml), 2-mM L-glutamine, 110-mg/liter sodium pyruvate supplemented with 5% fetal bovine serum (FBS) at 37°C and 5% CO_2. Cells are harvested by trypsinization when they reach 80% confluence. Cells are then sedimented by centrifugation, washed twice with phosphate buffered saline (PBS), and resuspended in 1-ml PBS per 40 million cells. Cell pellets should be frozen at −80°C until assayed. Immediately before the assay, cell pellets are thawed and homogenized by suspension in ice-cold 0.013% sodium cholate at a concentration of 20 million cells per ml and sonicated for 5 min on ice at 50% "duty cycle". The homogenate is then centrifuged for 10 min at 12,000 × *g* at 4°C to remove cell debris. The supernatants should be kept on ice until the assay.

2. Tissue Homogenates

Fresh tissues are minced in ice-cold homogenization buffer H containing 0.1-M Tris–HCl, 0.25-M sucrose, pH 7.4, 5-mM 2-mercaptoethanol, 0.5-μg/ml leupeptin, 0.7-μg/ml pepstatin A, and 100-μg/ml phenylmethylsulfonyl fluoride (PMSF). To remove red blood cells, minced tissue samples are washed twice with 5 volumes of 0.9% NaCl solution supplemented with 0.16-mg/ml heparin. Washed tissue samples are homogenized gently by hand (∼10 strokes) in 3 volumes of ice-cold buffer H using a Dounce-type glass body Teflon pestle homogenizer. Fibrous material and other tissue debris are eliminated by centrifugation of the tissue homogenate at 500 × *g* for 5 min at 4°C. The supernatants are used either as whole homogenates or are separated further into the subcellular fractions by differential centrifugation.

3. Isolated Mitochondria

Crude mitochondrial fractions are obtained from tissue homogenates by centrifugation at $14,000 \times g$ for 10 min at $4°C$. Pellets are washed in ice-cold buffer H, centrifuged at $14,000 \times g$ for 10 min at $4°C$, and resuspended in buffer H to a final concentration of 5- to 10-mg protein/ml. The pellets should be kept on ice until the assay.

B. SOD Activity Assay

SOD catalyzes the dismutation of superoxide anion $O_2^{•-}$ to molecular oxygen and H_2O_2:

$$2O_2^{•-} + 2H^+ \longrightarrow H_2O_2 + O_2 \tag{1}$$

In mammals, there are three forms of SOD: CuZn-SOD or SOD1, which is located in the cytoplasm and in the mitochondrial intermembrane space; Mn-SOD or SOD2, which is located in the mitochondrial matrix; and SOD3, which is extracellular. SOD1 is a dimer, while SOD2 and SOD3 are tetramers. SOD1 and SOD3 contain copper and zinc in their reactive centers, whereas SOD2 has manganese. The reaction catalyzed by all three SOD enzymes is a bimolecular reaction, with a catalytic cycle involving two distinct half-reactions: an oxidative half-reaction in which the substrate, $O_2^{•-}$, is oxidized to dioxygen, and a reductive half-reaction in which $O_2^{•-}$ is converted to H_2O_2.

1. SOD Assay with Cytochrome *c*

To date, several spectrophotometric SOD assays have been described employing different superoxide-detecting dyes, such as nitroblue tetrazolium (NBT; Beauchamp and Fridovich, 1971; Sun *et al.*, 1988), adrenaline (Bannister and Calabrese, 1987), or sulfanilamide (Elstner and Heupel, 1976). Luminometric (Bensinger and Johnson, 1981) and polarographic (Rigo and Rotilio, 1977) methods have also been described. A direct method for determining SOD activity based on recording H_2O_2 production has been developed as well (Segura-Aguilar, 1993). All of these assays have advantages and disadvantages, and to the best of our knowledge, a systematic comparison of these assays has not yet been performed. However, the most frequently used assay is that developed by (McCord and Fridovich, 1969). This assay is simple, reasonably reproducible, and inexpensive, and has been used successfully by us and others to measure SOD activity in cells and tissue homogenates.

This method employs the xanthine-xanthine oxidase system to generate $O_2^{•-}$, [Eq. (2)] and oxidized cytochrome *c* as a superoxide-trapping detection system.

The reduction rate of cytochrome c by $O_2^{\bullet-}$ is monitored spectrophotometrically at 550 nm [Eq. (3)]. The reduction of cytochrome c is inhibited when the SOD-containing sample is added because of enzymatic dismutation of superoxide anion $O_2^{\bullet-}$ [Eq. (4)]:

$$\text{Xanthine} + O_2 \longrightarrow \text{Uric acid} + O_2^{\bullet-} \text{ (catalyzed by XO)} \qquad (2)$$

$$\text{Cytochrome } c \text{ (Fe}^{3+}) + O_2^{\bullet-} \longrightarrow \text{Cytochrome } c \text{ (Fe}^{2+}) + 2O_2 \qquad (3)$$

$$2O_2^{\bullet-} + 2H^+ \longrightarrow H_2O_2 + O_2 \text{ (catalyzed by SOD)} \qquad (4)$$

The competition between reaction (3) and (4) is measured as a decrease in the observed rate of the reduction of cytochrome c, and this decrease is proportional to the total SOD activity present in the sample. The rate of the cytochrome c reduction is measured in the absence (total $\Delta A_{550\ nm}$) or in the presence (sample $\Delta A_{550\ nm}$) of a SOD-containing sample. SOD activity is expressed as:

$$\%\text{inhibition} = (\text{total } \Delta A_{550\ nm} - \text{sample } \Delta A_{550\ nm})/\text{total } \Delta A_{550\ nm}$$

Mn-SOD activity can be measured independently by adding 1-mM potassium cyanide (KCN) to the reaction mixture. KCN is a specific inhibitor of CuZn-SOD (Kasemset and Oberley, 1984). Therefore, activity detected in a KCN-treated sample is primarily Mn-SOD activity (see below). Moreover, CuZn-SOD activity can be determined by subtracting KCN-inhibited activity from the total SOD activity.

KCN inhibition also eliminates possible interference by cytochrome c oxidase (COX) and peroxidases. However, in samples that contain high cytochrome c oxidase or reductase activity, such as mitochondrial fractions or whole-tissue homogenates, use of native cytochrome c to detect SOD activity may produce erroneous results due to the reactions of native cytochrome c with reductases and oxidases in the sample. These unwanted reactions can be suppressed by chemically modifying cytochrome c protein, either by its acetylation or succinylation (Kuthan *et al.*, 1986). Partially acetylated cytochrome c is available commercially from several major distributors of biochemical reagents and is the best reagent for detecting SOD in mitochondria-enriched samples.

Procedure

Reaction mixture (3 ml total volume):

50-mM phosphate buffer, pH 7.8

0.1-mM Na_2-EDTA (to chelate-free copper that possess SOD activity)

1-mM NaN_3 (to inhibit endogenous CAT, COX, and peroxidase activities)

0.01-mM ferricytochrome c

0.05-mM xanthine

6-nM xanthine oxidase. (This amount of xanthine oxidase used results in reduction of ferricytochrome c at a rate of 0.025 absorbance units per minute. The optimum level of xanthine oxidase may vary with different preparations.)

To measure SOD activity, add 100 μl of sample (2–4 mg/ml of cell supernatant or 1–3 mg/ml of the tissue homogenate) or 100 μl of SOD standard sample to the

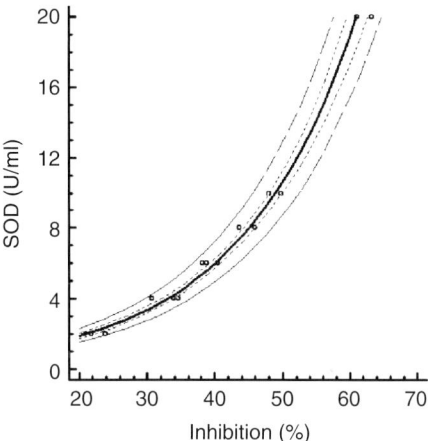

Fig. 1 Standard curve of SOD activity. The exponential regression analysis represents the correlation between SOD activity and percentage of cytochrome c inhibition [SOD (U/ml) = Exp (−0.5193 + 0.0577 × % inhibition]. Each point of SOD activity was performed by triplicate ($r = 0.9929$).

reaction mixture in the presence and absence of 1-mM cyanide (to measure Mn-SOD and Total-SOD activities, respectively). Monitor the reduction of cytochrome c spectrophotometrically by measuring the absorbance at 550 nm at 25 °C for 3–5 min after addition of the SOD sample. One unit of SOD activity is defined as the amount of SOD required to inhibit the rate of reduction of cytochrome c by 50% (i.e., to a rate of 0.0125 absorbance unit per minute).

A standard curve may be prepared by diluting a commercial SOD preparation [e.g., SOD from bovine erythrocytes (Sigma, St. Louis, MO) to 2–20 U/ml (0.2–2 U in the assay)] (Fig. 1). The assay is linear over a limited range of SOD concentrations. Therefore, the sample must be diluted so that the percentage of inhibition falls between 30% and 60%. Results are expressed as units of SOD per milligram of protein.

Table I shows results of SOD activity obtained using the described method in different studies in various cell lines [e.g., 143B-derived cybrids (King and Attardi, 1989), human fibroblasts, and mononuclear cells] and in muscle tissues. SOD activity values are expressed in U/mg protein.

2. SOD Assay with BXT-01050

The method introduced by Nebot *et al.* (1993) is a good alternative to the cytochrome c reduction assay described above. This assay has several technical advantages. It is based on a nonenzymatic reaction, and therefore is free from variations due to differences in commercially available xanthine oxidase and cytochrome c preparations. Also, it is a "positive" assay, that is, the rate of the SOD-detecting reaction increases with increasing SOD. As a result, the sensitivity and

Table I
SOD Activity (U/mg protein)[a]

	Total-SOD	CuZn-SOD	Mn-SOD	
Transmitochondrial cybrids				
MERRF CTL	1.99 ± 0.36	1.67 ± 0.37	0.32 ± 0.04	Vives-Bauza *et al.*, 2006
MERRF	3.07 ± 0.34	2.51 ± 0.28	0.55 ± 0.09	Vives-Bauza *et al.*, 2006
MELAS	3.29 ± 0.27	2.72 ± 0.20	0.55 ± 0.07	Vives-Bauza *et al.*, 2006
COI CTL	2.37 ± 0.55	1.70 ± 0.53	0.73 ± 0.12	Vives-Bauza *et al.*, 2006
COI	2.88 ± 0.75	2.06 ± 0.81	0.81 ± 0.20	Vives-Bauza *et al.*, 2006
Mononuclear cells	5.73 ± 0.15			Redon *et al.*, 2003
Fibroblasts				
Human controls		118 ± 24	116 ± 34	Lu *et al.*, 2003[b]
Patients with CPEO syndrome		149 ± 68	260 ± 24	
Muscle (vastus lateralis)				
Human young (22 ± 3.4 years)		10.8 ± 6.1	5.2 ± 1.8	Gianni *et al.*, 2004
Human old (71.8 ± 6.2 years)		11.2 ± 2.8	7.1 ± 0.8	Gianni *et al.*, 2004

[a]MERRF, Mitochondrial encephalomyopathy with ragged red fibers. Mutation A8344G in the mitochondrial DNA (mtDNA) tRNA^{Lys}. MELAS, mitochondrial encephalomyopathy with lactic acidosis and stroke-like episodes. Mutation A3243G in the mtDNA tRNA$^{Leu(UUR)}$. COI, mutation G6930A in the cytochrome *c* oxidase subunit I. CPEO, chronic progressive external opthalmoplegia. CTL, control.

[b]Uses NBT instead of cytochrome *c*.

the specificity of detection are improved, particularly when assaying a large number of samples. Another important advantage is that all required reagents and detailed instructions for this assay are commercially available from several sources (e.g., EMD Biosciences).

The method is based on the SOD-mediated increase in the rate of auto-oxidation of 5,6,6a,11b-tetrahydro-3,9,10-trihydrobenzo[*c*]fluorene (BXT-01050, Scheme 1) in aqueous alkaline solution. The BXT-01050 auto-oxidation yields a chromophore, which absorbs with maximal absorbance at 525 nm.

The main limitation of the assay is that BXT-01050 is subject to interference from redox-active compounds in the sample. The most common interfering agents, and the concentrations of these agents that interfere with the assay

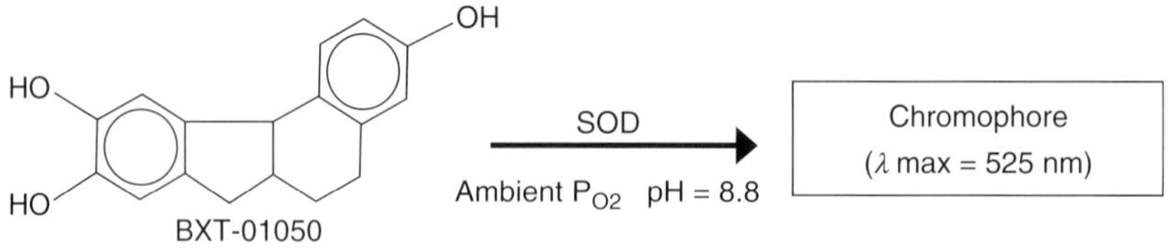

Scheme 1 SOD-mediated BXT-01050 chromophore formation.

(in parentheses), are hemoglobin (>0.5 μM), albumin (>0.1 μM), ascorbic acid (>0.5 μM), NADPH (10 μM), and butylated hydroxytoluene (BHT, >30 μM). These interfering compounds can be removed by protein extraction, dialysis, or gel filtration. However, the ethanol/chloroform extraction inactivates all SODs other than CuZn-SOD. Various biothiols that may be present in a sample (e.g., reduced glutathione) can also interfere with the reaction. Addition of a thiol scavenger, for example 1-methyl-2-vinylpyridinium (M2VP), to the sample can eliminate these sources of interference. Finally, cyanide interferes with the auto-oxidation of BXT-01050, and consequently cannot be used to distinguish CuZn-SOD activity from Mn-SOD activity.

Procedure and Reagents

Reagents:

> BXT-01050 solution: 0.66-mM BXT-01050 in 32-mM HCl containing 0.5-mM diethylenetriaminepentaacetic acid (DTPA) and 2.5% ethanol, pH 1.5. This solution can be stored at 2–8°C protected from light and air. Under these conditions, the compound is stable for up to 2 months.
>
> M2VP solution: 33.3-mM M2VP in 1-mM HCl.
>
> Reaction buffer: 50 mM of 2-amino-2-methyl-1,3-propanediol containing 3.3-mM boric acid and 0.11-mM DTPA, pH 8.8 at 37°C.

Procedure:

Mix by vortexing in a plastic cuvette 900 μl of reaction buffer with 40 μl of SOD-containing sample and 30 μl of M2VP solution. Incubate the mixture at 37°C for 1 min, and add 30 μl of BXT-01050 solution. Vortex and monitor the absorbance at 525 nm for 1 min. To measure the control rate of auto-oxidation, the SOD-containing samples are substituted with an equal volume of distilled water. Measure a minimum of four blank controls for each set of sample data.

SOD activity is determined from the ratio of the auto-oxidation rates measured in the presence (ΔA_{SOD}) or absence of SOD (ΔA_{Blank}). One SOD activity unit is defined as the activity that doubles the rate of the auto-oxidation background ($\Delta A_{SOD}/\Delta A_{Blank} = 2$). The relationship between the $\Delta A_{SOD}/\Delta A_{Blank}$ ratio and SOD activity follows a hyperbolic curve that is described as:

$$(\Delta A_{SOD}/\Delta A_{Blank}) = 1 + (SOD_{525\ nm})/[\alpha \times (SOD) + \beta]$$

where ΔA_{SOD} = rate of sample containing SOD, ΔA_{Blank} = average rate of blank samples, $SOD_{525\ nm}$ = SOD activity of the sample in SOD-525 nm arbitrary units, α = dimensionless coefficient = 0.073, and β = coefficient in SOD-525 units = 0.93.

3. In-gel Assay of SOD Activity

SOD activity can be detected semiquantitatively using an in-gel activity assay, which employs a redox-sensitive dye NBT as a $O_2^{\bullet -}$ detector, and a nonenzymatic superoxide-generating photochemical reaction, combined with polyacrylamide gel electrophoresis (Beauchamp and Fridovich, 1971). This section describes a modification of this method (Vijayvergiya *et al.*, 2005).

Procedure

Cell pellets or isolated mitochondria are solubilized in 0.5% Triton X-100 on ice for 15 min. Each sample (20–40 μg) is loaded on a 10% native polyacrylamide gel (Invitrogen, Carlsbad, CA) and subjected to electrophoresis. One unit of purified hSOD1 (Sigma, St Louis, MO) may be used as a positive control. The gel is then incubated in 25 ml of 1.23-mM NBT for 15 min and is washed briefly. The NBT-stained gel is then soaked in the dark in 30 ml of 100-mM potassium phosphate buffer, pH 7.0, containing 28-mM TEMED (*N,N,N′,N′*-tetramethylethylenediamine). Riboflavin is added at a final concentration of 2.8×10^{-2} mM and the gel is incubated for another 15 min in the dark.

To initiate the photochemical reaction, the gel is exposed to a fluorescent light. A few minutes later, the achromatic bands corresponding to SOD activity should appear on a dark blue background. SOD activity can be evaluated by densitometry of the achromatic bands.

C. GPx Activity Assay

GPxs catalyze the reduction of H_2O_2 and organic hydroperoxides (ROOH), using reduced glutathione (GSH) as the source of reducing equivalents. The GPx family includes at least five enzymes encoded in mammals by the genes GPx1, GPx2, GPx3, GPx4, and GPx5. These GPx isoforms have different subcellular localizations, and exist as cytoplasmic, mitochondrial, and extracellular forms. GPx activity is high in liver, kidney, and heart mitochondria and lower in brain and skeletal muscle mitochondria. Some reports have also suggested that GPx is present in the nucleus (Muse *et al.*, 1994). A monomeric form of GPx (GPx4) is associated with cellular membranes (Utsunomiya *et al.*, 1991). This enzyme reduces lipid peroxides in a reaction similar to that catalyzed by GPx1 (Brigelius-Flohe *et al.*, 1994) and is important in protecting the lipid membranes against lipid peroxidation. GPx1 contains a selenocysteine residue in the center of its active site, which is implicated in the catalysis of the enzyme. In contrast to CAT, the K_m of GPx1 for H_2O_2 is quite low, suggesting that this enzyme may function to fine-tune intracellular H_2O_2 concentrations at low levels.

1. Procedure for Measuring GPx Activity in Cells and Tissue Homogenates

GPx activity is measured by the method of Paglia and Valentine (1967), as modified by Lawrence and Burk (1976). In this method, oxidized gluthathione (GSSG), produced on reduction of organic peroxide by GPx, is recycled to its reduced state by glutathione reductase (GR) (Meister, 1988):

$$R\text{--}OOH + 2GSH \longrightarrow R\text{--}OH + GSSG + H_2O \quad \text{(catalyzed by GPx)} \quad (5)$$

$$GSSG + NADPH + H^+ \longrightarrow 2GSH + NADP^+ \quad \text{(catalyzed by GR)} \qquad (6)$$

The oxidation of NADPH is followed spectrophotometrically at 340 nm at 37°C. The choice of peroxide determines which GPx isoenzyme activity is detected. H_2O_2 or *tert*-butyl hydroperoxide are reduced only by the Se-dependent enzyme (Se-GPx). In contrast, cumene hydroperoxide is reduced by both the Se-GPx and the Non Se-GPx isoenzyme. Therefore, cumene hydroperoxide is used to estimate the total GPx activity, $GPx_{(total)}$.

Reaction mixture:

- 0.1 ml of 50-mM phosphate buffer, pH 7.0
- 0.1 ml of 1-mM EDTA
- 0.2 ml of 5-mM NaN_3
- 0.2 ml of 1-mM NADPH
- 6.4 μl GR (from a stock of 0.312 U/μl)
- 0.2 ml of 5-mM GSH
- 0.1 ml of sample-containing GPx (1–3 mg/ml of protein) (for the blank use 0.1 ml of phosphate buffer)

Substrates:

For total GPx: 0.2 ml of 1.5-mM cumene hydroperoxide

For Se-GPx: 0.2 ml of 0.25-mM H_2O_2 or 0.2 ml of 0.3-mM tert-butyl hydroperoxide. The H_2O_2 solution should be prepared fresh, immediately before use. Confirm the concentration of the H_2O_2 solution by checking the absorbance of the solution at 240 nm using a quartz cuvette. $\varepsilon_{240\ nm} = 0.0426$ mM^{-1} cm^{-1}.

The reaction assay is mixed by gentle agitation and incubated in the spectrophotometer at 37°C for 5 min in order to eliminate the nonspecific degradation of NADPH. The reaction is initiated by adding the substrate (either cumene hydroperoxide, H_2O_2, or *tert*-butyl hydroperoxide). The decay of NADPH is monitored for 5 min at 340 nm. The rate of decrease in the $\Delta A_{340\ nm}$ is directly proportional to the GPx activity in the sample. The linear rate is used to calculate the activity, which is reported in units per gram of protein:

$$\text{U/g prot} = (\Delta A/\text{min}_s - \Delta A/\text{min}_b) \times (V_t/V_s) \times 1/\varepsilon \times (1/\text{prot})$$

$$\text{U/g prot} = (\Delta A/\text{min}_s - \Delta A/\text{min}_b) \times 1/0.1 \times 1/6.22\ \mu\text{mol/ml} \times 1/\text{prot (g/ml)}$$

$$\text{U/g prot} = (\Delta A/\text{min}_s - \Delta A/\text{min}_b) \times 1.608/\text{prot}\ [\mu\text{mol/(min g prot)}]$$

where $\Delta A/\text{min}_s$ is the change in absorbance per minute for the sample; $\Delta A/\text{min}_b$ is the change of absorbance per minute for the blank; V_t is the total volume of the assay (expressed in ml); V_s is the sample volume (expressed in ml), and ε is the extinction coefficient (for NADPH, $\varepsilon = 6.22$ mM^{-1} cm^{-1}). One enzyme unit of GPx activity is defined as 1 μmol of NADPH oxidized per minute at 37°C.

Table II
GPx Activity (U/g protein)[a]

	Total-GPx	Se-GPx	Non-Se-GPx	
Transmitochondrial cybrids				
MERRF CTL	3.04 ± 1.11	2.32 ± 1.06	0.47 ± 0.13	Vives-Bauza *et al.*, 2006
MERRF	16.49 ± 2.00	13.77 ± 2.10	2.72 ± 0.40	Vives-Bauza *et al.*, 2006
MELAS	10.22 ± 1.50	9.06 ± 1.40	1.15 ± 0.20	Vives-Bauza *et al.*, 2006
COI CTL	4.02 ± 0.05	2.22 ± 0.16	1.80 ± 0.14	Vives-Bauza *et al.*, 2006
COI	3.76 ± 0.62	2.17 ± 0.62	1.59 ± 0.21	Vives-Bauza *et al.*, 2006
Mononuclear cells	57.18 ± 2.04			Redon *et al.*, 2003
Fibroblasts				
Human controls	65.9 ± 6.7			Lu *et al.*, 2003
Patients with CPEO syndrome	56.0 ± 15.0			Lu *et al.*, 2003

[a]MERRF, mitochondrial encephalomyopathy with ragged red fibers. Mutation A8344G in the mitochondrial DNA (mtDNA) tRNA[Lys]. MELAS, mitochondrial encephalomyopathy with lactic acidosis and stroke-like episodes. Mutation A3243G in the mtDNA tRNA[Leu(UUR)]. COI, mutation G6930A in the cytochrome *c* oxidase subunit I. CPEO, chronic progressive external opthalmoplegia. CTL, control.

Table II summarizes the GPx activity values obtained in different studies on cybrids, human fibroblasts, and mononuclear cells. For the cybrids and fibroblasts, control and pathogenic values are shown. GPx activity values are expressed in U/g protein.

Other considerations

i. Mercaptosuccinic acid is a specific inhibitor of GPx. It can be added to the control reaction mixture at 0.4 mM to improve the assay specificity toward the GPx.

ii. There is some overlap in activities of GPx and other enzymes, in particular with CAT, which catalyzes the reduction of H_2O_2, and the glutathione *S*-transferases that catalyze the reduction of lipid hydroperoxides. Therefore, for tissues that contain substantial amounts of glutathione *S*-transferases (such as, e.g., liver), only H_2O_2 should be used as substrate. In this case, it is also necessary to block the activity of CAT by adding 1-mM NaN_3.

2. Procedure for Measuring GPx1 Activity in Isolated Mitochondria

To measure the GPx1 activity in preparations of isolated mitochondria, the composition of the reaction assay and the procedure should be modified as follows.

Reaction mixture

0.9-ml 20-mM MOPS buffer, 0.15-mM NADPH, pH 7.1

2-μl (200-mM stock) GSH

2-μl (150–190 U/ml suspension in glycerol or ammonium sulfate) GR

100-μl (11.1 mM stock) H_2O_2

2–6 mg/ml of mitochondria

Procedure

Prepare 5 ml of reaction mixture containing 4.95 ml of 0.15-mM NADPH in 20-mM MOPS, pH 7.1 and 50 μl of 200-mM GSH. Resuspend the mitochondria in the reaction buffer at a concentration of 2–6 mg/ml. Add 2 ml of water to a cuvette and blank the spectrophotometer; remove the cuvette. Following, add 0.9 ml of reaction mixture to another clean cuvette. Thereafter, add 2 μl of GR (150–190 U/ml stock) and mix it by pipetting. Start the reaction by adding 100-μl H_2O_2 (11.1-mM stock). Follow the absorbance changes at 340 nm for 3 min. GPx activity is calculated from the absorbance changes taken at the linear part of the slope ($\Delta A_{340\,nm}$), which typically is from 1.6 to 2.6 min, and dividing it by the micromolar extinction coefficient for NADPH ($\varepsilon_{340\,nm} = 0.00622\ \mu M^{-1}\ cm^{-1}$) and by the content of mitochondria (in mg protein), to obtain the rate of reaction in nanomoles NADPH/min/mg prot = $(\Delta A_{340\,nm})/\varepsilon_{NADPH}$/mg prot.

D. Catalase Activity Assay

Catalase (CAT) is expressed ubiquitously in most tissues; it catalyzes the degradation of H_2O_2 into water and oxygen (Luck, 1954). CAT is specific for H_2O_2 and does not decompose organic peroxides. CAT is a peroxisomal enzyme present in almost all aerobic organisms. CAT is also present in rodent heart mitochondria, where it may comprise as much as 0.025% of all protein (Radi et al., 1991). The K_m of CAT is high, suggesting that it may function predominantly to facilitate bulk degradation of H_2O_2 (Chance et al., 1979).

The fine details of the catalytic mechanism of the enzyme still remains unclear. However, the most widely accepted reaction mechanism is that proposed by Deisseroth and Dounce (1970):

$$H_2O_2 + Fe^{2+}-E \rightarrow H_2O_2 + O{=}Fe^{3+}-E \tag{7}$$

$$H_2O_2 + O{=}Fe^{3+} \rightarrow H_2O + Fe^{2+}-E + O_2 \tag{8}$$

Fe–E represents the iron of the hematin prosthetic group bound to the rest of the enzyme. In the first stage, Eq. (7), one molecule of H_2O_2 enters in the heme prosthetic group. The Fe^{2+} heme reduces one molecule of H_2O_2 to water and generates a covalent oxyferryl species ($O{=}Fe^{3+}$; known as Compound I) plus a heme radical (Ivancich et al., 1997). The radical is quickly degraded in another electron transfer reaction, leaving the heme prosthetic group unaltered. During the second stage, Eq. (8), Compound I oxidizes the second molecule of H_2O_2 to give rise to the original enzyme ($Fe^{2+}-E$), plus another molecule of water and one molecule of oxygen.

CAT is inactivated at H_2O_2 concentrations above 0.1 M. As a result, it is not possible to saturate the enzyme with the substrate within a practically feasible concentration range, and the kinetics of CAT does not follow a normal pattern. Instead, the assay must be carried out at a relatively low concentration of H_2O_2 (10 mM), with short reaction times and relatively high enzyme concentrations. Under these conditions, the decay of H_2O_2 concentration due to the action of CAT is measured as a first-order reaction:

$$k = 1/\Delta t \times \ln(A_1/A_2) = 2.3/\Delta t \times \log(A_1/A_2)$$

$\Delta t = t_2 - t_1$ is measured time interval (in seconds)

$A_1 = $ absorbance at t_1; $A_2 = $ absorbance at t_2

A definition of CAT units (U) according to the recommendations of International Union of Biochemistry (IUB) is not feasible because of the abnormal kinetics of this enzyme. Instead, the rate of a first-order reaction (k) is used.

1. Assay Procedure

This section describes a modified direct method by Aebi (1984), which is based on the decomposition of H_2O_2 by CAT. The change in H_2O_2 concentration is monitored directly by following the decay in absorbance at 240 nm.

The reaction buffer is composed of a 10.3-mM H_2O_2 solution, which is prepared by diluting 30% commercial H_2O_2 stock ($\sim$9.4 M) in 50-mM phosphate buffer, pH 7.0. The reaction is carried in a 3-ml quartz or other UV-transparent cuvette. The reaction is started by adding 100 μl of the cell supernatant ($\sim$2–4 mg/ml of protein) to 2.9 ml of the reaction buffer. The absorbance reading is taken against a "blank" cuvette containing 2.9 ml of 50-mM phosphate buffer without H_2O_2, but with 100 μl of cell supernatant. The initial absorbance should be around 0.5. The decay in absorbance at 240 nm is monitored for 2 min. The linear rate is used to calculate the activity, which is presented in k per gram of protein:

$$k/\text{g prot} = (\text{VF}/\text{VM}) \times 2.3/\text{Dt} \times \log(A_1/A_2)/\text{prot g/ml} \times 1000$$

$$k/\text{g prot} = (3/0.1) \times 2.3/\text{Dt} \times \log(A_1/A_2)/\text{prot g/ml}$$

$$k/\text{g prot} = 69/\text{Dt} \times \log(A_1/A_2)/\text{prot g/ml}$$

CAT protein concentration can be determined based on its heme content, spectrophotometrically at 640–660 nm, $\varepsilon = 3.06$ mM^{-1} cm^{-1} (Sies *et al.*, 1973).

The same procedure can be utilized to measure CAT activity in tissue homogenates and in isolated mitochondria from rat heart. It is important to avoid using Triton X-100 and other detergents due to their typically high absorbance at 240 nm. Instead, sonication or homogenization can be used to unmask the full CAT activity.

Table III
Catalase Activity[a]

	Catalase	
Transmitochondrial cybrids (k/g protein)		
MERRF CTL	7.53 ± 0.34	Vives-Bauza et al., 2006
MERRF	16.40 ± 2.07	Vives-Bauza et al., 2006
MELAS	11.72 ± 1.17	Vives-Bauza et al., 2006
COI CTL	6.93 ± 0.28	Vives-Bauza et al., 2006
COI	6.04 ± 0.201	Vives-Bauza et al., 2006
Mononuclear cells (U/g protein)	276.9 ± 6.46	Redon et al., 2003
Fibroblasts (10^3 U/g protein)		
Human controls	6.6 ± 1.6	Lu et al., 2003
Patients with CPEO syndrome	6.2 ± 1.3	Lu et al., 2003
Muscle (vastus lateralis) (μmol/min mg protein)		
Human young (22 ± 3.4 years)	6.2 ± 2.4	Gianni et al., 2004
Human old (71.8 ± 6.2 years)	8.5 ± 2.0	Gianni et al., 2004

[a]MERRF, mitochondrial encephalomyopathy with ragged red fibers. Mutation A8344G in the mitochondrial DNA (mtDNA) tRNA[Lys]. MELAS, mitochondrial encephalomyopathy with lactic acidosis and stroke-like episodes. Mutation A3243G in the mtDNA tRNA[Leu(UUR)]. COI, mutation G6930A in the cytochrome c oxidase subunit I. CPEO, chronic progressive external opthalmoplegia. CTL, control.

Table III presents the typical CAT activities measured by this method in cells and tissue samples. The CAT units differ between the cited references and are mentioned within brackets.

Acknowledgments

This work has been supported by grants from the Spanish *Fondo de Investigaciones Sanitarias* (FIS PI 030168 and PI 050562).

References

Aebi, H. (1984). Catalase *in vitro*. *Methods Enzymol.* **105**, 121–126.

Andreyev, A. Y., Kushnareva, Y. E., and Starkov, A. A. (2005). Mitochondrial metabolism of reactive oxygen species. *Biochemistry* **70**, 200–214.

Bannister, J. V., and Calabrese, L. (1987). Assays for SOD. *Methods Biochem. Anal.* **32**, 279–312.

Beauchamp, C., and Fridovich, I. (1971). Superoxide Dismutase: Improved assays and an assay applicable to acrylamide gels. *Anal. Biochem.* **44**, 276–287.

Bensinger, R. E., and Johnson, C. M. (1981). Luminal assay for Superoxide dismutase. *Anal. Biochem.* **116**, 142–145.

Brigelius-Flohe, R., Aumann, K. D., Blocker, H., Gross, G., Kiess, M., Kloppel, K. D., Maiorino, M., Roveri, A., Schuckelt, R., and Usani, F. (1994). Phospholipid-hydroperoxide glutathione peroxidase. Genomic DNA, cDNA, and deduced amino acid sequence. *J. Biol. Chem.* **269**, 7342–7348.

Chance, B., Sies, H., and Boveris, A. (1979). Hydroperoxide metabolism in mammalian organs. *Physiol. Rev.* **59**, 527–605.

Deisseroth, A., and Dounce, A. L. (1970). Catalase: Physical and chemical properties, mechanism of catalysis, and physiological role. *Physiol. Rev.* **50**, 319–375.

Elstner, E. F., and Heupel, A. (1976). Inhibition of nitrite formation from hydroxylammoniumchloride: A simple assay for superoxide dismutase. *Anal. Biochem.* **70**, 616–620.

Gianni, P., Jan, K. J., Douglas, M. J., Stuart, P. M., and Tarnopolsky, M. A. (2004). Oxidative stress and the mitochondrial theory of aging in human skeletal muscle. *Exp. Gerontol.* **39**, 1391–1400.

Halliwell, B., and Gutteridge, J. M. (1990). Role of free radicals and catalytic metal ions in human disease: An overview. *Methods Enzymol.* **186**, 1–85.

Ivancich, A., Jouve, H. M., Sartor, B., and Gaillard, J. (1997). EPR investigation of compound I in Proteus mirabilis and bovine liver catalases: Formation of porphyrin and tyrosyl radical intermediates. *Biochemistry* **36**, 9356–9364.

Kasemset, D., and Oberley, L. W. (1984). Regulation of Mn-SOD activity in the mouse heart: Glucose effect. *Biochem. Biophys. Res. Commun.* **122**, 682–686.

King, M. P., and Attardi, G. (1989). Human cells lacking mtDNA: Repopulation with exogenous mitochondria by complementation. *Science* **246**, 500–503.

Kuthan, H., Haussmann, H. J., and Werringloer, J. (1986). A spectrophotometric assay for superoxide dismutase activities in crude tissue fractions. *Biochem. J.* **237**, 175–180.

Lawrence, R. A., and Burk, R. F. (1976). Glutathione peroxidase activity in selenium-deficient rat liver. *Biochem. Biophys. Res. Commun.* **71**, 952–958.

Lu, C. Y., Wang, E. K., Lee, H. C., Tsay, H. J., and Wei, Y. H. (2003). Increased expression of manganese-superoxide dismutase in fibroblasts of patients with CPEO sindrome. *Mol. Genet. Metab.* **80**, 321–329.

Luck, H. (1954). Quantitative determination of catalase activity of biological material. *Enzymologia* **17**, 31–40.

McCord, J. M., and Fridovich, I. (1969). Superoxide dismutase. An enzymic function for erythrocuprein (hemocuprein). *J. Biol. Chem.* **244**, 6049–6055.

Meister, A. (1988). Glutathione metabolism and its selective modification. *J. Biol. Chem.* **263**, 17205–17208.

Muse, K. E., Oberley, T. D., Sempf, J. M., and Oberley, L. W. (1994). Immunolocalization of antioxidant enzymes in adult hamster kidney. *Histochem. J.* **26**, 734–753.

Nebot, C., Moutet, M., Huet, P., Xu, J. Z., Yadan, J. C., and Chaudiere, J. (1993). Spectrophotometric assay of superoxide dismutase activity based on the activated autoxidation of a tetracyclic catechol. *Anal. Biochem.* **214**, 442–451.

Paglia, D. E., and Valentine, W. N. (1967). Studies on the quantitative and qualitative characterization of erythrocyte glutathione peroxidase. *J. Lab. Clin. Med.* **70**, 158–169.

Radi, R., Turrens, J. F., Chang, L. Y., Bush, K. M., Crapo, J. D., and Freeman, B. A. (1991). Detection of catalase in rat heart mitochondria. *J. Biol. Chem.* **266**, 22028–22034.

Redon, J., Oliva, M. R., Tormos, C., Giner, V., Chaves, J., Iradi, A., and Saez, G. T. (2003). Antioxidant Activities and Oxidative Stress Byproducts in Human Hipertension. *Hypertension* **41**, 1096–1101.

Rhee, S. G., Chae, H. Z., and Kim, K. (2005). Peroxiredoxins: A historical overview and speculative preview of novel mechanisms and emerging concepts in cell signaling. *Free Radic. Biol. Med.* **38**, 1543–1552.

Rigo, A., and Rotilio, G. (1977). Simultaneous determination of superoxide dismutase and catalase in biological materials by polarography. *Anal. Biochem.* **81**, 157–166.

Segura-Aguilar, J. (1993). A new direct method for determining superoxide dismutase activity by measuring hydrogen peroxide formation. *Chem. Biol. Interact.* **86**, 69–78.

Sies, H., Bucher, Th., Oshino, N., and Chance, B. (1973). Heme occupancy of catalase in hemoglobin-free perfused rat liver and of isolated rat liver catalase. *Arch. Biochem. Biophys.* **154**, 106–116.

Sun, Y., Oberley, L. W., and Li, Y. (1988). A simple method for clinical assay of superoxide dismutase. *Clin. Chem.* **34,** 497–500.

Utsunomiya, H., Komatsu, N., Yoshimura, S., Tsutsumi, Y., and Watanabe, K. (1991). Exact ultra-structural localization of glutathione peroxidase in normal rat hepatocytes: Advantages of microwave fixation. *J. Histochem. Cytochem.* **39,** 1167–1174.

Vijayvergiya, C., Beal, M. F., Buck, J., and Manfredi, G. (2005). Mutant superoxide dismutase 1 forms aggregates in the brain mitochondrial matrix of amyotrophic lateral sclerosis mice. *J. Neurosci.* **25,** 2463–2470.

Vives-Bauza, C., Gonzalo, R., Manfredi, G., García-Arumi, E., and Andreu, A. L. (2006). Enhanced ROS production and antioxidant defenses in cybrids harbouring mutations in mtDNA. *Neurosci. Lett.* **391,** 136–141.

CHAPTER 20

Methods for Measuring the Regulation of Respiration by Nitric Oxide

Sruti Shiva,[*] Paul S. Brookes,[†] and Victor M. Darley-Usmar[‡]

[*]Vascular Medicine Branch
NHLBI, NIH, Bethesda
Maryland 20892

[†]Department of Anesthesiology
University of Rochester Medical Center
Rochester, New York 14642

[‡]Cellular and Molecular Pathology
University of Alabama at Birmingham
Birmingham, Alabama 35294

METHODS IN CELL BIOLOGY, VOL. 80
0091-679X/07 $35.00
DOI: 10.1016/S0091-679X(06)80020-8

I. Introduction

It is now known that mitochondria are not only essential in maintaining the energy (ATP) levels of the cell, but are also involved in diverse cellular functions such as redox signaling and the regulation of apoptosis through the release of cytochrome c. Reactive nitrogen species have long been known to target mitochondrial respiratory proteins, altering their function, but the role of this interaction in physiology has remained unclear. The most well-characterized association of this type is the reversible binding of nitric oxide ($NO^\bullet$) to cytochrome a_3 at the active site of cytochrome c oxidase (complex IV), resulting in the inhibition of mitochondrial respiration. This phenomenon is part of a much larger signaling pathway, the nitric oxide ($NO^\bullet$)-cytochrome c oxidase signaling pathway, which has emerged as one of the most sensitive physiological mechanisms for the regulation of mitochondrial respiration (Brown, 2000; Cleeter *et al.*, 1994). Signaling is mediated through binding of $NO^\bullet$ to the binuclear Cu_B/heme$_{a_3}$ center in cytochrome c oxidase and has been characterized extensively (Giuffre *et al.*, 2000; Torres *et al.*, 1995). However, the implications for this binding in the control of mitochondrial respiration, ATP synthesis, and reactive oxygen species (ROS) generation have received less attention. It is important to understand the mechanisms underlying these phenomena, since we and others have shown that the degree of control by $NO^\bullet$ over mitochondrial function is perturbed in pathological situations (Brookes *et al.*, 2001; Dai *et al.*, 2001; Venkatraman *et al.*, 2003).

In this chapter, we outline basic methodologies for the measurement of $NO^\bullet$-dependent effects on mitochondrial respiration, including the measurement of respiratory thresholds in isolated mitochondria and cell systems. We describe the methodology used to measure respiration at low O_2 concentrations and the complications of using various $NO^\bullet$ donors. We also discuss the use of these methodologies to measure mitochondrial function in physiology and the change in function observed in pathology.

A. Basic Concepts of Metabolic Control

Like other complexes of the mitochondrial respiratory chain, cytochrome c oxidase (complex IV) is present in excess of the amount required to meet normal metabolic demand (Mazat *et al.*, 1997). One result of this property is that submaximal inhibition of the enzyme does not affect the rate of the overall metabolic pathway (i.e., respiration) until a threshold level of inhibition is reached. Beyond this threshold, respiratory inhibition occurs. This is illustrated in Fig. 1, which shows a "threshold curve" for cytochrome c oxidase of rat heart mitochondria. Implicit in the existence of such thresholds is the degree of control that an enzyme, in this case cytochrome c oxidase, has over respiration. Although a full discussion of metabolic control theory is beyond the scope of this chapter [see Mazat *et al.* (1997) for a more detailed discussion], it should be emphasized that in this context

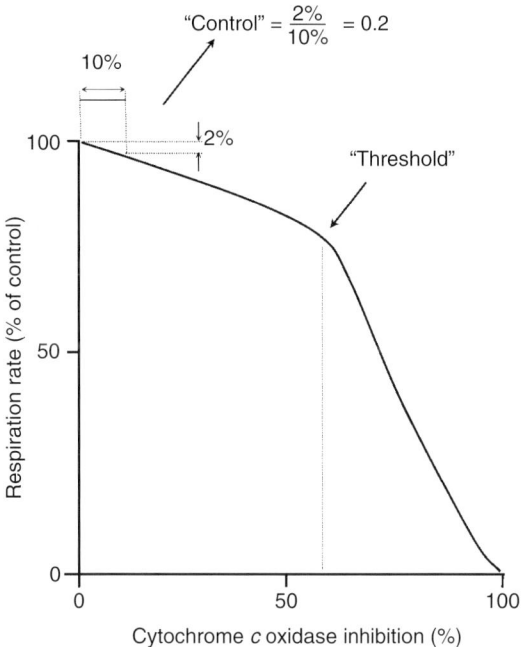

Fig. 1 Mitochondrial cytochrome c oxidase (complex IV) respiratory threshold. A representative trace for isolated rat heart mitochondria is shown in which respiration rate and complex IV activity were inhibited with cyanide. The specific definitions of the terms "threshold" and "control" in the context of this chapter are indicated.

the word "control" has a specific meaning, that is, the flux control coefficient. This represents the relationship between the fractional change in the metabolic pathway flux caused by a fractional change in the flux of the enzyme in question (Groen *et al.*, 1982). In other words, how does inhibition of a specific component in a metabolic pathway affect the overall activity of that pathway? In minimal terms, the control coefficient can be approximated from the slope of the initial portion of the threshold curve, as shown in Fig. 1.

Experimentally, respiratory thresholds and control coefficients are determined by measuring the effects of a specific inhibitor (e.g., cyanide for cytochrome c oxidase), both on the isolated activity of a respiratory complex and on the overall respiration rate (Brookes *et al.*, 2001; Mazat *et al.*, 1997; Venkatraman *et al.*, 2003). By combining the y-axes of these two graphs at shared concentrations of inhibitor, a threshold curve is constructed showing respiration as a function of the respiratory complex activity. This is shown theoretically in Fig. 2. For clarity, the x-axis of the threshold curve is often inverted to show respiration as a function of complex inhibition rather than of complex activity.

In this chapter, we outline developments to this methodology to measure the control over mitochondrial function by NO•. These methods are discussed in the

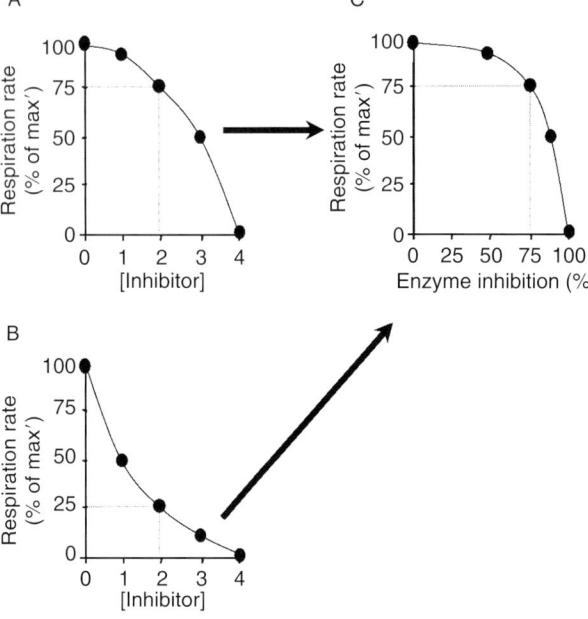

Fig. 2 Construction of threshold curves from inhibitor titration curves. The responses of respiration rate (panel A) and enzyme rate (panel B) to the inhibitor are determined. The *y*-axes of these graphs (i.e., enzyme rate and respiration rate) are then coplotted at corresponding values of inhibitor concentration to yield the threshold curve (panel C). Calculation of an example point on the curve is shown by the dotted lines: two units of inhibitor result in a 25% inhibition of respiration rate (A) and a 75% inhibition of the enzyme (B). Thus, 75% enzyme inhibition causes 25% respiration inhibition (C).

context of several important experimental considerations that arise when working with and measuring NO•, including the importance of measuring and controlling O_2 levels, when interpreting these data.

II. Apparatus

A. Classical Closed System Respirometer

The apparatus required to measure NO• effects on mitochondrial function consists of an O_2 electrode and an NO• electrode in a closed chamber (Fig. 3A). A cylindrical electrode chamber made of Perspex or Plexiglas holds a volume of 0.25–2.0 ml, and is water-jacketed to maintain the sample at 37 °C, via a circulating water bath. The contents are stirred by a magnetic or electrostatic stirring device with a glass- or Teflon-coated stir bar (Rank Brothers, Bottisham, Cambridge, UK). This apparatus is available commercially (e.g., WPI, Sarasota, FL). However, custom construction can also be used to accommodate specific experimental requirements. A segment of the water jacket is solid Perspex, with a hole drilled

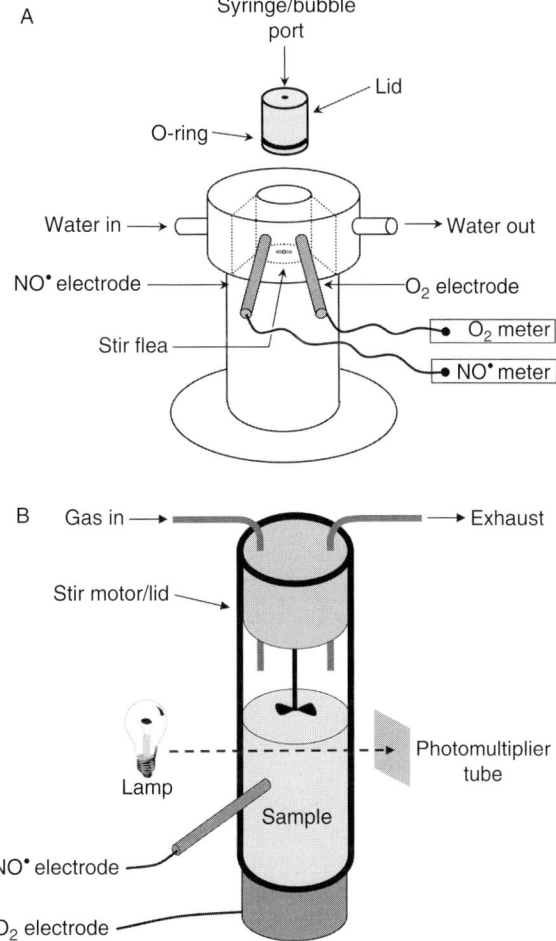

Fig. 3 Combined O_2 electrode and $NO^\bullet$ electrode chamber for measuring the effects of $NO^\bullet$ on mitochondrial and cellular respiration in a classical closed system (A) and open-flow system (B). For full description see text.

from the outside to the center of the chamber to house a Clark-type O_2 electrode (Instech, Plymouth Meeting, PA). An adjacent hole is also drilled to house a Clark-type $NO^\bullet$ electrode. Those made by either WPI or Innovative Instruments (Tampa, FL) offer good sensitivity and signal-to-noise ratio. The electrodes are positioned sufficiently high in the chamber to avoid contact with the magnetic stir bar. The $NO^\bullet$ and O_2 sensors lead to their respective meters, and data are recorded by a digital data-recording device (Dataq, Akron, OH) connected to a personal computer (PC). The chamber is capped by a Perspex lid containing an O-ring seal, which slides inside the chamber to meet the sample, leaving no headspace. The lid contains a port for the exit of air bubbles and for sample injection via Hamilton[TM] syringes. The underside of the lid is beveled to direct air bubbles upward toward the hole (not shown). If ambient electrical noise is high, the apparatus (minus the electrode meters and data acquisition device) should be housed in a well-grounded Faraday cage.

Owing to concerns regarding the potential diffusion of O_2 from plastic materials, such as Perspex and Teflon, high-resolution respiration systems have become available with stir bars made from PEEK, and the chamber and stoppers made from glass and/or titanium (e.g., Oxygraph 2K, Oroboros, Innsbrück, Austria). However, the high initial cost of such systems can be an obstacle to drilling holes for additional electrodes. Similarly, multichannel polarographic systems are available (e.g., WPI Apollo 4000) which can drive several electrodes ($NO^{\bullet}$, O_2, H_2O_2, H_2S), but such systems are often overcomplicated, being driven by an internal PC and requiring dedicated software. Often, simple chart-recorder-based systems, with each electrode driven by its own box, while more difficult to set up initially, are the more reliable and customizable solution.

B. Open-Flow Respirometer

In certain instances, it is necessary to measure the effects of $NO^{\bullet}$ on respiration while maintaining the chamber within a narrow range of O_2 tensions (detailed below). This can be accomplished by using an open-flow respirometer (Fig. 3B) in which the O_2 consumed by mitochondria is constantly replaced by O_2 diffusion into the liquid phase from a gaseous headspace (Brookes *et al.*, 2003; Cole *et al.*, 1982). The headspace gas, with a tightly controlled pO_2, is flowed over the surface of the liquid phase such that a steady-state $[O_2]$ results in the liquid phase, which is an equilibrium between O_2 consumption and delivery. This models the *in vivo* system in which real-time tissue O_2 tensions are a result of the balance between both O_2 delivery and consumption.

In practice, this is achieved by building a thermostated respiratory chamber with a polarographic O_2 sensor in its floor and an $NO^{\bullet}$ electrode in the side. The chamber contains 5 ml of solution and 5 ml of gas headspace. The lid of the chamber contains a port for injections by Hamilton[TM] syringe, and gas inflow and outlet tubes. Composition of the gas phase is regulated by mixing various ratios of dry air and O_2-free nitrogen gas, using digital mass flow controllers (e.g., Smart-Trak, Sierra Instruments, Monterey, CA). The gas is then humidified and passed through the headspace at a constant flow rate (usually 100 ml/min). Stirring is achieved by a propeller, which perturbs the liquid surface thereby maximizing the surface area for gas exchange. The use of open-flow systems is described later (Section V).

While not essential for this type of respirometer, the chamber can be modeled in a cylindrical quartz tube that serves as a cuvette for spectrophotometric measurements, as shown in Fig. 3. In this case described herein (Brookes *et al.*, 2003), the light path of the cuvette was 16mm, and the cuvette was fitted into the measuring chamber of a Cary model 14 recording spectrophotometer equipped

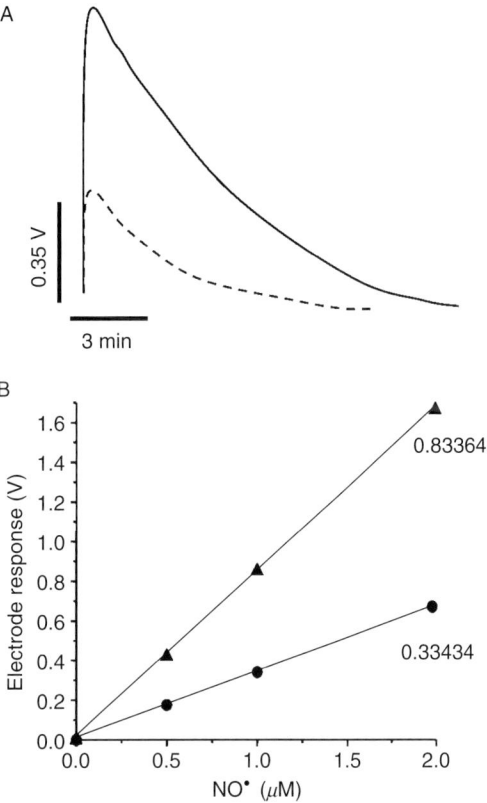

Fig. 4 Calibration of the NO$^\bullet$ electrode and effects of O$_2$. Panel A shows typical NO$^\bullet$ decay traces recorded in the NO$^\bullet$ electrode chamber containing mitochondrial respiration buffer (Table I), following bolus addition of 1-μM NO$^\bullet$ from a stock solution (see methods), with the lid closed. Solid line: at $\sim$20% O$_2$ saturation. Dotted line: at $\sim$80% O$_2$ saturation. The y-axis is the NO$^\bullet$ electrode response in volts. Panel B shows the electrode response (volts) as a function of added [NO$^\bullet$], at nominally high (80%, circles) and low (20%, triangles) O$_2$ levels. Numbers alongside the lines are slopes, in V/μM NO$^\bullet$. Traces and data are representative of at least four identical experiments.

The importance of a tight-fitting lid for the NO$^\bullet$ electrode chamber and effect of light/dark on measurements made in the chamber are emphasized by the data in Fig. 5. Figure 5A shows the decay of NO$^\bullet$ in mitochondrial respiration buffer with and without the lid on the chamber. The loss of NO$^\bullet$ to the headspace is thus a significant source of apparent NO$^\bullet$ consumption, and due care should be taken to eliminate air bubbles from the chamber, since even a small bubble ($\sim$10 μl) can represent a significant NO$^\bullet$ sink. Finally, it is important to calibrate the electrode frequently in this manner, since this relationship may change from experiment to experiment, depending on the age and condition of the electrode.

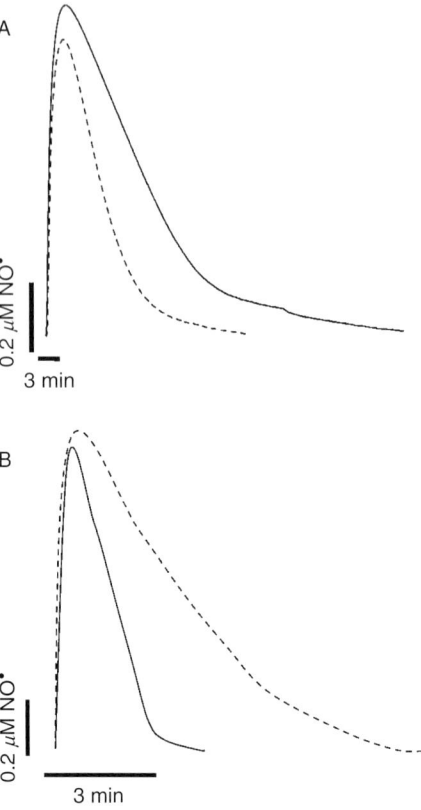

Fig. 5 Effects of chamber lid closure and light/dark on NO• decay in buffer. Panel A shows typical NO• decay traces recorded in the NO• electrode chamber containing mitochondrial respiration buffer (Table I), following bolus addition of 1-μM NO• from a stock solution (see methods), with the lid closed (solid line) or open to the atmosphere (dotted line). Panel B shows decay traces for 1-μM NO• recorded as above, in the NO• electrode chamber containing DMEM with the lid on, at ambient laboratory light levels (solid line) or in darkness (dotted line). Traces are representative of at least four identical experiments.

D. NO• Release from NO• Donor Compounds

One way to measure the effect of varying concentrations of NO• on both mitochondria and cells is by the use of NO• donor compounds. An example of a series of such compounds are the NONOates, which are stable at alkaline pH but degrade at neutral pH to release NO•, with half-lives of either seconds (PROLI-NONOate), minutes (DEA-NONOate), or hours (DETA-NONOate) (Keefer *et al.*, 1996). An important property of these compounds is that they release NO• independently of the biological system being studied. This is not the case for other NO• donors, for example, the *S*-nitrosothiols, which often require the presence of specific biological factors for NO• release to occur (Patel *et al.*, 1999). The concentration and exact type of NONOate used should be varied to produce an NO• release profile tailored to meet the experimental requirements. A typical NO• release profile from DPTA-NONOate is shown in Fig. 6, and consists of an initial positive slope followed by a flattening out to a steady-state plateau, where the rate of NO• consumption equals the rate of NO• production. Both the initial rate of NO• release and the steady-state level of NO• are dependent on the type and concentration of NONOate used, and of course on the pH of the buffer. Another advantage of NONOates is the ability to monitor their decomposition spectrophotometrically at 251 nm to obtain information on half-life and yields of NO• ($\varepsilon = 7680\,M^{-1}\,cm^{-1}$ for DETA-NONOate, and $7640\,M^{-1}\,cm^{-1}$ DPTA-NONOate).

Care should also be taken to ensure that the redox form of NO• released by the compound is the one desired (i.e., NO•, NO⁻, NO⁺). For example, Angelis' salt ($Na_2N_2O_3$) releases NO⁻ or HNO (Shiva *et al.*, 2004), whereas SIN-1 releases a mix of NO• and $O_2^{•-}$, resulting in formation of ONOO⁻. Furthermore, the other products formed on release of NO• from a donor may also have biological effects, which should be taken into consideration. For example, decay of GSNO forms GSH, and decay of spermine-NONOate forms spermine.

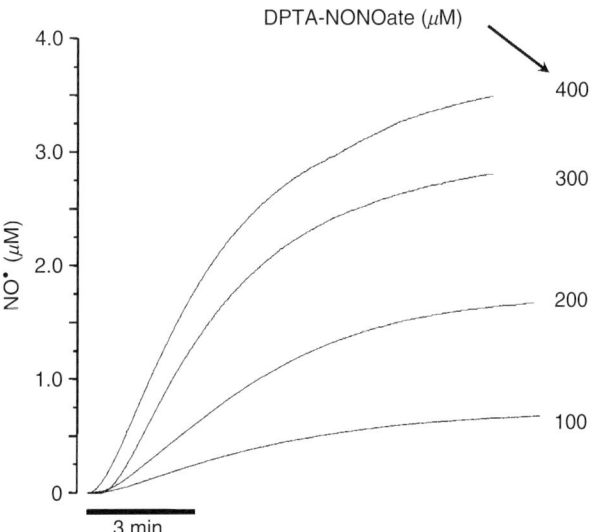

Fig. 6 Release of NO• from DPTA-NONOate in DMEM. Traces show NO• released from the indicated amounts of DPTA-NONOate, added from a 100-mM stock solution prepared in 10-mM NaOH. Recordings were made at 100% air saturation, and the electrode calibrated using authentic NO• as detailed in the text. Traces are representative of at least four identical experiments.

═══════ IV. NO• Threshold Measurement

A. Threshold Measurements in Isolated Mitochondria

As detailed in the introduction, the experimental derivation of mitochondrial respiratory thresholds consists essentially of two measurements: the effects of the inhibitor on an individual respiratory complex and the concomitant effects of the inhibitor on respiration. Procedures for measuring respiratory thresholds using commonly available pharmacological inhibitors are described in detail elsewhere (Brookes *et al.*, 2001; Mazat *et al.*, 1997). These include rotenone for complex I, thenoyltrifluoroacetone or malonate for complex II, myxothiazol or antimycin for complex III, and cyanide or azide for complex IV.

To measure the NO• threshold of cytochrome *c* oxidase, mitochondria are incubated in the electrode chamber in respiration buffer (Table I) at 1 mg protein/ml. Complex I-linked substrates (10-mM glutamate plus 2.5-mM malate) and ADP (200 μM) are added to initiate state 3 respiration. After a stable respiration rate is obtained and before O_2 saturation drops below 80% saturation, DETA-NONOate (0.3–1 mM) is added, resulting in release of NO• and a progressive inhibition of respiration. A typical combined O_2/NO• electrode trace is shown in Fig. 7A, with the addition of the NO• donor indicated by the arrow. The O_2 consumption rate (nmols O_2/min/mg protein) is determined at 1-min intervals along the trace. At these same time points, a line is dropped to the NO• trace, and [NO•] is determined, and in this manner a plot of [NO•] versus respiration can be constructed.

Since cytochrome *c* oxidase is the O_2-consuming enzyme of mitochondria, its activity can be assayed independently of the respiratory chain. This is achieved by replacing the respiratory substrates and ADP in the medium with ascorbate (1 mM), TMPD (50 μM), and FCCP (1 μM). Ascorbate plus TMPD is a reducing system that feeds electrons directly into cytochrome *c* oxidase (via cytochrome *c*), and FCCP is an uncoupler that dissipates the mitochondrial proton gradient, thereby allowing the H^+ pump of cytochrome *c* oxidase to function at maximal activity by removing feedback inhibition by the H^+ gradient. This effectively isolates cytochrome *c* oxidase from its respiratory chain environment, such that O_2 consumption rate = cytochrome *c* oxidase activity. DETA-NONOate is then added to the chamber as before, and both [NO•] and O_2 consumption rate (cytochrome *c* oxidase activity) are monitored. One potential complication of this system is the consumption of NO• by ascorbate due to auto-oxidation, which results in the measurement of lower-than-expected NO• concentrations by the electrode. However, we have found this ascorbate-dependent NO• consumption to be dependent on free metals, since it can be eliminated by addition of the chelator DTPA (100 μM) to the buffer.

In the same manner as for the respiration measurement, a plot of [NO•] versus cytochrome *c* oxidase activity can then be constructed. Finally, the threshold curve is constructed by combining the *y*-axes of the two NO• titration curves at corresponding concentrations of NO•, yielding a plot of cytochrome *c* oxidase

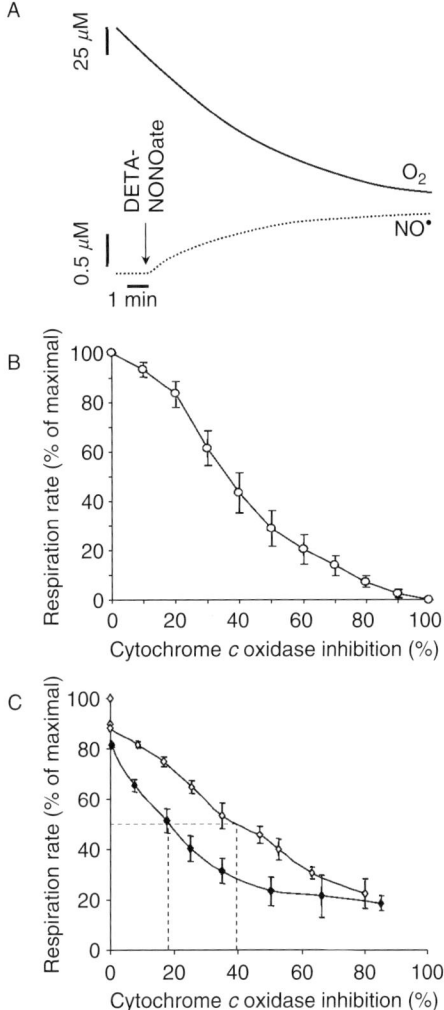

Fig. 7 Complex IV respiratory thresholds measured using NO•. Panel A: Rat heart mitochondria were incubated in the electrode chamber in respiration buffer at 1 mg/ml in the presence of glutamate, malate, and ADP (state 3 respiration, see text). At the indicated time, DETA-NONOate (0.2 mM) was added, and both NO• release and O_2 consumption monitored. Representative NO• and O_2 traces are shown. The experiment was then repeated in the presence of ascorbate/TMPD/FCCP (to measure isolated complex IV activity), and the threshold curve (Panel B) constructed from the combined NO• titration curves of respiration and enzyme activity. Panel C shows threshold curves for liver mitochondria from rats on a control (open) or ethanol (closed) diet. Dotted lines depict the IC_{50} for each group. Data are means ± SEM, $n = 6$.

activity versus respiration rate (Fig. 7B). The power of this technique is in the ability to construct a complete threshold curve from only two mitochondrial incubations, since a separate incubation is not required for each concentration of inhibitor (as would be the case, for example, if using cyanide as the inhibitor).

B. Altered Thresholds in Disease

This method can be applied to determine whether the regulation of mitochondrial function by $NO^\bullet$ or other mediators changes in models of disease. We have applied this technique to a rat model of chronic ethanol consumption, in which rats were fed a diet comprising 36% of calories derived from ethanol, or an isocaloric control diet with ethanol calories replaced by maltose-dextrin. After 6 weeks on this diet, liver mitochondria were isolated and threshold measurements were made (Venkatraman *et al.*, 2003). In this model, the respiration of mitochondria from ethanol-treated rats was more sensitive to $NO^\bullet$ than were control mitochondria. However, at low concentrations of $NO^\bullet$ there was no significant difference between the sensitivity of cytochrome *c* oxidase to inhibition by $NO^\bullet$. Thus, in this model, chronic ethanol consumption resulted in a decreased threshold for cytochrome *c* oxidase inhibition ($IC_{50} = 38 \pm 6\%$ for control vs $18 \pm 8\%$ for ethanol, Fig. 7C). In other words, the same degree of cytochrome *c* oxidase inhibition has a greater impact on overall respiration rates in the disease state (Venkatraman *et al.*, 2003). A similar effect was also noted in cardiac mitochondria from a rat aortic banding model of cardiac hypertrophy (Brookes *et al.*, 2001; Dai *et al.*, 2001).

Notably, in the model of cardiac hypertrophy, no alteration was seen in the maximal respiratory rate of mitochondria measured under optimal conditions (air-saturated buffer, high substrate concentrations, no inhibitors). Thus, simply measuring state 3 or 4 respiration under "normal" conditions can often lead to pathologic alterations in mitochondrial function being overlooked. Only by subjecting the isolated mitochondrial system to a physiological stress, such as $NO^\bullet$, can the bioenergetics defects be revealed.

C. Measurement of $NO^\bullet$ Thresholds in Cells

The inhibition of respiration by $NO^\bullet$ can also be measured in intact cells using the combined $NO^\bullet$ electrode/O_2 electrode/$NO^\bullet$ donor approach outlined above. While this is not a true respiratory complex threshold, the importance of measuring the dose response of cell respiration to $NO^\bullet$ should be stressed, since it can vary greatly under both physiological and pathological conditions (Thomas *et al.*, 2001). A physiological example is the respiratory state of mitochondria inside cells. In isolated mitochondria, it has been shown that $NO^\bullet$ is a more potent inhibitor of respiration in state 3 (ATP generating) than in state 4 (quiescent) (Borutaite and Brown, 1996). However, respiratory states do not strictly exist within cells, and mitochondria are thought to exist in "state 3.5." Thus, cellular activity and ATP consumption may have significant bearing on the control of respiration by $NO^{\bullet}$. In addition, variations in the number of mitochondria per cell, the amount and location of $NO^{\bullet}$ synthase isoforms, and the presence of species reactive toward $NO^{\bullet}$ (e.g., superoxide, myoglobin) may all affect $NO^{\bullet}$ delivery to the mitochondrion and thus affect the control of respiration by $NO^{\bullet}$ (Shiva et al., 2001). From a pathological perspective, we showed that the perturbation of the mitochondrial $NO^{\bullet}$-cytochrome c oxidase signaling pathway seen in cardiac hypertrophy (Brookes et al., 2001) is preserved at both the cell and whole-organ levels (Dai et al., 2001) such that cellular respiration and myocardial work output, respectively, are both more sensitive to $NO^{\bullet}$ in the hypertrophic state. Thus, profound defects in bioenergetics, ROS generation, and Ca^{2+} homeostasis can result from minor changes in respiratory thresholds and $NO^{\bullet}$ sensitivity.

When measuring $NO^{\bullet}$ effects in cells, several complications need to be considered. One example is the consumption of $NO^{\bullet}$ by components of the cell-culture media. Such consumption would decrease $NO^{\bullet}$ availability to the cells, and therefore calibration of the $NO^{\bullet}$ electrode should be performed in both the cell-culture medium and also in a non-$NO^{\bullet}$-consuming buffer, such as PBS, to quantify and correct for $NO^{\bullet}$ consumption by the medium. One mechanism by which cell-culture components may contribute to $NO^{\bullet}$ consumption is via the generation of ROS. It has previously been shown that this can be mediated by riboflavin (a common constituent in cell-culture media) in the presence of light (Grzelak et al., 2001). Therefore, it is important to conduct experiments under subdued light, as emphasized by the data in Fig. 5B showing that decay of $NO^{\bullet}$ is much steeper in ambient light versus dark conditions. This phenomenon is especially prevalent in, but not limited to, media containing albumin or proteins capable of forming nitrosothiols, since $NO^{\bullet}$ in its reactions with O_2 can form nitrosating agents leading to S-nitrosothiol formation, which then decay depending on experimental conditions. Since some of the most common mechanisms of $NO^{\bullet}$ consumption and S-nitrosothiol decay are dependent on metal ions, it is also recommended to include a metal chelator such as DTPA (100 μM) in the medium to limit such reactions (Patel et al., 1999). Similarly, addition of a wide-spectrum nitric oxide synthase (NOS) inhibitor (e.g., 1-mM l-nitroarginine) ensures that any effects of $NO^{\bullet}$ measured are from the exogenous known $NO^{\bullet}$, and not from endogenous sources such as eNOS.

Furthermore, since it is now becoming evident that classical $NO^{\bullet}$-dependent protein kinase signaling pathways (i.e., soluble guanylate cyclase/cGMP-dependent protein kinase) can impact on mitochondrial function, and possibly even phosphorylate mitochondrial proteins (Costa et al., 2005; Kim et al., 2004), it is important to test whether any effects of $NO^{\bullet}$ in the cellular system are mediated by such pathways. This can be accomplished either by trying to reproduce the mitochondrial effect using a cGMP analog (e.g., 8-Br-cGMP) or by seeing if the mitochondrial effect is sensitive to inhibitors of soluble guanylate cyclase (e.g., ODQ, 1H-[1,2,4]oxadiazole [4,3-a]quinoxalin-1-one) or cGMP-dependent protein kinase (e.g., KT5823).

The reaction between $NO^{\bullet}$ and O_2 also limits the ability to measure the effect of $NO^{\bullet}$ on cellular respiration (Liu et al., 1998a; Thomas et al., 2001). While physiological O_2 tension is estimated at 5–20 μM, most tissue-culture systems utilize atmospheric O_2 levels. It is not feasible to measure respiration at physiologically low O_2 tensions, since actively respiring cells would consume all of the O_2 in the respiration chamber within a short time period. One solution using the classical closed measurement system is to vary the level of NONOate donor added, to obtain inhibition of respiration at different O_2 tensions. As shown in Fig. 8, addition of 2-mM DETA-NONOate at 90% O_2 saturation results in a high rate of NO[•] release (trace iii), and an NO[•]–response curve that is right-shifted (Fig. 8B, open diamonds),

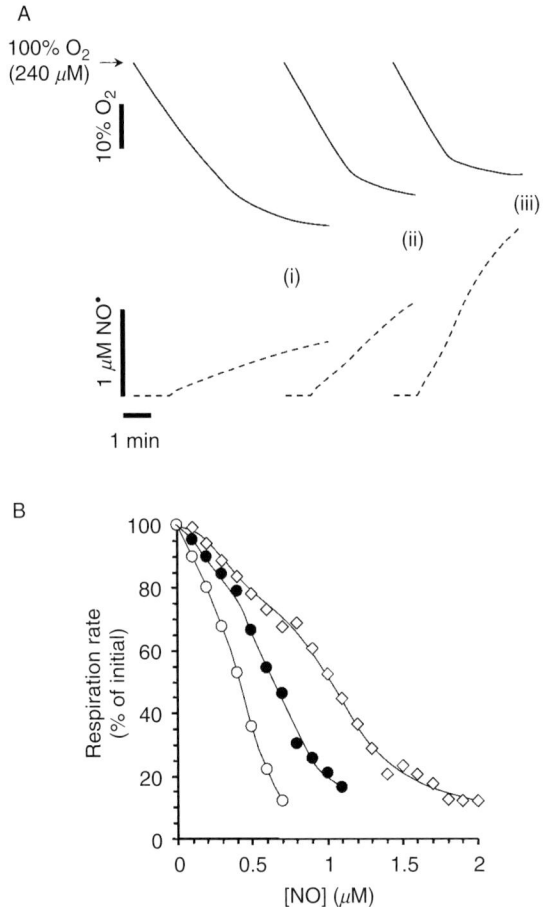

Fig. 8 Response of cell respiration to NO[•] and the effects of added NONOate level. Isolated adult rat ventricular cardiac myocytes were incubated in the combined electrode chamber in Krebs-Henseleit buffer plus 1-μM FCCP and 1-mM Ca^{2+}. Panel A shows representative O_2 consumption traces (solid lines) and NO[•] release traces (dotted lines) for (i) 0.6-mM, (ii) 1-mM, and (iii) 2-mM added DETA-NONOate. Panel B shows corresponding NO[•] titration curves calculated from the data in panel A. Open circles, 0.6-mM; closed circles, 1-mM; open diamonds, 2-mM DETA-NONOate. Traces and data are representative of at least five identical experiments.

since the inhibition takes place almost entirely at high O_2 tension where O_2 competes for $NO^\bullet$ binding at cytochrome c oxidase. By contrast, addition of 0.6-mM DETA-NONOate at the same 90% O_2 saturation level (trace i), results in a much slower rate of $NO^\bullet$ release such that the onset of inhibition of respiration is delayed until significant O_2 consumption has occurred. This results in an $NO^\bullet$–response curve that is left-shifted, since the inhibition takes place at a lower O_2 saturation, where-upon $NO^\bullet$ is a more potent inhibitor of respiration (Fig. 8B, open circles) (Brown, 2000). These results serve to highlight the effects of O_2 tension on $NO^\bullet$-inhibition of mitochondrial respiration (Cleeter et al., 1994; Koivisto et al., 1997). Another solution to the problem of measuring respiration at low oxygen tension is to utilize an open-flow type of apparatus (described below) in which a low O_2 tension is maintained by gas replenishment using mass flow controllers (Brookes et al., 2003; Cole et al., 1982).

V. Open–Flow Respirometry

In order to measure respiration in an open-flow system at controlled low O_2 tensions (Fig. 9), respiration buffer and mitochondria are added to the chamber, and then O_2 in the chamber is decreased, either by allowing mitochondria to respire or by purging with 100% N_2 gas. Once the liquid phase O_2 (measured by the O_2 electrode) is slightly higher than the desired O_2 tension, the headspace gas flow is adjusted to an empirically determined mixture of N_2/O_2 to maintain the desired O_2 tension. For example, in our system (Brookes et al., 2003), to maintain 1 mg of rat liver mitochondria respiring in state 3 (ADP turnover) on complex I substrates (glutamate + malate) at 5-μM O_2 required 39% air/61% N_2 in the headspace gas. While the mixture required will vary based on the respiratory rate of the mitochondria and the specific geometry of the system, a steady state will eventually be achieved at which O_2 consumption is equal to O_2 delivery. Once this steady state is maintained for 3–5 min, a bolus of $NO^\bullet$ is added to inhibit respiration, and both O_2 and NO traces are recorded until the $[NO^\bullet]$ returns to baseline (Fig. 9). Inhibition of respiration in this system results in an increase in the absolute concentration of O_2 in the chamber (since O_2 consumption is inhibited but O_2 delivery continues). The reversal of $NO^\bullet$-mediated inhibition then permits the return of $[O_2]$ to its original steady-state value.

In order to construct an inhibition curve from this type of trace, it is necessary to apply the following equation to calculate the respiration rate (Q) at various time points during the inhibition and recovery of respiration:

$$Q = m(C^* - C_1) - \frac{dC_1}{dt} \tag{1}$$

The principles behind this equation are rooted in the Fick equation (Sibitz, 1966), wherein the steady-state concentration of a species is the rate of its generation

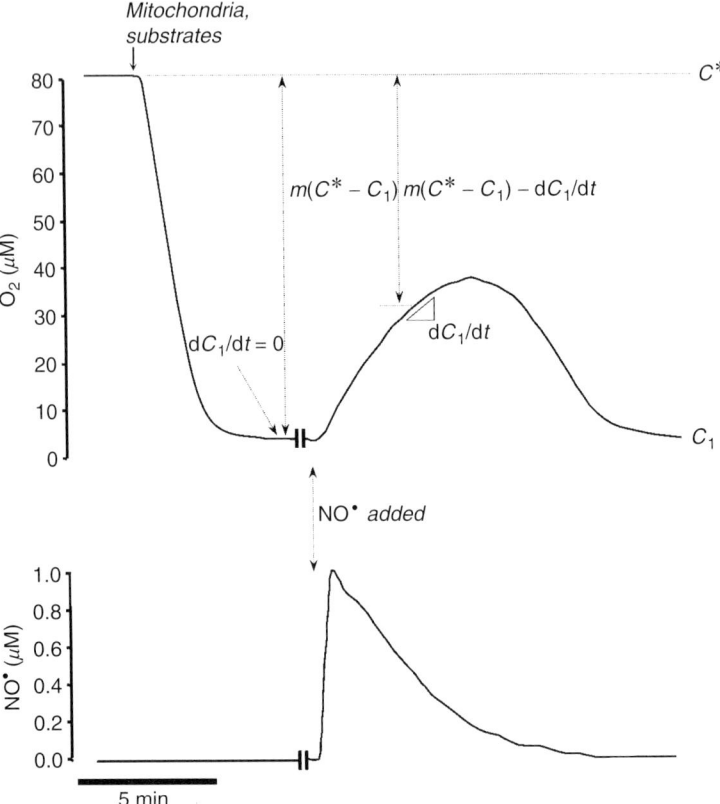

Fig. 9 Calculation of mitochondrial respiration rates using open-flow respirometry, at steady- and non-steady-state [O_2]. A typical set of traces for O_2 (upper panel) and NO$^•$ (lower panel) during state 3 respiration are shown. The predicted liquid phase [O_2] is termed C^*, and is based on the O_2 content of the headspace gas (dotted line across the top, see text). C_1 is the measured liquid phase [O_2], from the O_2 sensor (solid line). Addition of mitochondria and respiratory substrates (as shown by the arrow) results in establishment of a new steady-state value for C_1. At this steady state, the rate of mitochondrial respiration is calculated using the equation $Q = m(C^* - C_1)$. On inhibition of respiration by NO$^•$, C_1 rises transiently. During the NO$^•$ inhibition, dC_1/dt is factored into the rate calculation, as shown. Following disappearance of NO$^•$, respiration recovers and C_1 returns to baseline levels. See text for full explanation.

minus the rate of its decay. The full derivation of this equation is given in our previous publication on this topic (Brookes *et al.*, 2003). In Eq. (1), m is the mass transfer coefficient, which is the first-order rate constant for the transfer of O_2 between the gas and liquid phases. It will vary depending on the geometry of the open-flow system (e.g., stir speed, surface area, temperature), and therefore m must be determined empirically on the day of each experiment. This is done by equilibrating the liquid phase with a 100% O_2 gas phase, and then switching rapidly to 100% N_2. Plotting Ln[O_2] versus time (in seconds), the slope is m, with units of sec^{-1}.

C^* is the predicted liquid phase O_2 tension in the absence of mitochondria, and again is determined empirically for each headspace gas mixture. For example at sea level, mitochondrial respiration buffer at $37\,^\circ$C, equilibrated with room air (21% O_2), will have an $[O_2]$ of $\sim$195 μM. C_1 is the measured liquid phase O_2 at the exact time point the rate (Q) is being measured. dC_1/dt is the rate of change of C_1, that is, the slope of a tangent to the O_2 (C_1) trace, at the exact time point the rate (Q) is being measured. As seen in Fig. 9, at steady state, dC_1/dt is zero, and so cancels out from the equation, leaving the simplified equation $Q = m(C^*{-}C_1)$, which rearranges to give $C_1 = mC^*{-}Q$, that is, the simplified Fick equation wherein the steady-state concentration of O_2 (i.e., C_1) equals the rate of its generation (O_2 diffusion into the liquid phase, i.e., mC^*) minus the rate of its decay (respiration, i.e., Q).

By using the open-flow system, coupled with the equations described above, the rate of mitochondrial respiration in response to NO$^\bullet$ can be calculated at any O_2 tension, under tightly controlled conditions closely mimicking those *in vivo*. Using such a system, we were able to demonstrate that a previously documented (Borutaite and Brown, 1996) greater sensitivity of state 3 respiration to NO$^\bullet$ (vs state 4) finds its origins not in an experimental artifact related to the performance of these experiments at different O_2 tensions, but is rooted in the differential control that cytochrome c oxidase has over respiration in each respiratory state. Simply put, state 3 respiration is more sensitive to NO$^\bullet$ because the target of NO$^\bullet$ (cytochrome c oxidase) has more *control* in state 3 (Groen *et al.*, 1982).

VI. An *In Vitro* Model of Ischemia/Reperfusion Injury

A great deal of attention is now focused on the role of the mitochondrion during pathogenesis, and as outlined above, respirometry is an important tool for measuring changes in mitochondrial function in pathogenesis. However, in the case of ischemia/reperfusion (I/R) injury, the respirometer can provide an interesting *in vitro* model in which I/R-induced mitochondrial damage can be mimicked (Ozcan *et al.*, 2001). I/R-induced mitochondrial injury is characterized by the increased production of ROS, matrix Ca^{2+} overload, decreased ATP production and respiratory rate, and opening of the mitochondrial permeability transition (PT) pore [see Brookes *et al.* (2004) for review]. To induce these I/R-dependent changes *in vitro*, isolated mitochondria are suspended in the classical closed system respirometer and are supplied with sufficient substrate to respire in state 3. Once the chamber is sealed, the $[O_2]$ in the chamber decreases steadily due to respiration, until the chamber becomes anaerobic. The mitochondria are left in this anoxic condition for 30–45 min to model ischemia. After the desired anoxic period, the mitochondria are reoxygenated (to model reperfusion) by opening the lid, by directly introducing 100% oxygen into the chamber, or by centrifuging the mitochondria and resuspending them in fresh, oxygenated buffer. Sufficient substrate is then added to the reoxygenated mitochondria and the respiratory rate is measured once more (Fig. 10).

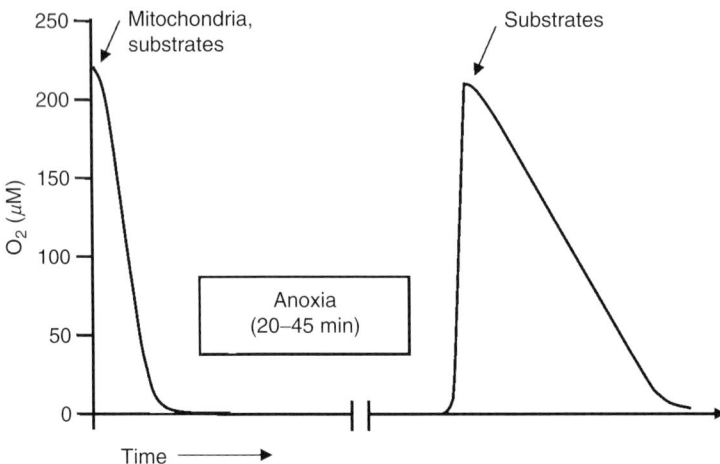

Fig. 10 *In vitro* model of mitochondrial ischemia/reperfusion. Typical O_2 trace showing preanoxic respiratory rate, 20–45 min of anoxia in the chamber, reoxygenation, and postanoxic respiratory rate. Data shown are for rat liver mitochondria.

Comparison of the respiratory rate before and after the anoxic period shows that mitochondrial respiration rate is substantially decreased after anoxia due to I/R-induced damage. This is a useful model to test treatments that potentially protect mitochondria from I/R injury, as such treatments given during the anoxic period would significantly increase postanoxic respiration rate in comparison to no treatment. For example, Terzic's group, previously used this model to test the effects of diazoxide, a mitochondrial K_{ATP}^{+} channel opener known to protect against I/R injury *in vivo* (Liu *et al.*, 1998b), on isolated mitochondrial function during I/R (Ozcan *et al.*, 2001). In control mitochondria after 20 min of anoxia, ADP-dependent respiration rate was 52% of the preanoxic rate. However, with 100-μM diazoxide present during the anoxic period, postanoxic respiration rate was significantly higher than in the control mitochondria ($\sim$87% of preanoxic respiration rate), suggesting protection of mitochondrial function. Clearly, given the growing realization that the mitochondrion is a target for some of the protective effects of NO• and its redox congeners, such as nitrite, during ischemia (Burwell *et al.*, 2006; Duranski *et al.*, 2005; Webb *et al.*, 2004), the use of such an isolated mitochondrial system can yield critical information regarding protective mechanisms.

VII. Conclusion

The methods described herein serve to illustrate not only the importance of measuring NO• thresholds in mitochondria and cells, but they also highlight many of the considerations that must be taken when making such measurements.

These include the presence of O_2, the source of $NO^\bullet$, the precise design of the apparatus, and experimental conditions such as buffer composition, light level, and temperature. When all these factors are considered, accurate interpretations of the role of $NO^\bullet$ in controlling mitochondrial function can be made. Such experiments can provide insight into intricate nature of the interaction between $NO^\bullet$ and mitochondria, and the variations in this system that can occur both under physiological and pathological conditions.

Acknowledgments

Support for the research described herein was provided by NIH grants AA13395 and HL070610 (to VDU), and HL071158 (to PSB). Certain parts of this article were adapted from an earlier chapter in Methods in Enzymology (Brookes *et al.*, 2002), with permission.

References

Borutaite, V., and Brown, G. C. (1996). Rapid reduction of nitric oxide by mitochondria, and reversible inhibition of mitochondrial respiration by nitric oxide. *Biochem. J.* **315**(Pt. 1), 295–299.

Brookes, P. S., Kraus, D. W., Shiva, S., Doeller, J. E., Barone, M. C., Patel, R. P., Lancaster, J. R., Jr., and Darley-Usmar, V. (2003). Control of mitochondrial respiration by $NO^\bullet$, effects of low oxygen and respiratory state. *J. Biol. Chem.* **278**, 31603–31609.

Brookes, P. S., Shiva, S., Patel, R. P., and Darley-Usmar, V. M. (2002). Measurement of mitochodrial respiratory thresholds and the control of respiration by nitric oxide. *Methods Enzymol.* **359**, 305–319.

Brookes, P. S., Yoon, Y., Robotham, J. L., Anders, M. W., and Sheu, S. S. (2004). Calcium, ATP, and ROS: A mitochondrial love-hate triangle. *Am. J. Physiol Cell Physiol.* **287**, C817–C833.

Brookes, P. S., Zhang, J., Dai, L., Zhou, F., Parks, D. A., Darley-Usmar, V. M., and Anderson, P. G. (2001). Increased sensitivity of mitochondrial respiration to inhibition by nitric oxide in cardiac hypertrophy. *J. Mol. Cell. Cardiol.* **33**, 69–82.

Brown, G. C. (2000). Nitric oxide as a competitive inhibitor of oxygen consumption in the mitochondrial respiratory chain. *Acta Physiol. Scand.* **168**, 667–674.

Burwell, L. S., Nadtochiy, S. M., Tompkins, A. J., Young, S., and Brookes, P. S. (2006). Direct evidence for S-nitrosation of mitochondrial complex I. *Biochem. J.* **394**, 627–634.

Cleeter, M. W., Cooper, J. M., Darley-Usmar, V. M., Moncada, S., and Schapira, A. H. (1994). Reversible inhibition of cytochrome c oxidase, the terminal enzyme of the mitochondrial respiratory chain, by nitric oxide. Implications for neurodegenerative diseases. *FEBS Lett.* **345**, 50–54.

Cole, R. P., Sukanek, P. C., Wittenberg, J. B., and Wittenberg, B. A. (1982). Mitochondrial function in the presence of myoglobin. *J. Appl. Physiol.* **53**, 1116–1124.

Costa, A. D., Garlid, K. D., West, I. C., Lincoln, T. M., Downey, J. M., Cohen, M. V., and Critz, S. D. (2005). Protein kinase G transmits the cardioprotective signal from cytosol to mitochondria. *Circ. Res.* **97**, 329–336.

Dai, L., Brookes, P. S., Darley-Usmar, V. M., and Anderson, P. G. (2001). Bioenergetics in cardiac hypertrophy: Mitochondrial respiration as a pathological target of NO. *Am. J. Physiol. Heart Circ. Physiol.* **281**, H2261–H2269.

Duranski, M. R., Greer, J. J., Dejam, A., Jaganmohan, S., Hogg, N., Langston, W., Patel, R. P., Yet, S. F., Wang, X., Kevil, C. G., Gladwin, M. T., and Lefer, D. J. (2005). Cytoprotective effects of nitrite during *in vivo* ischemia-reperfusion of the heart and liver. *J. Clin. Invest.* **115**, 1232–1240.

Giuffre, A., Barone, M. C., Mastronicola, D., D'Itri, E., Sarti, P., and Brunori, M. (2000). Reaction of nitric oxide with the turnover intermediates of cytochrome c oxidase: Reaction pathway and functional effects. *Biochemistry* **39**, 15446–15453.

Groen, A. K., Wanders, R. J., Westerhoff, H. V., Van der Meer, R., and Tager, J. M. (1982). Quantification of the contribution of various steps to the control of mitochondrial respiration. *J. Biol. Chem.* **257,** 2754–2757.

Grzelak, A., Rychlik, B., and Bartosz, G. (2001). Light-dependent generation of reactive oxygen species in cell culture media. *Free Radic. Biol. Med.* **30,** 1418–1425.

Keefer, L. K., Nims, R. W., Davies, K. M., and Wink, D. A. (1996). "NONOates" (1-substituted diazen-1-ium-1,2-diolates) as nitric oxide donors: Convenient nitric oxide dosage forms. *Methods Enzymol.* **268,** 281–293.

Kim, J. S., Ohshima, S., Pediaditakis, P., and Lemasters, J. J. (2004). Nitric oxide: A signaling molecule against mitochondrial permeability transition- and pH-dependent cell death after reperfusion. *Free Radic. Biol. Med.* **37,** 1943–1950.

Koivisto, A., Matthias, A., Bronnikov, G., and Nedergaard, J. (1997). Kinetics of the inhibition of mitochondrial respiration by NO. *FEBS Lett.* **417,** 75–80.

Liu, X., Miller, M. J. S., Joshi, M. S., Thomas, D. D., and Lancaster, J. R., Jr. (1998a). Accelerated reaction of nitric oxide with O_2 within the hydrophobic interior of biological membranes. *Proc. Natl. Acad. Sci. USA* **95,** 2175–2179.

Liu, Y., Sato, T., O'Rourke, B., and Marban, E. (1998b). Mitochondrial ATP-dependent potassium channels: Novel effectors of cardioprotection? *Circulation* **97,** 2463–2469.

Mazat, J. P., Letellier, T., Bedes, F., Malgat, M., Korzeniewski, B., Jouaville, L. S., and Morkuniene, R. (1997). Metabolic control analysis and threshold effect in oxidative phosphorylation: Implications for mitochondrial pathologies. *Mol. Cell. Biochem.* **174,** 143–148.

Muller, F. L., Liu, Y., and Van, R. H. (2004). Complex III releases superoxide to both sides of the inner mitochondrial membrane. *J. Biol. Chem.* **279,** 49064–49073.

Ozcan, C., Holmuhamedov, E. L., Jahangir, A., and Terzic, A. (2001). Diazoxide protects mitochondria from anoxic injury: Implications for myopreservation. *J. Thorac. Cardiovasc. Surg.* **121,** 298–306.

Patel, R. P., McAndrew, J., Sellak, H., White, C. R., Jo, H., Freeman, B. A., and Darley-Usmar, V. M. (1999). Biological aspects of reactive nitrogen species. *Biochim. Biophys. Acta* **1411,** 385–400.

Shiva, S., Brookes, P. S., Patel, R. P., Anderson, P. G., and Darley-Usmar, V. M. (2001). Nitric oxide partitioning into mitochondrial membranes and the control of respiration at cytochrome c oxidase. *Proc. Natl. Acad. Sci. USA* **98,** 7212–7217.

Shiva, S., Crawford, J. H., Ramachandran, A., Ceaser, E. K., Hillson, T., Brookes, P. S., Patel, R. P., and Darley-Usmar, V. M. (2004). Mechanisms of the interaction of nitroxyl with mitochondria. *Biochem. J.* **379,** 359–366.

Sibitz, G. R. (1966). Substance uptake in variable systems with special reference to the use of the Fick equation. *Respir. Physiol.* **2,** 118–128.

Stamler, J. S., and Feelisch, M. (1996). "Methods in Nitric Oxide Research." John Wiley and Sons, New York.

Thomas, D. D., Liu, X., Kantrow, S. P., and Lancaster, J. R., Jr. (2001). The biological lifetime of nitric oxide: Implications for the perivascular dynamics of NO and O_2. *Proc. Natl. Acad. Sci. USA* **98,** 355–360.

Torres, J., Darley-Usmar, V., and Wilson, M. T. (1995). Inhibition of cytochrome c oxidase in turnover by nitric oxide: Mechanism and implications for control of respiration. *Biochem. J.* **312**(Pt. 1), 169–173.

Venkatraman, A., Shiva, S., Davis, A. J., Bailey, S. M., Brookes, P. S., and Darley-Usmar, V. M. (2003). Chronic alcohol consumption increases the sensitivity of rat liver mitochondrial respiration to inhibition by nitric oxide. *Hepatology* **38,** 141–147.

Webb, A., Bond, R., McLean, P., Uppal, R., Benjamin, N., and Ahluwalia, A. (2004). Reduction of nitrite to nitric oxide during ischemia protects against myocardial ischemia-reperfusion damage. *Proc. Natl. Acad. Sci. USA* **101,** 13683–13688.

CHAPTER 21

Methods for Determining the Modification of Protein Thiols by Reactive Lipids

JooYeun Oh★,†, Michelle S. Johnson★,†, and Aimee Landar★,†

★Department of Pathology
University of Alabama at Birmingham
Birmingham, Alabama 35294

†Center for Free Radical Biology
University of Alabama at Birmingham
Birmingham, Alabama 35294

I. Introduction

A. Formation of Reactive Lipids in Physiology and Pathology

It is becoming increasingly evident that oxidized lipid mediators are an important class of small molecule second messengers in both physiological and pathological settings (Leitinger, 2005; Zmijewski *et al.*, 2005). These lipids are formed in response to a stimulus by processes that are ultimately controlled by metabolic or signal transduction proteins (Kagan *et al.*, 2006; Montine and Morrow, 2005; Shibata *et al.*, 2002).

The mechanisms of formation of lipid mediators can be grouped loosely into two categories: (1) lipids formed by enzymatic action and (2) lipids formed from nonspecific lipid peroxidation. Examples of enzymes, which oxidize lipids, include lipoxygenase and cyclooxygenase proteins which are often activated in response to inflammation. For example, the primary product of cyclooxygenase activity, prostaglandin H_2, is modified by downstream synthases to other products, some of which can undergo further oxidation to produce reactive lipids such as prostaglandins of the A and J series (Shibata *et al.*, 2002; Straus and Glass, 2001). Lipoxygenase 12/15 is also capable of generating lipid mediators (Kuhn and Borchert, 2002).

A number of proteins, including myeloperoxidase (Zhang *et al.*, 2002) and some cytochrome P450 isoforms (Caro and Cederbaum, 2006) have been shown to generate lipid oxidation products, although the biological function of these lipids is not as clear as for lipoxygenases and cyclooxygenases. An example of reactive lipids formed by nonspecific lipid peroxidation can be found in disease processes such as atherosclerosis and diabetes, where proteins such as nitric oxide synthase, NADPH oxidase, and dysfunctional mitochondrial respiratory complexes generate increased reactive oxygen and nitrogen species (ROS/RNS) (Niki *et al.*, 2005). These ROS/RNS can react with membrane lipids, causing lipid peroxidation, nitration, and the formation of a spectrum of lipids, some of which have biological activity (Coles *et al.*, 2002; Landar *et al.*, 2006c; Musiek *et al.*, 2005).

The oxidized lipids derived from either process may contain reactive functional groups which are capable of the posttranslational modification of proteins, and in this way, may affect protein function and elicit specific biological responses (Dickinson *et al.*, 2006; Landar *et al.*, 2006a). A major challenge in the field has been to develop methods to follow the fate of lipids and to determine the molecular mechanisms through which they mediate their biological effects. For example, the use of radioactivity to monitor these processes has been hampered by the high cost of custom radiolabeled lipid synthesis, as well as the requirement for specialized laboratory techniques to perform the experiments. However, over the last decade, avidin-biotin technology has proven to be an affordable and convenient alternative to radioactivity. Here, we describe some of the approaches we have recently developed to monitor and quantitate the formation of protein adducts with reactive lipids, using biotin-labeling strategies.

B. Electrophilic Lipids as Reactive Lipid Mediators

There are two major classes of reactive lipids that have been shown to covalently modify proteins and alter protein function in pathological situations: lipid aldehydes and electrophilic lipids. Of these reactive lipids, electrophilic lipids are particularly interesting because they can modify a number of different proteins,

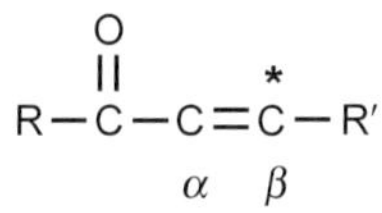

$$\overset{\displaystyle O}{\underset{\alpha \quad \beta}{\overset{\displaystyle \|}{R-C-C=C-R'}}}$$

Fig. 1 Structure of an α,β-unsaturated carbonyl group. Carbons in the α- and β-positions relative to the carbonyl are indicated. The electrophilic β-carbon is also marked with an asterisk. $R = H$ (aldehyde) or alkyl group (ketone); $R' = H$ or alkyl group.

producing a collection of modified proteins that have effects on cells, including adaptation to stress and apoptosis, and are referred to as the "electrophile responsive proteome" (Ceaser *et al.*, 2004). Electrophilic lipids contain an electrophilic carbon which is the β-carbon of an α,β-unsaturated carbonyl (Fig. 1). They are relatively stable *in vivo*, thereby allowing the diffusion of the lipid to a suitably reactive protein target.

C. Reactivity of Electrophilic Lipids with Protein Thiols and the Role of Mitochondria

Electrophilic lipids react with nucleophilic protein residues by Michael addition, forming a lipid–protein adduct. Amino acids containing nucleophilic side chains include cysteine, histidine, and lysine, where the nucleophiles are the deprotonated forms of the thiol, imidazole, and ε-amino group, respectively. The thiol group of cysteine is the most reactive of the nucleophilic groups because thiols having a low pK_a are deprotonated at physiological pH. Indeed, it has been hypothesized that electrophilic lipids interact primarily with this subset of cysteine residues. In this way, these lipids may control "thiol switches" which change the function of proteins in key cellular pathways. Our work using fluorescently tagged electrophilic lipids has demonstrated that they colocalize to the mitochondrion in endothelial cells, suggesting that mitochondrial thiols may be important targets for this class of reactive lipids (Landar *et al.*, 2006c). Here, we describe biotin-based techniques to measure protein thiols and adduct formation with electrophilic lipids.

II. Rationale

Biotin-based tagging techniques have also been applied with success to monitor the posttranslational modification of reactive protein thiols in cell signaling proteins and enzymes by ROS/RNS and electrophilic compounds (Dennehy *et al.*, 2006; Eaton *et al.*, 2003; Hao *et al.*, 2006; Jaffrey and Snyder, 2001; Kim *et al.*, 2000; Landar *et al.*, 2006b). One of the most versatile techniques is to attach biotin covalently to thiol-reactive molecules, and label thiol-containing proteins. After separation by proteomics methods, the biotin tag can be detected at a level of sensitivity in the picomole range, using Western blot analysis and horseradish peroxidase (HRP)-coupled streptavidin (Landar *et al.*, 2006b). In this case, the loss of the biotin signal is proportional to the degree of thiol modification. This method allows for detection and quantitation of the level of protein modification. Quantitation is important because the degree of posttranslational modification is generally an important indicator of the biological impact of regulation through this mechanism.

Another method to detect modification of thiols in proteins by reactive lipids is to label the lipid with biotin or another tag, incubate the labeled lipid with cells under different experimental conditions (e.g., high or low oxidative stress), and monitor the covalent attachment of labeled lipids into proteins using Western blot analysis. This method has been used for detection of adducts formed between electrophilic lipids, such as 15-deoxy-$\Delta^{12,14}$-prostaglandin J_2, and proteins in endothelial cells (Landar *et al.*, 2006c). Table I shows the lipids that we have labeled successfully. The tags generally contain a primary amine or hydrazide functional group and are designed to be used in a carbodiimide-mediated condensation reaction, which results in the formation of an amide bond with the loss of water.

In this chapter, we will describe in detail the techniques involved in the preparation, characterization, and quantitation of biotin-tagged lipids, as well as the use of biotin reagents to analyze thiols in biological samples. There are four major advantages of biotin: (1) the high affinity of avidin and streptavidin for biotin ($K_d \approx 10^{-15}$ M) (Green, 1975, 1990) makes this a very sensitive and specific way to detect the tag; (2) the binding of streptavidin, unlike an antibody, is not readily affected by flanking residues at the site of protein modification; (3) affinity resins are available for the purification of lipid–protein adducts; and (4) the biotin tag can be quantitated easily and accurately using biotinylated standards.

However, there are some caveats to using biotin as a tag which must also be considered. For example, in contrast to antibody-based methods, the exceptionally

Table I
Tagged Lipids Prepared in Our Laboratory

Lipid	Tag	Reference
15-Deoxy-$\Delta^{12,14}$-prostaglandin J_2	Biotin pentylamine[a]	Levonen *et al.*, 2004
	BODIPY-FL-EDA[b]	Landar *et al.*, 2006b
	Cy5 hydrazide[c]	NP[*]
15-A_2-Isoprostane	Biotin pentylamine[a]	Levonen *et al.*, 2004
Prostaglandin E_2	BODIPY-FL-EDA[b]	Landar *et al.*, 2006b
Arachidonic acid	Biotin pentylamine[a]	NP[*]
	BODIPY-FL-EDA[b]	NP[*]

Source of tagging reagents:
[a]Pierce.
[b]Molecular Probes, Inc.
[c]Amersham.
[*]NP: not published.

strong binding of streptavidin to biotin results in an association that is practically (though not formally) irreversible. The technical implications are that Western blots using enzyme-conjugated avidin/streptavidin as a means of detection cannot be stripped after development. In addition, affinity purification of biotinylated lipid–protein adducts requires extremely harsh conditions, which may not be suitable for downstream applications such as mass spectrometry. Since some proteins, such as carboxylases, have biotin covalently attached as an enzymatic cofactor, these proteins can give false positive results, and appropriate controls must be used to account for this during data interpretation and quantitation.

III. Methods

A. Biotin Labeling of Free Protein Thiols Using Biotinylated Iodoacetamide

Biotinylated iodoacetamide (BIAM) is a reagent that effectively labels free thiols on proteins with biotin. One of the original studies to use this approach in redox signaling examined proteins which had decreased labeling by BIAM in response to hydrogen peroxide (Kim *et al.*, 2000). BIAM only reacts with deprotonated thiols (thiolate anion), and depending on the pK_a of the thiol group, not all protein thiols will react with BIAM at physiological pH. We provide two protocols for labeling. The high-pH protocol will deprotonate and allow BIAM to label most free protein thiols. This is useful for global analysis and screening of a wide population of thiol containing proteins. The low-pH protocol is useful for analyzing the redox state of a subset of highly reactive protein thiols. Due to their low pK_a, they will remain deprotonated and will be susceptible to biotin tagging at low pH.

Materials

- Biological sample, such as isolated mitochondrial pellets (1 mg each).
- Low-pH labeling buffer (pH 6.5):

    50-mM 2-(*N*-morpholino)ethanesulfonic acid (MES), pH to 6.5 with 1-N NaOH (pH 6.5)

    0.5% Triton X-100 (v/v)

    1 tablet Complete Mini™ protease inhibitor cocktail (Roche Applied Science, Indianapolis, IN) per 10 ml of buffer (add immediately before use).

- High-pH labeling buffer (pH 8.5):

    10-mM Tris–HCl (pH 8.5)

    1% Triton X-100 (v/v)

    1 tablet Complete Mini™ protease inhibitor cocktail (Roche Applied Science) per 10 ml of buffer (add immediately before use).

- BIAM [*N*-(biotinoyl)-*N*-(iodoacetyl)ethylenediamine]: Prepare a stock solution by dissolving in dimethylformamide to 1 mM. Solution should be prepared immediately before use and protected from light.
- *β*-Mercaptoethanol: The concentration of undiluted *β*-mercaptoethanol is 14.3 M. Prepare a 500-mM stock solution in water.

Procedure

For a 1-mg mitochondrial pellet, add 0.5-ml labeling buffer with the desired pH. Incubate on ice for 10 min, vortexing two to three times during incubation. Add 55 μl of 1-mM stock solution of BIAM to mitochondria (100-μM final concentration of BIAM), vortex, and incubate 15 min at RT in the dark. Labeling is terminated by addition of 1/50 volumes of 500-mM stock solution of *β*-mercaptoethanol to a final concentration of 20 mM. Vortex and analyze labeling by one- or two-dimensional polyacrylamide gel electrophoresis (1D- or 2D-PAGE) followed by blotting to detect biotin tag (Section III.D). This procedure can be adapted for other types of samples such as cultured cells. In this case, phosphate buffered saline (PBS)-washed cells are lysed with the desired pH buffer immediately before adding the BIAM reagent.

B. Synthesis of a Biotinylated Electrophilic Lipid

We developed a protocol to conjugate biotin to the carboxylic acid group of the lipid, which allows for the analysis of covalent modification of proteins by electrophilic lipids. This method was adapted from two previously published methods which describe the biotinylation of prostaglandin A_2 (Parker, 1995) and 15-deoxy-$\Delta^{12,14}$-prostaglandin J_2 (15d-PGJ$_2$) (Cernuda-Morollon *et al.*, 2001).

1. Biotinylation of Lipid

Materials

- Stock solutions: All solvents should be of HPLC grade or better, and should be sparged with N_2 (or other inert gas) prior to use. To avoid oxidation of lipids, exposure to room air and light should be minimized. For example, fill headspace with argon during reactions, use brown glass vials, or wrap with foil, and so on.
- 1-mg vials of 15d-PGJ$_2$ (Cat. No. 18570, Cayman Chemical, Ann Arbor, MI). The solvent is evaporated under N_2 and the neat oil immediately resuspended in 600 μl of N_2-sparged acetonitrile.
- EZ-Link 5-(Biotinamido)pentylamine (Cat. No. 21345, Pierce, Rockford, IL). The stock solution is prepared to a concentration of 4 mg/ml in 77% acetonitrile:23% water (v/v).

- EDC (1-ethyl-3-[3-dimethylaminopropyl]carbodiimide hydrochloride) is prepared to a final concentration of 5 mg/ml in 100% acetonitrile. EDC should be stored under N_2 with precautions to avoid moisture. All solutions are prepared fresh and used immediately.

Procedure

In a brown glass vial combine in order:

600 μl of 15d-PGJ$_2$ (1 mg)
200 μl of EDC (1 mg)
200 μl EZ-Link biotin (0.8 mg)

Fill headspace with argon and close vial with Teflon-lined cap. Incubate reaction for 18 h at RT in the dark with constant gentle agitation.

2. HPLC Purification of Biotin–15d-PGJ$_2$

Materials

- HPLC system: BioCad Perfusion Chromatography system or other HPLC system equipped with ultraviolet/visible spectrophotometer and fraction collector.
- Preparative column: 10-μm Gemini C18 reversed phase column (Phenomenex, Sutter Creek, CA) with column dimensions of $250 \times 21.2 \, \text{mm}^2$ (column volume = 88.2 ml) using a flow rate of 20 ml/min.
- Mobile phase: Solvent [A] [10% acetonitrile/0.24% acetic acid/90% H_2O (v/v)]; Solvent [B] (100% acetonitrile). Acetonitrile should be HPLC grade.

Procedure

The column is equilibrated with Solvent A. The biotinylated lipid preparation is injected through a sample loop and 100% Solvent A is used to wash the column and flush the loop for 2 column volumes. Elution of lipid occurs over a linear gradient of 5% Solvent [A] to 95% Solvent [B] over 3 column volumes (Fig. 2A). Collect 5-ml fractions into glass tubes; fractions exhibiting absorbance at 306 nm are screened by electrospray mass spectrometry for the product (MW of biotin-15d-PGJ$_2$ is 626.90). In our experience, the product elutes in the second major peak (peak b, Fig. 2A and B).

3. Extraction of Biotinylated Lipid Product After HPLC

Materials

Acetonitrile (HPLC grade)
Methanol (HPLC grade)
Chloroform (HPLC grade, preserved with 0.75% ethanol)

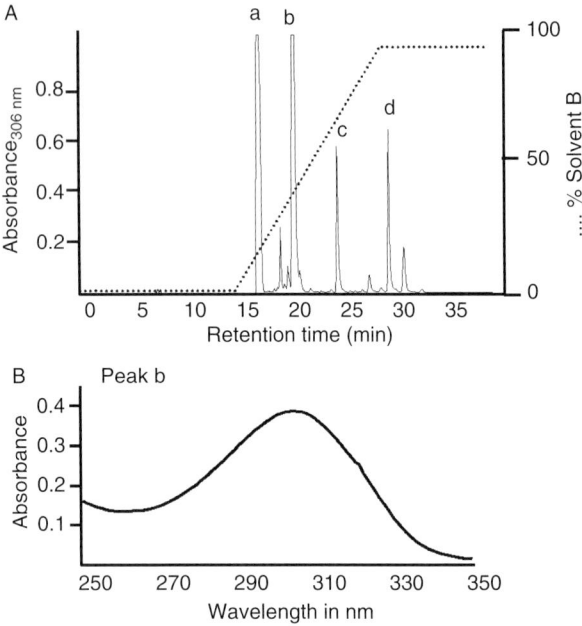

Fig. 2 HPLC purification of biotinylated 15d-PGJ$_2$. (A) HPLC chromatogram showing the elution of peaks exhibiting absorbance at 306-nm wavelength. (B) Wavelength scan of peak b containing the biotinylated 15d-PGJ$_2$. The molecular weight of the product was confirmed by mass spectrometry (not shown).

Acetic acid (certified ACS plus grade)

Ethanol (100%, molecular biology grade)

Procedure

To extract the biotinylated lipid, fractions containing the biotin-15d-PGJ$_2$ peak are combined and mixed with an equal volume of 1:1 methanol:chloroform and one drop of 1-M acetic acid. The mixture is vortexed until a uniform layer has formed. Thereafter, 1 volume each of 1-M NaCl and chloroform are added. Phases are allowed to equilibrate and the chloroform layer (bottom) containing the biotinylated lipid is isolated using a separation funnel. Solvent is evaporated under a steady stream of N$_2$. The purified product is reconstituted in dehydrated ethanol (200 μl for every milligram of 15d-PGJ$_2$ in the starting reaction).

Adduct concentration is measured by absorbance at 306 nm using the extinction coefficient 12,000 M^{-1} cm^{-1}, and is typically 10 mM. Store biotinylated lipids at $-80\,^{\circ}$C under N$_2$ for up to 3 months. The quality of stored materials should be verified periodically by spectrophotometry, mass spectrometry, and thin layer chromatography.

C. Preparation and Calibration of Biotinylated Cytochrome *c* and Other Proteins

The extent of posttranslational modification of a protein residue can be an important predictor of the biological impact. Thus, it is important to quantitate protein thiol modification accurately in order to relate these data to functional changes. The biotin signal resulting from biotinylated lipid–protein adduct formation or direct thiol labeling (BIAM) can be quantitated using biotinylated standard proteins containing known amounts of biotin per mole of protein (Landar *et al.*, 2006b).

1. Biotinylation of Proteins

Materials

- A stock solution of horse heart cytochrome *c* (cyt *c*, MW 12.4 kDa) is prepared by dissolving 10 mg in 1 ml PBS. In addition to cytochrome *c*, proteins of various molecular weights, including horse heart myoglobin (Mb, MW 17 kDa) and soybean trypsin inhibitor (SBTI, MW 21.5 kDa), are reacted with biotin as described above for cytochrome *c*. All proteins may be obtained from Sigma (St. Louis, MO). We do not recommend use of bovine serum albumin for these studies because the quantitative relationship between biotin content and signal is not consistent (J. Y. Oh, unpublished observations). This is likely due to direct interaction of HABA [2-(4′-hydroxyazobenzene) benzoic acid; see below] with albumin (Rutstein *et al.*, 1954).
- The biotin reagent is Sulfo-NHS-LC-biotin (Pierce) and is prepared by dissolving 5.5-mg biotin reagent in 1-ml PBS.
- PD-10 gel filtration column (8.3 ml) Sephadex G-25M, Amersham (Piscataway, NJ), equilibrated with PBS.
- Biotinylated standard proteins may be stored at 4°C with 1 tablet Complete Mini™ protease inhibitor cocktail (Roche Applied Science).
- Spectrophotometer.

Procedure

Combine 1-ml protein solution (10-mg/ml PBS) and 1-ml biotin solution (5.5 mg/ml). In our experiments, the optimal molar ratio of biotin:protein was 10:1, which, in the case of cytochrome *c*, resulted in the addition of an average of 4.6-mol biotin per mol protein. Allow reaction to proceed for 4 h on ice.

Unreacted biotin can be removed from the protein solution using a PD-10 gel filtration column. The column should be equilibrated with PBS (25 ml). After the PBS is drained to the level of the column bed, add biotinylated protein solution (in a volume of less than 2 ml), and allow solution to drain to the level of the column bed. Add 30 ml of PBS into column to elute the protein sample. Note: The elution behavior of colored proteins like cytochrome *c* can be monitored visually. As the band elutes from the column, begin collecting 1-ml fractions. For uncolored proteins, collect 0.5-ml fractions (4–5 drops) for 40 fractions and determine the protein elution

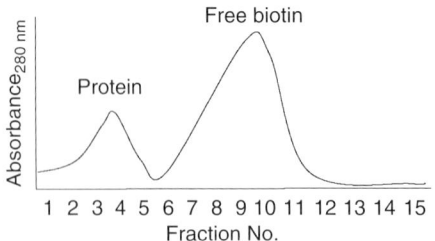

Fig. 3 Gel filtration chromatography of bt-cyt *c*. After the biotinylation reaction, free biotin was separated from cytochrome *c* using gel filtration chromatography. Fractions were collected and screened by monitoring absorbance at 280-nm wavelength.

behavior spectrophotometrically at 280 nm. To minimize contamination of eluted protein with free biotin, use the first 1–3 tubes from the first peak that contains protein (Fig. 3). Biotin incorporation is determined using the HABA assay (Section III.B.2).

2. Determination of the Biotin Incorporation in Standard Proteins (HABA Assay)

Biotin incorporation is determined using a colorimetric HABA dye displacement assay (Pierce). In this assay, a conjugate consisting of the colored dye HABA bound to avidin is added to the sample. The biotin that is present in the sample binds to avidin and displaces the HABA, which results in a decrease in the absorbance of HABA at 500 nm (Green, 1965). We have modified slightly the protocol recommended by Pierce in order to improve sensitivity.
Materials

- PBS
- HABA reagent (Pierce). Prepare a stock solution by dissolving 24.2 mg of HABA in 10 ml of 20-mM NaOH (10-mM final HABA concentration). Store at 4°C for several weeks
- 10 mg of ImmunoPureAvidin (Pierce). Prepare a stock solution by dissolving 10 mg of avidin in 19.4 ml of PBS
- Spectrophotometer

Procedure
Add 0.6 ml of 10-mM HABA solution in NaOH prepared above (HABA-avidin solution). Combine 0.5 ml of HABA-avidin solution with 0.5 ml of PBS in a 1-ml cuvette and blank the spectrophotometer using a wavelength scan from 300 to 700 nm. We prefer to measure the absorbance over a wavelength range to correct for light scatter. Add biotinylated protein to blanked HABA-avidin solution and measure absorbance as above (Fig. 4B). As can be seen in Fig. 5, increasing amounts of biotinylated protein leads to decreased absorbance at 500 nm. The light scatter in

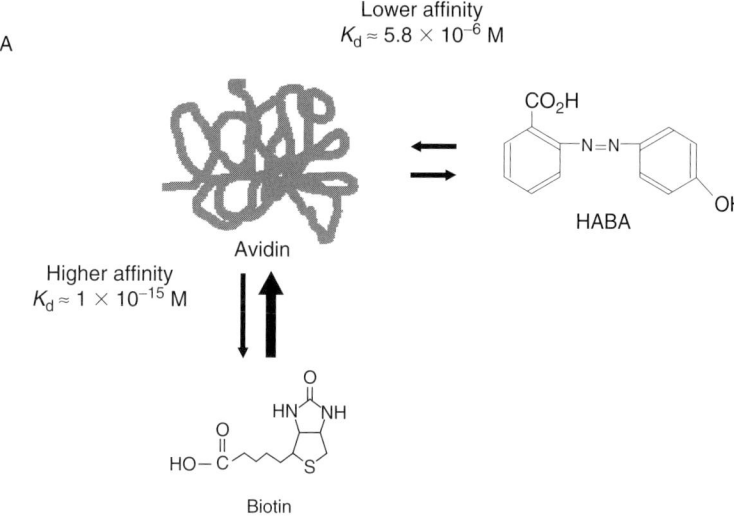

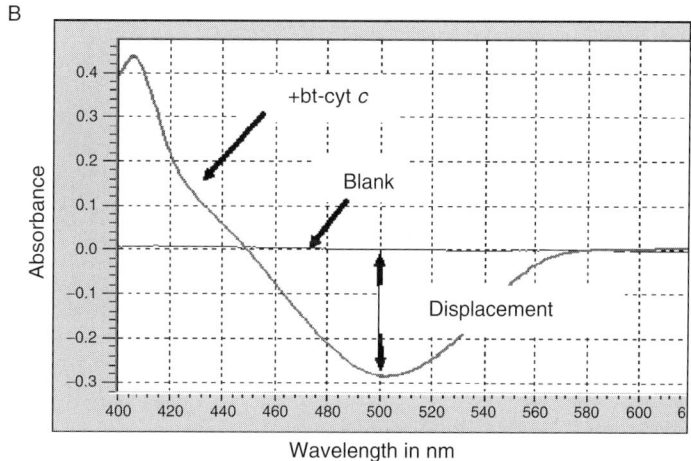

Fig. 4 HABA assay to determine biotin incorporation. (A) Diagram illustrating the HABA dye displacement from avidin by biotin due to the higher affinity of avidin-biotin binding. K_d of avidin for HABA is 5.8×10^{-6} M (Green, 1965); the K_d of avidin for biotin is $\sim 1 \times 10^{-15}$ M (Green, 1975). (B) Absorbance of diluted HABA-avidin solution alone (blank) or containing bt-cyt c. Note the change in absorbance at 500 nm on HABA displacement by biotin.

Sample f (Fig. 5A) is evident from absorbance at 560–600 nm that is greater than zero, and the difference can be used to correct for this phenomenon.

We have observed that the linear range of this assay, where an increase in biotin is represented by a proportional decrease in absorbance, is somewhat narrow (Fig. 5A). Therefore, it is necessary to repeat measurements using varying amounts

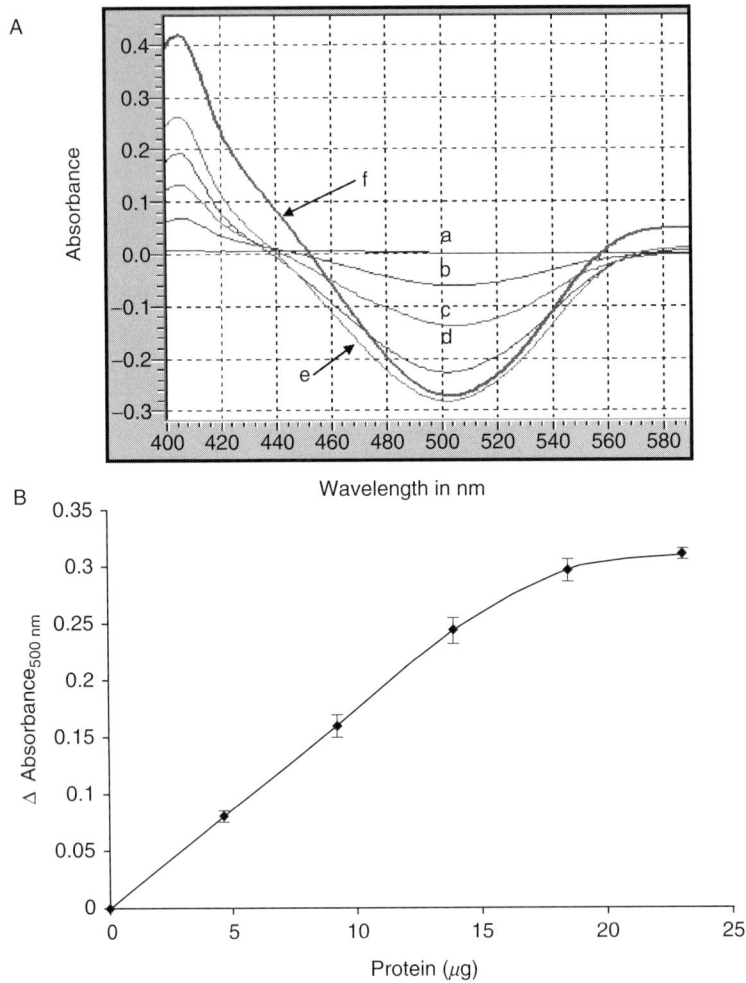

Fig. 5 Protein titration curve for HABA assay. (A) Increasing amounts of bt-cyt *c* were added to a cuvette containing diluted HABA-avidin solution and absorbance was measured over a wavelength range. a, 0 μg; b, 5 μg; c, 10 μg; d, 15 μg; e, 20 μg; and f, 25 μg. (B) The absorbance at 500 nm from panel A was replotted as a function of the amount of protein to illustrate the linear range of the assay.

of biotinylated protein, usually 1–20 μl. At this point, the absorbance at 500 nm may be plotted as a function of protein amount added (in micrograms) (Fig. 5B). In the linear range of the plot, quantitate the biotin incorporation in solution according to the extinction coefficient 34 mM^{-1} cm^{-1}. In the case of colored proteins, it may be necessary to correct for the absorbance of the protein itself by preparing a solution of protein in PBS, which does not contain HABA-avidin.

This absorbance should be subtracted from that of the HABA-avidin solution containing the protein.

3. Measurement of Protein Concentration

The cytochrome c concentration is measured most accurately using the spectral properties of the covalently bound heme prosthetic group in its reduced state. For other proteins, we use the Lowry protein assay method to determine protein concentration (Lowry *et al.*, 1951). The amount of biotin (moles) for a given amount of protein (moles) can be calculated from the concentrations of biotin (determined above by the HABA assay) and protein. We have also confirmed these calculations with detailed mass spectrometric analyses (Landar *et al.*, 2006b).

Materials

- PBS
- Sodium dithionite (for reducing cytochrome c)
- Biotinylated cytochrome c (bt-cyt c) prepared above

Procedure

Add 1 ml of PBS to cuvette and blank spectrophotometer using wavelength scan from 500 to 600 nm. Add the bt-cyt c into the PBS solution and add trace sodium dithionite (typically, a few grains of fresh powder) to reduce the heme group. Measure the absorbance by wavelength scan from 500 to 600 nm. Calculate the bt-cyt c concentration using the absorbance at 550 nm with an extinction coefficient of 27.6 mM^{-1} cm^{-1}.

D. Blotting and Image Analysis

High-resolution digital imaging techniques are essential for accurate quantitation of the biotin tag using Western blot analysis. X-ray film is not optimal, primarily because film saturates even at low exposures. As a result, quantitation obtained from an x-ray film image is often nonlinear and may underestimate quantified experimental changes. The dynamic range of digital camera imagers is far superior and is therefore recommended for Western blotting applications using biotin tags.

Materials

- A CCD camera imaging system such as the FluorChem 8000 (AlphaInnotech, San Leandro, CA).
- 1D- and 2D-polyacrylamide gel and Western blotting equipment. This is standard in most laboratories and will not be elaborated here. The aspects of the protocol that are critical to the biotin tags are explained in detail below.
- Sypro Ruby protein stain (Molecular Probes, Eugene, OR).
- Streptavidin, HRP conjugated (streptavidin-HRP) (Amersham)

- Blotting grade milk (Bio-Rad, Hercules, CA)
- Tris buffered saline containing 0.05% Tween-20 (TBS-T). PBS can be used interchangeably with TBS.
- SuperSignal Dura chemiluminescent substrate (Pierce). Note: The choice of substrate is very important. A chemiluminescent product that has a long duration is usually desirable, since camera imagers often require four to five times longer than film for exposure.

1. Blotting Protocol

Biotinylated protein samples are separated by 1D- or 2D-sodium dodecyl sulfate-polyacrylamide gel electrophoresis (SDS-PAGE). To measure biotin content in experimental samples, include biotinylated standard proteins on the gels. For 1D gels, a standard curve may be constructed using multiple biotinylated standard proteins in one lane, or alternatively, varying amounts of one biotinylated protein in a few lanes. For 2D gels, the biotinylated standard protein(s) must be run in a lane adjacent to the first dimension gel strip. Duplicate gels may be run so that proteins from one gel can be stained with Sypro Ruby and proteins from the other gel blotted for biotin detection. To quantitate moles biotin per mole protein, nonbiotinylated protein can be mixed with biotinylated protein, so that the standard can be used to calibrate both the stained gels and the Western blots (Landar *et al.*, 2006b).

To blot proteins, transfer to nitrocellulose membrane at 100 V for 2 h with cooling. Block nonspecific binding sites with 5% milk in TBS-T for 1 h. Wash membranes thoroughly to remove all milk. (Note: milk contains biotin, which is washed away in this step. However, since it can interfere with binding of streptavidin to biotinylated proteins, milk should not be included in subsequent steps.) Incubate blots with streptavidin-HRP (1:10,000 dilution in 10-ml TBS-T) for 1 h. Wash membranes three times for 10 min each. Add chemiluminescent substrate evenly to blot, ensuring coverage of the entire surface.

2. Imaging

Acquire an image of the Sypro Ruby-stained gel. Acquire a series of images from the Western blot using a CCD camera imager (AlphaInnotech). This can be done using the "movie" function, which integrates serial exposures. The result is a "movie strip" containing images of increasing intensity. Images should be saved as TIFF format files, which are used for subsequent analyses. Images containing saturated pixels should not be used for quantitation purposes.

3. Analysis

The amount of biotin (representing the amount of thiols) is quantitated by determining the density (in arbitrary units) of the selected area with the

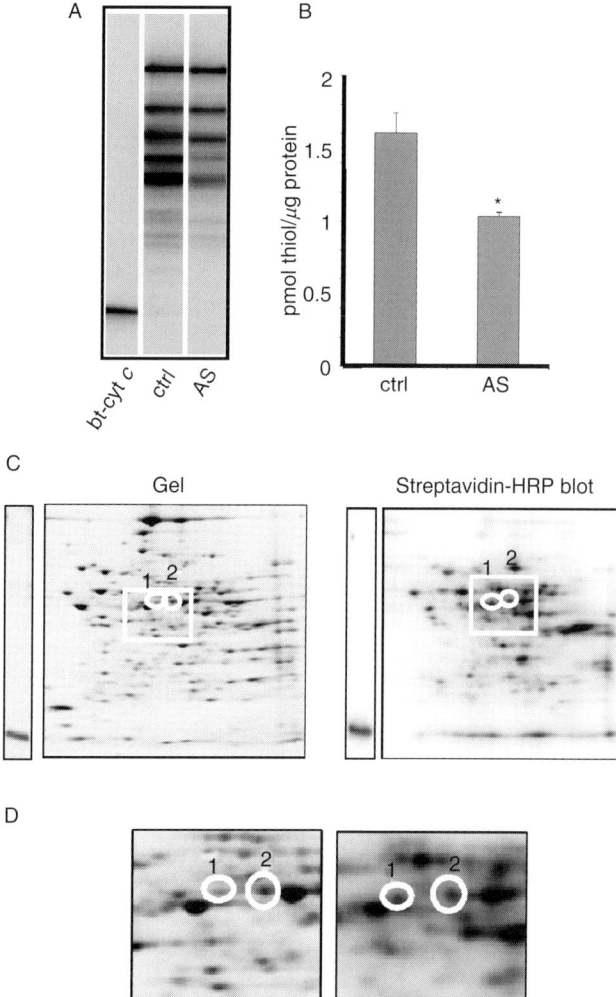

Fig. 6 Quantitation of thiol modification using a biotinylated internal proteomics standard. (A) Isolated mouse liver mitochondria were treated with vehicle (control) or 40-μM Angeli's salt (AS) to oxidize thiols, followed by labeling of free thiols using BIAM. Mitochondrial proteins (10 μg/lane) were separated by SDS-PAGE, including bt-cyt c as a standard, and biotin was detected by Western blot using streptavidin-conjugated HRP. (B) Biotin incorporation of the experiment show in panel A was determined using bt-cyt c as a standard. (C) Protein thiols were labeled with BIAM prior to separation by two-dimensional electrophoresis and Western blotting using streptavidin-conjugated HRP. A mixture of native cytochrome c (0.3 μg) and bt-cyt c (0.01 μg, 4.2 pmol biotin) was added to a lane alongside the samples. Total protein was visualized on one gel using Sypro Ruby. Protein thiols from an identical gel were detected using streptavidin-HRP. (D) A close-up image of the boxed regions from panel C.

AlphaEaseFC software version 4.0.0 supplied by the manufacturer of the camera. For 1D gels, the density of each lane containing experimental samples and biotinylated standard proteins (a "standard curve") is determined using the "1D-Multi" analysis tool. Figure 6A shows an example of a blot with one lane of the standard curve and two experimental samples containing BIAM-labeled endothelial cell proteins. Since the amount of biotin incorporated into biotin-standard proteins was determined independently by the HABA assay, the image can be calibrated by dividing the amount of biotin in the standard protein (in picomoles) by the density (arbitrary units). This value of pmol biotin (i.e., thiol) per unit density is then multiplied by the density in each experimental lane to give the pmol thiol for each lane. In the example in Fig. 6B, the pmol thiol determined for the experimental lanes was divided by the amount of protein loaded (in micrograms) to give pmol thiol/μg protein.

For 2D gels, density is measured using the "Spot Denso" analysis tool. Standard proteins are run in a well beside the isoelectric focusing strips as described in Section III.D.1, to provide both an internal protein and biotin standard. As with the 1D analysis, the images are calibrated by first dividing the picomole of biotin in the standard proteins by the density (arbitrary units). The amount of biotin (in picomoles) in spots of interest can be determined by multiplying this value by the spot density. If a duplicate gel is run and an unlabeled protein standard is included, the same procedure may be applied to calibrate the protein gel. In Fig. 6C, a duplicate blot and gel of BIAM-labeled endothelial cell proteins is shown, with lanes demonstrating the use of a standard protein mixture with known amounts of unlabeled cytochrome *c* and bt-cyt *c*. An absolute quantitation of thiol in a given spot (Fig. 6D) can be determined by dividing the amount of biotin (picomoles, representing thiols) in a spot by the amount of protein (in picomoles) in the corresponding gel spot. Thus, it is possible to follow the modification of thiols in a given protein on a molar basis. The panels in Fig. 6 were adapted from Landar *et al.* (2006b).

IV. Discussion

Modification of protein thiol groups by ROS/RNS or electrophilic lipids is important in redox cell signaling in physiological and pathological situations. Mitochondrial thiol modification may be particularly important in this process since (1) the mitochondrion is a source of ROS, and (2) specific mitochondrial thiols may be more susceptible to modification due to elevated intramitochondrial pH. Here, we have described biotin-based methods for the quantitation of protein thiols (BIAM) and adduct formation with electrophilic lipids (bt-15d-PGJ$_2$). These methods can be used in a variety of applications, including the assessment of modification of an individual protein. Furthermore, the basic protocols can be adapted easily for different reagents, such as thiol-reactive biotin-linked maleimide compounds, or to detect other modifications, such as lysine, using biotin-linked N-hydroxysuccinimide (NHS) esters. We have also adapted the basic method described here to monitor other reactive lipids and have added fluorescent tags for visualization as outlined in Table I. The increasing availability of new and sensitive methods for determining mechanisms of protein modification will facilitate the investigation of how proteins change function in response to stress. Since the mitochondrion is both a source and target or ROS/RNS, it is likely to be important in the refinement and development of these methods.

Acknowledgments

The authors are grateful to Victor Darley-Usmar for helpful discussions. A.L. is supported by a Scientist Development Grant from the American Heart Association and a Junior Faculty Development Award from the Comprehensive Cancer Center at the University of Alabama-Birmingham.

References

Caro, A. A., and Cederbaum, A. I. (2006). Role of cytochrome P450 in phospholipase A2- and arachidonic acid-mediated cytotoxicity. *Free Radic. Biol. Med.* **40,** 364–375.

Ceaser, E. K., Moellering, D. R., Shiva, S., Ramachandran, A., Landar, A., Venkartraman, A., Crawford, J., Patel, R., Dickinson, D. A., Ulasova, E., Ji, S., and Darley-Usmar, V. M. (2004). Mechanisms of signal transduction mediated by oxidized lipids: The role of the electrophile-responsive proteome. *Biochem. Soc. Trans.* **32,** 151–155.

Cernuda-Morollon, E., Pineda-Molina, E., Canada, F. J., and Perez-Sala, D. (2001). 15-Deoxy-Delta 12,14-prostaglandin J2 inhibition of NF-kappaB-DNA binding through covalent modification of the p50 subunit. *J. Biol. Chem.* **276,** 35530–35536.

Coles, B., Bloodsworth, A., Clark, S. R., Lewis, M. J., Cross, A. R., Freeman, B. A., and O'Donnell, V. B. (2002). Nitrolinoleate inhibits superoxide generation, degranulation, and integrin expression by human neutrophils: Novel antiinflammatory properties of nitric oxide-derived reactive species in vascular cells. *Circ. Res.* **91,** 375–381.

Dennehy, M. K., Richards, K. A., Wernke, G. R., Shyr, Y., and Liebler, D. C. (2006). Cytosolic and nuclear protein targets of thiol-reactive electrophiles. *Chem. Res. Toxicol.* **19,** 20–29.

Dickinson, D. A., Darley-Usmar, V. M., and Landar, A. (2006). The covalent advantage: A new paradigm for cell signaling by thiol reactive lipid oxidation products. *In* "Redox Proteomics: From Protein Modifications to Cellular Dysfunction and Diseases" (I. Dalle-Donne, A. Scalone, and D. A. Butterfield, eds.), pp. 345–367. John Wiley & Sons, Indianapolis.

Eaton, P., Jones, M. E., McGregor, E., Dunn, M. J., Leeds, N., Byers, H. L., Leung, K. Y., Ward, M. A., Pratt, J. R., and Shattock, M. J. (2003). Reversible cysteine-targeted oxidation of proteins during renal oxidative stress. *J. Am. Soc. Nephrol.* **14,** S290–S296.

Green, N. M. (1965). A Spectrophotometric assay for avidin and biotin based on binding of dyes by avidin. *Biochem. J.* **94,** 23C–24C.

Green, N. M. (1975). Avidin. *In* "Advances in Protein Chemistry" (C. B. Anfinsen, J. T. Edsall, and F. M. Richards, eds.), pp. 85–133. Academic Press, New York.

Green, N. M. (1990). Avidin and streptavidin. *Methods Enzymol.* **184,** 51–67.

Hao, G., Derakhshan, B., Shi, L., Campagne, F., and Gross, S. S. (2006). SNOSID, a proteomic method for identification of cysteine S-nitrosylation sites in complex protein mixtures. *Proc. Natl. Acad. Sci. USA* **103,** 1012–1017.

Jaffrey, S. R., and Snyder, S. H. (2001). The biotin switch method for the detection of S-nitrosylated proteins. *Sci STKE* **2001,** Pl1.

Kagan, V. E., Tyurina, Y. Y., Bayir, H., Chu, C. T., Kapralov, A. A., Vlasova, II, Belikova, N. A., Tyurin, V. A., Amoscato, A., Epperly, M., Greenberger, J., and Dekosky, S., *et al.* (2006). The "pro-apoptotic genies" get out of mitochondria: Oxidative lipidomics and redox activity of cytochrome c/cardiolipin complexes. *Chem. Biol. Interact.* **163**, 15–28.

Kim, J. R., Yoon, H. W., Kwon, K. S., Lee, S. R., and Rhee, S. G. (2000). Identification of proteins containing cysteine residues that are sensitive to oxidation by hydrogen peroxide at neutral pH. *Anal. Biochem.* **283**, 214–221.

Kuhn, H., and Borchert, A. (2002). Regulation of enzymatic lipid peroxidation: The interplay of peroxidizing and peroxide reducing enzymes. *Free Radic. Biol. Med.* **33**, 154–172.

Landar, A., Giles, N. M., Zmijewski, J. W., Watanabe, N., Oh, J. Y., and Darley-Usmar, V. M. (2006a). Modification of lipids by reactive oxygen and nitrogen species: The oxy-nitroxy-lipidome and its role in redox cell signaling. *Future Lipidology* **1**, 203–211.

Landar, A., Oh, J. Y., Giles, N. M., Isom, A., Kirk, M., Barnes, S., and Darley-Usmar, V. M. (2006b). A sensitive method for the quantitative measurement of protein thiol modification in response to oxidative stress. *Free Radic. Biol. Med.* **40**, 459–468.

Landar, A., Zmijewski, J. W., Dickinson, D. A., Le Goffe, C., Johnson, M. S., Milne, G. L., Zanoni, G., Vidari, G., Morrow, J. D., and Darley-Usmar, V. M. (2006c). Interaction of electrophilic lipid oxidation products with mitochondria in endothelial cells and formation of reactive oxygen species. *Am. J. Physiol. Heart Circ. Physiol.* **290**, H1777–H1787.

Leitinger, N. (2005). Oxidized phospholipids as triggers of inflammation in atherosclerosis. *Mol. Nutr. Food Res.* **49**, 1063–1071.

Levonen, A. L., Landar, A., Ramachandran, A., Ceaser, E. K., Dickinson, D. A., Zanoni, G., Morrow, J. D., and Darley-Usmar, V. M. (2004). Cellular mechanisms of redox cell signalling: Role of cysteine modification in controlling antioxidant defences in response to electrophilic lipid oxidation products. *Biochem. J.* **378**, 373–382.

Lowry, O. H., Rosebrough, N. J., Farr, A. L., and Randall, R. J. (1951). Protein measurement with the Folin phenol reagent. *J. Biol. Chem.* **193**, 265–275.

Montine, T. J., and Morrow, J. D. (2005). Fatty acid oxidation in the pathogenesis of Alzheimer's disease. *Am. J. Pathol.* **166**, 1283–1289.

Musiek, E. S., Gao, L., Milne, G. L., Han, W., Everhart, M. B., Wang, D., Backlund, M. G., Dubois, R. N., Zanoni, G., Vidari, G., Blackwell, T. S., and Morrow, J. D. (2005). Cyclopentenone isoprostanes inhibit the inflammatory response in macrophages. *J. Biol. Chem.* **280**, 35562–35570.

Niki, E., Yoshida, Y., Saito, Y., and Noguchi, N. (2005). Lipid peroxidation: Mechanisms, inhibition, and biological effects. *Biochem. Biophys. Res. Commun.* **338**, 668–676.

Parker, J. (1995). Prostaglandin A2 protein interactions and inhibition of cellular proliferation. *Prostaglandins* **50**, 359–375.

Rutstein, D. D., Ingenito, E. F., and Reynolds, W. E. (1954). The determination of albumin in human blood plasma and serum; a method based on the interaction of albumin with an anionic dye-2(4′-hydroxybenzeneazo) benzoic acid. *J. Clin. Invest.* **33**, 211–221.

Shibata, T., Kondo, M., Osawa, T., Shibata, N., Kobayashi, M., and Uchida, K. (2002). 15-Deoxy-delta 12,14-prostaglandin J2. A prostaglandin D2 metabolite generated during inflammatory processes. *J. Biol. Chem.* **277**, 10459–10466.

Straus, D. S., and Glass, C. K. (2001). Cyclopentenone prostaglandins: New insights on biological activities and cellular targets. *Med. Res. Rev.* **21**, 185–210.

Zhang, R., Brennan, M. L., Shen, Z., MacPherson, J. C., Schmitt, D., Molenda, C. E., and Hazen, S. L. (2002). Myeloperoxidase functions as a major enzymatic catalyst for initiation of lipid peroxidation at sites of inflammation. *J. Biol. Chem.* **277**, 46116–46122.

Zmijewski, J. W., Landar, A., Watanabe, N., Dickinson, D. A., Noguchi, N., and Darley-Usmar, V. M. (2005). Cell signalling by oxidized lipids and the role of reactive oxygen species in the endothelium. *Biochem. Soc. Trans.* **33**, 1385–1389.

PART V

Mitochondrial Genes and Gene Expression

CHAPTER 22

Detection of Mutations in mtDNA

Ali Naini and Sara Shanske

H. Houston Merritt Clinical Research Center for Muscular Dystrophy and Related disorders
Department of Neurology, College of Physicians and Surgeons
Columbia University, New York, New York 10032

I. Introduction

The human mitochondrial genome is a 16,569-bp (base pair) circle of double-stranded DNA. It contains 37 genes which specify 2 ribosomal RNAs (rRNAs), 22 transfer RNAs (tRNAs), and 13 polypeptides (Fig. 1). All 13 polypeptides encoded by mtDNA are components of the respiratory chain/oxidative phosphorylation system and include 7 subunits of complex I [NADH-coenzyme Q (CoQ) oxidoreductase], 1 subunit of complex III (CoQ-cytochrome c oxidoreductase), 3 subunits of complex IV (cytochrome c oxidase), and 2 subunits of complex V

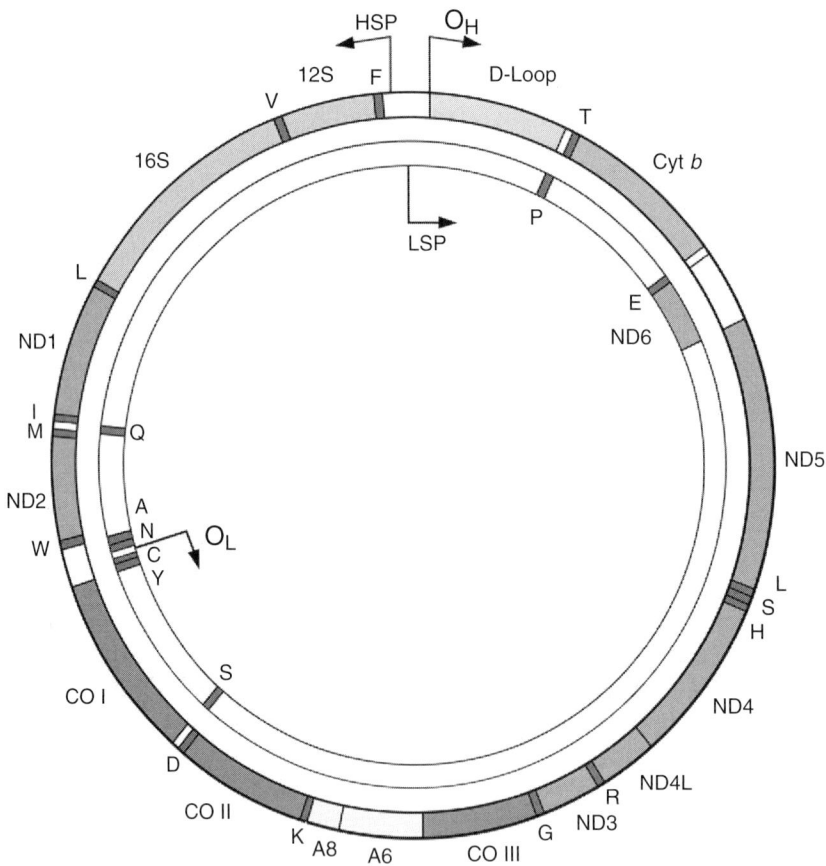

Fig. 1 The human mitochondrial genome encodes for 13 mRNAs, 22 tRNAs, and 2 rRNAs. The subunits of NADH-ubiquinone oxidoreductase (ND), cytochrome c oxidase (COX), cytochrome b (Cyt b), ATP synthase (A), 22 tRNAs, and two rRNAs (12S and 16S) are shown. The origins of heavy (O_H) and light (O_L) strand replication, and the heavy (HSP) and light (LSP) strand transcriptional promoters are also shown.

(ATP synthase). Each of these complexes also contains subunits encoded by nuclear genes, which are imported from the cytoplasm, assembled into holo-enzymes with the mtDNA-encoded subunits, and embedded in the mitochondrial inner membrane. Complex II (succinate dehydrogenase-CoQ oxidoreductase) is encoded entirely by nuclear genes.

There are several distinctive features of mtDNA that are relevant to the under-standing of mtDNA-related diseases. At fertilization, all mitochondria are con-tributed by the oocyte; thus, mtDNA is inherited only from the mother. Therefore, most mtDNA point mutations are maternally inherited: a woman carrying a mtDNA point mutation will transmit it to all her children, males as well as females, but only the daughters will transmit it to their progeny. However, while mtDNA itself is transmitted by the mother, disorders of mtDNA are not always inherited maternally. For example, large-scale single deletions in mtDNA are often sporadic and are not maternally inherited. Similarly, mtDNA depletion and deletion errors are often a result of mutations in nuclear genes that affect the integrity or quantity of mtDNA. As such, they are transmitted as autosomal recessive or autosomal dominant traits. In contrast to nuclear genes, each consisting of one maternal and one paternal allele, there are hundreds or even thousands of copies of mtDNA in every cell. Thus, when there is a deleterious mutation, both normal and mutated mtDNAs may coexist within a patient's tissues, a condition known as heteroplasmy. A critical number of mutated mtDNAs must be present before tissue dysfunction and clinical signs become apparent, the so-called threshold effect. Tissues with high requirements for oxidative energy metabolism, such as muscle, heart, eye, and brain, have relatively low thresholds and are particularly vulnerable to mtDNA mutations. At cell division, the proportion of mutant mtDNAs in daughter cells can shift, a phenomenon termed or known as mitotic segregation. If and when the pathogenic threshold for a particular tissue is exceeded, the phenotype can change. Thus, in a patient who is heteroplasmic for a pathogenic mutation, the clinical phenotype can change over the course of time.

Beginning in 1988, an ever-increasing number of mitochondrial diseases with distinct clinical phenotypes have been associated with mutations in mtDNA (reviewed in Schon, 2000), almost all of which resulted in neurological or neuromus-cular disorders. These errors fall into three major classes: (1) large-scale rearrange-ments of mtDNA, (2) depletion of mtDNA, or (3) point mutations in mtDNA.

II. Large-Scale Rearrangements of mtDNA

Large-scale single deletions of mtDNA are associated with three major clinical conditions: Kearns-Sayre syndrome (KSS), progressive external ophthalmoplegia (PEO), and Pearson syndrome (PS). Although dozens of different deletions have been described, each patient harbors only a single type of deletion (Fig. 2A), and the number of deleted genomes varies in different patients and in different tissues from the same patient (heteroplasmy). Most patients reported with deletions in

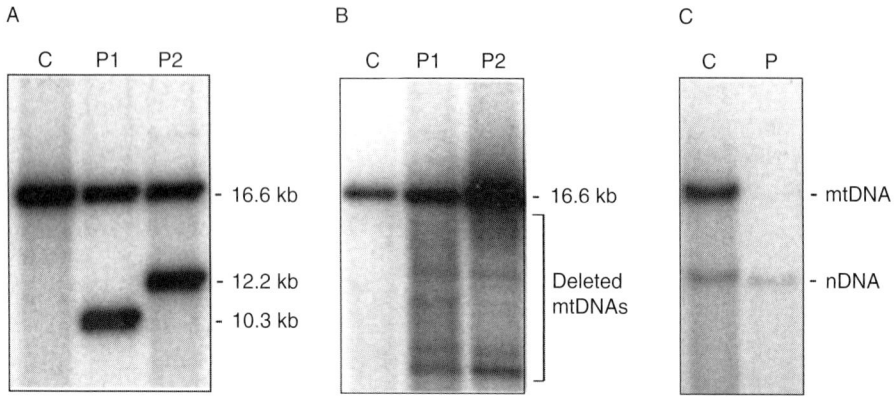

Fig. 2 Southern Blot analysis showing a typical single deletion (panel A), multiple deletions (panel B), and mtDNA depletion (panel C). C: control, P: patient. Adapted from Shanske and Wong (2004), with permission.

mtDNA have been sporadic; mothers of affected individuals and children of affected women are clinically normal and, when tested, have no detectable deletions (Graff *et al.*, 2000; Larsson *et al.*, 1992; Zeviani *et al.*, 1990). This suggested that deletions arose *de novo* early in embryogenesis or ovum. However, there have been several reports of deletions that appear to have been transmitted maternally (Bernes *et al.*, 1993; Boles *et al.*, 1997; Shanske *et al.*, 2002). Thus, one needs to be cautious when counseling. Some patients harbor more complex rearrangements in mtDNA, and duplications of mtDNA, both sporadic and maternally inherited, have been reported. In addition Mendelian-inherited disorders have been described in which affected family members harbor large quantities of multiple species of deletions of mtDNA in their tissues (Fig. 2B), which are apparently generated during the life span of the patient (Kaukonen *et al.*, 2000; Nishino *et al.*, 1999; Zeviani *et al.*, 1989). These disorders are a result of mutations in nuclear genes that either facilitate an intrinsic propensity of mtDNA to undergo rearrangements or impair recognition and elimination of spontaneously occurring rearrangements.

III. Detecting mtDNA Rearrangements

Single or multiple deletions in mtDNA are detected by Southern blot analysis and are best demonstrated by testing DNA isolated from muscle, since, in most patients, deletions are not detectable in DNA isolated from blood. To perform Southern blot analysis, total DNA is isolated, digested with restriction enzymes, separated on an agarose gel, transferred to a membrane (nitrocellulose or nylon), and incubated with a labeled mtDNA probe that will hybridize only to the mtDNA. The bands are detected by autoradiography, and their size(s) indicate whether only the normal-sized mtDNA (16.6 kb) is present or whether there is also a population of a smaller mtDNA, indicating a deletion. Figure 2A illustrates two patients, one having a hybridizing band at 12.2 kb (i.e., 4.4 kb deleted) and another with a band at 10.3 kb (i.e., 6.3 kb deleted). Depending on the restriction enzyme used, a band that is larger than 16.6 kb may indicate a duplicated species (see below). In patients with multiple deletions, Southern blots of DNA isolated from muscle show multiple bands representing species of mtDNA molecules harboring deletions of different sizes (Fig. 2B).

For Southern blot analysis, several issues are important to consider. The first is the tissue used for testing. As described above, deletions may only be detectable in DNA isolated from muscle. Another consideration is which restriction enzymes to use to digest the DNA. One usually uses at least one enzyme that has only a single cutting site in mtDNA, such as *Pvu*II (site at nt-2650), *Bam*HI (nt-14258), or *Eag*I (nt-2566), so that circular mtDNA is linearized. In some patients, a deleted species may be present in extremely low levels (i.e., less than 5%). If digestion with only one restriction enzyme is used, the intensity of the 16.6-kb wt-mtDNA fragment will often be so great that the signal will obscure any deleted fragment migrating on the gel below, but close to, the wild-type band. We have found that digestion with an enzyme that cuts two or three times in the mtDNA (e.g., *Eco*RI, *Hin*dIII, *Pst*I) can help to solve this problem. Following digestion with such an enzyme, the deleted fragment will often be larger than the wild-type fragment (owing to loss of one of the restriction sites in the deleted region). On electrophoresis and hybridization with an appropriate region-specific probe, the minority of deleted molecules will migrate above the wild-type band and can be identified and quantified relatively easily, with little background present.

In addition to single deletions, mtDNA in patients may also harbor duplications. In patients harboring mitochondrial genomes with deletions and duplications, the two rearranged species are topologically related; the duplicated mtDNA is composed of two mtDNA molecules—a wild-type mtDNA and a deleted mtDNA—arranged head to tail (Fig. 3A) (Tang *et al.*, 2000a,b). The only novel sequence in the duplicated mtDNA as compared to wild-type mtDNA is at the boundary of the duplicated region, which is the same as the boundary present in the corresponding deleted mtDNA (dashed line in Fig. 3A). Because the two molecules are related, a "standard" Southern blot analysis will not be able to distinguish between the two species.

Figure 3 shows an example of how these ambiguities can arise and how they can be circumvented. As shown in Fig. 3A, a patient with a 7813-bp deletion (Tang *et al.*, 2000a) harbored, in addition to the 16.6-kb wild-type mtDNA, a partially deleted genome 8.8-kb long (i.e., 7.8 kb was deleted) and a partially duplicated genome 25.3-kb long (i.e., 8.8 kb was inserted into the wild-type genome). Note that *Pvu*II cuts wild-type and deleted mtDNA once, but cuts duplicated mtDNA twice (Fig. 3A). Digestion with *Pvu*II is therefore expected to produce a 16.6-kb band corresponding to wild-type mtDNA, a smaller 8.8-kb band corresponding

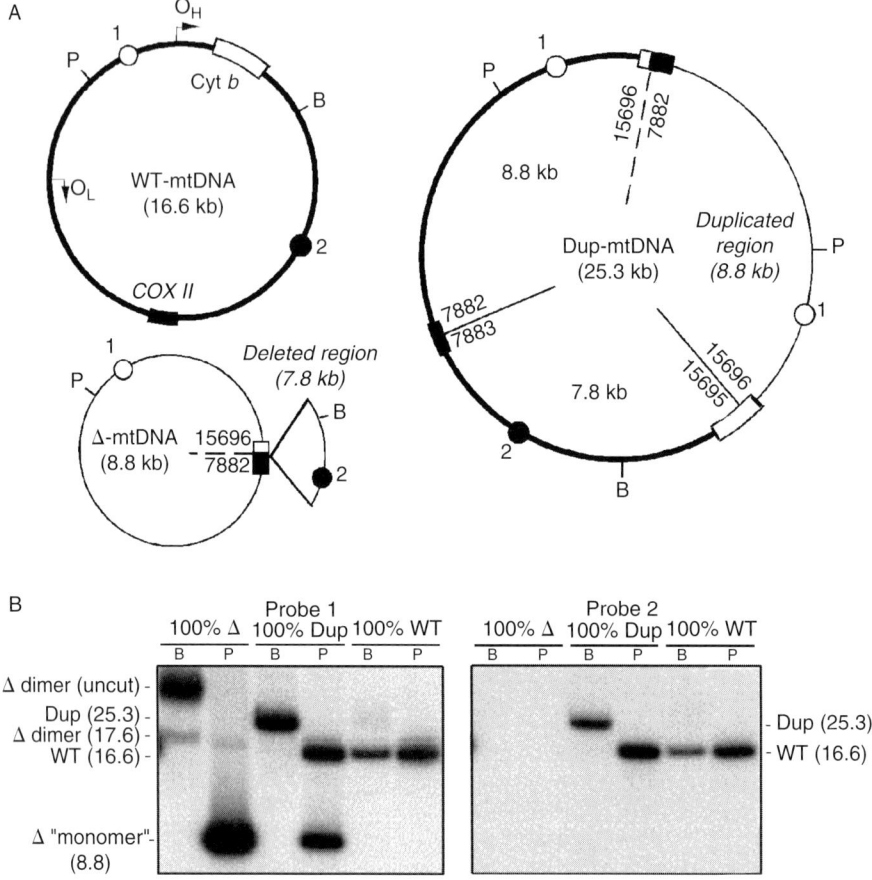

Fig. 3 A typical large-scale mtDNA rearrangement, showing the relationship among wild type, deleted, and duplicated species, using a patient with a 7.8-kb deletion (Tang *et al.*, 2000a) as an example. (A) The deletion (protruding "pie-shaped" segment) removes 7813 bp between the COX II [solid box] and Cyt *b* [open box] genes. The break point is indicated by a dashed line. The nucleotides associated with the rearrangement straddle the "pie slices." Numbers denote nucleotide positions. The locations of the *Pvu*II (P) and *Bam*HI (B) restriction sites, and probes 1 (open circle) and 2 (shaded circle) used in the Southern blot analyses (see text) are also shown. (B) Autoradiograms of Southern blot analyses of cybrid cell lines containing 100% wild type (WT), 100% duplicated (Dup), and 100% deleted (Δ) mtDNAs, following digestion with *Pvu*II (P) or *Bam*HI (B) and hybridization with probes 1 or 2. The identity of each hybridizing fragment and its size (in kb) is indicated at the sides. The 8.8-kb band represents either linearized full-length Δ-mtDNA (in deleted lines) or the duplicated region in dup-mtDNA (in duplicated lines).

to deleted mtDNA, and two bands derived from duplicated mtDNA, one of 16.6 kb (corresponding to the wild-type region of the molecule) and other of 8.8 kb (corresponding to the duplicated region of the molecule). When hybridized with a probe located outside the deleted region (i.e., probe 1), both the 16.6-kb and

8.8-kb fragments will hybridize, but in a heteroplasmic sample, it will not be clear from which of the three types of molecules these bands are derived. When hybridized with a probe located inside the deleted region (i.e., probe 2), there is no signal from the 8.8-kb band (derived from either the deleted or duplicated mtDNAs). Only a 16.6-kb fragment will hybridize, but it will not be clear whether this fragment derives from a wild-type molecule, from a duplicated molecule, or from both. Unlike *Pvu*II, *Bam*HI has no cutting site on the deleted mtDNA, leaving this species in nonlinearized circular forms. In addition, *Bam*HI cuts both wild-type mtDNA and duplicated mtDNA only once, yielding a 16.6- and 25.3-kb fragments, respectively (Fig. 3B). Thus, when *Bam*HI-digestion products are hybridized with probe 1, bands corresponding to all three species will be visualized: a 16.6-kb band corresponding to linearized wild-type mtDNA, a 25.3-kb band corresponding to linearized duplicated mtDNA, and one or more bands, migrating elsewhere on the gel, corresponding to the topological conformers of the uncut deleted mtDNA (e.g., supercoils and nicked circles). When hybridized with probe 2, however, none of the topological forms of the uncut deleted molecules will be visualized; only the 16.6- and 25.3-kb bands, corresponding to the wild-type and duplicated mtDNAs, will appear (Tang *et al.*, 2000a). Thus, by using a combination of the two restriction enzymes and the two regional probes, one can distinguish (and quantitate) unambiguously all three mtDNA species, even if they are present in a heteroplasmic mixture.

IV. Details of Southern Blot Hybridization Analysis

A. DNA Preparation

Prepare DNA from blood or solid tissue using Puregene DNA extraction kit and following manufacturer's recommended protocols, or according to the standard method for DNA mini-prep preparation, as described (Sambrook and Russell, 2001).

B. Probe Preparation

1. Amplify a 2.5-kb fragment of mtDNA in a total volume of 50 μl using the Expand High Fidelity PCR System kit (Roche Diagnostics, Indianapolis, IN) and primers.
   a. Forward primer: 5'-CCACTCCACCTTACTACCAGAC-3' (nt 1690–1711)
   b. Reverse primer: 5'-GTAATGCTAGGGTGAGTGGTAGG-3' (nt 4207–4185)
2. Load the entire polymerase chain reaction (PCR) product onto a 0.8% agarose gel in 1× TBE containing 0.5-μg/ml ethidium bromide (EtBr).
3. Cut the 2.5-kb fragment from the gel and purify using Gel Extraction Kit reagents (Eppendorf, Hamburg, Germany) and a recommended protocol.

Note: for analysis of mtDNA duplications, probes are prepared representing segments of mtDNA located "inside" and "outside" the deleted region.

C. Labeling of the Probe

1. Using the Random Primed DNA Labeling Kit (Boehringer Mannheim) to label the probe as follows: In a 500-μl microfuge tube, mix the 2-μl probe ($\sim$25 ng) with 8-μl water, and boil for 10 min. Chill immediately on ice.

2. To the above 10-μl solution, add the following:

   a. 1 μl of 0.5-mM dTTP

   b. 1 μl of 0.5-mM dCTP

   c. 1 μl of 0.5-mM dGTP

   d. 5 μl of [α-32P]-dATP (i.e., 50 μCi; 3000 Ci/mmol)

   e. 2 μl of 10$\times$ hexanucleotide random mixture

   f. 1 μl Klenow enzyme (2 units), mix, and incubate for 30 min at 37 °C

3. Add 2 μl of 0.2-M EDTA, pH 8.0, to stop the reaction.

4. Purify using G-50 microcolumns.

5. Add 1 μl probe to 10ml of Ultima Gold Scintillation liquid (Packard Bio-Science, Meriden, CT), mix, and count the radioactivity (should be $\sim 1 \times 10^6$ cpm, with a specific activity of $\sim 10^9$ cpm/μg).

D. Preparation of Restriction Enzyme Digest

1. In a 500-μl microfuge tube, prepare a digestion mix as follows:

   a. DNA 5–10 μg

   b. 1 μl of restriction enzyme (10 units/μl)

   c. 4 μl of buffer (10$\times$)

   d. 1 μl of RNAase (10 mg/ml)

   e. H$_2$O to make a total volume of 40 μl

   f. Incubate at 37 °C for 1–2 h

2. Check digestion

   a. Remove 4-μl solution from step 1 (continue to incubate the remaining mix).

   b. Add 6-μl H$_2$O and 2-μl 6$\times$ dye (bromphenol blue/xylene cyanol).

   c. Load a 0.8% or 1% agarose gel; run at $\sim$100 mA. Check digestion and amount of DNA.

   d. Add 1 μl of appropriate restriction enzyme (if not completely cut) or 1–2-μg DNA (if there is not enough DNA to detect on agarose gels), continue incubation as needed (overnight is alright).

E. Agarose Gel Electrophoresis

1. Prepare a 20-cm, 0.8% agarose gel containing 1-μg/ml ethidium bromide in 1× TAE buffer.
2. Load 20- to 40-μl digested DNA plus 4- to 8-μl 6× dye per well (try to apply similar amounts of DNA in each well); load markers (usually λ-digested with *Hin*dIII) in one well.
3. Run gel at ~140 mA for 8h or at ~80 mA overnight.
4. Stop when bottom dye is about three-fourths of the way down (e.g., 16 cm for a 20-cm gel).
5. Remove gel, place gel in glass dish on top of Saran Wrap, and photograph with a ultraviolet (UV)-visible ruler; cut one corner for orientation.
6. Rinse with double distilled (dd) H_2O for 5 min.
7. Add denaturing solution (1.5-M NaCl, 0.5-M NaOH), shake very gently for 30–40 min in a shaker.
8. Replace with neutralizing solution (1-M Tris, pH 7.4; 1.5-M NaCl), shake very gently for 15–30 min.

F. DNA Transfer

1. Set up transfer apparatus using "transfer buffer"—10× SSC and Whatman No. 3 filter paper.
2. Cut the membrane (Zeta-Probe [Bio-Rad]) to size of gel, wet in ddH_2O, then wet with "transfer buffer" right before applying.
3. Invert gel onto filter paper; remove bubbles.
4. Put wet membrane on gel; remove bubbles.
5. Put wet Whatman No. 3 filter on top of the membrane; remove bubbles.
6. Add several dry Whatman No. 3 filters and layers of paper towels; "seal" with Saran Wrap so that buffer does not evaporate.
7. After transfer, cut the membrane in one corner for orientation, carefully remove and put on filter paper and let it dry in oven at 80°C for ~45–60 min under vacuum. Alternatively, the DNA can be bound to the filter in a UV cross-linker (254-nm wavelength at 1.5 J/cm^2).

G. Hybridization

1. Roll membrane and place in a hybridization tube.
2. Prehybridize membrane in 7% SDS + 0.25-M Na_2HPO_4 for 15 min at 65°C in a hybridization oven. Rotate gently on a rotary shaker (speed ~5 rpm).
3. Boil the probe in water for 10 min to denature, and immediately chill on ice.

4. Add the probe [~0.5–1 × 10^6 cpm/ml (specific activity ~10^9 cpm/μg) in hybridization solution].

5. Hybridize overnight at 65 °C.

H. Washing the Membrane

1. Remove the probe (it can be kept at -20 °C and reused for up to 2 weeks).

2. Wash for 30 min in 0.02-M Na_2HPO_4 with 5% SDS, two times.

3. Wash for 30 min in 0.02-M Na_2HPO_4 with 1% SDS, two times.

4. Wash for 30 min in 0.02-M Na_2HPO_4 with 0.5% SDS.

I. Film Developing

1. Keep the membrane wet in case it is necessary to wash again to reduce background or to perform a second hybridization to increase signal.

2. Place filter between two pieces of Whatman 3MM paper to dry.

3. Place dry filter in sealable bag or place on a 20-cm × 20-cm sheet of Whatman 3MM and cover with plastic wrap (Saran Wrap is preferred).

4. Autoradiograph with Kodak XAR-5 film.

5. Expose overnight and develop.

Notes

1. It is advisable to hybridize the membrane first with probe 2 (the "inside" probe) and then with probe 1 (the "outside" probe). It is also advisable to strip the first probe off the membrane prior to hybridizing with the second probe.

2. In the example shown in Fig. 3, the Southern blots were performed on DNA isolated from cybrid cells containing homoplasmic levels of wild-type, deleted, and duplicated mtDNAs, to demonstrate more clearly how the two restriction enzymes and two regional probes can be used to distinguish among the three species. In a "real-life" situation, however, patient cells will almost certainly be heteroplasmic and will contain more than one species of mtDNA. In such a situation, the resulting Southern blot pattern will be more complicated, as it will be, in essence, a "superposition" of the two panels shown in Fig. 3B. Nevertheless, all three species can still be distinguished, but great care must be taken to keep track of the intensities among the relevant hybridizing fragments after hybridization with both the "inside" and "outside" probes. A change in relative intensity of the "deleted" band (8.8-kb band in Fig. 3B) following hybridization with the two probes is indicative of the presence of both deleted and duplicated species.

3. The example shown in Fig. 3 also highlights a second phenomenon often encountered in this analysis, namely the presence of multimeric forms of the various mtDNA species. The first lane in the left panel in Fig. 3B shows that the majority of

deleted mtDNA was present as a 17.6-kb dimer (uncut by *Bam*HI, but migrating on the gel as randomly linearized and nicked circles that were likely formed during the isolation of the DNA), rather than as an 8.8-kb monomer. However, in other samples from this patient (not shown), uncut supercoiled monomers were also found (Tang *et al.*, 2000a,b).

Deletions can also be detected by using polymerase chain reaction or PCR analysis. By employing sets of widely spaced primers, it is possible to detect both single deletions (where a specific small fragment is amplified) or multiple deletions (where there are many bands amplified representing deletions of various sizes). For example, the following primer pair flanks the 4977-bp "common" deletion located between nt 8483 and 13,460: forward primer from nt 8274 to 8305 and reverse primer from nt 13,720 to 13,692. In normal mtDNA, these primers would amplify a 5.5-kb fragment. In patients with the "common" deletion, 5-kb is deleted, leaving a fragment of ∼500 bp, which is detected after PCR amplification. One can also detect multiple mtDNA deletions by PCR; using widely spaced primers results in PCR products of varying sizes. However, when deleted molecules are present at very low levels and can be detected only by PCR, one needs to be cautious in interpretation, since deletions are normally present in older individuals and may not be pathogenic (DiMauro *et al.*, 2002).

V. Depletion of mtDNA

MtDNA depletion syndromes comprise a clinically heterogeneous group of disorders characterized by severe reduction in mtDNA copy number. Primary mtDNA depletion is inherited as an autosomal recessive trait and has been associated with mutations in nuclear-encoded genes responsible for mtDNA synthesis or maintenance of deoxynucleotide pools. MtDNA depletion may affect single organs, typically muscle or liver, or multiple tissues (Moraes *et al.*, 1991; Tritschler *et al.*, 1992). The mtDNA content can be measured by quantitative Southern blot analysis by taking the ratio of signal intensity of mtDNA to the signal intensity of a nuclear gene, usually the 18S rRNA gene (Fig. 2C) (Moraes *et al.*, 1991). However, this method has several inherent disadvantages, including a relatively long turnaround time (at least 2 days), the use of ^{32}P radio-labeled probes, and the low accuracy (semiquantitative). In contrast, quantitative real-time PCR offers several advantages: it is fast (less than 2 h), uses environmentally safe reagents, and has high accuracy. In Sections V.A and V.B, we describe both the Southern blot technique and a quantitative real-time PCR method.

A. Southern Blot Hybridization Analysis to Detect mtDNA Depletion

To quantify the mtDNA content in tissues, one can use a second probe on the Southern blot filter—a fragment containing a nuclear-encoded multicopy gene—as an internal control to correct for differences in the amount of DNA loaded in each lane (Moraes *et al.*, 1991). Total DNA is prepared as above, digested with *Pvu*II, subjected to electrophoresis using a 0.8% agarose gel, and transferred to nitrocellulose. The filters are then hybridized with two probes either sequentially or simultaneously. The first probe detects mtDNA, as above. The second probe detects the nuclear-encoded 18S ribosomal DNA genes which are present at about 400 copies per cell. We use clone pB (Wilson *et al.*, 1978) containing nuclear-encoded 18S rDNA sequences on a 5.8-kb *Eco*RI fragment. Digestion of the ribosomal DNA repeat regions containing the 18S rDNA genes with *Pvu*II will release hybridizable fragments ~12 kb in size. Thus, to quantitate the mtDNA and rDNA signals, we scan the nitrocellulose filters and count the mtDNA (16.6 kb) and rDNA (12.0 kb) signals (as counts per minute). The ratio mtDNA:rDNA (dimensionless) is calculated by dividing the mtDNA signal by the rDNA signal, after correcting for background signal on the filter.

Notes

In most standard Southern blot hybridizations, the probe is in excess and the quantitation of hybridizing bands is straightforward. In the case of blots for quantitating mtDNA, however, the probe may not be in excess, especially when multiple samples are present on one filter. It is therefore critical to be sure that the hybridization conditions, especially the ratio of probe to "hybridizable DNA" on the filter, allow for all hybridizable bands to be visualized in a quantitative manner. Moreover, quantitative results obtained from one blot cannot be compared to those obtained from a different blot unless great care is taken to be sure that the two, in fact, are indeed comparable. Minimally, the probe lengths and specific activities in the two blots must be known, so that the signals can be normalized. Furthermore, it is extremely useful to include, in each blot, "common samples", for example, age-matched normal tissue controls and at least one sample from a patient with known mtDNA depletion.

B. Quantitation of mtDNA by Real-Time PCR

The method uses two Taqman probes, one specific for 12S mitochondrial rRNA gene (mt-rDNA) and the other specific for the nuclear RNase P gene to be used as an internal standard. The amount of mtDNA relative to the amount of reference is calculated from their respective threshold cycles (Cts). Since the mtDNA content varies among different tissues and at different ages, a series of age-matched normal tissue controls are always included when mtDNA copy number is studied.

Reagents

1. Primers were designed using Primer Express version 2.0 software (Applied Biosystems, Foster City, CA). For the amplification of 12s mt-rDNA, use primers with the following sequences: 5′-ccacgggaaacagcagtgatt-3′ (forward) and 5′-ctatt-gacttgggttaatcgtgtga-3′ (reverse). Stock solutions (100 μM) of these primers are prepared in water and stored at $-80\,^{\circ}$C.

2. The probe used to quantitate accumulation of mt-rDNA by real-time PCR was also designed using Primer Express version 2.0 software (Applied Biosystems). It is a specific fluorescently labeled Taqman probe (FAM-5′-tgccagccaaccgcg-3′) that is used at a final concentration of 0.125 μM.

3. As for the internal standard and for the purpose of normalizing results, a nuclear gene (RNAse P) is concomitantly amplified. Reagents required for the amplification of RNAse P gene are available as a reagent kit (PDARs RNAse P, Cat. No. 4316844, Applied Biosystems).

4. TaqMan Universal Master Mix (Cat. No. 4304437, Applied Biosystems).

5. 96-Well optical reaction plate (Cat. No. 4306737, Applied Biosystems).

Protocol

Real-time PCR amplification is carried out in a 20-μl final volume containing:

2× Taqman Universal PCR Master Mix	10.0 μl
Forward primer (4.4 μM)	0.5 μl
Reverse primer (4.4 μM)	0.5 μl
TaqMan probe (5 μM)	0.5 μl
PDARS RNAse P (20×)	1.0 μl
Water	5.5 μl
DNA template (10–50 ng)	2.0 μl

The optimum thermal program for amplification is one cycle of 50 °C for 2 min, 95 °C for 10 min, followed by 40 cycles of 95 °C for 15 sec, and 60 °C for 1 min. In each run, a set of serially diluted genomic DNA containing 500, 250, 100, 50, 10, and 1ng DNA is included to construct the standard curve. The amount of mtDNA relative to RNAse P is calculated using the following formula: mtDNA/RNAseP $= 2^{-(Ct_{mtDNA}-Ct_{RNASeP})}$, where Ct is the threshold cycle.

Notes

1. All amplifications should be carried out in triplicates.

2. At least two no-template controls should be included with each run.

3. Each DNA specimen should be analyzed at least at two and preferably three different levels. This is necessary to confirm the linearity of the assay.

4. For accurate estimation of relative amount of mtDNA, the Ct values used for calculation should be within the linear exponential phase.

VI. Point Mutations in mtDNA

More than 150 pathogenic mtDNA point mutations have been documented and new ones are still being described (DiMauro and Schon, 2001; Servidei, 2003) (http://www.mitomap.org). These point mutations are located in all regions of the genome. About two-thirds of them are located in tRNA genes. The rest are located in the polypeptide-coding genes, with a few in rRNA genes. Some of these mutations have been reported in only one or a few families, while others appear to be more common. Two of the more common mtDNA point mutations in tRNA genes are associated with MERRF (myoclonus epilepsy and ragged red fibers) and MELAS syndromes (mitochondrial encephalomyopathy, lactic acidosis, and stroke-like episodes). The first point mutation in a tRNA gene to be reported was an A-to-G transition at nt-8344 (A8344G) in patients with MERRF (Shoffner et al., 1990). This mutation is present in ~90% of patients with MERRF, but other mutations have been reported (Servidei, 2003). The A-to-G transition at nucleotide (nt) position 3243 (A3243G) is the most frequently encountered mtDNA point mutation and it is associated with a spectrum of clinical presentations. It was originally described in patients with MELAS (Goto et al., 1990), but was subsequently observed in maternal relatives of patients with MELAS who were often oligosymptomatic, presenting with diabetes, short stature, migraine headaches, hearing loss, or other features, either alone or in various combinations (DiMauro and Schon, 2001). In addition, this same mutation has been described in some patients with maternally inherited PEO (Moraes et al., 1993) as well as diabetes and deafness (Kadowaki et al., 1995). Infantile presentation, while infrequent, has also been reported (Sue et al., 1999).

The more common mutations in protein-coding genes are associated with two disorders: LHON (Leber hereditary optic neuropathy) and NARP/MILS (neuropathy, ataxia, and retinitis pigmentosa/maternally inherited Leigh syndrome). The G-to-A transition at nt-11,778 in patients with LHON, a disorder characterized by acute or subacute onset of bilateral visual loss in young adults, was the first point mutation in mtDNA to be reported (Wallace et al., 1988). LHON is associated with three primary mutations in genes encoding subunits of complex I of the respiratory chain—G3460A, G11778A, and T14484C (Servidei, 2003)—and these mutations are usually homoplasmic.

A T-to-G mutation at nt-8993 (T8993G) in the mtDNA gene encoding ATPase 6 was initially described in patients with a maternally inherited multisystem disorder characterized by NARP (Holt et al., 1990). A family was subsequently reported in which an adult family member had the NARP syndrome and three children had MILS (Tatuch et al., 1992). This association has subsequently been confirmed. It is now established that, for this particular mutation, there is a good correlation between the percentage of mutant genomes (i.e., degree of heteroplasmy) and the severity of the clinical phenotype. Patients with low levels of mutation are unaffected or have only mild symptoms, those with intermediate levels present with NARP, whereas those with very high levels (more than 95%) of mutation are associated with MILS (White et al., 1999). In addition to the T-to-G mutation, patients with a T-to-C substitution at the same position (T8993C) have been reported; they have milder clinical manifestations (DiMauro and Bonilla, 2003).

Some patients with exercise intolerance often harbor mutations in protein-coding genes causing defects in complexes I, III, or IV (Andreu et al., 1999; DiMauro and Schon, 2001). These patients are almost always sporadic, with no evidence of maternal transmission. Additionally, in most cases, these mutations are present only in muscle and are not detectable in blood or fibroblasts. It is believed that these mutations are somatic, that is, they are spontaneous events that occurred in muscle and do not affect germ line cells.

Finally, pathogenic mutations in rRNA genes are rare. The most common one is in the 12S rRNA gene (A1555G) and has been reported in families with aminoglycoside-induced deafness or nonsyndromic hearing loss (Prezant et al., 1993).

A. Detecting mtDNA Point Mutations by PCR/RFLP Analysis

Known point mutations in mtDNA are screened for by amplifying an appropriate fragment of mtDNA by PCR, digesting with a diagnostic restriction enzyme, and analyzing the fragment sizes on a gel. This involves the following steps: (1) oligonucleotide primers are designed to allow amplification of mtDNA encompassing the mutated site, (2) appropriate mtDNA fragments are amplified by PCR amplification in the presence of ^{32}P-dATP, (3) PCR products are digested with a restriction enzyme that will cleave normal versus mutated sequences differentially (i.e., restriction fragment length polymorphism or RFLP), (4) digestion products are electrophoresed through nondenaturing polyacrylamide gels, and (5) fragment sizes are detected by autoradiography. PCR/RFLP analysis for the A3243G mutation is illustrated in Fig. 4. While more than 150 mtDNA point mutations have been reported to date (Schon et al., 2001; Servidei, 2001), fewer than two dozen are encountered with any frequency. The primer sequences, amplification conditions, and RFLP analyses required to detect these more common mutations are shown in Table I.

Tissue selection

It has become common practice to look for mtDNA point mutations in DNA isolated from blood. However, in certain cases, muscle DNA provides more secure molecular diagnoses and, in others, it is essential to do the analysis in DNA isolated from muscle. When looking for point mutations in tRNA genes, blood is usually an adequate source of tissue. However, some point mutations, typically sporadic point mutations causing myopathies (often in cytochrome *b*), are present only in muscle and not in blood. In addition, some mutations, including the common A3243G "MELAS" mutation, may have a low proportion of mutant genomes in blood, especially in maternal relatives of patients. Other accessible tissues that have higher mutation loads than blood that are useful for diagnosis include urine and cheek mucosa (Shanske et al., 2004). Cells from urinary sediment are obtained by collecting urine (about 10 ml, preferably from the first morning void). Cells from cheek mucosa are obtained by scraping the inside of the mouth with a cotton swab, dipping the swab in 3–5 ml of saline, and obtaining a pellet by low-speed centrifugation.

PCR amplification

1. Prepare the PCR cocktail as follows:
   a. ddH$_2$O   28.5 μl
   b. 10× Taq buffer   5 μl
   c. dNTP mix (2 mM each dNTP)   5 μl

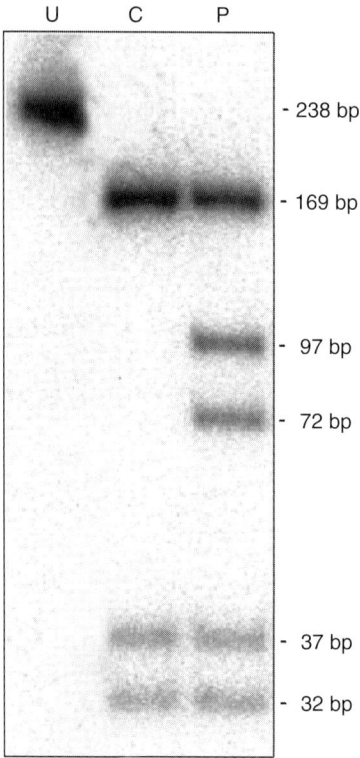

Fig. 4 PCR/RFLP analysis for the A3243G MELAS mutation. A 238-bp PCR product is digested with the restriction endonuclease *Hae*III. The A-to-G mutation at position 3243 creates an additional, diagnostic restriction site for *Hae*III, so that a 169-bp fragment that is present in normal DNA is cleaved into a 97-bp and a 72-bp fragment. Normal DNA has fragments of 169, 37, and 32 bp, whereas mutant DNA has fragments of 169, 97, 72, 37 and 32 bp. Adapted from Shanske and Wong (2004), with permission.

     d. Forward primer (2 pmol/μl)   5 μl
     e. Reverse primer (2 pmol/μl)   5 μl
     f. Template DNA   1 μl

2. Add 0.5-μl (2.5 units) Taq polymerase.
3. Set appropriate PCR program and run.
4. Remove PCR product, check 10 μl (+2 μl 6× loading dye) on a 0.8% agarose gel in 1× TAE buffer (40-mM Tris–acetate, 1-mM EDTA). Run the gel at 100 V.

Table I

Conditions for PCR/RFLP Analyses of the More Commonly Encountered Pathogenic mtDNA Point Mutations[a]

Mutation		Forward primer		Backward primer	
nt	Mut	Range	Sequence, 5'→3'	Range	Sequence, 5'→3'
1555	A→G	1009–1032	CACAAAATAGACTACGAAAGTGGC	1575–1556	ACTTACCATGTTACGACTGG
3243	A→G	3116–3134	CCTCCCTGTACCAAAGGAC	3353–3333	GCGATTAGAATGGGTACAATG
3256	C→T	3230–3255	GTTAAGATGGCAGCGCCGGTAAGCG	3353–3333	GCGATTAGAATGGGTACAATG
3271	T→C	3148–3169	CCTACTTCACAAAGCGCCTTCC	3295–3272	GAGGAATTGAACCTCGACTCTAA
3303	C→T	3277–3300	GTCAGAGGTTCAATTCCTCTTGTT	3353–3333	GCGATTAGAATGGGTACAATG
3460	G→A	3116–3134	CCTCCCTGTACCAAAGGAC	4349–4329	CCTCCCTGTACCAAAGGAC
4300	A→G	4281–4298	AGAGTTACTTTGATAGGG	4544–4523	GTGCGAGCTTAGCGCTGTGATG
4320	C→T	4185–4207	CCTACCACTCACCCTAGCATTAC	4544–4523	GTGCGAGCTTAGCGCTGTGATG
4409	T→C	389–4410	CACCCCATCCTAAAGTAACGTC	4544–4523	GTGCGAGCTTAGCGCTGTGATG
5537	InsT	5514–5536	AAATTTAGGTTAAATACAGACGA	5798–5777	GCAAATTCGAAGAAGCAGCTTC
5703	G→A	5472–5491	CTACTCCTACCTATCTCCCC	5798–5777	GCAAATTCGAAGAAGCAGCTTC
5814	T→C	5685–5709	CACTTAGTTAACAGCTAAGCACCC	5982–5960	CAGCTCATGCGCCGAATAATAGG
8344	A→G	8278–8296	CTACCCCCTCTAGAGCCCAC	8385–8345	GTAGTATTTAGTTGGGGCATTTCA-CTGTAAAGCCGTGTTGG
8356	T→C	8239–8263	CTTTGAAATAGGGCCCGTATTTACC	8380–8357	ATTTAGTTGGGGCATTTA
8363	G→A	8241–8361	AGAACCAACACCTCTTTACGG	8582–8561	GCGGGTAGGCCTAGGATTGTGG
8851	T→C	8829–8852	CCTAGCCATGGCCATCCCGTATG	9278–9256	GGCTGAGAGGGCCCCTGTTAGG
8993	T→G	8657–8685	CCACCCAACAATGACTAATCAAACTAACC	9278–9256	GGCTGAGAGGGCCCCTGTTAGG
8993	T→C	8657–8685	CCACCCAACAATGACTAATCAAACTAACC	9278–9256	GGCTGAGAGGGCCCCTGTTAGG
9176	T→C	8896–8914	GCCCTAGCCCACTTCTTAC	9278–9256	GTGTTGTCGTGCAGGTAGGGCTTCCT
9957	T→C	9933–9956	CATTTTGTAGATGTGGTTTGAGTA	10269–10238	GGGCAATTTCTAGATCAAATAATAAGAAGG
9997	T→C	9910–9931	CTTTGGCTTCGAAGCCGCCGCC	10269–10238	GGGCAATTTCTAGATCAAATAATAAGAAGG
10044	A→G	9910–9931	CTTTGGCTTCGAAGCCGCCGCC	10269–10238	GGGCAATTTCTAGATCAAATAATAAGAAGG
11084	A→G	1059–11083	CTCCCTACAAATCTCCTTAATTCTA	11272–11247	TAGGGTGTTGTGAGTGTAAATTAGTG
11778	G→A	12413–12393	CCAAACCCCTGAAGCTTCACGGCGCAG	12413–12393	GGGTTAACGAGGGTGGTAAGG
13513	G→A	13491–13512	CCTCACAGGTTTCTACTCCGAA	13610–13593	CGAGTGCTATAGGCGCTT
14459	G→A	14430–14458	ATGCCTCAGGATACTCCTCAATAGCCGTC	14874–14855	AGGATCAGGCAGCGCCAAG
14484	T→C	14463–14484	TAGTATATCCAAAGACAACGA	14810–14790	TTTGGGGGAGGTTATATGGG
14709	T→C	14688–14707	CTACAACCACGACCAATGACA	14967–14924	TTTACGTCTCGAGTGATGTGGGCG
15590	C→T	15803–15825	GTAGCATCCGTACTATACTTCAC	16014–15991	GTTTAAATTAGAATCTTAGCTCCG

(continues)

Table I *(continued)*

| Mutation | | PCR conditions | | | | | | RFLP analysis | |
nt	Mut	Cyc	Denat.	Anneal	Extend	Size	Enz	Normal	Mutant
1555	A→G	25	1 min @94	1 min@55	1 min@72	566	*Hae*III	455, **111**	455, **91**, **20**
3243	A→G	25	1 min @94	1 min@55	45 sec@72	238	*Hae*III	**169**, 37, 32	**97**, **72**, 37, 32
3256	C→T	25	1 min @94	1 min@55	1 min@72	123	*Hin*PI	**99**, 13, **11**	**110**, 13
3271	T→C	25	1 min @94	1 min@55	45 sec@72	147	*Dde*I	**103**, 44	**79**, 44, **24**
3303	C→T	25	1 min @94	1 min@55	1 min@72	76	*Hpa*I	**50**, **26**	**76**
3460	G→A	25	1 min @94	1 min@55	2 min@72	1233	*Acy*I	**889**, **344**	**1233**
4300	A→G	25	1 min @94	1 min@55	1 min@72	263	*Hph*I	**263**	**235**, **28**
4320	C→T	25	1 min @94	1 min@55	1 min@72	359	*Mnl*I	**282**, 77	**216**, 77, **66**
4409	T→C	30	1 min @94	1 min@60	1 min@72	155	*Mae*II	**137**, **18**	**155**
5537	InsT	25	1 min @94	1 min@60	1 min@72	284	*Mbo*II	**250**, **34**	**284**
5703	G→A	25	1 min @94	1 min@50	1 min@72	326	*Dde*I	**99**, 80, 60, 41, 34, **12**	**111**, 80, 60, 41, 34
5814	T→C	25	1 min @94	1 min@55	1 min@72	297	*Hph*I	**121**, 98, **78**	**199**, 98
8344	A→G	25	1.5 min @94	1.5 min@55	1 min@72	108	*Bgl*I	**108**	**73**, **35**
8356	T→C	25	1 min @94	1 min@55	1 min@72	141	*Dra*I	**119**, **22**	**141**
8363	G→A	25	1 min @94	1 min@55	1 min@72	240	*Hph*I	**113**, **53**, 44, **30**	**113**, **83**, 44
8851	T→C	25	1 min @94	1 min@60	1 min@72	450	*Rsa*I	280, **170**	280, **148**, **22**
8993	T→G	25	1 min @94	1 min@60	2 min@72	621	*Msp*I	**621**	**336**, **285**
8993	T→C	25	1 min @94	1 min@60	2 min@72	621	*Msp*I	**621**	**337**, **284**
9176	T→C	25	1 min @94	1 min@60	1 min@72	308	*Scr*FI	**211**, 97	**184**, 97, **27**
9957	T→C	25	1 min @94	1 min@55	1 min@72	336	*Rsa*I	259, 77	259, **54**, **23**
9997	T→C	25	1 min @94	1 min@55	1 min@72	367	*Bfa*I	162, **128**, 65, 12	162, **95**, 65, **33**, 12
10044	A→G	25	1 min @94	1 min@55	1 min@72	367	*Acl*I	**367**	**225**, **142**
11084	A→G	25	1 min @94	1 min@55	1 min@72	213	*Bfa*I	**161**, 52	**139**, 52, **22**
11778	G→A	25	1 min @94	1 min@60	1 min@72	745	*Sfa*NI	**634**, **111**	**745**
13513	G→A	30	1 min @94	1 min@60	1 min@72	119	*Bbs*I	**92**, **27**	**119**
14459	G→A	25	1 min @94	1 min@55	1 min@72	444	*Mae*III	**444**	**419**, **25**
14484	T→C	25	1 min @94	1 min@50	1 min@72	75	*Dpn*II	**54**, **21**	**75**
14709	T→C	25	1 min @94	1 min@55	1 min@72	279	*Nla*III	**124**, 70, 62, **23**	**147**, 70, 62
15590	C→T	25	1 min @94	1 min@50	1 min@72	211	*Msp*I	123, **66**, **22**	123, **88**

[a]Underlined letters denote mismatched nucleotides in primer. Underlined numbers indicate the PCR fragments associated with the RFLP.

Digestion and electrophoresis of PCR products

1. Perform "last-cycle-hot" labeling of PCR products. Add 0.5- to 1.0-μl [α-^{32}P]-dATP to 40 μl of PCR reaction, and perform one extension at 94 °C for 10 min.

2. Digest 10 μl of labeled PCR product with a diagnostic restriction enzyme at the appropriate temperature for ~2 h or overnight, as follows:

   a. PCR product   10 μl
   b. Buffer (10×)   4 μl
   c. Restriction enzyme   1 μl
   d. H$_2$O   25 μl

3. To 40 μl of restriction digest, add 8-μl dye, load ~20 μl on a 12% acrylamide gel (37.5:1 acrylamide:bis-acrylamide) in 1 × TBE (90-mM Tris–borate, 1-mM EDTA). Run at 300 V, 15W for about 1 h.

4. Remove the gel and place between two plastic sheet protectors.

5. Autoradiograph with Kodak XAR-5 film. Expose overnight and develop.

6. Detect and quantitate relevant fragments in a phosphorimager.

Notes

1. When designing a protocol for PCR/RFLP analysis, one should first look for a naturally occurring polymorphism at the site of the mutation. Often, such a convenient site is not available, in which case a "mismatched" primer can be synthesized to create an appropriate restriction site. In either case, it is better to perform the analysis so that the mutant allele "gains" the polymorphic site rather than "loses" it because a gain-of-site mutation is unambiguous—it generates a new fragment on the gel— whereas a loss-of-site cannot be distinguished from an uncut PCR fragment. In addition, it is useful to design the PCR/RFLP protocol so that at least two cutting sites are present on the amplified DNA. One site is present on both alleles and serves as an internal control for completeness of digestion and can assist in the quantitation of the percent heteroplasmy. The other site is present on either the wild-type or mutated allele but not on both, and is therefore diagnostic for the mutation.

2. During the amplification of a heteroplasmic population of mtDNAs, three species of products are formed after each cycle of denaturation/renaturation: wild-type homoduplexes, mutated homoduplexes, and heteroduplexes consisting of one wild-type strand annealed to one mutated strand. Cleavage of this mixture in the RFLP step will result in an underestimation of the proportion of mutation, because only the perfect homoduplexes will be cut. This problem can be circumvented by the addition of a labeled nucleotide in the last extension cycle of the PCR reaction, a procedure know as "last-cycle-hot" PCR (Moraes *et al.*, 1992; Petruzzella *et al.*, 1994). Under these conditions, the only species that are visualized (and cut) are those that have incorporated the label in a perfectly complementary "daughter" strand. In other words, as a result of last cycle labeling,

uncuttable heteroduplexes can be cut and visualized on the gel, because the heteroduplexes were denatured and converted to pairs of perfect duplexes in the final extension step.

3. If label was added at the beginning of the PCR reaction (i.e., "last-cycle-hot" PCR was not performed), it is prudent to estimate the "true" percent mutation by taking the square root of the raw proportion of the mutation. For example, if 64% of the alleles are calculated to be mutated, the true proportion of mutated molecules is actually 80% (i.e., $[0.64]^{1/2} = 0.8$).

4. If the proportion of heteroplasmy is calculated with unlabeled fragments (e.g., by densitometry of ethidium bromide-stained fragments on a gel), it is absolutely critical to amplify a set of serially diluted known samples as a standard. The standard can be DNA from homoplasmic cells or from templates subcloned into plasmids.

5. The use of RFLP analysis to detect point mutations may sometimes result in a false-positive result. This is especially true in the case of loss-of-site mutations, not only because of the problem of partial or failed digestion but also because a neutral mutation elsewhere within the restriction enzyme's recognition site will render the site uncleavable, even if the target mutation is absent. This problem was first uncovered in the case of the use of *Sfa*NI to detect mutations at nt-11778 in LHON patients (Johns and Neufeld, 1993; Mashima *et al.*, 1995; Yen *et al.*, 2000). Note that false positives may also exist in gain-of-site mutations when a new restriction site is generated by a polymorphism, especially when located in a region near the diagnostic RFLP site (Kirby *et al.*, 1998).

B. Quantification of Heteroplasmy Level by the Amplification–Refractory Mutation System and Quantitative Real–Time PCR

Amplification-refractory mutation system (ARMS), as described by Newton *et al.* (1989), is widely employed for detecting specific pathogenic mutations. Since a prerequisite for a successful DNA amplification during PCR is the presence of a perfect match at the $3'$ terminus of the primer, oligonucleotide primers can be designed to amplify target DNA sequences differing only by a single nucleotide in the region of interest. For this method, two reactions are set up: one for wild-type and one for mutant DNA. One of the pair of primers is common for both reactions and the other is specific for either wild type or mutant.

The principle of quantitative PCR (QPCR) is that, under an optimized conditions, the amount of amplicon will nearly double during each PCR cycle and this can be measured by the fluorescence intensity of a reporter dye present in the reaction. The measurement can be performed either at the end of the reaction cycles (end point QPCR) or while the amplification is progressing (real-time QPCR). However, more sensitive and reproducible results are obtained with real-time QPCR because the amplification efficiency tends to decrease during later cycles, due to the consumption of reagents and accumulation of inhibitors.

Fluorescent reporter dyes used in real-time QPCR can be a sequence-specific probe (an oligonucleotide labeled with a fluorescent molecule, such as FAM or TAMRA) or a DNA-binding fluorescent compound, such as SYBR green, which binds nonspecifically to any double-stranded DNA molecule. The fluorescence intensity of SYBR green increases about 1000-fold when it is bound to DNA. Therefore, by adding SYBR green to the PCR reagent, accumulation of newly synthesized DNA can be measured in real time by fluorescence intensity during the reaction.

Application of ARMS in combination with real-time QPCR methodology for mtDNA analysis has been described (Bai *et al.*, 2004). By employing this technique, it is now possible to reliably measure mitochondrial heteroplasmy levels below 5%, which cannot be obtained by RFLP and radiolabeling. The method described below is specific for the quantification of A3243G heteroplasmy level. However, by designing the appropriate primers and under optimized conditions, the method can be applied for the investigation of any other mtDNA mutation.

Reagents

1. Extract DNA from 0.3 ml of citrate- or EDTA-preserved blood using a DNA extraction kit (e.g., reagents and protocols supplied by Gentra Systems, Minneapolis, MN).

2. Olignucleotide sequences for wild type (3243A) and mutant (3243G) are as described previously (Bai *et al.*, 2004).

ARMS wild-type forward primer	5'-AGGGTTTGTTAAGATGGCtcA-3'
ARMS mutant forward primer	5'-AGGGTTTGTTAAGATGGCtcG-3'
ARMS revese primer	5'-TGGCCATGGGTATGTTGTTA-3'

Stock solutions of 100 μM are prepared in autoclaved deionized water and stored at $-20\,^{\circ}$C. Working solutions (5 μM) are prepared by diluting stock solutions 20 times with autoclaved deionized water, and stored at $-20\,^{\circ}$C.

3. PCR reagents: Taq DNA polymerase (Cat. No. 11192122), Taq buffer (Cat. No. 11192122), and dNTP (Cat. No. 11581295001) from Roche Diagnostics.

4. Real-time PCR reagents: Platinum® SYBR® Green (Cat. No. 11733-038; Invitrogen, Carlsbad, CA) and 96-well optical reaction plates (Cat. No. 4306737; Applied Biosystems).

5. Cloning reagents: for cloning PCR products we use pGEM® supplied by Promega (Madison, WI).

Instruments

Other than common laboratory equipment, a sensitive UV-spectrophotometer capable of taking measurements on small volumes and a real-time fluorescence reader are also required. It must be emphasized that because the calculation of mtDNA copy number during amplification is based on the initial number of mtDNA templates used, it is imperative that the initial concentration of DNA template is determined accurately. For this purpose, we use a NanoDrop® ND-100 spectrophotometer (NanoTechnologies, Wilmington, DE). This machine requires 1 μl of undiluted extracted or PCR-amplified DNA to determine the DNA concentration accurately. For the fluorescence reader, we use the ABI-Prism 7000 Sequence Detection System (Applied Biosystems). Obviously, similar instruments with the same capability manufactured by other companies can also be used.

Protocol

Construction of standard curve for the wild-type and mutant DNA:

Amplify 10-ng DNA extracted from 0.3-ml blood from a patient known to harbor the A3243G mutation. Two 50 μl-reaction mixes (one for the wild type and one for the mutant) contain 10-ng template DNA, 1× PCR buffer, 200-μM each dNTP, 200 nM each of forward and reverse primers, and 1 unit of Taq DNA polymerase. Amplification is carried out with the following thermal cycles: one cycle of initial denaturation at 95 °C, 5 min; followed by 30 cycles of denaturing at 95 °C, 40 sec; annealing at 55 °C, 30 sec; and extension at 72 °C, 30 sec. One cycle of 72 °C for 7 min is included for the final extension. The correct size (100bp) of both amplicons is verified on a 2% agarose gel. A 2-μl aliquot of PCR product is cloned directly following the manufacturer's suggested protocol.

Wild-type and mutant transformed colonies are grown in a 5-ml medium in culture tubes overnight and the plasmids are isolated and purified as described (Sambrook and Russell, 2001). It is important that plasmid preparations are free of RNA. Therefore, RNAse treatment of plasmid preparations are highly desirable. One can verify the correct sequence of wild type and mutant insert by direct sequencing of plasmids using the M13 forward (−20) primer. Measure the concentration of DNA and calculate the copy number of mtDNA molecules based on the molecular weight of plasmid, calculated as follows:

$m = [n] [1.096 \times 10^{-21} \text{ g/bp}]$, where n = genome size (bp plasmid + insert), and m = mass.

Make appropriate dilutions so that each 2-μl aliquot contains 106, 105, 104, 103, 102, 10, and 0 copies of either wild-type or mutant mtDNA.

Quantification:

Quantitative real-time PCR should be carried out in 96-well plates at least in duplicates and preferably in triplicates. In each run include two sets of standards, one for wild type and one for mutant mtDNA. For each reaction, mix the following:

2× Platinum® SYBR® Green mix	10.0 μl
Forward primer (ARMS wild type or mutant)	2.0 μl
Reverse primer (common for wild type and mutant)	2.0 μl
ROX dye (25 μM)	0.4 μl
Sterile deionized water	3.6 μl
Template DNA (2.5 ng/μl)	2.0 μl
Total volume	20.0 μl

Calculation of % heteroplasmy:

Since within the linear range of the standard curve the Ct is directly related to the initial copy number of template, the proportion of mutant mtDNA relative to wild type is estimated by their Cts using this formula:

$$\text{Proportion of mutant is} = \frac{1}{[1 + 1/2^{(Ct_{\text{wild type}} - Ct_{\text{mutant}})}]}$$

Notes

1. For convenience, make a master mix containing all of the above reagents except the DNA template. Calculate the amount of reagents needed by taking into account the number of two sets of standards and the number of DNA specimens to be analyzed. Using an automatic dispenser, add 18 μl of master mix into each well. Then add 2 μl of standard or DNA template samples and mix by repeated up-and-down pipetting.

2. For real-time amplification, use the following thermal preset program: one cycle of 50 °C, 2 min; followed by one cycle of 95 °C, 10 min; and 45 cycles of 95 °C, 15 sec, 63 °C, 1 min. At the end of the thermal cycles, analyze and review the results. Linearity of standard curve ($\log_{10}$ concentration versus threshold cycle) should be close to $R^2 = 1$ (values of 0.98 and above are acceptable). Amplification efficiency (E) should be close to 3.32, that is, the slope of standard curve should be -3.32 or lower. Finally, inspection of the dissociation curve for standard and specimen wells should show only a single curve representing a single melting product. Any well that shows several peaks with different melting points indicates multiple PCR products and Cts from these wells should not be trusted and must be repeated, because no meaningful quantification can be obtained from these data.

C. Detection of Unknown Mutations in mtDNA

When clinical features, maternal inheritance, muscle histochemistry, or biochemical results suggest an mtDNA-related disorder, an effort should be made to define the mutation. Guided by the clinical picture, one should first screen for the most common mutations associated with that particular syndrome, and then extend the search to the more rare mutations. However, in many patients the molecular defects remain unidentified.

Several methods for screening of unknown point mutations in mtDNA have been developed. These include single-strand conformation polymorphism (SSCP) (Suomalainen *et al.*, 1992), temporal temperature gradient gel electrophoresis (TTGE) (Chen *et al.*, 1999; Wong *et al.*, 2002), temperature gradient gel electrophoresis (TGGE) (Wartell *et al.*, 1990), denaturing gradient gel electrophoresis (DGGE) (Michikawa *et al.*, 1997; Sternberg *et al.*, 1998), and denaturing high performance liquid chromatography (dHPLC) (van Den Bosch *et al.*, 2000). Researchers at Johns Hopkins University and Affymetrix (Santa Clara, CA) have developed a mitochondrial Custom Reseq® microarray for rapid and high throughput analysis of mitochondrial DNA (Maitra *et al.*, 2004). The MitoChip® contains oligonucleotide probes and is able, according to the manufacturer, to sequence both strands of the entire human mitochondrial coding sequence with >99.99% accuracy in 48 h. With the availability of high throughput automatic nucleic acid sequencing technology, direct DNA sequencing is now often the preferred tool for mutation detection. References for some of these methodologies are included, but no details are provided here.

Regardless of the methodology employed, whenever a change from the reference sequence is detected, it is necessary to establish whether this is a benign polymorphism or a pathogenic mutation. Because mtDNA mutations are fixed in the population at a much higher rate than are nuclear DNA mutations, the mitochondrial genome harbors large numbers of neutral base changes that have no pathogenic significance (Wallace *et al.*, 1999). There are a number of criteria that are used to support the deleterious role of a novel mutation. Obviously, the mutation must not be a known neutral polymorphism, and it should be absent in controls from the same ethnic group. The base change must affect an evolutionarily conserved and functionally important site. Deleterious mutations are usually (but not always) heteroplasmic, while neutral changes are homoplasmic. The degree of heteroplasmy in different family members ought to be in rough agreement with the severity of symptoms. An elegant way to confirm pathogenicity is to correlate mutational load and functional abnormality by single-fiber PCR. In this protocol, the abundance of mutation is studied by PCR in individual abnormal muscle fibers (chosen on the basis of excessive mitochondrial proliferation, i.e., RRF) (Moraes and Schon, 1996). It is important to emphasize that while methodology is now available that allows for rather rapid and easy detection of base changes in mtDNA, this is but a first step. Simply describing changes in an individual's mtDNA that differ from a reference sequence does little to elucidate the basis for their disorder. It is essential that such findings be followed up with analyses that help establish whether a particular base change is pathogenic.

Acknowledgments

This chapter was adapted in large part from Schon *et al.* (2002).

References

Andreu, A. L., Hanna, M. G., Reichmann, H., Bruno, C., Penn, A. S., Tanji, K., Pallotti, F., Iwata, S., Bonilla, E., Lach, B., Morgan-Hughes, J., and DiMauro, S. (1999). Exercise intolerance due to mutations in the cytochrome *b* gene of mitochondrial DNA. *N. Engl. J. Med.* **341,** 1037–1044.

Bai, R. K., Perng, C. L., Hsu, C. H., and Wong, L. J. (2004). Quantitative PCR analysis of mitochondrial DNA content in patients with mitochondrial disease. *Ann. N. Y. Acad. Sci.* **1011,** 304–309.

Bernes, S. M., Bacino, C., Prezant, T. R., Pearson, M. A., Wood, T. S., Fournier, P., and Fischel-Ghodsian, N. (1993). Identical mitochondrial DNA deletion in mother with progressive external ophthalmoplegia and son with Pearson marrow-pancreas syndrome. *J. Pediatr.* **123,** 598–602.

Boles, R. G., Chun, N., Senadheera, D., and Wong, L. J. (1997). Cyclic vomiting syndrome and mitochondrial DNA mutations. *Lancet* **350,** 1299–1300.

Chen, T. J., Boles, R. G., and Wong, L. J. (1999). Detection of mitochondrial DNA mutations by temporal temperature gradient gel electrophoresis. *Clin. Chem.* **45,** 1162–1167.

DiMauro, S., and Bonilla, E. (2003). Mitochondrial disorders due to mutations in the mitochonrial genome. *In* "The Molecular and Genetic Basis of Eurologic and Psychiatric Disease" (R. N. Rosenberg, S. B. Prusiner, S. DiMauro, R. L. Barchi, and E. J. Nestler, eds.), Vol., pp. 189–196. Butterworth-Heinemann, Boston.

DiMauro, S., and Schon, E. A. (2001). Mitochondrial DNA mutations in human disease. *Am. J. Med. Genet.* **106,** 18–26.

DiMauro, S., Tanji, K., Bonilla, E., Pallotti, F., and Schon, E. A. (2002). Mitochondrial abnormalities in muscle and other aging cells: Classification, causes, and effects. *Muscle Nerve* **26,** 597–607.

Goto, Y., Nonaka, I., and Horai, S. (1990). A mutation in the tRNA(Leu)(UUR) gene associated with the MELAS subgroup of mitochondrial encephalomyopathies. *Nature* **348,** 651–653.

Graff, C., Wredenberg, A., Silva, J. P., Bui, T. H., Borg, K., and Larsson, N. G. (2000). Complex genetic counselling and prenatal analysis in a woman with external ophthalmoplegia and deleted mtDNA. *Prenat. Diagn.* **20,** 426–431.

Holt, I. J., Harding, A. E., Petty, R. K., and Morgan-Hughes, J. A. (1990). A new mitochondrial disease associated with mitochondrial DNA heteroplasmy. *Am. J. Hum. Genet.* **46,** 428–433.

Johns, D. R., and Neufeld, M. J. (1993). Pitfalls in the molecular genetic diagnosis of Leber hereditary optic neuropathy (LHON). *Am. J. Hum. Genet.* **53,** 916–920.

Kadowaki, T., Sakura, H., Otabe, S., Yasuda, K., Kadowaki, H., Mori, Y., Hagura, R., Akanuma, Y., and Yazaki, Y. (1995). A subtype of diabetes mellitus associated with a mutation in the mitochondrial gene. *Muscle Nerve* **3,** S137–S141.

Kaukonen, J., Juselius, J. K., Tiranti, V., Kyttala, A., Zeviani, M., Comi, G. P., Keranen, S., Peltonen, L., and Suomalainen, A. (2000). Role of adenine nucleotide translocator 1 in mtDNA maintenance. *Science* **289,** 782–785.

Kirby, D. M., Milovac, T., and Thorburn, D. R. (1998). A false-positive diagnosis for the common MELAS (A3243G) mutation caused by a novel variant (A3426G) in the ND1 gene of mitochondria DNA. *Mol. Diagn.* **3,** 211–215.

Larsson, N. G., Eiken, H. G., Boman, H., Holme, E., Oldfors, A., and Tulinius, M. H. (1992). Lack of transmission of deleted mtDNA from a woman with Kearns-Sayre syndrome to her child. *Am. J. Hum. Genet.* **50,** 360–363.

Maitra, A., Cohen, Y., Gillespie, S. E., Mambo, E., Fukushima, N., Hoque, M. O., Shah, N., Goggins, M., Califano, J., Sidransky, D., and Chakravarti, A. (2004). The Human MitoChip: A high-throughput sequencing microarray for mitochondrial mutation detection. *Genome Res.* **14,** 812–819.

Mashima, Y., Hiida, Y., Saga, M., Oguchi, Y., Kudoh, J., and Shimizu, N. (1995). Risk of false-positive molecular genetic diagnosis of Leber's hereditary optic neuropathy. *Am. J. Ophthalmol.* **119,** 245–246.

Michikawa, Y., Hofhaus, G., Lerman, L. S., and Attardi, G. (1997). Comprehensive, rapid and sensitive detection of sequence variants of human mitochondrial tRNA genes. *Nucleic Acids Res.* **25,** 2455–2463.

Moraes, C. T., Ciacci, F., Silvestri, G., Shanske, S., Sciacco, M., Hirano, M., Schon, E. A., Bonilla, E., and DiMauro, S. (1993). Atypical clinical presentations associated with the MELAS mutation at position 3243 of human mitochondrial DNA. *Neuromuscul. Disord.* **3,** 43–50.

Moraes, C. T., Ricci, E., Bonilla, E., DiMauro, S., and Schon, E. A. (1992). The mitochondrial tRNALeu(UUR) mutation in MELAS: Genetic, biochemical, and morphological correlations in skeletal muscle. *Am. J. Hum. Genet.* **50,** 934–949.

Moraes, C. T., and Schon, E. A. (1996). Detection and analysis of mitochondrial DNA and RNA in muscle by *in situ* hybridization and single-fiber PCR. *Methods Enzymol.* **264,** 522–540.

Moraes, C. T., Shanske, S., Tritschler, H-J, Aprille, J. R., Andreetta, F., Bonilla, E., Schon, E., and DiMauro, S. (1991). Mitochondrial DNA depletion with variable tissue expression: A novel genetic abnormality in mitochondrial diseases. . *Am. J. Hum. Genet.* **48,** 492–501.

Newton, C. R., Graham, A., Heptinstall, L. E., Powell, S. J., Summers, C., Kalsheker, N., Smith, J. C., and Markham, A. F. (1989). Analysis of any point mutation in DNA. The amplification refractory mutation system (ARMS). *Nucleic Acids Res.* **17,** 2503–2516.

Nishino, I., Spinazzola, A., and Hirano, M. (1999). Thymidine phosphorylase gene mutations in MNGIE, a human mitochondrial disorder. *Science* **283,** 689–692.

Petruzzella, V., Moraes, C. T., Sano, M. C., Bonilla, E., DiMauro, S., and Schon, E. A. (1994). Extremely high levels of mutant mtDNAs co-localize with cytochrome *c* oxidase-negative ragged-red fibers in patients harboring a point mutation at nt-3243. *Hum. Mol. Genet.* **3,** 449–454.

Prezant, T. R., Agapian, J. V., Bohlman, M. C., Bu, X., Oztas, S., Qiu, W. Q., Arnos, K. S., Cortopassi, G. A., Jaber, L., Rotter, J. I., *et al.* (1993). Mitochondrial ribosomal RNA mutation associated with both antibiotic-induced and non-syndromic deafness. *Nat. Genet.* **4,** 289–294.

Sambrook, J., and Russell, D. W. (2001). "Molecular Cloning, A Laboratory Manual," 3rd edn. Cold Spring Harbor Laboratory Press, Cold Spring Harbor.

Schon, E. A. (2000). Mitochondrial genetics and disease. *Trends Biochem. Sci.* **25,** 555–560.

Schon, E. A., Hirano, M., and DiMauro, S. (2001). Molecular genetic basis of the mitochondrial encephalomyopathies. *In* "Mitochondrial Disorders in Neurology II" (A. H. V. Schapira, and S. DiMauro, eds.), Vol., in press. Butterworth-Heinemann, Oxford.

Schon, E. A., Naini, A., and Shanske, S. (2002). Identification of mutations in mitochondrial DNA from patients suffering mitochondrial disease. *Methods Mol. Biol.* **197,** 55–74.

Servidei, S. (2001). Mitochondrial encephalopathies: Gene mutation. *Neuromuscul. Disord.* **11,** 121–130.

Servidei, S. (2003). Mitochondrial encephalomyopathies: Gene mutation. *Neuromuscul. Disord.* **13,** 685–690.

Shanske, S., Pancrudo, J., Kaufmann, P., Engelstad, K., Jhung, S., Lu, J., Naini, A., DiMauro, S., and De Vivo, D. C. (2004). Varying loads of the mitochondrial DNA A3243G mutation in different tissues: Implications for diagnosis. *Am. J. Med. Genet. A* **130,** 134–137.

Shanske, S., Tang, Y., Hirano, M., Nishigaki, Y., Tanji, K., Bonilla, E., Sue, C., Krishna, S., Carlo, J. R., Willner, J., Schon, E. A., and DiMauro, S. (2002). Identical mitochondrial DNA deletion in a woman with ocular myopathy and in her son with pearson syndrome. *Am. J. Hum. Genet.* **71,** 679–683.

Shanske, S., and Wong, L.-J. C. (2004). Molecular analysis of mtDNA disorders. *Mitochondrion* **4,** 403–415.

Shoffner, J. M., Lott, M. T., Lezza, A. M., Seibel, P., Ballinger, S. W., and Wallace, D. C. (1990). Myoclonic epilepsy and ragged-red fiber disease (MERRF) is associated with a mitochondrial DNA tRNA(Lys) mutation. *Cell* **61,** 931–937.

Sternberg, D., Danan, C., Lombes, A., Laforet, P., Girodon, E., Goossens, M., and Amselem, S. (1998). Exhaustive scanning approach to screen all the mitochondrial tRNA genes for mutations and its application to the investigation of 35 independent patients with mitochondrial disorders. *Hum. Mol. Genet.* **7,** 33–42.

Sue, C. M., Bruno, C., Andreu, A. L., Cargan, A., Mendell, J. R., Tsao, C. Y., Luquette, M., Paolicchi, J., Shanske, S., DiMauro, S., and De Vivo, D. C. (1999). Infantile encephalopathy associated with the MELAS A3243G mutation. *J. Pediatr.* **134,** 696–700.

Suomalainen, A., Ciafaloni, E., Koga, Y., Peltonen, L., DiMauro, S., and Schon, E. A. (1992). Use of single strand conformation polymorphism analysis to detect point mutations in human mitochondrial DNA. *J. Neurol. Sci.* **111,** 222–226.

Tang, Y., Manfredi, G., Hirano, M., and Schon, E. A. (2000a). Maintenance of human rearranged mtDNAs in long-term-cultured transmitochondrial cell lines. *Mol. Biol. Cell* **11,** 2349–2358.

Tang, Y., Schon, E. A., Wilichowski, E., Vazquez-Memije, M. E., Davidson, E., and King, M. P. (2000b). Rearrangements of human mitochondrial DNA (mtDNA): New insights into the regulation of mtDNA copy number and gene expression. *Mol. Biol. Cell* **11,** 1471–1485.

Tatuch, Y., Christodoulou, J., Feigenbaum, A., Clarke, J. T., Wherret, J., Smith, C., Rudd, N., Petrova-Benedict, R., and Robinson, B. H. (1992). Heteroplasmic mtDNA mutation (T—G) at 8993 can cause Leigh disease when the percentage of abnormal mtDNA is high. *Am. J. Hum. Genet.* **50,** 852–858.

Tritschler, H.-J., Andreetta, F., Moraes, C. T., Bonilla, E., Arnaudo, E., Danon, M. J., Glass, S., Zelaya, B. M., Vamos, E., Telerman-Toppet, N., Kadenbach, B., DiMauro, S., *et al.* (1992). Mitochondrial myopathy of childhood associated with depletion of mitochondrial DNA. *Neurology* **42**, 209–217.

van Den Bosch, B. J., de Coo, R. F., Scholte, H. R., Nijland, J. G., van Den Bogaard, R., deVisser, M., de Die-Smulders, C. E., and Smeets, H. J. (2000). Mutation analysis of the entire mitochondrial genome using denaturing high performance liquid chromatography. *Nucleic Acids Res.* **28**, E89.

Wallace, D. C., Brown, M. D., and Lott, M. T. (1999). Mitochondrial DNA variation in human evolution and disease. *Gene* **238**, 211–230.

Wallace, D. C., Singh, G., Lott, M. T., Hodge, J. A., Schurr, T. G., Lezza, A. M., Elsas, L. J., 2nd, and Nikoskelainen, E. K. (1988). Mitochondrial DNA mutation associated with Leber's hereditary optic neuropathy. *Science* **242**, 1427–1430.

Wartell, R. M., Hosseini, S. H., and Moran, C. P., Jr. (1990). Detecting base pair substitutions in DNA fragments by temperature-gradient gel electrophoresis. *Nucleic Acids Res.* **18**, 2699–2705.

White, S. L., Collins, V. R., Wolfe, R., Cleary, M. A., Shanske, S., DiMauro, S., Dahl, H. H., and Thorburn, D. R. (1999). Genetic counseling and prenatal diagnosis for the mitochondrial DNA mutations at nucleotide 8993. *Am. J. Hum. Genet.* **65**, 474–482.

Wilson, G., Hollar, B., Waterson, J., and Schmickel, R. (1978). Molecular analysis of cloned human 18S ribosomal DNA segments. *Proc. Natl. Acad. Sci. USA* **75**, 5367–5371.

Wong, L.-J. C., Liang, M.-H., Kwon, H., Park, J., Bai, R., and Tan, D. (2002). Comprehensive scanning of the whole mitochondrial genome for mutations. *Clin. Chem.* **48**, 1901–1912.

Yen, M. Y., Wang, A. G., Chang, W. L., Hsu, W. M., Liu, J. H., and Wei, Y. H. (2000). False positive molecular diagnosis of Leber's hereditary optic neuropathy. *Zhonghua Yi Xue Za Zhi (Taipei)* **63**, 864–868.

Zeviani, M., Gellera, C., Pannacci, M., Uziel, G., Prelle, A., Servidei, S., and DiDonato, S. (1990). Tissue distribution and transmission of mitochondrial DNA deletions in mitochondrial myopathies. *Ann. Neurol.* **28**, 94–97.

Zeviani, M., Servidei, S., Gellera, C., Bertini, E., DiMauro, S., and DiDonato, S. (1989). An autosomal dominant disorder with multiple deletions of mitochondrial DNA starting at the D-loop region. *Nature* **339**, 309–311.

CHAPTER 23

Diagnostic Assays for Defects in mtDNA Replication and Transcription in Yeast and Humans

Gerald S. Shadel and Bonnie L. Seidel-Rogol

Department of Biochemistry
Emory University School of Medicine
Atlanta, Georgia 30322

I. Introduction

Mitochondrial (mt) transcription and mtDNA replication are dependent on nucleus-encoded factors that are synthesized in the cytoplasm and imported into the organelle. These molecules include mitochondrial RNA polymerase and requisite transcription factors, mtDNA polymerase (pol γ) and accessory replication proteins, and RNA-processing enzymes (Shadel and Clayton, 1997). Studies of numerous model systems have elucidated basic machinery required for these processes, yet the full complement of necessary factors remains to be identified and characterized. In addition, how mitochondrial gene expression and mtDNA replication are regulated in accordance with metabolic, developmental, and tissue-specific demands remains largely undetermined. The yeast model system has provided a wealth

0091-679X/07 $35.00
DOI: 10.1016/S0091-679X(06)80023-3

of information regarding the structure and function of mitochondrial transcription and replication proteins and has also begun to serve as a convenient model for certain human mitochondrial disease processes. The utility of using the yeast *Saccaromyces cerevisiae* to study mitochondrial genetics comes from the ability of so-called petite mutants to survive on specific carbon sources in the absence of respiratory function. Thus, these strains are ideal for the study of nuclear gene products required for mitochondrial gene expression and mtDNA maintenance.

The involvement of mitochondrial dysfunction in human disease has long been appreciated, and the number of disease states either caused or compounded by the loss of mitochondrial activities continues to rise steadily. It is now clear that mtDNA mutations cause human disease through loss of expression of mitochondrial-encoded oxidative phosphorylation (OXPHOS) components. In addition, respiration defects also result from mutations in nuclear genes that encode proteins required for function or assembly of the OXPHOS system, or factors that participate in replication or expression of the mitochondrial genome. Examples of the latter include nuclear gene mutations responsible for mtDNA depletion syndrome and disorders that result in an increased incidence of multiple mtDNA deletions in patient cells. The essential nature of this organelle and its genome underscores the need for understanding how the nucleus and mitochondria communicate in order to regulate mitochondrial gene expression.

This chapter describes several simple assays for the determination of relative mtDNA copy number (as a measure of mtDNA replication and stability) and steady-state levels of mitochondrial transcripts (as a measure of an intact mitochondrial transcription apparatus) in human and yeast cells. These diagnostic assays allow potential defects in mtDNA replication and transcription to be identified in a time- and labor-efficient manner. Once a defect of this type is postulated based on the preliminary assays described here, we encourage use of more intensive *in vitro* and *in organello* methods to address the molecular nature of the defect more precisely (see Section IV).

II. Diagnosis of mtDNA Replication Defects in Yeast and Human Cells

A. Fundamental Aspects of mtDNA Replication

The process of mtDNA replication has been well characterized in mammalian cells, where it occurs by an unusual asynchronous mechanism involving two physically separated, unidirectional origins (termed origin of heavy-strand, O_H, and origin of light-strand, O_L) (Clayton, 1982). The human mtDNA molecule is a 16.5-kb, double-stranded circle that contains a major noncoding locus called the D-loop regulatory region. It is within this region where transcription of each of the two coding strands is initiated and transcripts derived from the L-strand promoter (LSP) are processed to prime H-strand mtDNA replication at O_H. Much progress has been made toward understanding this transcription-dependent mtDNA replication mechanism, and a wealth of data in vertebrate systems support a general model for the initiation of H-strand mtDNA replication, described by Shadel and Clayton (1997). In this model, RNA transcripts are initiated at the LSP by human mtRNA polymerase in the presence of the transcription factor h-mtTFA. Next, mtRNA polymerase traverses a series of conserved sequence blocks at O_H, where the elongating LSP transcript forms a stable RNA/DNA hybrid. This hybrid serves as a substrate for RNA-processing enzymes, such as RNase MRP (Chang and Clayton, 1987; Lee and Clayton, 1998), that cleave the RNA in the hybrid to generate mature RNA primers that are utilized by DNA polyγ to begin a productive mtDNA replication event. The degree of conservation of this mtDNA replication mechanism in nonvertebrates has been addressed in *S. cerevisiae*, and substantial evidence indicates that priming of wild-type mtDNA by transcription also occurs in yeast. Issues pertaining to the extent of similarity that exists between human and yeast mtDNA replication have been discussed (Shadel, 1999).

B. Methods for Detecting mtDNA and Measuring Copy Number

We routinely use two standard methods for detecting mtDNA in yeast and human cells: staining with nucleic acid-specific dyes (e.g., 4′,6-diamidino-2-phenylindole; DAPI) and Southern hybridization with mtDNA-specific probes. In recent years, polymerase chain reaction (PCR)-based protocols have also become a standard means to detect mtDNA. This section describes a quantitative competitive PCR method that allows mtDNA copy number to be assessed in human cells. Lack of detection of mtDNA or a reduced mtDNA signal in any of these assays can be taken as preliminary evidence of an mtDNA replication defect (see Section IV).

Because relic mtDNA sequences (mitochondrial-derived pseudogenes) are often present in the nuclear genome of eukaryotes (Wallace *et al.*, 1997), it should be emphasized that Southern hybridization or PCR-based methods performed on total cellular DNA, or inadequately purified mtDNA, are prone to artifactual, false-positive signals. To ensure that the observed signals in such assays are in fact derived from mtDNA, control cell lines with isogenic nuclear backgrounds that are devoid of mtDNA (rho°) should also be tested routinely for lack of a signal. In many cases, such cell lines can be generated by extended growth in the presence of ethidium bromide (King and Attardi, 1996). If generation of rho° lines is not possible, the test mtDNA should be isolated from mitochondria that have been purified to an extent that nuclear DNA does not contaminate the preparation.

1. Detection of mtDNA by Staining with DAPI

Staining mtDNA with nucleic acid-specific fluorescent dyes is a convenient means to assess quickly whether mtDNA is present in cells (Fig. 1). Our standard protocol for staining yeast mtDNA with DAPI is as follows. Yeast cells are grown to mid-exponential growth phase in synthetic dextrose medium (Bacto-yeast nitrogenous

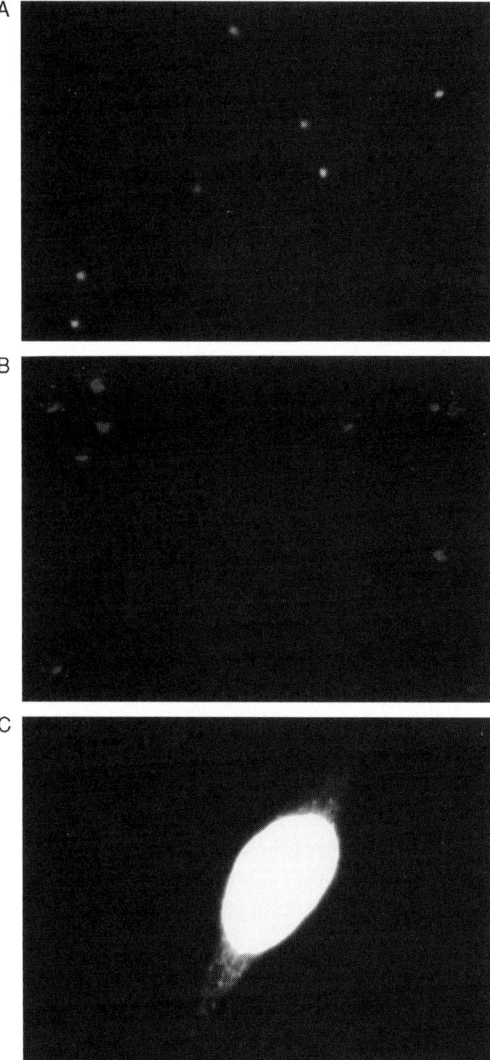

Fig. 1 Visualization of mtDNA in yeast and human cells using DAPI. The yeast strain GS125 (Wang and Shadel, 1999) was depleted of mtDNA after extensive growth in the presence of ethidium bromide to create the strain GS125 rho°. DAPI staining of GS125 rho° reveals only nuclear DNA staining (A), whereas the identically treated isogenic GS125 strain exhibits characteristic punctate cytoplasmic mtDNA staining in addition to the nuclear DNA signal (B). (C) A HeLa Tet-On cell stained with DAPI according to the procedure described in Spector *et al.* (1998). In order to visualize the punctate mtDNA signal throughout the cytoplasm in these cells, longer photographic exposures were required that "bleached out" the nuclear DNA signal (bright oval shape in the center of the cell).

base, 6.7 g/liter; dextrose, 20 g/liter) with required amino acids supplements as described (Sherman, 1991). Cells ($\sim$1.0 $\times$ 10^7) are harvested by centrifugation (30 s, 16,000 g) in a sterile 1.5-ml microcentrifuge tube, resuspended in 70% ethanol, and incubated for 10 min on ice. Cells are then washed twice with 1 ml of deionized water, resuspended in 100 μl of 50-ng/ml DAPI (Sigma Chemical; diluted from a 1-mg/ml stock that can be stored at $-20\,^\circ$C) in phosphate-buffered saline (PBS), and incubated for an additional 10 min. DAPI fluorescence is observed using a fluorescence microscope equipped with a UV filter set; we use an Olympus BX60 epifluorescence microscope. Both a nuclear DNA signal and an mtDNA signal are observed in wild-type cells (Fig. 1B). Alternatively, it is also possible to stain mtDNA in growing cell cultures by adding DAPI (1 μg/ml) to the growth medium. In our hands, the nuclear DNA signal is markedly reduced and variable under these conditions. In addition, DAPI can also be used to stain mtDNA in fixed human cells. We routinely use the protocol described in Spector *et al.* (1998) for this purpose (Fig. 1C).

2. Detection of mtDNA by Southern Hybridization

Because specific DNA sequences are detected by Southern hybridization, this technique can be used not only to detect and compare the relative amount of mtDNA in two samples, but to address whether the genome is intact. For example, the presence of mtDNA, as indicated by fluorescence microscopy, does not assess whether these genomes are full-length or mutated. This can be problematic, particularly in yeast, where mutant strains (*rho*$^-$ petite strains) arise spontaneously that propagate extensively deleted, repetitive mtDNA molecules. The following protocol is used for analyzing human mtDNA by Southern hybridization; yeast mtDNA can be analyzed in a similar manner using yeast mtDNA probes.

a. Southern Hybridization of mtDNA from Human Cell Lines

The following protocol was developed for HeLa cells, but should be generally applicable to any human cell line harboring an intact mtDNA genome. A HeLa Tet-On cell line was obtained from Clontech and cells were grown in monolayer in Dulbecco's Minimal Essential Medium (DMEM) (Mediatech Cellgro) containing 4.5-g/liter glucose, 10% fetal bovine serum, 2-mM l-glutamine, and 100-μg/ml G418 with subculturing every 3–4 days. Cells were grown in 75-cm^2 culture flasks to $\sim$80% confluency ($\sim$1 $\times$ 10^7 cells) and after trypsinization were harvested by centrifugation in a clinical centrifuge. Total cellular DNA was isolated using an RNA/DNA purification protocol exactly as described by the manufacturer (RNA/DNA Midi Kit, Qiagen Inc.). Five micrograms of total cellular DNA was digested to completion with *Nco*I, separated by electrophoresis on a 0.8% agarose gel, and transferred to a nylon membrane by capillary flow as described (Sambrook *et al.*, 1989). The DNA was then cross-linked to the membrane with UV light using a Stratalinker apparatus (Stratagene, Inc.). Nucleic acid hybridization was accomplished using Rapid-Hyb buffer exactly as described by the manufacturer (Amersham, Inc.) in a rolling-bottle hybridization oven (Techne Hybridiser, HB-1D). The mtDNA-specific probe was generated by PCR using total cellular DNA as a template (forward PCR primer: 5′-AGCAAACCACAGTTTCATGCCCA-3′; reverse primer: 5′-TAATGTTAG-TAAGGGTGGGGAAG-3′), which results in a 5456-bp product corresponding to nucleotides (nt) 8188–13,644 (according to Anderson *et al.*, 1981) of the mtDNA sequence. The PCR product was gel purified and radiolabeled with Ready-to-Go labeling beads according to the protocol provided by the manufacturer (Amersham, Inc.). A control nuclear DNA probe was generated in the same fashion, except that a 381-bp PCR product derived from the nuclear 28S rRNA gene locus was used as a template (forward PCR primer: 5′-GCCTAGCAGCCGACTTAGAACTGG-3′; reverse PCR primer: 5′-GGCCTTTCATTATTCTACACCTC-3′). After Southern hybridization, the mtDNA probe hybridized to a 7555-bp *Nco*I restriction fragment of human mtDNA and the 28S probe hybridized to a 2588-bp *Nco* I restriction fragment of nuclear DNA, corresponding to repeats from the ribosomal DNA locus. The mtDNA signal and the nuclear DNA signal were then quantitated by densitometry of exposed X-ray film (using a Bio-Rad Fluor-S MultiImager) or directly using a Phosphoimager system (FujiX BAS1000). The nuclear DNA signals were used to normalize the mtDNA signals for loading were differences, allowing the relative amounts of mtDNA in each sample to be compared.

3. Quantition of mtDNA by Competitive Polymerase Chain Reaction

A useful assay to determine more precisely the mtDNA copy number in cells is competitive PCR (e.g., see Zhang *et al.*, 1994). This method involves the generation of a control template that consists of an internally deleted version of the gene sequence that is to be analyzed in the experiment. This control template, which is amplified using the same primers as the test DNA, is added to the experimental DNA samples in known amounts, where it competes with the test template for primer binding and amplification. A series of competitive PCR reactions is generated consisting of several reactions with the same amount of test DNA, but differing in the amount of competitor template that is present. This series is amplified under identical PCR conditions, and the products are analyzed by agarose gel electrophoresis. Quantitation of the experimental and control products allows the calculation of an equivalence point (where the test and control signal are equal) that can be used to determine the concentration of test DNA in the sample (Freeman *et al.*, 1999).

We have applied this technique to assay mtDNA copy number in human cells. A 494-bp region of the mitochondrial ND2 gene can be amplified by PCR using the following primers: forward 5′-GGCCCAACCCGTCATCTAC-3′ and reverse 5′-CCACCTCAACTGCCTGCTATG-3′. To generate the competitive template for this assay, this ND2 fragment was first ligated into a pGEM-T vector (Promega, Inc.) and its nucleotide sequence was confirmed. We then generated an internal deletion by digesting the resulting plasmid with *Bst*EII, which cleaves a unique internal site in the ND2 gene fragment, and treating the linearized DNA with Bal31 as described (Sambrook *et al.*, 1989). After ligation of the deleted products, a control plasmid with an internal deletion of 204 bp of the ND2 gene (pND2Δ290) was isolated. Using ND2 gene-specific primers, this plasmid produces a 290-bp PCR product that, after electrophoresis, is easily distinguished from the 494-bp fragment generated from the wild-type ND2 gene (Fig. 2A).

Increasing amounts (0.026–2.6 ng) of pND2Δ290 are added to samples containing a constant amount (200 ng) of total HeLa Tet-On cellular DNA, and PCR amplification is performed (see legend of Fig. 2 for PCR conditions). The PCR products are precipitated in ethanol, separated on a 1% agarose gel, and visualized by staining with ethidium bromide (Fig. 2A). In our experiments, quantitation of

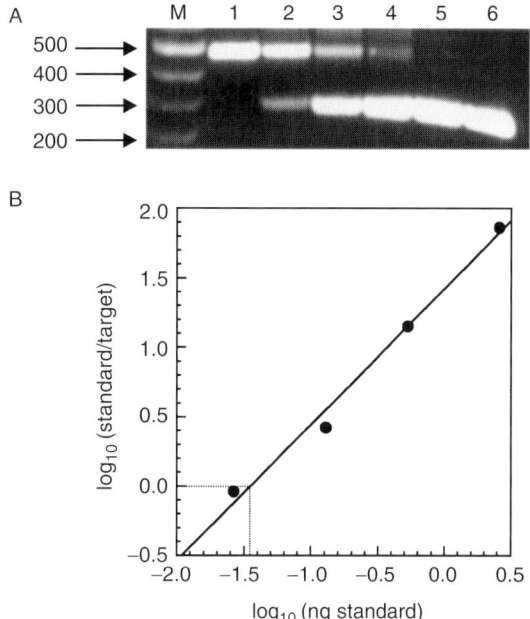

Fig. 2 Competitive PCR analysis of mtDNA from HeLa Tet-On cells. (A) Agarose gel electrophoretic analysis of competitive PCR products. Lanes 1–5 indicate products resulting from PCR reactions that contained 200 ng of total cellular DNA and differing amounts of external competitor DNA (pND2Δ290) as follows: lane 1, 0 ng; lane 2, 0.26 ng; lane 3, 0.39 ng; lane 4, 0.52 ng; and lane 5, 2.6 ng. Lane 6 contains 2.6 ng pND2Δ290 only. DNA molecular weight standards run with the PCR reactions are indicated on the left. PCR was performed in a final volume of $100\,\mu l$ containing 3.5 mM $MgCl_2$, 10 mM of each dNTP, 100 ng of each primer, and 2.5 units of *Taq* DNA polymerase (Bethesda Reseach Laboratories, Inc.). *Taq* DNA polymerase was added after a 5-min denaturation at 95°C, and amplification was for 25 cycles (1 min at 95°C, 1 min at 50°C, and 3 min at 72°C) followed by one 10-min elongation period at 72°C. (B) Quantitative analysis of competitive PCR products. The amount of each PCR product was determined by fluorescence using a Bio-Rad Fluor-S MultiImager, and these data are plotted as described by Freeman *et al.* (1999), which should yield a straight line. The equivalence point (indicated as dashed lines on the graph) is determined by calculating the value of X, where $Y = 0$ using the equation for this line. This value of X can be used to estimate the amount of target mtDNA in the sample. In this example, an estimate of 35 pg of mtDNA was obtained from a value of $X = -1.46$ (at $Y = 0$).

the fluorescent products is usually accomplished using a Bio-Rad Fluor-S Multi-Imager. These data are plotted and analyzed as described by Freeman *et al.* (1999) to determine the amount of mtDNA present in the sample (Fig. 2B). Knowledge of the total number of cells used for the DNA preparation allows a ready estimation of the average mtDNA copy number/per cell.

III. Analyzing Mitochondrial Transcripts *In Vivo*

A. Mitochondrial Transcription Initiation

Compared to other characterized transcription systems, including bacterial systems, the mitochondrial transcription apparatus is relatively simple in terms of protein requirements for initiation. This situation has allowed faithful *in vitro* transcription systems to be developed for several species, including yeast, *Xenopus,* mouse, and human, which have facilitated the study of protein–protein and protein–nucleic acid interactions required for transcription in the organelle. While interesting species-specific differences exist with regard to transcription (Shadel and Clayton, 1993), several basic principles are conserved from yeast to humans that are summarized here.

Perhaps the best understood mitochondrial transcription system is that from the yeast *S. cerevisiae*. Core yeast mitochondrial RNA polymerase (sc-mtRNA polymerase) is a 150-kDa protein encoded by the *RPO41* gene and is homologous to the single-subunit RNA polymerases of the bacteriophages T7, T3, and SP6 (Masters *et al.*, 1987). In addition, it has a unique amino-terminal extension of ~400 amino acids that harbors a mitochondrial matrix localization signal (aa 1–29) and comprises at least one additional functional domain required for mtDNA replication or stability (Wang and Shadel, 1999). Unlike bacteriophage RNA polymerases, which are themselves sufficient for transcription initiation, sc-mtRNA, polymerase requires a transcription factor, sc-mtTFB, to bind specifically to a simple nonanucleotide promoter ATATAAGTA(+1) and initiate transcription. Analysis of the nucleotide sequence of the *MTF1* gene, which encodes sc-mtTFB, revealed regions with apparent similarity to bacterial sigma factors (Jang and Jaehning, 1991). Indeed, certain properties of sc-mtTFB, observed during transcription *in vitro* (Mangus *et al.*, 1994), and aspects of its interaction with the core sc-mtRNA polymerase (Cliften *et al.*, 1997) suggest that sc-mtTFB functions in a manner analogous to this class of proteins. However, mutational analysis (Shadel and Clayton, 1995) and sequence comparisons of the three known yeast mtTFB homologs (Carrodeguas *et al.*, 1996) point to some potentially important differences between sigma factors and mtTFB in both structure and function. In addition to sc-mtTFB, yeast mitochondria also possess an abundant mtDNA-binding protein (sc-mtTFA) that contains two high mobility group (HMG) box DNA-binding domains (Diffley and Stillman, 1991). This protein, which is encoded by the *ABF2* gene, is part of the mtDNA nucleoid (Newman *et al.*, 1996), has been implicated in stabilizing recombination intermediates (MacAlpine *et al.*, 1998), and can stimulate or inhibit transcription *in vitro* depending on its concentration (Parisi *et al.*, 1993).

Mutations that disrupt the *RPO41* or *MTF1* genes, as well as the *ABF2* gene (under certain growth conditions), result in instability and ultimately loss of mtDNA, consistent with a role for each of these factors in mtDNA replication and/or stability as well as transcription.

As in yeast, core human mtRNA (h-mtRNA) polymerase is encoded by a nuclear gene (Tiranti *et al.*, 1997), is homologous to bacteriophage RNA polymerases, and requires at least one transcription factor, human mitochondrial transcription factor A (h-mtTFA). The h-mtTFA protein was identified in mitochondrial extracts through its ability to promote efficient specific transcription initiation by human mtRNA polymerase *in vitro* by binding to a specific site at human mtDNA promoters (Fisher and Clayton, 1988). Like its yeast homolog (sc-mtTFA), h-mtTFA contains two HMG-box DNA-binding domains, but in addition contains a basic C-terminal tail that imparts specific DNA-binding and transcriptional activation capabilities to the protein (Dairaghi *et al.*, 1995). The importance of mtTFA in mammals has been underscored through disruption of its gene in mice, which leads to early embryonic lethality, presumably through lack of mtDNA gene expression and replication (Larsson *et al.*, 1998). However, deciphering the precise role of mtTFA in transcription and mtDNA replication, as well as the molecular basis for the differential transcription factor requirements of yeast and humans (Shadel and Clayton, 1993), awaits establishment of mammalian *in vitro* transcription and replication systems that utilize recombinant mtRNA polymerase.

B. Standard Assays for Mitochondrial Transcription

In general, the methodologies for assaying mitochondrial transcript levels are similar, if not identical, to those utilized to analyze nucleus-derived transcripts. For example, Northern hybridization, nuclease protection (e.g., RNase protection), and reverse-transcriptase polymerase chain reaction (RT-PCR) assays have all been used successfully to monitor steady-state levels of specific mitochondrial transcripts in cultured human and yeast cells. Because these methods are standard in modern molecular biology laboratories and are well documented in other methods volumes (e.g., Sambrook, 1989), a detailed treatment of these common techniques will not be presented here. However, there are aspects of mitochondrial genome organization and expression that must be considered when these types of assays are performed on mitochondrial transcripts. For example, it should be appreciated that four different types of mature RNA molecules are synthesized in yeast and human mitochondria: mRNA, tRNA, rRNA, and origin-associated RNAs. These RNA molecules display a range of half-lives *in vivo*. In addition, with the exception of origin RNAs in yeast, many of these RNA species are synthesized as polycistronic RNA transcripts that must be processed into mature molecules. Thus, steady-state levels of these transcripts can be affected dramatically by alterations in RNA-processing activity as well as by transcription. Finally, the inherent and induced instability of the mitochondrial genome, especially in yeast, presents additional complications for transcript analysis *in vivo*. Although we outline assays in human and yeast cells in this chapter, the basic methodology should be applicable to any mitochondrial system where genome sequence information is available.

1. Assaying Mitochondrial Transcripts in Human Cells

Because human mtDNA has two major transcription units, one for each strand, and the H-strand transcript is subject to a site-specific termination event (Kruse *et al.*, 1989), we suggest employment of a set of at least four gene-specific probes to estimate whether mitochondrial transcription is normal in a human cell. The minimal set should consist of the following: one probe for the L-strand (e.g., ND6); two probes for the H-strand, one before (e.g., 12S or 16S rRNA) and one after (e.g., ND1 or COX I) the transcription termination site; and a control probe, corresponding to a nucleus-encoded RNA that does not change under the experimental conditions being tested (commonly, rRNA or glyceraldehyde 6-phosphate dehydrogenase, GAPDH). Because mtDNA transcripts are relatively abundant, whole cell RNA preparations can be used routinely, obviating the need to prepare RNA from purified mitochondria. However, when possible, control experiments should be carried out on cells lacking mtDNA (rho°), or with highly purified mitochondria, to ensure that the observed signals are of mitochondrial origin. To compare samples under different conditions, the mitochondrial and nuclear transcript signals are quantitated, allowing a ratio of mitochondrial to nuclear signal to be calculated, with the condition that the control signal does not change during the course of the experiment. In this manner, the relative steady-state amount of L-strand and H-strand transcripts can be determined and compared between samples. Comparison of the two H-strand signals may also provide information regarding the relative amount of transcription termination that is occurring at the termination site. For transcript analysis from cultured human cells, we generally use Qiagen RNA/DNA kits for RNA isolation. In our hands, these RNA preparations are suitable for analysis by Northern hybridization, RNase protection, and RT-PCR.

2. Assaying Mitochondrial Transcripts in Yeast

The strategy for analyzing transcript levels in yeast is virtually identical to that described earlier for human cells. However, the following considerations take into account unique features of the yeast mitochondrial genetic system, including the presence of ~20 independent transcription units and several transcripts that can harbor introns.

A conserved nonanucleotide promoter is associated with all the known yeast mitochondrial transcription units, suggesting that transcripts may respond similarly to changes in transcription initiation capacity. However, variability in nucleotides at +1 and +2 relative to the start site of transcription is known to affect promoter efficiency, making it possible that differential promoter regulation is operative under certain circumstances. Thus, estimation of whether transcription is occurring normally in yeast requires employment of a set of gene-specific probes that assay not only the different types of mature RNA molecules, but also different promoter contexts. A minimal set should include at least one of each the following: an mRNA (e.g., *COB1*, *COX1*, *COX2*, or *COX3*); an rRNA (e.g., 14S or 21S); a tRNA (e.g., tRNA^{Glu} or tRNA^{fMet}); and a control probe, corresponding to a nucleus-encoded RNA that does not change under the experimental conditions being tested (e.g., 18S rRNA). To be more thorough, one could also include probes for the *RPM1* gene, which encodes the RNA component of mitochondrial RNase P, and *VAR1*, which encodes a mitochondrial ribosomal protein. When designing these probes, care should be taken to avoid intron sequences often contained in the *COX1*, *COB1*, and the 21S rRNA genes. Examples of how this type of "global" transcript analysis has been used to assess changes in mitochondrial transcript levels that occur in yeast in response to glucose repression have been reported (Ulery, 1994; Mueller and Getz, 1986).

3. Assaying Mitochondrial Transcription in Yeast Under Condition of Wild-Type Genome Instability

Yeast mtDNA is inherently unstable and mutations in many nuclear genes can enhance this instability (Shadel, 1999). For example, disruption of genes encoding the yeast mitochondrial transcription apparatus (i.e., *RPO41* and *MTF1*) leads to mtDNA instability and eventual loss of the genome. This makes assaying mitochondrial transcript levels under such conditions problematic because the mtDNA either is not wild type or is absent altogether. We have utilized a class of mtDNA mutants called hypersuppressive (HS) *rho*$^-$ petites to circumvent this problem of template loss in order to assay transcription initiation capability *in vivo*. In particular, we have used the HS3324 *rho*$^-$ mitochondrial genome, which harbors a 966-bp, head-to-tail repeat of *ori5* (a putative yeast mitochondrial origin), that is stable in *RPO41* null backgrounds (Lorimer *et al.*, 1995). That is, the mtDNA present in this strain is able to replicate by a mechanism that does not require intact mtRNA polymerase activity. However, the *ori5* repeat does contain a nonanucleotide promoter that allows production of a specific RNA transcript to be assayed *in vivo*. The utility of such strains stems from the ability to introduce the HS3324 genome (or any HS genome) into strains of interest by a process known as cytoduction (Lancashire and Mattoon, 1979). In this manner, strains that cannot maintain wild-type mtDNA due to a nuclear DNA mutation can be assayed for the presence of an intact transcription apparatus. For example, we have used an *rpo41* null HS3324 strain to determine that mutated alleles of *RPO41* encode proteins that are capable of transcription *in vivo* (Fig. 3), but nonetheless are defective for wild-type mtDNA maintenance (Wang and Shadel, 1999). In principle, this method can be used to analyze mutants in other mitochondrial transcription components (e.g., sc-mtTFB) to determine their effects on transcription *in vivo* in the presence of a stable template. As is the case for mtRNA polymerase, additional functions for such factors may be elucidated in this manner.

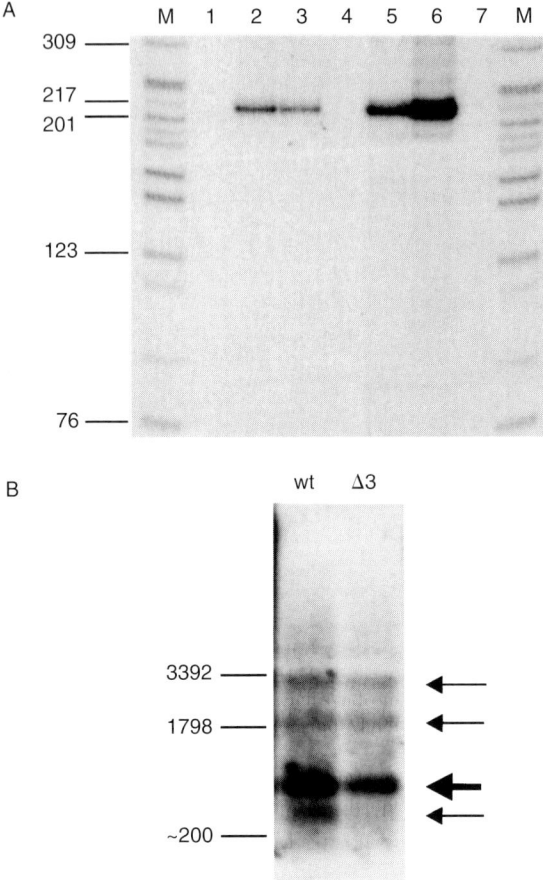

Fig. 3 Analysis of mitochondrial *ori5* transcripts in a hypersuppressive *rho⁻* yeast strain. The haploid yeast strain BS127 HS3324 ΔRPO41 contains an *ori5* repeat *rho⁻* mitochondrial genome and lacks mtRNA polymerase due to a chromosomal disruption of the *RPO41* locus (Lorimer *et al.*, 1995). This strain was transformed with plasmids that harbor either a wild-type *RPO41* gene (pGS348) or the temperature-sensitive allele *rpo41 Δ3* (pGS348Δ3). After growth at 30 and 37°C for ~20 generations, total cellular RNA was isolated from these strains as described (Schmitt *et al.*, 1990), as well as from the parental strain that lacks mtRNA polymerase activity. (A) Results of an S1 nuclease protection assay using an *ori5*-specific probe as described (Wang and Shadel, 1999). RNA analyzed from strains grown at 30°C (lanes 1–3) and from strains grown at 37°C (lanes 4–7) are shown. The parental strain (devoid of mtRNA polymerase) did not yield any *ori5* transcript signal (lanes 1 and 4), whereas strains harboring either the wild-type *RPO41*-containing plasmid (lanes 2 and 5) or an *rpo41 Δ3*-containing plasmid (lanes 3 and 6) each produced *ori5* transcripts that initiated at the predicted transcription start site. Lane 7 is a control lane that is the same as lane 6, except the sample was treated with RNase A prior to analysis, indicating that the observed signals were derived from RNA. M, *Hpa*II-digested pBR322 DNA markers (length in nucleotides is indicated on the left). (B) Northern analysis of *ori5* transcripts in *RPO41* (wt) and *rpo41 Δ3* (Δ3) plasmid-containing strains grown at 37°C performed as described (Wang and Shadel, 1999). The major *ori5* RNA transcript is indicated by the bold arrow. Two additional minor transcripts of ~2 and ~3 kb, corresponding to two and three *ori5* repeat lengths, are indicated by thin arrows, as is a subrepeat length transcript that was observed only in the wild-type strain. Indicated on the left is the position where known RNA species migrated.

IV. Additional Considerations

The assays described in this chapter provide an initial assessment of whether mtDNA replication or mitochondrial transcription is occurring normally in human or yeast cells. While these assays are helpful for initial diagnostic purposes, it is noted that a reduction in the steady-state level of mtDNA or mtRNA in a cell, which is the end point of all the assays described in this chapter, may be due to perturbations in the cell other than those that affect mtDNA replication or transcription per se. For example, a reduction in the steady-state level of mtDNA could, in principle, be caused by increased mtDNA instability. The situation in yeast is complicated further by the fact that mutations in a large number of genes, including those that abolish mitochondrial translation, result in mtDNA instability through mechanisms that remain unclear (Shadel, 1999). Thus, if a reduction in mtDNA copy number is detected in one of the assays described earlier, additional methods should be employed to determine if a DNA replication defect is the source of the problem. Examples of more direct assays for mtDNA replication include labeling of mtDNA with 5-bromo-2-deoxy uridine (Berk and Clayton, 1973; Davis and Clayton, 1996; Meeusen *et al.*, 1999) and ligation-mediated PCR to detect nascent mtDNA strands (Kang *et al.*, 1997).

Similar complications influence the interpretation of the transcription assays described herein. For example, a decrease in the steady-state level of a specific mtRNA species may be caused by a decrease in RNA stability or improper RNA processing rather than a reduction in transcription capacity. Again, additional assays have been developed that can be employed to address some of these issues more directly. For example, *in organello* methods for analyzing mitochondrial transcripts have been established for yeast (Poyton *et al.*, 1996) and vertebrate cells (Enriquez *et al.*, 1996; Gaines, 1996). By examining many transcripts simultaneously in this manner, such analyses could, in theory, differentiate between RNA processing and RNA transcription defects; the presence of incompletely processed precursor species indicates the former possibility. As already noted, faithful *in vitro* transcription assays have been developed for both human and yeast systems (Mangus and Jaehning, 1996; Shadel and Clayton, 1996). These systems can be employed to assay mitochondrial transcription initiation, elongation, or termination activity (Fernandez-Silva *et al.*, 1996).

Acknowledgment

This work was supported by Grant HL-59655 from the National Institutes of Health awarded to G.S.S.

References

Anderson, S., Bankier, A. T., Barrell, B. G., de Bruijn, M. H., Coulson, A. R., Drouin, J., Eperon, I. C., Nierlich, D. P., Roe, B. A., Sanger, F., Schreier, P. H., Smith, A. J., *et al.* (1981). Sequence and organization of the human mitochondrial genome. *Nature* **290,** 457–465.

Berk, A. J., and Clayton, D. A. (1973). A genetically distinct thymidine kinase in mammalian mito-chondria: Exclusive labeling of mitochondrial deoxyribonucleic acid. *J. Biol. Chem.* **248,** 2722–2729.

Carrodeguas, J. A., Yun, S., Shadel, G. S., Clayton, D. A., and Bogenhagen, D. F. (1996). Functional conservation of yeast mtTFB despite extensive sequence divergence. *Gene Expr.* **6,** 219–230.

Chang, D. D., and Clayton, D. A. (1987). A novel endoribonuclease cleaves at a priming site of mouse mitochondrial DNA replication. *EMBO J.* **6,** 409–417.

Clayton, D. A. (1982). Replication of animal mitochondrial DNA. *Cell* **28,** 693–705.

Cliften, P. F., Park, J. Y., Davis, B. P., Jang, S. H., and Jaehning, J. A. (1997). Identification of three regions essential for interaction between a sigma-like factor and core RNA polymerase. *Genes Dev.* **11,** 2897–2909.

Dairaghi, D. J., Shadel, G. S., and Clayton, D. A. (1995). Addition of a 29 residue carboxyl-terminal tail converts a simple HMG box-containing protein into a transcriptional activator. *J. Mol. Biol.* **249,** 11–28.

Davis, A. F., and Clayton, D. A. (1996). *In situ* localization of mitochondrial DNA replication in intact mammalian cells. *J. Cell Biol.* **135,** 883–893.

Diffley, J. F. X., and Stillman, B. (1991). A close relative of the nuclear, chromosomal high-mobility group protein HMG1 in yeast mitochondria. *Proc. Natl. Acad. Sci. USA* **88,** 7864–7868.

Enriquez, J. A., Perez-Martos, A., Lopez-Perez, M. J., and Montoya, J. (1996). In organello RNA synthesis system from mammalian liver and brain. *Methods Enzymol.* **264,** 50–57.

Fernandez-Silva, P., Micol, V., and Attardi, G. (1996). Mitochondrial DNA transcription initiation and termination using mitochondrial lysates from cultured human cells. *Methods Enzymol.* **264,** 129–139.

Fisher, R. P., and Clayton, D. A. (1988). Purification and characterization of human mitochondrial transcription factor 1. *Mol. Cell. Biol.* **8,** 3496–3509.

Freeman, W. M., Walker, S. J., and Vrana, K. E. (1999). Quantitative RT-PCR: Pitfalls and potential. *Biotechniques* **26,** 124–125.

Gaines, G. L., III (1996). In organello RNA synthesis system from HeLa cells. *Methods Enzymol.* **264,** 43–49.

Jang, S. H., and Jaehning, J. A. (1991). The yeast mitochondrial RNA polymerase specificity factor, MTF1, is similar to bacterial sigma factors. *J. Biol. Chem.* **266,** 22671–22677.

Kang, D., Miyako, K., Kai, Y., Irie, T., and Takeshige, K. (1997). *In vivo* determination of replication origins of human mtiochondrial DNA by ligation-mediated PCR. *J. Biol. Chem.* **272,** 15275–15279.

King, M. P., and Attardi, G. (1996). Isolation of human cell lines lacking mitochondrial DNA. *Methods Enzymol.* **264,** 304–313.

Kruse, B., Narasimhan, N., and Attardi, G. (1989). Termination of transcription in human mitochon-dria: Identification and purification of a DNA binding protein factor that promotes termination. *Cell* **58,** 391–397.

Lancashire, W. E., and Mattoon, J. R. (1979). Cytoduction: A tool for mitochondrial genetic studies in yeast. Utilization of the nuclear-fusion mutation kar 1-1 for transfer of drug r and mit genomes in *Saccharomyces cerevisiae. Mol. Gen. Genet.* **170,** 333–344.

Larsson, N. G., Wang, J., Wilhelmsson, H., Oldfors, A., Rustin, P., Lewandoski, M., Barsh, G. S., and Clayton, D. A. (1998). Mitochondrial transcription factor A is necessary for mtDNA maintenance and embryogenesis in mice. *Nature Genet.* **18,** 231–236.

Lee, D. Y., and Clayton, D. A. (1998). Initiation of mitochondrial DNA replication by transcription and R-loop processing. *J. Biol. Chem.* **273,** 30614–30621.

Lorimer, H. E., Brewer, B. J., and Fangman, W. L. (1995). A test of the transcription model for biased inheritance of yeast mitochondrial DNA. *Mol. Cell. Biol.* **15,** 4803–4809.

MacAlpine, D. M., Perlman, P. S., and Butow, R. A. (1998). The high mobility group protein Abf2p influences the level of yeast mitochondrial DNA recombination intermediates *in vivo. Proc. Natl. Acad. Sci. USA* **95,** 6739–6743.

Mangus, D. A., and Jaehning, J. A. (1996). Transcription *in vitro* with *Saccharomyces cerevisiae* mitochondrial RNA-polymerase. *Methods Enzymol.* **264,** 57–66.

Mangus, D. A., Jang, S. H., and Jaehning, J. A. (1994). Release of the yeast mitochondrial RNA polymerase specificity factor from transcription complexes. *J. Biol. Chem.* **269**, 26568–26574.

Masters, B. S., Stohl, L. L., and Clayton, D. A. (1987). Yeast mitochondrial RNA polymerase is homologous to those encoded by bacteriophages T3 and T7. *Cell* **51**, 89–99.

Meeusen, S., Tieu, Q., Wong, E., Weiss, E., Schieltz, D., Yates, J. R., and Nunnari, J. (1999). Mgm101p is a novel component of the mitochondrial nucleoid that binds DNA and is required for the repair of oxidatively damaged mitochondrial DNA. *J. Cell Biol.* **145**, 291–304.

Mueller, D. M., and Getz, G. S. (1986). Steady state analysis of mitochondrial RNA after growth of yeast *Saccharomyces cerevisiae* under catabolite repression and derepression. *J. Biol. Chem.* **261**, 11816–11822.

Newman, S. M., Zelenaya-Troitskaya, O., Perlman, P. S., and Butow, R. A. (1996). Analysis of mitochondrial DNA nucleoids in wild-type and a mutant strain of *Saccharomyces cerevisiae* that lacks the mitochondrial HMG box protein Abf2p. *Nucleic Acids Res.* **24**, 386–393.

Parisi, M. A., Xu, B., and Clayton, D. A. (1993). A human mitochondrial transcriptional activator can functionally replace a yeast mitochondrial HMG-box protein both *in vivo* and *in vitro. Mol. Cell. Biol.* **13**, 1951–1961.

Poyton, R. O., Bellus, G., McKee, E. E., Sevarino, K. A., and Goehring, B. (1996). In organello mitochondrial protein and RNA synthesis systems from *Saccharomyces cerevisiae. Methods Enzymol.* **264**, 36–42.

Sambrook, J., Fritsch, E. F., and Maniatis, T. (1989). ''Molecular Cloning: A Laboratory Manual,'' 2nd Ed. Cold Spring Harbor Press, Cold Spring Harbor, NY.

Schmitt, M. E., Brown, T. A., and Trumpower, B. L. (1990). A rapid and simple method for preparation of RNA from *Saccharomyces cerevisiae. Nucleic Acids Res.* **18**, 3091–3092.

Shadel, G. S. (1999). Yeast as a model for human mtDNA replication. *Am. J. Hum. Genet.* **65**, 1230–1237.

Shadel, G. S., and Clayton, D. A. (1993). Mitochondrial transcription initiation: Variation and conservation. *J. Biol. Chem.* **268**, 16083–16086.

Shadel, G. S., and Clayton, D. A. (1995). A *Saccharomyces cerevisiae* mitochondrial transcription factor, sc-mtTFB, shares features with sigma factors but is functionally distinct. *Mol. Cell. Biol.* **15**, 2101–2108.

Shadel, G. S., and Clayton, D. A. (1996). Mapping promoters in displacement-loop region of vertebrate mitochondrial DNA. *Methods Enzymol.* **264**, 139–148.

Shadel, G. S., and Clayton, D. A. (1997). Mitochondrial DNA maintenance in vertebrates. *Annu. Rev. Biochem.* **66**, 409–435.

Sherman, F. (1991). Getting started with yeast. *Methods Enzymol.* **194**, 3–21.

Spector, D. L., Goldman, R. D., and Leinwand, L. A. (1998). ''Cells, a Laboratory Manual,'' Vol. 3, p. 101.10. Cold Spring Harbor Press, Cold Spring Harbor, NY.

Tiranti, V., Savoia, A., Forti, F., D'Apolito, M. F., Centra, M., Rocchi, M., and Zeviani, M. (1997). Identification of the gene encoding the human mitochondrial RNA polymerase (h-mtRPOL) by cyberscreening of the expressed sequence tags database. *Hum. Mol. Genet.* **6**, 615–625.

Ulery, T. L., Jang, S. H., and Jaehning, J. A. (1994). Glucose repression of yeast mitochondrial transcription: Kinetics of derepression and role of nuclear genes. *Mol. Cell. Biol.* **14**, 1160–1170.

Wallace, D. C., Stugard, C., Murdock, D., Schurr, T., and Brown, M. D. (1997). Ancient mtDNA sequences in the human nuclear genome: A potential source of errors in identifying pathogenic mutations. *Proc. Natl. Acad. Sci. USA* **94**, 14900–14005.

Wang, Y., and Shadel, G. S. (1999). Stability of the mitochondrial genome requires an amino-terminal domain of yeast mitochondrial RNA polymerase. *Proc. Natl. Acad. Sci. USA* **96**, 8046–8051.

Zhang, H., Cooney, D. A., Sreenath, A., Zhan, Q., Agbaria, R., Stowe, E. E., Fornace, A. J., Jr., and Johns, D. G. (1994). Quantitation of mitochondrial DNA in human lymphoblasts by a competitive polymerase chain reaction method: Application to the study of inhibitors of mitochondrial DNA content. *Mol. Pharmacol.* **46**, 1063–1069.

CHAPTER 24

Microdissection and Analytical PCR for the Investigation of mtDNA Lesions

Sion L. Williams and Carlos T. Moraes

Department of Neurology
University of Miami Miller School of Medicine
Miami, Florida 33136

I. Introduction

The analysis of mtDNA from microdissected samples is becoming an increasingly important aspect of mitochondrial research. It enables the investigation of very specific subsets of cells and is providing insights into the biology of mtDNA that previously had been masked in studies of bulk DNA extracts. Over the past decade, analysis of microdissected samples has been used to investigate the molecular biology of mtDNA *in vivo* (Srivastava and Moraes, 2005; Zsurka *et al.*, 2005) and the role of mtDNA lesions in mitochondrial diseases (Moraes *et al.*, 1992, 1995; Giordano *et al.*, 2006; Pistilli *et al.*, 2003; Sciacco *et al.*, 1994), cancer (Chen *et al.*, 2002; Parr *et al.*, 2006; Polyak *et al.*, 1998; Rosson and Keshgegian, 2004; Wang *et al.*, 2005), Parkinson disease (Bender *et al.*, 2006; Cantuti-Castelvetri *et al.*, 2005; Kraytsberg *et al.*, 2006), HIV (Cossarizza *et al.*, 2003), and aging (He *et al.*, 2002; Khrapko *et al.*, 1999; Pak *et al.*, 2005; Wanagat *et al.*, 2001). It has also been used to examine both mouse models of mitochondrial disease (Tyynismaa *et al.*, 2005) and models of potential mtDNA gene therapy strategies (Bayona-Bafaluy *et al.*, 2005).

This chapter contains an overview of microdissection and analytical PCR techniques, and some example protocols applicable to the study of mtDNA. Obtaining sufficient DNA for analysis remains the most challenging aspect of working with microdissected samples. Where published techniques have been cited, brief details of samples sizes and amounts of template DNA used have been provided to help readers select protocols that may be applicable to their own research.

II. Microdissection and DNA Purification

A. Tissue Preparation and Sectioning

Paraffin embedding and cryosectioning are both compatible with microdissection. For ease of use however, many groups, including ours, favor cryosectioning. While tissues with a strong cellular architecture, such as skeletal muscle, can sometimes be analyzed without fixation, we generally fix samples in ten sample volumes of standard 4% paraformaldehyde PBS solution (pH 7.2) for 2 h. This is then followed by overnight storage in the same volume of 30% sucrose in PBS (plus a small amount of azide) to cryopreserve samples. We find this type of fixation compatible with PCR when used with either tissue lysis or commercial DNA extraction kits. To freeze tissues, we embed in Tissue Tek optimal cutting temperature (OCT) solution on an appropriate card or plastic support, and freeze either in isopentane in dry ice (<1 min) or nitrogen vapors above a beaker of liquid nitrogen (~3 min). Once frozen, tissue samples can be stored at −80 °C. For cryosectioning, we find that the best results are obtained with sections between 6–20 μm. When working with mtDNA, particularly mtDNA in cell extremities, care should be taken to avoid cutting the sections too thick. If sections are cut too thick, significant portions of contaminating cells can be present above or below the cells of interest.

Many histochemical stains can be applied to samples for microdissection. DNA is relatively robust and samples suitable for PCR amplification can be obtained from slides stained using a variety of dyes (Ehrig *et al.*, 2001). In terms of respiratory chain activity stains, standard cytochrome *c* oxidase (COX) activity stains, succinate dehydrogenase activity stains, and sequential COX and SDH activity double stains, are all compatible with mild fixation, and have little impact on DNA recovery following microdissection.

Immunofluorescent staining is also often used in microdissection. Although some reference manuals recommend short protocols, we find that standard staining procedures work well, provided that a suitable dehydration step is added following the staining (see below). As tissue sections for microdissection cannot be mounted like conventional slides, they often have poor optical qualities and suffer from high levels of autofluorescence. These issues must be addressed on a case-by-case basis, as increasing signal above background is dependent on sample autofluorescence, epitope abundance, and the quality of primary and secondary antibodies used. Most common fluorochromes are compatible with dehydration, although the signal is generally not as good as aqueous-mounted samples. Historically, we use Alexa dyes, which store fairly well at 4°C on dehydrated slides.

The most crucial modification to staining procedures for microdissection is the addition of a dehydration step at the end of staining. This is essential for laser microdissection, as even partially hydrated samples are almost impossible to cut. Many protocols recommend dehydrating stained slides using an ethanol/xylene series prior to air-drying. Generally, this takes the form of 30 sec in 70% ethanol, 30 sec in 95% ethanol, 30 sec in 100% ethanol, and 3 min in xylene. Following this, the excess xylene should be blotted off, and slides allowed to air-dry prior to storage in a sealed box at 4°C. Ethanol and xylene stocks should be changed regularly to ensure that they do not become hydrated. We find that dehydration can be greatly improved by storing stained slides with desiccant capsules. To do this, we store slides overnight at 4°C in a small slide box sealed with laboratory film containing five or six desiccant capsules (1 g each). Slides can be stored like this for up to 2 months without affecting microdissection or significant loss of immunofluorescent signal. For longer term storage, sections should be stored unstained at −80°C, and then stained and dehydrated just prior to microdissection.

B. Microdissection Techniques

Despite the recent ascendance of laser-based microdissection systems (described below), many different microdissection techniques have been described over the past 30 years [for review see Eltoum et al. (2002)]. Laser-based systems remain an expensive investment, and for many small projects, cheaper alternative techniques may be found. The only requirements are that the technique chosen is applicable to the tissue and cell types to be dissected, and that adequate care is taken to preserve the homogeneity of samples and to avoid contamination.

Needle microdissection is one of the most straightforward techniques. Using a 30-gauge needle and an inverted light microscope equipped with a 40× objective lens, tumor tissue can be dissected from 10-μm sections (Rabkin et al., 1997; Zhuang et al., 1995). Specialized tungsten needles mounted on micromanipulators have also been used, for instance, in the dissection individual colonic crypts from 7-μm paraffin sections submerged in proteinase K solution under an inverted microscope (Going and Lamb, 1996). Tissue sections can also be dissected using scalpel blades. Stratton and colleagues (Lakhani *et al.*, 1995a,b) used fine scalpel blades to dissect breast tumor tissue from 15-μm sections mounted on double-sided adhesive tape on glass microscope slides under a dissecting microscope.

Many groups have used variations on glass microcapillaries for microdissection. A combined microdissection and aspiration device, termed the "cytopicker," was conceived by Beltinger and Debatin (1998). It consists of a drawn glass micro-capillary mounted inside a steel cannula, and was originally applied to cytological specimens. The cannula is used to dissociate single cells in oil-submerged samples under an inverted microscope, followed by extension of the glass capillary, aspiration of the cell, and transfer to a sample tube. A glass microcapillary mounted on a micromanipulator was used by Zhou *et al.* (1997) to dissect and aspirate single neuronal somas from 35-μm sections bathed in Tris, EDTA (TE) buffer.

The fibrous nature of muscle makes it very amenable to microdissection. It is also a tissue that is of particular relevance to the study of mitochondrial diseases. An aerosol-protected mouth aspirator with a siliconized glass microcapillary was utilized by Moraes *et al.* (1992) to dissect single fibers from 30-μm transverse sections immersed in 50% ethanol under an inverted microscope. The same aspirator setup was used by Khrapko *et al.* (1999) to isolate individual cardiomyocytes. In a modification of the dissection procedure , they first used a collagenase solution to aid dissociation of single cells from 10-mg tissue sections prior to aspiration under an inverted microscope. Taylor *et al.* (2001) have also used the glass capillary procedure described by Zhou *et al.* (1997) above to dissect single muscle fibers from 20-μm transverse sections.

C. Laser Microdissection

There are currently three main laser microdissection technologies available [for review see Eltoum *et al.* (2002)]. Laser-capture microdissection (LCM) refers to a laser microdissection technique developed by the US National Cancer Institute and the National Institutes of Health, and commercialized by Arcturus and Molecular Devices. It is an inverted microscope system that uses a low-energy infrared laser to melt a thin polymer film onto cells of interest prior to separation and capture of the sample.

Laser microdissection and pressure catapulting (LMPC) dissecting microscopes are inverted UV laser-based systems where samples are cut from either membrane-coated or plain glass slides and then catapulted upward into collection caps by the induction of a high photon density beneath the sample (pressure catapulting). This technology is primarily marketed by Zeiss and PALM Microlaser Technologies.

An upright system, which uses a principle similar to LMPC, is marketed by Leica. In this system, sections are mounted on polymer membrane-coated slides, and a UV-laser is used to dissect samples that fall with gravity and pressure catapulting into collection caps below the stage. Although laser microdissection has been used since the 1970s (Isenberg *et al.*, 1976; Meier-Ruge *et al.*, 1976), it has only recently

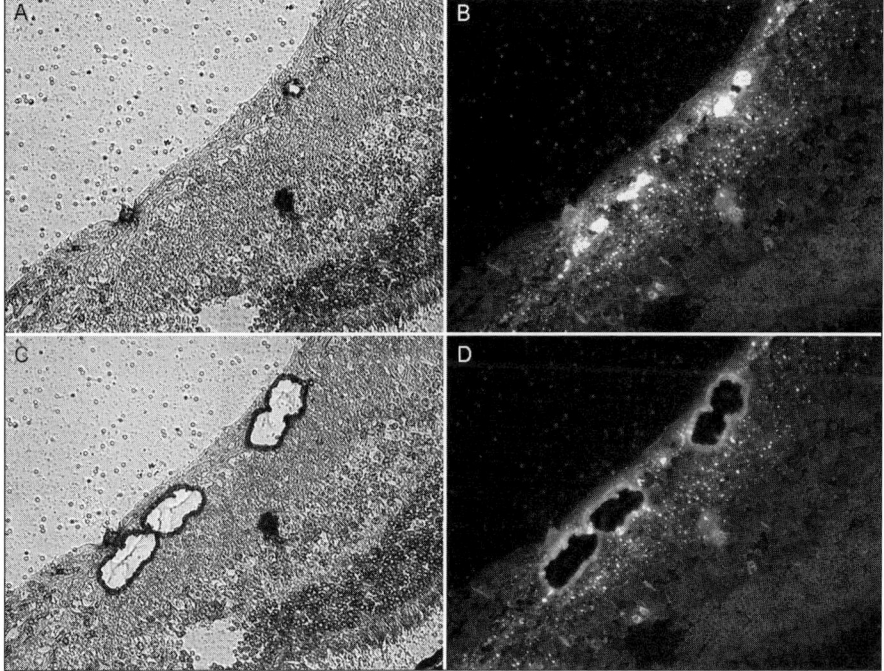

Fig. 1 Laser microdissection of single retinal ganglion cell somata using a Leica UV laser cutting microscope. Images show dissection of HA positive somata from an 8-μm section of mouse retina following injection of viral vector expressing an HA-tagged protein. Slide preparation and staining followed standard procedures are described in the text. (A) Differential interference contrast (DIC) image of intact retina. (B) Immunofluorescent image of HA positive cells, shown in white. (C) DIC image following laser microdissection. (D) Immunofluorescent image following laser microdissection. Samples were dissected at 40× magnification using default laser settings.

been commercialized, and as such, the technologies available are developing very rapidly. Combined systems that offer infrared lasers for LCM and UV lasers for laser-cutting have recently been released by Molecular Devices. In terms of their ability to dissect samples of various sizes for the analysis of mtDNA, there is really no difference between the three laser microdissection technologies and all have been successfully applied to mtDNA analysis, even at the single-cell level (Bender *et al.*, 2006; Cantuti-Castelvetri *et al.*, 2005; Zsurka *et al.*, 2005) (Fig. 1).

D. DNA Purification

Numerous DNA extraction techniques have been applied to microdissected samples for mtDNA analysis and any validated DNA extraction protocol applicable to microdissected samples should be suitable.

To ensure maximum DNA yield, many groups have used cell lysates when amplifying small samples. In general, samples are collected in microfuge tubes following microdissection, centrifuged briefly to ensure that the sample is in the bottom of the tube, and immersed in a small volume of an alkaline buffer containing detergent and proteinase K. Lysis buffers are well tolerated by PCR buffer systems, the only caveat being that care should be taken to inactive proteinase K completely.

A lysis solution initially used by Zhou *et al.* (1997) has been used in a number of studies. The buffer consists of 50-mM Tris–HCl, pH 8.5; 1-mM EDTA; 0.5% Tween-20; and 200-ng/ml proteinase K. Samples are incubated in the buffer for 2 h at 55°C followed by 10 min at 95°C to inactivate proteinase K. This buffer has been used for the lysis of (1) 20–30 neuronal somata sectioned at 35 μm in 50 μl, with 10 μl being added as template to each subsequent 50-μl PCR reaction; (2) individual colonic crypts sectioned at 20 μm in 10 μl, with 5 μl being used per subsequent 50-μl reaction (Taylor *et al.*, 2003); and (3) single transverse muscle fibers sectioned at 20 μm in 12 μl with identical template volumes as used for colonic crypts (He *et al.*, 2002; Taylor *et al.*, 2001). Published results indicate that this lysis buffer is compatible with real-time PCR buffer systems (He *et al.*, 2002; Wang *et al.*, 2005).

A slightly different lysis buffer system has been used by other groups. Khrapko *et al.* (1999) used a buffer composed of 10-mM EDTA, pH 8.0; 0.5% SDS; and 2-mg/ml proteinase K to lyse single dissociated cardiomyocytes in a volume of 1 μl. Samples were incubated at 37°C for 30 min and then diluted 1/10 with water. A 1-μl volume of lysate was then added to 5 μl of 1× PCR buffer and incubated at 95°C for 1 min to inactivate proteinase K. Samples were subsequently amplified in a final reaction volume of 20 μl. The same extraction protocol has also been used to lyse single muscle fibers sectioned at 10 μm (Cao *et al.*, 2001).

We have also used an alkaline lysis protocol developed by Cui *et al.* (1989) for genotyping single sperm, which avoids the use of proteinase K. Samples are lysed in 5 μl of buffer composed of 200-mM KOH, 50-mM DTT for 30 min at 65°C, and then neutralized by the addition of 5 μl of 900-mM Tris–HCl pH 8.3, 200-mM HCl. KOH solution should be made fresh weekly. This has been used successfully to liberate DNA from single transverse muscle fibers sectioned at 20 μm (Srivastava and Moraes, 2005; Bayona-Bafaluy *et al.*, 2005), and pooled CNS neurons sectioned at 20 μm (Bayona-Bafaluy *et al.*, 2005). Volumes of 1–2 μl of lysate are sufficient for amplification in 50-μl reaction volumes.

Commercially available DNA extraction kits optimized for microdissected samples are used by many groups. The QIAamp Micro kit from Qiagen has been used successfully to extract total DNA from single neuronal soma sectioned at 20–40 μm, the only adaptation being the use of 30 μl of water to elute DNA (Bender *et al.*, 2006). We routinely use this kit to extract DNA from pools of ≥3 retinal ganglion soma or retinal pigmented epithelium cells sectioned at 8 μm. For samples that were fixed previously using 4% paraformaldehyde, we digest for 16 h at 55°C and extract, using carrier RNA as recommended by the manufacturer. Our only modification is that we elute into 20 μl of 10-mM Tris, pH 8 (or Qiagen buffer EB) to avoid the use of an elution buffer containing EDTA that has the potential to inhibit subsequent amplification. The Picopure DNA extraction kit supplied by Arcturus is used by many groups using the Arcturus LCM system (Chen *et al.*, 2002), and the Ultraclean Bloodspin kit from MO BIO has been used to extract total DNA from microdissected pools of around 6000 prostate tumor cells (Parr *et al.*, 2006).

E. PCR

Due to the low template concentration in DNA extracts of microdissected samples, most groups use high-end commercial PCR kits for amplification. In particular, kits designed for long-range PCR, which have buffer and enzyme formulations optimized for highly efficient amplification, have proved reliable. Our group and others have had success using both the Expand Long Template PCR kit from Roche (Chen *et al.*, 2002) and the LA Taq polymerase kit from Takara (Khrapko *et al.*, 1999; Parr *et al.*, 2006). The Advantage HF 2 kit from Clontech has also been used with microdissected samples where high fidelity amplification was important (Cantuti-Castelvetri *et al.*, 2005).

III. Analytical PCR for Investigation of mtDNA in Microdissected Samples

The minute amounts of DNA recoverable from microdissected samples, typically in the pico- to nanogram range, necessitate use of specialized PCR-based protocols for analysis. Even within the spectrum of samples obtained using microdissection, mtDNA abundance can vary considerably from that extracted from a single cell to samples derived from multiple muscle fibers. Thus, not all examples given below will be applicable to all samples. Moreover, optimizing a microdissection and mtDNA analysis protocol will usually require some degree of experimentation. An important factor to bear in mind in relation to mtDNA analysis, is the potential for the occurrence of mtDNA depletion or ΔmtDNA species in tissues of interest which can also impact PCR efficiency relative to normal tissues.

A. Common Strategies for Improving PCR Signal Strength from Microdissected Samples

Problems caused by low abundance or poor quality template for PCR of microdissected samples can be overcome in a number of ways.

1. Increasing Template DNA Abundance and Recovery

One of the most obvious ways to increase the PCR efficiency of microdissected samples is to increase the amount of template DNA. This can be achieved simply by using more sample lysate in single reactions. However, this limits the analyses possible for each sample and is obviously risky when working with rare samples. In addition contaminants in lysates may also have an inhibitory effect on amplification, particularly when high template volumes are used. When working with purified DNA from microdissected samples we have managed to increase PCR yields by reducing total reaction volumes while maintaining template DNA volumes and component concentrations (Fig. 2). If applicable, sampling strategies can be altered to extract DNA from multiple pooled captures of a particular cell or tissue type. DNA recovery can sometimes be increased by altering extraction procedures or altering histological techniques. If problems are experienced in amplifying mtDNA from a sample set, increasing sample capture size prior to DNA extraction is the easiest way to troubleshoot a particular system. It should be remembered, however, that when optimizing a system, the samples used for optimization should be as close as possible to the test samples required for investigation. Microdissected DNA can be of poor quality and using diluted high-quality genomic or plasmid DNA will give a false indication of the suitability of a particular analytical technique for microdissected samples.

2. Increasing the Sensitivity of PCR Product Detection Using Radioactive Labeling

The sensitivity of radioactive detection methods is well recognized. Early work analyzing mtDNA from microdissected samples detected PCR products using [α-^{32}P]-dNTP incorporation throughout an entire PCR reaction (Sciacco et al., 1994) or for multiple cycles following an initial unlabeled reaction (Moraes

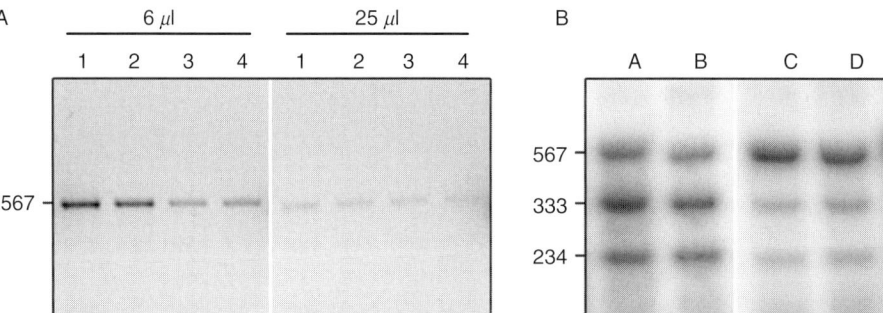

Fig. 2 (A) Advantages of small volume PCR reactions for amplification of DNA from microdissected samples. The left-hand panel shows entire 6-μl PCR reactions using 3 μl of template DNA to amplify a 567-bp fragment of mouse mtDNA. The right-hand panel shows entire 25-μl reactions amplifying the same fragment using identical 3-μl template volumes. Samples were separated using a 2.2% agarose gel and visualized using ethidium bromide. For samples 1 and 2 (different retinal regions) template volumes corresponded to around 110 cell-equivalents; samples 3 and 4 are 1/3 dilutions of samples 1 and 2, respectively. (B) Image from a storage phosphor imager of "last-cycle hot" PCR/RFLP of a 567-bp fragment of mtDNA amplified from heteroplasmic mouse retinal regions (lanes A–D). All procedures were carried out as described in the sample protocol. Each reaction used ~0.6 cell-equivalents as template; restriction enzyme cleavage resulted in fragments of 333 and 234 bp.

et al., 1992). For quantitative restriction fragment length polymorphism (RFLP) analysis, "last-cycle hot" PCR (Moraes *et al.*, 1992; Sciacco *et al.*, 1994) is the preferred method of labeling samples (see below). To reduce contamination and improve resolution, radiolabeled samples are conventionally run using polyacrylamide gel systems. Most laboratories have access to autoradiography cassettes and films suitable for imaging beta emitters, and the sensitivity of such systems exceeds that of standard fluorescent DNA stains. Storage phosphor systems such as the Perkin Elmer Cyclone or General Electric Storm offer even greater sensitivity, and most importantly, a linear response over a wide signal range. Systems such as these are useful for quantitation of relative signal intensities.

3. Increasing Amplification Efficiency Using Nested PCR

Nested PCR is often used to improve yields and reduce background signal when working with samples with low template abundance such as microdissected samples. It employs two sequential PCR reactions and is based on the principle that reducing template complexity improves amplification. The first round of amplification in nested PCR is carried out using primers external to the region of interest. Due to the extremely low abundance of specific template sequence relative to other oligonucleotides, such as genomic DNA, carrier RNA (when used in extraction), and even primers, this reaction mix is described as having a high template complexity. Resolution of the products of this reaction alone often results in a very weak and/or smeared signal; however, some amplification of template does occur, and relative to the initial template complexity the complexity of the reaction products is greatly reduced. This reduction in template complexity is harnessed in nested PCR by using an aliquot of the first reaction as template for a second amplification, using primers internal to those used in the first reaction. The use of alternative primers reduces the amplification of nonspecific products of the first reaction. The volumes of primary PCR used as template for secondary PCR vary between published protocols. Usually a <10% final volume is used, although if necessary, when using well-designed primers and small reaction volumes, the entire primary reaction can be used to seed a secondary reaction (Kraytsberg *et al.*, 2006). The number of cycles used in each reaction of nested PCR is usually kept low, at around 50 or fewer for both reactions in total. This helps to reduce amplification of nonspecific products, especially those originating from primer–primer interaction. The division of cycle numbers between the primary and secondary reactions varies between protocols, with some having a similar number of cycles in each reaction (Khrapko *et al.*, 1999) and others having more in the primary reaction (Kraytsberg *et al.*, 2006). Users may have to experiment with various protocols before finding one that works with a particular template and primer set. As with all PCR protocols, the success of nested PCR is dependent on primer design. Users may have to try a number of primer sets before finding suitable one. Where possible when working with microdissected samples, the advantage of using good *in silico*-designed primers cannot be overstated; many primers that work satisfactorily in conventional PCR will work poorly with microdissected samples.

One concern with nested PCR is the potential for cross-contamination of samples in sequential reactions. Strict adherence to standard precautions for avoiding PCR contamination should be followed. Examples include the use of different pipettes for PCR setup and analysis of amplicons, aliquoting all reagents, and, if possible, the use of separate bench areas or rooms for primary and secondary PCR setup. The use of partially nested primers, with one nested primer and one patient allele-specific primer, has been proposed as one way to help trace contaminated samples when working with mtDNA (Cantuti-Castelvetri et al., 2005) (see below). Numerous other variations of nested PCR have been used in mtDNA analysis of microdissected samples, some of which are outlined below.

B. RFLP of Known Polymorphisms and Mutations

RFLP analysis is one of the most straightforward techniques used to analyze polymorphisms in mtDNA populations, and has been applied to microdissected samples in many studies. The only adaptations to standard amplification/digestion RFLP protocols needed for microdissected samples are those that increase signal strength. As single nucleotide differences are generally assumed to have no significant influence on amplification efficiency, RFLP can be carried out subsequent to nested PCR. For accurate quantitation of polymorphism abundance, "last-cycle hot PCR" combined with storage phosphor-imager analysis is the preferred method of labeling and analyzing results (Bayona-Bafaluy et al., 2005; Moraes et al., 1992; Sciacco et al., 1994) (see below). As label is only added to the final cycle of amplification in "last-cycle hot PCR," the creation of labeled heteroduplex molecules that can interfere with restriction enzyme cleavage is minimized (Shoffner et al., 1990).

C. Sample Protocol

1. "Last–Cycle Hot" RFLP for mtDNA from Microdissected Samples

Slide preparation

1. Dissect tissue as appropriate.

2. Clean and wash tissue briefly in PBS and fix for up to 2 h in 4% PFA in PBS, pH 7.2. For sample sizes >100 mg, longer fixation may be necessary.

3. Transfer tissue to 30% sucrose and refrigerate at 4°C overnight until sample has sunk to the bottom of the tube.

4. To freeze, gently blot tissue on a paper towel and immerse in OCT in small plastic stand. We usually use pyramid-shaped corners cut from hexagonal weighing dishes.

5. Freeze over nitrogen vapors until OCT is solid (~2–3 min) and store at −80°C.

6. Cut 8-μm sections at an appropriate temperature dependent on water and lipid content, using a cryostat (e.g., brain, -10 to $-15\,^{\circ}$C; liver, -14 to $-18\,^{\circ}$C; eyecups -22 to $-24\,^{\circ}$C; muscle -20 to $-26\,^{\circ}$C).

7. Mount on appropriate slides (e.g., Leica PEN slides) warmed to $\sim 30\,^{\circ}$C and air-dry at RT for 20 min to 2 h.

Staining, dehydration, and microdissection

1. Stain as necessary (Fig. 1). Histochemical or immunofluorescent staining protocols generally do not require modification. Immunostains can be probed with primary antibody overnight at $4\,^{\circ}$C in humidified chambers without affecting microdissection. For COX activity staining, we incubate slides in a standard COX activity stain composed of 0.5-mg/ml diaminobenzidine (DAB), 1-mg/ml cytochrome *c*, and 0.1-mg/ml catalase in 50-mM sodium phosphate (NaPi) buffer, pH 7.4. For SDH staining, we stain in a solution of 5-mM EDTA, 1-mM KCN, 0.2-mM phenazine methasulfate, 50-mM succinic acid, and 1.5-mM nitro blue tetrazolium (NBT) in 50-mM NaPi buffer, pH 7.4. For COX-SDH double staining, we first stain for COX activity, wash slides briefly in NaPi buffer, and then stain for SDH activity. Both stains are incubated at $37\,^{\circ}$C in humidified chambers. Incubation times depend on tissue and fixation procedure. For instance, unfixed liver will develop visible stain in around 20min, whereas fixed liver may require up to 2 h.

2. Following staining, dehydrate slides by passing through 70%, 95%, and 100% ethanol for 30 sec each, and then clear in xylene for 3 min.

3. Blot off excess xylene and air-dry slides for about 5 min.

4. Store at $4\,^{\circ}$C in sealed, small volume slide boxes (e.g., $3'' \times 4''$, holding 50 slides) with six 1-g desiccant capsules. We find that storage at least overnight in this manner greatly improves dehydration and subsequent laser cutting.

5. Dissect cells of interest using a laser microdissection system and, if appropriate (e.g., when using the Leica system), spin down sample to the bottom of the collection tube before storage at $-20\,^{\circ}$C (for short term) or $-80\,^{\circ}$C (for over 7 days).

DNA extraction and amplification

1. Extract DNA using either cell lysis buffer or an appropriate extraction kit. For pooled samples of 3–12 cells, we use the QIAamp Micro kit from Qiagen, using 16-h sample digestions and 1 μg of carrier RNA per sample, as per manufacturer's instructions but eluting into 20 μl of 10-mM Tris–HCl, pH 8.5.

2. For PCR, we use the Expand long template PCR kit from Roche with either buffer 2 or 3, or the LA Taq polymerase kit from Takara. In both cases, we use 3 μl of Qiamp Micro kit DNA in a final reaction volume of 6 μl, with all other components as per manufacturer's standard recommendations (Fig. 2A). The small reaction volume increases initial template concentration leading to higher product

yields and combines well with last-cycle hot labeling and dilute RFLP (see below). PCR setup is carried out on ice and reactions are never stored prior to thermal cycling. We generally find that with samples obtained from <20 cells, signals from <600-bp fragments are best visualized with single PCR and last-cycle hot labeling or, if quantitation is not required, by nested PCR and ethidium bromide staining. For samples obtained from >100 cells, nested PCR is often unnecessary and radiolabeling can be avoided, unless results require quantitation.

3. For amplification of <600-bp fragments using a high-end commercial PCR kit, cycling is carried out in 0.2-μl thin-walled tubes as follows: 92 °C for 105 sec, then 32 cycles of 92 °C for 15 sec, T_m °C for 10 sec, 68 °C for 35 sec, followed by 7 min at 68 °C and hold at 4 °C. A hot lid set at 106 °C is used throughout heating cycles; oil is not required.

4. Samples can be stored at 4 °C for up to 72 h prior to "last-cycle hot" labeling, although we generally avoid this delay.

"Last-cycle hot" labeling

1. Make sufficient master mix for all samples containing, 2 μCi [α-^{32}P]-dCTP (generally 0.2 μl) and 1-U polymerase (generally 0.2 μl) in 2 μl of 1× PCR buffer per sample.

2. Add master mix to completed 6-μl PCR reactions on ice.

3. Cycle as follows: 92 °C for 3 min, 59 °C for 20 sec, 68 °C for 8 min and hold at 4 °C.

4. For standard amplified fragment length polymorphism (AFLP) such as three-primer PCR, add loading buffer and resolve samples using an appropriate poly-acrylamide gel electrophoresis (PAGE) system. Labeled samples can be frozen for around 24 h prior to electrophoresis without significant radioactive fragmentation.

Restriction digestion

1. Dilute 6 μl of labeled PCR reaction in a final volume of 24 μl of an appropriate digestion buffer containing 10-U restriction enzyme.

2. Incubate at 37 °C (or appropriate temperature) for 4–6 h, hold at 4 °C for convenience if necessary. Dilution of the PCR reaction and overdigestion removes the need for PCR cleanup. The high concentration of residual dNTPs, particularly dATP, in completed reactions can inhibit some restriction enzymes. In addition, avoiding PCR cleanup safeguards yield and reduces radioactive waste.

3. Resolve fragments using an appropriate nondenaturing PAGE system. We use 8% TBE-buffered gels to resolve fragments 200–600 bp in size.

4. Seal gels in heat-sealable plastic bags and wipe the outside of the bags with decontamination spray to reduce the chance of contamination.

5. Expose gels to storage-phosphor screen for an appropriate time. For 200- to 600-bp fragments from single PCR RFLP of DNA from 0.6–2.4 cell-equivalents

we generally expose for 4 days using the Perkin Elmer Cyclone system with standard screens (Fig. 2B). Such gels have counts of around 100–200 cpm at the start of exposure. Larger samples, such as muscle fibers or products of nested PCR reactions can be exposed in a matter of hours.

D. mtDNA Sequencing

Sequencing of mtDNA from microdissected samples has been achieved by a number of groups. Taylor *et al.* developed a tiled, nested strategy which is applicable to single muscle fibers (Taylor *et al.*, 2001), single colonic crypts (Taylor *et al.*, 2003), and single neurons (Bender *et al.*, 2006). In the first set of reactions, a series of nine overlapping fragments, each around 2 kb in size, that encompass the entire mtDNA genome are amplified. Following this, a second set of reactions, using nested M13-tagged primers and 2 μl of the first reaction as template per 50-μl reaction was used to amplify 600- to 700-bp fragments suitable for use as sequencing templates. mtDNA from whole lysates of single neuronal somata has also been sequenced using a partially nested PCR protocol. Cantuti-Castelvetri *et al.* (2005) used a single allele-specific primer paired with one of two nested primers within a 450- to 700-bp range in sequential PCR reactions. The allele-specific primers were unique to each patient investigated and were designed following sequencing of total mtDNA from tissue homogenates. The use of allele-specific primers was anticipated both to reduce contamination and to enable tracing of contaminated samples. This was an important precaution, as following amplification, fragments were cloned into TA vectors to allow for M13 sequencing of mtDNA.

Larger sample capture sizes enable sequencing of microdissected tissues without the use of nested PCR. DNA from 150 cell-equivalents was used as template for a single set of reactions using 34 primer pairs that was sufficient to generate sequencing templates encompassing the entire mtDNA genome (Parr *et al.*, 2006). Similarly, a single "touch-down" PCR reaction was sufficient to generate sequencing template from 100 cell-equivalents for partial mtDNA genome sequencing of hypervariable regions 1 and 2 in the D-loop (Chen *et al.*, 2002).

E. Quantification of mtDNA Content

Conventional Southern blot analysis is clearly not possible with microdissected samples, although mtDNA content can be measured using PCR-based approaches. A radioactive assay using standard thermal-cycling equipment was employed by Sciacco *et al.* (1994) to measure the mtDNA content of single muscle fibers in the <10 pg range. Incorporation of [α-^{32}P]-dATP into fragments amplified from test samples was measured relative to that incorporated in parallel reactions containing known concentrations of purified mtDNA template. Under the conditions used, a first-order correlation between initial template concentration and endpoint ^{32}P incorporation was observed in template standards.

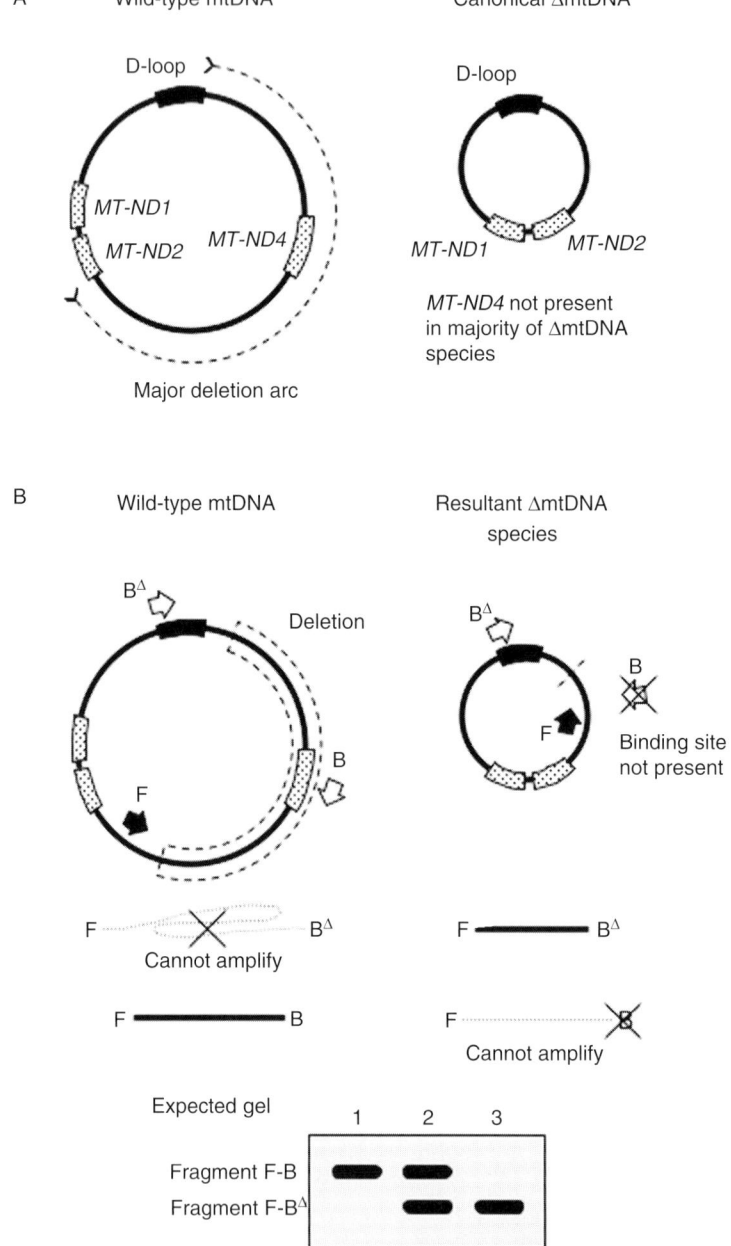

Fig. 3 The mtDNA major deletion arc. (A) Maps of wild-type and canonical ΔmtDNA. The wild-type map shows the position of the major deletion arc relative to the positions of *MT-ND1*, *MT-ND2*, *MT-ND4*, and the D-loop. The canonical ΔmtDNA map demonstrates the loss of *MT-ND4* in the majority of ΔmtDNA species. (B) The principle of three-primer PCR. The wild-type map shows the position of the

The endpoint ^{32}P incorporation in test samples was then read against this standard curve to give actual mtDNA content.

More recently, real-time PCR has become the method of choice for analysis of mtDNA abundance in microdissected samples. Giordano *et al.* (2006) adapted a dual color standard-curve method originally devised by Cossarizza *et al.* (2002) to investigate mtDNA copy number in microdissected samples of gastrointestinal tissue. They amplified an mtDNA fragment within *MT-ND2* (NADH dehydrogenase 2) and an nDNA fragment within *FASLG* and compared results from patient samples against a dilution series of standards Fas ligand, comparing results from patient samples with a single plasmid in which both fragments had been inserted in tandem. This enabled determination of the mtDNA copy-number per cell using a single set of standard reactions, as each cell contains *2n* copies of *FASLG*. Similarly, Wang *et al.* (2005) compared the abundance of a fragment within mtDNA region nt 3182–4216 to one within exon 2 of *HBB* (hemoglobin, beta), but used two separate sets of reactions. It should be noted that in both of these examples, researchers used mtDNA templates outside the major deletion arc (Samuels *et al.*, 2004) (Fig. 3A). This reduces the likelihood of underestimating mtDNA content by excluding ΔmtDNA species from analysis.

A real-time PCR technique that enables determination of the copy number of both wild-type and ΔmtDNA species has been developed by He and coworkers (He *et al.*, 2002). This method does not include an nDNA control and so provides information simply as a copy-number per sample basis rather than an nDNA: mtDNA ratio (see below).

F. Detection of ΔmtDNA Species

PCR is a potentially competitive reaction in which final product size and abundance are influenced by reaction conditions. As a result, the detection of ΔmtDNA species is not always quantitative. The simplest technique for detecting ΔmtDNA species is three-primer PCR (Moraes *et al.*, 1992; Sciacco *et al.*, 1994; Shoubridge *et al.*, 1990) (Fig. 3B). This technique uses three primers simultaneously: a standard primer pair outside the deletion boundary (F,B$^\Delta$) and a single primer within the deletion boundary (B). Using short extension times with mixed populations of wild type and ΔmtDNA as template, wild-type molecules are amplified by the F and B primers

---

hypothetical primers F, B, and B$^\Delta$. Using standard PCR conditions only the fragment between F and B is amplified from wild-type molecules, as F and B$^\Delta$ are too far apart. The position of a single hypothetical deletion is represented on the wild-type map using a dashed box. F and B$^\Delta$ are outside the deleted region, allowing for the amplification of the fragment between F and B$^\Delta$ from deleted molecules, whereas the fragment between F and B is not amplified, as there is no binding site for primer B. A cartoon of the expected electrophoresis pattern of the fragments described is shown underneath. In this example, fragment F-B$^\Delta$ is smaller than F-B, although this is entirely dependent on the primer design. Lane 1, homoplasmic wild-type cells; lane 2, heteroplasmic cells; and lane 3, homoplasmic mutant cells.

and ΔmtDNA by the F and B$^{\Delta}$ primers. Without optimization and a standard curve, this technique is nonquantitative, but allows for sensitive detection of ΔmtDNA. As with other PCR techniques for microdissected samples, signal detection can be increased by using radiolabeled or nested PCR (Srivastava and Moraes, 2005). Three-primer PCR is best applied to samples with known deletion boundaries or those suspected of harboring the common 4977-bp deletion (Schon et al., 1989).

Primer-shift PCR is a technique based on three-primer PCR that can be employed to search for ΔmtDNA species. Using this technique, reactions are set up using many different pairs of primers some of which will fall outside an unknown deletion boundary. These are essentially the same primers as described for three-primer PCR, except that multiple F and B/B$^{\Delta}$ primers are designed and used in separate reactions to scan large regions of template DNA. Primer shift can be used to screen total DNA from tissue homogenates for ΔmtDNA species. Primers that amplify ΔmtDNA successfully can then be used to investigate microdissected samples without the use of valuable microdissected samples for primer optimization. Combined with TA cloning, this approach has been used to investigated the role of double strand breaks on ΔmtDNA formation in single muscle fibers (Srivastava and Moraes, 2005).

Long-range PCR is often used for the detection of unknown mtDNA deletions, and many groups have used it to identify the clonal expansion of ΔmtDNA species. In the absence of substantial standardization, it is not generally considered a quantitative technique (Kajander et al., 1999), and is best suited to screening for ΔmtDNA species. Nested long-range PCR has been used for the detection of ΔmtDNA in single dissociated cardiomyocytes (Khrapko et al., 1999). In this study, a 0.1 cell-equivalent from a single cell lysate was amplified in two sequential 20-μl reactions of 25 and 20 cycles, respectively, using nested primers within the D-loop. A 1-μl volume from the primary reaction was used as template for the secondary reaction, and amplification was sufficient to enable visualization using ethidium bromide. Similar approaches have been used with samples derived from single muscle fibers from microdissected 10-μm transverse sections (Cao et al., 2001; Wanagat et al., 2001).

Nested long-range PCR can be combined with RFLP to localize deletion breakpoints. This approach has been used to map deletion breakpoints in single cardiomyocyte lysates (Bodyak et al., 2001; Khrapko et al., 1999). Following identification of clonal ΔmtDNA species in microdissected samples using nested long-range PCR, samples are digested with a combination of AvaI, DraI, and BclI, or with BstNI alone. The resulting fragments can be separated using 2% agarose gel electrophoresis and used to map of deletion breakpoints to a resolution of around 1 kb.

G. Quantification of ΔmtDNA Abundance

A number of techniques have been developed to quantitate the relative abundance of ΔmtDNA species in microdissected samples. When carefully standardized, three-primer PCR can be used to quantitate the abundance of ΔmtDNA

(Pistilli *et al.*, 2003; Sciacco *et al.*, 1994). Standardization is best achieved using radiolabeling of PCR products and densitometric or storage phosphor-imager analysis. First, PCR conditions need to be optimized to ensure that amplification is within a quasilinear range dependent on template concentration. Once this is achieved, a standard curve needs be prepared comparing results from three-primer PCR to those of Southern blotting over a known range of wild-type mtDNA: ΔmtDNA ratios. The resulting curve can be then used to correct readout from test PCR samples. Alternatively, some researchers have standardized three-primer PCR by simply comparing band intensities to those obtained over a known range of wild-type mtDNA:ΔmtDNA ratios at template concentrations similar to those obtained in microdissected samples (Khrapko *et al.*, 1999). Although standardization can be time consuming, three-primer PCR does not require specialized thermocyclers, and when properly applied, provides a level of accuracy similar to that obtained using real-time PCR (He *et al.*, 2002).

Fast, accurate quantitation of the abundance of ΔmtDNA species can be obtained using real-time PCR. He *et al.* (2002) developed and validated a strategy based on amplification of fragments inside and outside the major deletion arc that enables determination of wild-type and ΔmtDNA copy number on a per sample basis. Amplification of *MT-ND1* (NADH dehydrogenase 1) was used as a marker for regions outside the major arc and amplification of *MT-ND4* (NADH dehydrogenase 4) as a marker for regions inside the major arc (Fig. 3A). Results from each reaction were compared to a standard curves obtained using similar template DNA concentrations over a range of known mtDNA:ΔmtDNA ratios. This approach has been used to quantitate relative ΔmtDNA copy number in single muscle fibers using lysate from 0.5 fiber-equivalents microdissected from 20-μm transverse sections in each reaction (He *et al.*, 2002); in individual colonic crypts using lysate from 0.5 crypt-equivalents microdissected from 20-μm sections in each   reaction (Taylor *et al.*, 2003); and in somata from individual neurons using a 0.5 cell-equivalent per reaction (Bender *et al.*, 2006).

A unique quantal approach to assessing relative ΔmtDNA abundance has been developed by Kraytsberg and Khrapko (2005). It is based on a single-molecule PCR technique originally devised for investigation of minisatellite variation (Jeffreys *et al.*, 1990), also referred to as digital PCR (Pohl and Shih, 2004). For single-molecule PCR, multiple reactions are set up for a particular sample (typically >48), using template DNA diluted to the point at which around 50% of reactions do not receive amplifiable template. The assumption is that at this level of dilution, if amplification occurs it originates from a single intact template. Thus, in wells where amplification does occur, reaction products are ostensibly clonal. Depending on the PCR strategy used, final reaction products can serve as templates for sequencing reactions, RFLP, or AFLP such as duplex PCR (see below). Amplification is carried out using essentially standard conditions but with an increased concentration of polymerase. Due to the extremely low template concentration, visualization of single-molecule PCR reaction products may require the addition of a nested PCR amplification. The template simplicity of single-molecule PCR is advantageous because it reduces artifacts due to preferential amplification of ΔmtDNA species or the incidence of PCR jumping (Kraytsberg and Khrapko, 2005).

Using a nested, duplex, single-molecule PCR protocol, Kraytsberg *et al.* (2006) have analyzed mtDNA populations in single microdissected neuronal somata. Cells were microdissected from 10-μm sections and lysates were prepared for dilution of template DNA. Single-molecule templates were achieved using a 1/10,000 dilution of lysate per reaction. Two primer pairs were used simultaneously in each reaction (duplex PCR), one amplifying a region of *MT-ND1*, which is outside the major deletion arc, and one in a region of *MT-ND4*, which is inside the major deletion arc, similar to the real-time PCR strategy devised by He *et al.* (2002) (see above). The primary reaction was carried out in a final volume of 0.6 μl for 31 cycles and the entire reaction was used as template for a 3.6-μl, 15-cycle secondary reaction using nested primers. Reaction products were then resolved using agarose gel electrophoresis and could be visualized using ethidium bromide. The abundance of ΔmtDNA species could then be calculated from the number of lanes resolving only the *MT-ND1* fragment, representing major arc deletions, versus those resolving both the *MT-ND1* and *MT-ND4* fragments, representing wild-type molecules. Results were corrected by taking into account a slight bias in the amplification of ΔmtDNA species and accounting for the Poissonian probability of multiple molecules per well.

IV. Concluding Remarks

Analytical PCR of microdissected samples is emerging as a powerful tool for investigating the biology of mtDNA. Hopefully the examples described above have demonstrated that in its most fundamental form, this technique can be developed in any laboratory with access to basic molecular biology and microscopy equipment. Clearly, for complex analyses, access to specialized equipment is required, but as institutional investment in tools such as laser microdissection microscopes and real-time PCR machines increases, access within the mitochondrial research community will surely also increase. Application of these techniques to more diverse model systems and tissue archives can only benefit the field, both in terms of our understanding of mtDNA at the cellular level and in the development of improved analytical techniques.

Acknowledgments

This work was supported by grants from the National Institutes of Health (NINDS, NCI, and NEI) and the Muscular Dystrophy Association. We are grateful to Beata Frydel and Brigitte Shaw from the University of Miami Imaging Core for assistance with the Laser Capture Microscope.

References

Bayona-Bafaluy, M. P., Blits, B., Battersby, B. J., Shoubridge, E. A., and Moraes, C. T. (2005). Rapid directional shift of mitochondrial DNA heteroplasmy in animal tissues by a mitochondrially targeted restriction endonuclease. *Proc. Natl. Acad. Sci. USA* **102**(40), 14392–14397.

Beltinger, C. P., and Debatin, K. M. (1998). A simple combined microdissection and aspiration device for the rapid procurement of single cells from clinical peripheral blood smears. *Mol. Pathol.* **51**(4), 233–236.

Bender, A., Krishnan, K. J., Morris, C. M., Taylor, G. A., Reeve, A. K., Perry, R. H., Jaros, E., Hersheson, J. S., Betts, J., Klopstock, T., Taylor, R. W., and Turnbull, D. M. (2006). High levels of mitochondrial DNA deletions in substantia nigra neurons in aging and Parkinson disease. *Nat. Genet.* **38**(5), 515–517.

Bodyak, N. D., Nekhaeva, E., Wei, J. Y., and Khrapko, K. (2001). Quantification and sequencing of somatic deleted mtDNA in single cells: Evidence for partially duplicated mtDNA in aged human tissues. *Hum. Mol. Genet.* **10**(1), 17–24.

Cantuti-Castelvetri, I., Lin, M. T., Zheng, K., Keller-McGandy, C. E., Betensky, R. A., Johns, D. R., Beal, M. F., Standaert, D. G., and Simon, D. K. (2005). Somatic mitochondrial DNA mutations in single neurons and glia. *Neurobiol. Aging* **26**(10), 1343–1355.

Cao, Z., Wanagat, J., McKiernan, S. H., and Aiken, J. M. (2001). Mitochondrial DNA deletion mutations are concomitant with ragged red regions of individual, aged muscle fibers: Analysis by laser-capture microdissection. *Nucleic Acids Res.* **29**(21), 4502–4508.

Chen, J. Z., Gokden, N., Greene, G. F., Mukunyadzi, P., and Kadlubar, F. F. (2002). Extensive somatic mitochondrial mutations in primary prostate cancer using laser capture microdissection. *Cancer Res.* **62**(22), 6470–6474.

Cossarizza, A., Riva, A., Pinti, M., Ammannato, S., Fedeli, P., Mussini, C., Esposito, R., and Galli, M. (2002). Mitochondrial DNA content in CD4 and CD8 peripheral blood lymphocytes of HIV+ patients with lipodystrophy. *Antivir. Ther.* **7**(3), L44.

Cossarizza, A., Riva, A., Pinti, M., Ammannato, S., Fedeli, P., Mussini, C., Esposito, R., and Galli, M. (2003). Increased mitochondrial DNA content in peripheral blood lymphocytes from HIV-infected patients with lipodystrophy. *Antivir. Ther.* **8**(4), 315–321.

Cui, X. F., Li, H. H., Goradia, T. M., Lange, K., Kazazian, H. H., Jr., Galas, D., and Arnheim, N. (1989). Single-sperm typing: Determination of genetic distance between the G gamma-globin and parathyroid hormone loci by using the polymerase chain reaction and allele-specific oligomers. *Proc. Natl. Acad. Sci. USA* **86**(23), 9389–9393.

Ehrig, T., Abdulkadir, S. A., Dintzis, S. M., Milbrandt, J., and Watson, M. A. (2001). Quantitative amplification of genomic DNA from histological tissue sections after staining with nuclear dyes and laser capture microdissection. *J. Mol. Diagn.* **3**(1), 22–25.

Eltoum, I. A., Siegal, G. P., and Frost, A. R. (2002). Microdissection of histologic sections: Past, present, and future. *Adv. Anat. Pathol.* **9**(5), 316–322.

Giordano, C., Sebastiani, M., Plazzi, G., Travaglini, C., Sale, P., Pinti, M., Tancredi, A., Liguori, R., Montagna, P., Bellan, M., Valentino, M. L., Cossarizza, A., *et al.* (2006). Mitochondrial Neurogastrointestinal encephalomyopathy: Evidence of mitochondrial DNA depletion in the small intestine. *Gastroenterology* **130**(3), 893–901.

Going, J. J., and Lamb, R. F. (1996). Practical histological microdissection for PCR analysis. *J. Pathol.* **179**(1), 121–124.

He, L. P., Chinnery, P. F., Durham, S. E., Blakely, E. L., Wardell, T. M., Borthwick, G. M., Taylor, R. W., and Turnbull, D. M. (2002). Detection and quantification of mitochondrial DNA deletions in individual cells by real-time PCR. *Nucleic Acids Res.* **30**(14), e68.

Isenberg, G., Bielser, W., Meier-Ruge, W., and Remy, E. (1976). Cell surgery by laser micro-dissection: A preparative method. *J. Microsc.* **107**(1), 19–24.

Jeffreys, A. J., Neumann, R., and Wilson, V. (1990). Repeat unit sequence variation in minisatellites: A novel source of DNA polymorphism for studying variation and mutation by single molecule analysis. *Cell* **60**(3), 473–485.

Kajander, O. A., Poulton, J., Spelbrink, J. N., Rovio, A., Karhunen, P. J., and Jacobs, H. T. (1999). The dangers of extended PCR in the clinic. *Nat. Med.* **5**(9), 965–966.

Khrapko, K., Bodyak, N., Thilly, W. G., van Orsouw, N. J., Zhang, X., Coller, H. A., Perls, T. T., Upton, M., Vijg, J., and Wei, J. Y. (1999). Cell-by-cell scanning of whole mitochondrial genomes in aged human heart reveals a significant fraction of myocytes with clonally expanded deletions. *Nucleic Acids Res.* **27**(11), 2434–2441.

Kraytsberg, Y., and Khrapko, K. (2005). Single-molecule PCR: An artifact-free PCR approach for the analysis of somatic mutations. *Expert Rev. Mol. Diagn.* **5**(5), 809–815.

Kraytsberg, Y., Kudryavtseva, E., McKee, A. C., Geula, C., Kowall, N. W., and Khrapko, K. (2006). Mitochondrial DNA deletions are abundant and cause functional impairment in aged human substantia nigra neurons. *Nat. Genet.* **38**(5), 518–520.

Lakhani, S. R., Collins, N., Sloane, J. P., and Stratton, M. R. (1995a). Loss of heterozygosity in lobular carcinoma *in situ* of the breast. *Clin. Mol. Pathol.* **48**(2), M74–M78.

Lakhani, S. R., Collins, N., Stratton, M. R., and Sloane, J. P. (1995b). Atypical ductal hyperplasia of the breast: Clonal proliferation with loss of heterozygosity on chromosomes 16q and 17p. *J. Clin. Pathol.* **48**(7), 611–615.

Meier-Ruge, W., Bielser, W., Remy, E., Hillenkamp, F., Nitsche, R., and Unsold, R. (1976). The laser in the Lowry technique for microdissection of freeze-dried tissue slices. *Histochem. J.* **8**(4), 387–401.

Moraes, C. T., Ricci, E., Bonilla, E., DiMauro, S., and Schon, E. A. (1992). The mitochondrial tRNA (Leu(UUR)) mutation in mitochondrial encephalomyopathy, lactic-acidosis, and stroke-like episodes (MELAS), genetic, biochemical, and morphological correlations in skeletal-muscle. *Am. J. Hum. Genet.* **50**(5), 934–949.

Moraes, C. T., Sciacco, M., Ricci, E., Tengan, C. H., Hao, H., Bonilla, E., Schon, E. A., and DiMauro, S. (1995). Phenotype-genotype correlations in skeletal muscle of patients with mtDNA deletions. *Muscle Nerve* **3**, S150–S153.

Pak, J. W., Vang, F., Johnson, C., McKenzie, D., and Aiken, J. M. (2005). MtDNA point mutations are associated with deletion mutations in aged rat. *Exp. Gerontol.* **40**(3), 209–218.

Parr, R. L., Dakubo, G. D., Crandall, K. A., Maki, J., Reguly, B., Aguirre, A., Wittock, R., Robinson, K., Alexander, J. S., Birch-Machin, M. A., Abdel-Malak, M., Froberg, M. K., *et al.* (2006). Somatic mitochondrial DNA mutations in prostate cancer and normal appearing adjacent glands in comparison to age-matched prostate samples without malignant histology. *J. Mol. Diagn.* **8**(3), 312–319.

Pistilli, D., di Gioia, C. R., D'Amati, G., Sciacchitano, S., Quaglione, R., Quitadamo, R., Casali, C., Gallo, P., and Santorelli, F. M. (2003). Detection of deleted mitochondrial DNA in Kearns-Sayre syndrome using laser capture microdissection. *Hum. Pathol.* **34**(10), 1058–1061.

Pohl, G., and Shih, I. (2004). Principle and applications of digital PCR. *Expert Rev. Mol. Diagn.* **4**(1), 41–47.

Polyak, K., Li, Y., Zhu, H., Lengauer, C., Willson, J. K., Markowitz, S. D., Trush, M. A., Kinzler, K. W., and Vogelstein, B. (1998). Somatic mutations of the mitochondrial genome in human colorectal tumours. *Nat. Genet.* **20**(3), 291–293.

Rabkin, C. S., Janz, S., Lash, A., Coleman, A. E., Musaba, E., Liotta, L., Biggar, R. J., and Zhuang, Z. (1997). Monoclonal origin of multicentric Kaposi's sarcoma lesions. *N. Engl. J. Med.* **336**(14), 988–993.

Rosson, D., and Keshgegian, A. A. (2004). Frequent mutations in the mitochondrial control region DNA in breast tissue. *Cancer Lett.* **215**(1), 89–94.

Samuels, D. C., Schon, E. A., and Chinnery, P. F. (2004). Two direct repeats cause most human mtDNA deletions. *Trends Genet.* **20**(9), 393–398.

Schon, E. A., Rizzuto, R., Moraes, C. T., Nakase, H., Zeviani, M., and DiMauro, S. (1989). A direct repeat is a hotspot for large-scale deletion of human mitochondrial-DNA. *Science* **244**(4902), 346–349.

Sciacco, M., Bonilla, E., Schon, E. A., DiMauro, S., and Moraes, C. T. (1994). Distribution of wild-type and common deletion forms of mtDNA in normal and respiration-deficient muscle fibers from patients with mitochondrial myopathy. *Hum. Mol. Genet.* **3**(1), 13–19.

Shoffner, J. M., Lott, M. T., Lezza, A. M., Seibel, P., Ballinger, S. W., and Wallace, D. C. (1990). Myoclonic epilepsy and ragged-red fiber disease (MERRF) is associated with a mitochondrial DNA tRNA(Lys) mutation. *Cell* **61**(6), 931–937.

Shoubridge, E. A., Karpati, G., and Hastings, K. E. (1990). Deletion mutants are functionally dominant over wild-type mitochondrial genomes in skeletal muscle fiber segments in mitochondrial disease. *Cell* **62**(1), 43–49.

Srivastava, S., and Moraes, C. T. (2005). Double-strand breaks of mouse muscle mtDNA promote large deletions similar to multiple mtDNA deletions in humans. *Hum. Mol. Genet.* **14**(7), 893–902.

Taylor, R. W., Barron, M. J., Borthwick, G. M., Gospel, A., Chinnery, P. F., Samuels, D. C., Taylor, G. A., Plusa, S. M., Needham, S. J., Greaves, L. C., Kirkwood, T. B., and Turnbull, D. M. (2003). Mitochondrial DNA mutations in human colonic crypt stem cells. *J. Clin. Invest.* **112**(9), 1351–1360.

Taylor, R. W., Taylor, G. A., Durham, S. E., and Turnbull, D. M. (2001). The determination of complete human mitochondrial DNA sequences in single cells: Implications for the study of somatic mitochondrial DNA point mutations. *Nucleic Acids Res.* **29**(15), E74.

Tyynismaa, H., Mjosund, K. P., Wanrooij, S., Lappalainen, I., Ylikallio, E., Jalanko, A., Spelbrink, J. N., Paetau, A., and Suomalainen, A. (2005). Mutant mitochondrial helicase Twinkle causes multiple mtDNA deletions and a late-onset mitochondrial disease in mice. *Proc. Natl. Acad. Sci. USA* **102**(49), 17687–17692.

Wanagat, J., Cao, Z., Pathare, P., and Aiken, J. M. (2001). Mitochondrial DNA deletion mutations colocalize with segmental electron transport system abnormalities, muscle fiber atrophy, fiber splitting, and oxidative damage in sarcopenia. *FASEB J.* **15**(2), 322–332.

Wang, Y., Liu, V. W., Xue, W. C., Tsang, P. C., Cheung, A. N., and Ngan, H. Y. (2005). The increase of mitochondrial DNA content in endometrial adenocarcinoma cells: A quantitative study using laser-captured microdissected tissues. *Gynecol. Oncol.* **98**(1), 104–110.

Zhou, L., Chomyn, A., Attardi, G., and Miller, C. A. (1997). Myoclonic epilepsy and ragged red fibers (MERRF) syndrome: Selective vulnerability of CNS neurons does not correlate with the level of mitochondrial tRNAlys mutation in individual neuronal isolates. *J. Neurosci.* **17**(20), 7746–7753.

Zhuang, Z., Bertheau, P., Emmert-Buck, M. R., Liotta, L. A., Gnarra, J., Linehan, W. M., and Lubensky, I. A. (1995). A microdissection technique for archival DNA analysis of specific cell populations in lesions <1 mm in size. *Am. J. Pathol.* **146**(3), 620–625.

Zsurka, G., Kraytsberg, Y., Kudina, T., Kornblum, C., Elger, C. E., Khrapko, K., and Kunz, W. S. (2005). Recombination of mitochondrial DNA in skeletal muscle of individuals with multiple mitochondrial DNA heteroplasmy. *Nat. Genet.* **37**(8), 873–877.

CHAPTER 25

Transmitochondrial Technology in Animal Cells

Sandra R. Bacman and Carlos T. Moraes

Department of Neurology
University of Miami Miller School of Medicine
Miami, Florida 33136

I. Introduction

Studies of vertebrate mitochondrial DNA (mtDNA) maintenance and function have relied heavily on somatic cell experimentation in culture, because we are still unable to manipulate mtDNA sequences at will in animal cells. Initial patterns of mitochondrial segregation and species-specific compatibility were performed using somatic hybrid and cybrid cells. Cybrids, or cytoplasmatic hybrids, are eukaryotic cell lines produced by the fusion of a nuclear donor and mitochondria from a donor. Therefore, a cybrid is a cell formed by the nuclear genome from one source and mitochondrial genomes from a different one. As a result, in this model it is possible to dissociate the biochemical influence of the mitochondrial genome from its nuclear background. Hybrid or cybrid cells may be formed by several techniques, including cell fusion (Clayton *et al.*, 1971; De Francesco *et al.*, 1980; Giles *et al.*, 1980; Hayashi *et al.*, 1983; Ziegler and Davidson, 1981). *Rho-zero* cells (ρ^0) (cells devoid of mtDNA) can be repopulated easily with exogenous mtDNA, resulting in transmitochondrial cybrids (King and Attardi, 1989). A large number of pathogenic mtDNA mutations have been studied using this procedure, including rearrangements (Hayashi *et al.*, 1991; Lee *et al.*, 2005; Porteous *et al.*, 1998), point mutations in tRNA (Bacman *et al.*, 2003; Bornstein *et al.*, 2005; Chomyn *et al.*, 1991; Hao and Moraes, 1997; King *et al.*, 1992; Toompuu *et al.*, 1999), rRNA (Hsieh *et al.*, 2001; Inoue *et al.*, 1996), and protein-coding genes (Baracca *et al.*, 2005; Bruno *et al.*, 1999; Jun *et al.*, 1996; Trounce *et al.*, 1994). This technology has also been used to study evolutionary interactions between nuclear and mitochondrial genomes of closely related species (Dey *et al.*, 2000; Kenyon and Moraes, 1997; McKenzie and Trounce, 2000). This approach was used to generate transmitochondrial embryonic stem cells which were used successfully to produce transmitochondrial mice (Trounce *et al.*, 2004). Different tissue types and techniques have been employed for the generation of transmitochondrial cell lines in animals. Basically, a cell line devoid of mtDNA or with poisoned mitochondria works as the nuclear donor, while enucleated or fragmented cells function as the mtDNA donor (Fig. 1). On cell membrane fusions and appropriate selection, transmitochondrial cells are generated. In this chapter, we describe current protocols used to generate transmitochondrial animal cell lines in culture.

II. Generation of Nuclear Donors

A. Generation of ρ^0 Cells

DNA-intercalating agents, such as ethidium bromide (EtBr), have been shown to affect mtDNA replication in doses that are too low to affect nuclear DNA significantly (Nass, 1972; Smith *et al.*, 1971). Blocking mtDNA replication in culture conditions that allow exponential growth of cells will reduce the total amount of mtDNA by half every cell doubling. To keep EtBr-treated vertebrate cells growing exponentially, Morais *et al.* (1980) realized that the medium had to

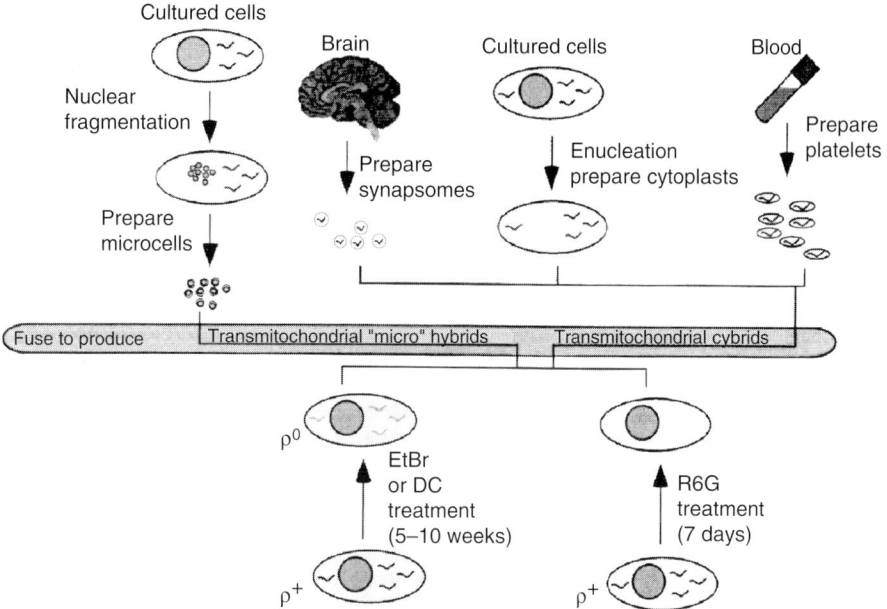

Fig. 1 Overview of methods to prepare transmitochondrial animal cell lines. Transmitochondrial cell lines are obtained by the fusion of cell lines devoid of mtDNA (ρ^0 cells or any other nuclear donors) and cellular fragments containing mitochondria (mtDNA donors). Different cells or tissues are suitable to prepare cytoplasmic bodies that can function as mtDNA donors. By using microcells and appropriated selection markers, one can introduce mtDNA and single chromosomes simultaneously into ρ^0 cells. Although ρ^0 cells are excellent nuclear donors, mitochondrial poisons, such as R6G, can also be used to eliminate mitochondria from cells that can then function as nuclear donors. All these procedures are described in this chapter.

be supplemented with pyrimidines. The reason for this requirement is related to the dependence of dihydroorotate dehydrogenase on a functional mitochondrial electron transport chain, and hence mtDNA, for the synthesis of pyrimidines (Gregoire *et al.*, 1984). Established cells made completely devoid of mtDNA by EtBr treatment (termed ρ^0 cells) were first described by Desjardins *et al.* (1986). King and Attardi (1989) isolated human ρ^0 cell lines by treating cells with 50-ng/ml EtBr and found that they were auxotrophic not only for pyrimidines but also for pyruvate. The dependence of pyruvate is not understood, but it is possibly due to an increased metabolism of pyruvate to lactate in ρ^0 cells. This would reduce the availability of pyruvate for the Krebs cycle (King and Attardi, 1989). Therefore, attempts to generate ρ^0 cell lines should be performed in high-glucose medium (4500 mg/ml) supplemented with Fetal Calf Serum (FCS 5–10%), 50-μg/ml uridine, and 1-mM sodium pyruvate.

Tiranti *et al.* (1998) have successfully generated mouse ρ^0 cells by treating them with 5-μg/ml EtBr. We have found that although this concentration can deplete mtDNA in some cells, most cell lines cannot grow in the presence of $\geq$500-ng/ml EtBr. As a general rule, one should try different concentrations to find the maximum concentration of EtBr that can be tolerated by the specific cell line. Bayona-Bafaluy *et al.* (2005b) produced ρ^0 cells from different primates by growing them in the presence of 150 ng/ml of EtBr for 3–4 months. The cells were used to generate cybrid fusions with mitochondrial donors from different apes.

For reasons that are not well understood, certain cell lines, such as mouse-derived ones, are refractory to complete mtDNA depletion by EtBr. Inoue *et al.* (1997a) found that a different intercalating agent, ditercalinium (DC), was more efficient than EtBr in generating mouse ρ^0 cells, as well as other ρ^0 cell lines. They used a concentration of 56 ng/ml–1.5 μg/ml, depending on the sensitivity to the drug. Okamaoto *et al.* (2003) found that ditercalinium chloride inhibits human DNA polymerase gamma activity as efficiently as does EtBr, and as expected, inhibits replication of mammalian mtDNA.

We found that DC is extremely toxic to most cells, and that relatively low doses (20–50 ng/ml) should be used to produce ρ^0 cells. It is important to maintain cells treated with DNA-intercalating agents in exponentially growing phase, as the mtDNA is diluted out rather than actively eliminated. By reducing their mtDNA levels cells became exclusively glycolytic, producing increased levels of lactate. This is easily observed by the fact that the phenol red from the medium turns yellow faster than usual. Therefore, during treatment, change the medium frequently and do not allow cells to reach confluency.

The duration of treatment varies from cell to cell. There is an initial rapid reduction in their mtDNA levels after EtBr treatment (King and Attardi, 1996; Moraes *et al.*, 1999). However, residual molecules are usually retained if the treatment lasts less than 25–30 days. If the DNA-intercalating agent is removed before complete depletion of mtDNA, cells repopulate with the residual genomes in a period that will depend on the size of the molecule (Moraes *et al.*, 1999) (Fig. 2). We find that smaller molecules (i.e., mtDNA with partial deletions) will repopulate cells five to eight times faster than do full-length genomes (Diaz *et al.*, 2002) (Fig. 2).

Typically, for ρ^0 generation, an exponentially growing cell line is treated for 2 months with EtBr or DC. EtBr is prepared as a filter-sterilized 10-mg/ml solution that is refrigerated and protected from light. Grow cells in EtBr or DC-containing medium for 2-month treatment. Freeze some cells for regrowth before the treatment is stopped in case the mtDNA depletion was not complete. After treatment, plate cells at low density to isolate single colonies. Grow cells in medium supplemented with uridine (Sigma U-3003) and pyruvate (Gibco 11360-070) for at least 15 days, before tests for the presence of mtDNA are performed. During this period, cells should repopulate themselves with mtDNA if the depletion was not complete. Test several individual colonies for mtDNA content. Southern blot and PCR assays are useful to assess the ρ^0 phenotype, but a more reliable assay is to test their auxotrophy to uridine. Cells harboring low levels of mtDNA will grow in the absence of added uridine. Table I describes some of the ρ^0 cells produced and used in transmitochondrial cybrid generation.

Because most ρ^0 cells are established with the purpose of being used in transmitochondrial cybrid generation, it is useful to choose a cell line that has a recessive

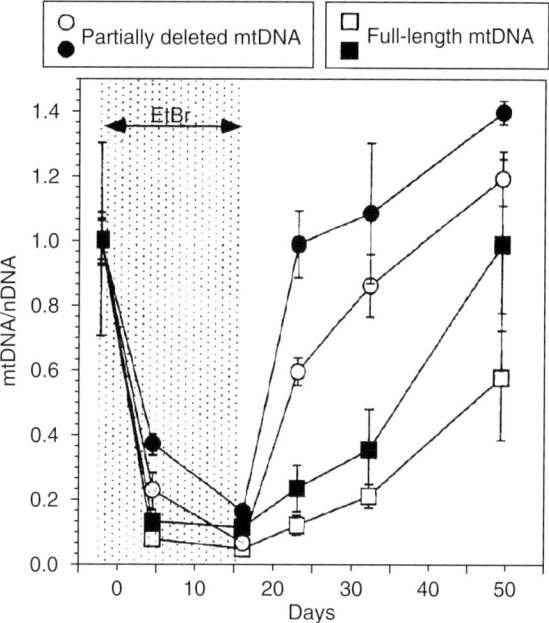

Fig. 2 mtDNA depletion and repopulation after treatment with EtBr. Cell lines essentially homo-plasmic for wild-type mtDNA (squares) or partially deleted mtDNA (7.5-kb deletion, circles) were treated with EtBr for 15 days (hatched time period), after which the drug was removed from the medium and the cells continued to grow for a total of 45 days. DNA was extracted from cell samples at different time points, and the ratio of mtDNA to the nuclear 18S rDNA gene determined by Southern blot, as described (Diaz *et al.*, 2002; Moraes *et al.*, 1999). Note that mtDNA repopulates cells at rates that are dependent on their genome sizes.

nuclear marker. Thymidine kinase (TK$^-$) or hypoxanthine guanine phosphorybosyl transferase (HPRT$^-$) deficiencies are good examples of recessive nuclear markers that can be obtained by chemical mutagenesis and selection in the presence of drugs that are toxic to wild-type cells (i.e., bromodeoxyuridine in the case of TK$^-$ cells and 8-azaguanine in the case of HPRT$^-$ cells). Chemical mutagenesis and selection proce-dures for recessive nuclear mutations have been described elsewhere (Anderson, 1995; Shapiro and Varshaver, 1975). Finally, because DNA-intercalating agents have the potential to cause nuclear DNA mutations, ρ^0 cells should be repopulated with a "wild-type" mtDNA to be used as control.

B. ρ^+ Cells Treated with Rhodamine 6G as Nuclear Donors

Ziegler and Davidson (1981) showed that pretreatment of cells with the mito-chondrial poison rhodamine 6G (R6G) increased the efficiency of interspecific hybrid formation. They found that parental cells that are treated with toxic levels

Table I
Examples of the Production and Initial Characterization of ρ^0 Cells Described in the Literature[a]

Cell line	Cell type	Species	Treatment	ρ^0 Phenotype confirmed	References
143B206	Osteosarcoma	Human	50-ng/ml EtBr	S/ Ur/Pyr	King and Attardi, 1989
Hela EB8	Cervical carcinoma	Human	50-ng/ml EtBr	S/N 6-thioguanine	Hayashi et al., 1991
A549	Lung carcinoma	Human	400-ng/ml EtBr	S/PCR Ur/Pyr	Bodnar et al., 1993
U937	Promonocytic leukemia	Human	5-ng/ml EtBr	PCR	Gamen et al., 1995
SY-SY5Y/ ρ^0 64/5	Neuroblastoma	Human	5-μg/ml EtBr	Biochemically, mitochondrial protein synthesis, Pyr	Miller et al., 1996
C2	Myoblast	Mouse	1.5-μg/ml DC	S/PCR Ur/Pyr	Inoue et al., 1997b
B82cap	Fibroblast	Mouse	1.5-μg/ml DC	S/PCR Ur/Pyr	Inoue et al., 1997a
NIH3T3	Fibroblast	Mouse	1.5-μg/ml DC	S/PCR Ur/Pyr	Inoue et al., 1997a
MIN 6	Pancreatic beta cell	Mouse	56-ng/ml DC	S Ur/Pyr	Inoue et al., 1997b
GM10611A-ρ^0	Hybrid	Hamster/ Human Chr.9	5-μg/ml EtBr	PCR/COX activity	Tiranti et al., 1998
L929	Connective tissue	Mouse	5-μg/ml EtBr	PCR/COX activity	Tiranti et al., 1998
Ntera2/D1 (NT2)	Teratocarcinoma	Human	25-ng/ml EtBr	PCR/COX activity Ur/Pyr	Binder et al., 2005; Swerdlow et al., 1997
ARPE-19 cells	Retinal pigment epithelial	Human	50-ng/ml EtBr	Quantitative RT-PCR Ur	Miceli and Jazwinski, 2005
MOLT-4	Lymphoblastoid T cells	Human	50-ng/ml EtBr	S/Quantitative PCR Ur/pyr	Armand et al., 2004
Primates ρ^0 cells	Fibroblasts	Chimpanzee (Ch), Gorilla (G), Baboon (B), Orangutan (O), and Rhesus (R) macaque	Ch, G, and B: 150-ng/ml, O and R: 500-ng/ml EtBr	S Ur, Gal	Bayona-Bafaluy et al., 2005b

[a]EtBr, ethidium bromide; DC, ditercalinium; S, Southern blot; N, Northern blot; Ur, uridine dependence; Pyr, pyruvate dependence; Gal, galactose.

of R6G could be rescued by fusion with the cytoplasm of untreated cells. Trounce and Wallace (1996) adapted the use of R6G for the production of transmitochondrial cybrids from a fusion of R6G-treated LMTK$^-$ mouse cells with enucleated mouse cells harboring a homoplasmic point mutation in the mtDNA 16S rRNA gene. Finally, Williams *et al.* (1999) fused cells from a patient with MERRF syndrome harboring the A8344G mtDNA mutation to enucleated HeLaCOT cells to produce cybrid clones with cytochrome *c* oxidase activity, demonstrating that R6G has some advantages over EtBr in removing the mitochondrial elements from cultured cells.

The procedure consists in treating the mtDNA recipient cell with 2- to 5-μg/ml R6G for 3–10 days. The sensitivity varies with cell type. The highest concentration that does not kill the cells in 3–10 days should be used. However, we find that several cell lines cannot survive for more than 48 h in R6G. For example, ES cells are very sensitive to R6G treatment and should be treated for short periods (Trounce *et al.*, 2004). After treatment, the cells should die even in complete medium supplemented with uridine, as they lose not only OXPHOS, but all mitochondrial functions. After R6G treatment of the mtDNA recipient, the medium is changed to R6G-free medium for $\sim$3–4 h. The R6G toxic effect is not reversible and dysfunctional mitochondria should not recover. After this "wash" period, treated cells can be fused to cytoplasts of the mtDNA donor. In theory, no nuclear selection is necessary, as R6G-treated cells should die in a few days after fusion. A control fusion without cytoplasts is recommended to assure that the R6G treatment is toxic enough to kill all cells that did not receive functional mitochondria.

We have had mixed results with this procedure. For xenomitochondrial cybrid production, we were frustrated to find that after several weeks, not all the mtDNA from the nuclear donor had been eliminated. Nevertheless, we had positive results with "same species" cybrid production. We produced homoplasmic xenomitochondrial mice using R6G treatment of mouse embryonic stem cells (McKenzie *et al.*, 2004). These animals were the first viable transmitochondrial mice with homoplasmic replacement of endogenous mtDNA, and were employed to confirm the feasibility of producing mitochondrial defects in mice using a xenomitochondrial approach.

Sample protocol

1. Prepare a stock solution (1 mg/ml) of R6G (Sigma Cat. No. R-4127) in sterile water and filter sterilize. Stock solutions can be kept for a month at 4°C.

2. Grow two T75 flasks of cells chosen to be mtDNA recipients in complete medium supplemented with 50-μg/ml uridine and 3-μg/ml R6G. The appropriate concentration of R6G will vary for each cell type. R6G-treated cells do not grow as fast as untreated cells, and trypsinization should be avoided.

3. Treat cells with R6G for 2–10 days (depending on cell type) and change the medium every 1–2 days.

4. After treatment, change the medium to complete medium without R6G. Leave the cells in this medium for 3–4 h before fusion.

═══════   ## III. Generation of mtDNA Donors

A. Cultured Cells as mtDNA Donors Enucleated with Cytochalasin B and Centrifugation (Cytoplasts)

Cultured cells (adherent or in suspension) are excellent mtDNA donors, but the enucleation and fusion procedures are different, depending on their growth substrate.

1. Adherent Cells

We commonly use the method described by King and Attardi (1989) to produce transmitochondrial cybrids using adherent cells as mtDNA donors. Enucleation efficiency varies depending on the cell type, and the conditions described here may not be able to produce cytoplasts from all cell lines. However, even relatively low-efficient enucleations can yield many transmitochondrial cybrids. True cybrids can only be produced from cells without nuclei.

Reagents and materials

Autoclave the following items:

> Long (30 cm) forceps wrapped in aluminum foil
> Two empty 250-ml centrifuge bottles with caps with sealing rings that are loosely capped
> Pack of paper towels wrapped in aluminum foil

Prepare enucleation medium:

> DMEM high glucose (Gibco BRL Cat. No. 11965) supplemented with 5% calf serum
> 10-μg/ml cytochalasin B (Sigma C-6762)
> Antibiotics (e.g., 20-μg/ml gentamicin)

Cell culture

Plate the cells to be enucleated in 35-mm dishes at least 24 h before enucleation (to ensure strong attachment to the surface). Plate more than one dish using different numbers of cells so that the most appropriate one(s) can be used for fusion. Place 35-mm dishes in 100-mm dishes to avoid contaminating the external surface of the 35-mm dish and facilitate handling during microscopic examination. Grow adherent cells to ~70% confluency.

Procedure

1. Prewarm a Sorvall GSA (Sorvall, Inc., Asheville, NC) or similar rotor to 35°C by centrifugation at 7000 rpm (8000 × g) without refrigeration for ~30 min prior to use.

2. Transfer 30 ml of prewarmed (37°C) enucleation medium to the sterile 250-ml centrifuge bottle taking care to keep the bottle/cap and forceps sterile.

3. Remove the lid of the 35-mm dish and insert the uncovered dish with cells upside down into the 250-ml centrifuge bottle using the sterile 30-cm forceps. The lid of the 35-mm dish is kept inside the original empty 100-mm dish for later use. It is almost impossible to remove all the air trapped inside the inverted 35-mm dish, but we have found that a relatively small "air bubble" will not interfere with the enucleation procedure.

4. Place cap on centrifuge bottle and centrifuge the prewarmed rotor (e.g., Sorvall GSA, Sorvall, Inc.) at 7000 rpm (8000 × g) for 25 min at 35°C.

5. Remove the 35-mm dish from the centrifuge bottle using long forceps and place it cell side up on top of the autoclaved paper towel in a sterile biological cabinet.

6. Wipe the outside of the 35-mm dish with sterile paper soaked with 70% ethanol. To maintain sterility, wear gloves and rinse hands and gloves thoroughly with 70% ethanol before handling the dish.

7. Place the clean medium-free dish back into the empty 100-mm dish and cover it with its sterile lid that was kept inside the same dish.

8. Quickly inspect the dish under the microscope to assess the enucleation. It should show the presence of cytoplasts, which resemble pieces of membrane. A good method to assess the enucleation efficiency is to stain a "test enucleation" dish with 1-μg/ml Hoechst 33342 (Sigma) and 200 nM of mitotracker (CmxRos, Molecular Probes, Invitrogen, Carlsbad, CA) in PBS.

2. Suspension Cells

Suspension cells have to be enucleated by centrifugation through a density gradient. Trounce and Wallace described two procedures that are efficient in enucleating suspension cells (Trounce et al., 1996).

1. Mix Percoll and medium 1:1 and preequilibrate overnight at 37°C in a 5% CO_2 incubator.

2. Transfer the Percoll solution to a sterile centrifuge tube (e.g., SS34) and add 100 μl of a 2-mg/ml cytochalasin B stock (in DMSO) (final concentration of 20 μg/ml).

3. Overlay the Percoll solution with 10^7 cells from a suspension culture in 10 ml of medium and centrifuge at 44,000 × g (e.g., 19,000 rpm in a fixed angle SS-34 rotor) for 70 min at 35–37°C.

4. At the end of the run, a distinct band can be identified one-third of the way up from the bottom. This band contains a mix of cytoplasts and karyoplasts that are appropriate for cybrid production if a nuclear recessive marker is used. Above this band there is a less well-defined band, which is enriched in cytoplasts (Trounce et al., 1996).

5. Transfer the defined band along with the hazy-band immediately above to a fresh centrifuge tube. Dilute the sample with 10 ml of medium and centrifuge at 650 × g at 15°C.

6. Resuspend the pellet obtained in 5 ml of medium. This material can be used in fusions using the methods described for fusion with "platelets as mtDNA donors" (see below). We also have used this procedure for adherent cells (after trypsinization) that could not be enucleated efficiently by the standard procedure.

B. Platelets as mtDNA Donors

Chomyn and colleagues (Chomyn, 1996; Chomyn *et al.*, 1994) described a simple procedure to obtain transmitochondrial cybrids using platelets as the mtDNA donor. Platelets are good mtDNA donors because they are nucleus-free fragments of megakaryocytes. Moreover, platelets are easy to obtain and can be frozen, like cultured cells, in DMSO-containing media, for future use in fusion experiments. The procedure has been described in detail by Chomyn (1996). It consists in isolated platelets (under sterile conditions) after concentrating red and white blood cells by low-speed centrifugation. The platelets obtained are washed in PBS and can be mixed with ρ^0 cells or other nuclear donor, as described above.

Protocol

1. Collect 7–20 ml of blood in heparin tubes (green top Vacutainer^{TM}) and mix with 0.1 volumes of 0.15-M NaCl and 0.1-M trisodium citrate (pH 7.0).

2. Centrifuge at $150 \times g$ for 15 min at RT. Transfer the top three-fourth of the platelet-rich plasma to a fresh centrifuge tube and centrifuge for 35 min at $2500 \times g$ at RT.

3. Resuspend the pellet obtained in 10 ml of 0.15-M NaCl, 0.015-M Tris–HCl (pH 7.4). Approximately $1–4 \times 10^7$ platelets are obtained by this procedure.

4. If necessary, aliquots can be frozen after centrifugation. To do so, resuspend the platelet pellet in cell culture freezing medium (e.g., DMEM supplemented with 30% FCS and 10% DMSO), and freeze in aliquots as done for cultured cells. These cells are ready to use in a fusion protocol.

C. Synaptosomes as mtDNA Donors

Inoue *et al.* (1997b) and Ito *et al.* (1999) described a procedure to use synaptosomes as mtDNA donors.

1. Wash a mouse brain in phosphate-buffered saline.

2. Homogenize the tissue in medium containing ice-cold 0.25-M sucrose; 50-mM HEPES, pH 7.5; and 0.1-mM EDTA using a Teflon-glass Potter-Elvehjem homogenizer.

3. Centrifuge the homogenate at $1000 \times g$ for 10 min at $4\,^{\circ}$C.

4. Transfer the supernatant to a fresh centrifuge tube and centrifuge at $17,000 \times g$ for 20 min at $4\,^{\circ}$C.

5. The pellet, which contains synaptosomes, can be used for the fusion with any nuclear donor.

D. Chemical Enucleation to Produce mtDNA Donors

In our laboratory, Bayona-Bafaluy *et al.* (2003) developed a chemical enucleation method for the production of transmitochondrial cybrids using actinomycin D (produced from *Streptomyces antibioticus*), one of the oldest chemotherapy drugs used in the clinic (Lurain, 2002). Actinomycin D acts as an inhibitor of nucleic acid synthesis by binding to DNA duplexes and interfering with the action of enzymes engaged in replication and transcription (Waring and Bailly, 1994). Since actinomycin D is toxic, cells treated with this agent will survive if they are fused to ρ^0 cells and selected for the presence of a functional oxidative phosphorylation system (Fig. 3).

Sample protocol

1. Plate cells (0.3×10^6) in a 35-mm dish and grow until they are ~70–80% confluent.

2. Treat cells with 0.5- to 5-μg/ml actinomycin D (Sigma A-9415) for different times. The concentration of the actinomycin D treatment needs to be determined for each cell type used. We recommend using a concentration of actinomycin D that kills cells after treatment for 1 week of treatment (Bayona-Bafaluy *et al.*, 2003) (Fig. 3).

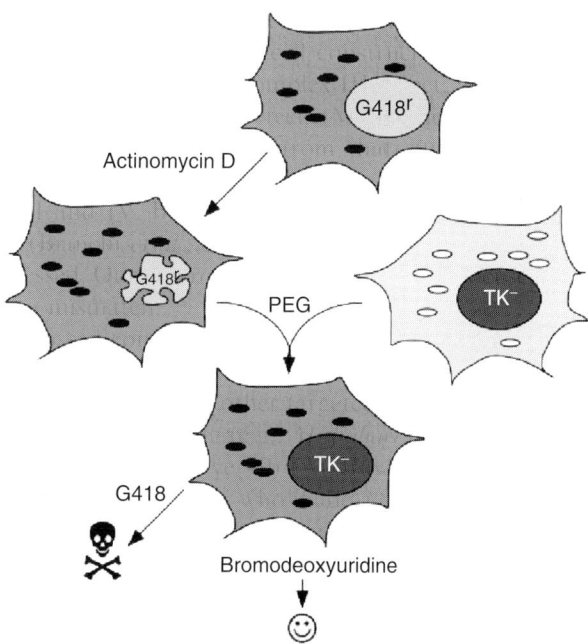

Fig. 3 Chemical enucleation and transfer of mtDNA to ρ^0 cells. Donor cells treated with actinomycin D are suitable mtDNA donors and can be fused to ρ^0 cells with PEG. Transmitochondrial cybrids clones can be obtained without nuclear selection (media lacking uridine to inhibit growth of ρ^0 cells) and can be checked later for loss of resistance to a dominant marker such as G418.

3. Wash treated cells with complete medium to remove the drug before using them as mtDNA donors for the fusion.

IV. Generation of Transmitochondrial Cybrids

We will describe a sample protocol in detail. Because many of the other protocols outlined in this chapter are similar, after nuclear and mitochondrial donors are prepared, the protocol described below can be followed for the other preparations.

For all of the protocols described, prepare the following fusion solution.

1. Autoclave a small glass bottle (e.g., 100 ml) containing 4.7 g of PEG 1450 (MW 1450, Sigma P-5402).

2. While the PEG is still liquid (i.e., warm but not hot), add 4 ml of DMEM (serum free) and 1 ml of DMSO. If the PEG hardens before adding the medium, warm up the bottle by slowly rotating it above a flame. Once the PEG is mixed with medium and DMSO, it will not solidify.

3. The pH of the final fusion solution is slightly acidic and yellow in color. The pH can be raised by the addition of ~1 μl of 10-N NaOH. Depending on the origin of the reagents, this amount may change, but one should be careful not to add too much NaOH. If this happens, the solution will turn purple due to high pH and should be discarded.

4. The diluted PEG solution should be used within 24 h.

A. Using Enucleated Cells as mtDNA Donors and ρ^0 Cells as Nuclear Donors

1. During the time that takes to enucleate the mtDNA donors, the ρ^0 cells should be trypsinized, counted, and placed in a Falcon tube at a concentration of 1.5×10^6 ρ^0 cells in 2.5 ml of complete medium supplemented with uridine and pyruvate.

2. Add 2.5 ml of the ρ^0 cells to the media-free 35-mm dish containing enucleated cells.

3. Incubate the 35-mm dish in a CO_2 incubator for 2.5–3 h at 37 °C. During this time, ρ^0 cells will attach to the dish and make contacts with the attached cytoplasts.

4. The mixture of cells and cytoplasts are washed three times with a serum-free DMEM medium. After the last wash has been aspirated, the dish should be placed at a 45° angle for 45–60 sec for the remaining medium to drain to the bottom of the dish, and removed by aspiration. For efficient PEG fusion, it is important that all of the medium is removed.

5. Add 2.5 ml of the final PEG solution to the dish, swirl for 10 sec. Let the solution remain in contact with the cells for exactly 60 sec, and remove the PEG solution by aspiration. Because of the high viscosity of the PEG solution, aspiration should start 55 sec after addition.

6. Wash the dish three times with DMEM supplemented with 10% DMSO, followed by one wash with DMEM.

7. Add 2.5 ml of complete medium supplemented with uridine to the dish, and incubate in a CO_2 incubator at $37°C$.

8. Change the medium on the following day. A considerable amount of dead cells will be evident. If fewer than 30% of the cells remain attached, the fusion conditions may have been too severe, and either the time or the concentration of PEG should be reduced (e.g., to 44%). As mentioned above, the sensitivity to PEG varies depending on the cell type. Fungizone (250-μg/ml amphotericin B) should not be used before, during, or immediately after the fusion, as we have observed that it increases the PEG toxicity significantly.

9. Selection for transmitochondrial cybrids starts between 16 and 30 h after fusion. Select for the presence of mtDNA by growth in the absence of uridine using dialyzed serum (regular FCS contains small levels of uridine). Select against nonenucleated mtDNA donors and hybrids using drugs that are not toxic to the specific ρ^0 cells (i.e., bromodeoxyuridine or 8-azaguanine).

10. Discrete clones (of $\sim$50–100 cells) should be observed by 10–14 days after selection starts. Good-size clones are usually observed by 15–20 days. We isolate clones using cloning rings (Bellco Glass Cat. No. 2090), sterile silicone vacuum grease (to keep the liquid inside the ring from leaking out), and trypsin. Clones can be isolated and, if desired, an aliquot of the cells can be immediately used for PCR typing, as described for single muscle fibers or small amounts of cells elsewhere in this volume (Chapter 24 by Williams and Moraes). We have found that BrdU-resistant cells may sometimes contain chromosomes from the mtDNA donor. Although these are the minority, it is advisable to test for the presence of additional nuclear markers (e.g., microsatellites) to assure that the BrdU-resistant clones are true cybrids.

B. Using Platelets as Mitochondrial Donors

Platelets from patients with neurodegenerative disorders, such as Parkinson and Alzheimer diseases (Ghosh et al., 1999; Gu et al., 1998; Ito et al., 1999; Sheehan et al., 1997; Swerdlow et al., 1997), and ALS (Gajewski et al., 2003), have been used extensively in the production of transmitochondrial cybrids. We follow the procedure described by Chomyn (1996).

Procedure

1. Mix $\sim$10^7 platelets with 10^6 ρ^0 cells and centrifuge at $200 \times g$ for 5 min at RT.

2. Aspirate all supernatant and resuspend the pellet obtained thoroughly with 100 μl of fusion solution (see above).

3. Incubate for 1 min at RT. Then, dilute the sample with complete medium (i.e., DMEM containing 10% FCS and 50-μg/ml uridine). If the ρ^0 cell has a nuclear marker, such as TK^-, BrdU can be added to the medium to eliminate any potential nucleated blood cells.

4. Distribute the cells into ten 100-mm dishes. If a large amount of clones is not required, discard half of the mix and plate the other half into five 100-mm dishes.

5. Three days later, replace the medium with DMEM supplemented with 10% dialyzed FCS and a drug that selects for a recessive nuclear marker of the ρ^0 mitochondrial recipient (e.g., BrdU). Do not add uridine.

6. Isolated clones should be visible 15 days after the fusion. In theory, the use of platelets as mtDNA donors makes the selection for nuclear markers unnecessary. Since white blood cells may contaminate the platelet preparation, analysis of the nuclear background is recommended.

C. Using ρ^+ Cells Treated with R6G as Nuclear Donors

Procedure

1. Grow the mtDNA donor cells in 35-mm dishes until they are ~70% confluent.

2. Trypsinize and count the R6G-treated cells.

3. Plate the R6G-treated cells (in 1.5 ml) on top of the enucleated cells and let them attach for 3 hours.

4. The remainder of the procedure is identical to that was described for fusions with ρ^0 cells.

5. Add 2.5 ml of complete medium supplemented with uridine to the dish and incubated in a CO_2 incubator.

6. Change the medium on the following day, when a considerable number of dead cells will be evident.

7. Selection for transmitochondrial cybrids starts ~30 h after fusion. Selection for the presence of mtDNA is not necessary, because fused R6G-treated cells should die in a few days. However, if a functional mtDNA is being transferred, remove uridine from the medium.

D. Using Synaptosomes as Mitochondrial Donors

The pellet containing synaptosomes is mixed with 5×10^6 mouse ρ^0 cells, and fusion is carried out in the presence of 50% (w/v) PEG 1450. The fusion mixture is cultivated in selective medium (without pyruvate and uridine). After a period of 14–20 days, the cybrid clones growing in the selection medium are isolated by the glass cylinder method. All the detailed conditions are as described in Section IV.A.

E. Using Cells Chemically Enucleated with Actinomycin D as Mitochondrial Donors

For this procedure it is useful to have a dominant nuclear marker (e.g., G418 resistance) in the cell treated with actinomycin D, as loss of the marker is a good indication of successful chemical enucleation after the fusion (Fig. 3).

Procedure

1. $1 \times 10^6 \rho^0$ cells (or any cell used as nuclear donor) are overlaid on the dish that contains actinomycin D-treated cells, and incubated for 3 h in a CO_2 incubator.

2. Add 3 ml of fusion solution (see above) and incubate for 60 sec.

3. Wash the plate three times with DMEM supplemented with 10% DMSO and once with DMEM.

4. Add 3 ml of complete medium to the dish and incubate in a CO_2 incubator for 48 h.

5. The cells are then trypsinized and plated onto ten 100-mm dishes (Bayona-Bafaluy *et al.*, 2003).

F. Generation of Transmitochondrial Hybrid Cells by Microcell–Mediated Chromosome and mtDNA Transfer

Microcell-mediated chromosome transfer has been used to transfer intact chromosomes from one mammalian cell to another (Ege and Ringertz, 1974; Fournier and Ruddle, 1977; McNeill and Brown, 1980). Using microcell-mediated chromosome transfer, Zhu *et al.* (1998) mapped the gene defect (*SURF1*) in Leigh Syndrome (LS) to chromosome 9q34 by complementation of the respiratory chain deficiency in patient fibroblasts. De Lonlay *et al.* (2002) also developed a functional complementation approach by transferring human chromosome fragments into respiratory chain-deficient fibroblasts and rescued the normal OXPHOS enzyme activity in deficient skin fibroblasts.

We have transferred a limited number of chromosomes and mtDNA simultaneously to mtDNA-free human cells (Barrientos and Moraes, 1998). The structure of mitochondria-containing microcells showed that they consisted of a single or a few micronuclei and a thin ring of cytoplasm surrounded by an intact plasma membrane. It has been shown that karyoplasts produced by cytochalasin B-induced enucleation contained ~11% of the mitochondrial volume of whole cells (Zorn *et al.*, 1979). The cytoplasmic region of microcells contains a variable number of mitochondria, and the composition of microcell preparations is very heterogeneous, and consists of small cytoplasts containing mitochondria, microcells containing mitochondria and a single micronucleus, microcells containing mitochondria and a few micronuclei, and in rare cases, microcells containing no mitochondria. The proportion of micronuclei-containing cell fragments is ~30% (Barrientos and Moraes, 1998). This feature should be taken into account when studies on restoration of normal maintenance and expression of mtDNA in defective cells are carried out using the microcell-mediated chromosome transfer system.

Only about 1% of the clones selected for the presence of mtDNA receive a particular chromosome. The number of clones selected for the presence of different chromosome markers was essentially the same whether selection for the presence of mtDNA was included, suggesting that microcells containing selected chromosomes also contained mtDNA. Using microcell-mediated chromosome transfer, hybrids are likely to receive donor chromosomes randomly, but it is possible that some chromosomes are transferred preferentially. Our experience suggests that different selectable chromosomes are transferred at a similar rate. Although most microcells contain one single chromosome, the presence of a few microcells containing several micronuclei suggests that clones containing more than one donor chromosome could also be generated during fusion. In some cases, selective pressure (i.e., no uridine) is necessary for repopulation of mtDNA introduced in cells using this system (Barrientos and Moraes, 1998).

Functional orangutan mtDNA was introduced and maintained in a human nuclear background using microcell-mediated chromosome and mitochondrial transfer (Barrientos *et al.*, 2000). Although partial oxidative phosphorylation function was restored only in the presence of most orangutan chromosomes, the human oxidative phosphorylation-related nuclear-coded genes were not able to complement many homologous orangutan ones. Microcell transfer was also used to restore respiratory chain-deficient cells, as was demonstrated by Seyda *et al.* (2001), who performed an analysis of microcell-mediated chromosome fusion between fibroblasts from patients with a mitochondrial disease and a panel of A9 mouse:human hybrids. Complementation was observed between the recipient cells and the mouse:human hybrid clone carrying human chromosome 2, showing that the underlying defect in the patients was under the control of a nuclear gene. The method described below was originally described by Fournier (1981).

1. Sample Protocol

1. Microcells are produced by treatment cells cultured in T25 flasks with 50-ng/ml colcemid (demecolcine) (Invitrogen D7499) (36–48 h, depending on cell type), which arrest the cells in mitosis and results in the formation of micronuclei (Fig. 3), and followed by treatment with 10-μg/ml cytochalasin B (20 min), which disrupts microfilaments.

2. Prewarm a GSA rotor (Sorvall, Inc.) to 35°C by centrifugation without refrigeration at 8000 rpm for 30 min. Add water to the rotor, as a cushion, and transfer T25 flasks to the rotor.

3. Enucleation is performed by centrifugation of the T25 flasks at 8000 $\times$ g (Sorvall GSA rotor, Sorvall, Inc.) for 60 min at 35°C. Not all commercial flasks are stable during centrifugation. As recommended by Fournier (1981), we used Costar flask No. 3025.

4. The microcell pellet obtained from the centrifugation is placed into 100 ml of DMEM and filtered sequentially through 12-, 8-, 5-, and 5-μm filters (Nucleopore filters, Costar, Cambridge, MA). Both the filter and filter holders (Costar) are autoclaved.

5. For hybridization experiments, centrifuge purified microcells at 2000 $\times$ g for 10 min at 4°C in an SS-34 rotor.

6. Resuspend the microcell pellet obtained in 3 ml of lectin solution (50-μg/ml phytohemaglutinin PHA-P; Sigma L1668). (This promotes adherence of microcells to recipient cells.)

7. A ρ^0 cell line, used as recipient cells, is grown in six 60-mm culture dishes. The recipient cells are washed in DMEM and incubated with the microcell solution (1 ml per dish) for 30 min at 37°C.

8. Cell fusion is performed by treatment with freshly prepared 46% polyethylene glycol 1500 solution (pH 7.5) for 1 min followed by 3 washes in DMEM, 5% DMSO.

9. Grow the cells overnight in complete medium, and select for the presence of appropriate markers in the next day, as described above.

V. Manipulating Heteroplasmy

A. Manipulating Heteroplasmy by the Use of EtBr

As shown in Fig. 2, mtDNA levels can be decreased transiently by EtBr treatment, providing an efficient approach to alter heteroplasmy. This procedure has been used to obtain homoplasmic-mutated clones harboring large deletions (King, 1996; Moraes et al., 1999) and point mutations (Hao and Moraes, 1997; Malik et al., 2002). Because partially deleted mtDNA repopulates cells faster than does full-length mtDNA (Moraes et al., 1999), it is easier to obtain homoplasmic deletion mutant clones compared to point mutations from heteroplasmic cells.

Cells should be treated with 50-ng/ml EtBr (in medium supplemented with uridine) for 15–22 days to reduce their mtDNA levels to very few molecules. EtBr solution should be filter-sterilized at high concentrations (10 mg/ml). Stocks should be protected from light and stored at 4°C. After the treatment, EtBr is removed from the medium and cells allowed to repopulate themselves with the residual mtDNA for 10–15 days. Depending on the number and type of mtDNA remaining in the cell, after repopulation, homoplasmic clones can be generated. When dealing with pathogenic mtDNA mutations, it is important to maintain uridine supplementation during the repopulation period. Although this procedure has worked successfully for some cell lines, it can also be very frustrating, as it is a very long procedure and at the end of 2 months one may find out that homoplasmic clones could not be obtained. Nevertheless, it is useful in cases where homoplasmic clones do not exist and studies addressing genotype–phenotype correlations are required.

B. Manipulating Heteroplasmy by Delivering Restriction Endonucleases to Mitochondria

Mutations in the mtDNA can cause a variety of human diseases. In most cases, such mutations are heteroplasmic (i.e., mutated and wild-type mtDNA coexist) and a small percentage of wild-type sequences can have a strong protective effect against a metabolic defect. A genetic approach to correct mtDNA mutations can be used to modulate heteroplasmy by the use of restriction endonucleases (RE) targeted to mitochondria, although this approach is limited by the presence of an appropriate restriction site. Srivastava and Moraes (2001) expressed *Pst*I RE in mammalian mitochondria, using a hybrid cell line containing both rat and mouse mtDNA (rat mtDNA contains no site for *Pst*I; mouse harbors two *Pst*I sites), and showed an increase of the rat mtDNA in the transfected cells, thereby allowing one to modulate mtDNA heteroplasmy in a predictable direction. Some cells expressing mitochondrial-targeted *Pst*I become ρ^0 (Srivastava and Moraes, 2001). Using a similar approach, Srivastava and Moraes (2005) targeted *Pst*I RE to the mitochondria of skeletal muscle in mouse, and because mouse mtDNA harbors two *Pst*I sites, transgenic founders developed a mitochondrial myopathy associated with mtDNA deletions and depletion. Tanaka *et al.* (2002) transfected cybrids containing the NARP T8993G mutation with a plasmid coding for a mitochondrial-targeted *Sma*I RE that recognizes this mutation, and induced complete elimination of the mutant DNA and restoration of the mitochondrial membrane potential and the ATP levels to WT levels. Bayona-Bafaluy *et al.* (2005a) could demonstrate a rapid directional shift of mtDNA heteroplasmy in a heteroplasmic mouse model that contains two polymorphic mtDNA sequence variants (NZB and Balb/C) by targeting *Apa*LI RE to mitochondria (there is a single *Apa*LI site in the Balb/C that is not present in the NZB variant).

Although manipulating heteroplasmy by an mtDNA-targeted RE can be a powerful approach to modulate mtDNA heteroplasmy, it is limited by the availability of unique RE cleavage site. The approach may be also useful in the context of multiple cleavage sites, if the expression of the RE targeted to mitochondria could be controlled.

Acknowledgments

This work was supported by grants from the National Institutes of Health (NINDS, NCI, and NEI) to C.T.M., and the Muscular Dystrophy Association. S.R.B. is supported by a minority supplement to PHS grant EY10804.

References

Anderson, P. (1995). Mutagenesis. *Methods Cell Biol.* **48,** 31–58.

Armand, R., Channon, J. Y., Kintner, J., White, K. A., Miselis, K. A., Perez, R. P., and Lewis, L. D. (2004). The effects of ethidium bromide induced loss of mitochondrial DNA on mitochondrial phenotype and ultrastructure in a human leukemia T-cell line (MOLT-4 cells). *Toxicol. Appl. Pharmacol.* **196,** 68–79.

Bacman, S. R., Atencio, D. P., and Moraes, C. T. (2003). Decreased mitochondrial tRNALys steady-state levels and aminoacylation are associated with the pathogenic G8313A mitochondrial DNA mutation. *Biochem. J.* **374,** 131–136.

Baracca, A., Solaini, G., Sgarbi, G., Lenaz, G., Baruzzi, A., Schapira, A. H., Martinuzzi, A., and Carelli, V. (2005). Severe impairment of complex I-driven adenosine triphosphate synthesis in leber hereditary optic neuropathy cybrids. *Arch. Neurol.* **62,** 730–736.

Barrientos, A., and Moraes, C. T. (1998). Simultaneous transfer of mitochondrial DNA and single chromosomes in somatic cells: A novel approach for the study of defects in nuclear-mitochondrial communication. *Hum. Mol. Genet.* **7,** 1801–1808.

Barrientos, A., Muller, S., Dey, R., Wienberg, J., and Moraes, C. T. (2000). Cytochrome c oxidase assembly in primates is sensitive to small evolutionary variations in amino acid sequence. *Mol. Biol. Evol.* **17,** 1508–1519.

Bayona-Bafaluy, M. P., Manfredi, G., and Moraes, C. T. (2003). A chemical enucleation method for the transfer of mitochondrial DNA to rho(o) cells. *Nucleic Acids Res.* **31,** e98.

Bayona-Bafaluy, M. P., Blits, B., Battersby, B. J., Shoubridge, E. A., and Moraes, C. T. (2005a). Rapid directional shift of mitochondrial DNA heteroplasmy in animal tissues by a mitochondrially targeted restriction endonuclease. *Proc. Natl. Acad. Sci. USA* **102,** 14392–14397.

Bayona-Bafaluy, M. P., Muller, S., and Moraes, C. T. (2005b). Fast adaptive coevolution of nuclear and mitochondrial subunits of ATP synthetase in orangutan. *Mol. Biol. Evol.* **22,** 716–724.

Binder, D. R., Dunn, W. H., Jr., and Swerdlow, R. H. (2005). Molecular characterization of mtDNA depleted and repleted NT2 cell lines. *Mitochondrion* **5,** 255–265.

Bodnar, A. G., Cooper, J. M., Holt, I. J., Leonard, J. V., and Schapira, A. H. (1993). Nuclear complementation restores mtDNA levels in cultured cells from a patient with mtDNA depletion. *Am. J. Hum. Genet.* **53,** 663–669.

Bornstein, B., Mas, J. A., Patrono, C., Fernandez-Moreno, M. A., Gonzalez-Vioque, E., Campos, Y., Carrozzo, R., Martin, M. A., del Hoyo, P., Santorelli, F. M., Arenas, J., and Garesse, R. (2005). Comparative analysis of the pathogenic mechanisms associated with the G8363A and A8296G mutations in the mitochondrial tRNA(Lys) gene. *Biochem. J.* **387,** 773–778.

Bruno, C., Martinuzzi, A., Tang, Y., Andreu, A. L., Pallotti, F., Bonilla, E., Shanske, S., Fu, J., Sue, C. M., Angelini, C., DiMauro, S., and Manfredi, G. (1999). A stop-codon mutation in the human mtDNA cytochrome c oxidase I gene disrupts the functional structure of complex IV. *Am. J. Hum. Genet.* **65,** 611–620.

Chomyn, A. (1996). Platelet-mediated transformation of human mitochondrial DNA-less cells. *Methods Enzymol.* **264,** 334–339.

Chomyn, A., Meola, G., Bresolin, N., Lai, S. T., Scarlato, G., and Attardi, G. (1991). *In vitro* genetic transfer of protein synthesis and respiration defects to mitochondrial DNA-less cells with myopathy-patient mitochondria. *Mol. Cell. Biol.* **11,** 2236–2244.

Chomyn, A., Lai, S. T., Shakeley, R., Bresolin, N., Scarlato, G., and Attardi, G. (1994). Platelet-mediated transformation of mtDNA-less human cells: Analysis of phenotypic variability among clones from normal individuals–and complementation behavior of the tRNALys mutation causing myoclonic epilepsy and ragged red fibers. *Am. J. Hum. Genet.* **54,** 966–974.

Clayton, D. A., Teplitz, R. L., Nabholz, M., Dovey, H., and Bodmer, W. (1971). Mitochondrial DNA of human-mouse cell hybrids. *Nature* **234,** 560–562.

De Francesco, L., Attardi, G., and Croce, C. M. (1980). Uniparental propagation of mitochondrial DNA in mouse-human cell hybrids. *Proc. Natl. Acad. Sci. USA* **77,** 4079–4083.

De Lonlay, P., Mugnier, C., Sanlaville, D., Chantrel-Groussard, K., Benit, P., Lebon, S., Chretien, D., Kadhom, N., Saker, S., Gyapay, G., Romana, S., Weissenbach, J., *et al.* (2002). Cell complementation using Genebridge 4 human:Rodent hybrids for physical mapping of novel mitochondrial respiratory chain deficiency genes. *Hum. Mol. Genet.* **11,** 3273–3281.

Desjardins, P., de Muys, J. M., and Morais, R. (1986). An established avian fibroblast cell line without mitochondrial DNA. *Somat. Cell Mol. Genet.* **12,** 133–139.

Dey, R., Barrientos, A., and Moraes, C. T. (2000). Functional constraints of nuclear-mitochondrial DNA interactions in xenomitochondrial rodent cell lines. *J. Biol. Chem.* **275,** 31520–31527.

Diaz, F., Bayona-Bafaluy, M. P., Rana, M., Mora, M., Hao, H., and Moraes, C. T. (2002). Human mitochondrial DNA with large deletions repopulates organelles faster than full-length genomes under relaxed copy number control. *Nucleic Acids Res.* **30,** 4626–4633.

Ege, T., and Ringertz, N. R. (1974). Preparation of microcells by enucleation of micronucleate cells. *Exp. Cell Res.* **87,** 378–382.

Fournier, R. E. (1981). A general high-efficiency procedure for production of microcell hybrids. *Proc. Natl. Acad. Sci. USA* **78,** 6349–6353.

Fournier, R. E., and Ruddle, F. H. (1977). Microcell-mediated transfer of murine chromosomes into mouse, Chinese hamster, and human somatic cells. *Proc. Natl. Acad. Sci. USA* **74,** 319–323.

Gajewski, C. D., Lin, M. T., Cudkowicz, M. E., Beal, M. F., and Manfredi, G. (2003). Mitochondrial DNA from platelets of sporadic ALS patients restores normal respiratory functions in rho(0) cells. *Exp. Neurol.* **179,** 229–235.

Gamen, S., Anel, A., Montoya, J., Marzo, I., Pineiro, A., and Naval, J. (1995). mtDNA-depleted U937 cells are sensitive to TNF and Fas-mediated cytotoxicity. *FEBS Lett.* **376,** 15–18.

Ghosh, S. S., Swerdlow, R. H., Miller, S. W., Sheeman, B., Parker, W. D., Jr., and Davis, R. E. (1999). Use of cytoplasmic hybrid cell lines for elucidating the role of mitochondrial dysfunction in Alzheimer's disease and Parkinson's disease. *Ann. N. Y. Acad. Sci.* **893,** 176–191.

Giles, R. E., Stroynowski, I., and Wallace, D. C. (1980). Characterization of mitochondrial DNA in chloramphenicol-resistant interspecific hybrids and a cybrid. *Somatic Cell Genet.* **6,** 543–554.

Gregoire, M., Morais, R., Quilliam, M. A., and Gravel, D. (1984). On auxotrophy for pyrimidines of respiration-deficient chick embryo cells. *Eur. J. Biochem.* **142,** 49–55.

Gu, M., Cooper, J. M., Taanman, J. W., and Schapira, A. H. (1998). Mitochondrial DNA transmission of the mitochondrial defect in Parkinson's disease. *Ann. Neurol.* **44,** 177–186.

Hao, H., and Moraes, C. T. (1997). A disease-associated G5703A mutation in human mitochondrial DNA causes a conformational change and a marked decrease in steady-state levels of mitochondrial tRNA(Asn). *Mol. Cell. Biol.* **17,** 6831–6837.

Hayashi, J., Tagashira, Y., Yoshida, M. C., Ajiro, K., and Sekiguchi, T. (1983). Two distinct types of mitochondrial DNA segregation in mouse-rat hybrid cells. Stochastic segregation and chromosome-dependent segregation. *Exp. Cell Res.* **147,** 51–61.

Hayashi, J., Ohta, S., Kikuchi, A., Takemitsu, M., Goto, Y., and Nonaka, I. (1991). Introduction of disease-related mitochondrial DNA deletions into HeLa cells lacking mitochondrial DNA results in mitochondrial dysfunction. *Proc. Natl. Acad. Sci. USA* **88,** 10614–10618.

Hsieh, R. H., Li, J. Y., Pang, C. Y., and Wei, Y. H. (2001). A novel mutation in the mitochondrial 16S rRNA gene in a patient with MELAS syndrome, diabetes mellitus, hyperthyroidism and cardiomyopathy. *J. Biomed. Sci.* **8,** 328–335.

Inoue, K., Takai, D., Soejima, A., Isobe, K., Yamasoba, T., Oka, Y., Goto, Y., and Hayashi, J. (1996). Mutant mtDNA at 1555 A to G in 12S rRNA gene and hypersusceptibility of mitochondrial translation to streptomycin can be co-transferred to rho 0 HeLa cells. *Biochem. Biophys. Res. Commun.* **223,** 496–501.

Inoue, K., Takai, D., Hosaka, H., Ito, S., Shitara, H., Isobe, K., LePecq, J. B., Segal-Bendirdjian, E., and Hayashi, J. (1997a). Isolation and characterization of mitochondrial DNA-less lines from various mammalian cell lines by application of an anticancer drug, ditercalinium. *Biochem. Biophys. Res. Commun.* **239,** 257–260.

Inoue, K., Ito, S., Takai, D., Soejima, A., Shisa, H., LePecq, J. B., Segal-Bendirdjian, E., Kagawa, Y., and Hayashi, J. I. (1997b). Isolation of mitochondrial DNA-less mouse cell lines and their application for trapping mouse synaptosomal mitochondrial DNA with deletion mutations. *J. Biol. Chem.* **272,** 15510–15515.

Ito, S., Ohta, S., Nishimaki, K., Kagawa, Y., Soma, R., Kuno, S. Y., Komatsuzaki, Y., Mizusawa, H., and Hayashi, J. (1999). Functional integrity of mitochondrial genomes in human platelets and autopsied brain tissues from elderly patients with Alzheimer's disease. *Proc. Natl. Acad. Sci. USA* **96,** 2099–2103.

Jun, A. S., Trounce, I. A., Brown, M. D., Shoffner, J. M., and Wallace, D. C. (1996). Use of transmitochondrial cybrids to assign a complex I defect to the mitochondrial DNA-encoded NADH dehydrogenase subunit 6 gene mutation at nucleotide pair 14459 that causes Leber hereditary optic neuropathy and dystonia. *Mol. Cell. Biol.* **16,** 771–777.

Kenyon, L., and Moraes, C. T. (1997). Expanding the functional human mitochondrial DNA database by the establishment of primate xenomitochondrial cybrids. *Proc. Natl. Acad. Sci. USA* **94,** 9131–9135.

King, M. P. (1996). Use of ethidium bromide to manipulate ratio of mutated and wild-type mitochondrial DNA in cultured cells. *Methods Enzymol.* **264,** 339–344.

King, M. P., and Attardi, G. (1989). Human cells lacking mtDNA: Repopulation with exogenous mitochondria by complementation. *Science* **246,** 500–503.

King, M. P., and Attardi, G. (1996). Mitochondria-mediated transformation of human rho(0) cells. *Methods Enzymol.* **264,** 313–334.

King, M. P., Koga, Y., Davidson, M., and Schon, E. A. (1992). Defects in mitochondrial protein synthesis and respiratory chain activity segregate with the tRNA(Leu(UUR)) mutation associated with mitochondrial myopathy, encephalopathy, lactic acidosis, and strokelike episodes. *Mol. Cell. Biol.* **12,** 480–490.

Lee, C. F., Liu, C. Y., Chen, S. M., Sikorska, M., Lin, C. Y., Chen, T. L., and Wei, Y. H. (2005). Attenuation of UV-induced apoptosis by coenzyme Q10 in human cells harboring large-scale deletion of mitochondrial DNA. *Ann. N. Y. Acad. Sci.* **1042,** 429–438.

Lurain, J. R. (2002). Advances in management of high-risk gestational trophoblastic tumors. *J. Reprod. Med.* **47,** 451–459.

Malik, S., Sudoyo, H., Pramoonjago, P., Sukarna, T., Darwis, D., and Marzuki, S. (2002). Evidence for the *de novo* regeneration of the pattern of the length heteroplasmy associated with the T16189C variant in the control (D-loop) region of mitochondrial DNA. *J. Hum. Genet.* **47,** 122–130.

McKenzie, M., and Trounce, I. (2000). Expression of Rattus norvegicus mtDNA in Mus musculus cells results in multiple respiratory chain defects. *J. Biol. Chem.* **275,** 31514–31519.

McKenzie, M., Trounce, I. A., Cassar, C. A., and Pinkert, C. A. (2004). Production of homoplasmic xenomitochondrial mice. *Proc. Natl. Acad. Sci. USA* **101,** 1685–1690.

McNeill, C. A., and Brown, R. L. (1980). Genetic manipulation by means of microcell-mediated transfer of normal human chromosomes into recipient mouse cells. *Proc. Natl. Acad. Sci. USA* **77,** 5394–5398.

Miceli, M. V., and Jazwinski, S. M. (2005). Nuclear gene expression changes due to mitochondrial dysfunction in ARPE-19 cells: Implications for age-related macular degeneration. *Invest. Ophthalmol. Vis. Sci.* **46,** 1765–1773.

Miller, S. W., Trimmer, P. A., Parker, W. D., Jr., and Davis, R. E. (1996). Creation and characterization of mitochondrial DNA-depleted cell lines with "neuronal-like" properties. *J. Neurochem.* **67,** 1897–1907.

Moraes, C. T., Kenyon, L., and Hao, H. (1999). Mechanisms of human mitochondrial DNA maintenance: The determining role of primary sequence and length over function. *Mol. Biol. Cell* **10,** 3345–3356.

Morais, R., Gregoire, M., Jeannotte, L., and Gravel, D. (1980). Chick embryo cells rendered respiration-deficient by chloramphenicol and ethidium bromide are auxotrophic for pyrimidines. *Biochem. Biophys. Res. Commun.* **94,** 71–77.

Nass, M. M. (1972). Differential effects of ethidium bromide on mitochondrial and nuclear DNA synthesis *in vivo* in cultured mammalian cells. *Exp. Cell Res.* **72,** 211–222.

Okamaoto, M., Ohsato, T., Nakada, K., Isobe, K., Spelbrink, J. N., Hayashi, J., Hamasaki, N., and Kang, D. (2003). Ditercalinium chloride, a pro-anticancer drug, intimately associates with mammalian mitochondrial DNA and inhibits its replication. *Curr. Genet.* **43,** 364–370.

Porteous, W. K., James, A. M., Sheard, P. W., Porteous, C. M., Packer, M. A., Hyslop, S. J., Melton, J. V., Pang, C. Y., Wei, Y. H., and Murphy, M. P. (1998). Bioenergetic consequences of accumulating the common 4977-bp mitochondrial DNA deletion. *Eur. J. Biochem.* **257,** 192–201.

Seyda, A., Newbold, R. F., Hudson, T. J., Verner, A., MacKay, N., Winter, S., Feigenbaum, A., Malaney, S., Gonzalez-Halphen, D., Cuthbert, A. P., and Robinson, B. H. (2001). A novel syndrome affecting multiple mitochondrial functions, located by microcell-mediated transfer to chromosome 2p14–2p13. *Am. J. Hum. Genet.* **68,** 386–396.

Shapiro, N. I., and Varshaver, N. B. (1975). Mutagenesis in cultured mammalian cells. *Methods Cell Biol.* **10,** 209–234.

Sheehan, J. P., Swerdlow, R. H., Miller, S. W., Davis, R. E., Parks, J. K., Parker, W. D., and Tuttle, J. B. (1997). Calcium homeostasis and reactive oxygen species production in cells transformed by mitochondria from individuals with sporadic Alzheimer's disease. *J. Neurosci.* **17,** 4612–4622.

Smith, C. A., Jordan, J. M., and Vinograd, J. (1971). *In vivo* effects of intercalating drugs on the superhelix density of mitochondrial DNA isolated from human and mouse cells in culture. *J. Mol. Biol.* **59,** 255–272.

Srivastava, S., and Moraes, C. T. (2001). Manipulating mitochondrial DNA heteroplasmy by a mitochondrially targeted restriction endonuclease. *Hum. Mol. Genet.* **10,** 3093–3099.

Srivastava, S., and Moraes, C. T. (2005). Double-strand breaks of mouse muscle mtDNA promote large deletions similar to multiple mtDNA deletions in humans. *Hum. Mol. Genet.* **14,** 893–902.

Swerdlow, R. H., Parks, J. K., Cassarino, D. S., Maguire, D. J., Maguire, R. S., Bennett, J. P., Jr., Davis, R. E., and Parker, W. D., Jr. (1997). Cybrids in Alzheimer's disease: A cellular model of the disease? [see comments]. *Neurology* **49,** 918–925.

Tanaka, M., Borgeld, H. J., Zhang, J., Muramatsu, S., Gong, J. S., Yoneda, M., Maruyama, W., Naoi, M., Ibi, T., Sahashi, K., Shamoto, M., Fuku, N., *et al.* (2002). Gene therapy for mitochondrial disease by delivering restriction endonuclease SmaI into mitochondria. *J. Biomed. Sci.* **9,** 534–541.

Tiranti, V., Hoertnagel, K., Carrozzo, R., Galimberti, C., Munaro, M., Granatiero, M., Zelante, L., Gasparini, P., Marzella, R., Rocchi, M., Bayona-Bafaluy, M. P., Enriquez, J. A., *et al.* (1998). Mutations of SURF-1 in Leigh disease associated with cytochrome c oxidase deficiency [see comments]. *Am. J. Hum. Genet.* **63,** 1609–1621.

Toompuu, M., Tiranti, V., Zeviani, M., and Jacobs, H. T. (1999). Molecular phenotype of the np 7472 deafness-associated mitochondrial mutation in osteosarcoma cell cybrids. *Hum. Mol. Genet.* **8,** 2275–2283.

Trounce, I., and Wallace, D. C. (1996). Production of transmitochondrial mouse cell lines by cybrid rescue of rhodamine-6G pre-treated L-cells. *Somat. Cell Mol. Genet.* **22,** 81–85.

Trounce, I., Neill, S., and Wallace, D. C. (1994). Cytoplasmic transfer of the mtDNA nt 8993 T → G (ATP6) point mutation associated with Leigh syndrome into mtDNA-less cells demonstrates cosegregation with a decrease in state III respiration and ADP/O ratio. *Proc. Natl. Acad. Sci. USA* **91,** 8334–8338.

Trounce, I. A., Kim, Y. L., Jun, A. S., and Wallace, D. C. (1996). Assessment of mitochondrial oxidative phosphorylation in patient muscle biopsies, lymphoblasts, and transmitochondrial cell lines. *Methods Enzymol.* **264,** 484–509.

Trounce, I. A., McKenzie, M., Cassar, C. A., Ingraham, C. A., Lerner, C. A., Dunn, D. A., Donegan, C. L., Takeda, K., Pogozelski, W. K., Howell, R. L., and Pinkert, C. A. (2004). Development and initial characterization of xenomitochondrial mice. *J. Bioenerg. Biomembr.* **36,** 421–427.

Waring, M. J., and Bailly, C. (1994). DNA recognition by intercalators and hybrid molecules. *J. Mol. Recognit.* **7,** 109–122.

Williams, A. J., Murrell, M., Brammah, S., Minchenko, J., and Christodoulou, J. (1999). A novel system for assigning the mode of inheritance in mitochondrial disorders using cybrids and rhodamine 6G. *Hum. Mol. Genet.* **8,** 1691–1697.

Zhu, Z., Yao, J., Johns, T., Fu, K., De Bie, I., Macmillan, C., Cuthbert, A. P., Newbold, R. F., Wang, J., Chevrette, M., Brown, G. K., Brown, R. M., *et al.* (1998). SURF1, encoding a factor involved in the biogenesis of cytochrome c oxidase, is mutated in Leigh syndrome. *Nat. Genet.* **20,** 337–343.

Ziegler, M. L., and Davidson, R. L. (1981). Elimination of mitochondrial elements and improved viability in hybrid cells. *Somatic Cell Genet.* **7,** 73–88.

Zorn, G. A., Lucas, J. J., and Kates, J. R. (1979). Purification and characterization of regenerating mouse L929 karyoplasts. *Cell* **18,** 659–672.

CHAPTER 26

Genetic Transformation of *Saccharomyces cerevisiae* and *Chlamydomonas reinhardtii* Mitochondria

Nathalie Bonnefoy,* Claire Remacle,† and Thomas D. Fox‡

*Centre de Génétique Moléculaire
CNRS UPR2167, Avenue de la Terrasse
91198 Gif-sur-Yvette cedex, France

†Laboratoire de Génétique des Microorganismes
Département des Sciences de la vie
Université de Liège, B-4000 Liège, Belgium

‡Department of Molecular Biology and Genetics
Cornell University, Ithaca
New York 14853

0091-679X/07 $35.00
DOI: 10.1016/S0091-679X(06)80026-9

I. Introduction

Methods by which genetic transformation of mitochondria can be used to generate a wide variety of defined alterations in mitochondrial DNA (mtDNA) have been worked out for only two species: the yeast *Saccharomyces cerevisiae* and very recently the green alga *Chlamydomonas reinhardtii*. DNA molecules can be delivered into mitochondria by microprojectile bombardment (biolistic transformation) and subsequently incorporated into mtDNA by the highly active homologous recombination machinery present in these organelles. In *Saccharomyces*, transformation frequencies are relatively low but the availability of strong selectable mitochondrial markers, both natural and synthetic, makes the isolation of transformants routine. In *Chlamydomonas*, the use of resistance mutations as well as recipient strains carrying mitochondrial deletions or point mutations allows for the selection of mitochondrial transformants. The strategies and procedures reviewed here allow the researcher to insert defined mutations into endogenous mitochondrial genes, and to insert new genes into mtDNA. These methods provide powerful *in vivo* tools for the study of mitochondrial biology in two highly divergent unicellular eukaryotes. A key feature of the yeast nuclear genetic system that has made it a preeminent tool for genetic and cell biological research is the fact that DNA transformed into nuclear chromosomes is incorporated into the genome only via homologous recombination (Hinnen *et al.*, 1978). This fact allows the researcher to add, subtract, and alter genetic information in a highly controlled fashion and essentially rewrite the yeast genome at will (Rothstein, 1991).

In *S. cerevisiae* and in the single-celled green alga *C. reinhardtii*, similar manipulations based on homologous recombination can be carried out on the mitochondrial genome. This chapter will focus largely on the more highly developed *S. cerevisiae* system to summarize some basic features of yeast mitochondrial genetics, describe current methods for delivery of DNA into the organelle, and outline strategies to create directed mutations in mitochondrial genes and insert new genes into mtDNA. In parallel, the general strategy and protocol for *Chlamydomonas* mitochondrial transformation will be also presented. *Chlamydomonas* mitochondrial transformation, and integration of wild-type DNA by homologous recombination, was first achieved 10 years ago (Boynton and Gillham, 1996) and repeated recently (Yamasaki *et al.*, 2005). However, the efficiency of the technique has now been considerably improved and has allowed the creation of a novel mutation in a gene encoding a complex I subunit (Remacle *et al.*, 2006). For more detailed discussions of the yeast mitochondrial genetics underlying the transformation strategies discussed here, the reader is referred to several previous reviews (Butow *et al.*, 1996; Dujon, 1981; Fox *et al.*, 1991; Perlman *et al.*, 1979; Pon and Schatz, 1991). General methods for yeast and *Chlamydomonas* genetics and manipulation have been compiled by Guthrie and Fink (2002) and Rochaix *et al.* (1998), respectively.

II. Important Features of *S. cerevisiae* and *C. reinhardtii* Mitochondrial Genetics

A. Phenotypes Associated with Mitochondrial Gene Expression

The common phenotype of mutations that affect yeast mitochondrial genes or their expression is the inability to grow on nonfermentable carbon sources. Wild-type *S. cerevisiae* strains grow well on complete medium containing nonfermentable carbon sources such as ethanol and glycerol [YPEG; 1% yeast extract, 2% peptone, 3% (v/v) ethanol plus 3% (v/v) glycerol]. Mutants which lack a functioning oxidative phosphorylation system cannot grow on such a nonfermentable medium, but grow relatively well on medium containing fermentable carbon sources such as glucose (YPD; 1% yeast extract, 2% peptone, 2% dextrose).

Chlamydomonas is a photosynthetic organism for which respiratory function is essential only when grown under heterotrophic conditions, that is, in the dark with acetate as a carbon source. Under these conditions, mitochondrial mutants inactivated in the *cox1* or *cob* gene are unable to grow, whereas mutants affecting the *nd* mitochondrial genes, encoding components of the complex I, exhibit slow growth. This growth difference can be used to create *nd* mutants by transforming the mitochondria of a *cob-* or *cox*-strain with a DNA that carries together a functional *cob* or *cox* gene and an *nd* mutation (Remacle *et al.*, 2006).

Respiratory growth of wild-type yeast can be impaired by several inhibitors of bacterial protein synthesis, and mutations conferring resistance can serve as genetic markers. Mutations in the mitochondrial ribosomal RNA genes can lead to resistance to chloramphenicol (Dujon, 1980), erythromycin (Sor and Fukuhara, 1982), and paromomycin (Li *et al.*, 1982). Mutations causing resistance to the ATP synthase inhibitor oligomycin (Ooi *et al.*, 1985; Sebald *et al.*, 1979) and the cytochrome *b* inhibitor diuron (di Rago *et al.*, 1986) also provide mitochondrial genetic markers. However, it is important to note that these drug resistance phenotypes arise spontaneously and can only be observed on nonfermentable medium in strains that respire. Thus, they are not ideal for use as selective markers in transformation experiments in yeast. In *Chlamydomonas*, the cytochrome *b* mutation *Mu* confers resistance to myxothiazol and mucidine (Bennoun *et al.*, 1991) and has been used to select mitochondrial transformants (Remacle *et al.*, 2006).

B. Novel Mitochondrial Genes Conferring Useful Mitochondrial Phenotypes in *S. cerevisiae*

Novel mitochondrial phenotypes have been generated by placing foreign genes into mtDNA. Phenotypes based on foreign genes can serve as mitochondrial genetic markers independently of respiratory function. One such phenotype is based on the fact that nuclear genes such as *URA3* and *TRP1* cannot be expressed when inserted into mtDNA, but will escape from mitochondria to the nucleus at high frequency leading to detectable growth phenotypes that can be scored on Petri plates (Thorsness and Fox, 1990, 1993).

At least some nuclear genes can be phenotypically expressed within mitochondria if they are rewritten in the *S. cerevisiae* mitochondrial genetic code (Fox, 1987), providing novel selectable markers. Expression of the synthetic gene *ARG8^m*, within mitochondria, allows nuclear *arg8* mutants to grow without arginine (Steele *et al.*, 1996). This protein, Arg8p, is normally imported into mitochondria from the cytoplasm, but also functions when synthesized within the organelle. Thus, Arg$^+$ prototrophy can become a phenotype dependent on mitochondrial gene expression. This new mitochondrial marker provides a convenient way to disrupt endogenous mitochondrial genes (Sanchirico *et al.*, 1998) as well as a useful reporter for studying mitochondrial gene expression (Bonnefoy and Fox, 2000; Bonnefoy *et al.*, 2001; Demlow and Fox, 2003; Fiori *et al.*, 2005; Green-Willms *et al.*, 2001; Perez-Martinez *et al.*, 2003; Steele *et al.*, 1996; Towpik *et al.*, 2004; Williams *et al.*, 2005) and genetic instability (Sia *et al.*, 2000; Strand and Copeland, 2002; Strand *et al.*, 2003). In addition, since *ARG8^m* specifies a soluble protein, translational fusions to endogenous mitochondrial genes can create chimeric proteins useful for the study of targeting and membrane translocation of mitochondrial translation products (Broadley *et al.*, 2001; He and Fox, 1997, 1999; Saracco and Fox, 2002).

Expression of novel mitochondrial genes can also be used to confer restoration of respiratory ability. A recoded version of the *RIP1* nuclear gene specifying the Rieske iron sulfur catalytic subunit of complex III, another protein normally located in mitochondria, has been successfully expressed within mitochondria to complement the corresponding nuclear mutation (Golik *et al.*, 2003). An approach to a novel selection system based on respiratory ability has been based on a mitochondrially expressed gene encoding the short protein Barstar, which specifically inhibits the toxic effects in the mitochondrial matrix of the barnase RNase targeted to mitochondria (Mireau *et al.*, 2003).

A visible reporter phenotype based on mitochondrial gene expression can also be generated by insertion into mtDNA of a synthetic gene, *GFP^m*, encoding the green fluorescent protein in the yeast mitochondrial code (Cohen and Fox, 2001; Demlow and Fox, 2003). Mitochondria in cells containing *GFP^m* inserted in place of *COX3* in mtDNA strains are long, tubular structures aligned along the mother bud axis, as in wild-type cells (Fig. 1).

Finally, transformation allows the fusion of tags onto mitochondrial endogenous mitochondrial genes: for example, the triple-HA tag has been fused to the

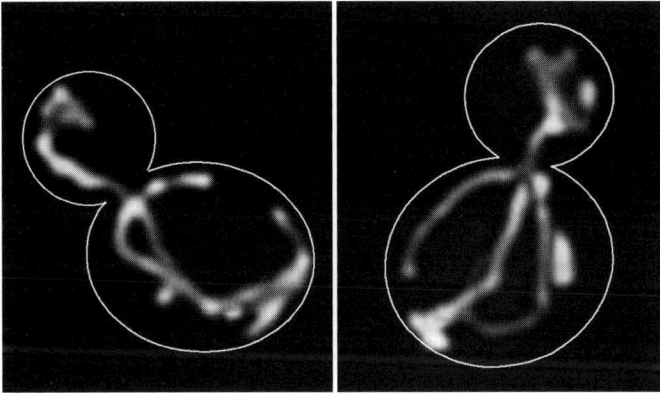

Fig. 1 Optical imaging (Anna Card Gay and Liza Pon, unpublished data) of a yeast strain JSC350X containing a mitochondrially expressed *cox3::GFP^m* reporter (Cohen and Fox, 2001). Images were obtained as a z-series of 20^+ optical sections through the entire cell. The optical z-series were then deconvoluted to remove out-of-focus fluorescence, and reconstructed in 3D. The images shown are 2D projections of 3D reconstructions. Phase images were also obtained to determine the cell boundaries (white outlines).

end of the Cox2 protein allowing easy detection of the protein with commercially available antibody (Saracco *et al.*, 2002).

C. Replication of mtDNA and Mitochondrial Deletion Mutants

Replication of yeast mtDNA is a complex and poorly understood process. Cells typically contain between 50 and 100 genome equivalents of mtDNA (Dujon, 1981) which are organized into a smaller number of nucleoid structures. The nucleoids, which are visible by fluorescence microscopy, are the genetic elements that are transmitted to daughter cells during cell division (Chen and Butow, 2005; Meeusen and Nunnari, 2003). Replication of complete, or *rho^+*, yeast mitochondrial genomes is thought to depend on a limited number of specific sites in the chromosome (de Zamaroczy *et al.*, 1984; Schmitt and Clayton, 1993) and, for unknown reasons, requires mitochondrial protein synthesis (Myers *et al.*, 1985).

The most frequent mutants in wild-type *S. cerevisiae* strains are the nonrespiring *rho^-* or cytoplasmic petite mutants. These strains have large deletions of mtDNA that destroy the organellar gene expression machinery by deleting components of its translation system (Dujon, 1981). The DNA sequences retained in *rho^-* mutants are typically reiterated such that the *rho^-* cell contains roughly the same amount of mtDNA as wild type. Replication of *rho^-* mtDNAs differs from that of *rho^+* mtDNAs in that it does not require mitochondrial protein synthesis.

Interestingly, the mtDNA sequences replicating in *rho⁻* strains can be derived from any portion of the chromosome, demonstrating that there is no clear requirement for a specific replication origin sequence in *rho⁻* mtDNAs. This is advantageous in creating mitochondrial transformants containing defined mtDNAs, since *rho⁰* yeast strains, entirely lacking mtDNA, can be transformed with bacterial plasmid DNAs that subsequently propagate as "synthetic" *rho⁻* molecules (Fox *et al.*, 1988).

In *Chlamydomonas*, the mitochondrial genome is a 15.8-kb linear molecule present at about 50 copies per cell. This genome has telomeric inverted repeats of about 500bp at each end that are thought to play a role in replication in conjunction with a reverse transcriptase-like protein (Vahrenholz *et al.*, 1993). Random mutagenesis with acriflavine has yielded point mutations in various mitochondrial genes, and deletion mutations that lack the left telomere and the neighboring *cob* gene, or *cob* and *nd4*. However, a mutant lacking both telomeres has never been isolated. The mtDNA present in the deletion mutants exists as both monomers and dimers arising from head-to-head fusions between monomers. The total amount of mtDNA in these mutants is generally lower than in wild type. The deletion mutants do not revert and can be used as recipients for mitochondrial transformation experiments.

D. Recombination and Segregation of mtDNA

Unlike the highly differentiated situation in animals and plants, there is true equality of the sexes in yeast mating. Haploid cells mate by fusion, cytoplasms are mixed, and the mitochondria of the haploid cells fuse to form an essentially continuous compartment (Azpiroz and Butow, 1993; Nunnari *et al.*, 1997). Homology-dependent recombination between the parental mtDNAs occurs at a high rate (Dujon, 1981) in the medial portion of the zygote (Azpiroz and Butow, 1993). Of great importance to the manipulation of yeast mtDNA is the fact that *rho⁻* mtDNA sequences readily recombine with complete *rho⁺* genomes. Thus, if a *rho⁻* mtDNA contains wild-type genetic information in a particular region, it can recombine with a *rho⁺* mtDNA bearing a mutation in that region. The result is that the mating of the nonrespiring *rho⁻* with the nonrespiring *rho⁺* mutant yields respiring recombinants at high frequency, whose growth can be selected on nonfermentable medium.

Yeast cells containing two different kinds of mtDNA can be created by mating or by mutation. Such heteroplasmic cells rapidly give rise to homoplasmic progeny, a phenomenon known as mitotic segregation (Dujon, 1981). However, exceptional cases of stable heteroplasmy in *S. cerevisiae* have been reported (Lewin *et al.*, 1979). *S. cerevisiae* differs in this regard from plant and animal cells, which can maintain heteroplasmic states for extended periods of growth (Hanson and Folkerts, 1992; Wallace, 1992). In this connection, it is important to note that, unlike most animal and plant cells, *S. cerevisiae* divides by a highly asymmetric budding process. New buds receive relatively few copies of mtDNA from mother cells (Zinn *et al.*, 1987), facilitating mitotic segregation. Heteroplasmic cells generated by transformation and subsequent homologous recombination also produce pure recombinant clones (Johnston *et al.*, 1988). Taken together, these features of the *S. cerevisiae* mitochondrial genetic system allow DNA introduced from outside the cell to be propagated within the organelle as a plasmid, and the plasmid-borne mitochondrial sequences to recombine homologously with complete *rho*$^+$ mtDNA (Fox *et al.*, 1988).

A drawback of the active mtDNA recombination system in yeast is that molecules with repeated sequences are often highly unstable. However, it has proven possible to construct mitochondrial genomes that contain all the normal endogenous genes plus new genes flanked by control regions necessary for their expression, despite the fact that the control regions are duplicated in these chromosomes. For example, the mitochondrially coded Barstar gene was inserted into a site upstream of *COX2*, flanked by 71 nucleotides encoding the *COX2* promoter and 5′-leader, and 119 nucleotides from the *COX2* 3′-untranslated region. These short repeats did not cause high instability, allowing maintenance of the Barstar gene in respiratory competent cells even without selection (Mireau *et al.*, 2003). The coding sequence for *COX1* has also been inserted into this ectopic "*COX2*" locus (Perez-Martinez *et al.*, 2003), as has *ARG8*^{m} (Duvezin-Caubet *et al.*, 2006). In the latter case, the function of the mitochondrial genome can be detected and/or selected either by respiratory growth or arginine prototrophy, or both. A different stable duplication of mitochondrial gene-flanking sequences was generated to express the recoded *RIP1* gene (Golik *et al.*, 2003). In this case, *RIP1*^{m} was inserted downstream *COX1*, surrounded by about 1 kb of *COX1* 5′-flanking sequence and 1 kb of *COX1* 3′-flanking sequence to allow its expression. The flanking sequences were derived from mtDNA of *S. douglasii*, which has several polymorphisms relative to the *S. cerevisiae* sequence. This sequence divergence presumably reduces the frequency of homologous recombination. Nevertheless, this ectopic gene was only stable in mtDNA derived from some *S. cerevisiae* strains, but unstable in others.

In *Chlamydomonas*, homologous mtDNA recombination is only detected after crosses between *mt*$^-$ and *mt*$^+$ strains in mitotic zygotes that do not undergo meiosis. In these zygotes and their mitotic progeny, mtDNA is transmitted by both parents and intermolecular recombinational events are frequent (Remacle and Matagne 1998). After 15–20 divisions, most of the diploid cells are homoplasmic for a mitochondrial genome, either recombinant or parental. Following mitochondrial transformation of haploid recipient cells, homologous recombination between the mutated mitochondrial genome of the recipient strain and exogenous DNA also occurs (Boynton and Gillham, 1996; Yamasaki *et al.*, 2005). Homologous sequences as short as 28 nucleotides can direct recombination (Remacle *et al.*, 2006). Most of the transformants recovered after selection in the dark are homoplasmic, but heteroplasmic clones could also be found (Remacle *et al.*, 2006).

═══════════ **III. Delivery of DNA to the Mitochondrial Compartment of rho^0 Cells and Detection of Mitochondrial Transformants**

A. Overview of Transformation Procedure

The standard device for microprojectile bombardment is the PDS-1000/He System, available from Bio-Rad, Inc. This instrument uses a helium shock wave in an evacuated chamber to accelerate microscopic metal particles toward a lawn of cells on a Petri plate. The shockwave is generated by rupture of a membrane at high pressure, and accelerates a second membrane (the macrocarrier or flying disk), carrying the metal particles, toward the plate. Some cells on the plate are penetrated by particles and survive. DNA precipitated on the particles is thus introduced into cells and is readily taken up by the nucleus. In addition, the mitochondria of a small fraction of such transformants also take up DNA. The PDS-1000 functions reproducibly for transformation of *S. cerevisiae* and *C. reinhardtii* mitochondria.

In a typical yeast mitochondrial transformation experiment, a large number of rho^0 cells are randomly bombarded by a large number of particles (Fig. 2). In the first step, cells that have been hit and that survived are allowed to make colonies on the Petri plates by selecting for a nuclear genetic marker that is included in the DNA precipitated on the particles. Mitochondrial transformants are identified among these colonies by genetic tests for the presence of new genetic information in the mitochondrial genome. This new information is typically a portion of wild-type mtDNA sequence that can rescue a known mitochondrial marker mutation by recombination, after the transformants are mated to an appropriate rho^+ tester strain, resulting in recombinants with a detectable growth phenotype. The new wild-type sequence may be an unaltered region of the gene of interest, or it may be another piece of wild-type mtDNA incorporated into a vector. Such marker rescue can work with as little as 50 bp of homologous sequence flanking the site of the mutation in the tester mtDNA. As shown in Fig. 2, transformation of rho^+ mutants can also be detected using this marker rescue strategy (or as discussed in Section V below, by directly selecting for a phenotypic change). Transformants can also be identified by scoring for expression of complete mitochondrial genes that function in *trans* (Butow *et al.*, 1996; Fox *et al.*, 1988).

Mitochondrial transformation of *C. reinhardtii* has been achieved to date only by transforming cells mutated in the *cob* or *cox1* genes with plasmids or polymerase chain reaction (PCR) fragments encompassing the mutation, so that transformants could be selected directly for heterotrophic growth on acetate medium in the dark. Employing this strategy with a plasmid containing the wild-type *cob* gene and a mutated complex I gene, *nd4*, has allowed the concomitant restoration of *cob* function and mutation of *nd4* by cointegration. In this case, colonies growing slowly in the dark were preferentially analyzed to increase the chance of finding an *nd4* mutant with a wild-type *cob* gene (Remacle *et al.*, 2006).

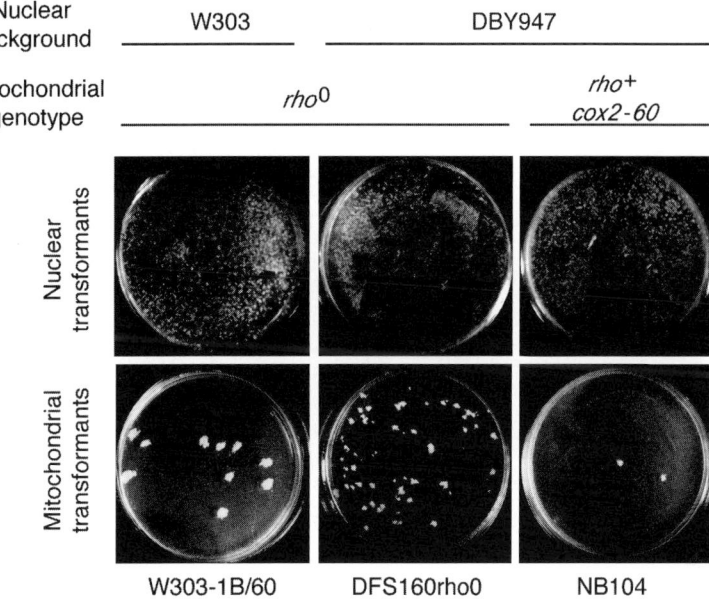

Fig. 2 Nuclear transformants and mitochondrial cotransformants obtained by bombardment of different yeast strains. The nuclear *LEU2* plasmid YEp351 (Hill *et al.*, 1986) and the *COX2* plasmid pNB69 (Bonnefoy and Fox, 1999) were precipitated together onto tungsten particles and bombarded on lawns of the *rho⁰* strains W303–1B/60 (*MATα, ade2-1, ura3-1, his3-11,15, trp1-1, leu2-3,112, can1-100 [rho⁰]*; a *rho⁰* derivative of W303–1B; Thomas and Rothstein, 1989) and DFS160 (*MATα ade2-101, leu2Δ, ura3-52, arg8Δ::URA3, kar1-1, [rho⁰]*; Steele *et al.*, 1996), or on lawns of the *rho⁺* strain NB104. W303–1B/60 was derived from W303 (ATCC 200060). DFS160 was derived from DBY947 (Neff *et al.*, 1983). NB104 *rho⁺* mtDNA carries a 129-bp deletion, *cox2-60*, located around *COX2* first codon (Bonnefoy and Fox, 1999), and is isonuclear to DFS160. The top plates correspond to minimal medium supplemented with sorbitol and lacking leucine. Typical plates showing about 3000 nuclear transformants for each strain have been presented. Nuclear transformants were crossed by replica plating to the nonrespiring tester strain (NB160), carrying a mutation of *COX2* initiation codon (Bonnefoy and Fox, 1999), and mitochondrial transformants (bottom plates) were detected by replica-plating the mated cells onto nonfermentable medium.

B. Experimental Details for Transformation and Identification of Mitochondrial Transformants

1. Strains

S. cerevisiae

Strain background is an important factor affecting the efficiency of transformation (Fig. 2). We have obtained the best results with strains in the S288c background, in particular those derived from DBY947 (Neff *et al.*, 1983). Strains derived from W303 (ATCC 200060) (Thomas and Rothstein, 1989) give lower but satisfactory efficiencies, while strains in the D273–10B (ATCC 24657) background are difficult to transform. Excellent hosts for mitochondrial transformation, derived from DBY947, can be obtained from the American Type Culture

Collection: MCC109rh0 (*MATα, ade2-101, ura3-52, kar1–1* [*rho*^{0}]) (Costanzo and Fox, 1993) and MCC123rho0, which is the identical strain with *MAT*a, available as ATCC 201440 and 201442, respectively.

C. reinhardtii

The mutant *dum11* (Dorthu *et al.*, 1992) (CC-4098 dum11 mt$^-$ at www.chlamy.org) has a 1.2-kb deletion including the left telomere and part of the *cob* gene, which causes the loss of complex III activity. *dum11* has proven to be a good recipient for mitochondrial transformation with DNA fragments containing at least the *cob* and *nd4* genes. Mitochondrial transformation has also been successful in a double mutant *dum19*, *dum25* (Remacle *et al.*, 2006) containing point mutations in *cox1* and *nd1*, respectively.

2. Preparation of Cells

S. cerevisiae

a. Grow the *rho*^{0} (or *rho*$^+$) strain to be bombarded for 2–3 days (stationary phase) at 30 °C with agitation in complete liquid medium (YP) containing either 2% raffinose or 2% galactose. These media may be supplemented with 0.1% glucose (to accelerate growth) and/or 100-mg/ml adenine (for Ade$^-$ auxotrophs).

b. Harvest cells and concentrate 40–100 times in liquid YPD medium to reach a cell density of $(1–5) \times 10^9$ cells/ml.

c. Spread 0.1 ml of cells onto minimal glucose medium (0.67% yeast nitrogen base, 5% glucose, 100-mg/ml adenine, 3.3% agar) containing 1-M sorbitol and supplemented to provide the appropriated prototrophic selection.

C. reinhardtii

a. Grow the recipient mutant cells for 2–3 days in the light (70 PAR) up to exponential phase [$(2–3) \times 10^6$ cells/ml] at 23 °C in liquid TAP medium (Tris-acetate-phosphate; Harris, 1989) with agitation. NH$_4$Cl 7.5 mM – MgSO$_4$ 0.41 mM – CaCl$_2$ 0.34 mM – K$_2$HPO$_4$ 0.62 mM – KH$_2$PO$_4$ 0.41 mM – Tris 20 mM pH 7.0 – Glacial acetic acid 0.1% – EDTA 171 μM – ZnSO$_4$ 31 μM – H$_3$BO$_3$ 184 μM – MnCl$_2$ 7.2 μM – FeSO$_4$ 18 μM – Co(NO)$_2$ 2.2 μM – CuSO$_4$ 6.3 μM – MoO$_3$ 4.9 μM KOH 552 μM.

b. Harvest cells and concentrate 500–300 times in liquid TAP medium to reach a cell density of 1×10^9 cells/ml.

c. Spread 0.1 ml of cells onto TAP medium [1.5% (w/v) agar]. Plates must be 1-cm thick to avoid damage to the agar by the combined action of the vacuum and the helium shock wave.

3. Preparation of Microprojectiles and Precipitation of DNA

We routinely use tungsten powder <1 μm 99.95% (metals basis) from Aesar/ Johnson Matthey, item 44210 CAS 7440–33–7, which is inexpensive and very effective. Tungsten powder 0.4 to 0.7 μm is also available from Bio-Rad (Cat. Nos. 165–2265 or 165–2266, respectively), as is gold powder 0.6 μm (Bio-Rad

Cat. No. 165–2262). The following describes our current procedure using tungsten particles. With the exception of the sterilization step described below, the same procedure gives comparable results for *S. cerevisiae* using gold particles. A slightly different procedure for preparation of gold particles has previously been described (Butow *et al.*, 1996), and gives equivalent results for *S. cerevisiae*. We have not tested gold particles for *Chlamydomonas* but other authors have reported very low transformation efficiency (Yamasaki *et al.*, 2005).

a. Sterilize up to 50 mg of tungsten particles by suspension in 1.5 ml of 70% ethanol in a microfuge tube, and incubation at room temperature for 10min. Wash the particles with 1.5 ml of sterile water and resuspend at 60 mg/ml in sterile freezer-stored 50% glycerol. Particles can be kept frozen for several months. Gold particles should be sterilized in 100% ethanol (Butow *et al.*, 1996).

b. For *S. cerevisiae* transformation, in a microfuge tube kept on ice, mix 5 μg of circular plasmid carrying the nuclear marker and a nuclear replication origin with 15–30 μg of circular plasmid carrying the mtDNA of interest in a total volume of 15–20 μl. For *Chlamydomonas*, use 15–30 μg of linearized plasmid (which is more efficient for mitochondrial transformation than circular plasmid), or 10 μg of PCR fragment in a total volume of 15–20 μl.

c. Add and mix 100 μl of tungsten particles, 4 μl of 1-M spermidine (as the free base), and 100 μl of 2.5-M CaCl$_2$ that was stored in the freezer. Incubate 10min on ice with occasional vortexing.

d. Spin briefly in a precooled centrifuge and remove the supernatant. Resuspend the particles thoroughly in 200 μl of 100% ethanol that was stored in the freezer, taking extreme care to break up aggregates of particles using the pipette tip. Repeat at least once until the particles resuspend easily.

e. Spin briefly, remove the supernatant, and add 50–60 μl of 100% ethanol. Distribute the resulting suspension evenly at the center of six macrocarriers (flying disks) placed in their holders, allowing the ethanol to evaporate (there is no need to prewash the macrocarriers nor to desiccate them after coating).

4. Bombardment

a. Carefully follow the manufacturer's instructions for use of the PDS-1000 apparatus. Place the rupture disk in its retaining cap and tighten using the torque wrench. Rupture disks of 1100–1350 psi can be use for efficient transformation of yeast and *Chlamydomonas*, although in our hands 1100-psi disks tend to give better results with tungsten or gold particles.

b. Load the macrocarrier in its holder into the assembly system. Interestingly, we have found that simply allowing the carrier disk to fly to the surface of the Petri plate, by not assembling the stopping screen, yields more yeast transformants than we observe if the stopping screen is employed. For *Chlamydomonas*, using a stopping screen mildly enhances the rate of transformation.

c. Place the open Petri plate carrying the lawn of cells at 5 cm (yeast) or 7 cm (*Chlamydomonas*) from the macrocarrier assembly. In *S. cerevisiae*, shorter distances result in very high colony densities in the center of the plate with few colonies at the periphery, while longer distances decrease the transformation efficiency.

d. Evacuate the vacuum chamber to a reading of 29–29.5 in. Hg on the PDS-1000's gauge. We have found that failure to draw the greatest vacuum possible reduces the transformation efficiency. Cell viability is not significantly affected by a prolonged stay under these vacuum conditions.

e. Fire.

f. Remove any fragments of the macrocarrier disk with a sterile forceps. Incubate the yeast plate at 30°C for 4–5 days until colonies appear (between 1000 and 10000 per plate for S288c related strains). Incubate the *Chlamydomonas* plate for one night in the light (70 PAR) at 23°C before transfer to the dark for 2 months.

g. Each investigator's experience will lead to modifications of these procedures. On the basis of our experience with mitochondrial transformation of yeast and *Chlamydomonas* we list our currently known optimal conditions in Table I.

Table I

Summary of the Factors Influencing Mitochondrial Transformation Efficiency in Yeast and Their Known Optima

Recipient strain		DNA precipitation		Biolistic parameters	
Parameter	Optimum	Parameter	Optimum	Parameter	Optimum
Genetic background		DNA -Size	5–6 kb	Rupture disks	1100 psi
-nuclear	DBY947	-Purity	Qiagen		
		-Concentration	>2 μg/μl	Stopping screens[a]	None
-mitochondrial[b]	rho^0	-Volume	<15–20 μl		
		-Quantity	5 μg (nuclear) 20–30 μg (mito)	Plate distance[c]	5 cm
		-Structure[d]	Circular		
Carbon source[e]	Raffinose			Vacuum	29–29.5
		Particles	Aesar 44210 Tungsten powder, <1 μm		
		Temperature	Ice cold		
		Precipitate	Finely dispersed		

Modifications for *Chlamydomonas* are the following:
[a]Yes.
[b]Left telomere deletions.
[c]7 cm.
[d]Linear.
[e]Acetate.

5. Identification of Mitochondrial Transformants

S. cerevisiae

a. During the incubation of the bombarded plates, set up a liquid YPD culture of an appropriate *rho*⁺ mutant (*mit*⁻) tester strain.

b. Replica plate the transformants onto a lawn of the tester strain freshly spread on a YPD plate.

c. Incubate at 30°C for 2 days to allow mating and recombination.

d. Print to YPEG medium (or another appropriate selection medium) to detect respiring diploids. In cases where a high number of nuclear transformants are present, it may be useful to also replicate the mated cells on medium that selects for the diploids, since comparison of the resulting plates may facilitate the identification of the desired transformant on the original bombarded plate.

e. Pick colonies off the bombarded plate that correspond to the position of respiring recombinants. Streak these colonies on YPD and repeat the marker rescue with the tester strain, as above. Such subcloning and retesting must usually be done three times before pure stable synthetic *rho*⁻ clones are obtained. (Cells usually lose the nuclear marker plasmid during these subcloning steps if no selection is applied for its maintenance.)

C. reinhardtii

a. After 4–6 weeks in the dark, microcolonies can be first detected with a stereoscopic microscope.

b. After a total incubation of 2 months in the dark, transfer the plates to dim light (5–10 PAR) for 1 week. Compared to control plates showing no colonies, bombarded plates contain typically 300–600 transformants for 3 µg, mostly at the periphery of the plate, of linearized plasmid and 200–400 transformants for 1.5 µg of PCR fragment (Fig. 3).

c. Patch the transformants on a new TAP plate and incubate 2 days in the light. If introducing an *nd* mutation while restoring *cob* or *cox1*, pick preferentially colonies growing slowly in the dark.

d. Test the restoration of respiratory proficiency by overlaying the transformants on plate with 0.5 mg/ml of 2,3,5-triphenyltetrazolium chloride (TTC) and 0.5% agar. Incubate 12 h in the dark at 23°C until a brown color develops (TTC+). This staining is due to the reduction of TTC to red formazan under anaerobic conditions by electrons from the respiratory chain when cytochrome *c* oxidase is active, showing that the respiratory deficiency is rescued. Generally, 95–100% of the transformants are TTC+, but the proportion of TTC+ clones can decrease to 80% if the plates are incubated in the dark for 5–6 weeks instead of 8 weeks, probably because recipient mutant mtDNA is still present in some clones when the selective pressure is removed too early.

e. Subclone the TTC+ transformants and carry out a molecular analysis of their mtDNA. In our hands, TTC+ transformants obtained after an 8-week

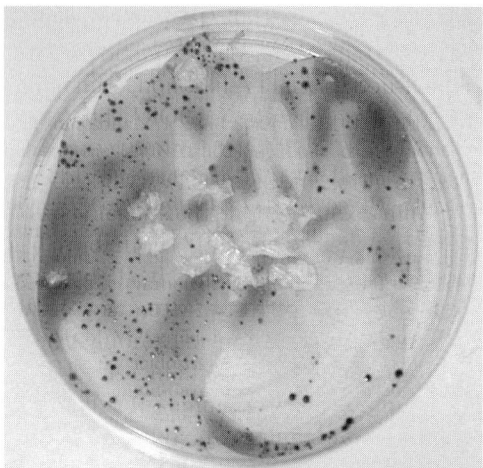

Fig. 3 Mitochondrial transformants of *Chlamydomonas*. The *dum11* mutant, lacking the left telomere and part of the *cob* gene (Dorthu *et al.*, 1992), was transformed with 3μg of a 6.6-kb PCR fragment containing the left telomere, *cob*, *nd4*, *nd5*, and *cox1* genes. Typically, mitochondrial transformants appear at the periphery of the TAP plate after 2 months of growth in the dark followed by a week of incubation under dim light. No stopping screen was used in this experiment.

incubation in the dark generally showed a homoplasmic recombined mtDNA. Three homoplasmic *nd4* mutants were recovered among 90 slow growth transformants in a typical experiment of simultaneous insertion of an *nd* mutation and restoration of wild-type *cox1* or *cob* (Remacle *et al.*, 2006).

IV. Strategies for Gene Replacement in *S. cerevisiae* mtDNA

In cases where the mitochondrial gene under study encodes an active RNA molecule, such as the mitochondrial RNase PRNA, it may be possible to assay the activity of wild-type and mutant genes in the primary synthetic *rho*⁻ transformants (Sulo *et al.*, 1995). More commonly however, mutations affecting protein-coding genes must be placed into *rho*⁺ mtDNA by a double recombination event.

A. Integration of Altered mtDNA Sequences by Homologous Double Crossovers

The most basic method for putting a mutant version of a mitochondrial gene, or a foreign piece of DNA flanked by mtDNA sequences, into the chromosome is to first introduce the altered sequence into a *rho*⁰ strain to create a synthetic *rho*⁻. This donor transformant (identified as described in Section III) is then, in a second step, mated with a wild-type *rho*⁺ recipient strain. As a result of this second mating, mitochondria from the two strains fuse and recombination between the two mtDNAs produces recombinant *rho*⁺ strains in which the new mtDNA sequence is integrated by double crossover events. Pure recombinant strains are generated by subsequent mitotic segregation. Since mtDNA recombination and segregation is so frequent, this simple procedure typically yields the desired integrants at frequencies between 1% and 50% of clones derived from zygotes.

If one of the strains in such a cross carries the karyogamy defective mutation *kar1-1* (Conde and Fink, 1976), which allows efficient mitochondrial fusion but greatly reduces nuclear fusion, haploid mitochondrial mutant cytoductants can be isolated after such a mating. This simple strategy has been successfully used, with variations specific to each study, in several laboratories (Boulanger *et al.*, 1995; Folley and Fox, 1991; Henke *et al.*, 1995; Mulero and Fox, 1994; Speno *et al.*, 1995; Szczepanek and Lazowska, 1996; Thorsness and Fox, 1993).

B. Experimental Details for Mating and Isolation of Recombinant Cytoductants

1. Mating

a. Grow cultures of the subcloned synthetic *rho⁻* strain and the recipient wild-type *rho⁺* strain overnight in liquid YPD. At least one of these two strains (usually the synthetic *rho⁻*) must carry the *kar1–1* mutation.

b. If the synthetic *rho⁻* donor and the *rho⁺* recipient strains share nuclear markers, and therefore cannot be distinguished selectively on glucose medium, mating mixtures should contain equal numbers of cells of both strains. If nuclear auxotrophic or drug resistance markers allow selection against the synthetic *rho⁻* donor strain, then the mating mixture should contain a fivefold excess of donor cells.

c. We have successfully used two different mating protocols for producing cytoductants.

Mix 0.5 ml of each parent (alternatively 1 ml of synthetic *rho⁻* and 0.2 ml of wild-type *rho⁺*) in a microfuge tube, spin, remove the supernatant, resuspend in residual liquid, and spread the mixture onto a YPD plate. Incubate at 30°C for 4–5 h. Check zygote formation microscopically. Scrape the mating cells from the plate and use them to inoculate fresh YPD liquid medium. Incubate at 30°C with agitation for a few hours to overnight.

Alternatively, mix both parents, in proportions as above, in 10 ml of liquid YPD and shake at 30°C for 3 h. Spin the culture in a tube and incubate the pellet at 30°C for 1 h without removing the medium. Resuspend by vortexing, transfer to a fresh flask, and incubate at 30°C with agitation for at least three additional hours.

2. Isolation of Cytoductants

a. Dilute the culture to obtain single colonies and plate on minimal medium, selecting for the recipient nuclear genotype and against the donor nuclear genotype, if possible. Alternatively, plate on YPD medium. Densities of 50–200 colonies per plate should be obtained.

b. Replica plate the colonies obtained to medium that will reveal the altered phenotype of the recipient strain as a result of integration of the mutant donor sequences into its mtDNA. For example, print to YPEG to identify clones that have acquired a mutation preventing respiratory growth.

c. Mate nonrespiring candidate clones to a rho^+ tester mutant, whose mitochondrial mutation is located outside of the region carried by the synthetic rho^-. The desired rho^+ recombinant cytoductants will produce respiring diploids after mating to this tester strain. (This step eliminates cytoductants that simply acquired the mtDNA of the donor strain.)

C. Streamlining the Integration of Multiple Mutations in a Short Region by Use of rho^+ Recipients Containing Defined Deletions

In situations where many nonfunctional mutations are to be placed in the same region of mtDNA, the strategy described above can be altered and made more efficient by using a rho^+ recipient that has a defined deletion in the region of interest (Costanzo and Fox, 1993; Mittelmeier and Dieckmann, 1993). It is, of course, often necessary to isolate such a recipient strain using the simple recombination method described above. However once the recipient is in hand, the nonrespiring recombinant cytoductants from crosses between nonrespiring synthetic rho^- strains and the nonrespiring recipient can be identified by a positive marker rescue screen employing an appropriate tester strain (Fig. 4). Following mating between the synthetic rho^- (carrying the experimental mutation "e" in Fig. 4) and the recipient, the cell population is plated on medium selecting for the recipient nuclear genotype. Recombinant cytoductants, unaltered recipient cells, and diploid cells will form colonies. However only the recombinant cytoductants will be able to form respiring diploids when mated to a rho^+ tester strain carrying a marker mutation ("m" in Fig. 4) located in the deleted region (and distinct from the experimentally induced mutation).

D. Experimental Details for Identification of Nonrespiring Cytoductants by Marker Rescue

The following steps assume that one already has in hand a rho^+ recipient with a deletion mutation in the region of interest, as well as a rho^- strain that carries wild-type information and can recombine with the recipient deletion mutant.

1. Carry out a mating between the synthetic rho^- donor and the rho^+ recipient containing a defined deletion to generate cytoductants as described in Section III.B.1 (at least one of these two strains must carry the *kar1-1* mutation) and spread dilutions of the cell mixture as described in Section III.B.2 , step a. Some of these colonies will be rho^+ recombinants that have the deleted region restored and contain the desired mutation (Fig. 4A).

2. Replica plate colonies from the mating mixture onto a YPD plate bearing a freshly spread lawn of a rho^+ tester strain that has a marker mutation within the recipient's deleted region but distinct from the new mutation to be introduced (Fig. 4B).

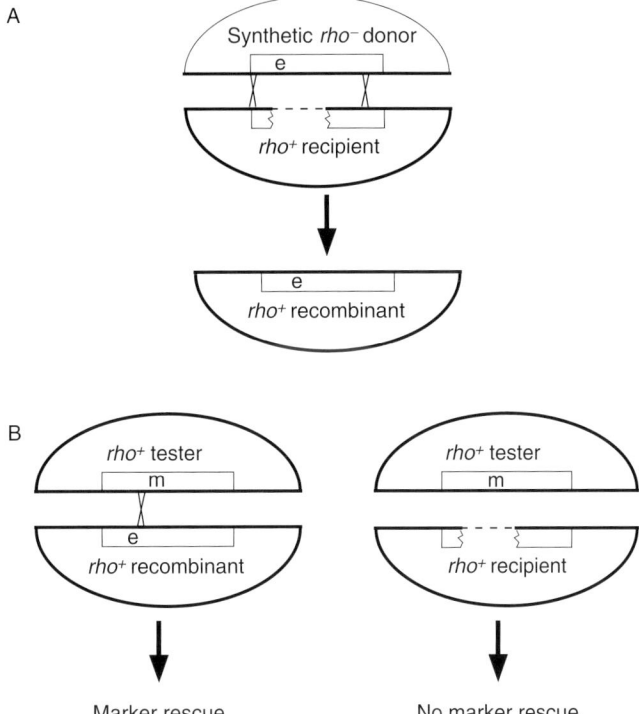

Fig. 4 Schematic diagram of recombination events that allow identification of nonrespiring recombinant cytoductants by marker rescue in yeast. Thick lines represent mtDNA sequences, thin lines represent vector DNA. The box represents a gene under study. (A) A karyogamy-defective (*kar1-1*) synthetic *rho⁻* donor containing an experimentally induced mutation "e" is mated to a *rho⁺* recipient strain with a deletion in the region of interest. The desired *rho⁺* recombinant cytoductants are present among the cells present in the mixture after mating. (B) To distinguish the desired *rho⁺* recombinant cytoductants from the unaltered recipient cells and other cell types present, clones derived from the mating mixture are mated to a *rho⁺* tester strain bearing the marker mutation "m." The desired *rho⁺* recombinant cytoductants can yield respiring recombinants when mated to this tester by a crossover between "e" and "m" (and a second resolving crossover anywhere else). The ability to produce respiring recombinants identifies the desired cytoductant clones. Unaltered recipient clones, and other cells types present, cannot yield such respiring recombinants.

3. After 2 days of incubation at 30°C, print the mated cells to YPEG. Identify haploid cytoductant clones on the plates from step 1 that correspond to respiring diploids in the mating of step 2.

4. Test the candidate cytoductant clones to be sure that they are *rho⁺* mutants by checking to see that they yield respiring diploids when mated to a *rho⁻* strain carrying wild-type information in the region deleted in the original recipient strain. This will eliminate cytoductants that simply contain the unrecombined mtDNA of the synthetic *rho⁻* donor.

V. Transformation of *rho*⁺ Cells with Plasmids or Linear DNA Fragments

The first demonstration of microprojectile bombardment's ability to deliver DNA into yeast mitochondria depended on integration of the transforming DNA directly into a mutant *rho*⁺ strain, converting a nonrespiring mutant into a respiring transformant (Johnston *et al.*, 1988). In addition to selection for restoration of respiratory growth, *rho*⁺ transformants can also be obtained with DNA sequences causing expression of *ARG8ᵐ* and selection for Arg⁺ growth. In cases where a mutated DNA sequence provides a function that can be selected phenotypically, direct transformation of *rho*⁺ strains bearing deletions in the region of interest can be used to integrate mutations into mtDNA (Bonnefoy and Fox, 2000).

In our hands, mitochondrial transformation is 10–20 times more efficient during bombardment of a *rho*⁰ strain than of an isogenic *rho*⁺ strain containing a small deletion in mtDNA (Fig. 2). This effect could be either due to physiological differences between the strains such as the properties of the mitochondrial inner membrane or due to an advantage in establishing an incoming DNA molecule in the absence of endogenous mtDNA. Effects consistent with the latter notion have been observed in comparisons of mtDNA behavior after *rho*⁺ × *rho*⁺ matings as opposed to *rho*⁺ × *rho*⁰ matings (Azpiroz and Butow, 1993).

Nevertheless, the relative inefficiency of transformation of *rho*⁺ hosts is offset in some situations by its convenience, since *rho*⁺ recombinants may be obtained more quickly. In addition, this strategy can be extended to transformation with linear DNA molecules obtained either from plasmid clones or PCR amplification. For example, we have found that linear DNA fragments having as little as 260bp of homologous sequence flanking each side of a deletion mutation in a *rho*⁺ recipient were able to yield respiring transformants at frequencies similar to those obtained with circular plasmids. The ability to use PCR-generated fragments to transform defined mtDNA deletion recipients can accelerate strain construction substantially.

Selection for transformation by DNA fragments with wild-type or near wild-type function is straightforward, by selecting first for nuclear transformants on glucose medium, as described above in Section III, and then screening for the mitochondrial phenotype by replica plating (Fig. 5) (Johnston *et al.*, 1988). Interestingly, two- to fivefold fewer transformants are detected by this phenotypic selection for integration of transforming DNA into *rho*⁺ mtDNA than are detected by mating of the *rho*⁺ colonies to a tester strain and scoring for respiring recombinants (as was done in the experiment of Fig. 2). Thus, it appears that only a fraction of the plasmids entering a *rho*⁺ mitochondrion during bombardment manage to integrate and become phenotypically expressed.

Selection of respiring transformants from nonrespiring host strains generally requires a period of outgrowth on a fermentable carbon source to allow establishment of the respiratory phenotype prior to selection for it. This phenotypic lag

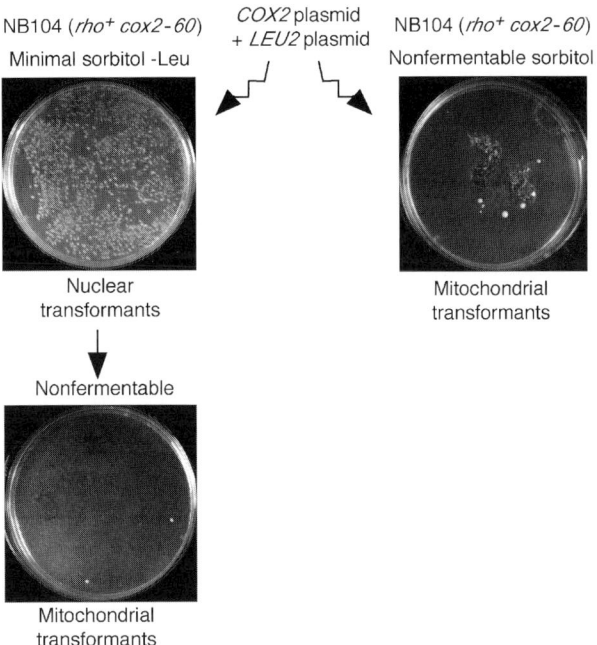

NB104 (*rho⁺ cox2-60*)   *COX2* plasmid   NB104 (*rho⁺ cox2-60*)
Minimal sorbitol -Leu   + *LEU2* plasmid   Nonfermentable sorbitol

Nuclear   Mitochondrial
transformants   transformants

Nonfermentable

Mitochondrial
transformants

Fig. 5 Selection of *rho⁺* mitochondrial transformants directly after yeast bombardment. The nuclear shuttle vector Yep351 (*LEU2*) and plasmid pNB69 (*COX2*) were bombarded together onto lawns of the *rho⁺ cox2-60* strain NB104 (see legend of Fig. 1). The bombarded lawns had been spread either on minimal medium supplemented with sorbitol but lacking leucine (top right), or on nonfermentable YPEG medium supplemented with sorbitol and 0.1% glucose (top left). Leu⁺ transformants were replica plated to nonfermentable medium (bottom plate) to select for mitochondrial transformants.

affects selection for respiring transformants with both nuclear (Müller and Fox, 1984) and mitochondrial genes (Johnston *et al.*, 1988). However, we have been able to successfully select respiring mitochondrial transformants in a single step by directly bombarding lawns of mutant *rho⁺* cells spread on YPEG plates supplemented with 0.1% glucose, to allow a brief period of outgrowth, and 1-M sorbitol (Fig. 5). We have also been able to select for *rho⁺* transformants expressing *ARG8ᵐ* by directly bombarding lawns spread on appropriate minimal glucose medium (see Section III.B.2, step c).

The detection of transformants of a *rho⁺* recipient that have incorporated a DNA sequence bearing a nonfunctional mutation is difficult, but possible, using essentially the same strategy as outlined in Fig. 4. If the recipient strain has a defined deletion, and integration of the transforming DNA restores the deleted sequence, then the desired integrants can be detected by their ability to rescue a known marker mutation in the deleted region.

VI. Concluding Remarks

The methods developed for manipulation of the *S. cerevisiae* mitochondrial genome have already been useful to set up an efficient technique for delivery of mtDNA into *Chlamydomonas* mitochondria and should also provide a useful model for other systems such as the petite-negative yeast *S. pombe*. So far, transformation of *Chlamydomonas* requires the selection of respiratory proficiency, and thus allows the creation of mutation in the *nd* genes, but not in *cob* or *cox1*. Although limited, this technique is of importance because this opens for the first time the way to *in vitro* mutagenesis of organelle genes for complex I, since *S. cerevisiae* does not contain a *bona fide* complex I. As appropriate selectable markers are developed for *Chlamydomonas* and other species, it seems likely that their full mitochondrial genomes will become amenable to *in vivo* experimental analysis, providing alternative model systems to *S. cerevisiae*.

Acknowledgments

We thank Anna Card Gay and Liza Pon for imaging analysis of the mitochondrial GFP yeast strain and David M. MacAlpine and Ronald A. Butow for advice on the use of gold particles. N.B. is supported by the Association Française contre les Myopathies, C.R. by Belgian Fonds de la Recherche Fondamentale Collective Grants 2.4587.04 and 2.4582.05, and T.D.F. by a grant from the US National Institutes of Health (GM29362). This work also benefits from a joint France-CGRI Wallonie/Bruxelles Tournesol grant to N.B. and C.R.

References

Azpiroz, R., and Butow, R. A. (1993). Patterns of mitochondrial sorting in yeast zygotes. *Mol. Biol. Cell* **4**, 21–36.

Bennoun, P., Delosme, M., and Kuck, U. (1991). Mitochondrial genetics of *Chlamydomonas reinhardtii*: Resistance mutations marking the cytochrome *b* gene. *Genetics* **127**, 335–343.

Bonnefoy, N., and Fox, T. D. (2000). *In vivo* analysis of mutated initiation codons in the mitochondrial *COX2* gene of *Saccharomyces cerevisiae* fused to the reporter gene *ARG8m* reveals lack of downstream reinitiation. *Mol. Gen. Genet.* **262**, 1036–1046.

Bonnefoy, N., Bsat, N., and Fox, T. D. (2001). Mitochondrial translation of *Saccharomyces cerevisiae* *COX2* mRNA is controlled by the nucleotide sequence specifying the pre-Cox2p leader peptide. *Mol. Cell. Biol.* **21**, 2359–2372.

Boulanger, S. C., Belcher, S. M., Schmidt, U., Dib-Hajj, S. D., Schmidt, T., and Perlman, P. S. (1995). Studies of point mutants define three essential paired nucleotides in the domain 5 substructure of a group II intron. *Mol. Cell. Biol.* **15**, 4479–4488.

Boynton, J. E., and Gillham, N. W. (1996). Genetics and transformation of mitochondria in the green alga *Chlamydomonas*. *Meth. Enzymol.* **264**, 279–296.

Broadley, S. A., Demlow, C. M., and Fox, T. D. (2001). Peripheral mitochondrial inner membrane protein, Mss2p, required for export of the mitochondrially coded Cox2p C tail in *Saccharomyces cerevisiae*. *Mol. Cell. Biol.* **21**, 7663–7672.

Butow, R. A., Henke, M., Moran, J. V., Belcher, S. M., and Perlman, P. S. (1996). Transformation of *Saccharomyces cerevisiae* mitochondria by the biolistic gun. *Meth. Enzymol.* **264**, 265–278.

Chen, X. J., and Butow, R. A. (2005). The organization and inheritance of the mitochondrial genome. *Nat. Rev. Genet.* **6,** 815–825.

Cohen, J. S., and Fox, T. D. (2001). Expression of green fluorescent protein from a recoded gene inserted into *Saccharomyces cerevisiae* mitochondrial DNA. *Mitochondrion* **1,** 181–189.

Conde, J., and Fink, G. R. (1976). A mutant of *S. cerevisiae* defective for nuclear fusion. *Proc. Natl. Acad. Sci. USA* **73,** 3651–3655.

Costanzo, M. C., and Fox, T. D. (1993). Suppression of a defect in the 5'-untranslated leader of the mitochondrial *COX3* mRNA by a mutation affecting an mRNA-specific translational activator protein. *Mol. Cell. Biol.* **13,** 4806–4813.

de Zamaroczy, M., Faugeron-Fonty, G., Baldacci, G., Goursot, R., and Bernardi, G. (1984). The ori sequences of the mitochondrial genome of a wild-type yeast strain: Number, location, orientation and structure. *Gene* **32,** 439–457.

Demlow, C. M., and Fox, T. D. (2003). Activity of mitochondrially synthesized reporter proteins is lower than that of imported proteins and is increased by lowering cAMP in glucose-grown *Saccharomyces cerevisiae* cells. *Genetics* **165,** 961–974.

di Rago, J. P., Perea, X., and Colson, A. M. (1986). DNA sequence analysis of diuron-resistant mutations in the mitochondrial cytochrome *b* gene of *Saccharomyces cerevisiae*. *FEBS Lett.* **208,** 208–210.

Dorthu, M. P., Remy, S., Michel-Wolwertz, M. R., Colleaux, L., Breyer, D., Beckers, M. C., Englebert, S., Duyckaerts, C., Sluse, F. E., and Matagne, R. F. (1992). Biochemical, genetic and molecular characterization of new respiratory-deficient mutants in *Chlamydomonas reinhardtii*. *Plant Mol. Biol.* **18,** 759–772.

Dujon, B. (1980). Sequence of the intron and flanking exons of the mitochondrial *21S* rRNA gene of yeast strains having different alleles at the *omega* and *rib-1* loci. *Cell* **20,** 185–197.

Dujon, B. (1981). Mitochondrial genetics and functions. *In* "The Molecular Biology of the Yeast *Saccharomyces*, Life Cycle and Inheritance" (J. N. Strathern, E. W. Jones, and J. R. Broach, eds.), pp. 505–635. Cold Spring Harbor Laboratory Press, Cold Spring Harbor, NY.

Duvezin-Caubet, S., Rak, M., Lefebvre-Legendre, L., Tetaud, E., Bonnefoy, N., and di Rago, J. P. (2006). A 'petite-obligate' mutant of *Saccharomyces cerevisiae*: Functional mtDNA is lethal in cells lacking the delta subunit of mitochondrial F1-ATPase. *J. Biol. Chem.* **281,** 16305–16313.

Fiori, A., Perez-Martinez, X., and Fox, T. D. (2005). Overexpression of the *COX2* translational activator, Pet111p, prevents translation of *COX1* mRNA and cytochrome *c* oxidase assembly in mitochondria of *Saccharomyces cerevisiae*. *Mol. Microbiol.* **56,** 1689–1704.

Folley, L. S., and Fox, T. D. (1991). Site-directed mutagenesis of a *Saccharomyces cerevisiae* mitochondrial translation initiation codon. *Genetics* **129,** 659–668.

Fox, T. D. (1987). Natural variation in the genetic code. *Annu. Rev. Genet.* **21,** 67–91.

Fox, T. D., Folley, L. S., Mulero, J. J., McMullin, T. W., Thorsness, P. E., Hedin, L. O., and Costanzo, M. C. (1991). Analysis and manipulation of yeast mitochondrial genes. *Meth. Enzymol.* **194,** 149–165.

Fox, T. D., Sanford, J. C., and McMullin, T. W. (1988). Plasmids can stably transform yeast mitochondria lacking endogenous mtDNA. *Proc. Natl. Acad. Sci. USA* **85,** 7288–7292.

Golik, P., Bonnefoy, N., Szczepanek, T., Saint-Georges, Y., and Lazowska, J. (2003). The Rieske FeS protein encoded and synthesized within mitochondria complements a deficiency in the nuclear gene. *Proc. Natl. Acad. Sci. USA* **100,** 8844–8849.

Green-Willms, N. S., Butler, C. A., Dunstan, H. M., and Fox, T. D. (2001). Pet111p, an inner membrane-bound translational activator that limits expression of the *Saccharomyces cerevisiae* mitochondrial gene *COX2*. *J. Biol. Chem.* **276,** 6392–6397.

Guthrie, C., and Fink, G. R. (eds.) (2002). Guide to yeast genetics and molecular and cell biology. *In* "Methods in enzymology," Vols. 350 and 351. Academic Press, San diego.

Hanson, M. R., and Folkerts, O. (1992). Structure and function of the higher plant mitochondrial genome. *Int. Rev. Cytol.* **141,** 129–172.

Harris, E. H. (1989). "The Chlamydomonas Sourcebook." Academic Press, San Diego.

He, S., and Fox, T. D. (1997). Membrane translocation of mitochondrially coded Cox2p: Distinct requirements for export of amino- and carboxy-termini, and dependence on the conserved protein Oxa1p. *Mol. Biol. Cell* **8**, 1449–1460.

He, S., and Fox, T. D. (1999). Mutations affecting a yeast mitochondrial inner membrane protein, Pnt1p, block export of a mitochondrially synthesized fusion protein from the matrix. *Mol. Cell. Biol.* **19**, 6598–6607.

Henke, R. M., Butow, R. A., and Perlman, P. S. (1995). Maturase and endonuclease functions depend on separate conserved domains of the bifunctional protein encoded by the group I intron *aI4* alpha of yeast mitochondrial DNA. *EMBO J.* **14**, 5094–5099.

Hill, J. E., Myers, A. M., Koerner, T. J., and Tzagoloff, A. (1986). Yeast/*E. coli* shuttle vectors with multiple unique restriction sites. *Yeast* **2**, 163–167.

Hinnen, A., Hicks, J. B., and Fink, G. R. (1978). Transformation of yeast. *Proc. Natl. Acad. Sci. USA* **75**, 1929–1933.

Johnston, S. A., Anziano, P. Q., Shark, K., Sanford, J. C., and Butow, R. A. (1988). Mitochondrial transformation in yeast by bombardment with microprojectiles. *Science* **240**, 1538–1541.

Lewin, A. S., Morimoto, R., and Rabinowitz, M. (1979). Stable heterogeneity of mitochondrial DNA in *grande* and *petite* strains of *S. cerevisiae*. *Plasmid* **2**, 474–484.

Li, M., Tzagoloff, A., Underbrink-Lyon, K., and Martin, N. C. (1982). Identification of the paromomycin-resistance mutation in the *15S* rRNA gene of yeast mitochondria. *J. Biol. Chem.* **257**, 5921–5928.

Meeusen, S., and Nunnari, J. (2003). Evidence for a two membrane-spanning autonomous mitochondrial DNA replisome. *J. Cell Biol.* **163**, 503–510.

Mireau, H., Arnal, N., and Fox, T. D. (2003). Expression of Barstar as a selectable marker in yeast mitochondria. *Mol. Genet. Genomics* **270**, 1–8.

Mittelmeier, T. M., and Dieckmann, C. L. (1993). *In vivo* analysis of sequences necessary for CBP1-dependent accumulation of cytochrome b transcripts in yeast mitochondria. *Mol. Cell. Biol.* **13**, 4203–4213.

Mulero, J. J., and Fox, T. D. (1994). Reduced but accurate translation from a mutant AUA initiation codon in the mitochondrial *COX2* mRNA of *Saccharomyces cerevisiae*. *Mol. Gen. Genet.* **242**, 383–390.

Müller, P. P., and Fox, T. D. (1984). Molecular cloning and genetic mapping of the *PET494* gene of *Saccharomyces cerevisiae*. *Mol. Gen. Genet.* **195**, 275–280.

Myers, A. M., Pape, L. K., and Tzagoloff, A. (1985). Mitochondrial protein synthesis is required for maintenance of intact mitochondrial genomes in *Saccharomyces cerevisiae*. *EMBO J.* **4**, 2087–2092.

Neff, N. F., Thomas, J. H., Grisafi, P., and Botstein, D. (1983). Isolation of the b-tubulin gene from yeast and demonstration of its essential function *in vivo*. *Cell* **33**, 211–219.

Nunnari, J., Marshall, W. F., Straight, A., Murray, A., Sedat, J. W., and Walter, P. (1997). Mitochondrial transmission during mating in *Saccharomyces cerevisiae* is determined by mitochondrial fusion and fission and the intramitochondrial segregation of mitochondrial DNA. *Mol. Biol. Cell* **8**, 1233–1242.

Ooi, B. G., Novitski, C. E., and Nagley, P. (1985). DNA sequence analysis of the *oli1* gene reveals amino acid changes in mitochondrial ATPase subunit 9 from oligomycin-resistant mutants of *Saccharomyces cerevisiae*. *Eur. J. Biochem.* **152**, 709–714.

Perez-Martinez, X., Broadley, S. A., and Fox, T. D. (2003). Mss51p promotes mitochondrial Cox1p synthesis and interacts with newly synthesized Cox1p. *EMBO J.* **22**, 5951–5961.

Perlman, P. S., Birky, C. W., Jr., and Strausberg, R. L. (1979). Segregation of mitochondrial markers in yeast. *Meth. Enzymol.* **56**, 139–154.

Pon, L., and Schatz, G. (1991). Biogenesis of yeast mitochondria. *In* "The Molecular and Cellular Biology of the Yeast *Saccharomyces*: Genome Dynamics, Protein Synthesis and Energetics" (J. R. Broach, J. R. Pringle, and E. W. Jones, eds.), Vol. 1, pp. 333–406 3 vols. Cold Spring Harbor Laboratory Press, Cold Spring Harbor.

Remacle, C., and Matagne, R. (1998). Mitochondrial genetics. *In* "The Molecular Biology of Chloroplasts and Mitochondria" (J. D. Rochaix, M. Goldschmidt-Clermont, and S. Merchant, eds.), pp. 661–674. Kluwer Academic Publishers, Dordrecht, The Netherlands.

Remacle, C., Cardol, P., Coosemans, N., Gaisne, M., and Bonnefoy, N. (2006). High-efficiency biolistic transformation of *Chlamydomonas* mitochondria can be used to insert mutations in complex I genes. *Proc. Natl. Acad. Sci. USA* **103**, 4771–4776.

Rochaix, J. D., Goldschmidt-Clermont, M., and Merchant, S. (eds.) (1998). *In* "The Molecular Biology of Chloroplasts and Mitochondria." Kluwer Academic Publishers, Dordrecht, The Netherlands.

Rothstein, R. (1991). Targeting, disruption, replacement and allele rescue: Integrative DNA transformation in yeast. *Meth. Enzymol.* **194**, 281–301.

Sanchirico, M. E., Fox, T. D., and Mason, T. L. (1998). Accumulation of mitochondrially synthesized *Saccharomyces cerevisiae* Cox2p and Cox3p depends on targeting information in untranslated portions of their mRNAs. *EMBO J.* **17**, 5796–5804.

Saracco, S. A., and Fox, T. D. (2002). Cox18p is required for export of the mitochondrially encoded *Saccharomyces cerevisiae* Cox2p C-tail and interacts with Pnt1p and Mss2p in the inner membrane. *Mol. Biol. Cell* **13**, 1122–1131.

Schmitt, M. E., and Clayton, D. A. (1993). Conserved features of yeast and mammalian mitochondrial DNA replication. *Curr. Opin. Genet. Dev.* **3**, 769–774.

Sebald, W., Wachter, E., and Tzagoloff, A. (1979). Identification of amino acid substitutions in the dicyclohexylcarbodiimide-binding subunit of the mitochondrial ATPase complex from oligomycin-resistant mutants of *Saccharomyces cerevisiae*. *Eur. J. Biochem.* **100**, 599–607.

Sia, E. A., Butler, C. A., Dominska, M., Greenwell, P., Fox, T. D., and Petes, T. D. (2000). Analysis of microsatellite mutations in the mitochondrial DNA of *Saccharomyces cerevisiae*. *Proc. Natl. Acad. Sci. USA* **97**, 250–255.

Sor, F., and Fukuhara, H. (1982). Identification of two erythromycin resistance mutations in the mitochondrial gene coding for the large ribosomal RNA in yeast. *Nucleic Acids Res.* **10**, 6571–6577.

Speno, H., Taheri, M. R., Sieburth, D., and Martin, C. T. (1995). Identification of essential amino acids within the proposed CuA binding site in subunit II of cytochrome *c* oxidase. *J. Biol. Chem.* **270**, 25363–25369.

Steele, D. F., Butler, C. A., and Fox, T. D. (1996). Expression of a recoded nuclear gene inserted into yeast mitochondrial DNA is limited by mRNA-specific translational activation. *Proc. Natl. Acad. Sci. USA* **93**, 5253–5257.

Strand, M. K., and Copeland, W. C. (2002). Measuring mtDNA mutation rates in *Saccharomyces cerevisiae* using the mtArg8 assay. *Methods Mol. Biol.* **197**, 151–157.

Strand, M. K., Stuart, G. R., Longley, M. J., Graziewicz, M. A., Dominick, O. C., and Copeland, W. C. (2003). *POS5* gene of *Saccharomyces cerevisiae* encodes a mitochondrial NADH kinase required for stability of mitochondrial DNA. *Eukaryot. Cell* **2**, 809–820.

Sulo, P., Groom, K. R., Wise, C., Steffen, M., and Martin, N. (1995). Successful transformation of yeast mitochondria with RPM1: An approach for in vivo studies of mitochondrial RNase P RNA structure, function and biosynthesis. *Nucleic Acids Res.* **23**, 856–860.

Szczepanek, T., and Lazowska, J. (1996). Replacement of two non-adjacent amino acids in the *S. cerevisiae bi2* intron-encoded RNA maturase is sufficient to gain a homing-endonuclease activity. *EMBO J.* **15**, 3758–3767.

Thomas, B. J., and Rothstein, R. (1989). Elevated recombination rates in transcriptionally active DNA. *Cell* **56**, 619–630.

Thorsness, P. E., and Fox, T. D. (1990). Escape of DNA from mitochondria to the nucleus in *Saccharomyces cerevisiae*. *Nature* **346**, 376–379.

Thorsness, P. E., and Fox, T. D. (1993). Nuclear mutations in *Saccharomyces cerevisiae* that affect the escape of DNA from mitochondria to the nucleus. *Genetics* **134**, 21–28.

Towpik, J., Chacinska, A., Ciesla, M., Ginalski, K., and Boguta, M. (2004). Mutations in the yeast *MRF1* gene encoding mitochondrial release factor inhibit translation on mitochondrial ribosomes. *J. Biol. Chem.* **279**, 14096–14103.

Vahrenholz, C., Riemen, G., Pratje, E., Dujon, B., and Michaelis, G. (1993). Mitochondrial DNA of *Chlamydomonas reinhardtii*: The structure of the ends of the linear 15.8-kb genome suggests mechanisms for DNA replication. *Curr. Genet.* **24,** 241–247.

Wallace, D. C. (1992). Diseases of the mitochondrial DNA. *Annu. Rev. Biochem.* **61,** 1175–1212.

Williams, E. H., Bsat, N., Bonnefoy, N., Butler, C. A., and Fox, T. D. (2005). Alteration of a novel dispensable mitochondrial ribosomal small-subunit protein, Rsm28p, allows translation of defective *COX2* mRNAs. *Eukaryot. Cell* **4,** 337–345.

Yamasaki, T., Kurokawa, S., Watanabe, K. I., Ikuta, K., and Ohama, T. (2005). Shared molecular characteristics of successfully transformed mitochondrial genomes in *Chlamydomonas reinhardtii*. *Plant Mol. Biol.* **58,** 515–527.

Zinn, A. R., Pohlman, J. K., Perlman, P. S., and Butow, R. A. (1987). Kinetic and segregational analysis of mitochondrial DNA recombination in yeast. *Plasmid* **17,** 248–256.

CHAPTER 27

Generation of Transmitochondrial Mice: Development of Xenomitochondrial Mice to Model Neurodegenerative Diseases

Carl A. Pinkert* and Ian A. Trounce[†]

*Department of Pathobiology
College of Veterinary Medicine
Auburn University
Auburn, Alabama 36849

[†]Centre for Neuroscience
University of Melbourne
Victoria, 3010 Australia

I. Introduction
II. Spontaneous and Induced Models of Mitochondrial Disease
III. Transgenic Models
IV. Introduction of Mutant mtDNA into Mitochondria
   A. Use of Transfected ρ^0 Cells as Intermediate Mitochondrial Carriers in the Production of Mouse Models
   B. Embryonic Stem Cell Technology
V. The First Transmitochondrial Mice
   A. mtDNA Injection Versus ES Cell-Derived Models
   B. Xenomitochondrial Mice
   C. Neurodegenerative Disorders
VI. Summary/Future Directions
   References

I. Introduction

Manipulation of the mitochondrial genome *in vitro* was technically feasible by the mid-1980s. Both human and murine mitochondrial genomes were cloned, sequenced, and available as constituents of plasmid constructs (Bigger *et al.*, 2000; Clayton, 1991).

METHODS IN CELL BIOLOGY, VOL. 80
Copyright 2007, Elsevier Inc. All rights reserved.

0091-679X/07 $35.00
DOI: 10.1016/S0091-679X(06)80027-0

The differences between nuclear and mitochondrial codon assignments were also identified, and the general features associated with mitochondrial targeting of nuclear-encoded mitochondrial proteins were described (Bibb *et al.*, 1981; Pfanner and Neupert, 1990a,b; Tapper *et al.*, 1983).

By the late 1990s, a handful of laboratories focused on methods to introduce foreign or altered mitochondrial genomes into somatic or germ cells, with a goal of employing genetically engineered mitochondria or mitochondrial genes for *in vivo* gene transfer or for gene therapy. Pioneering work in the development of cybrid technology, first using mtDNA mutants resistant to chloramphenicol (Wallace *et al.*, 1975), and then using mtDNA-less (ρ^0) cells, made the establishment of transmitochondrial mouse models inevitable (King and Attardi, 1989). A cybrid, by definition, can harbor more than one type of mtDNA. This coexistence of more than one form of mtDNA within a single cell or within cells of a tissue, organ, or individual organism is referred to as "heteroplasmy" (in contrast to maintenance of a single population of mtDNA genomes, referred to as "homoplasmy"). Early cybrid experiments demonstrated that multiple mitochondrial genotypes can be maintained within cells and that selective segregation and amplification of one genotype often occur (Hayashi *et al.*, 1986).

At present, there are scores of mutations of the mitochondrial genome that are known to be the underlying causes of various degenerative disorders. These mtDNA mutations, many of which exist in a heteroplasmic state, mainly affect tissues with high cellular energy requirements, such as brain, optic nerve, cardiac muscle, skeletal muscle, kidney, and endocrine organs (for review see Smeitink *et al.*, 2001). Mitochondrial function and mtDNA integrity are known to decline in human aging; increasing evidence points to the importance of mitochondrial failure in the common neurodegenerative diseases of aging (Howell *et al.*, 2005; Schon and Manfredi, 2003). The creation of transmitochondrial mice harboring introduced species-specific mitochondrial populations (our xenomitochondrial or "xenomice" model) represents a new model system that will provide a foundation to study mitochondrial function and dynamics in processes including aging and oxidative stress, and mitochondrial disease pathogenesis, leading to therapeutic strategies for human metabolic diseases affected by aberrations in mitochondrial function or mutation.

II. Spontaneous and Induced Models of Mitochondrial Disease

As late as 1996, animal models of heritable mitochondrial disease, including encephalomyopathy, were not available (Tracey *et al.*, 1997). Therefore, early studies employed models of mitochondrial dysfunction that were induced by treatment with respiratory chain inhibitors and other agents. For example, chronic administration of compounds, such as diphenylene iodonium, to rats produced phenotypes similar to those observed in known myopathies. Moreover, administration of germanium dioxide to rats produced hallmarks of mitochondrial disorders, including accumulation of ragged-red fibers with decreased cytochrome c oxidase (COX) activity (Cooper *et al.*, 1992; Higuchi *et al.*, 1991). Finally, neurodegeneration modeling and mitochondrial dysfunction have been observed following chronic exposure of mice to the pesticide MPTP$^+$ and rats to rotenone—both inhibitors of complex I (Betarbet *et al.*, 2000; Heikkila *et al.*, 1984). Here, specific inhibition of dopaminergic neurons was observed, resembling the Parkinson disease phenotype in humans, and reflective of epidemiological data related to exposure to environmental toxins.

A number of spontaneously arising rat models have been used to study nutritional regulation and mitochondrial function. The BHE rat was identified early as a unique non-obese model for diabetes mellitus (Berdanier, 1982, 1991; Lakshmanan *et al.*, 1977). Most rodents that develop insulin-independent diabetes mellitus spontaneously are obese. The BHE rat develops abnormal glucose tolerance and is lipemic with fatty liver, but is not obese (Berdanier, 1991). The BHE/Cdb rat and related models, which were studied in the 1990s, paved the way for current hypotheses related to nutritional regulation of mitochondrial function in mammals. This model is salt sensitive and carries two homoplasmic and three heteroplasmic mutations in the mitochondrial ATPase 6 gene, which result in impaired oxidative phosphorylation (OXPHOS) and glucose intolerance. The mutations responsible for these phenotypes were isolated and studied, leading to a molecular understanding of the structure–function relationships within the ATPase 6 gene (Everts and Berdanier, 2002; Mathews *et al.*, 2000).

There have also been a number of mouse strains with spontaneously arising mutations that affect mitochondrial functions. The copper-deficient *brindled* (*mottled-brindled*, *Mobr*, or *mo^{br}*; now *Atp7a^{Mo-br}*) mouse is normally considered a model of Menkes disease, with decreased COX enzymatic activity (Menkes *et al.*, 1962). COX and copper–zinc superoxide dismutase (SOD1) were compared in copper-deficient *brindled* and *blotchy* (*mottled-blotchy*, *Moblo*, or *Atp7a^{Mo-blo}*; also an osteoarthritis model) mouse mutants and normal mice (Phillips *et al.*, 1986). In *brindled* and *blotchy* mutants, enzyme deficiencies were identified in brain, heart, and skeletal muscle but not in liver, kidney, and lung, despite abnormal copper concentrations in unaffected tissues. However, COX activity was more severely affected than that of SOD1. Moreover, the observed COX deficiency correlated with decreased copper concentration and was comparable to that observed in nutritionally copper-deficient mice. Interestingly, administration of copper increased COX activity in all deficient tissues of *brindled* mice, but only in brain and heart of *blotchy* mice. However, skeletal–muscle COX activity in *blotchy* mutants did not respond to copper injection, indicative of differential developmental pathways.

In 1998, owing to the lack of existing animal models of mitochondrial disease, normal human myoblast cells were injected into muscle of SCID mice following induction of muscle necrosis to identify whether normal human cells would repopulate the impaired mouse tissue and restore function (Clark *et al.*, 1998). Post-injection, regenerated muscle fibers expressing human β-spectrin and subunit II of human COX were observed, illustrating a successful and localized reconstitution of human myoblast-derived mitochondria.

In the aggregate, a large data set has accumulated over the years that readily reflects mitochondrial dysfunction in animal models—and many of these models have been characterized at the molecular level following their initial description in the literature. All of the mutations in existing models affect proteins that are encoded in the nucleus, not the mtDNA genome. Indeed, there have been no reports of naturally occurring or spontaneously arising mtDNA mutants in laboratory mouse lines. The demonstration of methodologies to tailor mitochondrial genetics and mitochondrial:nuclear dynamics to the development of animal models was the next logical progression in efforts to study mitochondrial biology *in vivo*.

III. Transgenic Models

As described in a number of reports (Wallace, 2001), mouse knockout models of mutations in nuclear-encoded genes have been informative in studying pathogenesis of mitochondrial disease and critical pathways associated with mitochondrial function. There are hosts of characterized models, including those associated with ablation of the genes encoding the adenine nucleotide translocator-1 (Ant1), manganese superoxide dismutase (MnSOD; SOD2), mitochondrial transcription factor A (Tfam), glutathione peroxidase (GPx1), and uncoupling proteins 1, 2, and 3 (UCP1–3). As examples of the pathophysiology of knocking out key regulatory enzymes, Ant1-deficient mice exhibited severe mitochondrial myopathies. Two different models of MnSOD-deficient mice produced cardiomyopathy, neuronal degeneration, and liver dysfunction. Tfam-deficient mice exhibited a reduction in complexes I and IV activities. GPx1-deficient mice were viable but displayed growth retardation and liver-specific, but not heart, pathology. Lastly, UCP1, UCP2, or UCP3-deficient mice illustrated the nature of compensatory function when one or two of the three proteins were ablated, with phenotypes ranging from cold sensitivity to an increase in the generation of reactive oxygen species. With the potential interaction of literally a few thousand nuclear genes critical to metabolic function and normal mitochondrial activity, it is not surprising to see the large body of literature is developing from nuclear gene overexpression, knockout, knock-in, and conditional-expression models.

IV. Introduction of Mutant mtDNA into Mitochondria

To produce mice bearing mutant mitochondria or mtDNA, there is a fundamental need for efficient methods to introduce either (1) foreign or modified mtDNA or (2) intact foreign mitochondrial genomes into somatic or germ cells. However, outside of recapitulating a mitochondrial genomic deletion mutant (Irwin *et al.*, 2001), methods for the experimental manipulation of mtDNA genomes or for delivery of complete foreign mtDNA genomes into intact mitochondria in animal cells have not been developed (Collombet *et al.*, 1997; Irwin *et al.*, 2001).

The transfection of DNA into mitochondria presents some very formidable challenges. For example, transfection requires penetration of both the outer and inner mitochondrial membranes. Moreover, since the protein and lipid compositions of mitochondrial membranes are very different from that of the plasma membrane (Adachi *et al.*, 1994; de Kroon *et al.*, 1997; Hovius *et al.*, 1993; reviewed by Ellis and Reid, 1994), methods for introduction of DNA into cells may not be useful in introducing DNA into mitochondria. For example, liposome-mediated DNA delivery systems may not be useful for mtDNA transformation. Although *in vitro* fusion of inner mitochondrial membrane vesicles was reported, there have been no reports of fusion of outer membrane vesicles or whole mitochondria (Hackenbrock and Chazotte, 1986). Similarly, using mitochondrial-targeting peptides covalently attached to DNA molecules, Seibel *et al.* (1995) have successfully introduced a 322-bp DNA fragment into intact mitochondria via a protein import pathway. The import of a restriction endonuclease by a similar method, both *in vitro* and *in vivo*, has been shown to be feasible (Bayona-Bafaluy *et al.*, 2005). However, delivery of mitochondrial-specific DNA sequences and subsequent integration into the host mitochondrial genome by restriction endonuclease mediated integration (REMI) has not been reported, possibly because recombinations of mtDNA may be rare events or occur by mechanisms that are different from other DNA recombination events.

We have had some success with transfection of mitochondria by electroporation. This method was not successful for transfection with full length mtDNA molecules, and therefore may not be useful for models for many common human mtDNA-based diseases (e.g., LHON, MELAS, MERRF). However, using electroporation, we introduced a closed circular mtDNA molecule less than 7.2 kb, which contained O_H and O_L, into isolated mitochondria. Moreover, transfected mitochondria remained viable after transfer into recipient zygotes. We also used electroporation to introduce a 6.5-kb mouse mtDNA bearing a deletion, designated mtDNA-del, into isolated mitochondria. Transfected mitochondria were shown to be intact and coupled by closed-chamber respirometry, according to respiratory control ratio (RCR) and P/O ratio parameters. Unfortunately, when transfected mitochondria were introduced into zygotes and used for transmitochondrial mouse preparation, transfected mitochondria were not detectable in liveborn offspring (Irwin *et al.*, 2001, unpublished data).

A. Use of Transfected ρ^0 Cells as Intermediate Mitochondrial Carriers in the Production of Mouse Models

In the absence of methods for mtDNA transfection, cells bearing foreign mitochondria and mtDNA can be produced by microinjection of isolated mitochondria into cells of interest. To circumvent problems associated with variability and instability of mitochondrial populations in heteroplasmic cells, microinjection can be carried out in ρ^0 cells, which contain no mtDNA and require uridine and pyruvate in culture media for their survival (King and Attardi, 1989). Previous studies revealed that human cybrid cell lines containing mutant mtDNA for a number of mtDNA-related diseases could be produced by introduction of isolated patient mitochondria into ρ^0 cells (Chomyn *et al.*, 1991; Hayashi *et al.*, 1991, 1993; Trounce *et al.*, 1994, and in many reports since).

To develop transmitochondrial mice, we would take advantage of break-throughs in embryonic stem (ES) cell-based technologies and the work of Levy *et al.* (1999), where ES cells were used to transfer a foreign mitochondrial genome into mouse blastocysts and immortalized human and mouse ρ^0 cell lines (Bai and Attardi, 1998; King and Attardi, 1989; Trounce *et al.*, 2000). Here, isolated mitochondria were introduced into immortalized ρ^0 cells, by microinjection or cytoplast fusion, and cells bearing mitochondria were selected by their ability to grow in the absence of uridine and pyruvate in the culture medium. This selection can be employed prior to clonal expansion and subsequent analyses for mitochondrial mutants. In the case of the mtDNA deletion constructs, where complementation of respiratory chain activity would not be expected, the selection would not be used. Cells bearing mtDNA deletions can be identified by PCR analysis, as described by King and Attardi (1988). Such methodology allows for enhanced survival rates of transfected mitochondria, enhanced concentration of modified mitochondria, and serves as a culture system to propagate mitochondria and allow for controlled culture conditions to facilitate mitochondrial engineering.

An important advantage of this intermediate culture system is that heteroplasmic constructs can also be produced by mixing cells of interest in the enucleation step, then fusing the mixed cytoplasts (in different ratios, if desired), and genotyping many cybrid clones to select heteroplasmic clones. These clones can then be used in fusion with the R6G-treated ES cells, as described by McKenzie *et al.* (2004).

B. Embryonic Stem Cell Technology

ES cell transfer techniques have been used to produce random and targeted insertions; the latter have also been used for modification or ablation of discrete DNA sequences in the mouse genome. Compared to DNA microinjection for targeted insertion, the gene-targeting efficiency is extremely low (Brinster, 1993; Cappechi, 1989). The use of ES cell transfer into mouse embryos has been quite effective in allowing one to identify a specific genetic modification (a targeted event), via homologous recombination, at a precise chromosomal position. This selection capability has led to the production of mice that have incorporated a particular gene randomly within their genome, "knock-in" mice that carry modified endogenous genes and knockout models that lack a specific endogenous gene following targeting techniques. Indeed, technologies involving ES cells and primordial germ cells (PGCs) have been used to produce a host of useful and exciting mouse models (Brinster, 1993; Doetschman, 2002; Pinkert *et al.*, 2004; Rucker *et al.*, 2002).

Pluripotential ES cells are derived from early preimplantation embryos and are maintained in culture for a sufficient period for one to perform various *in vitro* manipulations (Figs. 1 and 2). The genome of ES cells can be manipulated *in vitro* by introducing foreign genes or foreign DNA sequences by techniques including electroporation, microinjection, precipitation reactions, transfection, or retroviral insertion (Nagy *et al.*, 2003; Pinkert, 2002). The use of ES cells to produce transgenic mice faced a number of procedural obstacles before it became competitive with DNA microinjection as a standard technique in mouse modeling. Within the last few years, the addition of coculture techniques involving tetraploid host embryos (four-cell stage to morulae) has resulted in founders that can be derived completely from the cocultured ES cells (Nagy *et al.*, 2003; Wood *et al.*, 1993). Hence, the founders are no longer chimeras, as all the cells come from the same progenitor cells, and the founder animals will breed truly (and transmit the genetic modification in the first generation offspring faithfully).

V. The First Transmitochondrial Mice

Early attempts to create transmitochondrial strains of mammalian species by introduction of foreign mitochondria into germ cells were not successful at generating heteroplasmic mice. However, they did illustrate important considerations in the development of a microinjection methodology (Ebert *et al.*, 1989). In experiments designed to mimic the transfer of mitochondria from spermatozoa during mammalian fertilization, introduction of ~120 intact mitochondria into fertilized murine zygotes did not result in detectable levels of foreign mtDNA in the offspring. Because these early experiments employed detection techniques that were less sensitive than PCR-based methods (e.g., Southern blot hybridization analyses), and because of the vast discrepancy in the number of foreign mitochondria microinjected versus the number known to be present in the average mouse ovum, it was quite possible that heteroplasmic or "transmitochondrial" offspring were indeed produced in these experiments. However, similar to sperm mitochondria that are programmed for destruction in fertilized oocytes, mitochondrial transfer efforts may have reflected a biological cascade of events critical to the fertilization/post-fertilization period (Cummins, 1998; Cummins *et al.*, 1999).

Techniques to create heteroplasmic mice then progressed via cytoplast fusion (Jenuth *et al.*, 1996, 1997) and by embryonic karyoplast transplantation (Meirelles and Smith, 1997). In these experiments, rapid segregation of mtDNA was possible within maternal lineages; however, specific manipulations were not readily controllable or quantifiable in first-generation animals. Generally, the recipient embryo appeared to dominate in terms of mitochondrial survival. The mechanisms for eliminating introduced mitochondria in eukaryotic organisms are extraordinarily diverse and mysterious. Nonetheless, any hint of heteroplasmy seems to be highly nonadaptive and is rapidly eliminated by selection processes (Cummins, 1999).

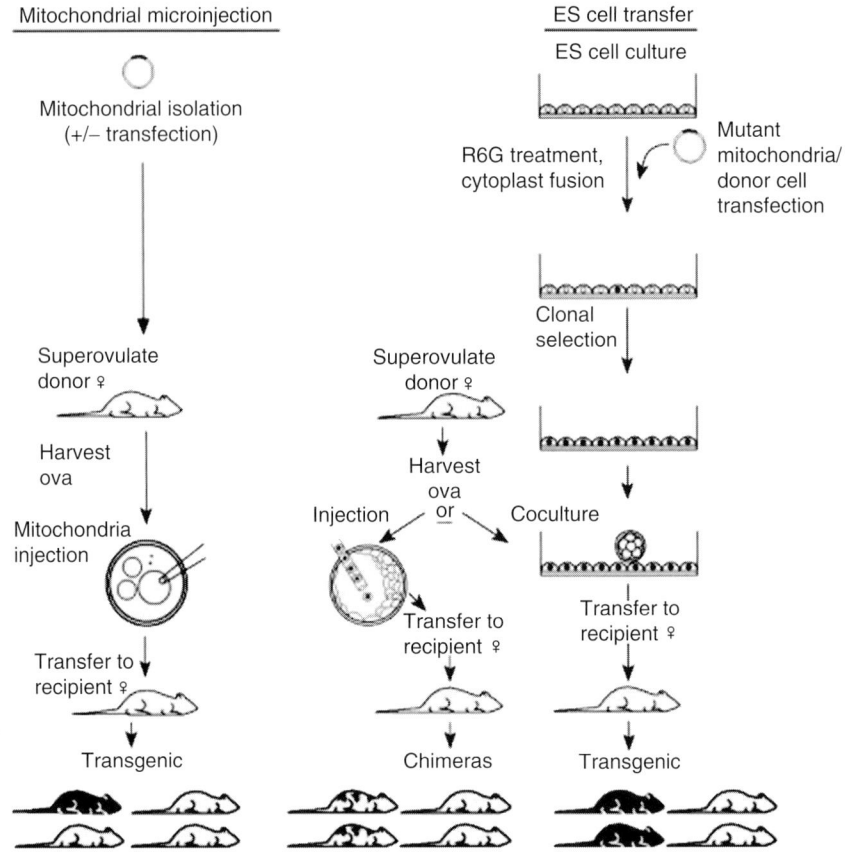

Fig. 1 Production of transmitochondrial mice. Mitochondrial injection/fusion and ES cell transfer technologies were used to produce transmitochondrial mice. A principal difference in the two techniques is the *in vitro* culture step allowing propagation of targeted clones using immortalized ES cells (right). On the left, for direct microinjection (or fusion) of intact mitochondria, the mitochondrial preparation is injected and/or electrofused into the embryo. Microinjection of mitochondria into the cytoplasm of the pronuclear unicellular fertilized ovum (zygote) causes expansion of the vitelline membrane/boundary, occasionally with the appearance of the extranuclear vacuole, as depicted. In general, using mitochondrial injection or fusion techniques, transgenic mice (or more specifically transmitochondrial mice), represented by the all-black mouse, possess heteroplasmic cells (harboring mutant and wild-type mitochondrial genomes). On the right, after clonal selection of transfected ES cells, one of two additional techniques is used for ES cell transfer. Ova are harvested between the eight-cell and blastocyst stages. R6G-treated and transfected ES cells are either injected directly into a host blastocyst (injection) or cocultured with eight-cell to morula stage ova, so that transfected ES cells are preferentially incorporated into the inner cell mass of the developing embryo (coculture). With blastocyst injection, transgenic offspring are termed chimeric, because some of their cells are derived from the host blastocyst and some from transfected ES cells (denoted by white mice with black patches). Using coculture and tetraploid embryos, one can obtain founder mice derived completely from the transfected ES cells (denoted as all-black mice) (Pinkert and Trounce, 2002).

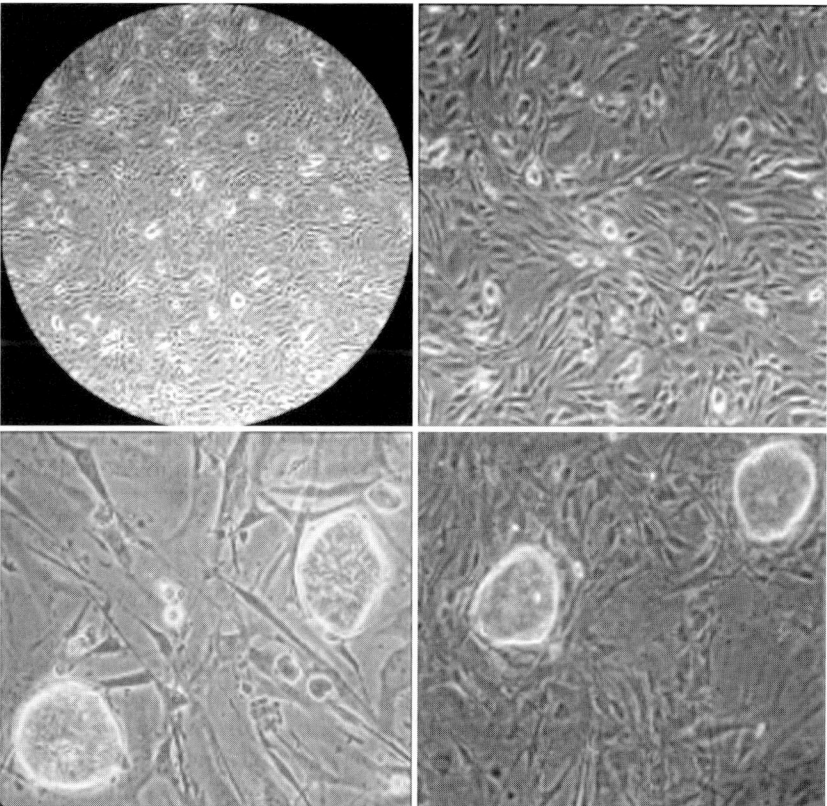

Fig. 2 ES cell culture. ES cells are cultured on gelatin and/or murine embryonic fibroblasts (MEFs) in high glucose containing DMEM supplemented with 15% fetal calf serum, L-glutamine, nonessential amino acids, β-mercaptoethanol, and leukemia inhibitory factor (Doetschman, 2002). ES cells form dense, rounded colonies with distinct edges that appear to "glow." The appearance of flattened colonies, fibroblast-like outgrowth, and irregular edges are evidence of differentiation. Upper panels: 40×, Lower panels: 200× (C. A. Cassar, J. L. Littleton, and C. A. Pinkert, unpublished data).

Interestingly, early reports on the development of cloned animals by nuclear transfer resulted in conflicting consequences when mitochondrial transmission was studied retrospectively (Evans *et al.*, 1999; Hiendleder *et al.*, 1999; Steinborn *et al.*, 2000; Takeda *et al.*, 1999). Similarly, using a human *in vitro* fertilization protocol, two heteroplasmic children may have been created inadvertently (Barritt *et al.*, 2001; Cohen *et al.*, 1997). As such, research independent of targeted mitochondrial genomic modifications may also help unlock mechanisms underlying the dynamics related to the persistence of foreign mitochondria and the maintenance of heteroplasmy in various cloning protocols. Indeed, specific culture-related conditions that influence the prevalence or development of heteroplasmy in these techniques may explain the range of results (heteroplasmy to homoplasmy) observed in nuclear transfer experimentation (Takeda *et al.*, 2002).

For *in vivo* modeling, a number of laboratories have described methodologies used to create transmitochondrial mouse models (Inoue *et al.*, 2000; Irwin *et al.*, 1999; Levy *et al.*, 1999; Marchington *et al.*, 1999; Pinkert *et al.*, 1997; Sligh *et al.*, 2000). All of these reports illustrated aspects of *in vivo* modeling of mitochondrial dynamics and human diseases. Interestingly, in some of the reports where developmental consequences of the genetic manipulations were observed, either aberrant or unexpected phenotypes were frequently encountered (Inoue *et al.*, 2000; Sligh *et al.*, 2000). Our efforts to devise a direct mitochondria-transfer technique offered a number of additional advantages (Fig. 1) compared to karyoplast and cytoplast methods, including the potential to engineer isolated mitochondria used for transmitochondrial mouse production (Irwin *et al.*, 2001).

A. mtDNA Injection Versus ES Cell-Derived Models

Early studies revealed that germ line transmission of foreign mitochondria that were injected directly into zygotes could occur (Irwin *et al.*, 1999; Pinkert *et al.*, 1997) using cytoplast injection and *in vivo* fusion studies (Laipis, 1996). Unfortunately, high levels of transmission of foreign mitochondria, detected as heteroplasmy, in founders or offspring, were not observed, perhaps reflecting the somatic cell origin of the transferred mitochondria. As a result, it was difficult to determine the fate of these foreign mitochondria in transmitochondrial mouse offspring during embryonic development and postnatal life (e.g., phenotypic changes associated with aging).

Subsequently, higher levels of foreign mitochondrial transmission and heteroplasmy were observed when ES cell were transferred into morula- or blastocyst-stage embryos. Levy *et al.* (1999) and Marchington *et al.* (1999) demonstrated independently that it was possible to fuse cytoplasts prepared from a chloramphenicol-resistant (CAP^R) cell line with mouse ES cells and then introduce the mutant cells into mouse blastocysts. Offspring exhibited respiratory defects expected from the CAP^R mutation (Levy *et al.*, 1999; Sligh *et al.*, 2000). Unexpectedly, chimeric founders exhibited ocular abnormalities and cataracts. Offspring of founder chimeras exhibited either heteroplasmy or homoplasmy for the introduced mutation, with striking developmental abnormalities (ranging from severe growth retardation to perinatal or *in utero* lethality).

Although the chloramphenicol selection method was an important development, this method was not sufficient to eliminate endogenous ES cell mitochondria from ES cell cybrids or the resulting transmitochondrial mice (Marchington *et al.*, 1999). An important innovation was in the use of rhodamine-6G (R6G, see Cannon *et al.*, 2004 for a review) to reduce transmission of endogenous ES cell mtDNAs, something that had been demonstrated previously only for somatic cells (Trounce and Wallace, 1996). While these initial studies did not involve a synthesized/engineered mitochondrial genome per se, they illustrated that such modeling would be feasible as a critical component in developing targeted animal models of human disease.

Finally, another limitation in development of mtDNA mutant mouse models is the lack of suitable mouse cell mutants. A handful of mutants were described over the past 20 years, all produced by *in vitro* mutagenesis followed by selection of drug-resistant mtDNA mutants (see Trounce and Pinkert, 2005). While some of these may prove interesting if successfully introduced into mice, Sligh *et al.* (2000) found that the CAP[R] mutation which results in mild impairment of respiration that were measurable using polarography (Levy *et al.*, 1999), resulted in the *in utero* or perinatal death of animals with a high prevalence of heteroplasmy. Data on respiratory chain function are lacking for many of the mouse cell mutants described, but when available, suggest that such mutant phenotypes may be severe (Bai and Attardi, 1998; Howell, 1990).

This suggested that mutations with mild respiratory chain impairment may be most useful in modeling human disorders. This was the basis for developing a "mitochondrial genomic gradient," by generating trans-species mitochondrial models based on relative amino acid divergence between related species of mtDNA-encoded proteins (Tables I and II). Preliminary analyses compared mtDNA sequences in published GenBank accessions and our analyses of the NADH dehydrogenase subunit 1 (ND1) of *Mus musculus domesticus*. Six substitutions were identified (Table II). As nine amino

Table I
Amino Acid Differences in Species-Specific mtDNA Translation Products[a]

		Estimated number of amino acid replacements compared with *M. musculus*				
mt gene product	No. of amino acids	*M. spretus*	*M. dunni*	*M. caroli*	*M. pahari*	*Rattus*
ND1	318	5	9	10	15	32
ND2	345	14	24	27	41	88
COX I	514	2	4	4	7	14
COX II	227	1	1	1	1	3
ATP8	67	2	4	4	7	14
ATP6	226	2	3	4	6	12
COX III	261	1	2	3	4	9
ND3	114	2	4	5	7	15
ND4L	97	2	4	4	7	14
ND4	459	9	16	18	27	59
ND5	607	21	36	41	61	133
ND6	172	5	9	11	16	34
Cyt *b*	**381**	**4**	**7**	**8**	**12**	**26**
Totals	3785	70	123	140	211	453

[a]Values for cytochrome *b*, shown in bold, are from complete sequences. Other values are extrapolations based on the relative divergence from *Rattus norvegicus* (*Rattus*). Full mtDNA sequences are required to define the exact amino acid differences (Trounce *et al.*, 2004).

Table II

M. *dunni* and M. *musculus* ND1 Amino Acid Sequence Variation[a]

Amino acid position	M. *dunni*	M. *musculus*
2	Y	F
157	S	N
175	I	L
176	I	L
311	L	T
317	H	Y

[a]Nucleic acid similarity to M. *m. domesticus* is 92%; amino acid similarity is 98% based on M. *m. domesticus* published nucleotide sequence data (accession AB042432). Amino acid information is derived from translated nucleotide sequence. A series of overlapping PCR fragments containing the complete ND1 gene amplified from M. *dunni* homoplasmic cell line DNA were sequenced using BigDye® Terminator chemistry (Applied Biosystems). Fragments were assembled and analyzed using Vector NTI Advance™ 9.0 (Informax™, Invitrogen™ software) (C. A. Cassar, W. K. Pogozelski, I. A. Trounce, and C. A. Pinkert, unpublished data).

acid changes were predicted for this subunit in M. *dunni* (Table I), our initial predictions were sound and reflective of the anticipated sequence divergence.

B. Xenomitochondrial Mice

Another approach to create mice with modified mitochondrial backgrounds is to introduce species-specific mitochondria into zygotes, that is, to develop xenomitochondrial cybrids for creation of chimeric mice. We found that *Mus spretus* mitochondria could be injected into wild-type mouse (M. *m. domesticus*) zygotes (Pinkert *et al.*, 1997). Subsequently, a different approach, based on interprimate transfer studies (Kenyon and Moraes, 1997), showed that mouse cells can be repopulated with rat mtDNA and that this produces severe respiratory defects (Dey *et al.*, 2000; McKenzie and Trounce, 2000).

The pioneering xenocybrid studies could be exploited, utilizing xenocybrids from several mouse species, to evaluate a graded respiratory impairment *in vivo* in mouse models (McKenzie *et al.*, 2004; Trounce *et al.*, 2004). Taking advantage of the enormous evolutionary diversity of *Muridae* species, it may be possible to create a series of cybrids with increasing levels of respiratory chain impairment resulting from the presence of mismatched nuclear-mtDNA subunits. Modeling of the defects was first done in a ρ^0 cybrid system, and lineages which exhibit the desired phenotype (e.g., a mild respiratory impairment) could be transferred readily into mouse female ES cells for production of chimeric mice.

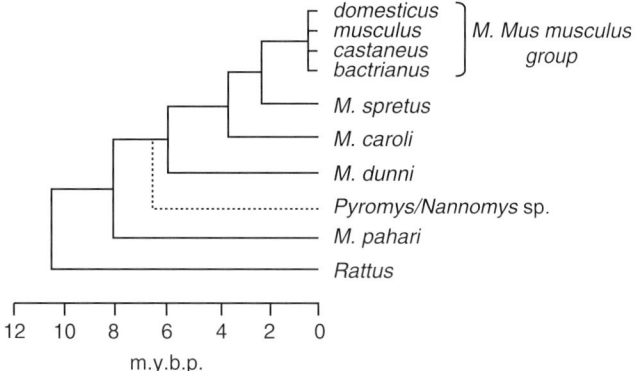

Fig. 3 Murinae phylogeny. Dendrogram illustrating evolutionary relationships of mouse species used in construction of xenomitochondrial cybrids (four underlined species), and related species of interest. The approximate divergence times are identified in million years before present (m.y.b.p.) (McKenzie and Trounce, 2000; Pinkert and Trounce, 2002).

Over the past 5 years, we have worked with collaborators to test the model that increasing evolutionary divergence (Fig. 3) results in an increase in the number of mismatched amino acids in the chimeric nuclear/mitochondria-encoded respiratory chain complexes I, III, and IV (McKenzie and Trounce, 2000). We showed that more divergent xenocybrid mouse L-cell constructs, carrying *Rattus* or *Otomys irroratus* mtDNA, express a severe complex III defect, while less diverged cybrids expressed mild complexes I and IV defects (McKenzie *et al.*, 2003). We have since targeted *Mus caroli* and *Mus pahari* from that study, which are most closely related to *M. musculus*. These species should display mild functional defects in both complexes I and IV, but since complex I exerts greater rate limitation on respiratory flux (Bianchi *et al.*, 2004), it is the complex I defects which are more likely to be expressed. Our *in vivo* modeling efforts have focused on the *Mus dunni: M. m. domesticus* mismatch.

While a backcrossed congenic strain harboring *M. spretus* mtDNA in a *M. m. domesticus* nuclear background was reported (Gyllensten *et al.*, 1985), note that, with the exception of *M. spretus*, other targeted mouse species exhibit karyotypic differences that preclude breeding (e.g., *M. pahari* has 48 chromosomes, while all sub-genera *Mus Mus* species have 40) (Auffray *et al.*, 2003). We also note the report of Roubertoux *et al.* (2003), who made a congenic backcross between mice with the common *M. musculus* laboratory strain mtDNA. While many inbred laboratory strains have mtDNA sequence identity and probably derive from a common female ancestor (Ferris *et al.*, 1982), New Zealand Black (NZB) mice exhibited the most divergent mtDNA among the laboratory strains.

Remarkably, differences between the congenic strain in several key brain morphological measures and behaviors were found, suggesting that increased numbers of mtDNA polymorphisms can have significant neuropsychiatric effects. The NZB mtDNA differs by only 15 amino acids compared with the common laboratory mouse mtDNA (Loveland *et al.*, 1990), whereas our constructs are predicted to differ by ~70 amino acids (*M. spretus*), 120 amino acids (*M. dunni*), and 210 amino acids (*M. pahari*) (see Table I).

We generated viable germ line mice with homoplasmic replacement of the *M. musculus* mtDNA with either *M. spretus* or *M dunni* mtDNA (McKenzie *et al.*, 2004). The major hurdle faced was the marginal germ line competency of the female ES cells used. We used a line derived by Alan Bradley (CC9.3.1), previously used by Wallace's group to produce transmitochondrial mice harboring a mutation in one mtDNA-encoded rRNA gene which confers chloramphenicol resistance (Sligh *et al.*, 2000). To date, this remains one of the most effective of the characterized female ES cell lines used to produce mice (Levy *et al.*, 1999; McKenzie *et al.*, 2004; Sligh *et al.*, 2000). Our efforts, following a large number of blastocyst injections using several cybrid clones, led to the production of over five dozen founder chimeras, with five germ line-competent founder females identified (McKenzie *et al.*, 2004; Trounce *et al.*, 2004, unpublished data).

In a follow-up report (Trounce *et al.*, 2004), we described our first female germ line animal that harbored *M. dunni* mtDNA in a *M. m. domesticus* nuclear background. Unfortunately, it is difficult to compare our results with those obtained in other studies because different ES cell lines were used. Most traditional mouse ES cell lines were derived from 129 strain mice, based on the strain's proclivity for generating ES cells that populate the germ line preferentially (Doetschman, 2002; Festing *et al.*, 1999; Nagy *et al.*, 2003; Threadgill *et al.*, 1997). However, for most ES cell transfer experiments, use of C57BL/6 embryos (or hybrids derived from C57BL/6) has been crucial to experimental success. While both 129 and C57BL/6 represent *M. m. domesticus* strains; strain-specific nuclear modifiers will account for phenotypic diversity (Engler *et al.*, 1991; Harris *et al.*, 1988; Pinkert, 2003). Consistent with this, analysis of learning and memory in C57BL/6J, 129S6, and DBA/2J mice strains revealed that differences in memory tasks among strains may reflect differences in exploratory behavior rather than reflecting a true spatial memory deficit. The results showed that differences existed when using the 129S6 and C57BL/6 strain backgrounds in specific testing regimens and supported the use of uniform strain backgrounds in studying memory deficits in mutant mice.

Our mice were derived from 129S6 ES cells injected into C57BL/6 blastocysts, then the D7 xenomouse lineage was backcrossed onto the C57BL/6 strain. Our xenomice demonstrated variable spatial memory deficits using a Barnes maze. The response variability was minimized, as the xenomouse lineage was further backcrossed onto the C57BL/6 inbred background through 10 generations. Although further studies are needed to confirm the presence of a true cognition impairment, it appeared that the early identification of a behavioral or cognitive deficiency based on Barnes Maze testing recapitulated the work by Holmes *et al.* (2002). Clearly, tools such as germ line-competent inbred female C57BL/6 ES cells (Ledermann and Bürki, 1991) would be a boon to standardization and to our mouse-modeling efforts.

C. Neurodegenerative Disorders

A large and growing body of evidence suggests that respiratory chain complex dysfunction may be central in human neurodegenerative disease pathogenesis (Beal, 2005; Schon and Manfredi, 2003). We have postulated that the production of mice with mild mtDNA-linked respiratory chain impairment using xenomitochondrial transfer technology will mirror the slow onset of age-related neurodegenerative disease. While one cannot control the locations of amino acid substitutions using xenomitochondrial transfer, we believe that important new insights into the patho-etiology of neurodegeneration processes will be gained. Thus, our "mitochondrial genomic gradient" approach (Fig. 4) will allow the superimposition, by crossbreeding, of increasing OXPHOS dysfunction onto neurodegenerative disease mice carrying, for example, nuclear genes linked to familial forms of Alzheimer's and Parkinson's diseases. Many efforts have been made to define mtDNA variants associated with altered risk of developing these diseases, with limited success (Elson *et al.*, 2006; van der Walt *et al.*, 2003). The absence of mtDNA-linked phenocopies of these diseases also argues against their causation by mtDNA mutations alone. Nonetheless, it is likely that human mtDNA polymorphisms may contribute to disease penetrance in concert with nuclear gene variants (Fig. 4).

The development and characterization of xenomouse lineages have been fraught with difficulty. In addition to the generation of the models in hand, much more groundwork is needed to establish baseline phenotypes and crosses/comparisons

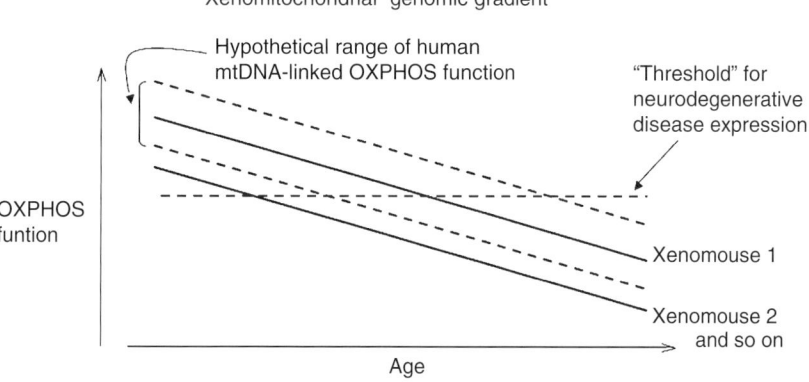

Fig. 4 Xenomitochondrial "genomic gradient." Our scheme illustrating how a xenomitochondrial genomic gradient may be used to explore mtDNA-linked functional OXPHOS variation. Human mtDNA is extensively polymorphic, potentially resulting in a range of OXPHOS function. This may eventually drop below a threshold for disease expression, when combined with the well-described decline in OXPHOS with human and mouse aging (exaggerated here). Since all inbred mouse strains harbor identical mtDNAs (see text), our production of "hyperpolymorphic" mtDNAs in the *M. m. domesticus* nuclear background is designed to model the range of human mtDNA polymorphism. By introducing successively more divergent mtDNAs, one can effectively "ratchet down" the OXPHOS function to model mtDNA effects in aging and neurodegeneration.

with existing neurodegeneration models to test proposed paradigms implicating mitochondrial mutation and subsequent functional impairment leading to or accelerating neurodegenerative phenomena. These efforts have provided a first glimpse of mitochondrial dynamics and human mtDNA disease in mouse models. This work is especially relevant to aging, mitochondrial disease pathogenesis, identification of nuclear compensatory mechanisms in OXPHOS deficiencies, and ultimately in developing therapeutic strategies for age-related neurodegenerative diseases affected by mitochondrial function or mutation.

VI. Summary/Future Directions

Since 1997, a number of laboratories have reported on methodologies used to create transmitochondrial animals. To date, methods for mitochondrial isolation and interspecific transfer of mitochondria have been reported, both in laboratory and domestic animal models (Hiendleder et al., 2003; McKenzie et al., 2004; Meirelles et al., 2001; Pinkert and Trounce, 2002). As discussed above, some reports on the development of cloned animals by nuclear transfer resulted in identification of induced heteroplasmy for mtDNA (Evans et al., 1999; Hiendleder et al., 1999, 2003; Steinborn et al., 2000; Takeda et al., 1999). Indeed, dependent on the specific methodology employed for nuclear and cytoplasm/ooplasm transfers to rescue embryos, additional models of heteroplasmy may have been characterized as a consequence of mitochondrial dysfunction. As such, research dependent and independent of targeted mitochondrial genomic modifications may also help unlock mechanisms underlying the dynamics related to persistence of foreign mitochondria and maintenance of heteroplasmy in various cloning protocols. In contrast, current ES cell approaches in mice hold significant promise in targeted modification of mitochondrial genes and in the study of nuclear–mitochondrial cross talk.

Xenomitochondrial mice promise to be valuable tools for understanding mechanisms of pathogenesis in mitochondrial diseases. Of equal significance will be the use of these mice in testing therapies for mitochondrial disease patients. Due to the unique features of mitochondrial genetics, nuclear gene knockout mice will not supersede such models, and the creation of mitochondrial mutants will shed additional light on mitochondrial function and dynamics of heteroplasmy. Such mice will also be of great interest in the study of Mendelian and sporadic neurodegenerative diseases where oxidative stress or secondary mitochondrial impairment is implicated.

Acknowledgments

We gratefully acknowledge the expertise, reagents, and assistance provided by R. B. Baggs, C. A. Berdanier, W. J. Bowers, A. Bradley, C. A. Cassar, M. V. Cannon, J. P. Corsetti, R. G. Cotton, P. J. Crouch, C. L. Donegan, J. A. Duce, D. A. Dunn, H. J. Federoff, M. K. Horne, R. L. Howell, C. A. Ingraham, C. A. Lerner, J. L. Littleton, M. McKenzie, M. P. Massett, C. L. Masters, W. K. Pogozelski, and K. Takeda.

References

Adachi, K., Matsuhashi, T., Nishizawa, Y., Usukura, J., Momota, M., Popinigis, J., and Wakabayashi, T. (1994). Further studies on physicochemical properties of mitochondrial membranes during formation of megamitochondria in the rat liver by hydrazine. *Exp. Mol. Pathol.* **61,** 134–151.

Auffray, J. C., Orth, A., Catalan, J., Gonzalez, J. P., Desmarais, E., and Bonhomme, F. (2003). Phylogenetic position and description of a new species of subgenus Mus (Rodentia, Mammalia) from Thailand. *Zool. Scr.* **32,** 119–127.

Bai, Y., and Attardi, G. (1998). The mtDNA-encoded ND6 subunit of mitochondrial NADH dehydrogenase is essential for the assembly of the membrane arm and the respiratory function of the enzyme. *EMBO J.* **17,** 4848–4858.

Barritt, J. A., Brenner, C. A., Malter, H., and Cohen, J. (2001). Mitochondria in human offspring derived from ooplasmic transplantation. *Hum. Reprod.* **16,** 513–516.

Bayona-Bafaluy, M. P., Blits, B., Battersby, B. J., Shoubridge, E. A., and Moraes, C. T. (2005). Rapid directional shift of mitochondrial DNA heteroplasmy in animal tissues by a mitochondrially targeted restriction endonuclease. *Proc. Natl. Acad. Sci. USA* **102,** 14392–14397.

Beal, M. F. (2005). Mitochondria take centre stage in aging and neurodegeneration. *Ann. Neurol.* **4,** 495–505.

Berdanier, C. D. (1982). Rat strain differences in gluconeogenesis by isolated hepatocytes. *Proc. Soc. Exp. Biol. Med.* **169,** 74–79.

Berdanier, C. D. (1991). The BHE rat: An animal model for the study of non-insulin-dependent diabetes mellitus. *FASEB J.* **5,** 2139–2144.

Betarbet, R., Sherer, T. B., MacKenzie, G., Garcia-Osuna, M., Panov, A. V., and Greenamyre, J. T. (2000). Chronic systemic pesticide exposure reproduces features of Parkinson's disease. *Nat. Neurosci.* **3,** 1301–1306.

Bianchi, C., Genova, M. L., Parenti Castelli, G., and Lenaz, G. (2004). The mitochondrial respiratory chain is partially organized in a supercomplex assembly: Kinetic evidence using flux control analysis. *J. Biol. Chem.* **279,** 36562–36569.

Bibb, M. J., Van Etten, R. A., Wright, C. T., Walberg, M. W., and Clayton, D. A. (1981). Sequence and gene organization of mouse mitochondrial DNA. *Cell* **26,** 167–180.

Bigger, B., Tolmachov, O., Collombet, J. M., and Coutelle, C. (2000). Introduction of chloramphenicol resistance into the modified mouse mitochondrial genome: Cloning of unstable sequences by passage through yeast. *Anal. Biochem.* **277,** 236–242.

Brinster, R. L. (1993). Stem cells and transgenic mice in the study of development. *Int. J. Dev. Biol.* **3,** 89–99.

Cannon, M. V., Pinkert, C. A., and Trounce, I. A. (2004). Xenomitochondrial stem cells and mice: Modeling human mitochondrial biology and disease. *Gene Ther. Regul.* **2,** 283–300.

Cappechi, M. R. (1989). Altering the genome by homologous recombination. *Science* **244,** 1288–1292.

Chomyn, A., Meola, G., Bresolin, N., Lai, S. T., Scarlato, G., and Attardi, G. (1991). *In vitro* genetic transfer of protein synthesis and respiratory defects to mitochondrial DNA-less cells with myopathy-patient mitochondria. *Mol. Cell. Biol.* **11,** 2236–2244.

Clark, K. M., Watt, D. J., Lightowlers, R. N., Johnson, M. A., Relvas, J. B., Taanman, J. W., and Turnbull, D. M. (1998). SCID mice containing muscle with human mitochondrial DNA mutations. An animal model for mitochondrial DNA defects. *J. Clin. Invest.* **15,** 2090–2095.

Clayton, D. A. (1991). Replication and transcription of vertebrate mitochondrial DNA. *Annu. Rev. Cell Biol.* **7,** 453–478.

Cohen, J., Scott, R., Schimmel, T., Levron, J., and Willadsen, S. (1997). Birth of infant after transfer of anucleate donor oocyte cytoplasm into recipient eggs. *Lancet* **350,** 186–187.

Collombet, J. M., Wheeler, V. C., Vogel, F., and Coutelle, C. (1997). Introduction of plasmid DNA into isolated mitochondria by electroporation. A novel approach toward gene correction for mitochondrial disorders. *J. Biol. Chem.* **272,** 5342–5347.

Cooper, J. M., Hayes, D. J., Challiss, R. A., Morgan-Hughes, J. A., and Clark, J. B. (1992). Treatment of experimental NADH ubiquinone reductase deficiency with menadione. *Brain* **115,** 991–1000.

Cummins, J. (1998). Mitochondrial DNA in mammalian reproduction. *Rev. Reprod.* **3**, 72–82.

Cummins, J. (1999). Elimination of the sperm mitochondrial DNA. *Embryo Mail News* comm. #1060, 30 September.

Cummins, J. M., Kishikawa, H., Mehmet, D., and Yanagimachi, R. (1999). Fate of genetically marked mitochondria DNA from spermatocytes microinjected into mouse zygotes. *Zygote* **7**, 151–156.

de Kroon, A. I., Dolis, D., Mayer, A., Lill, R., and de Kruijff, B. (1997). Phospholipid composition of highly purified mitochondrial outer membranes of rat liver and Neurospora crassa. Is cardiolipin present in the mitochondrial outer membrane? *Biochim. Biophys. Acta* **1325**, 108–116.

Dey, R., Barrientos, A., and Moraes, C. T. (2000). Functional constraints of nuclear-mitochondrial DNA interactions in xenomitochondrial rodent cell lines. *J. Biol. Chem.* **275**, 31520–31527.

Doetschman, T. (2002). Gene targeting in embryonic stem cells: I. History and methodology. *In* "Transgenic Animal Technology: A Laboratory Handbook" (C. A. Pinkert, ed.), 2nd edn. pp. 113–141. Academic Press, Inc., San Diego.

Ebert, K. M., Alcivar, A., Liem, B., Goggins, R., and Hecht, N. B. (1989). Mouse zygotes injected with mitochondria develop normally but the exogenous mitochondria are not detectable in the progeny. *Mol. Reprod. Dev.* **1**, 156–163.

Ellis, E. M., and Reid, G. A. (1994). Assembly of mitochondrial membranes. *Subcell. Biochem.* **22**, 151–181.

Elson, J. L., Herrnstadt, C., Preston, G., Thal, L., Morris, C. M., Edwardson, J. A., Beal, M. F., Turnbull, D. M., and Howell, N. (2006). Does the mitochondrial genome play a role in the etiology of Alzheimer's disease? *Hum. Genet.* **119**, 241–254.

Engler, P., Haasch, D., Pinkert, C. A., Doglio, L., Glymour, M., Brinster, R., and Storb, U. (1991). A strain-specific modifier on mouse chromosome 4 controls the methylation of independent transgene loci. *Cell* **65**, 939–947.

Evans, M. J., Gurer, C., Loike, J. D., Wilmut, I., Schnieke, A. E., and Schon, E. A. (1999). Mitochondrial DNA genotypes in nuclear transfer-derived cloned sheep. *Nat. Genet.* **23**, 90–93.

Everts, H. B., and Berdanier, C. D. (2002). Nutrient-gene interactions in mitochondrial function: Vitamin A needs are increased in BHE/Cdb rats. *IUBMB Life* **53**, 289–294.

Ferris, S. D., Sage, R. D., and Wilson, A. C. (1982). Evidence from mtDNA sequences that common laboratory strains of inbred mice are descended from a single female. *Nature* **295**, 163–165.

Festing, M. F., Simpson, E. M., Davisson, M. T., and Mobraaten, L. E. (1999). Revised nomenclature for strain 129 mice. *Mamm. Genome* **10**, 836.

Gyllensten, U., Wharton, D., and Wilson, A. C. (1985). Maternal inheritance of mitochondrial DNA during backcrossing of two species of mice. *J. Hered.* **76**, 321–324.

Hackenbrock, C. R., and Chazotte, B. (1986). Lipid enrichment and fusion of mitochondrial inner membranes. *Methods Enzymol.* **125**, 35–45.

Harris, A. W., Pinkert, C. A., Crawford, M., Langdon, W. Y., Brinster, R. L., and Adams, J. M. (1988). The Eµ-*myc* transgenic mouse: A model for high incidence spontaneous lymphoma and leukemia of early B cells. *J. Exp. Med.* **167**, 353–371.

Hayashi, J.-I., Ohta, S., Kikuchi, A., Takemitsu, M., Goto, Y.-I., and Nonaka, I. (1991). Introduction of disease-related mitochondrial DNA deletions into HeLa cells lacking mitochondrial DNA results in mitochondrial dysfunction. *Proc. Natl. Acad. Sci. USA* **88**, 10614–10618.

Hayashi, J., Ohta, S., Takai, D., Miyabayashi, S., Sakuta, R., Goto, Y., and Nonaka, I. (1993). Accumulation of mtDNA with a mutation at position 3271 in tRNA(Leu)(UUR) gene introduced from a MELAS patient to HeLa cells lacking mtDNA results in a progressive inhibition of mitochondrial respiratory function. *Biochem. Biophys. Res. Commun.* **197**, 1049–1055.

Hayashi, J., Werbin, H., and Shay, J. W. (1986). Effects of normal human fibroblast mitochondrial DNA on segregation of HeLaTG Mitochondrial DNA and on tumorigenicity of HeLaTG cells. *Cancer Res.* **46**, 4001–4006.

Heikkila, R. E., Manzino, L., Cabbat, F. S., and Duvoisin, R. C. (1984). Protection against the dopaminergic neurotoxicity of 1-methyl-4-phenyl-1,2,5,6-tetrahydropyridine by monoamine oxidase inhibitors. *Nature* **10**, 467–469.

Hiendleder, S., Schmutz, S. M., Erhardt, G., Green, R. D., and Plante, Y. (1999). Transmitochondrial differences and varying levels of heteroplasmy in nuclear transfer cloned cattle. *Mol. Reprod. Dev.* **54,** 24–31.

Hiendleder, S., Zakhartchenko, V., Wenigerkind, H., Reichenbach, H. D., Bruggerhoff, K., Prelle, K., Brem, G., Stojkovic, M., and Wolf, E. (2003). Heteroplasmy in bovine fetuses produced by intra- and inter-subspecific somatic cell nuclear transfer: Neutral segregation of nuclear donor mitochondrial DNA in various tissues and evidence for recipient cow mitochondria in fetal blood. *Biol. Reprod.* **68,** 159–166.

Higuchi, I., Takahashi, K., Nakahara, K., Izumo, S., Nakagawa, M., and Osame, M. (1991). Experimental germanium myopathy. *Acta Neuropathol.* **82,** 55–59.

Holmes, A., Wrenn, C. C., Harris, A. P., Thayer, K. E., and Crawley, J. N. (2002). Behavioral profiles of inbred strains on novel olfactory, spatial and emotional tests for reference memory in mice. *Genes Brain Behav.* **1,** 55–69.

Hovius, R., Thijssen, J., van der Linden, P., Nicolay, K., and De Kruijff, B. (1993). Phospholipid asymmetry of the outer membrane of rat liver mitochondria. Evidence for the presence of cardiolipin on the outside of the outer membrane. *FEBS Lett.* **330,** 71–76.

Howell, N. (1990). Glycine-231 residue of the mouse mitochondrial protonmotive cytochrome b: Mutation to aspartic acid deranges electron transport. *Biochemistry* **29,** 8970–8977.

Howell, N., Elson, J. L., Chinnery, P. F., and Turnbull, D. M. (2005). mtDNA mutations and common neurodegenerative disorders. *Trends Genet.* **11,** 583–586.

Inoue, K., Nakada, K., Ogura, A., Isobe, K., Goto, Y.-I., Nonaka, I., and Hayashi, J.-I. (2000). Generation of mice with mitochondrial dysfunction by introducing mouse mtDNA carrying a deletion into zygotes. *Nat. Genet.* **26,** 176–181.

Irwin, M. H., Johnson, L. W., and Pinkert, C. A. (1999). Isolation and microinjection of somatic cell-derived mitochondria and germline heteroplasmy in transmitochondrial mice. *Transgenic Res.* **8,** 119–123.

Irwin, M. H., Parrino, V., and Pinkert, C. A. (2001). Construction of a mutated mtDNA genome and transfection into isolated mitochondria by electroporation. *Adv. Reprod.* **5,** 59–66.

Jenuth, J. P., Peterson, A. C., Fu, K., and Shoubridge, E. A. (1996). Random genetic drift in the female germline explains the rapid segregation of mammalian mitochondrial DNA. *Nat. Genet.* **14,** 146–151.

Jenuth, J. P., Peterson, A. C., and Shoubridge, E. A. (1997). Tissue-specific selection for different mtDNA genotypes in heteroplasmic mice. *Nat. Genet.* **16,** 93–95.

Kenyon, L., and Moraes, C. T. (1997). Expanding the functional human mitochondrial DNA database by the establishment of primate xenomitochondrial cybrids. *Proc. Natl. Acad. Sci. USA* **94,** 9131–9135.

King, M. P., and Attardi, G. (1988). Injection of mitochondria into human cells leads to rapid replacement of the endogenous mitochondrial DNA. *Cell* **52,** 811–819.

King, M. P., and Attardi, G. (1989). Human cells lacking mtDNA: Repopulation with exogenous mitochondria by complementation. *Science* **246,** 500–503.

Laipis, P. J. (1996). Construction of heteroplasmic mice containing two mitochondrial DNA genotypes by micromanipulation of single-cell embryos. *Methods Enzymol.* **264,** 345–357.

Lakshmanan, M. K., Berdanier, C. D., and Veech, R. L. (1977). Comparative studies on lipogenesis and cholesterogenesis in lipemic BHE rats and normal Wistar rats. *Arch. Biochem. Biophys.* **183,** 355–360.

Ledermann, B., and Bürki, K. (1991). Establishment of a germ-line competent C57BL/6 embryonic stem cell line. *Exp. Cell Res.* **197,** 254–258.

Levy, S. E., Waymire, K. G., Kim, Y. L., MacGregor, G. R., and Wallace, D. C. (1999). Transfer of chloramphenicol-resistant mitochondrial DNA into the chimeric mouse. *Transgenic Res.* **8,** 137–145.

Loveland, B., Wang, C. R., Yonekawa, H., Hermel, E., and Lindahl, K. F. (1990). Maternally transmitted histocompatibility antigen of mice: A hydrophobic peptide of a mitochondrially encoded protein. *Cell* **60,** 971–980.

Marchington, D. R., Barlow, D., and Poulton, J. (1999). Transmitochondrial mice carrying resistance to chloramphenicol on mitochondrial DNA: Developing the first mouse model of mitochondrial DNA disease. *Nat. Med.* **5,** 957–960.

Mathews, C. E., Wickwire, K., Flatt, W. P., and Berdanier, C. D. (2000). Attenuation of circadian rhythms of food intake and respiration in aging diabetes-prone BHE/Cdb rats. *Am. J. Physiol. Regul. Integr. Comp. Physiol.* **279,** R230–R238.

McKenzie, M., Chiotis, M., Pinkert, C. A., and Trounce, I. A. (2003). Functional respiratory chain analyses in Murid xenomitochondrial cybrids expose coevolutionary constraints of cytochrome b and nuclear subunits of complex III. *Mol. Biol. Evol.* **20,** 1117–1124.

McKenzie, M., and Trounce, I. (2000). Expression of *Rattus norvegicus* mtDNA in *Mus musculus* cells results in multiple respiratory chain defects. *J. Biol. Chem.* **275,** 31514–31519.

McKenzie, M., Trounce, I. A., Cassar, C. A., and Pinkert, C. A. (2004). Production of homoplasmic xenomitochondrial mice. *Proc. Natl. Acad. Sci. USA* **101,** 1685–1690.

Meirelles, F. V., Bordignon, V., Watanabe, Y., Watanabe, M., Dayan, A., Lobo, R. B., Garcia, J. M., and Smith, L. C. (2001). Compete replacement of the mitochondrial genotype in a *Bos indicus* calf reconstructed by nuclear transfer to a *Bos taurus* oocyte. *Genetics* **158,** 351–356.

Meirelles, F. V., and Smith, L. C. (1997). Mitochondrial genotype segregation in a mouse heteroplasmic lineage produced by embryonic karyoplast transplantation. *Genetics* **145,** 445–451.

Menkes, J. H., Alter, M., Steigleder, G. K., Weakley, D. R., and Sung, J. H. (1962). A sex-linked recessive disorder with retardation of growth, peculiar hair, and focal cerebral and cerebellar degeneration. *Pediatrics* **29,** 764–779.

Nagy, A., Gertsenstein, M., Vintersten, K., and Behringer, R. (2003). "Manipulating the Mouse Embryo: A Laboratory Manual," 3rd edn. Cold Spring Harbor Press, New York.

Pfanner, N., and Neupert, W. (1990a). A mitochondrial machinery for membrane translocation of precursor proteins. *Biochem. Soc. Trans.* **18,** 513–515.

Pfanner, N., and Neupert, W. (1990b). The mitochondrial protein import apparatus. *Annu. Rev. Biochem.* **59,** 331–353.

Phillips, M., Camakaris, J., and Danks, D. M. (1986). Comparisons of copper deficiency states in the murine mutants blotchy and brindled. Changes in copper-dependent enzyme activity in 13-day old mice. *Biochem. J.* **238,** 177–183.

Pinkert, C. A. (2002). "Transgenic Animal Technology: A Laboratory Handbook," 2nd edn. Academic Press, San Diego.

Pinkert, C. A. (2003). Transgenic animal technology: Alternatives in genotyping and phenotyping. *Comp. Med.* **53,** 126–139.

Pinkert, C. A., Irwin, M. H., and Howell, R. L. (2004). Animal biotechnology and modeling. *In* "Encyclopedia of Molecular Cell Biology and Molecular Medicine" (R. A. Meyers, ed.), Vol. 1, pp. 209–240. Wiley-VCH, Weinheim.

Pinkert, C. A., Irwin, M. H., Johnson, L. W., and Moffatt, R. J. (1997). Mitochondria transfer into mouse ova by microinjection. *Transgenic Res.* **6,** 379–383.

Pinkert, C. A., and Trounce, I. A. (2002). Production of transmitochondrial mice. *Methods* **26,** 348–357.

Pinkert, C. A., and Trounce, I. A. (2005). Animal modeling: From transgenesis to transmitochondrial models. *In* "Mitochondria in Health and Disease. Oxidative Stress and Disease Series" (C. D. Berdanier, ed.), Vol. 15, pp. 559–580. CRC Press, Boca Raton.

Roubertoux, P. L., Sluyter, F., Carlier, M., Marcet, B., Maarouf-Veray, F., Chérif, C., Marican, C., Arrechi, P., Godin, F., Jamon, M., Verrier, B., and Cohen-Salmon, C. (2003). Mitochondrial DNA modifies cognition in interaction with the nuclear genome and age in mice. *Nat. Genet.* **35,** 65–69.

Rucker, E. B., III, Thomson, J. A., and Piedrahita, J. A. (2002). Gene targeting in embryonic stem cells: II. Conditional technologies. *In* "Transgenic Animal Technology: A Laboratory Handbook" (C. A. Pinkert, ed.), 2nd edn., pp. 144–171. Academic Press, Inc., San Diego.

Schon, E. A., and Manfredi, G. (2003). Neuronal degeneration and mitochondrial dysfunction. *J. Clin. Invest.* **111,** 303–312.

Seibel, P., Trappe, J., Villani, G., Klopstock, T., Papa, S., and Reichmann, H. (1995). Transfection of mitochondria: Strategy towards a gene therapy of mitochondrial DNA diseases. *Nucleic Acids Res.* **11,** 10–17.

Sligh, J. E., Levy, S. E., Waymire, K. G., Allard, P., Dillehay, D. L., Heckenlively, J. R., MacGregor, G. R., and Wallace, D. C. (2000). Maternal germ-line transmission of mutant mtDNAs from embryonic stem cell-derived chimeric mice. *Proc. Natl. Acad. Sci. USA* **97,** 14461–14466.

Smeitink, J., van den Heuvel, L., and DiMauro, S. (2001). The genetics and pathology of oxidative phosphorylation. *Nat. Rev. Genet.* **2,** 342–352.

Steinborn, R., Schinogl, P., Zakhartchenko, V., Achmann, R., Schernthaner, W., Stojkovic, M., Wolf, E., Muller, M., and Brem, G. (2000). Mitochondrial DNA heteroplasmy in cloned cattle produced by fetal and adult cell cloning. *Nat. Genet.* **25,** 255–257.

Takeda, K., Akagi, S., Takahashi, S., Onishi, A., Hanada, H., and Pinkert, C. A. (2002). Mitochondrial activity in response to serum starvation in bovine (*Bos taurus*) cell culture. *Cloning Stem Cells* **4,** 223–230.

Takeda, K., Takahashi, S., Onishi, A., Goto, Y., Miyazawa, A., and Imai, H. (1999). Dominant distribution of mitochondrial DNA from recipient oocytes in bovine embryos and offspring after nuclear transfer. *J. Reprod. Fertil.* **116,** 253–259.

Tapper, D. P., Van Etten, R. A., and Clayton, D. A. (1983). Isolation of the mammalian mitochondrial DNA and RNA and cloning of the mitochondrial genome. *Methods Enzymol.* **97,** 426–434.

Threadgill, D. W., Yee, D., Matin, A., Nadeau, J. H., and Magnuson, T. (1997). Genealogy of the 129 inbred strains: 129/SvJ is a contaminated inbred strain. *Mamm. Genome* **8,** 390–393.

Tracey, I., Dunn, J. F., and Radda, G. K. (1997). A ^{31}P-magnetic resonance spectroscopy and biochemical study of the mo^{vbr} mouse: Potential model for the mitochondrial encephalomyopathies. *Muscle Nerve* **20,** 1352–1359.

Trounce, I., Neill, S., and Wallace, D. C. (1994). Cytoplasmic transfer of the mitochondrial DNA 8993T → G (ATP6) point mutation associated with Leigh syndrome into mtDNA-less cells demonstrates co-segregation of decreased state III respiration and ADP/O ratio. *Proc. Natl. Acad. Sci. USA* **91,** 8334–8338.

Trounce, I., Schmiedel, J., Yen, H. C., Hosseini, S., Brown, M. D., Olson, J. J., and Wallace, D. C. (2000). Cloning of neuronal mtDNA variants in cultured cells by synaptosome fusion with mtDNA-less cells. *Nucleic Acids Res.* **28,** 2164–2170.

Trounce, I., and Wallace, D. C. (1996). Production of transmitochondrial mouse cell lines by cybrid rescue of rhodamine-6G pretreated L cells. *Somat. Cell Mol. Genet.* **22,** 81–85.

Trounce, I. A., McKenzie, M., Cassar, C. A., Ingraham, C. A., Lerner, C. A., Dunn, D. A., Donegan, C. L., Takeda, K., Pogozelski, W. K., Howell, R. L., and Pinkert, C. A. (2004). Development and initial characterization of xenomitochondrial mice. *J. Bioenerg. Biomembr.* **36,** 421–427.

Trounce, I. A., and Pinkert, C. A. (2005). Cybrids in the study of animal mitochondrial genetics and pathology. *In* "Mitochondria in Health and Disease. Oxidative Stress and Disease Series" (C. D. Berdanier, ed.), Vol. 15, pp. 539–558. CRC Press, Boca Raton.

van der Walt, J. M., Nicodemus, K. K., Martin, E. R., Scott, W. K., Nance, M. A., Watts, R. L., Hubble, J. P., Haines, J. L., Koller, W. C., Lyons, K., Pahwa, R., Stern, M. B., *et al.* (2003). Mitochondrial polymorphisms significantly reduce the risk of Parkinson disease. *Am. J. Hum. Genet.* **72,** 804–811.

Wallace, D. C. (2001). Mouse models for mitochondrial disease. *Am. J. Med. Genet.* **106,** 71–93.

Wallace, D. C., Bunn, C. L., and Eisenstadt, J. M. (1975). Cytoplasmic transfer of chloramphenicol resistance in human tissue culture cells. *J. Cell Biol.* **67,** 174–188.

Wood, S. A., Allen, N. D., Rossant, J., Auerbach, A., and Nagy, A. (1993). Non-injection methods for the production of embryonic stem cell-embryo chimaeras. *Nature* **365,** 87–89.

CHAPTER 28

In Vivo and In Organello Analyses of Mitochondrial Translation

P. Fernández-Silva, R. Acín-Pérez, E. Fernández-Vizarra, A. Pérez-Martos, and J. A. Enriquez

Departamento de Bioquímica y Biología Molecular y Celular, Universidad de Zaragoza
Zaragoza 50013, Spain

I. Introduction

From the point of view of their biogenesis, mitochondria are unique organelles in animals since they require the contribution of two physically separated genomes. The majority of mitochondrial proteins are encoded in the nucleus and are synthesized in the cytoplasm, usually as precursors that must be imported and processed inside the organelle. The human mitochondrial genome (mtDNA) contributes only 13 polypeptides that are components of four out of the five OXPHOS complexes located in the inner mitochondrial membrane. The correct expression of this small set of proteins is essential not only for energy production, but also for those processes in which mitochondria play a direct role (metabolic routes such as nucleotide or heme synthesis, apoptosis, ROS signaling, and so on). Thus, an

0091-679X/07 $35.00
DOI: 10.1016/S0091-679X(06)80028-2

increasing number of human diseases are associated with mutations in genes encoded not only by mtDNA but also by nuclear genes involved in mtDNA expression (Taylor and Turnbull, 2005).

The origin of mitochondria [most probably from α-purple bacteria that entered the eukaryotic cell ancestor as an endosymbiont around 2 thousand million years ago (Taanman, 1999)] and their evolution in a particular environment, along with the functional specialization of the organelles, explain most of the peculiarities of the mitochondrial genetic system. These peculiarities include a great economy in mtDNA organisation, the phenomenon of polyploidy, and exclusive maternal inheritance in mammals. The genetic code in mitochondria presents some deviations from the "universal" code, and even codon usage varies among different species. Concerning mitochondrial translation, we find that, reflecting its evolutionary origins, mitoribosomes show an antibiotic sensitivity resembling that of prokaryotic systems and use N-formylmethionine as the initiator amino acid, like prokaryotes. In addition, mitochondrial ribosomal RNAs (12S and 16S) are smaller than the bacterial ones, but have higher protein content (69% compared to 33% in bacterial ribosomes). The use of a simplified decoding system in mitochondria allows for the translation of the protein genes with only 22 tRNAs. These tRNAs are encoded in human mtDNA, but the aminoacyl-tRNA synthetases, as well as all mitochondrial ribosomal proteins and translation factors, are encoded in the nuclear genome. Furthermore, mitochondrial mRNAs either lack or have very short 5′ leader sequences, and do not contain the 7-methylguanylate cap structure that can guide ribosome binding. As a consequence, a large number of mitochondrial transcripts are required to provide an efficient translation (Taanman, 1999).

Two mitochondrial translation initiation factors have been identified to date: mtIF-2, which promotes fMet-tRNA binding to the 28S small ribosomal subunit (Liao and Spremulli, 1991; Spencer and Spremulli, 2005), and mtIF-3, which promotes formation of the initiation complex, probably by helping position the mRNA correctly onto the ribosome (Koc and Spremulli, 2002). Elongation factors in mitochondria (mtEF-Tu, mtEF-Ts, mtEF-G1, and mtEF-G2) also resemble their bacterial counterparts and have been cloned in humans and other species (Cai *et al.*, 2000; Hammarsund *et al*, 2001; Terasaki *et al.*, 2004). Finally, although less studied, a mitochondrial release factor has also been cloned both in humans (Zhang and Spremulli, 1998) and in yeast (Askarian-Amiri *et al.*, 2000).

Thus, it is assumed that mitochondrial translation, although with some peculiarities, resembles in general more the prokaryotic process than the eukaryotic one. However, contrary to the situation in bacteria, no efficient *in vitro* mitochondrial translation system capable of performing the synthesis of mtDNA-encoded products has yet been established. *In vitro* systems, in theory the most amenable to manipulation, have been used only for some applications. Thus, some systems composed of purified mammalian mitoribosomes and added elongation factors (EF-G and EF-Tu, either purified or recombinant) and directed by artificial templates (e.g., poly-U) have been developed. Such *in vitro* systems have been useful in determining the decoding capacity and translational efficiency of normal and modified or mutant tRNAs (Hanada *et al.*, 2001; Takemoto *et al.*, 1995; Yasukawa *et al.*, 2001), and have also been employed to assess the effects of ethanol consumption (Patel and Cunningham, 2002) and different antibiotics (Zhang *et al.*, 2005) on mitoribosome activity.

In organello approaches using isolated intact mitochondria represent an intermediate situation between the *in vitro* an *in vivo* ones, and allow some analyses that otherwise are not possible or very difficult to perform. These systems have been widely used with yeast (Poyton *et al.*, 1996; Yasuhara *et al.*, 1994), plant (Escobar-Galbis *et al.*, 1998; Giege *et al.*, 2005), and mammalian mitochondria. In mammals, *in organello* translation systems have been characterized in terms of isolation and incubation conditions for heart mitochondria (McKee *et al.*, 1990a,b) and have allowed for the analysis of different thyroid status of the animals (Leung and McKee, 1990). They have been also used to show that the stability of the labeled products is a function of membrane potential (Cote *et al.*, 1989) and to demonstrate a reduced activity of ribosomes in organelles from the liver of ethanol-consuming rats (Coleman and Cunningham, 1991). Furthermore, *in organello* analysis of translation has been applied in liver to study the effects of drugs such as clofibrate (Brass, 1992) and methapyrilene (Copple *et al.*, 1992), and in situations, such as mitochondrial differentiation after birth (Ostronoff *et al.*, 1996), that affect mitochondrial biogenesis and content.

On the other hand, the analysis of mitochondrial translation *in vivo* has been in use for several decades and has provided important basic as well as applied information. Thus, the ability to label mitochondrial translation products selectively has been crucial to identify some of the genes encoded in mtDNA (Chomyn, 1996). In addition, *in vivo* analyses have been very useful in understanding some of the molecular consequences of mutations in mtDNA, either identified in human patients or generated in cell culture systems (Acín-Pérez *et al.*, 2004; Chomyn *et al.*, 1991; Enriquez *et al.*, 1995).

In this chapter, we describe *in vivo* and *in organello* methods currently used in our laboratory to analyze the translation capacity of mitochondria from mammalian cells, both for basic research and as a tool to determine the molecular effects of mtDNA mutations.

II. Rationale

A variety of methods to analyze mitochondrial protein synthesis have been assayed in different organisms and at various levels: *in vivo*, *in vitro*, and *in organello*. As mentioned before, *in vitro* studies have been limited since mitochondrial transcripts are not translated in these systems. Even the more tractable mitochondria from *Saccharomyces cerevisiae* have not been readily amenable to the *in vitro* biochemical analysis of translation, probably because isolated ribosomes do not properly recognize the initiation codon of yeast mitochondrial transcripts (Poyton *et al.*, 1996). This fact, together with the lack of methods to transform mitochondria efficiently, has hampered the biochemical analysis of this process. Consequently, we focus in this chapter on the *in organello* and *in vivo* (cell culture) analyses of mitochondrial translation, particularly for animal cells, although the methods described herein could be adapted to other systems (for yeast translation analysis see Westermann *et al.*, 2001, this series).

The labeling of mtDNA-encoded proteins *in vivo* requires the use of inhibitors of cytoplasmic protein synthesis. Cycloheximide and emetine, as reversible or irreversible inhibitors, respectively, are most extensively used. Mitochondrial preparations used for *in organello* studies are very often contaminated with microsomes and other membrane fractions that can contain active cytoplasmic ribosomes (Fernández-Vizarra *et al.*, 2006). For this reason, the use of the inhibitors to reduce the background and to easily visualize the mitochondrial products is also recommended in this case. The small number of mitochondrial proteins (13 in mammals) and the high signal-to-noise ratio that can be obtained with appropriate labeling methods makes it easy to detect and quantify these translation products. As a consequence, these analyses have been widely used in the study of the molecular mechanisms of mtDNA mutations. Thus, analysis of total protein synthesis in cell models can detect (1) general defects in protein synthesis, such as those due to mutations in mitochondrial tRNAs and rRNAs (Chomyn *et al.*, 1991; Enriquez *et al.*, 1995; Guan *et al.*, 2000; Hao and Moraes, 1996), in nuclear genes for ribosomal proteins (Miller *et al.*, 2004) or in translation factors (Coenen *et al.*, 2004); (2) defects in single polypeptides such as those caused by point mutations in mtDNA (Bruno *et al.*, 1999; Hofhaus and Attardi, 1995; Tiranti *et al.*, 2000), deletions, and some mt-tRNA mutations (Limongelli *et al.*, 2004; Fig. 1, panels A and B); and (3) the appearance of new abnormal translation products, such as those caused by premature termination (Chomyn *et al.* 1991; Enriquez *et al.* 1995; Fig. 1, panel A). In addition, pulse-chase labeling experiments allow one to study the stability of the translation products (Bai *et al.*, 2000) and, when combined with blue native gel electrophoresis (BN-PAGE) analysis of respiratory complexes, it is possible to follow their assembly into higher order structures (Acín-Pérez *et al.*, 2004; Fig. 1, panels C and D). In this way, the effect of some mutations that do not alter the synthesis rate of the polypeptides but interfere with their stability and assembly (Acín-Pérez *et al.*, 2004), can also be detected.

In organello approaches are based on the fact that isolated intact mitochondria can retain their biogenetic capacity, reflecting the properties of the cells from which they are obtained (Enriquez *et al.*, 1996, 1999). These systems allow for easy manipulation of the incubation medium and can be useful to determine the effect on mitochondrial protein synthesis of metabolic changes (change in energetic substrates, effect of added drugs, and so on), to analyze differences in translation among differentiated cell types, or in cases in which there are no cultured cell lines available, for *in vivo* labeling. Critical for the validity of the *in organello* approach is the integrity of the mitochondrial membranes and their respiratory capacity (Enriquez *et al.*, 1996; Fernández-Vizarra *et al.*, 2006; McKee *et al.*, 1990a,b).

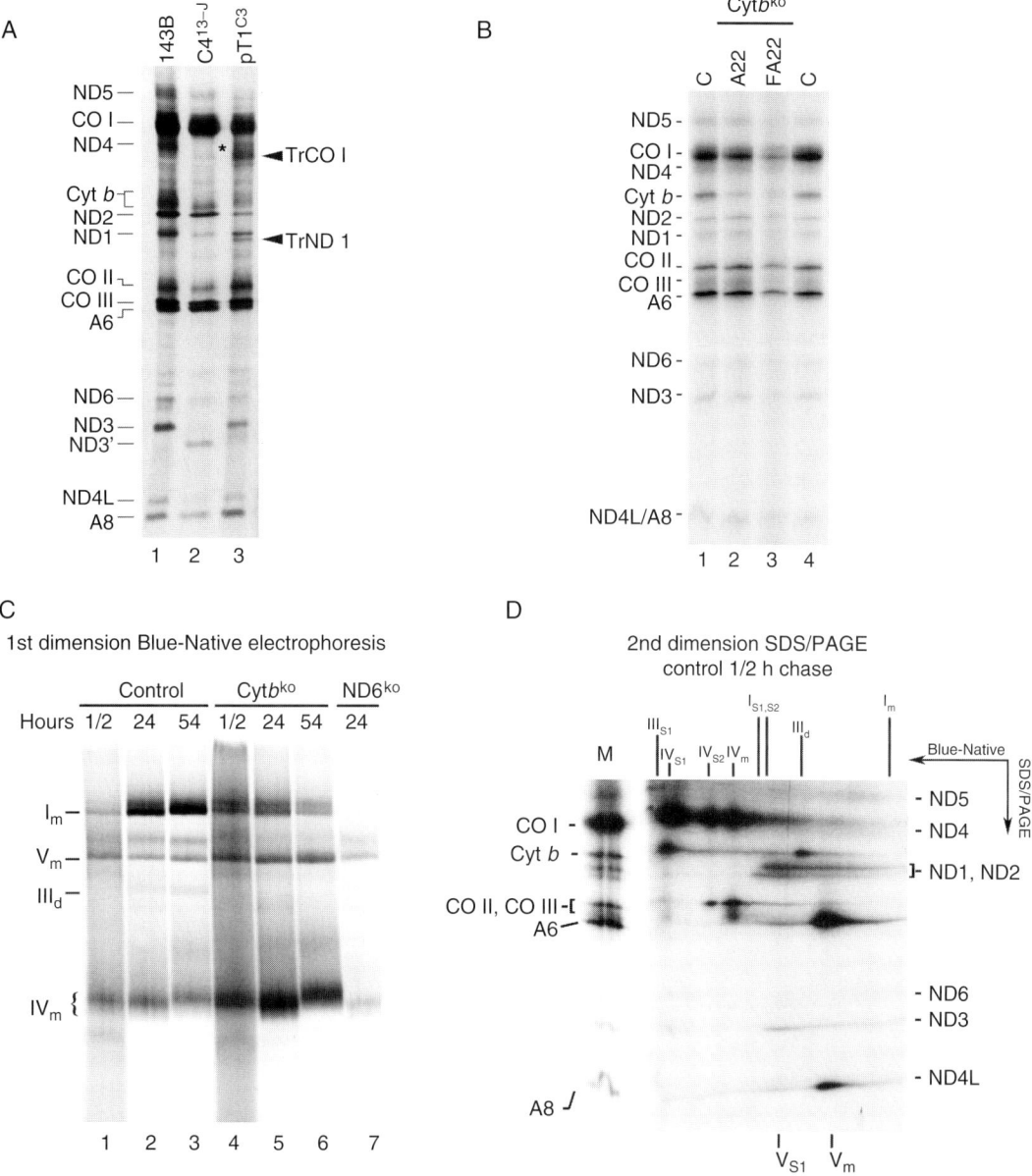

Fig. 1 *In vivo* protein synthesis analysis. (A) Pattern of pulse-labeled mitochondrial translation products corresponding to control (143B, lane 1) and mutant (C4^{13-J} and pT1b^{c3}, lanes 2 and 3) human cell lines. Cell line C4^{13-J} carries, in homoplasmic form, insertion of a C at position 10,947 in the ND4 gene, inducing a frameshift that causes the loss of this polypeptide (indicated by an asterisk, lane 2); cell line pT1b^{c3} carries a point mutation in the tRNA^{Lys} gene (A8344G, MERRF mutation), causing defects in protein synthesis, including the appearance of abnormal bands corresponding to

Other complementary analyses that can yield relevant information about the *in vivo* translation process in mitochondria include the study of tRNA aminoacylation (Enriquez and Attardi, 1996, Enriquez *et al.*, 1995) and the determination of polysome patterns (Chomyn *et al.*, 2000).

III. Methods

A. *In Vivo* Mitochondrial Protein Synthesis

1. General Considerations

Different types of cell lines can be used for the *in vivo* analysis of mitochondrial translation. Thus, transformed cells (both growing in suspension or attached to the substrate) (Chomyn, 1996), as well as primary cultures (Antonicka *et al.*, 2006; Clark *et al.*, 1999) and tissue-explanted cells (Attardi *et al.*, 1991), have been used successfully. The protocol described below is for the more commonly used anchorage-dependent transformed cells, and should be adapted for each particular case (e.g., cells growing in suspension do not need to be trypsinized but are concentrated by low-speed centrifugation, to collect them and/or to remove the growing medium and wash the cells prior to labeling).

Although the pattern of mitochondrial translation products in a given species is quite constant, some differences in the migration of a particular translation product (Chomyn, 1996), especially in the relative abundance of each polypeptide, are possible. Consequently, the availability of appropriate control lines of the same cell type, with the same nuclear background, is crucial, particularly when the effects of mtDNA

---

truncated polypeptides for COX I and for ND1 (TrCOI and TrND1, respectively, lane 3). ND1–6 plus ND4L are subunits of NADH dehydrogenase (complex I); CO I, CO II, and CO III refer to subunits of cytochrome *c* oxidase (complex IV); A6 and A8 are subunits of ATP synthetase (complex V). (B) Pattern of pulse-labeled mitochondrial translation products corresponding to control (E9 and FBalb/cJ, lanes 1 and 4, respectively) and cyt *b* mutant (A22 and FA22, lanes 2 and 3, respectively) mouse cell lines. A22 and FA22 (Cytb^{ko}) contain a homoplasmic missense mutation in the cyt *b* gene affecting the stability of cyt *b* protein (arrow) and complex III assembly. (C) Autoradiogram corresponding to the blue native analysis of pulse-chase labeled mitochondrial products in control (FBalb/cJ, lanes 1–3) and mutant (Cytb^{ko}, lanes 4–6, and ND6^{ko}, lane 7) cell lines. The ND6^{ko} cell line carries an insertion of a C at position 13,887 (C13,887), disrupting the synthesis of ND6 and affecting complex I assembly/stability. The cells were incubated for 2 h in the presence of ^{35}S-Methionine + cysteine and cycloheximide, followed by the indicated chase periods, revealing the differential stability of the complexes. (D) Analysis by second dimension electrophoresis of the mitochondrial translation products labeled as described in panel C and corresponding to the 1/2-h chase (lane 1 in panel C). The lane from the first dimension BN gel was excised, treated as described, and submitted to a second dimension denaturing (SDS-PAGE) electrophoresis. The position of the different polypeptide subunits and of the full complexes (I_m, V_m, and IV_m, represent monomer forms of complexes I, V, and IV, respectively; III_d is complex III in dimer form) as well as the proposed position of some subcomplexes (V_{S1}, $I_{S1,S2}$, IV_{S1}, and III_{S1}), is indicated. Panel A, reprinted from Enriquez *et al.* (2000) with permission; panels B and C, from Acín Pérez *et al.* (2004) with permission.

mutations are analyzed. See, for example, the alternative migration of the ND3 subunit, due to a polymorphism, in HeLa (Chomyn, 1996) and HeLa-derived C4[13−J] cells compared to 143B human cells (ND3 or ND3′ in Fig. 1A, lanes 1 and 2, respectively).

The amount of cells required for the labeling depends on their biogenetic activity and on the particular kind of analysis to be performed. With transformed cultures and for the detection of defects in total protein synthesis or in particular polypeptides, the labeling of one 6-cm or 10-cm Petri dish of subconfluent (70–80%) cells is usually sufficient.

Different pulse and chase times can be used, depending on the purpose of the analysis. Thus, if an analysis of the protein synthesis rate is to be performed, short-labeling times (pulse of 15–45 min) without chase are recommended. If a qualitative analysis of mitochondrial translation products is desired, to detect the absence of a particular polypeptide or the appearance of abnormal bands, a longer pulse of radioactivity (60–120 min) and no chase or very short chase (10–15 min) should be used. On the other hand, when analyzing the stability or assembly of mtDNA-encoded proteins, a long pulse (usually 2 h) followed by various times of chase (from 0.5 to 90 h) are required. When the assembly of proteins in the corresponding complexes has to be analyzed, either by complex inmunocapture or by blue-native electrophoresis, the previous purification of a mitochondrial fraction is necessary. To have a sufficient volume of cells for mitochondrial isolation, two options are available: either to scale up the labeling reaction (including several large culture plates) or to mix the labeled cells, after collecting them, with unlabeled ones. The objective is to have a starting volume of packed cells of at least 100 μl.

2. Labeling Procedure

a. Pulse–Labeling Experiments

1. Cells are cultured at 37°C in a 5% (v/v) CO_2 atmosphere, in the appropriate medium (usually DMEM supplemented with calf or fetal bovine serum at 5–10%). Fresh medium should be added 12–24 h before the labeling to ensure an optimal metabolic activity of the cells.

2. The growth medium is removed and the cells are rinsed with prewarmed methionine-free medium without serum, and incubated for 5–10 min at 37°C in this same medium to reduce the internal pool of methionine and increase the efficiency of labeling. Repeat the rinse and incubation steps.

3. Fresh methionine-free medium supplemented with 10% dialyzed serum and containing 100 μg/ml of emetine is added, and the cells are incubated at 37°C for 5–10 min to allow the antibiotic to block cytoplasmic protein synthesis. For a 6-cm or 10-cm culture dish, volumes of 2.5–3 ml and 4.5–6 ml of medium, respectively, are adequate.

4. The labeling is started by adding to the medium the radioactive aminoacid(s) (usually 0.2–1.0 mCi of Amersham's PRO-MIX^{TM}L-[^{35}S] (Barcelona, Spain) *in vitro* Cell Labeling Mix (>1000 Ci/mmol), containing [^{35}S]-labeled methionine and cysteine), mixing thoroughly, and incubating at 37 °C for the desired time.

5. The radioactive medium is removed carefully and the cells are rinsed once with prewarmed complete medium (containing methionine and nondyalized serum) and twice with PBS. The cells are collected by trypsinization, diluted with ~10 ml per plate of cold PBS containing 10% calf serum, and concentrated by centrifugation at 300 × g_{av} for 5 min.

6. The cell pellet is resuspended in 10 ml of cold PBS and the centrifugation step is repeated. Finally, the cells are resuspended in 1.5 ml of cold PBS, transferred to an Eppendorf tube and centrifuged in a microfuge at 500 × g_{av} for 2 min at 4 °C.

7. The final pellet of cells are resuspended in 50–100 μl of TE (10-mM Tris–HCl, pH 7.4; 1-mM EDTA) containing 1-μM PMSF, and are either stored at −80 °C until use or are lysed directly with SDS (0.1% final concentration) and then stored at −80 °C. Before use, 1/5 of the final volume of 5× loading buffer (4.25% glycerol; 125-mM Tris–HCl, pH 6.8; 5% SDS; 12-mM EDTA; 5% 2-mercaptoethanol; and 0.5-mg/ml bromophenol blue) is added to each sample. Repeated cycles of freezing and thawing with vigorous vortexing may be necessary to decrease the viscosity of the sample and facilitate the loading on the gel.

8. For the electrophoretic analysis of the pulse-labeled products, samples (50–75 μg of total protein) are loaded on a denaturing 15–20% acrylamide gradient gel (covered by a 3.6% stacking gel) as described (Chomyn, 1996) and run at 20–30 mA for 5–6 h. Representative results of this kind of analysis with control and mutant cell lines are shown in Fig. 1 (panels A and B). Other types of SDS-PAGE electrophoresis can also yield satisfactory results (Coenen *et al.*, 2004).

b. Pulse-Chase Experiments

1. Cells must be preincubated (8–15 h) in the presence of chloramphenicol (CAP) (40-μg/ml final concentration in the medium) prior to the addition of the radioisotope, to reversibly inhibit mitochondrial translation. This leads to the accumulation of a pool of cytosolically synthesized subunits of the OXPHOS complexes and facilitates the incorporation of the mitochondrial translation products labeled after removal of the drug (Costantino and Attardi, 1977).

2. After the incubation with CAP, the medium is removed, the cells are washed once with PBS, and normal growth medium is added to allow mitochondrial protein synthesis to resume. The cells are incubated in this medium for 30 min at 37 °C.

3. The labeling is performed as described in Section III.A.2.a, starting with the rinse and incubation steps (step 2) but substituting the antibiotic emetine by cycloheximide (100-μg/ml final concentration).

4. The cells are exposed for 2 h to [^{35}S] methionine and cysteine (pulse period) and different chases are performed (from 0.5 to 90 h). After the pulse, the cells are

washed twice with 5–10 ml of complete medium and incubated in this same medium for the desired chase period. In the case of long chase periods, trypsinization and replating of the cells may be needed to avoid reaching confluency of the culture. At the end of the chase, the cells are collected as described in Section III.A.2.a (step 5).

5. If purification of mitochondria is required for BN-PAGE analysis of assembled complexes, labeled cells are mixed with 100–200 μl of compact unlabeled cells, and mitochondria are isolated as described below.

c. Blue Native Analysis of Assembled Complexes

Estimation of the relative level of mitochondrial assembled complexes in the analyzed situations (control and mutant cell lines, different chase intervals, and so on) is performed by blue native electrophoresis (BN-PAGE) according to Schägger's protocol Schägger, 1996, 2001), with some modifications. Briefly, cells (50–70 million, corresponding to several plates of labeled cells or mixed labeled and unlabeled cells), are trypsinized, washed twice with PBS, and frozen at −80°C to increase cell breakage. The cells are allowed to thaw on ice for 5–10 min and are homogenized in a tightly fitting glass-Teflon homogenizer with about 7–8 cell pellet volumes of ice-cold Homogenizing buffer 1 (83-mM sucrose; 10-mM MOPS, pH 7.2). After adding an equal volume of ice cold Homogenizing buffer 2 (250-mM sucrose; 30-mM MOPS, pH 7.2), nuclei and unbroken cells are removed by centrifugation at $1000 \times g_{av}$ for 5 min at 4°C. The supernatant is transferred to Eppendorf tubes and mitochondria are collected by centrifugation at $12,000 \times g_{av}$ for 2 min and washed once under the same conditions with buffer A (320-mM sucrose; 1-mM EDTA; 10-mM Tris–HCl, pH 7.4). Mitochondrial pellets are suspended at a protein concentration of 5–10 μg/μl in buffer B (1-M 6-amiohexanoic acid; 50-mM Bis-Tris–HCl, pH 7.0) and the membrane proteins are solubilized by the addition of 1.6 g of dodecylmaltoside per gram of mitochondrial protein. After a 30-min centrifugation at $13,000 \times g_{av}$, the supernatant is collected and 10 μl of 5% Serva Blue G dye in 1-M 6-amiohexanoic acid are added to 30 μl of supernatant. A volume of 15–25 μl of sample ($\sim$100 μg of mitochondrial protein) is loaded in each lane and electrophoresed through 5–13% acrylamide BN-PAGE gels (Schägger, 1996). When the electrophoresis is finished, the gel is fixed, dried, and exposed for autoradiography. An example of the results obtained with the pulse labeling and BN analysis of assembled complexes is shown in Fig. 1C.

For the identification of the individual labeled polypeptides present in each complex or subcomplex, a second dimension gel can be performed. For that purpose, the first dimension lane from the BN gel is cut out and incubated in 1% SDS, 1% β-mercaptoethanol at RT for 1 h prior to running the second SDS-PAGE electrophoresis in 16.5% acrylamide in Tris–Tricine buffer (Schägger, 1996). Again, after the electrophoresis, the gel is fixed, dried, and exposed. A representative pattern of a second dimension gel obtained as described is shown in Fig. 1D.

The separated products, either from the first or from the second dimension gel, can be also transferred to nylon or PVDF membranes to perform western blot analysis using specific antibodies to confirm the identity of the observed complexes or the presence and position of a given polypeptide.

B. *In Organello* Protein Synthesis

1. General Considerations

Preparation of the mitochondrial fraction. As mentioned above, for *in organello* analysis of protein synthesis, the isolated mitochondria must be fully functional and able to perform coupled respiration. This requires a quick and gentle purification procedure. Depending on the source of mitochondria, different adaptations in the purification protocol may be needed. We routinely perform homogenization in isotonic buffers for tissue samples or in slightly hypotonic buffers that are rendered isotonic immediately after breakage of the cells for cell culture samples. The gentle homogenization step is followed by differential centrifugation and several washing steps so that the final mitochondrial fraction is relatively pure and highly functional. For a detailed description of the purification of mitochondria suitable for *in organello* analysis and from different sources see Fernández-Vizarra *et al.* (2006) (see also Chapter 1 by Pallotti and Lenaz and Chapter 4 by Kirby *et al.*, this volume).

Incubation time and conditions. In our experience, increasing the incubation times beyond 45 min does not increase the labeling of mitochondrial products under the selected incubation conditions. Other protocols in which the incubation temperature was decreased to 30 °C report a linear incorporation of radioactivity for up to 2 h (McKee *et al.*, 1990a). The inclusion of ADP and oxidizable substrates is essential for proper mitochondrial protein synthesis (McKee *et al.*, 1990b). It is also important to keep the mitochondria in suspension and well oxygenated during the assay, using, for example, a rotary wheel, to avoid the risk of anoxia due to their sedimentation to the bottom of the tube.

2. Labeling Procedure

1. Mitochondria isolated from cultured cells or from rat organs are suspended at a final concentration of 2 mg/ml of protein in 0.5 ml of Mitochondrial Incubation buffer (25-mM sucrose; 75-mM sorbitol; 100-mM KCl; 1-mM $MgCl_2$; 0.05-mM EDTA; 10-mM Tris–HCl; and 10-mM K_2HPO_4, pH 7.4) containing 10-mM glutamate, 2.5-mM malate, 1-mM ADP, and 1-mg/ml fatty acid-free bovine serum albumin (BSA).

2. For *in organello* protein synthesis assays, 100-μg/ml emetine, 100-μg/ml cycloheximide, and 10 μM each of the 20 L-amino acids (including methionine and cysteine) are added to the medium, and the mitochondria are incubated at 37 °C for 5 min with gentle shaking.

3. To start the labeling, >75 μCi of PRO-MIX™ L-[^{35}S] *in vitro* Cell Labeling Mix (>1000 Ci/mmol) (Amersham) are added to each sample, and incubation is performed in 1.5-ml Eppendorf tubes at 37 °C for 30 min in a rotary wheel (12 rpm).

4. After the incubation, the mitochondrial samples are concentrated by centrifugation at 13,000 × g_{av} for 1 min and the mitochondrial pellet is washed twice with 1 ml of cold buffer A (320-mM sucrose; 1-mM EDTA; 10-mM Tris–HCl, pH 7.4). The final mitochondrial pellet is resuspended in 40 μl of TE buffer containing 1-mM PMSF and lysed with 10 μl of 5× loading buffer.

5. The labeled mitochondrial proteins (150- to 300-μg total protein/lane) are separated by electrophoresis on denaturing gels, as described above for the *in vivo* pulse-labeled products. The gels produced are dried and exposed for X-ray film at −70 °C with a DuPont screen intensifier. Quantification of the amount of protein in the autoradiograms is carried out by standard procedures. Representative results of *in organello* protein synthesis analysis with organelles isolated from human cell lines and from different rat organs are shown in Fig. 2, panels A, B, and C.

C. Complementary Analyses: *In Organello* Aminoacylation Assay

1. General Considerations

Mitochondrial purification and incubation times and conditions. The same purification protocol that renders mitochondria suitable for *in organello* protein synthesis is optimal for analysis of aminoacylation (Enríquez and Attardi, 1996; Fernández-Vizarra *et al.*, 2006). Incubation is carried out under conditions similar to those described for protein synthesis, but for a shorter period (15 min).

2. Labeling Procedure

1. Mitochondria isolated from cultured cells or rat organs are suspended at a final concentration of ∼1 mg/ml of mitochondrial protein in 0.5 ml of Mitochondrial Incubation buffer (see above) containing 10-mM glutamate, 2.5-mM malate, 1-mM ADP, and 1-mg/ml fatty acid-free BSA.

2. For *in organello* aminoacylation assays, 10 μM of each of the 20 L-amino acids (except for the labeled one) are added to the medium. To start the labeling, >75 μCi of either L-[^{3}H] or L-[^{35}S]-labeled amino acid (1200 Ci/mmol) are added to each sample and incubated in 1.5-ml Eppendorf tubes at 37 °C for 15 min in a rotary wheel (12 rpm).

3. After the incubation, the mitochondrial samples are concentrated by centrifugation at 13,000 × g_{av} for 1 min at 4 °C, and washed twice with 1 ml of ice-cold buffer A (see above). The final mitochondrial pellet is resuspended in 10-mM Tris–HCl (pH 7.4), 0.15-M NaCl, 1-mM EDTA, and incubated for 10 min at 37 °C in the presence of 200-μg/μl proteinase K and 2.5% SDS. Total mitochondrial

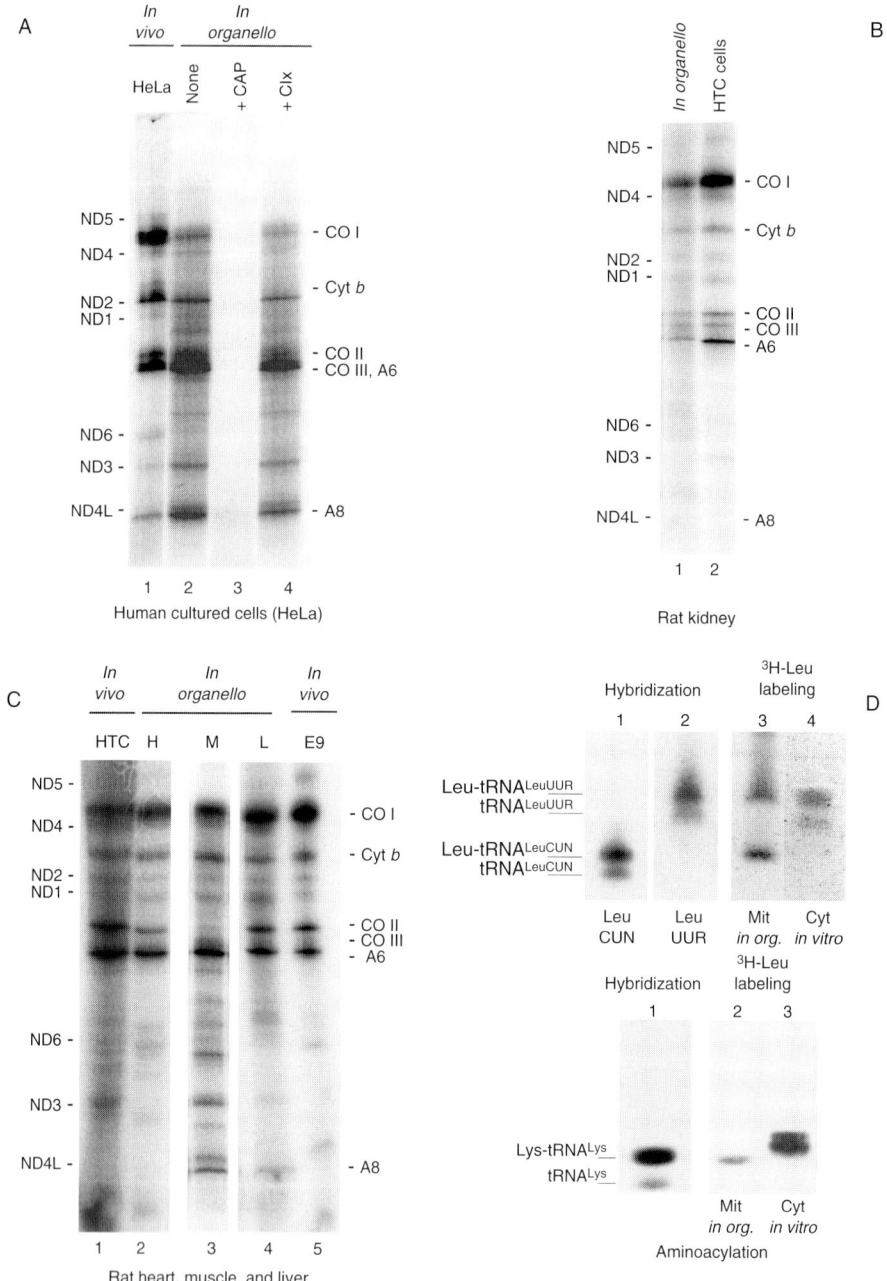

Fig. 2 *In organello* protein synthesis and aminoacylation. (A) Electrophoretic pattern of mitochondrial translation products derived from labeled intact human HeLa cells (*in vivo*, lane 1) or from isolated mitochondria from HeLa cells in the absence of antibiotics (None, lane 2) and in the presence of

nucleic acids, which contain radiolabeled mitochondrial tRNAs, are extracted with an equal volume of phenol (preequilibrated with 0.1-M sodium acetate, pH 4.7; 0.5-M NaCl; and 1% SDS). The aqueous phase is collected and nucleic acids are precipitated twice with ethanol at $-20\,^{\circ}$C. The samples are resuspended in sterile water, divided into aliquots, and stored at $-80\,^{\circ}$C.

4. To detect aminoacylation, the labeled mitochondrial tRNAs are separated by electrophoresis at $4\,^{\circ}$C, under acid conditions. The samples are mixed with 1.5 volumes of loading buffer (0.1-M sodium acetate, pH 5.0; 7-M urea; 0.05% bromophenol blue) and loaded on a 20-cm long, 1-mm thick 6.5% polyacrylamide/7-M urea gel. The gels are either electroblotted to a Zeta-probe membrane or dried and exposed for autoradiography at $-70\,^{\circ}$C with a DuPont screen intensifier. Quantification of the amount of aminoacyl-tRNA is carried out in the autoradiograms by standard procedures.

5. Although in the *in organello* assay, described here, aminoacylation of cytosolic tRNAs contaminating the mitochondrial fraction is almost negligible, it is recommended to run in parallel a sample of cytosolic tRNAs charged *in vitro* with the same amino acid, to verify the separation of the mitochondrial and the cytosolic tRNAs (Enriquez and Attardi, 1996; Enriquez *et al.*, 1995). Representative results of this kind of analysis for tRNA^{LeuUUR}, tRNA^{LeuCUN}, and tRNA^{Lys} are shown in Fig. 2D.

IV. Materials

Mouse control (E9 and Fbalb/cJ) and complex I knockout (ND6ko) cell lines were grown in DMEM medium supplemented with 5% calf or fetal bovine serum and 1-mM pyruvate. Complex III knockout cell lines (A22 and FA22) also require uridine, at a concentration of 50 μg/ml. Human control (143B and HeLa), complex I-mutant (C4–3-J), and tRNA^{Lys}-mutant (pT1C3) cell lines were grown in DMEM medium supplemented with 5% of bovine calf serum and 1-mM pyruvate. Rat control cell line HTC was grown in DMEM medium supplemented with 10% of bovine calf serum.

---

chloramphenicol (CAP, lane 3) and cyclohexymide (Clx, lane 4). (B) and (C) Comparison of the electrophoretic patterns of *in vivo* (cell cultures: rat cell line HTC, mouse cell line E9) and *in organello* labeled mitochondrial translation products from rat kidney (panel B, lane 1) and heart, skeletal muscle, and liver (panel C, lanes 2, 3, and 4, respectively). (D) *In organello* aminoacylation. Aminoacylation products anlayzed by acid polyacrylamide gel electrophoresis and blotting of L-[^{3}H]-Leu (upper panel, lane 3) or L-[^{3}H]-Lys (lower panel, lane 2) labeled mitochondria. The identity of the labeled tRNAs was established as tRNA^{LeuCUN} (upper panel, lane 1), tRNA^{LeuUUR} (upper panel, lane 2), and tRNA^{Lys} (lower panel, lane 1) by hybridization with appropriate probes. *In vitro* labeled cytoplasmic tRNAs (Cyt *in vitro*) were run in parallel to assess potential contamination (upper panel, lane 4; lower panel, lane 3). Panel A, from Fernández-Vizarra *et al.* (2002) with permission; panel D, from Enriquez and Attardi (1996) with permission.

The labeled amino acid radioisotopes, PRO-MIX^{TM} L-[^{35}S] *in vitro* Cell Labeling Mix (>1000 Ci/mmol), containing [^{35}S]-labeled methionine and cysteine; and ^{3}H-labeled L-Leucine and L-Lysisne (1200 Ci/mmol) were purchased from Amersham. *n*-Dodecyl-β-d-maltoside and Bis-Tris were purchased from Acros organics (Geel, Belgium); 6-amiohexanoic acid was from Fluka (Madrid, Spain), and the remaining chemicals and reagents were from Sigma (Madrid, Spain).

V. Discussion

Typical patterns of the mitochondrial polypeptides labeled in *in vivo* and *in organello* protein synthesis experiments are shown in Figs. 1 (panels A and B) and 2 (panels A and B), respectively. Normal exposure times both for *in vivo* pulse-labeling experiments and for *in organello* analysis range between 2 and 7 days, while for BN analysis of assembled complexes exposure times can be from 1 to 5 days and from 3 days to 2 weeks for the first and the second dimension gels, respectively. The most frequent problems with these types of analyses are a low signal in the labeling of mitochondrial products or a high background. In the case of low signal, the isotope concentration can be increased. If the background is the problem, care in the handling of samples and the use of protease inhibitors in the lysis buffer is recommended. In both cases the incubation time with the antibiotic and its activity, as well as the confluency of the cells should be checked. Partial degradation by protease activity tends to show as disappearance of the high–molecular-weight polypeptides and as an increase in background at the low-molecular-weight region of the gel.

The combination of pulse-chase labeling and BN-PAGE analysis allows for the detection of the OXPHOS complexes containing mtDNA-encoded subunits. Thus, as can be observed in Fig. 1C, complexes I, III, IV, and V are detectable in the first dimension BN gel in the control cell line. The cell line lacking cyt *b* (Cyt*b*^{ko}) shows the absence of assembled complex III and a decreased stability in complex I compared to the control, while the ND6 mutant (ND6^{ko}) does not assemble a full complex I. The assembly kinetics of each complex show clear differences and may require adjustment of the chase periods to be able to follow them (Fig. 1, panels C and D). In addition, changes on the stability of a given complex may affect the stability and turnover of other complexes (Acín-Pérez *et al.*, 2004; Fig. 1C).

Analysis by second dimension electrophoresis of the labeled mitochondrial translation products corresponding to the mouse control cell line (FBalb/cJ) after a 30-min chase is shown in Fig. 1D. By comparing with the first dimension gel from which the lane was excised, the position of each individual subunit can be followed and assigned to a full complex or to subcomplexes (see, e.g., the very conspicuous signals for COX I and ATPase6). Note that while most of the signal of complex V subunits (A6 and A8) appears in the area corresponding to the fully assembled complex (Vm), for complex I subunits (e.g., ND1 and ND2) the signal is more intense in areas corresponding to subcomplexes (I_{s1} and I_{s2}). Longer chase times would yield an increase in the proportion of signal located in the full complexes (Fig. 1C, lanes 2–3 and 5–6), and would provide information about complex stability (Acín-Pérez *et al.*, 2004).

Although the electrophoretic pattern of the *in organello* labeled products is very similar to the *in vivo* pattern (Fig. 2), additional bands of unknown nature usually appear in the region of low-molecular weight, making it difficult to identify some polypeptides such as ND3 and ND6. The differences in the pattern among the mitochondria from the various origins reflect their metabolic and biogenetic properties (Leung and McKee, 1990).

VI. Summary

Mitochondria possess their own translation system devoted to the synthesis of the mtDNA-encoded polypeptides. Analysis of the protein synthesis capacity of mitochondria can be a useful tool to study basic aspects of mtDNA expression, as well as to detect defects due to mutations in the mtDNA or in nuclear genes involved in this process. The combination of the different options for labeling and analysis of the translation products must be adapted for each particular application.

Acknowledgments

We thank Santiago Morales for his technical assistance. Our work was supported by the Spanish Ministry of Education (SAF2003-00103), the Instituto de Salud Carlos III (REDEMETH-G03/054, REDCIEN C03/06-Grupo RC-N34-3, and Research Project PI-042647), the EU (EUMITOCOMBAT-LSHM-CT-2004-503116), and by the Group of Excellence grant of the local government (DGA-B55). RA-P is supported by a fellowship from the Spanish Ministry of Education.

References

Acín-Pérez, R., Bayona-Bafaluy, M. P., Fernandez-Silva, P., Moreno-Loshuertos, R., Perez-Martos, A., Bruno, C., Moraes, C. T., and Enriquez, J. A. (2004). Respiratory complex III is required to maintain complex I in mammalian mitochondria. *Mol. Cells* **13**, 805–815.

Antonicka, H., Sasarman, F., Kennaway, N. G., and Shoubridge, E. A. (2006). The molecular basis for tissue specificity of the oxidative phosphorylation deficiencies in patients with mutations in the mitochondrial translation factor EFG1. *Hum. Mol. Genet.* **15**, 1835–1846.

Askarian-Amiri, M. E., Pel, H. J., Guevremont, D., McCaughan, K. K., Poole, E. S., Sumpter, V. G., and Tate, WP. (2000). Functional characterization of yeast mitochondrial release factor 1. *J. Biol. Chem.* **275**, 17241–17248.

Attardi, G., King, M. P., Chomyn, A., and Loguercio-Polosa, P. (1991). Novel genetic and molecular approaches to the study of mitochondrial biogenesis and mitochondrial diseases in human cells. *In* "Progress in Neuropathology" (T. Sato and S. DiMauro, eds.), pp. 75–92. Raven Press, New York.

Bai, Y., Shakeley, R. M., and Attardi, G. (2000). Tight control of respiration by NADH dehydrogenase ND5 subunit gene expression in mouse mitochondria. *Mol. Cell. Biol.* **20**, 805–815.

Brass, E. P. (1992). Translation rates of isolated liver mitochondria under conditions of hepatic mitochondrial proliferation. *Biochem. J.* **15**, 175–180.

Bruno, C., Martinuzzi, A., Tang, Y., Andreu, A. L., Pallotti, F., Bonilla, E., Shanske, S., Fu, J., Sue, C. M., Angelini, C., DiMauro, S., and Manfredi, G. (1999). A stop-codon mutation in the human mtDNA cytochrome c oxidase I gene disrupts the functional structure of complex IV. *Am. J. Hum. Genet.* **65** 611–620.

Cai, Y. C., Bullard, J. M., Thompson, N. L., and Spremulli, L. L. (2000). Interaction of mitochondrial elongation factor Tu with aminoacyl-tRNA and elongation factor Ts. *J. Biol. Chem.* **275**, 20308–20314.

Chomyn, A. (1996). *In vivo* labelling and analysis of human mitochondrial translation products. *In* "Methods in Enzymology" (G. Attardi, ed.), Vol. 264, pp. 197–210. Academic Press, San Diego, CA (USA).

Chomyn, A., Meola, G., Bresolin, N., Lai, S. T., Scarlato, G., and Attardi, G. (1991). *In vitro* genetic transfer of protein synthesis and respiration defects to mitochondrial DNA-less cells with myopathy-patient mitochondria. *Mol. Cell. Biol.* **11**, 2236–2244.

Chomyn, A., Enriquez, J. A., Micol, V., Fernandez-Silva, P., and Attardi, G. (2000). The mitochondrial myopathy, encephalopathy, lactic acidosis, and stroke-like episode syndrome-associated human mitochondrial tRNALeu(UUR) mutation causes aminoacylation deficiency and concomitant reduced association of mRNA with ribosomes. *J. Biol. Chem.* **275**, 19198–19209.

Clark, K. M., Taylor, R. W., Johnson, M. A., Chinnery, P. F., Chrzanowska-Lightowlers, Z. M., Andrews, R. M., Nelson, I. P., Wood, N. W., Lamont, P. J., Hanna, M. G., Lightowlers, R. N., and Turnbull, D. M. (1999). An mtDNA mutation in the initiation codon of the cytochrome C oxidase subunit II gene results in lower levels of the protein and a mitochondrial encephalomyopathy. *Am. J. Hum. Genet.* **64**, 1330–1339.

Coenen, M. J. H., Antonicka, H., Ugalde, C., Sasarman, F., Rossi, R., Heister, A., Newbold, R. F., Trijbels, F., van de Heuvel, L. P., Shoubridge, E. A., and Smeitink, J. A. M. (2004). Mutant mitochondrial elongation factor G1 and combined oxidative phosphorylation deficiency. *N. Engl. J. Med.* **351**, 2080–2086.

Coleman, W. B., and Cunningham, C. C. (1991). Effect of chronic ethanol consumption on hepatic mitochondrial transcription and translation. *Biochim. Biophys. Acta* **1058**, 178–186.

Copple, D. M., Rush, G. F., and Richardson, F. C. (1992). Effects of methapyrilene measured in mitochondria isolated from naïve and methapyrene-treated rat and mouse hepatocytes. *Toxicol. Appl. Pharmacol.* **116**, 10–16.

Costantino, P., and Attardi, G. (1977). Metabolic properties of the products of mitochondrial protein synthesis in HeLa cells. *J. Biol. Chem.* **252**, 1702–1711.

Cote, C., Poirier, J., and Boulet, D. (1989). Expression of the mammalian mitochondrial genome. Stability of mitochondrial translation products as a function of membrane potential. *J. Biol. Chem.* **264**, 8487–8490.

Enriquez, J. A., and Attardi, G. (1996). Analysis of aminoacylation of human mitochondrial tRNAs. *In* "Methods in Enzymology" (G. Attardi, ed.), Vol. 264, pp. 183–196. Academic Press, San Diego, CA (USA).

Enriquez, J. A., Chomyn, A., and Attardi, G. (1995). mtDNA mutation in MERRF syndrome causes defective aminoacylation of tRNA(Lys) and premature translation termination. *Nat. Genet.* **10**, 47–55.

Enriquez, J. A., Fernández-Silva, P., Perez-Martos, A., Lopez-Perez, M. J., and Montoya, J. (1996). The synthesis of RNA in isolated mitochondria can be mantained for several hours and is inhibited by high levels of ATP. *Eur. J. Biochem.* **237**, 601–610.

Enriquez, J. A., Fernández-Silva, P., Garrido-Pérez, N., Pérez-Martos, A., López Perez, M. J., and Montoya, J. (1999). Direct regulation of mitochondrial RNA synthesis by thyroid hormone. *Mol. Cell. Biol.* **19**, 657–670.

Enriquez, J. A., Cabezas-Herrera, J., Bayona-Bafaluy, M. P., and Attardi, G. (2000). Very rare complementation between mitochondria carrying different mitochondrial DNA mutations points to intrinsic genetic autonomy of the organelles in cultured human cells. *J. Biol. Chem.* **275**, 11207–11215.

Escobar-Galbis, M. L., Allen, J. F., and Hakansson, G. (1998). Protein synthesis by isolated pea mitochondria is dependent on the activity of respiratory complex II. *Curr. Genet.* **33**, 320–329.

Fernández-Vizarra, E., López-Pérez, M. J., and Enriquez, J. A. (2002). Isolation of biogenetically competent mitochondria from mammalian tissues and cultured cells. *Methods* **26**, 292–297.

Fernández-Vizarra, E., Fernández-Silva, P., and Enríquez, J. A. (2006). Isolation of mitochondria from mammalian tissues and cultured cells. *In* "Cell Biology: A Laboratory Handbook" (J. E. Celis, ed.), 3rd edn., Vol. 2, pp. 69–77. Elsevier-Academic Press, San Diego, CA (USA).

Giege, P., Sweetlove, L. J., Cognat, V., and Leaver, C. J. (2005). Coordination of nuclear and mitochondrial genome expression during mitochondrial biogenesis in Arabidopsis. *Plant Cell* **17**, 1497–1512.

Guan, M. X., Fischel-Ghodsian, N., and Attardi, G. (2000). A biochemical basis for the inherited susceptibility to aminoglycoside ototoxicity. *Hum. Mol. Genet.* **9**, 1787–1793.

Hammarsund, M., Wilson, W., Corcoran, M., Merup, M., Einhorn, S., Grander, D., and Sangfelt, O. (2001). Identification and characterization of two novel human mitochondrial elongation factor genes, hEFG2 and hEFG1, phylogenetically conserved through evolution. *Hum. Genet.* **109**, 542–550.

Hanada, T., Suzuki, T., Yokogawa, T., Takemoto-Hori, C., Sprinzl, M., and Watanabe, K. (2001). Translation ability of mitochondrial tRNAsSer with unusual secondary structures in an *in vitro* translation system of bovine mitochondria. *Genes Cells* **6**, 1019–1030.

Hao, H., and Moraes, C. T. (1996). Functional and molecular mitochondrial abnormalities associated with a C → T transition at position 3256 of the human mitochondrial genome. The effects of a pathogenic mitochondrial tRNA point mutation in organelle translation and RNA processing. *J. Biol. Chem.* **26**, 2347–2352.

Hofhaus, G., and Attardi, G. (1995). Efficient selection and characterization of mutants of a human cell line which are defective in mitochondrial DNA-encoded subunits of respiratory NADH dehydrogenase. *Mol. Cell. Biol.* **15**, 964–974.

Koc, E. C., and Spremulli, L. L. (2002). Identification of mammalian mitochondrial translational initiation factor 3 and examination of its role in initiation complex formation with natural mRNAs. *J. Biol. Chem.* **277**, 35541–35549.

Leung, A. C., and McKee, E. E. (1990). Mitochondrial protein synthesis during thyroxine-induced cardiac hypertrophy. *Am. J. Physiol.* **258**, E511–E518.

Liao, H.-X., and Spremulli, L. L. (1991). Initiation of protein synthesis in animal mitochondria: Purification and characterization of translational initiation factor 2. *J. Biol. Chem.* **266**, 20714–20719.

Limongelli, A., Schaefer, J., Jackson, S., Invernizzi, F., Kirino, Y., Suzuki, T., Reichmann, H., and Zeviani, M. (2004). Variable penetrance of a familial progressive necrotising encephalopathy due to a novel tRNA[Ile] mutation in the mitochondrial genome. *J. Med. Genet.* **41**, 342–349.

McKee, E. E., Grier, B. L., Thompson, G. S., and McCourt, J. D. (1990a). Isolation and incubation conditions to study heart mitochondrial protein synthesis. *Am. J. Physiol.* **258**, E492–E502.

McKee, E. E., Grier, B. L., Thompson, G. S., Leung, A. C. F., and McCourt, J. D. (1990b). Coupling mitochondrial metabolism and protein synthesis in heart mitochondria. *Am. J. Physiol.* **258**, E502–E510.

Miller, C., Saada, A., Shaul, N., Shabtai, N., Ben-Shalom, E., Shaag, A., Hershkovitz, E., and Elpeleg, O. (2004). Defective mitochondrial translation caused by a ribosomal protein (MRPS16) mutation. *Ann. Neurol.* **56**, 734–738.

Ostronoff, L. K., Izquierdo, J. M., Enriquez, J. A., Montoya, J., and Cuezva, J. M. (1996). Transient activation of mitochondrial translation regulates the expression of the mitochondrial genome during mammalian mitochondrial differentiation. *Biochem. J.* **316**, 183–191.

Patel, V. B., and Cunningham, C. C. (2002). Altered hepatic mitochondrial ribosome structure following chronic ethanol consumption. *Arch. Biochem. Biophys.* **398**, 41–50.

Poyton, R. O., Bellus, G., McKee, E. E., Sevarino, K. A., and Goehring, B. (1996). *In organello* mitochondrial protein and RNA synthesis systems from *Saccharomyces cerevisiae*. *In* "Methods in Enzymology" (G. Attardi, ed.), Vol. 264, pp. 36–42. Academic Press, San Diego, CA (USA).

Schägger, H. (1996). Electrophoretic techniques for isolation and quantification of oxidative phosphor-ylation complexes from human tissues. *In* "Methods in Enzymology" (G. Attardi, ed.), Vol. 264, pp. 555–566. Academic Press, San Diego, CA (USA).

Schägger, H. (2001). Blue-native gels to isolate protein complexes from mitochondria. *In* "Mitochondria-Methods in Cell Biology" (L. A. Pon, and E. A. Schon, eds.), Vol. 65, pp. 231–244. Elsevier, San Diego, CA (USA).

Spencer, A. C., and Spremulli, L. L. (2005). The interaction of mitochondrial translational initiation factor 2 with the small ribosomal subunit. *Biochim. Biophys. Acta* **1750,** 69–81.

Taanman, J.-W. (1999). The mitochondrial genome: Structure, transcription, translation and replica-tion. *Biochim. Biophys. Acta* **1410,** 103–123.

Takemoto, C., Koike, T., Yokogawa, T., Benkowski, L., Spremulli, L. L., Ueda, T. A., Nishikawa, K., and Watanabe, K. (1995). The ability of bovine mitochondrial transfer RNA^{Met} to decode AUG and AUA codons. *Biochimie* **77,** 104–108.

Taylor, R. W., and Turnbull, D. M. (2005). Mitochondrial DNA mutations in human disease. *Nat. Rev. Genet.* **6,** 389–402.

Terasaki, M., Suzuki, T., Hanada, T., and Watanabe, K. (2004). Functional compatibility of elongation factors between mammalian mitochondrial and bacterial ribosomes: Characterization of GTPase activity and translation elongation by hybrid ribosomes bearing heterologous L7/12 proteins. *J. Mol. Biol.* **336,** 331–342.

Tiranti, V., Corona, P., Greco, M., Taanman, J. W., Carrara, F., Lamantea, E., Nijtmans, L., Uziel, G., and Zeviani, M. (2000). A novel frameshift mutation of the mtDNA COIII gene leads to impaired assembly of cytochrome c oxidase in a patient affected by Leigh-like syndrome. *Hum. Mol. Genet.* **9,** 2733–2742.

Westermann, B., Herrmann, J. M., and Neupert, W. (2001). Analysis of mitochondrial translation products *in vivo* and *in organello* in Yeast. *In* "Mitochondria-Methods in Cell Biology" (L. A. Pon, and E. A. Schon, eds.), Vol. 65, pp. 429–438. Elsevier, San Diego, CA (USA).

Yasuhara, T., Mera, Y., Nakai, T., and Ohashi, A. (1994). ATP-dependent proteolysis in yeast mitochondria. *J. Biochem.* **15,** 1166–1171.

Yasukawa, T., Suzuki, T., Ishii, N., Ohta, S., and Watanabe, K. (2001). Wobble modification defect in tRNA disturbs codon-anticodon interaction in a mitochondrial disease. *EMBO J.* **20,** 4794–4802.

Zhang, L., Ging, N. C., Komoda, T., Hanada, T., Suzuki, T., and Watanabe, K. (2005). Antibiotic susceptibility of mammalian mitochondrial translation. *FEBS Lett.* **579,** 6423–6427.

Zhang, Y., and Spremulli, L. L. (1998). Identification and cloning of human mitochondrial translational release factor 1 and the ribosome recycling factor. *Biochim. Biophys. Acta* **1443,** 245–250.

PART VI

Assays for Mitochondrial Morphology and Motility

CHAPTER 29

Visualization of Mitochondria in Budding Yeast

Theresa C. Swayne, Anna C. Gay, and Liza A. Pon

Department of Anatomy and Cell Biology, College of Physicians and Surgeons
Columbia University, New York, New York 10032

I. Introduction

The small size, rounded shape, and rigid cell wall of fungi, including budding yeast, were once obstacles for optical imaging. However, it is now feasible for laboratories to acquire and analyze high-resolution multidimensional images of yeast and structures within yeast, including mitochondria. Since mitochondria cannot be detected in yeast using transmitted-light microscopy (phase-contrast or differential interference contrast), methods have been developed to visualize the organelle with vital fluorescent dyes, immunofluorescence, or targeted fluorescent proteins (FPs). This chapter describes labeling methods and optical approaches for visualizing yeast mitochondria using fluorescence microscopy. Applications and outcomes for each of the commonly used approaches are summarized in Table I and Fig. 1.

II. Use of Vital Dyes to Stain Yeast Mitochondria

Since vital dyes are commercially available and stain mitochondria with short incubation times, they are useful for rapidly assessing mitochondrial distribution, morphology, and dynamics in live cells. In addition, since vital dyes work by sensing membrane potential or by binding to DNA, they also provide information regarding mitochondrial integrity and function, and stain independently of the cell's ability to express and import foreign proteins into mitochondria. The main advantages of vital staining are speed and functional readout (Table I).

A. DNA-Binding Dyes

In yeast and other eukaryotes, mitochondrial DNA (mtDNA) assembles into punctate structures, called nucleoids, that are associated with the inner leaflet of

Table I
Recommendations for Imaging of Mitochondria in Budding Yeast[a]

Sample	Imaging method		
	Vital dye	Immunofluorescence	Targeted GFP
Fixed Cells			
Wild-type cells	*	✓	*
Cells with respiratory defects	*	*	*
Cells with import defects	*	*	*
Live Cells			
Wild-type cells	✓	x	✓
Cells with respiratory defects	*	x	✓
Cells with import defects	✓	x	*

[a]Choices for visualization. ✓: recommended; *: use with caveats (see text for discussion); x: not recommended/not possible.

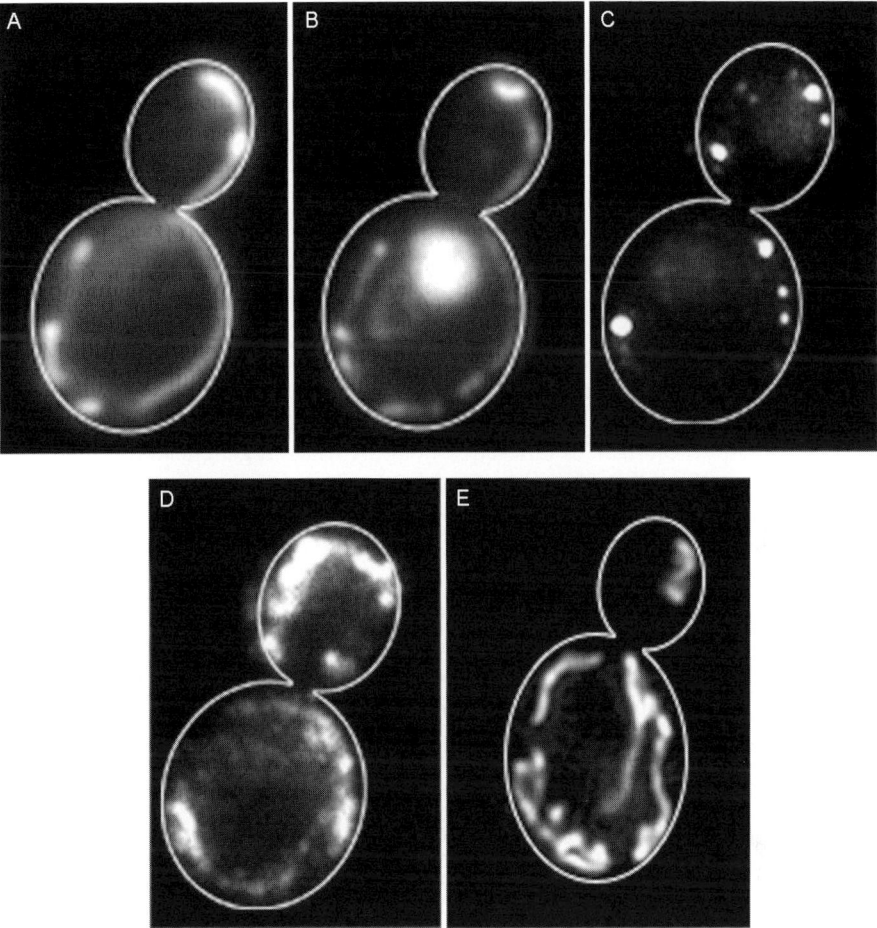

Fig. 1 Mitochondria visualized using different imaging approaches in fixed and living yeast cells. (A and B) Vital dyes. A living cell double-labeled with DiOC6(3) (A) and DAPI (B). (C and D) Immuno-fluorescence. A fixed cell double-labeled with antibodies against myc-tagged Mdm12p, an integral mito-chondrial outer membrane protein (C), and against mitochondrial outer membranes (D). (E) Genetically encoded live-cell tag. A living cell ectopically expressing GFP fused to the CIT1 signal sequence.

the inner mitochondrial membrane. mtDNA nucleoids contain multiple copies of mtDNA and proteins that contribute to the organization, replication, and expression of mtDNA. Since mitochondria are the only extranuclear organelles in animal cells and fungi that contain DNA, cytoplasmic DNA staining can be diagnostic for mitochondria. In yeast, a species in which mtDNA is dispensable, DNA-binding dyes are also used to determine if a strain is rho^0 (lacks mtDNA; also denoted ρ^0).

DAPI (4′,6′-diamidino-2-phenylindole) is the most common DNA-binding dye used in yeast. Upon binding to nucleic acids, DAPI fluorescence increases greatly, and the increase is more pronounced with DAPI binding to DNA than to RNA. These characteristics make DAPI a strong nuclear and mtDNA marker, with little cytoplasmic background staining. Another advantage is that it stains mtDNA independently of the metabolic state of the mitochondria. Consequently, it can be used in cells whose mitochondrial function may be impaired. Finally, DAPI stains DNA in both living and fixed cells, and produces a robust, persistent fluorescent signal.

There are two issues to bear in mind when working with DAPI. First, because DAPI also stains nuclear DNA, mtDNA nucleoids in close proximity to the nucleus are not well resolved. Second, because DAPI is visualized with UV illumination, sustained imaging of DAPI in live cells results in phototoxicity. Indeed, mitochondrial fragmentation or rupture can occur in DAPI-stained cells after 1–2 min of continuous illumination with conventional fluorescent light sources. Therefore, for prolonged or time-lapse imaging without genetic manipulation, we recommend the lipophilic dyes described below. For time-lapse imaging of mtDNA nucleoids, an FP should be fused to one of the nucleoid proteins (Section V).

B. Lipophilic Membrane Potential–Sensing Dyes

Membrane potential-sensing dyes are lipophilic, positively charged fluorophores that accumulate in cellular compartments that have a membrane potential. Because functioning mitochondria have the strongest membrane potential in the cell, these dyes accumulate in the mitochondria more readily than in any other compartment. Moreover, in contrast to the DNA-binding dyes, which stain punctate intramitochondrial structures, membrane potential-sensing dyes stain the entire mitochondrial membrane. As a result, they are excellent tools for investigating mitochondrial distribution and morphology. Finally, because of their potential-dependent accumulation, these dyes provide information regarding mitochondrial function and integrity.

The membrane potential-sensing dyes that work well in yeast are the carbocyanine $DiOC_6(3)$, the styryl dye DASPMI, the cationic rhodamine derivative rhodamine 123, and the fixable stains of the MitoTracker family (Table II). Of the dyes described here, rhodamine 123 is the least lipophilic and consequently the most sensitive to membrane potential. $DiOC_6(3)$ and various MitoTrackers are useful for double-labeling experiments because they have narrow excitation and emission spectra. Finally, the orange and red MitoTracker dyes persist in mitochondria after aldehyde fixation and permeabilization by acetone or Triton X-100, and thus are the only membrane potential-sensing dyes that can be used together with immunofluorescence staining.

These dyes are commercially available, well characterized, and stain with a high degree of specificity within very short incubation times. Therefore, visualization of mitochondria in living cells can be performed much more readily with membrane potential-sensing dyes compared to organelle-targeted FPs (see below). However,

Table II
Vital Dyes for Yeast Mitochondria

Dye	IUPAC name	λex	λem
DiOC$_6$(3)	3,3'-dihexyloxacarbocyanine iodide	484	501
DASPMI	4-(4-(dimethylamino)styryl)-N-methylpyridinium iodide (4-Di-1-ASP)	475	605[a]
Rhodamine 123	2-(6-Amino-3-imino-3H-xanthen-9-yl)benzoic acid methyl ester	505	534[a]
MitoTracker[b]	Various	Various	Various
DAPI	4',6-diamidino-2-phenylindole	358	461

[a]Broad emission range; not recommended for green/red double-label studies.
[b]Proprietary dyes from Invitrogen, Inc. (Carlsbad, CA).

we find that the fluorescence signal of these vital dyes is generally weaker and less persistent compared to targeted FPs.

C. Staining Protocol

1. Grow yeast cells to mid-log phase in liquid media.
2. Concentrate cells by centrifugation (5 sec at $\sim$13,000 $\times$ g) at RT.[1]
3. Wash cell pellet with of 1$\times$ PBS or growth media equal to the original volume of culture. Synthetic media is less autofluorescent than YPD. While PBS has even lower background fluorescence, it provides no energy and thus should be used only if cells will be observed for less than 30 min.
4. Resuspend cells to $\sim$2 $\times$ 10^8 cells/ml in PBS or growth media.[2]
5. Add appropriate volume of dye stock solution and mix thoroughly. Conditions that have been used to stain mitochondria in budding yeast are outlined in Table III.[3]
6. Incubate in the dark for the designated time.
7. Pipette gently to resuspend cells.[4]
8. Mount cells for short-term or long-term observation as described in Section VII.

Notes:

[1] Extended centrifugation can cause breakage of tubular mitochondria.

[2] A culture with OD$_{600}$ = 1.0 contains $\sim$10^7 cells/ml.

[3] The dye concentrations suggested in Table III should be considered only as a starting point. Specificity is highly dependent on concentration and mitochondrial membrane potential. When working with a new strain or mitochondrial mutant, titrate both dye concentration and incubation time. In general, keep the dye concentration as low as possible. In the case of membrane potential-sensing dyes, excessively high concentrations can alter mitochondrial morphology, produce defects in mitochondrial respiratory activity, and lead to staining of other cellular compartments (such as endoplasmic reticulum) that have weak membrane potentials.

[4] Washing cells is generally not necessary, as the fluorescence of most vital dyes is low until the dye binds to cell lipids. If high background is observed, cells can be washed once in media.

Table III
Use of Vital Dyes for Yeast Mitochondria

Dye	[Stock][a]	Staining conditions	
		[Dye]	Incubation
DiOC$_6$(3)[b]	10 mg/ml in ethanol	10–100 ng/ml	15–30 min RT
DASPMI[c]	1 mg/ml in ethanol	10–100 μg/ml	30 min RT
Rhodamine 123[d]	10 mg/ml in DMSO[e]	5–10 μg/ml	15–30 min RT
MitoTracker[f]	1 mM in DMSO[e]	25–500 nM	15–30 min RT
DAPI	1 mg/ml in H$_2$O	0.1 μg/ml	15 min RT

[a]All stock solutions should be stored in the dark.
[b]Simon *et al.*, 1995.
[c]McConnell *et al.*, 1990.
[d]Skowronek *et al.*, 1990; broad emission range is not recommended for green/red double-label studies.
[e]DMSO can be toxic and inhibits partitioning of the dye into the aqueous environment of the cells. If DMSO is the solvent, use a stock concentration that is at least 100×.
[f]Nunnari *et al.*, 1997.

III. Use of Immunostaining to Detect Mitochondria in Fixed Yeast Cells

For indirect immunofluorescence staining, yeast cells are typically grown to mid-log phase and fixed with paraformaldehyde. Since antibodies will not penetrate the yeast cell wall, the cell wall is removed from fixed cells enzymatically (e.g., with zymolyase or lyticase). Spheroplasts are then permeabilized using a nonionic detergent, and immobilized on a microscope coverslip. The sample-coated coverslip is then incubated with the primary antibody, which binds to the antigen of interest, and the secondary antibody, which binds to the invariant (Fc) region of the primary antibody and which is tagged with a fluorophore. Finally, the stained coverslip is applied to a microscope slide using mounting solution.

Several proteins can serve as markers for immunofluorescence visualization of mitochondria in *Saccharomyces cerevisiae*. These include proteins targeted to each of the submitochondrial compartments (Table IV). In addition to these antibodies to specific proteins, a polyclonal antibody raised against outer mitochondrial membranes has been used successfully for immunofluorescence (Riezman *et al.*, 1983; Smith *et al.*, 1995). Finally, with the advent of chromosomal tagging, proteins can be epitope-tagged and visualized by immunofluorescence using commercially available, well-characterized antibodies (Section V).

A. Pretreatment of Antibodies with Yeast Cell Walls

Rabbit antisera, even after affinity purification, often contain antibodies that recognize the yeast cell wall, bind to residual cell wall on spheroplasts, and generate background staining which may be punctate or uniformly distributed

Table IV
Marker Antigens for Yeast Mitochondria[a]

Protein	Location	References
Porin	OM	Mihara and Sato, 1985; Roeder *et al.*, 1998
Cytochrome oxidase subunit III	IM	Poot *et al.*, 1996; Taanman and Capaldi, 1993
Citrate synthase I	MAT	Okamoto *et al.*, 1998
OM14	OM	McConnell *et al.*, 1990; Riezman *et al.*, 1983
Abf2p	mtDNA	Diffley and Stillman, 1991

[a]Abbreviations: OM, mitochondrial outer membrane; IM, mitochondrial inner membrane; MAT, mitochondrial matrix; mtDNA, mitochondrial DNA.

over the surface of the spheroplast. Here, we describe a simple method for removing these contaminating antibodies by preadsorbing them to intact yeast cells. A batch of antibody may be pretreated in this way and stored for later use.

1. Reagents

PBS+
Make on day of use.
For 100 ml:

10× PBS	10 ml
Bovine serum albumin	1 g
10% NaN₃	1 ml

Bring volume to 100 ml with distilled H_2O.
Sterilize by passing through a 0.2 μm filter.

2. Protocol

1. Grow yeast to late log phase (24–30 h).
2. Dilute antiserum to 1/25 in PBS+.
3. Remove an aliquot of cells from the late log phase culture: 250 μl of cell culture/ml diluted antibody.
4. Concentrate cells by centrifugation (5 sec at ~13,000 × *g*) at RT. Resuspend cell pellet in 1 ml of PBS+. Repeat this wash procedure three times.
5. Resuspend washed cell pellet in diluted antibody and incubate with gentle mixing at 4°C for 2 h.
6. Concentrate cells by centrifugation, and transfer pretreated antibody to a fresh tube.

7. Carry out a second round of pretreatment. Repeat steps 3–4 and resuspend cell pellet in pretreated antibody recovered from step 6. Incubate with gentle mixing at 4 °C for 2 h.

8. Concentrate cells by centrifugation, and transfer pretreated antibody to a fresh tube. Store preadsorbed antibody in aliquots at –20 °C. Do not expose pretreated antibodies to multiple rounds of freezing and thawing.

B. Yeast Cell Growth, Fixation, and Cell Wall Removal

1. Growth Considerations

While it is sometimes possible to perform immunofluorescent staining on cells picked from colonies on solid media, internal structures, including mitochondria, of mid-log phase cells from liquid culture are better preserved and more reproducible. The choice of carbon source for growth affects the abundance and morphology of mitochondria (Damsky, 1976; Stevens, 1977; Visser *et al.*, 1995). Traditional glucose-based medium is not typically used because glucose represses mitochondrial bio-genesis and consequently makes the organelles more difficult to visualize. Lactate medium selects for cells with mtDNA and mitochondrial metabolic potential; it is also inexpensive and easy to prepare. For strains that cannot grow on nonfermen-table carbon sources, we use a raffinose-based medium. Another consideration in choice of medium is autofluorescence. Lactate and raffinose media as described below produce little general autofluorescence. However, in any carbon source, the red material that accumulates in vacuoles of *ade2⁻* cells is fluorescent and can interfere with microscopy in the red or green emission channels. Addition of extra adenine (two- to fivefold more than in normal media) can prevent this problem.

2. Fixation

There are two main classes of fixatives: cross-linkers and precipitants. The protocol described here uses paraformaldehyde, a cross-linking fixative which forms hydroxymethylene bridges between spatially adjacent amino acid residues. Methanol and acetone, a commonly used precipitating fixative mixture, is suitable for some immunofluorescence staining, including staining of the actin cytoskeleton. However, we find it to be a poor choice for mitochondria. It solubilizes many membranes and can cause extraction of antigens from mitochondria (Boldogh and Pon, 2001).

The fixation conditions can affect the quality of the immunofluorescence results. Common manipulations, including centrifugation and increasing or decreasing temperature, can alter internal structures in *S. cerevisiae* (Lillie and Brown, 1994). Therefore, we add the fixative directly to a liquid culture under growth conditions to minimize such disruptions. The other important variables in paraformaldehyde fixation are the concentration of fixative, duration of fixation, and pH of the fixative solution. High paraformaldehyde concentrations, low pH, and a long fixation period increase the number of cross-links formed and thereby improve structural preservation. Excessive cross-linking, however, can decrease antibody binding to antigens of interest and result in cross-linking of fluorescent substances in the media to cells, which increases background fluorescence surface. The optimal protocol strikes a balance between structural preservation and antigen accessibility.

3. Cell Wall Removal

The protocol described here uses zymolyase, a commercially available protein mixture derived from brewer's yeast. The optimum time for cell wall removal varies among strains and can be determined by visual inspection using phase-contrast or bright-field microscopy. Intact cells have prominently refractile edges, while spheroplasts, which do not have cell walls, are less refractile. Excessive zymolyase treatment can damage cellular structures, while insufficient treatment can result in background staining, as described above.

4. Reagents

Growth Media
Lactate medium
For 1 liter:

Yeast extract	3.0 g
Glucose	0.5 g
$CaCl_2$	0.5 g
NaCl	0.5 g
$MgCl_2$	0.6 g
KH_2PO_4	1 g
NH_4Cl	1 g
90% lactic acid	22 ml
NaOH	7.5 g

Dissolve ingredients in 800 ml distilled H_2O.
Adjust pH to 5.5 with NaOH.
Bring volume to 1 liter with distilled H_2O.
Optional: For growth of *ade2⁻* cells, add 0.1 mg/ml adenine.
Raffinose medium
For 1 liter:

Yeast extract	10 g
Bacto-peptone	20 g
Raffinose	20 g

Optional: For growth of *ade2⁻* cells, add 0.1 mg/ml adenine.

SE medium

For 1 liter:

Yeast extract	10 g
Bacto-peptone	20 g
Raffinose	20 g

Optional: For growth of *ade2⁻* cells, and 0.1 mg/ml adenine

SE medium

For 1 liter:

Glutamic acid, monosodium salt	1.0 g
Yeast nitrogen base *without* amino acids and ammonium sulfate	1.7 g
Dextrose	20.0 g

Amino acid supplements as needed.

Dissolve ingredients in 800 ml distilled H_2O.

Adjust pH to 5.5 with $NaHCO_3$.

Bring volume to 1 liter with distilled H_2O.

Optional: Supplement after autoclaving with 200 μg/ml G418.

NS

For 500 ml:

1 M Tris–HCl, pH 7.5	10 ml
Sucrose	21.4 g
0.5 M EDTA	1 ml
1 M $MgCl_2$	0.5 ml
1 M $ZnCl_2$	0.05 ml
0.5 M $CaCl_2$	0.05 ml

Adjust volume to 500 ml with distilled H_2O.

Sterilize by passing through a 0.2 μm filter and store in 40-ml aliquots at –20 °C.

NS⁺

On day of use, supplement 10 ml NS with the following:

200 mM phenylmethylsulfonylfluoride (PMSF) in ethanol	50 μl
10% NaN_3	500 μl

Paraformaldehyde[1]: EM Grade

Stock concentration: 16%

Source: Electron Microscopy Sciences, Hatfield, PA

Note:

[1] Paraformaldehyde is toxic and carcinogenic. It is absorbed through the skin and is irritating to the eyes, skin, and the respiratory tract. Always wear gloves and safety glasses when working with this fixative. When possible, work in a chemical hood.

Tris/DTT

Make on day of use.
For 100 ml:

1 M Tris–SO$_4$, pH 9.4	10 ml
1 M DTT	1 ml

Bring volume to 100 ml with distilled H$_2$O.

Wash Solution

Please refer to Section III.D.1.

Zymolyase

Make on day of use.
0.125-mg/ml Zymolyase 20T (Seikagaku, Inc., Tokyo, Japan) in wash solution.

5. Protocol

1. Grow a 5-ml culture to mid-log phase (10^6–10^7 cells/ml).
2. Add paraformaldehyde to the culture medium to a final concentration of 3.7%.
3. Incubate cells with fixative under normal culture conditions for 1 h.
4. Concentrate cells by centrifugation (5 sec at ~13,000 × g) and resuspend cell pellet in 1 ml Tris/DTT and incubate for 20 min at 30°C.
5. Concentrate cells by centrifugation. Resuspend cell pellet in 1 ml of zymolyase solution and incubate for 1.5 h at 30°C.
6. Concentrate cells by centrifugation and resuspend cell pellet in 1 ml of NS$^+$. Repeat this wash two times and resuspend the final cell pellet in 2 volumes NS$^+$.
7. Store fixed spheroplasts at 4°C for up to 1 week.

C. Preparation of a Staining Chamber

For all incubations, the sample-coated coverslip is placed on 20–40 μl of incubation solution on a platform of Parafilm sheets in a dark, humid chamber (Fig. 2). The staining chamber protects fluorescent dyes from unnecessary exposure to light and the sample from desiccation. To make the staining chamber, a platform consisting of ten 10 × 15 cm^2 sheets of Parafilm is placed on a stack (ca. 2 cm) of damp paper towels, and both are covered by an inverted opaque tray or pan. The Parafilm platform can be reused if it is washed after use. A small indentation can be made with forceps in the Parafilm next to each coverslip to aid in lifting it.

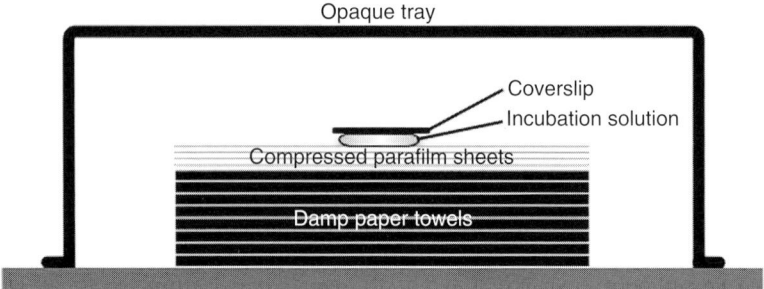

Fig. 2   Chamber for immunofluorescence staining. See text for description.

D. Immobilization of Fixed Spheroplasts on Coverslips and Incubation with Primary and Secondary Antibodies

In the method described, fixed spheroplasts are immobilized on a microscope coverslip and stained by incubation with primary and fluorochrome-coupled secondary antibodies. This is a modification of a procedure used to stain cultured animal cells. In other protocols, fixed spheroplasts are immobilized in the bottom of wells of multiwell microscope slides. This makes antibody incubations and washes less cumbersome for experiments involving large numbers of samples. However, we find that the coverslip method gives cleaner staining, and thus yields greater resolution.

1. Reagents

Mounting solution. The mounting solution used here is largely glycerol, which reduces spheroplast movement during imaging and prevents freezing of the samples during storage at –20 °C. It also contains *p*-phenylenediamine, an antiphotobleaching agent, which helps to prevent destruction of fluorophores by oxygen radicals generated during illumination.

For 100 ml:

PBS	10 ml
p-phenylenediamine	100 mg

Stir vigorously until dissolved.
Adjust pH to 9 with NaOH.
Add 90 ml glycerol and mix thoroughly.
Optional: For DAPI counterstaining of nuclear and mtDNA, add 100 μl of a stock solution of 1 mg/ml DAPI in distilled H_2O. Store at –20 °C in aliquots. DAPI is a possible carcinogen. It is harmful if inhaled, swallowed, or absorbed through the skin. Wear gloves, a face mask, and safety glasses when working with the dry compound.

NS and NS⁺

Please refer to Section III.B.4.

PBS⁺

Please refer to Section III.A.1.

PBT

1× PBS⁺

0.1% Triton X-100

Sterilize by passing through a 0.2 μm filter.

Polylysine (a polycation that adheres spheroplasts to coverslips)

Dissolve polylysine to a final concentration of 0.5 mg/ml in distilled H_2O.

Sterilize by passing through a 0.2 μm filter.

Store in aliquots at –20°C.

Wash solution

For 500 ml:

1 M potassium phosphate, pH 7.5	12.5 ml
1 M KCl	400 ml

Bring volume to 500 ml with distilled H_2O. Autoclave.

2. Protocol

1. Place a 20- to 40-μl drop of polylysine solution on the Parafilm platform.

2. Using forceps, lay a 22-mm² coverslip on the polylysine drop, taking care to avoid creating bubbles at the coverslip-slide interface.[1,2]

3. After at least 10 sec remove the coverslip from the polylysine. To do so, apply 200 to 300 μl distilled H_2O under one edge of the coverslip. This will float the coverslip off the Parafilm so it can be lifted off the drop with forceps.

4. Rinse by dipping the coverslip 5–10 times in a beaker of distilled H_2O, draining excess liquid with filter paper. Allow coverslip to air dry.

5. Mix 10 μl of the fixed spheroplast suspension with 100 μl PBS.

6. Place coated coverslip on Parafilm platform, coated side facing up. Apply the spheroplast mixture to the coated side of the coverslip and incubate for 30 min at room temperature in the staining chamber.

7. Remove unbound spheroplasts by dipping coverslip 5–10 times in a beaker of PBT. Drain excess liquid with filter paper. Do not overdry spheroplasts.

8. Place 20–40 μl of primary antibody diluted in PBT in a drop on the Parafilm platform. Lay the spheroplast-coated side of the coverslip on the drop of antibody and incubate for 2 h at room temperature in staining chamber.[3]

9. Lift the coverslip by introducing 200 to 300 μl PBT under the edge of the coverslip with a micropipette. Rinse the coverslip in PBT as for step 7. Drain excess liquid with filter paper.

10. Place the coverslip on a 20- to 40-μl drop of secondary antibody diluted in PBT and incubate 1 h at RT in the staining chamber.[3]

11. Lift the coverslip, rinse and drain, as for step 9.

12. Place 1 to 2 μl mounting solution on a microscope slide. Lower the coverslip, sample side down, onto the mounting solution, avoiding creation of bubbles. Dry any residual liquid from the edges of the coverslip. Seal the edges with clear nail polish and let dry.

13. Rinse the coverslip surface with distilled H_2O to remove residual salt deposits from staining and dry gently with a Kimwipe or Q-tip. View samples as soon as possible, and not more than a week after preparation.

Notes:

[1] The coverslip may be labeled with a fine-point marker in one corner of the uncoated side.

[2] Coverslip No. 1.5 is optimal for most applications because most objective lenses are corrected for the thickness of these coverslips (0.17 mm).

[3] Detection of two antigens simultaneously can be achieved by combining the two primary antibodies in the first step and the two secondary antibodies in the second step. For such an experiment, the primary antibodies must be raised in different species, and the fluorophores used for detection must have sufficiently separated excitation and emission spectra.

IV. Ectopic Expression of Mitochondria-Targeted Fluorescent Fusion Proteins

Immunofluorescence is a powerful tool. However, localization by immunofluorescence may be compromised by fixation and/or staining artifacts and relies on the availability of antibodies that are specific and can bind to antigens in fixed cells. The discovery that GFP could be used as a tool to visualize proteins in cells, development of GFP variants that have greater fluorescent output or a range of emission wavelengths, and development of methods to target FPs in living cells has revolutionized cell biology.

FPs can be targeted not just to mitochondria, but to any compartment within the organelle (the outer membrane, inner membrane, intermembrane space, and matrix), and to submitochondrial structures, including contact sites (where outer and inner membranes are closely apposed) and mtDNA nucleoids. For studies of morphology and dynamics of mitochondrial membranes, our laboratory uses FPs targeted to the matrix or the inner surface of the inner mitochondrial membrane. Appropriately targeted FPs can also be used to determine whether outer membrane, inner membrane, or both have fused (please refer to Chapter 32 by Ingerman *et al.*, this volume).

FPs can be targeted to mitochondria by two methods, both of which employ fusion proteins. One approach, described in this section, relies on ectopic expression of fusion proteins consisting of mitochondrial signal sequences fused to FPs. Section V describes the second method, fusion of FPs to endogenous mitochondrial proteins.

Signal sequences contain all of the information that is required for import of proteins from the cytosol into mitochondria. They may reside at the N-terminus, at the C-terminus, or within nuclear-encoded mitochondrial proteins. Mitochondrial signal sequences have been used to target FPs to mitochondria and to specific compartments within mitochondria. We have used several plasmid-borne targeted FPs to label yeast mitochondria (Table V). All of the targeted FPs used produce a robust fluorescent signal that is specific for mitochondria, and have no deleterious effect on cell growth or on mitochondrial morphology, motility, or respiratory activity.

A. Transformation of Yeast with Plasmid-Borne Targeted FPs

The lithium acetate method (Gietz *et al.*, 1995) is commonly used for yeast transformation. The following protocol is for one transformation reaction. A mock transformation containing no DNA should always be carried out in parallel.

1. Reagents

50% (w/v) Polyethylene glycol 3350
For 100 ml:
Polyethylene glycol, MW 3350 (Sigma, St. Louis, MO).
Add 35 ml distilled H_2O to 50 g of polyethylene glycol in a 150-ml beaker. Stir until dissolved. Bring volume to 100 ml with distilled H_2O. Autoclave. The concentration of this solution is critical. Cap the bottle tightly to prevent evaporation.
1 M Lithium acetate
For 1 liter:

Lithium acetate dihydrate	102 g

Bring volume to 1 liter with distilled H_2O.
Sterilize by passing through a 0.2-μm filter.
100 mM lithium acetate
Dilute 1 M lithium acetate 1/10 in distilled H_2O and sterilize by passing through a 0.2-μm filter.
Single-stranded calf-thymus DNA
Source: Sigma Cat. No. D-8661, diluted to 2.0 mg/ml in sterile distilled H_2O. Store at $-20\,^{\circ}$C.

2. Protocol

1. Grow yeast to mid-log phase.
2. Remove 10^8 cells from the culture.[1]
3. Wash cells with 100 mM lithium acetate[2] equal to the original volume of culture and resuspend in 50 μl of 100 mM lithium acetate.

Table V
Mitochondria–Targeted Signal Sequence–FP Fusion Proteins[a]

Site	Targeting	Promoter	Vector	FP	References
Matrix	*OLI1*[b] signal sequence	*ADH1*	2 μ pRS426 derivative	HcRed	Fehrenbacher *et al.*, 2004
	CIT1[c] signal sequence	*CIT1*	CEN-*URA3*	bGFP (F99S, M153T, V163A)	Okamoto *et al.*, 1998
	CIT1 signal sequence	*GAL1*	CEN-*URA3*	bGFP	Okamoto *et al.*, 1998
	OLI1	*GAL1*	CEN-*URA3*	DsRed	Mozdy *et al.*, 2000
OM	*TOM6* signal sequence	*GAL1*	CEN-*URA3*	bGFP	Okamoto *et al.*, 1998
IM	*YTA10*	*GAL1*	CEN-*URA3*	bGFP	Okamoto *et al.*, 1998
mtDNA	*ABF2*	*GAL1*	CEN-*URA3*	bGFP	Okamoto *et al.*, 1998

[a]For other mitochondria-targeted fusion proteins see Nunnari *et al.* (2002), and Westermann and Neupert (2000).
[b]F$_0$ATP synthase subunit 9.
[c]Citrate synthase 1.
Abbreviations: OM, mitochondrial outer membrane; IM, mitochondrial inner membrane; MAT, mitochondrial matrix; mtDNA, mitochondrial DNA.

4. Add the following reagents *in order, without mixing*:

   240 μl polyethylene glycol 3350 (50% w/v)

   36 μl 1.0 M lithium acetate

   25 μl single-stranded calf-thymus DNA, 2.0 mg/ml

   50 μl plasmid DNA diluted in distilled H_2O (>1.0 μg)

5. Vortex vigorously and incubate in 30°C water bath for 30 min.

6. Heat shock in 42°C water bath for 15–25 min.

7. Concentrate cells by mild centrifugation (30 sec, 7000 $\times$ g).

8. Resuspend cell pellet gently in 200 μl of sterile H_2O.

9. Plate 1 μl of the reaction on nonselective media (YPD).[3] Plate 10 μl, 50 μl, and the remainder of the reaction on media that selects for the plasmid marker.

Notes:

[3] When plating <100 μl, add 50 to 100 μl sterile H_2O to facilitate spreading.

[2] The lithium acetate wash should be as brief as possible, since this reagent is toxic to cells.

[3] When plating <100 μl, add 50 to 100 μl sterile H_2O to facilitate spreading.

B. Validation of Targeted FPs

Targeted FPs must be checked for toxicity, localization, and effect on mitochondrial morphology and function. Mislocalization of FP fusion markers may occur because targeting information is masked or insertion of the fusion protein into membranes is inhibited. For example, fusion proteins consisting of GFP fused to the C-terminus of Qrc1p (subunit 6 of the cytochrome bc_1 complex) or to porin localize to the cytosol (Okamoto *et al.*, 2001). Localization of FP fusion proteins may be assessed by visual inspection of the FP in cells in which mitochondria are counterstained, using vital dyes, or by biochemical methods for subcellular and submitochondrial fractionation. Possible effects of fusion proteins on mitochondrial morphology or respiratory activity should be assessed by visual inspection of mitochondria in fixed or living cells and by analysis of cell growth and growth rates on nonfermentable carbon sources.

V. Tagging Endogenous Proteins with Fluorescent Proteins

Tagging of the chromosomal copy of genes of interest with FPs is another commonly used method to (1) target FPs to mitochondria, (2) determine protein localization, and (3) analyze the dynamics of FP-tagged proteins and structures in living cells. Furthermore, since FP and epitope tags can be used in antibody-based techniques, including affinity purification, immunoprecipitation, Western blot analysis, and immunofluorescence, chromosomal-tagging technology facilitates biochemical characterization of proteins.

For chromosomal tagging, an insertion cassette, which consists of double-stranded linear DNA encoding the tag of interest plus a selectable marker, is inserted into a target site in the genome by homologous recombination. Tagging vectors have been developed for insertion of a variety of FPs (e.g., GFP, GFP color variants, and FPs that have been optimized for expression in yeast), epitopes (e.g., HA or myc), affinity tags (e.g., GST, TAP, or 6xHis), and various combinations of these tags. Some readily available tagging vectors are shown in Table VI.

Targeting of FPs to mitochondria requires more of a time investment compared to staining with vital dyes. However, plasmids for expression of mitochondria-targeted FPs and cassettes for insertion of FP genes into the yeast genome are readily available. Moreover, targeted FPs produce a signal that is stronger, more persistent, and more specific than that produced by vital dyes. Finally, the fluorescence of targeted FPs can persist after fixation, and targeted FPs can be detected using immunofluorescence with commercially available anti-FP antibodies.

A. Vectors for PCR–Mediated Tagging of Chromosomal Genes

Insertion cassettes are produced by PCR using tagging vectors as templates, and primers that hybridize both to the insertion cassette within the tagging vector and the target site within the yeast chromosome. The amplified DNA contains the desired tag and a selectable marker, flanked by DNA that is homologous to the desired insertion site (Fig. 3). The amplified DNA is transformed directly into yeast using a standard protocol (Gietz *et al.*, 1995). Recombinants that carry the inserted tag are identified using the selectable marker in the insertion cassette and characterized.

Tagging vectors are versatile, as different tags can be inserted into a target gene using a single set of primers and different cassettes from the same family. In addition, vectors are available for expression of tagged genes from their endogenous promoter or from the *GAL1* regulatable promoter. Finally, variations in the tagging cassettes have been developed in which the selectable markers can be excised from the tagged gene (Fig. 3B). As a result, tags can be inserted anywhere within the coding region of the gene of interest, and the tagged gene can be expressed at wild-type levels under control of the endogenous promoter. Moreover, the same selectable marker can be used for multiple rounds of insertion.

B. PCR Amplification of Insertion Cassette

Primers should be 60 or more bases long. The 5′-end of each primer should contain 40–45 bases of perfect homology to the target site. At the 3′-end of each primer should be 18–25 bases complementary to sequences encoding the epitope tag and the selectable marker from the plasmid (see also Fig. 3).

We find that varying the length of the linker region between the target gene and the GFP molecule can affect fluorescence output of the FP-tagged protein. For

Table VI
Yeast Tagging Cassette Vectors

Plasmid family	Tag position	Promoter	Tags	Markers
pFA6a[a]	C terminal	Endogenous	GFP(S65T)	*TRP1*
			3xHA	*kanMX6*
			13xMyc	*HIS3MX6*
			GST	
pFA6a-PGAL1[a]	N terminal or internal	*GAL1*	GFP(S65T)	*TRP1*
			3xHA	*kanMX6*
			GST	*HIS3MX6*
pUR[b]	C terminal	Endogenous	DsRed	*HIS3*
				URA3(K.l.)
pYM[c]	C terminal	Endogenous	yEGFP	*kanMX4*
			EGFP	*hphNT1*
			EBFP	*natNT2*
			ECFP	*HIS3MX6*
			EYFP	*klTRP1*
			DsRed, DsRedI	
			RedStar, RedStar2	
			eqFP611	
			FlAsH	
			1xHA, 3xHA, 6xHA	
			3xMyc, 9xMyc	
			1xMyc + 7xHis	
			TAP	
			Protein A	
pKT[d]	C terminal	Endogenous	yEGFP	*KanMX*
			yECFP	*SpHIS5*
			yEVenus	*CaURA3*
			yECitrine	
			yESapphire	
			yEmCFP[e]	
			yEmCitrine	
			tdimer2[f]	
			yECitrine + 3xHA	
			yECitrine + 13xMyc	
			yECFP + 3xHA	
			yECFP + 13xMyc	
pOM[g]	N terminal or internal	Endogenous[h]	yEGFP	*kanMX6*
			6xHA	*URA3(K.l.)*
			9xMyc	*LEU2(K.l.)*
			Protein A	
			TEV-ProteinA	
			TEV-GST-6xHis	
			TEV-ProteinA-7xHis	

[a]Longtine *et al.*, 1998.
[b]Rodrigues *et al.*, 2001.
[c]Janke *et al.*, 2004; available through EUROSCARF.
[d]Sheff and Thorn, 2004; available through EUROSCARF or Harvard University.
[e]Monomeric version.
[f]Tandem dimer of DsRed.
[g]Gauss *et al.*, 2005; available through EUROSCARF.
[h]After Cre-mediated removal of auxotrophic marker.

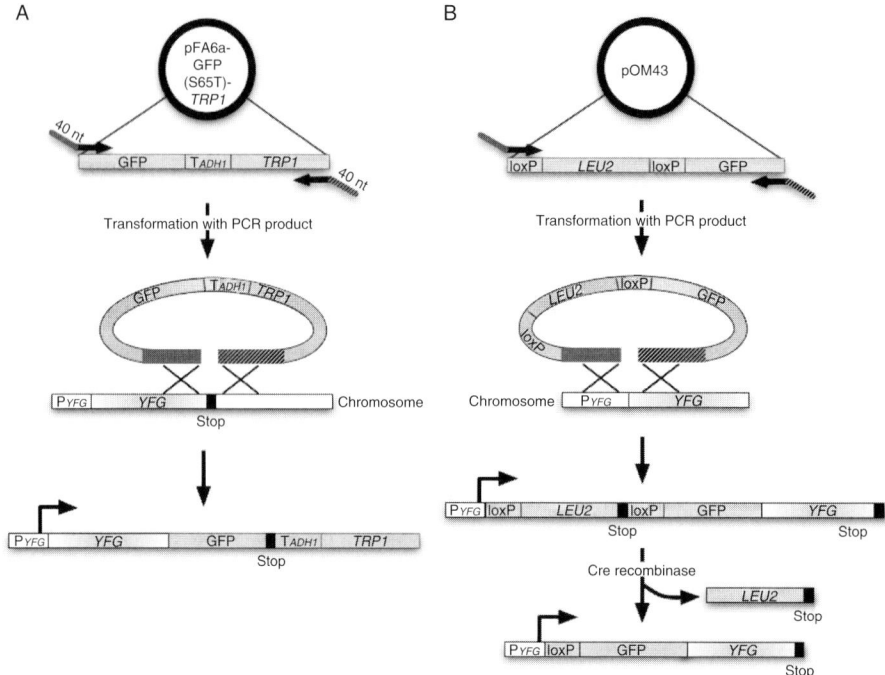

Fig. 3 Approaches for insertion of FPs or epitope tags into target sites in the yeast genome. (A) Traditional one-step gene tagging by homologous recombination with a PCR product. Primers are designed so as to amplify the tagging cassette (tag, terminator, and selectable marker) with 40-base-pair extensions derived from sequences flanking the site of insertion, immediately before the target gene's stop codon. Recombination inserts the tag at the C terminus of the chromosomal gene, followed by the selectable marker gene. (B) Tagging followed by Cre/Lox-mediated excision of the selectable marker. This strategy is used to preserve markers for future experiments, or to produce an N-terminal tagged protein expressed from its native promoter. For each technique, one of many available tagging vectors is shown. YFG: your favorite gene. P_{YFG}: promoter of YFG. T_{ADH1}: terminator of ADH1. Relevant stop codons are shown.

C-terminal tagging, this is accomplished by introducing DNA encoding extra amino acids between the 40 bases of DNA homologous to the target gene and the 20 bases corresponding to the plasmid template in the forward primer. Outcomes vary with different proteins. We suggest starting optimization studies with spacers that contain five alanines.

1. Reagents

> *PCR buffer*
> $3.3\times$ XL buffer II stock (Applied Biosystems), contains no Mg^{2+}
> *Mg(OAc)$_2$*
> 25 mM stock (Applied Biosystems)

dNTP mix
2.5 mM each dNTP (Applied Biosystems)
Primers
PAGE purified, 2×10^{-5} M
Vector template
Plasmid DNA, diluted to ~50 ng/μl
DNA polymerase
r*Tth* (Applied Biosystems)
The cassette is amplified using standard PCR conditions.

2. Protocol

Prepare the reaction mixture as described below.

Reagent	Volume (μl)
PCR buffer (3.3× stock, contains no Mg^{2+})	30
$Mg(OAc)_2$ (25-mM stock)[1]	8
dNTP mix (2.5 mM ea. dNTP)	8
Forward primer (2×10^{-5} M stock)[2]	1
Reverse primer (2×10^{-5} M stock)[2]	1
Vector template (~50 ng/μl stock)	1
r*Tth* DNA Polymerase	1
sterile distilled H_2O	50

[1]To optimize PCR efficiency, we suggest first titrating the Mg $(OAc)_2$ concentration (e.g., try four reactions with final $Mg(OAc)_2$ concentrations of 0.5, 1.0, 2.0, and 3.0 mM).
[2]To assure PCR specificity of untested primers, we suggest three control reactions: (i) no primers in the reaction, (ii) forward primer only, and (iii) reverse primer only.

Thermocycler conditions:
1 cycle of 94 °C (5 min)
30 cycles of: 94 °C (1 min), 55 °C (1 min), 72 °C (2–4 min, about 1 min per kb product length)
1 cycle of 72 °C (10 min)
Eight to ten 100-μl PCR reactions should be pooled to obtain DNA sufficient for a single yeast transformation (1 μg of DNA per 1×10^8 cells). Purify the PCR product by ethanol precipitation and 1% agarose gel electrophoresis to remove contaminants and undesired products. After cutting out the band containing the desired PCR product, we isolate and concentrate it using the Qiaex II Gel Extraction Kit (Qiagen).

C. Transformation of Yeast with the Amplified Insertion Cassette

Transform the purified PCR product into the target strain as described in Section IV.A.2. For transformations using auxotrophic markers, a single reaction typically yields 10–20 transformed clones for further characterization. For transformations using drug resistance markers (e.g., kanMX6), plate all cells from a single transformation on a nonselective (YPD) plate. Incubate for 2–3 days and replica plate onto a plate containing the drug (e.g., 200 mg/liter G418 for kanMX6). Incubate 2–3 days, and replica-plate onto a fresh plate containing the drug. Colonies can be picked from this plate for further analysis.

Note: Synthetic media containing ammonium sulfate interferes with G418 activity. To make drop-out plates containing G418, use SE (synthetic media with glutamate) rather than SC (see Section III.B.4 for protocols for media preparation).

D. Validation of Tagging and Gene Function

PCR is used to validate insertion of the tag into the target locus. This analysis uses genomic DNA isolated from transformants as a template and a pair of primers (20–25 bases) that hybridize on either side of the insertion. Optional PCR can be carried out using a third primer that anneals to a region within the integration cassette. Cells with correct tag integration at the target locus should be validated using Western blots to analyze protein expression. Finally, tagged constructs should be characterized for functionality given knowledge about the gene of interest. For example, if an inserted tag compromises function of a nonessential gene, then the tagged strain may show a phenotype resembling that of cells mutated in that gene. Some experimental outcomes for experiments testing an epitope-tagged gene are shown (Fig. 4).

E. Marker Excision by Cre Recombinase

The tagging vectors were first developed for C-terminal tag insertion. Similar vectors for one-step N-terminal tagging insert the selection marker and tag into the 5′ end of genes of interest. However, because the selection marker in these cassettes separates the tagged gene from its promoter, the tagged gene is expressed from a nonendogenous promoter included in the tagging cassette. These tagging vectors are used for regulated expression or overexpression studies. Until recently, tagging vectors were not available that could both (1) insert tags in the N-terminus or internal regions of proteins, and (2) express tagged proteins from their endogenous promoter (i.e., at wild-type levels).

The pOM family of tagging vectors, which were designed for removal of the selection marker after tag insertion, solved this problem (Fig. 3B). In this set of vectors, the selection marker is flanked by loxP sites, and can be removed using bacteriophage Cre recombinase (Gueldener *et al.*, 2002). Insertion and validation of the tagging cassette is carried out as described above. Yeast bearing integrated tags are then transformed with a plasmid that encodes Cre recombinase under control of the galactose-inducible promoter. To induce Cre expression, cells are grown in liquid media selecting for the plasmid, incubated for 4 h in media containing 1% raffinose + 1% galactose, and plated on nonselective (YPD) media. Colonies are then screened for loss of the tagging cassette marker by replica plating

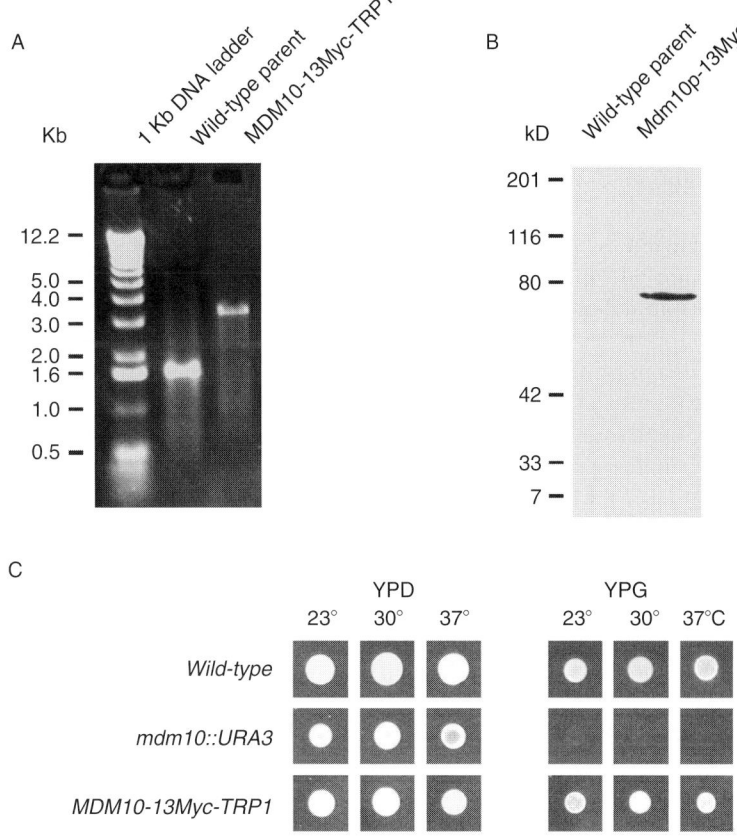

Fig. 4 Validation of tagged genes. After isolation of transformants on selective plates, clones are validated via PCR, protein expression, and functionality assays. (A) Integration of the cassette into the target locus is initially verified via PCR on isolated genomic DNA. In the above 1% agarose gel, PCR products from an untransformed parent wild-type haploid yeast, and yeast transformed with the 13Myc-*TRP1* cassette, are readily distinguished by size in wild-type *MDM10* and *MDM10*-13Myc-*TRP1* strains (*MDM10* was amplified using 20-base primers which correspond to chromosomal DNA 55 nucleotides upstream and downstream of the gene). (B) Clones positive for integration are verified for protein expression via Western Blot analysis. The above 10% SDS-PAGE gel of whole-cell extracts from wild-type and tagged strains was probed with a 9E10 αMyc monoclonal antibody. The Mdm10p-13Myc lane, but not the wild-type parent strain lane, shows expression of a single tagged protein of the expected molecular weight. (C) Tagged constructs can often be characterized for functionality given knowledge about the phenotype of the gene of interest. While deletion of the open reading frame of the *MDM10* gene results in yeast that is unable to grow on nonfermentable carbon sources (YPGlycerol), yeast bearing Mdm10p-13Myc behaves just like wild-type in that it retains the ability to grow on nonfermentable carbon sources at all temperatures.

or patching on selective plates. Further characterization is performed by PCR, Western blot analysis, and visualization, as described above. We use pSH47, which expresses Cre from a *GAL* promoter and contains the *URA3* selectable marker. We prefer *URA3* as a marker because it allows cells to be cured of the plasmid by counter-selection on 5-fluoroorotic acid (5-FOA) plates after marker excision.

VI. Strategies for Visualization of Yeast Mitochondria

The revolution in biological imaging that followed the development of GFP and the popularization of confocal microscopy has produced a powerful collection of fluorescence techniques: imaging modalities such as deconvolution, confocal and 2-photon excitation microscopy; manipulations of fluorescence such as photoactivation and photobleaching; and the use of fluorescence as a tool beyond simple imaging, for example as a molecular ruler using Förster resonance energy transfer (FRET). Further developments including 4π microscopy and structured illumination have only recently been commercialized but have been useful in some situations. Having more options is good, but the full arsenal of technology is not needed for every experiment. The investigator must determine which technology can best provide the information needed in a given study.

A. Defining Imaging Needs

Imaging is essentially a 2-dimensional process, while biological processes are intrinsically 3-dimensional. In addition, biological structures are dynamic and change position and morphology with time. As a result, a complete picture of a biological process requires at least 4 dimensions of information, which can be collected only by repeated imaging of the sample at different planes and time-points. However, each exposure to illumination diminishes the fluorescence output of probes (photobleaching) and can damage a living sample (phototoxicity). In addition, current technology allows for the collection of vast amounts of unnecessary information, often at great expense of money and time. Therefore, the investigator should define carefully the experimental questions to be asked before proceeding with imaging.

1. 2D Imaging

2D imaging is appropriate for rapid assessment of mitochondrial morphology and membrane potential. The round shape of the yeast cell can be a concern when taking single 2D images. However, mitochondria in budding yeast tend to lie near the cell cortex. As a result, a focal plane that captures that cortical region of the cell can provide interpretable images of mitochondria.

2. 3D Imaging

3D imaging is the method of choice for high-resolution imaging of mitochondrial morphology and distribution. As described above, the appearance of mitochondria in a normal yeast cell varies with focal plane. Since mitochondria are cortically distributed, a focal plane at the center of the cell will show cortical dots representing cross sections through mitochondrial tubules. Moreover, in yeast with abnormally aggregated mitochondria, the organelle may not be detectable in some optical planes. To fully characterize mitochondrial morphology and distribution, we collect images at a series of focal planes (a z-series) through the entire cell. This can be performed on living or fixed cells. The 3D image data can be reconstructed for viewing at any angle, and used for quantitative analysis of mitochondrial volume, morphology and distribution (Boldogh *et al.*, 2004; Fehrenbacher *et al.*, 2005; Yang *et al.*, 1999).

3D imaging is also essential for colocalization studies (Boldogh *et al.*, 2003; Fehrenbacher *et al.*, 2005; Hobbs *et al.*, 2001). With 2D imaging, two particles that are in different planes within a 3-dimensional cell may appear to colocalize. Thus, viewing structures of interest at all angles in 3D reconstructions is the only way to assess co-localization accurately.

3. Time-Lapse Imaging

This approach is used to study mitochondrial motility and plasticity. Like many other organelles, mitochondria are highly dynamic. They are motile (Simon *et al.*, 1995) and undergo fission and fusion (Nunnari *et al.*, 1997), and their overall morphology changes during processes such as sporulation, transition to stationary phase, and changes in carbon source (Egner *et al.*, 2002). The inheritance of mitochondria during budding and their mixing and resorting during mating (Azpiroz and Butow, 1993) are dynamic events in the life of mitochondria that have been followed with time-lapse fluorescence microscopy.

4. 4D Imaging

4D imaging (3D imaging over time) tracks mitochondria through multiple focal planes over time, and reveals the complete dynamic shape of the organelle as it evolves. However, optical sectioning increases sample illumination time and the resulting photodamage sustained by the cells. Thus, spatial resolution, temporal resolution, and label persistence/cell viability are in essential conflict.

5. Multicolor Imaging

Simultaneous imaging is a powerful method for defining the relationship between different proteins or organelles. One common application is determining whether a protein of interest colocalizes with a marker for a known compartment, such as mitochondria, or with another protein whose localization is known. Two-color imaging can also demonstrate functional relationships between proteins or organelles: does the disruption of one entity also disrupt the other?

For precise colocalization in optical microscopy, we recommend 3D imaging. As described above, objects in different focal planes may appear superimposed in a 2D image, but a z-series can reveal their true locations in 3 dimensions. However, because of unavoidable diffraction, the colocalization of two fluorescent probes by standard optical microscopy only places them within hundreds, or at best, several tens of nanometers of each other. Therefore, colocalization, even in 3D studies supports but does not prove that the proteins interact on a molecular level. An extension of fluorescence technology, Förster resonance energy transfer (FRET) imaging, can greatly improve the spatial resolution of two-color imaging, but has its own technical concerns that are beyond the scope of this chapter (please refer to Chapter 14 by Lemasters and Ramshesh, this volume).

As with multiple-label immunofluorescence, multiple-label live-cell imaging requires the selection of two or more fluorescent probes with different excitation and emission spectra, and appropriate filter sets to capture accurate images of each probe. Dynamic imaging of multiple probes requires specialized hardware, discussed below.

B. Imaging Technologies

There are 3 major strategies for collecting serial optical sections of fluorescent yeast mitochondria: wide-field microscopy with deconvolution, spinning-disk confocal microscopy, and scanning confocal microscopy.

1. Wide–Field Microscopy with Deconvolution

Deconvolution is a general term for computational techniques that increase the contrast and resolution of digital images. There are four primary sources for image degradation: noise, scatter, glare, and blur (Wallace *et al.*, 2001). Noise is semirandom image degradation produced by the signal or the digital imaging system. Scatter and glare are random disturbances of light produced by passage through areas with different refractive indices in the sample and through the lenses or filters of the imaging system. Finally, blur is the nonrandom spreading of light after passage through the lens. When a lens is used to view a 3D sample, some features in the image are in focus, and others are out of focus because they are at a different focal depth. Light from out-of-focus focal planes is the most significant cause of image blurring in fluorescence microscopy. One other source of degradation is diffraction. Although fluorescence emanates from point sources (individual fluorescent molecules), no optical system can resolve them perfectly because the diffraction of light waves blurs the image.

Since the spreading of blurred light is nonrandom, methods were developed to determine the point spread function (PSF) that describes the pattern of light spreading from a point source. A generic (theoretical) PSF may be calculated from data, such as the objective lens magnification and numerical aperture, dye emission wavelength, and camera pixel size. Alternatively, an empirical PSF can be determined from a z-series of images of subresolution ($<$0.2 μm diameter) fluorescent beads. We prefer the empirical PSF, but the theoretical PSF can also yield acceptable results. In a wide-field microscope, the PSF looks like a set of concentric rings in the x-y plane that extend from the original point source and like an hour-glass in the x-z or y-z planes with the point source at the crossover point in the hour glass.

Deconvolution algorithms apply the PSF to every point of light in a digital image to distinguish blurred light from the actual signal, and to eliminate or reduce blurred light from the acquired image. Nearly any image acquired by a digital fluorescence microscope can be deconvoluted. However, most users apply deconvolution to enhance images obtained using a wide-field fluorescence microscope, that is, a standard epifluorescence microscope equipped with a CCD (charge-coupled device) camera for digital image acquisition. For 3D and 4D studies, a z-series of wide-field images is collected. There are several algorithms for deconvolution of fluorescence images (Wallace *et al.*, 2001). The algorithm of choice depends on the sample, computation time, and software implementation. Our system (Volocity, Improvision, Inc.) uses a maximum-entropy iterative restoration algorithm.

2. Spinning-Disk Confocal Microscopy

Confocal microscopy and wide-field-deconvolution remove blur by opposite means. In confocal microscopy, pinholes in the light path prevent out-of-focus light from reaching the detector. The feature that distinguishes spinning-disk confocal microscopy from other forms of confocal microscopy is the simultaneous illumination of multiple points in the sample using a spinning disk that is perforated with strategically placed pinholes. Light delivery is enhanced using lasers, which provide high initial light flux, and using microlenses on each pinhole (Inoué and Inoué, 2002). Light capture is enhanced using increasingly sensitive cooled CCD cameras. As a result, large fields can be imaged rapidly, in some cases up to 30 frames per second (standard video rate). In practice, the actual temporal resolution depends on the sensitivity of the camera and the efficiency of light delivery through the pinholes. Nonetheless, because spinning disk confocal imaging allows for rapid acquisition of high-resolution images, many users are applying this approach to live cell imaging.

3. Point Scanning Confocal Microscopy

Scanning confocal imaging also uses pinholes in the excitation and emitted light paths to block detection of blurred light. However, in contrast to spinning disk methods, a single laser beam illuminates the sample and scans the field pixel by pixel. As a result of the long imaging acquisition times, this method is not widely used for imaging very rapid processes, such as mitochondrial motility. Moreover, the photomultiplier tubes (PMT) used to detect emitted light in current laser scanning systems are significantly less sensitive than CCD detectors. As a result, although scanning confocal microscopy can be used to image yeast mitochondria (Yang *et al.*, 1999), it requires a strong fluorescent signal. The lasers in scanning confocal systems can also be used for techniques that require high-intensity illumination of select regions, including FRAP (fluorescence recovery after photobleaching) and localized photoactivation. Thus, laser scanning systems are widely used to study protein and organelle dynamics (Lippincott-Schwartz *et al.*, 2003). However, we find that the signal emitted by yeast mitochondria or cytoskeletal structures with commonly used probes and detectors is too low to be usefully imaged by this method.

4. Line Scanning Confocal Microscopy

A variation on the scanning confocal microscope that has recently become commercially available is the line-scanning confocal microscope (Wolleschensky *et al.*, 2005). In this instrument, the area excited is a line rather than a point. By scanning more area simultaneously, the line-scanning confocal microscope can cover a field faster, and thus achieves much higher time resolution (100 frames of 512×512 pixels per second) than a point-scanning confocal microscope (a few frames per second). (These are the maximum frame rates; the available signal and detector noise usually reduces the practical frame rate by 50–90%.) In the future this may become a useful tool for studying mitochondrial dynamics, but with commonly used probes and detectors, the signal emitted by yeast mitochondria is too low to be imaged usefully by this method.

5. Alternative Excitation Methods: Multiphoton, 4π, and Others

The confocal imaging methods described above are all subtractive in nature: they achieve better resolution by excluding some of the emitted light from the detector. Because this unavoidably includes some in-focus, wanted light, these techniques entail both excessive exposure of the sample and loss of signal from the detector. One approach to reducing this inefficiency is to restrict the area of excitation. Nonlinear optical processes, such as multiphoton excitation, achieve this goal and can provide high spatial resolution without the use of pinholes. Basically, in conventional fluorescence the probability of a fluorophore being excited is linear with the light intensity. As a result, both focused and unfocused light excites fluorescence, producing a blurry image. In a nonlinear process the probability of the fluorophore being excited depends on the 2nd, 4th, or higher power of the light intensity, so that unfocused light is incapable of exciting fluorescence. Emission thus occurs only in a small focal volume, and blur is greatly reduced.

For yeast, two-photon excitation of probes such as GFP does not improve on wide-field or scanning confocal microscopy, because the wavelength required is much longer and thus the intrinsic spatial resolution of the image is reduced. However, another nonlinear technique, called 4π microscopy, has been used successfully to image living yeast cells with stunning 3D resolution (Egner *et al.*, 2002), albeit with poorer temporal resolution than wide-field methods. Other nonlinear techniques, such as STED (stimulated emission depletion), have been developed (Gustafsson, 2005) and may lead to new ways to break the diffraction barrier in microscopy.

C. Equipment for Imaging

The choice of microscopes, excitation sources, fluorescence filters and cameras for wide-field microscopy is discussed in Chapter 30 by De Vos and Sheetz, this volume. Here, we will discuss our particular setup for wide-field deconvolution microscopy, including some additional hardware that we use for rapid two-color imaging and for 4D (3D time-lapse) imaging studies. These methods have served us well in imaging of mitochondrial dynamics in yeast, and should also be applicable to larger cells, including mammalian cells.

Selection of an appropriate microscope and camera to detect fluorescence is essential for imaging mitochondria in living or fixed yeast cells. The small size of yeast cells requires a microscope equipped with a high-magnification objective lens and a high-resolution camera. Phototoxicity and photobleaching can be greatly reduced by image acquisition with a sensitive digital camera that can detect fluorescence with relatively weak, short illumination. Data processing speed and memory capacity of the computer linked to the digital camera can affect imaging speed and therefore should be tested thoroughly before committing to a particular imaging system.

An imaging system that works well for visualizing mitochondria in yeast is based on an upright epifluorescence microscope (e.g., Zeiss Axioskop 2 or Nikon E600) equipped with a Plan-Apochromat $100\times/1.4$ NA objective lens, and a cooled CCD camera such as the Orca ER (Hamamatsu). A software package (e.g., OpenLab from Improvision, Inc., or Metamorph from Universal Imaging) is used to control the hardware and capture images. (Iterative deconvolution is performed after image acquistion, on an independent computer workstation using Volocity software [Improvision].)

Fully motorized microscope stands are available, with integrated shutters, filter changers, and focus drives for automated multicolor, 3D, or 4D imaging. However, these functions can be added to an existing nonmotorized microscope. We have chosen the latter approach because it offers more flexibility, especially when optimizing for speed. For slower 3D imaging (time interval ≥ 1 min), a stepper motor coupled to the microscope fine focus drive (Ludl) is sufficient. This is equivalent to the built-in focus drive on motorized microscope stands. For fast

3D imaging (time interval 1 sec to 1 min), we use a piezoelectric focus motor (Physik Instrumente) on the objective lens. A shutter driver (Uniblitz D122, Vincent Associates) controls excitation light from the 100-W mercury arc lamp. We use two methods for multicolor imaging, with the choice again based on the time resolution required. The slower but more versatile method uses filter wheels (Ludl Electronic Products) in the excitation and emission light paths, with a double- or triple-line dichroic beamsplitter in the microscope turret. For fast imaging, we collect simultaneous images using an image splitter (DualView, Optical Insights Corp.). Finally, to control sample temperature (e.g., for studying temperature-sensitive mutants) we use a heating collar on the objective lens (Bioptechs).

VII. Methods for Imaging of Mitochondria in Living Yeast Cells

1. Reagents

Agarose Bed Growth Chamber

SC or lactate media[1]	5.0 ml
2% agarose (low melting)	0.1 g

Combine ingredients in 50-ml Falcon tube and boil ~5 min to ensure contents are fully dissolved. Dispense 200-μl aliquots[2] into microcentrifuge tubes. Store at RT, in the dark.

Notes:

[1] These media are less autofluorescent than YPD.
[2] Aliquots are recommended to avoid repeated boiling.

VALAP
Petrolatum (Vaseline)
Lanolin
Paraffin (hard)
Add ingredients in 1:1:1 ratio. Melt by submerging in a 70°C water bath. Aliquot melted VALAP into 60 × 15 mm^2 Petri dishes. Store at RT.

2. Time-Lapse Imaging Protocols

Time-lapse recordings of mitochondria yield useful data on the velocity and direction of mitochondrial movement, if the time interval between images is compatible with the rate of mitochondrial movement. For example, short-term recordings of mitochondrial movements in Latrunculin-A (Lat-A) treated cells, and in cells bearing temperature-sensitive mutations in actin or actin binding proteins, uncovered the requirement for an intact actin cytoskeleton for the control of mitochondrial organization and movement (Boldogh *et al.*, 1998; Simon *et al.*, 1995; Smith *et al.*, 1995).

a. Short-Term Imaging

Short-term mitochondrial visualization is carried out by adding cells directly to a glass microscope slide and imaging immediately for a period of no longer than 10 min. After 10 min, cells experience a significant decrease in viability and the use of a growth chamber becomes necessary. Protocol for the preparation of yeast and staining of mitochondria for short-term visualization is described below, followed by a protocol useful for those visualizations requiring longer imaging periods.

1. Grow yeast cells to mid-log phase in liquid medium.[1] Concentrate to density appropriate for visualization.[2]
2. Pipette 2.7–3.0 μl of sample on to a microscope slide and cover with a 22 × 22 mm^2 coverslip. Avoid creating bubbles between slide and coverslip.
3. View immediately without sealing.

Notes:

[1] Some media, such as YPD, autofluoresce. Therefore, we recommend growing cultures for visualization in either SC dropout media, or in lactate media.

[2] The volume used is important, since excess volume can cause cells to float, and too little volume can affect cell structure.

b. Long-Term Imaging

The growth chamber described below supports cell growth at wild-type levels for up to 5 h. The growth chamber also immobilizes cells in low-melting agarose, reducing cellular movement that can decrease image resolution. Finally, the chamber has minimal autofluorescence and remains transparent and thin enough for observation with oil-immersion lenses. For time-sensitive imaging trials, such as those involving the addition and removal of dyes or drugs, note that it takes roughly 5 min to make an agarose bed (Fig. 5).

1. Grow yeast cells to mid-log phase in liquid medium. Concentrate cells to appropriate density for imaging, as needed.
2. Melt an aliquot of agarose bed material in a boiling water bath (ca. 2 min).
3. Pipette 35 μl of agarose bed material onto a glass microscope slide. Cool slightly (5 sec).
4. Place a second microscope slide on top of agarose, applying light pressure to distribute the agarose bed until it is the size of a standard coverslip.
5. Let bed harden between microscope slides (~2 min).
6. Gently remove top slide by rotating the slide 90° and then sliding it past bottom slide.
7. Pipette 1.5 μl of imaging sample onto surface of the agarose bed.[1]
8. Cover with a 22 × 22 mm^2 coverslip and seal with VALAP. (To seal with VALAP, heat metal spatula in flame, dip into solid VALAP, and drag spatula along edges of coverslip.)

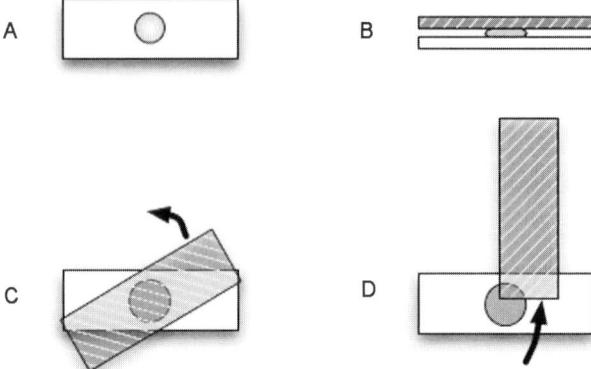

Fig. 5 Preparation of agarose bed for long-term live cell imaging. A. Agarose bed material (35 μl) on a microscope slide. B. Side view of agarose sandwiched between two microscope slides. C–D. Rotation and removal of top slide.

Note:

[1] If cells are not immobilized effectively, decrease the volume of sample added to agarose bed. Additionally, the water content of the agarose bed can be adjusted by letting the solidified bed stand uncovered for a few seconds before adding sample.

3. Optimizing Imaging Conditions and Preventing Toxicity

During live-cell imaging, photons react with cellular molecules to produce free radicals and reactive oxygen species. Mitochondrial health and cell viability can suffer as a result. We use the following specific criteria to confirm that our imaging conditions are not harmful to mitochondria:

1. Intact mitochondrial structure: intense or long excitation can cause cleavage or fragmentation of the organelle.
2. Mitochondrial movements: behavior of photo-damaged mitochondria has not been fully investigated; however, photodamage should be suspected if mitochondria show extremely low velocity-in wild-type cells.
3. If potential-sensitive dyes are used: maintenance of membrane potential (dye retention) throughout the imaging period.

If phototoxicity or photobleaching occurs, the following techniques (in order of preference) should be tried to reduce harmful light exposure:

1. Reduce excitation intensity (e.g., with a neutral density filter). Generally, a longer exposure time at lower intensity gives a comparable quality image with less photodamage.
2. Reduce exposure time.
3. Increase time interval and/or z sectioning interval.

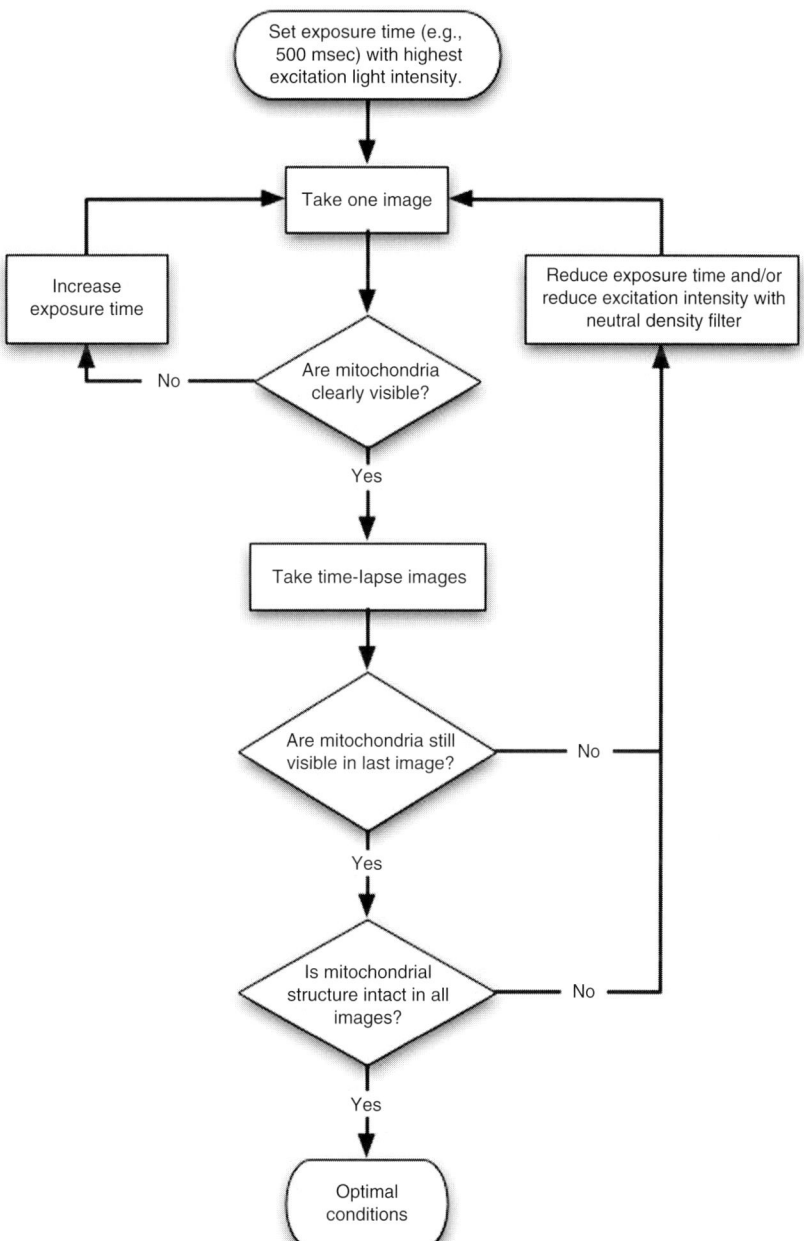

Fig. 6 A flow chart for setting optimum conditions for fluorescence image acquisition.

To increase signal-to-noise ratio without increasing light exposure:

1. Apply binning in the camera. With binning, the signal from a small array of detector pixels (usually 2×2 or 4×4) is pooled in one image pixel. The result is an image with fewer pixels, but with about fourfold greater signal, in the case of 2×2 binning. Noise is not appreciably increased. If necessary, preserve spatial resolution by adding a magnifying projection tube before the camera.

2. Increase gain in the camera. This will increase noise, but it preserves spatial resolution.

3. If these techniques fail, consider trying alternative filter sets with higher throughput: either a broader spectral window or more efficient coating processes.

We suggest a flow chart (Fig. 6) to help determine the optimum conditions for serial image acquisition. For methods to analyze and quantify mitochondrial motility, please refer to Chapter 30 by De Vos and Sheetz, this volume.

Acknowledgments

We thank the members of the Pon laboratory for the critical evaluation of the manuscript. The work was supported by research grants to L.A.P. from the National Institutes of Health (GM45735 and GM66307).

References

Azpiroz, R., and Butow, R. A. (1993). Patterns of mitochondrial sorting in yeast zygotes. *Mol. Biol. Cell* **4**, 21–36.

Boldogh, I., Vojtov, N., Karmon, S., and Pon, L. A. (1998). Interaction between mitochondria and the actin cytoskeleton in budding yeast requires two integral mitochondrial outer membrane proteins, Mmm1p and Mdm10p. *J. Cell Biol.* **141**, 1371–1381.

Boldogh, I. R., Nowakowski, D. W., Yang, H.-C., Chung, H., Karmon, S., Royes, P., and Pon, L. A. (2003). A protein complex containing Mdm10p, Mdm12p, and Mmm1p links mitochondrial membranes and DNA to the cytoskeleton-based segregation machinery. *Mol. Biol. Cell* **14**, 4618–4627.

Boldogh, I. R., and Pon, L. A. (2001). Assaying actin-binding activity of mitochondria in yeast. *Methods Cell Biol.* **65**, 159–173.

Boldogh, I. R., Ramcharan, S. L., Yang, H.-C., and Pon, L. A. (2004). A type V myosin (Myo2p) and a Rab-like G-protein (Ypt11p) are required for retention of newly inherited mitochondria in yeast cells during cell division. *Mol. Biol. Cell* **15**, 3994–4002.

Damsky, C. H. (1976). Environmentally induced changes in mitochondria and endoplasmic reticulum of *Saccharomyces carlsbergensis* yeast. *J. Cell Biol.* **71**, 123–135.

Diffley, J. F., and Stillman, B. (1991). A close relative of the nuclear, chromosomal high-mobility group protein HMG1 in yeast mitochondria. *Proc. Natl. Acad. Sci. USA* **88**, 7864–7868.

Egner, A., Jakobs, S., and Hell, S. W. (2002). Fast 100-nm resolution three-dimensional microscope reveals structural plasticity of mitochondria in live yeast. *Proc. Natl. Acad. Sci. USA* **99**, 3370–3375.

Fehrenbacher, K. L., Boldogh, I. R., and Pon, L. A. (2005). A role for Jsn1p in recruiting the Arp2/3 complex to mitochondria in budding yeast. *Mol. Biol. Cell* **16**, 5094–5102.

Fehrenbacher, K. L., Yang, H.-C., Gay, A. C., Huckaba, T. M., and Pon, L. A. (2004). Live cell imaging of mitochondrial movement along actin cables in budding yeast. *Curr. Biol.* **14**, 1996–2004.

Gauss, R., Trautwein, M., Sommer, T., and Spang, A. (2005). New modules for the repeated internal and N-terminal epitope tagging of genes in *Saccharomyces cerevisiae*. *Yeast* **22**, 1–12.

Gietz, R. D., Schiestl, R. H., Willems, A. R., and Woods, R. A. (1995). Studies on the transformation of intact yeast cells by the LiAc/SS-DNA/PEG procedure. *Yeast* **11**, 355–360.

Gueldener, U., Heinisch, J., Koehler, G. J., Voss, D., and Hegemann, J. H. (2002). A second set of loxP marker cassettes for Cre-mediated multiple gene knockouts in budding yeast. *Nucleic Acids Res.* **30**, e23.

Gustafsson, M. G. L. (2005). Nonlinear structured-illumination microscopy: Wide-field fluorescence imaging with theoretically unlimited resolution. *Proc. Natl. Acad. Sci. USA* **102**, 13081–13086.

Hobbs, A. E. A., Srinivasan, M., McCaffery, J. M., and Jensen, R. E. (2001). Mmm1p, a mitochondrial outer membrane protein, is connected to mitochondrial DNA (mtDNA) nucleoids and required for mtDNA stability. *J. Cell Biol.* **152**, 401–410.

Inoué, S., and Inoué, T. (2002). Direct-view high-speed confocal scanner: The CSU-10. *Methods Cell Biol.* **70**, 87–127.

Janke, C., Magiera, M. M., Rathfelder, N., Taxis, C., Reber, S., Maekawa, H., Moreno-Borchart, A., Doenges, G., Schwob, E., Schiebel, E., and Knop, M. (2004). A versatile toolbox for PCR-based tagging of yeast genes: New fluorescent proteins, more markers and promoter substitution cassettes. *Yeast* **21**, 947–962.

Lillie, S. H., and Brown, S. S. (1994). Immunofluorescence localization of the unconventional myosin, Myo2p, and the putative kinesin-related protein, Smy1p, to the same regions of polarized growth in *Saccharomyces cerevisiae*. *J. Cell Biol.* **125**, 825–842.

Lippincott-Schwartz, J., Altan-Bonnet, N., and Patterson, G. H. (2003). Photobleaching and photo-activation: Following protein dynamics in living cells. *Nat. Cell Biol.* Supplement: Imaging in Cell Biology, S7–S14.

Longtine, M. S., McKenzie, A., Demarini, D. J., Shah, N. G., Wach, A., Brachat, A., Philippsen, P., and Pringle, J. R. (1998). Additional modules for versatile and economical PCR-based gene deletion and modification in *Saccharomyces cerevisiae*. *Yeast* **14**, 953–961.

McConnell, S. J., Stewart, L. C., Talin, A., and Yaffe, M. P. (1990). Temperature-sensitive yeast mutants defective in mitochondrial inheritance. *J. Cell Biol.* **111**, 967–976.

Mihara, K., and Sato, R. (1985). Molecular cloning and sequencing of cDNA for yeast porin, an outer mitochondrial membrane protein: A search for targeting signal in the primary structure. *EMBO J.* **4**, 769–774.

Mozdy, A. D., McCaffery, J. M., and Shaw, J. M. (2000). Dnm1p GTPase-mediated mitochondrial fission is a multi-step process requiring the novel integral membrane component Fis1p. *J. Cell Biol.* **151**, 367–380.

Nunnari, J., Marshall, W. F., Straight, A., Murray, A., Sedat, J. W., and Walter, P. (1997). Mitochondrial transmission during mating in *Saccharomyces cerevisiae* is determined by mitochondrial fusion and fission and the intramitochondrial segregation of mitochondrial DNA. *Mol. Biol. Cell* **8**, 1233–1242.

Nunnari, J., Wong, E. D., Meeusen, S., and Wagner, J. A. (2002). Studying the behavior of mitochondria. *Meth. Enzymol.* **351**, 381–393.

Okamoto, K., Perlman, P. S., and Butow, R. A. (1998). The sorting of mitochondrial DNA and mitochondrial proteins in zygotes: Preferential transmission of mitochondrial DNA to the medial bud. *J. Cell Biol.* **142**, 613–623.

Okamoto, K., Perlman, P. S., and Butow, R. A. (2001). Targeting of green fluorescent protein to mitochondria. *Methods Cell Biol.* **65**, 277–283.

Poot, M., Zhang, Y. Z., Kramer, J. A., Wells, K. S., Jones, L. J., Hanzel, D. K., Lugade, A. G., Singer, V. L., and Haugland, R. P. (1996). Analysis of mitochondrial morphology and function with novel fixable fluorescent stains. *J. Histochem. Cytochem.* **44**, 1363–1372.

Riezman, H., Hase, T., Van Loon, A. P., Grivell, L. A., Suda, K., and Schatz, G. (1983). Import of proteins into mitochondria: A 70 kilodalton outer membrane protein with a large carboxy-terminal deletion is still transported to the outer membrane. *EMBO J.* **2**, 21261–22168.

Rodrigues, F., van Hemert, M., Steensma, H. Y., Corte-Real, M., and Leao, C. (2001). Red fluorescent protein (DsRed) as a reporter in *Saccharomyces cerevisiae*. *J. Bacteriol.* **183**, 3791–3794.

Roeder, A. D., Hermann, G. J., Keegan, B. R., Thatcher, S. A., and Shaw, J. M. (1998). Mitochondrial inheritance is delayed in *Saccharomyces cerevisiae* cells lacking the serine/threonine phosphatase PTC1. *Mol. Biol. Cell* **9**, 917–930.

Sheff, M. A., and Thorn, K. S. (2004). Optimized cassettes for fluorescent protein tagging in *Saccharomyces cerevisiae*. *Yeast* **21**, 661–670.

Simon, V. R., Swayne, T. C., and Pon, L. A. (1995). Actin-dependent mitochondrial motility in mitotic yeast and cell-free systems: Identification of a motor activity on the mitochondrial surface. *J. Cell Biol.* **130**, 345–354.

Skowronek, P., Krummeck, G., Haferkamp, O., and Rodel, G. (1990). Flow cytometry as a tool to discriminate respiratory-competent and respiratory-deficient yeast cells. *Curr. Genet.* **18**, 265–267.

Smith, M. G., Simon, V. R., O'Sullivan, H., and Pon, L. A. (1995). Organelle-cytoskeletal interactions: Actin mutations inhibit meiosis-dependent mitochondrial rearrangement in the budding yeast *Saccharomyces cerevisiae*. *Mol. Biol. Cell* **6**, 1381–1396.

Stevens, B. J. (1977). Variation in number and volume of the mitochondria in yeast according to growth conditions. A study based on serial sectioning and computer graphics reconstruction. *Biol. Cell.* **28**, 37–56.

Taanman, J. W., and Capaldi, R. A. (1993). Subunit VIa of yeast cytochrome *c* oxidase is not necessary for assembly of the enzyme complex but modulates the enzyme activity. Isolation and characterization of the nuclear-coded gene. *J. Biol. Chem.* **268**, 18754–18761.

Visser, W., van Spronsen, E. A., Nanninga, N., Pronk, J. T., Kuenen, J. G., and van Dijken, J. P. (1995). Effects of growth conditions on mitochondrial morphology in *Saccharomyces cerevisiae*. *Antonie Van Leeuwenhoek* **67**, 243–253.

Wallace, W., Schaefer, L. H., and Swedlow, J. R. (2001). A working person's guide to deconvolution in light microscopy. *BioTechniques* **31**, 1076–1078.

Westermann, B., and Neupert, W. (2000). Mitochondria-targeted green fluorescent proteins: Convenient tools for the study of organelle biogenesis in *Saccharomyces cerevisiae*. *Yeast* **16**, 1421–1427.

Wolleschensky, R., Zimmermann, B., Ankerhold, R., and Kempe, M. (2005). High-speed scanning confocal microscope for the life sciences. *Progr. Biomed. Opt. Imaging Proc. SPIE* **5860**, 58600N.

Yang, H.-C., Palazzo, A., Swayne, T. C., and Pon, L. A. (1999). A retention mechanism for distribution of mitochondria during cell division in budding yeast. *Curr. Biol.* **9**, 1111–1114.

CHAPTER 30

Visualization and Quantification of Mitochondrial Dynamics in Living Animal Cells

Kurt J. De Vos* and Michael P. Sheetz[†]

*Department of Neuroscience, MRC Centre for Neurodegeneration Research
The Institute of Psychiatry, King's College London, De Crespigny Park
Denmark Hill, London SE5 8AF, United Kingdom

[†]Department of Biological Sciences
Columbia University
New York, New York 10025

I. Introduction

Mitochondria have intrigued scientists since 1856, when Kölliker (1856) first described filamentous (mito) and grain (chondrium)-like structures in muscle cells that were later named mitochondria by Benda (1898). In the early twentieth century, Lewis and Lewis (1915) stained mitochondria with Janus green, a vital dye discovered by Michaelis (1899), and used hand drawing to document the behavior of mitochondria in living muscle cells for the first time. The Belgian researchers Frédéric and Chèvremont (1951, 1952) obtained the first moving pictures of mitochondrial movements, morphology changes, fusion, and fission in cultured fibroblasts (reviewed in Ernster and Lindberg, 1958). Later, Bereiter-Hahn (1978) described the remarkable variety of shapes and motile behaviors of mitochondria in living XTH cells. Since then, aided by the development of mitochondria-specific probes and advances in light microscopy, numerous groups have studied mitochondrial dynamics in living cells. In this chapter, we focus on methods to visualize mitochondria and quantify their dynamic properties in living mammalian cells. Furthermore, we discuss methods to investigate the interaction of mitochondria with the cytoskeleton in living animal cells.

Mitochondria display a variety of shapes, ranging from small and spherical or the classical "wormlike" shape to extended networks even within one cell (Bereiter-Hahn, 1990; Bereiter-Hahn and Voth, 1994; Collins *et al.*, 2002; De Vos *et al.*, 2005; Fig. 1). Mitochondria are also highly dynamic organelles; their dynamics have been subdivided into morphological dynamics and motility although several authors suggest that mitochondrial function, morphology, and motility are closely linked (Bereiter-Hahn, 1978; De Vos *et al.*, 2005; Hollenbeck *et al.*, 1985; Jendrach *et al.*, 2005; Miller and Sheetz, 2004; Morris and Hollenbeck, 1993). First, frequent transitions between different mitochondrial shapes can be observed, and include fusion, fission, and branching events (Bereiter-Hahn, 1990; Bereiter-Hahn and Voth, 1994). Second, mitochondria interact with and are transported along cytoskeleton tracks. Association of mitochondria with microtubules in mammalian cells was first reported more than 25 years ago (Chan and Bunt, 1978; Heggeness *et al.*, 1978; Raine *et al.*, 1971; Smith *et al.*, 1977). Around the same time, mitochondria were reported to interact with intermediate filaments (Mose-Larsen *et al.*, 1982; Toh *et al.*, 1980). Interaction of mitochondria with actin was reported in animal cells (Kuznetsov *et al.*, 1992, 1994;

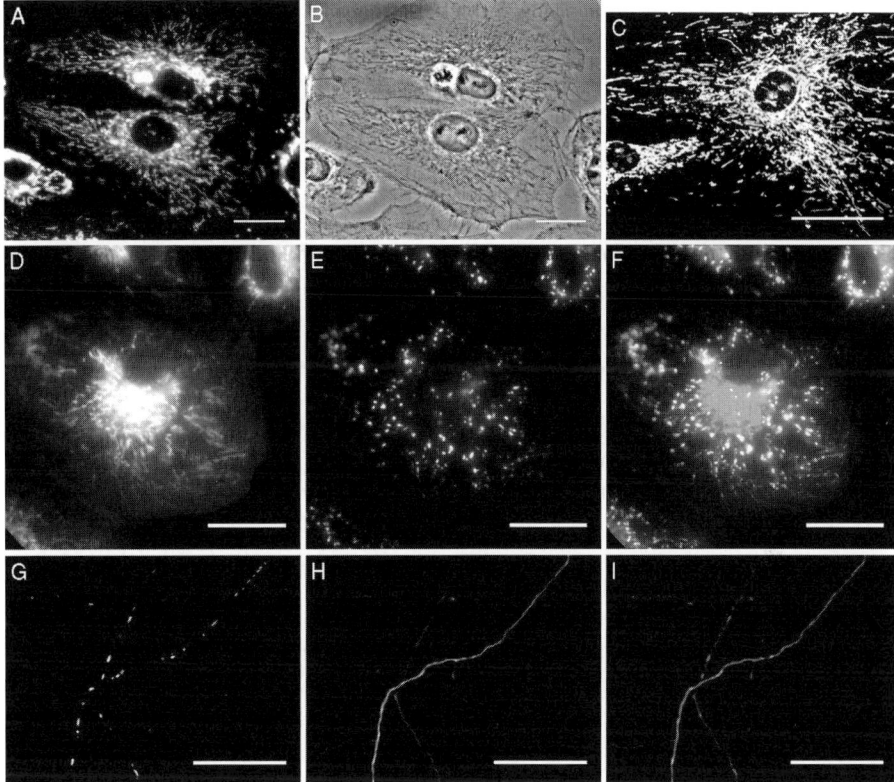

Fig. 1 Mitochondria in living cells. Mitochondria in the same CV-1 cell were visualized by mitochondrial expression of DsRed (A) and phase-contrast microscopy (B). In a primary mouse fibroblast, mitochondria were stained with 500-nM CMXROS (MitoTracker Red) for 30 min at 37°C (C). JC-1 staining of mitochondria (5 μM, 30 min, 37°C) in a CV-1 cell was recorded using fluorescein (D) and rhodamine (E) filters; an overlay is shown in (F). Mitochondria in the axon of primary rat cortical neurons were visualized by mitochondrial expression of DsRed (G); for easy identification of the axon, GFP was coexpressed (H); an overlay is shown in (I). Scale bars = 50 μm. (See Color Insert.)

Langford *et al.*, 1994; Ligon and Steward, 2000b; Morris and Hollenbeck, 1995). It is now well established that association of mitochondria with the cytoskeleton maintains mitochondrial distribution in animal cells and influences mitochondrial function.

In mammalian cells, the dispersion and positioning of mitochondria and other organelles is predominantly governed by the microtubule cytoskeleton (Soltys and Gupta, 1992). Early proof was obtained from pharmacological and morphological data, and was later strengthened by electron microscopy (Chan and Bunt, 1978; Smith *et al.*, 1977) and by *in vitro* motility assays (Brady *et al.*, 1982; Fahim *et al.*, 1985). After the discovery of the molecular motors conventional kinesin (Brady, 1985; Vale *et al.*, 1985) and cytoplasmic dynein (Paschal *et al.*, 1987), it became clear that mitochondria undergo bidirectional, long-distance motor-driven transport along microtubules. Conventional kinesin, a member of the kinesin-1 family, and the kinesin-3 family member KIF1Bβ associate with mitochondria and transport mitochondria toward the plus-end of microtubules, whereas cytoplasmic dynein mediates minus-end-directed transport (De Vos *et al.*, 1998, 2000; Jellali *et al.*, 1994; Leopold *et al.*, 1992; Nangaku *et al.*, 1994; Pilling *et al.*, 2006; Rodionov *et al.*, 1993; Tanaka *et al.*, 1998; Varadi *et al.*, 2004). This long-distance transport is the main mechanism to establish the dispersal of mitochondria in animal cells. In contrast to lower eukaryotes, such as yeast, where all transport of mitochondria occurs along filamentous actin (F-actin) (Boldogh and Pon, 2006; Lazzarino *et al.*, 1994; Simon *et al.*, 1995), short-range F-actin-based transport of mitochondria, probably by myosin family members, seems to fine-tune mitochondrial position after microtubule-based transport (Kuznetsov *et al.*, 1992, 1994; Langford *et al.*, 1994; Ligon and Steward, 2000b; Morris and Hollenbeck, 1995). Mitochondrial transport along intermediate filaments was not reported, but several studies showed that intermediate filaments are important to maintain the distribution of mitochondria in mammalian cells (Collier *et al.*, 1993; David-Ferreira and David-Ferreira, 1980; Eckert, 1986; Leterrier *et al.*, 1994; Linden *et al.*, 2001; Reipert *et al.*, 1999; Toh *et al.*, 1980). Indeed, deletion of the intermediate filament-associated protein, desmin, leads to dramatic changes in mitochondrial distribution and morphology in muscle cells (Linden *et al.*, 2001; Milner *et al.*, 2000; Reipert *et al.*, 1999; Thornell *et al.*, 1997). These findings raise the possibility that intermediate filaments may anchor mitochondria at specific sites in cells. However, evidence suggests that mitochondria might also dock on actin (Minin *et al.*, 2006).

Why are mitochondria so dynamic? Mitochondria are found in close alignment with the endoplasmic reticulum (ER), which is necessary for the ER-mediated mitochondrial sequestration of cytosolic calcium (Franke and Kartenbeck, 1971; Morre *et al.*, 1971; Wang *et al.*, 2000). During apoptosis mitochondria redistribute to the perinuclear region, possibly to facilitate delivery of mitochondrial proapoptotic proteins, such as apoptosis-inducing factor (Susin *et al.*, 1999), to their site of action, the nucleus (De Vos *et al.*, 1998, 2000). Beyond these specific cases, mitochondrial motility, which is driven by interactions with the cytoskeleton, is required to position the organelle at sites with high demands for energy, metabolites, and/or signaling molecules. For example, in neurons, in which the length of the axon poses the most dramatic transport problem of all cells, mitochondria are transported to and anchored at areas of intense ATP and metabolite consumption, including active growth cones and branches (Morris and Hollenbeck, 1993; Ruthel and Hollenbeck, 2003), synaptic buttons (Gotow *et al.*, 1991), myelination boundaries (Barron *et al.*, 2004; Bristow *et al.*, 2002; Yu Wai Man *et al.*, 2005), and nodes of Ranvier (Fabricius *et al.*, 1993). We showed that mitochondria are distributed evenly along the axon of cultured dorsal root ganglions, as if to ensure an even supply of ATP under basal activity conditions (Miller and Sheetz, 2004).

Several authors have proposed a correlation between mitochondrial morphology and function. For example, mitochondrial shape transitions can be developmentally regulated (Bate and Martinez Arias, 1993; Mignotte *et al.*, 1987), linked to disease (Arbustini *et al.*, 1998; Djaldetti, 1982; Nishino *et al.*, 1998; Tandler *et al.*, 2002), or cell death (Breckenridge *et al.*, 2003; Frank *et al.*, 2001), all conditions that influence mitochondrial function. Mitochondria were shown to elongate and fuse under hypoxic conditions, and disturbing mitochondrial function causes drastic mitochondrial shape changes (De Vos *et al.*, 2005).

Despite the mounting evidence for directed mitochondrial movement and the correlations between mitochondrial shape and function, the underlying molecular mechanisms and signaling pathways for these processes remain largely unknown. Several components of the mitochondrial transport machinery and of the fusion/fission machinery have been identified, for example molecular motors (kinesin-1, KIF1Bβ, and cytoplasmic dynein), motor receptors (Milton; Gorska-Andrzejak *et al.*, 2003; Stowers *et al.*, 2002, syntabulin; Cai *et al.*, 2005, and KBP; Wozniak *et al.*, 2005), fission factors (dynamin-related protein 1; Smirnova *et al.*, 1998, 2001, and hFis1; James *et al.*, 2003; Yoon *et al.*, 2003), and fusion factors (mitofusin 1 and 2; Legros *et al.*, 2002; Santel and Fuller, 2001). Similarly, studies have begun to elucidate the regulation of transport (Chada and Hollenbeck, 2004; De Vos *et al.*, 2003; Minin *et al.*, 2006) and mitochondrial shape transitions (De Vos *et al.*, 2005; Ishihara *et al.*, 2003; Varadi *et al.*, 2004). However, it is clear that more remain to be discovered. The methods described below to visualize mitochondria and quantify mitochondrial dynamics in living animal cells will facilitate elucidation of the regulation and molecular mechanisms of these fundamental, largely unexplored areas.

II. Labeling of Mitochondria in Living Animal Cells

Although Janus green was used to visualize mitochondria in living cells for the first time, it was later found to be quite toxic. A range of mitochondrial dyes, such as safranine, were used subsequently (Akerman and Wikstrom, 1976; Colnna *et al.*, 1973). However, the discovery of mitochondrion-specific fluorescent dyes, such as DASPMI (Bereiter-Hahn, 1976) and rhodamine 123 (Johnson *et al.*, 1980), paved the way for extensive studies of mitochondria in living cells. Numerous mitochondrial dyes are currently commercially available. The introduction of green fluorescent protein (GFP) and other fluorescent proteins as tags in living mammalian cells (Chalfie *et al.*, 1994) has allowed several researchers to target fluorescent proteins to mitochondria (Collins *et al.*, 2002; De Vos *et al.*, 2005; Minin *et al.*, 2006; Rizzuto *et al.*, 1995; Twig *et al.*, 2006).

A. Fluorescent Labeling of Mitochondria Using Mitochondrial Dyes and Fluorescent Proteins

1. Mitochondrial Dyes

Mitochondrial dyes can be subdivided in two groups based on their mechanism of accumulation in mitochondria. The first and largest group, the lipophilic cationic dyes, accumulate in mitochondria based on the mitochondrial membrane potential (MMP). Rhodamine, rosamine, carbocyanide, and styryl dyes are examples of these MMP-dependent dyes. The second group does not require MMP to accumulate in mitochondria, and includes dyes such as 10-*N*-nonyl acridine orange (NAO) and MitoTracker Green FM (see Table I for a list of dye properties and staining conditions).

a. MMP-Dependent, Lipophilic Cationic Mitochondrial Dyes

Mitochondria are unique among cellular organelles in that they have a large negative MMP (negative inside of ~180–220 mV). All lipophilic cationic mitochondrial dyes exploit the negative MMP: they cross membranes due to their lipophilicity and accumulate specifically in mitochondria because of the electrical attraction between their positive charge and the negative MMP. Accordingly, these dyes have been used to estimate MMP in living cells and to assess mitochondrial distribution and dynamics. However, MMP-sensing, lipophilic cationic dyes stain

Table I
Mitochondrial Dyes[a]

Dye	Abs./Em. (nm)	Usage	References
(a) MMP-dependent			
(1) Rhodamine dyes			
Rh123	507/529	10 μg/ml, 30 min, 37°C	Johnson *et al.*, 1980
TMRM—TMRE	549/573	0.5–1 μM, 10 min, 37°C	Zahrebelski *et al.*, 1995
(2) Rosamine dyes			
CMTMRos (MitoTracker Orange)	551/576	25–500 nM, 15–30 min, 37°C	Manufacturer
CMXRos (MitoTracker Red)	578/599	25–500 nM, 15–30 min, 37°C	Manufacturer
MitoTracker Red 580	581/644	25–500 nM, 15–30 min, 37°C	Manufacturer
MitoTracker Deep Red 633	644/665	25–500 nM, 15–30 min, 37°C	Manufacturer
(3) Carbocyanide dyes			
JC1	514/529; >0.1 μM: 485–585/590	10 μM, 10 min, 37°C	Smiley *et al.*, 1991
DiOC$_6$(3)	484/501	250 nM, 5 min, 37°C	Korchak *et al.*, 1982
(4) Styryl dyes			
DASPMI	475/605	0.1–5 μM, 15–30 min, 37°C	Bereiter-Hahn *et al.*, 1983
(b) MMP-independent			
NAO	495/519	0.1–5 μM, 10 min, 37°C	Septinus *et al.*, 1985
MitoTracker Green FM	490/516	20–200 nM, 15–30 min, 37°C	Manufacturer
MitoFluor Green	488/520	20–200 nM, 15–30 min, 37°C	Manufacturer
MitoFluor Red 589	588/622	25–500 nM, 15–30 min, 37°C	Manufacturer
MitoFluor Red 594	604/637	25–500 nM, 15–30 min, 37°C	Manufacturer

[a]Abbreviations: MMP, mitochondrial membrane potential; Abs., absorbance wavelength; Em., emission wavelength; Rh123, rhodamine123; TMRM, tetramethylrhodamine methyl ester; TMRE, tetramethylrhodamine ethyl ester; CMTMRos, chloromethyl tetramethyl rosamine; CMXRos, chloromethyl-X-rosamine; JC1, 5,5′,6,6′-tetrachloro-1,1′,3,3′-tetraethylbenzimidazolylcarbocyanine iodide; DiOC$_6$(3), 3,3′-dihexyloxacarbocyanine iodide; DASPMI, 4-(4-(dimethylamino)styryl)-*N*-methylpyridinium iodide; NAO, 10-*N*-nonyl acridine orange.

only mitochondria with respiratory activity. Furthermore, with the exception of the chloromethyl derivatives of rosamine (MitoTracker dyes) that covalently bind to mitochondrial proteins, lipophilic cationic dyes exhibit reversible accumulation in mitochondria and will leak out of the organelle if the MMP is lost. Therefore, one has to use these dyes with caution, especially when studying mitochondrial dynamics under conditions that might influence mitochondrial function and MMP. (Please refer to Chapter 29 by Swayne *et al.*, this volume for further discussion of MMP dyes.)

b. MMP-Independent Dyes

Mitochondrial dyes that accumulate in mitochondria independently of MMP include NAO, MitoTracker Green FM, and MitoFluor dyes (Jacobson *et al.*, 2002; Keij *et al.*, 2000; Pendergrass *et al.*, 2004) (Table I). The mechanism by which these accumulate specifically in mitochondria is not well understood. NAO can bind to the acidic glycerophospholipid cardiolipin (Erbrich *et al.*, 1984; Maftah *et al.*, 1990; Septinus *et al.*, 1985; Zinser *et al.*, 1991). Since cardiolipin is found primarily in the inner mitochondrial membrane, it was assumed that NAO is directed to mitochondria by its affinity for cardiolipin. However, since NAO stains mitochondria in cardiolipin-deficient yeast, binding of NAO to cardiolipin cannot be solely responsible for mitochondrial targeting (Gohil *et al.*, 2005). Indeed, some groups have challenged the MMP independency of mitochondrial accumulation of NAO, Mito-Tracker Green FM, and MitoFluor Green. Clearly caution is warranted when choosing these dyes to evaluate mitochondrial dynamics in living cells.

2. Mitochondria-Targeted Fluorescent Proteins

Although mitochondria have their own genome and protein translation machinery, most proteins found in mitochondria are nuclear encoded and are synthesized in the cytosol. These proteins contain a mitochondrial-targeting peptide that contains all information needed to target the protein to the correct submitochondrial location. Rizzuto *et al.* (1995) first exploited this mitochondrial-targeting system to direct fluorescent proteins to mitochondria by fusing GFP to the amino-terminal 31 amino acids of subunit VIII of cytochrome *c* oxidase, which form the mitochondrial-targeting sequence (see Chapter 15 by Pinton *et al.*, this volume). Mammalian expression vectors containing mitochondria-targeted fluorescent proteins are now commercially available (e.g., Living Colors from Clontech). In our experiments, we use *Discosoma* sp. red fluorescent protein (DsRed) coupled to the mitochondrial localization sequence of subunit VIII of cytochrome *c* oxidase (DsRed-mito, pDsRed1-mito or pDsRed2-mito, from Clontech, Palo Alto, CA; Fig. 1).

3. Choice of Mitochondrial Probe

As outlined above and in Table I, there is a rich choice of fluorescent mitochondrial probes to choose from. In our hands, most of the mitochondrial dyes show the same drawback for time-lapse imaging of mitochondria: photobleaching. In contrast, we have not observed significant photobleaching of GFP or DsRed targeted to mito-chondria. Several side effects of mitochondrial dyes have been reported. For example, rhodamine 123 can inhibit mitochondrial ATPase (Abou-Khalil *et al.*, 1985; Modica-Napolitano *et al.*, 1984), while $DiOC_6(3)$ also stains the ER (Terasaki *et al.*, 1984), inhibits complex I of the mitochondrial respiratory chain (Anderson *et al.*, 1993), and causes swelling of mitochondria when used at elevated levels (L.P., unpublished results). CMXROS (MitoTracker Red) photosensitizes mitochondria and can induce apoptosis on prolonged irradiation (Minamikawa *et al.*, 1999). Finally, CMTMROS (MitoTracker Orange) binds to the adenosine nucleotide translocator and affects mitochondrial function (Scorrano *et al.*, 1999). No side effects have been reported for mitochondria-targeted fluorescent proteins, although we find that expression of these proteins at high levels can affect mitochondrial structure and cell viability (K. D. V., unpublished results). In summary, when no simultaneous measurements of mitochondrial function and motility are required, we recommend using fluorescent proteins to image mitochondrial motility in living animal cells.

4. Phase–Contrast Microscopy

Despite the availability of a range of fluorescent mitochondrial probes, mitochon-dria are relatively large, phase-dense organelles that can be detected using phase-contrast microscopy, especially in thin cells, such as cultured fibroblasts or epithelial cells, and in neuronal processes (Bereiter-Hahn, 1978; Frédéric and Chèvremont, 1951, 1952; Overly *et al.*, 1996; Fig. 1). Imaging native, unlabeled mitochondria has several advantages. First, there are no artifacts associated with fluorescent labeling with dyes or by expression of fluorescent proteins. Furthermore, photo-bleaching, a major hurdle in fluorescence microscopy, is nonexistent. However, there are drawbacks, the most prominent of which is reliable identification of mitochon-dria. Although it is usually relatively easy to distinguish mitochondria from other cytoplasmic organelles on the basis of their size or filamentous morphology, it is difficult to distinguish punctate or reticular mitochondria from vesicles and other organelles (Fig. 1). Some experimenters have tried to circumvent this problem by using fluorescence methods to identify mitochondria before and after phase-contrast time lapse. Thus, although phase-contrast microscopy certainly is helpful, it is not used widely to visualize mitochondria.

III. Probing Interactions Between Mitochondria and the Cytoskeleton and Molecular Motors in Living Animal Cells

A. Visualization of the Cytoskeleton in Living Animal Cells

There are few cell-permeable fluorescent small molecule probes that can be used to visualize the cytoskeleton in living cells. One example is fluorescent Taxol (also called paclitaxel; Oregon Green 488 paclitaxel; Invitrogen, Molecular Probes, Eugene, OR). However, although fluorescent Taxol labels microtubules in living cells, it stabilizes microtubules and inhibits microtubule dynamics. Similar problems are encountered when fluorescently labeled phalloidin (Wulf *et al.*, 1979) is used to visualize actin in living cells. The most widely used technique to visualize the cytoskeleton in living animal cells is by introduction of fluorescently labeled cytoskeleton components. Several researchers microinjected purified fluorescently labeled cytoskeleton components such as tubulin, actin, and desmin (Keith *et al.*, 1981; Kreis and Birchmeier, 1982; Kreis *et al.*, 1979; Mittal *et al.*, 1989; Sammak and Borisy, 1988a,b; Saxton *et al.*, 1984; Taylor and Wang, 1978). After the discovery of GFP as a fluorescent protein tag in living animal cells (Chalfie *et al.*, 1994), many laboratories have fused GFP and other fluorescent proteins to cytoskeleton components and expressed them in living animal cells. Several mammalian expression vectors encoding actin or tubulin coupled to fluorescent proteins are available commercially (e.g., from Clontech). Fluorescent protein-tagged intermediate filament components, such as vimentin (Ho *et al.*, 1998), cytokeratins (Beil *et al.*, 2003), and neurofilaments (Ackerley *et al.*, 2000), have been reported. Figure 2 shows the distribution of mitochondria in living animal cells expressing tubulin, actin, or vimentin tagged with GFP. Note that overexpressing fluorescent protein-tagged cytoskeletal components sometimes disturbs the structure of the endogenous cytoskeleton, especially when expression levels are high; for example, we found that high levels of GFP-tagged vimentin disrupt the vimentin cytoskeleton in CV-1 cells (De Vos *et al.*, 2005). Comparing cytoskeletal structure in transfected and nontransfected controls by indirect immunofluorescence staining after fixation is necessary to exclude any artifacts due to overexpression.

B. Use of Cytoskeletal Inhibitors in Living Animal Cells

A plethora of commercially available chemicals disrupt or stabilize cytoskeletal components; their properties and usage are outlined in Table II. Most of these agents stabilize or induce depolymerization of F-actin or microtubules. Chemical inhibitors of intermediate filaments are less common and often less specific and effective. Therefore, microinjection of inhibitory antibodies (Pruss *et al.*, 1981) or expression of dominant-negative desmin (Schultheiss *et al.*, 1991; Yu *et al.*, 1994) are commonly used to disrupt intermediate filaments. We found that high overexpression of GFP-coupled vimentin can cause aggregation and disruption of vimentin intermediate filaments in CV-1 cells (De Vos *et al.*, 2005). These cytoskeleton inhibitors can be used to evaluate the role of the cytoskeleton in mitochondrial motility, distribution, and morphology in living cells using the methods described below.

C. Inhibition of Microtubule-Based Molecular Motors in Living Animal Cells

1. Chemical Inhibitors

Unfortunately, chemical inhibitors that specifically target mitochondrial microtubule-based motors are limited. No inhibitors of kinesin-1 or KIF1Bβ are currently available. Cytoplasmic dynein can be inhibited by vanadate or

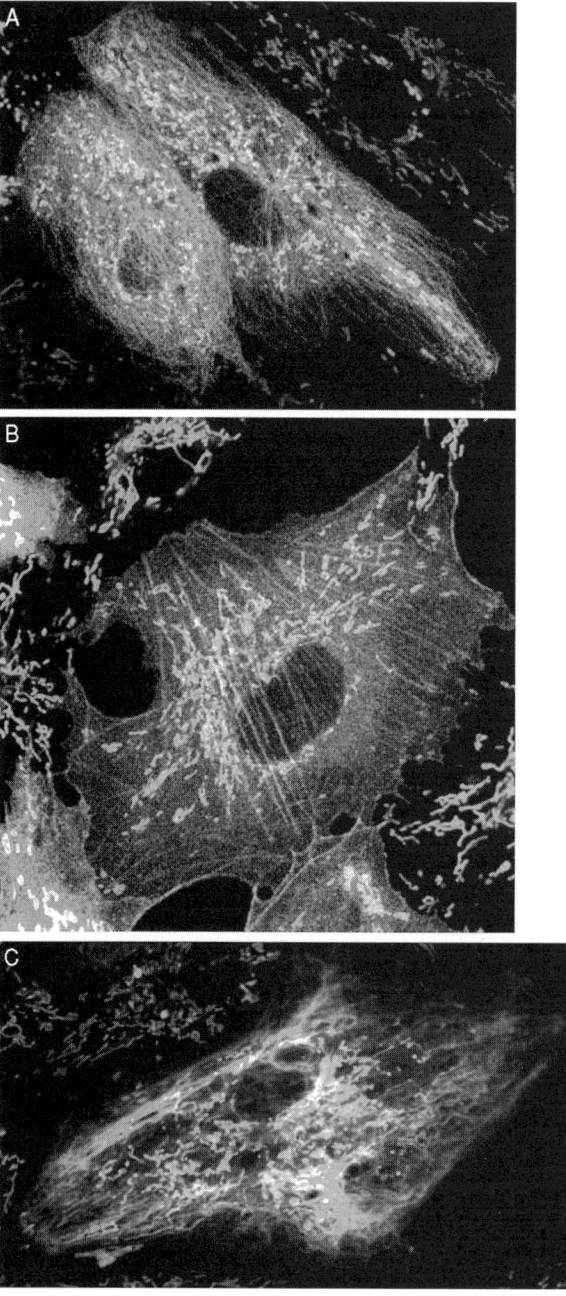

Fig. 2 Visualization of the cytoskeleton in living animal cells with GFP-tagged cytoskeleton components. CV-1 cells stably expressing mitochondrial DsRed (red) were transfected with tubulin (A), actin (B), or vimentin (C) tagged with EGFP (green), and observed by live confocal microscopy. (See Color Insert.)

Table II
Inhibition of the Cytoskeleton in Living Animal Cells

Inhibitor	Action mechanism and remarks	Usage[a]	References
Actin inhibitors			
Jasplakinolide	Inhibits F-actin disassembly; induces polymerization	1–2 μM, 1 h, 37°C	Bubb *et al.*, 1994; Cramer, 1999; Minin *et al.*, 2006
Cytochalasin	Induces repolymerization; several forms, e.g., B and D	1–2 μg/ml, 15–30 min, 37°C	Cooper, 1987; Flanagan and Lin, 1980; Forer *et al.*, 1972
Latrunculin A	Induces depolymerization	0.5–1 μM, 15–30 min, 37°C	Coue *et al.*, 1987; Spector *et al.*, 1983
Microtubule inhibitors[b]			
Taxol	Stabilizes microtubules	0.5–1 μM, 15–60 min, 37°C	De Brabander *et al.*, 1981; Saxton *et al.*, 1984
Cold/4°C	Inhibits polymerization; does not affect cold-stable microtubules		Baas and Heidemann, 1986
Nocodazole	Inhibits polymerization; fully reversible	0.1–5 μM, 15–60 min, 37°C	Baas and Heidemann, 1986; De Brabander *et al.*, 1977
Colchicine	Inhibits polymerization	0.1–5 μM, 15–60 min, 37°C	De Brabander *et al.*, 1977
Vinblastine	Inhibits polymerization	0.1–5 μM, 15–60 min, 37°C	De Brabander *et al.*, 1977
Intermediate filament inhibitors			
2,5-hexanedione	Disrupts	4 mM, 37°C	Durham *et al.*, 1983
Acrylamide	Disrupts	5 mM, 4 h, 37°C	Durham *et al.*, 1983; Eckert, 1985
Sphingosylphosphorylcholine	Induces perinuclear reorganization of intact keratin 8–18 filaments	10 μM, 30–60 min, 37°C	Beil *et al.*, 2003
Cycloheximide	Disrupts; also inhibits protein synthesis	10 μg/ml, 15–30 min, 37°C	Sharpe *et al.*, 1980
Expression of mutants	Disrupts desmin and vimentin		Schultheiss *et al.*, 1991; Yu *et al.*, 1994
High overexpression	Disrupts vimentin		De Vos *et al.*, 2005
Antibodies	Disrupt desmin and vimentin; not cell permeable		Pruss *et al.*, 1981

[a]The concentrations and incubation times of cytoskeletal inhibitors vary greatly in literature. The values given in the table are a compilation from multiple sources not all of which could be referenced. It is advisable to test the efficacy of cytoskeletal inhibitors in different cell types and applications, for example, by indirect immunofluorescence.

[b]Not all microtubule inhibitors are listed here; only the most commonly used in cell biology applications are listed.

erythro-9-[3-(2-hydroxynonyl)]adenine (EHNA) (Beckerle and Porter, 1982; Bouchard *et al.*, 1981; Gibbons *et al.*, 1978; Kobayashi *et al.*, 1978; Paschal *et al.*, 1987; Penningroth, 1986; Shpetner *et al.*, 1988). However, both inhibitors are rather unspecific.

2. Dominant–Negative Proteins

Kinesin-1 (conventional kinesin) is a tetramer consisting of two kinesin heavy chains and two kinesin light chains. The kinesin heavy chain harbors the motor domain, which is essential for microtubule binding and activity. Kinesin light chains are involved in cargo binding and regulation of kinesin heavy chain function (Vale and Fletterick, 1997). In addition, cargo binding involves kinesin receptors, such as kinectin (Kumar *et al.*, 1995; Toyoshima *et al.*, 1992), which links kinesin to the ER; Milton and syntabulin, which are required for the association kinesin-1 with mitochondria; and KBP (KIF1 binding protein), which is implicated in linking KIF1Bβ to mitochondria (Cai *et al.*, 2005; Gorska-Andrzejak *et al.*, 2003; Stowers *et al.*, 2002; Wozniak *et al.*, 2005).

Several groups have exploited the domain structure of kinesin-1 to design dominant-negative kinesin-1 mutants. Kinesin-1 function was successfully disrupted by expression of kinesin heavy chain constructs containing a point mutation in the ATP-binding motif (Nakata and Hirokawa, 1995). These mutated kinesin heavy chains dimerize with endogenous kinesin-1 and inhibit function. Alternatively, over-expression of the tetratricopeptide repeats (TPRs) of kinesin light chain has been successfully used to block kinesin-1, most likely by competition for binding sites on organelles (Muresan and Muresan, 2005; Semiz *et al.*, 2003; Verhey *et al.*, 2001). Similarly, expression of mutant KBP caused redistribution of mitochondria (Wozniak *et al.*, 2005).

Cytoplasmic dynein is a massive multisubunit complex (MW 1200 kDa) and is composed of two heavy chains, intermediate chains, light intermediate chains, and light chains. Dynein is associated with dynactin (i.e., dynein activator), p150Glued, dynamitin (p50), and actin-related protein 1 (ARP1) (Schroer, 1994, 2004). Cytoplasmic dynein function can be inhibited by overexpression of p50/dynamitin that disrupts the interaction of cytoplasmic dynein with the dynactin complex (Burkhardt *et al.*, 1997; Echeverri *et al.*, 1996).

3. Inhibitory Antibodies

Inhibitory antibodies have been used successfully to inhibit kinesin-1 and cyto-plasmic dynein function. Kinesin-1 was successfully inhibited in living animal cells by microinjection or syringe-loading delivery of antibodies directed against the motor domain of kinesin heavy chain, namely SUK4 and HD (De Vos *et al.*, 1998; Ingold *et al.*, 1988; Rodionov *et al.*, 1993); injecting cells with the well-characterized 70.1 mAb to the cytoplasmic dynein intermediate chains inhibits cytoplasmic dynein function (Burkhardt *et al.*, 1997; Dujardin *et al.*, 2003; Faulkner *et al.*, 2000; Yvon *et al.*, 2001).

4. Antisense RNA and RNA Interference

Antisense RNA and RNA interference (RNAi) technology has been successfully used to inhibit mitochondrial molecular motors. Antisense RNA was used to suppress kinesin-1 expression in hippocampal neurons (Bannai *et al.*, 2004; Feiguin *et al.*, 1994; Ferreira *et al.*, 1992; Kaether *et al.*, 2000; Severt *et al.*, 1999). We successfully used RNAi technology to deplete kinesin-1 from CHO cells (M.S., unpublished results). Depletion of KBP from cells using RNAi technology disrupted mitochondrial motility (Wozniak *et al.*, 2005). Finally, cytoplasmic dynein can be knocked down by RNAi technology (Harborth *et al.*, 2001; He *et al.*, 2005).

IV. Image Acquisition

A. Equipment

1. Temperature Control

Temperature control is a critical factor for reliable measurement of mitochondrial dynamics. In our experience, mitochondrial dynamics decrease dramatically with temperature. A decrease in temperature from 37 to 35 °C results in severe inhibition of mitochondrial motility in CV-1 cells and cortical neurons (K. D. V., unpublished results). A variety of temperature control units and observation chambers are commercially available. These are certainly adequate, but often quite expensive. The temperature control systems and observation chamber described here are relatively inexpensive, yet are sufficient for our imaging needs.

We use two methods to maintain the cells at 37 °C, which can be used separately or together. First, we use a heating collar (IntraCell, Cat. No. 150819) to heat the oil immersion objective. The contact maintained between the objective and the coverslip by the immersion oil transfers heat and keeps the cells at the desired temperature reliably. This method can only be applied when the contact surface between the objective and coverslip is large enough and the medium volume in the chamber is small ($\leq 400 \ \mu$l). We also use a forced air heater consisting of a hair dryer connected to a voltage controller (Staco Energy Products, Type 2PF1010; available from Fisher Scientific, Pittsburgh, PA). Warmed air is passed into a chamber created under the microscope stage with Plexiglas boarding. Using either or both methods, it is crucial to calibrate the temperature regularly and to allow the temperature to equilibrate before measurements.

2. Observation Chamber

We use a closed observation chamber that consists of a metal support with a square cutout (4 mm smaller than the size of the coverslips used) to which two coverslips are fixed to form the observation chamber (Fig. 3). A groove carved in the metal support allows medium changes on the microscope by capillary force. We have maintained cells in these chambers for up to 5 h without any noticeable

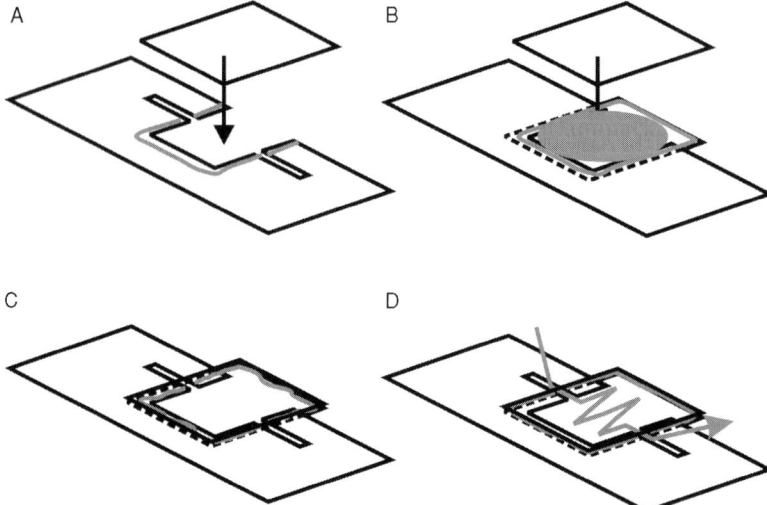

Fig. 3 Observation chamber assembly [see Section IV.A.2.a (Method 1)]. Fix a fresh coverslip to the metal support by pushing it down gently onto the greased side of the support (A). Turn over the support, apply grease, fill the chamber with culture medium, and close the chamber by gently pressing down the coverslip on which the cells were grown into the grease (B, C). Wipe off excess medium. Seal the chamber by applying VALAP to the coverslip-support boundaries. Medium in the chamber can be changed by capillary force (D).

decline in cell viability, and monitored mitochondrial motility for up to 3 hr. However, long-term (>3 h) measurements in closed observation chambers are prone to artifacts because of pH changes in the culture medium (mostly acidification) and oxygen depletion. To avoid these problems, we never maintain the cells longer than 2 h on the microscope stage. Moreover, we use CO_2-independent buffer systems, such as HEPES, to prevent pH changes of the culture medium by exposure to air and adapt the cells for growth in HEPES-buffered medium for several passages prior to imaging. Evaporation can be prevented by layering mineral oil on top of the culture medium. If longer measurements are required, it is advisable to invest in a complete environmental control system.

a. Method 1: Observation Chamber Assembly (See Fig. 3)
Materials

1. Metal support.
2. Glass coverslips, 22×22 mm^2, No. 1.
3. High vacuum grease (Dow Corning).
4. VALAP: Mix equal parts of Vaseline, lanolin, and paraffin. VALAP liquefies when moderately heated and rapidly solidifies on application. We do not use nail polish to seal the chamber because the solvents used in most kinds of nail

polish can be toxic for cells and can quench fluorescence, especially that of fluorescent proteins.

Procedure

1. Grow cells on a sterile coverslip (see below).

2. Apply a thin line of vacuum grease along the edges of the cutout at the grooved side of the support (Fig. 3A); keep the grooves free if media changes are required (Fig. 3D).

3. Fix a fresh coverslip to the metal support by pushing it down gently onto the greased side of the support (Fig. 3A).

4. Turn over the support.

5. Apply a thin line of grease along the edges of the cutout (Fig. 3B).

6. Apply a line of vacuum grease as thick as the support on the coverslip at the open end of the cutout (Fig. 3B).

7. Fill the chamber with culture medium (Fig. 3B); for most applications we use complete culture medium, however, when GFP or other green fluorescent probes are used it is advisable to use culture medium without phenol red to avoid high background fluorescence of phenol red.

8. Close the chamber by gently pressing down the coverslip on which the cells were grown into the grease (Fig. 3C).

9. Wipe off excess medium.

10. Seal the chamber by applying VALAP to the coverslip-support boundaries.

11. Medium changes are possible by wicking up the medium from one side of the chamber with filter paper while simultaneously adding new medium to the opposite side (Fig. 3D).

3. Microscope System

The main consideration when investing in a live imaging microscope system is the type of experiments one has in mind. For example, is two-dimensional or three-dimensional microscopy required? What is the desired time and spatial resolution? Is multicolor acquisition needed? Are low-light applications planned?

We use two microscope systems for time-lapse recording of mitochondrial dynamics in living cells. One is based on a Zeiss Axiovert 200 fluorescence microscope, the other is an Olympus FluoView 500 confocal laser scanning microscope (CLSM). The CLSM that we use is the standard Olympus configuration and is not described. The Axiovert 200 setup is described in detail below.

a. Choice of Microscope

Virtually any modern epifluorescence microscope is a suitable starting point to build a live imaging system. The first choice one has to make when investing in a new microscope is between an inverted (objective below sample) or upright (objective above sample) microscope. High numerical aperture (N.A.), water dipping, objectives now enable the use of open observation chambers on upright systems. Nonetheless, most live imaging systems are based on inverted microscopes because this configuration allows for use of open observation chambers and easy access to the sample for experiments involving perfusion, micromanipulators, electrodes, or microinjection.

b. Choice of Fluorescence Excitation Light Source

The standard excitation light source of most fluorescence microscopes is a 100-W mercury arc lamp that emits excitation light from ultraviolet (UV) to infrared. Mercury burners do not provide even intensity across the spectrum and are actually not nearly as bright in the blue excitation range used for excitation of fluorescein, GFP, or YFP compared to the near-UV or green excitation range (peaks at 313, 334, 365, 406, 435, 546, and 578 nm; Abramovitz, 1993). A xenon short-arc source provides a much more even intensity across the visual spectrum, but is less intense in the near-UV range. Generally, for the equivalent wattage, xenon arc sources are not as bright as mercury arc sources, but they are brighter in the 450- to 500-nm range where mercury sources do not perform as well. Xenon sources are preferred for quantitative analysis of fluorescence. As an alternative light source a tungsten-halogen lamp can be used, especially for blue or green excitation. Halogen sources are attractive for live imaging applications because they are less intense and therefore cause less phototoxicity. We use a xenon short-arc source.

c. Optical Filters

For epifluorescence, a combination of three filters is used: the excitation filter, a dichroic beam splitter, and an emission filter. Excitation filters transmit wavelength(s) used for fluorophore excitation and block all other wavelengths. Emission filters transmit the fluorescence emitted by the sample and block other wavelengths transmitted by the excitation filter. The dichroic beam splitter reflects the excitation light and transmits emitted fluorescence light. An excellent overview of different types of optical filters and their characteristics can be found in the *Handbook of Optical Filters for Fluorescence Microscopy from Chroma Technology Corp* (Reichman, 2000).

Depending on the application, different combinations of optical filters and hardware are used. For single-color applications standard filter cubes suffice; for multicolor applications multiple hardware configurations are possible. We use a combination of a multiband dichroic beam splitter with single-band emitter filters mounted in a filter wheel and a multiwavelength monochromator-based excitation setup. As an alternative for the monochromator-based excitation setup, single-band excitation filters can be mounted in a computer-controlled filter wheel. However, the monochromator allows wavelength switching in less than 3 ms compared to about 50 ms for a fast filter wheel. Speed is very important when imaging multiple fluorochromes simultaneously in time-lapse mode to minimize asynchronous recording. As an alternative for an emission filter wheel, dual-imaging systems can be used (please refer Chapter 29 by Swayne *et al.*, this volume). Additionally, to reduce photodamage and photobleaching during time-lapse recordings, it is crucial to block excitation light from reaching the sample in between recordings. This can be accomplished by inserting a shutter in the light path that is synchronized with the camera or by inserting a closed position in the excitation filter wheel. When a monochromator is used, no shutter is needed because the monochromator can be set to prevent any light coming through.

d. Objectives

For low-light applications such as time-lapse imaging of mitochondria (or other intracellular organelles), a high-magnification, high-N.A. oil immersion objective is desirable. It is especially important to choose an objective with high N.A., since N.A. determines the amount of light accepted by the objective, and the amount of light reaching the detector is proportional to the fourth power of the N.A. of the objective and inversely proportional to the magnification. The importance of high N.A. is also evident when one considers that only about 30% of the emitted fluorescence will actually reach the objective lens. Other characteristics to look out for are chromatic and spherical correction. We find that the Zeiss Plan Apochromat® or Plan NeoFluar® oil immersion objectives fit those criteria. Note that to achieve maximal spatial resolution, the objective magnification should be matched to the camera (see below).

e. Camera

Charge-coupled device (CCD) cameras are the most commonly used cameras for biological applications. In this section, we will briefly discuss the things to look out for when choosing a CCD camera for live imaging. Several camera vendors provide in depth information online (e.g., http://www.andor.com/; http://www.roperscientific.com/; http://www.ccd.com/).

There are two standard types of CCDs: front-illuminated and back-illuminated. Intensified CCDs (ICCDs) and electron-multiplying CCDs (EMCCDs) are modifications of these standard types. For live-cell imaging applications, the most important specifications to look out for are quantum efficiency (Q.E.), noise, and pixel size. Q.E. describes how effective the CCD converts incoming photons into electronic charge, or how efficiently the CCD detects light and is an important consideration for low-light applications such as live-cell imaging. Q.E. varies for different wavelengths of light. Therefore, it is important to evaluate Q.E. of the emission wavelengths used in your applications. Front-illuminated CCDs usually have a Q.E. between 30% and 40% in the light spectrum mostly used for live imaging (500–600 nm), whereas back-illuminated CCDs typically have a Q.E. between 70% and 90% for the same range.

The noise of a CCD camera is the combination of the readout noise of the CCD sensor and the circuitry and of thermal noise (black noise or current). The readout noise of a good CCD camera should be ≤ 15 electrons at 1-MHz readout speed. Note that the readout noise increases with the square root of the readout speed and with pixel size.

Black noise refers to charge building up in the CCD sensor even in the absence of light. Black noise depends on temperature and is reduced significantly in cooled CCD cameras. Note that readout noise and black current values found on camera specification sheets are usually average values and that peak values can be five times higher. In EMCCD cameras, additional noise comes from the amplification process (see below).

To achieve the maximal spatial resolution, one has to match the resolution of the microscope to the resolution of the camera. CCD pixel size plays an important role in making the correct match. The Nyquist limit states that the resolution of the detector has to be about two to three times smaller than the resolution of the system. In other words, the pixel size of the CCD should be two to three times smaller than the smallest structure to be detected. On the basis of Rayleigh criterion, the resolution of a diffraction-limited microscope is given by Eq. (1):

$$d = \frac{0.61 \times \lambda}{\text{N.A.}} \tag{1}$$

where d is the smallest distance between two structures that can be resolved and λ is the wavelength of light that is imaged. Assuming that we want to match the resolution of the microscope to 2.5 pixels on the CCD, we can calculate the optimal magnification according to Eq. (2):

$$Magnification = \frac{2.5 \times pixel\ size}{d} \tag{2}$$

where *Magnification* is the optimal magnification, *pixel size* is the pixel size of the CCD, and d is the smallest distance that can be resolved. For a CCD with 6.8-μm pixels, the optimal magnification equates to 55 for 525-nm light and a 1.3-N.A. objective. So in this case, a $40\times$ objective combined with a $1.3\times$ phototube would be ideal. Mitochondria are easily visible at this magnification.

EMCCDs—CCDs with a built-in electron multiplication structure (also called on-chip multiplication)—are becoming increasingly popular for live imaging applications. Without going in technical details, EMCCDs are capable of delivering high sensitivity (even detecting single photons) at high speed. EMCCDs come in the same varieties as other CCDs, thus the specifications discussed above apply. However, EMCCDs have an additional source of noise because of the on-chip amplification, and typical noise values are in the range of 50 electrons at 1-MHz readout speed.

f. Acquisition Software

Numerous acquisition (and analysis) software packages are available commercially for different operating systems. Widely used software packages are, for example, Openlab (OSX; Improvision Ltd., Coventry, UK), Volocity (OSX, Windows; Improvision Ltd., Coventry, UK), MetaMorph (Windows; Molecular Devices Corporation, Sunnyvale, CA), Simple PCI (Windows; Compix Inc., Sewickley, PA), and ISee (OSX, SGI, Windows; ISee Imaging Systems, Raleigh, NC), but others are available. Confocal microscope systems come with software included. All are up to the task, and mostly the choice will be based on pricing and user preference.

Our microscope setup

- Axiovert 200 (Carl Zeiss Ltd., Welwyn Garden City, Hertfordshire, UK)
- Polychrome IV monochromator connected by fiber optics to a Xenon short-arc source (Till Photonics, Gräfelfing, Germany)
- EGFP/DsRed filter set (Chroma Technology Corp., Rockingham, VT; Cat. No. 86007)
- 40× EC Plan-Neofluar® N.A. 1.3 (Carl Zeiss Ltd., Welwyn Garden City, Hertfordshire, UK)
- Objective heater (IntraCell, Shepreth, Royston Herts, UK; Cat. No. 150819)
- Lambda 10-2 filter wheel (Sutter Instrument Company, Novato, CA)
- Cameras: Hamamatsu C4880-80 CCD and Hamamatsu 9100-12 EMCCD (Hamamatsu Photonics, Welwyn Garden City, Hertfordshire, UK)
- Orbit II hardware controller (Improvision Ltd., Coventry, UK)
- Openlab software (Improvision Ltd., Coventry, UK)
- PowerPC G4 (Apple Computer, Cupertino, CA)

V. Preparation of Cells

In this section we describe the culture and transfection of our main model systems, primary cortical neurons, N2A neuroblastoma cells, and CV-1 fibroblasts. N2A cells and CV-1 cells will be discussed together. In principle, these protocols should be applicable for all cells that can be cultured on glass coverslips.

A. Isolation, Culture, and Transfection of Primary Cortical Neurons

1. Method 2: Isolation of Cortical Neurons

Materials

1. Hanks buffered saline solution (HBSS) w/o Ca^{2+}/Mg^{2+} ($HBSS^{-/-}$).
2. HBSS with Ca^{2+}/Mg^{2+} ($HBSS^{+/+}$).
3. Cortical neuron culture medium: Neurobasal medium supplemented with 2-mM L-glutamine, 200-U/ml streptomycin, 100-U/ml penicillin, and 1% B27 supplement (Invitrogen, Paisley, UK; GIBCO).
4. DNase solutions: DNase stock solution is prepared by dissolving 10-mg/ml DNase (Sigma Cat. No. D5025) in $HBSS^{+/+}$. Store the stock solution at $-20\,^{\circ}C$ in 50-μl aliquots. DNase working solution is prepared by adding 50 μl of DNase stock to 50 ml of $HBSS^{+/+}$. Store at $-20\,^{\circ}C$ in 5-ml aliquots after filter sterilization through a 0.2-μm filter.
5. Poly-L-lysine solution: Prepare the stock solution by dissolving 50-mg poly-L-lysine (MW 70,000–150,000) in 50 ml of sterile distilled water (0.1%).

6. Triturating solution is made up of 1% albumax (0.5 g; Invitrogen; GIBCO Cat. No. 11020-013), 25-mg soybean trypsin inhibitor (Sigma Cat. No. T9003), one 50-μl aliquot DNase stock solution in 50-ml HBSS$^{+/+}$, filter sterilized through a 0.2-μm filter, and stored at $-20\,^\circ$C in 1-ml aliquots.

7. Trypsin stock solution: 2.5% trypsin solution in saline (Sigma).

8. Glass coverslips, 22×22 mm^2, No. 1.

Procedure

1. Coverslips are coated overnight at 37 $^\circ$C in 0.01% poly-L-lysine solution and washed two times with distilled water and once with HBSS$^{-/-}$.

2. Dissect brains from E17 rat embryos or E15 mouse embryos, remove the meningus, and dissect cortex.

3. Store the dissected cortices in HBSS$^{-/-}$ during dissection and transfer in HBSS$^{-/-}$ to a 15-ml blue-cap tube.

4. After one wash with 10-ml HBSS$^{-/-}$, resuspend the cortices in 5-ml HBSS$^{-/-}$.

5. Start the digestion by adding 70 μl of the 2.5% trypsin stock solution.

6. Incubate 15 min at 37 $^\circ$C (gently invert the tube every 5 min).

7. Add 5-ml DNase working solution.

8. Incubate for 30 s at RT.

9. Aspirate the solution, leaving the cortices in the blue-cap tube.

10. Add 1 ml of triturating solution.

11. Triturate with three flame-polished, plugged, glass Pasteur pipettes with progressively smaller bore (triturate 15 strokes with each Pasteur pipette: one stroke = two passages through the pipette). Check for single-cell suspensions on a microscope. If a single-cell suspension is not yet achieved, continue triturating with the smallest bore pipette.

12. Add complete culture medium up to 5 ml and count the cells.

13. Plate the cortical neurons are at 10^6 cells per poly-L-lysine-coated 22×22 mm^2 coverslip in a six-well plate in 4-ml culture medium. This cell density is optimized for subsequent transfection.

14. Allow differentiating for at least 4–5 days prior to experiments.

Note: If fluorescent dyes are to be used, a two to three times lower density is advisable, followed by a 6–7 day differentiation period.

2. Method 3: Transfection of Primary Cortical Neurons

We routinely transfect primary cortical neurons using a modified calcium phosphate transfection protocol (Ackerley *et al.*, 2000; Nikolic *et al.*, 1996; Xia *et al.*, 1996). The following protocol describes transfection of primary cortical neurons plated on 22×22 mm^2 coverslips in a six-well plate. The protocol can be scaled up or down as necessary.

Materials

1. Promega Profection mammalian transfection system, calcium phosphate kit. This kit contains 3-M $CaCl_2$, nuclease-free H_2O, and 2× HEPES buffered saline (HBS). These solutions can be homemade, but using the kit enhances reproducibility of transfection.

2. Kynurenic acid stock solution (20×) is made up of 20-mM sodium kynureate, 10-mM $MgCl_2$, and 5-mM Na-HEPES. Adjust the pH to 7.5 with NaOH, aliquot, and store at –20 °C. Note that kynurenic acid is poorly soluble in H_2O and will only dissolve when neutral pH is reached.

3. Plasmid DNA mix: 10-μg plasmid DNA + (216 μl minus the volume of the plasmid) μl H_2O + 26 μl of 3-M $CaCl_2$. Add the $CaCl_2$ just before transfection.

Procedure

1. Collect the conditioned culture medium and keep it at 37 °C.

2. Add 1-ml prewarmed fresh culture medium containing 1× kynurenic acid per well and incubate for 30 min at 37 °C.

3. Just before transfection, prepare the calcium phosphate precipitates by adding the DNA mix dropwise to an equal volume of 2× HBS in a 15-ml blue-cap tube while vortex mixing at maximum speed. Usually the DNA mix we prepare contains 3 μg of pDsRed1-mito and 7 μg of EGFP-coding plasmid. The latter allows easy identification of transfected axons (Fig. 1).

4. Vortex for an additional 5–10 s and add the precipitates to the neurons (drop across the coverslip).

5. After 45 min incubation at 37 °C, remove the precipitate, and wash carefully: twice with fresh medium and once with conditioned medium.

6. Finally, add 4 ml of a 50/50 (v/v) mix of fresh and conditioned medium and return to 37 °C. The neurons will start express as early as 3–4 h after transfection, with maximal expression after 24–48 h. In our experiments, we image neurons 36 h after transfection.

B. Culture and Transfection of N2A Neuroblastoma Cells and CV-1 Cells

1. Method 4: Culture of N2A and CV-1 Cells

Materials

1. Culture medium: Dulbecco's Modified Eagles medium (DMEM) supplemented with 10% fetal calf serum, 2-mM L-glutamine, 4-mM sodium pyruvate, 200-U/ml streptomycin, and 100-U/ml penicillin (Invitrogen; GIBCO).

2. N2A differentiation medium: DMEM containing 1% N2 supplement (Invitrogen; GIBCO), 2-mM L-glutamine, 4-mM sodium pyruvate, 200-U/ml streptomycin, and 100-U/ml penicillin.

Procedure

1. N2A and CV-1 cells are routinely maintained in DMEM.

2. Prior to experiments, about 10^6 cells are seeded on poly-L-lysine-coated coverslips (see above).

3. Differentiation of N2A cells: Transfer N2A cells that were seeded on coverslips to differentiation medium for at least 24 h. Under these conditions, N2A cells extend processes, some of which are axonlike whereas others are neurites. We select the processes to analyze based on their length; axons are significantly longer than neurites in these cells (Yu *et al.*, 1997), and their morphology—in particular the transition from cell body to process—is better defined in the case of axonlike processes. The processes formed under these differentiation conditions that match our criteria are positive for the microtubule binding protein tau (present in both axons and neurites) and negative for the neurite-specific microtubule binding protein MAP2, confirming that they are axonlike (De Vos *et al.*, 2003).

2. Method 5: Transfection of N2A and CV-1 Cells

Both N2A and CV-1 cells are transfected easily by a variety of transfection reagents. We routinely use Lipofectamine Plus reagents (Invitrogen) or ExGen500 (MBI-Fermentas) according to the protocols supplied by the manufacturer. We advise against the use of Lipofectamine 2000 reagent (Invitrogen). Even though Lipofectamine 2000 transfects N2A and CV-1 cells efficiently, mitochondrial motility is severely reduced in Lipofectamine 2000 transfected cells, for reasons that are unclear. In case of N2A cells, transfecting in serum-free medium and then transferring to differentiation medium can combine transfection and differentiation. We typically perform experiments 24 h after transfection.

VI. Time–Lapse Recordings

Photobleaching and phototoxicity caused by repeatedly exposing fluorochromes and cells to excitation light are the main problems when observing mitochondria in living cells by fluorescence microscopy for prolonged periods. When imaging mitochondrial motility, one of the first indicators of phototoxicity in our experience is an abrupt reduction in motility. Appearance of membrane blebs indicates activation of apoptosis. In neurons or differentiated neuroblastoma cells, retraction of the growth cone is also observed when cells degenerate. Minimizing exposure time and excitation light intensity can reduce both photobleaching and phototoxicity. Inserting neutral density filters in the light path can attenuate the intensity of excitation light but leads to less fluorescence light and usually higher exposure times. Similarly, lower exposure times might compromise the quality of the recorded images. Using a high-sensitivity camera, such as an EMCCD or ICCD camera, can be helpful; also the right choice of mitochondrial probe is important (see above). Alternatively, increasing the time between recordings and/or shortening the total recording time can reduce the total exposure of the cells to excitation light over the course of an experiment. Clearly, there is a trade-off between the amount of data one collects and the well-being of the cells and mitochondria under observation. In the previous edition of this book, an excellent flow chart illustrates how to optimize conditions for live microscopy in yeast; this chart also can be applied for mammalian cells (Pon and Schon, 2001; Chapter 19, p. 347).

There is a vast variation in the measurement regime used to image mitochondrial motility. The time-lapse interval varies from less than a second to multiple minutes; the total recording time ranges from minutes to hours. There are no fixed rules that dictate the time-lapse interval or recording time, but some guidelines can be helpful to design a particular experiment.

In principle, measuring motility in real time, that is, at video rate or 24 frames/s is ideal; all movements will be recorded, and mitochondrial motility can be quantified fully. In practice, video rate recording is not possible when using fluorescence methods to visualize mitochondria because of photobleaching and phototoxicity. However, video rate recording can be performed using phase-contrast microscopy. When fluorescence methods are used, the measuring regime is a trade-off between photobleaching/phototoxicity and the collection of relevant data. To define the largest time interval that still allowed for accurate quantification of mitochondrial transport, we measured the average speed of mitochondria while varying the time-lapse interval stepwise. Measured at video rate, the average velocity of axonal microtubule-based mitochondrial transport is ~ 0.6 μm/s (Morris and Hollenbeck, 1995). The maximum time-lapse interval can be determined by increasing the time-lapse interval in 0.5-s intervals and quantifying average velocity; when the time interval becomes too large, the average velocity will drop below the average velocity measured at video rate, because stationary periods are included but are no longer detected. We find that a 3-s time interval yields reliable data on mitochondrial motility.

The minimum recording time can be estimated from statistical power analysis. For a two-tailed t-test with $\alpha = 0.05$, the sample size needs to be 210, or 105 per group, in order to detect medium-sized effects with 95% power. Assuming 20 mitochondria under observation, the maximal number of events that can be observed per time point is 20. Axonal mitochondria are stationary most of the time (De Vos $et\ al.$, 2003; Morris and Hollenbeck, 1995); therefore, assuming equal amounts of retrograde and anterograde transport and 90% stationary events, theoretically one will detect one retrograde and one anterograde motility event per time point. To collect 105 motility events per direction, 105 observations are needed. Using 3 s as a time-lapse interval, the total recording time needs to be minimally 5.25 min.

In our experiments, we record motility of mitochondria expressing DsRed1-Mito for 10–20 min with a 3-s time-lapse interval, using maximally 200-ms exposures. When higher time resolution or multicolor recording is required, we shorten the total recording time accordingly. Under these conditions we rarely encounter viability problems or photobleaching.

VII. Analysis of Mitochondrial Motility

In contrast to most other organelles, transport of mitochondria involves brief periods of rapid-directed movement followed by diffusion or stasis, that is, salta-tory movement (Bereiter-Hahn and Voth, 1994; De Vos et al., 2003; Hollenbeck, 1996; Hollenbeck and Saxton, 2005; Ligon and Steward, 2000a,b; Morris and Hollenbeck, 1995; Overly et al., 1996). Hence, a comprehensive quantitative description of mitochondrial motility needs to incorporate the relative frequency of directed movements, retrograde and anterograde, and pauses, and also needs to describe the characteristics of movements and pauses such as velocity and dura-tion. Here, we will discuss the quantitative analysis of mitochondrial motility from time-lapse recordings. We will focus mainly on the analysis of mitochondrial motility in neuronal axons, but these methods can be applied to any of cell type, with some modifications, which will be discussed where appropriate.

A. Software

We perform all image analysis with the free program ImageJ developed by Wayne Rasband (NIH, Bethesda; http://rsb.info.nih.gov/ij/; Abramoff et al., 2004). ImageJ is written in the Java programing language and therefore runs on any computer platform for which a Java virtual machine is available, including Windows, Mac OS9, Mac OSX, and Linux. ImageJ includes numerous image processing and analysis functions, and its functionality can be extended by plug-ins and macros.

B. Analysis of Total Mitochondrial Movement

A fast and easy way to evaluate the overall motility of mitochondria is by calculating the percentage of mitochondrial area that moved per time point. This is achieved by subtracting sequential images in a time series—mitochondria that remained stationary will disappear in the resulting image, whereas mitochon-dria that moved are retained—and quantifying the mitochondrial area before and after subtraction. From these areas, the overall mitochondrial motility per time point can be easily calculated according to Eq. (3):

$$Motility_t = \frac{Area_t^{moved}}{Area_t^{total}} \times 100 \qquad (3)$$

where $Motility_t$ is the total mitochondrial motility at time t as a percentage of total mitochondrial area, and $Area_t$ the total mitochondrial area (total) or the mito-chondrial area that moved (moved) at time t.

1. Method 6: Analysis of Total Mitochondrial Movement

Procedure

1. "Threshold" the time series and convert it to binary format. The threshold is chosen to include mitochondrial fluorescence and to exclude background fluorescence (*ImageJ: Image > Adjust... > Threshold*). After converting to binary format (*ImageJ: Threshold > Apply*), the mitochondria should show up white (pixel value = 255) and the background should be black (pixel value = 0). Converting to binary format has the advantage that fluctuations in fluorescence intensity that might influence the subtraction are avoided and facilitates the quantification of mitochondrial area. It might be necessary to invert the time series (*ImageJ: Edit > Invert*) after converting to binary format.

2. Now count the number of mitochondrial pixels per time point and determine the mitochondrial area, by dividing by the spatial calibration of the image according to Eq. (4):

$$Area_{mit} = \frac{Count_{pv = 255}}{Cal_{spatial}} \qquad (4)$$

where $Area_{mit}$ is the mitochondrial area in μm^2, $Count_{pv = 255}$ is the number of pixels with pixel value = 255, and $Cal_{spatial}$ is the spatial calibration in pixels/μm^2. Alternatively, the images can be calibrated beforehand (*ImageJ: Analyze > Set Scale...*). In ImageJ, the number white pixels in a binary image can be determined by generating an image histogram (*ImageJ: Analyze > Histogram*) and hovering the mouse over the right edge (pixel value = 255) of the histogram.

3. Generate a new time series showing only mitochondria that moved by sequentially subtracting the image at time $t–1$ from the image at time t. The resulting time series should now only show the mitochondrial area that moved in white.

4. Determine the mitochondrial area per time point in the subtracted time series as before in step 2.

5. Calculate the mitochondrial motility per time point by dividing the mitochondrial area from the subtracted image by the corresponding mitochondrial area from the original time series, according to Eq. (3). This yields the percentage of mitochondrial area moved per time point.

The source code for an ImageJ plug-in that automates this procedure can be found in Appendix A.

C. Kymograph Analysis

An alternative method to quantify mitochondrial motility extracts motility parameters from kymographs. In contrast to the analysis of total motility, this method analyzes individual mitochondria and yields additional information on the direction and velocity of mitochondrial motility.

1. Constructing a Kymograph

A kymograph is a two-dimensional (x-position vs time) representation of three-dimensional data (x, y-position vs time). First, the two-dimensional image data is reduced to an essentially one-dimensional image (x-dimension $\gg$ y-dimension). This is achieved by resampling the image along the path of movement of a chosen mitochondrion. As illustrated in Fig. 4, in the case of cells with similar x- and y-dimensions, such as fibroblasts or epithelial cells, the path of movement can be visualized by projecting the time series along the time axis (*ImageJ: Image > Stacks > Z Project...*; Fig. 4B, panel a). Subsequently, a one-pixel-wide line selection is made along the path of movement, which can then be used to resample the time series (Fig. 4B, panel b). Note that making a line selection will result in loss of some mitochondrial fluorescence data. In case of axonal transport, resample along the axon, which is already essentially one-dimensional (x-dimension $\gg$ y-dimension) (Figs. 1 and 4). Finally, the kymograph is constructed by reslicing the time series in the x, z-plane along the selection (*ImageJ: Image > Stacks >*

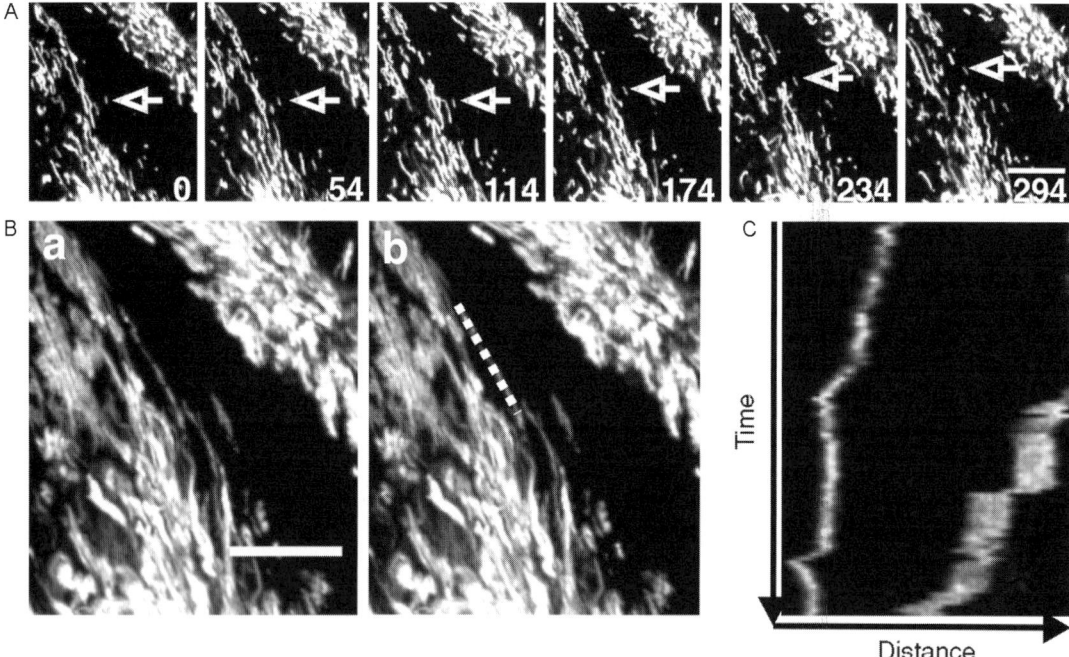

Fig. 4 Kymograph creation. Selected time points of a time series of CV-1 cells transfected with mitochondria-targeted DsRed; a moving mitochondrion is indicated by arrows (A). The path of movement of mitochondria can be visualized by projecting the time series along the time axis (B, panel a). Subsequently, a line selection is made along the path of movement (B, panel b), which can then be used to resample the time series by reslicing along the x-, z-panel (B, plane c). Scale bar = 10 μm.

Reslice [/]; Fig. 4B, panel c); this yields a single montage with successive time points beneath one another.

An alternative method to create kymographs, which retains all the original mitochondrial fluorescence data, is to straighten along a selection containing the whole mitochondron instead of a one-pixel-wide line, using the Straighten plug-in written by E. Kocsis (http://rsb.info.nih.gov/ij/plugins/straighten.html; Kocsis *et al.*, 1991; Fig. 5B and C). The resulting image stack is subsequently resliced along the *x*, *z*-plane (Fig. 5D). This results in a "side," *x–z* view, of the time series (Fig. 5E). Finally, the kymograph is created by a *z*-projection of the resliced time series (Fig. 5F). Distance along the path of movement is along the *x*-axis of a kymograph; time is along the *y*-axis.

The following ImageJ macro creates a kymograph starting from a straightened image using the reslice/projection method:

//kymograph macro
if (nSlices<2) exit("You need a stack")
run("Reslice [/]...", "input=1.000 output=1.000 start=Top");
run("Z Project...", "start=1 stop="+nSlices+"projection=[Average Intensity]");
//end of kymograph macro.

2. Analysis

Because a kymograph is a graph of position versus time, a line drawn through the positions of the mitochondria on the kymograph describes their motion (Fig. 6A). Counting the number of mitochondria that show retrograde or anterograde movements or remain stationary allows fast screening for changes in mitochondrial transport under different conditions (Fig. 6B). In addition, the slope of the line equals the average speed of mitochondrial motility [see Eq. (5); Fig. 6C]:

$$slope = \frac{\Delta x}{\Delta y} = \frac{(x_1 - x_0)}{(y_1 - y_0)} = \frac{(x_1 - x_0)}{(t_1 - t_0)} = \frac{\Delta x}{\Delta t} = v_{average} \tag{5}$$

An ImageJ plug-in to calculate average velocities from kymographs is available in Appendix B. Alternatively, one can use the Kymograph Analysis macros developed by J. Rietdorf and A. Seitz, available at http://www.embl-heidelberg.de/eamnet/html/body_kymograph.html. Most commercial image analysis packages now include a kymograph analysis function.

D. Full Quantitative Analysis of Mitochondrial Motility

In this analysis, mitochondrial motility is expressed in terms of (1) net displacement, (2) frequency, (3) velocity, and (4) persistence of mitochondrial motility. Additionally, we construct a motility histogram that shows the motility behavior of individual mitochondria. In addition to a quantitative description of motility, this analysis also yields quantitative data on pauses. The frequency of transport and pausing is an indicator of regulation of transport (activation vs inactivation).

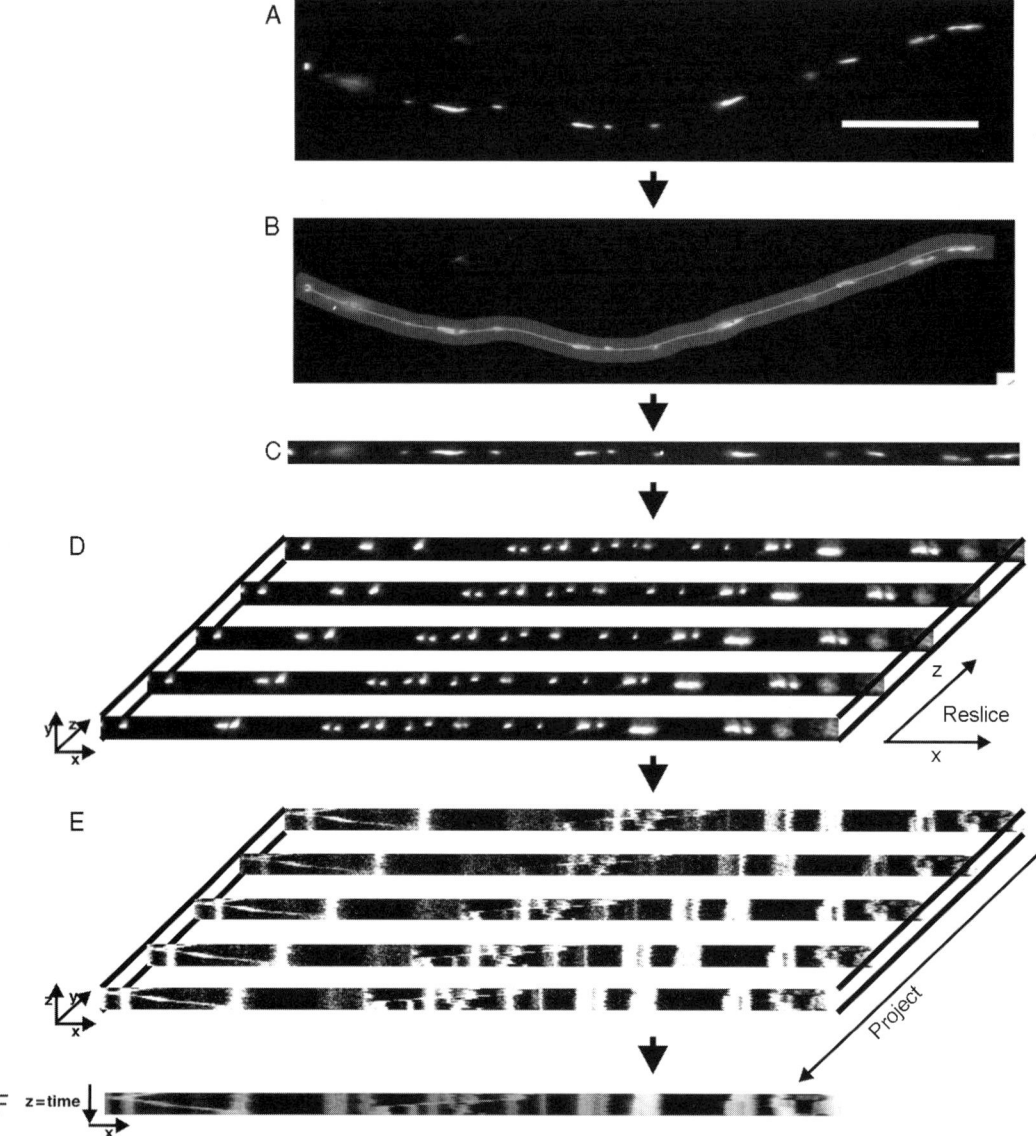

Fig. 5 Kymograph creation. The axon of a pDsRed1-mito-transfected cortical neuron was straightened using the Straighten plug-in (A–C). The resulting image stack was subsequently resliced along the x–z plane (D). This results in a "side," x–z view of the time series (E). Finally, the kymograph is created by a z-projection of the resliced time series (F). (See Color Insert.)

Velocity and persistence describe the physical properties of the molecular motors driving transport: the velocity is an indicator of the ATPase activity of the motor, whereas the persistence of movement indicates the active period of the motor.

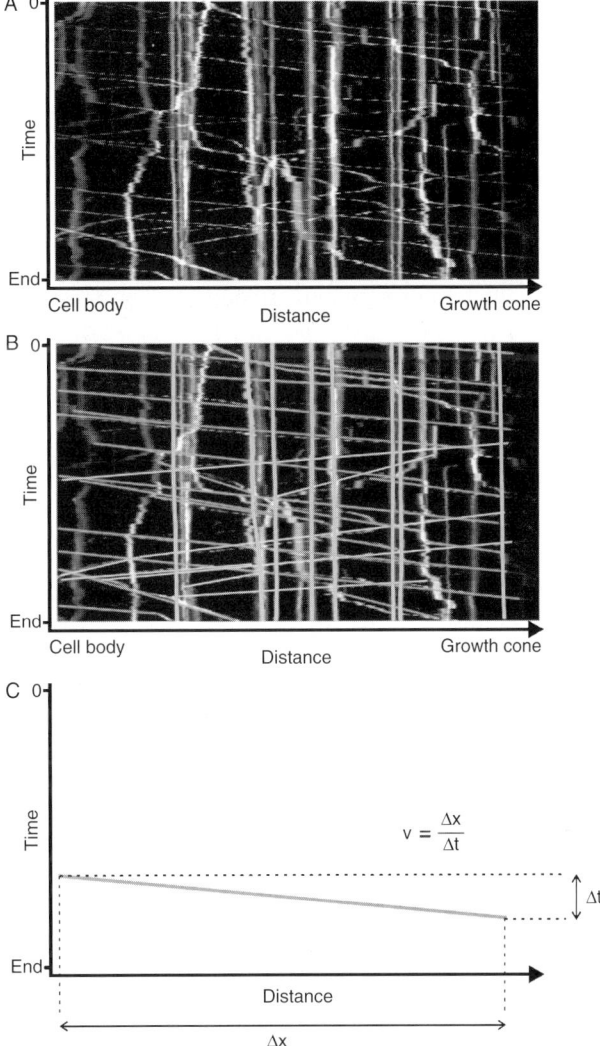

Fig. 6 Kymograph analysis. Cortical neurons were transfected with pDsRed1-mito, a time series was recorded (3-s interval, 20 min recording time), and a kymograph was constructed (A). Lines that describe the movement of individual mitochondria can be overlaid on the kymograph and the mitochondria classified according to their overall motility (B; retrograde, yellow; anterograde, green; stationary, red). From the path of movement, it is straightforward to calculate the mean velocity (C). (See Color Insert.)

In the following paragraphs, we focus on the analysis of axonal transport of mitochondria. Analysis of mitochondrial transport in other cell types follows the same principles as the analysis of axonal transport. However, because the polarity of microtubule array in nonneuronal cells is uncertain, reliable directional data cannot be obtained.

1. Straightening of Axons

Before proceeding with the quantitative analysis, it is important to straighten the time series along the path of movement because the analysis relies on the accurate measurement of distances between the positions of mitochondria at different time points, according to Pythagoras' theorem (see below). Image straightening can be performed using the Straighten plug-in (Fig. 5B and C).

2. Tracking of Mitochondria

The first and most laborious step in the analysis procedure is to determine the position (x–y coordinates) of each mitochondrion in the series of time-lapse frames. This can be done with ImageJ or any commercial image analysis package. We developed an ImageJ plug-in that assists in manual tracking of mitochondria. Some commercial packages also include autotracking modules, which will automatically track a selected mitochondrion through the time series. In our experience, however, none of these autotracker functions is completely reliable, insofar that the tracking routine often loses track of the mitochondrion being followed when it comes in close vicinity of other mitochondria. This is especially problematic when tracking axonal mitochondria or mitochondria in straightened image series because in these cases mitochondria cross each other frequently. Hence, user input is required to guide and validate the autotracking.

3. Direction of Mitochondrial Motility

Because of the highly organized architecture of the microtubule array in neuronal axons, with the plus-ends pointing away from the cell body toward the growth cone (Baas, 2002; Heidemann, 1996), we can include directional information in the analysis (in other cells one can define inward as toward the cell center, and outward movement as toward the cell periphery). Practically, we define a fixed reference point at the edge of the image in the direction of the cell body and compute the distance of the mitochondria to the reference point, according to Pythagoras' theorem per time point [see Eq. (6); Fig. 7]:

$$d_{ref} = \sqrt{(x_{mit} - x_{ref})^2 + (y_{mit} - y_{ref})^2} \tag{6}$$

where d_{ref} is the distance to the reference point and (x_{mit}, y_{mit}), and (x_{ref}, y_{ref}) are the coordinates of the mitochondrion and reference point, respectively. If the distance to the reference point increases between time points, there is anterograde (toward the growth cone) movement. Conversely, if the distance decreases, there is retrograde (toward the cell body) movement (Fig. 7). Once all positions and

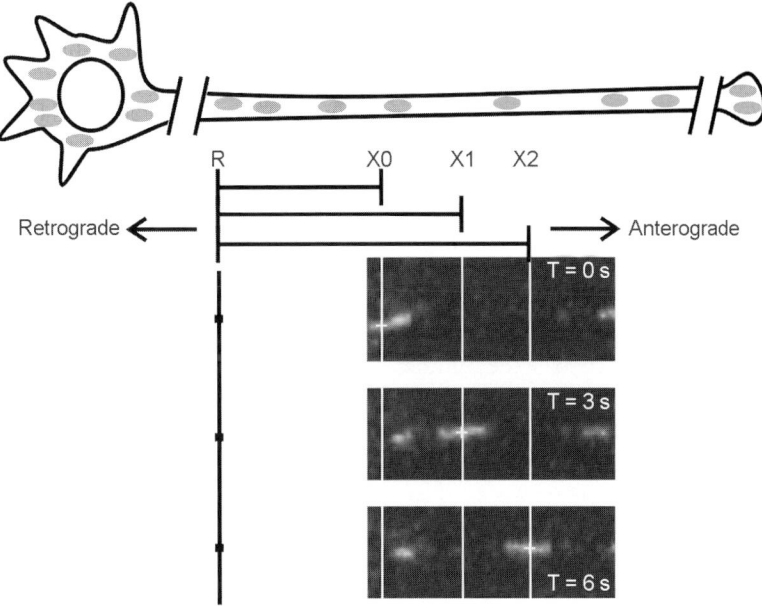

Fig. 7 Direction of mitochondrial transport. The direction of mitochondrial transport can be included in the analysis of mitochondrial motility by choosing a reference point (R) toward the cell body of the neuron, and calculating the distance between the reference point and the position of each mitochondrion per time point (X0, X1, and X2). If the distance to the reference point increases, there is anterograde motility; conversely, a distance decrease corresponds to retrograde transport.

distances to the reference point are determined, the motility parameters can be calculated easily in a spreadsheet.

4. Net Displacement of Mitochondria

First, we can analyze the net displacements of individual mitochondria. Net displacement is defined by comparing the position of each mitochondrion at the start and the end of recording, and analyzing their overall movement. Essentially, analysis of net displacement is equivalent to kymograph analysis, but the results are more accurate because the exact position of the mitochondrion is known. We calculate the distance between the start and end position by subtracting the corresponding distances to the reference point [Eq. (7)]:

$$d_{\text{net}} = d_{\text{ref}}^{t=\text{end}} - d_{\text{ref}}^{t=0} \tag{7}$$

where d_{net} is the net displacement, and $d_{\text{ref}}^{t=\text{end}}$ and $d_{\text{ref}}^{t=0}$ are the distance to the reference point at the end and the start of recording, respectively. To define meaningful net displacements, we usually apply a distance threshold of 10 times the average velocity. If the absolute value of the net displacement is greater than the distance threshold, the mitochondrion undergoes displacement. In contrast, in cases where the net displacement is smaller than the distance threshold, the mitochondrion is stationary. Tallying all mitochondria yields the number of motile and stationary mitochondria [see Eqs. (8) and (9)]:

$$M_{\text{displaced}} = \sum_{\text{mit}} (|d_{\text{net}}| > d_{\text{threshold}}) \tag{8}$$

$$M_{\text{stationary}} = \sum_{\text{mit}} (|d_{\text{net}}| < d_{\text{threshold}}) \tag{9}$$

where M is the number of mitochondria, d_{net} is the net displacement, and $d_{\text{threshold}}$ is the displacement threshold. If the net displacement has a positive value, the mitochondrion moved away from the cell body; conversely, a negative net displacement indicates retrograde movement [see Eqs. (10) and (11)]:

$$M_{\text{anterograde}} = \sum_{\text{mit}} [(|d_{\text{net}}| > d_{\text{threshold}}) \wedge (d_{\text{net}} > 0)] \tag{10}$$

$$M_{\text{retrograde}} = \sum_{\text{mit}} [(|d_{\text{net}}| > d_{\text{threshold}}) \wedge (d_{\text{net}} < 0)] \tag{11}$$

From these values, the percentage of stationary and motile (retrograde or antero-grade) mitochondria can be determined easily. The mean displacement length can also be determined from these data for both directions of transport by averaging the length of displacement per direction. To normalize different recordings, we express the mean displacement length as a percentage of the length of the axon in the field of view. Finally, as in case of kymograph analysis, the average velocity can be deter-mined according to Eq. (5). Note that the average velocity determined in this way is an underestimation because stationary periods are included, and that, for instance, in N2A cells mitochondria are stationary most of the time (De Vos et al., 2003).

5. Velocity of Mitochondrial Motility

We determine the absolute velocities per mitochondrion for each time point by calculating the distance each mitochondrion has moved relative to its position at the previous time point [see Eq. (12)] and dividing this distance by the time elapsed per frame, according to Eq. (13):

$$\Delta d = d_{\text{ref}}^{t} - d_{\text{ref}}^{t-1} \tag{12}$$

$$v_t = \frac{\Delta d}{t_{\text{interval}}} \tag{13}$$

where is Δd the distance traveled, d_{ref}^{t} is the distance to the reference point at time t, d_{ref}^{t-1} is the distance to the reference point at time $t-1$, is the velocity at time t, and

$t_{interval}$ is the time-lapse interval (time between recordings). This yields a positive velocity for anterograde movement and a negative velocity for retrograde movement. These absolute velocities can be used to construct velocity histograms or can be averaged to yield the mean velocity. In contrast to the average velocities obtained from kymographs and net displacement calculations, the velocities obtained in this way reflect the actual velocity of mitochondrial movement because the stationary periods can easily be excluded. To evaluate only microtubule-based transport, we apply a velocity threshold of 300 nm/s. This threshold excludes stationary periods and actin-based transport of axonal mitochondria, which is reportedly below 300 nm/s (D'Andrea *et al.*, 1994; Kuznetsov *et al.*, 1992; Morris and Hollenbeck, 1995).

6. Frequency of Mitochondrial Motility

Because analysis of net displacements is largely limited to unidirectionally moving mitochondria, and to further examine the transport activities underlying net mitochondrial displacement, we dissect the underlying individual events. We apply a threshold of absolute velocities to define motility events, 300 nm/s or two times the spatial resolution of the recorded images, whichever is greater, and define all motile and stationary events according to Eqs. (14) and (15):

$$e_{motile} = |v_t| > v_{threshold} \tag{14}$$

$$e_{stationary} = |v_t| < v_{threshold} \tag{15}$$

where e is a single event, v_t is the absolute velocity, and $v_{threshold}$ is the velocity threshold. This threshold is set to include only microtubule-based transport of mitochondria, since the reported velocity of actin-based transport of axonal mitochondria is below 300 nm/s (D'Andrea *et al.*, 1994; Kuznetsov *et al.*, 1992; Morris and Hollenbeck, 1995). Tallying these events per type allows us to determine the number of motile events and the number of stationary events per mitochondrion [see Eqs. (16) and (17)].

$$E_{motile} = \sum_{events} e_{motile} \tag{16}$$

$$E_{stationary} = \sum_{events} e_{stationary} \tag{17}$$

where E is the total number of events, e is a single event, and *events* is the total number of events. Averaging over the number of recorded frames yields the frequency of motile and stationary events for each mitochondrion [see Eqs. (18) and (19)]:

$$F_{motile} = \frac{E_{motile}}{frames} \tag{18}$$

$$F_{\text{stationary}} = \frac{E_{\text{stationary}}}{frames} \tag{19}$$

where F is the frequency, E is number of events, and *frames* is the number of frames recorded.

Further averaging over all mitochondria under observation yields the average frequencies per cell. Finally, the motile events can be subdivided in anterograde and retrograde motile events, and the average frequency of anterograde and retrograde transport can be calculated as above, keeping anterograde and retrograde events separate.

7. Persistence of Mitochondrial Motility

The persistence of mitochondrial motility is defined as the period of ongoing transport in one direction without pausing or reversal of direction. To determine the persistence of motility, we evaluate consecutive events and tally the number of motile events in the same direction that follow each other. Multiplying by the time interval yields the persistence of movement, that is, the duration of a continuous period of directed motility. The persistence of movement indicates the active period of the motor driving the transport. The duration of pausing can be calculated in the same way.

8. Motility Profile

Usually one finds three types of motile mitochondria in the axons: mitochondria that are transported predominantly in one direction, either anterograde or retrograde, and mitochondria that exhibit more balanced bidirectional transport. This transport behavior can be quantified by constructing a motility profile in the following way. First, tally anterograde and retrograde motility events per mitochondria according to Eqs. (16) and (17). Subsequently, calculate the relative percentage of anterograde and retrograde events per mitochondrion and use these percentages to construct a histogram using 10%-wide bins.

VIII. Analysis of Mitochondrial Morphology

Here we describe methods to analyze mitochondrial morphology dynamics.

A. Analysis of Morphology of Individual Mitochondria and Mitochondrial Fusion and Fission

To describe the shape of an object, several so-called "shape descriptors" can be determined. An excellent overview of shape descriptors can be found in *The Image Processing Handbook* by John Russ (Russ, 1992) or online in the *Image Processing*

Learning Resources (http://homepages.inf.ed.ac.uk/rbf/HIPR2/). Shape descriptors are combinations of size parameters that cancel out the dimensions so that the value of a shape descriptor does not change with the size of the object.

To determine the shape of individual mitochondria, we measure the mitochondrial area and the mitochondrial aspect ratio. The mitochondrial area indicates the size of mitochondria, and the aspect ratio is a measure of their elongation. Mitochondrial fusion and fission events can be quantified by counting the number of mitochondria per cell. To determine these parameters, we use the particle analysis function of ImageJ (*ImageJ: Analyze > Analyze Particles...*); area is obtained by counting the number of pixels that make up the mitochondrion. To determine the aspect ratio, an ellipse is fitted to the mitochondrion, and the major (longitudinal length) and minor axis (equatorial length) of the ellipse is used to calculate the aspect ratio [see Eq. (20)]:

$$AR = \frac{l_{major}}{l_{minor}} \tag{20}$$

where AR is the aspect ratio and l_{major} and l_{minor} are the lengths of the major and minor axes of the fitted ellipse, respectively. Because the shape of most mitochondria resembles an ellipsoid shape, calculation of the aspect ratio in this way yields a reliable approximation of the elongation of a given mitochondrion. This method only yields accurate results when applied to individual mitochondria. Mitochondria are often organized in large networks in many cell types, and the presence of a large mitochondrial reticulum will cause an overestimation of the mean area of individual mitochondria. Similarly, touching mitochondria will be regarded as a single organelle in the analysis, leading to an overestimation of mitochondrial area, an underestimation of the number of mitochondria, and a skewing of the aspect ratio. There are several ways to circumvent these possible artifacts: Mitochondrial networks are easily recognized and can be excluded from the measurements based on their size, while touching mitochondria can be separated by segmentation. To exclude mitochondrial networks from the analysis, a size threshold can be introduced in the particle analysis. Alternatively, the network can be deleted from the image before analysis.

A thorough description of segmentation techniques can be found in *The Image Processing Handbook* (Russ, 1992). An example of a strategy we often use to segment mitochondria in CV-1 cells is described below. Subsequently the obtained values for each parameter can be used to construct histograms or can be averaged to yield the mean values per time point. If possible we determine mitochondrial morphology before and after treatment in the same cell. In this case, the results can be normalized easily by calculating the percentage change compared to the mean value obtained from recording the same cells under untreated conditions, according to Eq. (21):

$$change = \left(\frac{value_t}{average_{pre}} \times 100 \right) - 100 \tag{21}$$

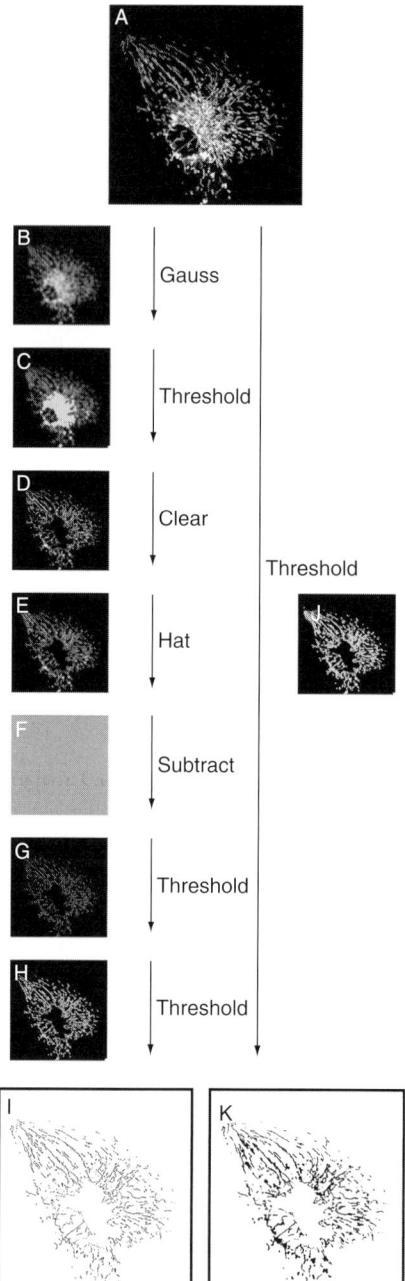

Fig. 8 Removal of mitochondrial networks and segmentation of mitochondria. An example segmentation procedure using a pDsRed1-mito transfected CV-1 cell (A) is shown. For a detailed description of the individual steps, see Section VIII.A.1 (Method 7). (B) Original image filtered with a 15×15 Gaussian

where *change* is the percentage change for a given parameter, $value_t$ is the value at time t after treatment, and $average_{pre}$ is the average value of the parameter before treatment.

1. Method 7: Analysis of Mitochondrial Morphology, Fission and Fusion

Procedure

This procedure is illustrated in Fig. 8. Note the difference between the binary image result after segmenting (Fig. 8I) and the equivalent image without segmenting routine (Fig. 8K).

Using ImageJ:

1. Optionally, remove networked mitochondria and segment touching mitochondria.

   • Removing networked mitochondria

      ○ Filter the original image with a 15 × 15 Gaussian filter [Eq. (22); *ImageJ: Process > Filters > Gaussian Blur... or Process > Filters > Convolve...*; Fig. 8B]:

$$\begin{matrix}
2 & 2 & 3 & 4 & 5 & 5 & 6 & 6 & 6 & 5 & 5 & 4 & 3 & 2 & 2 \\
2 & 3 & 4 & 5 & 7 & 7 & 8 & 8 & 8 & 7 & 7 & 5 & 4 & 3 & 2 \\
3 & 4 & 6 & 7 & 9 & 10 & 10 & 11 & 10 & 10 & 9 & 7 & 6 & 4 & 3 \\
4 & 5 & 7 & 9 & 10 & 12 & 13 & 13 & 13 & 12 & 10 & 9 & 7 & 5 & 4 \\
5 & 7 & 9 & 11 & 13 & 14 & 15 & 16 & 15 & 14 & 13 & 11 & 9 & 7 & 5 \\
5 & 7 & 10 & 12 & 14 & 16 & 17 & 18 & 17 & 16 & 14 & 12 & 10 & 7 & 5 \\
6 & 8 & 10 & 13 & 15 & 17 & 19 & 19 & 19 & 17 & 15 & 13 & 10 & 8 & 6 \\
6 & 8 & 11 & 13 & 16 & 18 & 19 & 20 & 19 & 18 & 16 & 13 & 11 & 8 & 6 \\
6 & 8 & 10 & 13 & 15 & 17 & 19 & 19 & 19 & 17 & 15 & 13 & 10 & 8 & 6 \\
5 & 7 & 10 & 12 & 14 & 16 & 17 & 18 & 17 & 16 & 14 & 12 & 10 & 7 & 5 \\
5 & 7 & 9 & 11 & 13 & 14 & 15 & 16 & 15 & 14 & 12 & 11 & 9 & 7 & 5 \\
4 & 5 & 7 & 9 & 10 & 12 & 13 & 13 & 13 & 12 & 10 & 9 & 7 & 5 & 4 \\
3 & 4 & 6 & 7 & 9 & 10 & 10 & 11 & 10 & 10 & 9 & 7 & 6 & 4 & 3 \\
2 & 3 & 4 & 5 & 7 & 7 & 8 & 8 & 8 & 7 & 7 & 5 & 4 & 3 & 2 \\
2 & 2 & 3 & 4 & 5 & 5 & 6 & 6 & 6 & 5 & 5 & 4 & 3 & 2 & 2
\end{matrix} \tag{22}$$

      ○ Threshold the resulting image to include the now blurred network (*ImageJ: Image > Adjust > Threshold...*; Fig. 8C).

      ○ Apply the threshold; mitochondria show up black.

---

filter. (C) The now-blurred network is thresholded (red) and (D) the networked mitochondria are removed. (E) The cleared image is filtered with a 7 × 7 Mexican Hat filter. (F) Result of subtracting the filtered image from the cleared image. (G) The resulting image was thresholded to include all mitochondria, converted to 8-bit, and (H) thresholded again, followed by conversion to the binary result (I). Note the difference between the binary image (I) after segmenting and the equivalent image (K) converted to binary using one threshold step (J) without the segmenting routine. (See Color Insert.)

- ○ Select the networked mitochondria using the wand tool.
- ○ Restore the selection on the original image (*ImageJ: Edit > Selection > Restore Selection*) and clear it (*ImageJ: Edit > Clear*; cleared image) (Fig. 8D).
- • Segment touching mitochondria
  - ○ Filter the cleared image with a 7×7 Mexican Hat filter [Eq. (23); *ImageJ: Process > Filters > Convolve...*; Fig. 8E; filtered image]:

$$
\begin{matrix}
0 & 0 & 0 & -1 & 0 & 0 & 0 \\
0 & 0 & -1 & -2 & -1 & 0 & 0 \\
0 & -1 & -2 & -3 & -2 & -1 & 0 \\
-1 & -2 & -3 & 16 & -3 & -2 & -1 \\
0 & -1 & -2 & -3 & -2 & -1 & 0 \\
0 & 0 & -1 & -2 & -1 & 0 & 0 \\
0 & 0 & 0 & -1 & 0 & 0 & 0
\end{matrix}
\tag{23}
$$

  - ○ Subtract the filtered image from the cleared image opting for a 32-bit result (*ImageJ: Process > Image Calculator...*; Fig. 8F).
  - ○ Threshold the resulting image to include all mitochondria and apply (Fig. 8G).
  - ○ Convert the thresholded image to 8-bit (*ImageJ: Image > Type > 8-bit*).
  - ○ Threshold the 8-bit image and apply (Fig. 8H).
2. Using the resulting binary image (Fig. 8I), count the number of mitochondria, and determine the area and the best fitting ellipse for each organelle per time point by particle analysis.
3. Calculations.

B. Analysis of Mitochondrial Branching and Networks

Dynamic mitochondrial morphology changes often include branching. To quantify mitochondrial branching, the number of branch points can be determined from skeletonized mitochondria. Binary skeletonization is achieved by performing binary image erosion using special rules. Fundamentally, erosion removes pixels that fit a chosen condition from an image. In most image analysis programs, the erosion function refers to the removal of any pixel that touches a background pixel. In case of skeletonization, pixels touching background pixels are removed except when doing so would disconnect a continuous region of pixels. Practically, this means that pixels are removed at the edges of objects until the objects are reduced to single-pixel-wide skeletons (Fig. 9C). From the skeleton, it is straightforward to determine branch points by counting nodes in the skeleton.

To estimate the extent of mitochondrial networks, counting the end points of the branches can be helpful. In the case of networking, the number of endpoints will be smaller than the number of branch points, whereas in the case of a branch point in an individual mitochondrion, the number of endpoints is larger than the

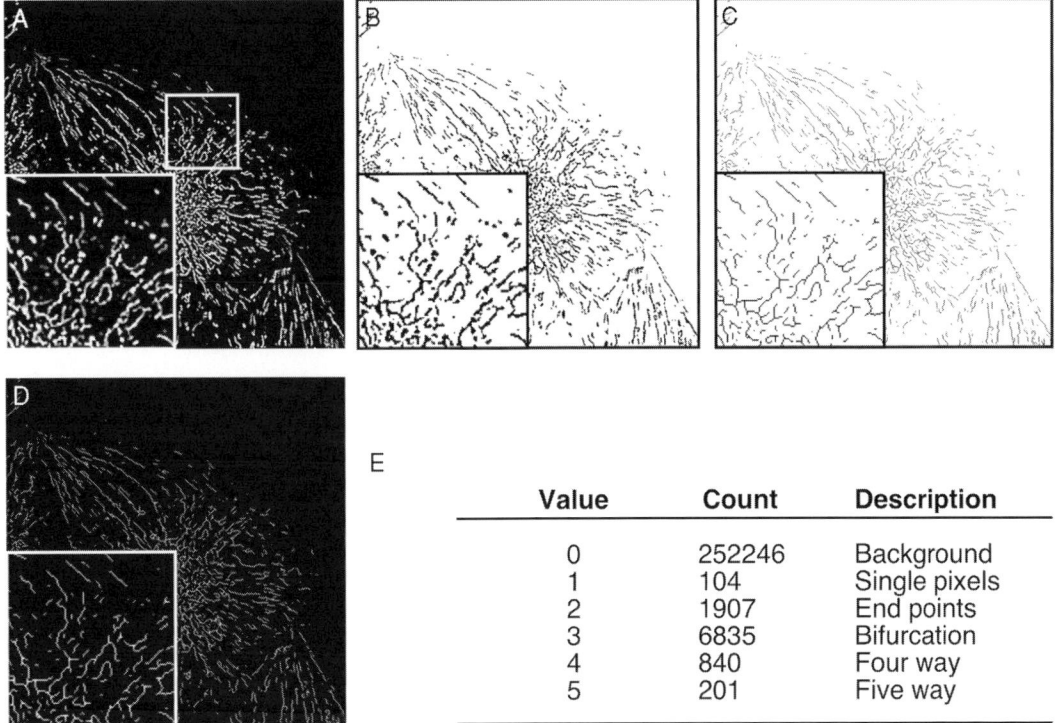

Value	Count	Description
0	252246	Background
1	104	Single pixels
2	1907	End points
3	6835	Bifurcation
4	840	Four way
5	201	Five way

Fig. 9 Analysis of mitochondrial branching and networks. An example of analysis of mitochondrial branching using a pDsRed1-mito-transfected CV-1 cell (A) is shown. For a description of the individual steps, see Section VIII.B.1 (Method 8). The original image was converted to binary format (B) and skeletonized (C). (D) Graphic output from the BinaryConnectivity plug-in and (E) the numeric results. The insets are zoomed two times.

number of branch points. In the simplest case of a bifurcation of a single mitochondrion, there is one branch point and three endpoints. Therefore, the ratio between endpoints and branch points is an estimation of the amount and extent of mitochondrial networking.

1. Method 8: Analysis of Mitochondrial Networks

Procedure

1. Convert the image to binary format (Fig. 9B).
2. Skeletonize (*ImageJ: Process > Binary > Skeletonize*; Fig. 9C).
3. Count branch points and endpoints using the BinaryConnectivity plug-in written by G. Landini (http://www.dentistry.bham.ac.uk/landinig/software/software.html) (Fig. 9D and E).

IX. Appendix A

A. Total_Motility

The Total Motility plug-in calculates the total motility of mitochondria from a time series. This plug-in was developed using Java version 1.5.0_06, and ImageJ 1.35 g. Copy the source code below in your favorite text editor and save the file as Total_Motility.java. Before the first use, compile the source code in ImageJ (*ImageJ: Plugins > Compile and run*). To perform the analysis: (1) open time series, (2) convert the image stack to binary format with the mitochondria showing up white, and (3) run the Total Motility plug-in.

```
//start of Total_Motility.java
import ij.IJ;
import ij.ImagePlus;
import ij.ImageStack;
import ij.WindowManager;
import ij.plugin.PlugIn;
import ij.process.Blitter;
import ij.process.ImageProcessor;
import ij.process.ImageStatistics;
import java.awt.image.ColorModel;

/*
 * Total_Motility.java
 *
 * Created on April 17, 2006, 7:50 PM
 *
 * @author Kurt De Vos © 2006
 *
 * This program is free software; you can redistribute it and/or
 * modify it under the terms of the GNU General Public License
 * as published by the Free Software Foundation (http://www.gnu.org/licenses/
   gpl.txt)
 *
 * This program is distributed in the hope that it will be useful,
 * but WITHOUT ANY WARRANTY; without even the implied warranty of
 * MERCHANTABILITY or FITNESS FOR A PARTICULAR PURPOSE.
 * See the GNU General Public License for more details.
 *
 * You should have received a copy of the GNU General Public License
```

```
* along with this program; if not, write to the Free Software
* Foundation, Inc., 59 Temple Place – Suite 330, Boston, MA 02111-1307,
  USA.
*/


public class Total_Motility implements PlugIn{
    ImagePlus img;
    ImageStatistics imStats;
    int stackSize;

    /**
     * Creates a new instance of Total_Motility
     */
    public Total_Motility() {
    }
    public void run(String arg){
        img = WindowManager.getCurrentImage();
        stackSize = img.getStackSize();
        if (img == null) {

            IJ.noImage();
            return;

        }
        if (img.getStackSize()==1){
            IJ.error("you need a stack");
            return;
        }
        if (!checkBinary()){
            IJ.error("you need a binary stack");
            return;
        }
        // First count the white pixels in the original stack...
        int[] originalCounts = binaryCount(img);
        //Do the subtraction
        ImagePlus subtractedImg = subtractStack(img);
        //Count the white pixels in the subtracted stack
        int[] subtractedCounts = binaryCount(subtractedImg);
```

```
      //Calculate the motility
      doCalculations(originalCounts,subtractedCounts);
    }

    // check if the image is in binary format
    public boolean checkBinary(){
      imStats = img.getStatistics();
      if (imStats.histogram[0]+imStats.histogram[255]==imStats.pixelCount){
        return true;
      }
      IJ.error("8-bit binary required – features in White!!");
      return false;
    }

    //Count white pixels
    public int[] binaryCount(ImagePlus img){
      int[] counts = new int[img.getStackSize()];
      for (int i=1;i<=img.getStackSize();i++){ //walks through the stack and
        counts
        IJ.showProgress((double)i/img.getStackSize());
        IJ.showStatus(i+"/"+img.getStackSize());
        img.setSlice(i);
        imStats = img.getStatistics();
        counts[i-1] = imStats.histogram[255]; // only the white pixels...
        System.out.println(img.getTitle()+" " +counts[i-1]);
      }
      img.setSlice(1);
      return counts;

    }

    //subtract
    public ImagePlus subtractStack(ImagePlus img){
      ImageStack stack = img.getStack();
      int stackSize = img.getStackSize();
      int currentSlice = img.getCurrentSlice();
      ColorModel colorMod = img.createLut().getColorModel();
      ImageStack newStack = new
ImageStack(stack.getWidth(),stack.getHeight(),colorMod);
```

```
            for(int i=1;i<=stackSize-1;i++) {

              IJ.showProgress((double)i/(double)(stackSize-1));
              IJ.showStatus(i+"/",+(stackSize-1));
            img.setSlice(i);
            ImageProcessor ip1 = img.getProcessor();
              ImageProcessor copyIp1 = ip1.duplicate();

            img.setSlice(i+1);
            ImageProcessor ip2 = img.getProcessor();
            ImageProcessor copyIp2 = ip2.duplicate();

              copyIp2.copyBits(copyIp1, 0, 0, Blitter.SUBTRACT);

              newStack.addSlice("", copyIp2);

            }

            ImagePlus sub = new ImagePlus(img.getTitle().substring(0, img.getTitle().
              lastIndexOf("."))+"_subtracted", newStack);
            sub.show();

            return sub;

          }


          // calculate and output
          public void doCalculations(int[] counts1, int[] counts2){

            IJ.setColumnHeadings("Slice\tpercent");
            for (int i=0; i<counts2.length; i++){

              int j = i+1;
              int slice = i+2;
              double percent = (double)counts2[i]/(double)counts1[j]*100;
              System.out.println("j="+j+" double "+ percent+" = "+counts1[j]+" / "+
                counts2[i]);
              IJ.write(slice+"\t"+IJ.d2s(percent));

            }

          }

        }
        //end of Total_Motility.java
```

X. Appendix B

A. SlopeToVelocity_

The SlopeToVelocity_ plug-in calculates the velocity of mitochondria from a kymograph. This plug-in was developed using Java version 1.5.0_06, and ImageJ 1.35 g. Copy the source code below in your favorite text editor and save the file as SlopeToVelocity_.java. Before the first use, compile the source code in ImageJ (*ImageJ: Plugins > Compile and run*). To perform the analysis: (1) create or open a kymograph; (2) make a line or segmented line selection, tracing the path of movement of a mitochondrion; and (3) run the SlopeToVelocity_ plug-in.

```
//start of SlopeToVelocity_ plugin

import ij.*;
import ij.plugin.*;
import ij.gui.*;
import java.awt.Rectangle;

/*
 * SlopeToVelocity_.java
 *
 * Created on April 17, 2006, 7:50 PM
 *
 * @author Kurt De Vos © 2006
 *
 * This program is free software; you can redistribute it and/or
 * modify it under the terms of the GNU General Public License
 * as published by the Free Software Foundation (http://www.gnu.org/licenses/
   gpl.txt)
 *
 * This program is distributed in the hope that it will be useful,
 * but WITHOUT ANY WARRANTY; without even the implied warranty of
 * MERCHANTABILITY or FITNESS FOR A PARTICULAR PURPOSE.
 *  See the GNU General Public License for more details.
 *
 * You should have received a copy of the GNU General Public License
 * along with this program; if not, write to the Free Software
 * Foundation, Inc., 59 Temple Place –Suite 330, Boston, MA 02111-1307, USA.
```

```
*
*
* This plugin calculates the velocity from line selections made on a kymograph
* Usage:
* Create or open a kymograph
* Make a line or segmented line selection, tracing the path of movement of an
  organelle
* Run this plugin
*
*/
public class SlopeToVelocity_ implements PlugIn {

  static final String headings = "Segment\tVelocity";

  public void run(String arg) {
    //get the image
    ImagePlus img = WindowManager.getCurrentImage();
    //check if there is an image
    if (img==null){
      IJ.noImage();
      return;
    }
    //get the selection
    Roi roi = img.getRoi();
    //Check if there is a selection and if so, if it is a line selection
    if (!(roi!=null && (roi.getType()==roi.LINE || roi.getType()==roi.POLY-
      LINE))){
      IJ.error("A straight line selection is required.");
      return;
    }
    //Check if the headings are set, if not do so
    if (!headings.equals(IJ.getTextPanel().getColumnHeadings())) {
      IJ.setColumnHeadings(headings);
    }

    double time =IJ.getNumber("Time Calibration in seconds/pixel",3);
    double scale=1.0;
    if (!img.getCalibration().scaled()){
      scale =IJ.getNumber("Spatial Calibration in pixels/μm",4.1);
```

```
      scale = 1/scale;
   }else{
      scale = img.getCalibration().pixelWidth;
   }
   double velocity = 0.0;
   if (roi.getType()==Roi.LINE) { //if it is a line selection
      Line line = (Line)roi;
      //calculate the velocity dx/dt
      velocity = ((double)line.x2 − (double)line.x1)/Math.abs((double)line.y1
         −(double)line.y2);
      //calibrate
      velocity = velocity*scale/time;
      //write the results
      IJ.write("Overall\t"+IJ.d2s(velocity));
   } else if (roi.getType()==Roi.POLYLINE){ // if it is a segemented line
      int segments = ((PolygonRoi)roi).getNCoordinates(); //get the number
         of line segments
      int[] x = convertXCoordinates((PolygonRoi)roi,((PolygonRoi)roi).
         getXCoordinates()); //get the x coordinates
      int[] y = convertYCoordinates((PolygonRoi)roi,((PolygonRoi)roi).
         getYCoordinates()); //get the y coordinates

      for (int i=0; i<segments−1; i++){ //for all segments do...
         //calculate the velocity dx/dt
         velocity = ((double)x[i+1] − (double)x[i])/Math.abs((double)y[i] −
            (double)y[i+1]);
         //calibrate
         velocity = velocity*scale/time;
         //write the results
         IJ.write((i+1)+"\t"+IJ.d2s(velocity));
      }
      //calculate the velocity over the whole selection
      velocity = ((double)x[segments−1] − (double)x[0])/Math.abs((double)y
         [0] − (double)y[segments−1]);
      //calibrate
      velocity = velocity*scale/time;
      //write the results
      IJ.write("Overall\t"+IJ.d2s(velocity));
   }
```

```
            IJ.register(SlopeToVelocity_.class);
    }

    /* We have to convert the coordinates to screen coordinates
     * because getXCoordinates() returns the ROI's X-coordinate relative
     * to the origin of the bounding box.
     */
    private int[] convertXCoordinates(PolygonRoi roi, int[] xCoordinates){
        int segments = roi.getNCoordinates();
        int[] convCoordinates = new int[segments];
        Rectangle rect = roi.getBounds();
        for (int i=0;i<segments;i++){
            convCoordinates[i] = xCoordinates[i]+rect.x;
        }
        return convCoordinates;
    }
    private int[] convertYCoordinates(PolygonRoi roi, int[] yCoordinates){
        int segments = roi.getNCoordinates();
        int[] convCoordinates = new int[segments];
        Rectangle rect = roi.getBounds();
        for (int i=0; i<segments; i++){
            convCoordinates[i] = yCoordinates[i]+rect.y;
        }
        return convCoordinates;
    }
}
//end of plugin SlopeToVelocity_
```

Acknowledgments

The authors want to thank R. Milner for critical reading of this manuscript. K. D. V. is supported by a BBRC grant (to A. Grierson, University of Sheffield) and an ALSA starter grant. Work in M. Sheetz' laboratory is supported by NIH grants NS23345 and GM23362.

References

Abou-Khalil, S., Abou-Khalil, W. H., Planas, L., Tapiero, H., and Lampidis, T. J. (1985). Interaction of rhodamine 123 with mitochondria isolated from drug-sensitive and -resistant Friend leukemia cells. *Biochem. Biophys. Res. Commun.* **127,** 1039–1044.

Abramoff, M. D., MagalHaes, P. J., and Ram, S. J. (2004). Image processing with ImageJ. *Biophotonics Int.* **11,** 36–42.

Abramovitz, M. (ed.) (1993). "Fluorescence Microscopy: The Essentials," Vol. 4. Olympus America, Inc., New York.

Ackerley, S., Grierson, A. J., Brownlees, J., Thornhill, P., Anderton, B. H., Leigh, P. N., Shaw, C. E., and Miller, C. C. (2000). Glutamate slows axonal transport of neurofilaments in transfected neurons. *J. Cell Biol.* **150**, 165–176.

Akerman, K. E., and Wikstrom, M. K. (1976). Safranine as a probe of the mitochondrial membrane potential. *FEBS Lett.* **68**, 191–197.

Anderson, W. M., Wood, J. M., and Anderson, A. C. (1993). Inhibition of mitochondrial and Paracoccus denitrificans NADH-ubiquinone reductase by oxacarbocyanine dyes. A structure-activity study. *Biochem. Pharmacol.* **45**, 2115–2122.

Arbustini, E., Diegoli, M., Fasani, R., Grasso, M., Morbini, P., Banchieri, N., Bellini, O., Dal Bello, B., Pilotto, A., Magrini, G., Campana, C., Fortina, P., *et al.* (1998). Mitochondrial DNA mutations and mitochondrial abnormalities in dilated cardiomyopathy. *Am. J. Pathol.* **153**, 1501–1510.

Baas, P. W. (2002). Microtubule transport in the axon. *Int. Rev. Cytol.* **212**, 41–62.

Baas, P. W., and Heidemann, S. R. (1986). Microtubule reassembly from nucleating fragments during the regrowth of amputated neurites. *J. Cell Biol.* **103**, 917–927.

Bannai, H., Inoue, T., Nakayama, T., Hattori, M., and Mikoshiba, K. (2004). Kinesin dependent, rapid, bi-directional transport of ER sub-compartment in dendrites of hippocampal neurons. *J. Cell Sci.* **117**, 163–175.

Barron, M. J., Griffiths, P., Turnbull, D. M., Bates, D., and Nichols, P. (2004). The distributions of mitochondria and sodium channels reflect the specific energy requirements and conduction properties of the human optic nerve head. *Br. J. Ophthalmol.* **88**, 286–290.

Bate, M., and Martinez Arias, A. (1993). "The Development of Drosophila Melanogaster." Cold Spring Harbor Laboratory Press, Plainview, NY.

Beckerle, M. C., and Porter, K. R. (1982). Inhibitors of dynein activity block intracellular transport in erythrophores. *Nature* **295**, 701–703.

Beil, M., Micoulet, A., von Wichert, G., Paschke, S., Walther, P., Omary, M. B., Van Veldhoven, P. P., Gern, U., Wolff-Hieber, E., Eggermann, J., Waltenberger, J., Adler, G., *et al.* (2003). Sphingosylphosphorylcholine regulates keratin network architecture and visco-elastic properties of human cancer cells. *Nat. Cell Biol.* **5**, 803–811.

Benda, C. (1898). *Arch. Anat. Physiol.* 393–398.

Bereiter-Hahn, J. (1976). Dimethylaminostyrylmethylpyridiniumiodine (daspmi) as a fluorescent probe for mitochondria *in situ. Biochim. Biophys. Acta* **423**, 1–14.

Bereiter-Hahn, J. (1978). Intracellular motility of mitochondria: Role of the inner compartment in migration and shape changes of mitochondria in XTH-cells. *J. Cell Sci.* **30**, 99–115.

Bereiter-Hahn, J. (1990). Behavior of mitochondria in the living cell. *Int. Rev. Cytol.* **122**, 1–63.

Bereiter-Hahn, J., Seipel, K. H., Voth, M., and Ploem, J. S. (1983). Fluorimetry of mitochondria in cells vitally stained with DASPMI or rhodamine 6 GO. *Cell Biochem. Funct.* **1**, 147–155.

Bereiter-Hahn, J., and Voth, M. (1994). Dynamics of mitochondria in living cells: Shape changes, dislocations, fusion, and fission of mitochondria. *Microsc. Res. Tech.* **27**, 198–219.

Boldogh, I. R., and Pon, L. A. (2006). Interactions of mitochondria with the actin cytoskeleton. *Biochim. Biophys. Acta* **1763**(5–6), 450–462.

Bouchard, P., Penningroth, S. M., Cheung, A., Gagnon, C., and Bardin, C. W. (1981). Erythro-9-[3-(2-hydroxynonyl)]adenine is an inhibitor of sperm motility that blocks dynein ATPase and protein carboxylmethylase activities. *Proc. Natl. Acad. Sci. USA* **78**, 1033–1036.

Brady, S. T. (1985). A novel brain ATPase with properties expected for the fast axonal transport motor. *Nature* **317**, 73–75.

Brady, S. T., Lasek, R. J., and Allen, R. D. (1982). Fast axonal transport in extruded axoplasm from squid giant axon. *Science* **218**, 1129–1131.

Breckenridge, D. G., Stojanovic, M., Marcellus, R. C., and Shore, G. C. (2003). Caspase cleavage product of BAP31 induces mitochondrial fission through endoplasmic reticulum calcium signals, enhancing cytochrome c release to the cytosol. *J. Cell Biol.* **160**, 1115–1127.

Bristow, E. A., Griffiths, P. G., Andrews, R. M., Johnson, M. A., and Turnbull, D. M. (2002). The distribution of mitochondrial activity in relation to optic nerve structure. *Arch. Ophthalmol.* **120**, 791–796.

Bubb, M. R., Senderowicz, A. M., Sausville, E. A., Duncan, K. L., and Korn, E. D. (1994). Jasplakinolide, a cytotoxic natural product, induces actin polymerization and competitively inhibits the binding of phalloidin to F-actin. *J. Biol. Chem.* **269**, 14869–14871.

Burkhardt, J. K., Echeverri, C. J., Nilsson, T., and Vallee, R. B. (1997). Overexpression of the dynamitin (p50) subunit of the dynactin complex disrupts dynein-dependent maintenance of membrane organelle distribution. *J. Cell Biol.* **139**, 469–484.

Cai, Q., Gerwin, C., and Sheng, Z. H. (2005). Syntabulin-mediated anterograde transport of mitochondria along neuronal processes. *J. Cell Biol.* **170**, 959–969.

Chada, S. R., and Hollenbeck, P. J. (2004). Nerve growth factor signaling regulates motility and docking of axonal mitochondria. *Curr. Biol.* **14**, 1272–1276.

Chalfie, M., Tu, Y., Euskirchen, G., Ward, W. W., and Prasher, D. C. (1994). Green fluorescent protein as a marker for gene expression. *Science* **263**, 802–805.

Chan, K. Y., and Bunt, A. H. (1978). An association between mitochondria and microtubules in synaptosomes and axon terminals of cerebral cortex. *J. Neurocytol.* **7**, 137–143.

Collier, N. C., Sheetz, M. P., and Schlesinger, M. J. (1993). Concomitant changes in mitochondria and intermediate filaments during heat shock and recovery of chicken embryo fibroblasts. *J. Cell. Biochem.* **52**, 297–307.

Collins, T. J., Berridge, M. J., Lipp, P., and Bootman, M. D. (2002). Mitochondria are morphologically and functionally heterogeneous within cells. *EMBO J.* **21**, 1616–1627.

Colnna, R., Massari, S., and Azzone, G. F. (1973). The problem of cation-binding sites in the energized membrane of intact mitochondria. *Eur. J. Biochem.* **34**, 577–585.

Cooper, J. A. (1987). Effects of cytochalasin and phalloidin on actin. *J. Cell Biol.* **105**, 1473–1478.

Coue, M., Brenner, S. L., Spector, I., and Korn, E. D. (1987). Inhibition of actin polymerization by latrunculin A. *FEBS Lett.* **213**, 316–318.

Cramer, L. P. (1999). Role of actin-filament disassembly in lamellipodium protrusion in motile cells revealed using the drug jasplakinolide. *Curr. Biol.* **9**, 1095–1105.

D'Andrea, L., Danon, M. A., Sgourdas, G. P., and Bonder, E. M. (1994). Identification of coelomocyte unconventional myosin and its association with *in vivo* particle/vesicle motility. *J. Cell Sci.* **107**, 2081–2094.

David-Ferreira, K. L., and David-Ferreira, J. F. (1980). Association between intermediate-sized filaments and mitochondria in rat Leydig cells. *Cell Biol. Int. Rep.* **4**, 655–662.

De Brabander, M., De May, J., Joniau, M., and Geuens, G. (1977). Ultrastructural immunocytochemical distribution of tubulin in cultured cells treated with microtubule inhibitors. *Cell Biol. Int. Rep.* **1**, 177–183.

De Brabander, M., Geuens, G., Nuydens, R., Willebrords, R., and De Mey, J. (1981). Taxol induces the assembly of free microtubules in living cells and blocks the organizing capacity of the centrosomes and kinetochores. *Proc. Natl. Acad. Sci. USA* **78**, 5608–5612.

De Vos, K., Goossens, V., Boone, E., Vercammen, D., Vancompernolle, K., Vandenabeele, P., Haegeman, G., Fiers, W., and Grooten, J. (1998). The 55-kDa tumor necrosis factor receptor induces clustering of mitochondria through its membrane-proximal region. *J. Biol. Chem.* **273**, 9673–9680.

De Vos, K., Severin, F., Van Herreweghe, F., Vancompernolle, K., Goossens, V., Hyman, A., and Grooten, J. (2000). Tumor necrosis factor induces hyperphosphorylation of kinesin light chain and inhibits kinesin-mediated transport of mitochondria. *J. Cell Biol.* **149**, 1207–1214.

De Vos, K. J., Sable, J., Miller, K. E., and Sheetz, M. P. (2003). Expression of phosphatidylinositol (4,5) bisphosphate-specific pleckstrin homology domains alters direction but not the level of axonal transport of mitochondria. *Mol. Biol. Cell* **14**, 3636–3649.

De Vos, K. J., Allan, V. J., Grierson, A. J., and Sheetz, M. P. (2005). Mitochondrial function and actin regulate dynamin-related protein 1-dependent mitochondrial fission. *Curr. Biol.* **15**, 678–683.

Djaldetti, M. (1982). Mitochondrial abnormalities in the cells of myeloma patients. *Acta Haematol.* **68**, 241–248.

Dujardin, D. L., Barnhart, L. E., Stehman, S. A., Gomes, E. R., Gundersen, G. G., and Vallee, R. B. (2003). A role for cytoplasmic dynein and LIS1 in directed cell movement. *J. Cell Biol.* **163,** 1205–1211.

Durham, H. D., Pena, S. D., and Carpenter, S. (1983). The neurotoxins 2,5-hexanedione and acrylamide promote aggregation of intermediate filaments in cultured fibroblasts. *Muscle Nerve* **6,** 631–637.

Echeverri, C. J., Paschal, B. M., Vaughan, K. T., and Vallee, R. B. (1996). Molecular characterization of the 50-kD subunit of dynactin reveals function for the complex in chromosome alignment and spindle organization during mitosis. *J. Cell Biol.* **132,** 617–633.

Eckert, B. S. (1985). Alteration of intermediate filament distribution in PtK1 cells by acrylamide. *Eur. J. Cell Biol.* **37,** 169–174.

Eckert, B. S. (1986). Alteration of the distribution of intermediate filaments in PtK1 cells by acrylamide. II: Effect on the organization of cytoplasmic organelles. *Cell Motil. Cytoskeleton* **6,** 15–24.

Erbrich, U., Septinus, M., Naujok, A., and Zimmermann, H. W. (1984). Hydrophobic acridine dyes for fluorescence staining of mitochondria in living cells. 2. Comparison of staining of living and fixed Hela-cells with NAO and DPPAO. *Histochemistry* **80,** 385–388.

Ernster, L., and Lindberg, O. (1958). Animal mitochondria. *Annu. Rev. Physiol.* **20,** 13–42.

Fabricius, C., Berthold, C. H., and Rydmark, M. (1993). Axoplasmic organelles at nodes of Ranvier. II. Occurrence and distribution in large myelinated spinal cord axons of the adult cat. *J. Neurocytol.* **22,** 941–954.

Fahim, M. A., Lasek, R. J., Brady, S. T., and Hodge, A. J. (1985). AVEC-DIC and electron microscopic analyses of axonally transported particles in cold-blocked squid giant axons. *J. Neurocytol.* **14,** 689–704.

Faulkner, N. E., Dujardin, D. L., Tai, C. Y., Vaughan, K. T., O'Connell, C. B., Wang, Y., and Vallee, R. B. (2000). A role for the lissencephaly gene LIS1 in mitosis and cytoplasmic dynein function. *Nat. Cell Biol.* **2,** 784–791.

Feiguin, F., Ferreira, A., Kosik, K. S., and Caceres, A. (1994). Kinesin-mediated organelle transloca-tion revealed by specific cellular manipulations. *J. Cell Biol.* **127,** 1021–1039.

Ferreira, A., Niclas, J., Vale, R. D., Banker, G., and Kosik, K. S. (1992). Suppression of kinesin expression in cultured hippocampal neurons using antisense oligonucleotides. *J. Cell Biol.* **117,** 595–606.

Flanagan, M. D., and Lin, S. (1980). Cytochalasins block actin filament elongation by binding to high affinity sites associated with F-actin. *J. Biol. Chem.* **255,** 835–838.

Forer, A., Emmersen, J., and Behnke, O. (1972). Cytochalasin B: Does it affect actin-like filaments? *Science* **175,** 774–776.

Frank, S., Gaume, B., Bergmann-Leitner, E. S., Leitner, W. W., Robert, E. G., Catez, F., Smith, C. L., and Youle, R. J. (2001). The role of dynamin-related protein 1, a mediator of mitochondrial fission, in apoptosis. *Dev. Cell* **1,** 515–525.

Franke, W. W., and Kartenbeck, J. (1971). Outer mitochondrial membrane continuous with endoplas-mic reticulum. *Protoplasma* **73,** 35–41.

Frédéric, J., and Chèvremont, M. (1951). [Research on chondriosomes in tissue culture by phase contrast microcinematography.]. *C. R. Seances Soc. Biol. Fil.* **145,** 1243–1244.

Frédéric, J., and Chèvremont, M. (1952). [Investigations on the chondriosomes of living cells by phase contrast microscopy and microcinematography.]. *Arch. Biol. (Liege)* **63,** 109–131.

Gibbons, I. R., Cosson, M. P., Evans, J. A., Gibbons, B. H., Houck, B., Martinson, K. H., Sale, W. S., and Tang, W. J. (1978). Potent inhibition of dynein adenosinetriphosphatase and of the motility of cilia and sperm flagella by vanadate. *Proc. Natl. Acad. Sci. USA* **75,** 2220–2224.

Gohil, V. M., Gvozdenovic-Jeremic, J., Schlame, M., and Greenberg, M. L. (2005). Binding of 10-N-nonyl acridine orange to cardiolipin-deficient yeast cells: Implications for assay of cardiolipin. *Anal. Biochem.* **343,** 350–352.

Gorska-Andrzejak, J., Stowers, R. S., Borycz, J., Kostyleva, R., Schwarz, T. L., and Meinertzhagen, I. A. (2003). Mitochondria are redistributed in *Drosophila* photoreceptors lacking milton, a kinesin-associated protein. *J. Comp. Neurol.* **463,** 372–388.

Gotow, T., Miyaguchi, K., and Hashimoto, P. H. (1991). Cytoplasmic architecture of the axon terminal: Filamentous strands specifically associated with synaptic vesicles. *Neuroscience* **40,** 587–598.

Harborth, J., Elbashir, S. M., Bechert, K., Tuschl, T., and Weber, K. (2001). Identification of essential genes in cultured mammalian cells using small interfering RNAs. *J. Cell Sci.* **114,** 4557–4565.

He, Y., Francis, F., Myers, K. A., Yu, W., Black, M. M., and Baas, P. W. (2005). Role of cytoplasmic dynein in the axonal transport of microtubules and neurofilaments. *J. Cell Biol.* **168,** 697–703.

Heggeness, M. H., Simon, M., and Singer, S. J. (1978). Association of mitochondria with microtubules in cultured cells. *Proc. Natl. Acad. Sci. USA* **75,** 3863–3866.

Heidemann, S. R. (1996). Cytoplasmic mechanisms of axonal and dendritic growth in neurons. *Int. Rev. Cytol.* **165,** 235–296.

Ho, C. L., Martys, J. L., Mikhailov, A., Gundersen, G. G., and Liem, R. K. (1998). Novel features of intermediate filament dynamics revealed by green fluorescent protein chimeras. *J. Cell Sci.* **111** (Pt. 13), 1767–1778.

Hollenbeck, P. J. (1996). The pattern and mechanism of mitochondrial transport in axons. *Front. Biosci.* **1,** d91–d102.

Hollenbeck, P. J., Bray, D., and Adams, R. J. (1985). Effects of the uncoupling agents FCCP and CCCP on the saltatory movements of cytoplasmic organelles. *Cell Biol. Int. Rep.* **9,** 193–199.

Hollenbeck, P. J., and Saxton, W. M. (2005). The axonal transport of mitochondria. *J. Cell Sci.* **118,** 5411–5419.

Ingold, A. L., Cohn, S. A., and Scholey, J. M. (1988). Inhibition of kinesin-driven microtubule motility by monoclonal antibodies to kinesin heavy chains. *J. Cell Biol.* **107,** 2657–2667.

Ishihara, N., Jofuku, A., Eura, Y., and Mihara, K. (2003). Regulation of mitochondrial morphology by membrane potential, and DRP1-dependent division and FZO1-dependent fusion reaction in mammalian cells. *Biochem. Biophys. Res. Commun.* **301,** 891–898.

Jacobson, J., Duchen, M. R., and Heales, S. J. (2002). Intracellular distribution of the fluorescent dye nonyl acridine orange responds to the mitochondrial membrane potential: Implications for assays of cardiolipin and mitochondrial mass. *J. Neurochem.* **82,** 224–233.

James, D. I., Parone, P. A., Mattenberger, Y., and Martinou, J. C. (2003). hFis1, a novel component of the mammalian mitochondrial fission machinery. *J. Biol. Chem.* **278,** 36373–36379.

Jellali, A., Metz-Boutigue, M. H., Surgucheva, I., Jancsik, V., Schwartz, C., Filliol, D., Gelfand, V. I., and Rendon, A. (1994). Structural and biochemical properties of kinesin heavy chain associated with rat brain mitochondria. *Cell Motil. Cytoskeleton* **28,** 79–93.

Jendrach, M., Pohl, S., Voth, M., Kowald, A., Hammerstein, P., and Bereiter-Hahn, J. (2005). Morpho-dynamic changes of mitochondria during ageing of human endothelial cells. *Mech. Ageing Dev.* **126,** 813–821.

Johnson, L. V., Walsh, M. L., and Chen, L. B. (1980). Localization of mitochondria in living cells with rhodamine 123. *Proc. Natl. Acad. Sci. USA* **77,** 990–994.

Kaether, C., Skehel, P., and Dotti, C. G. (2000). Axonal membrane proteins are transported in distinct carriers: A two-color video microscopy study in cultured hippocampal neurons. *Mol. Biol. Cell* **11,** 1213–1224.

Keij, J. F., Bell-Prince, C., and Steinkamp, J. A. (2000). Staining of mitochondrial membranes with 10-nonyl acridine orange, MitoFluor Green, and MitoTracker Green is affected by mitochondrial membrane potential altering drugs. *Cytometry* **39,** 203–210.

Keith, C. H., Feramisco, J. R., and Shelanski, M. (1981). Direct visualization of fluorescein-labeled microtubules in vitro and in microinjected fibroblasts. *J. Cell Biol.* **88,** 234–240.

Kobayashi, T., Martensen, T., Nath, J., and Flavin, M. (1978). Inhibition of dynein ATPase by vanadate, and its possible use as a probe for the role of dynein in cytoplasmic motility. *Biochem. Biophys. Res. Commun.* **81,** 1313–1318.

Kocsis, E., Trus, B. L., Steer, C. J., Bisher, M. E., and Steven, A. C. (1991). Image averaging of flexible fibrous macromolecules: The clathrin triskelion has an elastic proximal segment. *J. Struct. Biol.* **107,** 6–14.

Kölliker, A. (1856). *Z. Wiss. Zool.* **8,** 331–325.

Korchak, H. M., Rich, A. M., Wilkenfeld, C., Rutherford, L. E., and Weissmann, G. (1982). A carbocyanine dye, DiOC6(3), acts as a mitochondrial probe in human neutrophils. *Biochem. Biophys. Res. Commun.* **108**, 1495–1501.

Kreis, T. E., and Birchmeier, W. (1982). Microinjection of fluorescently labeled proteins into living cells with emphasis on cytoskeletal proteins. *Int. Rev. Cytol.* **75**, 209–214.

Kreis, T. E., Winterhalter, K. H., and Birchmeier, W. (1979). *In vivo* distribution and turnover of fluorescently labeled actin microinjected into human fibroblasts. *Proc. Natl. Acad. Sci. USA* **76**, 3814–3818.

Kumar, J., Yu, H., and Sheetz, M. P. (1995). Kinectin, an essential anchor for kinesin-driven vesicle motility. *Science* **267**, 1834–1837.

Kuznetsov, S. A., Langford, G. M., and Weiss, D. G. (1992). Actin-dependent organelle movement in squid axoplasm. *Nature* **356**, 722–725.

Kuznetsov, S. A., Rivera, D. T., Severin, F. F., Weiss, D. G., and Langford, G. M. (1994). Movement of axoplasmic organelles on actin filaments from skeletal muscle. *Cell Motil. Cytoskeleton* **28**, 231–242.

Langford, G. M., Kuznetsov, S. A., Johnson, D., Cohen, D. L., and Weiss, D. G. (1994). Movement of axoplasmic organelles on actin filaments assembled on acrosomal processes: Evidence for a barbed-end-directed organelle motor. *J. Cell Sci.* **107**(Pt. 8), 2291–2298.

Lazzarino, D. A., Boldogh, I., Smith, M. G., Rosand, J., and Pon, L. A. (1994). Yeast mitochondria contain ATP-sensitive, reversible actin-binding activity. *Mol. Biol. Cell* **5**, 807–818.

Legros, F., Lombes, A., Frachon, P., and Rojo, M. (2002). Mitochondrial fusion in human cells is efficient, requires the inner membrane potential, and is mediated by mitofusins. *Mol. Biol. Cell* **13**, 4343–4354.

Leopold, P. L., McDowall, A. W., Pfister, K. K., Bloom, G. S., and Brady, S. T. (1992). Association of kinesin with characterized membrane-bounded organelles. *Cell Motil. Cytoskeleton* **23**, 19–33.

Leterrier, J. F., Rusakov, D. A., Nelson, B. D., and Linden, M. (1994). Interactions between brain mitochondria and cytoskeleton: Evidence for specialized outer membrane domains involved in the association of cytoskeleton-associated proteins to mitochondria *in situ* and *in vitro*. *Microsc. Res. Tech.* **27**, 233–261.

Lewis, M. R., and Lewis, W. H. (1915). Mitochondria (and other cytoplasmic structures) in tissue cultures. *Am. J. Anat.* **17**, 339–401.

Ligon, L. A., and Steward, O. (2000a). Movement of mitochondria in the axons and dendrites of cultured hippocampal neurons. *J. Comp. Neurol.* **427**, 340–350.

Ligon, L. A., and Steward, O. (2000b). Role of microtubules and actin filaments in the movement of mitochondria in the axons and dendrites of cultured hippocampal neurons. *J. Comp. Neurol.* **427**, 351–361.

Linden, M., Li, Z., Paulin, D., Gotow, T., and Leterrier, J. F. (2001). Effects of desmin gene knockout on mice heart mitochondria. *J. Bioenerg. Biomembr.* **33**, 333–341.

Maftah, A., Petit, J. M., and Julien, R. (1990). Specific interaction of the new fluorescent dye 10-N-nonyl acridine orange with inner mitochondrial membrane. A lipid-mediated inhibition of oxidative phosphorylation. *FEBS Lett.* **260**, 236–240.

Michaelis, L. (1899). Die vitale Färbung, eine Darstellungsmethode der Zellgranula. *Arch. f. mikr. Anat.* Bd. **55**.

Mignotte, F., Tourte, M., and Mounolou, J. C. (1987). Segregation of mitochondria in the cytoplasm of Xenopus vitellogenic oocytes. *Biol. Cell* **60**, 97–102.

Miller, K. E., and Sheetz, M. P. (2004). Axonal mitochondrial transport and potential are correlated. *J. Cell Sci.* **117**, 2791–2804.

Milner, D. J., Mavroidis, M., Weisleder, N., and Capetanaki, Y. (2000). Desmin cytoskeleton linked to muscle mitochondrial distribution and respiratory function. *J. Cell Biol.* **150**, 1283–1298.

Minamikawa, T., Sriratana, A., Williams, D. A., Bowser, D. N., Hill, J. S., and Nagley, P. (1999). Chloromethyl-X-rosamine (MitoTracker Red) photosensitises mitochondria and induces apoptosis in intact human cells. *J. Cell Sci.* **112**(Pt. 14), 2419–2430.

Minin, A. A., Kulik, A. V., Gyoeva, F. K., Li, Y., Goshima, G., and Gelfand, V. I. (2006). Regulation of mitochondria distribution by RhoA and formins. *J. Cell Sci.* **119,** 659–670.

Mittal, B., Sanger, J. M., and Sanger, J. W. (1989). Visualization of intermediate filaments in living cells using fluorescently labeled desmin. *Cell Motil. Cytoskeleton* **12,** 127–138.

Modica-Napolitano, J. S., Weiss, M. J., Chen, L. B., and Aprille, J. R. (1984). Rhodamine 123 inhibits bioenergetic function in isolated rat liver mitochondria. *Biochem. Biophys. Res. Commun.* **118,** 717–723.

Morre, D. J., Merritt, W. D., and Lembi, C. A. (1971). Connections between mitochondria and endoplasmic reticulum in rat liver and onion stem. *Protoplasma* **73,** 43–49.

Morris, R. L., and Hollenbeck, P. J. (1993). The regulation of bidirectional mitochondrial transport is coordinated with axonal outgrowth. *J. Cell Sci.* **104**(Pt. 3), 917–927.

Morris, R. L., and Hollenbeck, P. J. (1995). Axonal transport of mitochondria along microtubules and F-actin in living vertebrate neurons. *J. Cell Biol.* **131,** 1315–1326.

Mose-Larsen, P., Bravo, R., Fey, S. J., Small, J. V., and Celis, J. E. (1982). Putative association of mitochondria with a subpopulation of intermediate-sized filaments in cultured human skin fibroblasts. *Cell* **31,** 681–692.

Muresan, Z., and Muresan, V. (2005). Coordinated transport of phosphorylated amyloid-beta precursor protein and c-Jun NH2-terminal kinase-interacting protein-1. *J. Cell Biol.* **171,** 615–625.

Nakata, T., and Hirokawa, N. (1995). Point mutation of adenosine triphosphate-binding motif generated rigor kinesin that selectively blocks anterograde lysosome membrane transport. *J. Cell Biol.* **131,** 1039–1053.

Nangaku, M., Sato-Yoshitake, R., Okada, Y., Noda, Y., Takemura, R., Yamazaki, H., and Hirokawa, N. (1994). KIF1B, a novel microtubule plus end-directed monomeric motor protein for transport of mitochondria. *Cell* **79,** 1209–1220.

Nikolic, M., Dudek, H., Kwon, Y. T., Ramos, Y. F., and Tsai, L. H. (1996). The cdk5/p35 kinase is essential for neurite outgrowth during neuronal differentiation. *Genes Dev.* **10,** 816–825.

Nishino, I., Kobayashi, O., Goto, Y., Kurihara, M., Kumagai, K., Fujita, T., Hashimoto, K., Horai, S., and Nonaka, I. (1998). A new congenital muscular dystrophy with mitochondrial structural abnormalities. *Muscle Nerve* **21,** 40–47.

Overly, C. C., Rieff, H. I., and Hollenbeck, P. J. (1996). Organelle motility and metabolism in axons vs dendrites of cultured hippocampal neurons. *J. Cell Sci.* **109**(Pt. 5), 971–980.

Paschal, B. M., Shpetner, H. S., and Vallee, R. B. (1987). MAP 1C is a microtubule-activated ATPase which translocates microtubules *in vitro* and has dynein-like properties. *J. Cell Biol.* **105,** 1273–1282.

Pendergrass, W., Wolf, N., and Poot, M. (2004). Efficacy of MitoTracker Green and CMXrosamine to measure changes in mitochondrial membrane potentials in living cells and tissues. *Cytometry A* **61,** 162–169.

Penningroth, S. M. (1986). Erythro-9-[3-(2-hydroxynonyl)]adenine and vanadate as probes for microtubule-based cytoskeletal mechanochemistry. *Methods Enzymol.* **134,** 477–487.

Pilling, A. D., Horiuchi, D., Lively, C. M., and Saxton, W. M. (2006). Kinesin-1 and Dynein are the primary motors for fast transport of mitochondria in *Drosophila* motor axons. *Mol. Biol. Cell* **17,** 2057–2068.

Pon, L. A., and Schon, E. A. (2001). "Mitochondria." Academic, San Diego, CA; London.

Pruss, R. M., Mirsky, R., Raff, M. C., Thorpe, R., Dowding, A. J., and Anderton, B. H. (1981). All classes of intermediate filaments share a common antigenic determinant defined by a monoclonal antibody. *Cell* **27,** 419–428.

Raine, C. S., Ghetti, B., and Shelanski, M. L. (1971). On the association between microtubules and mitochondria within axons. *Brain Res.* **34,** 389–393.

Reichman, J. (2000). "Handbook of Optical Filters for Fluorescence Microscopy." Chroma Technology Corp., Rockingham, VT, USA.

Reipert, S., Steinbock, F., Fischer, I., Bittner, R. E., Zeold, A., and Wiche, G. (1999). Association of mitochondria with plectin and desmin intermediate filaments in striated muscle. *Exp. Cell Res.* **252,** 479–491.

Rizzuto, R., Brini, M., Pizzo, P., Murgia, M., and Pozzan, T. (1995). Chimeric green fluorescent protein as a tool for visualizing subcellular organelles in living cells. *Curr. Biol.* **5**, 635–642.

Rodionov, V. I., Gyoeva, F. K., Tanaka, E., Bershadsky, A. D., Vasiliev, J. M., and Gelfand, V. I. (1993). Microtubule-dependent control of cell shape and pseudopodial activity is inhibited by the antibody to kinesin motor domain. *J. Cell Biol.* **123**, 1811–1820.

Russ, J. C. (1992). "The Image Processing Handbook." CRC Press, Boca Raton, FL.

Ruthel, G., and Hollenbeck, P. J. (2003). Response of mitochondrial traffic to axon determination and differential branch growth. *J. Neurosci.* **23**, 8618–8624.

Sammak, P. J., and Borisy, G. G. (1988a). Detection of single fluorescent microtubules and methods for determining their dynamics in living cells. *Cell Motil. Cytoskeleton* **10**, 237–245.

Sammak, P. J., and Borisy, G. G. (1988b). Direct observation of microtubule dynamics in living cells. *Nature* **332**, 724–726.

Santel, A., and Fuller, M. T. (2001). Control of mitochondrial morphology by a human mitofusin. *J. Cell Sci.* **114**, 867–874.

Saxton, W. M., Stemple, D. L., Leslie, R. J., Salmon, E. D., Zavortink, M., and McIntosh, J. R. (1984). Tubulin dynamics in cultured mammalian cells. *J. Cell Biol.* **99**, 2175–2186.

Schroer, T. A. (1994). Structure, function and regulation of cytoplasmic dynein. *Curr. Opin. Cell Biol.* **6**, 69–73.

Schroer, T. A. (2004). Dynactin. *Annu. Rev. Cell Dev. Biol.* **20**, 759–779.

Schultheiss, T., Lin, Z. X., Ishikawa, H., Zamir, I., Stoeckert, C. J., and Holtzer, H. (1991). Desmin/vimentin intermediate filaments are dispensable for many aspects of myogenesis. *J. Cell Biol.* **114**, 953–966.

Scorrano, L., Petronilli, V., Colonna, R., Di Lisa, F., and Bernardi, P. (1999). Chloromethyltetramethylrosamine (Mitotracker Orange) induces the mitochondrial permeability transition and inhibits respiratory complex I. Implications for the mechanism of cytochrome c release. *J. Biol. Chem.* **274**, 24657–24663.

Semiz, S., Park, J. G., Nicoloro, S. M., Furcinitti, P., Zhang, C., Chawla, A., Leszyk, J., and Czech, M. P. (2003). Conventional kinesin KIF5B mediates insulin-stimulated GLUT4 movements on microtubules. *EMBO J.* **22**, 2387–2399.

Septinus, M., Berthold, T., Naujok, A., and Zimmermann, H. W. (1985). Hydrophobic acridine dyes for fluorescent staining of mitochondria in living cells. 3. Specific accumulation of the fluorescent dye NAO on the mitochondrial membranes in HeLa cells by hydrophobic interaction. Depression of respiratory activity, changes in the ultrastructure of mitochondria due to NAO. Increase of fluorescence in vital stained mitochondria *in situ* by irradiation. *Histochemistry* **82**, 51–66.

Severt, W. L., Biber, T. U., Wu, X., Hecht, N. B., DeLorenzo, R. J., and Jakoi, E. R. (1999). The suppression of testis-brain RNA binding protein and kinesin heavy chain disrupts mRNA sorting in dendrites. *J. Cell Sci.* **112**(Pt. 21), 3691–3702.

Sharpe, A. H., Chen, L. B., Murphy, J. R., and Fields, B. N. (1980). Specific disruption of vimentin filament organization in monkey kidney CV-1 cells by diphtheria toxin, exotoxin A, and cycloheximide. *Proc. Natl. Acad. Sci. USA* **77**, 7267–7271.

Shpetner, H. S., Paschal, B. M., and Vallee, R. B. (1988). Characterization of the microtubule-activated ATPase of brain cytoplasmic dynein (MAP 1C). *J. Cell Biol.* **107**, 1001–1009.

Simon, V. R., Swayne, T. C., and Pon, L. A. (1995). Actin-dependent mitochondrial motility in mitotic yeast and cell-free systems: Identification of a motor activity on the mitochondrial surface. *J. Cell Biol.* **130**, 345–354.

Smiley, S. T., Reers, M., Mottola-Hartshorn, C., Lin, M., Chen, A., Smith, T. W., Steele, G. D., Jr., and Chen, L. B. (1991). Intracellular heterogeneity in mitochondrial membrane potentials revealed by a J-aggregate-forming lipophilic cation JC-1. *Proc. Natl. Acad. Sci. USA* **88**, 3671–3675.

Smirnova, E., Shurland, D. L., Ryazantsev, S. N., and van der Bliek, A. M. (1998). A human dynamin-related protein controls the distribution of mitochondria. *J. Cell Biol.* **143**, 351–358.

Smirnova, E., Griparic, L., Shurland, D. L., and van der Bliek, A. M. (2001). Dynamin-related protein Drp1 is required for mitochondrial division in mammalian cells. *Mol. Biol. Cell* **12**, 2245–2256.

Smith, D. S., Jarlfors, U., and Cayer, M. L. (1977). Structural cross-bridges between microtubules and mitochondria in central axons of an insect (Periplaneta americana). *J. Cell Sci.* **27**, 255–272.

Soltys, B. J., and Gupta, R. S. (1992). Interrelationships of endoplasmic reticulum, mitochondria, intermediate filaments, and microtubules—a quadruple fluorescence labeling study. *Biochem. Cell Biol.* **70**, 1174–1186.

Spector, I., Shochet, N. R., Kashman, Y., and Groweiss, A. (1983). Latrunculins: Novel marine toxins that disrupt microfilament organization in cultured cells. *Science* **219**, 493–495.

Stowers, R. S., Megeath, L. J., Gorska-Andrzejak, J., Meinertzhagen, I. A., and Schwarz, T. L. (2002). Axonal transport of mitochondria to synapses depends on milton, a novel *Drosophila* protein. *Neuron* **36**, 1063–1077.

Susin, S. A., Lorenzo, H. K., Zamzami, N., Marzo, I., Snow, B. E., Brothers, G. M., Mangion, J., Jacotot, E., Costantini, P., Loeffler, M., Larochette, N., Goodlett, D. R., *et al.* (1999). Molecular characterization of mitochondrial apoptosis-inducing factor. *Nature* **397**, 441–446.

Tanaka, Y., Kanai, Y., Okada, Y., Nonaka, S., Takeda, S., Harada, A., and Hirokawa, N. (1998). Targeted disruption of mouse conventional kinesin heavy chain, kif5B, results in abnormal perinuclear clustering of mitochondria. *Cell* **93**, 1147–1158.

Tandler, B., Dunlap, M., Hoppel, C. L., and Hassan, M. (2002). Giant mitochondria in a cardiomyopathic heart. *Ultrastruct. Pathol.* **26**, 177–183.

Taylor, D. L., and Wang, Y. L. (1978). Molecular cytochemistry: Incorporation of fluorescently labeled actin into living cells. *Proc. Natl. Acad. Sci. USA* **75**, 857–861.

Terasaki, M., Song, J., Wong, J. R., Weiss, M. J., and Chen, L. B. (1984). Localization of endoplasmic reticulum in living and glutaraldehyde-fixed cells with fluorescent dyes. *Cell* **38**, 101–108.

Thornell, L., Carlsson, L., Li, Z., Mericskay, M., and Paulin, D. (1997). Null mutation in the desmin gene gives rise to a cardiomyopathy. *J. Mol. Cell. Cardiol.* **29**, 2107–2124.

Toh, B. H., Lolait, S. J., Mathy, J. P., and Baum, R. (1980). Association of mitochondria with intermediate filaments and of polyribosomes with cytoplasmic actin. *Cell Tissue Res.* **211**, 163–169.

Toyoshima, I., Yu, H., Steuer, E. R., and Sheetz, M. P. (1992). Kinectin, a major kinesin-binding protein on ER. *J. Cell Biol.* **118**, 1121–1131.

Twig, G., Graf, S. A., Wikstrom, J. D., Mohamed, H., Haigh, S. E., Elorza, A. G., Deutsch, M., Zurgil, N., Reynolds, N., and Shirihai, O. S. (2006). Tagging and Tracking Individual Networks within a Complex Mitochondrial Web Using Photoactivatable GFP. *Am. J. Physiol. Cell Physiol.* **291**(1), C176–C184.

Vale, R. D., and Fletterick, R. J. (1997). The design plan of kinesin motors. *Annu. Rev. Cell Dev. Biol.* **13**, 745–777.

Vale, R. D., Reese, T. S., and Sheetz, M. P. (1985). Identification of a novel force-generating protein, kinesin, involved in microtubule-based motility. *Cell* **42**, 39–50.

Varadi, A., Johnson-Cadwell, L. I., Cirulli, V., Yoon, Y., Allan, V. J., and Rutter, G. A. (2004). Cytoplasmic dynein regulates the subcellular distribution of mitochondria by controlling the recruitment of the fission factor dynamin-related protein-1. *J. Cell Sci.* **117**, 4389–4400.

Verhey, K. J., Meyer, D., Deehan, R., Blenis, J., Schnapp, B. J., Rapoport, T. A., and Margolis, B. (2001). Cargo of kinesin identified as JIP scaffolding proteins and associated signaling molecules. *J. Cell Biol.* **152**, 959–970.

Wang, H. J., Guay, G., Pogan, L., Sauve, R., and Nabi, I. R. (2000). Calcium regulates the association between mitochondria and a smooth subdomain of the endoplasmic reticulum. *J. Cell Biol.* **150**, 1489–1498.

Wozniak, M. J., Melzer, M., Dorner, C., Haring, H. U., and Lammers, R. (2005). The novel protein KBP regulates mitochondria localization by interaction with a kinesin-like protein. *BMC Cell Biol.* **6**, 35.

Wulf, E., Deboben, A., Bautz, F. A., Faulstich, H., and Wieland, T. (1979). Fluorescent phallotoxin, a tool for the visualization of cellular actin. *Proc. Natl. Acad. Sci. USA* **76**, 4498–4502.

Xia, Z., Dudek, H., Miranti, C. K., and Greenberg, M. E. (1996). Calcium influx via the NMDA receptor induces immediate early gene transcription by a MAP kinase/ERK-dependent mechanism. *J. Neurosci.* **16,** 5425–5436.

Yoon, Y., Krueger, E. W., Oswald, B. J., and McNiven, M. A. (2003). The mitochondrial protein hFis1 regulates mitochondrial fission in mammalian cells through an interaction with the dynamin-like protein DLP1. *Mol. Cell Biol.* **23,** 5409–5420.

Yu, K. R., Hijikata, T., Lin, Z. X., Sweeney, H. L., Englander, S. W., and Holtzer, H. (1994). Truncated desmin in PtK2 cells induces desmin-vimentin-cytokeratin coprecipitation, involution of intermediate filament networks, and nuclear fragmentation: A model for many degenerative diseases. *Proc. Natl. Acad. Sci. USA* **91,** 2497–2501.

Yu, W., Sharp, D. J., Kuriyama, R., Mallik, P., and Baas, P. W. (1997). Inhibition of a mitotic motor compromises the formation of dendrite-like processes from neuroblastoma cells. *J. Cell Biol.* **136,** 659–668.

Yu Wai Man, C. Y., Chinnery, P. F., and Griffiths, P. G. (2005). Optic neuropathies—importance of spatial distribution of mitochondria as well as function. *Med. Hypotheses* **65,** 1038–1042.

Yvon, A. M., Gross, D. J., and Wadsworth, P. (2001). Antagonistic forces generated by myosin II and cytoplasmic dynein regulate microtubule turnover, movement, and organization in interphase cells. *Proc. Natl. Acad. Sci. USA* **98,** 8656–8661.

Zahrebelski, G., Nieminen, A. L., al-Ghoul, K., Qian, T., Herman, B., and Lemasters, J. J. (1995). Progression of subcellular changes during chemical hypoxia to cultured rat hepatocytes: A laser scanning confocal microscopic study. *Hepatology* **21,** 1361–1372.

Zinser, E., Sperka-Gottlieb, C. D., Fasch, E. V., Kohlwein, S. D., Paltauf, F., and Daum, G. (1991). Phospholipid synthesis and lipid composition of subcellular membranes in the unicellular eukaryote Saccharomyces cerevisiae. *J. Bacteriol.* **173,** 2026–2034.

CHAPTER 31

Cell-Free Assays for Mitochondria–Cytoskeleton Interactions

Istvan R. Boldogh,★ Liza A. Pon,★ Michael P. Sheetz,† and Kurt J. De Vos‡

★Department of Anatomy and Cell Biology, College of Physicians and Surgeons
Columbia University
New York, New York 10032

†Department of Biological Sciences
Columbia University, New York, New York 10025

‡Department of Neuroscience
MRC Centre for Neurodegeneration Research
The Institute of Psychiatry, King's College London, De Crespigny Park
Denmark Hill, London SE5 8AF, United Kingdom

I. Introduction

Mitochondria are essential organelles for a multitude of cellular activities, including aerobic energy production, metabolite biosynthesis, calcium regulation, and apoptosis. To carry out these functions, the organelle must localize to the sites where it is needed. Numerous observations have demonstrated that normal mitochondrial localization in the cell depends on interactions between mitochondria and the cytoskeleton (see also Chapter 29 by Swayne *et al.* and Chapter 30 by De Vos and Sheetz, this volume).

Several lines of evidence indicate that mitochondrial distribution in mammalian or *Drosophila* cells depends on the plus-end-directed microtubule motor, kinesin. For example, kinesin colocalizes with mitochondria, and loss of kinesin function results in abnormal clustering of mitochondria around the nucleus (De Vos *et al.*, 2000; Pereira *et al.*, 1997; Tanaka *et al.*, 1998). In the axons of neurons, where mitochondria travel great distances, both kinesins and the minus end-directed microtubule motors, dyneins, are implicated in mitochondrial movement, and most of the mitochondrial movement can be inhibited by the disruption of the microtubule network (Hollenbeck and Saxton, 2005; Ligon and Steward, 2000).

While many observations support a role for the microtubule cytoskeleton in mitochondrial transport in animal cells, a role for the actin cytoskeleton in mitochondrial motility and distribution has also emerged. First, the actin cytoskeleton supports short distance, cortical mitochondrial movements in axons (Ligon and Steward, 2000), plants, and several fungi (Boldogh *et al.*, 1998; Sheahan *et al.*, 2004; Suelmann and Fischer, 2000; Van Gestel *et al.*, 2002). Indeed, live recordings of budding yeast cells revealed that mitochondria can engage with and undergo bidirectional movement on actin cables (Fehrenbacher *et al.*, 2004). Although myosins have been implicated in actin-dependent cargo movement in animal cells and yeast, the Arp2/3 complex drives mitochondrial movement along actin cables in the budding yeast (Boldogh *et al.*, 2001; Fehrenbacher *et al.*, 2005). The actin cytoskeleton has also been implicated in anchorage of mitochondria at intracellular sites, including sites of high ATP utilization, in axons and in *Drosophila* (Chada and Hollenbeck, 2004; Minin *et al.*, 2006). In budding yeast, actin-dependent anchorage of mitochondria at the opposite poles of the cell is critical for equal distribution of the organelle during cell division (Boldogh *et al.*, 2004; Yang *et al.*, 1999).

Thus, both microtubules and the actin cytoskeleton contribute to mitochondrial motility and distribution. While studies using live-cell imaging are indispensable, numerous *in vitro* approaches have also extended our understanding of the mechanisms responsible for the interaction of mitochondria with the cytoskeleton. Below, we present *in vitro* motility assays for microtubule-based mitochondrial motility, using animal cell cultures and *in vitro* sedimentation assays to explore the interaction between mitochondria and the actin cytoskeleton in budding yeast.

II. *In Vitro* Reconstitution of Microtubule–Based Mitochondrial Transport

In this section, we describe two *in vitro* assays of mitochondrial motility: reconstitution of mitochondrial motility on (1) polarity-marked microtubules and (2) microtubule asters. Both assays can be used for live observation of mitochondrial motility. In addition, the microtubule aster assay is suitable to screen for factors that influence mitochondrial motility. The main components of *in vitro* assays for mitochondrial motility are (1) fluorescently labeled microtubules (labeled either polarity-marked or as centrosomal asters); (2) high-speed cytosol, the source of molecular motors and factors that support motility; and (3) isolated mitochondria.

In the following paragraphs we describe the necessary steps to reconstitute microtubule-based mitochondrial motility *in vitro*. The protocols described here were modified from the assays developed to reconstitute endosome motility *in vitro* (Nielsen *et al.*, 2001).

A. Preparation of Motility Assay Components

1. Observation Chambers

The observation chambers are made by sealing a glass coverslip to a glass microscope slide using grease (Fig. 1). Since these chambers are open at two sides, solutions can be easily exchanged by wicking up the solution at one side with a piece of filter paper while at the same time adding the new solution at the opposite side.

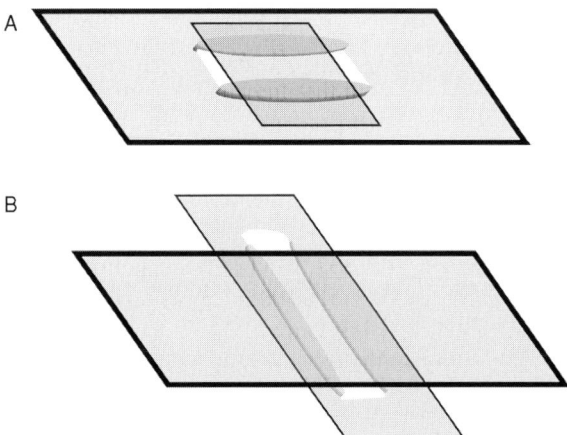

Fig. 1 Observation chambers. The observation chambers shown consist of coverslips that are attached to a microscope slide using two lines of vacuum grease. Sample areas are shown in white. Both chambers can be used for medium changes during measurements. The upper panel shows *Type 1*, the standard observation chamber that is used on an upright microscope. The lower panel shows *Type 2*, the chamber that is used on an inverted microscope.

Materials

1. Glass coverslips; 22 × 22 or 76 × 26 mm; No. 1 Gold Seal (Gold Seal, Electron Microscopy Sciences, Hatfield, PA).
2. Glass microscope slides; 76 × 26 × 1 mm, Select Micro Slides, washed (Chance Propper Ltd., Smethwick, Warley, UK, Cat. No. KTH 360).
3. High-vacuum grease (Dow Corning, Midland, MI).

Procedure

Depending on the microscope setup (inverted or upright), and if exchanging solutions on the microscope is desired, we prepare two different types of observation chamber, as illustrated in Fig. 1. Both are constructed similarly.

1. Apply two thin lines of vacuum grease on the microscope slide.
2. Fix a coverslip to a microscope slide by pushing it down gently onto the vacuum grease.
3. Perfuse the chamber with 10-μl buffer and adjust the chamber volume to 10 μl by gently pressing the glass coverslip down.

2. Preparation of Fluorescently Labeled Polarity–Marked Microtubules

Polarity-marked microtubules are prepared by polymerizing fluorescent microtubules using brightly fluorescent microtubule seeds as nucleators. Since microtubules grow faster at their plus-end compared to their minus-end, the brightly fluorescent seed will be toward the minus-end of the microtubule (Fig. 2). Therefore, it is straightforward to identify the polarity of these microtubules under the microscope. Alternatively, fluorescent microtubules can be polymerized without brightly fluorescent seeds, but in this case the microtubule polarity is unknown, and consequently the directionality of transport cannot be determined.

a. Synthesis of Polarity-Marked Microtubules
Materials

1. Purified tubulin (20 mg/ml); store at −80°C. Detailed protocols for the purification of tubulin from mammalian brain are described elsewhere (Kuznetsov and Gelfand, 2001).

2. Fluorescently labeled tubulin (5 mg/ml; Cytoskeleton, Inc., Denver, CO); store at −80°C.

3. Brightly fluorescent microtubule seeds. A detailed protocol to polymerize microtubule seeds was published before (Hyman, 1991; Nielsen *et al.*, 2001). This protocol uses GMP-CPP [guanylyl-(alpha, beta)methylenediphosphonate] to obtain short microtubule seeds. GMP-CPP is not available commercially and needs to be synthesized. Alternatively, we polymerize microtubule seeds from fluorescent tubulin, according to the procedure below, and obtain short seeds by shearing the microtubules.

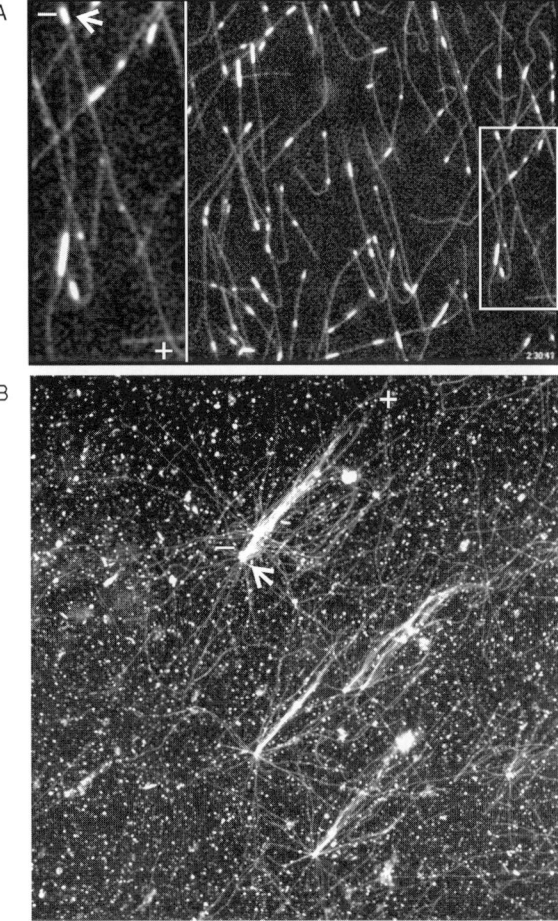

Fig. 2 Polarity-marked microtubules and microtubule asters. Fluorescent microtubules were poly-merized using brightly fluorescent microtubule seeds (A) or purified centrosomes (B). (A) The inset shown in the right panel is magnified two times at the left, and the seed (arrow) and plus (+) and minus (−)-ends of the microtubule are indicated. (B) A centrosome (arrow) and plus (+) and minus (−)-ends of the microtubules are indicated.

4. Glycerol.

5. BRB80 (Brinkley assembly buffer): 80-mM K-PIPES pH 6.8, 2-mM MgCl$_2$, and 1-mM EGTA; store at room temperature (RT).

6. 100-mM Taxol; solubilized in dimethyl sulfoxide (DMSO), and stored in aliquots at −20°C.

7. BRB80 + Taxol: BRB80 + 10-μM Taxol; prepare immediately before use.

8. 100-mM GTP. Dissolve in BRB80, readjust pH to 6.8, and store at −20°C.

Procedure

1. For polarity-marked microtubules, mix 2-μl microtubule seeds, 20-μl purified tubulin, and 2-μl fluorescent tubulin. For brightly fluorescent microtubule seeds, use 20-μl fluorescent tubulin. For moderately fluorescent, unmarked microtubules, mix 20-μl purified tubulin and 2-μl fluorescent tubulin.

2. Add GTP (to 1 mM) and glycerol (to 20%).

3. Incubate 30 min at 37°C in a warm water bath.

4. Bring volume to a final volume of 200 μl with BRB80 + Taxol buffer.

5. Sediment microtubules at 28 psi in an airfuge (Beckman) for 5 min, or at 100,000 × g in an ultracentrifuge at RT.

6. Gently resuspend the microtubule pellet in 100-μl BRB80 + Taxol, using care to avoid breaking the microtubules. For microtubule seeds, shear the microtubules by vigorous pipetting through a 100-μl yellow tip. Fluorescent microtubules can be kept 1 week at RT in the dark. Microtubule seeds are stored in aliquots at −80°C, after flash freezing in liquid nitrogen.

b. Adsorption of Polarity-Marked Microtubules to Observation Chambers
Materials

1. Observation chamber.
2. Taxol-stabilized, polarity-marked microtubules.
3. 0.01% Poly-L-ornithine (Sigma).
4. 5-mg/ml Casein in BRB80.

Procedure

1. Perfuse 10-μl 0.01% poly-L-ornithine in the observation chamber, adjust the chamber volume to 10 μl by gently pressing down the coverslip, and incubate 10 min at RT.

2. Wash 2× with 20-μl BRB80.

3. Perfuse 10-μl polarity-marked microtubules into the observation chamber and allow to bind for 5 min. Position the chamber with the coverslip facing down during incubation so that the microtubules preferentially attach to the coverslip.

4. Wash away unbound microtubules with casein solution and block the chamber with casein for 1 min.

3. Preparation of Microtubule Asters on Purified Centrosomes

In most interphase cells, microtubules are anchored with their minus-ends at the microtubule-organizing center (MTOC) in the center of the cell and radiate out with their plus-ends reaching for the cell periphery. Centrosomes, the main MTOC of mammalian cells, can be used as microtubule-nucleating centers *in vitro* to create microtubule asters. Since microtubules are attached to the centrosomes with their minus-ends and grow away from the centrosome, the polarity of microtubules in these asters is well defined (Fig. 2). Hence, *in vitro* motility of mitochondria on centrosomal arrays of microtubules can be used to define the direction of movement and to investigate changes in direction of motility under various regimens. Microtubule asters are synthesized *in situ* in the observation chamber.

a. Purification of Centrosomes from Human Lymphoblastic KE37 Cells

The purification of centrosomes has been described in detail before (Bornens and Moudjou, 1999; Bornens *et al.*, 1987; Mitchison and Kirschner, 1984, 1986). Our protocol is a combination and modification of those protocols.

Materials

1. Human lymphoblastic KE37 cells.
2. Culture medium: RPMI 1640, 10% fetal calf serum (FCS), 2-mM glutamine.
3. Nocodazole; 33-mM stock in DMSO, store at $-20\,^{\circ}$C.
4. Cytochalasin D; 10-mg/ml stock in DMSO, store at $-20\,^{\circ}$C.
5. Phosphate buffered saline (PBS).
6. 8% Sucrose in water (w/w).
7. 1/10 PBS + 8% (w/w) sucrose.
8. Lysis buffer: 1-mM Tris–HCl pH 8.0, 0.1% β-mercaptoethanol (β-ME), 0.5-mM MgCl$_2$, 1-mM phenylmethylsulphonyl fluoride (PMSF).
9. Lysis buffer + 1% NP40.
10. PIPES stock buffer (10×): 100-mM K-PIPES pH 7.2, 10-mM EDTA, and 1% β-ME.
11. 20% (w/w) Ficoll (MW 400,000) in 10-mM K-PIPES pH 7.2, 1-mM EDTA, 0.1% β-ME, and 0.1% NP40.
12. 70% (w/w) sucrose in 10-mM K-PIPES pH 7.2, 1-mM EDTA, 0.1% β-ME, and 0.1% Triton X-100.
13. 50% (w/w) sucrose in 10-mM K-PIPES pH 7.2, 1-mM EDTA, 0.1% β-ME, and 0.1% Triton X-100.
14. 40% (w/w) sucrose in 10-mM K-PIPES pH 7.2, 1-mM EDTA, 0.1% β-ME, and 0.1% Triton X-100.
15. MLS55 rotor and tabletop ultracentrifuge.
16. Refractometer.

Procedure

1. Grow 4 liters of KE37 cells in suspension culture ($\sim$10^6 cells/ml).
2. Before the start of centrosome isolation, incubate the cells with nocodazole (0.15 μM) and cytochalasin D (1 μg/ml) for 1.5 h at 37$\,^{\circ}$C.

3. The rest of the procedure is performed at 4°C.

4. Sediment cells by centrifugation (5 min, 1500 × *g* at 4°C). Resuspend the cell pellet in 200 ml of prechilled PBS, and transfer to four 50-ml tubes.

5. Wash cells once with 1/10 PBS + 8% sucrose, and once with 8% sucrose. The washing is performed by resuspending the cells in 50 ml of buffer and centrifugation for 3 min at 1500 × *g* at 4°C.

6. Gently resuspend the pellet to a final volume of 14 ml with ice-cold detergent-free lysis buffer, and divide the cell pellet in two equal parts in two 15-ml tubes.

7. It is important to perform the next steps as quickly as possible but without creating foam in the solution.

8. Add an equal volume of lysis buffer with 1% NP40, and lyse the cells by gently inverting the tubes until the solution clears.

9. Sediment unlysed cells, swollen nuclei, and chromatin by centrifugation (3 min, 1500 × *g* at 4°C).

10. Decant the supernatant into fresh tubes.

11. Add 1/10th volume of 10× PIPES stock buffer to the lysate.

12. Underlay 4 ml of lysate with 1.4 ml of a solution consisting of 20% Ficoll in 1× PIPES buffer, and 0.1% NP40 in an ultracentrifuge tube. (The volumes used are for the MLS55 rotor.)

13. Centrifuge for 15 min at 25,000 × *g* at 4°C.

14. Collect 1 ml at the interface between the two steps in the Ficoll gradient, and dilute with an equal volume of 1× PIPES buffer. This is the crude centrosome fraction.

15. Prepare a discontinuous sucrose gradient consisting, from the bottom up, of: 1-ml 70% sucrose, 0.5-ml 50% sucrose, and 1.5-ml 40% sucrose. Apply the crude centrosome fraction to the sucrose gradient and centrifuge at 100,000 × *g* for 1 h at 4°C.

16. Fractionate the sucrose gradient in ∼250-μl fractions.

17. Determine the sucrose content of the fractions using a refractometer. The centrosomes should be in the 50–65% sucrose fractions.

18. Test microtubule-nucleating activity of fractions obtained from the gradient, as described below.

c. In Situ *Polymerization of Microtubule Asters*
Materials

1. Observation chamber.
2. Purified centrosomes.
3. Purified tubulin (20 mg/ml).
4. Fluorescently labeled tubulin (5 mg/ml; Cytoskeleton, Inc.).

5. 100-mM GTP.

6. BRB80 (Section II.A.2.a).

7. BRB80 + Taxol: BRB80 + 10-μM Taxol; prepare immediately before use.

8. 5-mg/ml casein in BRB80.

9. Tubulin mix: mix 50-μl tubulin, 2-μl fluorescent tubulin, and 1-mM GTP, and adjust volume to 200 μl with BRB80.

Procedure

1. Flow 10 μl of 1-mM GTP in BRB80 into the observation chamber and adjust the chamber volume to 10 μl by gently pressing down the coverslip.

2. Flow 10 μl of centrosomes into the observation chamber and incubate 15 min at RT. Depending on the concentration of centrosomes they can be diluted in 1-mM GTP in BRB80.

3. To wash out unbound centrosomes, and to block the chamber, flow 20 μl of BRB80 + casein through the chamber.

4. Flow 10 μl of tubulin mix into the observation chamber, and incubate 1 h at 37°C.

5. To wash away the unpolymerized tubulin, flow 20 μl of BRB80 + Taxol into the observation chamber. Brisk washing will cause the centrosome-attached microtubules to orient themselves in the direction of flow (Fig. 2).

6. At this point the asters can be fixed by incubation with 1% gluturaldehyde in BRB80 for 5 min, followed by 20 μl of BRB80 + Taxol to wash out the gluturaldehyde.

4. Preparation of Microtubule–Associated Protein–Depleted High-Speed Cytosol

We use high-speed cytosol that has been depleted of microtubule-associated proteins (MAPs) as a source of motors and factors that stimulate transport of mitochondria along microtubules. In principle, high-speed cytosol can be obtained from any source of cells or tissue. However, for reasons that are not clear, some cytosol preparations support motility whereas others do not. We routinely test the activity of motors in cytosol preparations by performing microtubule-gliding assays (Carter and Cross, 2001). Although not strictly necessary, we remove MAPs by adsorption to microtubules, because it was shown that MAPs can influence mitochondrial transport along microtubules (Ebneth *et al.*, 1998; Lopez and Sheetz, 1993). The following protocol was used successfully to prepare active cytosol from L929 and N2A cells (De Vos *et al.*, 2000; K.D.V., unpublished results), and should be suitable for use with other cell lines.

a. Preparation of High-Speed Cytosol
Materials

1. N2A or L929 cells.

    2. PBS.

    3. PMEE buffer: 35-mM K-PIPES pH 7.4, 5-mM $MgSO_4$, 5-mM EGTA, 0.5-mM EDTA, 1-mM DTT, and 1-mM PMSF (Blocker *et al.*, 1997).

    4. Tissue grinder: pestle 19 (Kontes, Vineland, NJ), or a ball bearing cell cracker: 8.020-mm size, with a single 0.8002-mm ball bearing (HGM, Heidelberg, Germany).

Procedure

    1. Grow two to three 175-cm^2 flasks of cells to confluence, harvest the cells, and wash three times with ice-cold PBS. The remainder of the procedure is performed at 4°C.

    2. Resuspend the cell pellet in 0.9× the pellet volume using ice-cold PMEE.

    3. Homogenize the cell suspension with a tissue grinder (200 strokes) or a ball bearing cell cracker (5 strokes) on ice.

    4. Remove unbroken cells and nuclei by centrifugation at 3000 × g for 10 min at 4°C.

    5. Centrifuge the supernatant obtained at 200,000 × g for 30 min at 4°C. The supernatant obtained is high-speed cytosol, and can be stored up to 6 months at −80°C after snap freezing in liquid nitrogen. The activity of the samples will decline over time in storage.

b. Depletion of MAPs from High–Speed Cytosol by Microtubule Affinity
Materials

    1. High-speed cytosol.
    2. Purified tubulin (20 mg/ml).
    3. 100-mM GTP.
    4. Glycerol.
    5. 10-mM Taxol.
    6. BRB80.
    7. BRB80 + Taxol: BRB80 + 10-μM Taxol; prepare immediately before use.
    8. 100-mM Mg-ATP. Dissolve ATP in BRB80, readjust to pH 6.8, and add 100-mM $MgSO_4$. Store in aliquots at −20°C.

Procedures

    (i) *Synthesis of unlabeled Taxol-stabilized microtubules*:

    1. Polymerize microtubules from 50-μl purified, unlabeled tubulin by incubation for 30 min at 37°C in the presence of 1-mM GTP and 15-μl glycerol.

    2. After incubation, add BRB80-Taxol buffer to the polymerized microtubules until the final volume is 200 μl.

    3. Sediment polymerized microtubules by centrifugation at 28 psi in an airfuge for 5 min or at 100,000 × g in an ultracentrifuge at RT.

4. Gently resuspend the microtubule pellet in 50-μl BRB80-Taxol, using care to avoid breaking the microtubules. Microtubules can be kept for 1 week at RT, or stored at $-80\,^{\circ}$C after snap freezing in liquid nitrogen.

(ii) *MAP depletion of high-speed cytosol*:

1. Add Mg-ATP to 4 mM, Taxol to 10 μM, and GTP to 1 mM to the high-speed cytosol. (Adding ATP reduces binding of molecular motors to the microtubules.) Thereafter, add 20 μl of unlabeled, Taxol-stabilized microtubules. Alternatively, endogenous tubulin, which will polymerize after addition of Mg-ATP, Taxol, and GTP to the high-speed cytosol, may be sufficient to deplete MAPs.

2. Incubate the cytosol with microtubules for 15 min at $37\,^{\circ}$C.

3. Sediment the microtubules and microtubule-bound MAPs by centrifugation ($160{,}000 \times g$, 20 min at $20\,^{\circ}$C).

4. Keep the MAP-depleted cytosol on ice until use (not longer than 2 h) or snap freeze in liquid nitrogen and store at $-80\,^{\circ}$C up to 6 months. Activity of the MAP-depleted high-speed cytosol will decline over time.

5. Isolation of Mitochondria for *In Vitro* Motility Assays

We label the mitochondria with fluorescent proteins or dyes before isolation. As a result, it is easy to detect the mitochondria by fluorescence microscopy even if other nonlabeled organelles remain present. Therefore, this procedure is not designed to yield high purity mitochondria, but rather for speed of preparation. Speed is essential, as the suitability of mitochondria for *in vitro* motility assays declines rapidly after their isolation.

Materials

1. N2A, CV-1, or L929 cells.
2. 1-mM MitoTracker GreenFM; store at $-20\,^{\circ}$C (Invitrogen, Molecular Probes, Paisley, UK).
3. Mitochondrial isolation (MI) buffer: 10-mM HEPES pH 7.4, 220-mM mannitol, 70-mM sucrose, 1-mM EDTA, 1-mM DTT, and 1-mM PMSF.
4. Tissue grinder: pestle 19 (Kontes), or a ball bearing cell cracker: 8.020-mm size, with a 0.8002-mm ball bearing.

Procedure

1. Grow a 75-cm^2 flask of cells to confluence.
2. Label mitochondria with 1-μM MitoTracker GreenFM for 30 min under culture conditions. Alternatively, prepare mitochondria from cells expressing a mitochondria-targeted fluorescent protein.
3. Harvest the cells and transfer to $4\,^{\circ}$C for the rest of the procedure.
4. Wash twice with ice-cold PBS and once with ice-cold MI buffer.

5. Resuspend the washed cell pellet in 5 volumes of ice-cold MI buffer and homogenize with a tissue grinder (200 strokes) or a ball bearing cell cracker (5 strokes).

6. Sediment unbroken cells and nuclei by centrifugation (3000 × g, 10 min at 4 °C).

7. Centrifuge the supernatant obtained at 13,000 × g for 10 min at 4 °C.

8. Resuspend the pellet obtained, which is enriched in mitochondria, in 40-μl MI buffer and keep on ice. Mitochondria should be freshly isolated before each motility experiment and used within 2–3h after isolation.

B. Motility Assays

Materials

1. Observation chamber with absorbed polarity-marked microtubules or pre-formed microtubule asters.
2. MAP-depleted cytosol (2 μg/μl).
3. ATP/Taxol stock solution (5×): 25-mM ATP, 25-mM MgSO$_4$, and 50 μM Taxol in PMEE.
4. PMEE.

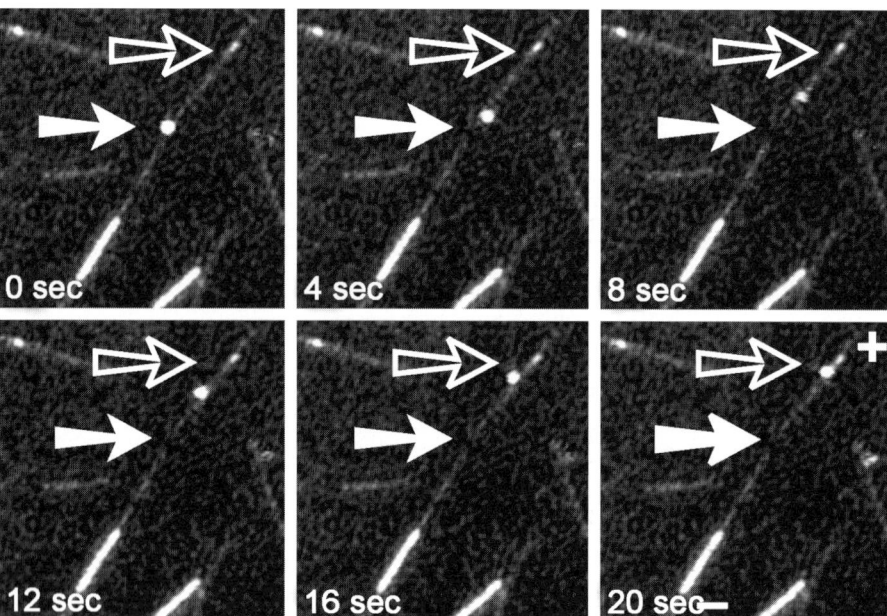

Fig. 3 Mitochondrial motility on polarity-marked microtubules. Still frames from a time-lapse series showing mitochondrial motility on a polarity-marked microtubule. The position of the mitochondria at $t = 0$ sec and $t = 20$ sec is indicated with a *closed* and *open arrow*, respectively. Although mitochondria are tubular structures in many cells, the organelle is fragmented into small spherical structures under conditions that are routinely used for cell homogenization and subcellular fractionation. The plus (+) and minus (−)-ends of the polarity-marked microtubule are indicated.

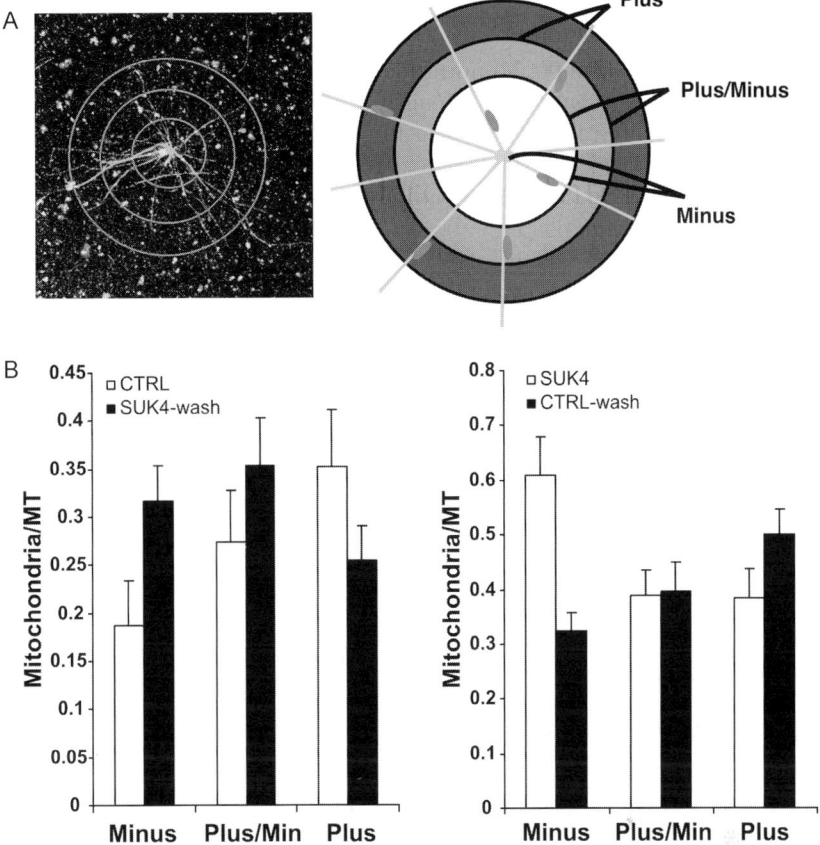

Fig. 4 Centrosomal assay of mitochondrial motility. (A) Experimental setup: Microtubules (green) polymerized from isolated centrosomes have a defined polarity, with their minus-ends at the centrosome and the plus-ends radiating outward. Fluorescently labeled, isolated mitochondria (red) were added in the presence of cytosol and ATP. After 1-h incubation at $37\,^{\circ}C$, the samples were observed by confocal laser scanning microscopy. For analysis, three radial zones (minus, plus/minus, and plus) were defined on the microtubule aster and the number of mitochondria in each zone was determined. (B) Reversible inhibition of plus-end-directed mitochondrial motility. Mitochondria were incubated with untreated, control cytosol (*left, open bars*). This resulted in a mitochondrial distribution biased toward the plus-end of the microtubules. Subsequent application of cytosol that was pretreated with kinesin-blocking SUK4 mAb resulted in a significant shift of the mitochondrial distribution to the minus-end of the micro-tubules (*left, filled bars*). In a complementary approach, the centrosome-based mitochondrial motility assay was performed using SUK4 mAb-treated cytosol (*right, open bars*). As expected, the majority of mitochondria were found at the minus-end of the microtubules. Replacement of SUK4 mAb-treated cytosol with control cytosol in the same sample restored plus-end motility (*right, filled bars*). (See Color Insert.)

Procedure

    1. Wash the observation chamber with 20-μl PMEE.

    2. Perfuse reaction mix containing 4-μl MAP-depleted cytosol, 4-μl isolated mitochondria, and 2-μl ATP/Taxol solution into the observation chamber.

    3. Record motility by time-lapse microscopy at RT or at 37°C. Alternatively, incubate 30–60 min at 37°C in an incubator, followed by fixation with 1% glutur-aldehyde.

C. Analysis of Mitochondrial Motility

    Both assays can be used to observe mitochondrial motility by time-lapse micros-copy (Figs. 3 and 5). In this case, analysis can be performed as discussed for mitochondrial motility in living animal cells (see Chapter 30 by De Vos and Sheetz, this volume). In addition, the distribution of mitochondria on the microtubule asters can be determined after fixation of the samples (Fig. 4). Furthermore, these assays can be used to evaluate binding of mitochondria to microtubules by quantifying the number of mitochondria per micrometer microtubule.

III. Sedimentation Assay for Binding of Actin to Mitochondria

    The sedimentation assay for binding of mitochondria with purified F-actin was originally used to document ATP-sensitive actin-binding activity on the surface of isolated yeast mitochondria (Lazzarino *et al.*, 1994). In addition, comparing actin-binding activities of wild-type mitochondria with those of mutant mitochon-dria revealed some of the molecular participants in mitochondria–actin interac-tions (Boldogh *et al.*, 1998). In a typical assay, isolated yeast mitochondria are incubated with phalloidin-stabilized yeast actin filaments in the presence or absence of ATP. Thereafter, mitochondria are separated from the reaction mix-ture by low-speed centrifugation through a sucrose cushion. Proteins recovered in low-speed pellets are identified by Western blot analysis, using antibodies raised against actin and mitochondrial marker proteins. Under these low-speed centrifu-gation conditions, all unbound F-actin is recovered in the supernatant. The level of porin, an integral mitochondrial outer membrane protein, is determined in the mitochondrial pellet to confirm equal recovery of mitochondria after low-speed centrifugation. The level of cytochrome b_2, a peripheral membrane protein that is on the intermembrane space side of the inner membrane, is determined in the mitochondrial pellet to confirm the integrity of the organelle. Here, any actin recov-ered in the mitochondrial pellet is bound to the organelle. A schematic outlining this assay is shown in Fig. 5A.

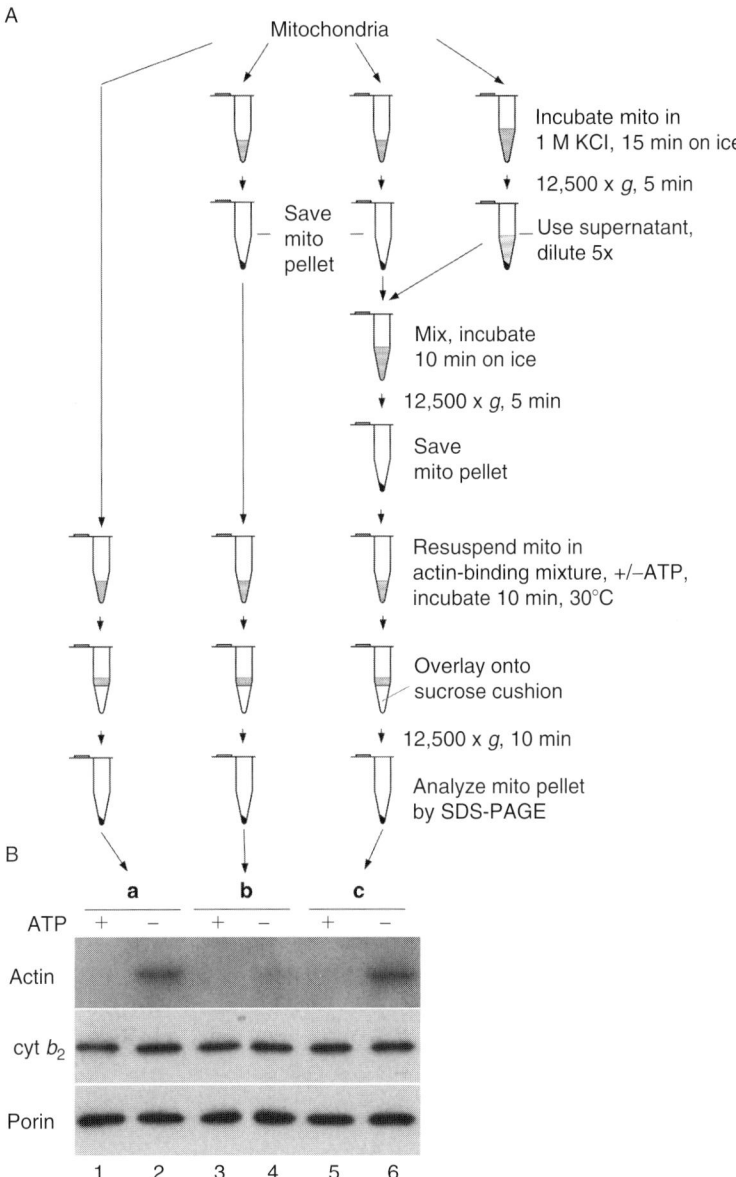

Fig. 5 Sedimentation assays of mitochondria isolated from D273-10B yeast with phalloidin-stabilized F-actin in the presence or absence of ATP. (A) Schematic illustration of sedimentation assay using untreated and salt-treated mitochondria. (B) SDS-PAGE and Western blot analysis of levels of actin and of two mitochondrial marker proteins (cytochrome b_2 and porin) in mitochondrial pellets of the sedimented reaction mixtures. Actin was detected using a monoclonal anti-actin antibody, c4d6 (Lessard, 1988). Recovery and integrity of mitochondria after sedimentation were determined with poly-clonal anti-cytochrome b_2 and anti-porin antibodies (gifts from G. Schatz). Lanes 1 and 2, untreated

A. Reagents

1. Stock Solutions

1. 1-M MgCl$_2$. Dissolve in water, autoclave and store at 4°C.
2. 1-M DTT. Dissolve in water and store in aliquots at −20°C.
3. 100-mM ATP. Dissolve in water, adjusted to pH 7.0 with KOH, and store in aliquots at −20°C.
4. 500-U/ml apyrase (Sigma, A6410). Dissolve in 20-mM HEPES-KOH pH 7.4, and store in aliquots at −20°C.
5. 5-mg/ml creatine kinase (Sigma, C3755). Dissolve in 50% glycerol, and store in aliquots at −20°C
6. 0.5-M creatine phosphate. Dissolve in water and store in aliquots at −20°C.
7. 2.4-M sorbitol. Autoclave and store at 4°C.
8. 2.5-M KCl. Autoclave and store at 4°C.
9. 50% sucrose. Dissolve in water and store at −20°C.
10. 1-M HEPES pH 7.4; pH is adjusted with KOH. Autoclave and store at 4°C.
11. 0.5-mg/ml phalloidin (Roche, Indianapolis, IN). Dissolve in ethanol. Store at −20°C.
12. 200-mM PMSF (Roche, Indianapolis, IN). Dissolve in ethanol and store in aliquots for up to several months at −20°C.
13. Protease inhibitor (PI) cocktails (all reagents are from Sigma, St. Louis, MO).
    (i) PI-1 (1000× stock):
       0.5-mg/ml pepstatin A (P4265).
       0.5-mg/ml chymostatin (C7268).
       0.5-mg/ml antipain (A6191).
       0.5-mg/ml leupeptin (L2884).
       0.5-mg/ml aprotinin (A1153).

Prepare 1000× stock from five starting solutions. Dissolve pepstatin A in DMSO, chymostatin in DMSO; antipain in water; leupeptin in water, and aprotinin in water, each at a final concentration of 10 mg/ml. Mix equal volumes of all five solutions and add water so that the final concentration of each PI in the mixture is 0.5 mg/ml.

    (ii) PI-2 (1000× stock):
       10-mM benzamidine-HCl (B6506).

---

mitochondria; lanes 3 and 4, salt-washed mitochondria; and lanes 5 and 6, salt-washed mitochondria pretreated with mitochondrial salt extract. Sample a, which is a control for salt washing, contains mitochondria that were incubated in buffer without KCl for 15 min on ice. Mitochondria in samples b and c were incubated in 1-M KCl. In all cases 170-µg mitochondria was loaded for actin detection and 16-µg mitochondria was loaded for cytochrome b_2 and porin detection.

1-mg/ml 1,10-phenanthroline (P9375).

Dissolve together in 100% ethanol.

Divide both PI-1 and PI-2 stock solutions into small aliquots. They can be stored for several months at $-20\,^{\circ}C$.

2. Working Solutions

1. RM0 buffer: 0.6-M sorbitol, 20-mM HEPES pH 7.4, 2-mM $MgCl_2$, $1\times$ PI-1, $1\times$ PI-2, 1-mM PMSF.

2. RM1 buffer: 0.6-M sorbitol, 20-mM HEPES pH 7.4, 2-mM $MgCl_2$, 1-M KCl, $1\times$ PI-1, $1\times$ PI-2, 1-mM PMSF.

3. RM buffer: 0.6-M sorbitol, 20-mM HEPES pH 7.4, 2-mM $MgCl_2$, 0.1-M KCl, 1-mM DTT, 1-mg/ml bovine serum albumin (fatty-acid free), $1\times$ PI-1, $1\times$ PI-2, 1-mM PMSF.

4. Sucrose cushion solution: 25% (w/v) sucrose, 20-mM HEPES-KOH pH 7.4, $1\times$ PI-1, $1\times$ PI-2, 1-mM PMSF.

5. $3\times$ SDS sample buffer: 0.19-M Tris–HCl, pH 6.8, 12% (v/w) SDS, 30% (v/v) glycerol, 15% (v/v) β-ME, 0.015% (v/w) bromophenol blue.

Note: All buffers containing PMSF are filtered through Whatman No. 4 filter paper, since adding PMSF to aqueous solution results in some level of precipitation, which might interfere with the sedimentation assay.

3. Mitochondria

For the mitochondria–actin-binding assay, we use Nycodenz-purified mitochondria that are flash frozen and stored in aliquots in liquid nitrogen. Mitochondria are stored at a concentration of 5- to 20-mg protein/ml in a buffer consisting of 0.6-M sorbitol, 20-mM HEPES-KOH, pH 7.4, 2-mM $MgCl_2$, 1-mM EGTA, pH 8.0, 200-mM PMSF, $1\times$ PI-1, $1\times$ PI-2, and 15–17% Nycodenz. A detailed description of the method used to isolate mitochondria from budding yeast is in Chapter 2 by Boldogh and Pon, this volume. Only freshly thawed and reisolated mitochondria should be used in the sedimentation assay, since prolonged storage of the organelle on ice leads to decreased actin-binding activity. All protein concentration measurements are done by BCA protein assay (Pierce, Rockford, IL).

Procedure

1. Thaw frozen mitochondrial suspension at $30\,^{\circ}C$.

2. Dilute the mitochondrial suspension with 5 volumes of ice-cold RM0 buffer. Concentrate mitochondria by centrifugation at $12,500 \times g$ for 5 min at $4\,^{\circ}C$.

3. Discard the supernatant obtained, and resuspend the mitochondrial pellet by pipetting gently in RM buffer to a final protein concentration of 10 mg/ml.

4. Preparation of Phalloidin–Stabilized F-actin

F-actin is purified from yeast as described by Boldogh and Pon (2001). The filamentous actin that is used in this assay is stabilized with phalloidin. We find that G-actin undergoes rapid polymerization, and that F-actin undergoes rapid depolymerization when incubated with mitochondria under the binding assay conditions. These actin dynamics are presumably due to Arp2/3 complex and other actin-binding proteins that are associated with yeast mitochondria. Therefore, treatment with F-actin with phalloidin is critical to maintain all added actin in the filamentous form and to obtain reproducible binding results.

Procedure

1. Prepare phalloidin-stabilized F-actin immediately before use in the binding assay.

2. Add 1/50 volume of 0.5-mg/ml phalloidin to 2-mg/ml F-actin solution (Section II), and incubate on ice for 30 min.

3. Dilute sample with 1 volume of $2\times$ RM buffer, and centrifuge the solution at $12,500 \times g$ for 10 min at $4°C$ to remove possible F-actin aggregates and to minimize nonspecific cosedimentation of F-actin with mitochondria.

4. Transfer supernatant obtained to a fresh tube and store on ice.

B. Binding Assay

A standard mitochondria-actin-binding assay is carried out in RM buffer in 80-μl final volume and contains mitochondria and phalloidin-stabilized F-actin. Since mitochondria exhibit ATP-sensitive actin-binding activity, ATP is depleted from samples using apyrase, an enzyme that catalyzes hydrolysis of ATP to ADP and of ADP to AMP. An ATP-treated sample is also included, as a control. In this sample, ATP levels are maintained at 2 mM by addition of ATP and an ATP-regenerating system consisting of creatine kinase and creatine phosphate.

In the assay described, F-actin is present at saturating levels, and mitochondria are present at levels that are not saturating. For binding mitochondria isolated from the yeast strain D273-10B to yeast F-actin, these concentrations are 200 and 15 μg, respectively. When working with a mutant or yeast strain for the first time, or for quantitative comparison of actin-binding activity of different mitochondria, we recommend time course experiments and/or titrating the level of mitochondria in the binding assay. Since the assay measures binding of mitochodria to actin polymers of variable and unknown length, it cannot be used to determine binding constants for mitochondrial-actin interactions. However, it can be used to identify actin-binding activity on mitochondria and mutations or treatments that inhibit this activity.

Procedure

The assay described has two samples: one incubated with ATP and another without (Fig. 5B, lanes 1–2). A mitochondria-free sample is an important control to confirm that the actin used does not sediment under the conditions used. An actin-free control is also useful to confirm that the actin in the mitochondrial pellet is added actin and not actin that was recovered with mitochondria on subcellular fractionation. Other controls and methods to evaluate the disposition of the actin-binding activity on mitochondrial membranes are described below.

Stock solution	Concentration in assay	Volume of stock solution for	
		Sample (+ATP)	Sample (−ATP)
2× RM buffer	1×	22.5 μl	22.5 μl
Water		17.7 μl	20.9 μl
ATP (100 mm)	2 mm	1.6 μl	–
Creatine kinase (5 mg/ml)	0.1 mg/ml	1.6 μl	–
Creatine phosphate (0.5 M)	10 mm	1.6 μl	–
Apyrase (500 U/ml)	12.5 U/ml	–	2 μl
F-actin (1 mg/ml)	188 μg/ml	15 μl	15 μl
Mitochondria (10 mg/ml)	2.5 mg/ml	20 μl	20 μl

1. Add all reagents listed, with the exception of mitochondria, to microfuge tubes. Vortex and let stand on ice.

2. Add mitochondria and incubate the samples for 10 min at 30°C.

3. Terminate the reaction by transferring the reaction mixtures to ice.

4. Overlay the reaction on 500 μl of ice-cold sucrose cushion in a fresh microfuge tube, and separate mitochondria from unbound actin by centrifugation at $12,500 \times g$ for 10 min at 4°C. (Sedimenting mitochondria through a sucrose cushion allows for the efficient separation of mitochondria and bound actin from unbound actin, which remains on the top of the cushion.)

5. Carefully remove the supernatant obtained.

6. Gently overlay 100 μl of ice-cold RM0 buffer onto the surface of mitochondrial pellet. This "wash" is then removed by aspiration without disturbing the integrity of the pellet.

7. Resuspend the mitochondrial pellet in 20 μl of RM0 buffer.

8. Add 10 μl of 3× SDS sample buffer, vortex, and heat sample to 95°C for 3 min.

9. Analyze solublized samples by SDS-PAGE and Western blots using antibodies against actin and the mitochondrial marker proteins porin and cytochrome b_2. Usually 25 μl of this mixture is used to detect actin and 2.5 μl is used to detect the mitochondrial markers.

C. Salt Treatment of Mitochondria

High ionic strength (1-M KCl) facilitates dissociation of loosely associated, peripheral membrane proteins from the mitochondrial membrane surface, but does not extract integral membrane proteins or affect the integrity of the organelle.

Therefore, the role of mitochondrial peripheral membrane proteins on actin binding to the organelle can be evaluated by testing the actin-binding activity of salt-washed mitochondria. Treatment of a 33-mg/ml mitochondria solution with 1-M KCl usually yields 0.5- to 0.6-mg/ml salt-extracted proteins in the supernatant. We find that salt washing results in loss of all mitochondrial–actin-binding activity (Fig. 5B, lanes 3 and 4). This activity, however, can be restored by incubation of salt-washed mitochondria with salt-extracted membrane proteins in a buffer containing 200-mM KCl. To regenerate the actin-binding activity of salt-washed mitochondria to the level of untreated mitochondria, salt-washed mitochondria must be incubated with 6–10 times the amount of salt extract as was removed by salt treatment (Fig. 5B lanes 5 and 6). Collectively, these findings indicate a peripheral protein component on the outer membrane in the mitochondria–actin interactions. We refer to this activity as mABP (mitochondria-actin-binding activity).

The mitochondria–actin-binding assay is also useful to identify integral protein components that are required for mitochondria–actin interactions. Isolated mitochondria from mutants of two integral outer membrane proteins, Mmm1p and Mdm10p, showed no actin-binding activity and no restoration of actin-binding activity to salt-washed mitochondria on incubation with salt extract from wild-type mitochondria. These properties of Mmm1p and Mdm10p mutants indicate that Mmm1p and Mdm10p might function as docking proteins for mABP (Boldogh *et al.*, 1998).

Protocols

(i) *Isolation of salt-washed mitochondria*

1. Thaw a frozen mitochondrial suspension at 30°C and dilute it with 5 volumes of RM0 buffer.

2. Remove two aliquots from the suspension, each containing 200 μg of mitochondrial protein. One aliquot will be the salt-washed mitochondria. The other will be retained as a control.

3. Concentrate mitochondria from each aliquot by centrifugation at 12,500 × g for 5 min at 4°C.

4. Resuspend the pellet of salt-washed mitochondria in ice-cold RM1 buffer to a final concentration of 10 mg/ml. Resuspend the other pellet in ice-cold RM buffer to a final concentration of 10 mg/ml.

5. Incubate these aliquots for 15 min on ice.

6. Separate mitochondria by centrifugation at 12,500 × g for 5 min at 4°C.

7. Discard the supernatant obtained, and resuspend the mitochondrial pellets in RM buffer to a final concentration of 10 mg/ml. These samples are used directly for the binding assay. If salt-washed mitochondria are to be regenerated with salt extract before the binding assay, resuspend the salt-washed mitochondrial pellet in RM0 buffer rather than RM buffer to a final protein concentration of 10 mg/ml.

(ii) *Preparation of mitochondrial salt-extractable membrane proteins*

1. Thaw a frozen mitochondrial suspension at 30 °C and dilute it with 5 volumes of RM0 buffer.
2. Concentrate mitochondria by centrifugation at 12,500 × *g* for 5 min at 4 °C.
3. Resuspend the mitochondrial pellet in RM1 buffer to a final concentration of 33 mg/ml.
4. Incubate for 15 min on ice.
5. Separate the salt extract from salt-washed mitochondria by centrifugation at 12,500 × *g* for 5 min at 4 °C. Save supernatant (salt extract).
6. Dilute the KCl concentration of supernatant to 0.2 M by adding 4 volumes of RM0 buffer. This solution is used for restoration of the actin-binding activity of salt-washed mitochondria.

(iii) *Restoration of actin-binding activity of salt-washed mitochondria*

1. Mix 300-μl diluted salt extract (see step ii.6) with 200-μg salt-washed mitochondria (see step i.6).
2. Incubate this mixture for 15 min on ice.
3. Concentrate mitochondria by centrifugation at 12,500 × *g* for 5 min, and resuspend the pellet obtained with ice-cold RM buffer to a final protein concentration of 10 mg/ml. This suspension is used for the binding assay.

D. Analysis of Mitochondria–Actin Binding

1. Image Densitometry

To quantitate the amount of actin and mitochondrial marker protein, the chemiluminescence of bands from Western blots is captured on a Kodak DS Image Station 440CF (Eastman Kodak Company, Rochester, NY). The images are analyzed using the Kodak 1D Image Analysis Software. To do so, an appropriate region of interest (ROI) is made that covers each of the individual bands and their mean gray values are measured. Signal from an area that is the same size as the ROI and is either above or below the band of interest is measured as an indicator of background gray level. The background gray value is subtracted from the gray value of the corresponding band. There are other methods for defining background for ROIs: for example, the mean gray levels of the perimeter of ROI can be used as background. The choice of method depends on how even or uneven the staining is in the background.

For quantitation of actin binding, the density of the band of actin recovered in the mitochondrial pellet is compared with the densities of known amounts of actin standards. Our standards range from 50- to 500-ng actin per lane. We find that usually 200 μg of mitochondria bind to 100- to 150-ng actin under our standard assay conditions. For accuracy, it is important to have actin standards with the samples on the same gel and to use only the data points that are in the linear range of the calibration curve. This usually requires loading different amounts of the same samples.

2. Establishing Correlation Between Actin–Binding Activity and Mitochondrial Purity

We use Nycodenz-purified mitochondria for our assay. Low levels of membrane contaminants are present in the preparation. For this reason, it is important to compare the fractionation pattern of mitochondria–actin-binding activity with that of mitochondria and other membrane compartments. In different cellular fractions, the levels of selected membrane marker proteins are evaluated by Western blot analysis and compared with actin–binding activity (Fig. 6). Comparison of the amount of marker proteins and actin binding revealed that enrichment of the mitochondrial marker and removal of the membrane markers

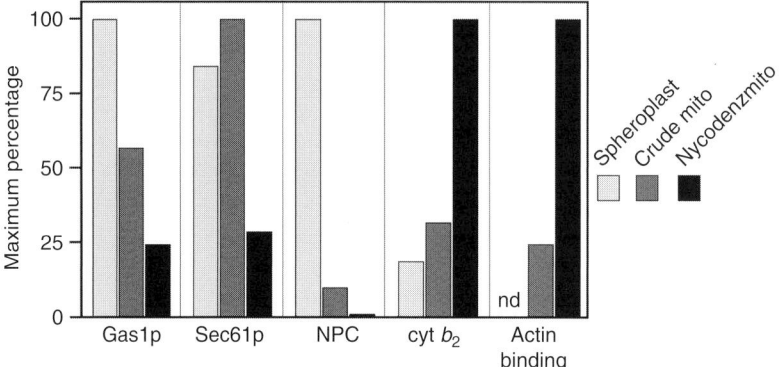

Fig. 6 Relative levels of membrane marker proteins and actin-binding activity in spheroplasts, crude mitochondria, and gradient-purified mitochondria. Spheroplasts, crude mitochondria, and gradient-purified mitochondria fractions were subjected to centrifugation at $12,500 \times g$ for 5 min at $4\,^{\circ}$C. The pellets were resuspended in RMO buffer and the protein concentrations of the samples were determined with the BCA assay (Pierce, Rockford, IL) using bovine serum albumin as a standard. Since no purified membrane marker proteins were available for standard curves, only relative levels of marker proteins were determined in the three fractions. To do so, three different amounts of each fraction were analyzed by SDS-PAGE and Western blot. The gray values of the bands that reflected the loading differences in all three fractions on the same blot were used to normalize the relative amount of membrane markers in these fractions. For each marker protein, 100% was defined as the maximum amount of marker per milligram of total protein. The marker proteins used were: Gas1p, a glycosyphophatidylinositol-anchored protein that is transported through the endoplasmic reticulum (ER), Golgi apparatus, and secretory vesicles prior to insertion into the plasma membrane (Nuoffer *et al.*, 1993); Sec61p, an ER integral membrane protein (Stirling *et al.*, 1992); NPC, nuclear pore complex proteins (Davis and Blobel, 1986); and cytochrome b_2 (cyt b_2), a mitochondrial inner membrane protein. The relative actin-binding activity per milligram protein was determined in the crude and gradient-purified mitochondria. Hundred percent was defined as the amount of binding activity in the sample containing the highest specific activity. nd: not determined.

examined in Nycodenz-purified mitochondria compared to those in crude mitochondria and in whole-cell extracts (spheroplasts). This increase in the purity of mitochondria correlates with an enrichment in the ATP-sensitive actin-binding activity. Thus, the ATP-sensitive actin-binding activity detected in the isolated organelles resides in mitochondria and not in membrane contaminants within the mitochondrial preparation.

Acknowledgments

We thank members of the Pon laboratory and R. Milner for critical evaluation of the manuscript. Work on these protocols was supported by research grants from the National Institutes of Health to L.P. (GM45735 and GM66307) and M.S. (GM036277). K.D.V. is supported by a BBRC grant (to A. Grierson, University of Sheffield) and by an ALSA starter grant.

References

Blocker, A., Severin, F. F., Burkhardt, J. K., Bingham, J. B., Yu, H., Olivo, J. C., Schroer, T. A., Hyman, A. A., and Griffiths, G. (1997). Molecular requirements for bi-directional movement of phagosomes along microtubules. *J. Cell Biol.* **137,** 113–129.

Boldogh, I., Vojtov, N., Karmon, S., and Pon, L. A. (1998). Interaction between mitochondria and the actin cytoskeleton in budding yeast requires two integral mitochondrial outer membrane proteins, Mmm1p and Mdm10p. *J. Cell Biol.* **141,** 1371–1381.

Boldogh, I. R., and Pon, L. A. (2001). Assaying actin-binding activity of mitochondria in yeast. *Methods Cell Biol.* **65,** 159–173.

Boldogh, I. R., Ramcharan, S. L., Yang, H. C., and Pon, L. A. (2004). A type V myosin (Myo2p) and a Rab-like G-protein (Ypt11p) are required for retention of newly inherited mitochondria in yeast cells during cell division. *Mol. Biol. Cell.* **15,** 3994–4002.

Boldogh, I. R., Yang, H. C., Nowakowski, W. D., Karmon, S. L., Hays, L. G., Yates, J. R., III, and Pon, L. A. (2001). Arp2/3 complex and actin dynamics are required for actin-based mitochondrial motility in yeast. *Proc. Natl. Acad. Sci. USA* **98,** 3162–3167.

Bornens, M., and Moudjou, M. (1999). Studying the composition and function of centrosomes in vertebrates. *Methods Cell Biol.* **61,** 13–34.

Bornens, M., Paintrand, M., Berges, J., Marty, M. C., and Karsenti, E. (1987). Structural and chemical characterization of isolated centrosomes. *Cell Motil. Cytoskeleton* **8,** 238–249.

Carter, N., and Cross, R. (2001). An improved microscope for bead and surface-based motility assasy. *In* "Kinesin Protocols" (I. Vernos, ed.), Vol. 164, pp. 73–89. Humana Press, Totowa, NJ.

Chada, S. R., and Hollenbeck, P. J. (2004). Nerve growth factor signaling regulates motility and docking of axonal mitochondria. *Curr. Biol.* **14,** 1272–1276.

Davis, L. I., and Blobel, G. (1986). Identification and characterization of a nuclear pore complex protein. *Cell* **45,** 699–709.

De Vos, K., Severin, F., Van Herreweghe, F., Vancompernolle, K., Goossens, V., Hyman, A., and Grooten, J. (2000). Tumor necrosis factor induces hyperphosphorylation of kinesin light chain and inhibits kinesin-mediated transport of mitochondria. *J. Cell Biol.* **149,** 1207–1214.

Ebneth, A., Godemann, R., Stamer, K., Illenberger, S., Trinczek, B., and Mandelkow, E. (1998). Overexpression of tau protein inhibits kinesin-dependent trafficking of vesicles, mitochondria, and endoplasmic reticulum: Implications for Alzheimer's disease. *J. Cell Biol.* **143,** 777–794.

Fehrenbacher, K. L., Boldogh, I. R., and Pon, L. A. (2005). A role for jsn1p in recruiting the arp2/3 complex to mitochondria in budding yeast. *Mol. Biol. Cell* **16,** 5094–5102.

Fehrenbacher, K. L., Yang, H. C., Gay, A. C., Huckaba, T. M., and Pon, L. A. (2004). Live cell imaging of mitochondrial movement along actin cables in budding yeast. *Curr. Biol.* **14,** 1996–2004.

Hollenbeck, P. J., and Saxton, W. M. (2005). The axonal transport of mitochondria. *J. Cell Sci.* **118,** 5411–5419.

Hyman, A. A. (1991). Preparation of marked microtubules for the assay of the polarity of microtubule-based motors by fluorescence. *J. Cell Sci. Suppl.* **14,** 125–127.

Kuznetsov, S. A., and Gelfand, V. I. (2001). Purification of kinesin from brain. *In* "Kinesin Protocols" (I. Vernos, ed.), Vol. 164, pp. 1–7. Humana Press, Totowa, NJ.

Lazzarino, D. A., Boldogh, I., Smith, M. G., Rosand, J., and Pon, L. A. (1994). Yeast mitochondria contain ATP-sensitive, reversible actin-binding activity. *Mol. Biol. Cell* **5,** 807–818.

Lessard, J. L. (1988). Two monoclonal antibodies to actin: One muscle selective and one generally reactive. *Cell Motil. Cytoskeleton* **10,** 349–362.

Ligon, L. A., and Steward, O. (2000). Role of microtubules and actin filaments in the movement of mitochondria in the axons and dendrites of cultured hippocampal neurons. *J. Comp. Neurol.* **427,** 351–361.

Lopez, L. A., and Sheetz, M. P. (1993). Steric inhibition of cytoplasmic dynein and kinesin motility by MAP2. *Cell Motil. Cytoskeleton* **24,** 1–16.

Minin, A. A., Kulik, A. V., Gyoeva, F. K., Li, Y., Goshima, G., and Gelfand, V. I. (2006). Regulation of mitochondria distribution by RhoA and formins. *J. Cell Sci.* **119,** 659–670.

Mitchison, T., and Kirschner, M. (1984). Microtubule assembly nucleated by isolated centrosomes. *Nature* **312,** 232–237.

Mitchison, T. J., and Kirschner, M. W. (1986). Isolation of mammalian centrosomes. *Methods Enzymol.* **134,** 261–268.

Nielsen, E., Severin, F., Hyman, A., and Zerial, M. (2001). *In vitro* reconstitution of endosome motiltiy along microtubules. *In* "Kinesin Protocols" (I. Vernos, ed.), Vol. 164, pp. 133–146. Humana Press, Totowa, NJ.

Nuoffer, C., Horvath, A., and Riezman, H. (1993). Analysis of the sequence requirements for glycosyl-phosphatidylinositol anchoring of *Saccharomyces cerevisiae* Gas1 protein. *J. Biol. Chem.* **268,** 10558–10563.

Pereira, A. J., Dalby, B., Stewart, R. J., Doxsey, S. J., and Goldstein, L. S. (1997). Mitochondrial association of a plus end-directed microtubule motor expressed during mitosis in *Drosophila. J. Cell Biol.* **136,** 1081–1090.

Sheahan, M. B., Rose, R. J., and McCurdy, D. W. (2004). Organelle inheritance in plant cell division: The actin cytoskeleton is required for unbiased inheritance of chloroplasts, mitochondria and endoplasmic reticulum in dividing protoplasts. *Plant J.* **37,** 379–390.

Stirling, C. J., Rothblatt, J., Hosobuchi, M., Deshaies, R., and Schekman, R. (1992). Protein transloca-tion mutants defective in the insertion of integral membrane proteins into the endoplasmic reticulum. *Mol. Biol. Cell* **3,** 129–142.

Suelmann, R., and Fischer, R. (2000). Mitochondrial movement and morphology depend on an intact actin cytoskeleton in *Aspergillus nidulans. Cell Motil. Cytoskeleton* **45,** 42–50.

Tanaka, Y., Kanai, Y., Okada, Y., Nonaka, S., Takeda, S., Harada, A., and Hirokawa, N. (1998). Targeted disruption of mouse conventional kinesin heavy chain, kif5B, results in abnormal peri-nuclear clustering of mitochondria. *Cell* **93,** 1147–1158.

Van Gestel, K., Kohler, R. H., and Verbelen, J. P. (2002). Plant mitochondria move on F-actin, but their positioning in the cortical cytoplasm depends on both F-actin and microtubules. *J. Exp. Bot.* **53,** 659–667.

Yang, H. C., Palazzo, A., Swayne, T. C., and Pon, L. A. (1999). A retention mechanism for distribution of mitochondria during cell division in budding yeast. *Curr. Biol.* **9,** 1111–1114.

CHAPTER 32

In Vitro Assays for Mitochondrial Fusion and Division

Elena Ingerman,[1] Shelly Meeusen,[1] Rachel DeVay, and Jodi Nunnari

Department of Molecular and Cellular Biology
University of California Davis
Davis, California 95616

I. Introduction

Most of our knowledge concerning the mechanisms that mediate mitochondrial division and fusion has come from genetic studies in *Saccharomyces cerevisiaie* and from the direct biochemical and cytological observations of division and

[1] Equal contributors.

fusion component behaviors *in vivo* (Meeusen and Nunnari, 2005). These approaches have generated good working models for their roles in mitochondrial dynamics and have revealed that, remarkably, both mitochondrial fusion and division depend on highly conserved dynamin-related proteins (DRPs), which are large self-assembling GTPases that regulate membrane dynamics in a variety of cellular processes. However, the lack of *in vitro* systems for studying these processes has prevented a deeper investigation of mechanism.

This chapter outlines two methods that have proved instrumental in recapitulating events *in vitro* that occur during mitochondrial division and fusion, and have led to more mechanistic models of these processes. First, we outline the methodology behind an assay developed to recapitulate the process of mitochondrial fusion *in vitro* (Meeusen *et al.*, 2004). Using this *in vitro* fusion assay, we have separated outer mitochondrial membrane fusion from inner mitochondrial membrane fusion, and have begun to understand the molecular machines that mediate each of these separate events. Second, we outline the methodology of a coupled, continuous GTPase assay that has allowed us to study accurately the kinetic properties of the dynamin-related GTPase, Dnm1, which is the master regulator of mitochondrial division. This assay will be useful for future mechanistic studies of mitochondrial division and also for the study of mitochondrial fusion, which requires the activity of the dynamin-related GTPases, Fzo1/Mfn1/2 and Mgm1/OPA1.

II. Mitochondrial Fusion *In Vitro*

The importance of understanding the fundamental mechanism of mitochondrial fusion is underscored by the fact that defects in fusion have recently been linked to the onset of neurodegenerative diseases. Mutations in MFN2 (mammalian ortholog of yeast Fzo1) and OPA1 (mammalian ortholog of yeast Mgm1) are associated with the neurodegenerative disorders Charcot Marie Tooth Neuropathy, Type 2 and Autosomal Dominant Optic Atrophy, respectively (Alexander *et al.*, 2000; Delettre *et al.*, 2000; Zuchner *et al.*, 2004). Thus, understanding the roles of these proteins in mitochondrial fusion will likely aid in the development of potential therapeutic agents.

The mechanism of mitochondrial fusion is unique. No paradigm exists for fusion of a double-membraned system. In addition, heterotypic fusion between mitochondria and other membranes has not been observed, and SNARES are not required. The majority of the known core mitochondrial fusion components belong uniquely to the dynamin-related GTPase family, whose members are large GTPases that regulate membrane dynamics through self-assembly (Danino and Hinshaw, 2001; Osteryoung and Nunnari, 2003; Praefcke and McMahon, 2004). How these large GTPases control mitochondrial outer and inner membrane fusions is an outstanding question. Thus, the mechanism underlying mitochondrial fusion is distinct from other membrane fusion events and has evolved independently.

The development of a fluorescent-based *in vitro* fusion assay has given the field an important tool to unravel the mechanics of mitochondrial fusion (Meeusen *et al.*, 2004). This *in vitro* assay is relatively simple, but requires that mitochondria be brought into contact for a critical duration of time (by centrifugation and incubation) and resuspended in a reaction buffer (RB) containing exogenous GTP and an energy regeneration system (Meeusen *et al.*, 2004). Interestingly, under these conditions, cytosol is not essential for fusion. In addition, by varying energetic conditions, mitochondrial outer membrane fusion can be resolved from inner membrane fusion, allowing for the analysis of productive fusion intermediates.

Fusion of outer and inner mitochondrial membranes *in vitro* can be observed directly using either confocal microscopy or wide-field deconvolution microscopy of fluorescent proteins that have been targeted *in vivo* to specific mitochondrial compartments and/or membranes (Meeusen *et al.*, 2004). Specifically, outer and inner membrane fusions can be quantified distinctly by mixing mitochondria isolated from cells expressing an outer membrane-targeted GFP (om-GFP) and a matrix-targeted dsRED (m-dsRED) with mitochondria isolated from cells expressing a matrix-targeted blue fluorescent protein, or BFP (m-BFP). Fusion of outer and inner membranes results in a single continuous green outer membrane encircling a single colocalized blue and red matrix compartments. In the event that outer membrane fusion occurs but inner membrane fusion is blocked, two tightly juxtaposed distinct red and blue matrix compartments encircled by a single continuous green outer membrane are observed.

Due to the low quantum fluorescence yield and sensitivity to photo bleaching of m-BFP, detection can be difficult on most conventional microscopes equipped with CCD cameras, especially after m-BFP becomes more dilute by matrix content mixing following fusion. To address this experimental problem, inner membrane fusion can be assessed in a separate assay by using mitochondria from cells expressing m-dsRED with mitochondria from cells expressing matrix-targeted GFP (m-GFP, Fig. 1). On inner membrane fusion, mitochondria whose matrix contents contain colocalized red and green fluorophores are observed. The ability to dissect and observe directly the process of mitochondrial fusion *in vitro* will continue to contribute to our mechanistic understanding of this complex and unique process.

The efficiency of mitochondrial fusion *in vitro* is sensitive to a number of different parameters. This methods section explains in detail the protocols used for isolating fusogenic mitochondria, conducting fusion *in vitro*, and analyzing completed fusion reactions. We also use complementary electron microscopic analysis of mitochondrial fusion reactions, which is not reviewed here (Meeusen *et al.*, 2004).

A. Plasmids

Plasmids used to generate fluorescent mitochondria in yeast for use in *in vitro* fusion assays are as follows: m-GFP (pVT100UGFP) and m-BFP (pYES-mtBFP) were gifts from Dr. B. Westermann and were created as described by Westermann

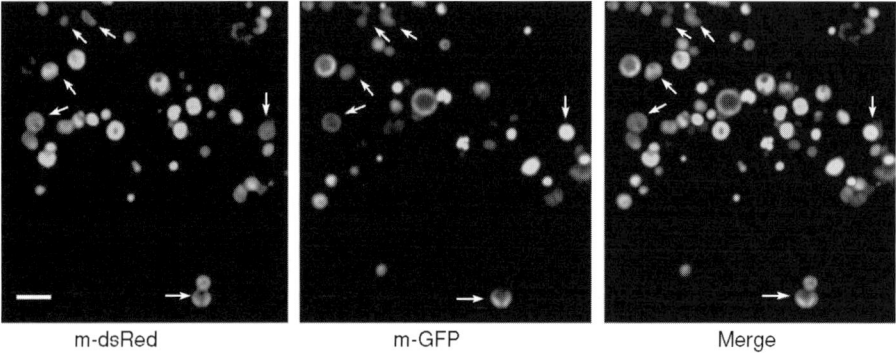

| m-dsRed | m-GFP | Merge |

Fig. 1 *In vitro* fusion of mitochondria. A representative field of mitochondria that have been subjected to fusion assay conditions. Mitochondria isolated from a strain expressing matrix-targeted GFP (m-GFP, middle panel) mixed with mitochondria isolated from a strain expressing m-dsRed (left panel). Fused mitochondria are identified by colocalization of m-GFP and m-dsRed, and are indicated by arrows in the right panel. Scale bar = 2 µm. (See Color Insert.)

and Neupert (2000); m-dsRED was created by replacing GFP in pVT100U-mtGFP with dsRED2 (a gift from Dr. B. Glick, University of Chicago, Chicago, IL) (Meeusen *et al.*, 2004); and om-GFP (pOM45-GFP) was a gift from Dr. R. Jensen, Johns Hopkins University (Sesaki *et al.*, 2003).

B. Cell Culturing and Preparation of Fusion Competent Mitochondria

Two days prior to the *in vitro* assay, two 100-ml cultures are inoculated with *S. cerevisiaie* W303 cells containing plasmids that encode different mitochondrial-targeted fluorophores. Yeast are inoculated into synthetic minimal complete media composed of complete minimal dextrose or complete minimal galactose media lacking either uracil, leucine, or both, as necessary, which are used to maintain episomal plasmids, and grown at 30 °C. The choice of plasmids depends on whether one wants to assay outer or inner membrane fusion. When isolating mitochondria from temperature-sensitive mutants, it may be important to conduct all steps at lower temperatures to prevent the manifestation of mutant phenotypes prior to the *in vitro* assay.

One day after inoculation, two 2-liter Erlenmeyer flasks containing 1 liter of rich YPGlycerol (YPG) media are inoculated with cells at OD_{600} 0.1 and placed on a rotary shaker at 250 rpm and grown to log phase. It is critical that the cultures are shaken so that they are aerated sufficiently, because the *in vivo* respiratory status of the cells affects the fusion competence of isolated mitochondria *in vitro*. In addition, we have observed that less aeration impedes growth rates, contributing to thicker cell walls, and ultimately less cell lysis, diminishing yield from the mitochondrial preparation.

Cells between OD_{600} 0.6–1.0 are harvested by centrifugation at $1500 \times g$ for 5 min and washed once in deionized H_2O. From this step until isolated mitochondria are placed in RB, all manipulations must be conducted as quickly as possible; the ability of mitochondria to fuse *in vitro* is compromised by long preparation times. The entire preparation time should not exceed 4 h. Resuspend cells at 20 OD_{600}/ml in Tris, β-ME buffer (0.1-M Tris, pH 9.4; 50-mM β-mercaptoethanol) and incubate at 30°C for 20 min to begin cell wall lysis. Cells are collected by centrifugation at $1500 \times g$ for 5 min, washed in 100 ml of 1.2-M sorbitol, and collected as above. Cells are then resuspended at 50 OD_{600}/ml in 1.2-M sorbitol containing 3-mg/ml yeast lytic enzyme (dissolve 3-mg/ml ICN *Arthrobacter luteus* yeast lytic enzyme into 1.2-M Sorbitol immediately prior to use), and incubated at <100 rpm at 30°C for 30 min to generate spheroplasts. Spheroplasts are then collected by centrifugation at $1500 \times g$ for 5 min, washed in 100 ml of 1.2-M sorbitol, and collected as above. All residual 1.2-M sorbitol is removed from cell pellets using a micropipetman, and the spheroplast pellet is resuspended in a minimal volume of ice-cold NMIB (0.6-M sorbitol, 5-mM $MgCl_2$, 50-mM KCl, 100-mM KOAc, 20-mM HEPES, pH 7.4). For 1000 OD_{600}, 4 ml of NMIB is sufficient. The NMIB solution should not contain serine protease inhibitors; these and other protease inhibitors affect fusion negatively. Thus, all steps between the cell lysis and fusion steps must be conducted on ice to minimize protease activity. Spheroplast resuspensions are transferred immediately to chilled tight dounces (Wheaton) that are prewashed with cold NMIB. Spheroplast resuspensions are then dounced a minimum of 100 times until $\sim$60% of the cells are dark when visualized using phase microscopy.

Extracts are diluted threefold in cold NMIB and centrifuged at $3000 \times g$ for 5 min at 4°C to remove unlysed cells and large cellular debris. The supernatant fraction is then centrifuged at $10,170 \times g$ for 10 min at 4°C to yield a mitochondrial-enriched pellet. Mitochondrial pellets are resuspended immediately in a minimal volume of cold NMIB to a final protein concentration of 5–10 mg/ml. For example, mitochondrial pellets from 1000 OD_{600} of cells should be resuspended in $\sim$0.25-ml NMIB and quantified using a Bradford assay.

C. Fusion Reactions

For fusion reactions, equal amounts of freshly prepared, differentially labeled mitochondria are mixed thoroughly by pipetting. The total mitochondrial protein content should equal 0.5 mg per reaction. Mixed mitochondria are concentrated by centrifugation at $10,170 \times g$ for 10 min at 4°C. Centrifuged reactions are then incubated on ice for 10 min. This step likely promotes associations between partner outer mitochondrial membranes. The supernatant is removed by aspiration and mitochondrial pellets are suspended immediately in 40 μl of either Stage 1 reaction mix (20-mM Pipes–KOH, pH 6.8; 150-mM KOAc; 5-mM $Mg(OAc)_2$; 0.6-M sorbitol) or Stage 2 reaction mix (20-mM Pipes–KOH, pH 6.8; 150-mM KOAc; 5-mM $Mg(OAc)_2$; 0.6-M Sorbitol; 0.2-mg/ml creatine phosphokinase; 40-mM creatine phosphate; 1-mM ATP; 0.5-mM GTP) (see below). It is important to note that creatine phosphokinase (Sigma, St. Louis, MO) is unstable and its activity is critical for efficient mitochondrial fusion. When opening a new bottle of lyophilized enzyme, make dry aliquots containing 2–4 mg per tube and dessicate. Immediately prior to use, resuspend in RB. One molar of creatine phosphate (Sigma) can be resuspended in RB, aliquoted, and stored at −80 °C. GTP (0.5 M) and 1-mM ATP (Sigma) can be resuspended in RB, aliquoted, and stored at −80 °C. The GTP stock may need to be adjusted to pH 7.0.

Stage 1 reaction mix supports only outer membrane fusion. Stage 2 reaction mix supports the fusion of outer and inner mitochondria membranes. Fusion reactions are usually carried out at 22 °C for up to 1 h.

D. Analysis of Fusion *In Vitro*

Aliquots of Stages 1 and 2 reactions are fixed by resuspension in 2× volume of 8% paraformaldehyde in PBS, pH 7.4, for a minimum of 20 min. Aliquots of fixed reactions are immobilized on microscope slides by mixing with an equal volume of 4% low-melting point agarose in NMIB. Three-dimensional data sets should be collected on a fluorescence microscope to distinguish between outer and inner mitochondrial membrane fusion events. We collect data using an Olympus IX70 Deltavision Microscope using a 60× objective (Olympus), a 100-W Mercury lamp (Applied Precision, Inc.), and an integrated, cooled CCD-based Princeton Micromax Camera equipped with a Sony Interline Chip. Other microscopes, such as laser scanning or spinning disk confocal, may also be used to assess fusion. Given that the average diameter of isolated mitochondria is 0.5 μm, a total distance of 3 μm in the z-direction is sufficient to analyze fusion intermediates and fused structures. It is important to collect data from areas of the slide where mitochondria are in a dispersed monolayer. Imaging highly concentrated mitochondria can interfere with the quantification of outer membrane tethering and other early steps of fusion. Fusion efficiency is quantified by dividing the number of fused mitochondria (mitochondria with colocalized fluorophores) by the total number of mitochondria in a given field. Fusion efficiency averages ~7% for Stage 1 outer membrane fusion and 15% for Stage 2 inner membrane fusion.

III. Analysis of Dynamin–Related GTPase Activity *In Vitro*

Dynamin-related GTPases (DRPs) have evolved in eukaryotic cells to function in such diverse processes as membrane trafficking, organelle division, and resistance to viral infection (Danino and Hinshaw, 2001; Osteryoung and Nunnari, 2003; Praefcke and McMahon, 2004; Song and Schmid, 2003). The *S. cerevisiae* genome encodes three DRPs: Dnm1 and Mgm1, which function in the division and fusion of mitochondria, respectively, and Vps1, which functions in the biogenesis of Golgi-derived vesicles and in the division of peroxisomes (Ekena *et al.*, 1993;

Gurunathan *et al.*, 2002; Hoepfner *et al.*, 2001; Osteryoung and Nunnari, 2003; Vater *et al.*, 1992). Mammalian cells possess homologues of Dnm1 (DRP1, also denoted DNM1L) and Mgm1 (OPA1) that also function in the regulation of apoptosis, in addition to other DRPs, such as dynamin-1, which mediates the scission of clathrin-coated pits from the plasma membrane during endocytosis and which is the prototype of the DRP superfamily (Hinshaw, 2000; Osteryoung and Nunnari, 2003; Schmid, 1997; Youle, 2005).

Recently determined x-ray crystal structures indicate that DRPs possess a core fold similar to that of all regulatory GTPases, thus establishing them as members of the GTPase superfamily (Klockow *et al.*, 2002; Niemann *et al.*, 2001; Prakash *et al.*, 2000). However, in addition to their relatively large mass, the kinetic properties of DRPs are unique among other classes of GTPases, such as the Ras and heterotrimeric Gα families (Song and Schmid, 2003). Besides a GTPase domain, all DRPs possess three other hallmark domains: a middle domain of 150 amino acids, a divergent sequence called insert B, and a C-terminal assembly or GTPase-effector domain (GED) (van der Bliek, 1999). These domains cooperate via both intra- and intermolecular interactions to promote the self-assembly of higher order DRP structures.

Most of the information concerning DRP function has come from the analysis of dynamin-1. *In vitro*, dynamin self-assembles into spiral-like filaments (Hinshaw and Schmid, 1995). Self-assembly stimulates dynamin's GTPase activity and is required for dynamin's ability to remodel membranes during endocytosis (Damke *et al.*, 2001a; Song *et al.*, 2004; Warnock *et al.*, 1996). In DRPs, GED/GED and GED/GTPase domain interactions are required for self-assembly and assembly-driven GTPase activity (Sever *et al.*, 1999; Smirnova *et al.*, 1999). However, the molecular mechanism of GED-stimulated GTPase activity is currently unknown. The relative affinities of dynamin for GTP and GDP are lower than those of Ras and Gα, and the rate of GTP hydrolysis of dynamin is significantly higher in the assembled versus the unassembled state (Damke *et al.*, 2001a; Eccleston *et al.*, 2002; Warnock *et al.*, 1996). In addition, the dissociation rate of GDP from dynamin is fast in comparison to that of Ras, indicating that GDP release is not likely to be a rate-limiting step in the dynamin GTPase cycle, and obviating the need for a nucleotide exchange factor *in vivo* (Damke *et al.*, 2001a; Eccleston *et al.*, 2002).

On the basis of its activities and kinetic properties, dynamin has been postulated to play a mechanochemical role in severing endocytic vesicles from the plasma membrane (Hinshaw and Schmid, 1995). However, findings suggest that dynamin and other DRPs function as classical-signaling GTPases, which, in their GTP-bound form, recruit downstream effectors, such as endophilin, which are responsible for membrane severing (Fukushima *et al.*, 2001; Newmyer *et al.*, 2003; Sever, 2002; Sever *et al.*, 1999, 2000a,b). The *in vitro* kinetic properties of dynamin predict that it would exist *in vivo* predominantly in the GTP-bound form (Song and Schmid, 2003). Thus, to function as a classical switch GTPase, other factors, such as interacting proteins, would be predicted to modulate significantly the kinetic properties of DRPs such that production of the GTP-bound form would be rate limiting (Song and Schmid, 2003). Thus, while our knowledge of the properties of DRPs is considerable, the exact mechanistic roles played by DRPs in membrane-remodeling events are unresolved at present.

Methods for the detailed and accurate kinetic analysis of GTPase activity are essential tools for unraveling the exact molecular role of assembling DRPs in membrane-remodeling events (Damke *et al.*, 2001b). Two fundamentally different types of GTPase assays are available for measuring activity: fixed-time-point and continuous assays. To date, only fixed-time-point assays have been used to characterize DRP GTPase activity. Specifically, these include a radioactive [^{32}P]GTP assay and a nonradioactive malachite green colorimetric assay that measures inorganic phosphate release (Lanzetta *et al.*, 1979). In fixed-time-point assays, we have found that the relatively high rates of GTP hydrolysis by DRPs under self-assembling conditions cause substrate depletion, resulting in inaccurate estimates of K_m, V_{max}, and other kinetic parameters. In addition, fixed-time-point assays do not have the time resolution required to detect rapid changes in the rates of GTP hydrolysis over time, as might occur on DRP self-assembly. For example, a lag phase would be overlooked in a fixed-time-point assay, resulting in inaccurately low estimates of GTP hydrolysis rates.

To overcome these problems, we have employed a continuous GTPase assay, which includes a GTP-regenerating system, for the kinetic analysis of Dnm1 GTPase activity. Interestingly, this assay has revealed two unique kinetic parameters that were not detected using standard fixed-time-point analyses of GTPase activity (Ingerman *et al.*, 2005). First, we found that both Dnm1 and dynamin-1 exhibit significant lags (5 min) before reaching a steady-state GTPase activity. In a fixed-time-point assay, this lag was not detected, and thus, the measured rate of Dnm1's GTP hydrolysis was underestimated. In other self-assembling systems, such as F-actin and mictrotubules, a kinetic lag is indicative of the rate-limiting formation of a structure that nucleates self-assembly, which we now postulate to be the case for DRPs. Second, we found that Dnm1 exhibits significant cooperativity with respect to GTP concentration during steady-state GTPase activity (Hill coefficient = 2.8–1.4), suggesting that an intrinsic GTP-dependent switch is functioning to stimulate self-assembly. Here, we describe this assay, which also will be of general use in the characterization of other GTPases possessing relatively high turnover rates for GTP hydrolysis.

A. Continuous GTPase Assay

Our continuous, coupled substrate-regenerating assay has been modified from an assay used previously for measuring the enzymatic activities of ATPases (Renosto *et al.*, 1984). We have adapted this assay for measuring the activity of the dynamin-related GTPase, Dnm1, in 96-well plate format (Kiianitsa *et al.*, 2003). This assay also has been used to characterize the activity of the self-assembling bacterial tubulin-like GTPase, FtsZ (Margalit *et al.*, 2004).

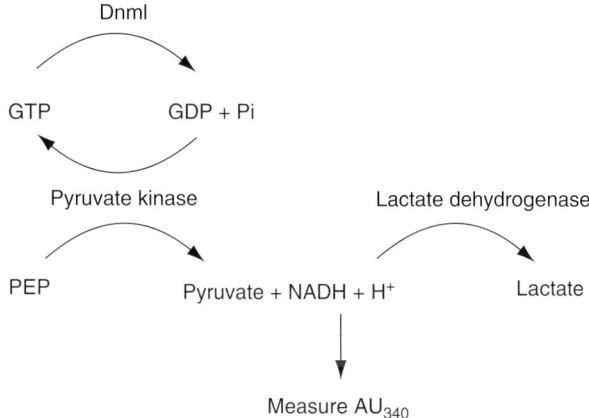

Fig. 2 Schematic representation of the coupled GTP-regenerating assay for GTP hydrolysis. See text for a detailed description of each reaction in the assay.

In the continuous, substrate-regenerating assay, Dnm1 hydrolyzes GTP to GDP and inorganic phosphate. As shown in Fig. 2, GTP is regenerated from GDP and phospho(enol)pyruvate (PEP) by pyruvate kinase. The pyruvate produced by this regenerative reaction is reduced to lactate by the enzyme lactate dehydrogenase, using NADH as a cosubstrate. NADH depletion is measured by monitoring continuously a decrease in absorbance at 340 nm over time, and is directly proportional to GTP hydrolysis.

B. Protein Purification and Storage

Dnm1 is purified from baculovirus-infected Hi5 cells in 25-mM HEPES; 25-mM PIPES, pH 7.0; 500-mM NaCl; 500-mM imidazole, pH 7.4, as described (Ingerman *et al.*, 2005). For maximal protein stability, dimethyl sulfoxide (DMSO) is added to a final concentration (v/v) of 20% to purified protein (0.8-mg/ml Dnm1), rendering the final freezing buffer concentrations 20-mM HEPES, 20-mM PIPES, 400-mM NaCl, 400-mM imidazole, 20% DMSO. Aliquots of protein are flash frozen in liquid nitrogen and stored at −80 °C.

C. Reagents

RB (20×), containing 25-mM HEPES; 25-mM PIPES, pH 7.0; 100-mM MgCl$_2$; 20-mM PEP, and 150-mM KCl, is prepared in advance and stored at −80 °C in aliquots. Similarly, stock solutions of 50-mM GTP (in H$_2$O, pH adjusted to 7 with 0.5-M NaOH) and 20-mM NADH (in 25-mM HEPES, 25-mM PIPES, pH 7.0) are also made in advance and stored at −80 °C in aliquots. We routinely verify the concentration of prepared GTP stock solutions spectrophotometrically, by measuring the absorbance of the prepared stock solution at 253 nm (molar absorbtivity of GTP at 253 nm = 13,700 $M^{-1} cm^{-1}$).

Each GTPase assay reaction contains ~25-mM HEPES; 25-mM PIPES, pH 7.0; 150-mM NaCl; 7.5-mM KCl; ~20 U/ml of pyruvate kinase/lactate dehydrogenase mixture (Sigma P0294); 5-mM $MgCl_2$; 1-mM PEP (Sigma P7002); 600-μM NADH (Sigma N8129); variable amounts of GTP (Sigma G8877); and 0.06 mg/ml of Dnm1. Each reaction also contains 30-mM imidazole and 1.5% DMSO, two additional components of the freezing buffer.

D. Procedure

For a complete kinetic analysis of Dnm1, we measure the GTPase activity of Dnm1 at the following GTP concentrations: 10, 25, 50, 100, 150, 200, 300, 400, 500, 600, 750, and 1000 μM.

We prepare 10× GTP stock solutions by diluting the 50-mM GTP stock in 25-mM HEPES, 25-mM PIPES, pH 7.0 to 0.1, 0.25, 0.5, 1, 1.5, 2, 3, 4, 5, 6, 7.5, and 10 mM. The use of 10× GTP stock solutions allows us to place a uniform volume of GTP into each reaction tube, resulting in better reproducibility. Each 200-μl GTPase assay reaction contains:

10 μl of 20× RB

6 μl of 20-mM NADH

12 μl of 2-M NaCl (this amount accounts for the 15 μl of 400-mM NaCl contained in the Dnm1-freezing buffer and yields a final NaCl concentration of 150 mM)

132 μl of 25-mM HEPES, 25-mM PIPES, pH 7.0

5-μl Pyruvate kinase/Lactate dehydrogenase mixture

20 μl of 10× GTP stock solution

15-μl Dnm1 (0.8-mg/ml Dnm1 in freezing buffer).

For greater accuracy and reproducibility, a master mix containing 20× RB, NADH, NaCl, HEPES/PIPES, and pyruvate kinase/lactate dehydrogenase is prepared. Thus, for every reaction we add 20 μl of 10× GTP stock and 165 μl of master mix. To start the reaction, 15 μl of Dnm1 are added and the contents of the tube are mixed quickly and gently using a pipettor. For better reproducibility and precision, 150 μl of the 200-μl reaction are aliquoted immediately into a 96-well plate.

Absorbance measurements are recorded in a SpectraMAX Pro (Molecular Devices, Sunnyvale, CA) 96-well plate reader that has been preheated to 30°C. We collect absorbance readings at 340 nm (λ_{max} for NADH) at 20-sec intervals for 40 min. To assess the sensitivity of the assay and to account for autohydrolysis/nonspecific GTPase activity, one of the reactions in the 96-well plate should be a "blank," in which 15 μl of freezing buffer is used in place of Dnm1.

The values for absorbance versus time are dependent on the concentration of protein being assayed and its GTP turnover rate. At 10-μM GTP, using Dnm1 (MW 87,000 Da) at a final assay concentration of 0.057-mg/ml, Dnm1 produces a slope of 2.6×10^{-5} AU$_{340}$/sec, which translates into an activity of 1 min^{-1}. At 1-mM GTP, a slope of 1.3×10^{-3} AU$_{340}$/sec is typically observed, which translates into an activity of 52 min^{-1}.

Determining the precise path length is essential for correctly converting data in the form of AU$_{340}$ versus time to an accurate GTPase activity. To measure the path length of the 96-well plate, we first accurately measure the true concentration of NADH in a GTP-free reaction solution (containing 25-mM HEPES; 25-mM PIPES, pH 7.0; 150-mM NaCl; 7.5-mM KCl; 7~20 U/ml of pyruvate kinase/lactate dehydrogenase mixture; 5-mM MgCl$_2$; 1-mM PEP; ~100-μM NADH; 30-mM imidazole; 1.5% DMSO; and 0.06-mg/ml of Dnm1) using a quartz cuvette with a known path length of 1cm and a standard spectrophotometer. A corresponding NADH-free solution is used as a reference blank. An aliquot of 150 μl of the NADH-free and the NADH-containing solution is placed into separate wells of a 96-well plate. The average AU$_{340}$ measurement of six NADH-free and six NADH-containing wells is determined and the average absorbance of the NADH-free wells is subtracted from the NADH-containing wells. (Net absorbance values are typically 0.235.) Using the experimentally determined concentration of NADH and the Beer–Lambert law ($A = \varepsilon \cdot c \cdot l$, we calculate the path length of a 150-μl well volume, which, under our assay conditions, is 0.38 cm.

1. Notes on the Assay

Excessive amounts of imidazole (>40 mM) inhibit the GTPase activity of Dnm1, as well as that of other DRPs. DMSO, in amounts up to 10%, has a slight stimulatory effect on GTPase activity, likely via an effect on self-assembly. 150-mM NaCl was chosen as a physiological salt concentration. In practice, DRPs should be assayed using a range of NaCl concentrations (20–500 mM), to assess their kinetic properties under both assembly and nonassembly conditions.

NADH is not regenerated in this assay because its depletion is required for monitoring protein activity by the coupled assay. Thus, the initial concentration of NADH in the assay, while not affecting the activity of a protein directly, may need to be increased for DRPs with especially high turnovers (>50 min^{-1}), to allow continuous measurement of protein activity over longer periods (>30 min).

E. Data Analysis

Using Microsoft Excel, we plot absorbance data versus time. Because Dnm1 exhibits a lag in reaching steady-state GTP hydrolysis, we do not utilize data from early time points in our calculation of steady-state GTPase velocities. Instead, we determine the hydrolysis rate using absorbance data that are linear with respect to time. Using the "add trendline" function of Microsoft Excel, we fit a line to each plot of absorbance versus time that is obtained for each GTP concentration. The slope of this line is multiplied by 60 (sec/min) and divided by the molar absorbitivity of NADH (6220 M^{-1} cm^{-1}) and by the previously determined path length (e.g., 0.38 cm). This operation converts the slope of the plot of absorbance versus time (AU_{340}/sec) to the velocity of GTP hydrolysis (M/min). The velocity is then converted to a protein activity by multiplying the velocity by the molecular weight of Dnm1 and dividing by the concentration of Dnm1 protein in the assay. Activity at each GTP concentration is plotted as a function of GTP concentration and the Genfit function of Mathcad (MathSoft, Inc., Cambridge, MA), which is an algorithm for fitting data to a curve using nonlinear least squares regression, is used to determine k_{cat}, K_m, or $K_{0.5}$, and the Hill coefficient.

Acknowledgments

This work was supported by grants from the NSF (MCB-0110899) and NIH (R01-GM62942A) to J.N.

References

Alexander, C., Votruba, M., Pesch, U. E. A., Thiselton, D. L., Mayer, S., Moore, A., Rodriguez, M., Kellner, U., Leo-Kottler, B., Auburger, G., Bhattacharya, S. S., and Wissinger, B. (2000). OPA1, encoding a dynamin-related GTPase, is mutated in autosomal dominant optic atrophy linked to chromosome 3q28. *Nat. Genet.* **26,** 211–215.

Damke, H., Binns, D. D., Ueda, H., Schmid, S. L., and Baba, T. (2001a). Dynamin GTPase domain mutants block endocytic vesicle formation at morphologically distinct stages. *Mol. Biol. Cell* **12,** 2578–2589.

Damke, H., Muhlberg, A. B., Sever, S., Sholly, S., Warnock, D. E., and Schmid, S. L. (2001b). Expression, purification, and functional assays for self-association of dynamin-1. *Methods Enzymol.* **329,** 447–457.

Danino, D., and Hinshaw, J. E. (2001). Dynamin family of mechanoenzymes. *Curr. Opin. Cell Biol.* **13,** 454–460.

Delettre, C., Lenaers, G., Griffoin, J.-M., Gigarel, N., Lorenzo, C., Belenguer, P., Pelloquin, L., Grosgeorge, J., Turc-Carel, C., Perret, E., Astarie-Dequeker, C., Lasquellec, L.,., *et al.* (2000). Nuclear gene OPA1, encoding a mitochondrial dynamin-related protein, is mutated in dominant optic atrophy. *Nat. Genet.* **26,** 207–210.

Eccleston, J. F., Binns, D. D., Davis, C. T., Albanesi, J. P., and Jameson, D. M. (2002). Oligomerization and kinetic mechanism of the dynamin GTPase. *Eur. Biophys. J.* **31,** 275–282.

Ekena, K., Vater, C. A., Raymond, C. K., and Stevens, T. H. (1993). The VPS1 protein is a dynamin-like GTPase required for sorting proteins to the yeast vacuole. *Ciba Found Symp.* **176,** 198–211; discussion 211–214.

Fukushima, N. H., Brisch, E., Keegan, B. R., Bleazard, W., and Shaw, J. M. (2001). The AH/GED sequence of the dnm1p GTPase regulates self-assembly and controls a rate-limiting step in mitochondrial fission. *Mol. Biol. Cell* **12,** 2756–2766.

Gurunathan, S., David, D., and Gerst, J. E. (2002). Dynamin and clathrin are required for the biogenesis of a distinct class of secretory vesicles in yeast. *EMBO J.* **21,** 602–614.

Hinshaw, J. E. (2000). Dynamin and its role in membrane fission. *Annu. Rev. Cell Dev. Biol.* **16,** 483–519.

Hinshaw, J. E., and Schmid, S. L. (1995). Dynamin self-assembles into rings suggesting a mechanism for coated vesicle budding. *Nature* **374,** 190–192.

Hoepfner, D., van den Berg, M., Philippsen, P., Tabak, H. F., and Hettema, E. H. (2001). A role for Vps1p, actin, and the Myo2p motor in peroxisome abundance and inheritance in *Saccharomyces cerevisiae. J. Cell Biol.* **155,** 979–990.

Ingerman, E., Perkins, E. M., Marino, M., Mears, J. A., McCaffery, M., Hinshaw, J., and Nunnari, J. (2005). Dnm1 forms spirals that are structurally tailored to fit mitochondria. *J. Cell. Biol.* **170**(7), 1021–1027.

Kiianitsa, K., Solinger, J. A., and Heyer, W. D. (2003). NADH-coupled microplate photometric assay for kinetic studies of ATP-hydrolyzing enzymes with low and high specific activities. *Anal. Biochem.* **321,** 266–271.

Klockow, B., Tichelaar, W., Madden, D. R., Niemann, H. H., Akiba, T., Hirose, K., and Manstein, D. J. (2002). The dynamin A ring complex: Molecular organization and nucleotide-dependent conformational changes. *EMBO J.* **21,** 240–250.

Lanzetta, P., Alvarez, L., Reinach, P., and Candia, O. (1979). An improved assay for nanomole amounts of inorganic phosphate. *Anal. Biochem.* **100,** 95–97.

Margalit, D. N., Romberg, L., Mets, R. B., Hebert, A. M., Mitchison, T. J., Kirschner, M. W., and Ray Chaudhuri, D. (2004). Targeting cell division: Small-molecule inhibitors of FtsZ GTPase perturb cytokinetic ring assembly and induce bacterial lethality. *Proc. Natl. Acad. Sci. USA* **101,** 11821–11826.

Meeusen, S., McCaffery, J. M., and Nunnari, J. (2004). Mitochondrial fusion intermediates revealed *in vitro. Science* **305,** 1747–1752.

Meeusen, S. L., and Nunnari, J. (2005). How mitochondria fuse. *Curr. Opin. Cell Biol.* **17**(4), 389–394.

Newmyer, S. L., Christensen, A., and Sever, S. (2003). Auxilin-dynamin interactions link the uncoating ATPase chaperone machinery with vesicle formation. *Dev. Cell* **4,** 929–940.

Niemann, H. H., Knetsch, M. L., Scherer, A., Manstein, D. J., and Kull, F. J. (2001). Crystal structure of a dynamin GTPase domain in both nucleotide-free and GDP-bound forms. *EMBO J.* **20,** 5813–5821.

Osteryoung, K. W., and Nunnari, J. (2003). The division of endosymbiotic organelles. *Science* **302,** 1698–1704.

Praefcke, G. J., and McMahon, H. T. (2004). The dynamin superfamily: Universal membrane tubulation and fission molecules? *Nat. Rev. Mol. Cell Biol.* **5,** 133–147.

Prakash, B., Praefcke, G. J., Renault, L., Wittinghofer, A., and Herrmann, C. (2000). Structure of human guanylate binding protein 1 representing a unique class of GTP binding proteins. *Nature* **403,** 567–571.

Renosto, F., Seubert, P. A., and Segel, I. H. (1984). Adenosine 50-phosphosulfate kinase from *Penicillium chrysogenum.* Purification and kinetic characterization. *J. Biol. Chem.* **259,** 2113–2123.

Schmid, S. L. (1997). Clathrin-coated vesicle formation and protein sorting: An integrated process. *Annu. Rev. Biochem.* **66,** 511–548.

Sesaki, H., Southard, S. M., Yaffe, M. P., and Jensen, R. E. (2003). Mgm1p, a dynamin-related GTPase, is essential for fusion of the mitochondrial outer membrane. *Mol. Biol. Cell.* **14,** 2342–2356.

Sever, S. (2002). Dynamin and endocytosis. *Curr. Opin. Cell Biol.* **14,** 463–467.

Sever, S., Damke, H., and Schmid, S. L. (2000a). Dynamin: GTP controls the formation of constricted coated pits, the rate limiting step in clathrin-mediated endocytosis. *J. Cell Biol.* **150,** 1137–1147.

Sever, S., Damke, H., and Schmid, S. L. (2000b). Garrotes, springs, ratchets, and whips: Putting dynamin models to the test. *Traffic* **1,** 385–392.

Sever, S., Muhlberg, A. B., and Schmid, S. L. (1999). Impairment of dynamin's GAP domain stimulates receptor-mediated endocytosis. *Nature* **398,** 481–486.

Smirnova, E., Shurland, D.-L., Newman-Smith, E. D., Pishvaee, B., and van der Bliek, A. M. (1999). A model for dynamin self-assembly based on binding between three different protein domains. *J. Biol. Chem.* **274,** 14942–14947.

Song, B. D., and Schmid, S. L. (2003). A molecular motor or a regulator? Dynamin's in a class of its own. *Biochemistry* **42,** 1369–1376.

Song, B. D., Yarar, D., and Schmid, S. L. (2004). An assembly-incompetent mutant establishes a requirement for dynamin self-assembly in clathrin-mediated endocytosis *in vivo*. *Mol. Biol. Cell* **15**, 2243–2252.

van der Bliek, A. M. (1999). Functional diversity in the dynamin family. *Trends Cell Biol.* **9**, 96–102.

Vater, C. A., Raymond, C. K., Ekena, K., Howald-Stevenson, I., and Stevens, T. H. (1992). The VPS1 protein, a homolog of dynamin required for vacuolar protein sorting in *Saccharomyces cerevisiae*, is a GTPase with two functionally separable domains. *J. Cell Biol.* **119**, 773–786.

Warnock, D. E., Hinshaw, J. E., and Schmid, S. L. (1996). Dynamin self-assembly stimulates its GTPase activity. *J. Biol. Chem.* **271**, 22310–22314.

Westermann, B., and Neupert, W. (2000). Mitochondria-targeted green fluorescent proteins: Convenient tools for the study of organelle biogenesis in *Saccharomyces cerevisiae*. *Yeast* **16**, 1421–1427.

Youle, R. J. (2005). Morphology of mitochondria during apoptosis: Worms-to-beetles in worms. *Dev. Cell* **8**, 298–299.

Zuchner, S., Mersiyanova, I. V., Muglia, M., Bissar-Tadmouri, N., Rochelle, J., Dadali, E. L., Zappia, M., Nelis, E., Patitucci, A., Senderek, J., Parman, Y., Evgrafov, O., *et al.* (2004). Mutations in the mitochondrial GTPase mitofusin 2 cause Charcot-Marie-Tooth neuropathy type 2A. *Nat. Genet.* **36**, 449–451. Epub 2004 Apr 4.

PART VII

Methods to Determine Protein Localization to Mitochondria

CHAPTER 33

Electrophoretic Methods to Isolate Protein Complexes from Mitochondria

Ilka Wittig and Hermann Schägger

Molekulare Bioenergetik, ZBC
Universitätsklinikum Frankfurt
60590 Frankfurt am Main, Germany

I. Introduction

Blue native electrophoresis (BNE), which was initially named blue native polyacrylamide gel electrophoresis (BN-PAGE), was first described in 1991 (Schägger and von Jagow, 1991). The original protocols for BNE and related techniques have been improved and expanded considerably, as summarized here. As a basic modification, imidazole buffer has replaced Bis-Tris buffer, because it does not interfere with protein determinations. The basic principles underlying the native electrophoretic techniques used, however, remained unchanged. Mild nonionic detergents are used for solubilization of biological membranes for

BNE. After solubilization, an anionic dye (Coomassie blue G-250) is added which binds to the surface of all membrane proteins and many water-soluble proteins.

This binding of a large number of dye molecules to solubilized proteins produces a number of effects that are advantageous for BNE. First, the isoelectric points (pI) of the proteins are shifted to the acidic range. Therefore, all proteins migrate to the anode and can be separated on the basis of their migration distance in gradient polyacrylamide gels. Second, the negative charge of the dye on the surface of proteins reduces their aggregation and converts membrane proteins into water-soluble proteins. Therefore, once Coomassie dye is bound, any detergent can be omitted from the gels. This minimizes the risk of denaturation of detergent-sensitive membrane protein complexes. Third, proteins that bind Coomassie dye are visible during electrophoresis and migrate as blue bands. This facilitates excision of bands and recovery of proteins and complexes in a native state by native electroelution.

Novel applications of clear native electrophoresis (CNE; Wittig and Schägger, 2005) are also described. CNE differs from BNE in that Coomassie dye is not used in CNE. Therefore, proteins migrate in an electrical field according to two parameters: the intrinsic pI of the protein and the actual running pH in the gel. In CNE, which uses the imidazole buffer system, a pH of 7.5 develops in the running gel during electrophoresis. As a result, this type of CNE can only separate proteins with pI <7.5. An alternative Tris buffer system (running at pH 8.7) can be used to separate proteins with higher pI. CNE has lower resolution compared to BNE but is the method of choice in cases where Coomassie dye interferes with techniques required to further analyze native complexes, for example, in-gel determination of catalytic activities, fluorescence resonance energy transfer (FRET) analyses, and in-gel detection of fluorescent markers in general (Gavin *et al.*, 2003, 2005). Careful choice of detergent and appropriate detergent/protein ratio has helped to preserve some native protein–protein interactions during BNE, including the dimeric state of mitochondrial ATP synthase (Arnold *et al.*, 1998; Schägger and Pfeiffer, 2000). Using identical solubilization conditions but CNE instead of BNE, even higher oligomeric states of ATP synthase, such as tetramers, hexamers, and octamers, have been identified and separated (Wittig and Schägger, 2005).

Second dimension BNE (modified by adding detergent to the blue cathode buffer) can follow first dimension CNE or first dimension BNE whenever gentle dissociation of supramolecular assemblies into their constituent complexes is desired (Schägger and Pfeiffer, 2000). SDS-PAGE or doubled SDS-PAGE (dSDS-PAGE; i.e., two orthogonal SDS-gels with strongly differing gel types for first and second dimensions) usually follow as final steps to separate the subunits of complexes and to identify very hydrophobic membrane proteins by mass spectrometry (Rais *et al.*, 2004).

II. Materials and Methods

A. Chemicals

Dodecyl-β-D-maltoside, Triton X-100, digitonin (Cat. No. 37006, >50% purity, used without recrystallization), and 6-aminohexanoic acid are from Fluka (Buchs, CH, Switzerland). Acrylamide, bis-acrylamide (the commercial, twice-crystallized products), and Serva Blue G (Coomassie blue G-250) are from Serva (Heidelberg, Germany). All other chemicals are from Sigma (Buchs, CH, Switzerland).

B. First Dimension: Blue Native Electrophoresis

1. Detergents, Stock Solutions, and Buffers

In principle, any nonionic detergent or mild anionic detergent, such as cholic acid derivatives, can be used for the solubilization of biological membranes for BNE, as long as the detergent can solubilize the desired protein and keep it in the native state. We prefer to use dodecyl-β-D-maltoside, Triton X-100, and digitonin, all of which are stored as 1–20% stock solutions in water. Stock solutions of Tricine (1M), 6-aminohexanoic acid (2M), imidazole (1M), and imidazole/HCl (1M, pH 7.0) are stored at 7°C. Buffers for BNE are summarized in Table I.

Table I
Buffers for BNE[a]

Solution	Composition
Cathode buffer B[b] (deep blue)	50-mM Tricine, 7.5-mM imidazole (the resulting pH is around 7.0), plus 0.02% Coomassie blue G-250
Cathode buffer B/10[c] (slightly blue)	As above but lower dye concentration (0.002%)
Gel buffer 3× (triple concentrated)	75-mM Imidazole/HCl (pH 7.0), 1.5-M 6-aminohexanoic acid[d]
Anode buffer	25-mM Imidazole/HCl (pH 7.0)
5% Coomassie blue	Suspended in 500-mM 6-aminohexanoic acid
AB mix (49.5% T, 3% C acrylamide-bisacrylamide mixture)[e]	48-g Acrylamide and 1.5-g bisacrylamide per 100 ml

[a]Imidazole is introduced instead of Bis-Tris because Bis-Tris interferes with many commonly used protein determinations.
[b]Cathode buffer B is stirred for several hours before use and stored at room temperature, since Coomassie dye can form aggregates at low temperature that can prevent proteins from entering the gel.
[c]Cathode buffer B/10 and all other solutions can be stored at 7°C.
[d]6-Aminohexanoic acid is not essential for BNE but it improves protein solubility and is an efficient and inexpensive serine protease inhibitor. BN-gels can be stored at 4°C for several days. We did not observe protease degradation under these conditions.
[e]T, total concentration of acrylamide and bisacrylamide monomers; % C, percentage of cross-linker to total monomer.

2. Gel Types

Gels that resolve specific molecular mass ranges are listed in Table II. Each of these gels are 0.16×14×14 cm, and are used either as "analytical gels" (with 0.5- or 1.0-cm sample wells) or as "preparative gels" (with one 14-cm sample well). Gradient gel preparation is exemplified in Table III.

3. Sample Preparation

a. How Much Detergent Is Required for Complete Solubilization of Membrane Proteins from Biological Membranes?

As a general rule of thumb, bacterial membranes require Triton X-100 or dodecylmaltoside at a detergent/protein ratio of 1 g/g for solubilization, but 2–3 g/g is necessary for solubilization of mitochondrial membranes. Using digitonin, the

Table II
Gel Types for BNE

Mass range (kDa)[a]	Sample gel (% T)	Gradient gel (% T)
10–10,000	3.0	3→13
10–3000	3.5	4→13
10–1000	4.0	5→13
10–500	4.0	6→18

[a]Resolution in the 10- to 100-kDa range is low in BNE with all gel types.

Table III
Gradient Gel Preparation[a]

	Sample gel	Gradient separation gel	
	4% T	5% T	13% T
AB mix	0.5 ml	1.9 ml	3.9 ml
Gel buffer 3×	2 ml	6 ml	5 ml
Glycerol	–	–	3 g
Water	3.5 ml	10 ml	3 ml
10% APS[b]	50 μl	100 μl	75 μl
TEMED	5 μl	10 μl	7.5 μl
Total volume	6 ml	18 ml	15 ml

[a]Volumes for one gel (0.16×14×14 cm). Linear gradient separation gels are cast at 4°C and maintained at room temperature for polymerization. The volume of the 5% T solution is greater than that of the 13% T solution containing glycerol. This assures that the two solutions initially are not mixed when the connecting tube is opened. The sample gel is cast at room temperature. After removal of the combs, gels are overlaid with gel buffer 1× and stored at 4°C.

[b]10% Aqueous ammonium persulfate solution, freshly prepared.

detergent/protein ratio is doubled (e.g., 4–6 g/g for mitochondrial membranes) to achieve comparable solubilization, since the mass of digitonin is about twice the mass of Triton X-100 and dodecylmaltoside.

For solubilization pilot studies, use a series of detergent/protein ratios, for example, 0.5, 1.0, 1.5, and 2.0 g dodecylmaltoside per gram of protein (for bacterial membranes). This titration will reveal the efficiency of solubilization, and whether the enzyme of interest is detergent sensitive with respect to its physiological aggregation state, catalytic activity, or subunit composition. For example, the enzyme might be dimeric (active) using low-detergent but monomeric (inactive) using high-detergent conditions. Retention or loss of detergent-labile subunits can be analyzed by second dimension SDS-PAGE.

b. What Is the Maximal and Minimal Protein Load for BN–PAGE?

As a general rule of thumb, the maximum protein load depends more on the DNA content of the solubilized sample than on the protein quantity. Using isolated mitochondria (low DNA), the maximum protein load for application to a 0.16×1 cm sample well is 200–400 μg of total protein. Using bacterial membranes directly (without removing DNA), the protein load should be reduced to around 50–100 μg. DNA probably blocks the pores of the sample gel, which prevents proteins from entering the gel. Application of DNases does not help. In our hands, deep blue artifact bands appeared in gels of DNase-treated samples. The best way to remove DNA is by mild disruption of cells and differential centrifugation.

The minimal load for BNE depends on the sensitivity of the protein detection method. When specific antibodies are available there is essentially no minimal protein load for BNE. However, the detergent/protein ratio used for solubilization of small and large amounts of protein must be the same, and the detergent concentration thereby must be clearly above the critical micelle concentration (cmc). The latter condition can be kept using low final volumes for low-protein amounts.

c. General Scheme for Solubilizing Biological Membranes

Biological membranes usually are suspended with small volumes of carbohydrate- or glycerol-containing buffers (e.g., 250-mM sucrose, 400-mM sorbitol, or 10% glycerol). Aliquots of these samples can be shock-frozen in liquid nitrogen and stored at $-80\,^{\circ}$C without dissociating multiprotein complexes. The salt concentration in these suspensions should be low (0- to 50-mM NaCl). Potassium and divalent cations should be avoided because they can cause aggregation of Coomassie dye and precipitation of Coomassie-associated proteins. The preferred buffer is 50-mM imidazole/HCl, pH 7.0. Other buffers, for example, Na-phosphate or Na-MOPS, can be used at 5- to 20-mM concentrations. Concentrated protein suspensions (>10 mg/ml) can be used directly for solubilization. Organelle or vesicle suspensions with low-protein concentrations should be diluted and then concentrated by centrifugation. Finally, all solubilization steps are carried out on ice.

Procedure

1. Solubilize membrane pellets (50- to 400-μg sedimented protein) by adding 10–40 μl of solubilization buffer (50-mM NaCl, 50-mM imidazole/HCl, pH 7.0, usually containing 5-mM 6-aminohexanoic acid and 1-mM EDTA) and detergent (from 1% to 20% stock solutions) at predetermined detergent/protein ratios (see Section II.B.3.a). Higher salt concentrations should be avoided, since high salt can lead to stacking of proteins in the sample wells and highly concentrated membrane proteins tend to aggregate. Solubilization is complete within several minutes.

2. Centrifuge the sample for 10–30 min at 100,000 $\times$ g at 4 °C, depending on the sample volume. In cases where large proteins complexes (e.g., >5-MDa complexes) are analyzed, centrifuge solubilized samples at 20,000 $\times$ g for 20 min at 4 °C. The supernatant obtained from this centrifugation is applied to the BN-gel.

3. Prior to sample application, rinse the BN-gel wells to remove gel storage buffer.

4. Add 5% glycerol to the supernatants to facilitate sample loading.

5. Shortly before loading the sample, add Coomassie dye from a 5% suspension in 500-mM 6-aminohexanoic acid. The amount of added dye depends on the amount of detergent used. The optimum dye/detergent ratio is in the range from 1:4 to 1:10 g/g.

Note: Careful determination of the correct amount of dye to add is particularly important with samples that require high-detergent concentrations, for example, 2–4%, for solubilization. Excess lipid/detergent micelles incorporate the anionic dye and therefore migrate very fast. This removal of lipid/detergent/dye micelles improves protein resolution. A mixture of nonionic detergent and anionic dye can mimic some properties of an anionic detergent. For most multiprotein complexes, this "anionic detergent" does not dissociate labile protein subunits, because the extracted membrane protein complexes are sufficiently shielded by boundary lipid.

However, there are situations where addition of Coomassie dye to the sample is not advisable, and clear samples or samples containing the red dye Ponceau S, as in CNE, are preferred. For example, membrane protein complexes that are purified by chromatographic protocols often contain reduced amounts of boundary lipid. Detergent-labile subunits can dissociate from these partially delipidated samples on addition of Coomassie dye.

4. Running Conditions

1. BNE usually is performed at 4–7 °C, as broadening of bands was observed at RT, and cathode buffer B (Table I) is commonly used.

2. 100 V is applied until the sample has entered the gel. Thereafter, the current is limited to 15 mA and voltage is limited to 500 V.

3. After about one-third of the total running distance, cathode buffer B is removed, and the run is continued using cathode buffer B/10. (This buffer exchange improves detection of faint protein bands, helps with native blotting, as less Coomassie dye competes with protein binding to PVDF membranes, and improves the performance of SDS electrophoresis in the second dimension.)

4. Run times are typically 2–5 h.

5. Electroelution of Native Proteins

Two points are essential for the electroelution of native proteins. First, a band of the desired protein (or at least a band of a marker protein in the vicinity) must be detectable at the end of the run to facilitate excision of gel bands and electroelution of native proteins. Second, since proteins stop migrating completely when they approach a mass-specific pore size limit, it is essential that BNE is terminated early, for example, after half of the normal running distance. At this stage, the proteins are still mobile in the BN-gel, and can be efficiently extracted from the gel using H-shaped elutor vessels built according to Hunkapiller *et al.* (1983).

Procedure

1. Seal both lower ends of the elutor vessel with dialysis membranes with a low-cutoff value, for example 2 kDa. (We find that low-cutoff dialysis membranes are more mechanically stable compared to other dialysis membranes (Schägger, 2003a,b).

2. Excise blue protein bands from the gel, mash the gel by several passages through 1-ml syringes, and transfer the mashed gel to the cathodic arm of the H-shaped elutor vessel.

3. Fill both arms of the chamber and the horizontal connecting tube with electrode buffer (25-mM Tricine; 3.75-mM imidazole, pH 7.0; and 5-mM 6-amino-hexanoic acid (as a protease inhibitor).

4. Extract for several hours using 500 V with current that is limited to 2 mA per elutor vessel (to prevent damage if a high-salt buffer was used erroneously). Partially aggregated proteins are collected as a blue layer on the anodic dialysis membrane. To avoid protein aggregation, the voltage can be reduced to 200 V.

6. Semidry Electroblotting of Native Proteins

Similar to native electroelution, native electroblotting also requires short runs of BNE for the efficient transfer of proteins (see earlier discussion). Use cathode buffer B/10 and not cathode buffer B as the BNE cathode buffer before electroblotting, because Coomassie dye binds to PVDF membranes, reducing their protein-binding capacity. Do not use nitrocellulose membranes because they cannot be destained using the conditions described below.

Table IV
Buffers for CNE (Running pH Around 7.5)

Solution	Composition
Cathode buffer	50-mM Tricine, 7.5-mM imidazole (resulting pH is around 7.0 at 4°C)
Gel buffer 3× (triple concentrated)	75-mM Imidazole/HCl (pH 7.0; 4°C), 1.5-M 6-aminohexanoic acid
Anode buffer	25-mM Imidazole/HCl (pH 7.0; 4°C)
Glycerol/Ponceau S solution	50% Glycerol, 0.1% Ponceau S
AB mix (49.5% T, 3% C)	48-g Acrylamide and 1.5-g bisacrylamide per 100 ml

Procedure

1. Place a stack of 4 sheets of Whatman 17 CHR filter papers (total width is 3 millimeters) that has been wetted with electrode buffer on the lower electrode of the semi-dry blotting apparatus (the cathode in this arrangement). The electrode buffer is the cathode buffer for CNE, Table IV; 50-mM Tricine, 7.5 mM imidazole; the resulting pH is around 7.0.

2. Place the gel and then the PVDF membrane, for example, Immobilon P® (wetted with methanol and soaked in electrode buffer) on the filter paper stack. Then put another stack of wetted sheets of filter papers on top, mount the anode of the electroblotting apparatus, and finally place a 5 kg load on top.

3. The transfer is at 4°C (which is recommended) or at RT, using voltage that is set to 20 V (actual voltage around 7 V) and current that is limited to 0.5 mA/cm^2 (80 mA for a 12×14 cm^2 gel area). The transfer is usually complete after 3 h.

4. To destain the background, incubate membrane in 25% methanol, 10% acetic acid.

C. First Dimension: Clear Native Electrophoresis

A basic version of CNE, originally termed colorless native PAGE (CN-PAGE; Schägger *et al.*, 1994), was described shortly after the development of BNE (Schägger and von Jagow, 1991). Advantages and limitations of CNE have been discussed (Wittig and Schägger, 2005). The resolution of CNE is clearly lower compared to BNE (Fig. 2A). However, CNE is the method of choice when Coomassie dye used in BNE interferes with techniques required to further analyze native proteins.

The first step in CNE and BNE is identical, that is, biological membranes are solubilized under the same ionic strength, pH, and detergent conditions. Dodecylmaltoside is the common detergent for the separation of individual complexes, and digitonin is used for the isolation of larger physiological assemblies. Following clarification of solubilized sample by centrifugation, add 1 volume of a

Table V
Buffers for CNE (Running pH Around 8.7)

Solution	Composition
Cathode buffer	100-mM glycine, 15-mM Tris (resulting pH is around 8.7 at 4 °C)
Gel buffer 3× (triple concentrated)	75-mM Tris–HCl (pH 8.5; 4 °C), 1.5-M 6-aminohexanoic acid
Anode buffer	25-mM Tris–HCl (pH 8.4; 4 °C)
Glycerol/Ponceau S solution	50% Glycerol, 0.1% Ponceau S
AB mix (49.5% T, 3% C)	48-g Acrylamide and 1.5-g bisacrylamide per 100 ml

solution consisting of 50% glycerol and 0.1% Ponceau S to 9 volumes of solubilized sample. The red dye Ponceau S facilitates loading the sample (around 20 μl per 0.16 × 0.5 cm sample well) and marks the running front but does not bind to proteins.

Buffers required for CNE are summarized in Tables IV and V. Using the imidazole–Tricine system (Table IV), the running pH is around 7.5. This CNE system is useful for proteins with pI < 7.5 that will migrate toward the anode. The alternative Tris–glycine system (Table V), where the running pH is around 8.7, may be used for proteins with pI < 8.7. We have not tried to increase the running pH beyond 8.7 because multiprotein complexes may dissociate when exposed to high pH.

Gel types for BNE (Table II) can also be used for CNE with minor modifications. (1) When dodecylmaltoside is used for membrane solubilization, 0.03% dodecylmaltoside should be added to the gradient gel mixtures (Table III), since dissociation of protein-bound detergent during electrophoresis can lead to protein aggregation. In contrast, dissociation of digitonin from the lipid/detergent annulus around membrane proteins seems less important. Therefore, when digitonin is used for membrane solubilization there is no need to add detergent to the gel. (2) Electrophoretic mobility of proteins usually is low in CNE, since there is no anionic dye to shift the charge of solubilized proteins. Therefore, low-percentage acrylamide gels should be used to assure a certain running distance, for example, 3→13% acrylamide-gradient gels.

The running conditions are the same as described for BNE. At present, no suitable protocols for electroelution and electroblotting of proteins from CN-gels are available. However, current investigations promise that CNE can be optimized to reach the resolution of BNE, and electroblotting of CN-gels can work satisfactorily (Wittig and Schägger, manuscript in preparation).

D. Second Dimension: Modified Blue Native Electrophoresis

Two-dimensional native electrophoresis using first dimension CNE and second dimension BNE has been used to compare the different separation principles of CNE and BNE (Schägger *et al.*, 1994). In another application, we tested whether supramolecular assemblies of several multiprotein complexes can be dissociated into the constitutive individual complexes by two-dimensional native electrophoresis (Schägger and Pfeiffer, 2000). We found that BNE containing low levels of detergent in the cathode buffer B (0.02% dodecylmaltoside or 0.03% Triton X-100) can dissociate supramolecular structures but retain the structure of the individual complexes. Since CNE can separate supramolecular protein complexes, we have used combinations of these native techniques, including CNE/modified BNE (Fig. 2B) and BNE/modified BNE (not shown).

Procedure

1. For 2D native gels, the first dimension gel should be terminated before proteins reach their mass-specific pore size limit and stop migrating. This is necessary to transfer the protein from the first dimension to the second.

2. Strips of 1 cm excised from the first dimension gel are dipped into water for 1 sec, and placed on glass plates at the position usually occupied by the stacking gel. Spacers (slightly thinner than for first dimension gels, e.g., 1.4 mm) are placed onto the glass plate. A second glass plate is placed on top of the spacers and held in place with clamps.

3. Drain excess water before bringing the gel in an upright position.

4. Introduce acrylamide solution that will be used for the second dimension separation at the gaps between 1D gel strip and spacers, and overlay the acrylamide solution with water.

5. Following polymerization, overlay with more water to cover the 1D native gel strip, and push the strip on the top of the separating gel using plastic cards. Remove the water and fill the gaps between gel strip and spacers with a 10% acrylamide native gel (prepared by analogy with gels in Table III).

6. Add either 0.02% dodecylmaltoside or 0.03% Triton X-100 to cathode buffer B (Table I) and start 2D BNE under the running conditions for 1D BNE (current limited to 15 mA; voltage increases gradually during the run from about 200 to 500 V).

7. In contrast to 1D gels, the 2D gels should be terminated late to concentrate protein into sharp spots. Under these conditions, the intense band of free Coomassie dye may elute from the gel front.

E. Second or Third Dimension: SDS-PAGE

Although the Tris–Glycine system (Laemmli, 1970) offers excellent resolution in 1D SDS-PAGE, Tricine-SDS-PAGE (Schägger, 2003a; Schägger and von Jagow, 1987) is recommended for the resolution of subunits of complexes in the second dimension (Fig. 1B). Similarly, Tricine-SDS-PAGE can follow two-dimensional native electrophoresis (2D CNE/BNE; Fig. 2B) for third dimension resolution (not shown). Finally, since lower acrylamide concentrations can be used in Tricine-SDS-PAGE, transfer of proteins to blotting membranes from Tricine SDS gels is more efficient than from Tris–Glycine SDS gels.

Procedure

1. Excise 0.5-cm lanes from 1D native or 2D native gels.

2. Wet gel slices with 1% SDS for 15 min. Solutions containing thiol reagents are not recommended, except in some rare cases where cleavage of disulfide bridges is important. In those cases, the strips are wetted for up to 2 h in 1% SDS, 1% mercaptoethanol, and then briefly rinsed with water, since mercaptoethanol is an efficient inhibitor of acrylamide polymerization.

3. Place the equilibrated gel slice on a glass plate at the position usually occupied by the stacking gel. Apply spacers and place the second glass plate on top of the spacers. Using spacers that can be thinner that the native gel (e.g., 0.7 mm for 1.4-mm native gels), 0.5-cm lanes of the native gel are compressed to a width of about 1 cm and will not move when the glass plates are brought to a vertical position.

4. Pour the separating gel mixture between the glass plates, usually a 10% acrylamide gel mixture for Tricine-SDS-PAGE (for the 5- to 100-kDa mass range and efficient electroblotting) or a 16% acrylamide gel mixture (for the 5- to 100-kDa mass range; for sharper bands), until there is a 5-mm gap between the SDS gel and the native gel strip. Overlay with water to fill the gap.

5. Following polymerization, the native gel strip is gently pushed onto the separating gel using a 0.6-mm plastic card and residual water is removed.

6. Fill the gaps to the left and right of the native gel strip with a 10% acrylamide native gel mixture (Table III).

7. Second- or third-dimension SDS-PAGEs using the Tricine-SDS-system and gel dimensions $0.07 \times 14 \times 14$ cm are started at RT, with a maximal voltage of 200 V and current limited to 50 mA. When the current falls below 50 mA, the voltage can be increased with the current still limited to 50 mA. The run times for 10% and 16% gels are 3–4 h and 5–6 h, respectively.

In some situations, another SDS-gel may follow to complete three- or four-dimensional electrophoresis (1D native or 2D native electrophoresis are followed by dSDS-PAGE (the successive application of two orthogonal SDS-gels). dSDS-PAGE is especially useful when 1D SDS-PAGE is not sufficient to resolve a given protein mixture properly (Rais *et al.*, 2004). It is also useful when nonreducing and reducing conditions must be applied sequentially, for example, following use of chemical cross-linkers containing cleavable disulfide bonds.

III. Applications

There are several principal ways to use native electrophoretic techniques. It is possible to use 1D native gels for analytical purposes, including in-gel catalytic activity assays (Wittig and Schägger, 2005; Zerbetto *et al.*, 1997), colorimetric quantification of protein amounts (Schägger, 1995a, 1996), estimation of native masses and oligomeric states (Schägger *et al.*, 1994), and—following native electroblotting—for immunological detection and quantification of separated proteins.

Electroblotting of native gels followed by immunodetection is prone to false positive detection. For Western blot analysis, subunits of complexes should be separated by 2D SDS-PAGE, and then identified by specific antibodies (Carrozzo *et al.*, 2006). Reliability of immunodetection on electroblotted 2D SDS-gels is much higher, since background signals and cross-reactions can be easily identified whenever a signal is not detected in the column of protein subunits of the complex of interest, or the assigned mass is not compatible with the mass of the specific subunit. Similarly, quantification of protein amounts on Coomassie-stained 2D SDS-gels is preferable to quantification on 1D BN-gels, since only true subunits of complexes are used for quantification, whereas protein bands in 1D BN-gels contain unknown amounts of background protein.

BNE is also used for preparative purposes, applying up to 3-mg protein samples to a 14-cm wide native gel. Bands that are visible during the run can be excised and the proteins can be electroeluted, as described in Section II.B.5. Proteins isolated by preparative BNE have been used for antibody production, protein characterization, and proteomic studies (Fandino *et al.*, 2005; Rais *et al.*, 2004).

BNE was developed originally as a means to isolate and analyze the five oxidative phosphorylation complexes from mammalian mitochondria, which are rather stable membrane protein complexes (Schägger and von Jagow, 1991; Schägger *et al.*, 2004). It has also been used to analyze protein complexes from yeast and plant mitochondria (Arnold *et al.*, 1998; Eubel *et al.*, 2003), chloroplasts (Kügler *et al.*, 1997; Rexroth *et al.*, 2004), and chromaffin granules. Specific proteins, including vacuolar ATP synthase (Ludwig *et al.*, 1998), bacterial respiratory supercomplexes (Stroh *et al.*, 2004), and the mitochondrial protein import machinery in yeast and plants (Dietmeyer *et al.*, 1997; Jänsch *et al.*, 1998), have been studied using this method. Finally, BNE has been used to study neurotransmitter assembly (Griffon *et al.*, 1999), apoptosis (Vahsen *et al.*, 2004), and mitochondrial encephalomyopathies (Bentlage *et al.*, 1995; Carrozzo *et al.*, 2006; Schägger *et al.*, 2004).

A. Protein Complexes from Mitochondria and Chloroplasts

1. Isolation of Detergent-Stable Mitochondrial Complexes

Protein complexes that can be solubilized by mild nonionic detergents, such as dodecylmaltoside or Triton X-100, and separated by BNE are defined here as detergent-stable complexes. The largest detergent-stable multiprotein complexes that have been isolated from mammalian mitochondria are the oxoglutarate dehydrogenase complex (OGDC; native mass around 2.5 MDa) and the pyruvate dehydrogenase complex (PDC; native mass around 10 MDa), as shown in

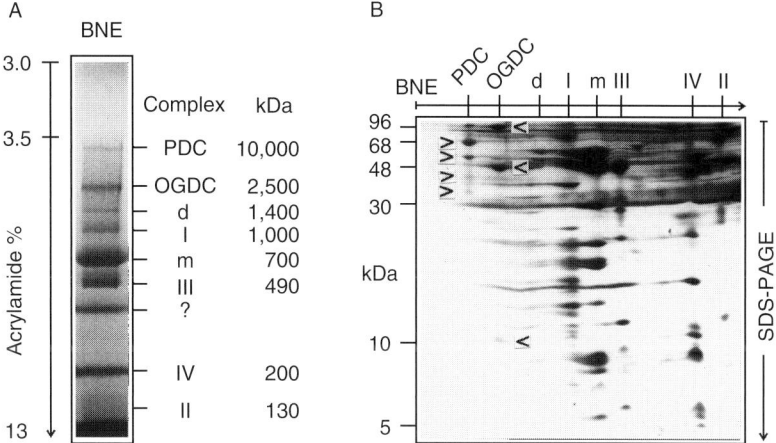

Fig. 1 Isolation of detergent-stable mitochondrial complexes by blue native electrophoresis using Triton X-100 for solubilization. (A) Solubilized bovine heart mitochondrial complexes I–IV (I, II, III, IV) were separated using a linear 3.5%→13% acrylamide gradient gel for BNE, overlaid by a 3% acrylamide sample gel. The mitochondrial ATP synthase (also called complex V) was identified as a major monomeric form (m) and a minor dimeric form (d). OGDC, oxoglutarate dehydrogenase complex; PDC, pyruvate dehydrogenase complex. (B) Subunits of the native complexes were separated by Tricine-SDS-PAGE using a 16.5% T, 3% C gel type. The identity of PDC and OGDC was confirmed by amino-terminal protein sequencing of the 68-kDa subunit E2 of PDC and the 96- and 48-kDa subunits E1 and E2, respectively, of OGDC.

Fig. 1. Concentrations of 3% are the lowest acrylamide concentration that can be handled routinely in gradient gels. As a result, the size of PDC marks the upper mass limit of BNE. Prokaryotic ribosomes (around 2–3 MDa) and the larger eukaryotic ribosomes (6–10 MDa) could be isolated by BNE under special conditions (Schägger, unpublished). However, individual mitochondrial ribosomes (similar in size to prokaryotic ribosomes) have not been isolated by BNE, potentially because they are present in polysomes and are not recovered in the soluble phase using conventional solubilization conditions.

2. Isolation of Detergent–Labile Supramolecular Assemblies

a. Dimeric and Oligomeric ATP Synthases

The structural organization of enzymes in biological membranes can be very complex. An example is ATP synthase, which is usually isolated in monomeric form. However, electron microscopy revealed that this protein exists as a polymeric structure that winds as helical double row of particles around tubular cristae of the mitochondrial inner membrane (Allen *et al.*, 1989).

BNE has been used to isolate ATP synthase as dimers or oligomers. This protein was first isolated as a dimer from yeast mitochondria, using Triton X-100 at very low detergent/protein ratio and BNE (Arnold *et al.*, 1998). Dimeric and even oligomeric ATP synthases were later isolated using digitonin from yeast mitochondria (Paumard *et al.*, 2002; Pfeiffer *et al.*, 2003). ATP synthases in the mass range 1.5–6 MDa have been isolated from mammalian mitochondria (Wittig and Schägger, 2005). For a discussion of the functional role(s) of oligomeric ATP synthase, see Wittig *et al.* (2006).

b. Respirasomes

Respirasomes, stoichiometric assemblies of respiratory complexes in the mitochondrial membrane, were originally identified when supercomplexes consisting of complexes I, III, and IV were isolated from bovine heart using BNE (Schägger and

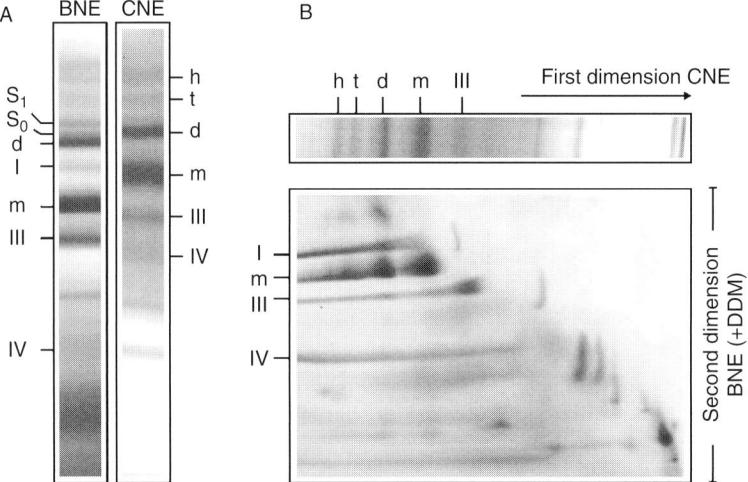

Fig. 2 Analysis of supramolecular assemblies of oxidative phosphorylation complexes using one- and two-dimensional native electrophoresis techniques. (A) Isolation of supramolecular assemblies of complexes using digitonin for solubilization and blue native electrophoresis (BNE) or clear native electrophoresis (CNE) for separation. I, III, and IV, respiratory complexes I, III, and IV, respectively. S_0 and S_1, respiratory supercomplexes or respirasomes containing monomeric complex I, dimeric complex III, and no (0) or one (1) copy of complex IV, respectively. m, d, t, and h, monomeric, dimeric, tetrameric, and hexameric forms of mitochondrial complex V, respectively. Resolution of BNE is higher compared to CNE but oligomeric forms of complex V are better preserved in CNE. (B, upper panel) First dimension CNE (1D CNE) is similar to that in panel A, right lane, but digitonin amounts for solubilization were reduced by 75%. (B, lower panel) Oligomeric forms of complex V were then dissociated into monomeric complex V (m) during second dimension BNE, which was modified by adding dodecylmaltoside [2D BNE (+DDM)].

Pfeiffer, 2000) (Fig. 2). Moreover, analyses of muscle tissue and cell lines of patients with mitochondrial disorders indicated that association of respiratory complexes into respirasomes is essential for complex I stability (Acin-Perez *et al.*, 2004; Schägger *et al.*, 2004). Assembly of respiratory complexes into respirasomes may also facilitate channeling of the electron carrier quinone between complexes I and III, which makes electron transfer rates essentially independent of the midpoint potential of the quinone pool (Bianchi *et al.*, 2004). Interestingly, recent evidence suggested that respirasomes are not the largest assemblies of respiratory complexes in the membrane. Specifically, electron microscopy studies have revealed that large particles are arranged at regular intervals on mitochondrial inner membranes (Allen, 1995; Allen *et al.*, 1989). Therefore, it is possible that respirasomes, with masses up to 2.3 MDa in mammalian mitochondria, associate in a larger linear structure, which we refer to as the "respiratory string" (Wittig *et al.*, 2006).

3. Determination of Molecular Masses of Native Protein Complexes

As shown in Fig. 1, using dodecylmaltoside or Triton X-100 for solubilization BNE can separate proteins according to their molecular masses. Many commercially available water-soluble proteins and the bovine mitochondrial complexes have been shown to fit a general calibration curve (Schägger, 2003b; Schägger *et al.*, 1994). With the exception of some extremely basic membrane proteins and some water-soluble proteins that do not bind Coomassie dye, the native masses of membrane proteins can be determined with a maximal error of 20%.

Using digitonin instead of dodecylmaltoside or Triton X-100 for BNE, the migration of membrane protein in general is slightly reduced. Since digitonin delipidates membrane proteins less than does dodecylmaltoside or Triton X-100, the lower migration of complexes solubilized in digitonin compared to other detergents may be due to altered protein/lipid/detergent/Coomassie ratios within solubilized complexes. Therefore, for mass determination using BNE, solubilize membrane proteins in the same detergent as the protein of interest. Estimation of native masses in CN-gels is more complicated, as described by Wittig and Schägger (2005).

4. In-Gel Catalytic Activity Measurements

Coomassie dye can adversely affect catalytic activity assays (Schägger and von Jagow, 1991). In spite of this, Zerbetto *et al.* (1997) have measured activity of oxidative phosphorylation complexes separated in BN-gels. Not surprisingly, the same assays can be used for CN-gels. In CN-gels, the assays can be much faster and inhibitor sensitive (Pfeiffer *et al.*, 2003; Wittig and Schägger, 2005).

B. Isolation of Protein Complexes from Tissue Homogenates and Cell Lines

Mitochondrial protein represents about 1%, 5%, and 10% of the total cellular protein in cell lines, skeletal muscle, and heart muscle, respectively. This corresponds to less than 10 μg total mitochondrial protein in 10 mg of sedimented cells (wet weight), or less than 100 μg in 10-mg heart tissue. The lower protein amounts were sufficient for in-gel activity assays and silver staining. The larger protein amounts even allowed for Coomassie staining of 2D gels and densitometric quantification of the subunits of respiratory complexes. Protocols for processing different tissues (Bentlage *et al.*, 1995; Schägger and Ohm, 1995) and cell lines (Schägger *et al.*, 1996) have been described elsewhere, and will not be repeated here.

C. Purification of Partially Purified Protein Complexes Using BN–PAGE

Purification of membrane proteins and complexes is often difficult, and only partial purification may be achieved by conventional isolation protocols. BNE is a valuable means for final purification of these problematic proteins and has been used as a final purification step for antibody production and protein chemistry studies. One example of this type of purification is described in Schägger (1995b). About 1–2 mg of a partially purified mitochondrial complex could be loaded on each preparative gel and recovered by native electroelution. Three general rules may help to avoid dissociation of prepurified complexes during final purification.

1. The salt concentration in the sample should be low, but sufficient to keep proteins in solution, for example, 25- to 50-mM NaCl. High salt causes protein concentration and aggregation within the sample well during electrophoresis.

2. Keep detergent concentrations for protein elution from chromatographic columns low (e.g., 0.05% Triton X-100 or 0.1% dodecylmaltoside), because the combination of a neutral detergent and the anionic Coomassie dye used subsequently in BNE exhibits some properties of a mild anionic detergent.

3. Do not add Coomassie dye to the sample if the desired protein complex contains detergent-labile subunits. The Coomassie dye that is required in BNE is supplied with cathode buffer B. Cathode buffer B/10 can be used directly when the detergent concentration in the sample is as low as 0.2%.

IV. Outlook

BNE and CNE preserve some physiological protein–protein interactions, and allow for the isolation of special supramolecular protein assemblies. However, weak and dynamic interactions in a cell or organelle may not be detected using this method. Chemical cross-linking often is required to fix two interacting proteins together. Multidimensional electrophoresis, using native conditions first, and dSDS-PAGE last, can then be used to prepare samples for the mass spectrometric identification of the physiologically associated proteins. This application may become an essential alternative to two-hybrid systems, or an experimental assay to verify or dismiss predicted interactions by two-hybrid systems.

Acknowledgment

This work was supported by the Deutsche Forschungsgemeinschaft, Sonderforschungsbereich 628, Project P13 (H.S.).

References

Acin-Perez, R., Bayona-Bafaluy, M. P., Fernandez-Silva, P., Moreno-Loshuertos, R., Perez-Martoz, A., Bruno, C., Moraes, C. T., and Enriquez, J. A. (2004). Respiratory complex III is required to maintain complex I in mammalian mitochondria. *Mol. Cell* **13,** 805–815.

Allen, R. D. (1995). Membrane tubulation and proton pumps. *Protoplasma* **189,** 1–8.

Allen, R. D., Schroeder, C. C., and Fok, A. K. (1989). An investigation of mitochondrial inner membranes by rapid-freeze deep-etch techniques. *J. Cell Biol.* **108,** 2233–2240.

Arnold, I., Pfeiffer, K., Neupert, W., Stuart, R. A., and Schägger, H. (1998). Yeast F_1F_0-ATP synthase exists as a dimer: Identification of three dimer specific subunits. *EMBO J.* **17,** 7170–7178.

Bentlage, H., De Coo, R., Ter Laak, H., Sengers, R., Trijbels, F., Ruitenbeek, W., Schlote, W., Pfeiffer, K., Gencic, S., von Jagow, G., and Schägger, H. (1995). Human diseases with defects in oxidative phosphorylation. 1. Decreased amounts of assembled oxidative phosphorylation complexes in mitochondrial encephalomyopathies. *Eur. J. Biochem.* **227,** 909–915.

Bianchi, C., Genova, M. L., Parenti Castelli, G., and Lenaz, G. (2004). The mitochondrial respiratory chain is partially organized in a supercomplex assembly: Kinetic evidence using flux control analysis. *J. Biol. Chem.* **279,** 36562–36569.

Carrozzo, R., Wittig, I., Santorelli, F. M., Bertini, E., Hofmann, S., Brandt, U., and Schägger, H. (2006). Subcomplexes of human ATP synthase mark mitochondrial biosynthesis disorders. *Ann. Neurol.* **59,** 265–275.

Dietmeyer, K., Hönlinger, A., Bömer, U., Dekker, P. J. T., Eckerskorn, C., Lottspeich, F., Kübrich, M., and Pfanner, N. (1997). Tom 5 functionally links mitochondrial preprotein receptors to the general import pore. *Nature* **388,** 195–200.

Eubel, H., Jänsch, L., and Braun, H.-P. (2003). New insights into the respiratory chain of plant mitochondria: Supercomplexes and a unique composition of complex II. *Plant Physiol.* **133,** 274–286.

Fandino, A. S., Rais, I., Vollmer, M., Elgass, H., Schägger, H., and Karas, M. (2005). LC-nanospray-MS/MS analysis of hydrophobic proteins from membrane protein complexes isolated by blue-native electrophoresis. *J. Mass Spectrom.* **40,** 1223–1231.

Gavin, P. D., Devenish, R. J., and Prescott, M. (2003). FRET reveals changes in the F_1-stator stalk interaction during activity of F_1F_0-ATP synthase. *Biochim. Biophys. Acta* **1607,** 167–179.

Gavin, P. D., Prescott, M., and Devenish, R. J. (2005). Yeast F_1F_0-ATP synthase complex interactions *in vivo* can occur in the absence of the dimer specific subunit e. *J. Bioenerg. Biomembr.* **37,** 55–66.

Griffon, N., Büttner, C., Nicke, A., Kuhse, J., Schmalzing, G., and Betz, H. (1999). Molecular determinants of glycine receptor subunit assembly. *EMBO J.* **18,** 4711–4721.

Hunkapillar, M. W., Lujan, E., Ostrander, F., and Hood, L. E. (1983). Isolation of microgram quantities of proteins from polyacrylamide gels for amino acid sequence analysis. *Methods Enzymol.* **91,** 227–236.

Jänsch, L., Kruft, V., Schmitz, U. K., and Braun, H.-P. (1998). Unique composition of the preprotein translocase of the outer mitochondrial membrane from plants. *J. Biol. Chem.* **273,** 17251–17257.

Kügler, M., Jänsch, L., Kruft, V., Schmitz, U. K., and Braun, H.-P. (1997). Analysis of the chloroplast protein complexes by blue-native polyacrylamide gel electrophoresis. *Photosynth. Res.* **53,** 35–44.

Laemmli, U. K. (1970). Cleavage of structural proteins during the assembly of the head of bacteriophage T4. *Nature* **227,** 680–685.

Ludwig, J., Kerscher, S., Brandt, U., Pfeiffer, K., Getlawi, F., Apps, D. K., and Schägger, H. (1998). Identification and characterization of a novel 9.2 kDa membrane sector associated protein of vacuolar proton-ATPase from chromaffin granules. *J. Biol. Chem.* **273,** 10939–10947.

Paumard, P., Vaillier, J., Coulary, B., Schaeffer, J., Soubannier, V., Mueller, D. M., Brethes, D., di Rago, J.-P., and Velours, J. (2002). The ATP synthase is involved in generating mitochondrial cristae morphology. *EMBO J.* **21,** 221–230.

Pfeiffer, K., Gohil, V., Stuart, R. A., Hunte, C., Brandt, U., Greenberg, M. L., and Schägger, H. (2003). Cardiolipin stabilizes respiratory chain supercomplexes. *J. Biol. Chem.* **278,** 52873–52880.

Rais, I., Karas, M., and Schägger, H. (2004). Two-dimensional electrophoresis for the isolation of integral membrane proteins and mass spectrometric identification. *Proteomics* **4,** 2567–2571.

Rexroth, S., Meyer zu Tittingdorf, J. M. W., Schwassmann, H. J., Krause, F., Seelert, H., and Dencher, N. A. (2004). Dimeric H^+-ATP synthase in the chloroplast of *Chlamydomonas reinhardtii. Biochim. Biophys. Acta* **1658,** 202–211.

Schägger, H. (1995a). Quantification of oxidative phosphorylation enzymes after blue native electrophoresis and two-dimensional resolution: Normal complex I protein amounts in Parkinson's disease conflict with reduced catalytic activities. *Electrophoresis* **16,** 763–770.

Schägger, H. (1995b). Native electrophoresis for isolation of mitochondrial oxidative phosphorylation protein complexes. *Methods Enzymol.* **260,** 190–202.

Schägger, H. (1996). Electrophoretic techniques for isolation and quantitation of oxidative phosphorylation complexes from human tissues. *Methods Enzymol.* **264,** 555–566.

Schägger, H. (2003a). SDS electrophoresis techniques. *In* "Membrane Protein Purification and Crystallization" (C. Hunte, G. von Jagow, and H. Schägger, eds.), pp. 85–103. Academic Press, San Diego.

Schägger, H. (2003b). Blue native electrophoresis. *In* "Membrane Protein Purification and Crystallization" (C. Hunte, G. von Jagow, and H. Schägger, eds.), pp. 105–130. Academic Press, San Diego.

Schägger, H., and von Jagow, G. (1987). Tricine-sodium dodecyl sulfate polyacrylamide gel electrophoresis for the separation of proteins in the range from 1–100 kDalton. *Anal. Biochem.* **166,** 368–379.

Schägger, H., Bentlage, H., Ruitenbeek, W., Pfeiffer, K., Rotter, S., Rother, C., Böttcher-Purkl, A., and Lodemann, E. (1996). Electrophoretic separation of multiprotein complexes from blood platelets and cell lines: Technique for the analysis of diseases with defects in oxidative phosphorylation. *Electrophoresis* **17,** 709–714.

Schägger, H., Cramer, W. A., and von Jagow, G. (1994). Analysis of molecular masses and oligomeric states of protein complexes by blue native electrophoresis and isolation of membrane protein complexes by two-dimensional native electrophoresis. *Anal. Biochem.* **217,** 220–230.

Schägger, H., De Coo, R., Bauer, M. F., Hofmann, S., Godinot, C., and Brandt, U. (2004). Significance of respirasomes for the assembly/stability of human respiratory chain complex I. *J. Biol. Chem.* **279,** 36349–36353.

Schägger, H., and Ohm, T. (1995). Human diseases with defects in oxidative phosphorylation: II. F_1F_0 ATP-synthase defects in Alzheimer's disease revealed by blue native polyacrylamide gel electrophoresis. *Eur. J. Biochem.* **227,** 916–921.

Schägger, H., and Pfeiffer, K. (2000). Supercomplexes in the respiratory chains of yeast and mammalian mitochondria. *EMBO J.* **19,** 1777–1783.

Schägger, H., and von Jagow, G. (1991). Blue native electrophoresis for isolation of membrane protein complexes in enzymatically active form. *Anal. Biochem.* **199,** 223–231.

Stroh, A., Anderka, O., Pfeiffer, K., Yagi, T., Finel, M., Ludwig, B., and Schägger, H. (2004). Assembly of respiratory chain complexes I, III, and IV into NADH oxidase supercomplex stabilizes Complex I in *Paracoccus denitrificans. J. Biol. Chem.* **279,** 5000–5007.

Vahsen, N., Candé, C., Brière, J.-J., Bénit, P., Joza, N., Mastroberardino, P. G., Pequignot, M. O., Casares, N., Larochette, N., Métivier, D., Feraud, O., Debili, N., *et al.* (2004). AIF deficiency compromises oxidative phosphorylation. *EMBO J.* **23,** 4679–4689.

Wittig, I., Carrozzo, R., Santorelli, F. M., and Schägger, H. (2006). Supercomplexes and subcomplexes of mitochondrial oxidative phosphorylation. *Biochim. Biophys. Acta* **1757,** 1066–1072.

Wittig, I., and Schägger, H. (2005). Advantages and limitations of clear native polyacrylamide gel electrophoresis. *Proteomics* **5,** 4338–4346.

Zerbetto, E., Vergani, L., and Dabbeni-Sala, F. (1997). Quantification of muscle mitochondrial oxidative phosphorylation enzymes via histochemical staining of blue native polyacrylamide gels. *Electrophoresis* **18,** 2059–2064.

CHAPTER 34

Analysis of Protein–Protein Interactions in Mitochondria

Johannes M. Herrmann[*] and Benedikt Westermann[†]

[*]Institut für Zellbiologie
Universität Kaiserslautern
67663 Kaiserslautern, Germany

[†]Institut für Zellbiologie
Universität Bayreuth
95440 Bayreuth, Germany

I. Introduction

Many mitochondrial proteins are subunits of large oligomeric structures. Examples are the respiratory chain complexes and the F_0F_1-ATPase in the inner membrane, the mitochondrial ribosome in the matrix, the protein translocation complexes, and the molecular machinery mediating fusion and fission of the mitochondrial membranes. The analysis of protein–protein interactions is therefore of particular importance for the study of the biogenesis and function of mitochondrial proteins. In this chapter, we present a number of techniques that have been used successfully to analyze protein–protein interactions within cells and organelles. Among these, coimmunoprecipitation, copurification and chemical cross-linking have proven to be particularly useful to study various aspects of

0091-679X/07 $35.00
DOI: 10.1016/S0091-679X(06)80034-8

mitochondrial biogenesis. These techniques are powerful methods for the analysis of the sequential interactions of precursor proteins with the various components of the translocation machineries in the mitochondrial membranes. Moreover, they are useful for the investigation of protein–protein interactions during mitochondrial protein synthesis, folding and degradation, and for the determination of the molecular composition of oligomeric protein complexes, as for example the respiratory chain complexes or protein translocation machineries.

Here, we outline the rationale behind the analysis of protein–protein interactions in mitochondria. Relevant methods are described and illustrated with specific examples. This chapter is a revised and updated version of an article that has been published in an earlier volume of this series (Herrmann et al., 2001).

II. Rationale

Several points have to be considered before starting an analysis of protein–protein interactions. (1) Protein–protein interactions are often weak and unstable in vitro. For example, in the case of membrane protein complexes, solubilization with detergents may destabilize oligomers, thereby preventing their analysis in extracts. (2) Interactions can be transient. This may prevent coisolation of the interaction partners, since their physical interaction might not last during the isolation procedure. (3) Proteins often interact with several different binding partners. For example, during import into mitochondria a precursor protein contacts several proteins of the translocation machinery in a sequential manner. (4) Many proteins tend to bind to other proteins upon extraction. For example, membrane proteins and nonassembled subunits of complexes often expose hydrophobic surfaces, causing them to stick to other proteins. These nonnative interactions can be surprisingly specific and strong, and it is often difficult to distinguish them from authentic endogenous interactions.

For these reasons, it is crucial that experiments addressing protein–protein interactions are well controlled and backed up by the use of different approaches that should consistently reveal the same interactions. Various methods have been employed to monitor interactions of mitochondrial proteins. Copurification procedures are the standard assays to detect protein–protein interactions, since they typically reflect in vivo interactions reliably and allow for the isolation of previously unknown binding partners. Chemical cross-linkers can stabilize weak or short-lived interactions and therefore have been used extensively to monitor transient interactions. However, cross-linking often is rather inefficient, and the isolation and identification of yet unknown interaction partners is therefore difficult. Advantages and limitations of both techniques are discussed below.

Besides copurification and chemical cross-linking, several other methods are commonly employed to assess protein–protein interactions. These include (1) blue native gel electrophoresis, which is described in detail in Chapter 33 by Wittig and Schägger, this volume, and (2) compositional analysis of mitochondrial protein complexes using size exclusion chromatography or gradient centrifugation. While both techniques give information on the sizes of native protein complexes, they typically do not allow for the identification of interaction partners of given proteins.

Several biophysical methods can be used to study interactions of purified proteins. For example, plasmon resonance was used to measure the binding affinities of preproteins and receptors of the mitochondrial outer membrane (Iwata and Nakai, 1998; Okamoto et al., 2002; Vergnolle et al., 2005). Interactions of proteins can also be tested by two-hybrid analysis in the yeast Saccharomyces cerevisiae (Fields, 1993). This method has been used in several cases to track down interacting domains of mitochondrial proteins (Armstrong et al., 1999; Brown et al., 1994; Cerveny and Jensen, 2003; Yano et al., 2003). Large-scale analyses have begun to systematically test each of the ca. 6300 yeast genes for their two-hybrid interactions in arrays covering the entire yeast genome (Ito et al., 2001; Phizicky et al., 2003; Uetz et al., 2000). It is not difficult to predict that these ca. 40 million combinations will reveal a number of new protein–protein interactions of mitochondrial proteins. However, as protein–protein interactions in the yeast two-hybrid system occur in the nucleus, they always have to be confirmed by assaying the interaction partners in their natural environment. Advantages and limitations of different approaches for the analysis of protein–protein interactions have been discussed in a number of excellent reviews (Gietz, 2006; Phizicky and Fields, 1995; Uetz, 2002).

In copurification or pull-down assays, a known protein serves as a bait for coisolation with its binding partners. Innumerable versions of copurification approaches have been used to identify protein–protein interactions. Two different experimental setups are possible. First, the bait protein or peptide can be immobilized on a resin to which a protein extract is applied. Interacting proteins will then associate with the bait protein and unbound material can be washed off. Here, the interaction occurs after generation of the extracts, for example after lysis of mitochondria. This method is typically employed to assess binding parameters such as affinities. It can also be used for the isolation of larger quantities of binding partners. This type of copurification is generally referred to as affinity chromatography. Alternatively, the bait can be purified from an extract and interacting proteins are thereby coisolated. Here, the protein complexes are formed before preparation of the extracts and therefore reflect the authentic interactions that are present in the cell. This type is typically used to monitor in vivo interactions. The most commonly used procedures to isolate protein complexes are coimmunoprecipitation and copurification using affinity tags (see below). In this chapter, we illustrate the techniques of coimmunoprecipitation and copurification, as well as chemical cross-linking, and present typical protocols and examples.

III. Methods

The methods described here have been developed for yeast as a model organism; however, they can be readily applied to other organisms. Preferably, isolated mitochondria should be used as starting material in order to reduce the background,

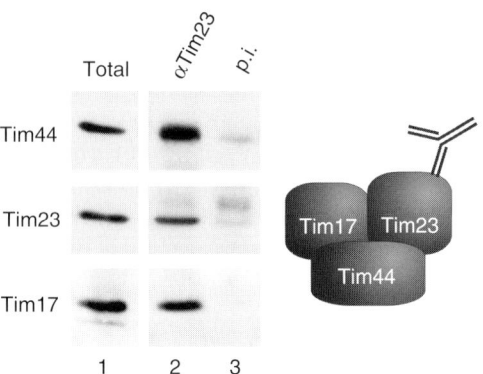

Fig. 1 Coimmunoprecipitation of subunits of the TIM23 complex with Tim23-specific antibodies. Tim23-specific antibodies or antibodies from preimmune serum (p.i.) were coupled to protein A-Sepharose. The resin was incubated with mitochondrial extracts. Proteins bound to the resin were analyzed by Western blotting with sera against the three subunits of the TIM23 complex: Tim44, Tim23, and Tim17. See protocol 1 for details. The cartoon on the right shows the composition of the coimmunoprecipitated complex. For simplicity, additional subunits of the TIM23 complex are not depicted.

which might be rather high when starting from whole cells (see Chapter 1 by Pallotti and Giorgio, Chapter 2 by Boldogh and Pon, and Chapter 3 by Millar *et al.* for isolation of mitochondria). Mitochondria can be used directly (as shown in Fig. 1) or following various manipulations, including analysis after import of radiolabeled precursor proteins (Fig. 2).

A. Coimmunoprecipitation

The procedure of coimmunoprecipitation is illustrated by two examples. First, coimmunoprecipitation is used to assess the composition of the translocase of the mitochondrial inner membrane, the TIM23 complex (protocol 1). In brief, mitochondria are gently lysed using a mild detergent, and the extract is cleared by centrifugation. Tim23-specific antibodies and unspecific control antibodies are coupled to protein A-Sepharose CL-4B beads (Amersham Biosciences Europe, Freiburg, Germany). The beads are then incubated with the mitochondrial extract, washed, and bound proteins are analyzed by Western blotting.

In the second example (protocol 2), a radiolabeled mitochondrial precursor protein is arrested in the TIM23 translocase during translocation. By coimmunoprecipitation, this import intermediate is precipitated with Tim23-specific antibodies and visualized by autoradiography.

In addition, recrystallization of digitonin (protocol 3) and the covalent binding of antibodies to protein A (protocol 4) are described, as both procedures can be very useful for coimmunoprecipitation experiments.

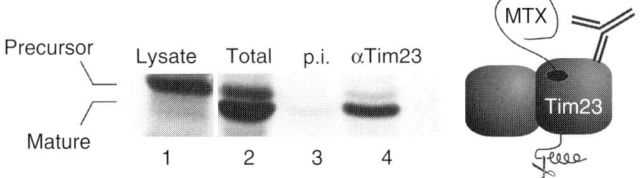

Fig. 2 Coimmunoprecipitation of radiolabeled Cox5a-DHFR with the TIM23 complex. [^{35}S]-radiolabeled Cox5a-DHFR was preincubated with methotrexate (MTX) to stabilize folding of the DHFR domain. The preprotein was kept untreated (lane 1) or incubated with mitochondria for 1 min at 25°C. The mitochondria were lysed and either loaded directly on the gel (lane 2), or the resulting extract was used for coimmunoprecipitation with preimmune serum (p.i., lane 3) or serum against Tim23 (αTim23, lane 4). Total samples correspond to 2% (lane 1) or 20% (lane 2) of the material used for immunoprecipitation. See protocol 2 for details. The cartoon on the right shows the arrested precursor protein in the Tim23 channel, which is recognized by specific antibodies. Scissors indicate processing of the precursor by the matrix-localized mitochondrial processing peptidase.

1. Protocol 1: Coimmunoprecipitation of TIM23 Complex Subunits

a. Lysis of Mitochondria

Purified yeast mitochondria (300 μg) are lysed by gentle agitation for 15 min at 4°C in 50-μl lysis buffer [1% (w/v) digitonin, 50-mM NaCl, 2-mM EDTA, 1-mM phenylmethylsulfonyl fluoride, 10-mM HEPES/KOH, pH 7.4]. Washing buffer [1900 μl; 0.1% (w/v) digitonin, 50-mM NaCl, 2-mM EDTA, 1-mM phenylmethylsulfonyl fluoride, 10-mM HEPES/KOH, pH 7.4] is added. The extract is cleared by centrifugation for 30 min at 30,000 × g at 4°C. The resulting supernatant is the extract used for coimmunoprecipitation.

b. Coupling of Antibodies to Protein A–Sepharose Beads

For coupling of the antibodies to the beads, 40-μl protein A-Sepharose CL-4B slurry (50% suspension) is washed twice in washing buffer and incubated for 30 min at room temperature either with 10-μg affinity-purified polyclonal antibodies against Tim23 or with 5-μl preimmune serum. The beads are washed three times in washing buffer.

c. Coimmunoprecipitation

One milliliter of the extract is added to the beads loaded with antibodies against Tim23 or preimmune antibodies, respectively. After incubation for 2 h at 4°C, the beads are washed three times with 1-ml washing buffer, once with 10-mM HEPES/KOH, pH 7.4, and resuspended in gel loading buffer containing SDS. The suspension is incubated for 3 min at 96°C. Isolated proteins are separated by SDS-PAGE and analyzed by Western blotting using antisera against three subunits of the TIM23 complex (Tim44, Tim23, and Tim17). The result of this experiment is shown in Fig. 1.

2. Protocol 2: Coimmunoprecipitation of Import Intermediates with the TIM23 Complex

a. Arrest of a Radiolabeled Preprotein in the Mitochondrial Import Machinery

For this import experiment, a fusion protein of a mitochondrial preprotein (yeast Cox5a) and mouse dihydrofolate reductase (DHFR) is used (Meier *et al.*, 2005). The preprotein (Cox5a-DHFR) is synthesized in reticulocyte lysate and imported into isolated yeast mitochondria essentially as described in Chapter 36 by Stojanovski *et al.*, this volume. Before addition of the preprotein to the mitochondria, the reticulocyte lysate is incubated with 1-μM methotrexate and 1-mM NADPH in order to stabilize the folded DHFR domain. The import reaction is carried out with 300-μg mitochondria for 1 min at 25 °C. This results in formation of a membrane spanning import intermediate which accumulates in the TIM23 translocase (see sketch on the right of Fig. 2).

b. Lysis

Mitochondria are reisolated and lysed by gentle agitation for 15 min at 4 °C in 100-μl ice-cold lysis buffer [1% (w/v) digitonin, 50-mM NaCl, 2-mM EDTA, 1-mM phenylmethylsulfonyl fluoride, 30-mM HEPES/KOH, pH 7.4]. The extract is diluted with 1900-μl washing buffer [0.1% (w/v) digitonin, 50-mM NaCl, 2-mM EDTA, 1-mM phenylmethylsulfonyl fluoride, 30-mM HEPES/KOH, pH 7.4] and cleared by centrifugation for 30 min at 47,000 $\times$ g at 4 °C.

c. Coupling of Antibodies to Protein A–Sepharose Beads

For coupling of the antibodies to the beads, 40-μl protein A-Sepharose CL-4B slurry (50% suspension) are washed twice in washing buffer and incubated either with 10-μg purified polyclonal Tim23-specific antibodies or with 5 μl of preimmune serum for 30 min at room temperature. The beads are washed three times in washing buffer.

d. Coimmunoprecipitation

One milliliter of the extract containing lysed mitochondria is added to the beads loaded with antiserum against Tim23 or preimmune serum, respectively. The samples are incubated for 2 h at 4 °C with gentle agitation. The beads are then washed twice in washing buffer, once in 30-mM Tris–HCl, pH 7.4 and resuspended in gel loading buffer containing SDS as described above. Isolated proteins are separated by SDS-PAGE and visualized by autoradiography. The result of this experiment is shown in Fig. 2.

Comments

1. The mild buffer conditions may cause a problem known as "post-lysis binding," meaning that the interaction of the bait with the coisolated proteins occurs only in the extract after solubilization of the membranes and therefore does not reflect an *in vivo* interaction. This may be a serious problem, which necessitates careful controls. Typical controls are: (i) Competition experiments: Here, an excess of potential binding partners of the bait is added after lysis. This should

not influence the result of the coimmunoprecipitation. Potential binding partners might be identical to the coisolated protein. Alternatively, other known interaction partners of the bait can be added. (ii) Dilution experiments: If the binding of the bait to its binding partner occurred before lysis, the protein concentration in the lysate should not play any significant role. In contrast, post-lysis interactions should be more pronounced at higher protein concentrations in the lysate. (iii) Mixing experiments: In this case, bait and potential substrate proteins are combined after lysis. No coimmunoprecipitation should be observed in this case.

2. A useful control for the specificity of a coimmunoprecipitation can be the use of an extract that does not contain the antigen, for example extracts obtained from deletion or depletion strains.

3. Affinity purification of the antibodies often significantly reduces the background obtained with antiserum [see Harlow and Lane (1998) for protocols].

4. Instead of prebinding the antibodies to the protein A-Sepharose, the serum (or purified antibodies) and the protein A-Sepharose can be added directly to the extract. However, in this case the amount of serum required has to be carefully determined because it is important that all antibodies bind to the protein A-Sepharose during the incubation.

5. The sometimes inefficient immunoprecipitation of epitope-tagged proteins by monoclonal antibodies can be significantly improved by the addition of several copies of the tag onto the protein, thereby increasing the avidity of binding of the antibody to the antigen (e.g., three or more epitopes in a row).

6. Instead of protein A-Sepharose, protein G-Sepharose can be used, which is often advisable when antibodies obtained from mice, rats, or goats are used.

7. The rather high amount of IgG proteins often leads to an inferior quality of the SDS-PAGE. This can be a problem especially for proteins migrating in a range of 50–60 kDa, which is the size of the IgG heavy chains. This problem may be solved by omitting the boiling of the samples before loading. When the samples are instead mixed in loading buffer at room temperature for 10 min, the disulfide bridges between the heavy and the light chains remain largely intact even in the presence of 2.5% β-mercaptoethanol, and the IgG "blob" therefore is shifted to about 90 kDa. Alternatively, the antibodies can be cross-linked to the protein A-Sepharose to prevent their release into the gel loading buffer containing SDS (Harlow and Lane, 1998). This procedure is described in protocol 3.

8. It is often useful to check whether the immunoprecipitation is quantitative. To do this, the supernatant after sedimentation of the protein A-Sepharose/ antibody/antigen conjugates can be analyzed either by Western blotting or by reimmunoprecipitation. After quantitative immunoprecipitation the supernatant will be completely depleted of the antigen.

9. The solubilization conditions are critical and have to be carefully optimized. The use of the appropriate detergent is crucial, especially in the case of membrane proteins. Typically, nonionic detergents are used such as Triton X-100, octyl

glucopyranoside, dodecyl maltoside, or digitonin (all can be obtained from Sigma, St. Louis, MO). Triton X-100 is easy to handle, highly soluble in water, stable (in the dark), inexpensive, and the solubilization of membrane protein complexes with Triton X-100 is highly reproducible. However, some complexes are not stable in Triton X-100. In the case of fragile complexes, digitonin is often the detergent of choice. The quality of digitonin varies a lot between different batches and suppliers. It is essential to recrystallize digitonin before use (see protocol 4). In addition to the type of detergent, the ionic strength and the pH of the solubilization buffer can be critical. Both have to be tested and adapted for each protein complex. We routinely use buffers with 50- to 150-mM NaCl and pH 7.0–8.5. Lower pH values are not recommended since they interfere with the binding of the antibodies.

3. Protocol 3: Cross-Linking of Antibodies to Protein A–Sepharose

a. Binding of Antibodies to Protein A–Sepharose Beads

Antiserum (250 μl) is incubated with 500-μl PBS (8-g/l NaCl, 0.2-g/l KCl, 1.44-g/l Na_2HPO_4, 0.24-g/l NaH_2PO_4) and 200-μl protein A-Sepharose slurry (50% in PBS) for 1 h at 25°C.

b. Cross-Linking

The beads are washed three times in 0.2-M $NaBO_4$, pH 9.0, and resuspended in 1.8 ml of this buffer with 10.4-mg dimethyl pimelidate (DMP; Pierce, Rockford, IL).

c. Quenching

After incubation for 30 min at 30°C, the beads are washed and then incubated in 0.2-M ethanolamine for 2 h at 25°C to quench the cross-linker. Finally, they are resuspended in 150-μl PBS/20-mM NaN_3. This solution can be stored at 4°C for several weeks.

Comments

1. In some cases cross-linking may interfere with the binding efficiency of the antibodies to the antigen. This has to be tested for each serum.

2. The extent of the immobilization of antibodies on the protein A-Sepharose varies significantly. However, in most cases cross-linking of the IgGs to protein A-Sepharose markedly improves the quality of the gels.

4. Protocol 4: Recrystallization of Digitonin

One gram of digitonin is dissolved in 20-ml ethanol and boiled in a water bath using a stir bar. The solution is then very slowly (several hours) cooled to −20°C. It is critical that the digitonin solution is cooled down along a very shallow temperature gradient. This can be achieved by placing the boiled solution in a closed styrofoam box which is incubated subsequently at room temperature, in the cold room, and then in the freezer over a period of several hours. Depending on the quality of the starting material, about a quarter to a third of the original digitonin powder is typically lost by recrystallization. After incubation for 16 h at $-20\,^{\circ}C$, the solution is centrifuged for 15 min at $12,000 \times g$ at $4\,^{\circ}C$, the supernatant removed, and the pelleted digitonin crystals are dried under vacuum for several hours. The resulting powder can be stored at $4\,^{\circ}C$.

Comments

1. Good digitonin is well soluble in water up to 2–4% (depending on the buffer composition). Impurities reduce the solubility of digitonin.

2. Digitonin tends to form precipitates in solution. These precipitates can have detrimental effects when used in chromatography as they might clog columns irreversibly.

B. Copurification Using Affinity Tags

Affinity tags are typically short peptide sequences which bind efficiently and with high specificity to binding sites on affinity matrices. Using recombinant gene technology, these tags can be fused to any protein of interest and used for purification of the protein. The use of affinity tags has become more and more popular, since the expression of fusion proteins in most biological systems is relatively easy and this technique makes the generation of protein-specific antibodies dispensable. Affinity tags can be epitopes which allow for immunoprecipitation with commercially available antibodies. Widely used examples are HA (hemagglutinin), myc, or flag tags. These epitopes typically comprise about 5–12 amino acid residues and are often used in several copies to increase the avidity of the antibodies. In addition, an increasing number of affinity tags are used which do not rely on an antibody–antigen interaction. These include hexahistidine tags (see below), *tandem affinity purification* tags (TAP, consisting of a calmodulin binding peptide, a tobacco etch virus (TEV) protease cleavage site, and the IgG binding domain of protein A) (Puig *et al.*, 2001), protein A tags (Graslund *et al.*, 2002), or streptavidin tags (Skerra and Schmidt, 2000). A description of a great variety of different tags and their use can be found in recently published excellent reviews (Harlow and Lane, 1998; Jarvik and Telmer, 1998; Lichty *et al.*, 2005). For yeast, the availability of comprehensive libraries of tagged proteins facilitates the rapid analysis of a great number of protein–protein interactions. Recently constructed collections include a library containing more than 3600 yeast strains carrying a $3\times$ HA tag as a mini-transposon insertion (Ross-Macdonald *et al.*, 1999), a yeast strain collection encompassing 98% of all yeast open reading frames (ORFs) fused to a C-terminal TAP tag (Ghaemmaghami *et al.*, 2003), a strain collection containing 6029 yeast strains with chromosomally GFP-tagged ORFs (Huh *et al.*, 2003), and a yeast expression plasmid collection encompassing 5854 yeast ORFs fused to a tag consisting of a hexahistidine tag, an HA epitope, a 3C protease site, and a protein A domain (Gelperin *et al.*, 2005).

Hexahistidine tags have been particularly useful for the isolation of mitochondrial protein complexes by affinity chromatography on matrices charged with metal ions (Bauer *et al.*, 1996; Berthold *et al.*, 1995; Chacinska *et al.*, 2003; Künkele *et al.*, 1998; Nargang *et al.*, 2002; Waizenegger *et al.*, 2004). Several affinity matrices are commercially available carrying side groups with immobilized nickel ions, for example nickel nitrilotriacetic acid (NiNTA). Background binding can be reduced by addition of imidazole to the buffers, which competes with proteins for the binding to the resin. Higher concentrations of imidazole (150–500 mM) or EDTA (5–25 mM) are typically used to elute bound proteins from the resin.

As an example, we describe here the purification of the Mia40–Erv1 complex (Mesecke *et al.*, 2005). In this experiment, a yeast strain is used which expresses Mia40 with a C-terminal hexahistidine tag. *In vivo*, Mia40 and Erv1 are bound to each other covalently via disulfide bridges, which remain intact even under stringent salt conditions. Specific binding of the hexahistidine tag with high affinity to metal ions, such as nickel, permits the isolation of the tagged protein together with its interaction partner.

1. Protocol 5: Copurification of Erv1 with a Hexahistidine-Tagged Variant of Mia40

a. Lysis of Mitochondria

Mitochondria are isolated from a yeast strain expressing the Mia40 protein with a C-terminal hexahistidine tag (Mesecke *et al.*, 2005). Mitochondria (400 μg) are resuspended in lysis buffer (1% Triton X-100, 300-mM NaCl, 10-mM imidazole, 50-mM sodium phosphate, pH 8) and incubated for 30 min at 4°C. DTT (10 mM) is added during lysis to samples in which disulfide bonds are to be reduced. The resulting extract is cleared by centrifugation at $100,000 \times g$ for 10 min at 4°C.

b. Binding of the Hexahistidine-Tagged Mia40 Protein to NiNTA Sepharose

NiNTA beads (50 μl) are washed three times in lysis buffer and incubated with the mitochondrial extract for 1 h at 4°C under agitation. Then, the beads are washed extensively in washing buffer (0.05% Triton X-100, 300-mM NaCl, 30-mM imidazole, 50-mM sodium phosphate, pH 8). Bound proteins are released with SDS-PAGE loading buffer containing 300-mM imidazole and 5% (v/v) β-mercaptoethanol. The result of this experiment is shown in Fig. 3.

Comments

1. Chelating agents like EDTA must be omitted from all buffers as they remove the metal ligands from the binding resins.

2. The binding buffers should contain at least 50- to 100-mM salt to reduce nonspecific interactions of proteins with the resin. Neutral to basic buffers are recommended as acidic buffers lead to protonation of the histidine residues, which interfere with the binding of the tags to the NiNTA resin.

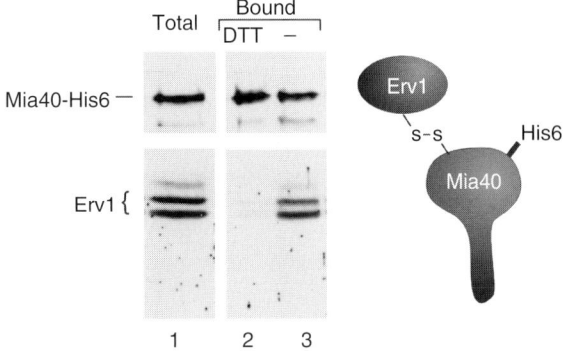

Fig. 3 Purification of the Mia40-Erv1 complex on NiNTA Sepharose. Mitochondria (800 μg) were purified from a strain expressing a Mia40 variant with a C-terminal hexahistidine tag (Mia40-His6). Mitochondria were lysed in lysis buffer and split into two aliquots. One aliquot was incubated with (lane 2) and one without (lane 3) DTT. After a clarifying spin, the mitochondrial extracts were incubated with NiNTA Sepharose for 1 h. Following three washing steps, bound proteins were eluted with SDS-PAGE loading buffer and detected by Western blotting with antisera against Mia40 and Erv1. For control, 40 μg of mitochondria were directly applied to the gel (lane 1). See protocol 5 for details. The cartoon shows a schematic drawing of Erv1 bound to hexahistidine-tagged Mia40 by a disulfide bond.

3. The addition of tags may interfere with the functionality of proteins. Alternatively, due to steric hindrance, the tag of the fusion protein might not be accessible and, hence, the fusion protein may not bind to the affinity resin. Testing both N- and C-terminal fusions and placing spacer sequences of several amino acid residues between the tag and the protein of interest can overcome this problem.

4. The NiNTA Sepharose can be used as bulk material or in columns. The latter is advisable for larger amounts of material, for example for preparative purifications. A large variety of different columns and nickel chelating matrices is available.

C. Cross-Linking

Chemical cross-linkers introduce covalent bonds within proteins or between proteins that are in close proximity to each other. This makes cross-linking a powerful method for the detection of transient or weak interactions that might not be detected by coimmunoprecipitation. Cross-linking has been used extensively to assess interactions between mitochondrial proteins. A limitation of cross-linking is the dependence on specific amino acid side chains in the proteinaceous interaction partners. Obviously, these side chains need to be in a certain distance and orientation to allow cross-linking. Cross-linking is usually not very efficient: in the case of stable interactions cross-linking yields might exceed 10%; however, in the case of transient interactions, they are typically less than 1%. Therefore, a sensitive method is required to detect the cross-linked products. For stable interactions of abundant binding partners (e.g., subunits of the respiratory chain complexes), cross-linked products can be detected by Western blotting. Transient interactions are usually assessed using radiolabeled proteins. Synchronization of the interactions is often required to increase the yield of specific cross-links. In the case of experiments addressing protein translocation across mitochondrial membranes, radiolabeled precursor proteins containing a C-terminal fusion to DHFR can be used. DHFR can be folded stably by addition of the substrate analog methotrexate, which arrests the protein as a translocation intermediate of defined length. This has been used in many cases to cross-link precursor proteins to components of the mitochondrial protein import machineries (Berthold *et al.*, 1995; Vestweber *et al.*, 1989a).

There are two different approaches using cross-linking reagents: (1) (Photoreactive) cross-linkers can be incorporated into proteins, either co- or posttranslationally. Then, these proteins are allowed to interact with other proteins and the cross-linking reaction is triggered by a light flash (Brunner, 1993, 1996). This type of cross-linking led to the identification of Tom40 as a component of the translocation machinery in the mitochondrial outer membrane (Vestweber *et al.*, 1989a,b). (2) The second type uses bifunctional cross-linking reagents to connect preexisting complexes. This method is versatile and can be used to investigate many different processes. For example, cross-linking of subunits of the respiratory chain complexes has been used extensively to identify nearest neighbors within these multisubunit oligomers (Briggs and Capaldi, 1977; Smith *et al.*, 1978; Todd and Douglas, 1981). It has also been used in several cases to screen for components in the proximity of a precursor protein during mitochondrial import or folding (Curran *et al.*, 2002; Mesecke *et al.*, 2005; Mokranjac *et al.*, 2003).

A large variety of bifunctional cross-linkers are commercially available (e.g., from Pierce, Rockford IL or Molecular Probes, Eugene OR). Cross-linking reagents differ in several properties:

1. Depending on their reactive groups cross-linkers differ in their specificity. Targets can be primary amines (the amino terminus and lysine residues), sulfhydryls (cysteine residues), carboxy groups (aspartate and glutamate residues and the carboxy terminus), or carbohydrates. Some photoreactive reagents, like phenyl azides, are more or less nonselective. Both homo- and heterobifunctional cross-linkers are available.

2. Some cross-linkers have to be activated by a light flash which allows the trapping of protein–protein interactions at a specific time point.

3. Cross-linking reagents differ in the length of the spacer arm, which is usually in the range of 0.4–1.2 nm.

4. Some cross-linkers allow the cleavage of the cross-bridge to release the adduct from the bait.

5. Some cross-linkers can be radiolabeled. This can be used, for example, to specifically iodinate binding partners to enable their identification after cleavage of the cross-bridge.

6. Several cross-linkers are available in membrane-permeable and membrane-impermeable versions.

Since the yields of cross-links between interacting proteins are not predictable, several cross-linkers (e.g., differing in the length of the spacer arm) have to be titrated and the conditions for each interaction have to be optimized. Cross-linking reagents that have been used recently to detect interactions of mitochondrial proteins include *m*-maleimidobenzoyl-*N*-hydroxysuccinimide ester (MBS), 1,5-difluoro-2,4-dinitrobenzene (DFDNB), dithiobis (succinimidyl propionate) (DSP), disuccinimidyl glutarate (DSG), disuccinimidyl suberate (DSS), and 1,4-di (3'-2'-pyridyldithio-propionamido)butane (DPDPB).

In the example described below (protocol 6), a radiolabeled precursor protein was arrested as a translocation intermediate during an *in vitro* import reaction and treated with the cross-linker DSS. By immunoprecipitation, one cross-linking adduct was identified as Tim17, a component of the translocation machinery in the inner membrane.

1. Protocol 6: Cross-Linking of Radiolabeled Precursor Proteins to Subunits of the TIM23 Complex

a. Pretreatment of Mitochondria to Arrest a Radiolabeled Precursor Protein in the Import Channel

Mitochondria (150 μg) are preincubated in 600-μl import buffer (220-mM sucrose, 80-mM KCl, 10-mM MOPS/KOH, pH 7.2, 3% bovine serum albumin, 25-mM potassium phosphate, 5-mM MgOAc, and 1-mM MnCl$_2$) in the presence of 1-μM methotrexate for 5 min at 4°C. Then, 20-μl reticulocyte lysate containing [^{35}S]-radiolabeled pSu9(1–112)-DHFR is added (Ungermann *et al.*, 1994). After incubation for 5 min on ice to allow binding of methotrexate to DHFR, the sample is incubated for 15 min at 25°C.

b. Cross-Linking

The sample is split into a 50-μl aliquot for mock treatment (total, without cross-linker) and a 550-μl aliquot. To the first aliquot, 0.5-μl DMSO, and to the latter 5.5-μl DSS (30-mM stock in DMSO) are added. After incubation for 30 min at 25°C, Tris–HCl, pH 8.0 is added to 100-mM final concentration to quench unreacted cross-linker.

c. Immunoprecipitation

Fifty-five microliters of the cross-linked sample are removed (total, with cross-linker). From all samples, mitochondria are reisolated by centrifugation at 12,000 × *g* for 5 min at 4°C and washed in buffer containing 0.6-M sorbitol, 250-mM KCl, 10-mM HEPES, pH 7.4 and either dissolved in SDS-PAGE loading buffer (for totals) or in 80-μl 1% SDS, 100-mM Tris–HCl, pH 7.4 (for immunoprecipitation). The latter sample is vortexed, boiled for 3 min, diluted with 1-ml lysis buffer

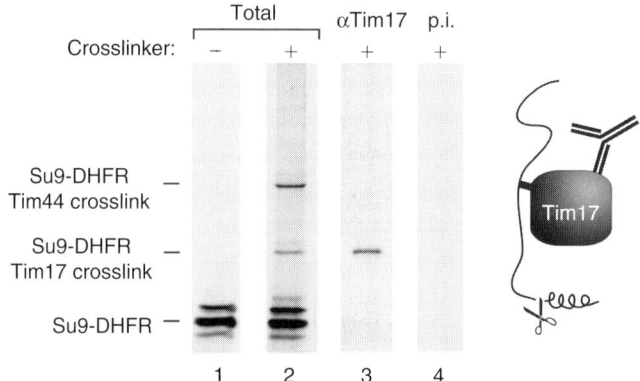

Fig. 4 Cross-linking of a mitochondrial preprotein to Tim17. [^{35}S]-radiolabeled Su9(1–112)-DHFR was arrested in the mitochondrial import channel prior to incubation in the absence (lane 1) or presence of the cross-linking reagent DSS (lanes 2–4). Mitochondria were lysed and either loaded directly on the gel (lanes 1 and 2), or the resulting extract was used for immunoprecipitation with serum against Tim17 (αTim17, lane 3) or with preimmune serum (p.i., lane 4). Samples loaded in lanes 1 and 2 (total) correspond to 20% of the material used for immunoprecipitation. See protocol 6 for details. The cartoon shows a schematic representation of immunprecipitated Tim17 cross-linked to radiolabeled precursor protein that has been processed by the mitochondrial processing peptidase.

[1% Triton X-100 (w/v), 300-mM NaCl, 5-mM EDTA, 1-mM phenylmethylsulfonyl fluoride, 10-mM Tris–HCl, pH 7.4], and centrifuged for 10 min at 30,000 × g at 4°C. The resulting supernatant is split and used for immunoprecipitation with a serum against Tim17 or with a preimmune serum as control. The result of this experiment is shown in Fig. 4.

Comments

1. DSS belongs to *N*-hydroxysuccinimide esters (NHS esters) which are the most commonly used cross-linking reagents. They react with unprotonated primary amines which excludes the use of Tris- or glycine-containing buffers during cross-linking. Instead, Tris or glycine (pH 7.5) are often used as quenching reagents following cross-linking. For NHS esters, it is important to keep the pH above 7–7.5 to reduce protonation of the amino groups. These reagents are rather instable in water especially at a pH above 8. To prevent hydrolysis of the cross-linkers during storage, they have to be carefully kept dry. It is very important to warm the vials containing the cross-linkers to room temperature before opening to protect them against condensing water. The cross-linker solutions in DMSO should be prepared fresh before use.

2. Quenching reagents should be present during all steps after the cross-linking reaction.

3. We did not observe any effects of addition of up to 2–4% DMSO (final concentration) on protein import into isolated mitochondria. However, higher concentrations can have harmful effects on membrane integrity.

4. ATP-depletion, dissipation of the membrane potential, or low temperature may be used to arrest translocating chains instead of using DHFR/methotrexate.

Acknowledgments

The authors would like to thank Drs. Albrecht Gruhler, Stephan Meier, and Christian Kozany for providing data shown in the figures.

References

Armstrong, L. C., Saenz, A. J., and Bornstein, P. (1999). Metaxin 1 interacts with metaxin 2, a novel related protein associated with the mammalian mitochondrial outer membrane. *J. Cell Biochem.* **74,** 11–22.

Bauer, M. F., Sirrenberg, C., Neupert, W., and Brunner, M. (1996). Role of Tim23 as voltage sensor and presequence receptor in protein import into mitochondria. *Cell* **87,** 33–41.

Berthold, J., Bauer, M. F., Schneider, H.-C., Klaus, C., Dietmeier, K., Neupert, W., and Brunner, M. (1995). The MIM complex mediates preprotein translocation across the mitochondrial inner membrane and couples it to the mt-Hsp70/ATP driving system. *Cell* **81,** 1085–1093.

Briggs, M. M., and Capaldi, R. A. (1977). Near-neighbor relationships of the subunits of cytochrome c oxidase. *Biochemistry* **16,** 73–77.

Brown, N. G., Constanzo, M. C., and Fox, T. D. (1994). Interactions among three proteins that specifically activate translation of the mitochondrial *COX3* mRNA in *Saccharomyces cerevisiae*. *Mol. Cell. Biol.* **14,** 1045–1053.

Brunner, J. (1993). New photolabeling and crosslinking methods. *Annu. Rev. Biochem.* **62,** 483–514.

Brunner, J. (1996). Use of photocrosslinkers in cell biology. *Trends Cell Biol.* **6,** 154–157.

Cerveny, K. L., and Jensen, R. E. (2003). The WD-repeats of Net2p interact with Dnm1p and Fis1p to regulate division of mitochondria. *Mol. Biol. Cell* **14,** 4126–4139.

Chacinska, A., Rehling, P., Guiard, B., Frazier, A. E., Schulze-Specking, A., Pfanner, N., Voos, W., and Meisinger, C. (2003). Mitochondrial translocation contact sites: Separation of dynamic and stabilizing elements in formation of a TOM-TIM-preprotein supercomplex. *EMBO J.* **22,** 5370–5381.

Curran, S. P., Leuenberger, D., Schmidt, E., and Koehler, C. M. (2002). The role of the Tim8p-Tim13p complex in a conserved import pathway for mitochondrial polytopic inner membrane proteins. *J. Cell Biol.* **158,** 1017–1027.

Fields, S. (1993). The two-hybrid system to detect protein-protein interactions. *Methods Enzymol.* **5,** 116–124.

Gelperin, D. M., White, M. A., Wilkinson, M. L., Kon, Y., Kung, L. A., Wise, K. J., Lopez-Hoyo, N., Jiang, L., Piccirillo, S., Yu, H., Gerstein, M., Dumont, M. E., *et al.* (2005). Biochemical and genetic analysis of the yeast proteome with a movable ORF collection. *Genes Dev.* **19,** 2816–2826.

Ghaemmaghami, S., Huh, W. K., Bower, K., Howson, R. W., Belle, A., Dephoure, N., O'Shea, E. K., and Weissman, J. S. (2003). Global analysis of protein expression in yeast. *Nature* **425,** 737–741.

Gietz, R. D. (2006). Yeast two-hybrid system screening. *Methods Mol. Biol.* **313,** 345–371.

Graslund, S., Eklund, M., Falk, R., Uhlen, M., Nygren, P. A., and Stahl, S. (2002). A novel affinity gene fusion system allowing protein A-based recovery of non-immunoglobulin gene products. *J. Biotechnol.* **99,** 41–50.

Harlow, E., and Lane, D. (1998). "Using Antibodies: A Laboratory Manual." Cold Spring Harbor Laboratory Press, Cold Spring Harbor, NY.

Herrmann, J. M., Westermann, B., and Neupert, W. (2001). Analysis of protein-protein interaction in mitochondria by co-immunoprecipitation and chemial cross-linking. *In* "Mitochondria: Methods in Cell Biology" (L. A. Pon, and E. A. Schon, eds.), Vol. 65, pp. 217–230. Academic Press, San Diego.

Huh, W. K., Falvo, J. V., Gerke, L. C., Carroll, A. S., Howson, R. W., Weissman, J. S., and O'Shea, E. K. (2003). Global analysis of protein localization in budding yeast. *Nature* **425**, 686–691.

Ito, T., Chiba, T., Ozawa, R., Yoshida, M., Hattori, M., and Sakaki, Y. (2001). A comprehensive two-hybrid analysis to explore the yeast protein interactome. *Proc. Natl. Acad. Sci. USA* **98**, 4569–4574.

Iwata, K., and Nakai, M. (1998). Interaction between mitochondrial precursor proteins and cytosolic soluble domains of mitochondrial import receptors, Tom20 and Tom70, measured by surface plasmon resonance. *Biochem. Biophys. Res. Commun.* **253**, 648–652.

Jarvik, J. W., and Telmer, C. A. (1998). Epitope tagging. *Annu. Rev. Genet.* **32**, 601–618.

Künkele, K.-P., Heins, S., Dembowski, M., Nargang, F. E., Benz, R., Thieffry, M., Walz, J., Lill, R., Nussberger, S., and Neupert, W. (1998). The preprotein translocation channel of the outer membrane of mitochondria. *Cell* **93**, 1009–1019.

Lichty, J. J., Malecki, J. L., Agnew, H. D., Michelson-Horowitz, D. J., and Tan, S. (2005). Comparison of affinity tags for protein purification. *Protein Expr. Purif.* **41**, 98–105.

Meier, S., Neupert, W., and Herrmann, J. M. (2005). Proline residues of transmembrane domains determine the sorting of inner membrane proteins in mitochondria. *J. Cell Biol.* **170**, 881–888.

Mesecke, N., Terziyska, N., Kozany, C., Baumann, F., Neupert, W., Hell, K., and Herrmann, J. M. (2005). A disulfide relay system in the intermembrane space of mitochondria that mediates protein import. *Cell* **121**, 1059–1069.

Mokranjac, D., Paschen, S. A., Kozany, C., Prokisch, H., Hoppins, S. C., Nargang, F. E., Neupert, W., and Hell, K. (2003). Tim50, a novel component of the TIM23 preprotein translocase of mitochondria. *EMBO J.* **22**, 816–825.

Nargang, F. E., Preuss, M., Neupert, W., and Herrmann, J. M. (2002). The Oxa1 protein forms a homooligomeric complex and is an essential part of the mitochondrial export translocase in Neurospora crassa. *J. Biol. Chem.* **277**, 12846–12853.

Okamoto, K., Brinker, A., Paschen, S. A., Moarefi, I., Hayer-Hartl, M., Neupert, W., and Brunner, M. (2002). The protein import motor of mitochondria: A targeted molecular ratchet driving unfolding and translocation. *EMBO J.* **21**, 3659–3671.

Phizicky, E., Bastiaens, P. I., Zhu, H., Snyder, M., and Fields, S. (2003). Protein analysis on a proteomic scale. *Nature* **422**, 208–215.

Phizicky, E. M., and Fields, S. (1995). Protein-protein interactions: Methods for detection and analysis. *Microbiol. Rev.* **59**, 94–123.

Puig, O., Caspary, F., Rigaut, G., Rutz, B., Bouveret, E., Bragado-Nilsson, E., Wilm, M., and Seraphin, B. (2001). The tandem affinity purification (TAP) method: A general procedure of protein complex purification. *Methods* **24**, 218–229.

Ross-Macdonald, P., Coelho, P. S., Roemer, T., Agarwal, S., Kumar, A., Jansen, R., Cheung, K.-H., Sheehan, A., Symoniatis, D., Umansky, L., Heidtman, M., Nelson, K., *et al.* (1999). Large-scale analysis of the yeast genome by transposon tagging and gene disruption. *Nature* **402**, 413–418.

Skerra, A., and Schmidt, T. G. (2000). Use of the Strep-Tag and streptavidin for detection and purification of recombinant proteins. *Methods Enzymol.* **326**, 271–304.

Smith, R. J., Capaldi, R. A., Muchmore, D., and Dahlquist, F. (1978). Cross-linking of ubiquinone cytochrome *c* reductase (complex III) with periodate-cleavable bifunctional reagents. *Biochemistry* **17**, 3719.

Todd, R. D., and Douglas, M. G. (1981). A model for the structure of the yeast mitochondrial adenosine triphosphatase complex. *J. Biol. Chem.* **256**, 6984–6989.

Uetz, P. (2002). Two-hybrid arrays. *Curr. Opin. Chem. Biol.* **6**, 57–62.

Uetz, P., Giot, L., Cagney, G., Mansfield, T. A., Judson, R. S., Knight, J. R., Lockshon, D., Narayan, V., Srinivasan, M., Pochart, P., Qureshi-Emili, A., Li, Y., *et al.* (2000). A comprehensive analysis of protein-protein interactions in *Saccharomyces cerevisiae*. *Nature* **403**, 623–627.

Ungermann, C., Neupert, W., and Cyr, D. M. (1994). The role of Hsp70 in conferring unidirectionality on protein translocation into mitochondria. *Science* **266**, 1250–1253.

Vergnolle, M. A., Baud, C., Golovanov, A. P., Alcock, F., Luciano, P., Lian, L. Y., and Tokatlidis, K. (2005). Distinct domains of small Tims involved in subunit interaction and substrate recognition. *J. Mol. Biol.* **351,** 839–849.

Vestweber, D., Brunner, J., Baker, A., and Schatz, G. (1989a). A 42K outer-membrane protein is a component of the yeast mitochondrial protein import site. *Nature* **341,** 205–209.

Vestweber, D., Brunner, J., and Schatz, G. (1989b). Modified precursor proteins as tools to study protein import into mitochondria. *Biochem. Soc. Trans.* **17,** 827–828.

Waizenegger, T., Habib, S. J., Lech, M., Mokranjac, D., Paschen, S. A., Hell, K., Neupert, W., and Rapaport, D. (2004). Tob38, a novel essential component in the biogenesis of beta-barrel proteins of mitochondria. *EMBO Rep.* **5,** 704–709.

Yano, M., Terada, K., and Mori, M. (2003). AIP is a mitochondrial import mediator that binds to both import receptor Tom20 and preproteins. *J. Cell. Biol.* **163,** 45–56.

CHAPTER 35

Analysis and Prediction of Mitochondrial Targeting Signals

Shukry J. Habib, Walter Neupert, and Doron Rapaport

Institut für Physiologische Chemie, Universität München
D-81377 Munich, Germany

I. Introduction

Eukaryotic cells are subdivided into functionally distinct, membrane-bound compartments. These include the nucleus, mitochondria, chloroplasts, the endo-membrane system [including the endoplasmic reticulum (ER), the Golgi complex,

0091-679X/07 $35.00
DOI: 10.1016/S0091-679X(06)80035-X

endo- and exocytic vesicles and lysosomes], and peroxisomes. Each compartment harbors a variety of proteins that carry out various biochemical reactions necessary for viability of the cell. With the exception of a few proteins that are synthesized within mitochondria and chloroplasts, all cellular proteins are nuclear-encoded and synthesized on cytosolic ribosomes. A significant portion of the proteins in a eukaryotic cell is targeted to a specific organelle. Thus, a fundamental problem in molecular cell biology is to understand how these latter proteins reach the intracellular location at which they exert their functions. This process, usually called protein sorting or protein trafficking, involves information encoded in the protein sequence itself as well as the cellular machinery that decodes this information and delivers the protein to its correct location (Blobel, 1975).

About half of the mitochondrial precursor proteins, especially those destined for the matrix, are synthesized with an N-terminal extension, the so-called presequence (also known as matrix-targeting sequence).

Presequences do not share sequence homology. They are ca. 10–80 residues long and rich in positively charged amino acid residues. Presequences have the potential to form amphiphilic α-helices. In contrast to matrix-destined proteins, all proteins of the mitochondrial outer membrane and some of the proteins destined to the inner membrane and the intermembrane space (IMS) are devoid of a typical presequence. The targeting information in these proteins is rather contained in the protein sequence itself. This chapter describes the known mitochondrial-targeting signals, their analysis and available methods to predict the presence of such signals in eukaryotic proteins.

The mitochondrial machineries that recognize these signals and mediate correct submitochondrial sorting will be mentioned only briefly as they are beyond the scope of this chapter.

II. Overview on Protein Translocation into Mitochondria

Mitochondria are made up of the outer and inner membranes, which separate the IMS and the matrix from the cytosol. Mitochondria have been estimated to contain over 1000 different proteins in mammalian cells and about 700–800 proteins in yeast (Sickmann *et al.*, 2003; Taylor *et al.*, 2003). As only 1–2% of mitochondrial proteins are encoded by the mitochondrial genome and synthesized within the organelle itself, biogenesis of mitochondria depends on uptake of newly synthesized proteins into the organelle. Therefore, importing precursor proteins into the organelle and sorting them into the correct submitochondrial compartment are essential processes for mitochondrial biogenesis and thereby for eukaryotic cell viability.

Precursor proteins are recognized at the surface of mitochondria by the *trans*locase of the *outer mitochondrial* membrane (TOM complex) (Fig. 1). The TOM complex has the capacity to insert proteins into the outer membrane and to translocate all other precursor proteins across the outer membrane (for recent

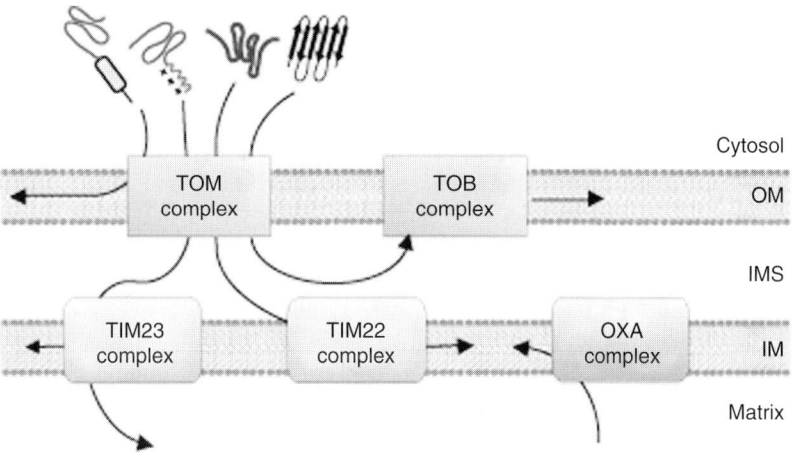

Fig. 1 Import pathways of mitochondrial precursor proteins. The TOM complex mediates the insertion of proteins into or translocation of mitochondrial precursor proteins across the outer membrane. After crossing the outer membrane (OM), presequence-containing precursor proteins are engaging the TIM23 complex in the inner membrane (IM), whereas polytopic inner membrane proteins lacking presequences engage the TIM22 complex for membrane insertion. β-barrel precursors cross the outer membrane via the TOM complex and are then inserted into the outer membrane via the TOB complex. The OXA complex mediates the insertion into the inner membrane of proteins coming from the matrix side.

reviews see Paschen and Neupert, 2001; Pfanner and Geissler, 2001). A number of membrane-embedded β-barrel proteins made up from antiparallel β-sheets constitute a distinct group of mitochondrial outer membrane proteins (Gabriel *et al.*, 2001; Rapaport, 2003). Precursors of β-barrel proteins are initially recognized by the receptor components of the TOM complex (Tom20 and Tom70). They are then translocated through the import pore of the TOM complex (Krimmer *et al.*, 2001; Model *et al.*, 2001; Rapaport, 2002; Rapaport and Neupert, 1999; Schleiff *et al.*, 1999). From the TOM complex β-barrel precursors are transferred to the TOB/ SAM complex which mediates their insertion into the outer membrane (Fig. 1) (Kozjak *et al.*, 2003; Paschen *et al.*, 2003; Wiedemann *et al.*, 2003).

Import of mitochondrial precursor proteins into the inner membrane, matrix, and in some cases the IMS requires the additional action of *t*ranslocases of the *i*nner *m*itochondrial membrane (TIMs) (Fig. 1). Precursor proteins harboring a presequence are transferred from the TOM complex to the TIM23 machinery. In contrast, the TIM22 import pathway mediates the import and insertion into the inner membrane of proteins spanning the inner membrane several times and belonging to two subclasses, the family of the mitochondrial solute carrier proteins and the import components Tim17, Tim22, and Tim23 (Koehler *et al.*, 1999; Rehling *et al.*, 2003).

III. Mitochondrial-Targeting Signals

A. Matrix-Targeting Signal (Presequence)

The best-characterized mitochondrial-targeting signal is the matrix-targeting signal, also called the presequence. About half of the 522 yeast mitochondrial reference proteins currently listed in the MitoP2 database (http://www.mitop.de) have a predicted presequence (Prokisch *et al.*, 2006). Presequence are cleavable N-terminal extensions of about 10–80 amino acid residues which were shown to be necessary and sufficient to direct proteins into the mitochondrial matrix (Hurt *et al.*, 1984; Neupert, 1997). Presequences are rich in positively charged and hydroxylated amino acid residues (von Heijne, 1986a). Plant mitochondrial presequences differ from other eukaryotic mitochondrial presequences in being usually longer and also richer in serine (Glaser *et al.*, 1998; Lister *et al.*, 2003).

Analysis of a large number of mitochondrial presequences suggested that most of them have the potential to form a positively charged amphiphilic α-helix in which segregation of positively charged and hydrophobic residues on opposite faces of the helix occurs (Fig. 2) (Roise and Schatz, 1988; von Heijne, 1986a). Two-dimensional nuclear magnetic resonance (NMR), fluorescence methods and circular dichroism measurements demonstrated the ability of targeting sequences to form amphipathic α-helices in membranes or membrane-like environments, whereas in aqueous solution they are essentially unstructured (Bruch and Hoyt, 1992;

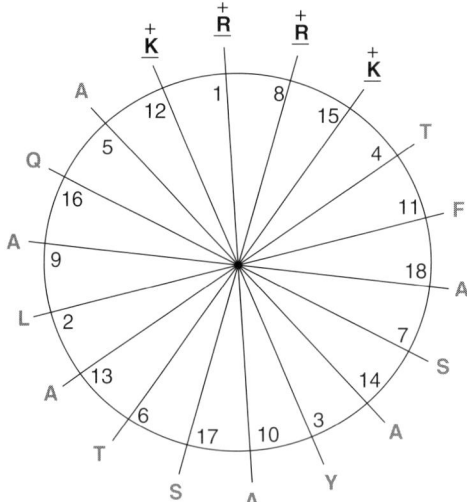

Fig. 2 Mitochondrial presequences have the potential to form amphiphilic α-helices. Amino acid residues 5–22 of yeast β-subunit of ATPase are plotted on a helical wheel. The numbers represent the positions of the amino acids within the helix. Positively charged residues are in bold and underlined whereas other residues are in gray.

Hammen *et al.*, 1994; Karslake *et al.*, 1990; Lemire *et al.*, 1989; von Heijne, 1986a,b). The importance of amphiphilicity to the function of presequences is illustrated by the observation that synthetic positively charged amphipathic peptides composed only of amino acid residues leucine, serine, and arginine are able to function as presequences (Roise *et al.*, 1986).

1. Interactions of Presequences on the Surface of Mitochondria

Interaction of presequences with lipid bilayers has been observed in several studies. For example, the presequence of yeast cytochrome oxidase subunit IV was not only able to insert into monolayers but also to facilitate contact between bilayers of unilamellar vesicles (Leenhouts *et al.*, 1993, 1994; Torok *et al.*, 1994). The positively charged residues in the presequence were previously assumed to be crucial for its recognition by Tom20 receptor (Schatz, 1997), whereas the amphipathic property was proposed to facilitate the membrane insertion. The first part of this assumption had to be reconsidered after the NMR structure of rat Tom20 cytosolic domain together with the presequence of rat aldehyde dehydrogenase was determined (Abe *et al.*, 2000). Surprisingly, the recognition of the presequence by the receptor turned out to be mediated by hydrophobic interactions rather than ionic ones. Notably, the amphiphilic helix of the presequence was present in the absence of a membrane environment, bound in a hydrophobic groove of Tom20. Thus, the interaction of the presequence with the membrane is not a prerequisite for formation of the amphiphilic helix. Furthermore, detergent-purified TOM complex can interact in the absence of lipid membranes with presequence-containing precursor proteins. Hence, lipid membranes are not essential for a productive recognition of a presequence by the TOM complex (Stan *et al.*, 2000).

2. Processing of Presequences by Mitochondrial Peptidases

Cleavable presequences are processed in the matrix by the mitochondrial processing peptidase (MPP). The MPPs in all organisms identified to date consist of two subunits termed α-MPP and β-MPP, both of which are required for the processing activity (Emtage and Jensen, 1993; Geli *et al.*, 1990; Glaser and Dessi, 1999; Hawlitschek *et al.*, 1988; Kalousek *et al.*, 1993). Due to the high degree of sequence variation in presequences, it is difficult to predict MPP cleavage sites by statistical or other methods. In many cases, the presence of an arginine residue at position –2 within the presequence is important for cleavage, the so-called "R-2 rule" (Hendrick *et al.*, 1989; von Heijne *et al.*, 1989). Four motifs of cleavage sites were identified: xRx↓x(S/x) (R-2 motif); xRx(Y/x)↓(S/A/x)x (R-3 motif); Rx↓ (F/L/I)xx(T/S/G)xxxx↓ (R-10 motif which is discussed in detail below); and xx↓x (S/x) (R-none motif) (Gakh *et al.*, 2002; Gavel and von Heijne, 1990). However, in a later analysis of 71 mitochondrial presequences from yeast, R-2, R-3, or R-10 motifs were found in only 65% of them (Branda and Isaya, 1995). A large body of mutational studies led to the conclusion that structural elements, mainly in the presequence but probably also in the mature portion of the protein, represent the most important determinants for cleavage of any precursor by MPP (Gakh *et al.*, 2002).

Following an initial cleavage by MPP, several precursors require a second cleavage carried out by mitochondrial intermediate peptidase (MIP) (Isaya *et al.*, 1991; Kleiber *et al.*, 1990). This dual cleavage is characterized by the presence of the octapeptide motif Rx↓(F/L/I)xx(T/S/G)xxxx↓ at the C-terminus of the presequence (Gavel and von Heijne, 1990; Hendrick *et al.*, 1989). MPP first cleaves the motif two peptide bonds from the arginine residue, yielding a processing intermediate with the octapeptide at its N-terminus that is then cleaved by MIP, yielding the mature protein.

3. Matrix–Destined Precursors Devoid of Cleavable N–Terminal Presequence

Although mitochondrial matrix proteins and membrane proteins that are imported by the TIM23 complex usually contain a cleavable presequence at their N-terminus, exceptional cases exist. Some precursor proteins with an ostensible N-terminal presequences, such as chaperonin10 and 3-oxo-acyl-CoA thiolase, are not cleaved after import into the matrix (Hammen *et al.*, 1996; Jarvis *et al.*, 1995; Rospert *et al.*, 1993; Waltner and Weiner, 1995). Another special case is the DNA helicase Hmil, where the cleavable presequence is present at its C-terminus (Lee *et al.*, 1999). A C- to N-terminal direction of import of this precursor was demonstrated, as well as for constructs that had a normally N-terminal presequence at the C-terminus (Fölsch *et al.*, 1998). Furthermore, the yeast mitochondrial transcription factor MTF1 protein, which is located in the matrix, is an example of a protein that lacks a conventional matrix-targeting signal. Remarkably, its import is independent of membrane receptors, membrane potential, and ATP hydrolysis (Sanyal and Getz, 1995). It was proposed that the import of MFT1 relies on a favorable import-competent conformation of the entire protein (Biswas and Getz, 2002).

4. Proteins Containing a Presequence in Combination with a Sorting Signal

Bipartite sorting signals consist of a typical mitochondrial presequence which is followed by a hydrophobic sorting domain (Fig. 3). The mitochondrial presequence is removed by an initial processing event catalyzed by MPP. Then, the remaining protein is inserted by a stop-transfer mechanism into the lipid core of the inner membrane. In some cases, the protein is processed further to release the mature soluble form into the IMS. This second cleavage event occurs within or at the outer surface of the inner membrane. In the cases of the yeast enzyme l-lactate dehydrogenase (cytochrome b_2) and the proteins cytochrome c_1 and cytochrome b_5 reducatase (Mcr1), the second cleavage is mediated by the inner membrane protease (IMP; Glick *et al.*, 1992; Hahne *et al.*, 1994; Hartl *et al.*, 1987).

Many proapoptotic proteins, such as endonuclease G, apoptosis-inducing factor (AIF), the serine protease HtrA2/Omi, and Smac/DIABLO, reach the IMS via

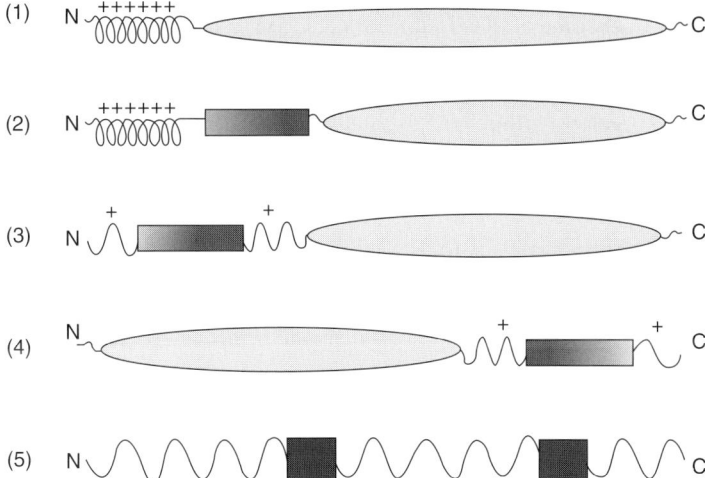

Fig. 3 Mitochondrial-targeting signals. Proteins destined to the matrix usually contain a matrix-targeting signal (presequence) at their N-terminus (1). Some proteins residing in the inner membrane or in the IMS are targeted to the mitochondria by a presequence which is followed by a hydrophobic sorting domain (2). The targeting signal of signal-anchored and tail-anchored proteins of the outer membrane is composed of a hydrophobic segment with positively charged residues at its flanking regions. The targeting signal is located at the N-terminus of signal-anchored proteins (3) and at the C-terminus of tail-anchored proteins (4). Many polytopic proteins which reside in the outer and the inner membrane contain internal targeting signals that are spread within the protein sequence (5). Gray boxes represent hydrophobic segments whereas black squares represent internal targeting signals.

bipartite sorting signals. Although most of these signals are characterized, the enzymes that catalyze the release of the mature proteins into the IMS have been identified only in a few cases. For example, Smac/DIABLO was found to be processed in the mitochondrial IMS by the IMP protease (Burri *et al.*, 2005).

Other proteins contain more complicated targeting and sorting signals. An interesting example is Mgm1, a dynamin-related GTPase that regulates mitochondrial morphology. It has a bipartite sorting signal and an additional hydrophobic sorting domain. Arrest of the first hydrophobic domain in the inner membrane by a stop-transfer results in the formation of a long N-terminally anchored isoform. Alternatively, further translocation of this domain and insertion of the second hydrophobic domain into the membrane allows for proteolytic cleavage mediated by the mitochondrial rhomboid protease, Pcp1. Consequently, a short isoform is released to the IMS (Herlan *et al.*, 2003, 2004; McQuibban *et al.*, 2003). The yeast cytochrome *c* oxidase (CCPO), which is involved in degradation of reactive oxygen species, is also a substrate of Pcp1. However, the presequence in this case is not cleaved by MPP. Rather it is cleaved by mAAA protease of the inner membrane (Esser *et al.*, 2002).

Studies on the internal targeting signal and the sorting mechanism of cytochrome oxidase subunit Va, CoxVa, revealed interesting insights into the targeting and import properties of this protein. CoxVa has a typical presequence and a hydrophobic segment toward the C-terminal part of the protein that is required for directing the polypeptide into the mitochondrial inner membrane, where it serves as a transmembrane anchor (Glaser and Cumsky, 1990). Deletion of the presequence impaired import of the protein into mitochondria; however, overexpression of this truncated version permitted import. One interpretation of this observation is that CoxVa contains a second weaker and internal targeting signal that normally does not play a role but can be recruited to mediate import (Dircks and Poyton, 1990). Such a cryptic targeting signal was found in the mature part of the β-subunit of mitochondrial ATPase (Bedwell *et al.*, 1987). In addition to its roles in mitochondrial targeting and translocation, the internal F1β signal was found to maintain the precursor in an import-competent conformation prior to import (Hajek *et al.*, 1997).

B. Internal Targeting Signals

All proteins residing in the outer membrane and some of the proteins located in the inner membrane and the IMS are targeted to mitochondria via internal targeting signals.

1. Internal Targeting Sequences of Inner Membrane Proteins

a. Precursors Sorted by the TIM22 Complex

The inner membrane contains a large family of metabolite transporters, of which the ADP/ATP carrier (AAC) is the most prominent member. Like the rest of the carrier family, AAC lacks a cleavable presequence and consists of three structurally related modules. Each module is composed of two membrane-spanning segments connected by a loop and a conserved signature motif PX(DE)XX(RK) (Nelson *et al.*, 1998). Each of the three modules contains information for mitochondrial targeting (Endres *et al.*, 1999). A peptide library screen was applied to study the interaction of various sequences of the yeast phosphate carrier with TOM receptors components. It was shown that peptides (about 10 amino acids residues) in various regions within the protein could interact with different receptors. The observed interactions are weak, suggesting that these regions can act individually but most likely in a cooperative manner (Brix *et al.*, 1997). Some precursors of carrier proteins have N-terminal cleavable sequence (Murcha *et al.*, 2005; Winning *et al.*, 1992; Zara *et al.*, 1992). In the case of the mammalian phosphate carrier, these sequences are not required for targeting although they enhance and promote specificity to the import process. The sequence itself could target a passenger protein to mitochondria, although with a low efficiency.

b. Precursors Sorted by the TIM23 Complex

A paradigm for an internal targeting signal is provided by BCS1, a protein involved in the biogenesis of Rieske iron–sulfur proteins. BCS1 is anchored to the inner membrane via a hydrophobic stretch, with the N-terminus exposed to the IMS and the large C-terminal domain to the matrix (Fölsch et al., 1996; Nobrega et al., 1992). The targeting and sorting information are encoded within the first 126 amino acid residues. This region contains three distinct targeting/sorting elements (1) a conserved transmembrane domain (TMD) ca. 40 residues from the N-terminus, directly followed by (2) a positively-charged segment that has the characteristics of a classical matrix-targeting signal, and (3) an adjacent import auxiliary region. While the TMD is an essential element for import and internal sorting, deletion of the N-terminal segment does not affect the import efficiency. When the N-terminal segment and the TMD were deleted, the protein was efficiently targeted to the matrix, and processed by MPP at an MPP cleavage site that is not used upon import of the full length protein. In contrast, deletion of the transmembrane region alone led to a complete inability of the resulting variant to become imported (Fölsch et al., 1996). This underlines the general observation that presequences present in the interior of polypeptide chains are inactive. It was demonstrated, by employing a peptide library scan, that the TOM receptors (Tom70, Tom20, and Tom22) bind with highest affinity to the auxiliary presequence. Therefore, the initial recognition of BCS1 precursor at the surface of the organelle was proposed to depend mainly on the auxiliary region (Stan et al., 2003).

2. Targeting Sequences of Intermembrane Space Proteins

Proteins targeted to the IMS can be classified into three groups according to their sorting pathway and energetic requirements: (1) soluble IMS protein with a bipartite targeting signal, (2) soluble IMS proteins that are trapped by folding in the IMS and (3) proteins that bind to membrane proteins in the IMS (Herrmann and Hell, 2005). As the first group was discussed above, we will discuss IMS proteins that do not contain a presequence, that is, classes (2) and (3).

Two important features seem to be required for the targeting and import of proteins of the second class. First, these IMS proteins are of low molecular mass. Second, they contain conserved patterns of cysteine (and histidine) residues in their sequence (Herrmann and Hell, 2005). It seems that the low molecular mass of these proteins allows the unfolded precursors to diffuse through the general import pore of the TOM complex. Once in the IMS, the precursor becomes associated to factors residing in this subcompartment. These factors can act as receptors and are able also to catalyze the folding of the polypeptide by triggering cofactor-binding events and/or by oxidation reaction of thiol residues (Herrmann and Hell, 2005; Wiedemann et al., 2006). Consequently, the protein becomes folded and trapped in the IMS. The small Tim proteins are examples of this class. These proteins have a conserved twin Cx_3C motif in which two cysteine residues are separated by three amino acids. Spacing between each Cx_3C varies from 11 to 16 amino acids (Koehler, 2004). This signature is crucial for mitochondrial targeting, import, and assembly (Allen *et al.*, 2003; Lutz *et al.*, 2003). Mutations in the twin Cx_3C motif of the human homologue of Tim8p, DDP1 (deafness dystonia polypeptide 1), causes the inherited X-linked disease Mohr-Tranebjaerg syndrome, a progressive neurodegenerative disorder that leads to deafness, blindness, dystonia, and mental deterioration (Jin *et al.*, 1996).

The metallochaperone Cox17 is an 8-kDa copper-binding protein located in the IMS and cytosol. It contains six conserved cysteine residues; among them the last four form the twin Cx_9C motif which is required for targeting and import. Notably, the cysteine residues that coordinate the copper factor are distinct from the residues required for mitochondrial targeting (Heaton *et al.*, 2000). Several other proteins in the IMS, like Cox19, Cox23, and Som1 have twin Cx_9C motifs, pointing to possible common targeting and import mechanisms (Herrmann and Hell, 2005).

Cytochrome *c* presents a unique case of targeting and import into the IMS. It is synthesized in the cytosol as apocytochrome *c* precursor that lacks an N-terminal targeting signal. Mutagenesis studies suggest that apocytochrome *c* has redundant structural targeting information located in both termini of the protein (Jordi *et al.*, 1989; Nye and Scarpulla, 1990; Sprinkle *et al.*, 1990). The TOM complex appears to decode this structural information by a yet-unknown mechanism. Apocytochrome *c* interacts in the IMS with cytochrome *c* heme lyase (CCHL), an inner membrane-associated protein. The latter incorporates the heme cofactor into apocytochrome *c* by formation of two thioether bonds, thereby triggering both folding and the release of holocytochrome *c* into the IMS (Dumont *et al.*, 1988; Nargang *et al.*, 1988). Folding of the protein may trap it in the IMS.

The CCHL, mentioned above, belongs to the third class of proteins residing in the IMS. The precursor of CCHL lacks a matrix-targeting signal but contains a topogenic signal with two highly conserved motifs that are a characteristic signature of all known mitochondrial heme lyases (Diekert *et al.*, 1999). The signal, which is about 60 amino acid residue long, is found in the third quarter of these proteins and is sufficient for translocation into the IMS. It is highly hydrophilic containing a similar number of positive and negative charges distributed over the entire sequence element. Secondary structure prediction suggests the presence of two α-helices and one extended structure. Insertion of this signal into the cytosolic protein DHFR enabled the latter to be translocated into the mitochondrial IMS (Diekert *et al.*, 1999). Thus, this signal is sufficient for targeting to and import into mitochondria. Interestingly, when attached to either terminus of DHFR, the topogenic signal will not support import. These findings might point to the requirement for a looped structure for import of heme lyase into the IMS.

3. Internal Targeting Signals of Outer Membrane Proteins

The outer membrane of mitochondria contains a diverse set of proteins with different topologies. They can be classified as follows: (1) Signal anchor proteins like Tom70, Tom20, and Tom45. These proteins expose a large domain to the cytosol and only a small N-terminal segment is present on the IMS side (Fig. 3). (2) Tail-anchored proteins, such as Tom5, Bcl-2, and Fis1, have a single TMD at the C-terminus and a large N-terminal region is exposed to the cytosol (Fig. 3). (3) Fzo1 (a component of the mitochondrial fusion machinery) spans the outer membrane twice, exposing a small loop in the IMS. (4) β-Barrel proteins like porins and Tom40 are predicted to traverse the outer membrane in a series of antiparallel β-strands (Rapaport, 2003; Waizenegger *et al.*, 2004). (5) Peripherally associated membrane proteins, like Mas37 and Tob38, which are present at the cytosolic side of the outer membrane (Gratzer *et al.*, 1995; Paschen *et al.*, 2005; Wiedemann *et al.*, 2003). The targeting signals of the first two groups were analyzed in some detail. In contrast to proteins that span the outer membrane once, the targeting information in β-barrel proteins appears to be encoded in a structural element that involves different regions rather than a contiguous linear sequence (Rapaport, 2003).

a. The Targeting Sequences of Signal–Anchored Proteins

N-terminally anchored proteins are also known as "signal-anchored" proteins because their TMD and its flanking regions function both as an intracellular sorting signal and as an anchor to the membrane (Shore *et al.*, 1995). However, these proteins do not share any sequence similarity in their signal-anchor domains. Therefore, the targeting information for these proteins is probably encoded in structural elements rather than in a specific primary sequence.

Analysis of the signal-anchored segments in both mammalian and yeast cells revealed that the TMD is required for both mitochondrial targeting and membrane anchoring of the protein, whereas the positively charged residues in the flanking regions are required to enhance the import rate (McBride *et al.*, 1992; Waizenegger *et al.*, 2003). The moderate hydrophobicity of the TMD, and to a various extent its length, may be critical for targeting to mitochondria (Kanaji *et al.*, 2000). Recently, it was also shown in yeast that the moderate hydrophobicity of the TMD is the most important requirement for targeting and anchoring. Furthermore, anchor domains of outer membrane proteins were shown to be functionally interchangeable. Hence, they seem to play only a minor role in the specific function of these proteins, but have a decisive role in topogenic signaling (Waizenegger *et al.*, 2003).

b. The Targeting and Sorting Sequences of Tail–Anchored Proteins

Like the N-terminally anchored proteins, tail-anchored proteins do not share any sequence conservation in their tail-region. Therefore, the mitochondrial-targeting information must be encoded in the structural features of this part of the protein. The importance of the positive charges in the TMD-flanking regions was recognized with several proteins. Proteins of the Bcl-2 family are central regulators of apoptosis. Bcl-X_L is specifically targeted to the mitochondrial outer membrane, whereas Bcl-2 is present on several intracellular membranes. The TMD of both proteins have the same length and hydrophobicity. However, the signal in Bcl-X_L contains two basic amino acids at both ends of the TMD. The signal in Bcl-2 contains only one basic residue on either side, which is presumably why it is inserted into different membranes in an apparently nonspecific manner (Kaufmann *et al.*, 2003).

Cytochrome b_5 and VAMP1 are other examples of dual-membrane localization. Both proteins exist in two isoforms: one in the ER membrane and the other in the mitochondrial outer membrane. Proper targeting and insertion of these proteins into the mitochondrial outer membrane requires a short TMD within the tail domain and presence of positively charged amino acids in the region that flanks the TMD at its C-terminus region (Borgese *et al.*, 2003; Isenmann *et al.*, 1998).

Individual residues within the TMD were also shown to play a role in the targeting of tail-anchored proteins such as the components of the TOM core complex: Tom5, Tom6, Tom7, and Tom22. The TMD of these proteins harbors a conserved proline residue. In the case of Tom7, this residue was shown to be important for efficient targeting (Allen *et al.*, 2002). The proline residue, which is known as an α-helical destabilizer, may introduce flexibility within the TMD. This flexibility might help the tail domain to be anchored in the membrane.

Further information was obtained by studying yeast Tom5 in a mammalian system. Moderate length of the TMD, positive charges in the C-segment, and the distance between, or the context of the TMD and C-segment appear critical for targeting (Horie *et al.*, 2002). The importance of a positive net charge in the C-terminal segment for correct mitochondrial targeting has been demonstrated also in the case of Fis1 (Habib *et al.*, 2003; Horie *et al.*, 2003). In summary, moderately hydrophobic and relatively short TMD together with positively charged residues at its flanking regions are the crucial features of targeting signals of tail-anchored proteins.

IV. Bioinformatics Tools to Predict Mitochondrial Targeting Signals

Since the internal targeting signals are highly variable and not sufficiently characterized, there are currently no algorithms available to predict these signals. In contrast, several algorithms have been designed to predict mitochondrial pre-sequences. The prediction provided by these programs is based mainly on (1) physicochemical parameters such as the abundance of certain amino acids and the hydrophobicity in certain regions and/or (2) analysis of the plain residue patterns in (part of) the amino acid sequences. Below, we provide a short overview of some widely used programs. Links to some of these programs together with additional information on the localization of proteins to mitochondria can be found in the database MitoP2 (http://www.mitop.de). MitoP2 provides information about mitochondrial proteins from budding yeast, *Neurospora crassa*, human, and mouse.

A. TargetP

TargetP (http://www.cbs.dtu.dk/services/TargetP/) is a neural network-based tool that uses only N-terminal sequence information to discriminate among proteins destined to mitochondria, chloroplasts, secretory pathway, or "other" localization (Emanuelsson *et al.*, 2000). It can also predict a potential cleavage site for presequence processing. A nonplant version of TargetP that distinguishes only between mitochondrial presequences, signal peptides, and other signal sequences has also been constructed. One drawback of neural networks, however, is that it is generally difficult to understand and interpret how and why they make such predictions.

B. PSORT II

PSORT II (http://www.psort.org/) requests a full-length amino acid sequence and its origin (Gram-negative bacteria, Gram-positive bacteria, yeasts, animals, and plants). The program then calculates the values of 22 feature variables that reflect various characteristics of the sequence (Nakai and Horton, 1999). Accordingly, it estimates the likelihood of the protein being sorted to each candidate site. In the case of mitochondrial presequence, the criterion is based on amino acid composition of the N-terminal 20 residues and some weak cleavage-site consensus.

C. MITOPRED

MITOPRED (http://bioinformatics.albany.edu/~mitopred/) predicts nuclear-encoded mitochondrial proteins from all eukaryotic species, including plants. Prediction is based on the occurrence patterns of Pfam domains (version 16.0) in different cellular locations, amino acid composition, and pI value differences between mitochondrial and nonmitochondrial proteins (Sonnhammer *et al.*, 1997).

D. MitoProt II

MitoProt II (http://ihg.gsf.de/ihg/mitoprot.html) predicts the N-terminal protein region that can represent a matrix-targeting sequence (presequence) and the cleavage site. A complete description of the method to make the prediction is available in Claros and Vincens (1996).

E. Predotar

Predotar (http://urgi.infobiogen.fr/predotar/predotar.html) is a neural network-based approach (Predotar—*Pred*iction of *O*rganelle *Tar*geting sequences) which recognizes the N-terminal targeting sequences of classically targeted precursor proteins. For each protein sequence, Predotar provides a probability estimate as to whether the sequence contains a mitochondrial, plastid, or ER targeting sequence (Small *et al.*, 2004).

In summary, prediction programs are very useful tools to obtain initial information on a possible localization of an unknown protein to mitochondria. As the different programs are based on different criteria, a combination of results from some of them may provide a better prediction. In Section V, we discuss how experimental methods can provide the required support for these theoretical predictions.

V. Experimental Analysis of Mitochondrial–Targeting Signals

There are several experimental approaches to analyze a putative mitochondrial-targeting signal.

A. Construction of Hybrid Proteins

To confirm the targeting capacity of a sequence domain, this sequence can be fused to a passenger protein and the ability of the resulting hybrid protein to be imported into mitochondria *in vivo* and *in vitro* is then investigated. In many cases, fusion of mitochondrial presequences to a nonmitochondrial protein (passenger proteins) was reported to guide the passenger protein into the organelle (Horwich *et al.*, 1985; Hurt *et al.*, 1984, 1985). For example, pSu9-DHFR, a hybrid protein consisting of the first 69 amino acids of the precursor of subunit 9 of the mito-chondrial F_0-ATPase (pSu9) fused to the amino terminus of mouse dihydrofolate reductase (DHFR), is rapidly imported into isolated mitochondria (Pfanner *et al.*, 1987). Import of precursor proteins can be analyzed by the protection of the imported precursors against externally added proteases and/or by the processing of the presequence by MPP.

Furthermore, the subcellular location of a hybrid protein composed of a putative mitochondrial-targeting signal and the green fluorescent protein (GFP) can also be determined *in vivo* by fluorescence microscopy or subcellular fractionation followed by immunodecoration with antibodies against GFP (Chapter 1 by Pallotti and Lenaz, Chapter 2 by Boldogh and Pon, and Chapter 3 by Millar *et al.*). Transformation of cells with such constructs and using fluorescence microscopy yielded brightly fluorescent mitochondria (De Giorgi *et al.*, 1999; Westermann and Neupert, 2000).

B. Mutagenesis Approach

Analysis of precursor proteins with systematic mutations and deletions is a valuable approach to identify and characterize targeting signal(s). The targeting ability of the mutated protein can be examined *in vivo* and/or *in vitro*. For the *in vivo* studies, the mutated protein can be inserted as a GFP fusion protein into the appropriate expression vector and its localization to mitochondria can be analyzed by fluorescence microscopy. Alternatively, subcellular fractionation and immuno-decoration can be applied. Another *in vivo* method is performing a functional complementation assay in the yeast *Saccharomyces cerevisiae*. This method can be used either when the gene is essential for cell viability or when the gene is not an essential one but its deletion has a definite phenotype (for examples see

Waizenegger *et al.*, 2003). In this approach, the mutated construct is transformed into a deletion strain of the corresponding gene and the ability of the mutated protein to restore growth is tested [in the case of essential genes the "plasmid-shuffling" method is applied (Sikorski and Hieter, 1989)]. It is important to be aware that some mutations can affect the function of the protein and not the targeting information. Thus, a combination of several assays provides a better interpretation of the results. For the *in vitro* analysis, the establishment of a reliable *in vitro* import assay for the wild-type protein is prerequisite for studying the mutated variants.

C. Peptide Library Scan

Targeting signals are usually deciphered by components of the mitochondrial translocation machineries. Detection of interaction between these components and certain sequences within the investigated protein may provide information about targeting signals. Peptide library scans, have been used to identify mitochondrial internal targeting signals (Brix *et al.*, 1997; Stan *et al.*, 2003). For this procedure, cellulose-bound peptide libraries are prepared by automated spot synthesis (Frank, 1992; Kramer and Schneider-Mergener, 1998). Peptides of 13 amino acid residues with an overlap of 10 residues cover the sequence of interest. The membranes are incubated with a recombinant-purified component of the mitochondrial translocation machinery (e.g., the cytosolic domain of Tom20, Tom22, or Tom70). After a washing step, the bound protein is transferred to a polyvinylidene difluoride (PVDF) membrane, followed by detection with antibodies against the corresponding TOM component. The spots obtained can be quantified and thus the analysis of the intensity of the various spots provides information about the affinity of the import receptor to various regions within the sequence of the precursor protein. This method is applicable only when the targeting signal is built up from a continuous sequence and would not provide reliable data in cases where the targeting information is encoded in a structural element composed from separate segments within the precursor protein.

D. Replacing of the Putative Signal with Known Mitochondrial-Targeting Sequences

In order to verify the targeting information of a sequence, one can replace the putative signal by a known mitochondrial-targeting sequence which can target the protein to the same submitochondrial compartment. Correct localization can be verified by subcellular fractionation and immunodecoration of the hybrid protein or by using the complementation assay mentioned above. For example, the signal-anchor domains of some outer membrane proteins were exchanged among proteins from this family without significant effects on the functions of these proteins. This confirmed the targeting capacity of these segments (Waizenegger *et al.*, 2003). A similar approach was also employed successfully also with outer membrane tail-anchored proteins (Habib *et al.*, 2003). Of note, a complementation assay cannot be applied when the targeting signal is part of the functional domain of the protein.

VI. Summary

The targeting of mitochondrial precursor proteins from the cytosol to the organelle and the subsequent intramitochondrial sorting depend on specific targeting and sorting information within the precursor sequence. In the case of the cleavable presequences, the characteristics of this signal are well understood and various bioinformatics and experimental tools are available to analyze them. The challenge for the coming years is to achieve a similar level of characterization and prediction ability for the various internal targeting and sorting signals.

Acknowledgments

We thank Dušan Popov-Čeleketić for helpful discussions and Igor Siwanowicz for the help in preparing the figures.

References

Abe, Y., Shodai, T., Muto, T., Mihara, K., Torii, H., Nishikawa, S., Endo, T., and Kohda, D. (2000). Structural basis of presequence recognition by the mitochondrial protein import receptor Tom20. *Cell* **100,** 551–560.

Allen, R., Egan, B., Gabriel, K., Beilharz, T., and Lithgow, T. (2002). A conserved proline residue is present in the transmembrane-spanning domain of Tom7 and other tail-anchored protein subunits of the TOM translocase. *FEBS Lett.* **514,** 347–350.

Allen, S., Lu, H., Thornton, D., and Tokatlidis, K. (2003). Juxtaposition of the two distal CX3C motifs via intrachain disulfide bonding is essential for the folding of Tim10. *J. Biol. Chem.* **278,** 38505–38513.

Bedwell, D. M., Klionsky, D. J., and Emr, S. D. (1987). The yeast F1-ATPase beta subunit precursor contains functionally redundant mitochondrial protein import information. *Mol. Cell Biol.* **7,** 4038–4047.

Biswas, T. K., and Getz, G. S. (2002). Import of yeast mitochondrial transcription factor (Mtf1p) via a nonconventional pathway. *J. Biol. Chem.* **277,** 45704–45714.

Blobel, G. B. (1975). Transfer of proteins across membranes. II. Reconstitution of functional rough microsomes from heterologous components. *J. Cell Biol.* **67,** 852–862.

Borgese, N., Colombo, S., and Pedrazzini, E. (2003). The tale of tail-anchored proteins: Coming from the cytosol and looking for a membrane. *J. Cell Biol.* **161,** 1013–1019.

Branda, S. S., and Isaya, G. (1995). Prediction and identification of new natural substrates of the yeast mitochondrial intermediate peptidase. *J. Biol. Chem.* **270,** 27366–27373.

Brix, J., Dietmeier, K., and Pfanner, N. (1997). Differential recognition of preproteins by the purified cytosolic domains of the mitochondrial import receptors Tom20, Tom22 and Tom70. *J. Biol. Chem.* **272,** 20730–20735.

Bruch, M. D., and Hoyt, D. W. (1992). Conformational analysis of a mitochondrial presequence derived from the F1-ATPase beta-subunit by CD and NMR spectroscopy. *Biochim. Biophys. Acta* **1159,** 81–93.

Burri, L., Strahm, Y., Hawkins, C. J., Gentle, I. E., Puryer, M. A., Verhagen, A., Callus, B., Vaux, D., and Lithgow, T. (2005). Mature DIABLO/Smac is produced by the IMP protease complex on the mitochondrial inner membrane. *Mol. Biol. Cell* **16,** 2926–2933.

Claros, M. G., and Vincens, P. (1996). Computational method to predict mitochondrially imported proteins and their targeting sequences. *Eur. J. Biochem.* **241,** 779–786.

De Giorgi, F., Ahmed, Z., Bastianutto, C., Brini, M., Jouaville, L. S., Marsault, R., Murgia, M., Pinton, P., Pozzan, T., and Rizzuto, R. (1999). Targeting GFP to organelles. *Methods Cell Biol.* **58,** 75–85.

Diekert, K., Kispal, G., Guiard, B., and Lill, R. (1999). An internal targeting signal directing proteins into the mitochondrial intermembrane space. *Proc. Natl. Acad. Sci. USA* **96,** 11752–11757.

Dircks, L. K., and Poyton, R. O. (1990). Overexpression of a leaderless form of yeast cytochrome *c* oxidase subunit V_a circumvents the requirement for a leader peptide in mitochondrial import. *Mol. Cell. Biol.* **10,** 4984–4986.

Dumont, M. E., Ernst, J. F., and Sherman, F. (1988). Coupling of heme attachment to import of cytochrome *c* into yeast mitochondria. *J. Biol. Chem.* **263,** 15928–15937.

Emanuelsson, O., Nielsen, H., Brunak, S., and von Heijne, G. (2000). Predicting subcellular localization of proteins based on their N-terminal amino acid sequence. *J. Mol. Biol.* **300,** 1005–1016.

Emtage, J. L., and Jensen, R. E. (1993). MAS6 encodes an essential inner membrane component of the yeast mitochondrial protein import pathway. *J. Cell Biol.* **122,** 1003–1012.

Endres, M., Neupert, W., and Brunner, M. (1999). Transport of the ADP/ATP carrier of mitochondria from the TOM complex to the TIM22.54 complex. *EMBO J.* **18,** 3214–3221.

Esser, K., Tursun, B., Ingenhoven, M., Michaelis, G., and Pratje, E. (2002). A novel two-step mechanism for removal of a mitochondrial signal sequence involves the mAAA complex and the putative rhomboid protease Pcp1. *J. Mol. Biol.* **323,** 835–843.

Fölsch, H., Gaume, B., Brunner, M., Neupert, W., and Stuart, R. A. (1998). C- to N-terminal translocation of preproteins into mitochondria. *EMBO J.* **17,** 6508–6515.

Fölsch, H., Guiard, B., Neupert, W., and Stuart, R. A. (1996). Internal targeting signal of the BCS1 protein: A novel mechanism of import into mitochondria. *EMBO J.* **15,** 479–487.

Frank, R. (1992). Spot synthesis: An easy technique for the positionally addressable, parallel chemical synthesis on a membrane support. *Tetrahedron* **48,** 9217–9232.

Gabriel, K., Buchanan, S. K., and Lithgow, T. (2001). The alpha and beta: Protein translocation across mitochondrial and plastid outer membranes. *Trends Biochem. Sci.* **26,** 36–40.

Gakh, O., Cavadini, P., and Isaya, G. (2002). Mitochondrial processing peptidases. *Biochim. Biophys. Acta* **1592,** 63–77.

Gavel, Y., and von Heijne, G. (1990). Cleavage-site motifs in mitochondrial targeting peptides. *Protein Eng.* **4,** 33–37.

Geli, V., Yang, M., Suda, K., Lustig, A., and Schatz, G. (1990). The MAS-encoded processing protease of yeast mitochondria: Overproduction and characterization of its two nonidentical subunits. *J. Biol. Chem.* **31,** 19216–19222.

Glaser, E., and Dessi, P. (1999). Integration of the mitochondrial-processing peptidase into the cytochrome bc_1 complex in plants. *J. Bioenerg. Biomembr.* **31,** 259–274.

Glaser, E., Sjoling, S., Tanudji, M., and Whelan, J. (1998). Mitochondrial protein import in plants. Signals, sorting, targeting, processing and regulation. *Plant. Mol. Biol.* **38,** 311–338.

Glaser, S., and Cumsky, M. (1990). Localization of a synthetic presequence that blocks protein import into yeast mitochondria. *J. Biol. Chem.* **265,** 8817–8822.

Glick, B. S., Brandt, A., Cunningham, K., Muller, S., Hallberg, R. L., and Schatz, G. (1992). Cytochromes c_1 and b_2 are sorted to the intermembrane space of yeast mitochondria by a stop-transfer mechanism. *Cell* **69,** 809–822.

Gratzer, S., Lithgow, T., Bauer, R. E., Lamping, E., Paltauf, F., Kohlwein, S. D., Haucke, V., Junne, T., Schatz, G., and Horst, M. (1995). Mas37p, a novel receptor subunit for protein import into mitochondria. *J. Cell Biol.* **129,** 25–34.

Habib, S. J., Vasiljev, A., Neupert, W., and Rapaport, D. (2003). Multiple functions of tail-anchor domains of mitochondrial outer membrane proteins. *FEBS Lett.* **555,** 511–515.

Hahne, K., Haucke, V., Ramage, L., and Schatz, G. (1994). Incomplete arrest in the outer membrane sorts NADH-cytochrome b_5 reductase to two different submitochondrial compartments. *Cell* **79,** 829–839.

Hajek, P., Koh, J. Y., Jones, L., and Bedwell, D. M. (1997). The amino terminus of the F1-ATPase beta-subunit precursor functions as an intramolecular chaperone to facilitate mitochondrial protein import. *Mol. Cell. Biol.* **17,** 7169–7177.

Hammen, P. K., Gorenstein, D. G., and Weiner, H. (1994). Structure of the signal sequences for two mitochondrial matrix proteins that are not proteolytically processed upon import. *Biochemistry* **33,** 8610–8617.

Hammen, P. K., Gorenstein, D. G., and Weiner, K. (1996). Amphiphilicity determines binding properties of three mitochondrial presequences to lipid surface. *Biochemistry* **35,** 3772–3781.

Hartl, F.-U., Ostermann, J., Guiard, B., and Neupert, W. (1987). Successive translocation into and out of the mitochondrial matrix: Targeting of proteins to the intermembrane space by a bipartite signal peptide. *Cell* **51,** 1027–1037.

Hawlitschek, G., Schneider, H., Schmidt, B., Tropschug, M., Hartl, F. U., and Neupert, W. (1988). Mitochondrial protein import: Identification of processing peptidase and of PEP, a processing enhancing protein. *Cell* **53,** 795–806.

Heaton, D., Nittis, T., Srinivasan, C., and Winge, D. R. (2000). Mutational analysis of the mitochondrial copper metallochaperone Cox17. *J. Biol. Chem.* **275,** 37582–37587.

Hendrick, J. P., Hodges, P. E., and Rosenberg, L. E. (1989). Survey of amino-terminal proteolytic cleavage sites in mitochondrial precursor proteins: Leader peptides cleaved by two matrix proteases share a three-amino acid motif. *Proc. Natl. Acad. Sci. USA* **86,** 4056–4060.

Herlan, M., Bornhovd, C., Hell, K., Neupert, W., and Reichert, A. S. (2004). Alternative topogenesis of Mgm1 and mitochondrial morphology depend on ATP and a functional import motor. *J. Cell Biol.* **165,** 167–173.

Herlan, M., Vogel, F., Bornhovd, C., Neupert, W., and Reichert, A. S. (2003). Processing of Mgm1 by the rhomboid-type protease Pcp1 is required for maintenance of mitochondrial morphology and of mitochondrial DNA. *J. Biol. Chem.* **278,** 27781–27788.

Herrmann, J. M., and Hell, K. (2005). Chopped, trapped or tacked-protein translocation into the IMS of mitochondria. *Trends Biochem. Sci.* **30,** 205–211.

Horie, C., Suzuki, H., Sakaguchi, M., and Mihara, K. (2002). Characterization of signal that directs C-tail-anchored proteins to mammalian mitochondrial outer membrane. *Mol. Biol. Cell* **13,** 1615–1625.

Horie, C., Suzuki, H., Sakaguchi, M., and Mihara, K. (2003). Targeting and assembly of mitochondrial tail-anchored protein Tom5 to the TOM complex depend on a signal distinct from that of tail-anchored proteins dispersed in the membrane. *J. Biol. Chem.* **278,** 41462–41471.

Horwich, A. L., Kalousek, F., Mellmann, I., and Rosenberg, L. E. (1985). A leader peptide is sufficient to direct mitochondrial import of a chimeric protein. *EMBO J.* **4,** 1129–1135.

Hurt, E. C., Pesold-Hurt, B., and Schatz, G. (1984). The cleavable prepiece of an imported mitochondrial protein is sufficient to direct cytosolic dihydrofolate reductase into the mitochondrial matrix. *FEBS Lett.* **178,** 306–310.

Hurt, E. C., Pesold-Hurt, B., Suda, K., Opplinger, W., and Schatz, G. (1985). The first twelve aminoacids (less than half of the presequence) of an imported mitochondrial protein can direct mouse cytosolic dihydrofolate reductase into the yeast mitochondrial matrix. *EMBO J.* **4,** 2061–2068.

Isaya, G., Kalousek, F., Fenton, W. A., and Rosenberg, L. E. (1991). Cleavage of precursors by the mitochondrial processing peptidase requires a compatible mature protein or an intermediate octapeptide. *J. Cell. Biol.* **113,** 65–76.

Isenmann, S., Khew-Goodall, Y., Gamble, J., Vadas, M., and Wattenberg, B. W. (1998). A splice-isoform of vesicle-associated membrane protein-1 (VAMP-1) contains a mitochondrial targeting signal. *Mol. Biol. Cell* **9,** 1649–1660.

Jarvis, J. A., Ryan, M. T., Hoogenraad, N. J., Craik, D. J., and Hoj, R. B. (1995). Solution structure of the acetylated and noncleavable mitochondrial targeting signal of rat chaperonin 10. *J. Biol. Chem.* **270,** 1323–1331.

Jin, H., May, M., Tranebjaerg, L., Kendall, E., Fontan, G., Jackson, J., Subramony, S. H., Arena, F., Lubs, H., Smith, S., Stevenson, R., Schwartz, C., *et al.* (1996). A novel X-linked gene, DDP, shows mutations in families with deafness (DFN-1), dystonia, mental deficiency and blindness. *Nat. Genet.* **14,** 177–180.

Jordi, W., Li-Xin, Z., Pilon, M., Demel, R. A., and de Kruijff, B. (1989). The importance of the aminoterminus of the mitochondrial precursor protein apocytochrome *c* for translocation across model membranes. *J. Biol. Chem.* **264,** 2292–2301.

Kalousek, F., Neupert, W., Omura, T., Schatz, G., and Schmitz, U. K. (1993). Uniform nomenclature for the mitochondrial peptidases cleaving precursors of mitochondrial proteins. *Trends Biochem. Sci.* **18,** 249.

Kanaji, S., Iwahashi, J., Kida, Y., Sakaguchi, M., and Mihara, K. (2000). Characterization of the signal that directs Tom20 to the mitochondrial outer membrane. *J. Cell Biol.* **151,** 277–288.

Karslake, C., Piotto, M. E., Pak, Y. K., Weiner, H., and Gorenstein, D. G. (1990). 2D NMR and structural model for a mitochondrial signal peptide bound to a micelle. *Biochemistry* **29,** 9872–9878.

Kaufmann, T., Schlipf, S., Sanz, J., Neubert, K., Stein, R., and Borner, C. (2003). Characterization of the signal that directs Bcl-xL, but not Bcl-2, to the mitochondrial outer membrane. *J. Cell Biol.* **160,** 53–64.

Kleiber, J., Kalousek, F., Swaroop, M., and Rosenberg, L. E. (1990). The general mitochondrial matrix processing protease from rat liver: Structural characterization of the catalytic subunit. *Proc. Natl. Acad. Sci. USA* **87,** 7978–7982.

Koehler, C. M. (2004). The small Tim proteins and the twin Cx3C motif. *Trends Biochem. Sci.* **29,** 1–4.

Koehler, C. M., Merchant, S., and Schatz, G. (1999). How membrane proteins travel across the mitochondrial intermembrane space. *Trends Biochem. Sci.* **24,** 428–432.

Kozjak, V., Wiedemann, N., Milenkovic, D., Lohaus, C., Meyer, H. E., Guiard, B., Meisinger, C., and Pfanner, N. (2003). An essential role of Sam50 in the protein sorting and assembly machinery of the mitochondrial outer membrane. *J. Biol. Chem.* **278,** 48520–48523.

Kramer, A., and Schneider-Mergener, J. (1998). Synthesis and screening of peptide libraries on continuous cellulose membrane supports. *Methods Mol. Biol.* **87,** 25–39.

Krimmer, T., Rapaport, D., Ryan, M. T., Meisinger, C., Kassenbrock, C. K., Blachly-Dyson, E., Forte, M., Douglas, M. G., Neupert, W., Nargang, F. E., and Pfanner, N. (2001). Biogenesis of the major mitochondrial outer membrane protein porin involves a complex import pathway via receptors and the general import pore. *J. Cell Biol.* **152,** 289–300.

Lee, C. M., Sedman, J., Neupert, W., and Stuart, R. A. (1999). The DNA helicase, Hmi1p, is transported into mitochondria by a C-terminal cleavable targeting signal. *J. Biol. Chem.* **274,** 20937–20942.

Leenhouts, J. M., de Gier, J., and de Kruijff, B. (1993). A novel property of a mitochondrial presequence. Its ability to induce cardiolipin-specific interbilayer contacts which are dissociated by a transmembrane potential. *FEBS Lett.* **327,** 172–176.

Leenhouts, J. M., Torok, Z., Demel, R. A., de, G. J., and de, K. B. (1994). The full length of a mitochondrial presequence is required for efficient monolayer insertion and interbilayer contact formation. *Mol. Membr. Biol.* **11,** 159–164.

Lemire, B. D., Fankhauser, C., Baker, A., and Schatz, G. (1989). The mitochondrial targeting function of randomly generated peptide sequences correlates with predicted helical amphiphilicity. *J. Biol. Chem.* **264,** 20206–20215.

Lister, R., Murcha, M. W., and Whelan, J. (2003). The mitochondrial protein import machinery of plants (MPIMP) database. *Nucleic Acids Res.* **31,** 325–327.

Lutz, T., Neupert, W., and Herrmann, J. M. (2003). Import of small Tim proteins into the mitochondrial intermembrane space. *EMBO J.* **22,** 4400–4408.

McBride, H. M., Millar, D. G., Li, J. M., and Shore, G. C. (1992). A signal-anchor sequence selective for the mitochondrial outer membrane. *J. Cell Biol.* **119,** 1451–1457.

McQuibban, G. A., Saurya, S., and Freeman, M. (2003). Mitochondrial membrane remodelling regulated by a conserved rhomboid protease. *Nature* **423,** 537–541.

Model, K., Meisinger, C., Prinz, T., Wiedemann, N., Truscott, K. N., Pfanner, N., and Ryan, M. T. (2001). Multistep assembly of the protein import channel of the mitochondrial outer membrane. *Nat. Struct. Biol.* **8,** 361–370.

Murcha, M. W., Millar, A. H., and Whelan, J. (2005). The N-terminal cleavable extension of plant carrier proteins is responsible for efficient insertion into the inner mitochondrial membrane. *J. Mol. Biol.* **351,** 16–25.

Nakai, K., and Horton, P. (1999). PSORT: A program for detecting sorting signals in proteins and predicting their subcellular localization. *Trends Biochem. Sci.* **24,** 34–36.

Nargang, F. E., Drygas, M. E., Kwong, P. L., Nicholson, D. W., and Neupert, W. (1988). A mutant of *Neurospora crassa* deficient in cytochrome c heme lyase activity cannot import cytochrome *c* into mitochondria. *J. Biol. Chem.* **263,** 9388–9394.

Nelson, D. R., Felix, C. M., and Swanson, J. M. (1998). Highly conserved charge-pair networks in the mitochondrial carrier family. *J. Mol. Biol.* **277,** 285–308.

Neupert, W. (1997). Protein import into mitochondria. *Annu. Rev. Biochem.* **66,** 863–917.

Nobrega, F. G., Mobrega, M. P., and Tzagoloff, A. (1992). BSC1, a novel gene required for the expression of functional Rieske iron-sulfur protein in *Saccharomyces cerevisiae. EMBO J.* **11,** 3821–3829.

Nye, S. H., and Scarpulla, R. C. (1990). Mitochondrial targeting of yeast apoiso-1-cytochrome *c* is mediated through functionally independent structural domains. *Mol. Cell. Biol.* **10,** 5763–5771.

Paschen, S. A., and Neupert, W. (2001). Protein import into mitochondria. *IUBMB Life* **52,** 101–112.

Paschen, S. A., Neupert, W., and Rapaport, D. (2005). Biogenesis of beta-barrel membrane proteins of mitochondria. *Trends Biochem. Sci.* **30,** 575–582.

Paschen, S. A., Waizenegger, T., Stan, T., Preuss, M., Cyrklaff, M., Hell, K., Rapaport, D., and Neupert, W. (2003). Evolutionary conservation of biogenesis of β-barrel membrane proteins. *Nature* **426,** 862–866.

Pfanner, N., and Geissler, A. (2001). Versatility of the mitochondrial protein import machinery. *Nat. Rev. Mol. Cell Biol.* **2,** 339–349.

Pfanner, N., Tropschug, M., and Neupert, W. (1987). Mitochondrial protein import: Nucleoside triphosphates are involved in conferring import-competence to precursors. *Cell* **49,** 815–823.

Prokisch, H., Andreoli, C., Ahting, U., Heiss, K., Ruepp, A., Scharfe, C., and Meitinger, T. (2006). MitoP2: The mitochondrial proteome database—now including mouse data. *Nucleic Acids Res.* **34,** D705–D711.

Rapaport, D. (2002). Biogenesis of the mitochondrial TOM complex. *Trends Biochem. Sci.* **27,** 191–197.

Rapaport, D. (2003). How to find the right organelle—targeting signals in mitochondrial outer membrane proteins. *EMBO Rep.* **4,** 948–952.

Rapaport, D., and Neupert, W. (1999). Biogenesis of Tom40, core component of the TOM complex of mitochondria. *J. Cell Biol.* **146,** 321–331.

Rehling, P., Pfanner, N., and Meisinger, C. (2003). Insertion of hydrophobic membrane proteins into the inner mitochondrial membrane—a guided tour. *J. Mol. Biol.* **326,** 639–657.

Roise, D., Horvath, S. J., Tomich, J. M., Richards, J. H., and Schatz, G. (1986). A chemically synthesized pre-sequence of an imported mitochondrial protein can form an amphiphilic helix and perturb natural and artificial phospholipid bilayers. *EMBO J.* **5,** 1327–1334.

Roise, D., and Schatz, G. (1988). Mitochondrial presequences. *J. Biol. Chem.* **263,** 4509–4511.

Rospert, S., Junne, T., Glick, B. S., and Schatz, G. (1993). Cloning and disruption of the gene encoding yeast mitochondrial chaperonin 10, the homolog of *E. coli* groES. *FEBS Lett.* **335,** 358–360.

Sanyal, A., and Getz, G. S. (1995). Import of transcription factor MTF1 into the yeast mitochondria takes place through an unusual pathway. *J. Biol. Chem.* **270,** 11970–11976.

Schatz, G. (1997). Just follow the acid chain. *Nature* **388,** 121–122.

Schleiff, E., Silvius, J. R., and Shore, G. C. (1999). Direct membrane insertion of voltage-dependent anion-selective channel protein catalyzed by mitochondrial Tom20. *J. Cell Biol.* **145,** 973–978.

Shore, G. C., McBride, H. M., Millar, D. G., Steenaart, N. A. E., and Nguyen, M. (1995). Import and insertion of proteins into the mitochondrial outer membrane. *Eur. J. Biochem.* **227,** 9–18.

Sickmann, A., Reinders, J., Wagner, Y., Joppich, C., Zahedi, R., Meyer, H. E., Schonfisch, B., Perschil, I., Chacinska, A., Guiard, B., Rehling, P., Pfanner, N., *et al.* (2003). The proteome of *Saccharomyces cerevisiae* mitochondria. *Proc. Natl. Acad. Sci. USA* **100,** 13207–13212.

Sikorski, R. S., and Hieter, P. (1989). A system of shuttle vectors and host strains designed for efficient manipulation of DNA in *Saccharomyces cerevisiae. Genetics* **122,** 19–27.

Small, I., Peeters, N., Legeai, F., and Lurin, C. (2004). Predotar: A tool for rapidly screening proteomes for N-terminal targeting sequences. *Proteomics* **4,** 1581–1590.

Sonnhammer, E. L., Eddy, S. R., and Durbin, R. (1997). Pfam: A comprehensive database of protein domain families based on seed alignments. *Proteins* **28**, 405–420.

Sprinkle, J. R., Hakvoort, T. B., Koshy, T. I., Miller, D. D., and Margoliash, E. (1990). Amino acid sequence requirements for the association of apocytochrome *c* with mitochondria. *Proc. Natl. Acad. Sci. USA* **87**, 5729–5733.

Stan, T., Ahting, U., Dembowski, M., Künkele, K.-P., Nussberger, S., Neupert, S., and Rapaport, D. (2000). Recognition of preproteins by the isolated TOM complex of mitochondria. *EMBO J.* **19**, 4895–4902.

Stan, T., Brix, J., Schneider-Mergener, J., Pfanner, N., Neupert, W., and Rapaport, D. (2003). Mitochondrial protein import: Recognition of internal import signals of BCS1 by the TOM complex. *Mol. Cell. Biol.* **23**, 2239–2250.

Taylor, S. W., Fahy, E., Zhang, B., Glenn, G. M., Warnock, D. E., Wiley, S., Murphy, A. N., Gaucher, S. P., Capaldi, R. A., Gibson, B. W., and Ghosh, S. S. (2003). Characterization of the human heart mitochondrial proteome. *Nat. Biotechnol.* **21**, 281–286.

Torok, Z., Demel, R. A., Leenhouts, J. M., and de, K. B. (1994). Presequence-mediated intermembrane contact formation and lipid flow. A model membrane study. *Biochemistry* **33**, 5589–5594.

von Heijne, G. (1986a). Mitochondrial targeting sequences may form amphiphilic helices. *EMBO J.* **5**, 1335–1342.

von Heijne, G. (1986b). Towards a comparative anatomy of N-terminal topogenic protein sequences. *J. Mol. Biol.* **189**, 239–242.

von Heijne, G., Stepphun, J., and Herrmann, R. G. (1989). Domain structure of mitochondrial and chloroplast targeting peptides. *Eur. J. Biochem.* **180**, 535–545.

Waizenegger, T., Habib, S. J., Lech, M., Mokranjac, D., Paschen, S. A., Hell, K., Neupert, W., and Rapaport, D. (2004). Tob38, a novel essential component in the biogenesis of beta-barrel proteins of mitochondria. *EMBO Rep.* **5**, 704–709.

Waizenegger, T., Stan, T., Neupert, W., and Rapaport, D. (2003). Signal-anchor domains of proteins of the outer membrane of mitochondria: Structural and functional characteristics. *J. Biol. Chem.* **278**, 42064–42071.

Waltner, M., and Weiner, H. (1995). Conversion of a nonprocessed mitochondrial precursor protein into one that is processed by the mitochondrial processing peptidase. *J. Biol. Chem.* **270**, 26311–26317.

Westermann, B., and Neupert, W. (2000). Mitochondria-targeted green fluorescent proteins: Convenient tools for the study of organelle biogenesis in *Saccharomyces cerevisiae*. *Yeast* **16**, 1421–1427.

Wiedemann, N., Kozjak, V., Chacinska, A., Schönfish, B., Rospert, S., Ryan, M. T., Pfanner, N., and Meisinger, C. (2003). Machinery for protein sorting and assembly in the mitochondrial outer membrane. *Nature* **424**, 565–571.

Wiedemann, N., Pfanner, N., and Chacinska, A. (2006). Chaperoning through the mitochondrial intermembrane space. *Mol. Cell* **21**, 145–148.

Winning, B. M., Sarah, C. J., Purdue, P. E., Day, C. D., and Leaver, C. J. (1992). The adenine nucleotide translocator of higher plants is synthesized as a large precursor that is processed upon import into mitochondria. *Plant J.* **2**, 763–773.

Zara, V., Palmieri, F., Mahlke, K., and Pfanner, N. (1992). The cleavable presequnce is not essential for zimport and assembly of the phosphate carrier of mammalian mitochondria but enhances the specificity and efficiency of import. *J. Biol. Chem.* **267**, 12077–12081.

CHAPTER 36

Import of Proteins into Mitochondria

Diana Stojanovski, Nikolaus Pfanner, and Nils Wiedemann

Institut für Biochemie und Molekularbiologie
Zentrum für Biochemie und Molekulare Zellforschung
Universität Freiburg, Hermann-Herder-Straße 7
D-79104 Freiburg, Germany

I. Introduction

Mitochondria are prominent and essential members of the eukaryotic cytoplasm, which have puzzled researchers for more than a century. Most notably, mitochondria represent the primary site for ATP synthesis via oxidative phosphorylation, but in recent years the organelle has been placed at the forefront of much attention, as its involvement in many processes essential for cell viability is being realized. These include its role in Fe–S biosynthesis (Kispal *et al.*, 1999; Lange *et al.*, 2000; Li *et al.*, 2001; Lill and Mühlenhoff, 2005; Lill *et al.*, 1999), cellular Ca^{2+} buffering (Leo *et al.*, 2005; Pinton *et al.*, 1998; Rizzuto *et al.*, 2000), and its pivotal role in programed cell death (Desagher and Martinou, 2000; Green and Reed, 1998; Youle and Karbowski, 2005). The majority of mitochondrial precursors, ~1000 in the Baker's yeast *Saccharomyces cerevisae*, are encoded by the nuclear genome and are synthesized on cytosolic ribosomes (Schatz and Dobberstein, 1996; Sickmann *et al.*, 2003; Wiedemann *et al.*, 2004a). Proper organelle functioning and ultimately cellular survival relies heavily on the successful delivery and integration of these nuclear-encoded precursors into the organelle.

Individual- or multiple-targeting elements within the primary sequence of nuclear-encoded mitochondrial precursors not only govern their initial direction to the organelle, but also dictate their final intramitochondrial destination. Specialized translocation machineries within the organelle's outer and inner membranes, in addition to translocation mediators within the organelle's soluble regions—the mitochondrial intermembrane space (IMS) and mitochondrial matrix—take on the task of sorting these nuclear-encoded precursors to their correct mitochondrial subcompartments. Briefly, the translocase of the outer mitochondrial membrane (TOM complex) represents the recognition site and central entry gate for the import of essentially all nuclear-encoded mitochondrial proteins. Precursors referred to as simple outer membrane proteins, defined by the presence of a single transmembrane anchor, engage with the TOM complex, on which their release or insertion into the lipid bilayer can be facilitated by an as-yet-undefined mechanism (Pfanner *et al.*, 2004; Rapaport, 2003). Precursors of the outer membrane with more complex topologies, such as porin and Tom40, which possess a β-barrel structural fold, require the further action of the sorting and assembly machinery (SAM complex) of the outer membrane for their successful integration and assembly into functional complexes (Gentle *et al.*, 2004; Ishikawa *et al.*, 2004; Kozjak *et al.*, 2003; Milenkovic *et al.*, 2004; Paschen *et al.*, 2003; Waizenegger *et al.*, 2004; Wiedemann *et al.*, 2003, 2004b).

The "classical" import pathway into mitochondria is defined by the presence of a cleavable presequence at the amino-terminus of preproteins, and is exploited by the majority of matrix residents and some proteins of the mitochondrial inner membrane and IMS. These members are translocated through the TOM complex and are delivered to the inner membrane presequence translocase (TIM23 complex). The membrane potential across the inner membrane activates the Tim23 channel and drives the translocation of the amino-terminal presequence by an electrophoretic mechanism (Wiedemann *et al.*, 2004a). The TIM23 complex can exist in two alternative states influenced by the presence or absence of the inner membrane constituent, Tim21 (Chacinska *et al.*, 2005). The absence of Tim21 permits association of the ATP-driven presequence translocase-associated motor (PAM complex) with the TIM23 complex, which is now competent for protein translocation into the matrix, on which presequence removal mediated by the mitochondrial processing peptidase (MPP) can take place (Chacinska *et al.*, 2005). Alternatively, a Tim21-bound and PAM-free TIM23 complex is competent for the sorting of inner membrane precursors (Chacinska *et al.*, 2005). The progression of such precursors into the mitochondrial matrix is halted by the presence of a hydrophobic-sorting anchor, or stop transfer signal, which typically follows the positively charged presequence (Glick *et al.*, 1992). On arrest within the TIM23 translocon, inner membrane precursors are released laterally into the inner membrane. After release into the inner membrane, some precursors are exposed to a processing event on the IMS side of the inner membrane, which results in the release of a soluble IMS protein (Burri *et al.*, 2005; Glick *et al.*, 1992).

Other IMS proteins, which are typically small in size and contain characteristic cysteine residues, exploit a recently defined specific mitochondrial intermemembrane space import and assembly machinery, consisting of the IMS components Mia40 and Erv1 (Allen *et al.*, 2005; Chacinska *et al.*, 2004; Mesecke *et al.*, 2005; Naoé *et al.*, 2004; Rissler *et al.*, 2005). The abundant class of inner membrane carrier proteins employ an alternative import pathway that is dictated by internal cryptic-targeting information. These precursors are guided to an alternative translocase of the inner membrane, the TIM22 complex, by small Tim chaperone complexes of the IMS after their translocation through the TOM complex (Curran *et al.,* 2002; Koehler, 2004; Rehling *et al.*, 2003; Truscott *et al.*, 2002; Wiedemann *et al.*, 2004a). The TIM22 complex facilitates the insertion of carrier proteins into the inner membrane in a $\Delta\Psi$-dependent manner.

The characterization of these alternative import pathways, by which the organelle upholds and maintains its integrity, has been greatly facilitated through the use of *in vitro* import studies. Through the incubation of isolated mitochondria under alternative conditions with *in vitro*-translated ^{35}S-labeled precursor proteins, one can delineate the import pathway and import requirements for the precursor of interest. The *in vitro* mitochondrial import assay is not only a valuable tool in the analysis of the mitochondrial import machinery, but can provide much insight into the biogenesis of different mitochondrial precursors. In particular, such an assay is of great benefit in the early characterization of proteins, when tools like antibodies may not be available. We describe here the basis of the *in vitro* mitochondrial import assay and the various questions that can be addressed by exploiting this technique.

II. *In Vitro* Synthesis of Precursor Proteins

A. RNA Preparation and *In Vitro* Transcription

The success of *in vitro* import experiments is markedly influenced by the quality of the *in vitro*-translated precursor, which in turn is largely influenced by the quality of the RNA employed in the reaction. To achieve good yields and high quality *in vitro* generated RNA transcripts, pure template DNA free of contaminants is required, in addition to employing a high fidelity RNA promoter. For the *in vitro* transcription reaction, two template options are available: (1) the use of plasmid constructs encoding the open-reading frame (ORF) of interest or (2) DNA templates generated by the polymerase chain reaction (PCR) (Ryan *et al.*, 2001).

The use of plasmid constructs entails cloning of the desired ORF into a suitable vector, which for such applications include Promega's (Madison, WI) pGEM series, Stratagenes's (La Jolla, CA) pBluescript, and Ambion's (Austin, TX) pDP. These often contain two bacteriophage polymerase promoters, such as SP6 and T7, typically on either side of the multiple-cloning site (MCS). The ORF must be cloned in the correct orientation (depending on the polymerase to be utilized in subsequent transcription), and the initiation codon should be relatively close to the promoter. It is often suggested that prior to its application in the transcription reaction, plasmid DNA be linearized just downstream of the stop codon or the transcription termination site. This ensures that the generated RNA transcripts are of a defined length and sequence. Although in many instances this may not be necessary; this is an important variable that one can decipher with relative ease in early experimental trials. Conversely, the use of PCR-generated DNA templates entails the amplification of the desired ORF from genomic DNA (most *S. cerevisiae* genes are intron free) or cDNA. This approach involves the use of a 5′ primer that incorporates an additional RNA polymerase promoter site, such as an SP6 site, upstream of the ORF-encoding sequence. The purified PCR product can subsequently be used directly in the transcription reaction. This option is useful for quick screens, as it eliminates additional cloning procedures.

The transcription reaction is essentially performed as outlined in the instructions provided from the manufacturer of the appropriate RNA polymerase. The procedure presented here is the method we routinely adopt in-house. However, high-quality and high-yield capped RNA can also be obtained by exploiting the simplified reaction format of RNA kits, such as the Ambion mMESSAGE mMACHINE kits.

1. Reagents and Methodology

a. Reagents

$10\times$ Transcription buffer
    400-mM HEPES-KOH, pH 7.4
    60-mM $Mg(OAc)_2$
    20-mM Spermidine in RNAse-free water
    Store in 400-μl aliquots at $-20\,^{\circ}C$
Transcription premix
    400-μl Transcription buffer

> 20 μl of 20-mg/ml Fatty acid-free bovine serum albumin (BSA)
>
> 40 μl of 1-M Dithiothreitol (DTT)
>
> 20 μl of 0.1-M ATP
>
> 20 μl of 0.1-M CTP
>
> 20 μl of 0.1-M UTP
>
> 2 μl of 0.1-M GTP
>
> 2.67-ml RNAse-free H_2O
>
> Filter sterilize and store in 120-μl aliquots at $-20\,^{\circ}$C

1-mM Cap

Make a 1-mM m^7G(5$'$)ppp(5$'$)G (GE Healthcare, Amersham Biosciences, Little Chalfont, Buckinghamshire, UK) stock solution in RNase-free H_2O. Store in 10-μl aliquots at $-20\,^{\circ}$C. Some RNA transcripts may not require Cap for efficient translation. In these instances, the transcription premix should be adjusted to incorporate an equal amount of GTP relative to the other nucleotides.

Ribonuclease inhibitor (e.g., RNasin from Promega or SUPERase-In from Ambion)

RNA polymerase: SP6 or T7 RNA polymerase (e.g., Stratagene)

Template DNA (20-μg plasmid DNA or 2-μg purified PCR product)

10-M LiCl in RNAse-free H_2O

Absolute ethanol

b. Procedure

Combine template DNA, 120 μl of transcription premix, 200-U RNase inhibitor, 10 μl of 1-mM CAP, and 100 U of the appropriate RNA polymerase, and bring solution to a final volume of 200 μl with sterile RNase-free H_2O. The sample is incubated at 37$\,^{\circ}$C for 1 h (incubation for 2 h is recommended to obtain higher yields, particularly for SP6 reactions that tend to be slower than T7 or T3 reactions). After incubation, precipitate RNA by adding 10 μl of 10-M LiCl and 600-μl absolute ethanol to each 200-μl transcription reaction. Mix the solution and allow precipitation to precede for 0.5–16 h at $-20\,^{\circ}$C. Concentrate precipitated RNA by centrifugation at 12,000 $\times$ g for 30 min at 4$\,^{\circ}$C. Discard the supernatant obtained and subject the pellet to a second round of centrifugation to facilitate removal of ethanolic supernatant. The pellet obtained is then air-dried at room temperature (RT) for 5–10 min, and resuspended in 100-μl sterile RNase-free H_2O containing 40 U of RNasin. Alternatively, RNA purification can be undertaken with readily available RNA purification kits, such as the Ambion MEGAclear Kit or the QIAGEN RNeasy MinElute Cleanup Kit.

RNA should be separated into aliquots, which contain amounts of RNA that are sufficient for a single translation reaction, and stored at $-80\,^{\circ}$C. The concentration of RNA can be determined spectrophotometrically (1 A_{260nm} RNA = 40 μg/ml).

To do so, one must remove the template DNA by DNase treatment before RNA purification. It is also possible to determine the RNA yield and integrity by agarose gel electrophoresis. This latter approach is valuable when working with problematic translation products or when optimizing translation conditions.

Titrations should be performed to determine the amount of RNA needed for cell-free translation. The concentration of RNA used for titration should be in the range that is recommended by the manufacturer of the translation system (0.5–2μg/ 50-μl translation mix for GE Healthcare rabbit reticulocyte lysate system detailed below).

B. *In Vitro* Translation

The most commonly used *in vitro* translation system consists of extracts from rabbit reticulocytes and contains all the necessary macromolecular components required for the translation of exogenous RNA. Although lysate systems from wheat germ and *Escherichia coli* extracts are available, in many instances these appear nonproductive for *in vitro* import reactions into yeast mitochondria, possibly due to different chaperone compositions. There are two alternative approaches one can employ in the generation of *in vitro* proteins, depending on the starting genetic material, that is DNA or RNA. Most standard systems require the addition of exogenous RNA as a template, but a number of kits offer a coupled transcription–translation reaction by using DNA as template, such as the Promega TNT Quick Coupled Transcription/Translation. The appropriate system to exploit, that is the single or coupled reaction, is something that needs to be established empirically by the investigator for their specific precursor(s), since quite distinct differences in translation efficiency and quality can be obtained with the alternative systems for individual precursors. The coupled system offers the convenience of a single-tube reaction, but as acknowledged by the manufacturer, in many instances this system is not suitable for use with PCR products as template. Thus, the system we detail below exploits the GE Healthcare (formerly Amersham) Rabbit Reticulocyte System, and the addition of exogenous RNA prepared as detailed in Section II.A.

1. Reagents and Methodology

a. Reagents

Rabbit Reticulocyte Lysate System [GE Healthcare; constituents required for this application include: Rabbit reticulocyte lysate, translation mix minus methionine, 2.5-M KCl, and 25-mM Mg(OAc)$_2$]

7.51-μCi/ml [^{35}S]-methionine (70%)/cysteine (30%) (Pro-mix-L[^{35}S] *in vitro* cell-labeling mix; GE Healthcare)

200-mM Cold methionine

1.5-M Sucrose

b. Procedure

The procedure described below is our standard translation reaction. We routinely optimize RNA and salt concentrations, and try alternative salts, as needed, for different translation reactions. For optimization, our goal is to minimize the translation products produced by internal initiations, and obtain high yields of correctly translated proteins. Additionally, the protocol below utilizes a large amount of reticulocyte lysate. The amount of lysate should be adjusted as needed for specific applications.

For cell-free translation, prepare the following mixture:

20-μl Translation mix minus methionine

10-μl of 2.5-M KOAc

5-μl of 25-mM Mg(OAc)$_2$

20-μl ^{35}S-methionine/cysteine

100-μl Rabbit reticulocyte lysate

Bring volume up to 250 μl with RNase-free H$_2$O

Add RNA to the mixture and incubate at 30 °C for 1 h (translation reactions performed using commercial lysate, which have low levels of protease activity, can be incubated for up to 2 h). Thereafter, add cold methionine (200-mM stock) to a final concentration of 5 mM, and centrifuge at 125,000 $\times$ g for 30 min at 4 °C. The supernatant obtained is removed and made isotonic to import buffer by the addition of sucrose (1.5-M stock) to a final concentration of 250-mM. The pellet obtained from the centrifugation contains ribosomes and should be discarded. Translation products should be stored in aliquots at −80 °C.

To assess the efficacy of the translation reaction and quality of the translated product, subject a small fraction (1 μl) of the translation reaction to SDS-PAGE and subsequent digital autoradiography. Apply radioactive ink over the protein standards, which run in parallel, for a direct assessment of the molecular weight of the translated species.

III. The Import Reaction

A. Mitochondrial Isolation

Mitochondria are isolated as detailed in Chapter 2 (Boldogh and Pon, this volume). Mitochondria isolated from budding yeast retain import activity after freezing and storage at −80 °C in SEM buffer (250-mM sucrose, 1-mM EDTA, 10-mM MOPS-KOH, pH 7.2). However, for import into mitochondria from *Neurospora crassa*, assays must be performed on organelles immediately after isolation. Mammalian mitochondria are also generally used for import studies immediately after isolation. However, at least one study has reported import activity in mammalian mitochondria after freezing and storage at −80°C in a buffer consisting of 500-mM sucrose and 10-mM HEPES-KOH, pH 7.4 (Johnston *et al.*, 2002). In cases where the *in vitro* import reaction is carried using mitochondria isolated from different yeast strains or cell lines, mitochondria should be isolated in parallel, under identical conditions, and assayed at equivalent protein levels.

Regardless of the source of mitochondria, membrane integrity and membrane potential must be assessed in isolated mitochondria prior to use in import assays. The integrity of the mitochondrial outer membrane can be evaluated by assessing whether protease-sensitive IMS proteins remain inaccessible to externally added protease in isolated mitochondria (Section IV.A). The membrane potential of isolated mitochondria can be determined by measuring the fluorescent quenching of the potential-sensitive fluorescent dye $DiSC_3(5)$ (3,3′-dipropylthiadicarbocyanine iodide) (Gärtner *et al.*, 1995). These controls are crucial to maintain confidence that any effects seen in the subsequent import reactions are not a consequence of variability among batches of mitochondria.

B. Standard Import Reaction into Isolated Yeast Mitochondria

1. Reagents and Methodology

a. Reagents

SEM buffer:
  250-mM Sucrose
  1-mM EDTA
  10-mM MOPS-KOH, pH 7.2
Import buffer:
  3% (w/v) Fatty acid-free BSA
  250-mM Sucrose
  80-mM KCl
  5-mM $MgCl_2$
  2-mM KH_2PO_4
  5-mM Methionine
  10-mM MOPS-KOH, pH 7.2 (store in aliquots at −20°C)
200-mM ATP (prepared in H_2O, titrated to pH 7.2 with KOH) (store in aliquots at −20°C)
200-mM NADH (prepared fresh in import buffer)
500-mM Creatine phosphate (prepared in H_2O and stored in aliquots at −20°C)
10-mg/ml Creatine kinase (prepared fresh in import buffer)

b. Procedure

The import buffer for the import reaction is a MOPS-based buffer at pH 7.2, which contains sucrose at a final concentration of 250 mM for organelle stability during the import assay. Additionally, 5-mM methionine is included in the import buffer to prevent nonspecific binding of unincorporated radiolabeled methionine from the cell-free translation lysate to mitochondrial proteins, and to reduce background noise in subsequent autoradiography. We typically perform single import reactions in a final volume of 100 μl. Each import reaction consists of import buffer supplemented with NADH, which serves as a substrate for mitochondrial energy production. In many cases, we include 1-mM ATP and an ATP-regenerating system, which consists of 5-mM creatine phosphate and 100-μg/ml creatine kinase. These constituents are added to a microfuge tube, mixed, and maintained on ice. Thereafter, mitochondria (25- to 50-μg protein/import reaction) are added and the mixture is incubated at 25°C for 2 min to allow the sample to equilibrate at the temperature used for the import reaction. After this brief incubation, the reticulocyte lysate containing the radiolabeled precursor protein is added at a concentration of 1–20% (v/v). Addition of precursor marks the initiation of the import reaction. The import is carried our for 2–60 min. To stop import of proteins dependent on a mitochondrial membrane potential, import is terminated by dissipation of the membrane potential (Section III.C). For import of other proteins, the reaction is terminated by transferring the reaction to ice, followed by centrifugation at 12,000 × g for 5 min at 4°C. Mitochondria which are recovered in the pellet after centrifugation are then separated from unincorporated precursors in the supernatant. Methods used for further manipulation of mitochondria after import will vary, depending on the question being addressed, as described below.

C. Time Course Experiment

The rate of import of proteins into mitochondria is related linearly to the time of import. When performing *in vitro* import studies, in particular with a new, uncharacterized, precursor protein, it is important to analyze the precursor's import kinetics. Analysis of import kinetics is also critical when comparing import efficiencies under different conditions, or when comparing mitochondrial preparations from different strains. The time course assay is simple and entails the incubation of radiolabeled precursor with mitochondria, as is done for a standard import reaction (Section III.B). However, import is terminated at various times throughout the reaction, thus allowing for an analysis of import kinetics.

The procedure described below is the classical example of a time course analysis utilizing Su9-DHFR (presequence of *N. crassa* F_0-ATPase subunit 9 fused to the passenger protein dihydrofolate reductase) as a substrate. Import of presequence-containing preproteins is monitored by several criteria. Proteolytic removal of the presequence, and liberation of the mature protein, can be monitored by the increased mobility of the shortened protein on SDS-PAGE. Translocation of proteins across mitochondrial membranes can be monitored by testing the sensitivity of imported proteins to degradation by externally added protease. The protocol described below also includes a valuable control in which import is inhibited by dissipating the mitochondrial membrane potential.

1. Reagents and Methodology

a. Reagents

Reagents as described in Section III.B

AVO mix (make up in ethanol and store at $-20\,^{\circ}$C):

0.8-mM *A*ntimycin A from *Streptomyces* sp. (inhibits electron transfer at complex III of the respiratory chain; Sigma, St. Louis, MO)

0.1-mM *V*alinomycin (potassium ionophore that uncouples oxidative phosphorylation; Sigma)

2-mM *O*ligomycin from *Streptomyces diastatochromogenes* (blocks F_0/F_1 ATPase; Sigma)

Table I shows components of a time course experiment. Tube 1 ($+\Delta\Psi$) is an import reaction that contains enough sample for four time points. Although the kinetic analysis described consists only of three time points, we routinely carry out import reactions in larger volumes to correct for potential pipetting inaccuracies. Tube II ($-\Delta\Psi$) is the control in which the membrane potential is dissipated and consists of enough sample for a single time point.

The components of the import reaction are added to labeled microfuge tubes and are maintained on ice. Mitochondria are added and the samples are incubated at $25\,^{\circ}$C for 2 min. The precursor protein-containing lysate [e.g., 2.5% (v/v)] is added at $t = 0$ for the time course. At 2, 5, and 10 min after initiation of import, 200 μl of the import reaction is removed from tube 1 and transferred to a fresh

Table I
Assembly of Import Reaction for Time Course Analysis

Import mix	$+\Delta\Psi$ (μl)	$-\Delta\Psi$ (μl)
Import buffer	728	186
0.2-M NADH	16	–
0.2-M ATP	8	2
AVO mix	–	2
Ethanol	8	–
Mitochondria (10 mg/ml)	20	5
Mix and incubate at $25\,^{\circ}$C for 2 min		
^{35}S-precursor	20	5
Mix and import at $25\,^{\circ}$C		

tube containing 2-μl AVO mix. These samples are mixed gently and are stored on ice. At 10 min, tube II ($-\Delta\Psi$) receives 2-μl of ethanol and is also placed on ice. The four samples (2, 5, 10 min + $\Delta\Psi$, and 10 min $-\Delta\Psi$) are each split further into two equal-sized aliquots. One aliquot is retained as an untreated control, while the other aliquot is subject to proteinase K treatment (Section IV.A) to degrade unincorporated protein. After treatment with protease, the mixture is centrifuged at 12,000 $\times$ g for 5 min at 4°C, washed with SEM buffer, resuspended in SDS-PAGE-loading dye (5 min on a vortex shaker), incubated at 95°C for 5 min, and loaded onto a denaturing SDS gel. An aliquot of the precursor (1-μl reticulocyte lysate) is loaded on the gel to serve as a standard to assess precursor processing and import efficiency (Fig. 1).

In the time course experiment described above, import reactions were initiated at the same time and terminated at different times. To assess the import kinetics of $\Delta\Psi$-independent precursors, a reverse kinetics approach can be employed. In such an experiment, import reactions for each time point are carried out in separate microfuge tubes, and import is initiated at different times but is terminated at the same time. For example, a four time point experiment of 5, 10, 15, and 30 min

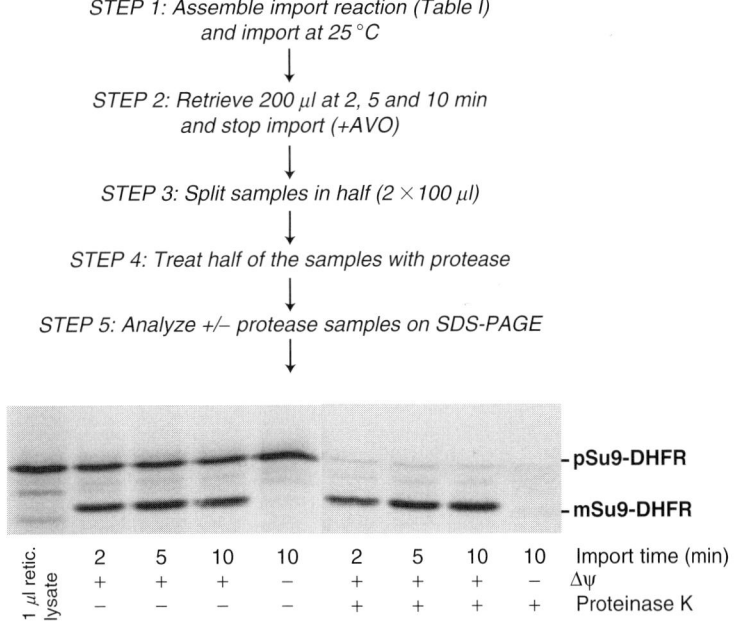

Fig. 1 Import of Su9-DHFR into isolated yeast mitochondria. *In vitro*-translated ^{35}S-labeled pSu9-DHFR (precursor) was incubated with isolated yeast mitochondria at 25°C and treated as displayed in steps 1–5 and as defined in Section III.C. Lane 1 represents 20% of the input signal for pSu9-DHFR in each import reaction. Import of the precursor into the mitochondrial matrix is monitored by generation of mSu9-DHFR (mature, processed form), as denoted in the figure.

would involve initiation of the independent import reactions at 0, 15, 20, and 25 min, and terminating import at 30 min by placing the samples on ice, followed by centrifugation to separate mitochondria from unincorporated material.

IV. Steps to Resolve the Location of Imported Mitochondrial Precursors

In the previous section, we described procedures to assess import into or across the mitochondrial membranes and proteolytic removal of presequences from precursor proteins. In this section, we introduce procedures that can be used to manipulate mitochondria after import, and describe how these methods can assist in disclosing the disposition of imported proteins within the organelle. Figure 2 provides a diagrammatic representation of the approaches and techniques used to determine submitochondrial location of a newly imported mitochondrial protein.

A. Protease Protection Assays

The translocation or import of precursors through the mitochondrial outer membrane delivers them to a proteolytically protected location within the organelle, that is, to a location where they cannot be degraded by externally added proteases. Such protease-protection assays provide the first clue that the precursor is

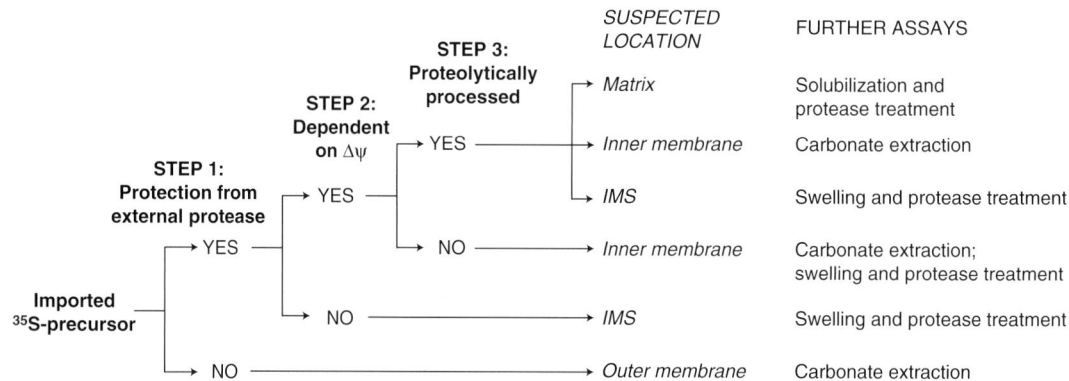

Fig. 2 Flow diagram of the steps taken to determine a precursor's submitochondrial location. For analyses of the submitochondrial location of an imported protein, steps 1–3 are performed first, that is (1) assessing the precursor's accessibility to externally added protease, (2) assessing the precursor's dependence on a membrane potential ($\Delta\Psi$) for import, and (3) assessing if the precursor is proteolytically processed after import. These three assays provide the basic insight into the potential location of the precursor, which is then subsequently analyzed employing the alternative techniques described (far right).

located within the organelle. Following the import reaction, mitochondria are isolated, resuspended in SEM buffer ($\sim$200 μl), and the sample is divided into two equal-sized aliquots. Proteinase K is added to one sample to a final concentration of 10–50 μg/ml from a 1-mg/ml stock made immediately before use in SEM buffer. An equivalent volume of SEM buffer is added to the other aliquot. Samples are then incubated on ice for 15 min. Thereafter, protease activity is inhibited by the addition of phenylmethylsulfonyl fluoride (PMSF; 200-mM stock in isopropanol) to a final concentration of 2 mM and incubation on ice for 10 min. In some instances, alternative proteases, such as trypsin, may be necessary. For most applications, trypsin is used at a final concentration of 10–50μg/ml, and inactivated by the addition of soybean trypsin inhibitor in 30-fold excess to trypsin. After termination of the protease treatment, mitochondria are isolated by centrifugation at 12,000 $\times$ g for 5 min at 4°C, washed with SEM buffer, solubilized in SDS-PAGE-loading dye, and subjected to SDS-PAGE.

B. Mitochondrial Swelling

As described above, protease-protection assays are used to monitor translocation of proteins across mitochondrial membranes. To determine the submitochondrial location of newly imported proteins, the protease sensitivity of imported proteins is assessed in mitoplasts (mitochondria with ruptured outer membranes but intact inner membranes). Incubation of mitochondria in hypotonic media results in mitochondrial swelling, rupture of the outer membrane, and conversion of mitochondria to mitoplasts. This, in turn, renders proteins in the IMS or on the IMS-side of the inner membrane accessible to added protease.

After the import reaction, mitochondria are washed in SEM buffer, isolated by centrifugation, resuspended in SEM buffer (40 μl), and divided into two equal-sized aliquots. One aliquot is maintained as untreated intact mitochondria, and the other is incubated in hypotonic buffer to induce swelling, outer membrane rupture, and mitoplast formation. SEM buffer (180 μl) is added to the untreated control, while 180 μl of swelling buffer (10-mM MOPS-KOH, pH 7.2) is added to the mitoplast sample. Pipette the mitoplast sample up and down 10 times, and incubate both samples on ice for 15 min. Thereafter, samples are divided into two equal-sized aliquots for further treatment in the presence or absence of proteinase K, as described in Section IV.A.

The efficiency of the swelling reaction should be monitored by immunodecoration with proteins that only become accessible to protease after swelling (e.g., cytochrome c heme lyase). Additionally, the integrity of the inner membrane and matrix compartment should be assessed by comparing the protease resistance of selected matrix proteins (e.g., citrate synthase) in mitochondria and mitoplasts by Western blot analysis.

C. Mitochondrial Solubilization for Protease Treatment

Resistance to protease in mitoplasts can imply one of two things: the precursor is either a resident of the mitochondrial inner membrane or matrix, or it is an aggregate that cosedimentants nonspecifically with mitochondria. To eliminate the latter, an import experiment is performed in the absence of mitochondria. Treatment of mitochondria with detergent and subsequent protease treatment provide further evidence for localization of an imported protein to the inner membrane or matrix compartments. To test the protease-sensitivity of imported proteins in detergent solubilized mitochondria, mitochondria from *in vitro* import assays are isolated by centrifugation and are resuspended in SEM (100–200 μl). The sample is divided into two equal-sized aliquots and the mitochondria of one sample is solubilized by the addition of Triton X-100 (10% (v/v) stock) to a final concentration of 0.5% (v/v). Both samples are subsequently split again into two equal-sized aliquots, and one aliquot of mitochondria and detergent-treated mitochondria are treated with proteinase K as described in Section IV.A. Samples are then TCA precipitated, as described in Section IV.D, and analyzed by SDS-PAGE and Western blots.

D. Is the Newly Imported Mitochondrial Protein Soluble, Peripherally Membrane Associated, or Integrated into Mitochondrial Membranes?

Two approaches are commonly used to address this question. The first method involves a sonication of mitochondrial extracts, which results in separation of soluble IMS and matrix proteins from those that are membrane bound. Since ionic interactions are salt sensitive and hydrophobic interactions are salt resistant, performing the sonication in increasing salt concentrations (0–500 mM) allows one to assess how tightly associated a protein is with the membrane. The alternative procedure is carbonate extraction (treatment of mitochondria under alkali conditions), which extracts soluble and peripheral membrane proteins but does not extract integral membrane proteins. Carbonate-extracted and -unextracted material can be separated by centrifugation. Samples from the both sonication and carbonate extraction procedures are analyzed on SDS-PAGE and Western blots, and the authenticity of the data should be confirmed by assessing the behavior of mitochondrial marker proteins.

For extraction by sonication of mitochondrial membranes, the mitochondria are isolated after import by centrifugation, and are washed in SEM buffer. The mitochondrial pellet is then resuspended in an extraction buffer consisting of 10-mM MOPS–KOH, pH 7.2; 500-mM NaCl (500 μl); and the sample is subjected to sonication on ice, either in a sonicating ice bath for 5 min or in a prechilled sonication vessel using a microtip sonifier (3 × 6 pulses; 40% duty cycle; microtip output setting 5 for Branson Sonifier 250). For separation of membrane proteins from soluble proteins, the extract is centrifuged at 100,000 × g for 1 h at 4 °C. The pellet obtained contains membranes and membrane-associated proteins, and the supernatant obtained contains liberated soluble proteins. The supernatant fraction is subjected to a trichloroacetic acid (TCA) precipitation for concentration, as described below. Finally, the pellet, which contains membrane-associated proteins and the soluble TCA-precipitated proteins, is solublilized in SDS-PAGE-loading dye.

To determine if an imported protein is an integral membrane protein, mitochondria from two import reactions (50- to 100-μg protein) are divided into two equal-sized aliquots, and are isolated by centrifugation at 12,000 $\times$ g for 5 min at 4°C. The mitochondrial pellets are washed in SEM buffer (100 μl). Thereafter, one pellet is treated for analysis by SDS-PAGE by solubilization in SDS-PAGE-loading dye. The other pellet is resuspended in 100 μl of ice-cold 100-mM Na_2CO_3 (prepared fresh in water), and incubated on ice for 30 min. The mixture is centrifuged at 100,000 $\times$ g for 1 h at 4°C. The pellet from this centrifugation contains membranes and integral membrane proteins, while the supernatant contains extracted soluble and peripheral membrane proteins. Proteins in the supernatant are precipitated with TCA, and both the TCA-precipitated proteins and the pellet are solubilized in SDS-PAGE-loading dye.

Both of the procedures described above rely on the successful concentration of liberated soluble mitochondrial proteins by TCA precipitation. Briefly, sodium deoxycholate (which serves as a carrier to enhance protein precipitation) is added to the isolated supernatant to a final concentration of 0.0125%. Thereafter, 72% (w/v) TCA is added at one-fifth of the final sample volume, and the sample is incubated on ice for 30 min. Precipitated species are then isolated by centrifugation at 18,000 $\times$ g for 30 min at 4°C. The pellet is washed with ice-cold acetone, air-dried briefly (5–10 min), and resuspended in SDS-PAGE-loading dye. For solubilization of protein pellets in Laemmli buffer, the samples should be mixed vigorously on a vortex shaker until the pellets are completely solubilized, prior to heating at 95°C for 5 min. Some degree of protein loss occurs with TCA precipitation, particularly for samples with low-protein levels. One alternative to TCA precipitation is to concentrate proteins using commercially available products such as StrataClean[TM] resin (Stratagene).

V. Assaying Protein Complex Assembly by BN–PAGE

Once the location of the mitochondrial precursor has been elucidated, steps to clarify its function may begin. The assembly of mitochondrial precursors into higher-ordered macromolecular complexes, and the elucidation of complex constituents, can provide much insight into the respective function(s) of the unknown candidate. These processes can be analyzed by blue native page electrophoresis (BN-PAGE) (Schägger and von Jagow, 1991), which is detailed in Chapter 33 (Wittig and Schäger, this volume). This method has served as a powerful tool in analyzing the assembly of the TOM (Dekker et al., 1998; Dietmeier et al., 1997; Model et al., 2001; Rapaport and Neupert, 1999) and TIM machineries (Dekker et al., 1997; Kurz et al., 1999).

A key aspect in the use of BN-PAGE is finding a detergent that can disrupt lipid–lipid interactions and separate a protein complex from unrelated proteins without disturbing interactions between protein components within the complex. Detergents used routinely for the solubilization of mitochondrial protein complexes include *n*-dodecylmaltoside, Triton X-100, and digitonin. Comparative testing of alternative detergents and of suitable detergent-to-protein ratios are necessary to determine the optimal detergent and conditions for solubilization of the desired complex(es). We routinely use digitonin for the solubilization of yeast mitochondria, and this protocol is detailed below.

Following import, mitochondria are washed by resuspension in SEM buffer and centrifugation. Mitochondria (25- to 50-μg protein) recovered in the pellet are solubilized in 50 μl of ice-cold digitonin buffer [1% (w/v) digitonin; 20-mM Tris–Cl, pH 7.4; 0.1-mM EDTA; 50-mM NaCl; 10% (v/v) glycerol; 1-mM PMSF]. It is vital that the solubilization conditions be kept constant between samples (e.g., that each mitochondrial pellet is pipetted the same number of times) to ensure that variability among samples is not a consequence of variability in solubilization conditions. Samples are incubated on ice for 10 min, and clarified by centrifugation at $12,000 \times g$ for 5 min at $4\,^\circ$C. The supernatant, which contains extracted proteins and protein complexes, is transferred to a microfuge tube containing 5 μl of blue native sample buffer [5% (w/v) Coomassie brilliant blue G-250; 100-mM Bis-Tris, pH 7.0; 500-mM 6-aminocaproic acid]. The sample is then loaded onto a blue native polyacrylamide gel of the desired gradient, and subjected to electrophoresis at $4\,^\circ$C.

It may be necessary to perform a second dimension SDS-PAGE with protein complexes separated on BN-PAGE (Schägger and von Jogow, 1991; Schägger *et al.*, 1994). For the second dimension, the desired lane from the BN-PAGE is excised and placed horizontally between the glass plates in the initial assembly of the gel. Thus, the BN strip is in position during the casting of the gel. The separating gel is poured and allowed to polymerize. Thereafter, a stacking gel is poured carefully to prevent bubbles being trapped beneath the BN strip.

Proteins separated by BN-PAGE can be also subjected to Western blot analysis (Schägger, 2001). This type of analysis is very informative in ascertaining steady-state levels of import machineries, such as the TOM complex, in mitochondria from mutant yeast strains with defects in mitochondrial protein import.

VI. Resolving Protein Complex Constituents

A. Antibody Shift and Antibody Depletion Assays

Antibody shift assays can be used in conjunction with BN-PAGE to determine whether radiolabeled imported proteins are associated with specific protein complexes, as long as antibodies to known complex components are available. This approach served as a powerful tool to identify the composition of protein complexes (Johnston *et al.*, 2002; Truscott *et al.*, 2002) and when working with translocation intermediates, which are often present in small quantities that can only be detected using radioactively labeled proteins.

For complexes where the antigen is exposed to the cytoplasm, the mitochondrial pellet is isolated after import, resuspended in SEM buffer (100 μl), and incubated with the desired antiserum. For assays of antigens with IMS-exposed domains, antibodies are added after rupturing of the outer membrane of mitochondria after import, as described in Section IV.B (Chacinska *et al.*, 2004). In either case, antiserum (1–10 μl) or affinity-purified antibodies in SEM buffer (10–40 μl) are added to the desired sample, and incubated on ice for 30 min. Mitochondria or mitoplasts are then isolated and washed once in SEM buffer, followed by solubilization in 1% digitonin (or the detergent of choice) and analyzed by BN-PAGE (Fig. 3).

The alternative to the antibody shift assay is an antibody depletion assay. In this approach, mitochondria are first solubilized in the detergent of choice. Antibody is then added, and the samples are incubated on ice for 30 min with occasional mixing. We find that purified freeze-dried IgGs give the best results for antibody depletion because they do not significantly change the concentration or detergent in the sample. Antibody-bound complexes are then depleted by the addition of Protein A-Sepharose (bed volume, 20 μl) that has been preequilibrated with detergent buffer. Samples are incubated for 30 min on ice with IgG bound to Sepharose beads, and are concentrated by low-speed centrifugation at $2500 \times g$ for 2 min at $4\,^{\circ}$C. The supernatant obtained from this centrifugation is carefully removed and analyzed by BN-PAGE. Depletion of the desired complex(es) can be monitored by autoradiography or immunodecoration. This approach has been used successfully to deplete endogenous TOM complex for studies of mitochondrial protein import (Truscott *et al.*, 2002).

B. Cross-Linking After Import

Cross-linking experiments have been used in conjunction with *in vitro* import assays to study direct physical interactions of radiolabeled proteins during their import into mitochondria (Kozany *et al.*, 2004; Mokranjac *et al.*, 2003; Söllner *et al.*, 1992; Wiedemann *et al.*, 2001). The choice of cross-linker is highly dependent on the application. In each circumstance, alternative cross-linker and optimal cross-linker concentrations need to be established. Here, we provide the standard protocol for cross-linking after import of a radiolabeled precursor with bifunctional membrane permeable cross-linkers such as EGS (ethylene glycol bis [succinimidyl succinate]), MBS (*m*-maleimidobenzoyl-*N*-hydroxysuccinimide ester) and DSG (disuccinimidyl glutarate) (Pierce, Rockford, IL).

Mitochondria are isolated after import and thoroughly washed to remove unincorporated precursor, BSA that is present in the import buffer, and free radiolabeled methionine from the lysate. For thorough washing, the import reaction is applied to the top of a 0.5-M sucrose cushion (650 μl, consisting of 1-mM EDTA,

STEP 1: Isolate mitochondrial pellet after import

↓

STEP 2: Resuspended in 100 μl SEM buffer

↓

STEP 3: Add desired antiserum and mix

↓

STEP 4: Incubate on ice for 30 min

↓

STEP 5: Isolate and wash mitochondria in SEM

↓

STEP 6: Analyze solubilized mitochondria on BN-PAGE

↓

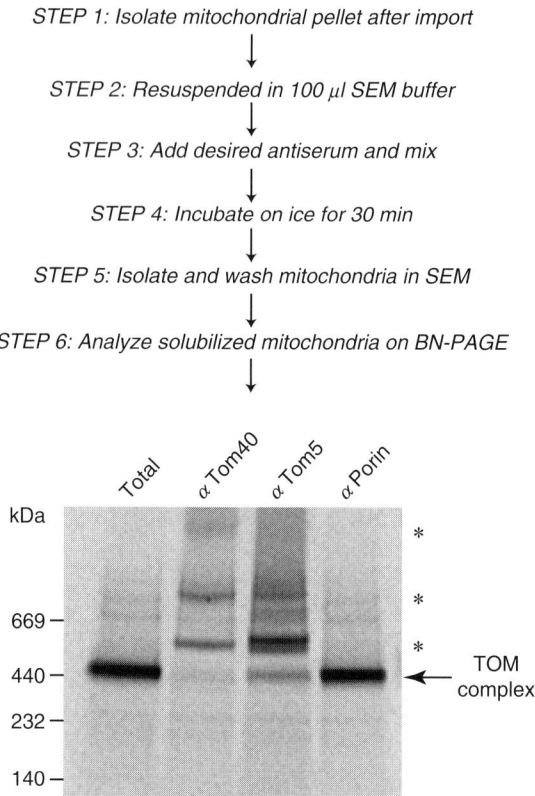

Fig. 3 Unraveling protein complex constituents by antibody shift BN-PAGE analysis. *In vitro*-translated ^{35}S-labeled Tom22 was incubated with isolated mitochondria at 25 °C for 30 min and treated as shown in steps 1–6 and as defined in Section VI.A. After incubation with antisera against the outer membrane (Tom40, Tom5, and porin), which are accessible on the surface of the organelle, samples were separated by BN-PAGE and analyzed by digital autoradiography. Lane 1 (Total) represents an untreated control, and illustrates assembly of Tom22 into the mature TOM complex, as indicated. Antibody-bound and antibody-shifted complexes are indicated by asterisks (*). Lanes 2–3 illustrate the mobility shift of the Tom22-containing complex by treatment with antibodies raised against Tom40 and Tom5. Thus, Tom22, Tom5, and Tom40 are present in the same complex. Lane 4 is the negative control. Here, treatment with α-porin antibodies has no effect on the mobility of the Tom complex.

10-mM MOPS–KOH, pH 7.2). The sample is then centrifuged at 12,000 × *g* for 10 min at 4 °C. The mitochondrial pellet is resuspended in the cross-linking buffer of choice (100 μl); in our hands, SEM buffer has worked well. Cross-linkers are made fresh, immediately prior to use, and are dissolved in the appropriate solution. Water is the solvent for water-soluble and membrane-impermeable cross-linkers such as Sulfo-MBS. DMSO is used as the solvent for water-insoluble and membrane-permeable cross-linkers such as MBS. After addition of cross-linker to the mitochondrial sample (final concentration between 50 μM and 2 mM), cross-linking is performed on ice for 30–60 min. Cross-linking is terminated by the addition of an appropriate quenching agent and incubation for 15 min on ice. The quenching agent for a homo-bifunctional amino-reactive cross-linker, such as EGS, is Tris–Cl, pH 7.2 at a final concentration of 50 mM. For a hetero-bifunctional cross-linker, such as MBS, which reacts with both amines and sulfhydryls, cysteine at a final concentration of 50 mM is added in addition to Tris. Mitochondria are then isolated by centrifugation, washed in SEM buffer, solubilized in SDS-PAGE-loading dye, and subjected to SDS-PAGE. The apparent molecular weight of the cross-linked species may give some indication of the interacting component's identity, but solid evidence is provided through immunoprecipitation of the radiolabeled species with antibodies against the candidate interacting protein. As an additional control, the formation of the cross-linked species can be assessed in mitochondria from deletion strains or from tagged strains of the potential-interacting protein, if these are available.

VII. Assaying Protein Import in Yeast Mutants

Gene disruption or mutation is a fundamental tool in the unraveling of gene function with high confidence. The relative simplicity of performing genetic manipulations in the yeast *Saccharomyces cerevisiae* has contributed considerably to the delineation of alternative import pathways and import mechanisms in this organism. The early events associated with import of a nuclear-encoded precursor can be determined by analysis of import into mitochondria isolated from yeast strains bearing deletions in nonessential outer membrane receptors (Tom5, Tom6, Tom7, Tom20, and Tom70).

Given the identification of a deletion that produces a defect in mitochondrial protein import, one concern that must be addressed is whether the deleted gene has direct or indirect effects on the import machinery. This point is illustrated by the analysis of yeast bearing a deletion in *TOM20*. This deletion strain displays severe defects in respiration and import into mitochondria. Deletion of *TOM20* results in a reduction in the steady-state levels of Tom22 (Hönlinger *et al.*, 1995). Interestingly, restoration of Tom22 levels via overexpression in the *TOM20* deletion strain resulted in restoration of mitochondrial respiration and import activity. Thus, the phenotypes observed on deletion of *TOM20* are due to secondary effects on the steady-state levels of Tom22 (Hönlinger *et al.*, 1995).

In assessing precursor dependence on protein components essential for yeast viability, two different strategies can be employed: the use of conditional mutants or regulated gene expression. The use of temperature-sensitive mutants has proved very useful in the analysis of import components from all four compartments of the organelle, including Tom40 (Gabriel *et al.*, 2003; Kassenbrock *et al.*, 1993; Schleiff *et al.*, 1999), Sam50 (Kozjak *et al.*, 2003), and Sam35 (Milenkovic *et al.*, 2004) of the outer membrane; Tim10, Tim12 (Koehler *et al.*, 1998), Mia40

(Chacinska *et al.*, 2004), and Erv1 (Rissler *et al.*, 2005) of the IMS; Tim23 (Dekker *et al.*, 1997), Pam18 (Truscott *et al.*, 2003), Pam16 (Frazier *et al.*, 2004), Tim50 (Chacinska *et al.*, 2005), and Tim44 (Bömer *et al.*, 1998) of the inner membrane, and Mge1 (Laloraya *et al.*, 1995) and mtHsp70 (Gambill *et al.*, 1993; Voos *et al.*, 1993) of the mitochondrial matrix. The behavior of individual temperature-sensitive mutants can be quite distinct, and thus requires the careful characterization. Some mutants may possess specific defects when grown at permissive temperatures, whereas others may display no effect. However, in the case of the latter, in some instances a brief *in vitro* heat shock to isolated mitochondria (15 min at 37 °C) before commencing the import assay may be useful in eliciting abnormal activity.

The alternative to temperature-sensitive mutants is to place the essential gene of interest under the control of a galactose-inducible promoter. The shifting of yeast cells to growth medium lacking galactose leads to repression of the gene. If the protein has a rapid half-life, down-regulation of gene expression can deplete the essential gene product from mitochondria. This approach has also been used to characterize alternative mitochondrial protein import pathways (Geissler *et al.*, 2002; Ishikawa *et al.*, 2004; Kozany *et al.*, 2004; Mesecke *et al.*, 2005; Mokranjac *et al.*, 2003; Naoé *et al.*, 2004; Waizenegger *et al.*, 2004).

Regardless of the method employed, that is, use of temperature-sensitive mutants or strains in which protein levels have been down regulated or completely abolished due to gene disruption, a number of vital controls must be employed to determine whether observed defects are primary or secondary effects of the mutation of interest. As described in Section III.A, wild-type and mutant mitochondria should always be isolated in parallel, and steady-state protein levels in all four compartments assessed. It is also wise to assess the import kinetics of substrates from all four mitochondrial compartments, as described in Section III.C. To identify effects on membrane potential, the $\Delta\Psi$ of mitochondria should be evaluated. It should also be noted that the recent advances in RNA interference technology now make it possible to perform this type of analysis in mammalian systems. Such an approach has been exploited to reveal a requirement for the import of human Tom40 of the human Sam50 homologue (Humphries *et al.*, 2005).

VIII. Concluding Remarks

The ability to reconstitute the process of mitochondrial import *in vitro* has provided a very successful and straightforward means by which the submitochondrial residence and import pathway of most mitochondrial proteins can be classified. In the past 4–5 years, we have witnessed the emergence not only of new translocation components but also the elucidation of novel translocation machineries. These include the sorting and assembly machinery of the outer membrane (SAM complex) and the IMS-specific import machinery consisting of Mia40 and

Erv1. The discovery of novel subunits of the mitochondrial import motor has also given much insight into the dynamic function of this molecular machine, now known as the PAM complex. The identification of these novel components has been greatly facilitated by determination of the proteome of *S. cerevisae* mitochondria (Sickmann *et al.*, 2003) and development of affinity purification approaches. However, the *in vitro* mitochondrial import assay continues to be the primary assay for unraveling the function of these novel import components and pathways. This assay will continue to play a pivotal role in defining the means and mechanisms by which protein import and integration into the organelle are executed.

Acknowledgments

Work of the authors' laboratory was supported by the Deutsche Forschungsgemeinschaft, Sonderforschungsbereich 388, Max Planck Research Award, the Alexander von Humboldt Foundation, Bundesministerium für Bildung und Forschung, Gottfried Wilhelm Leibniz-Program, and the Fonds der Chemischen Industrie. D.S. is a recipient of an Alexander von Humboldt Research Fellowship.

References

Allen, S., Balabanidou, V., Sideris, D. P., Lisowsky, T., and Tokatlidis, K. (2005). Erv1 mediates the Mia40-dependent protein import pathway and provides a functional link to the respiratory chain by shuttling electrons to cytochrome c. *J. Mol. Biol.* **353**, 937–944.

Bömer, U., Maarse, A. C., Martin, F., Geissler, A., Merlin, A., Schönfisch, B., Meijer, M., Pfanner, N., and Rassow, J. (1998). Separation of structural and dynamic functions of the mitochondrial translocase: Tim44 is crucial for the inner membrane import sites in translocation of tightly folded domains, but not of loosely folded preproteins. *EMBO J.* **17**, 4226–4237.

Burri, L., Strahm, Y., Hawkins, C. J., Gentle, I. E., Puryer, M. A., Verhagen, A., Callus, B., Vaux, D., and Lithgow, T. (2005). Mature DIABLO/Smac is produced by the IMP protease complex on the mitochondrial inner membrane. *Mol. Biol. Cell* **16**, 2926–2933.

Chacinska, A., Pfannschmidt, S., Wiedemann, N., Kozjak, V., Sanjuán-Szklarz, L. K., Schulze-Specking, A., Truscott, K. N., Guiard, B., Meisinger, C., and Pfanner, N. (2004). Essential role of Mia40 in import and assembly of mitochondrial intermembrane space proteins. *EMBO J.* **23**, 3735–3746.

Chacinska, A., Lind, M., Frazier, A. E., Dudek, J., Meisinger, C., Geissler, A., Sickmann, A., Meyer, H. E., Truscott, K. N., Guiard, B., Pfanner, N., and Rehling, P. (2005). Mitochondrial presequence translocase: Switching between TOM tethering and motor recruitment involves Tim21 and Tim17. *Cell* **120**, 817–829.

Curran, S. P., Leuenberger, D., Oppliger, W., and Koehler, C. M. (2002). The Tim9p-Tim10p complex binds to the transmembrane domains of the ADP/ATP carrier. *EMBO J.* **21**, 942–953.

Dekker, P. J. T., Martin, F., Maarse, A. C., Bömer, U., Müller, H., Guiard, B., Meijer, M., Rassow, J., and Pfanner, N. (1997). The Tim core complex defines the number of mitochondrial translocation sites and can hold arrested preproteins in the absence of matrix Hsp70-Tim44. *EMBO J.* **16**, 5408–5419.

Dekker, P. J. T., Ryan, M. T., Brix, J., Müller, H., Hönlinger, A., and Pfanner, N. (1998). Preprotein translocase of the outer mitochondrial membrane: Molecular dissection and assembly of the general import pore complex. *Mol. Cell. Biol.* **18**, 6515–6525.

Desagher, S., and Martinou, J. C. (2000). Mitochondria as the central control point of apoptosis. *Trends Cell Biol.* **10**, 369–377.

Dietmeier, K., Hönlinger, A., Bömer, U., Dekker, P. J. T., Eckerskorn, C., Lottspeich, F., Kübrich, M., and Pfanner, N. (1997). Tom5 functionally links mitochondrial preproteins receptors to the general import pore. *Nature* **388,** 195–200.

Frazier, A. E., Dudek, J., Guiard, B., Voos, W., Li, Y., Lind, M., Meisinger, C., Geissler, A., Sickmann, A., Meyer, H. E., Bilanchone, V., Cumsky, M. G., *et al.* (2004). Pam16 has an essential role in the mitochondrial protein import motor. *Nat. Struc. Mol. Biol.* **11,** 226–233.

Gabriel, K., Egan, B., and Lithgow, T. (2003). Tom40, the import channel of the mitochondrial outer membrane, plays an active role in sorting imported proteins. *EMBO J.* **22,** 2380–2386.

Gambill, B. D., Voos, W., Kang, P. J., Miao, B., Langer, T., Craig, E. A., and Pfanner, N. (1993). A dual role for mitochondrial heat shock protein 70 in membrane translocation of preproteins. *J. Cell Biol.* **123,** 109–117.

Gärtner, F., Voos, W., Querol, A., Miller, B. R., Craig, E. A., Cumsky, M. G., and Pfanner, N. (1995). Mitochondrial import of subunit Va of cytochrome c oxidase characterized with yeast mutants. *J. Biol. Chem.* **270,** 3788–3795.

Geissler, A., Chacinska, A., Truscott, K. N., Wiedemann, N., Bradner, K., Sickmann, A., Meyer, H. E., Meisinger, C., Pfanner, N., and Rehling, P. (2002). The mitochondrial presequence translocase: An essential role of Tim50 in directing preproteins to the import channel. *Cell.* **111,** 507–518.

Gentle, I. E., Gabriel, K., Beech, P., Waller, R., and Lithgow, T. (2004). The Omp85 family of proteins is essential for outer membrane biogenesis in mitochondria and bacteria. *J. Cell Biol.* **164,** 19–24.

Glick, B. S., Brandt, A., Cunningham, K., Müller, S., Hallberg, R. L., and Schatz, G. (1992). Cytochromes c_1 and b_2 are sorted to the intermembrane space of yeast mitochondria by a stop-transfer mechanism. *Cell* **69,** 809–822.

Green, D. R., and Reed, J. C. (1998). Mitochondria and apoptosis. *Science* **281,** 1309–1312.

Hönlinger, A., Kübrich, M., Moczko, M., Gärtner, F., Mallet, L., Bussereau, F., Eckerskorn, C., Lottspeich, F., Dietmeier, K., Jacquet, M., and Pfanner, N. (1995). The mitochondrial receptor complex: Mom22 is essential for cell viability and directly interacts with preproteins. *Mol. Cell. Biol.* **15,** 3382–3389.

Humphries, A. D., Streimann, I. C., Stojanovski, D., Johnston, A. J., Yano, M., Hoogenraad, N. J., and Ryan, M. T. (2005). Dissection of the mitochondrial import and assembly pathway for human Tom40. *J. Biol. Chem.* **280,** 11535–11543.

Ishikawa, D., Yamamoto, H., Tamura, Y., Moritoh, K., and Endo, T. (2004). Two novel proteins in the mitochondrial outer membrane mediate beta-barrel protein assembly. *J. Cell Biol.* **166,** 621–627.

Johnston, A. J., Hoogenraad, J., Dougan, D. A., Truscott, K. N., Yano, M., Mori, M., Hoogenraad, N. J., and Ryan, M. T. (2002). Insertion and assembly of human Tom7 into the preprotein translocase complex of the outer mitochondrial membrane. *J. Biol. Chem.* **277,** 42197–42204.

Kassenbrock, C. K., Cao, W., and Douglas, M. G. (1993). Genetic and biochemical characterization of ISP6, a small mitochondrial outer membrane protein associated with the protein translocation complex. *EMBO J.* **8,** 3023–3034.

Kispal, G., Csere, P., Prohl, C., and Lill, R. (1999). The mitochondrial proteins Atm1p and Nsf1p are essential for biogenesis of cytosolic Fe/S proteins. *EMBO J.* **18,** 3981–3989.

Koehler, C. M. (2004). The small Tim proteins and the twin Cx_3C motif. *Trends Biochem. Sci.* **29,** 1–4.

Koehler, C. M., Jarosch, E., Tokatlidis, K., Schmid, K., Schweyen, R. J., and Schatz, G. (1998). Import of mitochondrial carriers mediated by essential proteins of the intermembrane space. *Science.* **279,** 369–373.

Kozany, C., Mokranjac, D., Sichting, M., Neupert, W., and Hell, K. (2004). The J domain-related cochaperone Tim16 is a constituent of the mitochondrial TIM23 preprotein translocase. *Nat. Struc. Mol. Biol.* **11,** 234–241.

Kozjak, V., Wiedemann, N., Milenkovic, D., Lohaus, C., Meyer, H. E., Guiard, B., Meisinger, C., and Pfanner, N. (2003). An essential role of Sam50 in the protein sorting and assembly machinery of the mitochondrial outer membrane. *J. Biol. Chem.* **278,** 48520–48523.

Kurz, M., Martin, H., Rassow, J., Pfanner, N., and Ryan, M. T. (1999). Biogenesis of Tim proteins of the mitochondrial carrier import pathway: Differential targeting mechanisms and crossing over with the main import pathway. *Mol. Biol. Cell.* **10,** 2461–2474.

Laloraya, S., Dekker, P. J. T., Voos, W., Craig, E. A., and Pfanner, N. (1995). Mitochondrial GrpE modulates the function of matrix Hsp70 in translocation and maturation of preproteins. *Mol. Cell. Biol.* **15,** 7098–7105.

Lange, H., Kaut, A., Kispal, G., and Lill, R. (2000). A mitochondrial ferredoxin is essential for biogenesis of cellular iron–sulfur proteins. *Proc. Natl. Acad. Sci. USA.* **97,** 1050–1055.

Leo, S., Bianchi, K., Brini, M., and Rizzuto, R. (2005). Mitochondrial calcium signalling in cell death. *FEBS J.* **272,** 4013–4022.

Li, J., Saxena, S., Pain, D., and Dancis, A. (2001). Adrenodoxin reductase homolog (Arh1p) of yeast mitochondria required for iron homeostasis. *J. Biol. Chem.* **276,** 1503–1509.

Lill, R., and Mühlenhoff, U. (2005). Iron-sulfur-protein biogenesis in eukaryotes. *Trends Biochem. Sci.* **30,** 133–141.

Lill, R., Diekert, K., Kaut, A., Lange, H., Pelzer, W., Prohl, C., and Kispal, G. (1999). The essential role of mitochondria in the biogenesis of cellular iron–sulfur proteins. *Biol. Chem.* **380,** 1157–1166.

Mesecke, N., Terziyska, N., Kozany, C., Baumann, F., Neupert, W., Hell, K., and Hermann, J. M. (2005). A disulfide relay system in the intermembrane space of mitochondria that mediates protein import. *Cell* **121,** 1059–1069.

Milenkovic, D., Kozjak, V., Wiedemann, N., Lohaus, C., Meyer, H. E., Guiard, B., Pfanner, N., and Meisinger, C. (2004). Sam35 of the mitochondrial protein sorting and assembly machinery is a peripheral outer membrane protein essential for cell viability. *J. Biol. Chem.* **279,** 22781–22785.

Model, K., Meisinger, C., Prinz, T., Wiedemann, N., Truscott, K. N., Pfanner, N., and Ryan, M. T. (2001). Multistep assembly of the protein import channel of the mitochondrial outer membrane. *Nat. Struc. Biol.* **8,** 361–370.

Mokranjac, D., Paschen, S. A., Kozany, C., Prokisch, H., Hoppins, S. C., Nargang, F. E., Neupert, W., and Hell, K. (2003). Tim50, a novel component of the TIM23 preprotein translocase of mitochondria. *EMBO J.* **22,** 816–825.

Naoé, M., Ohwa, Y., Ishikawa, D., Ohshima, C., Nishikawa, S., Yamamoto, H., and Endo, T. (2004). Identification of Tim40 that mediates protein sorting to the mitochondrial intermembrane space. *J. Biol. Chem.* **279,** 47815–47821.

Paschen, S. A., Waizenegger, T., Stan, T., Preuss, M., Cyrklaff, M., Hell, K., Rapaport, D., and Neupert, W. (2003). Evolutionary conservation of biogenesis of beta-barrel membrane proteins. *Nature* **426,** 862–866.

Pfanner, N., Wiedemann, N., Meisinger, C., and Lithgow, T. (2004). Assembling the mitochondrial outer membrane. *Nat. Struct. Mol. Biol.* **11,** 1044–1048.

Pinton, P., Brini, M., Bastianutto, C., Tuft, R. A., Pozzan, T., and Rizzuto, R. (1998). New light on mitochondrial calcium. *Biofactors* **8,** 243–253.

Rapaport, D. (2003). Finding the right organelle. Targeting signals in mitochondrial outer membrane proteins. *EMBO Rep.* **4,** 948–952.

Rapaport, D., and Neupert, W. (1999). Biogenesis of Tom40, core component of the TOM complex of mitochondria. *J. Cell Biol.* **146,** 321–331.

Rehling, P., Model, K., Bradner, K., Kovermann, P., Sickmann, A., Meyer, H. E., Kühlbrandt, W., Wagner, R., Truscott, K. N., and Pfanner, N. (2003). Protein insertion into the mitochondrial inner membrane by a twin-pore translocase. *Science* **299,** 1747–1751.

Rissler, M., Wiedemann, N., Pfannschmidt, S., Gabriel, K., Guiard, B., Pfanner, N., and Chacinska, A. (2005). The essential mitochondrial protein Erv1 cooperates with Mia40 in biogenesis of intermembrane space proteins. *J. Mol. Biol.* **353,** 485–492.

Rizzuto, R., Bernardi, P., and Pozzan, T. (2000). Mitochondria as all-round players of the calcium game. *J. Physiol.* **529,** 37–47.

Ryan, M. T., Voos, W., and Pfanner, N. (2001). Assaying protein import into mitochondria. *In* "Methods in Cell Biology" (L. A. Pon and E. A. Schon, eds.), Vol. 65, pp. 189–215. Elsevier Academic Press, New York.

Schatz, G., and Dobberstein, B. (1996). Common principles of protein translocation across membranes. *Science* **271,** 1519–1526.

Schägger, H. (2001). Blue-native gels to isolate protein complexes from mitochondria. *In* "Methods in Cell Biology" (L. A. Pon and E. A. Schon, eds.), Vol. 65, pp. 231–244. Elsevier Academic Press, New York.

Schägger, H., and von Jagow, G. (1991). Blue native electrophoresis for isolation of membrane protein complexes in enzymatically active form. *Anal. Biochem.* **199**, 223–231.

Schägger, H., Cramer, W. A., and von Jagow, G. (1994). Analysis of molecular masses and oligomeric states of protein complexes by blue native electrophoresis and isolation of membrane complexes by two-dimensional native electrophoresis. *Anal. Biochem.* **217**, 220–230.

Schleiff, E., Silvius, J. R., and Shore, G. C. (1999). Direct membrane insertion of the volrage-dependent anion selective channel protein catalyzed by mitochondrial Tom20. *J. Cell Biol.* **145**, 973–978.

Sickmann, A., Reinders, J., Wagner, Y., Joppich, C., Zahedi, R., Meyer, H. E., Schönfisch, B., Perschil, I., Chacinska, A., Guiard, B., Rehling, P., Pfanner, N., *et al.* (2003). The proteome of *Saccharomyces cerevisiae* mitochondria. *Proc. Natl. Acad. Sci. USA* **100**, 13207–13212.

Söllner, T., Rassow, J., Wiedmann, M., Schlossmann, J., Keil, P., Neupert, W., and Pfanner, N. (1992). Mapping of the protein import machinery in the mitochondrial outer membrane by crosslinking of translocation intermediates. *Nature* **355**, 84–87.

Truscott, K. N., Wiedemann, N., Rehling, P., Müller, H., Meisinger, C., Pfanner, N., and Guiard, B. (2002). Mitochondrial import of the ADP/ATP carrier: The essential TIM complex of the intermembrane space is required for precursor release from the TOM complex. *Mol. Cell. Biol.* **22**, 7780–7789.

Truscott, K. N., Voos, W., Frazier, A. E., Lind, M., Li, Y., Geissler, A., Dudek, J., Müller, H., Sickmann, A., Meyer, H. E., Meisinger, C., Guiard, B., *et al.* (2003). A J-protein is an essential subunit of the presequence translocase-associated protein import motor of mitochondria. *J. Cell Biol.* **163**, 707–713.

Voos, W., Gambill, D., Guiard, B., Pfanner, N., and Craig, E. A. (1993). Presequence and mature part of preproteins strongly influence the dependence of mitochondrial protein import in heat shock protein 70 in the matrix. *J. Cell Biol.* **123**, 119–126.

Waizenegger, T., Habib, S. J., Lech, M., Mokranjac, D., Paschen, S. A., Hell, K., Neupert, W., and Rapaport, D. (2004). Tob38, a novel essential component in the biogenesis of beta-barrel proteins of mitochondria. *EMBO Rep.* **5**, 704–709.

Wiedemann, N., Pfanner, N., and Ryan, M. T. (2001). The three modules of the ADP/ATP carrier cooperate in receptor recruitment and translocation into mitochondria. *EMBO J.* **20**, 951–960.

Wiedemann, N., Kozjak, V., Chacinska, A., Schönfisch, B., Rospert, S., Ryan, M. T., Pfanner, N., and Meisinger, C. (2003). Machinery for protein sorting and assembly in the mitochondrial outer membrane. *Nature* **424**, 565–571.

Wiedemann, N., Frazier, A. E., and Pfanner, N. (2004a). The protein import machinery of mitochondria. *J. Biol. Chem.* **279**, 14473–14476.

Wiedemann, N., Truscott, K. N., Pfannschmidt, S., Guiard, B., Meisinger, C., and Pfanner, N. (2004b). Biogenesis of the protein import channel Tom40 of the mitochondrial outer membrane: Intermembrane space components are involved in an early stage of the assembly pathway. *J. Biol. Chem.* **279**, 18188–18194.

Youle, R. J., and Karbowski, M. (2005). Mitochondrial fission in apoptosis. *Nat. Rev. Mol. Cell Biol.* **6**, 622–643.

PART VIII

Appendices

APPENDIX 1

Basic Properties of Mitochondria

Luis J. García-Rodríguez

Department of Anatomy and Cell Biology, Columbia University
New York, New York 10032

0091-679X/07 $35.00
DOI: 10.1016/S0091-679X(06)80040-3

Organism	Mitochondria per cell	% of total cell volume	Genomes per organelle	Genomes per cell	Genome size (kb)	Size (μm)	Volume (μm³)
Yeast (*Saccharomyces cerevisiae*)	1(A*) 10; 22 (C*) <10; 30–50 (B*)	12 (C**) 12; 4 (B**)	2–50 (D) 4 (C)	44; 88 (C*) 50; 100 (E*)	75 (KK)	0.16 (C*)	0.14 (C*)
Yeast (*Neurospora crassa*)	50–100 (F)	11.9 (G*)	7 μg/mg mito protein	~2.45 mg/g dry cell wt	65 (RR)	–	0.131 (H)
Penicilum marneffei					35 (RR)		
Aspergillus nidulans					33 (KK)		
Arabidopsis thaliana					367 (KK)		
Drosophila melanogaster					19 (KK)		
Rana temporaria oocyte	>100 × 10⁶ (MM)	0.86 ± 0.11 (MM**)					
Bufo bufo oocyte	>100 × 10⁶ (MM)	0.64 ± 0.05 (MM**)					
Xenopus oocyte	480–390,000 (I*) 16 ± 2 × 10⁶ (J*)		5–10 (D) 12 (J***)	1.92 × 10⁸ (K)		0.44 ± 0.08 (NN)	0.30 (J*)
Mouse oocyte	14–29 (L*)	5.29–7.16 (L*/**)		2.6 × 10⁸ (OO)	16.3 (PP)		avg 0.21–0.26 (L*^)
Mouse L cell	250 ± 70 (M)		2–6 (M,N)	1800 (O)		0.36 diameter (P*)	–
Mouse fertilized embryo				119,000 (HH)			
Rat liver cell	1400–2200 (Q*) 2601 (R*)	28 (S) 18.4 (T*)	5–10 (D)	7000–22,000 (U) 13,005–26,010 (V)			3.83–0.64 (Q*)
Pig oocyte	0.1/μm³ (W*/**)	6 (W**)		49,000–404,000 (QQ)			–
Human HeLa cell	383–882 (X*)	9.4–11 (X*)		8800 (O)	16.6 (LL)	0.62 (W**)	0.27–0.31 (X*)
Human liver cell	2,200 (Y)	17.6 (Y)					0.84 (Y)
Human osteosarcoma cell 143B				9100 ± 1600 (Z)			7.0 picoliters (Z)
Human fibrosarcoma cell HT1080	–		–	9700 ± 1800 (Z)			6.1 picoliters (Z)
Human melanoma cell VA₂-B/CAP23	–			10,300 ± 2100 (Z)			4.1 picoliters (Z)
Human HeLa S3 cell	–		–	7900 ± 1700 (Z)			4.9 picoliters (Z)
Human cultured A2780 cells	avg 107 (AA)		1–15, avg 4.6 (AA)	500 (AA)			–
Human fibroblast cell lines (6)				2400 ± 6000 (BB)			–
Human transformed cell lines (4)				1600 ± 7200 (BB)			–
Human squeletal muscle				3650 ± 620 (SS*)			
Human cardiac muscle				6970 ± 920 (SS*)			
Human platelets	4 (CC)		1 (CC)	4 (CC)			–
Human oocytes				130,000 ± 80,000 (DD)			–
Human sperm				100 (EE)			–
Bovine oocytes	10,000 (FF)			100,000 (FF)			–
Bovine sperm				75 (GG)			–
Chicken embryo fibroblast (CEF)	–			604 ± 134 (II)			–
Chicken liver tumor DU24	–	3.8 (JJ)	–	360 ± 80 (JJ)			–
Quail cell line LSCC-H32	–	4.5 (JJ)	–	276 ± 88 (JJ)			–

References:

(A) Hoffmann and Avers (1973). *Science* **181**, 749–751.
(B) Stevens (1977). *Cell. Biol.* **28**, 37–56.
(C) Grimes et al. (1974). *J. Cell. Biol.* **61**, 565–574.
(D) Alberts et al. (1994) *Molecular Biology of the Cell*, 3rd Ed., Garland, New York, pp. 553 and 706.
(E) Williamson (1970). *Cold Spr. Harb. Sym. Quant. Biol.* **24**, 247–276.
(F) Prokisch, H. (1999) Institut für Physiol. Chemie, Univ. of Munich. Personal communication.
(G) Hawley and Wagner (1967). *J. Cell Biol.* **35**, 489–499.
(H) Luck and Reich (1962). *Proc. Natl. Acad. Sci. USA* **52**, 931–938.
(I) Marinos and Billett (1981). *J. Embryol. Exp. Morphol.* **62**, 395–409.
(J) Marinos (1985). *Cell Diff.* **16**, 139–143.
(K) Based on 1.6 × 10⁷ organelles/cell and 12 mtDNAs/organelle (ref J)
(L) Nogawa et al. (1988). *J. Morph.* **195**, 225–234.
(M) Nass (1969). *J. Mol. Biol.* **42**, 521–528.
(N) Nass (1966). *Proc. Natl. Acad. Sci. USA* **56**, 1215–1232.
(O) Bogenhagen and Clayton (1974). *J. Biol. Chem.* **249**, 7991–7995.
(P) King et al. (1972). *J. Cell Biol.* **53**, 127–142.
(Q) David (1979). *Exp. Pathol.* **17**, 359–373.
(R) Pieri et al. (1975). *Exp. Gerontol.* **10**, 291–304.
(S) Blouin et al. (1977). *J. Cell. Biol.* **72**, 441–455.
(T) Loud (1962). *J. Cell. Biol.* **15**, 481–487.
(U) Based on a range of 5–10 (ref D) to 1,400–2,200 (ref Q) mtDNAs/organelle.
(V) Based on a range of 5–10 (ref D) to 2,601 (ref R) organelles/cell.
(W) Cran (1985). *J. Reprod. Fertil.* **74**, 237–245.

(X) Posakony et al. (1977). *J. Cell Biol.* **74**, 468–491.
(Y) Rohr et al. (1976). *In* "Progress in Liver Diseases" (Popper and Schaffner, eds.), Vol. 5, pp. 24–34. Grune and Stratton, New York.
(Z) King and Attardi (1989). *Science* **246**, 500–503.
(AA) Satoh and Kuroiwa (1991). *Exp. Cell. Res.* **196**, 137–40.
(BB) Shmookler Reis and Goldstein (1983). *J. Biol. Chem.* **258**, 9078–9085.
(CC) Shuster et al. (1988). *Biochem. Biophys. Res. Commun.* **155**, 1360–1365.
(DD) Chen et al. (1995). *Am. J. Hum. Genet.* **57**, 239–247.
(EE) Birky (1983). *Science* **222**, 468–475.
(FF) Michaels and Hauswirth (1982). *Dev. Biol.* **94**, 246–251.
(GG) Bahr and Engler (1970). *Exp. Cell. Res.* **60**, 338–340.
(HH) Piko and Taylor (1987). *Dev. Biol.* **123**, 364–374.
(II) Desjardins et al. (1985). *Mol. Cell. Biol.* **5**, 1163–1169.
(JJ) Morais et al. (1988). *In Vitro Cell. Dev. Biol.* **24**, 649–658.
(KK) Brown (2002). Genomes 2. BIOS Scientific publishers.
(LL) Shealan et al. (2005). *Plant J.* **44**, 744–755.
(MM) Romek and Krzysztofowicz (2005). *Fol. Histochem. et Biol.* **1**, 57–63.
(NN) Marchant et al. (2002). *Am. J. Physiol. Cell Physiol.* **282**, 1374–1386.
(OO) Pico and Taylor (1987). *Dev. Biol.* **123**, 364–374.
(PP) Larson and Clayton (1995). *Annu. Rev. Gent.* **29**, 151–178.
(QQ) Shourbagy et al. (2006). *Reproduction* **131**, 233–245.
(RR) Woo et al. (2003). *FEBS Lett.* **555**, 469–477.
(SS) Miller et al. (2003). *Nucl. Acid Res.* **31**–11, e61.

Notes:

^ Stereologic and morphometric analysis allows for 3-dimensional calculation from 2-dimensional measurements. Areas and diameters in table are precursors to these kinds of calculations, but cannot give the same information without necessary coefficients and parameters.

A* Values are from analysis of a diploid iso-N strain grown in liquid nutrient containing glucose or ethanol as energy sources.

B* Values are the number of mitochondria in an exponentially growing culture in glycerol or glucose, and in a stationary culture, respectively.

B** Values represent the total volume of mitochondria per cell volume in derepressed respiring cells and in glucose-repressed cells, respectively.

C* Values are based on haploid and diploid cells, respectively, growing exponentially on lactate medium (without catabolite repression).

C** Values are constant regardless of ploidy.

E* Values are in haploid and diploid cells, respectively.

G* Value is the average of time points taken over 30 hours of growth, and is the % mitochondrial volume/cytoplasmic volume, assuming sphericity.

I* Calculation based on average number of mitochondria profiles/100 mm² section of EM and calculated size of mitochondria in previtellogenic oocytes. Mitochondrial number was only calculated in oocytes with diameters between 50–250 mm. Range correlates with increase in oocyte size during development.

J* Calculation based on mature oocyte and a standard ellipsoid mitochondrion diameter of 0.2–0.4 mm.

J** Value based on an average number of 2 × 108 mtDNAs in a mature oocyte, and 25 × 104 ng mtDNA/mitochondrion in a mature oocyte.

L* Value is mean number of mitochondria per section, based on 2-dimensional analysis of single EM sections (i.e., not 3-dimensionally extrapolated).

L** Value is the ratio of the area of mitochondria to the area of cytoplasm in a 2-dimensional EM section.

P*   Value based on 2-dimensional analysis of single EM sections (i.e., not 3-dimensionally extrapolated).
Q*   Values are a range of mean calculations from postnatal rats, 2 to 6 months of age, respectively.
R*   Value is number of mitochondria per cell after 27 months of age.
T*   Value is % total cytoplasmic area occupied by the mitochondria of an average normal rat liver cell.
W*   Value is mean number of mitochondria/μm^3 of oocyte volume, based on 110 μm oocyte diameter and 0.62 μm mean mitochondrial diameter.
W** Values based on mature oocytes 40 hours after stimulation with hCG, after which fertilization can be achieved.
X*   Values are a range over entire cell cycle (beginning with time 0 after mitosis) of HeLa F-315 cells. Values are based only on 2-dimensional analysis.
MM* Volume densities of mitochondria in relation to the whole cytoplasm of the oocyte in stage 1 of cleavage.
SS* Value is mean number of mtDNA copy number per diploid nuclear genome.

APPENDIX 2

Direct and Indirect Inhibitors of Mitochondrial ATP Synthesis

Nanette Orme-Johnson

Department of Biochemistry, Tufts University School of Medicine
Boston, Massachusetts 02111

A few important notes about the tables in this appendix. The information in the tables is in no sense complete with respect to the compounds listed or the references cited for their use. I have tried to include the more commonly used inhibitors and others that are of either historical interest or that have been found recently and may become more widely used. The frequency of use of a compound may be assessed by the number of references indicated, that is, *(fewer than 10 references), **(10–100 references), or ***(greater than 100 references). Additionally, when possible, I have tried to cite mainly recent references, thus, if the latest reference for a compound is, for example, 1975, you may conclude that this compound is probably not used frequently now. I have also tried to cite papers using a given inhibitor in a number of experimental settings, for example, tissue or cells (T/C), whole mitochondria (M), and so on. This is intended to indicate the concentration of inhibitor that will cause the effect in that context. An interesting review by Wallace (1) on mitochondria as targets of drug toxicity considers how the drugs inhibit various reactions needed for the synthesis of ATP. I have not considered here compounds, for example, many of the HIV therapeutic drugs (2), that decrease mitochondrial ATP production indirectly by inhibiting the synthesis of mitochondrial DNA thus reducing the levels of proteins needed for ATP synthesis.

METHODS IN CELL BIOLOGY, VOL. 80
Copyright 2007, Elsevier Inc. All rights reserved.

0091-679X/07 $35.00
DOI: 10.1016/S0091-679X(06)80041-5

Table 2a

Inhibitors of Electron Transport Chain

The family of complex I inhibitors is large and structurally diverse. A number of excellent review articles are available. The review by Degli Esposti (3) lists complex I inhibitors with references to their use in beef heart mitochondria and/or submitochondrial particles, citing the concentrations used in that system and the efficacy of the inhibitors compared to that of rotenone. The adjacent review by Miyoshi (4) discusses the relation of the structure of complex I inhibitors to their efficacy. A subsequent review by Miyoshi (5) discusses the acetogenins specifically and the observation that the correlation of efficacy with structures is not always obvious. The review by Alali *et al.* (6) and the papers by Landolt *et al.* (7), by Zafra-Polo *et al.* (8), and by Miyoshi *et al.* (9) discuss in detail the annonaceous acetogenins. Reference (10) by Miyoshi *et al.* and references therein discuss pyridinium-type inhibitors. Certain quinone antagonists may inhibit both complex I and complex III (11) but at very different concentrations; the inhibition that occurs at lower inhibitor concentration is cited in this table. A number of inhibitors of the respiratory chain are potential or actual antibiotics, antiparasitics, chemotherapeutic agents, and/or insecticides. The review by Lümmen (12) discusses complex I inhibitors as insecticides and acaricides.

Inhibitor	Site[a]	References[b]	Notes[c]
Amiodarone*	Complex I	**M:** Hamster lung and liver (13); mouse liver (14); rat liver (15)	(a) Also inhibits complex V (16) and carnitine palmitoyltransferase-1 in rat heart mitochondria (17)
Amytal** (Amobarbitol)	Complex I	**T/C:** Cardiomyocytes (18); ROC-1 cells (19). **M:** *C. parapsilosis* (20); digitonin-treated homogenates of the mouse brain (21); rat liver (22); potato tuber (23). **SMP:** Rat heart (22)	
Annonaceous acetogenins** Includes: Annonin, annonacin, asimicin, bullatacin, gigantetrocins, molvazarin, otivarin, rolliniastatins, squamocin, trilobactin	Complex I	**T/C:** Insect Sf9 (Bullatacin and analogues) (24); solid tumor cell lines, see reference (6) for a review. **M:** Beef heart and *N. crassa* (Annonin VI) (25); European corn borer (26); rat liver and midgut of *Manduca sexta* larvae (bullatacin and analogues) (24); rat liver (20 compounds including annonacin, asimicin, gigantetrocins, bullatacin, molvazarin, trilobactin) (7). **SMP:** Beef heart (molvazarin, otivarin, rolliniastatin-1 and -2, squamocin) (27), (cherimolin-1, rolliniastatin-2) (28), (11 compounds including bullatacin, gigantetrocin, squamocins) (9), (six compounds including guanacone and derivatives, motrillin, rolliniastatin-2) (29), (bullatacin) (30), (six compounds including annonacin, corossolin, rolliniastatin-1) (31), crossolin, rolliniastatin-1 and -2 (32); blowfly flight muscle (33). **IC:** Beef heart (bullatacin and analogues) (24); (molvazarin, otivarin, tolliniastatin-1 and -2, squamocin) (27)	(a) Very large class of natural products (tetrahydrofuran acetogenins) isolated from *Annonaceae* plants. Those compounds tested inhibit complex I. A much larger number of these compounds have been found to be toxic to transformed cells, presumably, at least in part, due to this inhibition (b) Structure–function studies are reviewed in (6–9)
Antimycin A***	Complex III	**T/C:** *G. graminis* mycelia (also inhibits AOX) (34); mouse fibrosarcoma (35); mouse C2C12 myocytes (36); *S. cerevisiae* (37); transgenic cultured tobacco cells (38); *P. yoelii* in intact erythrocytes (39); tachyzoites (permeabilized) of *T. gondii* (40); *T. brucei brucei* (41). **M:** *G. graminis* (34); *L. mexicana* promastigote (42); potato tuber (43); lugworm (44); rat brain (45); rat liver (46); soy bean-seedling roots (47); tobacco (38); *T. brucei brucei* (41). **SMP:** Beef heart (48–49); rat liver (50). **IC:** Beef heart and *N. crassa* (51); beef Heart (crystal structure of inhibitor complex) (52)	(a) Center N inhibitor (53)

(continues)

Table 2a   *(continued)*

Inhibitor	Site[a]	References[b]	Notes[c]
Atovaquone* (Mepron)	Complex III	**T/C:** *P. yoelii* in intact erythrocytes (39); *P. carinii* (54); tachyzoites (permeabilized) of *T. gondii* (40). **M:** *P. falciparum* and *P. yoelii* (55); rat liver (40); rat and human livers (56)	(a) Inhibits mammalian as well as parasite dihydroorotate dehydrogenase, however the mammalian enzyme is much less sensitive than the parasite enzyme
Aureothin*	Complex I	**M:** Beef heart and *N. crassa* (25); rat liver (57). **IC:** *N. crassa* (25)	(a) Formerly known as mycolutien (58)
Azide***	Complex IV	**T/C:** *C. albicans* (59); CRI-G1 insulinoma cell line (60); crayfish neurons (61); human kidney cells (62); mouse C2C12 myocytes (36); rat hepatocytes (63); *S. cerevisiae* (37); SY5Y cells (64). **M:** Digitonin-treated homogenates of the mouse brain (21); rat liver (65); *L. mexicana* promastigote (42). **SMP:** Beef heart (66)	(a) Also inhibits complex V
β-Carboline derivatives**	Complex I	**T/C:** *T. cruzi* epimastigotes (67). **M:** Beef heart (β-carbolines and β-carboliniums) (68); rat liver (four compounds) (69), (seven compounds) (70) and (TaClo) (69)	(a) 5*H*-Pyrido[4,3-β]indoles (b) Inhibits the synthesis of 1MeTIQ (71)
β-Methoxyacrylate derivatives** Includes: Melithiazol, myxothiazol, MOA stilbene, oudemansin, pterulone, strobilurin	Complex III	**T/C:** Cardiomyocytes (myxothiazol) (18, 72); *P. yoelii* in intact erythrocytes (myxothiazol) (39); rat hepatocytes (myxothiazol) (73). **M:** *L. mexicana* promastigote (myxothiazol) (42); soy bean-seedling roots (myxothiazol) (47). **SMP:** Beef heart (74), (MOA stilbene or myxothiazol) (48, 75), (melithiazols) (76); rat liver (myxothiazol) (77). **IC:** Beef heart (myxothiazol) (51, 78–80), (crystal structure for inhibitor complex with MOA stilbene or myxothiazol) (52); *N. crassa* (myxothiazol) (51); *S. cerevisiae* mutants (MOA stilbene or myxothiazol) (81)	(a) Also called MOAs (b) Binds at Q(o) site (79) (c) Reference (82) discusses an MOA photoaffinity label (d) Melithiazol and myxothiazol are products of myxobacteria (e) Strobilurin is also called mucidin (83)
Capsaicin*	Complex I	**M:** Beef heart and potato tuber (84); rat liver (85). **SMP:** Beef heart (86). **IC:** Beef heart (also several analogues) (87)	(a) Isolated from *Capsicum* (b) The capsaicin site overlaps the rotenone site but not the piericidin A site (86)
Cyanide***	Complex IV	**T/C:** *G. graminis* mycelia (34); *P. yoelii* in intact erythrocytes (39); *S. cerevisiae* (37); *T. gondii* tachyzoites (permeabilized) (40); *T. brucei brucei* (41). **M:** *G. graminis* (34); *L. mexicana* promastigote (42); lugworm (44); rat brain (45); soy bean-seedling roots (47); potato tuber (43); soy beans (88). **SMP:** Rat liver (50)	
Diphenylene-Iodonium**	Complex I (AOX)	**T/C:** Cultured chick cardiomyocytes (18); cultured fibroblasts from normals and from LHON patients (89); tobacco cells (AOX) (90). **M:** Beef heart (91); potato tuber and cuckoo pint spadix (92). **SMP:** Potato tuber and pea leaves (93). **IC:** Beef heart (94)	
Ethoxyformic anhydride*	Complex III	**M:** Pigeon heart (95). **SMP:** Beef heart (48, 96); rat liver (50)	(a) Modifies histidine residues in Rieske Fe–S protein
Fenazaquin*	Complex I	**T/C:** Insect Sf9 cells (97); MCF-7 human breast cancer cells (98). **SMP:** Beef heart (86); rat liver (97). **IC:** Beef heart (97)	(a) Reference (99) discusses quinazolines photoaffinity labels
Fenpyroximate*	Complex I	**M:** Beef heart and *N. crassa* (25); rat liver and spider mites (100). **SMP:** Beef heart (86)	

(continues)

Table 2a *(continued)*

Inhibitor	Site[a]	References[b]	Notes[c]
Flavonoids** Includes: Flavone	Complex I	**T/C:** Complex I-deficient Chinese hamster CCL 16-B2 cells transfected with NDI1 of *S. cerevisiae* (flavone) (101). **M:** Beef heart (structure–function studies) (102–103); *L. mexicana* promastigote (flavone) (42); rat brain (flavone) (104); rat liver (8 compounds) (105). **SMP:** Beef heart (11 compounds) (106). **IC:** *S. cerevisiae* (flavone) (107)	(a) See also rotenoids (b) Some are free radical scavengers and antioxidants (c) Some inhibit complex V (e.g., quercetin)
Funiculosin**	Complex III	**M:** Rat liver (108). **SMP:** Beef heart (49). **IC:** *N. crassa* and beef heart (51)	(a) Fish mitochondria are resistant (109) (b) Binds at Q_i site (49); center N inhibitor (53)
HQNO**	Complex III	**T/C:** *B. bovis* (110); mouse LA9 (111); U937 cells (112). **M:** *L. mexicana* promastigote (42); rat liver (113). **SMP:** Beef heart (114)	(a) Binds at Q_i site (49, 112)
Isoquinolines or Isoquinoliniums**	Complex I	**T/C:** PC12 and SK-N-MC dopaminergic lines (isoquinoline and TIQ) (115); cultured endothelial cells (salsolinol) (116). **M:** Rat brain (TIQ) (117); mouse brain (N-Me-TIQ and N-Me-IQ$^+$) (118); rat liver (15 isoquinoline derivatives structurally related to MPTP) (119); mouse liver (MIQ$^+$) (120). **SMP:** Beef heart (nine *N*-methyl quinolinium analogues) (121); rat forebrain (11 isoquinolines, 2 dihydroisoquinolines, salsolinol, and 8 other 1,2,3,4-tetrahydroisoquinolines) (122)	(a) Inhibits α-ketoglutartate but not glutamate dehydrogenase (six isoquinoline derivatives) (123) (b) Some prevent the synthesis of 1MeTIQ (71) (c) Inhibits monoamine oxidase from human brain synaptosomal mitochondria (124)
Malonate***	Complex II	**T/C:** PC12 cells (125); rat marrow stroma (126). **M:** Beef heart, potato tuber and rat liver (uncoupled) (127); *L. mexicana* promastigote (42); rat brain (45, 128); rat heart (129); rat liver (130); *T. brucei brucei* (41). **SMP:** Beef heart (131); rat liver (50)	(a) Protects against covalent modification by ethoxyformic anhydride and rose bengal (132)
MPP$^+$***	Complex I	**T/C:** Human SH-SY5Y neuroblastoma cells (133); rat cerebellar granule cells (134). **M:** Beef heart and rat liver (135); rat brain and liver (136); rat brain and NB69 cells (138); synaptosomal cells from different rat brain areas (139). **SMP:** Beef heart (139–140), (MPP$^+$, six *N*-methyl pyridinium analogues and nine *N*-methyl quinolinium analogues) (121) and (MPP$^+$, five 4′-MPP$^+$ analogues and five 4′-phenylpyridine analogues) (141); beef heart (MPP$^+$, also used structural analogues of MPP$^+$) (142); fragments from rat forebrain (MPP$^+$) (122); rat liver inverted membrane preparations (MPP$^+$) (143). **IC:** Beef heart (MPP$^+$, five 4′-MPP$^+$ analogues and five 4′-phenylpyridine analogues) (141)	(a) Neurotoxic metabolite, produced by the action of monoamine oxidase B on MPTP (144–145) (b) Inhibits α-ketoglutarate dehydrogenase (123, 146–147) (c) Inhibits the synthesis of 1MeTIQ (71) (d) Inhibits synthesis of mitochondrial DNA (148) not by inhibiting DNA polymerase but by destabilizing the D-loop (149)
Myxobacterial products*** Includes: Aurachins, melithiazol, myxalamid, myxothiazol, phenalamid, phenoxan, stigmatellin, thiagazole	Complex III	**M:** Cuckoo pint spadices (aurachin C and analogues) (also inhibits AOX) (150); beef heart and *N. crassa* (aurachins A and B, myxalamid, phenalamid, phenoxan, thiagazole) (25). **SMP:** Beef heart (myxothiazol) (49) and (stigmatellin) (48, 114, 151); rat liver (myxothiazol) (77). **IC:** Beef heart (myxothiazol) (51, 78–80); *N. crassa* (myxothiazol) (51), and (aurachins) (25); beef heart (crystal structure for inhibitor complex with stigmatellin) (52)	(a) Isolated from myxobacterium (b) Aurachin binds at Q_i^3 site (150) (c) Myxothiazol and stigmatellin bind at Q_o site (49) (d) Melithiazol and myxothiazol are β-methoxyacrylates and listed as such

(continues)

Table 2a   (*continued*)

Inhibitor	Site[a]	References[b]	Notes[c]
3-Nitropropionic acid**	Complex II	**T/C:** Murine embryonal carcinoma cells (152); PC12 cells (125); rat cerebellar granule cells and astrocytes (153); rat hippocampal slices (154) and neurons (155); rat isolated, beating atria (156). **M:** PC6 cells and TNF-treated neurons (157); rat heart (156)	
Piericidin**	Complex I	**M:** Beef heart and *N. crassa* (piericidin A) (25); rat liver (158); *Y. lipolytica* (159). **SMP:** Beef heart (9, 27, 86, 140). **IC:** *N. crassa* (25)	(a) The piericidin A site overlaps the rotenone site but not the capsaicin site (86) (b) Reference (160) studied structural analogues
n-Propylgallate**	AOX	**T/C:** Harpin-treated tobacco cells (90); several types of marine phytoplankton and *C. reinhardtii* (161). **M:** Cuckoo pint spadices (also *n*-propyl gallate analogues) (150); soybean cotyledon (88, 150)	
Pyridaben*	Complex I	**T/C:** MCF-7 human breast cancer cells (98). **M:** Rat liver and tobacco hook worm midgut (97). **SMP:** Beef heart (30, 106). **IC:** Beef heart (98)	(a) (Trifluoromethyl) diazirinyl[³H]pyridaben ([³H]TDP) is a photoaffinity label (30) (b) Inhibits TPA-induced ornithine decarboxylase activity in cultured cells (98)
Rhein**	Complex I	**T/C:** Rat hepatocytes (162). **M:** Rat liver (163–164). **SMP:** Beef heart (165)	
Rotenoids*** Includes: Deguelin, rotenone	Complex I	**T/C:** Astrocytes (rotenone) (166); cultured mouse epidermal cells (deguelin) (167); *T. brucei brucei* (rotenone) (41). **M:** Beef heart (rotenone) (27); beef heart and *N. crassa* (rotenone) (25); *L. mexicana* promastigote (rotenone) (42); lugworm (rotenone) (44); rat brain (rotenone) (45); *T. brucei brucei* (rotenone) (41). **SMP:** Beef heart (rotenone) (140); beef heart and potato tuber (rotenone and 16 analogues) (168); beef heart (rotenone and 9 analogues, deguelin and 18 analogues) (106); rat heart (rotenone) (22). **IC:** *N. crassa* (rotenone) (25)	(a) Isoflavonoids from isolated from *Leguminosae* plants (b) Structural features of rotenoids needed for inhibition of complex I are similar to those for blocking phorbol ester-induced ornithine decarboxylase (98, 106, 169) (c) The rotenone site overlaps both the piericidin A and the capsaicin site, the latter two sites do not overlap (86)
Salicylhydroxamic acid**	AOX	**T/C:** *G. graminis* mycelia (34); tea leaf discs (43); several types of marine phytoplankton and *C. reinhardtii* (161); *T. brucei brucei* (41). **M:** Lugworm (44); potato tuber (43)	(a) Also called SHAM
Tannins*	Complex I	**M:** Rat liver (170). **IC:** Beef heart (171)	
TTFA**	Complex II	**T/C:** Germinating seeds of wheat and maize (172); human cervical carcinoma C33A cells (173); mouse fibrosarcoma (35); rat renal proximal tubular cells (174). **M:** Bone marrow-derived cell line, 32D (175); rat brain [v3]nonsynaptic tissue (176); *L. mexicana* promastigote (42); rat brain (177). **SMP:** Beef heart (178); *S. ratti* (179)	(a) Thenoyltrifluoroacetone
UHDBT**	Complex III	**T/C:** Trypomastigote forms of *T. brucei* (180). **M:** Jerusalem artichoke tuber and cuckoo pint spandices (also inhibits AOX) (181); rat liver (182). **IC:** Beef heart (183); chicken heart (79); *S. cerevisiae* (184); beef heart (crystal structure of inhibitor complex) (52)	(a) Synthetic analogue of ubiquinone (b) Binds at Q(o) site (78–79)

Table 2b
Inhibitors of F_0F_1-ATPase (Complex V) or the Adenine Nucleotide Translocator

Inhibitor	Site[a]	References[b]	Notes[c]
Atractyloside***	ANT	**T/C:** Neuronally differentiated PC12 cells (147); permeabilized guinea pig heart muscle (185); rat hepatocytes (73); sperm of the telost fishes (185). **M:** *L. mexicana* promastigotes (42); rat brain (136); rat heart (187); rat liver (188); rat liver and skeletal muscle (189); *S. cerevisiae* (190). **SMP:** Beef heart (191). **IC:** Beef heart (192); rat heart mitochondria (136); *S. cerevisiae* (193)	(a) The difference in binding of atractylate and carboxy-atractylate to rat liver mitochondria is discussed in reference (194)
Aurovertin**	Complex V	**T/C:** AS-30D rat hepatoma cells (195); ρ^0 HeLa S3 cells (196). **M:** Beef heart (197); *L. mexicana* promastigotes (42); pea leaves (198). **SMP:** Beef heart (199). **IC:** Beef heart in reconstituted liposomes (199); human F_1-ATPase (196); *T. cruzi* (200)	
Azide***	Complex V	**T/C:** ρ^0 Human tumor cells (201); ρ^0 HeLa S3 cells (196). **SMP:** Beef heart (66). **IC:** Human F_1-ATPase (196)	(a) Also inhibits complex IV
Bongkrekic***	ANT	**T/C:** Cultured cerebrocortical neurons (202); ρ^0 HeLa S3 cells (196); HepG2cells and rat primary hepatocytes (203); Jurkat cells (204); murine pre-B cell line BAF3 (205); ρ^0 HeLa S3 cells (196). **M:** Rat heart (206); rat liver (207); rat liver and skeletal muscle (189); *S. cerevisiae* (190). **SMP:** Beef heart (199, 208). **IC:** Beef heart (192, 209); *S. cerevisiae* (193)	(a) Produced by *Burkholderia cocovenenans*
Carboxy-atractyloside***	ANT	**M:** Beef heart (210); rat heart from cold exposed and normal rats (211); rat liver (212); rat liver and skeletal muscle (189); *S. cerevisiae* (190). **SMP:** Beef heart (208). **IC:** Beef heart (192, 209); *S. cerevisiae* (193)	
DCCD***	Complex V	**M:** Rat liver (213); rat liver (from rats infected with *F. hepatica*) (214); *S. cerevisiae* (215). **SMP:** Beef heart (216); inner membrane (217); *H. diminuta* (218). **IC:** Beef heart (219); spinach leaves (220)	(a) Modifies carboxylic groups of glutamic or aspartic residues (b) Also inhibits complex III (221) and complex IV (222–223) (c) Binds to Asp-160 of yeast cyt *b* (224) (d) Inhibits purified mitochondrial VDAC channel (225)

(continues)

Table 2b   (*continued*)

Inhibitor	Site[a]	References[b]	Notes[c]
Efrapeptin**	Complex V	**T/C:** *A. castellanii* (226); CHO cells (227). **M:** Beef heart (228–229). **SMP:** Beef heart (230); rat testis (231). **IC:** Beef heart (232); *T. cruzi* (200)	(a) Blocks exocytic but not endocytic trafficking of proteins (233) (b) Binds to a unique site in the central cavity of F_1-ATPase
Leucinostatins**	Complex V	**T/C:** CHO cell lines (227). **M:** Rat liver (213)	(a) Uncouples at higher concentrations (213)
Oligomycin***	Complex V	**T/C:** *A. castellanii* (226); AS-30D rat hepatoma cells (195); CHO cell lines (227); neuronally differentiated PC12 cells (147); rat hepatocytes (73); *S. cerevisiae* (37); Jurkat T cells (234). **M:** Rat brain (45); *L. mexicana* promastigotes (42); rat heart (211); rat liver (212). **SMP:** Beef heart (151, 216); rat heart (235). **IC:** Human F_1-ATPase (196); spinach leaves (220)	
Venturicidin**	Complex V	**M:** Pea leaves (198); potato tuber (236); rat liver (213). **SMP:** Beef heart (237–238). **IC:** Beef heart reconstituted in lipid bilayer (239); beef heart F_0 (240)	

Table 2c
Uncouplers

These compounds decrease the yield of ATP by decreasing one or both of the two terms that contribute to the protomotive force ($\Delta\mu_{H^+}$). Thus, they decrease either the membrane potential ($\Delta\Psi$), the pH gradient (ΔpH), or both.

Inhibitor	References[b]	Notes[c]
A23187**	**T/C:** Rabbit thoracic aorta rings (241); rat pituitary (GH3, GC, and GH3B6), adrenal (PC12), and mast (RBL-1) cell lines (242); rat hepatocytes (243). **M:** Beef heart (244); rat liver (245–246). **SMP:** *L. braziliensis* promastigotes (247)	(a) Principally a divalent cation ionophore; preference $Ca^{++} > Mg^{++}$ (b) Also called calcimycin
CCCP***	**T/C:** Bovine sperm (248); filarial nematodes (249); human cervical carcinoma C33A (173); Jurkat and CEM (250); mouse C2C12 myocytes (36); rat anterior pituitary GH (3) (251); rat cerebellar neurons (252); spinach leaves (253). **M:** *H. diminuta* (218); mouse C2C12 myocyte (36); rat brain (128, 254); rat forebrain (128); rat heart and liver (255); spinach leaves (253). **SMP:** Beef heart (249, 256). **IC:** Beef heart complex III (257)	(a) Protonophore

(*continues*)

Table 2c   (*continued*)

Inhibitor	References[b]	Notes[c]
DNP***	**T/C:** Jurkat and CEM (250); sperm of the telost fishes (186); *S. cerevisiae* (258). **M:** Beef liver and pea stem (259); rat heart, liver and skeletal muscle (255); rat liver (260); rat liver and skeletal muscle (189). **SMP:** Rat heart (255)	(a) Protonophore (b) Alkylated dinitrophenols (2-alkyl-4,6-dinitrophenol) inhibits complex II (261–262)
FCCP***	**T/C:** *T. brucei brucei* (263); hamster brown fat (264); rat hepatocytes (73); rat renal proximal tubule cells (265); Jurkat T cells (234). **M:** Beef liver and pea stem (259); hamster brown fat (266); *L. mexicana* (42); rat heart and liver (255); rat liver (267); rat liver and skeletal muscle (189); rodent brown adipose (268). **SMP:** Beef heart (74); *H. diminuta* (218); *L. braziliensis* promastigotes (247)	(a) Protonophore
Gramicidin**	**T/C:** Bloodstream forms of *T. brucei brucei* (269); rat cerebellar neurons (252); rat hepatocytes (73); *S. cerevisiae* (270). **M:** Rat heart and liver (255); rat liver and skeletal muscle (189). **SMP:** Beef heart (271)	(a) Pore forming (b) Polypeptide antibiotic (c) Reference (272) studies pore forming by gramicidin analogues
Monensin**	**T/C:** Rat brain synaptosomes (273); rat hepatocytes and liver and lung slices (274); rat renal proximal tubule cells (265). **M:** Hamster brown fat (266); rat liver and skeletal muscle (189); rat liver (275)	(a) Causes disintegration of the ER-Golgi system in spinach leaves (253) (b) Electroneutral monovalent antiporter, especially Na^+/H^+ (c) Complex polyether antibiotic structurally related to nigericin
Nigericin**	**T/C:** Bloodstream forms of *T. brucei brucei* (289); rat renal proximal tubule cells (265). **M:** Hamster brown fat (266); rat liver (275)	(a) Electroneutral monovalent antiporter, especially K^+/H^+ (b) Complex polyether antibiotic, structurally related to monensin
SF6847**	**M:** Cell line overexpressing human Bcl-2 or from livers of Bcl-2 transgenic mice (276); rat heart and liver (255); rat heart from cold exposed and normal rats (211). **SMP:** Beef heart (277)	(a) Protonophore
Valinomycin***	**T/C:** Bloodstream forms of *T. brucei brucei* (269); mouse C2C12 myocytes (36); murine pre-B cell line BAF3 (205); rat cerebellar neurons (252); rat renal proximal tubule cells (265); *S. cerevisiae* (270). **M:** Hamster brown fat (266); rat heart (278); rat liver (275). **SMP:** Beef heart (74). **IC:** Beef heart complex III in asolectin liposomes (279) and in potassium-loaded phospholipid vesicles (221)	(a) Monovalent uniporter for K^+

(*continues*)

Compounds

Common name or abbreviation	Chemical name
1MeTIQ	1-Methyl-1,2,3,4-tetrahydroisoquinoline
A23187	6S-[6α92S*, 3S*), 8β(R*), 9β, 11α]-5-(methylamino)-2-([3,9,11-trimethyl-8-[1-methyl-2-oxo-2-(1H-pyrrol-2-yl)ethyl]-1,7-dioxaspiro[5,5]undec-2-yl]methyl)-4-benzoxazole-carboxylic acid
Amiodarone or AMD	(2-Butyl-3-benzofuranyl)[4-[2-(diethylamino)ethoxy]-3,5-diiodophenyl]methanone
Amobarbital or Amytal	5-Ethyl-5-(3-methylbutyl)-2,4,6-(1H,3H,5H)-pyrimidinetrione
Antimycin	3-Methylbutanoic acid 3-([3-(formylamino)-2-hydroxybenzoyl]amino)-8-hexyl-2,6-dimethyl-4,9-dioxo-1,5-dioxonan-7-yl ester
Arylazido-β-alanine ADP-ribose	N-(4-Azido-2-nitrophenyl) β alanine
Atovaquone	2-[trans-4-(4'-Chlorophenyl)cyclohexyl]-3-hydroxy-1,4-naphthoquinone
Aureothin	2-Methoxy-3,5-dimethyl-6-(tetrahydro-4-[2-methyl-3-(4-nitrophenyl)-2-propenylidene]-2-furanyl)-4H-pyran-4-one
β-Carboline alkaloids	9H-Pyrido-[3,4-b]-indole derivatives
CCCP	Carbonyl cyanide m-chlorophenylhydrazone
DCCD	Dicyclohexylcarbodiimide
DNP	Dinitrophenol
FCCP	Carbonyl cyanide p-trifluoromethoxyphenylhydrazone
Fenazaquin	2-Decyl-4-quinazolinylamine
Fenpyroximate	4(cis-4-tert-Butylcyclohexylamino)5-chloro-6-ethylpyrimidine
Flavone	2-Phenyl-4H-1-benzopyran-4-one
HQNO	n-Heptyl-4-hydroxyquinoline-N-oxide
MIQ$^+$	1-Methyl-isoquinolone
MOA stilbene	3-Methoxy-2(2-styrylphenyl)propenic acid-methylester
MPTP	1-Methyl-4-phenyl-1,2,3,6-tetrahydropyridine
MPP$^+$	1-Methyl-4-phenylpyridinium
N-Me-IQ$^+$	N-Methyl-isoquinolinium ion
N-Me-TIQ	N-Methyl-1,2,3,4-tetrahydroisoquinoline
Piercidin	2,3-Dimethoxy-4-hydroxy-5-methyl-6-polyprenyl-pyridine
Rhein	4,5-Dihydroxyanthraquinone-2-carboxylic acid
Salsolinol	1-Methyl-6,7-dihydroxy-1,2,3,4-tetrahydroisoquinoline
TaClo	1-Trichloromethyl-1,2,3,4-tetrahydro-β-carboline
THIQ	1,2,3,4-Tetrahydroisoquinoline, also called TIQ
TIQ	1,2,3,4-Tetrahydroisoquinoline, also called THIQ
TTFA	4,4,4-Trifluoro-1-(2-thienyl)-butane-1,3-dione
TTFB	4,5,6,7-Tetrachloro-2-trifluoromethylbenzimidazole
UHDBT	5-n-Undecyl-6-hydroxy-4,7-dioxobenzothiazole

Organisms

Abbreviation	Full name of organism	Abbreviation	Full name of organism
A. castellanii	Acanthamoeba castellanii	N. crassa	Neurospora crassa
B. bovis	Babesia bovis	P. falciparum	Plasmodium falciparum
C. albicans	Candida albicans	P. yoelii	Plasmodium yoelii
C. parapsilosis	Candida parapsilosis	P. carinii	Pneumocystis carinii
C. reinhardtii	Chlamydomonas reinhardtii	S. cerevisiae	Saccharomyces cerevisiae
F. hepatica	Fasciola hepatica	S. ratti	Strongyloides ratti
G. graminis	Gaeumannomyces graminis	T. gondii	Toxoplasma gondii
H. diminuta	Hymenolepis diminuta	T. brucei brucei	Trypanosoma brucei brucei
L. braziliensis	Leishmania braziliensis	T. cruzi	Trypanosoma cruzi
L. mexicana	Leishmania mexicana	Y. lipolytica	Yarrowia lipolytica

(continues)

[a]ANT, Adenine nucleotide translocator; AOX, Alternative oxidase; Complex I, NADH:ubiquinone reductase; Complex II, Succinate:ubiquinone reductase; Complex III, Ubiquinol:cytochrome c reductase; Complex IV, Cytochrome oxidase; Complex V, F_0F_1-ATPase; UC, Uncoupler.

[b]Abbreviations used: **T/C**, tissue of cells; **M**, mitochondria; **SMP**, submitochondrial particles; and **IC**, isolated complex; for each the source is indicated.

[c]Center N (equivalently, the Qi, Q(i), or Q_i site) is the ubiquinone reduction site in complex III. Center P (equivalently, the Qo, Q(o), or Q_o site) is the ubiquinol oxidation site in complex III. Center N inhibitors include: n-heptyl-4-hydroxyquinoline-N-oxide (HQNO), antimycins, and aurachins. Center P inhibitors include: 5-n-undecyl-6-hydroxy-4,7-dioxobenzothiazole (UHDBT), β-methoxyacrylates, stigmatellin, and myxothiazol.

References Listed in Table

1. Wallace, K. B. *et al.* (2000). *Annu. Rev. Pharmacol. Toxicol.* **40,** 353–388.
2. Walker, U. A. (2003). *J. HIV Ther.* **8,** 32–35.
3. Degli Esposti, M. (1998). *Biochim. Biophys. Acta* **1364,** 222–235.
4. Miyoshi, H. (1998). *Biochim. Biophys. Acta* **1364,** 236–244.
5. Miyoshi, H. (2001). *J. Bioenerg. Biomembr.* **33,** 223–231.
6. Alali, F. Q. *et al.* (1999). *J. Nat. Prod.* **62,** 504–540.
7. Landolt, J. L. *et al.* (1995). *Chem. Biol. Interact.* **98,** 1–13.
8. Zafra-Polo, M. C. *et al.* (1996). *Phytochemistry* **42,** 253–271.
9. Miyoshi, H. *et al.* (1998). *Biochim. Biophys. Acta* **1365,** 443–452.
10. Miyoshi, H. *et al.* (1998). *J. Biol. Chem.* **273,** 17368–17374.
11. Degli Esposti, M. *et al.* (1993). *Biochem. Biophys. Res. Commun.* **190,** 1090–1096.
12. Lummen, P. (1998). *Biochim. Biophys. Acta* **1364,** 287–296.
13. Card, J. W. *et al.* (1998). *Toxicol. Lett.* **98,** 41–50.
14. Fromenty, B. *et al.* (1990). *J. Pharmacol. Exp. Ther.* **255,** 1377–1384.
15. Ribeiro, S. M. *et al.* (1997). *Cell Biochem. Funct.* **15,** 145–152.
16. Chen, C. L. *et al.* (1994). *Res. Commun. Mol. Pathol. Pharmacol.* **85,** 193–208.
17. Kennedy, J. A. *et al.* (1996). *Biochem. Pharmacol.* **52,** 273–280.
18. Becker, L. B. *et al.* (1999). *Am. J. Physiol.* **277,** H2240–H2246.
19. Jurkowitz, M. S. *et al.* (1998). *J. Neurochem.* **71,** 535–548.
20. Guerin, M. *et al.* (1989). *Biochimie* **71,** 887–902.
21. Kunz, W. S. *et al.* (1999). *J. Neurochem.* **72,** 1580–1585.
22. Wyatt, K. M. *et al.* (1995). *Biochem. Pharmacol.* **50,** 1599–1606.
23. Raison, J. K. *et al.* (1973). *J. Bioenerg.* **4,** 409–422.
24. Ahammadsahib, K. I. *et al.* (1993). *Life Sci.* **53,** 1113–1120.
25. Friedrich, T. *et al.* (1994). *Eur. J. Biochem.* **219,** 691–698.
26. Lewis, M. A. *et al.* (1993). *Pestic. Biochem. Physiol.* **45,** 15–23.
27. Degli Esposti, M. *et al.* (1994). *Biochem. J.* **301,** 161–167.
28. Estornell, E. *et al.* (1997). *Biochem. Biophys. Res. Commun.* **240,** 234–238.
29. Gallardo, T. *et al.* (1998). *J. Nat. Prod.* **61,** 1001–1005.
30. Schuler, F. *et al.* (1999). *Proc. Natl. Acad. Sci. USA* **96,** 4149–4153.
31. Tormo, J. R. *et al.* (1999). *Chem. Biol. Interact.* **122,** 171–183.
32. Tormo, J. R. *et al.* (1999). *Arch. Biochem. Biophys.* **369,** 119–126.
33. Londershausen, M. *et al.* (1991). *Pestic. Sci.* **33,** 427–438.
34. Joseph-Horne, T. *et al.* (1998). *J. Biol. Chem.* **273,** 11127–11133.
35. Schulze-Osthoff, K. *et al.* (1992). *J. Biol. Chem.* **267,** 5317–5323.
36. Biswas, G. *et al.* (1999). *EMBO J.* **18,** 522–533.
37. Grant, C. M. *et al.* (1997). *FEBS Lett.* **410,** 219–222.
38. Maxwell, D. P. *et al.* (1999). *Proc. Natl. Acad. Sci. USA* **96,** 8271–8276.
39. Srivastava, I. K. *et al.* (1997). *J. Biol. Chem.* **272,** 3961–3966.
40. Vercesi, A. E. *et al.* (1998). *J. Biol. Chem.* **273,** 31040–31047.
41. Beattie, D. S. *et al.* (1996). *Eur. J. Biochem.* **241,** 888–894.
42. Bermudez, R. *et al.* (1997). *Mol. Biochem. Parasitol.* **90,** 43–54.
43. Kumar, S. *et al.* (1999). *Anal. Biochem.* **268,** 89–93.
44. Volkel, S. *et al.* (1996). *Eur. J. Biochem.* **235,** 231–237.
45. Zini, R. *et al.* (1999). *Drugs Exp. Clin. Res.* **25,** 87–97.
46. Ohkohchi, N. *et al.* (1999). *Transplantation* **67,** 1173–1177.
47. Millar, A. H. *et al.* (1993). *FEBS Lett.* **329,** 259–262.

48. Matsuno-Yagi, A. *et al.* (1999). *J. Biol. Chem.* **274,** 9283–9288.

49. Rich, P. R. *et al.* (1990). *Biochim. Biophys. Acta* **1018,** 29–40.

50. Moreno-Sanchez, R. *et al.* (1999). *Eur. J. Biochem.* **264,** 427–433.

51. Bechmann, G. *et al.* (1992). *Eur. J. Biochem.* **208,** 315–325.

52. Yu, C. A. *et al.* (1998). *Biochim. Biophys. Acta* **1365,** 151–158.

53. Brasseur, G. *et al.* (1994). *FEBS Lett.* **354,** 23–29.

54. Cushion, M. T. *et al.* (2000). *Antimicrob. Agents Chemother.* **44,** 713–719.

55. Fry, M. *et al.* (1992). *Biochem. Pharmacol.* **43,** 1545–1553.

56. Knecht, W. *et al.* (2000). *FEBS Lett.* **467,** 27–30.

57. Magae, J. *et al.* (1993). *Biosci. Biotechnol. Biochem.* **57,** 1628–1631.

58. Schwartz, J. L. *et al.* (1976). *J. Antibiot. (Tokyo)* **29,** 236–241.

59. Maesaki, S. *et al.* (1998). *J. Antimicrob. Chemother.* **42,** 747–753.

60. Harvey, J. *et al.* (1999). *Br. J. Pharmacol.* **126,** 51–60.

61. Nguyen, P. V. *et al.* (1997). *J. Neurophysiol.* **78,** 281–294.

62. Zager, R. A. *et al.* (1997). *Kidney Int.* **51,** 728–738.

63. Nakagawa, Y. *et al.* (1996). *Toxicology* **114,** 135–145.

64. Cassarino, D. S. *et al.* (1998). *Biochem. Biophys. Res. Commun.* **248,** 168–173.

65. Starkov, A. A. *et al.* (1997). *Biochim. Biophys. Acta* **1318,** 173–183.

66. Galkin, M. A. *et al.* (1999). *FEBS Lett.* **448,** 123–126.

67. Rivas, P. *et al.* (1999). *Comp. Biochem. Physiol. C Pharmacol. Toxicol. Endocrinol.* **122,** 27–31.

68. Krueger, M. J. *et al.* (1993). *Biochem. J.* **291,** 673–676.

69. Janetzky, B. *et al.* (1995). *J. Neural. Transm. Suppl.* **46,** 265–273.

70. Albores, R. *et al.* (1990). *Proc. Natl. Acad. Sci. USA* **87,** 9368–9372.

71. Yamakawa, T. *et al.* (1999). *Neurosci. Lett.* **259,** 157–160.

72. Yao, Z. *et al.* (1999). *Am. J. Physiol.* **277,** H2504–H2509.

73. Chandel, N. S. *et al.* (1997). *J. Biol. Chem.* **272,** 18808–18816.

74. Galkin, A. S. *et al.* (1999). *FEBS Lett.* **451,** 157–161.

75. Junemann, S. *et al.* (1998). *J. Biol. Chem.* **273,** 21603–21607.

76. Sasse, F. *et al.* (1999). *J. Antibiot. (Tokyo)* **52,** 721–729.

77. Kozlov, A. V. *et al.* (1999). *FEBS Lett.* **454,** 127–130.

78. Kim, H. *et al.* (1998). *Proc. Natl. Acad. Sci. USA* **95,** 8026–8033.

79. Crofts, A. R. *et al.* (1999). *Biochemistry* **38,** 15807–15826.

80. Crofts, A. R. *et al.* (1999). *Proc. Natl. Acad. Sci. USA* **96,** 10021–10026.

81. Geier, B. M. *et al.* (1992). *Eur. J. Biochem.* **208,** 375–380.

82. Mansfield, R. W. *et al.* (1990). *Biochim. Biophys. Acta* **1015,** 109–115.

83. Von Jagow, G. *et al.* (1986). *Biochemistry* **25,** 775–780.

84. Satoh, T. *et al.* (1996). *Biochim. Biophys. Acta* **1273,** 21–30.

85. Chudapongse, P. *et al.* (1981). *Biochem. Pharmacol.* **30,** 735–740.

86. Okun, J. G. *et al.* (1999). *J. Biol. Chem.* **274,** 2625–2630.

87. Shimomura, Y. *et al.* (1989). *Arch. Biochem. Biophys.* **270,** 573–577.

88. Hoefnagel, M. H. *et al.* (1995). *Arch. Biochem. Biophys.* **318,** 394–400.

89. Cock, H. R. *et al.* (1999). *J. Neurol. Sci.* **165,** 10–17.

90. Xie, Z. *et al.* (2000). *Mol. Plant. Microbe Interact.* **13,** 183–190.

91. Ragan, C. I. *et al.* (1977). *Biochem. J.* **163,** 605–615.

92. Roberts, T. H. *et al.* (1995). *FEBS Lett.* **373,** 307–309.

93. Bykova, N. V. *et al.* (1999). *Biochem. Biophys. Res. Commun.* **265,** 106–111.

94. Majander, A. *et al.* (1994). *J. Biol. Chem.* **269,** 21037–21042.

95. Herrero, A. *et al.* (1997). *Mech. Ageing Dev.* **98,** 95–111.

96. Anderson, W. M. *et al.* (1995). *Biochim. Biophys. Acta* **1230,** 186–193.

97. Hollingworth, R. M. *et al.* (1994). *Biochem. Soc. Trans.* **22,** 230–233.

98. Rowlands, J. C. *et al.* (1998). *Pharmacol. Toxicol.* **83,** 214–219.

99. Latli, B. *et al.* (1996). *Chem. Res. Toxicol.* **9,** 445–450.

100. Obata, T. *et al.* (1992). *Pestic. Sci.* **34,** 133–138.

101. Seo, B. B. *et al.* (1998). *Proc. Natl. Acad. Sci. USA* **95,** 9167–9171.

102. Hodnick, W. F. *et al.* (1988). *Biochem. Pharmacol.* **37,** 2607–2611.

103. Hodnick, W. F. *et al.* (1994). *Biochem. Pharmacol.* **47,** 573–580.

104. Ratty, A. K. *et al.* (1988). *Biochem. Med. Metab. Biol.* **39,** 69–79.

105. Santos, A. C. *et al.* (1998). *Free Radic. Biol. Med.* **24,** 1455–1461.

106. Fang, N. *et al.* (1998). *Proc. Natl. Acad. Sci. USA* **95,** 3380–3384.

107. de Vries, S. *et al.* (1988). *Eur. J. Biochem.* **176,** 377–384.

108. Pietrobon, D. *et al.* (1981). *Eur. J. Biochem.* **117,** 389–394.

109. Degli Esposti, M. *et al.* (1992). *Arch. Biochem. Biophys.* **295,** 198–204.

110. Gozar, M. M. *et al.* (1992). *Int. J. Parasitol.* **22,** 165–171.

111. Howell, N. *et al.* (1988). *J. Mol. Biol.* **203,** 607–618.

112. Brambilla, L. *et al.* (1998). *FEBS Lett.* **431,** 245–249.

113. Halestrap, A. P. (1987). *Biochim. Biophys. Acta* **927,** 280–290.

114. Matsuno-Yagi, A. *et al.* (1996). *J. Biol. Chem.* **271,** 6164–6171.

115. Seaton, T. A. *et al.* (1997). *Brain Res.* **777,** 110–118.

116. Melzig, M. F. *et al.* (1993). *Neurochem. Res.* **18,** 689–693.

117. Suzuki, K. *et al.* (1989). *Biochem. Biophys. Res. Commun.* **162,** 1541–1545.

118. Suzuki, K. *et al.* (1992). *J. Neurol. Sci.* **109,** 219–223.

119. McNaught, K. S. *et al.* (1996). *Biochem. Pharmacol.* **51,** 1503–1511.

120. Aiuchi, T. *et al.* (1996). *Neurochem. Int.* **28,** 319–323.

121. Miyoshi, H. *et al.* (1997). *J. Biol. Chem.* **272,** 16176–16183.

122. McNaught, K. S. *et al.* (1995). *Biochem. Pharmacol.* **50,** 1903–1911.

123. McNaught, K. S. *et al.* (1995). *Neuroreport* **6,** 1105–1108.

124. Minami, M. *et al.* (1993). *J. Neural. Transm. Gen. Sect.* **92,** 125–135.

125. Keller, J. N. *et al.* (1998). *J. Neurosci.* **18,** 4439–4450.

126. Klein, B. Y. *et al.* (1996). *J. Cell. Biochem.* **60,** 139–147.

127. Brand, M. D. *et al.* (1994). *Eur. J. Biochem.* **226,** 819–829.

128. Vogel, R. *et al.* (1999). *Neurosci. Lett.* **275,** 97–100.

129. Korshunov, S. S. *et al.* (1997). *FEBS Lett.* **416,** 15–18.

130. Chien, L. F. *et al.* (1996). *Biochem. J.* **320,** 837–845.

131. Catia Sorgato, M. *et al.* (1985). *FEBS Lett.* **181,** 323–327.

132. Hederstedt, L. *et al.* (1986). *Arch. Biochem. Biophys.* **247,** 346–354.

133. Fall, C. P. *et al.* (1999). *J. Neurosci. Res.* **55,** 620–628.

134. Camins, A. *et al.* (1997). *J. Neural. Transm.* **104,** 569–577.

135. Murphy, M. P. *et al.* (1995). *Biochem. J.* **306,** 359–365.

136. Cassarino, D. S. *et al.* (1999). *Biochim. Biophys. Acta* **1453,** 49–62.

137. Pardo, B. *et al.* (1995). *J. Neurochem.* **64,** 576–582.

138. Bougria, M. *et al.* (1995). *Eur. J. Pharmacol.* **291,** 407–415.

139. Pecci, L. *et al.* (1994). *Biochem. Biophys. Res. Commun.* **199,** 755–760.

140. Singer, T. P. *et al.* (1994). *Biochim. Biophys. Acta* **1187,** 198–202.

141. Gluck, M. R. *et al.* (1994). *J. Biol. Chem.* **269,** 3167–3174.

142. Hasegawa, E. *et al.* (1997). *Arch. Biochem. Biophys.* **337,** 69–74.

143. Sablin, S. O. *et al.* (1996). *J. Biochem. Toxicol.* **11,** 33–43.

144. Castagnoli, N. Jr. *et al.* (1985). *Life Sci.* **36,** 225–230.

145. Heikkila, R. E. *et al.* (1985). *Neurosci. Lett.* **62,** 389–394.

146. Mizuno, Y. *et al.* (1988). *J. Neurol. Sci.* **86,** 97–110.

147. Chalmers-Redman, R. M. *et al.* (1999). *Biochem. Biophys. Res. Commun.* **257,** 440–447.

148. Miyako, K. *et al.* (1999). *Eur. J. Biochem.* **259,** 412–418.

149. Umeda, S. *et al.* (2000). *Eur. J. Biochem.* **267,** 200–206.

150. Hoefnagel, M. H. *et al.* (1995). *Eur. J. Biochem.* **233,** 531–537.

151. Matsuno-Yagi, A. *et al.* (1997). *J. Biol. Chem.* **272,** 16928–16933.

152. Pass, M. A. *et al.* (1994). *J. Nat. Toxins* **2,** 386–394.

153. Olsen, C. *et al.* (1999). *Brain Res.* **850,** 144–149.

154. Riepe, M. W. *et al.* (1996). *Exp. Neurol.* **138,** 15–21.

155. Brorson, J. R. *et al.* (1999). *J. Neurosci.* **19,** 147–158.

156. Lopez, P. S. *et al.* (1998). *Toxicol. Appl. Pharmacol.* **148,** 1–6.

157. Bruce-Keller, A. J. *et al.* (1999). *J. Neuroimmunol.* **93,** 53–71.

158. Bhuvaneswaran, C. *et al.* (1972). *Experientia* **28,** 777–778.

159. Kerscher, S. J. *et al.* (1999). *J. Cell Sci.* **112,** 2347–2354.

160. Chung, K. H. *et al.* (1989). *Zeit Natur for Sec C. J. Biosci.* **44,** 609–616.

161. Eriksen, N. T. *et al.* (1999). *Aquatic Microbial. Ecol.* **17,** 145–152.

162. Bironaite, D. *et al.* (1997). *Chem. Biol. Interact.* **103**, 35–50.

163. Miccadei, S. *et al.* (1993). *Anticancer Res.* **13**, 1507–1510.

164. Floridi, A. *et al.* (1994). *Biochem. Pharmacol.* **47**, 1781–1788.

165. Glinn, M. A. *et al.* (1997). *Biochim. Biophys. Acta* **1318**, 246–254.

166. Di Monte, D. A. *et al.* (1999). *Toxicol. Appl. Pharmacol.* **158**, 296–302.

167. Gerhauser, C. *et al.* (1997). *Cancer Res.* **57**, 3429–3435.

168. Ueno, H. *et al.* (1996). *Biochim. Biophys. Acta* **1276**, 195–202.

169. Udeani, G. O. *et al.* (1997). *Cancer Res.* **57**, 3424–3428.

170. Konishi, K. *et al.* (1993). *Biol. Pharm. Bull.* **16**, 716–718.

171. Konishi, K. *et al.* (1999). *Biol. Pharm. Bull.* **22**, 240–243.

172. Igamberdiev, A. U. *et al.* (1995). *FEBS Lett.* **367**, 287–290.

173. Suzuki, S. *et al.* (1999). *Oncogene* **18**, 6380–6387.

174. van de Water, B. *et al.* (1995). *Mol. Pharmacol.* **48**, 928–937.

175. Berridge, M. V. *et al.* (1993). *Arch. Biochem. Biophys.* **303**, 474–482.

176. Barja, G. *et al.* (1998). *J. Bioenerg. Biomembr.* **30**, 235–243.

177. Suno, M. *et al.* (1989). *Arch. Gerontol. Geriatr.* **8**, 291–297.

178. Yu, L. *et al.* (1987). *J. Biol. Chem.* **262**, 1137–1143.

179. Armson, A. *et al.* (1995). *Int. J. Parasitol.* **25**, 257–260.

180. Turrens, J. F. *et al.* (1986). *Mol. Biochem. Parasitol.* **19**, 259–264.

181. Cook, N. D. *et al.* (1985). *Arch. Biochem. Biophys.* **240**, 9–14.

182. Trumpower, B. L. *et al.* (1980). *J. Bioenerg. Biomembr.* **12**, 151–164.

183. Zhang, L. *et al.* (1999). *FEBS Lett.* **460**, 349–352.

184. Ljungdahl, P. O. *et al.* (1989). *J. Biol. Chem.* **264**, 3723–3731.

185. Kohnke, D. *et al.* (1997). *Mol. Cell Biochem.* **174**, 101–113.

186. Lahnsteiner, F. *et al.* (1999). *J. Exp. Zool.* **284**, 454–465.

187. Mildaziene, V. *et al.* (1995). *Arch. Biochem. Biophys.* **324**, 130–134.

188. Elimadi, A. *et al.* (1997). *Br. J. Pharmacol.* **121**, 1295–1300.

189. Andreyev, A. *et al.* (1989). *Eur. J. Biochem.* **182**, 585–592.

190. Roucou, X. *et al.* (1997). *Biochim. Biophys. Acta* **1324**, 120–132.

191. Ziegler, M. *et al.* (1993). *J. Biol. Chem.* **268**, 25320–25328.

192. Brandolin, G. *et al.* (1993). *J. Bioenerg. Biomembr.* **25**, 459–472.

193. Fiore, C. *et al.* (2000). *Protein Expr. Purif.* **19**, 57–65.

194. Klingenberg, M. *et al.* (1975). *Eur. J. Biochem.* **52**, 351–363.

195. Nakashima, R. A. *et al.* (1984). *Cancer Res.* **44**, 5702–5706.

196. Buchet, K. *et al.* (1998). *J. Biol. Chem.* **273**, 22983–22989.

197. Vazquez-Laslop, N. *et al.* (1990). *J. Biol. Chem.* **265**, 19002–19006.

198. Valerio, M. *et al.* (1994). *Eur. J. Biochem.* **221**, 1071–1078.

199. Persson, B. *et al.* (1987). *Biochim. Biophys. Acta* **894**, 239–251.

200. Cataldi de Flombaum, M. A. *et al.* (1981). *Mol. Biochem. Parasitol.* **3**, 143–155.

201. Appleby, R. D. *et al.* (1999). *Eur. J. Biochem.* **262**, 108–116.

202. Budd, S. L. *et al.* (2000). *Proc. Natl. Acad. Sci. USA* **97**, 6161–6166.

203. Pastorino, J. G. *et al.* (2000). *Hepatology* **31**, 1141–1152.

204. Stridh, H. *et al.* (1999). *Chem. Res. Toxicol.* **12**, 874–882.

205. Furlong, I. J. *et al.* (1998). *Cell Death Differ.* **5**, 214–221.

206. Griffiths, E. J. *et al.* (1993). *Biochem. J.* **290**, 489–495.

207. Hermesh, O. *et al.* (2000). *Biochim. Biophys. Acta* **1457**, 166–174.

208. Majima, E. *et al.* (1995). *J. Biol. Chem.* **270**, 29548–29554.

209. Brustovetsky, N. *et al.* (2000). *J. Neurosci.* **20**, 103–113.

210. Huber, T. *et al.* (1999). *Biochemistry* **38**, 762–769.

211. Simonyan, R. A. *et al.* (1998). *FEBS Lett.* **436**, 81–84.

212. Valenti, D. *et al.* (1999). *FEBS Lett.* **444**, 291–295.

213. Shima, A. *et al.* (1990). *Cell Struct. Funct.* **15**, 53–58.

214. Lenton, L. M. *et al.* (1994). *Biochim. Biophys. Acta* **1186**, 237–242.

215. Guerin, B. *et al.* (1994). *J. Biol. Chem.* **269**, 25406–25410.

216. Gaballo, A. *et al.* (1998). *Biochemistry* **37**, 17519–17526.

217. Hekman, C. *et al.* (1991). *J. Biol. Chem.* **266**, 13564–13571.

218. Mercer, N. A. *et al.* (1999). *Exp. Parasitol.* **91**, 52–58.

219. Walker, J. E. *et al.* (1991). *Biochemistry* **30**, 5369–5378.

220. Hamasur, B. *et al.* (1992). *Eur. J. Biochem.* **205,** 409–416.

221. Miki, T. *et al.* (1994). *J. Biol. Chem.* **269,** 1827–1833.

222. Wikstrom, M. *et al.* (1985). *J. Inorg. Biochem.* **23,** 327–334.

223. Prochaska, L. J. *et al.* (1986). *Biochemistry* **25,** 781–787.

224. Beattie, D. S. (1993). *J. Bioenerg. Biomembr.* **25,** 233–244.

225. Shoshan-Barmatz, V. *et al.* (1996). *FEBS Lett.* **386,** 205–210.

226. Edwards, S. W. *et al.* (1982). *Biochem. J.* **202,** 453–458.

227. Simmons, W. A. *et al.* (1983). *Somatic Cell Genet.* **9,** 549–566.

228. Lardy, H. *et al.* (1975). *Fed. Proc.* **34,** 1707–1710.

229. Bossard, M. J. *et al.* (1981). *J. Biol. Chem.* **256,** 1518–1521.

230. Cross, R. L. *et al.* (1978). *J. Biol. Chem.* **253,** 4865–4873.

231. Vazquez-Memije, M. E. *et al.* (1984). *Arch. Biochem. Biophys.* **232,** 441–449.

232. Abrahams, J. P. *et al.* (1996). *Proc. Natl. Acad. Sci. USA* **93,** 9420–9424.

233. Muroi, M. *et al.* (1996). *Biochem. Biophys. Res. Commun.* **227,** 800–809.

234. Tirosh, O. *et al.* (2003). *Biochem. Pharmacol.* **66,** 1331–1334.

235. Panov, A. V. *et al.* (1995). *Arch. Biochem. Biophys.* **316,** 815–820.

236. Valerio, M. *et al.* (1993). *FEBS Lett.* **336,** 83–86.

237. Matsuno-Yagi, A. *et al.* (1993). *J. Biol. Chem.* **268,** 6168–6173.

238. Matsuno-Yagi, A. *et al.* (1993). *J. Biol. Chem.* **268,** 1539–1545.

239. Miedema, H. *et al.* (1994). *Biochem. Biophys. Res. Commun.* **203,** 1005–1012.

240. Collinson, I. R. *et al.* (1994). *Biochemistry* **33,** 7971–7978.

241. Cappelli-Bigazzi, M. *et al.* (1997). *J. Mol. Cell Cardiol.* **29,** 871–879.

242. Pizzo, P. *et al.* (1997). *J. Cell Biol.* **136,** 355–366.

243. Marsh, D. C. *et al.* (1993). *Hepatology* **17,** 91–98.

244. Jung, D. W. *et al.* (1996). *Arch. Biochem. Biophys.* **332,** 19–29.

245. Panov, A. *et al.* (1996). *Biochemistry* **35,** 12849–12856.

246. Meinicke, A. R. *et al.* (1998). *Arch. Biochem. Biophys.* **349,** 275–280.

247. Benaim, G. *et al.* (1990). *Mol. Biochem. Parasitol.* **39,** 61–68.

248. Breitbart, H. *et al.* (1990). *Biochim. Biophys. Acta* **1026,** 57–63.

249. Hayes, D. J. *et al.* (1990). *Mol. Biochem. Parasitol.* **38,** 159–168.

250. Linsinger, G. *et al.* (1999). *Mol. Cell Biol.* **19,** 3299–3311.

251. Wu, S. N. *et al.* (1999). *J. Pharmacol. Exp. Ther.* **290,** 998–1005.

252. Sureda, F. X. *et al.* (1997). *Cytometry* **28,** 74–80.

253. Wanke, M. *et al.* (2000). *Biochim. Biophys. Acta* **1463,** 188–194.

254. Curti, C. *et al.* (1999). *Mol. Cell Biochem.* **199,** 103–109.

255. Starkov, A. A. *et al.* (1997). *Biochim. Biophys. Acta* **1318,** 159–172.

256. Tran, T. V. *et al.* (1991). *Biochim. Biophys. Acta* **1059,** 265–274.

257. Cocco, T. *et al.* (1992). *Eur. J. Biochem.* **209,** 475–481.

258. Oyedotun, K. S. *et al.* (1999). *J. Biol. Chem.* **274,** 23956–23962.

259. Vianello, A. *et al.* (1995). *FEBS Lett.* **365,** 7–9.

260. Colleoni, M. *et al.* (1996). *Pharmacol. Toxicol.* **78,** 69–76.

261. Tan, A. K. *et al.* (1993). *J. Biol. Chem.* **268,** 19328–19333.

262. Yankovskaya, V. *et al.* (1996). *J. Biol. Chem.* **271,** 21020–21024.

263. Xiong, Z. H. *et al.* (1997). *J. Biol. Chem.* **272,** 31022–31028.

264. Mohell, N. *et al.* (1987). *Am. J. Physiol.* **253,** C301–C308.

265. Rodeheaver, D. P. *et al.* (1993). *J. Pharmacol. Exp. Ther.* **265,** 1355–1360.

266. Jezek, P. *et al.* (1998). *Am. J. Physiol.* **275,** C496–C504.

267. Catisti, R. *et al.* (1999). *FEBS Lett.* **464,** 97–101.

268. Rial, E. *et al.* (1999). *EMBO J.* **18,** 5827–5833.

269. Ruben, L. *et al.* (1991). *J. Biol. Chem.* **266,** 24351–24358.

270. Haass-Mannle, H. *et al.* (1997). *J. Photochem. Photobiol. B* **41,** 90–102.

271. Kotlyar, A. B. *et al.* (1998). *Biochim. Biophys. Acta* **1365,** 53–59.

272. Rottenberg, H. *et al.* (1989). *Biochemistry* **28,** 4355–4360.

273. Erecinska, M. *et al.* (1996). *Brain Res.* **726,** 153–159.

274. Kawanishi, T. *et al.* (1991). *J. Biol. Chem.* **266,** 20062–20069.

275. Grijalba, M. T. *et al.* (1998). *Free Radic. Res.* **28,** 301–318.

276. Shimizu, S. *et al.* (1998). *Proc. Natl. Acad. Sci. USA* **95,** 1455–1459.

277. Matsuno-Yagi, A. *et al.* (1989). *Biochemistry* **28,** 4367–4374.

278. Holmuhamedov, E. L. *et al.* (1999). *J. Physiol. (Lond.)* **519**(Pt. 2), 347–360.

279. Tolkatchev, D. *et al.* (1996). *J. Biol. Chem.* **271,** 12356–12363.

APPENDIX 3

Linearized Maps of Circular Mitochondrial Genomes from Representative Organisms

Compiled by Eric A. Schon

Department of Neurology and Department of Genetics and Development
Columbia University, New York, New York 10032

METHODS IN CELL BIOLOGY, VOL. 80
Copyright 2007, Elsevier Inc. All rights reserved.

0091-679X/07 $35.00
DOI: 10.1016/S0091-679X(06)80042-7

828

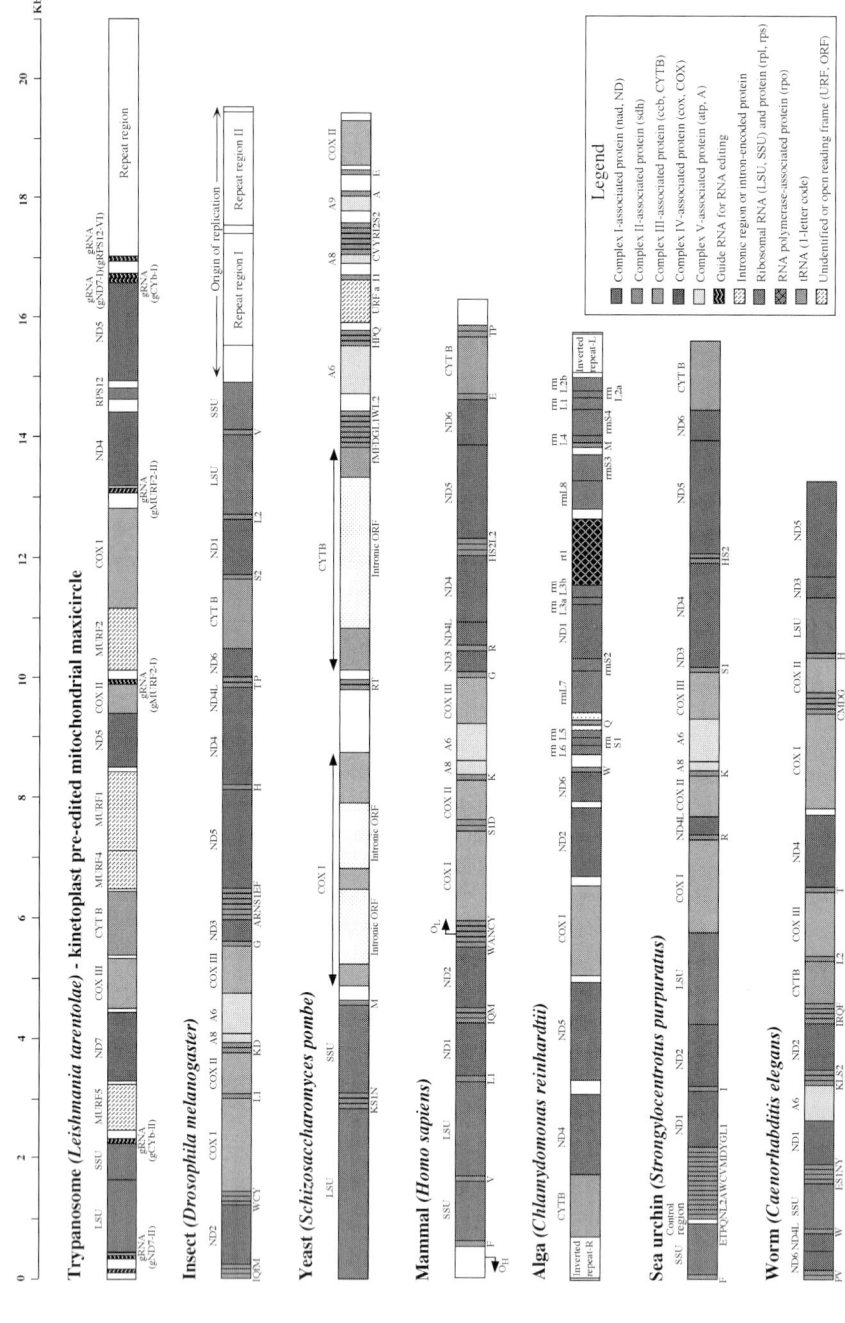

Trypanosome (*Leishmania tarentolae*) - kinetoplast pre-edited mitochondrial maxicircle

Insect (*Drosophila melanogaster*)

Yeast (*Schizosaccharomyces pombe*)

Mammal (*Homo sapiens*)

Alga (*Chlamydomonas reinhardtii*)

Sea urchin (*Strongylocentrotus purpuratus*)

Worm (*Caenorhabditis elegans*)

Legend

Complex I-associated protein (nad, ND)
Complex II-associated protein (sdh)
Complex III-associated protein (ccb, CYTB)
Complex IV-associated protein (cox, COX)
Complex V-associated protein (atp, A)
Guide RNA for RNA editing
Intronic region or intron-encoded protein
Ribosomal RNA (LSU, SSU) and protein (rpl, rps)
RNA polymerase-associated protein (rpo)
tRNA (1-letter code)
Unidentified or open reading frame (URF, ORF)

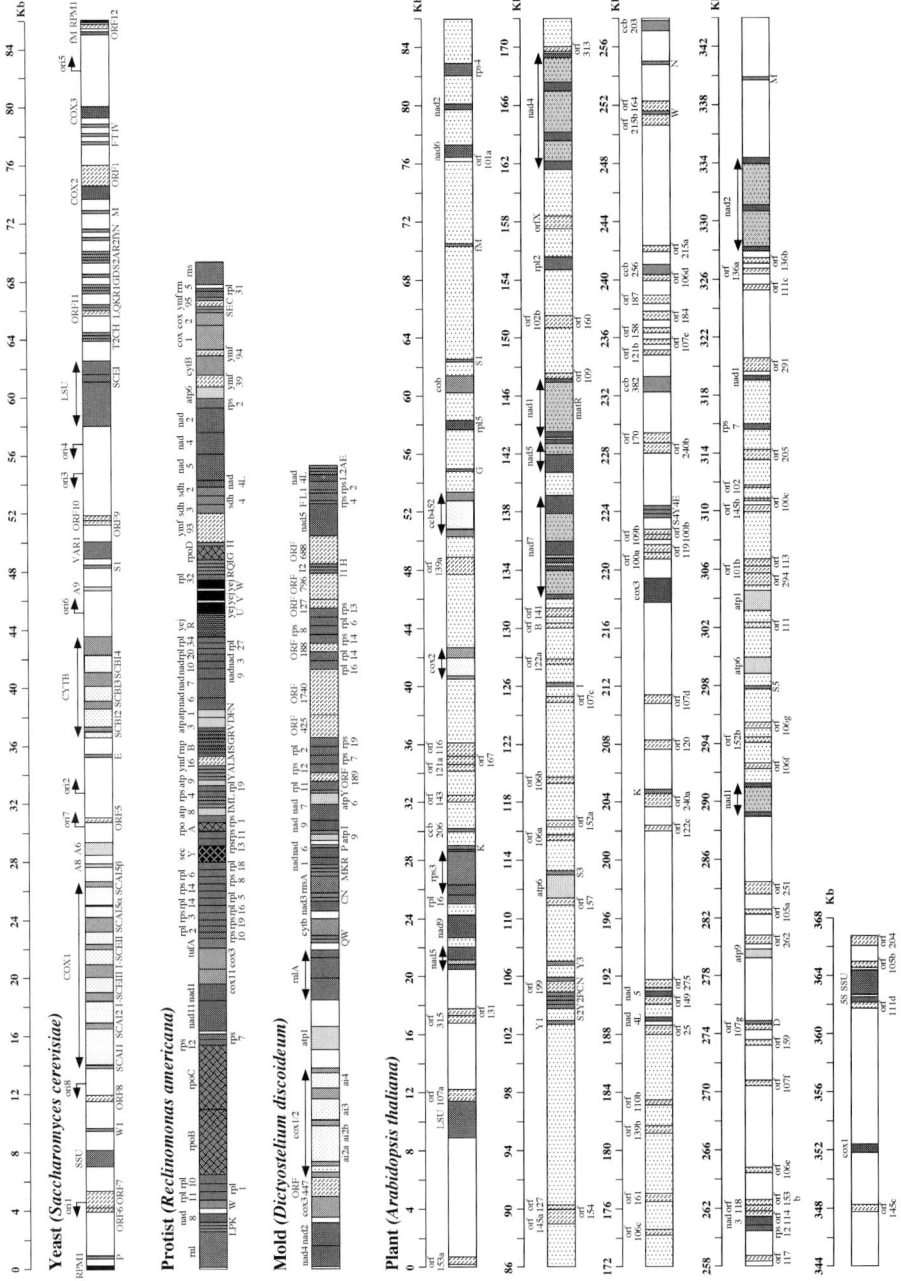

829

APPENDIX 4

Mitochondrial Genetic Codes in Various Organisms

Compiled by Eric A. Schon

Department of Neurology and Department of Genetics and Development
Columbia University, New York, New York 10032

METHODS IN CELL BIOLOGY, VOL. 80

0091-679X/07 $35.00
DOI: 10.1016/S0091-679X(06)80043-9

The standard genetic code

UUU	F	Phe	UCU	S	Ser	UAU	Y	Tyr	UGU	C	Cys
UUC	F	Phe	UCC	S	Ser	UAC	Y	Tyr	UGC	C	Cys
UUA	L	Leu	UCA	S	Ser	UAA	*	Ter	UGA	*	Ter
UUG	L	Leu	UCG	S	Ser	UAG	*	Ter	UGG	W	Trp
CUU	L	Leu	CCU	P	Pro	CAU	H	His	CGU	R	Arg
CUC	L	Leu	CCC	P	Pro	CAC	H	His	CGC	R	Arg
CUA	L	Leu	CCA	P	Pro	CAA	Q	Gln	CGA	R	Arg
CUG	L	Leu	CCG	P	Pro	CAG	Q	Gln	CGG	R	Arg
AUU	I	Ile	ACU	T	Thr	AAU	N	Asn	AGU	S	Ser
AUC	I	Ile	ACC	T	Thr	AAC	N	Asn	AGC	S	Ser
AUA	I	Ile	ACA	T	Thr	AAA	K	Lys	AGA	R	Arg
AUG	M	Met	ACG	T	Thr	AAG	K	Lys	AGG	R	Arg
GUU	V	Val	GCU	A	Ala	GAU	D	Asp	GGU	G	Gly
GUC	V	Val	GCC	A	Ala	GAC	D	Asp	GGC	G	Gly
GUA	V	Val	GCA	A	Ala	GAA	E	Glu	GGA	G	Gly
GUG	V	Val	GCG	A	Ala	GAG	E	Glu	GGG	G	Gly

Mitochondrial genetic codes: Deviations from the standard code

	Standard	(a)	(b)	(c)	(d)	(e)	(f)	(g)	(h)	(i)	(j)	(k)	(l)	(m)
UAG	Stop								Ala	Leu	Leu			
UGA	Stop	Trp	Trp	Trp	Trp	Trp	Trp		Trp			Cys		Trp
UCA	Ser										Stop			
UUA	Leu												Stop	
CUU	Leu		Thr											
CUC	Leu		Thr											
CUA	Leu		Thr											
CUG	Leu		Thr											
AUA	Ile	Met	Met		Met		Met							Met
AAA	Lys					Asn		Asn	Asn					Asn
AGA	Arg	Stop			Ser	Ser	Gly		Ser					Ser
AGG	Arg	Stop			Ser	Ser	Gly		Ser					Ser

Note: The standard code is present in the mitochondria of land plants and green algae.

(a) Includes all known vertebrates.

(b) Includes Fungi *(Candida, Hansenula, Kluveromyces, Saccharomyces, Schizosaccharomyces).*

(c) Includes Fungi *(Ascobolus, Aspergillus, Candida, Neurospora, Podospora);* Coelenterata; Mycoplasma; Protozoa *(Leishmania, Paramecia, Tetrahymena, Trypanosoma);* Spiroplasma; also a red alga *(Chondrus).*

(d) Includes Inverebrata, such as Crustaceae *(Artemia);* Insecta *(Apis, Drosophila, Locusta);* Mollusca; Nematoda *(Ascaris, Caenorhabditis);* Acoelomorpha and Catenulida (related to flatworms).

(e) Includes Echinodermata, such as Asterozoa (starfish) and Echinozoa (sea urchins).

(f) Includes Ascidiae (sea squirts).

(g) Includes Rhabditophora (flatworms); see Telford *et al.* (2000) *Proc. Natl. Acad. Sci. USA* **97**, 11359–11364.

(h) Includes Chlorophycean algae *(Hydrodictyon, Pediastrum, Tetraedron);* see Hayashi-Ishimaru *et al.* (1996). *Curr. Genet.* **30**, 29–33.

(i) Includes Chlorophycean algae *(Coelastrum, Scenedesmus).*

(j) Includes a green alga *(Scenedesmus obliquus).*

(k) Includes Ciliata *(Euplotes).*

(l) Includes a protist *(Thraustochytrium aureum).*

(m) Includes Trematoda.

		The standard genetic code												
	Standard	(a)	(b)	(c)	(d)	(e)	(f)	(g)	(h)	(i)	(j)	(k)	(l)	(m)
UAG	Stop								Ala	Leu	Leu			
UGA	Stop	Trp	Trp	Trp	Trp	Trp	Trp	Trp	Trp			Cys		Trp
UAA	Stop							Tyr						
UCA	Ser										Stop			
UUA	Leu												Stop	
CUU	Leu		Thr											
CUC	Leu		Thr											
CUA	Leu		Thr											
CUG	Leu		Thr											
AUA	Ile	Met	Met		Met		Met							Met
AAA	Lys					Asn		Asn	Asn					Asn
AGA	Arg	Stop			Ser	Ser	Gly	Ser	Ser					Ser
AGG	Arg	Stop			Ser	Ser	Gly	Ser	Ser					Ser

Note: The standard code is present in the mitochondria of land plants and green algae.

Data derived from a compilation of genetic codes by A. Elzanowski and J. Ostell, National Center for Biotechnology Information (NCBI), (http://www.ncbi.nlm.nih.gov/Taxonomy/Utils/wprintgc.cgi?mode=c#SG4).

(a) Includes all known vertebrates.

(b) Includes Fungi *(Candida, Hansenula, Kluveromyces, Saccharomyces, Schizosaccharomyces).*

(c) Includes Fungi *(Ascobolus, Aspergillus, Candida, Neurospora, Podospora);* Coelenterata; Mycoplasma; Protozoa *(Leishmania, Paramecia,Tetrahymena, Trypanosoma);* Spiroplasma; also a red alga *(Chondrus).*

(d) Includes Inverebrata, such as Crustaceae *(Artemia);* Insecta *(Apis, Drosophila, Locusta);* Mollusca; Nematoda *(Ascaris, Caenorhabditis);* Acoelomorpha and Catenulida (related to flatworms).

(e) Includes Echinodermata, such as Asterozoa (starfish) and Echinozoa (sea urchins).

(f) Includes Ascidiae (sea squirts).

(g) Includes Rhabditophora (flatworms).

(h) Includes Chlorophycean algae *(Hydrodictyon, Pediastrum, Tetraedron).*

(i) Includes Chlorophycean algae *(Coelastrum, Scenedesmus).*

(j) Includes a green alga *(Scenedesmus obliquus).*

(k) Includes Ciliata *(Euplotes).*

(l) Includes a protist *(Thraustochytrium aureum).*

(m) Includes Trematoda.

Gene Products Present in Mitochondria of Yeast and Animal Cells

Compiled by Eric A. Schon

Department of Neurology and Department of Genetics and Development
Columbia University, New York, New York 10032

METHODS IN CELL BIOLOGY, VOL. 80
Copyright 2007, Elsevier Inc. All rights reserved.

0091-679X/07 $35.00
DOI: 10.1016/S0091-679X(06)80044-0

AMINO ACID AND NITROGEN METABOLISM

Gene product	Function; comment	Location	E.C. #	Fungal		Animal		
				Symbol	GenBank	Symbol	GenBank	Human Chr.
Acetohydroxy acid reductoisomerase	Ile/Leu/Val syn; nucleoid "parsing"	MAT?	1.1.1.86	ILV5	X04969.1			
Acetolactate synthase, catalytic subunit	Ile/Leu/Val syn	MAT?	4.1.3.18	ILV2	X02549.1	ILVBL	BF310206	19p13.12
Acetolactate synthase, regulatory subunit	Ile/Leu/Val syn	MAT?	4.1.3.18	ILV6	X59720.1			
Acetylglutamate kinase/acetylglutamyl phosphate reductase	Arg/ornithine syn; transcription factor	MAT?	2.7.2.8	ARG5,6	X57017.1			
Acetylornithine acetyltransferase bifunctional protein	Arg/ornithine syn	MAT?	2.3.1.35	ARG7	U90438.1			
Acetylornithine aminotransferase	Arg/ornithine syn	MAT?	2.6.1.11	ARG8	X84036.1			
Acyl-CoA dehydrogenase-8 (isobutyryl-CoA dehydrogenase)	Val catab; br chain beta oxidation	MAT?	1.3.99.–	NCU06543.1 (Nc)	XM_326397.1 (Nc)	ACAD8	AF126245.1	11q25
Agmatinase (arginase family) (S. pombe SPAC11D3.09 may be cytosolic)	Arg deg; agmatine syn	MAT	3.5.3.11			AGMAT	AK027037.1	1p36.13
Alanine aminotransferase	Ala metab	MAT?	2.6.1.2	ALT1	Z73261.1	AAT1	D10355.1	8q24.3
Alanine aminotransferase 2, potentially mitochondrial	Nitrogen metab	NC	2.6.1.2			GPT2	AY029173.2	16q11.2
Alanine:glyoxylate aminotransferase [1]	Gly syn	MAT?	2.6.1.44/51	AGX1	D50617.1	AGXT2	AJ292204.1	5p13.2
Aldehyde dehydrogenase (NAD+); glycerol trinitrate reductase	Nitroglycerine metab; in yeast nucleoid	MAT	1.2.1.3	ALD4	Z75282.1	ALDH2	Y00109.1	12q24.12
Alpha-aminoadipic semialdehyde synthase [3]	Lys deg, 1st 2 steps	ND	1.5.1.8/9			AASS	AJ007714.2	7q31.32
Arginase II	Arg deg/NOS regulation	MAT?	3.5.3.1			ARG2	D86724.1	14q24.1
Arginine decarboxylase (ODC-P), may be mitochondrial (but no obvious MTS)	Arg deg; agmatine syn	NC	4.1.1.19			ADC	AY050634.1	1p35.1
Aspartate aminotransferase	Asp/Glu transamination	MAT	2.6.1.1	AAT1	X68052.1	GOT2	M22632.1	16q21
Betaine aldehyde dehydrogenase (aldehyde dehydrogenase E3)	Gly syn	MAT	1.2.1.3	YGL059w	Z72581.1	ALDH9A1	U34252.1	1q24.1
Branched chain alpha-ketoacid dehydrogenase (BCKDH) kinase	Leu/Ile/Val deg	MAT?	2.7.11.4			BCKDK	AF026548.1	16p11.2
Branched chain amino acid aminotransferase	Leu/Ile/Val deg	MAT?	2.6.1.42	BAT1	X78961.1	BCAT2	U68418.1	19q13.33
Carbamoyl phosphate synthetase 1 (Ura2) [1]	Urea cycle	MAT?	6.3.4.16	URA2	M27174.1	CPS1	Y15793.1	2q35
Choline dehydrogenase	Gly syn	MIM	1.1.99.1			CHDH	BC034502.1	3p21.2
Cysteine synthase [1] [7]	Cysteine syn	MOM	2.5.1.47	YGR012w	Z72797.1			
2-Dehydro-3-deoxyphosphoheptonate aldolase (DAHPS)	Aromatic aa syn, 1st step	NC	4.1.2.15	ARO3	X13514.1			
Delta-1-pyrroline-5-carboxylate dehydrogenase	Pro/Glu metab	MAT	1.5.1.12	PUT2	M10029.1	ALDH4	U24266.1	1p36
Diamine acetyltransferase 1 (spermidine/spermine N1-acetyltransferase)	Polyamine catabolism	ND	2.3.1.57	AN2479.2 (An)	XM_406616.1 (An)	SAT	M77693.1	Xp22.11
Dihydroxy acid dehydratase (DAD)	Ile/Leu/Val syn	MAT?	4.2.1.9	ILV3	Z49516.1			
Dimethylglycine dehydrogenase (Me2GlyDH)	Gly metab; binds folate	MAT	1.5.99.2	AN8654.2 (An)	XM_412791.1 (An)	DMGDH	AF111858.1	5q13.3
FAD-dependent oxidoreductase domain containing 1 [8]	Amino acid transp/metab?	ND				FOXRED1	BC013902.2	11q24.2
Gamma-aminobutyrate aminotransferase (GABA-T)	Glu metab; GABA shunt; ROS in plants	MAT	2.6.1.19	gatA (An)	X15647.1 (An)	ABAT	L32961.1	16p13.2
Gamma-glutamyl phosphate reductase	Pro syn (first 2 steps in human)	MIM	1.2.1.41	PRO2	U43565.1	PYCS	X94453.1	10q24.3
Glutamate dehydrogenase 1 (cytosolic in yeast)	Glu metab	MAT	1.4.1.3			GLUD1	M20867.1	10q23.3
Glutamate dehydrogenase 2 (NAD[P]+) (cytosolic in yeast)	Glu metab	MAT	1.4.1.3			GLUD2	U08997.1	Xq24
Glutaminase, kidney isoform (also alt spliced isoforms) (cytosolic in yeast)	Glu metab	MIM	3.5.1.2			GLS	AB020645.1	2q32.3
Glutaminase, liver isoform (also alt spliced isoform); nuclear in brain (cyto in yeast)	Glu metab	MIM	3.5.1.2			GA	AF223944.1	12q13.3
Glutamine transaminase K (cytoplasmic cysteine conjugate-beta lyase)	Glu metab?	MOM?	2.6.1.64	BNA3	Z49335.1	CCBL1	X82224.1	9q34.11
Glutaryl-CoA dehydrogenase (GCD)	Lys/Trp metab; br chain beta-oxidation	MAT	1.3.99.7	NCU02291.1 (Nc)	XM_331066.1 (Nc)	GCDH	U69141.1	19p13.13
Glycine C-acetyltransferase	Gly metab	MAT?	2.3.1.29	HEM1	M26329.1	GCAT	AF077740.1	22q13.1
Glycine amidinotransferase (GATM)	Creatine syn	MIM	2.1.4.1			AGAT	S68805.1	15q21.1
Glycine decarboxylase H protein (aminomethyltransferase)	Gly deg	MAT	1.4.4.2	GCV3	U12980.1	GCSH	M69175.1	16q23.1
Glycine decarboxylase L protein (dihydrolipoyl dehydrogenase)	Gly deg	MAT	1.8.1.4	LPD1	M20880.1	DLD	J03490.1	7q31.1
Glycine decarboxylase P protein (dehydrogenase)	Gly deg	MAT	1.4.4.2	GCV2	U20641.1	GLDC	D90239.1	9p24.1
Glycine decarboxylase T protein (aminomethyltransferase)	Gly degr; folate/C1 metab	MAT	2.1.2.10	GCV1	L41522.1	GCST	D13811.1	3p21.31
Glycine-N-acyltransferase (aralkyl acyl-CoA:amino acid N-acyltransferase) [3] [12]	Gly metab	MIM	2.3.1.13	NCU05419.1 (Nc)	XM_325273.1 (Nc)	GLYAT	AF023466.1	11q12.1
Homoaconitase	Lys syn	MAT	4.2.1.36	LYS4	X93502.1			
Homoisocitrate dehydrogenase	Lys syn	MAT	1.1.1.155	LYSI2	Z46728.1			
3-Hydroxyisobutyryl-CoA hydrolase	Val deg	MAT?	3.1.2.4	EHD3	Z74332.1	HIBCH	U66669.1	2q32.2
3-Hydroxymethyl-3-methylglutaryl-CoA lyase	Leu deg	MAT	4.1.3.4	NCU05419.1 (Nc)	XM_325273.1 (Nc)	HMGCL	L07033.1	1p36.12

(continues)

Protein	Function	Loc	EC	Fungal gene	Accession	Accession	Human gene	Accession	Chromosome
4-Hydroxyphenylpyruvate dioxygenase, potential [6]	Aromatic aa metab?	ND	1.13.11.27	NCU01830.1 (Nc)	M12893.1	BX284754.1 (Nc)	GLOXD1	BC007293.1	1p34.1
Isomerase, putative, potentially mito [1] [3] [4] (no introns in human)	Amino acid deg?	NC	5.3.3.–	YNL168c	Z75016.1	Z71444.1	FAHD1	BC063017.1	16p3.3
2-Isopropylmalate synthase I	Leu syn	MAT	4.1.3.12	LEU4	—		—	—	—
2-Isopropylmalate synthase II, probably mitochondrial	Leu syn	NC	4.1.3.12	LEU9	—		—	—	—
Isovaleryl-CoA dehydrogenase	Leu deg; br chain beta oxidation	MAT	1.3.99.10	NCU02126.1 (Nc)	XM_959191.1 (Nc)		IVD	M34192.1	15q15.1
Kynurenine/alpha-aminoadipate aminotransferase	Lys syn	MAT?	2.6.1.7.39	—	—		AADAT	AF097994.1	4q33
Kynurenine 3-monooxygenase (kynurenine 3-hydroxylase) [7]	Trp deg; NAD syn	MOM	1.14.13.9	BNA4	Z35859.1		KMO	Y13153.1	1q43
Maleylacetoacetate isomerase	Phe/Tyr catab; glutathione cofactor	NC	5.2.1.2	MAIA (En)	AJ001837.1 (En)		GSTZ1	U86529.1	14q24.3
3-Methylglutaconyl-CoA hydratase (AU-specific RNA-binding protein)	Leu metab; may direct mt-RNA deg	ND	4.2.1.18	EHD3	Z74332.1		AUH	X79888.1	9q22.2
Methylcrotonyl-CoA carboxylase, alpha subunit (MCCA)	Leu metab	MIM	6.4.1.4	NCU00591.1 (Nc)	XM_960638.1 (Nc)		MCCC1	AF310339.1	3q27.1
Methylcrotonyl-CoA carboxylase, beta subunit (MCCB)	Leu metab	MIM	6.4.1.4	NCU02127.1 (Nc)	XM_959192.1 (Nc)		MCCC2	AF310971.1	5q13.2
Methylenetetrahydrofolate bifunctional enzyme (NAD)	Met metab	MAT	1.5.1.15	—	—		MTHFD2	X16396.1	2p13.1
Methylenetetrahydrofolate reductase	Met metab; folate/C1 metab	MAT?	1.5.1.20	MET12	U36624.1		—	—	—
Methylenetetrahydrofolate reductase	Met metab; folate/C1 metab	MAT?	1.5.1.20	MET13	Z72647.1		—	—	—
Methylenetetrahydrofolate trifunctional enzyme (NADP)	Met metab	MAT?	1.5.1.5	MIS1	J03724.1		FTHFSDC1	AL117452.1	6q25.1
Methylmalonate-semialdehyde dehydrogenase	Val/pyrimidine metab	MAT?	1.2.1.27	ALD5	U56605.1		ALDH6A1	AF148505.1	14q24.3
Mitochondrial genome maintenance protein (human UK114)	Ile syn; mtDNA maintenance	ND	—	MMF1	Z38060.1		HRSP12	X95384.1	8q22.2
Monoamine oxidase A (peroxisomal in yeast)	Amine neurotransmitter deg	MOM	1.4.3.4	—	—		MAOA	M68840.1	Xp11.23
Monoamine oxidase B (peroxisomal in yeast)	Amine neurotransmitter deg	MOM	1.4.3.4	—	—		MAOB	M69177.1	Xp11.23
N-acetylglutamate synthase	Arg/ornithine syn	MAT?	2.3.1.1	ARG2	Z49346.1		NAGS	AY116538.1	17q21.31
NADH kinase	Arg syn, 3rd step; antioxidant protection	MAT	2.7.1.86	POS5	Z73544.1		—	—	—
Ornithine aminotransferase (cytosolic in yeast)	Arg/ornithine syn	MAT?	2.6.1.13	—	—		OAT	M12267.1	10q26.13
Ornithine carbamoyltransferase	Arg/ornithine syn	MAT	2.1.3.3	ARG3	M11946.1		OTC	K02100.1	Xp21.1
Proline dehydrogenase (proline oxidase)	Pro/Glu metab	MAT	1.5.–.–	PUT1	M18107.1		PRODH	U79754.1	22q11.21
Propionyl-CoA carboxylase, alpha chain	Met deg	MAT	6.4.1.3	CNC01910 (Cn)	XM_569564.1 (Cn)		PCCA	X14608.1	13q32
Propionyl-CoA carboxylase, beta chain	Met deg	MAT	6.4.1.3	CNC01820 (Cn)	XM_569537.1 (Cn)		PCCB	X73424.1	3q22.3
Pyrroline-5-carboxylate reductase, potentially mitochondrial	Pro syn, 3rd (last) step	ND	1.5.1.2	PRO3	M57886.1		PYCRL	BC007993.1	8q24.3
Sarcosine dehydrogenase	Gly metab	MAT	1.5.99.1	—	—		SARDH	AJ223317.1	9q34
Serine hydroxymethyltransferase	Ser/gly metab; folate/C1 metab	MAT	2.1.2.1	SHM1	Z36131.1		SHMT2	U23143.1	12q13
Serine/threonine deaminase	Ser/Thr metab	NC	4.2.1.13	CHA1	M85194.1		SDS	J05037.1	12q24.13
Threonine deaminase	Ile syn	MAT?	4.2.1.16	ILV1	M36383.1		—	—	—
Threonine dehydrogenase	Thr deg	MAT	1.1.1.103	—	—		TDH	AY101186.1	8p23.1
APOPTOSIS									
"Intrinsic" apoptotic agents									
Activator of apoptosis harakiri (neuronal death protein DP5/BID3)	Pro-apoptotic protein; binds p32 (C1QBP)	MAT?	—	—	—		HRK	U76376.1	12q24.22
AIF-homologous mitochondrion-associated inducer of death (AMID); p53 homolog	Pro-apoptotic protein	MOM?	1.6.5.3	NDI1	X61590.1		AMID	AK027403.1	10q22.1
AIF-homologous mitochondrion-associated inducer of death (AMID); p53 homolog	Pro-apoptotic protein	MOM?	—	NDE2	Z74133.1		AMID	AK027403.1	10q22.1
AIF-like protein (AIFL)	Pro-apoptotic protein	MIM	—	—	—		AIFL	BC032485.1	22q11.21
APOP-1 (cyclophilin D-dependent but Bax/Bak-independent apoptosis	Pro-apoptotic protein	ND	—	—	—		C14orf153	AK223092.1	14q32.33
Apoptosis-associated speck-like protein (CARD5), 3 alt-spl forms	Adaptor protein for BAX translocation	MOM?	—	—	—		ASC	AF384465.1	16p11.2
Apoptosis inducing factor (AIF); alt spl forms	Pro-apoptotic protein; mito morphology	IMS	—	NCU05850.1 (Nc)	XM_325704.1 (Nc)		PDCD8	AF100928.1	Xq25
Apoptosis signal-regulating kinase 1 (ASK1)	Inhibited by TRX1/TRX2 in cyto/mito	MOM?	2.7.1.–	—	—		MAP3K5	U67156.1	6q23.3
Autophagy protein 5, 24-kDa N-term cleavage product	Pro-apoptotic protein; translocates to mitos	MOM?	—	—	—		ATG5	BC002699.2	6q21
B-cell lymphoma 2 protein (Bcl-2) (2 splice variants)	Anti-apoptotic protein	MIM	—	—	—		BCL2	M13994.1	18q21.3
Bcl2/adeno E1B 19-kD-interacting protein 3 (Bnip3)	Pro-apoptotic protein; transloc fr nucl->mitos	MOM?	—	—	—		BNIP3	U15174.1	10q26.3
Bcl2/adeno E1B 19-kD-interacting protein 3-like (Bnip3L)	Pro-apoptotic protein; translocates to mitos	MOM?	—	—	—		BNIP3L	AF079221.1	8p21.2
Bcl2-antagonist/killer (Bak), 2 isoforms	Binds VDAC2 to prevent apoptosis	MOM?	—	—	—		BAK1	U16811.1	6p21.31
Bcl2-antagonist of cell death (Bad)	Binds Bcl-x/Bcl-2	MOM?	—	—	—		BAD	AF031523.1	11q13.2

Appendix 5 (*continued*)

Gene product	Function; comment	Location	E.C. #	Fungal Symbol	Fungal GenBank	Animal Symbol	Animal GenBank	Human Chr.
Bcl2-associated athanogene 1 (Bag-1), short form	Bcl-2 binding protein	ND	—	—	—	BAG1	Z35491.1	9p12
Bcl2-associated X protein, alpha isoform (Bax alpha)	Pro-apoptotic protein	MOM	—	—	—	BAX	L22473.1	19q13.32
Bcl2 binding component 3 (p53 upregulated modulator of apoptosis [PUMA])	Pro-apoptotic protein	MOM?	—	—	—	BBC3	AF354654.1	19p13
Bcl2 inhibitor of transcription (Bit1; peptidyl-tRNA hydrolase 2 (Pth2))	Pro-apoptotic; translation	IMS?	3.1.1.29	PTH2	Z35818.1	PTH2	BC006807.1	17q23.2
Bcl2-interacting killer, apoptosis-inducing (Bik)	Pro-apoptotic protein	MOM?	—	—	—	BIK	U34584.1	22q13.31
Bcl2 interacting mediator of cell death (Bim), BimEL isoform (anoikis)	Translocates to mitos; induces anoikis	MOM?	—	—	—	BCL2L11	AF032458.1	2q13
Bcl2-like 1 (Bcl-x) (at least 3 isoforms, including Bcl-xL, Bcl-xS)	Anti-apoptotic protein	MOM	—	—	—	BCL2L1	Z23115.1	20q11.2
Bcl2-like protein 10 (Boo/Diva)	Translocates to mitos	NC	—	—	—	BCL2L10	AF285092.1	15q21.2
Bcl2-related ovarian killer, long/short splice forms	Pro-apoptotic protein	NC	—	—	—	BOK	AF174487.1	2q37.3
Bcl2-related protein A1 (Bfl-1) (hematopoetic-specific), mito alt spl isoform	Anti-apoptotic protein	MOM?	—	—	—	BCL2A1	U27467.1	15q24.3
Bcl-Rambo (also encodes MIL1 internally)	Pro-apoptotic protein	ND	—	—	—	BCL2L13	AF325209.1	22q11.21
Bcl-w apoptosis regulator	Anti-apoptotic protein	ND	—	—	—	BCLW	U59747.1	14q11.2
BH3 interacting domain death agonist, both full-length and truncated (tBID) forms	Pro-apoptotic protein; binds cardiolipin	MOM	—	—	—	BID	AF042083.1	22q11.2
Bifunctional apoptosis regulator (BAR)	Anti-apoptotic protein	ND	—	—	—	BAR	AF173003.1	16p13.12
Brain protein 44, sim to USMG5 and YGR243w (Fmp43) [1]	Function unknown	ND	—	YGR243w	Z73028.1	BRP44	AL110297.1	1q24.2
Brain protein 44, sim to USMG5 and YHR162w, potentially mito [5]	Function unknown	ND	—	YHR162w	U00027.2	BRP44	AL110297.1	1q24.2
Brain protein 44-like (apoptosis-regulating basic protein) [1] [4]	Function unknown	MIM?	—	YGL080w	Z72602.1	BRP44L	AF151887.1	6q27
BRCC2	Pro-apoptotic protein	ND	—	—	—	BRCC2	AF303179.1	11q24.1
c-Abl tyrosine kinase (4 alt spl isoforms)	Translocates to mitos from ER	ND	2.7.1.112	—	—	ABL1	X16416.1	9q34.2
Caspase 3, activated (forms from 277-aa precursor (p12/p17 isoforms)	Pro-apoptotic protein	IMS	3.4.22.–	—	—	CASP3	U13737.1	4q35.1
Caspase 8, activated form	Pro-apoptotic protein	MOM	3.4.22.–	—	—	CASP8	AF102146.1	2q33
Caspase 9, activated forms from 416-aa precursor (p10/p35 isoforms)	Pro-apoptotic protein	IMS	3.4.22.–	—	—	CASP9	AB019205.1	1p36.21
Cell death activator EGL-1	Pro-apoptotic protein	ND	—	—	—	egl-1 (w)	AF057309.1 (w)	—
Cell death effector (CED-4) (human homolog APAF-1 is cytosolic)	Pro-apoptotic protein	ND	—	—	—	ced-4 (w)	X69016.1 (w)	12q23.1
Cell death inducing DFF45-like effector A (CIDE-A) (expressed in brown fat)	Pro-apoptotic protein; Binds UCP1	ND	—	—	—	CIDEA	AY364639.1	18p11.21
Cell death inducing DFF45-like effector B (CIDE-B)	Pro-apoptotic protein	ND	—	—	—	CIDEB	AF218586.1	14p11
Chloride intracellular channel p64H1	Apoptotic target; translocates to nucleus	ND	—	—	—	CLIC4	AF097330.1	9p21.3
c-Jun NH2-terminal kinase 1	Inactivates Bcl-xL	NC	2.7.1.–	—	—	JNK1	L26318.1	10q11.22
c-Jun NH2-terminal kinase 2	Inactivates Bcl-xL	NC	2.7.1.–	—	—	JNK2	L31951.1	5q35
Coflin (actin depolymerizing factor) (Yeast Cof1 is cytosolic)	Translocates to nucleus	ND	—	—	—	CFL1	D00682.1	11q13.1
Cyclin dependent kinase 11 (CDK11), N-terminal p60 fragment	Translocates fr nucl->mitos; pro-apoptotic	ND	2.7.1.37	—	—	CDC2L1	AF067512.1	1p36.33
DAP kinase-interacting protein 1 ("Mind bomb")	Binds DAPK1; ubiquitin ligase	ND	—	—	—	MIB	AY149908.1	18q11.2
Death-associated protein kinase 1	Pro-apoptotic protein	ND	2.7.1.–	—	—	DAPK1	X76104.1	9q21.33
Death-inducing protein DIP	Pro-apoptotic protein; target of E2F1	ND	—	—	—	DIP	BC080189.1	22q13.31
Direct IAP binding protein with low PI (Smac/DIABLO)	Translocates to nucleus	IMS	—	—	—	DIABLO	BC011909.1	12q24.31
DNA endonuclease G	ER-mito communication; apoptosis via Bid	ER-MAM	3.1.30.–	—	—	ENDOG	X79444.1	9q34.13
Early secretory pathway sorting connector	Phosphorylates Bcl-2	MOM	2.7.1.–	—	—	PACS2	AY320284.1	14q32.33
Extracellular signal-related kinase 1 (Erk1)	Phosphorylates Bcl-2	MOM	2.7.1.–	—	—	ERK1	X60188.1	16p12.2
Extracellular signal-related kinase 2 (Erk2)	Binds Bcl-xL (BCL2L1)	MOM	2.7.1.–	—	—	ERK2	M84489.1	22q11.2
Fas-activated serine/threonine kinase (FAST)	Anchors Bcl-2/Bcl-xL to mitos	MOM?	—	—	—	FASTK	X86779.1	7q36.1
FK506-binding protein 8, 38-kDa [10]	Pro-apoptotic protein	MOM?	—	FKB1	M57967.1	FKBP8	AY225339.1	19p13.11
Galectin-3 internal gene, ORF2 (Mitogaligin); alt splicing of LGALS3		ND	—	—	—	GALIG	AF266280.1	14q22.3
Ganglioside GD3 synthase, potentially mitochondrial	Ceramide-induced apoptosis	NC	2.4.99.8	—	—	SIAT8A	X77922.1	12p12.3

(continues)

Protein	Function	Localization	EC no.	Homolog	Homolog acc.	Gene symbol	Accession	Chromosome
Ganglioside D3 (GD3)	Translocates to mitos	NC						—
Gelsolin (actin regulatory protein)	Anti-apoptotic protein	NC				GSN	X04412.1	9q34.11
GRIM-19 (B16.6 subunit of Complex I)	Pro-apoptotic protein	MIM?	1.6.5.3			NDUFA13	AF286697.1	19p13.11
Hid protein	Interacts with Bcl-xL	ND		NCU09299.1 (Nc)	XM_331990.1 (Nc)	HID (f)	U31226.1 (f)	
Histidine triad nucleotide binding protein 2	Apoptotic sensitizer; hydrolyzes AMP-aa's	ND		HNT1	Z74173.1	HINT2	AF356515.1	9p13.3
Histone deacetylase 7a (HDAC7)	Pro-apoptotic protein	IMS				HDAC7A	AY321367.1	12q13.11
Immuno-associated nucleotide (IAN) family protein IAN-4, T/B cell-specific	Reg mito integrity; T cell apoptosis	MOM				IAN4L1	AK002158.1	7q36.1
Integral membrane protein 2B, short form [ITM2B(S)]	Induces apoptosis	ND				ITM2B	AB030204.1	13q14.2
Interferon-induced protein 6-16 precursor (Ifi-6-16)	Anti-apoptotic protein; inhibits caspase 3	ND				G1P3	X02492.1	1p36.11
DnaJ homolog subfamily A member 3, long isoform (hTid-1L); homolog of yeast Mdj1	Increases apoptosis	MAT		MDJ1	Z28336.1	DNAJA3	AF061749.1	16p13.3
DnaJ homolog subfamily A member 3, short isoform (hTid-1S); homolog of yeast Mdj1	Suppresses apoptosis	MAT		MDJ1	Z28336.1	DNAJA3	AF061749.1	16p13.3
Metastasis suppressor for carcinoma	Pro-apoptotic protein	NC				CC3	U69161.1	11p15.1
Metastasis suppressor for carcinoma, alt spliced product (TC3)	Anti-apoptotic protein	NC				CC3	AF092095.1	11p15.1
Mitochondrial acidic matrix protein (human P32 [gC1q-R]), hyaluronan-binding protein	Ca binding? Binds HRK, DP5, rubella capsid	MAT		MAM33	Z38060.1	C1QBP	L04636.1	17p13.3
Modulator of apoptosis-1 (MAP-1)	Pro-apoptotic protein; target of Bax	ND				MOAP1	BC015044.1	14q32.13
Mucin 1 carcinoma-associated protein, C-terminal subunit	Anti-apoptotic protein	MOM				MUC1	J05582.1	1q22
Myeloid cell leukemia sequence 1 (Mcl-1)	Anti-apoptotic protein	MOM				MCL1	AF118124.1	1q21.3
Nascent polypeptide-associated complex beta subunit (betaNAC) (Sc Egd1 NC)	Anti-apoptotic; caspase substrate?	MOM?				BTF3	X53280.1	5q13.2
Noxa ("damage", mediator of p53-dependent apoptosis	Pro-apoptotic protein	ND				PMAIP1	D90070.1	18q21.32
p21-activated protein kinase 2, cleaved product	Cleaved product translocates to mitos	MOM?	2.7.1.-			PAK2	U24153.1	3q29
p21-activated protein kinase 5 (PAK5; also called PAK7), brain-specific	Anti-apoptotic protein	MOM?	2.7.1.-			PAK7	AB040812.1	20p12.2
p53 tumor suppressor protein	Pro-apoptotic protein; binds Bcl-xL	ND				TP53	M14695.1	17p13.1
p53-regulated apoptosis-inducing protein 1 (3 isoforms)	Pro-apoptotic protein	ND				P53AIP1	AB045830.1	11q24
p53 binding protein 2 (53BP2) (Apoptosis stimulating of p53 protein 2 [AAPP2])	Pro-apoptotic protein	ND				TP53BP2	AJ318888.2	1q42.11
Peptidyl-prolyl cis-trans isomerase NIMA-interacting 1	Proapoptotic protein; binds JIP3, p-BIM-EL	ND	5.2.1.8			PIN1	U49070.1	19p13.2
Phosphatidylinositol-3,4,5-trisphosphate 3-phosphatase PTEN	Proapoptotic protein; translocates to mitos	ND	3.1.3.67			PTEN	U93051.1	10q23.31
Protein kinase Cdelta	Induces apoptosis; transl to Sc mitos w. farnesol	MOM	2.7.1.-	PKC1	Z35866.1	PRKCD	D10495.1	3p21.31
Protein phosphatase 2A, catalytic subunit	Dephosphorylates Bcl-2	ND	3.1.3.16			PPP2CA	X12646.1	5q31.1
Protein phosphatase 2A, regulatory subunit (subunit B), form B56alpha	Dephosphorylates Bcl-2 @ Ser-70	NC	3.1.3.16			PPP2R5A	L42373.1	1q32.3
Raf-1 oncogene, cytosolic serine/threonine kinase	Targeted to mitos by Bcl-2	MOM	2.7.1.-			RAF1	X03484.1	3p25
Reaper (D. melanogaster; no human homolog found)	Pro-apoptotic protein	MOM				RPR (f)	L31631.1 (f)	
Ribosomal protein YmL27	Pro-apoptotic protein; binds Bcl2	MAT		MRPL27	Z36151.1	MRPL41	BE326258.1	9q34.3
Ribosomal protein S29	Pro-apoptotic protein	MAT?		RSM23	D10263.1	DAP3	X83544.1	1q23.1
Ribosomal protein S30 (programmed cell death 9 protein PDCD9)	Pro-apoptotic protein	MAT?				MRPS30	AF146192.2	5p12
RIP-like protein kinase RIP3	Pro-apoptotic protein	ND				RIP3	AF156884.1	14q11.2
Septin 4 (PNUTL2; ARTS; H5, alt spl mito isoform (M-septin) (binds XIAP)	Pro-apoptotic protein; transloc to nucleus	IMS?				SEPT4	AF176379.1	17q23.2
Serine protease HtrA2/Omi (Smac-like inhibitor of XIAP) (binds XIAP)	Pro-apoptotic protein; degrades IAP's	IMS	3.4.21.-			OMI	AF020760.2	2p13.1
SIVA-1 (CD27-binding protein, long isoform)	Anti-apoptotic protein; binds CD27, Bcl-XL	MOM?				SIVA	U82938.1	14q32.33
Hypoxia inducible factor prolyl hydroxylase 3 (no obvious MTS in human EGLN3)	Pro-apoptotic protein in rat	ND				Egln3 (r)	U06713.1 (r)	14q13.1
Stress-activated protein kinase (SAPK)	Translocates to mitos/Bcl-xL	MOM?	2.7.1.-			MAPK9	L31951.1	5q35.3
Survivin (IAP-related inhibitor of apoptosis)	Anti-apoptotic protein	ND				BIRC5	U75285.1	17q25.3
Target of Rapamycin (only human protein determined to be mitochondrial)	Kinase; relocates fr mito->nucleus	MOM	2.7.1.137	TOR1	X74857.1	FRAP1	L34075.1	1p36.22
Target of Rapamycin	Kinase; relocates fr mito->nucleus	MOM	2.7.1.137	TOR2	X71416.1	FRAP1	L34075.1	1p36.22
Thyroid hormone nuclear receptor subfamily 4, group A, member 1 (NGFI-B) (TR3)	Pro-apoptotic; transloc to mitos	MOM?				NR4A1	L13740.1	12q13.13

839

Appendix 5 *(continued)*

Gene product	Function; comment	Location	E.C. #	Fungal		Animal		
				Symbol	GenBank	Symbol	GenBank	Human Chr.
WW domain-containing oxidoreductase (WOX1); translocates from mitos->nucleus	Pro-apoptotic protein; tumor suppressor	ND	—	—	—	WWOX	AF187015.1	16q23.2
Yeast suicide protein 1 [5]	Mitochondrial fragmentation	ND	—	YSP1	U10397.1	—	—	—
"Extrinsic" apoptotic agents								
Cytomegalus virus-encoded vMIA protein fr unspliced exon 1 UL37 mRNA, N-term frag	Inhibitor of apoptosis; binds ANT	MOM	—	—	—	HCMVUL37	X17403.1	—
E. Coli EspF protein (fk15)	Pro-apoptotic protein	ND	—	—	—	EspF (Ec)	AF022236.1 (Ec)	—
E. Coli mitochondrial associated protein (Map(ORF19), type III secreted protein	Function unknown	MAT	—	—	—	Map (Ec)	AF022236.1 (Ec)	—
Hepatitis B virus transactivating protein X	Binds VDAC3	MOM?	—	—	—	HBX	X69798.1 (Hb)	—
Hepatitis C virus core protein (1st 191 aa of 3010-aa polyprotein, cut @ aa-179)	Modulate apoptosis or lipid transfer	MOM	—	—	—	None	AF165045.1	—
Hepatitis C virus non-structural protein 4A (NS4A)	Promotes apoptosis	MOM	—	—	—	None	AJ278830.1	—
HIV transactivating regulatory protein Tat	Microtubule polymerization; pro-apoptotic	ND	—	—	—	TAT	M11840.1 (Hiv)	—
HIV viral protein R (at least 2 isoforms)	Induces apoptosis	MOM?	—	—	—	VPR	U81843.1 (Hiv)	—
HTLV-I p13(II) protein coded by the x-II ORF	Pro-apoptotic protein	MIM	—	—	—	XII	L08433.1	—
Influenza A virus PB1-F2 protein, alt out-of-frame ORF of PB1 mRNA	Pro-apoptotic protein	MOM/MIM	—	—	—	PB1	M25932.1	—
Kaposi's sarcoma-associated herpesvirus (KSHV) K7 protein	Anti-apoptotic protein	MOM?	—	—	—	[ORF K7]	U93872.2	—
Kaposi's sarcoma-associated herpesvirus (KSHV) K15 protein	Binds HAX1	MOM?	—	—	—	[ORF K15]	U75698.1	—
Myxoma virus integral membrane protein type I-alpha (M11L)	Anti-apoptotic protein; binds MPT & Bak	MOM	—	—	—	M11L	M93049.1	—
Outer membrane protein A (OMP-38) of Acinetobacter baumannii (Ab)	Pro-apoptotic	ND	—	—	—	OMPA (Ab)	AY485227.1 (Ab)	—
Papillomavirus (HPV) E1-E4 fusion protein (E1^E4)	Pro-apoptotic	ND	—	—	—	E1^E4	D00735.1	—
Porin (PorB) from Neisseria meningitidis (Nm)	Anti-apoptotic protein	NC	—	—	—	PorB (Nm)	X65530.1 (Nm)	—
Pseudorabies virus (swine herpesvirus 1) Ser/Thr kinase us3, long isoform us3a	Anti-apoptotic protein?	ND	—	—	—	US3	DAA02204.1	—
Translocated intimin receptor of Escherichia coli (Ec)	Pro-apoptotic protein	MOM?	—	—	—	TIR (Ec)	AB026719.1	—
Vacuolating cytotoxin VacA of Helicobacter pylori (Hp)	Pro-apoptotic protein	ND	—	—	—	VACA (Hp)	U29401.1 (Hp)	—
Vaccinia virus inhibitor of apoptosis	Anti-apoptotic protein (C-term MTS)	ND	—	—	—	F1L (Vv)	U94848.1 (Vv)	—
Verotoxin II (Shiga-like toxin 2), subunit A of Escherichia coli (Ec)	Binds Bcl2 BH1 domain	MOM?	3.2.2.22	—	—	SLT-IIA (Ec)	X65949.1 (Ec)	—
Vesicular stomatitis virus (Vsv) matrix (M) protein	Loss of mito membrane permeability	MOM	—	—	—	M (Vsv)	J02428.1 (Vsv)	—
Walleye dermal sarcoma virus Orf C	Pro-apoptotic protein	ND	—	—	—	None	L41838.1	—
CARRIERS AND TRANSPORTERS								
ABC protein, sub-family C (CFTR/MRP), member 6, short form	Function unknown	NC	—	—	—	ABCC6	X95715.1	16p13.1
ADP/ATP carrier (adenine nucleotide translocator) isoform 1 (human ANT1)	ADP/ATP import/export; in nucleoid	MIM	—	AAC1	M12514.1	SLC25A4	J02966.1	4q35
ADP/ATP carrier (adenine nucleotide translocator) isoform 2 (human ANT2)	ADP/ATP import/export	MIM	—	AAC2	J04021.1	SLC25A5	M57424.1	Xq24
ADP/ATP carrier (adenine nucleotide translocator) isoform 3 (human ANT3)	ADP/ATP import/export	MIM	—	AAC3	M34076.1	SLC25A6	BC014775.1	Xp22.33
ADP/ATP carrier (adenine nucleotide translocator) isoform 4	ADP/ATP import/export	MIM	—	AAC3	M34076.1	SLC25A31	AL136857.1	4q28.1
Anion exchange protein 1 (AE1) [2] (yeast Bor1 NC)	Anion transp	ND	—	BOR1	Z71551.1	SLC4A1	M27819.1	17q21.31
Aquaporin 8	Water transp	MIM	—	—	—	AQP8	AB013456.1	16p12.1
Aquaporin 9, spl mitochondrial isoform	Water transp; found in brain mitos	MIM	—	—	—	AQP9	NM_020980.2	15q21.3
Aspartate/glutamate carrier 1 (AGC1) (Aralar1)	Asp/Glu transp	MIM?	—	AGC1	Z71255.1	SLC25A12	Y14494.1	2q31
Aspartate/glutamate carrier 2 (AGC2) (citrin [calcium-dependent mito carrier])	Asp/Glu transp, esp in liver	MIM	—	NCU01241.1 (Nc)	XM_326733.1 (Nc)	SLC25A13	AF118838.1	7q21.3
ATP-Mg/Pi) carrier, Ca-binding, isoform 1 (SCaMC-1/APC1)	Adenine nucleotide transporter	MIM	—	SAL1	Z71359.1	SLC25A24	AF123303.1 (i)	1p13.3
ATP-Mg/Pi) carrier, Ca-binding, isoform 2 (SCaMC-3/APC2)	Adenine nucleotide transporter	MIM	—	SAL1	Z71359.1	SLC25A23	AJ619962.1	19p13.3
ATP-Mg/Pi) carrier, Ca-binding, isoform 3 (SCaMC-2/APC3)	Adenine nucleotide transporter	MIM	—	SAL1	Z71359.1	SLC25A25	AJ619963.1	9q34.11

(continues)

Description	Function	EC no.	Localization	Yeast gene	Yeast GenBank	Human gene	Human GenBank	Chromosome
ATP-Mg-P(i) carrier, Ca-binding, isoform 4 (SCaMC-2/APC4)? (MGC34725)	Adenine nucleotide transporter?	—	MIM?	SAL1	Z71359.1	SLC25A41	BC031671.1	19p13.3
Calsequestrin, skeletal muscle isoform (calmine)	Calcium binding protein		MIM			CASQ1	S73775.1	1q21
Carnitine/acylcarnitine translocase (CACT)	Ac-carnitine transp to mitos		MIM	CRC1	Z75008.1	SLC25A20	Y10319.1	3p21.31
Carnitine/acylcarnitine translocase-like protein (CACL)	Ac-carnitine transp to mitos		MIM?	CRC1	Z75008.1	SLC25A29	BX247983.1	14q32.2
Carnitine O-acetyltransferase	Fatty acid transp	2.3.1.7	MIM	CAT2	Z14021.1	CRAT	X78706.1	9q34.1
Carnitine O-acetyltransferase	Fatty acid transp	2.3.1.7	MIM	YAT1	X74553.1			
Carnitine O-acetyltransferase, similar to Yat1p	Fatty acid transp		ND	YAT2	U18778.1			
Carnitine palmitoyltransferase I, liver type	Fatty acid transp	2.3.1.21	MOM			CPT1A	L39211.1	11q13.3
Carnitine palmitoyltransferase I, muscle type	Fatty acid transp	2.3.1.21	MOM			CPT1B	U62733.1	22q13.3
Carnitine palmitoyltransferase IC, brain/testis-specific	Does not transport FA; reg energy metab	2.3.1.21	MOM			CPT1C	AF357970.1	19q13.33
Carnitine palmitoyltransferase II (carnitine O-acetyltransferase 2)	Fatty acid transp	2.3.1.21	MIM			CPT2	U09648.1	1p32
Cationic amino acid transporter-2A, potentially mitochondrial	Cationic aa transp		NC			ATRC2	U76368.1	8p22
Chloride ion channel 27 [2]	Chloride transp?		NC			CLIC1	U93205.1	6p21.33
Chloride intracellular channel p64H1	Chloride transp; translocates to nucleus		ND			CLIC1	AF097330.1	9p21.3
Cobalamin transporter (CblA complementation group)	Translocates cobalamin into mitos		MIM			MMAA	AF524846.1	4q31.21
Cobalt transporter	Co transp		MIM	COT1	M88252.1			
Coenzyme A transporter (Graves disease carrier; GDC)	CoA transp		MIM	LEU5	U10555.1	SLC25A16	M31659.1	10q21.3
Copper transporting ATPase, 140-kDa isoform	Cu transp	3.6.1.36	ND			ATP7B	U11700.1	13q14.3
Cyclin M4 (Ancient Conserved Domain Protein family 4 [ACDP4]), potentially mito	Binds Cox11; transporter chaperone?		NC	MAM3	Z74802.1	CNNM4	BC063295.1	2q11.2
Diazepam binding inhibitor (acyl-CoA binding protein [ACBP]) (long/short forms)	Acyl-CoA ester transp		MOM			DBI	M14200.1	2q14.2
Dicarboxylate carrier (DIC)	Malate-for-citrate transp		MIM	DIC1	U79459.1	SLC25A10	AJ131613.1	17q25.3
Equilibrative nucleoside transporter 1 (ENT1) (mito in human but not in rat/mouse)	Purine and pyrimidine transp		ND	FUN26	L05146.2	SLC29A1	U81375.1	6p21.1
Excitatory amino acid transporter 1 (EAAT1)	Glu/Asp transp		MIM			SLC1A3	U03504.1	5p13.2
Folate transporter; FAD carrier protein (MCF protein)	Folate/FAD (flavin) export; may transp NAD+		MIM	FLX1	L41168.1	SLC25A32	AF283645.1	8q22.3
Fatty acid translocase FAT/CD36 (platelet glycoprotein IV)	Long chain F.A. transp; binds CPT-1		ND			CD36	M24795.1	7q21.11
Folate transporter (reduced folate carrier), 60-kDa isoform	Folate transp		ND			SLC19A1	AL163302.2	21q22.3
General amino acid permease AGP2	Also transports carnitine		ND	AGP2	Z36001.1			
Glucose transporter, type 3 (brain isoform), potentially mitochondrial	Glucose transp		ND			GLUT3	M20681	12p13.31
Glutamate carrier, isoform 1 (GC1)	Glutamate/H(+) symporter		NC	NCU01241.1 (Nc)	XM_326733.1 (Nc)	SLC25A22	AK023106.1	11p15.5
Glutamate carrier, isoform 2 (GC2)	Glutamate/H(+) symporter		NC	NCU01241.1 (Nc)	XM_326733.1 (Nc)	SLC25A18	AY008285.1	22q11.21
GTP/GDP carrier (Yhm1)	GTP/GDP transp		MIM	GGC1	Z74246.1			
Hypothetical protein YDR438w (THI74)	May be involved in thiamine transp		ND	YDR438w	Z49808.1			
Iron accumulation protein	Fe transp?		MIM	MMT1	Z37580.1			
Iron accumulation protein	Fe transp?		MIM	MMT2	X56445.1			
Iron transport protein; RNA splicing protein (MCF)	Splicing of mt-rRNAs; Fe/Cu uptake		MIM	MRS3	Z28277.1	SLC25A28	AK056782.1	10q24.2
Iron transport protein; RNA splicing protein (MCF)	Splicing of mt-rRNAs; Fe/Cu uptake		MIM	MRS4	Z28277.1	SLC25A37	AF223466.1	8p21.2
Iron transport protein, sim to yeast Mrs3/Mrs4 and to mouse Mscp (alt spl forms)	Fe import for heme syn; in erythroid cells		MIM	MRS4		SLC25A37	AF223466.1	8p21.2
Porphyrin transporter, ATP-binding cassette (ABC); mitochondrial ABC transporter 3	Porphyrin transp; heme syn		MOM	ATM1	X82612.1	ABCB6	AF076775.1	2q35
Iron transporter homolog similar to Atm1p (ABC)	Fe homeostasis; heme syn; binds ferrochelatase		ND	ATM1	X82612.1	ABCB7	AF038950.1	Xq13.3
Kanadaptin (Cl-/HCO3- anion exchanger 1-binding protein), esp in kidney	Anion transp; nucl/mito localization		ND			SLC4A1AP	AK001486.1	2p23.3
Kidney Mitochondrial Carrier Protein 1 (KMCP1)	Similar to UCP5/BMCP1		ND			SLC25A30	AL627107.32	13q14.13
Magnesium transporter	Mg transp		MAT?	MRS2	M82916.1	MRS2L	AF28828.1	6p22.2
Magnesium transporter, homolog of Mrs2p (LPE10)	Mg transp		MIM	LPE10	U39205.1			
Manganese trafficking factor (MCF protein CGI-69; LOC51629)	Activation of Mn-SOD		MIM	MTM1	Z73042.1	SLC25A39	BC009330.2	17q21.31
Manganese transporter	Mn transp		ND	SMF1	U15929.1			
Manganese transporter, probable	Mn transp		MIM	SMF2	U00062.1			
MCF protein FLJ20551	Function unknown, probable		MIM	YDL119c	Z74167.1	SLC25A38	BC013194.1	3p22.1
MCF protein	Function unknown; sim to Age1 and Pet8		MIM	YMR166c	Z49705.1			

Appendix 5 *(continued)*

Gene product	Function; comment	Location	E.C. #	Fungal Symbol	Fungal GenBank	Animal Symbol	Animal GenBank	Human Chr.
MCF protein, homolog of Ymc2, putatively mito in human	Oleate transp?; Glu syn? Related to SLC25A29	MIM		YMC1	X67122.1	LOC153328	BC025747.1	5q31.1
MCF protein (hepatocellular carcinoma-downregulated mito carrier)	Oleate transp?; Glu syn? Related to SLC25A29	MIM		YMC2	Z35973.1	HDMCP	AY569438.1	14q32.2
MCF protein, intronless expressed retrogene derived from MGC4399	Function unknown	NC		–	–	None	AA757281.1 (I)	11p12
MCF protein LOC283130, homolog of Ymc2	Oleate transp?; Glu syn?	MIM		YMC1	X67122.1	SLC25A45	BC041100.1	11q13.1
MCF protein, similar to Mtm1/CGI-69 (MCFP)	Function unknown	MIM?		MTM1	Z73042.1	SLC25A40	BC027322.1	7q21.12
MCF protein, similar to Rim2; nucleoid binding protein	Function unknown; mtDNA attachment?	MIM		YHM2	Z48756.1			
Metallothionein, isoform 1A	Zn transp	IMS		–	–	MT1A	K01383.1	16q13
Metallothionein, isoform 2A	Zn transp	IMS		–	–	MT2A	X97260.1	4q13.3
Metallothionein, isoform 3	Zn transp	IMS		–	–	MT3	X89604.1	16q13
Mitochondrial carrier homolog 1 (a and b isoforms)	Function unknown; brain-specific	MOM		–	–	MTCH1	AF176006.3	6p21.2
Mitochondrial carrier protein KIAA0446	Function unknown; potential glutamine carrier	MIM?		–	–	SLC25A44	BC008843.2	1q22
Mitochondrial carrier protein LOC91137, some similarity to Sfc1 in C-term	Function unknown	MIM?		–	–	SLC25A46	BC017169.1	5q22.1
Mitochondrial carrier protein MGC4399	Function unknown	MIM		YPR011c	Z71255.1	SLC25A33	BC004991.1	1p36.22
Mitochondrial carrier protein MGC26694, sim to SLC25A16 and YPR011c [12]	ANT-related; function unknown	MIM				SLC25A42	BC045598.1	19p13.11
Mitochondrial solute carrier protein LOC203427, similar to SLC25A16 and Leu5	Function unknown	NC				SLC25A43	BC019584.1	Xq24
Monocarboxylate transporter (MCT1)	Lactate/pyruvate transp	MIM		ESBP6	X84903.1 (a)	SLC16A1	L31801.1	1p13.1
Multi-drug resistance-like transporter 1 (antigen/erythroid transporter ABC-me)	m-AAA peptide/heme export	MIM		MDL1	L16958.1	ABCB10	AB013380.1	1q42.13
Multi-drug resistance-like transporter 1 (Mdl1) homolog	Function unknown	ND		MDL1	L16958.1	ABCB8	AF047690.1	7q36.1
Multi-drug resistance-like transporter 2	FeS cluster transp	ND		MDL2	L16959.1	ABCB2	L21204.1	6p21.3
Nicotinamide dinucleotide transporter (YIL006w)	NAD+ transp	MIM		NDT1	Z38113.1	MFTC	AF283645.1	8q22.3
Nicotinamide dinucleotide transporter (YEL006w)	NAD+ transp	MIM		NDT2	U18530.1	MFTC	AF283645.1	8q22.3
NADPH oxidase homolog 1 (NOH1)	Proton channel; reg pH; makes superoxide	ND				NOX1	AF166327.1	Xq22.1
Organic cation transporter (OCTN1)	Organic cation/carnitine transp	ND		HXT9	Z49494.1	SLC22A4	AB007448.1	5q23.3
Organic cation transporter (OCTN2)	Maternofetal carnitine transport	ND				SLC22A5	AF057164.1	5q23.3
Ornithine decarboxylase antizyme (frameshift translation), 29-kDa mito isoform	Regulates polyamine transp; degrades OTC	ND				OAZ1	D78361.1	19p13.3
Ornithine transporter 1 (ORNT1 in human)	Ornithine, citrulline, Lys, Arg carrier	MIM		ORT1	X87414.1	SLC25A15	AF112968.1	13q14
Ornithine transporter 2, functional retrogene (no introns) (ORNT2)	Ornithine, citrulline, Lys, His, Arg carrier	MIM		ORT1	X87414.1	SLC25A2	AF378119.1	5q31.3
Oxaloacetate/sulfate carrier	Oxaloacetate/[thio]sulfate/malonate transp	MIM		OAC1	AJ238698.1	SLC25A35	AK097536.1	17p13.1
Oxaloacetate/sulfate carrier homolog LOC2847237, sim to Oac1	Function unknown	MIM		OAC1	AJ238698.1	SLC25A34	BC027998.1	1p36.21
Oxodicarboxylate carrier, isoform 1 (ODC1)	Oxoglutarate/oxoadipate/oleate transp	MIM		ODC1	U43703.1	SLC25A21	AJ278148.1	14q11.2
Oxodicarboxylate carrier, isoform 2 (ODC2), MCART1 potentially mito	Oxoglutarate/oxoadipate/oleate transp	MIM		ODC2	U75130.1	MCART1	BC008500.1	9p13.2
Oxoglutarate/malate carrier (MCF protein) (OGC)	Oxoglutarate/malate exch; porphyrin transp	MIM		DIC1	U79459.1	SLC25A11	AF070548.1	17p13.3
Peripheral benzodiazepine receptor associated protein (PAP7)	Cholesterol transp	MOM?				BZRP	U12421.1	22q13.31
Peripheral benzodiazepine receptor	Cholesterol transp w PKA-RIalpha	NC				ACBD3	BU153320.1	1q42.12
Peripheral benzodiazepine receptor interacting protein	Function unknown	ND				BZRAP1	AF039571.1	17q23.2
Phosphate carrier (MCF protein) (import receptor p32) (PHC)	Phosphate carrier (PIC)	MIM		MIR1	X57478.1	SLC25A3	X77337.1	12q23
Phosphate carrier (MCF), similar to Mir1	Phosphate carrier (PIC)	MIM		PIC2	U18796.1			
Phospholipid scramblase 3	Phospholipid/cardiolipin transfer	ND		YJR100c	Z49600.1	PLSCR3	AF159442.1	17p13.1
Potassium channel (Kir4.1), putatively mitochondrial	ATP-sensitive K+ channel	ND				KCNJ10	U73192.1	1q23.2
Potassium channel (Kir6.1), putatively mitochondrial	ATP-sensitive K+ channel	ND				KCNJ8	D50312.1	12p11.23
Potassium channel (Kir6.2), putatively mitochondrial	ATP-sensitive K+ channel	ND				KCNJ11	D50582.1	11p15.1
Potassium channel, Ca2+-activated BK-type (BCKAalpha)	K+ transp; interacts w. KCNMB1-4	MIM				KCNMA1	U23767.1	10q22.3
Potassium channel, Ca2+-activated BK-type (BKbeta 1)	K+ transp; interacts w. KCNMA1	MIM?				KCNMB1	U25138.1	5q35.1
Potassium channel regulatory subunit (sulfonylurea receptor 2, isoform A)	Assoc. with Kir6.1/Kir6.2	ND				SUR2	AF061323.1	12p12.1

(continues)

Description	Function	EC	Location	Yeast gene	Yeast accession	Human gene	Human accession	Locus
Potassium voltage-gated channel subfamily A member 3 (Kv1.3)	K+ transp		MIM	—		KCNA3	L23499.1	1p13.3
Pyrimidine nucleotide transporter (Pyt1)	Pyrimidine nucleotide transp		MIM	RIM2	Z36061.1	SLC25A36	AK001480.1	3q23
Ryanodine receptor 1 (skeletal muscle calcium release channel)	Ca release channel		MIM	—		RYR1	J05200.1	19q13.2
S-adenosylmethionine carrier (SAMC)	S-adenosylmethionine transp		MIM	PET8	U02536.1	SLC25A26	AK092495.1	3p14.1
Sarcoplasmic/endoplasmic reticulum calcium ATPase 1 (SERCA1) [13]	In brown adipose tissue (BAT) mitos		MIM	—		ATP2A1	U96781.1	16p11.2
Sideroflexin 1, potential alpha-isopropylmalate carrier	Leu syn?		MIM	FSF1	Z75179.1	SFXN1	AF327346.1	5q35.2
Sideroflexin 2, putative Fe transporter, may be mitochondrial	Mito Fe metab?		NC	—		SFXN2	BC022091.1	10q24.32
Sideroflexin 3, putative Fe transporter, may be mitochondrial	Mito Fe metab?		NC	—		SFXN3	BC000124.1	10q24.31
Sideroflexin 4, putative Fe transporter, may be mitochondrial	Mito Fe metab?		NC	—		SFXN4	BC063562.1	10q26.11
Sideroflexin 5 (rat tricarboxylate carrier [BBG-TCC] is brain-specific)	Citrate transp?		MIM?	—		SFXN5	AY044437.1	2p13.2
Sodium/calcium exchanger (NCX1), potentially mitochondrial	Ca++/H+ antiporter		ND	—		SLC8A1	M91368.1	2p22.1
Sodium/hydrogen exchanger (NHE1), potentially mitochondrial	Na+/H+ antiporter		ND	NHX1	U33007.1	SLC9A1	S68616.1	1p36.11
Sodium/potassium/calcium exchanger (NCKX1), potentially mitochondrial	Na+/Ca++, K+ antiporter		ND	—		SLC24A1	AF026132.1	15q22.32
Steroidogenic acute regulatory protein (StAR)	Sterol transp		MOM	—		STAR	U17280.1	8p11.2
Steroidogenic acute regulatory protein-like (CAB1, MLN64)	Lysosome->mito cholesterol transp		MOM	—		STARD3	X80198.1	17q12
Steroidogenic acute regulatory protein-related lipid transfer domain containing protein 13	Rho GTPase protein-related lipid transfer		ND	—		STARD13	AL049801.1	13q13.1
Sterol carrier protein 2	Lysosome -> mito sterol transfer	2.3.1.16	ND	—		SCP2	M75883.1	1p32.3
Succinate/fumarate carrier (antiporter)	Succinate-fumarate transp		MIM	SFC1	Z25485.1	—	—	—
Succinate/fumarate transporter, potentially mitochondrial	Succinate-fumarate transp?		MIM	YFR045w	AF419345.1	—	—	—
Superoxide dismutase 1 copper chaperone	Delivers Cu to Cu,Zn-SOD		IMS	CCS1	U17378.1	CCS	AF002210.1	11q13.3
Thiamine pyrophosphate carrier (not a deoxynucleotide carrier [DNC] in humans)	ThPP/ThMP transp		MIM	TPC1	Z72881.1	SLC25A19	AF182404.1	17q25.1
Thiamine transporter (THT1), putatively mitochondrial	Thiamine transp		ND	THI72	Z75100.1	SLC19A2	AF135488.1	1q24.2
Tricarboxylate transporter (citrate carrier; CIC)	Citrate transp		MIM	CTP1	U17503.1	SLC25A1	U25147.1	22q11.21
Uncoupling protein 1 (brown fat; thermogenin) (UCP1)	Proton carrier; binds CIDEA		MIM	—		SLC25A7	U28480.1	4q31.1
Uncoupling protein 2 (UCP2)	Proton carrier		MIM	—		SLC25A8	U76367.1	11q13.4
Uncoupling protein 3 (UCP3)	Proton carrier		MIM	—		SLC25A9	U84763.1	11q13.4
Uncoupling protein 4 (UCP4)	Function unknown		MOM	—		SLC25A27	AF110532.1	6p12.3
Uncoupling protein 5 (UCP5) (3 isoforms) (also called BMCP1)	Function unknown		MOM	—		SLC25A14	AF078544.1	Xq26.1
Vanilloid receptor type 1 (VR1/TRPV1), claimed to be mitochondrial	Reg of chlorhydropeptic secretion		MIM	—		TRPV1	AJ277028.1	17p13.2
Vesicular inhibitory amino acid transporter, potentially mitochondrial	GABA transp		NC	—		VIAAT	AY044836.1	20q11.23
Voltage-dependent anion channel 1 (Porin)	Anion import		MOM	POR1	M34907.1	VDAC1	L06132.1	5q31.1
Voltage-dependent anion channel 2 (Porin), at least 4 alt spl isoforms	Binds BAK to prevent apoptosis		MOM	POR2	Z38125.1	VDAC2	L08666.1	10q22.2
Voltage-dependent anion channel 3 (Porin), at least 2 alt spl isoforms	Anion import		MIM	—		VDAC3	AF038962.1	8p11.2
INTERMEDIATE METABOLISM								
Abhydrolase domain containing 10 [6]	Aromatic compound metabolism		ND	—		ABHD10	AK002204.1	3q13.2
Acetate non-utilizing protein	Carbon assimilation		IMS	ACN9	U33057.1	DC11	AF201933.1	7q21.3
Acetyl-CoA hydrolase (human CACH1 is cytosolic)	Hydrolyzes acetyl-CoA	3.1.2.1	MAT	ACH1	M31036.1			
Acetyl-CoA synthetase 2, AMP-forming, mitochondrial isoform (confusing nomenclature)	TCA cycle-associated	6.2.1.1	MAT	ACS1	X66425.1	ACSS1	BC039261.1	20p11.21
Aconitate hydratase (aconitase)	TCA cycle	4.2.1.3	MAT	ACO1	M33131.1	ACO2	U80040.1	22q13.2
Aconitase homolog [1]	Function unknown	4.2.1.3	NC	ACO2	Z49475.1	—	—	—
Alcohol dehydrogenase III	Ethanol syn	1.1.1.1	MAT	ADH3	K03292.1	—	—	—
Alcohol dehydrogenase IV, potentially mitochondrial	Ethanol syn	1.1.1.1	NC	ADH4	Z72778.1	ADHFE1	AY033237.1	8q13.1
Aldehyde dehydrogenase 2 (NAD+)	Ethanol metab	1.2.1.3	MAT	ALD2	Z17314.1	LDHIA2	AB015226.1	15q21.3
Aldehyde dehydrogenase (NAD+); glycerol trinitrate reductase	Ethanol metab; nitroglycerine metab	1.2.1.3	MAT	ALD4	Z75282.1	ALDH2	Y00109.1	12q24.12
Aldehyde dehdrogenase ([1],[7])	Ethanol metab	1.2.1.5	MOM	HFD1	Z49702.1	ALDH3A1	M74542.1	17p11.2
Aldehyde dehydrogenase 5 (NAD+)	Ethanol metab	1.2.1.3	MAT	ALD5	U56605.1	ALDH5B1	M63967.1	9p13.2
Aldose reductase, potentially mitochondrial [2]	Ethanol metab	1.1.1.21	NC	YPR1	X80642.1	AKR1B1	X15414.1	7q33

Appendix 5 (continued)

Gene product	Function; comment	Location	E.C. #	Fungal Symbol	Fungal GenBank	Animal Symbol	Animal GenBank	Human Chr.
Alpha-ketoglutarate dehydrogenase complex E1 component	TCA cycle-related	MAT	1.2.4.2	KGD1	M26390.1	OGDH	D10523.1	7p13
Alpha-ketoglutarate dehydrogenase complex E2 component	Keto acid decarboxylation	MAT	2.3.1.61	KGD2	M34531.1	DLST	D16373.1	14q24.3
Alpha-ketoglutarate dehydrogenase complex E3 component	Dihydrolipoamide dehydrogenase	MAT	1.8.1.4	LPD1	M20880.1	DLD	J03690.1	7q31.1
Arabinono-1,4-lactone oxidase [7] (pseudogene w. no introns in primates)	Ascorbic acid syn	MOM	1.1.3.37	ALO1	U40390.1	GULO (r)	D12754.1 (r)	—
Beta,beta-carotene-9',10'-dioxygenase, putative [2]	Cleavage of provitamin A	ND	—	—	—	BCDO2	AJ290393.1	11q23.1
Biotin apo-protein ligase (yeast Bpl1 is cytosolic)	Biotinylates carboxylases	ND	6.3.4.—	—	—	HLCS	D23672.1	21q22.1
Biotinidase, 48-kDa alt spl isoform	Function in mitos not clear	MAT	3.5.1.12	—	—	BTD	U03274.1	3p25.1
Biotin synthase, putatively mitochondrial	Biotin syn; requires Pet8	NC	2.8.1.6	BIO2	Z73071.1	—	—	—
Branched-chain keto acid dehydrogenase complex, E1-alpha subunit	Br chain keto acid metab	MAT	1.2.4.4	AN1726.2 (An)	XM_405863.1 (An)	BCKDHA	Z14093.1	19q13.13
Branched-chain keto acid dehydrogenase complex, E1-beta subunit	Br chain keto acid metab	MAT	1.2.4.4	AN8559.2 (An)	XM_412696.1 (An)	BCKDHB	M55575.1	6q14.2
Branched-chain keto acid dehydrogenase complex, E2 subunit	Br chain keto acid metab; in nucleoid	MAT	2.3.1.—	B12J7.120 (Nc)	BX842635.1 (Nc)	DBT	X66785.1	1p31
Carbonic anhydrase VA (also called V)	C-compound metab	MAT?	4.2.1.1	—	—	CA5A	L19297.1	16q24.3
Carbonic anhydrase VB	C-compound metab	MAT?	4.2.1.1	—	—	CA5B	AB021660.1	Xp22.1
Carbonyl reductase 2 (dicarbonyl/L-xylulose reductase) [2]	Carbonyl cpd metab.	MAT?	1.1.1.10	—	—	DCXR	AF113123.1	17q25.3
Carbonyl reductase 3 (NADPH), potentially mitochondrial	Carbonyl cpd metab	NC	1.1.1.184	—	—	CBR3	AB004854.1	21q22.12
Citrate synthase	TCA cycle	MAT	4.1.3.7	CIT1	Z23259.1	CS	AF047042.1	19p13.12
Citrate synthase 2 (peroxisomal), mitochondrial (has cryptic MTS)	TCA cycle?	MAT?	4.1.3.7	CIT2	M14686.1	CS	AF047042.1	19p13.12
Citrate synthase 3	TCA cycle	MAT	4.1.3.7	CIT3	X88846.1	CS	AF047042.1	19p13.12
CoA synthase (bifunctional protein), ubiquitous (alpha) & brain-specific (beta) isoforms	CoA syn, 4th and 5th steps	MOM	2.7.7.3/1.24	YDR196c	Z48784.1	COASY	BC067254.1	17q21.2
Cob(I)alamin adenosyltransferase (CblB complementation group)	Cobalamin metab	ND	2.5.1.17	—	—	MMAB	BC011831.2	12q24.11
Cytochrome b2 (L-lactate cytochrome c oxidoreductase) (peroxisomal in human)	Lactate oxdn	IMS	1.1.2.3	CYB2	X03215.1	—	—	—
Dehydrogenase/reductase SDR family member 4 [3]	Oxidoreductase	ND	1.1.1.184	—	—	DHRS4	AK001870.1	14q11.2
Diacylglycerol O-acyltransferase homolog (ACAT-related protein 1) (ACAT)	Triacylglycerol syn, last step	ER-MAM?	2.3.1.20	DGA1	Z75153.1	DGAT1	AF059202.1	8q24.3
Dihydrolipoamide S-acetyltransferase (PDHC-E2) (Pda2)	TCA cycle-related	MAT	2.3.1.12	LAT1	J04096.1	DLAT	J03866.1	11q23.1
Dihydropteroate synthetase	Dihydrofolate syn, 1st 2 steps	ND	4.1.2.25	FOL1	Z71532.1	—	—	—
Dolichyl-phosphate mannose synthase	Glycoprotein syn	MOM	2.4.1.109	PMT1	L19169.1	—	—	—
DRAP deaminase (Rib2) homolog [3]	Riboflavin syn?	NC	4.2.1.70	YGR021w	Z72806.1	CCDC44	BC005049.1	17q23.3
Folylpolyglutamate synthase	Folate metab	MIM	6.3.2.17	—	—	FPGS	M98045.1	9q34.13
Fumarate hydratase (fumarase)	TCA metab	MAT	4.2.1.2	FUM1	J02802.1	FH	U59309.1	1q42.1
Fumarate reductase	TCA cycle-related	MAT	1.3.99.1	OSM1	Z49551.1	—	—	—
Gamma-butyrobetaine hydroxylase (Fmp12) [1]	Carnitine biosyn, 4th step; BBOX1 NC	ND	1.14.11.1	FMP12	U11582.1	BBOX1	AF082868.1	11p14.2
Glucokinase (hexokinase IV) (3 alt spl isoforms in human)	Glycolysis	MOM	2.7.1.1	HXK2	M11181.1	GCK	M90299.1	7p13
Glucokinase regulatory protein	Glycolysis; binds GCK	MOM?	—	Afu1g17490 (Af)	XM_748022.1 (Af)	GCKR	Z48475.1	2p23.3
Glyceraldehyde-3-phosphate dehydrogenase, potentially mitochondrial	Glycolysis, second phase, 1st step	ND	1.2.1.12	TDH3	V01300.1	GAPD	M17851.1	12p13.31
Glycerate kinase, mitochondrial alt spl isoform (GLYCTK1)	Glyoxylate catabolism?; ser deg	ND	2.7.1.31?	—	—	GLYCTK	AF448855.1	3p21.1
Glycerol kinase (3 isoforms in human)	Glycerol metab	ND	2.7.1.30	GUT1	X69049.1	GK	L13943.1	Xp21.2
Glycerol kinase, testis-specific isoform 1 (no introns)	Glycerol metab	ND	2.7.1.30	GUT1	X69049.1	GKP3	X78711.1	4q32.3
Glycerol kinase, testis-specific isoform 2 (no introns)	Glycerol metab	ND	2.7.1.30	GUT1	X69049.1	GKP2	X78712.1	4q21.21
Glycerol-3-phosphate dehydrogenase (FAD) [2]	Glycerol metab	MIM	1.1.99.5	GUT2	X71660.1	GPD2	U12424.1	2q24.1
Glycerol-3-phosphate dehydrogenase 2 (NAD+) [2]	Glycerol metab	ND	1.1.1.8	GPD2	Z74801.1	GPD1L	BC028726.1	3p24.1
Glycogen phosphorylase, muscle form [2]	Carbohydrate metab	ND	2.4.1.1	—	—	PYGM	M32598.1	11q13.1
Glycogen phosphorylase, brain form [2]	Carbohydrate metab	ND	2.4.1.1	—	—	PYGB	U47025.1	20p11.21
Glycosyltransferase 1 domain containing 1, potentially mitochondrial	Possible hexosyltransferase	NC	2.4.1.—	—	—	GLT1D1	AK056540.1	12q24.32
Glyoxylate reductase/hydroxypyruvate reductase [1][4]	Glyoxylate metab	ND	1.1.1.79/26	YNL274c	Z71550.1	GRHPR	AF134895.1	9p13.2

(continues)

Protein	Function	Gene	EC	Loc	Accession	Symbol	Accession	Location
Hexokinase I	Glycolysis	HXK1	2.7.1.1	MOM?	M14410.1	HK1	M75126.1	10q22.1
Hexokinase II	Glycolysis	HXK2	2.7.1.1	MOM?	M11181.1	HK2	Z46376.1	2p12
3-Hydroxybutyrate dehydrogenase	Ketone body metab	—	1.1.1.30	MAT	BX908809.1 (Nc)	BDH	BC005844.2	3q29
3-Hydroxyisobutyrate dehydrogenase	Br chain oxo-acid metab	[29E8.120] (Nc)	1.1.1.31	MAT	XM_325084.1 (Nc)	HIBADH	AAF19251.1	7p15.2
Hypothetical protein [9] (B13C5.090)	Potential amidase	NCU00905.1 (Nc)		MOM				
Hypothetical protein, potentially mitochondrial [5][6]	Phosphatase/hydrolase? Carbohyd metab?	YKR070w		ND	Z28295.1	CECR5	AF273271.1	22q11.1
Hypothetical protein MGC5352; Bcl-XL-binding protein v68 [6]	Similar to phosphoglycerate mutase			ND		PGAM5	BC008196.1	13q24.33
Isocitrate dehydrogenase 3 (NAD+) subunit 1 [6]	TCA cycle	IDH1	1.1.1.41	MAT	M95203.1	IDH3G	Z68907.1	Xq28
Isocitrate dehydrogenase 3 (NAD+) subunit alpha in human)	TCA cycle	IDH2	1.1.1.41	MAT	M74131.1	IDH3A	U07681.1	15q25.1
Isocitrate dehydrogenase 3 (NAD+) subunit beta, isoform B	TCA cycle	IDH1	1.1.1.41	MAT	M95203.1	IDH3B	U49283.1	20p13
Isocitrate dehydrogenase 2 (NADP+)	TCA cycle	IDP1	1.1.1.42	MAT	M57229.1	IDH2	X69433.1	15q26.1
Isocitrate lyase (may be inactive)	Propionyl-CoA/Thr metab	ICL2	4.1.3.1	MAT	Z48951.1			
L-2-hydroxyglutarate dehydrogenase	Synthesis of alpha-ketoglutarate	NCU08048.1 (Nc)	1.2.99.2	MAT?	XM_328753.1 (Nc)	L2HGDH	AY757363.1	14q21.3
Lipoate protein ligase	Lipoate biosynthesis	LIP2		NC	U19027.1	LIPT1	AB017567.1	2q11.2
Lipoate protein ligase A homolog, potentially mitochondrial	Lipoate syn/transp?	Y1L046w		NC	Z49321.1			
Lipoic acid synthase	Lipoate syn	LIP5	2.8.1.–	MAT?	Z75104.1	LIAS	BC023635.2	4p14
D-lactate dehydrogenase (cytochrome)	Lactate/pyruvate metab	DLD1	1.1.2.4	MIM	X66052.1	LDHD	NM_194436.1	16q22.3
D-lactate dehydrogenase (actin-interacting protein) (yeast Aip2)	Function unclear	DLD2	1.1.2.4	MAT	Z74226.1	D2HGDH	BC036604.2	2q37.3
L-lactate dehydrogenase A-like, testis-specific, potentially mitochondrial	Lactate -> pyruvate		1.1.1.27	NC		LDHL	AY009108.1	15q22.2
L-lactate dehydrogenase C chain, sperm-specific isozyme C4	Lactate/pyruvate metab		1.1.1.27	ND		LDHC	U13680.1	11p15.1
Lysine-ketoglutarate reductase /saccharopine dehydrogenase [3]	Oxidoreductase			ND		AASS	AJ007714.2	7q31.32
Malate dehydrogenase, mitochondrial	Malate-aspartate shuttle	MDH1	1.1.1.37	MAT	J02841.1	MDH2	AF047470.1	7q11.23
Malic enzyme, NADP(+)-dependent	TCA cycle-related	MAE1	1.1.1.38	MAT	Z28029.1	ME2	M55905.1	18q21.1
Malic enzyme. NADP(+)-dependent	TCA cycle-related	MAE1	1.1.1.40	MAT	Z28029.1	ME3	X79440.1	11q14.2
5,10-Methenyltetrahydrofolate synthetase, probably mitochondrial	Folate metab	FAU1	6.3.3.2	NC	U18922.1	MTHFS	L38928.1	15q25.1
Methionine synthase reductase, isoform A	Regeneration of cob(I)alamin		2.1.1.135	NC		MTRR	AF121213.1	5p15.31
Methylmalonyl-CoA epimerase (racemase)	Propionyl-CoA metab		5.1.99.1	ND		MCEE	AF364547.1	2p13.3
Methylmalonyl-CoA mutase	TCA cycle-related		5.4.99.2	MAT		MUT	M65131.1	6p21
3-Methyl-2-oxobutanoate hydroxymethyltransferase [1]	Pantothenate syn, 1st branch, 1st step	ECM31	2.1.2.11	ND	Z36045.1			
Molybdenum cofactor biosynthesis protein A/C, potentially mitochondrial	Molybdenum cofactor syn, step 1	AN0947.2 (An)		NC	XM_405084.1 (An)	MOCS1	AF034374.1	6p21.2
O-linked N-acetylglucosamine (GlcNAc) transferase, alt spl mito isoform	Glycosylation?		2.4.1.–	MIM		OGT	U77413.1	Xq13.1
Pantothenate kinase, long isoform	CoA syn		2.7.1.33	MAT?		PANK2	AF494409.1	20p13
Phenacrylic acid decarboxylase, probably mitochondrial	Cinnamic acid resistance	PAD1	4.1.1.–	NC	L09263.1			
Phosphoenolpyruvate carboxykinase 2 (mitochondrial isoform)	Gluconeogenesis		4.1.1.32	MAT?		PCK2	X92720.1	14q11.2
Pterin-4-alpha-carbinolamine dehydratase, putative, potentially mito	Tetrahydrobiopterin syn?	YHL018w	4.2.1.96	NC	U11582.1	DCOHM	AF499009.1	5q31.1
Nicotinamide mononucleotide adenylyl-transferase, type 3 (NC if mito in yeast)	NAD syn	NMA1	2.7.7.1	ND	U20618.1	NMNAT3	AF345564.1	3q23
Pyruvate carboxylase	TCA cycle-related		6.4.1.1	MAT		PC	S72370.1	11q13.3
Pyruvate dehydrogenase (lipoamide) E1-alpha subunit	TCA cycle-related	PDA1	1.2.4.1	MAT	M29582.1	PDHA1	M24848.1	Xp22.13
Pyruvate dehydrogenase (lipoamide) E1-alpha subunit 2, testis-specific	TCA cycle-related	PDA1	1.2.4.1	MAT	M29582.1	PDHA2	M86808.1	4q22.3
Pyruvate dehydrogenase (lipoamide) E1-beta subunit	TCA cycle-related	PDB1	1.2.4.1	MAT		PDHB	M98476.1	3p14.3
Pyruvate dehydrogenase complex, protein X	Binds E3 to E2 core	PDX1		MAT	Z72978.1	PDX1	AF001437.1	11p13
Pyruvate dehydrogenase kinase isozyme 1	Regulation of PDHC	PKP1	2.7.11.1	MAT	Z46861.1	PDK1	L42450.1	2q31.1
Pyruvate dehydrogenase kinase isozyme 2	Regulation of PDHC	PKP1	2.7.11.2	MAT	Z46861.1	PDK2	L42451.1	17q21.33
Pyruvate dehydrogenase kinase isozyme 3	Regulation of PDHC	PKP1	2.7.11.2	MAT	Z46861.1	PDK3	L42452.1	Xp22.11
Pyruvate dehydrogenase kinase isozyme 4	Regulation of PDHC; phosphorylates Pda1	PKP1	2.7.11.2	MAT	Z46861.1	PDK4	U54617.1	7q21.3
Pyruvate dehydrogenase phosphatase, cat subunit, isoform 1	Regulation of PDHC; PDHC phosphatase	PTC5	3.1.3.43	MAT	Z74998.1	PDP1	AF155661.1	8q22.1
Pyruvate dehydrogenase phosphatase, cat subunit, isoform 2	Regulation of PDHC; PDHC phosphatase	PTC5	3.1.3.43	MAT	Z74998.1	PDP2	BC028030.1	Xp22.13
Pyruvate dehydrogenase phosphatase, regulatory subunit	Regulation of PDHC	AN8654.2 (An)	3.1.3.43	MAT	XN_412791.1 (An)	PDPR	BX538155.1	16q22.1
Riboflavin kinase (flavokinase)	FAD syn: flavoprotein biogenesis	FMN1	2.7.1.26	MIM	Z49701.1	RFK	AK002011.1	9q21.13
Succinic semialdehyde dehydrogenase, NAD-dependent	GABA deg; GABA shunt; ROS in plants		1.2.1.24	MAT?		ALDH5A1	Y11192.1	6p22.2

Appendix 5 *(continued)*

Gene product	Function; comment	Location	E.C. #	Fungal		Animal		
				Symbol	GenBank	Symbol	GenBank	Human Chr.
Succinyl CoA:3-oxoacid CoA transferase, somatic isoform (SCOT-s)	Ketone body metab	MAT	2.8.3.5	NCU06881.1 (Nc)	XM_327166.1 (Nc)	OXCT1	U62961.1	5p13.1
Succinyl CoA:3-oxoacid CoA transferase, testis-specific isoform (SCOT-t)	Ketone body metab	MAT	2.8.3.5	NCU06881.1 (Nc)	XM_327166.1 (Nc)	OXCT2	AB050193.1	1p34.2
Succinyl-CoA synthetase, subunit alpha	TCA cycle	MAT	6.2.1.4	LSC1	Z75050.1	SUCLA1	AF104921.1	2p11.2
Succinyl-CoA synthetase, subunit beta (ATP-specific in human)	TCA cycle; binds ALAS-E	MAT	6.2.1.4	LSC2	Z73029.1	SUCLA2	AB035863.1	13q14.2
Succinyl CoA synthetase, subunit beta (GTP-specific in human)	TCA cycle	MAT	6.2.1.5	LSC2	Z73029.1	SUCLG2	AF058954.1	3p14.1
Trimethyllysine hydroxylase (trimethyllysine dioxygenase)	Carnitine syn, 1st step	MAT	1.14.11.8	cbs–1 (Nc)	AJ421151.1 (Nc)	TMLHE	AF373407.1	Xq28

LIPID METABOLISM
Cholesterol, steroid, and xenobiotic metabolism

Gene product	Function; comment	Location	E.C. #	Fungal Symbol	Fungal GenBank	Animal Symbol	Animal GenBank	Human Chr.
Acetoacetyl-CoA thiolase (acetyl-CoA C-acetyltransferase 1) (cyto in yeast)	Mevalonate/sterol syn, 1st step	MAT?	2.3.1.9	UM03571.1 (Um)	XN_401186.1 (Um)	ACAT1	D90228.1	11q22.3
Adrenodoxin (ferredoxin 1)cholesterol side-chain-cleavage system)	Steroid and FeS cluster syn	MAT	—	YAH1	Z73608.1	FDX1	J03548.1	11q22
Adrenodoxin/ferredoxin reductase (alt spl in human)	Pregnenolone syn	MAT	1.18.1.2	ARH1	U38689.1	FDXR	J03826.1	17q25.1
3-Beta hydroxy-5-ene steroid dehydrogenase type I	Steroid syn	MIM?	1.1.1.145			HSD3B1	M27137.1	1p12
3-Beta hydroxy-5-ene steroid dehydrogenase type II	Steroid syn	MIM?	1.1.1.145			HSD3B2	M67466.1	1p12
3-Beta hydroxy-5-ene steroid dehydrogenase type III	Steroid syn	MIM?	1.1.1.145			Hsd3b3 (m)	M77015.1 (m)	—
3-Beta hydroxy-5-ene steroid dehydrogenase type IV	Steroid syn	MIM?	1.1.1.145			Hsd3b4 (m)	L16919.1 (m)	—
3-Beta hydroxy-5-ene steroid dehydrogenase type V	Steroid syn	MIM?	1.1.1.145			Hsd3b5 (m)	L41519.1 (m)	—
3-Beta hydroxy-5-ene steroid dehydrogenase type VI	Steroid syn	MIM?	1.1.1.145			Hsd3b6 (m)	AF031170.1 (m)	—
Cytochrome P450 1A1 (derived from cytP4501A1), alt spl mito/nucl variant	Xenobiotic metab; trans fr ER->mitos	ND	1.14.14.1			CYP1A1	K03191.1	15q24.3
Cytochrome P450 2B1	Xenobiotic metab; trans fr ER->mitos	ND	1.14.14.1			CYP2B	M29873.1	19q13.2
Cytochrome P450 2B2	Xenobiotic metab; trans fr ER->mitos	ND	1.14.14.1			CYP2B6	M29874.1	19q13.2
Cytochrome P450 2E1	Xenobiotic metab; trans fr ER->mitos	ND	1.14.14.1			CYP2E1	J02625.1	10q26.3
Cytochrome P450 3A1	Xenobiotic metab	ND	1.14.14.1			CYP3A1 (r)	M10161.1 (r)	—
Cytochrome P450 3A2	Xenobiotic metab	ND	1.14.14.1			CYP3A2 (r)	M13646.1 (r)	—
Cytochrome P450 3A4	Xenobiotic metab	NC	1.14.14.1			CYP3A4	J04449.1	7q22.1
Cytochrome P450 11B1 (11 beta, polypeptide 1)	Steroid 11-beta-hydroxylase	ND	1.14.15.4			CYP11B1	X55764.1	8q21
Cytochrome P450 11B2 (11 beta, polypeptide 2)	Steroid 11-beta-hydroxylase	ND	1.14.15.4			CYP11B2	X54741.1	8q21
Cytochrome P450 25-hydroxy vitamin D-1 alpha hydroxylase	Vitamin D metab	ND	1.14.-.-			CYP27B1	AB005038.1	12q14
Cytochrome P450 C-22 hydroxylase (Sad)	Ecdysone syn	ND	1.14.-.-			CYP315A1 (f)	AY079170.1 (f)	—
Cytochrome P450 C-22 hydroxylase (Dib)	Ecdysone syn	ND	1.14.-.-			CYP302A1 (f)	NM_080071.2 (f)	—
Cytochrome P450-SCC side chain cleavage enzyme	Cholesterol monooxygenase	ND	1.14.15.6			CYP11A1	M14565.1	15q24.2
Cytochrome P450 sterol 27-hydroxylase	Vitamin D metab	ND	1.14.14.-			CYP27A1	M62401.1	2q35
Cytochrome P450 vitamin D(3) 24-hydroxylase (CC24)	Vitamin D metab	ND	1.14.-.-			CYP24A1	L13286.1	20q13.2
Epoxide hydrolase, potentially mitochondrial [5]	Aromatic hydrocarbon catab	NC	3.3.2.3	YGR031w	Z72816.1	WBSCR21	AF412030.1	7q11.23
Estradiol 17 beta-dehydrogenase 8	Estrogen syn	ND	1.1.1.62			HSD17B8	AL031228.1	6p21.32
3-Hydroxyacyl-CoA dehydrogenase, type II (SCHAD) (17beta-HSD10)	Steroid metab; binds A4	MAT	1.1.1.35			HADH2	AF037438.1	Xp11.22
Hydroxymethylglutaryl-CoA synthase (HMG-CoA synthase)	Mevalonate syn	ND	4.1.3.5	ERG13	X96617.1	HMGCS2	X83618.1	1p11.2
NADP-cytochrome P450 reductase [1]	Ergosterol biosyn	ND	1.6.2.4	NCP1	D13788.1	POR	AF258341.1	7q11.23
Oxysterol binding protein 2, potentially mitochondrial	Binds 7-ketocholesterol	NC	—	HES1	U03914.1	OSBP2	AF323731.1	22q12.2
S-adenosyl-methionine:delta-24-c-sterol methyltransferase	Ergosterol syn	MOM	2.1.1.41	ERG6	X74249.1			—
Sigma receptor (SR-31747 binding protein 1) (Erg2p claimed to be in ER)	Sterol syn; Ca reg; other unknown	MOM	—	ERG2	M74037.1	SR-BP1	U75283.1	9p13.3

Fatty acid metabolism

Gene product	Function; comment	Location	E.C. #	Fungal Symbol	Fungal GenBank	Animal Symbol	Animal GenBank	Human Chr.
ACP synthase (phosphopantetheine protein transferase)	Pantetheinylation of ACP	MAT?	2.7.8.7	PPT2	Y16253.1			—
Acetyl Co-A carboxylase 2 (ACC2), mitochondrial form [1]	Provides malonyl-CoA for lipoic acid syn	ND	6.4.1.2	HFA1	Z22558.1	ACACB	U89344.1	12q24.11
Acetyl-CoA synthetase, medium chain (Sa protein)	Beta oxidation of fatty acids?	MAT	6.2.1.1	ACS1	X66425.1	SAH	D16350.1	16p12.3

(continues)

Description	Localization	EC	Gene (Nc/Um)	Accession (Nc/Um)	Human gene	Accession	Chromosome
Acety-CoA synthetase, medium chain (Ac-CoA synthetase 3; MACS1); prefers butyrate. Beta oxidation of fatty acids	MAT	6.2.1.2	ACS1	X66425.1	BUCS1	AB059429.1	16p12.3
Acyl carrier protein (ACP) (also in mammalian complex 1). Component of type II fatty acid synthase	MAT	1.6.5.3	ACP1	Z28192.1	NDUFAB1	AF087660.1	16p12.2
Acyl-CoA synthetase, medium chain, olfactory-specific in rat (O-MACS; LOC341392). Beta oxidn of fatty acids? Ensembl ORF	ND	6.2.1.2	ACS1	X66425.1	[OMACS]	XM_937749.2	12p13.31
Acyl-CoA dehydrogenase, long chain-specific (LCAD). Beta-oxidation of linear fatty acids	MAT	1.3.99.13	NCU08924.1 (Nc)	BX294024.1 (Nc)	ACADL	M74096.1	2q34
Acyl-CoA dehydrogenase, medium chain-specific (MCAD). Beta-oxidation of linear fatty acids	MAT	1.3.99.3	NCU02126.1 (Nc)	XM_959191.1 (Nc)	ACADM	M91432.1	1p31
Acyl-CoA dehydrogenase, short/branched chain fatty acids (SBCAD). Beta-oxdn of branched chain fatty acids	MAT	1.3.99.–	NCU06543.1 (Nc)	XM_326397.1 (Nc)	ACADSB	U12778.1	10q26.13
Acyl-CoA dehydrogenase, short chain-specific (SCAD). Beta-oxidation of linear fatty acids	MAT	1.3.99.2	NCU02126.1 (Nc)	XM_959191.1 (Nc)	ACADS	M26393.1	12q24.31
Acyl-CoA dehydrogenase, very long chain-specific (VLCAD). Beta-oxidation of linear fatty acids	MIM	1.3.99.–	NCU06543.1 (Nc)	XM_326397.1 (Nc)	ACADVL	D43682.1	17p13.1
Acyl-CoA dehydrogenase-8 (isobutyryl-CoA dehydrogenase). Val catabolism; br chain beta oxidation	MAT?	1.3.99.–	NCU06543.1 (Nc)	XM_326397.1 (Nc)	ACAD8	AF126245.1	11q25
Acyl-CoA dehydrogenase-9, similar to VLCAD. Beta-oxidation of fatty acids	MIM?	1.3.9.–	NCU06543.1 (Nc)	XM_326397.1 (Nc)	ACAD9	AF27351.1	3q21.3
Acyl-CoA dehydrogenase-10 (3 alt spl forms). Beta-oxidation of fatty acids	MAT?				ACAD10	AY323912.1	12q24.12
Acyl-CoA desaturase 1 (no E. coli homolog). Relocalizes from ER->mitochondria	ER-MAM	1.14.99.5	OLE1	J05676.1	SCD	AB032261.1	10q24.31
Acyl-CoA synthetase, very long chain (= fly Bubblegum/lipidosin) [7]. Beta-oxidation of fatty acids	MOM	6.2.1.3	FAA1	X66194.1	BG1	AF179481.1	15q25.1
Acyl-CoA thioesterase. Activation of fatty acids	ND	3.1.2.2	UM06364.1 (Um)	XM_403979.1 (Um)	ACATE2	AF132950.1	Xp22.11
Acyl-CoA thioesterase, very long chain-specific (MTE1). Beta-oxidation of fatty acids	MAT	3.1.2.2	TES1	AF124265.1	PTE2A	AK001939.1	14q24.3
Acyl-CoA thioesterase, type II, mitochondrial alt spl isoforms. Lipid metabolism	ND	3.1.2.2	UM06364.1 (Um)	XM_403979.1 (Um)	BACH	AB074417.1	1p36.31
Acyl-CoA thioesterase, brown fat-inducible (BFIT), potentially mitochondrial. Acyl-CoA thioesterase activity	NC	3.1.2.–			THEA	AF416921.1	1p32.3
Alcohol acyl transferase [7]. Medium-chain fatty acid ethyl ester syn	MOM		EHT1	AB012577.1			
Alpha-methylacyl-CoA racemase (peroxisomal/mitochondrial alt spl forms). Beta-oxidation of fatty acids	ND	5.1.99.4			AMACR	AK022765.1	5p13.2
Beta-ketoacyl-ACP synthase (3-oxoacyl-ACP synthase). Component of type II fatty acid synthase	MAT?	2.3.1.41	CEM1	X73488.1	OXSM	BC008202.1	3p24.2
Cytochrome b5, mitochondrial isoform, targeted via N-myristoylation. Fatty acid/sterol syn; androgen syn	MOM		CYB5	L22494.1	CYB5M	AB009282.1	16q22.1
Cytochrome b5 reductase (diaphorase). Desaturation of fatty acids	MOM	1.6.2.2	CBR1	Z28365.1	DIA1	Y09501.1	22q13.2
Cytochrome b5 reductase, IMS form, 32 kDa. Electron transfer to cyt c	IMS	1.6.2.2	MCR1	X81474.1			
Cytochrome b5 reductase, MOM form, 34 kDa. Reduces cytochrome b5	MOM	1.6.2.2	MCR1	X81474.1			
Delta3,5-delta2,4-dienoyl-CoA isomerase (HPXEL). Beta-oxidation of fatty acids	MAT?	5.3.3.–	NCU06647.1 (Nc)	XM_326932.1 (Nc)	ECH1	U16660.1	19q13
2,4-Dienoyl-CoA reductase. Beta-oxidation of unsaturated fatty acids	MAT?	1.3.1.34	UM01935.1 (Um)	XM_399550.1 (Um)	DECR1	L26050.1	8q21.3
Dodecenoyl-CoA delta isomerase (3,2-trans-enoyl-CoA isomerase). Beta-oxidation of unsaturated fatty acids	MAT	5.3.3.8			DCI	Z25820.1	16p13.3
Electron transfer flavoprotein, subunit alpha (ETF-alpha). Electron transfer from FAD	MAT		YPR004c	Z71255.1	ETFA	J04058.1	15q24.3
Electron transfer flavoprotein, subunit beta. Electron transfer from FAD	MAT		ETF-beta	Z72992.1	ETFB	X71129.1	19q13.3
Enoyl CoA hydratase, short chain 1. Beta-oxidation of fatty acids	MAT	4.2.1.17	EHD3	Z74332.1	ECHS1	D13900.1	10q26.3
Enoyl-CoA hydratase domain containing 1, potentially mito. Function unknown	NC		UM01935.1 (Um)	XM_399550.1 (Um)	ECHDC1	BC003549.1	6q22.33
Enoyl-CoA hydratase domain containing 2, potentially mito. Function unknown	NC		NCU09058.1 (Nc)	XM_331449.1 (Nc)	ECHDC2	BC044574.1	1p32.3
Fatty acid amide hydrolase, similar to chicken mitochondrial VDHAP [9]. Amidase; deg bioactive fatty acids	MOM		NCU08356.1 (Nc)	XM_329401.1 (Nc)	FAAH	U82535.1	1p33
Fatty acid-binding protein, adipocyte. Fatty acid/retinoic acid transp	ND				FABP4	J02874.1	8q21.13
Fatty acid-binding protein, muscle. Fatty acid transp	ND				FABP3	X56549.1	1p35.2
Fatty acid-CoA ligase, long-chain 1 (FACL1/2) [12]. Beta-oxidation of fatty acids	MIM	6.2.1.3	FAA2	X77783.1	ACSL1	D10040.1	4q35.1
Fatty acid-CoA ligase, long-chain 3 (FACL3). Beta-oxidation of fatty acids	MOM	6.2.1.3	NCU02291.1 (Nc)	XM_331066.1 (Nc)	ACSL3	D89053.1	2q35
Fatty acid-CoA ligase, long-chain 4 (FACL4) (alt spl forms) [12]. Beta-oxidation of fatty acids	ER-MAM	6.2.1.3	NCU08354.1 (Nc)	XM_329399.1 (Nc)	ACSL4	AF030555.1	Xq23
Fatty acid-CoA ligase, long-chain 5 (FACL5) (alt spl forms). Beta-oxidation of fatty acids	MOM	6.2.1.3	NCU09058.1 (Nc)	XM_331449.1 (Nc)	ACSL5	AB033899.1	10q25.2
Fatty acid-CoA ligase, long-chain 6 (FACL6) (alt spl forms). Beta-oxidation of fatty acids	MOM	6.2.1.3	NCU06448.1 (Nc)	XM_326302.1 (Nc)	ACSL6	AF099740.1	5q31.1
Glutaryl-CoA dehydrogenase (GCD). Lys/Trp metab; br chain beta-oxidation	MAT	1.3.99.7			GCDH	U69141.1	19p13.2
Hypothetical protein [9]. Short chain dehydrogenase family	MOM						
Hypothetical protein, sim to EHHADH, aa 50-270, potentially mito. Beta-oxidation of fatty acids?	NC	4.2.1.17			EHHADH	L07077.1	3q27.2
Hypothetical protein, sim to EHHADH, aa 10-190, potentially mito. Beta-oxidation of fatty acids?	NC	4.2.1.17	NCU09553.1 (Nc)	XM_329911.1 (Nc)	EHHADH	L07077.1	3q27.2
Hypothetical protein, sim to EHHADH, aa 290-460, potentially mito. Beta-oxidation of fatty acids?	NC				EHHADH	L07077.1	3q27.2
2-Enoyl thioester reductase (enoyl reductase). Component of type II fatty acid synthase	MAT		ETR1	D26606.1	MECR	BC001419.1	1p35.2

Appendix 5 *(continued)*

Gene product	Function; comment	Location	E.C. #	Fungal		Animal		
				Symbol	GenBank	Symbol	GenBank	Human Chr.
3-Hydroxyacyl-CoA dehydrogenase, short chain (SCHAD)	Beta-oxidation of fatty acids	MAT	1.1.1.35	NCU08058.1 (Nc)	XM_960096.1 (Nc)	HADHSC	AF095703.1	4q25
3-Hydroxyacyl-thioester dehydratase [5]	Fatty acid biosynthesis	ND		HTD2	P38790			
Isovaleryl-CoA dehydrogenase (IVD)	Leu deg; br chain beta oxidation	MAT	1.3.99.10	NCU02126.1 (Nc)	XM_959191.1 (Nc)	IVD	M34192.1	15q15.1
3-Ketoacyl-CoA thiolase (acetyl-CoA acyltransferase 2)	Beta-oxidation of fatty acids	MIM	2.3.1.16			ACAA2	D16294.1	18q21.1
Malonyl-CoA:acyl carrier protein transacylase (malonyl transferase)	Component of type II fatty acid synthase	ND	2.3.1.39	MCT1	Z75129.1	MT	AL359401.1	22q13.2
Malonyl-CoA decarboxylase	Regulates fatty acid metab	MAT?	4.1.1.9			MLYCD	AF090834.1	16q24
Molybdenum cofactor sulfurase C-term domain containing protein 2 (HMCS-CT)	Hydroxylamine metab w Cytb5/Cytb5 reduct	MOM?				MOSC2	BC016859.1	1q41
Oxidoreductase, hypothetical, ER, also mito	Short-chain dehydrogenases/reductases	ND		YBR159w	Z36028.1			
3-Oxoacyl-[ACP] reductase (beta-ketoacyl reductase)	Component of type II fatty acid synthase	MAT?	1.-.-.-	OAR1	Z28055.1	HEP27	U31875.1	14q11.2
Thioesterase-like protein [3] (06l00l2H03Rik) [3]	Function unknown	ND				None (m)	AK002603.1 (m)	11p13
Triacylglycerol lipase (lipase 2)	Triglyceride lipolysis	ND	3.1.1.3	TGL2	X98000.1			
Trifunctional protein, alpha subunit (3-hydroxyacyl-CoA dehydrogenase [HCAD])	Beta-oxidation of fatty acids; in nucleoid	MAT?	1.1.1.35			HADHA	D16480.1	2p23.3
Trifunctional protein, beta subunit (3-ketoacyl-CoA thiolase)	Beta-oxidation of fatty acids	MAT?	2.3.1.16			HADHB	D16481.1	2p23
Phospholipid metabolism								
Acylglycerol kinase (AGK; MULK [multiple substrate lipid kinase])	Lysophosphatidic acid (LPA) and PA syn	ND				MULK	AJ278150.1	7q34
Cardiolipin synthase	Cardiolipin syn	MIM	2.7.8.-	CRD1	Z74190.1	CRLS1	AL035461.11	20p12.3
CDP-diacylglycerol synthase	Phospholipid syn	MIM/MOM	2.7.7.41	CDS1	Z35898.1	CDS2	Y16521.1	20p13
CDP-diacylglycerol synthase (highly expressed in retina)	Phospholipid syn	MIM/MOM	2.7.7.41			CDS1	U60808.1	4q21.1
Diacylglycerol cholinephosphotransferase	Phospholipid syn	MOM?	2.7.8.2	CPT1	J05203.1			
Glycerol-3-phosphate acyltransferase (GPAT)	Phospholipid syn	ND	2.3.1.15			GPAM	BC030783.1	10q25.2
Glycerophosphate acyltransferase beta	Phospholipid syn	ND	2.3.1.51	YBR042c	Z35911.1	AGPAT2	AF000237.1	9q34.3
Glycerophosphate acyltransferase epsilon, potentially mito	Phospholipid synthesis	NC	2.3.1.51	YOR022c	Z74930.1	AGPAT5	AK002072.1	8p23.1
Hypothetical protein, potentially mitochondrial [5]	Phospholipase activity	ND		SCS3	D21200.1			
Inositol phospholipid synthesis protein, putatively mitochondrial	Function unknown	ND		NCU06460.1 (Nc)	XM_952258.1 (Nc)			
Lysophosphatidic acid (LPA) phosphatase (LPAP)	Triacylglycerol metab	ND				LPAP	BC009965.1	1q21.1
Lysophospholipase 1 (not clear if YLR118c is mitochondrial)	Phospholipid metab	ND	3.1.1.5	YLR118c	X89514.1	LYPLA1	AF081281.1	8q11.23
NADPH-dependent 1-acyl dihydroxyacetone phosphate reductase [7]	Phosphatidic acid biosynthesis from DHAP	MOM	1.1.1.101	AYR1	Z46833.1			
Phosphatidylcholine transfer protein (mito translocation in endothelial cells)	Phosphatidylcholine transp	ND				PCTP	AF151638.1	17q22
Phosphatidylethanolamine binding protein (not clear if Tfs1 is mito)	Inhibits RAF1	ND		TFS1	X62105.1	PBP	D16111.1	12q24.23
Phosphatidylethanolamine N-methyltransferase, isoform 2 (PEMT2)	Phosphatidylcholine syn (also in ER)	ER-MAM		PEM1	M16987.1	PEMT	NM_148173.1	17p11.2
Phosphatidylglycerolphosphate (PG-P) synthase	Cardiolipin syn	MOM	2.7.8.8	PGS1	AJ012047.1	PGS1	BC025951.1	17q25.3
Phosphatidylinositol synthase	Phospholipid syn	MOM	2.7.8.11	PIS1	J02697.1	PIS1	AF014807.1	16p11.2
Phosphatidylserine decarboxylase 1	Phospholipid syn	MIM	4.1.1.65	PSD1	L20973.1	PISD	AL050371.1	22q12.2
Phosphatidylserine synthase	Phospholipid syn	ER-MAM	2.7.8.8	CHO1	X05944.1	PTDSS1	D16494.1	9q22.3
Phosphatidylserine synthase 2	Phospholipid syn	ER-MAM	2.7.8.-			PTDSS2	BC001210.1	11p15.5
Phospholipase A2, group IB	Phospholipid metab	MIM	3.1.1.4			PLA2G1B	M21054.1	12q24.23
Phospholipase A2, group IIA	Phospholipid metab	MIM	3.1.1.4			PLA2G2A	M22430.1	1p36.13
Phospholipase A2, group IVb (phospholipase A2 beta), mito isoform PLA2beta3	Phospholipid metab	ND	3.1.1.4			PLA2G4B	AF065215.1	15q15.1
Phospholipase A2, group IVC	Phospholipid metab	ND	3.1.1.4			PLA2G4C	AF058921.1	19q13.32
Phospholipase A2, group V	Phospholipid metab	MIM	3.1.1.4			PLA2G5	U03090.1	1p36.13
Phospholipase A2, calcium-independent	Phospholipid metab	ND	3.1.1.4			PLA2G6	AF064594.1	22q13.1
Phospholipase C; (delta)1 (not sure if yeast Plc1 is mito)	Phospholipid metab; reg mtCa uptake	ND	3.1.4.11	PLC1	D12738.1	PLCD1	U09117.1	3p22.3
Phospholipase D (CG12314) (MitoPLD)	Required for mito fusion; hydrolyzes cardiolipin	MOM				LOC201164	BC031263.1	17p11.2
Tafazzin (mutated in Barth syndrome)	Cardiolipin phospholipid acyltransferase	MOM+MIM		TAZ1	U40829.1	TAZ	X92763.1	Xq28

Sphingolipid/glycolipid metabolism

Name	Function	Ortholog	EC	Loc	Ortholog acc.	Gene	Accession	Chr
Alpha-4-galactosyltransferase, potentially mitochondrial	Glycosphingolipid syn	—	—	NC		A14GALT	AB037883.1	22q13.31
Beta 1,3 N-acetylglucosaminyltransferase 5, potentially mitochondrial	Glycolipid syn	—	—	NC		B3GNT5	AB045278.1	3q27.1
Beta-1,4 N-acetylgalactosaminyltransferase (GALGT)	Ganglioside glycosylation	—	2.4.1.92	ER-MAM		SIAT2	M83651.1	12q13.3
Ceramide glucosyltransferase (glucosylceramide synthase)	Glycosphingolipid syn, 1st step	—	2.4.1.80	ER-MAM		UGCG	D50840.1	9q31.3
Galactoside alpha(2-3) sialyltransferase	Ganglioside glycosylation	—	2.4.99.4	ER-MAM		SIAT4	L29555.1	8q24.22
Galactoside alpha(2-6) sialyltransferase	Ganglioside glycosylation	—	2.4.99.1	ER-MAM		SIAT1	X17247.1	3q27.2
Ganglioside GD3 synthase, potentially mitochondrial	Ceramide-induced apoptosis	—	2.4.99.8	NC		SIAT8A	X77922.1	12p12.3
Ganglioside D3 (GD3)	Translocates to mitos	—	—	NC				
Inositol phosphosphingolipid phospholipase C (neutral sphingomyelinase)	Translocates fr ER to mitos in log phase	ISC1	3.1.4.—	ND	U18778.1	SMPD2	AJ222801.1	6q21
Lactosylceramide synthase (glucosyl-ceramide-galactosyltransferase)	Glycosphingolipid syn, 2nd step	—	2.4.1.—	ER-MAM		B4GALT6	AF038664.1	18q12.1
Sialidase 4 (N-acetyl-alpha-neuraminidase 4)	Glycosphingolipid metab	—	3.2.1.18	ND		NEU4	AJ277883.1	2q37.3
UDP-glucose:dolichylmonophosphate glucosyltransferase	Glycosylation	—	—	MOM		UGCGL1	BC041098.1	2q14.3

NUCLEIC ACID METABOLISM

Nucleotide and phosphate metabolism

Name	Function	Ortholog	EC	Loc	Ortholog acc.	Gene	Accession	Chr
Adenine deaminase (dihydropyrimidinase related protein-3)	Adenine metab	—	3.5.4.2	ND		DPYSL3	D78014.1	5q32
Adenylate kinase 2 (ATP:AMP phosphotransferase)	Adenine nucleotide metab	ADK2	2.7.4.3	IMS	X65126.1	AK2	U34371.1	1p35.1
Adenylate kinase 3 (GTP:AMP phosphotransferase)	Function unknown	ADK2	2.7.4.10	MAT	X65126.1	AK3	AB021870.1	9p24.1
Adenylate kinase 4 (GTP:AMP phosphotransferase)	Adenine nucleotide metab	ADK2	2.7.4.10	MAT	X65126.1	AK4	X60673.1	1p31.3
ADP-ribose pyrophosphatase (nudix hydrolase)	Hydrolyzes ADP ribose	YSA1	3.6.1.13	NC	Z35980.1	NUDT9	AY026252.1	4q22
NUDT9-binding protein C17orf25, alt spl of CGI-150 (similar to glyoxalase)	Function unknown	—	—	ND		CGI-150	AF177342.1	17p13.3
Creatine kinase, mitochondrial 1 (ubiquitous)	High-energy phosphate transfer	—	2.7.3.2	IMS		CKMT1	J04469.1	15q15
Creatine kinase, mitochondrial 2 (sarcomeric)	High-energy phosphate transfer	—	2.7.3.2	IMS		CKMT2	J05401.1	5q13.3
2',3'-cyclic nucleotide-3'-phosphodiesterase (CNP), mito isoform (CNP2)	Linearizes cyclic nucleosides	—	3.1.4.37	ND		CNP	BC001362.2	17q21.2
Deoxyguanosine kinase	Purine syn	—	2.7.1.113	MAT		DGUOK	U41668.1	2p13.1
5'(3')-Deoxyribonucleotidase (dNT2)	Thymine/uracil metab	—	3.1.3.5	ND		NT5M	AF210652.1	17p11.2
Dihydroorotate dehydrogenase (cyto in S. cerevisiae; mito in S. pombe)	Pyrimidine syn	URA3 (Sp)	1.3.3.1	MIM	X65114.1 (Sp)	DHODH	M94065.1 (i)	16q22.2
Exopolyphosphatase, mitochondrial isoform (Ppx1a)	Polyphosphate deg	PPX1	3.6.1.11	MAT	L28711.1			
Hypothetical protein [9], sim to Sc Adk1/YDR226w	Adenylate kinase	NCU01550.1 (Nc)	—	MOM	XM_327988.1 (Nc)			
Inorganic pyrophosphatase	Phosphate metab	PPA2	3.6.1.1	MIM?	M81880.1	PPA2	AF217187.1	4q24
NAD(P) transhydrogenase (nicotinamide nucleotide transhydrogenase)	Redox of NAD/NADP	NCU01140.1 (Nc)	1.6.1.1	MIM	XP_326633.1 (Nc)	NNT	U40490.1	5p12
Nucleoside diphosphate kinase D (nm23-H4)	Non-ATP NTP syn	YNK1	2.7.4.6	MIM	D13562.1	NME4	Y07604.1	16p13.3
Nucleoside diphosphate kinase 6 (nm23-H6)	Non-ATP NTP syn	YNK1	2.7.4.6	IMS?	D13562.1	NME6	AF051941.1	3p21.3
Nucleoside diphosphate-linked moiety X motif 6, alt sp form, potentially mito	Nudix hydrolase	—	—	NC		NUDT6	BC009842.2	4q27
Nucleoside monomphosphate kinase (UMP-CMP kinase), potential mito form	Pyrimidine nucleoside mono-P kinase	—	2.7.4.14	ND		UCK	AK025258.1	1p33
Regulator of thymidylate synthase (rTS), gamma (mitochondrial) isoform	Down-regulation of thymidylate synthase	—	—	ND		ENOSF1	BC001285.1	18p11.32
Taurocyamine kinase (Arenicola brasiliensis [Ab])	Phosphotaurocyamine syn in invertebrates	—	2.7.3.4	ND		TK (Ab)	AB186412 (Ab)	—
Thymidine kinase	dTTP syn	—	2.7.1.21	MAT?		TK2	Y10498.3	16q22.1

DNA replication

Name	Function	Ortholog	EC	Loc	Ortholog acc.	Gene	Accession	Chr
Acetohydroxy acid reductoisomerase	Nucleoid "parsing"; also Ile/Val syn	ILV5	1.1.1.86	MAT?	X04969.1			
Cruciform cutting endonuclease	Holliday junction resolvase	CCE1	3.1.—.—	MIM	M65275.1			
DNA exonuclease, 3'-5'	DNA editing	REX2	—	MAT	Z73231.1	SFN	AL110239.1	11q23.2
DNA helicase	mtDNA concatamerization	HMI1	3.6.1.—	MIM	Z74837.1			
DNA helicase	DNA unwinding	PIF1	—	MAT	X05342.1	PIF1	DQ437529.1	15q22.31
DNA helicase/primase (Twinkle) (alt spliced Twinkle)	5'-3' helicase; unwinds DNA	—	—	MAT		PEO1	AF292004.1	10q24.31

(continues)

Appendix 5 *(continued)*

Gene product	Function; comment	Location	E.C. #	Fungal Symbol	Fungal GenBank	Animal Symbol	Animal GenBank	Human Chr.
DNA ligase I (yeast)/DNA ligase III (human)	Ligates free DNA ends (BER)	MAT	6.5.1.1	CDC9	X03246.1	LIG3	X84740.1	17q12
DNA polymerase delta interacting protein 38 [3]	Binds DNA Pol delta p50 subunit/PCNA	ND	—	—	—	PDIP38	AF077203.1	17q11.2
DNA polymerase gamma, accessory subunit	Increases processivity	MAT?	2.7.7.7	—	—	POLG2	AF177201.1	17q24.2
DNA polymerase gamma, catalytic subunit	Replication of mtDNA	MAT?	2.7.7.7	MIP1	J05117.1	POLG	U60325.1	15q26.1
DNA polymerase gamma binding protein Sed1	Function unknown	ND	—	SED1	X66838.1	—	—	—
Single-stranded DNA-binding protein	Coats ss-mtDNA	MAT	—	RIM1	S43128.1	SSBP1	M94556.1	7q34
Topoisomerase I	Breaks and rejoins ss-mtDNA	MAT?	—	—	—	TOP1MT	AF349017.1	8q24.3
Topoisomerase II beta-1 (insert 5-aa after Val-23), potentially mitochondrial	Decatenates circles	MAT	5.99.1.3	—	—	TOP2B	X68060.1	3p24.2
Topoisomerase III alpha	Reduces negative supercoils	MAT?	5.99.1.2	—	—	TOP3A	U43431.1	17p11.2
DNA plasticity, recombination, and repair								
3-Alkyladenine DNA glycosylase	Excises 3-meA/7-meG	NC	3.2.2.21	—	—	MPG	M74905.1	16p13.3
Apurinic/apyrimidinic endonuclease (similar to *E. coli* Exo III)	Phosphatase (BER)	MAT?	3.1.11.2	—	—	APEX	M80261.1	14q11.2
Apurinic/apyrimidinic endonuclease (similar to *E. coli* Nfo [Endo IV])	Mismatch repair; binds Pir1p	MAT?	4.2.99.18	APN1	M33667.1	APEX	M80261.1	14q11.2
Breast cancer type 1 susceptibility protein BRCA1, 220-kDa phosphorylated isoform	DNA repair	ND	—	—	—	BRCA1	U14680.1	17q21.31
Deoxycytidyl transferase	Error-prone translesion synthesis (TLS)	ND	2.7.7.-	REV1	M22222.1	—	—	—
DNA endonuclease (mtDNA-encoded)	Helicase	MAT?	—	SCA15-alpha	AJ011856.1	—	—	—
DNA endonuclease I-SceI (21S rRNA) (mtDNA-encoded)	Transposase; intron homing	MAT	—	SCEY	M11280.1	—	—	—
DNA endonuclease I-SceII (aI4) (mtDNA-encoded)	COX1 aI4 intron homing	MAT?	—	SCAI4	V00694.1	—	—	—
DNA endonuclease I-SceIII (aI3) (mtDNA-encoded)	COX1 aI3 intron homing	MAT?	—	SCAI3	V00694.1	—	—	—
DNA endonuclease III (*E. coli* Nth [Endo III])	Removes thymine glycols (BER)	MAT	4.2.99.18	NTG1	L05146.1	NTH1	AC005600.1	16p13.3
DNA endonuclease VIII-like 1	Creates abasic sites (BER)	ND	—	—	—	NEIL1	AB079068.1	15q24.2
DNA endonuclease G-like protein 1, probably mitochondrial	DNA endo/exonuclease	NC	3.1.30.-	NUC1	X06670.1	ENDOGL1	AB020523.1	3p22.1
DNA polymerase zeta catalytic subunit Rev3	Error-prone translesion synthesis (TLS)	ND	2.7.7.7	REV3	M29683.1	—	—	—
DNA polymerase zeta subunit Rev7	Error-prone translesion synthesis (TLS)	ND	—	REV7	U07228.1	—	—	—
DNA-repair protein XRCC9 (Fanconi anemia group G protein)	Binds PRDX3	NC	—	—	—	FANCG	U70310.1	9p13.3
DNA repair protein for interstrand crosslinks, potentially mito	Incises crosslinks	NC	—	PSO2	X64004.1	—	—	—
dUTP pyrophosphatase	Prevents U incorp. in DNA (BER)	MAT?	3.6.1.23	DUT1	X74263.1	DUT	U90223.1	15q21.1
Histone-like HMG protein (*Physarum polycephalum* [*Pp*])	Condenses mtDNA; in nucleoid	MAT	—	ABF2	M73753.1	GLOM (Pp)	AB049419.1 (Pp)	—
Increased Recombination Centers protein (YDR332w), potentially mitochondrial [5]	Similar to DEAD box helicase	ND	—	IRC3	U32517.2	—	—	—
KU80 autoantigen, C-terminal truncated isoform, 68 kDa	DNA end-binding activity	ND	—	—	—	XRCC5	M30938.1	2q35
Maturase/reverse transcriptase aI1 (mtDNA-encoded)	COX1 aI1 intron homing	MAT?	—	SCAI1	L36897.1	—	—	—
Maturase/reverse transcriptase aI2 (mtDNA-encoded)	COX1 aI2 intron homing	MAT?	—	SCAI2	V00694.1	—	—	—
Mitochondrial genome maintenance protein	Function unknown	NC	—	PET18	X59720.1	—	—	—
Mitochondrial genome maintenance protein (in mito nucleoid)	Repairs oxidative damage	MAT	—	MGM101	X68482.1	—	—	—
Mitochondrial homologous recombination protein	Promotes heteroduplex pairing	MAT?	—	MHR1	AB016430.1	—	—	—
Mitochondrial inner membrane nuclease (Exo I)	DNA damage-inducible gene	MIM	—	DIN7	X90707.1	—	—	—
MutM homolog (8 mito isoforms)	8-oxo-G DNA glycosylase (BER)	MAT?	—	MSH1	M84169.1	OGG1	AB019528.1	3p26.2
MutS homolog	Not involved in DNA repair?	MAT?	—	—	—	—	—	—
MutT homolog, p26 isoform	8-oxo-dGTPase (BER)	MIM	3.1.6.-	—	—	MTH1	AB025240.1	7p22
MutY homolog (3 mito isoforms)	A and 2-OH-A glycosylase	MAT	—	—	—	MUTYH	AB032920.1	1p34.1
Photolyase-like protein (cryptochrome 2)	Cuts cyclobutane dimers	MAT?	4.1.99.3	PHR1	X03183.1	CRY2	BC041814.1	11p11.2
Poly(ADP-ribosyl)transferase, mito isoform not clear	Induced by DNA breaks	MIM	2.4.2.30	—	—	PARP1	M18112.1	1q42.12
Protein associated w. mito biogenesis/cell cycle reg	mtDNA maintenance?	MAT?	—	ERV2	Z68111.1	—	—	—
RecA protein	Recombinase; strand exchange	ND	2.7.7.-	—	—	recA (s)	AC116979.3 (s)	—
Telomerase reverse transcriptase (catalytic subunit)	Function unknown	ND	—	—	—	TERT	AF015950.1	5p15.33
Tumor suppressor protein p53	Participates in BER	ND	—	—	—	TP53	M14695.1	17p13.1
Uracil-DNA glycosylase	Removes U from DNA (BER)	MAT?	3.2.2.-	UNG1	J04470.1	UNG1	X15653.1	12q24.13
UV-damaged DNA endonuclease	UV-dependent excision repair	ND	—	UVDE (Sp)	D78571.1 (Sp)	—	—	—

(continues)

Kinetoplast-associated proteins

Protein	Function	Localization	EC	Gene	Accession	Gene	Accession	Chrom.
Cdc2p-related protein kinase 1 (*T. cruzi [Tc]*)	Binds mammalian cyclins	Kinetoplast	—			CRK1 (Tz)	U74762.1 (Tz)	—
DNA ligase kalpha (*C. fasciculata [Cf]*)	Ligates kinetoplast nicks, everywhere	Kinetoplast	—			LIGk alpha	AY380335.2 (Cf)	—
DNA ligase kbeta (*C. fasciculata [Cf]*)	Ligates kinetoplast nicks, antipodal	Kinetoplast	—			LIGk beta	AY380335.1 (Cf)	—
DNA polymerase beta (*C. fasciculata [Cf]*)	Replication of mtDNA	Kinetoplast	2.7.7.7			None	U19912.2 (Cf)	—
Guide RNA binding protein, 27 kDa (*C. fasciculata [Cf]*)	mRNA editing in kinetoplast	Kinetoplast	—			gBP27 (Cf)	AF157559.1 (Cf)	—
Guide RNA binding protein, 29 kDa (*C. fasciculata [Cf]*)	mRNA editing in kinetoplast	Kinetoplast	—			gBP29 (Cf)	AF156855.1 (Cf)	—
Kinetoplast-associated protein p16 (*C. fasciculata [Cf]*)	Histone H1-like; binds kDNA	Kinetoplast	—			KAP4 (Cf)	S56498.1 (Cf,i)	—
Kinetoplast-associated protein p17 (*C. fasciculata [Cf]*)	Histone H1-like; binds kDNA	Kinetoplast	—			KAP3 (Cf)	S56496.1 (Cf,i)	—
Kinetoplast-associated protein p18 (*C. fasciculata [Cf]*)	Histone H1-like; binds kDNA	Kinetoplast	—			KAP2 (Cf)	AF008944.1 (Cf)	—
Kinetoplast-associated protein p21 (*C. fasciculata [Cf]*)	Histone H1-like; binds kDNA	Kinetoplast	—			KAP1 (Cf)	AF03451.1 (Cf)	—
Mitochondrial DEAD BOX protein (*T. brucei [Tb]*)	mRNA editing in kinetoplast	Kinetoplast	—			MHEL61 (Tb)	U86382.1 (Tb)	—
Mitochondrial oligo_U binding protein TBRGG1 (*Trypanosoma cruzi [Tc]*)	RNA editing in kinetoplast	Kinetoplast	—			[TRBGG1] (Tc)	XM_804443.1(Tc)	—
RNA editing ligase (*T. brucei [Tb]*)	RNA editing in kinetoplast	Kinetoplast	6.5.1.3			MP48 (Tb)	AY009111.1 (Tb)	—
RNA editing protein, 16 kDa (*T. brucei [Tb]*)	RNA editing and stability; binds Tb p22	Kinetoplast	—			RBP16 (Tb)	AF042492.1 (Tb)	—
RNA editing protein, 18 kDa (*T. brucei [Tb]*)	RNA editing in kinetoplast	Kinetoplast	—			MP18 (Tb)	AF382336.1 (Tb)	—
RNA editing protein, 24 kDa (*T. brucei [Tb]*)	RNA editing in kinetoplast	Kinetoplast	—			MP24 (Tb)	AY228168.1 (Tb)	—
RNA editing protein, 42 kDa (*T. brucei [Tb]*)	RNA editing in kinetoplast	Kinetoplast	—			MP42 (Tb)	AF382335.1 (Tb)	—
RNA editing protein, 44 kDa (*T. brucei [Tb]*)	RNA editing in kinetoplast	Kinetoplast	—			MP44 (Tb)	AY228170.1 (Tb)	—
RNA editing protein, 46 kDa (*T. brucei [Tb]*)	RNA editing in kinetoplast	Kinetoplast	—			MP46 (Tb)	AY228169.1 (Tb)	—
RNA editing protein, 48 kDa (REL2) (*T. brucei [Tb]*)	RNA editing in kinetoplast	Kinetoplast	—			MP48 (Tb)	AY009111.1 (Tb)	—
RNA editing protein, 52 kDa (REL1) (*T. brucei [Tb]*)	RNA editing in kinetoplast; ligase	Kinetoplast	6.5.1.3			MP52 (Tb)	AY009110.1 (Tb)	—
RNA editing protein, 57 kDa (RET1) (*T. brucei [Tb]*)	RNA editing in kinetoplast	Kinetoplast	—			MP57 (Tb)	AY228173.1 (Tb)	—
RNA editing protein, 61 kDa (*T. brucei [Tb]*)	RNA editing in kinetoplast	Kinetoplast	—			MP61 (Tb)	AY228167.1 (Tb)	—
RNA editing protein, 63 kDa (*T. brucei [Tb]*)	RNA editing in kinetoplast	Kinetoplast	—			MP63 (Tb)	AF382334.1 (Tb)	—
RNA editing protein, 67 kDa (*T. brucei [Tb]*)	RNA editing in kinetoplast	Kinetoplast	—			MP67 (Tb)	AY228166.1 (Tb)	—
RNA editing protein, 81 kDa (*T. brucei [Tb]*)	RNA editing in kinetoplast	Kinetoplast	—			MP81 (Tb)	AF382333.1 (Tb)	—
RNA editing protein, 90 kDa (*T. brucei [Tb]*)	RNA editing in kinetoplast	Kinetoplast	—			MP90 (Tb)	AY228165.1 (Tb)	—
RNA editing protein, 99 kDa (*T. brucei [Tb]*)	RNA editing in kinetoplast	Kinetoplast	—			MP99 (Tb)	AY228172.1 (Tb)	—
RNA editing protein, 100 kDa (*T. brucei [Tb]*)	RNA editing in kinetoplast	Kinetoplast	—			MP100 (Tb)	AY228171.1 (Tb)	—
Structure-specific endonuclease 1 (*C. fasciculata [Cf]*)	Binds kDNA	Kinetoplast	—			SSE1 (Cf)	AF124228.1 (Cf)	—
Terminal uridylyl transferase, 121 kDa (RET1) (*T. brucei [Tb]*)	3' TUTase: gRNA 3' end maturation	Kinetoplast	—			3' TUTase (Tb)	AY029070.1 (Tb)	—
Terminal uridylyl transferase, 57 kDa (RET2) (*T. brucei [Tb]*)	3' TUTase: U-insertion	Kinetoplast	—			MP57 (Tb)	AY228173.1 (Tb)	—
Topoisomerase II (TOP2) (*T. brucei [Tb]*)	Decatenates circles	Kinetoplast	5.99.1.3			[TbTOP2mt] (Tb)	M26803.1 (Tb)	—
Universal minicircle sequence binding protein (*C. fasciculata [Cf]*)	Binds kDNA	Kinetoplast	—			UMSBP (Cf)	AF045246.1 (Cf)	—

RNA transcription, processing, and maturation

Protein	Function	Localization	EC	Gene	Accession	Gene	Accession	Chrom.
Acetylglutamate kinase/acetylglutamyl phosphate reductase	Arg/ornithine syn; transcription factor	MAT?	2.7.2.8	ARG5,6	X57017.1			—
COX1 mRNA splicing protein	Splices COX1 intron al5-beta	MAT?	—	MSS18	X07650.1			—
Cyt b pre-mRNA processing protein	Cyt b mRNA processing	MAT?	—	CBT1	Z28208.1			—
Cyt b pre-mRNA processing protein 1	Cyt b mRNA processing	MIM	—	CBP1	K02647.1			—
Cyt b pre-mRNA processing protein 2	Splices COB al5 intron	MAT?	—	CBP2	K00138.1			—
Dihydrouridine synthase	Modify U on tRNA-Asp (+ others?)	ND	—	SMM1	X91816.1	DUS2L	AK000406.1	16q22.1
G-rich RNA sequence binding factor 1 [2]	Poly(A)+ mRNA binding protein	NC	—			GRSF1	U07231.1	4q13.3
Heterogeneous nuclear ribonucleoprotein (hnRNP) K protein, alt spl forms	Binds poly(C) in mRNAs	MAT	—			HNRPK	S74678.1	9q21.32
Maturation protein of pre-rRNA	RNA processing	MIM	—	YME2	S92205			—
3-Methylglutaconyl-CoA hydratase (AU-specific RNA-binding protein)	Leu metab; may direct mt-RNA deg	ND	4.2.1.18	EHD3	Z74332.1	AUH	X79888.1	9q22.2
Mitochondrial protein al5-beta (mtDNA-encoded)	Function unknown	MAT?	—	SCA15-beta	AJ011856.1			—
mRNA maturase bi2 for Cyt b intron (mtDNA-encoded)	Splices COB bi2 intron	MAT	—	SCB12	AJ011856.1			—
mRNA maturase bi3 for Cyt b intron (mtDNA-encoded)	Splices COB bi3 intron	MAT	—	SCB13	AJ011856.1			—
mRNA maturase bi4 for Cyt b intron (mtDNA-encoded)	Splices COB bi4 intron	MAT	—	SCB14	AJ011856.1			—
Nuclear cleavage/polyadenylation factor I, component	pre-mRNA 3' end processing	ND	—	RNA14	M73461.1			—
Poly(A) binding protein 2, potentially mitochondrial	Polyadenylates mt-RNAs?	NC	—			PABP2	AF026029.1	14q11.2

Appendix 5 (continued)

				Fungal		Animal		
Gene product	Function; comment	Location	E.C. #	Symbol	GenBank	Symbol	GenBank	Human Chr.
Poly(A) polymerase	Adds poly(A) to mt-mRNAs; for stability	MAT?	—	—	—	PAPD1	AK022188.1	10p11.23
Polynucleotide phosphorylase	3′-5′ exoribonuclease; stabilizes mt-mRNAs	IMS	2.7.7.8	—	—	PNPT1	AJ458465.1	2p16.1
Pre-mRNA branch site protein p14, potentially mitochondrial	May bind RNA	NC	—	—	—	SF3B14	AF151868.1	2p23.3
Ribonuclease II	RNA degradosome	MAT?	2.7.7.8	MSU1	U15461.1	MGC4562	AK095407.1	15q22.31
Ribonuclease UK114 (heat-responsive protein 12) [3]	Inhib trans by cleaving mRNA	ND	—	MMF1	Z38060.1	HRSP12	X95384.1	8q22.2
RNA end formation protein 2	mRNA 3′ end maturation	ND, Nuc?	—	REF2	U20261.1	—	—	—
RNA helicase (DEAD box) (ATP-dependent)	RNA degradosome	MAT	—	SUV3	M91167.1	SUV3	AF042169.1	10q22.1
RNA helicase (DEAD box)	Group II intron splicing in yeast	MAT	—	MSS116	Z48784.1	DDX18	AK001467.1	2q14.1
RNA helicase (DEAD box), potentially mitochondrial	Function unknown	NC	—	—	—	DDX46	AF106680.1	5q31.1
RNA helicase (DEAD box)	Function unknown	ND	—	—	—	DDX28	AF329821.1	16q22.1
RNA helicase (DEAD box)	RNA metab	MAT?	—	MRH4	Z72586.1	—	—	—
RNA polymerase II	Transcription	MAT?	2.7.7.6	RPO41	M17539.1	POLRMT	U75370.1	19p13.3
RNase H [3]	RNA dg	NC	3.1.26.4	—	—	RNASEH1	AF039652.1	2p25.3
RNase L (2-5A-dependent ribonuclease)	Reg mRNA stability	NC	3.1.26.-	—	—	RNASEL	L10381.1	1q25.3
RNase MRP processing protein, 15.8 kDa, potentially mitochondrial	Not clear	NC	—	POP7	Z36036.1	—	—	—
RNase MRP processing protein, 22.6 kDa, potentially mitochondrial	Not clear	NC	—	POP3	X95844.1	—	—	—
RNase MRP processing protein, 100 kDa, potentially mitochondrial	In RNase P and RNase MRP	NC	3.1.26.5	POP1	X80358.1	POP1	D31765.1	8q22.2
RNase MRP, protein component	Endoribonuclease	MAT	—	SNM1	Z37982.1	—	—	—
RNase MRP, RNA component	Endoribonuclease	MAT	—	NME1	Z14231.1	RMRP	X51867.1	9p13.2
RNase P, protein component	5′ processing of mt-tRNAs	MAT?	3.1.26.5	RPM2	L06209.1	RPP38	U77664.1	10p13
RNase P, RNA component (mtDNA-encoded 9S RNA in yeast)	5′ processing of mt-tRNAs	MAT?	—	RPM1	U46211.1	[H1 RNA]	X15624.1	14q11.2
RNase Z (tRNA 3′ processing endonuclease, potentially mito (Sc Trz1 prob not mito)	Processing of 3′ ends of mt-tRNAs	NC	—	SPAC1D4.10 (Sp)	Z69239 (Sp)	ELAC2	BC004158.1	17p12
RNA splicing protein	Splices COB intron bI3	MAT?	—	MRS1	X05509.1	—	—	—
rRNA (guanosine-2′-O)-methyltransferase	Methylates G-2270 in 21S rRNA	MAT?	2.1.1.-	PET56	L19947.1	MRM1	BC072411.1	17q12
rRNA methyltransferase	Methylates U-2791 of 21S rRNA	MAT	2.1.1.-	MRM2	Z72658.1	FJH1	AF093415.1	7p22.2
rRNA pseudouridine synthase	Pseudouridine-2819 in 21S rRNA	MAT?	4.2.1.70	PUS5	U51921.1	RPUSD2	XM_031554.5	15q15.1
tRNA splicing endonuclease, Sen2 subunit (catalytic)	Processing of cytosolic tRNAs	MOM (CTS)	3.1.27.9	SEN2	M32336.1	—	—	—
tRNA splicing endonuclease, Sen54 subunit	Processing of cytosolic tRNAs	MOM (CTS)	3.1.27.9	SEN54	U41849.1	—	—	—
Transcription factor A (TFAM: MTTFA)	Mito transcription initiation	MAT	—	ABF2	M73753.1	TFAM	M62810.1	10q21.1
Transcription factor B, isoform 1	12S rRNA adenine methyltransferase	MAT	—	MTF1	X13513.1	TFB1M	AF151833.1	6q25.3
Transcription factor B, isoform 2	Mito transcription initiation	MAT	—	MTF1	X13513.1	TFB2M	AK026835.1	1q44
Transcription factor 2 (NAM1)	Mito transcript stability	MAT	—	MTF2	X14719.1	—	—	—
Transcription termination factor	Terminates at 16S rRNA	MAT	—	—	—	MTERF	Y09615.1	7q21.2
Transcription termination factor-like protein (MTERFL; LOC80298)	Function unknown	ND	—	—	—	MTERFD3	AY008301.1	12q23.3
ORGANELLAR MORPHOLOGY AND INHERITANCE								
Actin-associated proteins								
Actin	Rescues mdm20 mutations	MOM?	—	ACT1	V01288.1	ACTB	M10277.1	7p22.1
Actin-related protein Arp2 of Arp2/3 complex	Dynamic actin filament assembly	MOM?	—	ARP2	X61502.1	ARP2	AF006082.1	2p14
Actin-related protein Arc15 of Arp2/3 complex, 17 kDa	Mito movement on actin cables	MOM?	—	ARC15	Z38060.1	ARPC5L	BC000798.1	9q34.11
Actin-related protein Arc18 of Arp2/3 complex, 20 kDa	Dynamic actin filament assembly	MOM?	—	ARC18	U19103.1	ARPC3	AF006086.1	12q24.11
Actin-related protein Puf1 of Arp2/3 actin-organizing complex	Actin assembly and function	MOM	—	JSN1	L43493.1	—	—	—
Actin-related protein Puf3 of Arp2/3 actin-organizing complex	Binds cyto mt-mRNAs; links Arp2/3 to mitochore	MOM	—	PUF3	Z73118.1	—	—	—
GTP-binding protein Ypt11	Mito distribution to bud; binds Myo2	MOM?	—	YPT11	Z71580.1	—	—	—
Mitochondrial distribution and morphology protein Mdm10	Binds mitos to actin w. Mdm12/Mmm1	MOM	—	MDM10	X80874.1	—	—	—
Mitochondrial distribution and morphology protein Mdm12	Binds mitos to actin w. Mdm10/Mmm1	MOM	—	MDM12	U62252.1	—	—	—
Mitochondrial inheritance factor Mmr1	Mito distribution to bud; binds Myo2	MOM?	—	MMR1	U17246.1	—	—	—
Myosin, class V	Mito distribution to bud; binds Ypt11	MOM?	—	MYO2	M35532.1	MYO5A	Y07759.1	15q21.2

Protein	Function	Yeast gene	Accession	Loc	Human gene	Accession	Locus
Myosin, class VI (unconventional myosin)	Molecular motor protein	MLP1	L01992.1	NC	MYO6	U90236.2	6q14.1
Nuclear envelope spectrin repeat protein 2 (Nesprin 2); 9 isoforms	Linking network between organelles			ND	SYNE2	AF435011.1	14q23.2
	Binds mitos at contact sites to actin w. Mdm10/12	MMM1	L32793.1	MOM+MIM		—	—
Protein associated w. mito shape and structure (Mmm2)	Req. for nucleoids	MDM34	Z72741.1	MOM		—	—
Reduced viability upon starvation protein 167	Binds actin	RVS167	M92092.1	ND		—	—
Autophagy-related proteins							
Autophagy related protein 9 (cytoplasm to vacuole targeting protein 7)	Cycles betw mitos & autophagosomes	ATG9	Z74197.1	ND	APG9L1	BX537984.1	2q35
Autophagy related protein 9 (cytoplasm to vacuole targeting protein 7)	Cycles betw mitos & autophagosomes	ATG9	Z74197.1	ND	NOS3AS	AY515311.1	7q36.1
Autophagy protein 5, 24-kDa N-term cleavage product	Pro-apoptotic protein; translocates to mitos			MOM?	ATG5	BC002699.2	6q21
Cyclin-dependent kinase inhibitor 2A (p19ARF), short mitochondrial form (smARF)	Autophagy & caspase-independent cell death			MIM?	CDKN2A	U26727.1	9p21.3
Protein phosphatase 2C-like protein	Mitochondrial autophagy (mitophagy)	PTC6	X59720.2	IMS	PPM1K	BC037552.1	4q22.1
"Youth" protein Uth1p	Autophagic degradation of mitochodria	UTH1	Z28267.1	MOM		—	—
Microtubule-associated proteins							
ADP-ribosylation-like factor 2 GTPase	Binds ANT1/BART1/PP2A	CIN4	L36669.1	IMS	ARL2	L13687.1	11q13.1
Binder of ARL2, arf-like 2 binding protein BART1	Binds mouse ANT1/ARL2			IMS	BART1	AF126062.1	16q13
Dynactin 1	Dynein-dynactin complex			ND	DCTN1	AF064205.1	2p11.3
Dynein heavy chain (MDHC7) (ciliary dynein, axonemal, heavy chain 7 [DNHAC7])	Attaches to mitos	USO1	X54378.1	NC	DNAH7	AF327442.1	2q32.3
Dynein light chain (Tctex1)	Binds VDAC1			MOM?	TCTEL1	D50663.1	6q25.3
Gamma-SNAP (only a fraction is associated w. mitos)	Vesicular transport			MOM	NAPG	U78107.1	18p11.22
Gamma-SNAP associated factor-1 (GAF1; RIP11A; RAB11 family interacting protein 5)	Binds gamma-SNAP/gamma-tubulin			MOM	RAB11FIP5	AF334812.1	2p13.2
HIV transactivating regulatory protein Tat	Microtubule polymerization; pro-apoptotic			ND	TAT	M11840.1 (Hiv)	—
KIF1 binding protein (KFB)	Binds KIF1Balpha; regulates mito loc			ND	KIAA1279	BC012180.1	10q22.1
Kinectin (kinesin receptor), 120-kDa alt spl mitochondrial isoform	Binds kinesin			MOM?	KTN1	AY264265.1	14q22.3
Kinesin-associated mito adaptor (trafficking kinesin-binding protein 1) (Milton1)	Binds kinesin and mitos; binds Miro, 2 isoforms			MOM?	TRAK1	BC015922.1	3p22.1
Kinesin-associated mito adaptor (trafficking kinesin-binding protein 2) (Milton2; GRIF1)	Binds kinesin and mitos; binds Miro			MOM?	TRAK2	BC048093.1	2q33.1
Kinesin heavy chain KIF5B (kinesin-1 family)	Microtubule motor protein			MOM	KIF5B	X65873.1	10p11.23
Kinesin light chain 2-like, isoform b (sim to rat kinesin light chain 3)	Binds spermatid mitos in rat			MOM?	KLC2L	BC07841.1	19q13.32
Kinesin light chain, isoform C (= rat isoform B), mito-specific alt spl isoform	Microtubule motor protein			MOM	KNS2	L04733.1	14q32.33
Kinesin-like protein KIF1B, alpha isoform (kinesin-3 family)	Microtubule motor protein			MOM	KIF1B	AX039604.1	1p36.22
Kinesin-like protein KIF1C (NcKin3) (kinesin-3 family)	Binds mitos; dimeric non-processive motor	1TE5.250 (Nc)	AL513467.1 (Nc)	MOM?		—	—
Kinesin-like protein Cin8[1]	Assembly of mitotic spindle	CIN8	Z11859.1	ND		—	—
Kinesin-like protein Kip2[1]	Assembly of mitotic spindle; nucl migr	KIP2	Z11963.1	ND		—	—
Leucine zipper putative tumor suppressor 1 (FEZ1), potentially mitochondrial	Function unknown			NC	LZTS1	AF123659.1	8p21.3
Microtubule and mitochondria interacting protein (Mmi1; YKL056c) (NC if TP1 is mito)	Translocates from cyto->mito	MMI1	Z28056.1	MOM	TPT1	X16064.1	13q14.12
Microtubule-associated protein 1B	Reg mito axonal retrograde transp in devel			MOM?	MAP1B	L06237.1	5q13.2
Microtubule interacting protein 1S/BPY2 interacting protein 1 (C19orf5)	Binds microtubules/dsDNA/RASSF1/LRPPRC			NC	BPY2IP1	BC080547.1	19p13.11
Mitochondrial rho GTP-binding protein Miro-1 (Arht1)	Anterograde mito axonal transp; binds Milton	GEM1	U12980.1	MOM	RHOT1	AJ517412.1	17q11.2
Mitochondrial rho GTP-binding protein Miro-2 (Arht2)	Anterograde mito axonal transp; binds Milton	GEM1	U12980.1	MOM	RHOT2	AJ517413.1	16p13.3
Ras association (RALGDS/AF-6) domain family 1 protein, isoforms A and C	Binds microtubules, LRPPRC, BPY2IP1			NC	RASSF1	AF132675.1	3p21.31
Syntabulin (STB) (microtubule-associated protein (FLJ20366)	Links mitos to kinesin-1; syntaxin transp			MOM	SYBU	AY705385.1	8q23.2
Spartin	Binds microtubules			MOM?	SPG20	AY123329.1	13q13.3
Tubulin-alpha	Binds VDAC	TUB3	M28428.1	MOM?	TUBA1	K00558.1	12q13.12

(continues)

Appendix 5   (continued)

Gene product	Function; comment	Location	E.C. #	Fungal		Animal		
				Symbol	GenBank	Symbol	GenBank	Human Chr.
Tubulin-beta	Binds VDAC	MOM?	–	TUB2	V01296.1	TUBB1	J00314.1	20q13.32
Tubulin-gamma1	Binds GAF-1 and gamma-SNAP	MOM	–	–		TUBG1	BC000619.1	17q21.1
Tubulin-gamma2	Binds GAF-1 and gamma-SNAP	MOM	–	–		TUBG2	AF225971.1	17q21.1
Mitochondrial shape and structure								
Apoptosis inducing factor (AIF); alt spl forms	Pro-apoptotic protein; mito morphology	IMS	–	–		PDCD8	AF100928.1	Xq25
ATPase subunit e (synonym ATP21 in yeast)	Cristae morphology	MIM	3.6.1.14	TIM11	U32517.1	ATP5I	D50371.1	4p16.3
ATPase subunit g	Cristae morphology	MIM	3.6.1.14	ATP20	Z49919.1	ATP5IG	AF092124.1	11q23.3
Cell surface glycoprotein 115–120 kDa	Glycolipid anchored surface protein	ND	–	GAS1	X53424.1	–		–
Cell surface protein Gas4, homolog of Gas1	Cell surface protein?	ND	–	GAS4	Z74874.1	–		–
Cell wall component Sun4p	Cell wall function	MAT	–	SUN4	Z71342.1	–		–
Dynamin-related GTPase, long and short isoforms	MOM/MIM fusion w. Fzo1 and Ugo1	IMS	–	MGM1	X62834.1	OPA1	AB011139.1	3q29
Dynamin-related protein (called DLP1/DRP1/dymple in mammals)	Mito fission with Fis1; GTPase	MOM	–	DNM1	L40588.1	DNM1L	AF061795.1	12p11.21
Ganglioside-induced differentiation-associated protein 1	Regulates mitochondrial network w. MFN's	MOM	–	–		GDAP1	Y17849.1	8q21.11
Endophilin B1	Fatty acyl transferase; reg MOM-MIM coupling	MOM?	2.3.1.-	–		SH3GLB1	AF263293.1	1p22.3
GTP-binding protein RAN	Formation of contact sites?	MOM	–	–		RAN	BC004272.1	12q24.33
GTP-binding protein G(t), alpha-1 subunit	Formation of contact sites?	MOM	–	–		GNAI1	AF055013.1	7q21.11
GTP-binding protein G(S), alpha subunit	Formation of contact sites?	MOM	–	GPA2	J03609.1	GNAS1	U29343.1	20q13.32
Hyaladherin (receptor for hyaluronan-mediated motility), 75-kDa isoform	Cell motility	ND	–	–		HMMR	U29343.1	5q34
Latent transforming growth factor-beta 1 (TGF-beta 1)	Growth/differentiation factor	ND	–	–		TGFB1	X02812.1	19q13.13
Latent TGF-beta binding protein 1 (LTBP-1)	Growth/differentiation factor	ND	–	–		LTBP1	M34057 (i)	2p23.3
Membrane-associated ring finger (C3HC4) 5 (RNF153); ubiquitin protein ligase (MITOL)	Affects mito fission; binds MFN2/DNM1L	MOM	–	–		MARCH5	NM_017824.4	10q23.32
Misato (Drosophila) homolog Dml1	Organelle partitioning?	MOM	–	DML1	Z49809.1	MSTO1	BC002535.1	1q22
Mitochondrial Distribution and Morphology protein Mdm30 (F-box protein)	Regulates mito morphology (with Mfb1)	MOM	–	MDM30	U19103.1	–		–
Mitochondrial Distribution and Morphology protein Mfb1 (F-box protein)	Regulates mito morphology (with Mdm30)	MOM	–	MFB1	Z48612.1	–		–
Mitochondrial Distribution and Morphology protein Mdm33	Fission of the inner membrane	MIM	–	MDM33	U32274.1	–		–
Mitochondrial Distribution and Morphology protein Mdm35	Function unknown; MIA	IMS	–	MDM35	Z28052.1	–		–
Mitochondrial Distribution and Morphology protein Mdm36	Assoc w. fission; links mitos to cytoskeleton	MOM?	–	MDM36	U51033.1	–		–
Mitochondrial division protein	Binds Dnm1 and Fis1	MOM	–	MDV1	Z49387.1	–		–
Mitochondrial division protein FtsZA	Mitochondrial division	ND	–	–		FtsZA (s)	AF304356.1 (s)	–
Mitochondrial division protein FtsZB	Mitochondrial division	ND	–	–		FtsZB (s)	AF30441.1 (s)	–
Mitochondrial fission-associated protein (tetratricopeptide repeat protein 11)	Binds Dnm1 and Mdv1	MOM	–	FIS1	Z38060.1	TTC11	BC009428.1	7q22.1
Mitochondrial fusion protein Ugo1	Binds Fzo1 and Mgm1; MCF motifs	MOM	–	UGO1	U33050.2	–		–
Mitochondrial protein 18 kDa (MTP18) (MGC5762), subset of HSPC242	Req for mito fission	ND	–	NCU06587.1 (Nc)	XM_32644I.1 (Nc)	HSPC242	AAH46132.1	22q12.2
Mitofusin 1 ("fuzzy onions" protein)	GTPase for mito fusion	MOM	–	FZO1	Z36048.1	MFN1	AF329637.1	3q26.32
Mitofusin 2 ("fuzzy onions" protein)	GTPase for mito fusion	MOM	–	FZO1	Z36048.1	MFN2	NM_014874.2	1p36.22
Mitofusin-binding protein (MIB) (synaptic vesicle membrane protein VAT-1, cyto/mito	Binds MFN1/2	MOM?	1.-.-.-	–		VAT1	BT007369.1	17q21.31
OPA3 protein, potentially mitochondrial	Function unknown	MIM	–	–		OPA3	AK025840.1	19q13.32
ORF-640 of linear mito plasmid mF (*Physarum polycephalum* [*Pp*]) (= ScTID3?)	Mito fusion protein in Pp	MOM	–	SPBC1703.11 (Sp)	AL136516.1 (Sp)	URF6 (Pp)	D29637.1 (Pp)	–
Phospholipase D (CG12314) (MitoPLD)	Required for mito fusion; hydrolyzes cardiolipin	MOM	–	–		LOC201164	BC031263.1	17p11.2
WD40 protein Caf4 [1], similar to Mdv1	Mito fission; recruits Dnm1 to mitos	ND	–	CAF4	Z28261.1	–		–
Mitochondrial "germline-specific" proteins								
Alpha-gustducin (guanine nucleotide-binding protein G(t), alpha-3 subunit)	G protein subunit; In spermatozoa	ND	–	–		GNAT3	XM_294370.5	7q21.11

854

Protein	Localization	Function	EC	Gene (other)	Accession (other)	Gene symbol	Accession	Chromosome
Asparaginase-like protein	ND	Sperm autoantigen				ASRGL1	BC021295.2	11q12.1
Basonuclin, isoform 1a	ND	Spermatid mitochondria				BNC	L03427.1	15q25.3
Don Juan protein, sperm-specific	ND	Mitos in sperm flagella	3.5.1.1			DJ (f)	U90537.1 (f)	—
Fatty acid desaturase ("Degenerative spermatocyte")	ND	Meiosis in spermatogenesis				DEGS	BC000961.2	1q42.12
Mitochondria-associated cysteine-rich protein SMCP	MOM	Sperm mito structure				MCSP	X89960.1	1q21
Ooplasm-specific protein 1 (maternal antigen that embryos require [MATER])	ND	Function unknown				NALP5	AY054986.1	19q13.43
Sperm associated antigen 1 (testis and skin-type isoforms in mouse [t-Tpis, s-Tpis])	MOM	Elimination of paternal mitos		STI1	M28486.1	SPAG1	AF311312.1	8q22.2
Spermatogenic specific-gene 1 (Spergen-1)	ND	Induces mito aggregation in sperm				SPAS1	AA398848.1 (i)	11q25
Spermatogenesis associated factor (SPAF), potentially mitochondrial	NC	AAA family protease				SPAF	AF361489.1	4q28.1
Testis-expressed protein of 22 kDa, homolog of mouse Tep22? (LOC647310)	NC	Mito sheath of midpiece?				[TEX22]	XM_930381.2	14q32.33
Other "morphology-related" proteins								
Annexin 1	ND	Function unknown				ANX1	X05908.1	9q21.11
Annexin V [2]	NC	Function unknown				ANX5	X12454.1	4q27
Annexin VI	ND	Function unknown				ANX6	Y00097.1	5q33.1
CDC6-related protein, potentially mitochondrial	NC	Function unknown				CDC6	U77949.1	17q21.1
Connexin 43 gap junction alpha-1 protein	MIM	Translocates to mitos				GJA1	X52947.1	6q22.31
Dihydropyrimidinase related protein-5/CRMP5-associated molecule (CRAM)	MOM?	Translocates to mitos via M-septin				DPYSL5	AF157634.1	2p23.3
Golgi phosphoprotein 3 (GPP34; MIDAS [mitochondrial DNA absence sensitive factor])	IMS	Regulates mitochondrial mass				GOLPH3	AJ296152.1	5p13.3
Mitochondrial distribution and morphology protein Mdm31	MIM	Mito shape/mtDNA w Mmm1,2/Mdm10,12		MDM31	NC_001140.2			—
Mitochondrial distribution and morphology protein Mdm32	MIM	Mito shape/mtDNA w Mmm1,2/Mdm10,12		MDM32	Z75055.1			—
Mitochondrial protein X (MIX) (Leishmania major [Lm])	MIM	Reg morphology, segregation, virulence		NCU00894.1 (Nc)	XM_325073.1 (Nc)	MIX (Lm)	CA02308.1 (Lm)	—
Mitofilin, heart motor protein (HMP)	MIM	Controls cristae morphology				IMMT	D21094.1	2p11.2
Myelin-associated oligodendrocytic basic protein, 155-aa mito isoform (MOBP155)	ND	Function unknown				MOBP	AF120475.1 (m)	3p22.2
Nuclear migration protein Num1 [1]	ND	Affects mito morphology		NUM1	X61236.1			—
Parasitophorous vacuole membrane protein ROP2 (Toxoplasma gondii [Tg])	MOM?	Fuses PVM to mitos (and ER)				ROP2 (Tg)	Z36906.1 (Tg)	—
Parasitophorous vacuole membrane protein ROP8 (Toxoplasma gondii [Tg])	MOM?	Fuses PVM to mitos				ROP8 (Tg)	AF011377.1 (Tg)	—
Plectin 1b ("cytolinker" protein) alt spl mitochondrial isoform	MOM	Binds intermediate filaments				PLEC1	Z54367.1	8q24.3
Protein associated w. filamentous growth	NC	Function unknown		DFG16	Z74938.1			—
Septin 4 (PNUTL2; ARTS; H5), alt spl mito isoform (M-septin)	ND	Axonal guidance; sperm differentiation				SEPT4	AB078010.1	17q23.2
Slit (Drosophila) homolog 3	ND	Cell migration in development?				SLIT3	AB017169.1	5q34
Synaptojanin 2A (same as synaptojanin 2)	MOM	Mito clustering; binds SYNJ2BP	3.1.3.36			SYNJ2	AF318616.1	6q25.3
Synaptojanin 2 binding protein (outer membrane protein OMP25)	MOM	Mito clustering; binds SYNJ2				SYNJ2BP	AK002133.1	14q24.2
Syntaphilin, brain-specific	MOM	Synaptic maturation; in mitos and PM				SNPH	AF187733.1	20p13
Tortoise protein	MAT?	Chemotaxis				TorA (s)	AF303221.1 (s)	—
Tumor suppressor DICE1 (mito in worm; DICE1 is nuclear in human)	MIM	Remodeling cristae				dic-1	NM_001027946.1 (w)	—
Vesicle associated membrane protein-1B, mitochondrial form (VAMP-1B)	MOM	Fuses transport vesicles				VAMP1	AF060538.1	12p13.31
Vimentin-like protein associated w mito morphology and inheritance	NC	Intermediate filament formation?		MDM1	X66371.1	VIM	M14144.1	10p12.33
Yeast suicide protein 1 [5]	ND	Mitochondrial fragmentation		YSP1	U10397.1			—
PROTEIN SORTING								
Outer membrane components								
Cytochrome c import factor (maturation of c-type cytochromes)	ND	Cyc1 Cyc7 imp; binds CCHL; reduces heme		CYC2	L28428.1			—
Disrupted in schizophrenia 1 protein (DISC1), many alt spl isoforms	ND	Function unknown; binds NUDEL				DISC1	AK025293.1	1q42.2
DnaJ homolog subfamily A member 1	MOM?	Co-chaperone of Hsc70		API	Z71353.1	HSJ2	L08069.1	9p21.1

(continues)

Appendix 5 (*continued*)

Gene product	Function; comment	Location	E.C. #	Fungal Symbol	Fungal GenBank	Animal Symbol	Animal GenBank	Human Chr.
Heat shock cognate 71-kDa protein HSC70	Binds TOM70	MOM		SSA1	X12926.1	HSC70	Y00371.1	11q24.1
Heat shock protein HSP90-alpha	Binds TOM70	MOM		HSP82	K01387.1	HSP90A	X15183.1	14q32.31
Heat shock cognate protein HSC82 [1]	Function unknown	ND		HSC82	M26644.1	HSPCB	M16660.1	6p21.1
Metaxin 1	Translocase, binds metaxin 2	MOM		MSP1	X68055.1	MTX1	U46920.1	1q21
Metaxin 2	Translocase, binds metaxin 1	MOM				MTX2	AF053551.1	2q31.2
Mitochondrial distribution and morphology protein Mdm10	MOM assembly (SAM complex)	MOM		MDM10	X80874.1			
Mitochondrial fusion targeting protein	Targeting factor for import	NC		MFT1	S57517.1			
Mitochondrial import stimulating factor (MSF) L subunit (14-3-3 protein)	Targets precursor to mitos in human	MOM		BMH1	X66206.1	YWHAE	U28936.1	17q11.2
Mitochondrial import stimulating factor (MSF) S1 subunit (14-3-3 protein)	Targets precursor to mitos in human	MOM		BMH2	U01883.1	YWHAZ	U28964.1	8q22.3
NUDEL, endooligopeptidase A (LIS1-interacting protein)	Binds YWHAE and DISC1	ND				NDEL1	BC026101.1	17p13.1
Outer membrane sorting/assembly protein OMP85/TOB55 [1] [4]	Beta-barrel protein assembly (SAM/TOB compl)	MOM		SAM50	Z71302.1	SAMM50	AK001087.1	22q13.31
Outer membrane import receptor subunit, 5 kDa	Subunit of GIP	MOM		TOM5	U40829.1			
Outer membrane import receptor subunit, 6 kDa	Assembly of GIP complex	MOM		TOM6	Z22815.1			
Outer membrane import receptor subunit, 7 kDa	Dissociation of GIP complex	MOM		TOM7	Z71346.1	TOMM7	BC001732.1	7p15.3
Outer membrane import receptor subunit, 13 kDa	Beta-barrel protein assembly (SAM/TOB compl)	MOM		TOM13	Z74768.1			
Outer membrane import receptor subunit, 20 kDa	Translocase receptor	MOM		TOM20	X75319.1	TOMM20	D13641.1	1q42.3
Outer membrane import receptor subunit, 22 kDa	Translocase receptor	MOM		TOM22	X82405.1	TOMM22	AB040119.1	22q13.1
Outer membrane import receptor subunit, 38 kDa	Beta-barrel protein assembly (SAM/TOB compl)	MOM		SAM35	U10556.1	AIP	U78521.1	11q13.2
Outer membrane import receptor subunit, 40 kDa	Translocase channel	MOM		TOM40	X56885.1	TOMM40	AF043250.1	19q13.32
Outer membrane import receptor subunit, 70 kDa	Translocase receptor	MOM		TOM70	X05585.1	TOMM70A	BC003633.1	3q12.2
Outer membrane import receptor subunit, 71 kDa	Function unknown	MOM		TOM71	U00059.1	TOMM70A	BC003633.1	3q12.2
Outer membrane sorting and assembly machinery subunit, 37 kDa (MAS37)	Sorts MOM proteins fr GIP to MOM	MOM		TOM37	Z49703.1	TOMM34	U58970.1	20q13.12
Phosphofurin acidic cluster sorting protein 2 (early secretory pathway sorting connector)	ER-mito communication; apoptosis via Bid	ER-MAM				PACS2	AY320284.1	14q32.33
Intermembrane space components								
Import factor Mia40/Tim40 (Fmp15) [1]	IMS import via disulfide relay, w. Erv1; not MIA	IMS		MIA40	Z28195.1	CHCHD4	BC017082.1	3p25.1
Inner membrane import receptor subunit, 8 kDa	Binds Tim9/Tim10	IMS		TIM8	Z49636.1	TIMM8A	U66035.1	Xq22
Inner membrane import receptor subunit, 8 kDa	Mediates import across IMS	IMS		TIM8	Z49636.1	TIMM8B	AF152350.1	11q23
Inner membrane import receptor subunit, 9 kDa, putative	Mediates import across IMS	IMS?				TIMM9B	AF150105.1	11p15.4
Inner membrane import receptor subunit TIMM13, 13 kDa, a and b isoforms	Binds Tim9/Tim10; import across IMS; MIA	IMS		TIM13	Z72966.1	TIMM13	AF152351.1	19p13.3
Small TIM complex assembly factor HOT13	Assembles Tim8-13; MIA	IMS		HOT13	Z28084.1			
Subunit of the TIM22 complex	Binds Tim10	IMS		TIM9	AF093244.1	TIMM9	AF152353.1	14q23.1
Subunit of the TIM22 complex (also called Mrs11p)	Binds Tim9	IMS		TIM10	Z80875.1	TIMM10	AF152354.1	11q12.1
Subunit of the TIM22 complex (also called Mrs5p)	Binds Tim22/Tim54	IMS/MIM		TIM12	M90689.1			
Sulfhydryl oxidase, FAD-linked	IMS import via disulfide relay sys, w. Mia40; MIA	IMS		ERV1	X60722.1	ALR	U31176.1	16p13.3
Inner membrane components								
Import factor Mia40/Tim40 (Fmp15) [1]	IMS import via disulfide relay sys, w. Mia40	IMS		MIA40	Z28195.1	CHCHD4	BC017082.1	3p25.1
Inner membrane peptidase (IMP), subunit 1	Cuts signal seq in IMS	MIM	3.4.99.-	IMP1	S55518.1	IMMP1L	AK057788.1	11p13
Inner membrane peptidase (IMP), subunit 2	Cuts signal seq in IMS	MIM	3.4.99.-	IMP2	Z49213.1	IMMP2L	AF359563.1	7q31.1
Inner membrane peptidase (IMP), putative subunit 3	Req for Imp1 function	MIM		SOM1	X90459.1			
Inner membrane protease (human NOGO-interacting mitochondrial protein)	Processing/export from MAT	MIM		YIM1	Z47071.1	NIMP	AF336861.1	6q21
Membrane insertion protein	Inserts proteins into MIM	MIM		OXA1	X77558.1	OXA1L	X80695.1	14q11.2
Membrane insertion protein	Oxa1-indep insertion of Atp6 and Cyt b	MIM		MDM38	X74769.1	LETM1	AF061025.1	4p16.3
Membrane insertion protein	Inserts proteins into MIM	MIM		MBA1	Z36054.1			
Membrane insertion protein (pentamidine-resistance protein)	Protein export, w. Cox18/Mss2	MIM		PNT1	Z75174.1			
Methyltransferase-like protein OMS1 [5]	Suppresses Oxa1p mutations	MIM		OMS1	U28374.1			

Subunit of the TIM22 complex (inner membrane translocase, 22 kDa)	In Tim54-Tim22 complex	—	MIM	TIM18	Z75205.1	—	—	—
Subunit of the TIM22 complex (inner membrane translocase, 22 kDa)	In translocation channel	—	MIM	TIM22	Z74265.1	TIMM22	AF155330.1	17p13.3
Subunit of the TIM22 complex (inner membrane translocase, 54 kDa)	In translocation channel	—	MIM	TIM54	Z49329.1	—	—	—
Subunit of the TIM23 complex (inner membrane translocase, 15 kDa) (Mdj2)	Protein folding chaperone; binds Tim16	—	MIM	MDJ2	Z71604.1	—	—	—
Subunit of the TIM23 complex (inner membrane translocase, 16 kDa) (Tim16/Pam16)	In protein import motor (PAM); MMC	—	NC	PAM16	Z49379.1	MAGMAS	AF349455.1	16p13.3
Subunit of the TIM23 complex (inner membrane translocase, 16.5 kDa)	Binds Tim23; directs preproteins to MAT	—	MOM+MIM	TIM17	X77796.1	TIMM17A	AF106622.1	1q32.1
Subunit of the TIM23 complex (inner membrane translocase, 16.5 kDa)	Binds Tim23; directs preproteins to MAT	—	MOM+MIM	TIM17	X77796.1	TIMM17B	AF034790.1	Xp11.23
Subunit of the TIM23 complex (inner membrane translocase, 17 kDa) (Pam17)	In protein import motor (PAM); MMC	—	MIM	PAM17	Z28290.1	—	—	—
Subunit of the TIM23 complex (inner membrane translocase, 18 kDa) (Tim14/Pam18)	In protein import motor (PAM)	—	MIM	PAM18	Z73180.1	DNAJD1	AF126743.1	13q14.11
Subunit of the TIM23 complex (inner membrane translocase, 18 kDa) (Tim14/Pam18)	In protein import motor (PAM); MMC	—	MIM	PAM18	Z73180.1	DNAJC19	BC009702.1	3q26.33
Subunit of the TIM23 complex (inner membrane translocase, 21 kDa) (Tim21)	Binds Tim23/Tom22; directs preproteins to MIM	—	ND	TIM21	Z72818.1	—	—	—
Subunit of the TIM23 complex (inner membrane translocase, 23 kDa)	Connects MOM and MIM	—	MOM+MIM	TIM23	X74161.1	TIMM23	AF030162.1	10q11.23
Subunit of the TIM23 complex (inner membrane translocase, 44 kDa)	In protein import motor (PAM); MMC	—	MIM	TIM44	X67276.1	TIMM44	AF041254.1	19p13.2
Subunit of the TIM23 complex (inner membrane translocase, 55 kDa)	Maintains TIM23 permeability barrier	—	MIM	TIM50	U39205.1	TIMM50	AY551341.1	19q13.2
Matrix components								
Heat shock protein 10	Associates w Hsp60	—	MAT	HSP10	X75754.1	HSPE1	U07550.1	2q33.1
Heat shock protein 60	Associates w Hsp10	—	MAT	HSP60	M33301.1	HSPD1	M34664.1	2q33.1
Heat shock protein 70 family member HSP70 (human mortalin-2)	In protein import motor (PAM); MMC	—	MAT	SSC1	M27229.1	HSPA9B	L15189.1	5q31.2
Heat shock protein 70 family member (Ssc3)	Import protein; in yeast nucleoid	—	MAT	ECM10	U18530.1	HSPA9B	L15189.1	5q31.2
Heat shock protein 70 family member, similar to human mortalin-2	FeS cluster syn/assembly, binds Isu1	—	MAT?	SSQ1	U19103.1	HSPA9B	L15189.1	5q31.2
Heat shock protein 70 interacting protein (Hep1/Tim15/Zim17)	Prevents aggregation of HSP70s	—	MAT	ZIM17	Z71586.1	DNAJA3	AF061749.1	16p13.3
Heat shock protein Mdj1	In protein import motor (PAM); MMC	—	MAT	MDJ1	Z28336.1	C3orf31	BC015088.2	3p25.2
Mitochondrial matrix protein, 37 kDa (MMP37)	Protein import across MIM	—	MAT	TAM41	Z72831.1	—	—	—
Mitochondrial processing peptidase (MPP), alpha subunit	Cleaves MTS	3.4.24.64	MAT	MAS2	X14105.1	MPPA	D50913.1	9q34.3
Mitochondrial processing peptidase (MPP), beta subunit	Cleaves MTS	3.4.24.64	MAT	MAS1	X07649.1	PMPCB	AF054182.1	7q22.1
Mitochondrial intermediate peptidase (MIP)	Cleaves MTS	3.4.24.59	MAT	OCT1	U10243.1	MIPEP	U80034.1	13q12
Nucleotide exchange/release factor, isoform 1 (GrpE protein homolog 1)	In protein import motor (PAM); binds HSP70	—	MAT	MGE1	Z75140.1	GRPEL1	AF298592.1	4p16.1
Nucleotide exchange/release factor, isoform 2 (GrpE protein homolog 2)	In protein import motor (PAM); binds HSP70	—	MAT	MGE1	Z75140.1	GRPEL2	AK074293.1	5q35
Peptidyl-prolyl cis-trans isomerase (cyclophilin D in human)	Cyclosporin-sensitive	5.2.1.8	MAT	CPR3	M84758.1	PPIF	M80254.1	10q22.3
Protein associated w. intramitochondrial sorting (YLR168c; CGI-107)	Homolog of fly sfmo; function unknown	—	ND	MSF1′	X70279.1	C20orf45	AF151865.1	20q13.32
Protein associated w. intramitochondrial sorting, similar to Msf1	Homolog of fly retm; function unknown	—	ND	YDR185c	Z46727.1	SEC14L5	XM_032693.3	16p13.3
Protein associated w. intramitochondrial sorting, similar to Msf1	Processing of Mgm1/OPA1	—	IMS	UPS1	U14913.1	PREL1	AF201925.1	5q35.3
PROTEIN TRANSLATION AND STABILITY **Ribosomal components – large subunit**								
Ribosomal protein L1 (homolog of E. coli L1)	Component of large subunit	—	MAT	MRPL1	Z48758.1	MRPL1	AB049474.1	4q21.21
Ribosomal protein L2 (homolog of E. coli L2)	Component of large subunit	—	MAT	RML2	U18179.1	MRPL2	AF132956.1	6p21.1
Ribosomal protein L3 (homolog of E. coli L3)	Component of large subunit	—	MAT	MRPL9	X65014.1	MRPL3	X06323.1	3q22.1
Ribosomal protein L4 (homolog of E. coli L4)	Component of large subunit	—	MAT	YML6	Z46659.1	MRPL4	AB049635.1	19p13.2

Gene product	Function; comment	Location	E.C. #	Fungal		Animal		
				Symbol	GenBank	Symbol	GenBank	Human Chr.
Ribosomal protein YmL5/YmL7 (homolog of E. coli L5)	Component of large subunit	MAT		MRPL7	Z49701.1	–	AB049636.1	1q21.3
Ribosomal protein YmL16 (homolog of E. coli L6)	Component of large subunit	MAT		MRPL6	X69480.1	–	BC015904.1	17q21.32
Ribosomal protein L9 (homolog of E. coli L9)	Component of large subunit	MAT				MRPL9	AF151871.1	11q13.2
Ribosomal protein L10 (homolog of E. coli L10)	Component of large subunit	MAT		MRPL11	Z74250.1	MRPL10	X79865.1	17q25
Ribosomal protein L11 (homolog of E. coli L11)	Component of large subunit	MAT		MRPL19	Z71461.1	MRPL11	AB049640.1	8q24.12
Ribosomal protein L7/L12 (homolog of E. coli L7/L12)	Component of large subunit; in Sc nucleoid	MAT		MNP1	Z72591.1	MRPL12	NM_032111.1	6p21.1
Ribosomal protein L13 (homolog of E. coli L13)	Component of large subunit	MAT		MRPL23	Z75058.1	MRPL13	BC000891.1	8q11.23
Ribosomal protein L14 (homolog of E. coli L14)	Component of large subunit	MAT		MRPL38	Z28169.1	MRPL14	AB049642.1	11q12.1
Ribosomal protein L15 (homolog of E. coli L15)	Component of large subunit	MAT		MRPL10	Z71560.1	MRPL15	AF164797.1	11p15.4
Ribosomal protein L16 (homolog of E. coli L16)	Component of large subunit	MAT		MRPL16	Z35799.1	MRPL16	AL136633.1	6q26
Ribosomal protein L17 (homolog of E. coli L17)	Component of large subunit	MAT		MRPL8	X53841.1	MRPL17	D14660.1	2p12
Ribosomal protein L18 (homolog of E. coli L18)	Component of large subunit	MAT				MRPL18	AB049644.1	1p36.33
Ribosomal protein L19 (no E. coli homolog found)	Component of large subunit	MAT		IMG1	X59720.1	MRPL19	BC055088.1	11q13.3
Ribosomal protein L20 (homolog of E. coli L20)	Component of large subunit	MAT		MRPL49	Z49371.1	MRPL20	AF161507.1	5q33.2
Ribosomal protein L21 (homolog of E. coli L21)	Component of large subunit	MAT		MRPL22	Z71453.1	MRPL21	U26596.1	11p15.5
Ribosomal protein L22 (homolog of E. coli L22)	Component of large subunit	MAT		MRPL41	M81696.1	MRPL22	BC016700.1	1q23.1
Ribosomal protein L23 (homolog of E. coli L23)	Component of large subunit	MAT				MRPL23	AB049647.1	17q21.33
Ribosomal protein L24 (homolog of E. coli L24)	Component of large subunit	MAT		MRP7	Z71281.1	MRPL24	BC000990.2	16p13.3
Ribosomal protein L27 (homolog of E. coli L27)	Component of large subunit	MAT		MRPL24	Z47815.1	MAAT1	AF151083.1	2q11.2
Ribosomal protein YmL14/YmL24 (homolog of E. coli L28)	Component of large subunit	MAT		MRPL33	D90217.1	MRPL30	AB049649.1	7p14.1
Ribosomal protein L30 (homolog of E. coli L30)	Component of large subunit	MAT		MRPL32	AY693214.1	MRPL32	AF047440.1	2p23.2
Ribosomal protein L32 (homolog of E. coli L32)	Component of large subunit	MAT		MRPL39	Z49810.1	MRPL33	AB049652.1	19p13.3
Ribosomal protein L33 (homolog of E. coli L33)	Component of large subunit	MAT		YDR115w	Z48758.1			
Ribosomal protein L34 (homolog of E. coli L34)	Component of large subunit	MAT				MRPL34	L38941.1	4q25
Ribosomal protein L34, potentially mitochondrial	Cytoplasmic ribosomal protein	NC		RPL34A	U18813.1	RPL34	AF208849.1	2p11.2
Ribosomal protein L35 (homolog of E. coli L35, alt spl isoforms a and b	Component of large subunit	MAT				MRPL35	AF155653.1	5p15.33
Ribosomal protein L36 (homolog of E. coli L36.B)	Component of large subunit	MAT		GON5	Z73539.1	MRPL36	AL554345 (i)	21q21.3
Ribosomal protein L39, L5 family, ubiquitous isoform MRP-L5 (alt spl)	Component of large subunit	MAT				MRPL39	AF270511.1	21q21.3
Ribosomal protein L39, L5 family, heart-specific isoform MRP-LSV1 (alt spl)	Component of large subunit	MAT				MRPL39	BC000041.1	1p32.3
Ribosomal protein L36 (no E. coli homolog found)	Component of large subunit	MAT				MRPL37	BE326258.1	9q34.3
Ribosomal protein YmL27 (no E. coli homolog found)	Component of large subunit; binds Bcl2	MAT		MRPL27	Z36151.1	MRPL41	AB049656.1	10q24.31
Ribosomal protein (CI-B8 family) (no E. coli homolog found)	Component of large subunit	MAT		MRPL51	U32445.1	MRPL43	AK022763.1	2q36.1
Ribosomal protein YmL3 (no E. coli homolog found)	Component of large subunit	MAT		MRPL3	Z49211.1	MRPL44	BC006235.1	17q12
Ribosomal protein (no E. coli homolog found)	Component of large subunit	MAT				MRPL45	AF205435.1	15q25.3
Ribosomal protein YmL17/YmL30 (no E. coli homolog found)	Component of large subunit	MAT		MRPL30	Z71528.1	MRPL46	AF285120.1	3q26.33
Ribosomal protein YmL4 (no E. coli homolog found)	Component of large subunit	MAT		MRPL4	Z30582.1	MRPL48	AF151876.1	11q13.4
Ribosomal protein (no E. coli homolog found)	Component of large subunit	MAT				MRPL49	U39400.1	11q13.1
Ribosomal protein (no E. coli homolog found)	Component of large subunit	MAT		IMG2	X59720.1	MRPL50	AK000500.1	9q31.1
Ribosomal protein (no E. coli homolog found)	Component of large subunit	MAT				MRPL51	AF151075.1	12p13.31
Ribosomal protein (no E. coli homolog found)	Component of large subunit	MAT				MRPL52	BC068070.1	14q11.2
Ribosomal protein (no E. coli homolog found)	Component of large subunit	MAT				MRPL53	BC012163.1	2p13.1
Ribosomal protein YmL37 (no E. coli homolog found)	Component of large subunit	MAT		MRPL37	Z36137.1	MRPL54	BC065273.1	19p13.3
Ribosomal protein (no E. coli homolog found)	Component of large subunit	MAT				MRPL55	BC052806.1	1q42.13
Ribosomal protein L56? (may actually be a serine beta-lactamase-like protein)	Component of large subunit?	MAT				LACTB	AK027808.1	15q22.2
Ribosomal protein YmL13 (no E. coli homolog found)	Component of large subunit	MAT		MRPL13	X73673.1			
Ribosomal protein YmL15 (no E. coli homolog found)	Component of large subunit	MAT		MRPL15	U20618.1			
Ribosomal protein YmL20 (no E. coli homolog found)	Component of large subunit	MAT		MRPL20	X55840.1			
Ribosomal protein YmL25 (no E. coli homolog found)	Component of large subunit	MAT		MRPL25	X56106.1			
Ribosomal protein YmL28 (no E. coli homolog found)	Component of large subunit	MAT		MRPL28	M88597.1	NLVCF	AJ295637.1	22q11.21

Ribosomal protein YmL31 (no E. coli homolog found)	Component of large subunit	MAT	—	MRPL31	X15099.1	MRPL38	BC013311.1	17q25.1
Ribosomal protein YmL35 (no E. coli homolog found)	Component of large subunit	MAT	—	MRPL35	U32517.1	—	—	—
Ribosomal protein YmL36 (homolog of E. coli L31 [RpmE])	Component of large subunit	MAT	—	MRPL36	Z35991.1	—	—	—
Ribosomal protein YmL40 (no E. coli homolog found)	Component of large subunit	MAT	—	MRPL40	Z73529.1	—	—	—
Ribosomal protein YML44 (no E. coli homolog found)	Component of large subunit	MAT	—	MRPL44	X17552.1	—	—	—
Ribosomal protein (no E. coli homolog found, but similar to L9)	Component of large subunit	MAT	—	MRPL50	Z71637.1	—	—	—
Ribosomal RNA, large subunit (mtDNA-encoded)	Component of large subunit	MAT	—	rRNA21S	AJ011856.1	MTRNR2	J01415.1	mtDNA
Ribosomal components – small subunit								
Ribosomal protein (no E. coli homolog found)	Component of small subunit	MAT	—	MRP8	Z28142.1	MRP63	BC000002.2	13q12.11
Ribosomal protein (no E. coli homolog found)	Component of small subunit	MAT	—	MRP10	Z74093.1	—	—	—
Ribosomal protein (no E. coli homolog found)	Component of small subunit	MAT	—	MRP51	U43503.1	—	—	—
Ribosomal protein (no E. coli homolog found)	Component of small subunit	MAT	—	MRPS35	Z72950.1	—	—	—
Ribosomal protein (no E. coli homolog found), same as NDUFA6	Component of small subunit	MAT	—	NUO-14.8 (Nc)	X76344.1 (Nc)	NDUFA6	AF047182.1	22q13.2
Ribosomal protein (E. Coli RPS3) (mtDNA-encoded)	Component of small subunit	MAT	—	VAR1	V00705.1	—	—	—
Ribosomal protein MRP1 (no E. coli homolog found)	Component of small subunit	MAT	—	MRP1	M15160.1	—	—	—
Ribosomal protein MRP13 (no E. coli homolog found)	Component of small subunit	MAT	—	MRP13	Z72869.1	—	—	—
Ribosomal protein NAM9 (E. Coli S4 [RpsD])	Component of small subunit	MAT	—	NAM9	M60730.1	—	—	—
Ribosomal protein PET123 (no E. coli homolog found)	Component of small subunit	MAT-MIM	—	PET123	X52362.1	—	—	—
Ribosomal protein RSM28 (no E. coli homolog found)	Component of small subunit	MAT	—	RSM28	AF459095.1	—	—	—
Ribosomal protein S2 (E. Coli RPS2 [RpsB])	Component of small subunit	MAT	—	MRP4	M82841.1	MRPS2	AF151849.1	9q34.3
Ribosomal protein S3a [2]	Component of small subunit?	MAT?	—	RPS1B	Z38114.1	RPS3A	M77234.1	4q31.3
Ribosomal protein S5 (homolog of E. coli S5 [RpsE])	Component of small subunit	MAT	—	MRPS5	Z36120.1	MRPS5	AB049940.1	2q11.1
Ribosomal protein S6 (homolog of E. coli S6)	Component of small subunit	MAT	—	MRP17	X58362.1	MRPS6	AB049942.1	21q22.12
Ribosomal protein S7 (homolog of E. coli S7)	Component of small subunit	MAT	—	RSM7	Z49613.1	MRPS7	AF077042.1	17q25.1
Ribosomal protein (homolog of E. coli S8)	Component of small subunit	MAT	—	MRPS8	Z49705.1	—	—	—
Ribosomal protein S9 (homolog of E. coli S9)	Component of small subunit	MAT	—	MRPS9	Z36015.1	MRPS9	BC057240.1	2q12.1
Ribosomal protein S10 (homolog of E. coli S10), putative	Component of small subunit	MAT	—	RSM10	Z54075.1	MRPS10	AK001429.1	6p21.1
Ribosomal protein S11 (homolog of E. coli S11)	Component of small subunit	MAT	—	MRPS18	Z71582.1	MRPS11	AB049944.1	15q26.1
Ribosomal protein S12 (homolog of E. coli S12)	Component of small subunit	MAT	—	MRPS12	Z71651.1	MRPS12	Y11168.1	19q13.2
Ribosomal protein S13 (homolog of E. coli S13)	Component of small subunit	MAT	—	YNR036c	Z71357.1	—	—	—
Ribosomal protein S14 (homolog of E. coli S14)	Component of small subunit	MAT	—	SWS2	M15161.1	MRPS14	AL049705.1	1q25.1
Ribosomal protein S15 (homolog of E. coli S15)	Component of small subunit	MAT	—	MRP2	X55977.1	MRPS15	AB049946.1	1p34.3
Ribosomal protein S16 (homolog of E. coli S16)	Component of small subunit	MAT	—	MRPS28	U33335.1	MRPS16	AF151890.1	10q22.2
Ribosomal protein S17 (homolog of E. coli S17)	Component of small subunit	MAT	—	MRPS16	Z49808.1	MRPS17	AF077035.1	7p11.2
Ribosomal protein S17-like (homolog of E. coli S17), potentially mito	Component of small subunit?	NC	—	MRPS17	Z49702.1	—	—	—
	Component of small subunit?	NC	—	YMR118c	—	—	—	—
Ribosomal protein S18 (homolog of E. coli S13) (shown to be mito in human)	Translation initiation; binds fMET	MAT	—	RPS18A	U33007.1	RPS18	X69150.1	6p21.3
Ribosomal protein S18 (homolog of E. coli S13) (shown to be mito in human)	Translation initiation; binds fMET	MAT	—	RPS18B	Z46659.1	RPS18	X69150.1	6p21.3
Ribosomal protein S18-A (homolog of E. coli S18)	Component of small subunit	MAT	—	RSM18	U18796.1	MRPS18A	AB049952.1	6p21.1
Ribosomal protein S18-B (homolog of E. coli S18)	Component of small subunit	MAT	—	RSM18	U18796.1	MRPS18B	AF100761.1	6p21.33
Ribosomal protein S18-C (homolog of E. coli S18)	Component of small subunit	MAT	—	RSM18	U18796.1	MRPS18C	AF151892.1	4q21.23
Ribosomal protein S19 (homolog of E. coli S19)	Component of small subunit	MAT	—	RSM19	Z71652.1	—	—	—
Ribosomal protein S21 (homolog of E. coli S21)	Component of small subunit	MAT?	—	MRP21	Z58851.1	MRPS21	AF182417.1	1q21.3
Ribosomal protein S22 (no E. coli homolog found)	Component of small subunit	MAT?	—	—	—	MRPS22	AF226045.1	3q23
Ribosomal protein S23 (no E. coli homolog) (CGI-138)	Component of small subunit	MAT?	—	RSM25	Z46728.1	MRPS23	AF151896.1	17q23.2
Ribosomal protein S24 (no E. coli homolog found)	Component of small subunit	MAT	—	—	—	MRPS24	BC012167.1	7p13
Ribosomal protein S25 (no E. coli homolog found)	Component of small subunit	MAT?	—	MRP49	Z28167.1	MRPS25	AK024702.1	3p25.1
Ribosomal protein S26 (no E. coli homolog found)	Component of small subunit	MAT?	—	—	—	MRPS26	BC013018.2	20p13
Ribosomal protein S27 (no E. coli homolog found)	Component of small subunit	MAT?	—	—	—	MRPS27	NM_015084.1	5q13.2
Ribosomal protein S28 (no E. coli homolog found)	Component of small subunit	MAT?	—	RSM24	Z46727.1	MRPS35	AF182422.1	12p11.22
Ribosomal protein S29 (no E. coli homolog found) (also pro-apoptotic protein)	Component of small subunit	MAT?	—	RSM23	D10263.1	DAP3	X83544.1	1q23.1

(continues)

Appendix 5 *(continued)*

Gene product	Function; comment	Location	E.C. #	Fungal Symbol	Fungal GenBank	Animal Symbol	Animal GenBank	Human Chr.
Ribosomal protein S30 (no *E. coli* homolog found) (also pro-apoptotic protein)	Component of small subunit	MAT?	—	—	—	PDCD9	AF146192.2	5p12
Ribosomal protein S31 (no *E. coli* homolog found)	Component of small subunit	MAT?				MRPS31	Z68747.1	13q14.11
Ribosomal protein S32 (no *E. coli* homolog found)	Component of small subunit	MAT?				MRPS32	AF135160.1	12q22
Ribosomal protein S33 (no *E. coli* homolog found)	Component of small subunit	MAT?				MRPS33	AF078858.1	7q34
Ribosomal protein S34 (no *E. coli* homolog found)	Component of small subunit; binds DLG1	MAT?				MRPS34	BF027127.1	16p13.3
Ribosomal protein S35 (no *E. coli* homolog found)	Component of small subunit	MAT				MRPS28	BC010150.1	8q21.13
Ribosomal protein S36 (no *E. coli* homolog found)	Component of small subunit	MAT		YMR31	X17540.1	MRPS36	BE785097.1	5q33.2
Ribosomal protein S50 (no *E. coli* homolog found)	Not clear; may actually be a methyltransferase	MAT?		RSM22	Z28155.1	—	—	—
Ribosomal protein S54 (no *E. coli* homolog found)	Component of small subunit	MAT?		RSM26	Z49661.1	—	—	—
Ribosomal protein S55 (no *E. coli* homolog found)	Component of small subunit	MAT		RSM27	Z73000.1	—	—	—
Ribosomal protein YmS2 (no *E. coli* homolog found) (also protein phosphatase?)	Component of small subunit	MAT		PPE1	U10556.1	PME1	BC003046.1	11q13.14
Ribosomal RNA, small subunit (mtDNA-encoded)	Component of small subunit	MAT		rRNA15S	AJ011856.1	MTRNR1	J01415.1	mtDNA
Ribosomal RNA, 5S	Function unknown	MAT?				[5SRNA]	X71799.1	1q42.13
Aminoacyl-tRNA synthetases								
Alanyl-tRNA synthetase	Charging of tRNA-Ala	MAT?	6.1.1.7	ALA1	U18672.1	AARSL	AB033096.1	6p21.1
Arginyl-tRNA synthetase	Charging of tRNA-Arg	MAT	6.1.1.19	MSR1	L39019.1	RARSL	AK023550.1	6q15
Asparaginyl-tRNA synthetase	Charging of tRNA-Asn	MAT	6.1.1.22	PMP1	X59720.1	NARS2	AK056980.1	11q14.1
Aspartyl-tRNA synthetase	Charging of tRNA-Asp	MAT	6.1.1.12	MSD1	M26020.1	DARS2	BC045173.2	1q25.1
Cysteinyl-tRNA synthetase, potentially mitochondrial	Charging of tRNA-Cys	NC	6.1.1.16	YNL247w	Z71523.1	FLJ12118	AK022180.1	13q34
Glutaminyl-tRNA synthetase	Charging of tRNA-Gln	MAT?	6.1.1.18	GLN4	Z75076.1	—	—	—
Glutamyl-tRNA synthetase	Charging of tRNA-Glu	MAT	6.1.1.17	MSE1	L39015.1	EARS2	AL832489.1	16p12.1
Glycyl-tRNA synthetase	Charging of tRNA-Gly	MAT?	6.1.1.14	GRS1	Z35990.1	GARS	D30658.1	7p14.3
Histidyl-tRNA synthetase	Charging of tRNA-His	MAT	6.1.1.21	HTS1	M14048.1	HARSR	U18937.1	5q31.3
Isoleucyl-tRNA synthetase	Charging of tRNA-Ile	MAT	6.1.1.5	ISM1	L38957.1	IARS2	D28500.1	1q41
Leucyl tRNA synthetase	Charging of tRNA-Leu	MAT	6.1.1.4	NAM2	J03495.1	LARS2	D21851.1	3p21.31
Lysyl-tRNA synthetase	Charging of tRNA-Lys	MAT	6.1.1.6	MSK1	X57360.1	KARS	AF285758.1	16q23
Methionyl-tRNA synthetase	Charging of tRNA-Met	MAT	6.1.1.10	MSM1	X14629.1	MARS2	BC009115.1	2q33.1
Phenylalanyl-tRNA synthetase alpha chain	Charging of tRNA-Phe	MAT	6.1.1.20	MSF1	J02691.1	FARS1	AF097441.1	6p25.1
Prolyl-tRNA synthetase	Charging of tRNA-Pro	MAT	6.1.1.15	YER087w	U18839.1	PARS2	BC011758.1	1p32.2
Seryl-tRNA synthetase, potentially mitochondrial	Charging of tRNA-Ser	NC	6.1.1.11	DIA4	U10400.1	SARS2	AB029948.1	19q13.13
Threonyl-tRNA synthetase	Charging of tRNA-Thr	MAT	6.1.1.3	MST1	M12087.1	TARSL1	BC000541.1	1q21.2
Tryptophanyl-tRNA synthetase	Charging of tRNA-Trp	MAT	6.1.1.2	MSW1	M12081.1	WARS2	AJ242739.1	1p12
Tyrosyl-tRNA synthetase	Charging of tRNA-Tyr	MAT	6.1.1.1	MSY1	L42333.1	YARS2	AK024057.1	12p11.21
Valyl-tRNA synthetase	Charging of tRNA-Val	MAT	6.1.1.9	VAS1	J02719.1	VARS2L	AK000511.1	6p21.33
Transfer RNAs								
Cysteine desulfurase	Thio-modification of tRNAs	MAT	—	NFS1	M98808.1	NIFS	AF097025.1	20q11.22
Dimethylguanine tRNA methyltransferase	Methylates G34 in tRNAs	MIM	2.1.1.32	TRM1	M17193.1	TRMT1	AF194479.1	19p13.13
Glutamyl-tRNA(Gln) amidotransferase, subunit A	Charging Gln-tRNA(Gln)	MIM?	6.3.5.–	YMR293c	X80836.1	QRSL1	AL136679.1	6q21
Glutamyl-tRNA(Gln) amidotransferase, subunit B	Charging Gln-tRNA(Gln)	MIM	6.3.5.–	PET112	L22072.1	PET112L	AF026851.1	4q31.3
Hypothetical protein 15E1.2 [6]	tRNA-Gln amidotransferase?	ND	—			15E1.2	AL021546.1	12q24.31
Hypothetical protein LOC51250 [8]	Contains pseudouridylate synthase motif	ND	—			C6orf203	BC010899.2	6q21
Methionyl-tRNA formyltransferase	Formylation of tRNA-Met; C1 metab	MAT	2.1.2.9	FMT1	Z35774.1	FMT	BC033687.1	15q22.31
Peptidyl-tRNA hydrolase 1	Releases aa's from aa-tRNAs	ND	3.1.1.29	PTH1	U00030.1	PTRH1	BC047012.1	9q34.11
Peptidyl-tRNA hydrolase 2 (Pth2; Bcl2 inhibitor of transcription (Bit1)	Releases aa's from aa-tRNAs; pro-apoptotic	IMS?	3.1.1.29	PTH2	Z35818.1	PTRH2	BC006807.1	17q23.2
Pseudouridine synthase 1, isoform 1, potentially mitochondrial	Pseudouridine-27,28 in tRNAs	NC	5.4.99.12	—	—	PUS1	AF318369.1	12q24.33
Pseudouridine synthase (Deg1)	Pseudouridine-38,39 in tRNAs	ND	5.4.99.–	PUS3	D44600.1			—
Pseudouridine synthase	Pseudouridine-55 in tRNAs	ND	4.2.1.70	PUS4	Z71568.1			—
Pseudouridine synthase	Pseudouridine-31 in tRNAs	NC	4.2.1.70	PUS6	Z72954.1	RPUSD2	BC016967.1	15q15.1
Pseudouridine synthase	Pseudouridine-32 in tRNAs	ND	4.2.1.70	PUS9	Z74084.1	RPUSD2	BC016967.1	15q15.1

(continues)

Description	Function	Loc	EC	Gene (yeast)	Accession	Gene (human)	Accession	Location
Translational accuracy factor	tRNA modification @ U34; binds Mss1	MAT?	—	MTO1	Z72758.1	MTO1	AF132937.1	6q13
tRNA modification GTPase (alt spl in human)	tRNA modification @ U34; binds Mto1	MAT?	—	MSS1	X69481.1	GTPBP3	AF360742.1	19p13.12
tRNA isopentenyltransferase	Isopentenylation of mt-tRNAs	MAT	2.5.1.8	MOD5	M15991.1	IPT	AY303390.1	1p34.2
tRNA 5-methylaminomethyl-2-thiouridylate-methyltransferase (2-thiouridylase 1)	tRNA modification @ wobble base (MTU1)	MAT?	2.1.1.61	TRMU	Z74081.1	TRMU	AY062123.1	22q13.31
tRNA nucleotidyltransferase	Adds CCA to 3′ end of tRNAs	MAT	2.7.7.25	CCA1	M59870.1	TRNT1	AB063105.1	3p26.2
tRNA-Ala (mtDNA-encoded)	Transfer RNA - Ala	MAT	—	tA(TGC)Q	AJ011856.1	MTTA	J01415.1	mtDNA
tRNA-Arg1 (mtDNA-encoded) (codon AGR in yeast)	Transfer RNA - Arg (TCT)	MAT	—	tR(TCT)Q1	AJ011856.1	MTTR	J01415.1	mtDNA
tRNA-Arg2 (mtDNA-encoded) (codon CGN in yeast)	Transfer RNA - Arg (ACG)	MAT	—	tR(ACG)Q2	AJ011856.1	—	—	—
tRNA-Asn (mtDNA-encoded)	Transfer RNA - Asn	MAT	—	tN(GTT)Q	AJ011856.1	MTTN	J01415.1	mtDNA
tRNA-Asp (mtDNA-encoded)	Transfer RNA - Asp	MAT	—	tD(GTC)Q	AJ011856.1	MTTD	J01415.1	mtDNA
tRNA-Cys (mtDNA-encoded)	Transfer RNA - Cys	MAT	—	tC(GCA)Q	AJ011856.1	MTTC	J01415.1	mtDNA
tRNA-iMet (mtDNA-encoded)	Transfer RNA - fMet	MAT?	—	tM(CAT)Q2	M29654.1	—	—	—
tRNA-Gln[CUG] (CDC65) (nucleus-encoded)	Transfer RNA - Gln	MAT?	—	tQ(CUG)M	Z74331.1	—	—	—
tRNA-Gln (mtDNA-encoded)	Transfer RNA - Gln	MAT	—	tQ(TTG)Q	AJ011856.1	MTTQ	J01415.1	mtDNA
tRNA-Glu (mtDNA-encoded)	Transfer RNA - Glu	MAT	—	tE(TTC)Q	AJ011856.1	MTTE	J01415.1	mtDNA
tRNA-Gly (mtDNA-encoded)	Transfer RNA - Gly	MAT	—	tG(TCC)Q	AJ011856.1	MTTG	J01415.1	mtDNA
tRNA-His (mtDNA-encoded)	Transfer RNA - His	MAT	—	tH(GTG)Q	AJ011856.1	MTTH	J01415.1	mtDNA
tRNA-Ile (mtDNA-encoded)	Transfer RNA - Ile	MAT	—	tI(GAT)Q	AJ011856.1	MTTI	J01415.1	mtDNA
tRNA-Leu1 (mtDNA-encoded) (codon UUR in human)	Transfer RNA - Leu	MAT	—	tL(TAA)Q	AJ011856.1	MTTL1	J01415.1	mtDNA
tRNA-Leu2 (mtDNA-encoded) (codon CUN in human)	Transfer RNA - Leu (CUN)	MAT	—		AJ011856.1	MTTL2	J01415.1	mtDNA
tRNA-Lys (mtDNA-encoded)	Transfer RNA - Lys	MAT	—	tK(TTT)Q	AJ011856.1	MTTK	J01415.1	mtDNA
tRNA-Lys[CUU] (nucleus-encoded)	Transfer RNA - Lys	MAT	—	tRNA-K1	K00286.1	—	—	—
tRNA-Met (mtDNA-encoded)	Transfer RNA - Met	MAT	—	tM(CAT)Q1	AJ011856.1	MTTM	J01415.1	mtDNA
tRNA-Phe (mtDNA-encoded)	Transfer RNA - Phe	MAT	—	tF(GAA)Q	AJ011856.1	MTTF	J01415.1	mtDNA
tRNA-Pro (mtDNA-encoded)	Transfer RNA - Pro	MAT	—	tP(TGG)Q	AJ011856.1	MTTP	J01415.1	mtDNA
tRNA-Ser1 (mtDNA-encoded) (codon UCN in human)	Transfer RNA - Ser (UCN)	MAT	—	tS(TGA)Q2	AJ011856.1	MTTS1	J01415.1	mtDNA
tRNA-Ser2 (mtDNA-encoded) (codon AGY in human)	Transfer RNA - Ser (AGY)	MAT	—	tS(GCT)Q1	AJ011856.1	MTTS2	J01415.1	mtDNA
tRNA-Thr1 (mtDNA-encoded) (codon XXX)	Transfer RNA - Thr (XXX)	MAT	—	tT(xxx)Q2	AJ011856.1	MTTT	J01415.1	mtDNA
tRNA-Thr2 (mtDNA-encoded) (codon ACN in yeast)	Transfer RNA - Thr (ACA)	MAT	—	tT(TGT)Q1	AJ011856.1	—	—	—
tRNA-Trp (mtDNA-encoded)	Transfer RNA - Trp	MAT	—	tW(TCA)Q	AJ011856.1	MTTW	J01415.1	mtDNA
tRNA-Tyr (mtDNA-encoded)	Transfer RNA - Tyr	MAT	—	tY(GTA)Q	AJ011856.1	MTTY	J01415.1	mtDNA
tRNA-Val (mtDNA-encoded)	Transfer RNA - Val	MAT	—	tV(TAC)Q	AJ011856.1	MTTV	J01415.1	mtDNA
Translation factors								
Hypothetical protein Rmd9 [1]	Processing mt-mRNAs; rescues *Oxa1*	MAT	—	RMD9	Z72629.1	—	—	—
Hypothetical protein, potentially mitochondrial [5]	Translation release factor?	ND	—	YLR281c	U17243.1	FLJ38663	AF061733.1	12q24.31
Methionine aminopeptidase (N-terminal methionine excision [NME] pathway)	Removes N-terminal Met	MAT?	3.5.11.18			MAP1D	AY374142.1	2q31.1
Mitochondrial GTPase 2	Mito ribosome biogenesis	MAT	—	MTG2	U00027.2	GTPBP5	NM_015666.2	20q13.33
Peptide deformylase (N-terminal methionine excision [NME] pathway)	Removes formyl group from f-Met	NC	3.5.1.88			PDF	AF239156.1	16q22.1
Protein involved in mitochondrial metabolism, 73 kDa	Transcription-coupled translation	ND	—	SLS1	Z48452.1	—	—	—
Protein required for protein synthesis	Ribosome assembly?	MIM	—	MTG1	Z49807.1	MTG1	BC026039.1	10q26.3
Protein required for protein synthesis	Function unknown	MAT?	—	PET130	U37712.1	—	—	—
Regulator of protein synthesis	Interacts w. transl apparatus	ND	—	MBR1	M63309.1	—	—	—
Regulator of protein synthesis	Interacts with Mbr1	ND	—	ISF1	X72671.1	—	—	—
Ribosome recycling factor	Recycling of ribosomes	MAT	—	RRF1	AB016033.1	MRRF	BC013049.1	9q33.2
RNA 3′-terminal phosphate cyclase-like, potentially mitochondrial	Ribosome maturation	NC	—	RCL1	Z74752.1	RNAC	AK022904.1	9p24.1
Translation elongation factor 1 alpha 2	Function unknown	ND	—	TEF1	X00779.1	EEF1A2	X70940.1	20q13.33
Translation elongation factor 1 alpha 2	Function unknown	ND	—	TEF2	M15667.1	EEF1A2	X70940.1	20q13.33
Translational elongation factor G	Translocation from A to P site	MAT	—	MEF1	X58378.1	GFM1	AF367998.1	3q25.32
Translational elongation factor G2	Translocation from A to P site	MAT	—	MEF2	L25088.1	GFM2	AF367997.1	5q13.3
Translational elongation factor Ts	GDP release from EF-Tu	MAT	—	SPBC800.07c (Sp)	AL391034.1 (Sp)	TSFM	BC022862.1	12q13.3
Translational elongation factor Tu	Delivers tRNAs to A site	MAT	—	TUF1	Z75095.1	TUFM	L38995.1	16p11.2

Gene product	Function; comment	Location	E.C. #	Fungal		Animal		
				Symbol	GenBank	Symbol	GenBank	Human Chr.
Translational elongation factor 4 (LepA family GTP binding protein), potentially mito	Back-translates the ribosome	NC	—	GUF1	U22360.1	GUF1	BC036768.1	4p12
Translational initiation factor 2	Forms initiation complex	MAT	—	IFM1	X58379.1	MTIF2	L34600.1	2p16.1
Translational initiation factor 3	Initiation of protein syn	MAT?	—	YGL159w	Z72681.1	MTIF3	AF410851.1	13q12.2
Translation initiation factor eIF3, 135-kDa subunit	Mito clustering	NC	—	CLU1	AF004911.1	KIAA0664	AK023003.1	17p13.3
Translation release factor 1	Releases polypeptides	MAT?	—	MRF1	X60381.1	MTRF1	AF072934.1	13q14.11
Protein degradation, modification, and stability								
Aminopeptidase II (Ysc II) [1]	Leucyl aminopeptidase	ND	3.4.11.—	APE2	X63998.1	NPEPPS	AK096709.1	17q21.32
Anaphase promoting complex (APC) subunit 10 [1]	Cell cycle-regulated ubiquitin ligase	ND	—	DOC1	Z72762.1			
Aspartyl protease, putative, potentially mitochondrial [5]	Peptidase A1 activity?	ND	—	YPS3	Z73293.1			
Basigin (EMMPRIN/CD147) [2]	In gamma-secretase complex	ND	—			BSG	D45131.1	19p13.3
Calpain-10 (calcium-dependent cysteine protease)	Protease	MAT	3.4.22.—			CAPN10	AF089088.1	2q37.3
Cathepsin B, alt spliced form	Protease	ND	3.4.22.1			CTSB	L16510.1	8p23.1
Cathepsin D [2] [11]	Protease	ND	3.4.23.5			CTSD	M11233.1	11p15.5
Clp protease, proteolytic subunit	General cleavage protease	MAT	3.4.21.92	CNN01350 (Cn)	XM_568594.1 (Cn)	CLPP	Z50853.1	19p13.3
ClpX protease homolog	Chaperone/proteasome subunit	MAT	—	MCX1	Z36096.1	CLPX	AJ006267.1	15q22.32
Cysteine proteinase 1 (bleomycin hydrolase) [1]	General aminopeptidase	ND	3.4.22.40	YCP1	M97910.1			
Cytosol aminopeptidase, mitochondrial alt spl isoform [6]	Leucine/proline aminopeptidase	ND	3.4.11.—	SPAC13A11.05 (Sp)	Z54096.1	LAP3	AF061738.1	4p15.32
Heat shock protein 78 (*E. coli* ClpB/HSP104)	Component of mito proteolysis system	MAT		HSP78	L16533.1	SKD3	AL136909.1	11q13.4
Histone deacetylase, NAD-dependent (Sirtuin, type 3) (yeast Sir2p is nuclear)	Protein deacetylase	MAT				SIRT3	AF083108.2	11p15.5
Histone deacetylase, NAD-dependent (Sirtuin, type 4), potentially mito (Hst2p is nuc)	Protein deacetylase	NC				SIRT4	AF083109.1	12q24.31
Histone deacetylase, NAD-dependent (Sirtuin, type 5), potentially mito	Protein deacetylase	NC	3.5.1.—			SIRT5	AF083110.1	6p23
Hypothetical protein [1][3]	Carbon-nitrogen hydrolase motif	ND		NIT3	U19102.1	NIT2	AF260334.1	3q12.2
Hypothetical protein, potentially mitochondrial [5]	HSP40 chaperone	ND		JID1	Z49219.1	DNAJA2	AJ001309.1	16q11.2
Hypothetical protein [9], sim to Sc Ste24	Probable zinc metallo-protease	MOM		NCU03637.1 (Nc)	XM_322938.1 (Nc)			
Insulysin (insulin degrading enzyme), mitochondrial isoforms	Degrades insulin and Abeta peptide	ND	3.4.24.56			IDE	NM_004969.1	10q23.33
LON protease (alt spl in human)	ATP-dependent protease	MAT	3.4.21.—	PIM1	X74544.1	PRSS15	U02389.1	19p13.3
Matrix metalloproteinase-1 (interstitial collagenase)	Assoc w mitos during apoptosis	MOM?	3.4.24.7			MMP1	X54925.1	11q22.2
Metalloprotease, homolog of presequence protease Pre-P/Zn-MP in *Arabidopsis*	General oligopeptidase (Cym 1 is in IMS)	MAT		CYM1	U33007.1	PITRM1	AF061243.1	10p15.2
Metalloprotease of the AAA family [1] [8]	Function unknown	ND		AFG1	U18779.1	LACE1	AF520418.1	6q21
Metalloprotease of the AAA family (paraplegin-related protein) [12]	Human AFG3L1 mRNA is not translated	MIM	3.4.24.—	AFG3	X76643.1	Afg3l1 (m)	AF329695.1 (m)	16q24.3
Metalloprotease of the AAA family (m-AAA) (paraplegin-related protein) (Yta12)	MIM protein deg, with Yta12	MIM	3.4.24.—	AFG3	X76643.1	AFG3L2	Y18314.1	18p11
Metalloprotease of the AAA family (m-AAA) (paraplegin)	MIM protein deg, with Afg3	MIM	3.4.24.—	YTA12	U09358.1	SPG7	Y16010.1	16q24.3
Metalloprotease of the AAA family (i-AAA) (paraplegin-like)	MIM protein deg?	MIM	3.4.24.—	YME1	L14616.1	YME1L1	AJ132637.1	10p14
Metalloprotease of the AAA family, TOB3, isoform A [4]	In mammalian nucleoid	MIM	3.4.24.—	YME1	L14616.1	ATAD3A	BC007803.2	1p36.33
Metalloprotease of the AAA family, TOB3, isoform B: long and short alt spl forms	In mammalian nucleoid	MIM	3.4.24.—	YME1	L14616.1	ATAD3B	BC002542.2	1p36.33
Metalloprotease of the AAA family (i-AAA) accessory component YCL044c	Protein turnover; binds Yme1	MIM		MGR1	X59720.2	MPRP1	BC012915.1	1p32.1
Metalloprotease of the AAA family, metallopeptidase	Cleaves Oxa1 (not obligately)	MIM		OMA1	Z28312.1			
Methionine-S-sulfoxide reductase A	Repairs oxidized Met in proteins	MAT	1.8.4.6	MXR1	U18796.1	MSRA	AJ242973.1	8p23.1
Methionine-R-sulfoxide reductase B, mitochondrial form (MSRB3B)	Repairs oxidized Met in proteins	NC	1.8.4.6	YCL033c	X59720.1	MSRB3	BC040053.1	12q14.3
Methionine-R-sulfoxide reductase (pilin-like transcription factor) (MSRB2; CBS-1)	Repairs oxidized Met in proteins	NC	1.8.4.6	YCL033c	X59720.1	PILB	AF122004.1	10p12.2
Nardilysin, potentially mitochondrial	Protease @ Arg residues	NC	3.4.24.61			NRD1	U64898.1	1p32.2
N-terminal acetyltransferase 2 [1]	N-terminal acetylation of proteins	ND	2.3.1.88	NAT2	L25608.1			

O-sialoglycoprotein endopeptidase, potentially mitochondrial	Hydrolyzes O-sialoglycoproteins	NC	QR17	3.4.24.57	Z74152.1	OSGEPL1	AJ295148.1	2q32.2
Parkin (at least 3 alt spl isoforms) [10]	Has E3 ubiquitin ligase activity	MOM	—	—	—	PARK2	AB009973.1	6q26
Peptidyl-prolyl cis-trans isomerase [2] [10]	Protein folding; Modifies RYR-1	ND	FKB1	5.2.1.8	M57967.1	FKBP1A	M34539.1	20p13
Peptidyl-prolyl cis-trans isomerase (Cyclophilin)	Protein folding (cyclosporin-sensitive)	ND	CPH1	5.2.1.8	X17505.1	PPIA	Y00052.1	21q21.1
Peptidyl-prolyl cis-trans isomerase NIMA-interacting 1	Proapoptotic protein; binds JIP3, p-BIM-EL	ND	—	5.2.1.8	—	PIN1	U49070.1	19p13.2
Peptidyl-prolyl cis-trans isomerase NIMA-interacting 4 (parvulin 14; EPVH)	Rotamase; chaperone?	MAT	—	5.2.1.8	—	PIN4	AB009690.1	Xq13.1
Presenilin-1	In gamma-secretase complex	MIM	—	—	—	PSEN1	U40379.1	14q24.2
Presenilin-binding protein APH1A (Presenilin stabilization factor)	In gamma-secretase complex	ND	—	—	—	APH1A	AY113699.1	1q21.2
Presenilin-binding protein APH2 (Nicastrin; NCT)	In gamma-secretase complex	ND	—	—	—	NCSTN	AF240468.1	1q23.2
Presenilin-binding protein PEN2 (Presenilin enhancer protein 2)	In gamma-secretase complex	ND	—	—	—	PEN2	AF220053.1	19q13.12
Prohibitin complex, subunit 1	Inhibits m-AAA protease	MIM	PHB1	—	U16737.1	PHB	S85655.1	17q11.33
Prohibitin complex, subunit 2	Inhibits m-AAA protease; in nucleoid	MIM	PHB2	—	Z73016.1	REA	AF150962.1	12p13.31
Proline aminopeptidase, putative, potentially mitochondrial	Function unknown	NC	YER078c	3.4.—	U18839.1	LOC63929	BC004989.1	22q13.2
Protease, rhomboid-like (PARL)	Signal peptidase w. Yta10/Yta12	MIM	PCP1	3.4.21.—	Z72886.1	PARL	AF197937.1	3q27.1
Proteasome (26S) regulatory subunit, possibly mitochondrial	Protein deg	NC	RPN11	—	X79561.1	POH1	U86782.1	2q24.2
Protein disulfide isomerase [2]	Rearranges S-S bonds	ER->MOM	PDI1	5.3.4.1	X52313.1	PDIA3	D83485.1	15q15
Protein disulfide isomerase-related protein P5 [2]	Rearranges S-S bonds	NC	PDI1	5.3.4.1	X52313.1	PDIA6	D49489.1	2p25.1
Protein disulfide isomerase/prolyl 4-hydroxylase [2]	Rearranges S-S bonds	NC	PDI1	5.3.4.1	X52313.1	P4HB	M22806.1	17q25.3
Protein farnesyltransferase beta subunit Ram1	Beta subunit of farnesyltransferase	ND	RAM1	2.5.1.58	Z74138.1	—	—	—
Protein N-terminal amidase, potentially mitochondrial [5]	In N-end rule pathway	ND	NTA1	3.5.1.—	Z49562.1	—	—	—
Protein required for ATP-dependent degradation of apo-iso-1-cyt c (apo-Cyc1p)	Function unknown	IMS?	SUE1	—	U40829.1	—	—	—
Renin, alt spl mito-specific isoform (exclusive isoform in heart)	Neutral aspartyl protease	ND	—	3.4.23.15	—	REN	M10152.1	1q32.1
Saccharopepsin (vacuolar proteinase A) [1]	Peptidase of Prb1, Prc1, Pep4	ND	PEP4	3.4.23.25	M13358.1	—	—	—
Serine protease HtrA2/Omi (Smac-like inhibitor of XIAP) [1]	Pro-apoptotic protein; degrades IAP's	IMS	—	3.4.21.—	—	HTRA2	AF141305.1	2p13.1
Stomatin-like protease 2 (SLP2) [3] [4] [6]	Prohibitin-like chaperone; binds MFN2	MIM	—	—	—	STOML2	AF282596.1	9p13.3
Sulfite oxidase	Degrades S-containing aa's	IMS	NCU05633.1 (Nc) FG07500.1 (Gz)	1.8.3.1	XM_955019.1 (Nc) XM_387676.1 (Gz)	SUOX	L31573.1	12q13.2
Thimet oligopeptidase (yeast saccharolysin; human neurolysin)	ATP-independent metalloendopeptidase	IMS	PRD1	3.4.24.37	X76504.1	NLN	AJ300837.1	5q12.3
Transmembrane protease, serine 5 (Adrenal mito protease [AmP], alt spl mito isoform	Function unknown	NC	—	—	—	TMPRSS5	AB028140.1	11q23.2
Ubiquitin carboxyl-terminal hydrolase 4 (deubiquitinating enzyme)	Deubiquitinating enzyme	ND	UBP4	3.1.2.15	U02518.1	USP6	X63547.1	17p13.2
Ubiquitin carboxyl-terminal hydrolase 16 (no evidence yet that USP16 is mito)	Function unknown; deubiquitination?	MOM	UBP16	3.2.1.15	U41849.1	USP16	AF126736.1	21q21.3
Ubiquitin-like protein SBT2, dendritic cell-derived (DC-UbP)	Function unknown	ND	—	—	—	DC-UbP	AF251700.1	5q35.1
Ubiquitin-like protein SUMO-1 conjugating enzyme type UBC9 (Ubcp NC)	Binds DNM1L at mito fission sites	ND	UBC9	6.3.2.19	X82538.1	UBC9	X96427.1	16p13.3
Ubiquitin-like protein SMT3C (small ubiquitin-like modifier 1 SUMO1) (Smt3p NC)	Binds DNM1L at mito fission sites	ND	SMT3	—	U27233.1	UBL1	U67122.1	2q33.1
Ubiquitin-like-specific protease 2, potentially mitochondrial [5]	SUMO-specific protease activity	ND	ULP2	3.4.22.—	Z46861.1	—	—	—
Ubiquitin protein ligase (nedd4-like protein)	Binds Dnm1, Act1, Sla2	NC	RSP5	6.3.2.—	U18916.2	NEDD4L	DQ181796.1	18q21.31
Ubiquitin protein ligase binding protein	Binds Rsp5	NC	BUL1	—	D50083.1	—	—	—
Ubiquitin protein ligase, homolog of Goliath, ring finger protein 130	Ring-H2 zinc finger protein	NC	—	—	—	RNF130	AY083998.1	5q35.3
Ubiquitin protein ligase (MITOL); membrane-associated ring finger (C3HC4) 5 (RNF153)	Affects mito fission; binds MFN2 & DNM1L	MOM	—	—	—	MARCH5	NM_017824.4	10q23.32
VHL-interacting deubiquitinating enzyme 1, type I, potentially mitochondrial	Ubiquitin-specific protease	NC	—	—	—	VDU1	AF383172.1	1p31.1
Von Hippel-Lindau tumor suppressor, 18-kDa isoform	Ubiquitin ligase E3 component?	ND	—	—	—	VHL	AF010238.1	3p25.2
X-prolyl aminopeptidase P homolog	Releases Pro-linked aa's	ND	YLL029w	3.4.11.9	Z73134.1	XPNPEPL	X95762.1	10q25.3

RESPIRATORY CHAIN/OXIDATIVE PHOSPHORYLATION
Complex 1 (NADH dehydrogenase, ubiquinone [NDU])

Hypothetical protein (Fmp36) [1]	Complex I subunit I? Sim to NDUFB9	ND	—	—	U33050.4	LYRM3	BC047079.1	5q23.3
LYR motif-containing protein 2, potentially mitochondrial	Complex I subunit I? Sim to NDUFB9	NC	YDR493w	—	—	LYRM2	BC009782.1	6q15
Hypothetical protein, potentially mitochondrial [10]	Function unknown; similar to NDUFA9	ND	YLR290c	1.6.5.3/ 1.6.99.3	U17243.1	—	—	—

(continues)

Gene product	Function; comment	Location	E.C. #	Fungal		Animal		
				Symbol	GenBank	Symbol	GenBank	Human Chr.
Mitochondrial Respiration generator of Truncated DLST (MIRTD), translated fr exon 8	Assembly of complexes I and IV	IMS	2.3.1.61	KGD2	M34531.1	DLST	D16373.1	14q24.3
NADH dehydrogenase, external, rotenone-insensitive	Oxidizes cyto NADH	MIM	—	NDE1	Z47071.1	AMID	—	10q22.1
NADH dehydrogenase, internal, rotenone-insensitive	Oxidizes cyto NADH in MAT	MIM	1.6.5.3	NDI1	X61590.1	AMID	AK027403.1	10q22.1
NADH dehydrogenase, rotenone-insensitive	Oxidizes cyto NADH in IMS	MIM	—	NDE2	Z74133.1	AMID	AK027403.1	10q22.1
NDU assembly protein	Function unknown	MIM?	—	CIA30 (Nc)	AJ001726.1 (Nc)	NDUFAF1	AF151823.1	15q15.1
NDU assembly protein	Function unknown	MIM?	—	CIA84 (Nc)	AJ001712.1 (Nc)			
NDU assembly protein B17.2L (mimitin; paralog of structural subunit NDUFA12)	Assembly protein	MOM	—	NCU00278.1 (Nc)	XM_322363.1 (Nc)	NDUFA12L	AB183433.1	5q12.1
NDU subunit 1 (mtDNA-encoded)	Structural subunit	MIM	1.6.5.3	ND1 (Nc)	AY101381.1 (Cn)	MTND1	J01415.1	mtDNA
NDU subunit 2 (mtDNA-encoded)	Structural subunit	MIM	1.6.5.3	ND2 (Cn)	AY101381.1 (Cn)	MTND2	J01415.1	mtDNA
NDU subunit 3 (mtDNA-encoded)	Structural subunit	MIM	1.6.5.3	ND3 (Cp)	AY101381.1 (Cn)	MTND3	J01415.1	mtDNA
NDU subunit 4 (mtDNA-encoded)	Structural subunit	MIM	1.6.5.3	ND4 (Cn)	AY101381.1 (Cn)	MTND4	J01415.1	mtDNA
NDU subunit 4L (mtDNA-encoded)	Structural subunit	MIM	1.6.5.3	nd4L (Yl)	AJ307410.1 (Yl)	MTND4L	J01415.1	mtDNA
NDU subunit 5 (mtDNA-encoded)	Structural subunit	MIM	1.6.5.3	ND5 (Cn)	AY101381.1 (Cn)	MTND5	J01415.1	mtDNA
NDU subunit 6 (mtDNA-encoded)	Structural subunit	MIM	1.6.5.3	ND6 (Cn)	AY101381.1 (Cn)	MTND6	J01415.1	mtDNA
NDU subunit NDUFA1, 7.5 kDa	Structural subunit	MIM	1.6.5.3	NCU04781.1 (Nc)	XM_324137.1 (Nc)	NDUFA1	U54993.1	Xq24
NDU subunit NDUFA2, 8 kDa	Structural subunit	MIM	1.6.5.3	NUO-10.5 (Nc)	X69929.1 (Nc)	NDUFA2	AF047185.1	5q31.2
NDU subunit NDUFA3, 9 kDa	Structural subunit	MIM	1.6.5.3	None (Bc)	AL112163.1 (Bc)	NDUFA3	AF044955.1	19q13.42
NDU subunit NDUFA4, 9 kDa	Structural subunit	MIM	1.6.5.3	UM02800.1 (Um)	XM_400415.1 (Um)	NDUFA4	U94586.1	7p21.3
NDU subunit NDUFA5, 13 kDa	Structural subunit	MIM	1.6.5.3	NUO-32 (Nc)	X56237.1 (Nc)	NDUFA5	U53468.1	7q32
NDU subunit NDUFA6, 14 kDa (mt-small ribosomal protein C1-B14)	Structural subunit?	MIM	1.6.5.3	NUO-14.8 (Nc)	X76344.1 (Nc)	NDUFA6	AF047182.1	22q13.1
NDU subunit NDUFA7, 14.5 kDa (B14.5a)	Structural subunit; phosphorylated @ Ser-95	MIM	1.6.5.3			NDUFA7	AF054178.1	19p13.2
NDU subunit NDUFA8, 19 kDa	Structural subunit	MIM	1.6.5.3	NCU02472.1 (Nc)	XM_331670.1 (Nc)	NDUFA8	BC001016.1	9q33.2
NDU subunit NDUFA9, 39 kDa	Structural subunit	MAT	1.6.5.3	NUO-40 (Nc)	X56238.1 (Nc)	NDUFA9	L04490.1	12p13.32
NDU subunit NDUFA10, 42 kDa	Structural subunit; phosphorylated @ Ser-59	MAT	1.6.5.3			NDUFA10	AF087661.1	2q37.3
NDU subunit NDUFA11, 14.7 kDa (no MTS) (LOC126328)	Structural subunit	NC	1.6.5.3			NDUFA11	XM_050030.3	19p13.3
NDU subunit NDUFA12, 17.2 kDa (B17.2; DAP13)	Structural subunit	MIM	1.6.5.3	CNE03960 (Cn)	XM_571023.1 (Cn)	NDUFA12	AF217092.1	12q22?
NDU subunit NDUFA13, 16.7 kDa (same as GRIM19 pro-apoptotic protein)	Structural subunit	MIM	1.6.5.3	NCU09299.1 (Nc)	XM_331990.1 (Nc)	NDUFA13	AF286697.1	19p13.11
NDU subunit NDUFAB1, 8 kDa (acyl carrier protein [ACP])	Structural subunit	MIM	1.6.5.3	NCU05008.1 (Nc)	X59258.1 (Nc)	NDUFAB1	AF087660.1	16p12.2
NDU subunit NDUFB1, 7 kDa	Structural subunit	MIM	1.6.5.3			NDUFB1	AF054181.1	14q32.12
NDU subunit NDUFB2, 8 kDa	Structural subunit	MIM	1.6.5.3			NDUFB2	AF050639.1	7q34
NDU subunit NDUFB3, 12 kDa	Structural subunit	MIM	1.6.5.3	NCU09002.1 (Nc)	XM_331393.1 (Nc)	NDUFB3	AF035839.1	2q33.1
NDU subunit NDUFB4, 15 kDa	Structural subunit	MIM	1.6.5.3			NDUFB4	AF044957.1	3q13.33
NDU subunit NDUFB5, 16 kDa	Structural subunit	MIM	1.6.5.3			NDUFB5	AF047181.1	3q26.33
NDU subunit NDUFB6, 17 kDa	Structural subunit	MIM	1.6.5.3			NDUFB6	AF035840.1	9p13.3
NDU subunit NDUFB7 (B18)	Structural subunit	MIM	1.6.5.3			NDUFB7	M33374.1	19p13.12
NDU subunit NDUFB8, 19 kDa	Structural subunit	MIM	1.6.5.3	NCU09460.1 (Nc)	XM_332151.1 (Nc)	NDUFB8	AF044958.1	10q24.31
NDU subunit NDUFB9, 22 kDa	Structural subunit	MIM	1.6.5.3	UM05625.1 (Um)	XM_403240.1 (Um)	NDUFB9	AF044956.1	8q13.3
NDU subunit NDUFB10, 22 kDa	Structural subunit	MIM	1.6.5.3			NDUFB10	AF044954.1	16p13.3
NDU subunit NCUFB11, 17.3 kDa (P17.3)	Structural subunit	NC	1.6.5.3			NDUFB11	AF044213.1	Xp11.23
NDU subunit NDUFC1, 6 kDa	Structural subunit	MIM	1.6.5.3			NDUFC1	AF047184.1	4q31.1
NDU subunit NDUFC2, 14.5 kDa	Structural subunit	MIM	1.6.5.3			NDUFC2	AF087659.1	11q14.1
NDU subunit NDUFS1, 75 kDa; caspase substrate in apoptosis	Structural subunit; in FeS component	MIM	1.6.5.3	NUO-78 (Nc)	X57602.1 (Nc)	NDUFS1	X61100.1	2q33.3
NDU subunit NDUFS2, 49 kDa	Structural subunit; in FeS component	MIM	1.6.5.3	NCU02534.1 (Nc)	XM_331732.1 (Nc)	NDUFS2	AF031360.1	1q23.3
NDU subunit NDUFS3, 30 kDa	Structural subunit; in FeS component	MIM	1.6.5.3	NCU04074.1 (Nc)	XM_323392.1 (Nc)	NDUFS3	AF067139.1	11p11.2
NDU subunit NDUFS4, 18 kDa	Structural subunit; in FeS component	MIM	1.6.5.3	NUO-21 (Nc)	X78082.1 (Nc)	NDUFS4	AF020351.1	5q11.1
NDU subunit NDUFS5, 15 kDa	Structural subunit; in FeS component	MIM	1.6.5.3			NDUFS5	AF047434.1	1p34.3
NDU subunit NDUFS6, 13 kDa	Structural subunit; in FeS component	MIM	1.6.5.3	NCU00484.1 (Nc)	XM_322569.1 (Nc)	NDUFS6	AF044959.1	5p15.33
NDU subunit NDUFS7, 20 kDa	Structural subunit; in FeS component	MIM	1.6.5.3	NCU03953.1 (Nc)	XM_323271.1 (Nc)	NDUFS7	AF060512.1	19p13.3
NDU subunit NDUFS8, 23 kDa	Structural subunit; in FeS component	MIM	1.6.5.3	NCU05009.1 (Nc)	X95547.1 (Nc)	NDUFS8	U65579.1	11q13.3
NDU subunit NDUFV1, 51 kDa	Structural subunit; in Fp component	MIM	1.6.5.3	NCU04044.1 (Nc)	XM_323362.1 (Nc)	NDUFV1	AF053070.1	11q13.3

Protein	Function	Localization	EC	S. cerevisiae / N. crassa	Accession	Human gene	GenBank	Locus
NDU subunit NDUFV2, 24 kDa	Structural subunit; in Fp component	MIM	1.6.5.3	NUO-24 (Nc)	X78083.1 (Nc)	NDUFV2	M22538.1	18p11.22
NDU subunit NDUFV3, 9 kDa	Structural subunit; in Fp component	MIM	1.6.5.3	—	—	NDUFV3	X99726.1	21q22.3
Complex II (succinate dehydrogenase [SDH])								
SDH assembly protein (similar to Hsp60p)	SDH-specific chaperone	MIM		TCM62	Z35913.1			—
SDH cytochrome b large subunit (CII-3)	Anchor w Sdh4; binds Ub	MIM	1.3.5.1	SDH3	X73884.1	SDHC	D49737.1	1q23.3
SDH cytochrome b small subunit (CII-4)	Anchor w Sdh3; reduces Ub	MIM	1.3.5.1	SDH4	L26333.1	SDHD	AB006202.1	11q23.1
SDH flavoprotein (Fp) subunit, liver isoform (Type 1) (Tyr-629/Val-658)	Catalytic subunit with Sdh2	MIM	1.3.5.1	SDH1	M86746.1	SDHA	D30648.1	5p15.33
SDH flavoprotein (Fp) subunit, heart isoform (Type II) (Phe-629/Ile-658)	Catalytic subunit with Sdh2	MIM	1.3.5.1	SDH1	M86746.1	SDHA	L21936.1	5p15.33
SDH flavoprotein (Fp) subunit homolog (SDH1b), centromeric member of gene pair	Function unknown	MIM	1.3.5.1	YJL045w	Z49320.1	—		
SDH iron-sulfur protein (Ip) subunit	Catalytic subunit w Sdh1	MIM	1.3.5.1	SDH2	Z73146.1	SDHB	U17248.1	1p36.13
Complex III (ubiquinone cytochrome c reductase [UCR])								
Alternate oxidase	Oxid ubiquinol; bypasses compl III/IV	MIM?		aod-1 (Nc)	AY140653.1 (Nc)			—
Alternate oxidase	Oxid ubiquinol; bypasses compl III/IV	MIM?		[aod-3] (Nc)	XM_324230.1 (Nc)			—
Translational activator of COB mRNA	Cyt b mRNA processing	MAT		CBP6	M10154.1			—
Translational activator of COB mRNA, 23 kDa	Ribosome positioning?	MIM		CBS1	X15650.1			—
Translational activator of COB mRNA, 45 kDa	Ribosome positioning?	MIM		CBS2	X13523.1			—
UCR assembly protein	Iron chaperone; assembles Rieske ISP	MIM		BCS1	S47190.1	BCS1L	AF026849.1	2q35
UCR assembly protein, 20 kDa	Function unknown	MIM		CBP4	U10700.1			—
UCR assembly protein, 34.5 kDa (BZFB [β])	Function unknown	MIM		CBP3	J04830.1	BZFB	AK001712.1	20q11.22
UCR assembly-like protein similar to Abc1, potentially mitochondrial	Function unknown	NC		YLR253w	U20865.1	ADCK5	AK092773.1	8q24.3
UCR subunit 1, 47 kDa in human	Core protein 1	MIM	1.10.2.2	COR1	J02636.1	UQCRC1	L16842.1	3p21.3
UCR subunit 2, 45 kDa in human	Core protein 2	MIM	1.10.2.2	QCR2	X05120.1	UQCRC2	J04973.1	16p12
UCR subunit 3, 35 kDa in human (COB) (mtDNA-encoded)	Cytochrome b	MIM	1.10.2.2	CYTB	A0011856.1	MTCYB	J01415.1	mtDNA
UCR subunit 4, 28 kDa in human	Cytochrome c1	MIM	1.10.2.2	CYT1	X00791.1	CYC1	J04444.1	8q24.3
UCR subunit 5, 22 kDa in human (196-aa mature polypeptide)	Rieske iron-sulfur protein	MIM	1.10.2.2	RIP1	M23316.1	UQCRFS1	L32977.1	19q12
UCR subunit 6, 13.4 kDa in human (also called subunit 8)	Cyt b-associated protein	MIM	1.10.2.2	QCR7	X00256.1	UQCRB	M22348.1	8q22.1
UCR subunit 7, 9.5 kDa in human	Core-associated protein	MIM	1.10.2.2	QCR8	X05550.1	QP-C	D50369.1	5q23.3
UCR subunit 8, 9.2 kDa in human (also called subunit 6)	Hinge protein	MIM	1.10.2.2	QCR6	X00551.1	UQCRH	M36647.1	1p33
UCR subunit 9, 8.0 kDa in human (78-aa MTS from subunit 5)	Rieske ISP targeting peptide	MIM	1.10.2.2			UQCRFS1	L32977.1	19q12
UCR subunit 10, 7.2 kDa in human	Cyt c1-associated protein	MIM	1.10.2.2	QCR9	M59797.1	UCRC	AF112217.1	22q12.2
UCR subunit 11, 6.4 kDa in human	Rieske ISP-associated protein	MIM	1.10.2.2	QCR10	U07275.1	UQCR	D55636.1	19p13.3
Complex IV (cytochrome c oxidase [COX])								
COX assembly protein Cox10	Heme A farnesyltransferase	MIM	2.5.1.29	COX10	M55566.1	COX10	NM_001303.2	17p12
COX assembly protein Cox11	Inserts Cu into CuB?	MIM		COX11	X55731.1	COX11	NM_004375.2	17q22
COX assembly protein Cox14	Function unknown	MIM		COX14	U15040.1			
COX assembly protein Cox 15 (2 spl variants in human)	Heme O hydroxylase w Yah1/Arh1	MIM		COX15	L38643.1	COX15	NM_078470.2	10q24.2
COX assembly protein Cox16	Function unknown	MIM		COX16	Z49278.1	C14orf112	NM_016468.3	14q24.2
COX assembly protein Cox17	Copper transf (to COX11 and SCO?); MIA	Cyto/IMS		COX17	L75948.1	COX17	NM_005694.1	3q13.33
COX assembly protein Cox18 (OXA1L2), 2 alt spl isoforms	Exports Cox2 C-term to IMS	MIM?		COX18	U59742.1	COX18	AK096310.2	4q13.3
COX assembly protein Cox19	Copper chaperone? MIA	Cyto/IMS		COX19	Z73123.1	COX19	AY957566.1	7p22.3
COX assembly protein Cox20	Cox2 maturation/assembly	MIM		COX20	Z48612.1	FAM36A	BC018519.1	1q44
COX assembly protein Cox23	Copper chaperone; C-term sim to Cox17	Cyto/IMS		COX23	U00059.1	CHCHD7	BC002546.2	8q12.1
COX assembly protein Cox24 (Qri5)	Processing and translation of COX1 mRNA	MAT?		QRI5	S43721.1			
COX assembly protein Mss2	Exports Cox2 to MIM, w. Cox18/Pnt1	MAT?		MSS2	X81477.1			
COX assembly protein Pet100	Assembles subunits VII, VIIa, VIII	ND		PET100	U91943.1		NM_020536.2	20p11.23
COX assembly protein Pet117	Function unknown; in 5′ end of CSRP2BP	NC		PET117	L06066.1	[PET117]	BC047722.1	2q11.2
COX assembly protein Pet191	Function unknown	NC		PET191	L06067.1	MGC52110	AF026852.1	17p13.1
COX assembly protein Sco1	Cu transp? Redox sensor?	MIM		SCO1	X17441.1	SCO1		
COX assembly protein Sco2	Cu transp? Redox sensor?	MIM		SCO2	Z35893.1	SCO2	NM_005138.1	22q13.33
COX assembly protein Shy1	Maturation of heme a3-CuB center	MIM		SHY1	Z72897.1	SURF1	BC028314.1	9q34.2

(continues)

Appendix 5 (continued)

Gene product	Function; comment	Location	E.C. #	Fungal		Animal		
				Symbol	GenBank	Symbol	GenBank	Human Chr.
COX1 RNA-associated protein (leucine-rich pentatricopeptide repeat containing protein)	Stability/translation of COX1 mRNA	MIM	—	PET309	L06072.1	LRPPRC	AY289212.1	2p21
COX subunit I (mtDNA-encoded)	Structural subunit	MIM	1.9.3.1	COX1	AJ011856.1	MTCO1	J01415.1	mtDNA
COX subunit II (mtDNA-encoded)	Structural subunit	MIM	1.9.3.1	COX2	AJ011856.1	MTCO2	J01415.1	mtDNA
COX subunit III (mtDNA-encoded)	Structural subunit	MIM	1.9.3.1	COX3	AJ011856.1	MTCO3	J01415.1	mtDNA
COX subunit IV in yeast (subunit Vb in human)	Structural subunit	MIM	1.9.3.1	COX4	X01418.1	COX5B	NM_001862.2	2q11.2
COX subunit Va in yeast (subunit IV, isoform 1 in human)	Structural subunit	MIM	1.9.3.1	COX5A	X02561.1	COX4I1	NM_001861.2	16q24.1
COX subunit Vb in yeast (subunit IV, isoform 2 in human)	Structural subunit	MIM	1.9.3.1	COX5B	M17799.1	COX4I2	NM_032609.2	20q11.21
COX subunit VI in yeast (subunit Va in human)	Structural subunit	MIM	1.9.3.1	COX6	M10138.1	COX5A	NM_004255.2	15q24.1
COX subunit VIa in yeast (subunit VIa-L in human)	Structural subunit	MIM	1.9.3.1	COX13	X72970.1	COX6A1	NM_004373.2	12q24.31
COX subunit VIa in yeast (subunit VIa-H in human)	Structural subunit	MIM	1.9.3.1	COX13	X72970.1	COX6A2	NM_005205.2	16p11.2
COX subunit VIb in yeast (subunit VIb in human)	Structural subunit	IMS	1.9.3.1	COX12	M98332.1	COX6B1	NM_001863.3	19q13.12
COX subunit VIb, isoform 2 (testis-specific)	Structural subunit	MIM	1.9.3.1	COX12	M98332.1	COX6B2	NM_144613.4	19q13.42
COX subunit VIb, isoform 3 [2]	Function unknown	NC	1.9.3.1?	COX12	M98332.1	C1orf31	BC025793.1	1q42.2
COX subunit VII in yeast (subunit VIIa-H in human)	Structural subunit	MIM	1.9.3.1	COX7	X51506.1	COX7A1	NM_001864.2	19q13.12
COX subunit VIIa in yeast (subunit VIc in human)	Structural subunit	MIM	1.9.3.1	COX9	J02633.1	COX6C	NM_004374.2	8q22.2
COX subunit VIIa-L in human	Structural subunit	MIM	1.9.3.1			COX7A2	NM_001865.2	6q14.1
COX subunit VIIa-related protein in human	Function unknown	NC				COX7A2L	NM_004718.2	2p21
COX subunit VIIb in human	Structural subunit	MIM	1.9.3.1			COX7B	Z14244.1	Xq21.1
COX subunit VIIb, isoform 2	Function unknown	MIM?	1.9.3.1			COX7B2	BC035923.1	4p12
COX subunit VIII in yeast (subunit VIIc in human)	Structural subunit	MIM	1.9.3.1	COX8	J02634.1	COX7C	X16560.1	5q14.3
COX subunit VIIIa-H in mouse (COX8-1)	Structural subunit	MIM	1.9.3.1			Cox8b (m)	U15541.1 (m)	11q13.1
COX subunit VIII (VIIIa-L in mouse) (COX8-2)	Structural subunit	MIM	1.9.3.1			COX8A	J04823.1	11q13.1
COX subunit VIII, isoform III (COX8-3)	Complex IV subunit VIII-3	MIM	1.9.3.1			COX8C	AY161004.1	14q32.13
Mitochondrial Respiration generator of Truncated DLST (MIRTD)	Assembly of complexes 1 and 4	IMS	2.3.1.61			DLST	D16373.1	14q24.3
Translational activator of COX1 mRNA	Translation of COX1 mRNA	MIM		MSS51	J01487.1			
Translational activator of COX2 mRNA	Translation of COX2 mRNA	MIM		PET111	M17143.1			
Translational activator of COX3 mRNA	Translation of COX3 mRNA	MIM		PET54	X13427.1			
Translational activator of COX3 mRNA	Translation of COX3 mRNA	MIM		PET122	X07558.1			
Translational activator of COX3 mRNA	Translation of COX3 mRNA	MIM		PET127	Z74925.1			
Translational activator of COX3 mRNA, may bind Pet309	Translation of COX3 mRNA	MIM		PET494	K03520.1			
Complex V (ATP synthase [ATPase])								
ATPase assembly protein	Assembly of Atp6 in F0 (A6 chaperone)	MIM		ATP10	J05463.1			
ATPase assembly protein	Interacts w. subunit beta	MAT		ATP11	M87006.1	ATPAF1	AK026004.1	1p33
ATPase assembly protein	Interacts w. subunit alpha	MAT		ATP12	M61773.1	ATPAF2	AF052185.1	17p11.2
ATPase assembly protein	Assembly of F0; translation of Atp6	MIM		ATP22	U32306.1			
ATPase assembly protein (metalloprotease) (ATP23)	Processing of Atp6 w Atp10; binds Phb1/2	MIM		YNR020c	Z71635.1			
ATPase coupling factor B	Structural subunit of F0	MIM	3.6.1.14			ATPW	AY052377.1	14q21.3
ATPase expression protein	Assoc w Atp6/Atp8 expression	MOM		NCA2	AAB68052.1			
ATPase expression protein	Assoc w Atp6/Atp8 expression	ND		NCA3	L20786.1			
ATPase expression protein (sim to Rice PPR8-1 pentatricopeptide repeat)	Atp6/8 mRNA processing	MIM		AEP3	Z71255.1	PTCD3	BC011832.2	2p11.2
ATPase inhibitor protein, 10 kDa (= C-terminus of E. coli subunit epsilon)	ATPase inhibitor	MIM		INH1	D00443.1	ATPIF1	AB029042.1	1p35.3
ATPase interacting protein	Affects expression of Atp2	MIM?		YJR120w	Z49620.1			
ATPase proteolipid 68 MP homolog, 6.8 kDa	Function unknown	ND				MP68	AF054175.1	14q32.33
ATPase 9 RNA-associated protein, 59 kDa	Req. for translation of Atp9	ND		AEP1	M80615.1			
ATPase 9 RNA-associated protein, 67.5 kDa	Accum of ATP9 mRNA	ND		AEP2	M59860.1			
ATPase stabilizing factor, 9 kDa	Function unknown	MIM		STF1	D00347.1			
ATPase stabilizing factor, 15 kDa	Function unknown	MIM		STF2	D00444.1			
ATPase stabilizing factor, 18 kDa	Stabilizes/folds Atp12	MAT		FMC1	Z38125.1			
ATPase subunit alpha	Structural subunit of F1	MIM	3.6.1.14	ATP1	J02603.1	ATP5A1	D14710.1	18q21.1
ATPase subunit beta	Structural subunit of F1	MIM	3.6.1.14	ATP2	M12082.1	ATP5B	M27132.1	12p13.3

Name	Function	Loc	EC	Yeast gene	Yeast acc.	Human gene	Human acc.	Location
ATPase subunit gamma	Structural subunit of F1	MIM	3.6.1.14	ATP3	U08318.1	ATP5C1	D16561.1	10p14
ATPase subunit delta (= E. coli subunit epsilon)	Structural subunit of F1; couples c to gamma	MIM	3.6.1.14	ATP16	Z21857.1	ATP5D	X63422.1	19p13.3
ATPase subunit epsilon (not E. coli subunit epsilon)	Structural subunit of F1	MIM	3.6.1.14	ATP15	X64767.1	ATP5E	BC001690.1	20q13.31
ATPase subunit 4 (human subunit b)	Structural subunit of F0	MIM	3.6.1.14	ATP4	X06732.1	ATP5F1	X60221.1	1p13.2
ATPase subunit 5 (OSCP)	Structural subunit of F0	MIM	3.6.1.14	ATP5	X12356.1	ATP5O	X83218.1	21q22.12
ATPase subunit 6 (subunit a in E. coli) (mtDNA-encoded)	Structural subunit of F0	MIM	3.6.1.14	ATP6	AJ011856.1	MTATP6	J01415.1	mtDNA
ATPase subunit 7 (human subunit d)	Structural subunit of F0	MIM	3.6.1.14	ATP7	Z28016.1	ATP5JD	AF087135.1	17q25.1
ATPase subunit 8 (mtDNA-encoded)	Structural subunit of F0	MIM	3.6.1.14	ATP8	AJ011856.1	MTATP8	J01415.1	mtDNA
ATPase subunit 9 (human subunit c), P1 isoform (mtDNA-encoded in yeast)	Structural subunit of F0	MIM	3.6.1.14	ATP9	AJ011856.1	ATP5G1	D13118.1	17q21.32
ATPase subunit 9 (human subunit c), P2 isoform (mtDNA-encoded in yeast)	Structural subunit of F0	MIM	3.6.1.14	ATP9	AJ011856.1	ATP5G2	D13119.1	12q13.13
ATPase subunit 9 (human subunit c), P3 isoform (mtDNA-encoded in yeast)	Structural subunit of F0	MIM	3.6.1.14	ATP9	AJ011856.1	ATP5G3	U09813.1	2q31.1
ATPase subunit e (synonym ATP21 in yeast)	Structural subunit of F0	MIM	3.6.1.14	TIM11	U32517.1	ATP5I	D50371.1	4p16.3
ATPase subunit f	Structural subunit of F0	MIM	3.6.1.14	ATP17	U72652.1	ATP5J2	AF088918.1	7q22.1
ATPase subunit g	Structural subunit of F0	MIM	3.6.1.14	ATP20	Z49919.1	ATP5L	AF092124.1	11q23.3
ATPase subunit h (human subunit F6 [coupling factor 6])	Structural subunit of F0	MIM	3.6.1.14	ATP14	U51673.1	ATP5J	M37104.1	21q21.3
ATPase subunit i (also called subunit j)	Dimerizes ATPase (via F0?)	MIM	3.6.1.14	ATP18	AF073791.1	—	—	—
ATPase subunit k (ORF YOL077w-a)	Dimerizes w subunits e g	MIM	3.6.1.14	ATP19	Z74820.1	—	—	—
Heme/cytochrome/FeS group metabolism								
5-Aminolevulinate synthase, non-erythroid specific (ALAS-N)	Heme syn	MAT	2.3.1.37	HEM1	M26329.1	ALAS1	X56351.1	3p21.1
5-Aminolevulinate synthase 2, erythroid-specific (ALAS-E)	Heme syn	MAT	2.3.1.37	HEM1	M26329.1	ALAS2	X56352.1	Xp11.21
Biliverdin reductase A	Heme deg	ND	1.3.1.24	—	—	BLVRA	X93086.1	7p13
Chaperone, homolog of E. coli Hsc20/DnaJ	FeS cluster syn/assembly; binds Ssq1	ND	—	JAC1	Z72540.1	HSC20	AY191719.1	22q12.1
Coproporphyrinogen III oxidase, long isoform (yeast Hem13 is cytosolic)	Heme syn	IMS	1.3.3.3	—	—	CPOX	BC023551.2	3q12.1
Cysteine desulfurase	Provides S to FeS clusters	MAT		NFS1	M98808.1	NIFS	AF097025.1	20q11.22
Cytochrome c, isoform 1 (somatic isoform in humans and rodents)	Electron transfer	IMS		CYC1	V01298.1	CYCS	BT006946.1	7p15.2
Cytochrome c, isoform 1 (testis-specific isoform in rodents)	Electron transfer	IMS	—	CYC1	V01298.1	CYCT (r)	M20623.1 (r)	
Cytochrome c, isoform 2	Electron transfer	IMS	—	CYC7	V01299.1	ZC116.2 (w)	Z74046.1 (w)	
Cytochrome c heme lyase (CCHL)	Cytochrome c syn	MIM	4.4.1.17	CYC3	X04776.1	HCCS	U36787.1	Xp22.22
Cytochrome c1 heme lyase (CC1HL)	Cytochrome c1 syn	MIM	4.4.1.—	CYT2	X67017.1			
Ferritin, mitochondrial form (intronless gene)	Fe transp	MAT	1.16.3.1			FTMT	BC034419.1	5q23.1
Ferrochelatase	Heme syn	MIM	4.99.1.1	HEM15	J05395.1	FECH	D00726.1	18q21.3
Frataxin	Heme and Fes syn w. Hem15	MAT?	—	YFH1	Z74168.1	FRDA	U43747.1	9q13
Glutaredoxin 5	FeS cluster syn/assembly	MAT	—	GRX5	U39205.1	GLRX5	BC047680.1	14q32.13
Heat shock protein 70 family member, similar to human mortalin-2	FeS cluster syn/assembly; binds Isu1	MAT?	—	SSQ1	U19103.1	HSPA9B	L15189.1	5q31.2
Heme oxygenase-1 (in ER and mitochondria) (Sc Hmx1 may not have HO activity)	Heme deg	MIM	1.4.99.3	HMX1 (Ca)	XM_713840.1 (Ca)	HMOX1	X06985.1	22q12.3
Iron metabolism protein	FeS cluster syn/assembly; binds Ssq1	MAT	—	ISU1	U43703.1	ISCU	AV705042.1	12q23.3
Iron metabolism protein	FeS cluster syn/assembly	MAT	—	ISU2	Z75133.1	ISCU	AV705042.1	12q23.3
Iron metabolism protein NFU, isoform 1 (mitochondrial)	FeS cluster syn/assembly	ND	—	NFU1	Z28040.1	HIRIP5	AY286306.1	2p13.3
Iron-sulfur assembly protein	FeS cluster syn/assembly	MAT	—	ISA1	Z73132.1	HBLD2	BC002675.1	9q31.33
Iron-sulfur assembly protein	FeS cluster syn/assembly	MAT	—	ISA2	Z71255.1	HBLD1	BC015771.1	14q24.3
Iron-sulfur assembly protein; binds Nfs1	FeS cluster syn/assembly; binds Nfs1	MAT	—	ISD11	AY558455.1	LYRM4	BC009552.2	6p25.1
Molybdenum co-factor (MOCO) sulphurase C-terminal domain containing 1	Hypothetical FeS protein; function unknown	MIM	—			MOSC1	BC010619.1	1q41
Porphobilinogen deaminase, alt spl form, potentially mito (yeast Hem3 is cyto)	Heme biosynthesis, 3rd step	NC	2.5.1.61			HMBS	M95623.1	11q23.3
Protoporphyrinogen oxidase	Heme syn	MIM	1.3.3.4	HEM14	Z71381.1	PPOX	D38537.1	1q23.3
Sulfhydryl oxidase, FAD-linked	FeS protein maturation in cytosol	IMS	—	ERV1	X60722.1	ALR	U31176.1	16p13.3
Ubiquinone metabolism								
Electron transfer flavoprotein dehydrogenase	Reduction of ubiquinone	MIM	1.5.5.1	YOR356w	Z75264.1	ETFDH	S69232.1	4q32.1
Farnesyl diphosphate synthase homolog (KIAA1293)	Isoprenoid/mevalonate syn	ND	2.5.1.10	—	—	FDPS	NM_002004.2	1q22

(continues)

Gene product	Function; comment	Location	E.C. #	Fungal Symbol	Fungal GenBank	Animal Symbol	Animal GenBank	Human Chr.
Geranylgeranyl diphosphate synthase [5] (human NC)	Polyprenyl syn; ubiquinone syn?	ND	2.5.1.-	BTS1	U31632.1	GGPS1	NM_004837.3	1q42.3
Decaprenyl-diphosphate synthase subunit 1 (transprenyltransferase) (Sc COQ1)	Ubiquinone syn	MIM	2.5.1.-	dps1 (Sp)	D84311.1 (Sp)	PDSS1	NM_014317.3	10p12.1
Decaprenyl-diphosphate synthase subunit 2 (transprenyltransferase) (Sc COQ1)	Ubiquinone syn	MIM	2.5.1.-	dlp1 (Sp)	AB118853.1 (Sp)	PDSS2	AB210839.1	6q21
p-Hydroxybenzoate:polyprenyltransferase	Ubiquinone syn	MIM	2.5.1.39	COQ2	M81698.1	COQ2	NM_015697.6	4q21.23
Hexaprenyl dihydroxybenzoate methyltransferase	Ubiquinone syn	MIM	2.1.1.114	COQ3	M73270.1	COQ3	NM_017421.2	6q16.3
Polyprenyl-hydroxybenzoate decarboxylase?, potentially mito	Ubiquinone syn?	NC		FG04669.1 (Gz)	XM_384845.1 (Gz)			—
Polyprenyl-hydroxybenzoate decarboxylase?, potentially mito	Ubiquinone syn?	NC		FG06434.1 (Gz)	XM_386610.1 (Gz)			—
Ubiquinone biosynthesis protein COQ4	Ubiquinone syn; function unknown	MIM	2.5.1.-	COQ4	AF005742.1	COQ4	NM_016035.1	9q34.13
Ubiquinone C-methyltransferase	Ubiquinone syn	MIM	2.1.1.-	COQ5	Z97052.2	COQ5	NM_032314.2	12q24.31
Ubiquinone monooxygenase, flavin-dependent	Ubiquinone syn	ND	1.14.13.-	COQ6	AF003698.1	COQ6	NM_182476.1	14q24.3
Ubiquinone monooxygenase or hydroxylase (yeast Cat5)	Ubiquinone syn	MIM		COQ7	X82930.1	COQ7	NM_016138.3	16p12.3
Ubiquinone biosynthesis protein (same as Coq8) [12]	Ubiquinone syn, early step	MIM		ABC1	X59027.1	CABC1	NM_020247.4	1q42.13
Ubiquinone biosynthesis protein (YLR201c) [6]	Ubiquinone syn	ND		COQ9	U14913.1	COQ9	NM_020312.1	16q13
Ubiquinone biosynthesis protein, potential	May bind Q6 in yeast	MIM		COQ10	Z74750.1	COQ10A	NM_144576.2	12q13.3
Ubiquinone biosynthesis protein, potential	Oligoketide cyclase/lipid transport protein?	NC		COQ10	Z74750.1	COQ10B	NM_025147.2	2q33.1
Ubiquinone biosynthesis protein AARF, putative, potentially mito	Ubiquinone syn?	NC		YPL109c	U43503.1	ADCK2	NM_052853.3	7q34
SIGNAL TRANSDUCTION								
A-kinase anchoring protein 1, 149-kDa (AKAP121) (84-kDa in spermatids [s-AKAP84])	Binds protein kinase PKAII	MIM				AKAP1	U34074.1	17q22
A-kinase anchoring protein 10, 74 kDa	Binds protein kinase	MOM?				AKAP10	AF037439.1	17p11.2
A-kinase anchoring protein, 25 kDa	Binds type II regulatory subunit of PKA	MOM?				RAB32	AF498958.1	6q24.3
A-kinase anchoring protein WAVE-1 (WASP family 1)	In sperm mito sheath	MOM?				WASF1	AF134303.1	6q21
AMAP-1	Binds AMY-1	MOM?				AMAP-1	AB083068.1	17q21.33
AMY-1 (c-Myc binding protein)	Binds AKAP149 and s-AKAP84	MOM?				AMY1	D50692.1	1p34.3
AMY-1-associating protein expressed in testis 1 (C3orf15), alpha isoform (testis-specific)	Binds AMY-1/AKAP84 PKA RII subunit	MOM?				AAT1	AF440403.1	3q13.33
Androgen receptor, mitochondrial loc in sperm	Binds androgen, HIPK3, RANBP9	ND				AR	M20132.1	Xq12
cAMP-dependent protein kinase, catalytic subunit	Reg mito enzymatic content	ND	2.7.1.37	TPK3	M17074.1	PKX1	X85545.1	Xp22.3
cAMP-dependent protein kinase, regulatory subunit RII alpha	Binds AKAP1	MOM				PRKAR2A	X14968.1	3p21.31
cAMP-dependent protein kinase type II-beta regulatory subunit	Binds AKAP1	MIM				PRKAR2B	M31158.1	7q22.3
cAMP-responsive element binding protein (CREB)	Signalling; binds Cre site in D-loop	MAT?				CREB1	X55545.1	2q33.3
Cellular retinoic acid-binding protein, type I	Retinoid metab?	ND				CRABP1	S74445.1	15q25.1
C-fos proto-oncogene	To mitos after kainate?	ND				FOS	K00650.1	14q24.3
Discs large tumor suppressor protein (Drosophila), human homolog	Binds MRPS34	ND				DLG1	U13897.1	3q29
Downstream of tyrosine kinase 4 (docking protein 4)	Recruits c-Src to mitos	ND				DOK4	BC003541.1	16q13
Estrogen receptor alpha, mitochondrial loc in uterus and sperm	Binds 17beta-estradiol	MAT				ESR1	X03635.1	6q25.3
Estrogen receptor beta, mitochondrial loc in uterus and sperm	Binds 17beta-estradiol	MAT				ESR2	AF051427.1	14q23.2
ETS domain-containing protein Elk-1 (transcription factor)	Binds permeability transition pore	MIM				ELK1	M25269.1	Xp11.23
Glucocorticoid receptor (GRL), alpha isoform	In retinal Muller cells	ND				NR3C1	M10901.1	5q31.3
Glycogen synthase kinase 3beta (2 isoforms) (also cyto/nuclear)	Binds PSI, AKAP220; phosphorylates KLC's	ND	2.7.1.37	RIM11	U03280.1	GSK3B	L33801.1	3q13.33
G-protein gamma-12 subunit, potentially mitochondrial [2]	Signalling	ND				GNG12	AF188181.1	1p31.2
Growth arrest and DNA-damage-inducible proteins-interacting protein 1	Inhibits CDKs; regulates G-S progression	MAT				GADD45GIP1	BC069200.1	19p13.2
Growth factor receptor-bound protein 10, zeta isoform	Binds RAF1	MOM				GRB10	AF000017.1	7p12.2
GTP-binding protein, ERA homolog, potentially mitochondrial	Signalling; cell cycle?	ND				ERAL1	AF082657.1	17q11.2
GTPase-activating protein Gyp8 [5]	Function unknown	ND		GYP8	D50617.1			—
GTP-binding protein, potentially mitochondrial	Function unknown	NC		YLF2	Z29089.1	PTD004	AF078859.1	2q31.1
Guanine nucleotide binding protein alpha 12 [2]	Signalling	ND				GNA12	L01694.1	7p22.3
Hepatocellular carcinoma suppressor 1 (HCCS1) (yeast Vps53 is in Golgi)	Putative human tumor suppressor	ND				VPS53	AF246287.1	17p13.3

(continues)

Name	Comments	Localization	EC	Yeast gene	Yeast accession	Symbol	Accession	Chromosome
HS1-associated protein	Binds tyrosine kinase	MOM?	—			HAX1	U68566.1	1q21.3
Human cervical cancer oncogene HCCR-1; BRI3-binding protein	Negative regulator of p53; tumor suppressor	ND	—			BRI3BP	AF284094.1	12q24.31
I-kappaB kinase, subunit alpha (inhibitor of NF-kappa-B kinase) (IKKA)	Phosphorylates I-kappaB alpha	MAT?	2.7.11.10			CHUK	AF012890.1	10q24.31
I-kappaB kinase, subunit beta (inhibitor of NF-kappa-B kinase) (IKKB)	Phosphorylates inhibitors of I-kappaB	MAT?	2.7.11.10			IKBKB	AF080158.1	8p11.21
I-kappaB kinase, subunit gamma (inhibitor of NF-kappa-B kinase)	Phosphorylates I-kappaB alpha	MAT?	2.7.1.—			IKBKG	AF074382.1	Xq28
I-kappa-B kinase epsilon subunit (Inhibitor of NF kappaB kinase)	Phosphorylates inhibitors of I-kappaB	MOM	2.7.11.1			IKBKE	AF241789.1	1q32.1
Inositol 1,4,5-triphosphate receptor, type 1	Calcium signaling	ER-MAM	—			ITPR1	D26070.1	3p26.2
Leucine-rich repeat kinase 2 (cyto/mito)	Mixed-lineage kinase	MOM	—			LRRK2	AY792511.1	12q12
Long chain base-responsive inhibitor of protein kinases Pkh1p and Pkh2p	With Pil1, regulates Pkc1/Ypk1 pathways	ND	—	LSP1	Z71255.1			—
Long chain base-responsive inhibitor of protein kinases Pkh1p and Pkh2p	With Lsp1, regulates Pkc1/Ypk1 pathways	ND	—	PIL1	Z72871.1			—
MAVS cleavage protease NS3/4A of heptitis c virus polyprotein	Cleaves MAVS at Cys-508	MOM	—			None	AJ242653.1	
Mitochondrial antiviral signaling protein (MAVS)/Virus-induced signaling adapter (VISA)	Activates NF-kappaB/IRF3; involved in immunity	MOM	—			VISA	AB097003.1	20p13
Mitochondrial carrier homolog 2, long form (Met-induced mitochondrial protein [MIMP])	Met-HGF/SF signalling; binds tBID	MOM	—			MTCH2	AF176008.1	11p11.2
Mitochondrial tumor suppressor MTSG1 (AT2 receptor-interacting protein 1), mito form	Transcription factor MTSG1	ND	—			MTUS1	AF121259.1	8p22
Mothers against decapentaplegic homolog 5 (SMAD5)	Translocates to nucleus	ND	—			MADH5	U59913.1	5q31.1
Mothers against decapentaplegic homolog 6 (SMAD6), potentially mitochondrial	Translocates to nucleus?	NC	—			MADH6	U59914.1	15q22.31
Mothers against decapentaplegic homolog 7 (SMAD7), potentially mitochondrial	Translocates to nucleus?	NC	—			MADH7	AF010193.1	18q21.1
Mitogen-activated protein kinase 12	Binds OMP25	ND	—			MAPK12	Y10487.1	22q13.33
Myocilin (trabecular inducible glucocorticoid response protein)	Mutated in glaucoma	ND	—			MYOC	U85257.1	1q24.3
Myotonin-protein kinase, isoforms C (+VSGGG) and D (-VSGGG)	Lys/Arg-directed kinase	MOM	2.7.1.—			DMPK	L00727.1	19q13.32
Nerve growth factor receptor associated protein BEX3	Binds perinuclear mitochondria	ND	—			NGFRAP1	BC003190.1	Xq22.2
Neurofibromin	Stimulates RAS GTPase	ND	—			NF1	M82814.1	17q11.2
Neurotrophin receptor TrkB (BDNF/NT-3 growth factors receptor), alt spl isoforms	Receptor for BDNF, NT3, and NT4/5	ND	2.7.1.112			NTRK2	U12140.1	9q21.33
NF-kappaB inhibitor alpha (I-kappaB alpha)	Binds ANT	IMS	—			NFKB1	M69043.1	14q13.2
NF-kappaB subunit p50 alt spl product from p100 precursor	Neg reg of mtRNAs	MAT	—			NFKB2	X61498.1	10q24.32
NF-kappaB inhibitory subunit p65	Transcr. fact.; binds ANT; reg mtRNAs	IMS	—			NFKB3	M62399.1	11q13.3
Nitric oxide synthase, neuronal isoform (nNOS, also called mtNOS), alt spl forms	NO syn; translocates to mitos after T3	MIM	1.14.13.39			NOS1	U17327.1	12q24.2
Nitric oxide synthase, candidate, putatively mitochondrial, sim to AtNOS1	Has MTS	MIM	—			C4orf14	AK074953.1	4q12
Phosphatidylinositol 4-kinase 230	Intracellular signalling	MOM	2.7.1.67			PI4K230	AF012872.1	22q11.21
PPARgamma2, 45-kDa isoform (mt-PPAR)	Binds D-loop w. THRA1 and mt-RXR	MAT	—			PPARG	U79012.1	3p25.2
Prostaglandin G/H synthase 2 (cyclooxygenase-2)	Arachidonic acid -> prostaglandin H2	ND	—			PTGS2	M90100.1	1q31.1
Prostaglandin E synthase 2 [6]	Prostaglandin syn	ND	—			PTGES2	AK057049.1	9q34.11
Protein kinase A-RAF	MAP kinase signaling pathway	IMS, MAT	2.7.1.—			ARAF1	X04790.1	Xp11.3
Protein kinase Calpha	Phosphorylates Bcl-2	ND	2.7.1.—			PRKCA	X52479.1	17q24.1
Protein kinase Cdelta	Induces apoptosis; transl to Sc mitos w. farnesol	MOM	2.7.1.—	PKC1	Z35866.1	PRKCD	D10495.1	3p21.31
Protein kinase Cepsilon	Translocates to mitos; binds COX IV	MOM?	2.7.1.37			PRKCE	X65293.1	2p21
Protein phosphatase 1K	Function unknown	ND	3.1.3.16			PPM1K	BC037552.1	4q22.1
Protein phosphatase 2C	Function unknown	NC	3.1.3.16	PTC7	U10556.1	TA-PP2C	AF385435.1	12q24.11
Protein phosphatase 2A, regulatory subunit B, mito form Bbeta2	Signalling; neuron-specific	MOM?	—			PPP2R2B	BC031790.1	5q32
Protein phosphatase methylesterase-1 (also small ribosomal protein)	Protein carboxymethylation	MAT	—	PPE1	U10556.1	PME1	AF157028.1	11q13.14
Phosphatidylinositol-3,4,5-trisphosphate 3-phosphatase PTEN	Proapoptotic protein; translocates to mitos	ND	3.1.3.67			PTEN	U93051.1	10q23.31
PTEN induced putative kinase 1 (PINK1)	Serine-threonine protein kinase	ND	—			PINK1	BC028215.1	1p36.12

Gene product	Function; comment	Location	E.C. #	Fungal Symbol	Fungal GenBank	Animal Symbol	Animal GenBank	Human Chr.
RAN-binding protein 2 (Nuclear pore complex protein Nup358)	Binds COX11 and HK1	ND	–	–	–	RANBP2	D42063.1	2q13
RAP guanine-nucleotide-exchange factor 3	Activates RAP1; in mitosis?	ND	–	–	–	EPAC	BC017728.1	12q13.11
Ras-cyclic AMP pathway, inhibitory regulator [1]	Attenuates Ras1	ND	–	IRA2	M33779.1	–	–	–
Ras GTPase, Harvey isoform (H-ras)	GTPase	ND	–	–	–	HRAS	AF493916.1	11p15.5
RAS GTPase, Kirsten isoform (K-ras), isoform 2B/KRAS 4B, translocates to mitos	GTPase; mito form farnesylated on S181	MOM	–	–	–	KRAS	M54968.1	12p12.1
RAS GTPase, N-ras	GTPase	MOM+MIM	–	–	–	NRAS	X07251.1	1p13.2
Ras-related protein RAB40A	GTP-binding protein?	MOM	–	YPT1	X00209.1	RAB40A	Z95624.1	Xq22.1
Receptor expression-enhancing protein 5 (REEP5); Deleted in Polyposis 1 (DP1)	HCCR-1 binding protein	ND	–	–	–	REEP5	AY562243.1	5q22.2
Respiration factor 1 (cis acting mtDNA-binding protein)	Binds both nuclear and mtDNA UAS	ND	–	RSF1	Z49213.1	–	–	–
Retinoblastoma-like protein 2, E1a-associated (p130)	May regulate cell cycle	ND	–	–	–	RBL2	X76061.1	16q12.2
Retinoic acid receptor alpha, 44-kDa isoform (mt-RXR)	Binds D-loop w. THRA1 and mt-PPAR	MAT	–	–	–	RXRA	X52773.1	9q34.2
Rho-GTPase-activating protein 8, alt sp form, potentially mito	Activates Rho, Rac, Cdc42	NC	–	–	–	ARHGAP8	AL351192.1	22q13.31
Serine/threonine protein kinase AKT (protein kinase B)	Translocates to mitos after PI3K activation	MOM/MIM	2.7.1.37	–	–	AKT1	M63167.1	14q32.33
Serine/threonine protein kinase 11 (LKB1), nuclear/mitochondrial	Tumor suppressor	ND	2.7.1.37	SNF1	M13971.1	STK11	AF035625.1	19p13.3
Serine/threonine protein kinase [1]	Signal transduction?	ND	2.7.1.–	YGR052w	Z72837.1	–	–	–
Serine/threonine protein kinase, potentially mitochondrial	Signal transduction?	NC	2.7.1.–	KIN82	X59720.1	RSK3	X85106.1	6q26
Serine/threonine protein kinase PCTAIRE-2	Binds TRAP; signal transduction?	ND	2.7.1.37	–	–	PCTK2	X66360.1	12q23.1
Serine/threonine-protein kinase VRK2 (vaccinia-related kinase), isoform A (VRK2A)	Phosphorylates p53 (mito/ER isoform)	ND	2.7.11.1	–	–	VRK2	AB000450.1	2p16.1
Serine threonine protein phosphatase	Reg glucose repression	ND	3.1.3.16	SIT4	M24395.1	PPP6C	X92972.2	9q33.3
Serine/threonine protein phosphatase 2C gamma	Dephosphorylates NDUFS4	ND	3.1.3.16	–	–	PPM1G	Y13936.1	2p23.2
Serum glucocorticoid-regulated kinase-1 (SGK1) (Serine/threonine-protein kinase)	Activates Na, K, and Cl channels	ND	2.7.11.1	–	–	SGK	BC001263.1	6q23.2
SH3 binding protein SAB	BTK- and JNK-interacting protein	NC	–	–	–	SH3BP5	AB005047.1	3p25.1
Signal transducer and activator of transcription 3	Affects mito transcription; binds GRIM19	NC	–	–	–	STAT3	BC000627.1	17q21.2
Signaling intermediate in Toll pathway, evolutionarily conserved (SITPEC) [8]	Function unknown	ND	–	–	–	SITPEC	BC008279.1	19p13.2
Src homology and collagene (Shc), 66-kDa isoform (p66Shc)	Binds HSP70; generates H2O2	IMS	–	–	–	SHC1	U73377.1	1q22
Stanniocalcin-1 (50-kDa homodimer; STC50)	Hormone regulating Ca transp	MAT	–	–	–	STC1	U25997.1	8p21.2
Steroid receptor RNA activator (SRA) stem-loop interacting RNA binding Protein	Recruited to promoters	ND	–	–	–	SLIRP	BC017895	14q24.3
Transcriptional activator (Cu-sensing), potentially mitochondrial	Similar to Grisea (*Podospora anserina*)	NC	–	MAC1	X74551.1	–	–	–
Transcription factor AP1-like, Yap1, nucl/cyto, also mito	Regulates oxidative stress/redox genes	ND	–	YAP1	X58693.1	–	–	–
Transcription factor Mlx	Activator of glycolysis with MondoA	MOM	–	–	–	MLX	AF203978.1	17q21.31
Transcription factor MondoA	Activator of glycolysis with Mlx	MOM	–	–	–	MLXIP	AF312918.1	12q24.31
Transforming protein RhoA [1] [2]	Function unknown	NC	–	RHO1	M51189.1	RHOA	L25080.1	3p21.31
Triiodothyronine (T3) receptor c-ErbAalpha1-related protein p43 (NR1A1)	Binds D-loop w. mt-PPAR and mt-RXR	MAT	–	–	–	THRA1	X55005.1	17q11.2
Tuberin (Tuberous sclerosis 2 protein), mito-tageted form (internal MTS)	W. hamartin (TSC1), represses mTOR	ND	–	–	–	TSC2	X75621.1	16p13.3
Tudor protein	Signal transduction?	ND	–	–	–	TUD (f)	X62420.1 (f)	6p12.3
Tudor repeat associator with PCTAIRE 2	Binds PCTAIRE2; signal transduction?	ND	–	–	–	TDRD7	BC028694.1	9q22.33
Tumor necrosis factor type 1 receptor associated protein TRAP-1	HSP-90 homolog	MAT	–	–	–	TRAP1	AF154108.1	16p13.3
Tyrosine kinase Csk (c-src tyrosine kinase)	Signal transduction?	IMS	–	–	–	CSK	X60114.1	15q24.1
Tyrosine kinase ErbB1 (EGFR), phosphorylated form (Tyr845)	Translocates to mitos; binds COX II	ND	2.7.1.112	–	–	EGFR	X00588.1	7p11.2
Tyrosine kinase ErbB4, C-terminal intracellular domain (4ICD)	Signal transduction?	ND	2.7.1.112	–	–	ERBB4	L07868.1	2q34
Tyrosine kinase Fyn	Signal transduction?	IMS	–	–	–	FYN	M14676.1	6q21
Tyrosine kinase Lyn	Signal transduction?	IMS	–	–	–	LYN	M16038.1	8q12.1
Tyrosine kinase Src (v-src avian sarcoma)	Binds AKAP121 and PTPN21	IMS	–	–	–	SRC	BC011566.1	20q11.22
Tyrosine phosphatase, dual specificity, mitochondrial 1	Protein-tyrosine-phosphatase	MIM	–	–	–	PTPMT1	BC020242.1	11p11.2
Tyrosine phosphatase, potentially mitochondrial [5]	Protein-tyrosine-phosphatase	ND	3.1.3.48	SDP1	Z38125.1	DUSP5	U16996.1	10q25.2

Tyrosine phosphatase, non-receptor type 1	In brain, but not heart, muscle, liver mitos	ND	3.1.3.48	PTP1	M64062.1	PTPN1	M31724.1	20q13.13
Tyrosine phosphatase, non-receptor type 11	Protein-tyrosine-phosphatase	ND	3.1.3.48	PTP1	M64062.1	PTPN11	X70766.1	12q24.13
Tyrosine phosphatase, non-receptor type 21, potentially mitochondrial	Binds AKAP121 and SRC	NC	3.1.3.48	PTP1	M64062.1	PTPN21	X79510.1	14q31.3
Tyrosine phosphatase, receptor-type, H precursor	Non-catalytic domain	MOM	3.1.3.48	PTP1	M64062.1	PTPRH	D15049.1	19q13.42
Tyrosine phosphatase interacting protein 51 (PTPIP51)	May be pro-apoptotic	ND				FAM2C	NM 018145.1	15q15.1
WNT13, alt spl mitochondrial isoform WNT13B WNT2B1	Binds frizzled/LRP5/LRP6; aff mito morphology	MIM				WNT2B	Z71621.1	1p13.2
WW domain-containing oxidoreductase (WOX1); translocates from mitos->nucleus	Tumor suppressor; binds p53/JNK1	ND				WWOX	AF211943.1	16q23.2
Zeocin resistance protein 1	Zeocin resistance: reg Pkc1/Slt2/Mpk1 paths	ND		ZEO1	AAT93226.1		—	—
STRESS RESPONSE								
Apoptosis inducing factor (AIF)	DNA laddering; H2O2 scavenger	IMS				PDCD8	AF100928.1	Xq25
Azurocidin (heparin-binding protein)	Neutrophils->endothelial mitos	ND				AZU1	M96326.1	2p11.2
Catalase (dual targeting to peroxisomes and mitochondria)	H2O2 deg	MAT	1.11.1.6	CTA1	X13028.1	CAT	X04076.1	11p13
Crystallin, alpha B (Drosophila heat shock protein 22 homolog)	Protect against oxidative stress?	MAT				CRYAB	S45630.1	11q23.1
Cytochrome c peroxidase	H2O2 deg; oxidative stress signalling	IMS	1.11.1.5	CCP1	X62422.1			
Disulfide isomerase (hypothetical protein FLJ22625), potentially mitochondrial	Function unknown	NC				C5orf14	BC001615.1	5q31.1
Flavohemoglobin (nitric oxide dioxygenase)	Oxidative/nitrosidative stress response	MAT	1.14.12.17	YHB1	Z73019.1			
Glutaredoxin 2	Thioltransferase	NC		TTR1	U33057.1	GLRX2	AF132495.2	1q31.2
Glutathione peroxidase 1	Oxidizes glutathione	MAT?	1.11.1.9			GPX1	X13709.1	3p21.3
Glutathione peroxidase 4	Oxidizes glutathione	MAT?	1.11.1.9			GPX4	X71973.1	19p13.3
Glutathione reductase	Reduces glutathione	MAT?	1.6.4.2	GLR1	L35342.1	GSR	X15722.1	8p21.1
Glutathione S-transferase	Glutathione conjugation	MOM	2.5.1.18			MGST1	J03746.1	12p12.3
Glutathione S-transferase 3, microsomal [2]	Glutathione conjugation	ND	2.5.1.18			MGST3	AF026977.1	1q24.1
Glutathione S-transferase, class alpha 1 (glutathione S-transferase A1)	Glutathione conjugation	ND	2.5.1.18			GSTA1	BC053578.1	6p12.1
Glutathione S-transferase, class alpha 2 (glutathione S-transferase A2)	Glutathione conjugation	ND	2.5.1.18			GSTA2	BC002895.2	6p12.1
Glutathione S-transferase, class alpha 4 (glutathione S-transferase A4)	Glutathione conjugation	ND	2.5.1.18			GSTA4	Y13047.1	6p12.2
Glutathione S-transferase, class kappa (subunit 13)	Glutathione conjugation	MAT?	2.5.1.18			GSTK1	AF070657.1	7q34
Glutathione S-transferase, class pi	Glutathione conjugation	NC	2.5.1.18			GSTP1	M24485.1	11q13.3
Glutathione S-transferase, class zeta (also maleylacetoacetate isomerase)	Phe/Tyr catab; glutathione cofactor	NC	5.2.1.2	MAIA (En)	AJ001837.1 (En)	GSTZ1	U86529.1	14q24.3
Glyoxalase II (hydroxyacyl glutathione hydrolase)	Glutathione reduction	MAT	3.1.2.6	GLO4	Z74948.1	HAGH	BC000840.1	16p13.3
Heat shock protein 5 (glucose-regulated protein, 78-kDa)	Protein assembly in ER; to mitos under stress	ND				HSPA5	X87949.1	9q33.3
Histatin 5 (HSN-5) antifungal salivary peptide	Binds Candida albicans mitos	ND				HTN3	M26665.1	4q13.3
Hypothetical protein YKR049c [1]	Redox protein with a thioredoxin fold	MIM		FMP46	Z28274.1			
Hypothetical protein YLR251w (human MPV17 is not peroxisomal [8])	Ethanol metab and heat shock	MIM		SYM1	AY558211.1	MPV17	NM 002437.4	2p23.3
Hypothetical protein YLR251w (human FKSG24 [8])	Ethanol metab and heat shock?	MIM?		SYM1	AY558211.1	FKSG24	BC093008.1	19p13.11
Hypoglycemia/hypoxia Inducible Mitochondrial Protein (HIMP1a)	Hypoxia inducible gene 1a: stress protection	MIM?				HIGD1A	AF077034.1	3p22.1
Hypoglycemia/hypoxia Inducible Mitochondrial Protein (HIMP1b)	Hypoxia inducible gene 1b: stress protection	MIM?				HIGD1B	BC020667.1	17q21.31
Hypoxia induced protein? [1] (human not clear)	Hypoxia induced protein?	ND		YML030w	Z46659.1	HIGD2A	BC000587.1	5q35.2
Mercaptopyruvate sulfurtransferase	Cyanide detoxification	ND	2.8.1.2		Z73544.1	TST2	X59434.1	22q13.1
NADH kinase	Antioxidant protection; Arg syn	MAT		POS5				
NADPH oxidase homolog 1 (NOH1)	Proton channel; reg cell pH; makes superoxide	ND				NOX1	AF166327.1	Xq22.1
Oxidation resistance protein 1	Oxidative repair	ND		OXR1	Z73552.1	OXR1	BC032710.2	8q23.1
Peroxiredoxin 3 (thioredoxin-dependent peroxide reductase)	Thiol-specific antioxidant?	ND	1.11.1.-	PRX1	Z35825.1	PRDX3	D49396.1	10q26.11
Peroxiredoxin 4 (thioredoxin peroxidase AO372) [2]	Antioxidant protein	NC	1.11.1.-			PRDX4	U25182.1	Xp22.11
Peroxiredoxin 5 (yeast Ahp1 is peroxisomal)	Peroxidase	ND				PRDX5	AF110731.1	11q13.2
Stannin	Confers sensitivity to trimethyltin	ND				SNN	AF030196.1	16p13.13
Sulfhydryl oxidase, putative, potentially mitochondrial	Function unknown	NC				SOXN	AJ318051.1	9q34.3
Sulfide dehydrogenase (human CGI-44 may be cytosolic)	Detoxifies cadmium	ND		HMT2 (Sp)	AF042283.1 (Sp)	SQRDL	AAD34039.1	15q21.1
Superoxide dismutase 1 (Cu,Zn-SOD)	Scavenges free radicals	IMS	1.15.1.1	SOD1	J03279.1	SOD1	X02317.1	21q22.1

(continues)

Appendix 5 (continued)

Gene product	Function; comment	Location	E.C. #	Fungal Symbol	Fungal GenBank	Animal Symbol	Animal GenBank	Human Chr.
Superoxide dismutase 2 (Mn-SOD)	Scavenges free radicals	MAT	1.15.1.1	SOD2	X02156.1	SOD2	M36693.1	6q25.3
Superoxide dismutase (Fe-SOD) (Toxoplasma gondii (Tg))	Unusual MTS	ND	1.15.1.1	–	–	SODB2 (Tg)	AY176062.1 (Tg)	–
Thioredoxin 2 (TRX2)	Reduces disulfides; inhibits MAP3K5	MAT?	–	TRX3	X59720.1	TXN2	U78678.1	22q12.3
Thioredoxin reductase 2	Reduces disulfides	MAT?	1.8.1.9	TRR2	U00059.1	TXNRD2	AF044212.1	22q11.21
Thioredoxin-dependent peroxide reductase (AOP2) [2]	Antioxidant protein	NC	1.11.1.7	–	–	PRDX6	D14662.1	1q25.1
Thiosulfate sulfurtransferase (rhodanese)	Cyanide detoxification	MAT	2.8.1.1	YOR251c	Z75159.1	TST	D87292.1	22q13.1
Thyroid sulfurtransferase (rhodanese)-like KAT protein [1]	Rhodanese-related protein	ND	–	YOR286w	Z75194.1	KAT	AL591806.16	1q23.3
Thyroid peroxidase (TPO) (NADH/NADPH thyroid oxidase p138-tox)	Generates hydrogen peroxide	ND	1.11.1.8	–	–	DUOX2	AF181972.1	15q21.1
Tryparedoxin peroxidase, trypanothione-dependent (T. brucei [Tb])	Peroxide detoxification	ND	–	–	–	TRYP2 (Tb)	AF196570 (Tb)	–
MISCELLANEOUS								
Aflatoxin B1 aldehyde reductase member 2 [3]	Aldo/keto reductase 2 family	ND	1.1.1.–	AAD10	Z49655.1	AKR7A2	BK000395.1	1p36.13
Amyloid beta A4 precursor protein (APP695)	Cell surface receptor; PM and mitos	MIM	–	–	–	APP	Y00264.1	21q21.3
Ankyrin repeat domain 29, alt spl isoforms, potentially mitochondrial	Function unknown	NC	–	–	–	ANKRD29	BC030622.1	18q11.2
Antiquitin (aldehyde dehydrogenase family) [2]	Function unknown	ND	–	–	–	ALDH7A1	S74728.1	5q23.2
Apolipoprotein H (beta-2-glycoprotein I)	Binds phospholipids	NC	–	–	–	APOH	X58100.1	17q24.1
Beta propeller protein, putative (T. cruzi [Tc])	WD repeat protein; function unknown	ND	–	–	–	BPP1 (Tc)	CAE12294.1 (Tc)	–
Biphenyl hydrolase-related protein (Valacyclovir hydrolase) [3]	Serine/biphenyl hydrolase	ND	–	–	–	BPHL	AJ617684.1	6p25.2
BMP/Retinoic acid-inducible neural-specific protein-3 (BRINP3; DBCCR1L1)	Upreg in gonadotropinoma	ND	–	–	–	FAM5C	AB111893.1	1q31.1
C21orf2, 21.3-kDa protein	Function unknown	ND	–	–	–	C21orf2	Y11392.1	21q22.3
Calcipressin-1 (Down syndrome critical region gene 1), sim to fly Nebula, may be mito	Binds calcineurin A	NC	–	–	–	DSCR1	U85266.2	21q22.12
Calcium binding atopy-related autoantigen 1, potentially mitochondrial [8] [12]	Function unknown	MIM	–	–	–	CBARA1	BC004216.2	10q22.1
Caveolin-1, alpha and beta isoforms (mainly in lung)	Caveolae trafficking	ND	–	–	–	CAV1	Z18951.1	7q31.2
Coiled-coil domain containing protein 2 (CCDC2), potentially mitochondrial [8]	Function unknown; slight similarity to Yme2	NC	–	–	–	IFT74	AY040325.1	9p21.2
Coiled-coil domain containing 54, potentially mitochondrial [8]	Function unknown	NC	–	–	–	CCDC54	BC030780.2	3q13.12
Coiled-coil domain containing protein 109A, potentially mitochondrial [12]	Function unknown	MIM	–	–	–	CCDC109A	BC034235.1	10q22.1
Coiled-coil-helix-coiled-coil-helix domain containing 3 [3] [4] [6] [12]	Function unknown	MIM	–	–	–	CHCHD3	BC014839.1	7q32.3
CSA-Conditional, T-cell Activation-Dependent gene (CSTAD), long/short isoforms	Function unknown; mouse-specific	MOM	–	–	–	Cstad (m)	AK020589.1 (m)	2qB (m)
C-terminal modulator protein [1]	Function unknown	ND	–	FMP10	U18922.1	CTMP	AJ313515.1	1q21.3
Cytokeratin type II	Intermediate filament protein	MIM	–	–	–	K6HF	Y17282.1	12q13
DJ-1 protein [2]	RNA-binding protein regulatory subunit	MAT/IMS	–	–	–	PARK7	D61380.1	1p36.23
Electron transport protein homolog	Function unknown	NC	–	YTP1	U22109.1	–	–	–
ES-1 protein homolog, KNP-Ia (long form) and KNP-Ib (short form)	Function unknown	ND	–	–	–	HES1	D86061.1	21q22.3
Ethylmalonic encephalopathy protein (formerly HSCO)	Metallo-beta-lactamase/glyoxalase	MAT	–	–	–	ETHE1	D83198.2	19q13.31
Extracellular matrix protein 1, nucleus, also mito	Function unknown	ND	–	ECM1	U12980.2	–	–	–
Four-and-a-half LIM-only protein 1 [2]	Cytoskeletal remodeling	ND	–	–	–	FHL1	U60115.1	Xq26.3
Four-and-a-half LIM-only protein 2 (in skeletal muscle: heart)	Cytoskeletal remodeling	ND	–	–	–	FHL2	U29332.1	2q12.1
Four-and-a-half LIM-only protein 3 (in skeletal muscle)	Cytoskeletal remodeling	ND	–	–	–	FHL3	U60116.2	1p35.1
Galectin (beta-galactosidase binding lectin), homolog of chicken CG-16	Binds beta-galactoside	ND	–	–	–	LGALS1	J04456.1	22q13.1
Growth-related gene, short isoform (EEG-1S) (also called C1QDC1)	Inhibitor of growth	ND	–	–	–	EEG1	AY074491.1	12p11.21
Heat shock protein, 27 kD protein 2 (HSPB2), possibly mitochondrial	Function unknown (binds DMPK)	MOM?	–	–	–	HSPB2	U75898.1	11q23.1

(continues)

Description	Function	Localization	ORF / ID	Accession	Human homolog	Accession	Location
Hepatitis C virus NS5A-transactivated protein 9 (KIAA0101)	Function unknown	ND			NS5ATP9	BC005832.2	15q22.31
High mobility group protein HMG-I/HMG-Y (A1)	Binds D-loop (nucl/cyto/mito)	MAT?			HMGA1	X14957.1	6p21.31
Hit family protein 2, potentially mitochondrial [5]	Function unknown	ND	HNT2	U28374.1	KIAA1522	BO933230.1	1p35.1
Homolog of mouse C77080, potentially mitochondrial	Function unknown	NC					
Huntingtin	Function unknown	MOM			HD	L12392.1	4p16.3
Huntingtin-associated protein 1 (HAP1)	Binds huntingtin	ND			HAP1	AJ224877.1	1p34.2
Hypothetical protein [9]	Function unknown	MOM	NCU06247.1 (Nc)	XM_326101.1 (Nc)			
Hypothetical protein [9] (G11A3.010)	Function unknown	MOM	NCU01401.1 (Nc)	XM_326893.1 (Nc)			
Hypothetical protein [9]	Serine hydrolase domain	MOM	NCU04946.1 (Nc)	XM_954301.1 (Nc)			
Hypothetical protein [1]	Function unknown	MAT	NCU03158.1 (Nc)	XM_330593.1 (Nc)			
Hypothetical protein [1]	Function unknown	ND	EM15	Z74813.1	FLJ20487	BC002331.1	11q12.2
Hypothetical protein [1]	Function unknown	MAT	FUN14	L22015.1			
Hypothetical protein [1]	Req for G1 phase of cell cycle	ND	GIF1	Z38061.1			
Hypothetical protein [1]	Function unknown	ND	MSC6	Z75262.1			
Hypothetical protein [1]	Propionate metab?	ND	PDH1	Z71255.1			
Hypothetical protein [1]	RNA/DNA helicase family?	ND	SOV1	Z48952.1			
Hypothetical protein [1]	Function unknown	ND	YDL157c	Z74205.1			
Hypothetical protein [1]	Function unknown	ND	YFR011c	D50617.1			
Hypothetical protein [1]	Function unknown	ND	YGR235c	Z73020.1			
Hypothetical protein [1]	Function unknown	ND	YIL087c	Z46728.1			
Hypothetical protein [1]	Function unknown	ND	YLR091w	Z73263.1			
Hypothetical protein [1]	Function unknown	ND	YNL100w	Z71376.1			
Hypothetical protein [1]	Function unknown	ND	YNR018w	Z71633.1			
Hypothetical protein [1]	Function unknown	ND	YNR040w	Z71655.1			
Hypothetical protein (Fmp16) [1]	Molybdopterin/thiamine syn?	MOM	YOR215c	Z75123.1			
Hypothetical protein (Fmp21) [1]	Molybdopterin/thiamine syn?	MOM	YHR003c	U10555.1	UBA3	AB012190.1	3p14.1
Hypothetical protein (Fmp22) [1]	Function unknown	MAT	YKL027w	Z28027.1	UBA3	AB012190.1	3p14.1
Hypothetical protein (Fmp23) [1]	Function unknown	ND	YDR070c	Z74366.1			
Hypothetical protein (Fmp24) [1]	Function unknown	ND	YBR269c	Z36138.1			
Hypothetical protein (Fmp25) [1] (human KIAA0032 prob not mito)	Function unknown	ND	YHR198c	U00030.1			
Hypothetical protein (Fmp29) [1]	Function unknown	ND	YBR047w	Z35916.1			
Hypothetical protein (Fmp30) [1]	Function unknown	ND	YMR115w	Z49702.1			
Hypothetical protein (Fmp32) [1] [7]	Function unknown	ND	YLR077w	Z73249.1			
Hypothetical protein (Fmp33) [1]	Function unknown	ND	YER080w	U18839.1			
Hypothetical protein (Fmp34) [1]	Function unknown	ND	YPL103c	U43281.1			
Hypothetical protein (Fmp38) [1]	Function unknown	ND	YFL046w	D50617.1			
Hypothetical protein (Fmp39) [1]	Function unknown	ND	YIL161w	Z49436.1			
Hypothetical protein (Fmp40) [1], human selenoprotein O, potentially mitochondrial	Function unknown	MOM	YHR199c	U00030.1	SELO	AY324823.1	22q13.33
Hypothetical protein (Fmp51) [1]	Function unknown	ND	YOR205c	Z75113.1	C6orf79	BC016850.1	6p23
Hypothetical protein (Fmp52) [1] [7]	Function unknown	ND	YMR157c	Z49705.1	HTATIP2	BC002439.2	11p15.1
Hypothetical protein Scm4 [1]	Effector of Cdc4 function	NC	YPL222w	Z73578.1			
Hypothetical protein, potentially mitochondrial	Some similarity to ferredoxin	MIM?	YBR262c	Z36131.1	MGC19604	BC063460.1	19p13.2
Hypothetical protein, potentially mitochondrial	Function unknown	NC	YER004w	U18778.1			
Hypothetical protein, potentially mitochondrial	Function unknown	NC	SCM4	X69566.1			
Hypothetical protein, potentially mitochondrial	Function unknown	NC	CNG00160 (Cn)	XM_571743.1 (Cn)			
Hypothetical protein, potentially mitochondrial	Function unknown	NC	YER182w	U18922.1			
Hypothetical protein, potentially mitochondrial	Function unknown	NC	YGL226w	Z72748.1			
Hypothetical protein, potentially mitochondrial	Function unknown	NC	YGR265w	Y07777.1			
Hypothetical protein, potentially mitochondrial	Function unknown	NC	YJL067w	Z49342.1			
Hypothetical protein, potentially mitochondrial	Function unknown	NC	MSC1	Z50178.1	C7orf30	BC012331.1	7p15.3
Hypothetical protein, potentially mitochondrial	Function unknown	NC	YNL184c	Z71461.1	KIAA0391	NM_014672.2	14q13.2
Hypothetical protein, potentially mitochondrial [8]	Function unknown	NC	YAL10B09845g (Yl)	CR382128.1 (Yl)	FLJ44955	AK126903.1	6q24.3
Hypothetical protein, potentially mitochondrial [5]	Function unknown	ND	FYV4	U00061.1	C1orf69	NM_001010867.1	1q42.13
			CAF17	Z49622.1			
			DEM1	Z36032.1			

Appendix 5 (*continued*)

Gene product	Function; comment	Location	E.C. #	Fungal Symbol	Fungal GenBank	Animal Symbol	Animal GenBank	Human Chr.
Hypothetical protein, potentially mitochondrial [5]	Function unknown	ND	-	ECM18	Z48758.1	-	-	-
Hypothetical protein, potentially mitochondrial (cyto/mito [5])	Function unknown	ND	-	GPB2	U12980.2	-	-	-
Hypothetical protein, potentially mitochondrial [5]	Function unknown	ND	-	ECM19	U19729.2	-	-	-
Hypothetical protein, potentially mitochondrial [5]	Function unknown	ND	-	MNE1	Z75258.1	-	-	-
Hypothetical protein, potentially mitochondrial [5]	Function unknown	ND	-	NGL1	Z74784.1	-	-	-
Hypothetical protein, potentially mitochondrial [5]	Function unknown	ND	-	NCS6	Z72733.1	ATPBD3	BC009037.2	19q13.41
Hypothetical protein, potentially mitochondrial [5]	Function unknown	ND	-	YBL059w	Z35820.1	-	-	-
Hypothetical protein, potentially mitochondrial [5]	Function unknown	ND	-	YBL095w	Z35857.1	-	-	-
Hypothetical protein, potentially mitochondrial [5]	ABC transporter activity	ND	-	YDR061w	Z74357.1	-	-	-
Hypothetical protein, potentially mitochondrial [5]	Function unknown	ND	-	YDR065w	Z74361.1	-	-	-
Hypothetical protein, potentially mitochondrial [5]	Function unknown	ND	-	YDR379c-a	NC_001136.5	-	-	-
Hypothetical protein, potentially mitochondrial [5]	Function unknown	ND	-	YER077c	U18839.1	-	-	-
Hypothetical protein, potentially mitochondrial [5]	Function unknown	ND	-	YGL057c	Z72579.1	-	-	-
Hypothetical protein, potentially mitochondrial [5]	Function unknown	ND	-	YGL085w	NC_001139.4	-	-	-
Hypothetical protein, potentially mitochondrial [5]	Function unknown	ND	-	YGR015c	Z72800.1	-	-	-
Hypothetical protein, potentially mitochondrial [5]	Function unknown	ND	-	YGR102c	Z72887.1	-	-	-
Hypothetical protein, potentially mitochondrial [5]	Function unknown	ND	-	YIL077c	Z37997.1	-	-	-
Hypothetical protein, potentially mitochondrial [5]	Function unknown	ND	-	YIL043w	Z49318.1	-	-	-
Hypothetical protein, potentially mitochondrial [5]	Function unknown	ND	-	YIL062w-a	NC_001142.5	-	-	-
Hypothetical protein, potentially mitochondrial [5]	Function unknown	ND	-	YIL131c	Z49406.1	-	-	-
Hypothetical protein, potentially mitochondrial [5]	Function unknown	ND	-	YJL147c	Z49422.1	-	-	-
Hypothetical protein, potentially mitochondrial [5]	Function unknown	ND	-	YJR003c	Z49503.1	-	-	-
Hypothetical protein, potentially mitochondrial [5]	Function unknown	ND	-	YJR080c	Z49578.1	-	-	-
Hypothetical protein, potentially mitochondrial [5]	Function unknown	ND	-	YJR111c	Z49611.1	-	-	-
Hypothetical protein, potentially mitochondrial [5]	Function unknown	ND	-	YKL162c	Z28162.1	-	-	-
Hypothetical protein, potentially mitochondrial [5]	Function unknown	ND	-	YLR132c	Z73304.1	-	-	-
Hypothetical protein, potentially mitochondrial [5]	Function unknown	ND	-	YLR283w	U17243.1	-	-	-
Hypothetical protein, potentially mitochondrial [5]	Function unknown	ND	-	YLR346c	U19028.1	-	-	-
Hypothetical protein, potentially mitochondrial [5]	Function unknown; functionally sim to Scm4	ND	-	YLR356w	U19102.1	-	-	-
Hypothetical protein, potentially mitochondrial [5]	Function unknown	ND	-	YMR098c	Z49807.1	-	-	-
Hypothetical protein, potentially mitochondrial [5]	Function unknown	ND	-	YMR252c	Z48639.1	-	-	-
Hypothetical protein, potentially mitochondrial [5]	Function unknown	ND	-	YNL122c	Z71398.1	-	-	-
Hypothetical protein, potentially mitochondrial [5]	Function unknown	ND	-	YNL199c	Z71475.1	-	-	-
Hypothetical protein, potentially mitochondrial [5]	Function unknown	ND	-	YNL211c	Z71487.1	-	-	-
Hypothetical protein, potentially mitochondrial [5]	Function unknown	ND	-	YOL150c	Z74893.1	-	-	-
Hypothetical protein, potentially mitochondrial [5]	Function unknown	ND	-	YOR228c	Z75136.1	-	-	-
Hypothetical protein, potentially mitochondrial [5]	Function unknown	ND	-	YOR305w	Z75213.1	-	-	-
Hypothetical protein, potentially mitochondrial [5]	Function unknown	ND	-	YPL107w	U43281.1	-	-	-
Hypothetical protein, potentially mitochondrial [5]	Function unknown	ND	-	YPL168w	Z73524.1	-	-	-
Hypothetical protein, potentially mitochondrial [5]	Function unknown	ND	-	YPR116w	U32445.1	-	-	-
Hypothetical protein, potentially mitochondrial [5]	Function unknown	ND	-	YDR514c	U33057.1	-	-	-
Hypothetical protein, potentially mitochondrial [5]	Function unknown	ND	-	YLR072w	Z73244.1	-	-	-
Hypothetical protein, potentially mitochondrial [5] [7]	Function unknown	MOM	-	YPR098c	U32445.1	-	-	-
Hypothetical protein, potentially mitochondrial [10]	Function unknown	ND	-	YDL027c	Z74075.1	-	-	-
Hypothetical protein, potentially mitochondrial [7] [10]	Function unknown	MOM	-	YDR381c-a	U28373.1	-	-	-
Hypothetical protein, potentially mitochondrial, alt spl isoforms	Function unknown	NC	-			C14orf159	AY358516.1	14q32.12
Hypothetical protein, potentially mitochondrial, alt spl isoforms	Function unknown	NC	-			C21orf57	AK128875.1	21q22.3
Hypothetical protein, potentially mitochondrial, alt spl isoforms	Function unknown	NC	-			TBRG4	AL833840.1	7p13
Hypothetical protein, probably mitochondrial	Function unknown	ND	-	YNL213c	Z71489.1	-	-	-
Hypothetical protein, probably mitochondrial	Function unknown; contains TPR repeats	ND	-			KIAA0141	D50911.1	5q31.3
Hypothetical protein, similar to Rmd9 [5]	Function unknown	ND	-	YBR238c	Z36107.1	-	-	-
Hypothetical protein FLJ10737 [4] [10]	DnaJ homolog?	MIM	-	XDJ1	X76343.1	DNAJC11	AK001599.1	1p36.31
Hypothetical protein FLJ25853, potentially mitochondrial	Function unknown	NC	-			CCDC70	AK098719.1	13q14.3
Hypothetical protein LOC91689 [8]	Function unknown	ND	-			C22orf32	BC024237.2	22q13.2
Hypothetical protein LOC79133, potentially mitochondrial	SAM-dependent methyltransferase activity	NC	-			C20orf7	NM_024120.3	20p12.1

(continues)

Protein	Function	Loc.	EC	Yeast/fungal gene	Ortholog accession	Human symbol	Accession	Locus
Hypothetical protein TMEM70 (transmembrane protein 70) [8]	Function unknown	ND	—	YLH47	U48418.1	TMEM70	BC002748.2	8q21.11
Hypothetical protein YPR125w/Ylh47, similar to Mdm38	Assoc w. ribosomes; suppresses MRS2-1	MIM	—	YDR282c	U51030.1	LETM1	AF061025.1	4p16.3
Hypothetical protein YDR282c, potentially mitochondrial (C6orf96 [8])	Function unknown	ND	—	—	—	C6orf96	BC106065.1	6q25.1
Leucine repeat-rich protein	Function unknown	NC	—	—	—	LRRC10	AK123908.1	12q15
MDR1 and Mitochondrial Taxol Resistance Associated Gene (MM-TRAG)	Taxol and doxorubicin resistance	ND	—	—	—	C7orf23	BT006973.1	7q21.12
Methyltransferase 11 domain containing 1, potentially mitochondrial	rRNA methyltransferase?	NC	—	RSM22	Z28155.1	METT11D1	BC005053.1	14q11.2
Methyltransferase, SAM-dependent, potentially mitochondrial	Function unknown	NC	2.1.1.–	MTQ1	Z71339.1	HEMK1	AK025973.1	3p21.31
Mitocalcin (EF-hand domain family, member DI; swiprosin-2)	Calcium-binding protein; function unknown	MIM	—	—	—	EFHD1	BC002449.2	2q37.1
Mitochondrial acidic matrix protein (human P32 [gC1q-R])	Ca binding? Binds Hrk/DP5	MAT	—	MAM33	Z38060.1	C1QBP	L04636.1	17p13.3
Mitochondrial coiled-coil domain 1 (MCCD1)	Function unknown	ND	—	—	—	MCCD1	BN000141.1	6p21.33
Mitochondrial intermembrane space cysteine motif protein of 14 kDa	Function unknown; MIA	IMS	—	MIC14	Z68196.1	CHCHD5	BC004498.2	2q13
Mitochondrial intermembrane space cysteine motif protein of 17 kDa [3]	Function unknown; induced by NR4A1; MIA	ND	—	MIC17	Z48613.1	CHCHD2	BC015639.2	7p11.2
Mitochondrial peculiar membrane protein 1	Function unknown	ND	—	MPM1	Z49342.1	—	—	—
Mitochondrial protein	Function unknown	ND	—	YIL157c	Z38059.1	—	—	—
Myeloma overexpressed gene protein, alt spl short form, potentially mitochondrial	Function unknown	NC	—	—	—	MYEOV	AJ223366.1	11q13.3
Neighbor of COX4 (NOC4)	Function unknown	ND	—	—	—	COX4AL	AF005888.1	16q24.1
Neuropeptide Y, may be mito via alt translation initiation of preprotein @ aa-45	Function unknown	ND	—	—	—	NPY	K01911.1	7p15.3
4-Nitrophenylphosphatase domain and non-neuronal SNAP25-like protein 1 [3][4][12]	May be in mammalian nucleoid	MIM	—	NCU08092.1 (Nc)	XM_328797.1 (Nc)	NIPSNAP1	AJ001258.1	22q12.2
4-Nitrophenylphosphatase domain and non-neuronal SNAP25-like protein 3A [6]	Function unknown	ND	—	NCU08092.1 (Nc)	XM_328797.1 (Nc)	NIPSNAP3A	AK024015.1	9q31.1
Nuclear matrix transcription factor 4 (Zinc finger protein 384), alt spl mito isoform	Transcription factor	ND	—	—	—	ZNF384	AB070238.1	12p13.31
Nudix (nucleoside diphosphate-linked moiety X)-type motif 19 [3]	Androgen-regulated protein RP2	ND	—	FG00801.1 (Gz)	XM_380977.1 (Gz)	NUDT19	XM_939641.2	19q13.11
O-Acetyltransferase 1, potentially mitochondrial	Protein glycosylation?	NC	—	—	—	CAS1	AF397424.2	7q21.3
ORF Q0010 (mtDNA-encoded)	Function unknown	ND	—	[ORF6]	AJ011856.1	—	—	—
ORF Q0017 (mtDNA-encoded)	Function unknown	ND	—	[ORF7]	AJ011856.1	—	—	—
ORF Q0032 (mtDNA-encoded)	Function unknown	ND	—	[ORF8]	AJ011856.1	—	—	—
ORF Q0092 (mtDNA-encoded)	Function unknown	ND	—	[ORF5]	AJ011856.1	—	—	—
ORF Q0142 (mtDNA-encoded)	Function unknown	ND	—	[ORF9]	AJ011856.1	—	—	—
ORF Q0143 (mtDNA-encoded)	Function unknown	ND	—	[ORF10]	AJ011856.1	—	—	—
ORF Q0144 (mtDNA-encoded)	Function unknown	ND	—	[Q0144]	AJ011856.1	—	—	—
ORF Q0167 (mtDNA-encoded)	Function unknown	ND	—	RF2	AJ011856.1	—	—	—
ORF Q0182 (mtDNA-encoded)	Function unknown	ND	—	[ORF11]	AJ011856.1	—	—	—
ORF Q0255 (mtDNA-encoded)	Function unknown	ND	—	[ORF1]	AJ011856.1	—	—	—
ORF Q0297 (mtDNA-encoded)	Function unknown	ND	—	[ORF12]	AJ011856.1	—	—	—
Outer Membrane Protein of 14 kDa (YBR230c) [1][7]	Function unknown	MOM	—	OM14	Z36099.1	—	—	—
Ovary-specific acidic protein-1 (corneal endothelium-specific protein 1 [CESP-1])	Function unknown	ND	—	—	—	OSAP	AF484960.1	4q31.1
Pentatricopeptide repeat protein 1, potentially mitochondrial Pet20	Function unknown; sim to Sue1,Cox14, NDUFA6	NC	—	AN2310_2 (An) / PET20	XM_654822.1 (An) / Z73515.1	PTCD1	BC080580.1	7q22.1
Poly(A) binding protein interacting protein [1]	Binds poly(A) binding protein	ND	—	PBP1	Z72963.1	—	—	—
Proline rich protein 5 (3 variants), potentially mitochondrial, alt spl isoforms	Function unknown	NC	—	—	—	PRR5	BK005639.1	22q13.31
Protein C20orf52 [3] (not clear if Mgr2 is mito)	Mgr2 is req by cells lacking mtDNA	ND	—	MGR2	U43281.1	C20orf52	BC008488.1	20q11.22
Protein FLJ14668 [3]	Function unknown	ND	—	—	Z82341.1	FLJ14668	AK027574.1	2p13.3
Protein involved in respiration	Function unknown	MIM	—	YKR016w	U18262.1	—	—	—
Protein involved w nuclear control of mitochondria [1]	Function unknown; may also be nuclear	ND	—	GDS1	U01883.1	—	—	—
Protein kinase C inhibitor protein 1 (14-3-3 protein gamma)	Binds human Hsp60, PrPC	ND	—	BMH2	M31796.1	YWHAG	AF142498.1	7q11.23
Protein of the outer mitochondrial membrane, 45 kDa	Function unknown	MOM	—	OM45	—	—	—	—

Gene product	Function; comment	Location	E.C. #	Fungal		Animal		
				Symbol	GenBank	Symbol	GenBank	Human Chr.
Protein required for respiratory growth, potentially mitochondrial	Function unknown	NC	–	YKL137w	Z28137.1	–	–	–
Protoplast secreted protein 2 [1]	Function unknown	ND	–	PST2	Z74328.1	–	–	–
Receptor expression-enhancing protein 1	Function unknown	ND	–	–	–	REEP1	AY562239.1	2p11.2
Stretch responsive muscle protein, potentially mitochondrial	Proprioception in sk. muscle	NC	–	–	–	SMPX	AJ250584.1	Xp22.11
SWR1-complex protein 3, alt initiation [1]	Subunit of SWR1 chromatin remodeling complex	ND	–	SWC3	L05146.1	–	–	–
T cell receptor gamma chain (TCRG) alternate reading frame protein (TARP)	Androgen regulated protein, prostate-specific	MOM	–	–	–	TARP	AF151103.1	7p14.1
Transcript Antisense to Ribosomal RNA	In 25S rDNA locus	ND	–	TAR1	AF479964.1	–	–	–
Transmembrane protein 14C [13]	Function unknown	ND	–	–	–	TMEM14C	BC002496.2	6p24.2
Upregulated during skeletal muscle growth 5 (HCVFTP2) [4][13]	Function unknown	MIM?	–	–	–	USMG5	BC007087.1	10q24.33
Yeast suicide protein 2	Function unknown	ND	–	YSP2	U32517.2	–	–	–
Zinc finger CCHC domain containing protein 12, potentially mitochondrial	Function unknown	NC	–	–	–	ZCCHC12	BC031241.1	Xq24
Zinc finger CDGSH domain-containing protein 1 ("mito-NEET", MDS029, C10orf70) [3]	Binds thiazolidinedione	ND	–	BUD20	Z73246.1	ZCD1	BC005962.1	10q21.1
Zinc finger protein T86, potentially mitochondrial (alt trans init in human?)	Transcription?	NC	–	–	–	ZNF593	BI824424.1	1p36.11
Zuotin (ribosome-associated complex [RAC]) [1]	Function unknown	NC	–	ZUO1	X63612.1	ZRF1	X98260.1 (i)	7q22.1
TOTAL NUMBER OF GENE LOCI				996		1317		

Notes: Names of gene products usually follow the yeast nomenclature. For those animal gene products that have alternatively-spliced isoforms, only one representative isoform is usually indicated. Gene symbols within brackets are unofficial. Most yeast loci are from *Saccharomyces cerevisiae*, but some gene products are absent from *S. cerevisiae* (Sc) but present in other fungi, such as *Aspergillus fumigatus* (Af), *Aspergillus nidulans* (An), *Botrytis cinerea* (Bc), *Candida albicans* (Ca), *Cryptococcus neoformans* (Cn), *Emericella nidulans* (En), *Gibberella zeae* (Gz), *Magnaporthe grisea* (Mg), *Neurospora crassa* (Nc), *Pneumocystis carinii* (Pc), *Schizosaccharomyces pombe* (Sp), *Ustilago maydis* (Um), and *Yarrowia lipolytica* (Yl), as indicated. Absence of a gene in *S. cerevisiae* means that it is either missing in the genome, or if present, is apparently not mitochondrial. All animal loci are from human, unless indicated otherwise: b, bovine; c, chicken; f, fly (*D. melanogaster*); h, Chinese hamster; m, mouse; p, pig; r, rat; s, slime mold (*D. discoideum*); u, sea urchin; w, worm (*C. elegans*). Other abbreviations: a, alternative reading frame or alternative AUG start; ABC, ATP-binding cassette; ANT, adenine nucleotide translocator; BER, base excision repair; deg, degradation; CTS, C-terminal targeting signal; ER-MAM, endoplasmic reticulum mitochondrial-associated membrane; GIP, general import pore; i, incomplete partial sequence; IMS, intermembrane space; ISP, iron-sulfur (FeS) protein; ITS, internal targeting signal; LOC, location in mitochondria; MAT, matrix; metab, metabolism; MCF, mitochondrial carrier family; MIA, mitochondrial import and assembly pathway (Cx3C and Cx9C proteins); MIM, mitochondrial inner membrane; MMC, mitochondrial motor and chaperone; MOM, mitochondrial outer membrane; MTS, mitochondrial [N-terminal] targeting signal; NC, mitochondrial localization not checked; ND, localized to mitochondria, but specific compartment not determined; NOS, nitric oxide synthase; PAM, presequence translocase-associated motor; SAM, sorting and assembly machinery of MOM; syn, synthesis; TOB, topogenesis of MOM beta-barrel proteins; transp, transport; ?, presumed mitochondrial localization within mitochondria, but not verified experimentally.

Note also: (1) The function in yeast may not correspond to the function in animals. (2) Some gene products have multiple functions and may have been listed more than once. (3) Absence of a yeast homolog to an animal gene, or vice versa, does not necessarily mean that the homolog does not exist, as it may exist but not be mitochondrial. (4) Total number of animal gene products does not include (i) kinetoplast-associated proteins or (ii) non-animal gene products (e.g. viral, bacterial) that are targeted to animal mitochondria following infection. (5) Finally, this is simply a manually-curated list generated by the author to assist in understanding the biology of mitochondria. **Disclaimer: No claim of accuracy of this list should be implied, for example, by the presence or absence of a gene product, by the indicated relationship(s) between a fungal gene product and its animal homolog, by the assignment of any specific function to a gene product, or by the assignment of a gene to a chromosomal locus. Caveat emptor!**

References: Gene products for which no literature reference is available are included if there is a credible proteome study supporting a mitochondrial localization.
[1] Sickmann et al. (2003) PNAS 100:13207 (found in yeast mitos);
[2] Taylor et al. (2003) Nat. Biotechnol. 21:281 (found in human mito 2-D gels);
[3] Mootha et al. (2003) Cell 115:629 (found in mouse mitos);
[4] Da Cruz et al. (2003) JBC 278:41566 (found in isolated mouse MIM);
[5] Huh et al. (2003) Nature 425:686 (mito GFP in yeast);
[6] Fountoulakis et al. (2003) Electrophoresis 24:260 (found in human IMR-32 mitos);
[7] Zahedi et al. (2006) MBC 17:1436 (found in yeast MOM);
[8] Calvo et al. (2006) Nat. Genet. 38:576 (found in mouse by "Maestro");
[9] Schmitt et al. (2006) Proteomics 6:72 (found in N. crassa isolated MOM);
[10] Reinders et al. (2006) J. Proteome. Res. 5:1543 (found in yeast mito proteome);
[11] Johnson (2006) Am. J. Physiol. Cell Physiol. in press (found in rat mitos);
[12] McDonald et al. (2006) Mol. Cell Proteomics in press (found in isolated rat liver MIM);
[13] Carroll et al. (2006) PNAS 103:16170 (found in bovine mitochondrial membranes).

APPENDIX 6

Changes in the Mitochondrial Transcriptome and Proteome Under Various Stresses and Growth Conditions

Compiled by Eric A. Schon

Department of Neurology and Department of Genetics and Development
Columbia University, New York, New York 10032

METHODS IN CELL BIOLOGY, VOL. 80

0091-679X/07 $35.00
DOI: 10.1016/S0091-679X(06)80045-2

Expression of Mitochondrial-Targeted Gene Products in ρ^+ and ρ^0 Yeast

Table 1
Genes with Induced Expression in ρ^0 Cells

Gene	Fold change	Description
Mitochondrial biogenesis and function		
NAM1	1.6	Mitochondrial splicing and protein synthesis
PET494	1.6	Translational activator of COX3 mRNA
MEF1	2.8	Mitochondrial translation elongation factor G
MSF1	1.7	Mitochondrial phenylalanyl-tRNA synthetase
MRS3	1.6	Mitochondrial carrier family; splicing protein
PET117	1.6	Cytochrome c oxidase assembly factor
COX14	1.5	Cytochrome c oxidase assembly factor
COX15	1.9	Cytochrome c oxidase assembly factor
PET100	1.6	Cytochrome c oxidase assembly factor
COX11	1.9	Heme a synthesis
COX10	1.6	Heme a synthesis
CBP3	1.8	Cytochrome c reductase assembly factor
CBP4	1.7	Cytochrome c reductase assembly factor
ATP12	1.7	F1-ATP synthase assembly protein
ATP11	2	F1-ATP synthase assembly protein
CYC1	2	Cytochrome c isoform I
CYC7	2	Cytochrome c isoform II
CYB2	2.4	Cytochrome b2
COX5B	1.7	Cytochrome c oxidase chain Vb
COQ5	1.6	Ubiquinone biosynthesis
COQ3	1.7	Ubiquinone biosynthesis
TOM6	1.5	Mitochondrial protein import
TOM20	1.6	Mitochondrial import receptor
TIM17	1.9	Mitochondrial protein import
PHB2	1.7	Prohibitin
PHB1	2.3	Prohibitin
HSF1	1.6	Heat shock transcription factor
HSP78	1.6	Mitochondrial heat shock protein
HSP60	2.2	Mitochondrial chaperonin
HSP10	2.4	Mitochondrial chaperonin
SSC1	1.9	Mitochondrial chaperonin
MAS1	2	Subunit of mitochondrial processing peptidase
MSF1	1.7	Intramitochondrial sorting
YPR011C	1.6	Mitochondrial carrier family
YMR166C	1.6	Mitochondrial carrier family
YER053C	2	Mitochondrial carrier family
HAP4	1.8	Transcriptional activator; HAP2/3/4/5 complex
ISU1	2.2	Iron metabolism
NFS1	1.5	Iron metabolism
COX17	1.7	Copper metabolism
DLD1	1.7	D-Lactate ferricytochrome c oxidoreductase
NDE1	2	Mitochondrial NADH dehydrogenase
MAM33	2.4	Required for respiratory growth
YJR101W	1.9	Required for respiratory growth
YKL137W	1.5	Required for respiratory growth

(continues)

Table 1 (*continued*)

Gene	Fold change	Description
Genes repressed by the cAMP-PKA pathway and/or induced by Msn2/Msn4		
HXK1	2.1	Hexokinase I
HOR2	2	DL-Glycerol phosphate phosphatase
GPD2	1.8	Glycerol-3-phosphate dehydrogenase
HSP78	1.6	Mitochondrial heat shock protein
UBI4	1.5	Ubiquitin polyprotein
GSY2	2	Glycogen synthetase isoform 2
TPS1	1.8	Trehalose-6-phosphate synthase
TPS2	2.1	Trehalose-6-phosphate synthase
TPS3	1.6	Trehalose-6-phosphate synthase
TSL1	1.8	Trehalose-6-phosphate synthase
GLK1	1.8	Glucokinase
ARA1	1.5	Subunit of NADP+-dependent D-arabinose dehydrogenase
MDH1	1.6	Malate dehydrogenase, mitochondrial
MDH2	1.5	Malate dehydrogenase, cytosolic
YOR173W	1.8	Unknown
YML128C	1.6	Unknown
YNL200C	1.6	Unknown
Genes regulated by the cell wall integrity signaling pathway		
HSP150	2	Secreted O-glycosylated protein
PIR1	1.8	Protein similar to Hsp150 and Pir3
PIR3	2.1	Protein similar to Hsp150 and Pir1
CWP1	2.5	Cell wall mannoprotein
PST1	2	Similar to the Sps2-Ecm33-YCL048 family
SED1	2.1	Cell surface glycoprotein
MPK1	2	Serine/threonine protein kinase
Other genes related to cell stress		
PEP4	1.8	Proteinase A
YKL137W	1.5	Respiratory growth; resistance to NaCl and H2O2
YJR101W	1.9	Respiratory growth; stress resistance
YGL157W	1.6	Related to GRE2
Other metabolic enzymes		
SCS7	2.3	Ceramide hydrolase
CAR1	1.6	Arginase
AAT2	2.5	Aspartate aminotransferase
THI13	1.5	Pyrimidine biosynthesis
XKS1	1.5	Xylulokinase
FBP26	1.8	Fructose-2,6-bisphosphatase
YDR316W	2.4	Putative SAM-dependent methyltransferases
YPR140W	1.6	Putative acyltransferase
YFR047C	1.5	Putative quinolinicacid phosphoribosyltransferase
YAL061W	1.7	Similar to alcohol/sorbitol dehydrogenase
YAL060W	2.4	Similar to alcohol/sorbitol dehydrogenase
YDR516C	1.8	Similar to GLK1 and HXK1

(*continues*)

Table 1 (*continued*)

Gene	Fold change	Description	Gene	Fold change	Description
Glyoxylate and tricarboxylic acid cycle			**Other**		
MDH2	1.6	Malate dehydrogenase, cytosolic	AFR1	1.9	Morphogenesis of the mating projection
MDH1	1.5	Malate dehydrogenase, mitochondrial	RME1	1.7	Repressor of IME1 transcription
CIT1	1.5	Citrate synthase, mitochondrial	MSC1	1.6	Insertional mutant causes dominant meiotic lethality
ACO1	4	Aconitase	MSC6	2	Meiotic recombination
IDH1	3.6	Isocitrate dehydrogenase subunit I	AUT1	1.5	Autophagocytosis
			AUT7	1.7	Autophagocytosis
IDH2	5	Isocitrate dehydrogenase subunit II	HXT3	1.8	Low affinity hexose transporter
Glycolysis			YPS5	1.7	Yapsin 5
PGI1	2.3	Glucose-6-phosphate isomerase	TPK1	1.6	Protein kinase A subunit
			EX21	1.9	Peroxisomal protein targeting
HXK2	1.6	Hexokinase II			
HXK1	2.1	Hexokinase I	TUB4	1.6	Gamma-tubulin
GLD2	1.9	Glyceraldehyde-3-phosphate dehydrogenase 2	TUB1	1.7	Alpha-tubulin
			ECM13	1.9	Possibly involved in cell wall structure
ADH1	1.7	Alcohol dehydrogenase I			
GPP1	1.5	DL-Glycerol phosphate phosphatase	CAF17	1.7	Component of the CCR4 transcriptional complex
HOR2	2	DL-Glycerol phosphate phosphatase	YBR022W	1.5	Hydrolase involved in RNA splicing
Genes induced by Pdr1/Pdr3 overexpression			YGL064C	1.8	Member of the DEAD-box RNA helicase family
PDR5	3	ABC transporter/multidrug resistance			
LPG20	1.7	Putative aryl-alcohol dehydrogenase			
HXK1	2.1	Hexokinase I			
YOR049C	6	Transporter of unknown substrate			
YGR035C	2.2	Unknown			
YOR152C	2	Unknown			
YDR516C	1.8	Similar to GLK1 and HXK1			

Mitochondrial ribosomal proteins (1.6–2.6-fold induced)

MRP2, MRP4, MRP7, MRP13, MRP17, MRP21, MRP49, MRPL3, MRPL6, MRPL7, MRPL10, MRPL13, MRPL16, MRPL17, MRPL23, MRPL24, MRPL25, MRPL27, MRPL31, MRPL32, MRPL35, MRPL36, MRPL37, MRPL38, MRPL49, MRPS5, MRPS9, MRPS28, NAM9, PET123, RML2, YDR115W, YDR116C, YGL068W, YGL129C, YMR158W, YMR188C.

Unknown genes (1.5–2.8-fold induced)

YBR262C, YCR004C, YDR134C, YDR175C, YDR384C, YDR494W, YDR511W, YER119C, YER182W, YFR047C, YGL066W, YGL069C, YGL107C, YGR035C, YGR165W, YGR219W, YHL021C, YHR087W, YIL157C, YKL169C, YKL195W, YLR205C, YLR463C, YMR102C, YMR157C, YMR245W, YMR291W, YMR293C, YMR316W, YNL177C, YNL200C, YOL083W, YOR052C, YOR135C, YOR152C, YOR173W, YOR286W, YPL067C, YPL196W, YPR100W, YPR148C, YPR151C.

Table 2
Genes with Decreased Expression in ρ^0 Cells

Gene	Fold change	Description	Gene	Fold change	Description
Mitochondrial biogenesis and function			**Other**		
INH1	1.4	Inhibitor of the mitochondrial ATPase	HOM3	1.4	Threonine and methionine biosynthesis
YHM2	1.4	Mitochondrial carrier family	HIS1	1.6	Histidine biosynthesis
ATP20	1.4	F1-F0-ATP synthase complex subunit	TRP5	1.4	Tryptophan synthase
			SAM3	1.4	S-Adenosylmethionine permease
ATP5	1.5	F0-ATP synthase subunit V	PHO89	2.2	Na^+-dependent phosphate transporter
YMC2	1.4	Mitochondrial carrier family			
QCR2	1.4	Ubiquinol cytochrome c reductase complex	ZRT1	1.4	High affinity zinc transport protein
COX4	1.4	Cytochrome c oxidase subunit IV	FUR1	1.4	Uracil phosphor:bosyltransferase
			GIT1	2.3	Transporter involved in inositol metabolism
Ribosomal proteins (1.4-fold reduced)			UBP15	1.4	Putative ubiquitin-specific protease
RPL22A, RPS16B, RPS18B					
Unknown genes (1.4–1.7-fold reduced)			ATR1	1.7	Multidrug resistance transporter family
YCR102C, YDR209C, YDR380W, YLR432W					
			YGR138C	1.5	Multidrug resistance transporter family

Reference: Traven *et al.* (2001). Interorganellar Communication. Altered nuclear gene expression profiles in a yeast mitochondrial DNA mutant. *J. Biol. Chem.* **276,** 4020–4027. Data adapted from Tables 1 and 2, with permission.

Expression of Mitochondrial-Targeted Gene Products in ρ^+ and ρ^0 Breast Cancer Cells

Genes up-regulated in ρ^0 cells ($\geq$3 fold)		Genes down-regulated in ρ^0 cells ($\geq$3 fold)	
Gene product	Gene**	Gene product	Gene**
Transcription		**Transcription**	
v-maf protein G	MAFG	Dachshund (Drosophila) Homolog	DACH1
General transcription factor IIH polypeptide 3	GTF2H3	**Cell signalling**	
IFN consensus sequence binding protein 1	?	LIM binding domain 2	LDB2
Nuclear autoantigen	?	Phospholipase A2 Group IB	PLA2G1B
Runt-related transcription factor 1	RUNX1	Phospholipase C, epsilon	PLCE1
(AML-1 Oncogene)		Lipopolysaccharide response-68 protein	C14orf43
Cell signalling		Agouti related protein	AGRP
Slit (Drosophila) homolog 3	SLIT3	PKC, gamma	PRKCG
Phospholipase C, epsilon	PLCE1	Protein Tyrosine Phosphatase, Receptor Type c,	PTPRCAP
Synaptojanin 2*	SYNJ2	Polypeptide-Associated Protein	
HLA DR beta 5	HLA-DRB5	Neurotensin Receptor I	NTSR1
Cell architecture		Phosphodiesterase 1A, calmodulin dependent	PDE1A
Tubulin specific chaperone D	TBCD	**Cell architecture**	
Hyaluronoglucosaminidase 1*	HYAL1	Adaptor related protein complex 4 mu 1 subunit	AP4M1
Collagen Type IV	?	Synuclein gamma	SNCG
Gap Junction Protein 43	GJA1	Thymopoietin	TMPO
Metabolism		**Metabolism**	
Apolipoprotein D	APPOD	PIBF1	C13orf24
Arylacetamide deacetylase (esterase)*	AADAC	Carboxypeptidase B2 (plasma)	CPB2
Neuromedin U	NMU	Cytochrome P450 subfamily VIIB, polypeptide 1	CYP7B1
ATP5A1	ATP5A1	Protective protein for beta-galactosidase	PPGB
UDP-galactose:N-acetylglucosamine	B4GALT1	**Cell growth and differentiation**	
beta-1,4-galactosyltransferase I		Cyclin-dependent kinase inhibitor P19	CDKN2D
Cysteine dioxygenase type I	CDO1	GAP-43	GAP43
Apoptosis		**Intracellular protein degradation**	
TNF superfamily member 6, (Fas Ligand)*	FASLG	Huntingtin interacting protein B	HYPB
Ectodermal dysplasia 1, anhidrotic	EDA	Ubiquitin-like 4	UBL4A
Unknown			
FBR89	SEMA6A ?		
FLJ23231	ZC3H12A		
ESTs, Weakly similar to AAB47496	PRRT1		
NG5 [H. sapiens]			
EST	?		
FLJ13164	OSBPL11		
Uncharacterized Bone marrow protein, BM033	C6orf35		
FJ 10407 (FLJ10407?*)	TMEM48?		

*Correction of presumed spelling error in Delsite et al.

**Gene symbols were not provided by Delsite et al.; the indicated gene symbols are presumed to be correct, but may be in error.

?Denotes that the gene cannot be deduced unequivocally from the description given.

Reference: Delsite R, et al. (2002). Nuclear genes involved in mitochondria-to-nucleus communication in breast cancer cells. *Mol. Cancer* **1**, 6–15. Data adapted from Tables 1 and 2, with permission. Human breast cancer line MDA-MB-435 rho-plus cells vs MDA-MB-435 rho-zero cells.

Alterations in the Mitochondrial Proteome of ρ^+ vs ρ^0 Osteosarcoma Cells

SW-PROT	Gene name	Gene	x-Fold decrease ρ^0 vs. ρ^+
Respiratory complexes subunits and assembly proteins			
P28331	NADH-ubiquinone oxidoreductase 75 kDa subunit	NDUFS1	6
O75489	NADH-ubiquinone oxidoreductase 30 kDa subunit	NDUFS3	1.6
P19404	NADH-ubiquinone oxidoreductase 24 kDa subunit	NDUFV2	1.24
O00217	NADH-ubiquinone oxidoreductase 23 kDa subunit	NDUFS8	21
Q9P0J0	NADH-ubiquinone oxidoreductase B16.6 subunit	NDUFA13	5.4
P47985	Ubiquinol-cytochrome c reductase iron-sulfur subunit	UQCRFS1	9
P31930	Ubiquinol-cytochrome c reductase complex core protein 1	UQCRC1	1.8
P22695	Ubiquinol-cytochrome c reductase complex core protein 2	UQCRC2	1.7
Q8TB65	Cytochrome c oxidase subunit Va	COX5A	0.75
P10606	Cytochrome c oxidase polypeptide Vb	COX5B	6
P25705	ATP synthase alpha chain	ATP5A1	1.1
P06576	ATP synthase beta chain	ATP5B	1.1
P30049	ATP synthase delta chain	ATP5D	0.9
O75947	ATP synthase D chain	ATP5H	1.1
O75880	SCO1 protein homolog	SCO1	3.7
Mitochondrial translation apparatus			
Q9BYD6	Mitochondrial ribosomal protein L1	MRPL1	4.5
P52815	39S ribosomal protein L12	MRPL12	10
Q96GC5	Mitochondrial ribosomal protein L48	MRPL48	4
Q9Y399	Mitochondrial 28S ribosomal protein S2	MRPS2	20
P82650	Mitochondrial 28S ribosomal protein S22	MRPS22	3
Q9Y3D9	Mitochondrial ribosomal protein S23	MRPS23	2.3
Q92552	Mitochondrial 28S ribosomal protein S27	MRPS27	2
P51398	Mitochondrial 28S ribosomal protein S29	DAP3	2.9
Q9NP92	Mitochondrial 28S ribosomal protein S30	MRPS30	10.5
Q9UGM6	Tryptophanyl-tRNA synthetase	WARS2	2.9
Q96Q11	tRNA-nucleotidyltransferase 1 (CCA addition)	TRNT1	5.5
Mitochondrial transport systems			
O75431	Metaxin 2	MTX2	2.1
P57105	Mitochondrial outer membrane protein 25	SYNJ2BP	2.5
Q9P129	Calcium-binding transporter	SLC25A24	3.5
Q9Y5J7	Mito. Import inner mb. Translocase subunit TIM9 A	TIMM9	5
Q9NS69	Mitochondrial import receptor subunit TOM22	TOMM22	2.3
Q9UJZ1	Stomatin-like protein 2	STOML2	1.9
Energy systems			
P13804	Electron transfer flavoprotein alpha-subunit	ETFA	2.4
P38117	Electron transfer flavoprotein beta-subunit	ETFB	2.3
Q16134	Electron transfer flavoprotein-ubiquinone oxidoreductase	ETFDH	2.3
Q16836	Short chain 3-hydroxyacyl-CoA dehydrogenase	HADHSC	3.5
Q99714	3-Hydroxyacyl-CoA dehydrogenase type II	HADH2	2.3
P49748	Acyl-CoA dehydrogenase, very-long-chain specific	ACADVL	1.74
P36957	Dihydrolipoamide succinyltransf. (2-oxoglutarate DH cplx)	DLST	2.1
P09622	Dihydrolipoamide dehydrogenase	DLD	6.1
Q9P2R7	Succinyl-CoA ligase (ADP-forming) beta-chain	SUCLA2	8.6
Q9HCC0	Methylcrotonoyl-CoA carboxylase beta chain	MCCC2	5.8
Q8WVX0	Putative enoyl-CoA hydratase	ECH1	1.7
P78540	Arginase II	ARG	1.9
Miscellaneous			
O00165	Hax-1	HAX1	2.8
Q9NR28	Smac protein	DIABLO	1.8
Q16762	Rhodanese	TST	1.8
Q6YN16	Hydroxysteroid dehydrogenase	HSDL2	4.45
Q6P587	fum. Acetoacetate lyase domain cont. Protein	FAHD1	3
Q9BSH4	UPF0082 protein PRO0477	CCDC44	2.2
Q96EH3	Chromosome 7 ORF 30	C7orf30	7

Reference: Chevallet *et al.* (2006). Alterations of the mitochondrial proteome caused by the absence of mitochondrial DNA: A proteomic view. *Electrophoresis* **27**, 1574–1583. Data adapted from Table 1, with permission. Human 143B rho-plus vs 143B206 rho-zero cells.

Effect of Complex I Deficiency on the Expression of Human Mitochondrial-Targeted Gene Products

| Gene symbol | Accession | Control | | | | | NDUFS4 | | NDUFS2 | NDUFV1 | | NDUFS7 | NDUFS8 |
		C1	C2	C3	C4	C5	P1	P2	P3	P4	P5	P6	P7
HIBADH	N77326	−1.4	−1.3	1.9	Nd	−1.4	Nd	−5	Nd	−2	−1.7	−3.3	Nd
MTHFD2	X16396	−1.3	1.9	Nd	Nd	Nd	−3.3	−2.5	Nd	−1.3	−2	−2.5	2.3
PDK1	AI026814	−1.4	−2	Nd	Nd	Nd	−5	Nd	Nd	−1.3	−2	−5	1.4
ADH4	AA007395	1.2	−1.4	Nd	Nd	Nd	Nd	Nd	1.8	−1.4	−5	−1.7	3.9
ALDH6	AA455235	Nd	−5	Nd	Nd	1.2	Nd	9.4	Nd	−3.3	−1.4	−5	2.5
LDHA	AA489611	−1.7	1	0.7	−3.3	−1.7	1.3	1	−5	−2.5	−1.1	1	−3.3
ADH6	AI244615	−1.4	−2.5	2.4	Nd	1.6	−2	−2.5	1.3	1.3	−3.3	−1.7	−1.1
ALDH2	R93551	−1.7	−1.7	0.9	Nd	Nd	−10	1	−1.7	−1.4	−1.4	−3.3	−1.7
BPGM	AA678065	−1.7	Nd	Nd	Nd	Nd	Nd	−3.3	Nd	1.4	−1.7	−2.5	2
c11beta-HSD1	AA150918	Nd	Nd	Nd	Nd	Nd	Nd	Nd	Nd	Nd	3.5	1.4	3.7
ACAD8	AA282273	−1.4	−1.7	1.4	Nd	2.2	−1.3	−1.3	−1.1	1	−1.7	−3.3	−1.7
STAR	T77120	−1.1	−1.1	1.1	−2.5	−1.1	2.9	−1.7	−2	1	−1.4	−1.4	−2.5
CYP1B1	AA448157	1.3	1.8	1	Nd	Nd	Nd	Nd	Nd	−2	2.7	−2.5	−2
CYP2 C	N53136	−1.7	−1.7	1.3	Nd	3.2	Nd	−1.7	Nd	1.1	−2.5	−3.3	Nd
ALAS2	AA010004	−2	−1.7	0.8	−1.3	−1.4	1.5	1.3	1	1.8	−1.4	−1.4	−1.1
CS	N67639	1.7	2.1	0.5	Nd	Nd	−2	1.2	Nd	1.4	1.6	2.6	1.5
ATP5 B	AA708298	1.2	2.6	0.9	Nd	2.3	2.4	1.5	−1.4	1.1	2.6	2.8	1.2
MTNDL4	MTDNA	1.7	1	1.2	−1.3	−1.7	2.2	1	−5	1.4	2.5	2.3	−3.3
MTCO1	MTDNA	1.5	−1.4	0.9	−2	−1.4	2	1.4	−4	1.3	1.9	1.3	−1.4
MTCO2	MTDNA	2.5	1.7	0.9	−1.4	1	1.9	1.4	−6	1.3	3.7	2.1	−2.5
MTCO3	MTDNA	3.1	1.4	0.9	1.2	−1.1	1.3	1.1	−8	1.2	1.9	1.6	−5
MTND4	MTDNA	2.4	1.8	1.4	−1.1	−1.3	1.6	1.4	−3.3	1.3	4.6	3.8	−1.7
MTND5	MTDNA	2.3	1.4	1.2	−1.3	−1.7	1	1.2	−3.3	1.2	4.8	3.6	−1.7
MTCYB	MTDNA	1.6	1.1	1	−1.3	−1.1	1.4	−1.3	−10	−1.1	2.4	1.5	−10
MTND1	MTDNA	1.3	1	1.5	−1.1	−1.7	1.3	−1.1	−2.5	1.1	1.3	−1.1	−5
CYB5	R92281	1.9	2.8	0.7	Nd	Nd	−2	1.2	Nd	1	2.7	3.1	2.4
MTATP8	MTDNA	3.2	1.2	1.3	1.2	1	1	1.4	−5	1.1	2.5	2	−3.3
MTATP6	MTDNA	2.1	−1.4	1.2	1.2	−1.1	1.5	1	−5	1.2	1.1	−1.4	−5
PGD	AA598759	2.1	3	1	Nd	−2.5	1.8	1.2	2.2	2.3	1.6	1.9	2.9
PRKAR2 B	AA181500	Nd	Nd	Nd	Nd	Nd	Nd	Nd	Nd	Nd	2.6	1.4	Nd
GNAT1	W92431	1	−1.1	0.7	−2.5	Nd	2.1	−1.1	−2	1.1	−1.3	−1.3	−2.5
AK3	AA007279	1.1	Nd	Nd	Nd	1.1	Nd	−1.7	Nd	−1.7	−1.1	−2.5	−1.3
MSP1	AI796091	−1.4	−1.1	1.1	Nd	Nd	Nd	Nd	Nd	−1.3	−2	−2.5	Nd
POLRMT	AA521239	−1.4	−1.1	1.1	Nd	Nd	1	−1.9	2.2	−1.3	−1.9	−1.9	−1.6
MRS3	H40449	1.4	−1.3	1.2	Nd	−1.6	−2.5	Nd	4.8	−1.2	1.2	1.4	Nd
AZF	AI637719	1.4	1.3	1	Nd	−1.3	2.4	1.4	3.2	1.3	1.8	2	1.2
AAC1	AI184139	−1.1	−1.1	1	−2.5	−2	2.1	1.3	1	1.4	−1.1	−1.4	3
FXYD1	H57136	2	1.6	1.4	Nd	1.1	−1.3	1.1	1.3	−1.5	1.7	2.2	4.1
NBC1	AA452278	−1.1	Nd	1.9	Nd	5.1	Nd	Nd	−5	−1.1	−2.5	−2.5	Nd
MT1E	AA872383	2	−3.3	2.3	Nd	3.4	7.2	6.9	6.5	2.1	5.6	7.4	9.8
MT2	AI289110	1.5	−3.3	2.4	1.7	1.5	11.7	5.5	5.2	2.3	3.8	5.3	10.5
MT1 G	H53340	1.4	−5	2.6	1.3	1.4	8.3	6.9	6.2	2.4	4.9	4.9	10.8
MT1 B	H72722	2	−2.5	2.9	Nd	Nd	10.8	9.5	7.5	2.2	6.4	10.1	21
MT1 H	H77766	1.9	−1.7	1.9	1.2	1.2	4.8	7	3.7	2.1	4.8	5.9	14
MT1 F	N55459	1.7	−3.3	2.5	2.1	1.6	6.8	8.2	6.5	2.2	4.8	4.8	15.1
MT1 L	N80129	1.8	−2.5	2.4	1.6	1.6	9.3	9.4	7.1	2.4	6.5	7.5	17.5
ATP6E	AA702541	1.5	2.3	1	Nd	Nd	−1.3	1.2	0.9	1.6	2.1	1.9	−1.3
ATP1G1	AA775899	1.2	−1.7	1.6	−1.1	−1.3	5.8	2.4	3.2	1.9	2.1	2.6	7.4

Reference: van der Westhuizen FH, *et al.* (2003). Human mitochondrial complex I deficiency: investigating transcriptional responses by microarray. *Neuropediatrics* **34**, 14–22. Adapted from Table 2, with permission. Genes showing differential expression in control (C) and patient (P) fibroblast cell lines due to glucose/galactose transition as carbon energy source. The values indicate relative increase (positive value) or decrease (negative value) when glucose-containing medium were substituted with galactose-containing medium. Mutated complex I gene indicated above each patient.

Effect of Exercise on the Expression of Human Mitochondrial-Targeted Gene Products

Gene name	Accession #	3 hr	48 hr	Potential relevant function
Metabolism and mitochondria				
Forkhead transcription factor O1A	AA134749	**5.2 ± 2.6**	1.1 ± 0.4	• Activates PDK4 and PGC1a
*Pyruvate dehydrogenase kinase 4	AA169469	**3.5 ± 1.2**	1.3 ± 0.5	• Negatively regulates pyruvate dehydrogenase
Mitochondrial ribosomal protein L2	N94366	**3.4 ± 1.4**	5.5 ± 2.8	• Involved in mitochondrial translation
IL-6 receptor	T52330	**3.1 ± 0.5**	2.3 ± 0.7	• Subunit for IL-6 receptor complex
Ras-related associated with Diabetes	W84445	**2.9 ± 0.8**	2.6 ± 2.1	• A role in glucose metabolism
*PPAR γ coactivator 1α	N89673	**2.9 ± 0.8**	0.6 ± 0.1	• Regulates mitochondrial biogenesis
*PPAR	AA088517	**2.7 ± 0.7**	1.4 ± 0.6	• Positively regulates fat metabolism
Nuclear receptor binding protein 2	N30573	**2.6 ± 0.4**	**3.8 ± 0.9**	• Binds to and co-modulates PPARa
Aminolevulinate δ syntetase 2	AA699919	**2.3 ± 0.4**	1.5 ± 0.1	• Catalyzes first step in the heme biosynthesis
Interferon regulatory factor 1	AA478043	**2.1 ± 0.3**	1.0 ± 0.1	• Transcription factor for iNOS expression
IL-6 signal transducer (gp130)	T61343	**2.0 ± 0.1**	1.5 ± 0.1	• Component for IL-6 receptor complex
†PPAR δ	n/a	**2.6 ± 0.6**	1.1 ± 0.1	• Positively regulates fat metabolism
†PPAR α	n/a	**1.7 ± 0.1**	1.6 ± 0.4	• Positively regulates fat metabolism
Oxidant stress and signaling				
Metallothionein 1G	H53340	**7.5 ± 0.6**	2.4 ± 0.9	• Metal ion homeostasis and detoxification,
Metallothionein 1H	H77766	**7.0 ± 1.0**	2.0 ± 0.5	protection against oxidative stress, cell
Metallothionein 1F	N55459	**6.6 ± 1.8**	2.3 ± 1.0	proliferation and apoptosis; responds
Metallothionein 3	AI362950	**6.1 ± 0.9**	1.6 ± 0.8	transiently to most forms of stress or injury
Metallothionein 1B	H72722	**5.3 ± 0.8**	2.0 ± 0.6	providing cytoprotective action, particularly
Metallothionein 2A	AA872383	**4.7 ± 0.5**	2.2 ± 1.0	oxidative injury
Metallothionein 1L	AI289110	**4.2 ± 0.4**	2.2 ± 0.5	
Tyrosyl-DNA phosphodiesterase 1	AI215965	**6.7 ± 4.1**	**8.0 ± 4.0**	• Repairs free-radical mediated DNA damage
*JunB	N94468	**4.9 ± 0.4**	n/a	• Part of AP-1 complex
Interferon regulatory factor 1	AA478043	**2.1 ± 0.3**	1.0 ± 0.1	• O_2 mediated transcription factor
Electrolyte transport				
NMDA receptor	R88267	n/a	**25 ± 16**	• Role in synaptic plasticity
*Ca^{2+} ATPase (SERCA 3)	AA857542	**4.5 ± 0.8**	1.6 ± 0.6	• Pumps Ca^{2+} into SR
Solute carrier 17 (1)	N73241	**3.9 ± 14**	2.7 ± 0.3	• Transports phosphate into cells
Na^+/K^+ ATPase (b3)	AA489275	**2.7 ± 0.2**	n/a	• Regulatory component of Na^+/K^+ ATPase pump
Chloride channel 4	AA019316	**2.4 ± 0.3**	0.6 ± 0.2	• Regulates cell volume and intracellular pH
Solute carrier 22 (3)	AA460012	**2.4 ± 0.5**	1.9 ± 0.5	• Mediates transport of organic cations
GABA receptor	R40790	**2.2 ± 0.3**	**2.3 ± 0.3**	• Chloride channel; inhibitory neurotransmitter

*Differential expression confirmed with real-time RT-PCR.

†Not measured on microarrays; differential expression determined using real-time RT-PCR.

Bold, Statistically significant change (FDR < 5% on SAM or P < 0.05 on ANOVA). Note: Not all differentially expressed genes are shown in the tables. For a complete listing of the data, please visit www.ncbi.nlm.nih.gov/geo/

Reference: Mahoney DJ, *et al.* (2005) Analysis of global mRNA expression in human skeletal muscle during recovery from endurance exercise. *FASEB J.* **19**, 1498–1500. Data adapted from Table 1. Differentially increased gene expression after exercise.

Expression of Gene Products in Muscle from Caloric–Restricted Rats

Change		Gene name	Function	Gene name
+	3.7	Mitochondrial COX subunits I, II, III	Energy metabolism	Mtco1-Mtco3
+	3.2	COX subunit IV	Energy metabolism	Cox4i1?
+	3.1	Mitochondrial cytochrome B	Energy metabolism	Mtcytb
+	2.8	Cu-Zn superoxide dismutase	Free radical scavenger	Sod1
+	2.6	Polyubiquitin	Protein metabolism	Ubc
+	2.3	Cyclin G	Cell cycle	Ccng1
+	2.2	Vascular cell adhesion molecule-1	Cell growth/adhesion	Vcam1
+	2.1	Glutathione peroxidase I	Free radical scavenger	Gpx1
+	2.1	Mn superoxide dismutase	Free radical scavenger	Sod2
+	2.0	Mitochondrial NADH dehydrogenase	Energy metabolism	?
+	2.1	COX subunit VIII	Energy metabolism	Cox8?
+	2.1	Mitochondrial COX subunit Va	Energy metabolism	Cox5a
+	2.0	Alpha-crystallin B chain	Extracellular matrix	Cryab
−	2.0	Lipoprotein lipase	Lipid metabolism	Lpl
−	2.1	Cyclophilin B	Drug metabolism	Ppib
−	2.1	Guanosine monophosphate reductase	Purine metabolism	Gmpr
−	2.1	Ras-related protein	Signal transduction	?
−	2.1	Heat shock protein 70	Stress response/chaperone	Hspa1a
−	2.2	Alpha-tubulin	Structure/contractile	Tuba1
−	2.2	Stress-inducible chaperone mtGrpE	Stress response/chaperone	Grpel1
−	2.3	Collagen alpha-1 type I	Extracellular matrix	Col1a1
−	2.3	Aldose reductase	Glucose metabolism	Akr1b1
−	2.3	Vascular endothelial cell growth factor	Cell growth/adhesion	Vegf
−	2.4	Proteasome subunit RC7 II	Protein metabolism	?
−	2.5	Beta-actin	Structure/contractile	Actb
−	2.5	ST1B1	Unknown	Sult1b1
−	2.5	Cyclin-dependent kinase 4	Cell cycle	Cdk4
−	2.7	Phospholipase C form II	Lipid metabolism	?
−	2.7	Porphobilinogen deaminase	Heme synthesis	Hmbs
−	2.9	c-ras-H-1	Unknown	Hras1
−	3.3	Nap1 protein	Proteinase	Nckap1
−	3.6	ATP-sensitive potassium channel-1	Ion channel	Kcnj1
−	3.7	Cyclin D1	Cell cycle	Ccnd1
−	3.8	Chaperonin 60	Stress response/chaperone	Hsp60
−	4.7	Pro alpha-1 collagen type III	Extracellular matrix	Col3a1

Reference: Sreekumar *et al.* (2002). Effects of caloric restriction on mitochondrial function and gene transcripts in rat muscle. *Am. J. Physiol. Endocrinol. Metab.* **283,** E38–E43. Data adapted from Table 1, with permission. Expression of genes in muscle from caloric-restricted vs control rats.

Effect of Cold on the Expression of Mitochondrial-Targeted Gene Products in Carp

Gene ID	Name	Blast E-value	Acc. #
252 cold-induced genes, of which at least 40 encode mitochondrial proteins			
539-1	ADP,ATP carrier protein, fibroblast isoform (ADP/ATP translocase 2)	e-152	CA965413
539-2	ADP,ATP carrier protein, isoform T2 (ADP/ATP translocase 3)	9e-77	CA966975
747-1	Electron transfer flavoprotein beta-subunit (Beta-ETF)	2e-83	CA967122
312-1	Succinate dehydrogenase [ubiquinone] cytochrome B small subunit	1e-40	CA965167
04h21	Cytochrome c, iso-1 and iso-2	4e-42	CA965344
590-1	ATP-binding cassette, sub-family F, member 2 (Iron inhibited ABC)	e-129	CA965822
05g12	Electron transfer flavoprotein alpha-subunit, mitochondrial	1e-25	CA967527
24-1	ATP synthase alpha chain heart isoform, mitochondrial precursor	0.0	CA967441
17e15	Phosphate carrier protein, mitochondrial precursor (PTP)	3e-96	CA964942
948-1	ATP synthase gamma chain, mitochondrial precursor	e-101	CA965223
696-1	ATP synthase beta chain, mitochondrial precursor	0.0	CA966721
28a24	Mitochondrial import receptor subunit TOM20 homolog (Mitochondrial)	3e-59	CA965079
751-2	ATP synthase lipid-binding protein, mitochondrial precursor	3e-32	CA969711
751-3	ATP synthase lipid-binding protein, mitochondrial precursor	1e-39	CA965268
16e05	ATP synthase delta chain, mitochondrial precursor	1e-55	CA966367
729-1	Cytochrome c1, heme protein, mitochondrial precursor (Cytochrome)	2e-75	CA966996
896-1	Ubiquinol-cytochrome C reductase complex core protein I	9e-75	CA968888
19k18	NADH-ubiquinone oxidoreductase 49 kDa subunit (Complex I-49KD)	e-128	CA965052
28d20	NADH-ubiquinone oxidoreductase 75 kDa subunit, mitochondrial	e-126	CF662979
12c16	Cytochrome c oxidase polypeptide VB (VI)	8e-37	CA966760
290-1	Vacuolar ATP synthase subunit G 1 (V-ATPase G subunit 1) (Vacuolar)	8e-26	CA969859
05o19	Cytochrome c oxidase polypeptide VIb (AED)	2e-30	CA965512
11a12	NADH-ubiquinone oxidoreductase chain 3	8e-23	CF662979
16c12	Ornithine decarboxylase antizyme, long isoform (ODC-Az-L)	7e-47	CA966991
20o23	ATP-dependent CLP protease ATP-binding subunit ClpX-like	2e-95	CA967005
730-2	Stress-70 protein, mitochondrial precursor (75 kDa glucose)	3e-97	CA967001
730-1	Stress-70 protein, mitochondrial precursor (75 kDa glucose regulated)	3e-88	CA969061
35h23	Stress-70 protein, mitochondrial precursor (75 kDa glucose)	7e-49	CF661512
114-1	12.6 kDa FK506-binding protein (FKBP-12.6) (Peptidyl-prolyl)	7e-47	CA968035
283-1	Superoxide dismutase [Mn], mitochondrial precursor	7e-91	CA969803
879-1	Isocitrate dehydrogenase [NADP], mitochondrial precursor	e-148	CF662883
727-2	Pyruvate dehydrogenase E1 component alpha subunit, somatic form	4e-10	CF662449
727-1	Pyruvate dehydrogenase E1 component alpha subunit, somatic form	9e-97	CA966982
30-2	Trifunctional enzyme beta subunit, mitochondrial precursor	3e-38	CA964076
28g10	3-Ketoacyl-CoA thiolase, mitochondrial (Beta-ketothiolase)	1e-85	CA965105
06n21	5-Aminolevulinic acid synthase, erythroid-specific, mitochondrial	e-101	CF662529
26e24	14-3-3 protein epsilon (Mitochondrial import stimulation factor L)	1e-09	CA970320
397-2	ARP2/3 complex 21 kDa subunit (P21-ARC) (Actin-related protein 2/3)	5e-93	CA967899
739-1	Basigin precursor (Blood-brain barrier HT7 antigen) (Neurothelin)	6e-20	CA967050
24i08	Cytochrome c oxidase assembly protein COX15	3e-16	CA969874
8 cold-repressed genes, none of which appear to encode mitochondrial proteins			

Reference: Gracey AY, *et al.* (2006) Coping with cold: An integrative, multitissue analysis of the transcriptome of a poikilothermic vertebrate. *Proc. Natl. Acad. Sci. USA* **101,** 16970–16975. Data adapted from Table 1, with permission. The common response genes. A response threshold test was employed to identify genes that exhibited a similar response to cold in all tissues (see S*upporting Materials and Methods*). Gene ID corresponds to the unique identifier given to clones following alignment of their 5′ ESTs (see *Supporting Materials and Methods*). Clones that did not share sequence similarity with any other clone were assigned an ID that corresponds to their location in the 384-well microtiter plate in which they were found. For example, Gene ID 26c22 was found in plate 26, at coordinates c22. Clones that shared sequence similarity in the first round of alignment are grouped and assigned the same first numerical identifier. They are then assigned a second identifier that indicates whether they share similarity with another clone in the group. For example, Gene IDs 262-1 and 262-2 were grouped in cluster 262 in the first round of alignment, but were then split into two separate clusters by more rigorous alignment. Accession numbers of ESTs that were aligned to identify each putative gene are indicated.

INDEX

VOLUMES IN SERIES

Founding Series Editor
DAVID M. PRESCOTT

Volume 10 (1975)
Methods in Cell Biology
Edited by David M. Prescott

Volume 11 (1975)
Yeast Cells
Edited by David M. Prescott

Volume 12 (1975)
Yeast Cells
Edited by David M. Prescott

Volume 13 (1976)
Methods in Cell Biology
Edited by David M. Prescott

Volume 14 (1976)
Methods in Cell Biology
Edited by David M. Prescott

Volume 15 (1977)
Methods in Cell Biology
Edited by David M. Prescott

Volume 16 (1977)
Chromatin and Chromosomal Protein Research I
Edited by Gary Stein, Janet Stein, and Lewis J. Kleinsmith

Volume 17 (1978)
Chromatin and Chromosomal Protein Research II
Edited by Gary Stein, Janet Stein, and Lewis J. Kleinsmith

Volume 18 (1978)
Chromatin and Chromosomal Protein Research III
Edited by Gary Stein, Janet Stein, and Lewis J. Kleinsmith

Volume 19 (1978)
Chromatin and Chromosomal Protein Research IV
Edited by Gary Stein, Janet Stein, and Lewis J. Kleinsmith

Volume 20 (1978)
Methods in Cell Biology
Edited by David M. Prescott

Advisory Board Chairman
KEITH R. PORTER

Volume 21A (1980)
Normal Human Tissue and Cell Culture, Part A: Respiratory, Cardiovascular, and Integumentary Systems
Edited by Curtis C. Harris, Benjamin F. Trump, and Gary D. Stoner

Volume 21B (1980)
Normal Human Tissue and Cell Culture, Part B: Endocrine, Urogenital, and Gastrointestinal Systems
Edited by Curtis C. Harris, Benjamin F. Trump, and Gray D. Stoner

Volume 22 (1981)
Three-Dimensional Ultrastructure in Biology
Edited by James N. Turner

Volume 23 (1981)
Basic Mechanisms of Cellular Secretion
Edited by Arthur R. Hand and Constance Oliver

Volume 24 (1982)
The Cytoskeleton, Part A: Cytoskeletal Proteins, Isolation and Characterization
Edited by Leslie Wilson

Volume 25 (1982)
The Cytoskeleton, Part B: Biological Systems and *in Vitro* Models
Edited by Leslie Wilson

Volume 26 (1982)
Prenatal Diagnosis: Cell Biological Approaches
Edited by Samuel A. Latt and Gretchen J. Darlington

Series Editor
LESLIE WILSON

Volume 27 (1986)
Echinoderm Gametes and Embryos
Edited by Thomas E. Schroeder

Volume 28 (1987)
***Dictyostelium discoideum:* Molecular Approaches to Cell Biology**
Edited by James A. Spudich

Volume 50 (1995)
Methods in Plant Cell Biology, Part B
Edited by David W. Galbraith, Don P. Bourque, and Hans J. Bohnert

Volume 51 (1996)
Methods in Avian Embryology
Edited by Marianne Bronner-Fraser

Volume 52 (1997)
Methods in Muscle Biology
Edited by Charles P. Emerson, Jr. and H. Lee Sweeney

Volume 53 (1997)
Nuclear Structure and Function
Edited by Miguel Berrios

Volume 54 (1997)
Cumulative Index

Volume 55 (1997)
Laser Tweezers in Cell Biology
Edited by Michael P. Sheetz

Volume 56 (1998)
Video Microscopy
Edited by Greenfield Sluder and David E. Wolf

Volume 57 (1998)
Animal Cell Culture Methods
Edited by Jennie P. Mather and David Barnes

Volume 58 (1998)
Green Fluorescent Protein
Edited by Kevin F. Sullivan and Steve A. Kay

Volume 59 (1998)
The Zebrafish: Biology
Edited by H. William Detrich III, Monte Westerfield, and Leonard I. Zon

Volume 60 (1998)
The Zebrafish: Genetics and Genomics
Edited by H. William Detrich III, Monte Westerfield, and Leonard I. Zon

Volume 61 (1998)
Mitosis and Meiosis
Edited by Conly L. Rieder

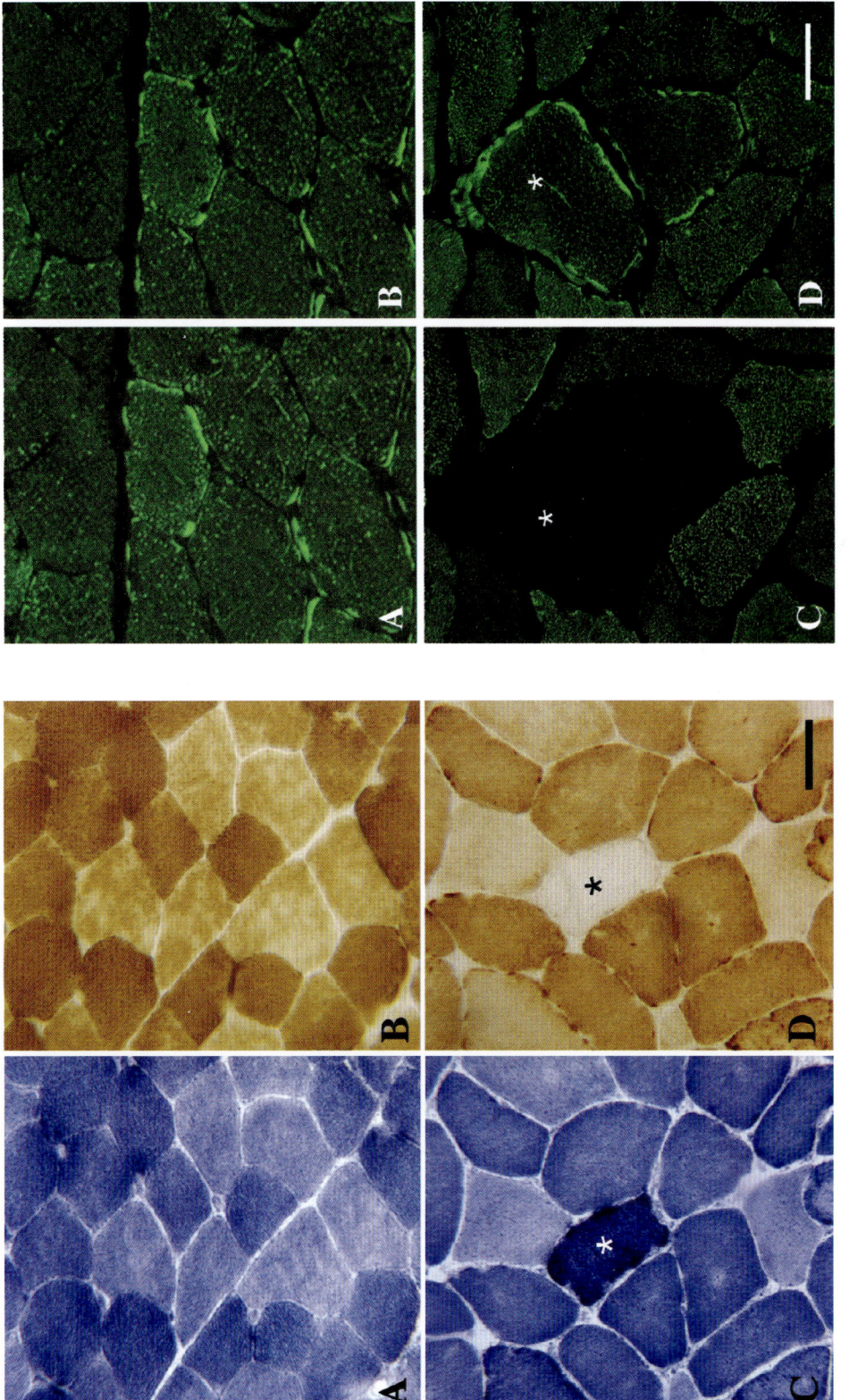

Chapter 6, Fig. 2

Chapter 6, Fig. 1

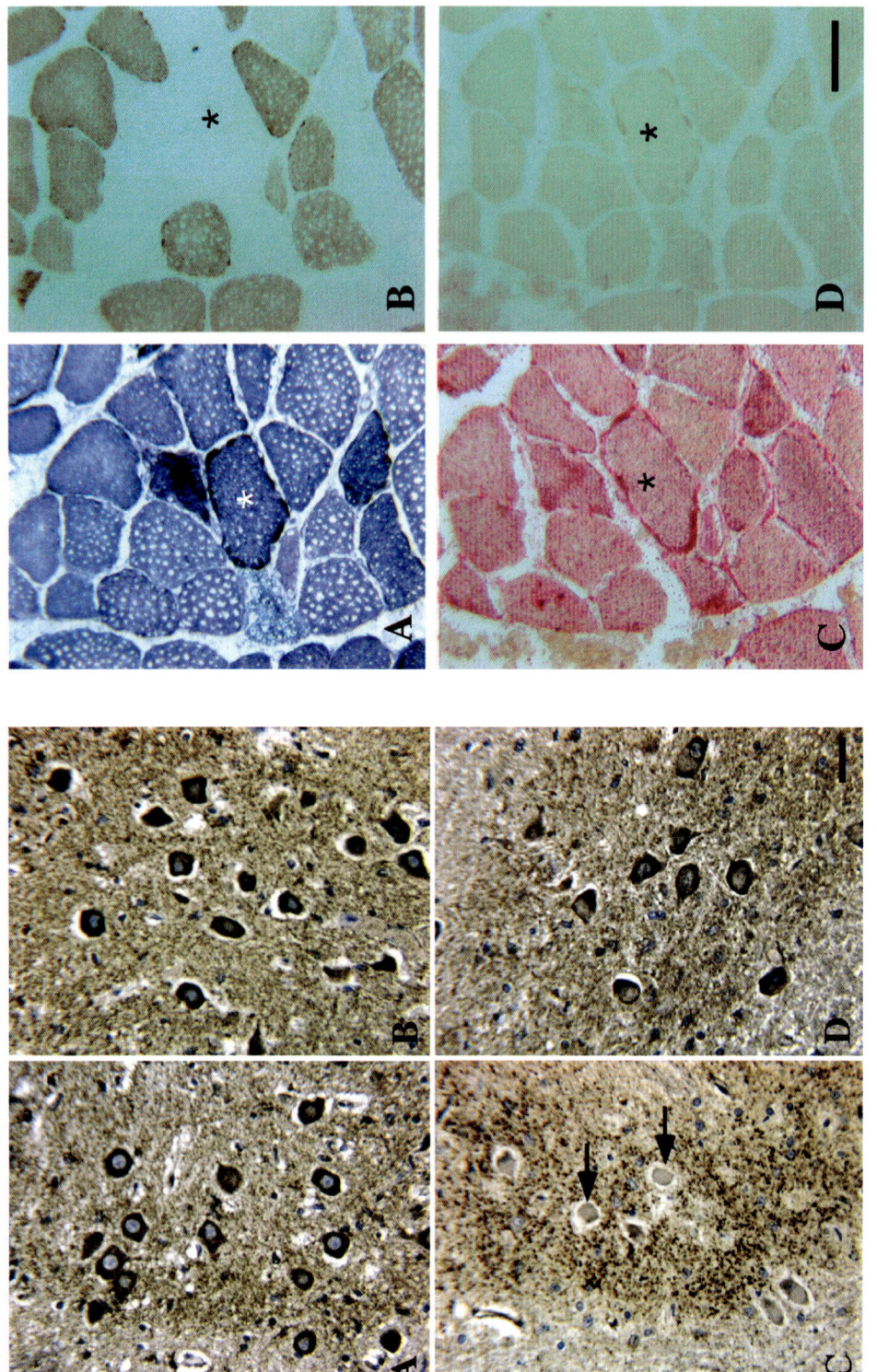

Chapter 6, Fig. 4

Chapter 6, Fig. 3

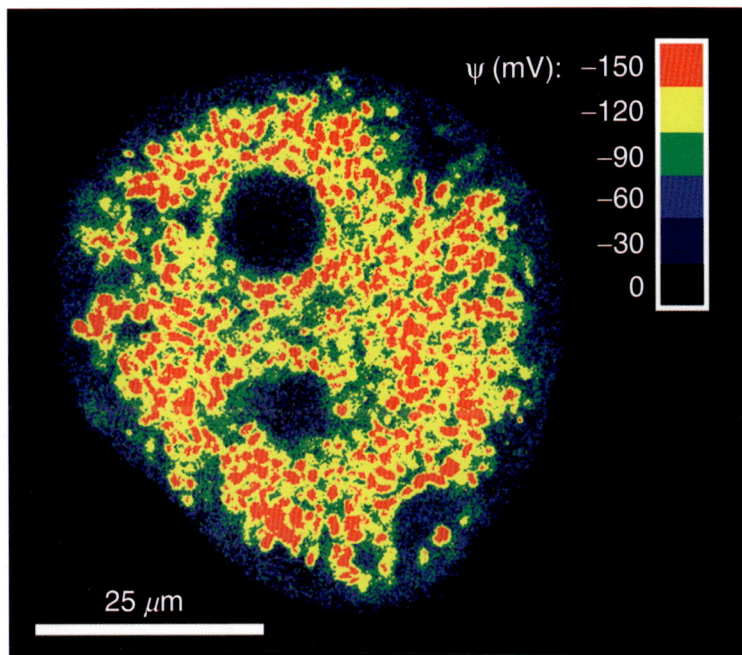

ψ (mV): −150
−120
−90
−60
−30
0

25 μm

Chapter 14, Fig. 1

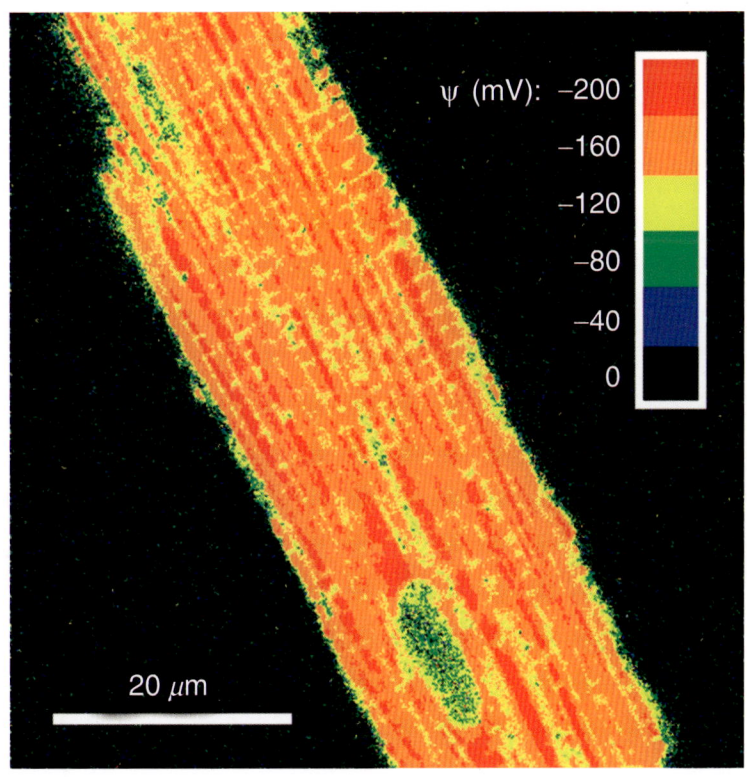

ψ (mV): −200
−160
−120
−80
−40
0

20 μm

Chapter 14, Fig. 2

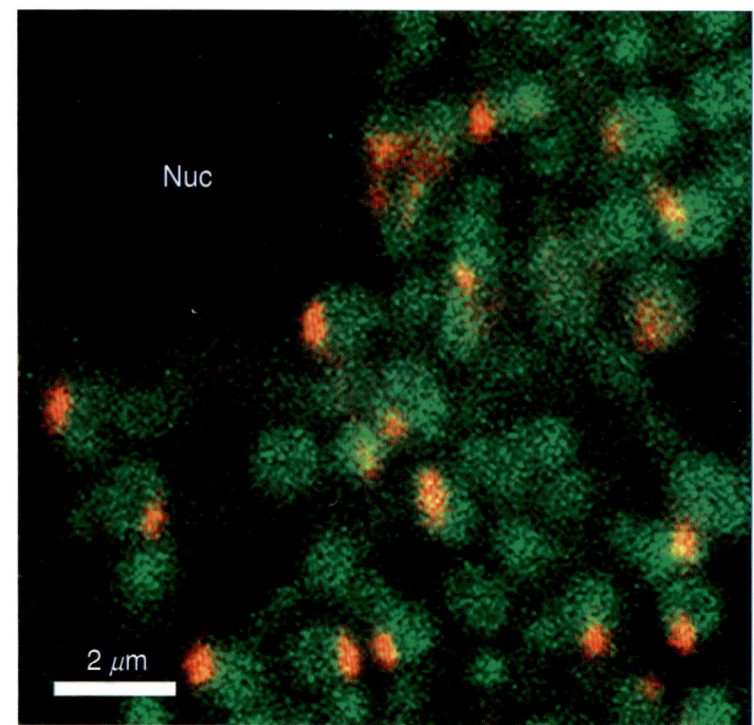

Chapter 14, Fig. 3

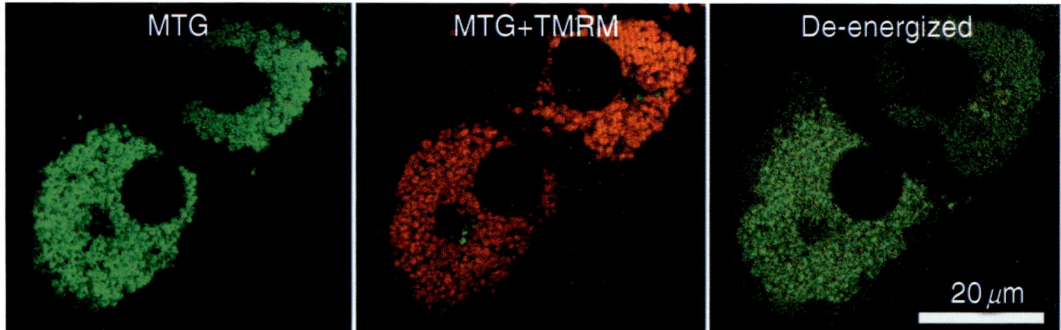

Chapter 14, Fig. 4

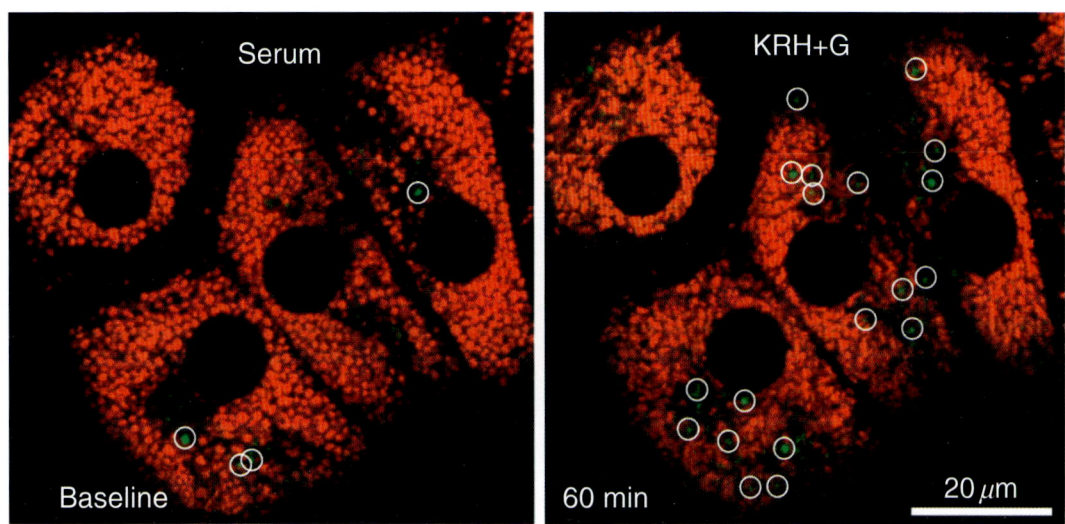

Chapter 14, Fig. 5

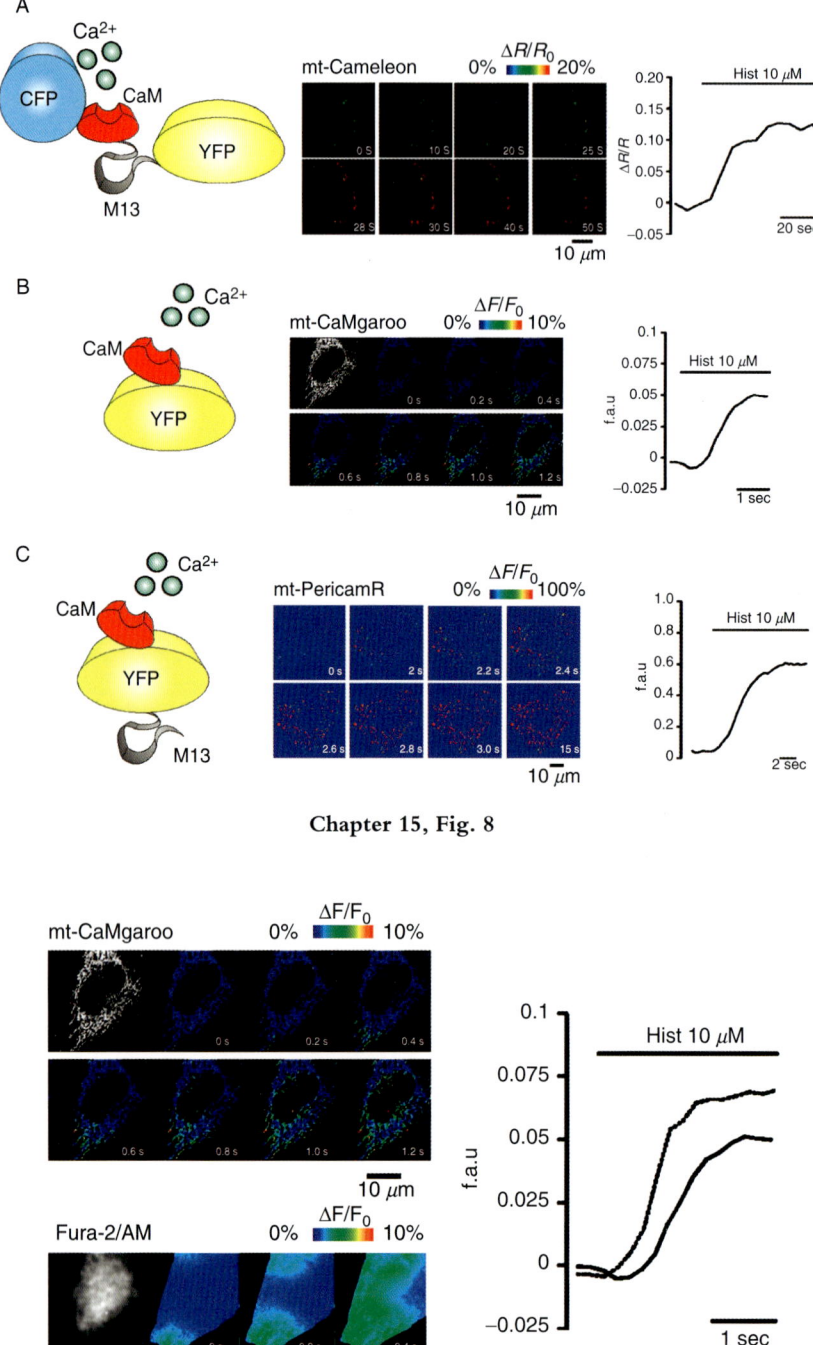

Chapter 15, Fig. 8

Chapter 15, Fig. 9

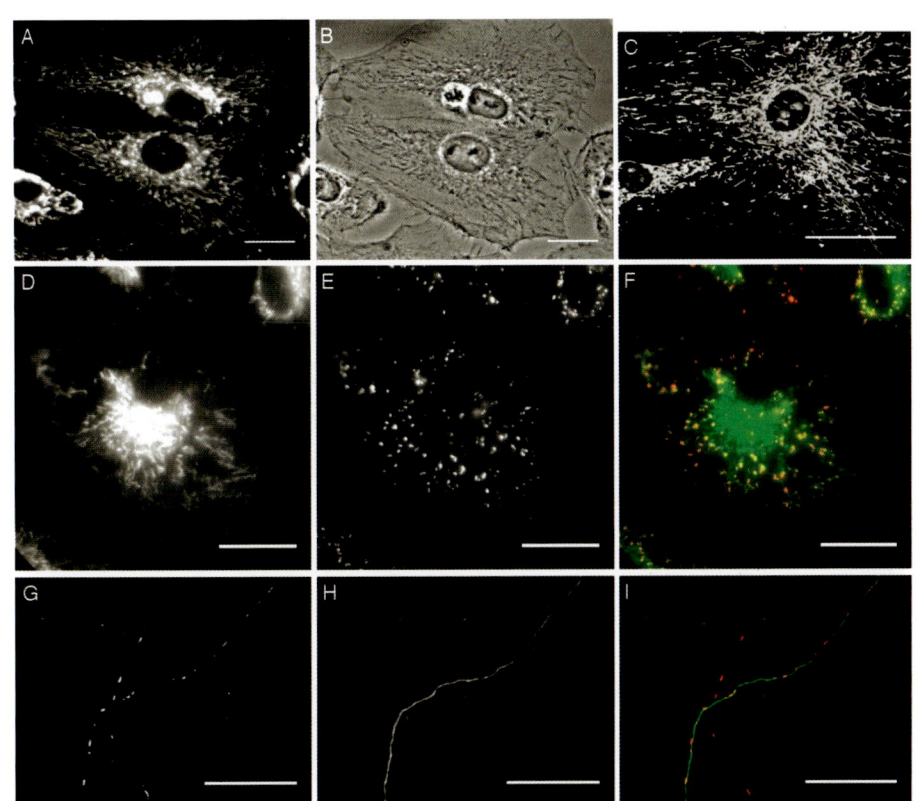

Chapter 30, Fig. 1

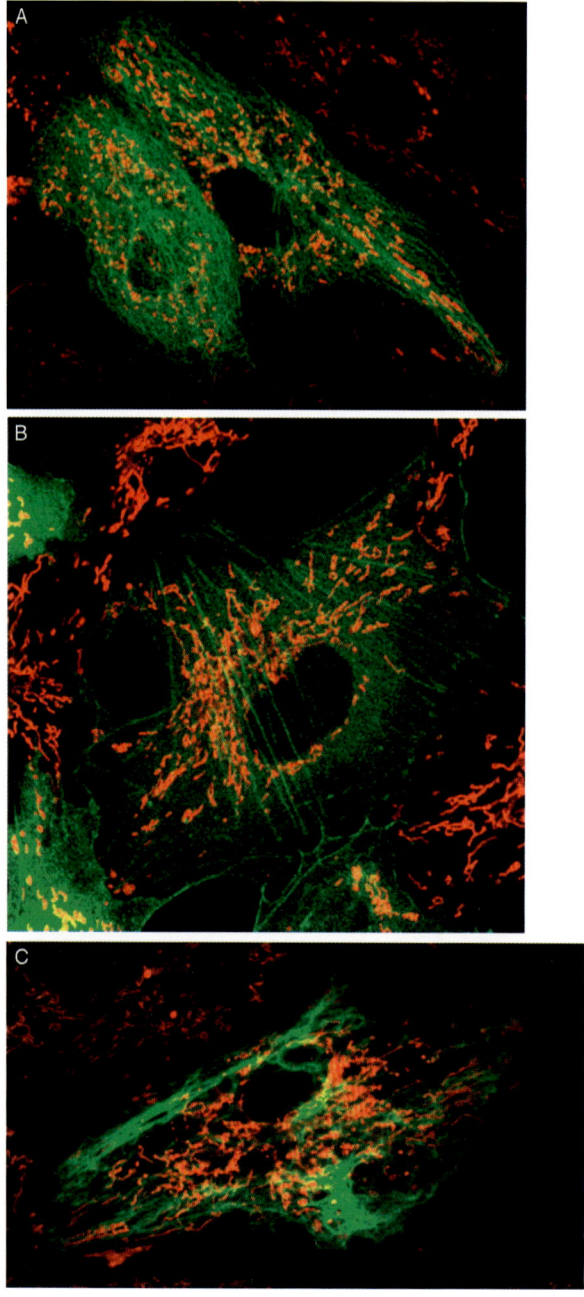

Chapter 30, Fig. 2

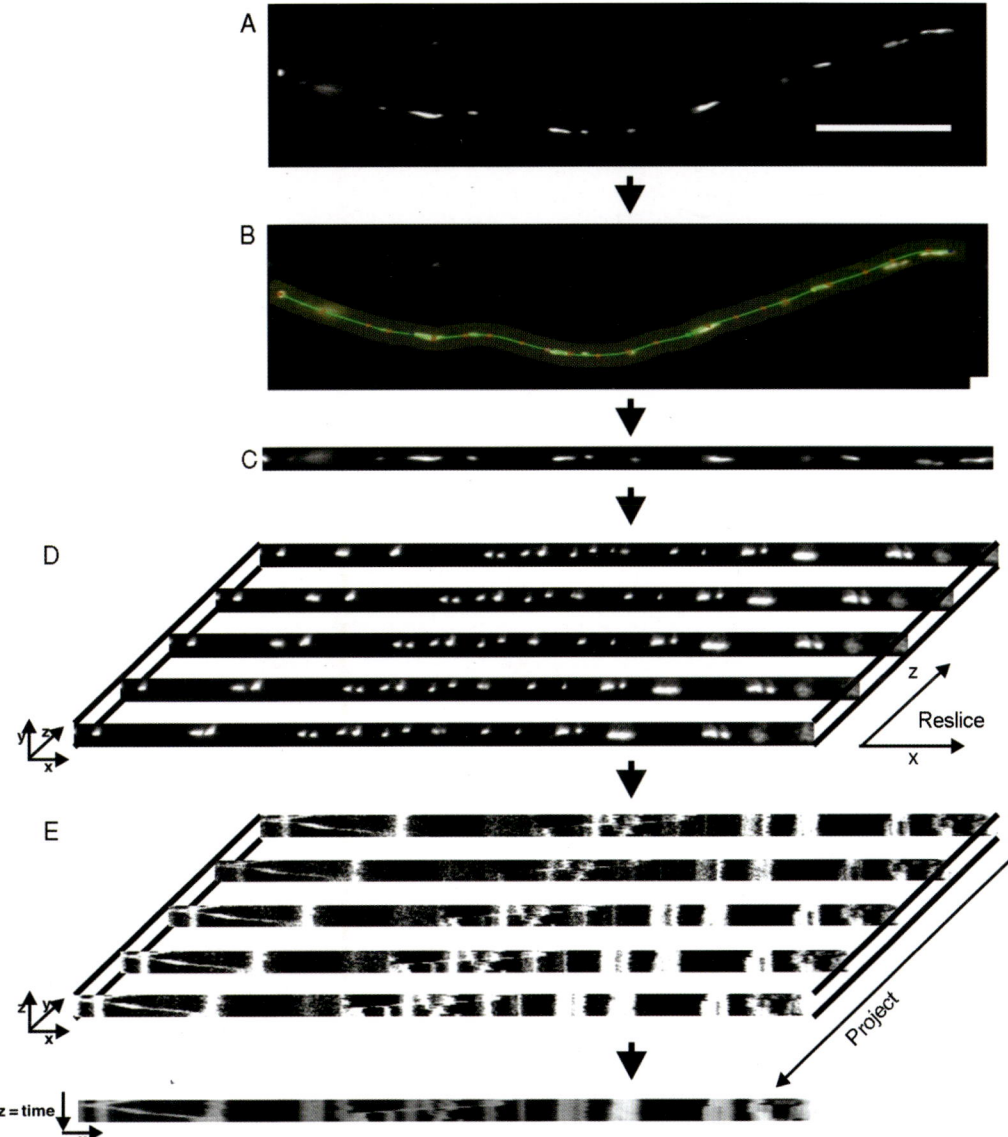

Chapter 30, Fig. 5

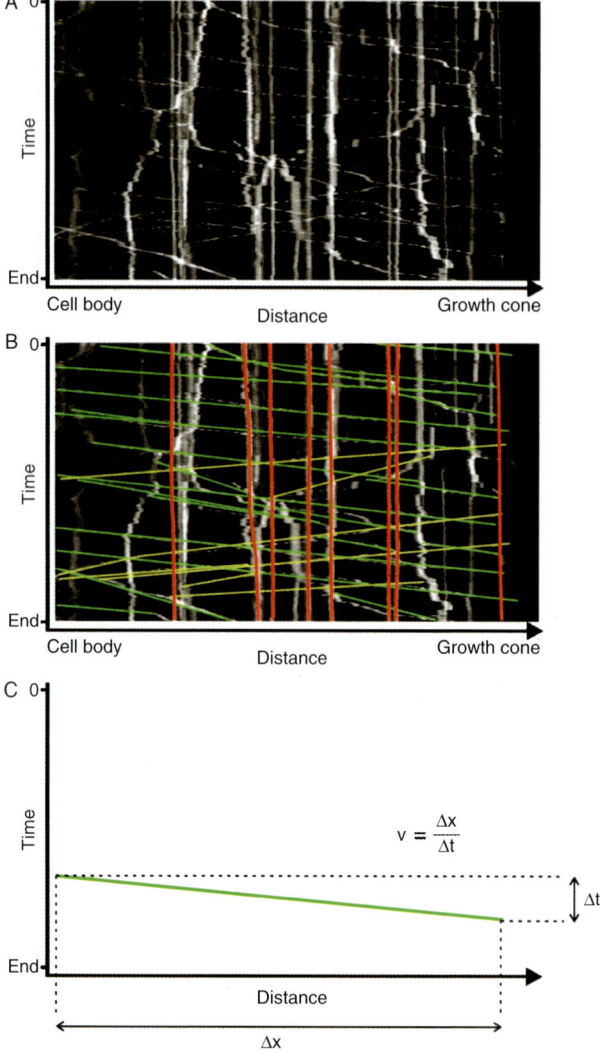

Chapter 30, Fig. 6

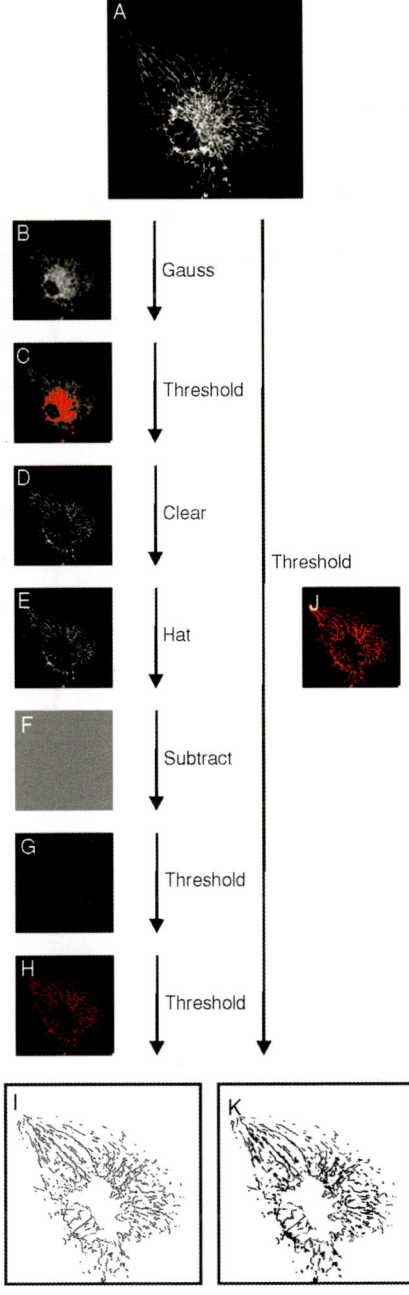

Chapter 30, Fig. 8

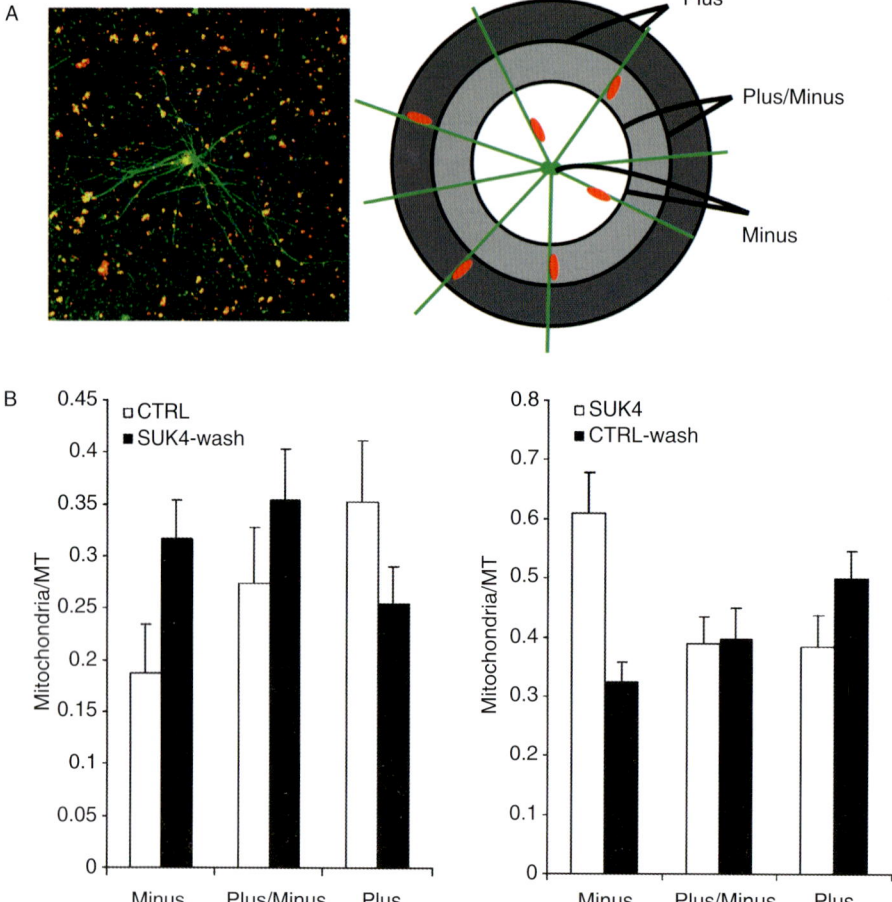

Chapter 31, Fig. 4

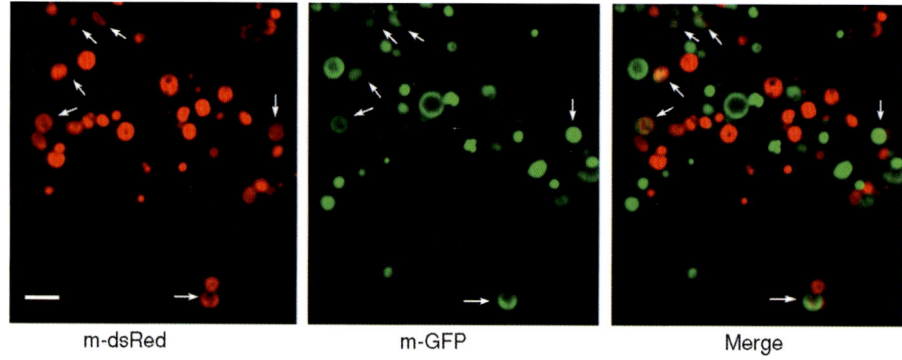

m-dsRed

m-GFP

Merge

Chapter 32, Fig. 1